Pesticide Science and Biotechnology

International Union of Pure and Applied Chemistry
in conjunction with
The Chemical Institute of Canada

ORGANIZING COMMITTEE

Chairman	H. V. Morley
Vice-Chairman	R. Greenhalgh
Secretary	A. Alper

Members

J. W. ApSimon, D. B. Carlisle, W. P. Cochrane, C. A. Franklin,
J. Hollebone, J. K. King

SCIENTIFIC PROGRAMME COMMITTEE

Chairman	R. Greenhalgh
Vice-Chairman	J. Miyamoto

Members

J. W. ApSimon, A. Ambrus, N. A. Ashford, D. B. Carlisle, W. P. Cochrane,
H. O. Esser, C. A. Franklin, R. J. Hemingway, R. M. Hollingworth,
W. Klein, D. W. Lamb, J. J. Menn, K. Mori, J. R. Roberts,
T. R. Roberts, G. Synnatschke, C. F. Wilkinson

International Union of Pure and Applied Chemistry

Pesticide Science and Biotechnology

Proceedings of the
Sixth International Congress of Pesticide Chemistry
held in Ottawa, Canada, 10–15 August 1986

Edited by

R. GREENHALGH

Plant Research Centre, Agriculture Canada
Ottawa, Ontario K1A 0C6, Canada

T. R. ROBERTS

Shell Research Ltd,
Sittingbourne Research Centre
Sittingbourne, Kent ME9 8AG, UK

Blackwell Scientific Publications

OXFORD LONDON EDINBURGH
BOSTON PALO ALTO MELBOURNE

Blackwell Scientific Publications
Editorial offices:
Osney Mead, Oxford OX2 0EL
8 John Street, London WC1N 2ES
23 Ainslie Place, Edinburgh EH3 6AJ
52 Beacon Street, Boston, Massachusetts 02108, USA
667 Lytton Avenue, Palo Alto, California 94301, USA
107 Barry Street, Carlton, Victoria 3053, Australia

First published 1987

Printed in Great Britain at the Alden Press, Oxford

DISTRIBUTORS

USA and Canada
Blackwell Scientific Publications Inc
PO Box 50009, Palo Alto
California 94303

Australia
Blackwell Scientific Publications
(Australia) Pty Ltd
107 Barry Street, Carlton, Victoria 3053

British Library
Cataloguing in Publication Data
International Congress of Pesticide
Chemistry (*6th* : *1986* : *Ottawa*)
Pesticide science and biotechnology:
proceedings of the 6th International Congress
of Pesticide Chemistry held in Ottawa,
Canada, 10–15th August 1986.
1. Pesticides
I. Title II. Greenhalgh, R. III. Roberts,
T. R. IV. International Union of Pure and
Applied Chemistry
668′.65 SB951
ISBN 0-632-01618-3

Library of Congress
Cataloging in Publication Data
International IUPAC Congress of Pesticide Chemistry
(6th : 1986 : Ottawa, Ont.)
Pesticide science and biotechnology.

Sponsored by IUPAC in conjunction with Chemical
Institute of Canada.
Includes index.
1. Pesticides—Congresses. I. Greenhalgh, R. (Roy)
II. Roberts, T. R. (Terence Robert), 1943–
III. International Union of Pure and Applied
Chemistry. IV. Chemical Institute of Canada.
V. Title
TP248.P47158 1986 632′.95 86-33410
ISBN 0-632-01618-3

Contents

Environmental fates and effects

Metabolism in plants and animals

Pesticide toxicology and human health

Preface

The International Union of Pure and Applied Chemistry (IUPAC) has sponsored a series of international congresses of Pesticide Chemistry, and the Sixth Congress was held in Ottawa from 10–15 August, 1986. These Congresses are the only international scientific meetings that deal with all aspects of pesticide science and toxicology, notably synthesis, mode of action, formulation, fate and effects.

As with earlier Congresses, therefore, the scientific programme for the Ottawa meeting encompassed all the major areas of pesticide chemistry and related areas such as toxicology and regulatory issues. However, greater emphasis was given to the fate and effects of pesticides in the environment and to the current status of those aspects of biotechnology relevant to pesticide science.

In his plenary lecture, Dr Geissbühler stressed the importance of the biorational approach in the design of new bioactive molecules. He also felt that biotechnology will feature as an important component of future research approaches. The potential for biotechnology as an alternative strategy for crop protection was also the theme of a special symposium held as part of the Congress.

The Congress programme comprised both symposia and poster/workshop presentations, the latter being chosen as the main medium for communication. The invited speakers in the symposia were asked not only to review current progress but also to prognosticate on the future of their specific field of research. The posters were intentionally separated from the workshops in an endeavour to stimulate an active discussion involving the participants. The invited speakers in the workshop sessions were asked to concentrate on specific problems associated with each subtopic.

The Scientific Programme Committee decided that the Proceedings of this Congress should be published as a single volume, with emphasis on the themes of pesticides in agriculture and the environment. A selection was made of the presentations, generally two from each sub-topic representative of the sub-topic themes. These, together with all the symposium presentations and the plenary talks, make up the Proceedings.

We wish to acknowledge the assistance of Janice Hayes (Ottawa) and Judith Smith (Sittingbourne) in the preparation of the final manuscripts for publication.

October 1986

R. Greenhalgh
T. R. Roberts

Plenary lecture

Reflections on the future of chemical plant protection research

Hans Geissbühler*, Christian d'Hondt, Erich Kunz, Robert Nyfeler and Klaus Pfister

Agricultural Division, CIBA-GEIGY Ltd., CH-4002 Basle, Switzerland

Abstract - The environment in which future chemical plant protection research will be conducted is envisaged to be dominated by four major factors: a) by rapid further advances in crop and pest biochemistry and molecular biology; b) by a continuing decrease of the success rate of conventional research approaches (such as mass screening of random chemistry for its toxicity against pests); c) by the extensive use of electronic information, data processing and storage procedures and equipment, not only in research, but also in agricultural production; and d) by an increasing economic and ecological pressure on agricultural inputs. This twofold pressure will, in turn lead to modified or new crop production technologies, and to even more restrictive regulatory requirements.

The major consequences of this environment are predicted to be the following:

- Biotechnology will be an important component of plant protection strategy and research.

- Bio(rational) approaches in the design of new bioactive molecules in terms of their biological performance, toxicological properties and environmental behaviour will become increasingly feasible.

- More sophisticated evaluation and testing procedures in the areas of chemistry, biology and safety will be required to characterize and define the mode of action, practical use, economic benefits, and environmental acceptability of new and existing compounds.

- More efficient and safer, i.e. more target-oriented delivery systems will be needed, a need which calls for commensurate efforts and adaptations in formulation and application research and techniques.

INTRODUCTION

Whilst it would be tempting to dwell for a moment on past achievements and on the essential contribution of chemical plant protection agents in safeguarding the productivity of modern agriculture, we have decided to take the forward approach.

Some professional analysts are inclined to call today's chemical plant protection market maturing or mature. Whilst this may be true in strictly economic terms for certain highly developed or industrialized areas, this opinion is not acceptable from the scientific/-technical view point, for the following three reasons:

- pest insect, disease and, to some extent, even weed populations exhibit a high degree of biological dynamism which allows them to adapt quickly to changing environmental conditions of their respective crop ecosystems. Therefore, plant protection measures also require continuous major adaptations, including the introduction of new compounds.

- questions are being raised as to whether a number of chemical plant protection agents used today really do meet modern standards of regulatory acceptability. For example, there are at this moment between 10 and 15 active ingredients under so called "Special Review" by the U.S. Environmental Protection Agency. Whatever the merits of such questions for any particular compound or case may be, we must realize and accept the urgency of developing improved solutions.

- both the technical progress and the accentuating environmental constraints to which agriculture is exposed, call for continuous improvements and changes in production technology, which, in turn necessitate corresponding modifications in plant protection.

In our reflections on the future of chemical plant protection research, we shall first attempt to describe the future research environment over a time horizon of 20-25 years. We shall then deduce the major consequences of this environment and illustrate each one with appropriate examples. Some of these examples will be conceptual rather than scientific, technical or experimental.

THE FUTURE RESEARCH ENVIRONMENT

We believe that our future research environment will be dominated by four major factors:

- first; by further rapid, if not spectacular, advances in our knowledge on crop and pest biochemistry and molecular biology.
- second; by a decreasing success rate of conventional research approaches such as mass screening of random chemistry for its toxicity against pests.
- third; by the extensive use of electronic information, data processing and storage procedures and equipment, not only in research but also in agricultural production.
- fourth; by an increasing economic and ecological pressure on agricultural inputs including pesticides. This twofold pressure will, in turn, lead to modified or new crop production technologies, and to even more restrictive regulatory requirements. It is evident that the two latter developments are bound to affect our R&D strategies and approaches in plant protection.

In looking at this future research environment, we foresee the following major consequences:

- Biotechnology will be an important component of plant protection strategy and research.
- (Bio)rational approaches in the design of new bioactive molecules in terms of their biological performance, toxicological properties and environmental behaviour will become increasingly feasible.
- More sophisticated evaluation and testing procedures in the areas of chemistry, biology and safety will be required to characterize and define the mode of action, practical use, economic benefits, and environmental acceptability of new and existing compounds.
- More efficient and safer, i.e. more target-oriented delivery systems will be needed; a need which calls for commensurate efforts and adaptations in formulation and application research and techniques.

Before elaborating on the three latter consequences in more detail, let us briefly discuss the mentioned economic/ecological pressure on inputs and the need for adaptations in crop production techniques.

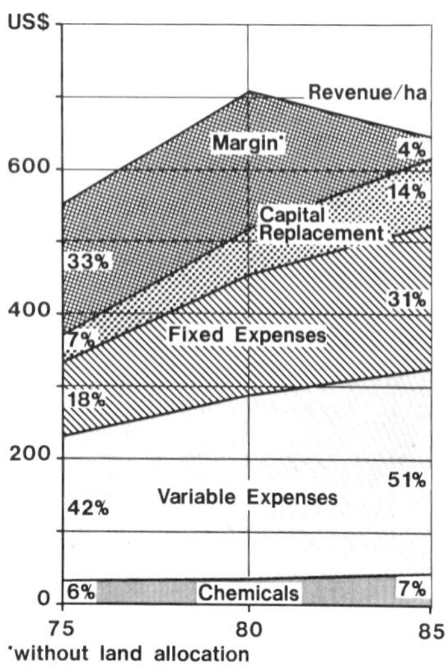

Fig. 1. Average U.S. corn production costs and profit margins per hectare for the period 1975 to 1985 (ref. 1).

It is an accepted fact, that, for a number of reasons, and with few exceptions, the worldwide farm community is, or continues to be, in an economic bottleneck. This is illustrated in Fig. 1 which (as just one typical example) shows the cost/profit picture of the average American corn farmer, over the last ten years (ref. 1). Whilst his variable, fixed, and capital replacement expenses have risen continously, his profit margin has taken a significant dip in recent years. Predictions are that the farm community will remain under economic pressure for quite some time. The message for us, plant protection researchers, is the following: Much more than in the past, we shall have to demonstrate to the farmer and his advisers not only the technical but also the economic benefits of our proposed solutions taking into account his particular political and social conditions.

A number of long-recognized and definitely accentuating ecological and economic constraints, such as

- the erosion and degradation of agricultural soils in a number of geographical areas
- the decreasing availability of water, even in some temperate zones
- the long-term shortage and corresponding increase in cost of fossil energy
- the real and potential contamination of soil, water and air - major components of our environment
- the continuing climatic vulnerability of many agricultural areas
- the growing volumes of agricultural wastes

must be increasingly counteracted by appropriate adaptations in crop production techniques.

The following partial list of such techniques attempts to relate them to their respective key demands on plant protection:

For the different modes of conservation tillage, effective herbicides are indispensible and the appropriate control of soil pests and diseases becomes more prominent (refs. 2,3).

Efficient exploitation of the various mixed cropping techniques, which are aimed at either preserving soil or nutrients, requires growth inhibitors to manage certain fast growing crops. Also, the often critical selectivity of herbicides can be improved by the use of safeners. On the other hand, residues in the edible parts of mixed crops may be critical and need careful observation and control (refs. 4,5).

Minimum (i.e. drip or trickle) irrigation offers the opportunity of more directed pesticide placement through their delivery in irrigation water (i.e. chemigation). It is obvious that products used in this way must meet physico-chemical requirements, which prevent losses upon transport and which confer systemic properties (refs. 6,7).

The further improvement of integrated pest management techniques would seem to depend upon amelioration of two key elements - the establishment of extensive and reliable threshold data bases and of accurate and real-time forecasting procedures (refs. 8,9).

The rapidly expanding variations of "glasshouse cropping", including the use of plastic foil cover-schemes, offer further economic opportunities for delivery via irrigation and also for a rational approach to exploiting controlled evaporation of active ingredients (ref. 10).

Electronic monitoring of key environmental parameters and computerized control of inputs will allow more rational timing of pesticide applications, and more differentiated, localized placement (refs. 11,12).

Efforts to reduce agricultural wastes by recycling must include the search for improved or new solutions for the safe disposal of chemical plant protection remnants (ref. 13).

While many of these opportunities are within the domain of agronomists and biologists, the chemists can make substantial contributions by paying increased attention to the physico-chemical features of their active ingredients and formulations. This will eventually allow us to optimize the desired effects more rationally and effectively instead of relying on the trial and error approach we usually take now when faced with plant protection problems in new production systems.

(BIO)RATIONAL APPROACHES

(Bio)rational approaches in the design of new bioactive molecules should not be seen as an isolated activity but as part of a more comprehensive synthetic chemistry concept, which will continue to include random, analog and natural products chemistry (ref. 14). Its main distinction is that it will follow a different testing sequence, which is practically opposite to that normally used for the classical approaches (i.e. primary testing is conducted on in vitro biochemical systems rather than on whole plants). It is this testing sequence which allows the systematic application of the various modern, computerized molecular modelling procedures and thus a rational determination of a compound's intrinsic activity features (Fig. 2).

Chemistry	Biol. Testing	Support
Random-	Morphological	-
Analog-	↓ Physiological	QSAR
Natural Products	↓ Biochemical	-
Biorational	Biochemical ▼ (Physiological) ▼ Morphological	Molecular Modelling

Fig. 2. Synthetic chemistry approaches as related to biological testing procedures and physical/chemical support methods.

Type of Activity	Testing System	Observed Effect	Support
Ergosterol biosynthesis	intact fungi	reduced ergosterol content	
C14-demethylation of 24-methylen-dihydro-lanosterol	micro-somal system	increase of 14-alpha-methyl-sterols	x-ray analysis quantum mechanics conformation analysis molecular graphics
Interaction with cyto-chrome P-450	pure cyto-chrome P-450	specific change of UV-spectrum	

Fig. 3. Physiological, biochemical and molecular levels for the biorational approach to inhibiting biosynthesis of ergosterol.

Let us briefly demonstrate this by two practical examples:

It has been observed that a number of recent fungicide classes, including the imidazoles, 1,2,4-triazoles, pyrimidines and pyridines inhibit the normal functioning of fungal cell membranes by blocking the biosynthesis of ergosterol from lanosterol (ref. 15). This can be and is exploited for designing new molecules which are active at the same site. Some information can already be generated by activity testing at the physiological level (i.e. by measuring the ergosterol content of intact, treated fungi). However, to fully exploit the modern support methods, such as X-ray analysis, quantum mechanics, conformation analysis and molecular graphics, testing at the biochemical or at the molecular level is indispensible. This can be achieved, for example, by following the inhibition of C14-demethylation of 24-methylene-dihydrolanosterol in a microsomal system or by investigating the interactions of candidate compounds with the coenzyme cytochrome P-450 (Fig. 3).

The application of molecular graphics is illustrated in Fig. 4 which shows the computerized model of the interaction between the triazole fungicide propiconazole and the heme portion of cytochrome P-450. In. Fig. 5 the Van der Waals volumes of an additional number of active compounds have been combined to demonstrate the essential steric requirements which have to be met by novel active molecules. This example represents the indirect modelling approach which relies on structural information provided by a series of active and inactive compounds and which can lead to the spatial description of the receptor (ref. 16).

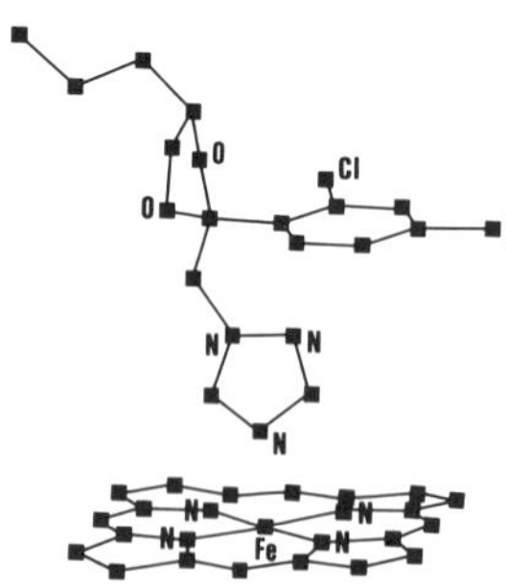

Fig. 4. Computer model of the fungicide propioconazole, (±)-1-[2-(2,4-dichlorphenyl)-4-propyl-1,3 dioxolan-2-ylmethyl]-1H-1,2,4-triazole, and the heme-portion of cytochrome P-450 (coordinates derived from force field calculations and a porphyrin cristal structure, respectively).

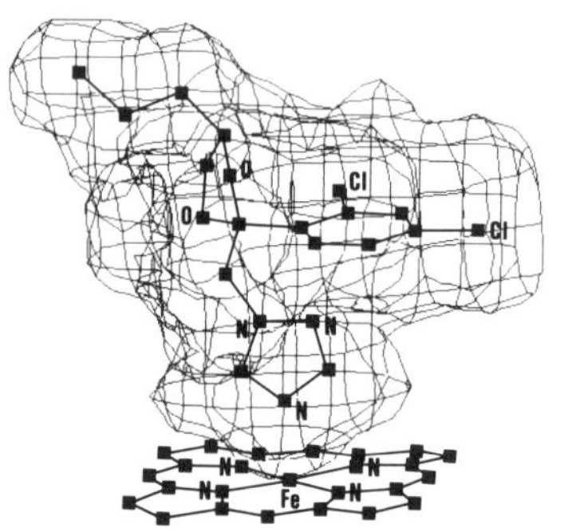

Fig. 5. Extension of the model in Fig. 4 including the Van der Waals volume common to a number of C14-demethylation inhibitors. (The models in Figs. 4 and 5 were produced using the CHEMX-program, distributed by CDL, Oxford, UK).

The alternative approach is to fist establish, by x-ray analysis, a defined model of the biological receptor and then deduce chemical structures which fit this particular target (ref. 17). As a relevant example for agrochemistry, the steric configuration of the receptor protein for the herbicide terbutryn in a photosynthetic bacterium has recently been determined (ref. 18). This bacterial system can be considered to be a first model for future studies in higher plants.

It is obvious that the success of biorational approaches will depend primarily on the availability of an appropriate number of specific and sufficiently well-defined target sites. We have examined the availability of such potential target sites in two areas, where, we believe, new and structurally different active ingredients are required. The corresponding compilations are far from being exhaustive. They are meant to be sufficiently illustrative to convey some basic reflections.

In insect control (Fig. 6), where our present structural armament of insecticides is too restricted, we believe that those target sites of the insect nervous system which have already been identified such as the post-synaptic receptor systems and the voltage-dependant inorganic ion channels offer as yet unexplored opportunities for being inhibited, blocked, accelerated or antagonized (ref. 19). Lead structures of varying activity for the mentioned target sites are available. However, for some of these systems, including the glutamate and octopamine specific receptors and the calcium-channel, additional biochemical and molecular information on compound/site interaction will be required before rational approaches can be initiated. For acetylcholine-esterase we suggest exploiting the enzyme of resistant species in order to generate alternative structures to the present classical inhibitors. Outside the insect nervous system, inhibitors of the enzymes involved in cuticle formation and deposition, and of the enzymes regulating juvenile hormone biosynthesis and degradation appear to be promising targets for preventing normal insect growth and development.

Target site(s)	Anticipated effect	Identified, active compounds
Postsynapt. Receptors (acetylcholine, glutamate, octopamine)	agonism, antagonism of transmission	nicotine, cartap spider toxins formamidines
Voltage-dependent $Na^{+}(Ca^{++})$channels	disturbance of axonal transmission	pyrethroids benzimidazoles
Acetylcholine-esterase (resistant)	overexcitation	N-propyl-carbamates
Chitin synthase	inhibition of cuticle formation	benzoylureas
Enzymes of JH-biosynthesis, degradation	inhibition (interference with JH-titer)	fluomevalonate CF_3-ketones

Fig. 6. Potential targets and corresponding lead structures for the biorational approach to insect control agents.

In plant growth regulation, the conventional approaches have, so far, not been very successful and it appears that earlier, exaggerated expectations in this area have been reduced to more practical levels (ref. 20). Although some of the original objectives may now be more conveniently reached by biotechnology, especially if the envisaged target mechanisms involve the regulation and expression of specific genes, we believe that chemical plant growth regulators still have a future and that the search for them can be made more productive by biorational experimentation. Fig. 7 summarizes the availability of biochemical information on a number of potential target sites in four major areas of plant growth: plant hormones (ref 21), assimilate transport and distribution (ref. 22), regulatory plant systems (refs. 23,24), and resistance to stress (ref. 25). Fig. 7 demonstrates that, whilst we have quite extensive knowledge on the identity of the molecules involved in these processes and on their biosynthetic pathways, our information on the mechanism of action at the molecular level continues to be seriously deficient. This leaves us with three options for pursuing biorational approaches:

- to interfere with the biosynthetic pathway of any of the listed plant growth components

- to investigate in much more detail than at present, the mode of action of the current commercial and developmental plant growth regulators in order to identify and describe suitable target sites, or

- to generate much more extensive information on the mechanism of action at the biochemical and molecular levels of the listed growth components. We plead for a forceful and more extended contribution by academic institutions to increase the promise of the last option.

Whichever option is chosen, we believe that, at the start, we should not focus unduly on any particular practical objective, but rather search for strong intrinsic activity in a key growth-regulating mechanism and then explore it further at the physiological and eventually whole plant level.

Potential Targets	Involved molecules	Biosyn-thesis	Molecular mechanism	Compounds comm./exp.
Hormones				
Gibberellins	+	+	-	+
Auxins	+	+	-	+
Cytokinins	+	+	-	+
Abscisic Acid	+	+	-	(+)
Ethylene	+	+	-	+
Assimilate Trans.				
source	+	+*	+	-
transport	(+)		-	-
sink	(+)	(+)	-	(+)
Regulatory syst.				
Calmodulin	+	(+)	-	+
Phytochrome	+	+	-	-
Stress resist.	(-)		-	-

*Photosynthesis

Fig. 7. Availability of biochemical information on potential targets for biorational approaches to plant growth regulation (+ = sufficient information; - = insufficient or no information). Last column lists commercial or developmental plant growth regulators which appear to interfere with corresponding target mechanisms.

To achieve useful biological performance in practice, sufficient activity at the target site is indispensible. However, equally important is the transfer of an adequate quantity or dose from the site of application to the target (ref. 26). During this transfer, a molecule has to cope with such phenomena as evaporation, penetration, translocation and degradation. Opponents of the biorational approach postulate that it is exactly this dosage transfer which can hardly ever be rationalized. However, we believe that this area continues to be unduly neglected and that we ought to pay just as much attention to it as do our colleagues in pharmaceutical research. By systematically determining a number of key physico-chemical parameters, such as vapor pressure, water solubility, partitioning coefficients and dissociation, hydrolysis and photolysis constants, and by subjecting them to systematic computerized simulation and sensitivity analyses, we ought to be able to produce a small set of auxiliary models, which allow prediction of dosage transfer in plants within a given chemical class of bioactive agents.

Fortunately, we are somewhat more advanced in exploiting physico-chemical knowledge for computerized modelling and predicting the environmental behaviour of pesticides (refs. 27,28,29). This is illustrated with a model paddy rice ecosystem which is made up of the components air, water, fish and sediments Fig. 8). The equilibrium distribution among these components depends on a limited number of physico-chemical parameters which are listed in Fig. 9.

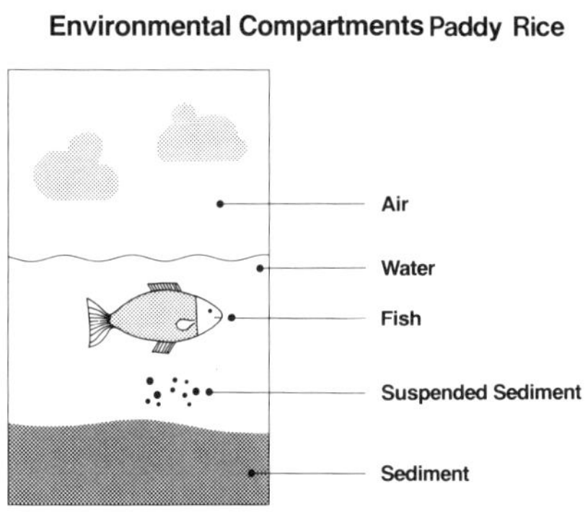

Fig. 8. Model paddy rice ecosystem comprising as components air, water, fish and sediments.

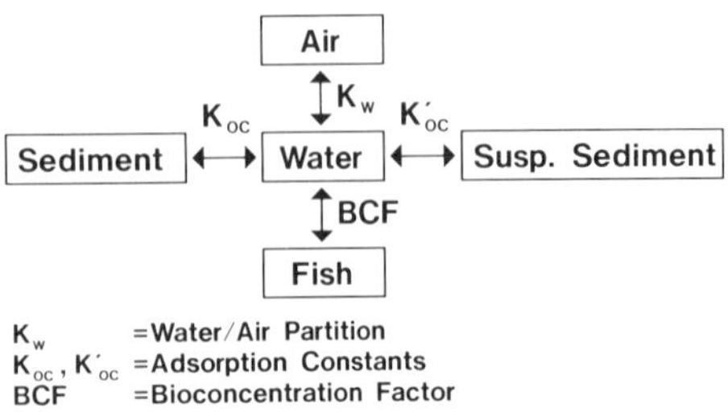

Fig. 9. Scheme of physico-chemical parameters involved in equilibrium distribution of chemicals in paddy rice ecosystem, as presented in Fig. 8.

For illustration, the three related insecticidal (thio)methyl-carbamates carbofuran (2,3-dihydro-2,2-dimethylbenzofuran-7-yl methylcarbamate), furathiocarb (butyl-2,3-dihydro-2,2-dimethylbenzofuran-7-yl N,N-dimethyl-N,N-thiodicarbamate) and carbosulfan (2,3-dihydro-2,2-dimethylbenzofuran-7-yl[dibutylaminothio]-methylcarbamate), have been subjected to the described modelling procedure.

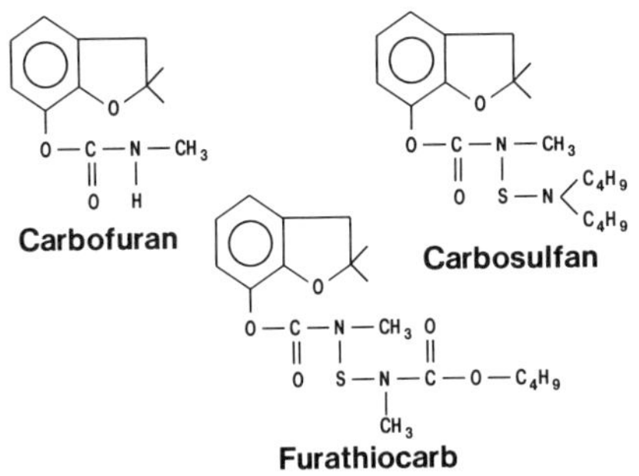

Though these compounds are closely related in terms of their structure and biological activity, their physico-chemical properties, such as water solubility partition coefficient and adsorption constant exhibit significant differences (Table 1). As a consequence and as Table 2 demonstrates, their equilibrium distribution among the compartments is indeed quite different, when evaluated under identical small plot conditions.

TABLE 1. Physico-chemical parameters of methylcarbamate insecticides subjected to modelling in paddy rice ecosystem.

	carbofuran	furathiocarb	carbosulfan
water solubility ppm	700	10	0.3
partition coefficient log P	1.5	4.8	6.7
adsorption constant Koc	19	393	2200

TABLE 2. Equilibrium distribution of methylcarbamate insecticides among compartments of paddy rice ecosystem (% of compound).

compartment	carbofuran	furathiocarb	carbosulfan
water	96.85	59.66	20.26
sediment	3.10	39.57	75.23
susp. sediment	0.04	0.47	0.9
fish	0.0007	0.28	3.6
air	0.0002	0.003	0.008

In essence (bio)rational approaches depend not only on the availability of sophisticated information which still has to be generated, but also and equally on a more sophisticated and intelligent use of the vast amount of information which we already generate in the normal course of our work.

SOPHISTICATED TESTING PROCEDURES

We shall demonstrate the need for more sophisticated evaluation and testing procedures in the areas of chemistry, biology and safety with five examples:

In chemistry where increasingly elaborate molecules are synthesized, more attention must be given to the existence, separation, identification and separate biological testing of optical and diastereo-isomers. Provided the economics are justified and appropriate technical procedures are available, this will not only allow a substantial lowering of the rate of application in many instances, but can also reduce or even eliminate certain undesirable side effects, such as phytotoxicity. This is illustrated in Fig. 10 which compares the two enantiomers of an acylalanine structure with regard to their respective fungicidal and herbicidal activities (ref. 30). While this may be an extreme example, the differences in biological activity of the two enantiomers are striking enough to indicate a fit to different biochemical target sites.

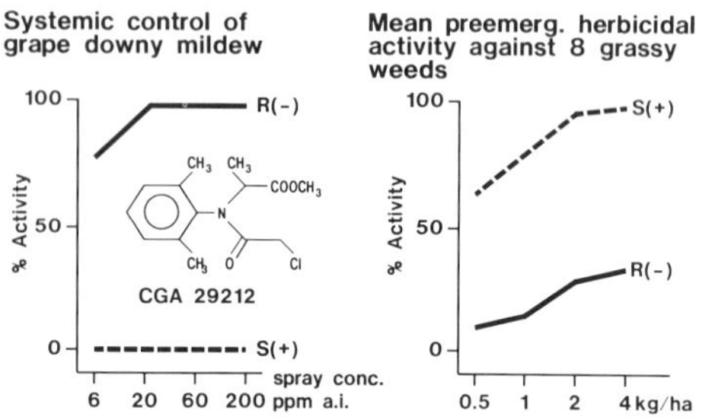

Fig. 10. Fungicidal and herbicidal activity of the two enantiomers of CGA 29212, methyl N-(2-chloroacetyl)-N-(2,6-xylyl)-DL-alaninate. Adapted from ref. 30.

In modern plant protection, the concept of not necessarily killing, but rather manipulating or managing pest populations to keep them below economic threshold levels must be and is receiving increasing attention. This calls for considerable modifications in our screening and field testing systems.

Thus, if potential insect control agents are to be screened for their antifeeding and/or repellent activity, new, rather sophisticated test systems must be developed and implemented. A model which allows testing of a large number of chemicals is shown in Fig. 11. It gives the test insect a dual choice for feeding when exposed to a half-treated leaf surface and then determines the activity of a compound by registering the feeding traces on the exposed leaf (ref. 31).

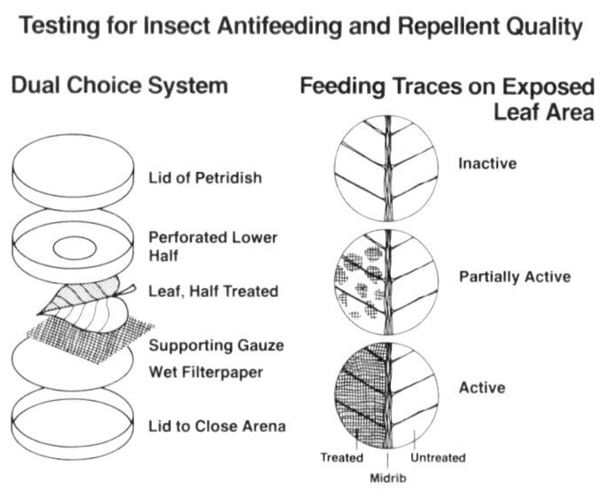

Fig. 11. Testing procedure for determining antifeeding and/or repellent activity of chemicals against insects. Adapted from ref. 31.

In field testing, we must place more emphasis on demonstrating not only the technical feasibility and/or advantages of a particular control solution, but also its economic value to the farmer. In essence, this means determining crop yields in relation to different control parameters and in relation to the ensuing cost/benefit ratio. For example, in weed control, yields have to be related to the control of particular weed species, to the density of the weed canopy, the timing of control measures, etc. For illustration Fig. 12 summarizes the effects of black-grass and broad-leaved weed control on yields of English winter cereals in three different soils (ref. 32). The data demonstrate that, whilst adequate control of black grass was indispensible to achieve significant yield increases, the control of broad-leaved weeds caused less response in two of the three described conditions. Although such single-season results have to be verified over a period of several years, they suggest that broad-leaved weed populations might remain at or below the economic threshold level in a number of instances.

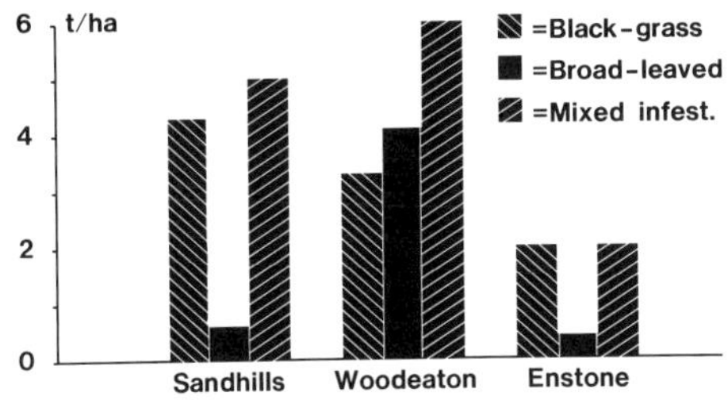

Fig. 12. Yield responses of English winter cereals to the control of black grass (A.myosuorides), broad-leaved weeds and mixed infestations in three different soils. Adapted from ref. 32.

Computerized modelling procedures would again appear to be the means of choice for evaluating multi-factor economic implications of pest control measures in practice.

To make progress in delaying and eventually preventing resistance phenomena in disease and insect control, efforts are required in two directions. First, those technical procedures which have already been recognized and established, i.e. the mixing and/or alternation of active ingredients and the exploitation of those IPM-components which reduce selection pressure, have to be implemented more systematically and consistently. Second, increasing attention must be given to the initiation of compound-specific experiments, such as selection trials, mode of action studies (to identify the vulnerability of target sites), mode of resistance investigations and experiments on population genetics, before a compound is actually introduced into practice (ref. 33).

In toxicology, a more mechanistic and less empirical approach towards safety testing seems to be indicated for the following three reasons:

- to identify and possibly eliminate suspicious compounds at an early stage of development. This would avoid false promotions and subsequent lengthy and costly development activities, such as chronic toxicity testing, process development, etc.
- to reduce, we hope, the number of empirical subchronic and chronic toxicity experiments which are now required for regulatory purposes.
- to allow, as with biological efficacy, a more rational optimisation of the toxicity features of a particular structural class of active compounds.

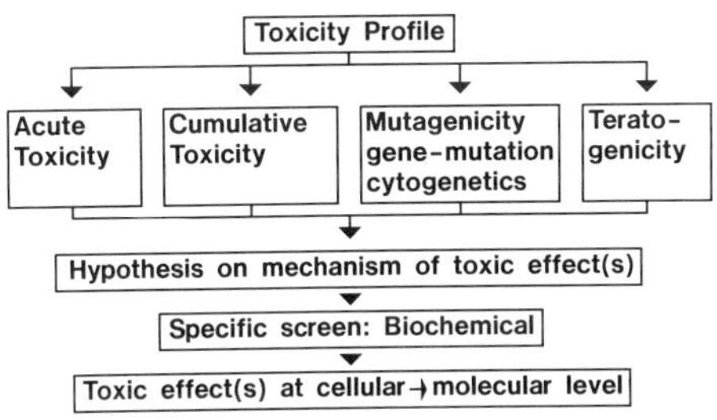

Fig. 13. Proposed toxicity testing scheme for establishing toxicity profile and for describing toxic effects of chemicals at cellular, biochemical and/or molecular levels.

The scheme in Fig. 13 proposes the early establishment of a toxicity profile based on four short-term testing procedures: Acute toxicity, cummulative toxicity, mutagenicity (both gene mutation and cytogenetics) and teratogenicity. The results of these tests permit the formulation of a first hypothesis on the mechanism of the observed toxic effect(s). The hypothesis is subsequently verified by a tailor-made, specific screen consisting of selected cellular and biochemical assays including enzyme assays. As a result of these endeavours, toxicity effects would be characterized in mechanistic terms at the cellular, biochemical or even molecular levels.

DELIVERY SYSTEMS

It is a well documented and generally acknowledged fact that current delivery systems for plant protection agents are inefficient (ref. 26). Though it would appear that innumerable opportunities for economic, ecological and safety improvements exist, the corresponding progress is too slow. Instead of presenting a comprehensive catalogue of suggestions, we shall concentrate on a few systems where, we believe, chemists can make substantial contributions.

Seeds as carriers for chemicals have long been recognized as being efficient and economic delivery systems and the use of this technique continues to expand (refs. 35,36). However, it seems that the potential is far from being fully exploited. The key chemical issues are the current deficiencies in system design and a lack of active ingredients which meet the specific physico-chemical criteria required for this application.

The release of a chemical from a seed system must be manipulated in such a way that it fits the location, the timing and the concentration required for controlling a particular pest population. The release can be controlled efficiently as shown in Fig. 14 which demonstrates the rate of release of the two insecticides diazinon and furathiocarb from coated sorghum seed (ref. 37). For diazinon the release was decelerated by microencapsulation whereas for furathiocarb it was modified by a change in the polymer component of the carrier. Provided the release is adequately controlled, rates of application can be substantially reduced. This is demonstrated in Table 3 which compares the level of control of the corn wireworm and the crop yields with granules and coated seeds respectively. The same crop yield can be achieved with rates of application via the seed system 6 times lower than with granules.

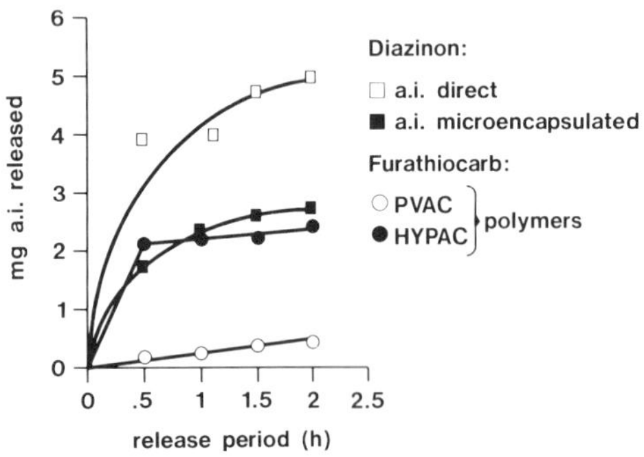

Fig. 14. Rates of release of the insecticides diazinon and furathiocarb from different coated sorghum seeds.

TABLE 3. Control of corn wireworm (Agriotes sp.) by granules and by a seed system containing furathiocarb. Mean of 5 trials (ref. 37).

Delivery system	Application rate g/ha	Level of control*	Yield t/ha
Granules	600	78	5.7
Seed	90	69	5.7
Untreated	-	47	3.6

* percent undamaged plants 60 days after planting

Electrostatic spraying devices are being designed to reduce application rates and to prevent drift of spray clouds (refs. 38,39). To make the practical use of this system more consistent and sucessful, more specific formulations than currently available are required and the corresponding quality control parameters will need to meet more stringent criteria. Thus, for solvent-based systems, this means keeping electrical conductance properties within strictly defined limits.

More extended acceptance of controlled droplet or spinning disk devices, which are designed to reduce spray volumes and to provide pre-determined and defined spray deposits, will depend on improved biological performance and equipment reliability (ref. 40). Formulations to be used in these systems can contribute to such improvements, for instance, by careful adjustment of their rheological properties (including viscosity) which influence filament break off.

Though not necessarily a major task for the chemists, closed filling systems will increasingly be needed to assure the applicator's safety and to allow more convenient cleaning and disposal of containers.

CONCLUSION

The major objective of our presentation was to demonstrate that chemical plant protection research continues to be a highly dynamic and challenging activity. It offers innumerable opportunities to chemists and biochemists to apply their ingenuity and perseverance in order to discover more economic, more efficient, more stable and safer practical solutions. If we succeed in these endeavours, chemicals will remain an indispensible component of crop protection in the foreseeable future.

REFERENCES

1. Anonymous, Chase Econometrics USA, Nov. 2 (1985).
2. M.R. Gebhardt, T.C. Daniel, E.E. Schweizer and R.R. Allmaras, Science, 230, 625-630 (1985).
3. H.M. Young, Jr., No-Tillage-Farming, No-Till Farmer, Brookfield USA (1982).
4. B.N. Okigbo, Opportunities For Increasing Crop Yields, p. 233, Pitman, Boston, London, Melbourne (1980).
5. W.C. Beets, Multiple Cropping and Tropical Farming Systems, Gower, Aldershot UK/Westview Press, Boulder USA (1982).
6. Z. Gerstl and N. Abbasel, Field Distribution of Pesticides Applied via a Drip Irrigation System, Inst. Soils Water, Bet Dagan, Israel (1984).
7. B. Hanke, Bewässerungstechnik, VEB Deutscher Landwirtschaftsverlag, Berlin DDR (1982).
8. R.E. Frisbie, Consortium for Integrated Pest Management, Texas A&M Univ. USA (1983).
9. M.L. Flint and R. van den Bosch, Introduction to Integrated Pest Management, Plenum Press New York and London (1981).
10. Frost and Sullivan, Commercial Greenhouses and Associated Equipment and Agrochemicals Market, Frost and Sullivan, Fulton, New York (1983)
11. W.O. Rasmussen et al., Computer Applications in Agriculture, Westview Press, Boulder USA (1985).
12. J.M. McKinion and H.E. Lemmon, Computer and Electronics in Agric., 1, 31-40 (1985).
13. FOA: Guidelines for the Disposal of Waste Pesticide and Pesticide Containers on the Farm, Food and Agriculture Organization of the United Nations, Rome (1985).
14. H. Geissbühler, U. Müller, J.P. Pachlatko and H.R. Waespe, Chemistry and World Food Supplies: The New Frontiers, (CHEMRAWN II, L.W. Shemilt, ed.), p.643, Pergamon Press Oxford (1983).
15. T. Kato, Chemistry in Plant Protection. Sterol-Biosynthesis Inhibitors and Antifeeding Compounds (G. Haug et al., eds.), p. 1, Springer, Berlin (1986).
16. N.C. Cohen, Adv. Drug Res., 14, 41-145 (1985).
17. P.J. Goodford, J. Med. Chem., 27, 557-564 (1984).
18. H. Michel, O. Epp and J. Deisenhofer, EMBO J. (in press).
19. A.C. Baillie, Approaches to New Leads for Insecticides, H.C. von Keyserlingk et al., eds.), p. 9, Springer, Berlin (1985).
20. W. Draber, Pflanzenschutz und Schädlingsbekämpfung, (K.H. Büchel, ed.), p. 190, Thieme, Stuttgart (1977).
21. P.F. Wareing (ed.), Plant Growth Substances, Academic Press, London (1982).
22. H.W. Heldt and U.I. Flügge, Naturwiss., 73, 1-7 (1986).
23. J.C. Lagarias, Photochem., Photobiol., 42, 811-820 (1985).
24. C.B. Klee and T.C. Vanaman, Adv. Protein Chem., 35, 213-321 (1982).
25. J. Key and T. Kosuge (eds.), Cellular and Molecular Biology of Plant Stress, A.R. Liss, New York (1985).
26. I.J. Graham-Bryce, Pesticide Chemistry: Human Welfare and Environment (J. Miyamoto and P.C. Kearney, eds.), p.21, Pergamon Press, Oxford (1983).
27. R.L. Swann and A. Eschenroeder (eds.), Fate of Chemicals in the Environment, ACS Symp. Series 225, Washington D.C. (1983).
28. P.J. McCall, D.A. Laskowski, R.L. Swann, and H.J. Dishburger, Residue Rev., 85, 231-244 (1983).
29. M. Leistra, Pestic. Sci., 17, 256-264 (1986).
30. A. Hubele, W. Kunz, W. Eckhardt and E. Sturm, Pesticide Chemistry: Human Welfare and Environment (J. Miyamoto and P.C. Kearney, eds.), p. 233, Pergamon Press, Oxford (1983)
31. M. Schwinger, Umschau, Heft 3, 170-173 (1986).
32. B.J. Wilson, M.E. Thornton and P.J.W. Lutman, Aspects Appl. Biol., 9, 41-48 (1985).
33. T. Staub and D. Sozzi, Plant Disease, 68 , 1026-1031 (1984).
34. J.A. Timbrell, Principles of Biochemical Toxicology, Taylor & Francis, London (1982).
35. K.A. Jeffs (ed.), Seed Treatment, CIPAC Monograph No. 2, Coll. Int. Pesticide Anal. Council Ldt., Harpenden, Herts. (1978).

36. P.J. Baughan, A.J. Biddle, J.A. Blackett and A.M. Toms, Symp. on Application and Biology, BCPC Monographs No. 28, p. 97, Croydon (1985).
37. N. Burkhard and D.J. Nevill, CIBA-GEIGY Ltd., Basle, unpubl. (1986).
38. S. Law, IEEE Trans., IA, 19, 160-168 (1983).
39. E.C. Hislop, Proc. 1983, British Crop Protection Conf., 2, 469-477 (1983).
40. M.R. Gebhardt, C.L. Weber and L.F. Bouse, Trans. ASAE, 382-392 (1985).

Synthesis of pesticides and bioregulators

Enantioselective synthesis of bioregulators

Dieter Enders* and Beatrice E.M. Rendenbach

Institut für Organische Chemie der RWTH Aachen, Professor-Pirlet-Str.1, 5100 Aachen, Federal Republic of Germany

Abstract - Metalated hydrazones (B) based on (S)- or (R)-1-amino-2-methoxymethyl-pyrrolidine (SAMP or RAMP) (A) have proved to be very useful chiral enolate equivalents (C), allowing electrophilic substitutions α to the carbonyl group of aldehydes and ketones with excellent, up to virtually complete overall enantioselectivities. The application of this methodology in the synthesis of highly enantiomerically enriched "bioregulators", such as insect pheromones, defensive substances etc. is described.

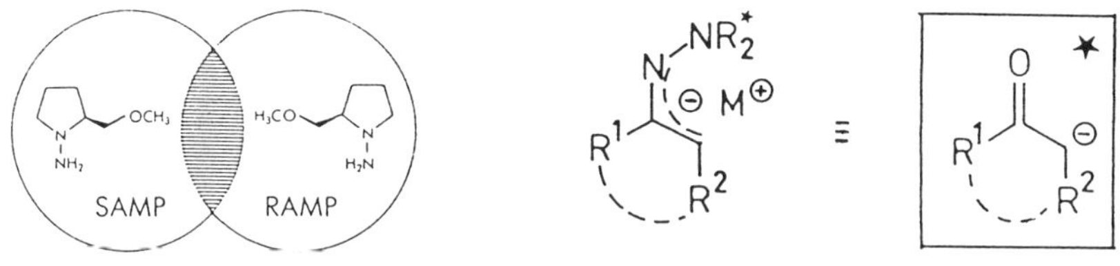

A B C

Since it was discovered in the early seventies that the biological acticity of the enantiomers of intra and interspecifically operating "chemical signals" is usually different, the enantioselective synthesis of "bioregulators" has been studied extensively. Among the various techniques, which are at our disposal to prepare enantiomerically pure compounds, asymmetric synthesis is, in principle, the most elegant and economic possibility (ref.1). As is shown in fig. 1, our group was able to develop a general and highly efficient method for diastereo- and enantioselective carbon-carbon bond formation α to the carbonyl group of aldehydes and ketones via their corresponding metalated chiral hydrazones (ref.2). This paper summarizes some of the applications of our SAMP-/RAMP-hydrazone method in highly stereoselective syntheses of insect constituents, such as various pheromones and defensive substances, and important key intermediates for bioregulator synthesis. As key steps α-alkylations of acyclic ketones (electrophile = alkylhalides) and Michael additions (electrophile = α,β-unsaturated esters) with virtually complete asymmetric induction are used.

Both enantiomers of the enantiomerically pure hydrazine used as chiral auxiliary in our method, (S)- and (R)-1-amino-2-methoxymethyl-pyrrolidine (SAMP and RAMP, ee≥99.9% ref.3), are easily prepared from the amino acids (S)-proline (4 steps, 50% yield) (ref.4) and (R)-glutamic acid (6 steps, 35% yield) (ref.5). Because in these preparations hazardous nitrosamines are involved as intermediates, an alternative, nitrosamine-free route to SAMP and RAMP was developed recently (ref.6). As is shown in fig. 2, the new procedure involves an N-Amination via Hofmann degradation as a key step. Thus, (S)- or (R)-2-methoxymethyl-pyrrolidine is transformed into its crystalline urea derivative (step 5), followed by treatment with a concentrated hypochlorite solution (step 6). In this manner, the auxiliaries are obtained in good overall yields (57%) by simple "water chemistry" and from readily available starting materials. SAMP and RAMP are also commercially available from all major chemical suppliers.

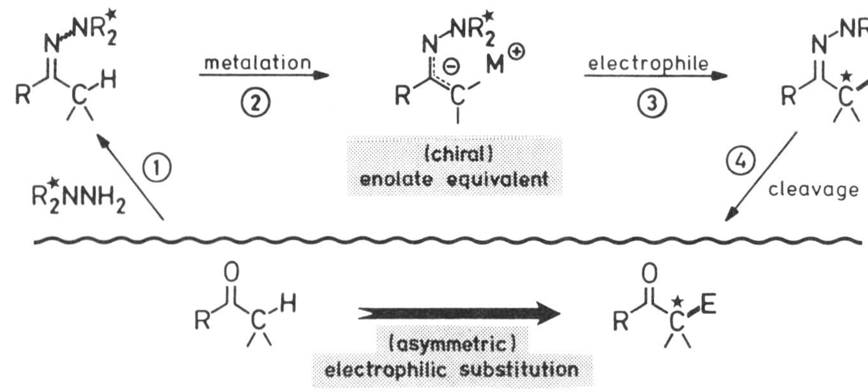

Fig. 1. Eletrophilic substitutions α to the carbonyl group of aldehydes and ketones via metalated chiral hydrazones.

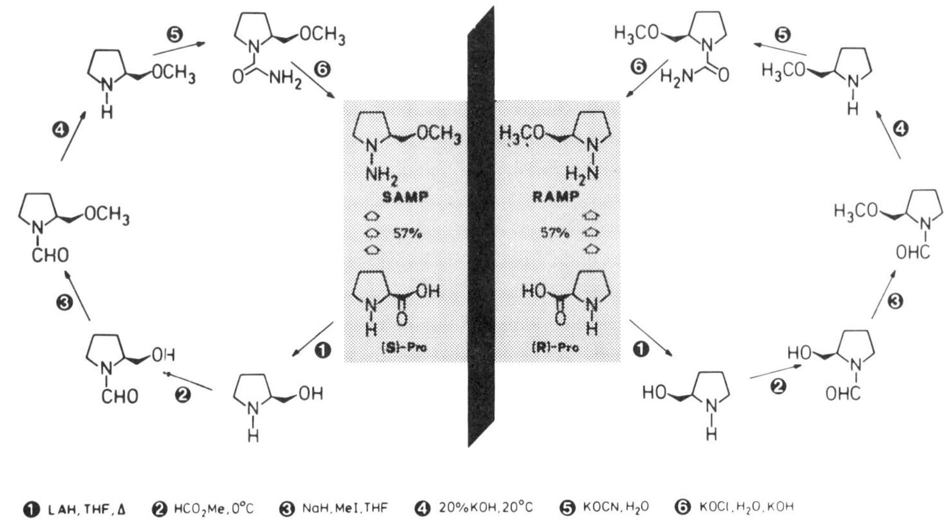

Fig. 2. "Nitrosamine-free" preparations of SAMP and RAMP via Hofmann degradation.

(S)-(+)-4-Methyl-3-heptanone (S)-4 is the principle alarm pheromone of the leaf cutting ant *Atta texana* , which is 400 times more active than its optical antipode. As is shown in fig. 3, it is easily prepared in a three step synthesis starting from diethylketone 1 and n-propyliodide. According to polarimetry, this classical ketone alkylation occurs with practically complete asymmetric induction (ee=99.5%).

(S)-4 has also been identified as an alarm pheromone in three other ant genera of the subfamily *Myrmicinae* , as a component of the defensive secretion of the "daddy longleg" *Leiobunum vittatum* (opiliones) and as being produced by the elm bark beetles *Scolytus scolytus* (F.) and *S. multistriatus* (ref.7).

Another remarkable example for a highly enantioselective acyclic ketone alkylation is the synthesis of the defensive substance of "daddy longlegs" *Leiobunum vittatum* and *L. calcar* , which belong to the class of spiders. When attached, the animals visibly wet themselves with secretion and release a powerful odor. Since only a few microliters of this secretion are available from nature and the absolute configuration of this enone 6 is unknown, the asymmetric synthesis of both enantiomers was undertaken.

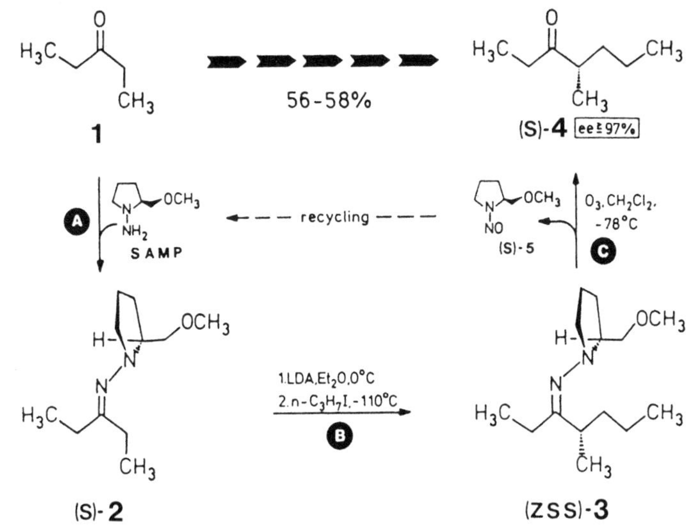

Fig. 3. Enantioselective synthesis of the alarm pheromone of the leaf cutting ant *Atta texana* .

As is indicated in fig.4, the retrosynthetic analysis leads to diethylketone and the bromide 7. Employing the SAMP-/RAMP-methodology the (S)- or (R)-enantiomer of 6 could be prepared in an overall chemical yield of 70% and in very high enantiomeric purity ($ee \geq 95\%$), respectively. This means that almost complete stereoselection in the deprotonation and alkylation steps have been achieved and virtually no racemization occurred during hydrazone cleavage (ref.8). In this way, g-amounts of both optical antipodes, as well as racemic 6 (via the corresponding dimethylhydrazone), were prepared. However, in collaboration with Prof. Meinwald (Cornell) the appropriate bioassays did not show much difference for both enantiomers.

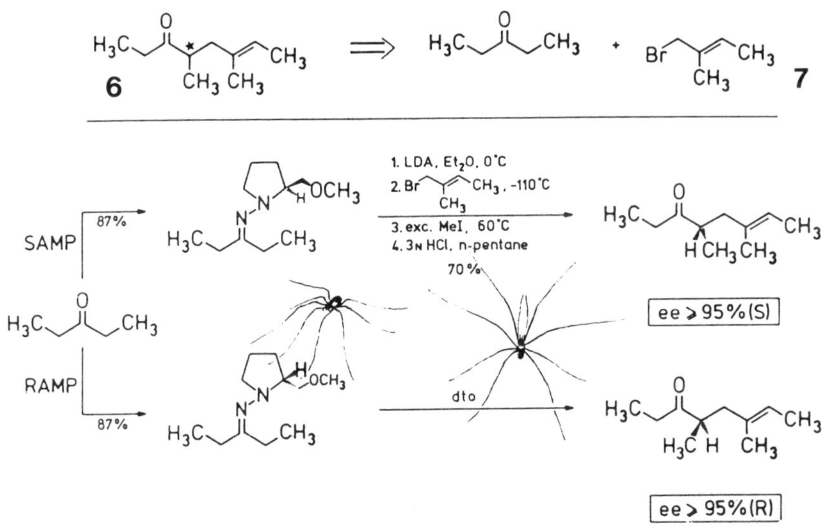

Fig. 4. Asymmetric synthesis of both enantiomers of the defense substance of "daddy longlegs" *Leiobunum vittatum* and *L. calcar* (opilionids).

At this point the elegant serricornin synthesis of Mori et al. (ref.9) should be mentioned. In the key step of this synthesis of the cigarette beetle pheromone (SSS)-8,again a virtually complete asymmetric induction could be achieved via the hydrazone method. Both absolute configurations at the new stereocenter were obtained at will, by simply changing from SAMP to RAMP (fig.5).

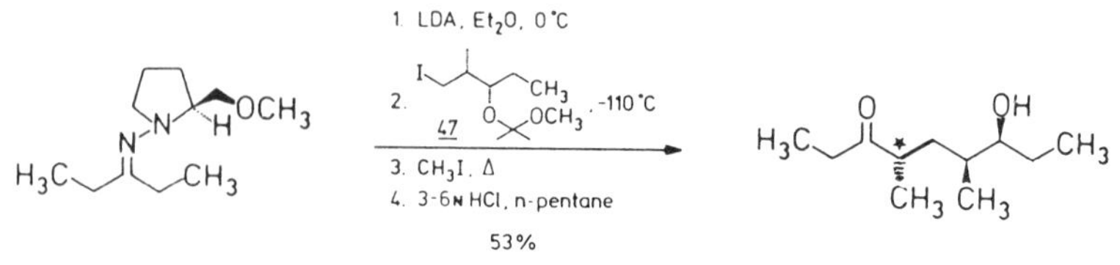

(SSS)-**8** (ee=99%)

Fig. 5. Serricornin synthesis (cigarette beetle pheromone) of Mori et al. (ref.9) using the SAMP-/RAMP-hydrazone method.

Recently we discovered that lithiated SAMP-/RAMP-hydrazones of methylketones undergo conjugate addtion to α,β-unsaturated esters in good overall yields and with practically complete asymmetric induction (ref.10). In this way, a number of β-substituted δ-ketoesters (R)- or (S)-9 could be obtained in enantiomerically pure form (fig.6).

In the remaining part of this paper the successful application of these novel asymmetric Michael-additions in the synthesis of a variety of insect constituents will be described (ref.11).

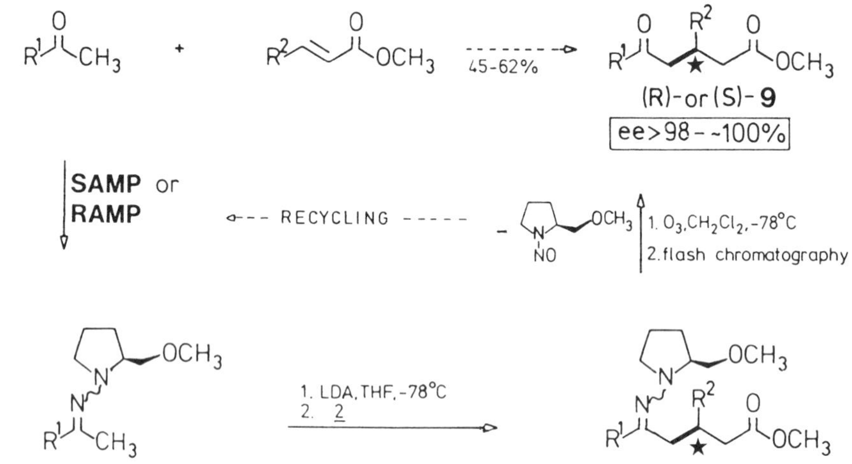

R^1= Me,Et,n-Pr,i-Pr,n-Pent,n-Hex,Ph ; R^2=Me,Ph (12 examples)

Fig. 6. Enantioselective synthesis of β-substituted δ-ketoesters via SAMP-/RAMP-hydrazone-1.4-additions to enoates.

The chiral aliphatic aldehyde (4R,8R)-11 is the aggregation pheromone of the flour beetles *Tribolium castaneum* and *T. confusum* (ref.12). In their synthesis Mori et al. used the esteraldehyde (R)-10 as a key intermediate. As is evident from fig.7, (R)-10 can now easily be prepared in three steps by asymmetric Michael-addition of acetaldehyde to methylcrotonate using SAMP as chiral auxiliary.

The enantiomer (S)-10, available in the same way using RAMP, is also an important intermediate in the synthesis of demethyldysidenin 12, a metabolite isolated from the sponge *Dysidea herbacea* (ref.13).

In addition, the new efficient preparation of the Michael-adduct (S)-10 opens an alternative entry to (R)-5-benzyloxy-3-methyl-pentanal (R)-13, again used by Mori et al. (ref.14) in their synthesis of the sex pheromone of the yellow scale (*Aonidiella citrinal*) (S)-14. Protection of the aldehyde function, LAH-reduction and O-benzylation, followed by deprotection back to the aldehyde, furnished the key intermediate in 66% yield and high enantiomeric purity (ee ≧ 95%)(fig.8).

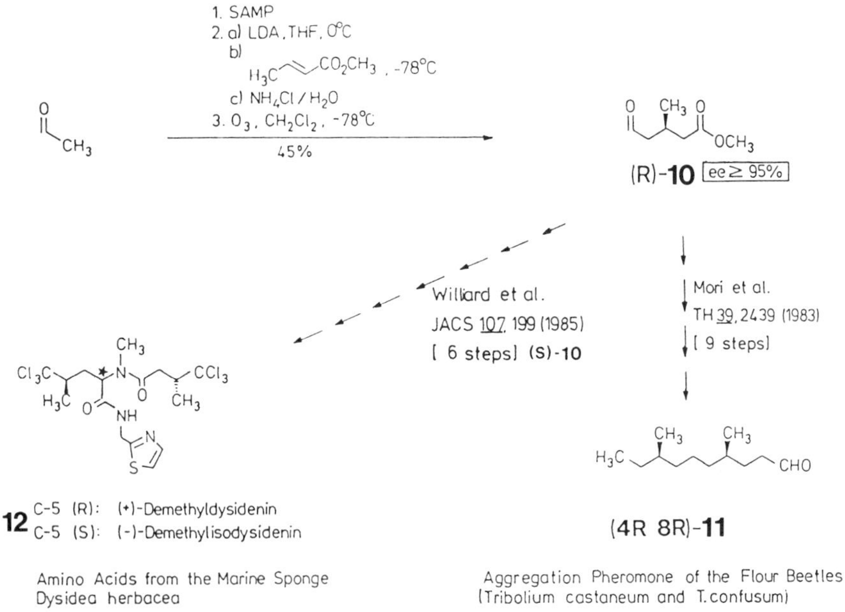

Fig. 7. Short enantioselective synthesis of (R)-10, the key intermediate in the synthesis of the aggregation pheromone of the flour beetles.

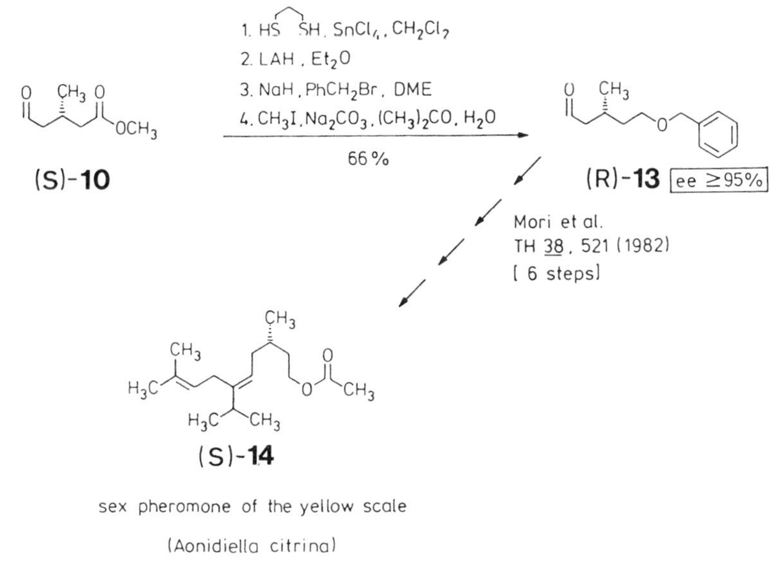

Fig. 8. Enantioselective synthesis of (R)-13, key intermediate of the sex pheromone syntheses of the yellow scale.

Recently, we were able to extend our new asymmetric Michael-addition technique to the synthesis of β,γ-disubstituted δ-ketoesters in excellent anti-diastereo- and enantioselectivities (ref. 15). Having in hand such a short and stereochemically efficient route to these important chiral building blocks, it seemed attractive to apply the new method in the synthesis of biologically active compounds.

In fact, it was only a matter of a few weeks to prepare all known chiral pheromone constituents of the small forest ant *Formica polyctena* and *F. rufa* in excellent enantiomeric purity, g-amounts and as both enantiomers, respectively (ref. 11, 16).

As is shown in fig. 9, n-propanal can be added to simple enoates in a conjugate fashion affording anti-3.4-disubstituted aldehydeesters (SS)-15 (de, ee≥ 96%), if SAMP is used as auxiliary. Alternatively, the optical antipode is available via RAMP. A recycling of the chirality information is possible by LAH-reduction of the nitroso intermediate (S)-5, generated during ozonolysis, the last step of the sequence.

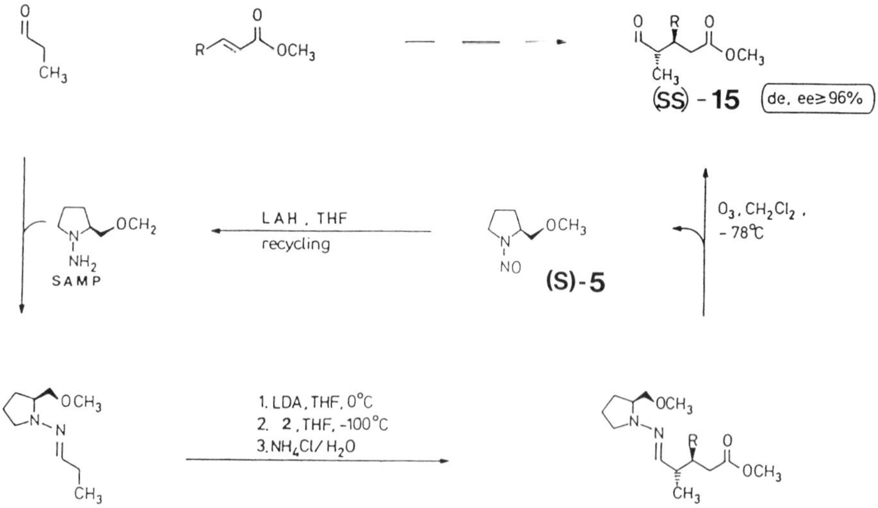

Fig. 9. Anti-diastereo- and enantioselective synthesis of 3.4-disubstituted-5-oxo-alkanoates using the SAMP-/RAMP-hydrazone methode.

Starting from (SS)-15 ($R=CH_3,C_2H_5$) the ant constituents could be obtained in a straight forward and convergent manner, employing routine chemical operations (fig.10). The simple esters and alcohols (SS)-17, (S)-20 and (SS)-21 have been isolated by Francke et al (ref.17) from heads of workers and in higher concentrations from heads of old queens. In laboratory bioassays the ester (SS)-17 ($R=C_2H_5$), which was also found in *Formica pratensis*, showed a strong aggression inhibiting effect(note a).

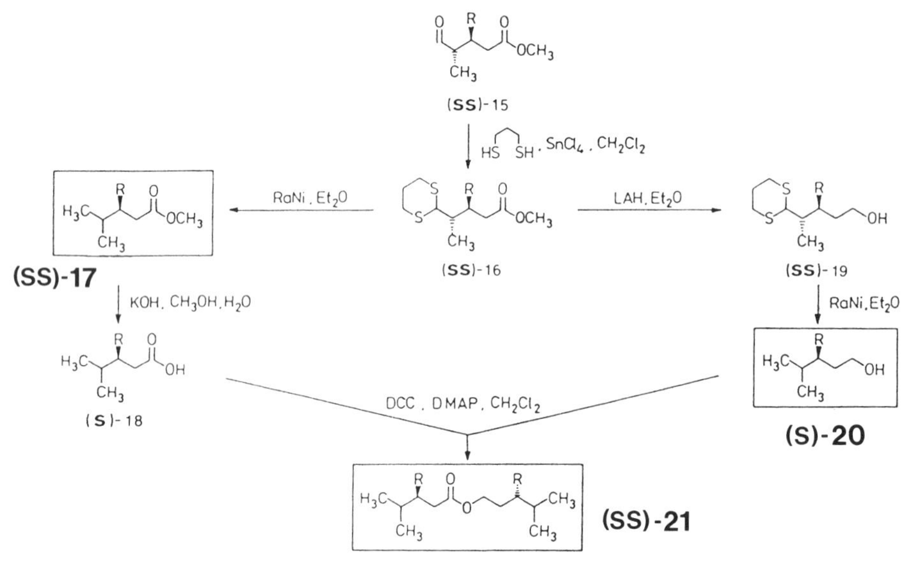

Fig. 10. Enantioselective synthesis of the pheromones of the small forest ant *Formica polyctena*, *F. rufa* and *F. pratensis*.

---

a Absolute configurations predicted.

All ant pheromones prepared in this way in highly enantiomeric purity are summarized in fig. 11, showing number of synthetic steps and overall chemical yield (not optimized). Appropriate bioassays are now being planned in collaboration with Prof. Francke, Hamburg, to establish the absolute configurations of the natural compounds.

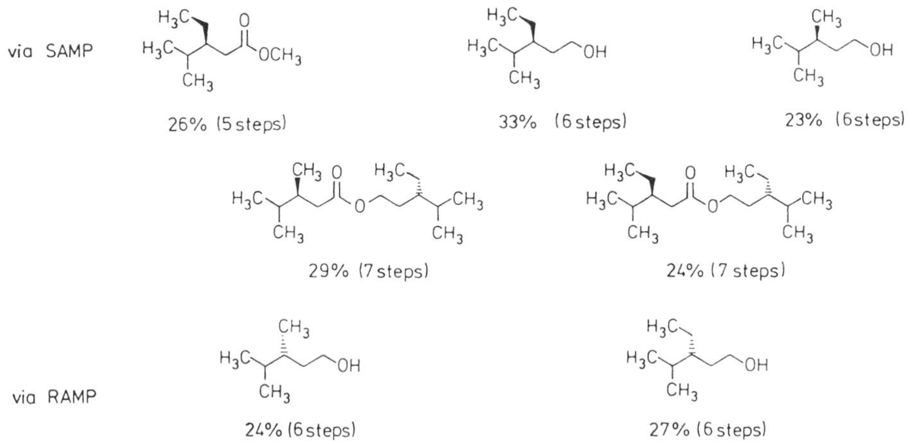

Fig. 11. Enantiomers of ant pheromones prepared via SAMP-/RAMP-hydrazone-Michael-additions.

Finally, we could demonstrate the synthetic utility of our novel asymmetric Michael-reaction in a short, highly diastereo- and enantioselective synthesis of the δ-lactone (RR)-23, which is a key intermediate of Mori's (ref.18) and Baker's (ref. 19, racemic) faranal synthesis (3S,4R)-24, the trail pheromone of pharaoh's ant (*Monomorium pharaonis*) (fig. 12).

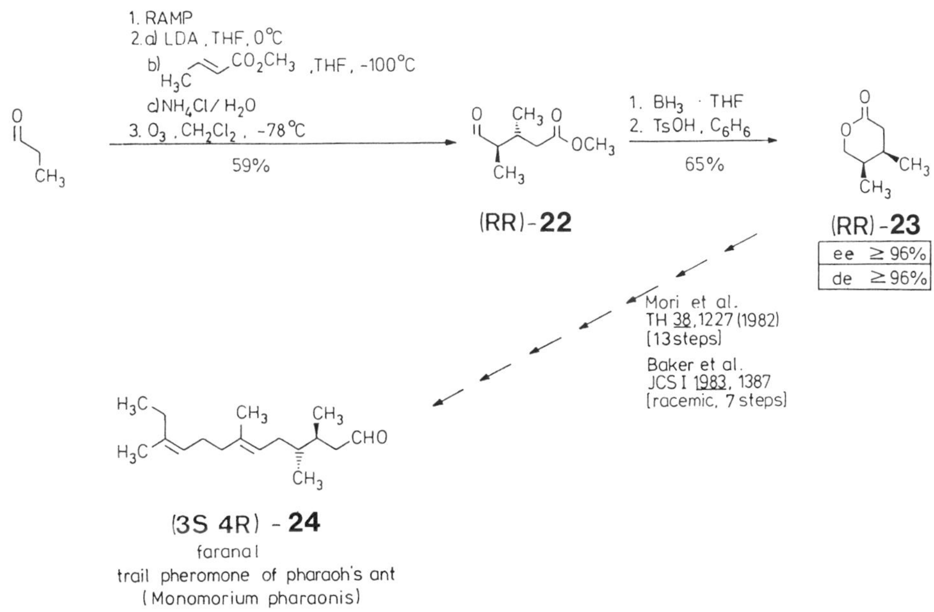

Fig. 12. Diastereo- and enantioselective synthesis of (RR)-23, key intermediate of the faranal synthesis.

In summary, the SAMP-/RAMP-hydrazone method seems to be a useful tool in the ever growing chemical arsenal for stereoselective synthesis.

ACKNOWLEDGEMENTS

This work was supported by the Deutsche Forschungsgemeinschaft, the Fonds der chemischen Industrie, the Minister für Wissenschaft und Forschung des Landes Nordrhein-Westfalen and the companies Degussa AG, BASF AG, Bayer AG and Hoechst AG. The author expresses his sincere gratitude to his co-workers Dr. H. Eichenauer, Dr. P. Fey, H. Kipphardt, Dr. U. Baus and Dr. K. Papadopoulos for their skill and hard work.

REFERENCES

1. Most recent review: D. Enders and R.W. Hoffmann, Chem.i.u.Zeit, 19, 177 (1985).
2. a) D. Enders, "Alkylation of Chiral Hydrazones" in J.D. Morrison: Asymmetric Synthesis, Vol. 3, p. 275, Academic Press, New York (1984).
   b) D. Enders, Chemtech. 11, 504 (1981).
   c) D. Enders, "Regio-, Diastereo- and Enantioselective C-C-Coupling Reactions Using Metalated Hydrazones, Formamides, Allylamines and Aminonitrils" in H. Nozaki: Current Trends in Organic Synthesis, p. 151, Pergamon Press, Oxford (1983).
   d) D. Enders, "Asymmetric Synthesis of Carbonyl Compounds and Primary Amines" in W. Bartmann, B.M. Trost: Selectivity - a Goal for Synthetic Efficiency, p. 65, Verlag Chemie, Weinheim (1984).
   e) D. Enders, "Recent Developments in Asymmetric Synthesis via Metalated Chiral Hydrazones" in S. Gronowitz: Asymmetric Organic Synthesis, Nobel Symposium 60, Chemica Scripta 25, 139 (1985).
3. K.Günther, J. Martens and M. Messerschmidt, J.Chromatogr., 288, 203(1984).
4. D.Enders and H. Eichenauer, Chem.Ber., 112, 2933 (1979).
5. D.Enders, H. Eichenauer and R. Pieter, Chem. Ber., 112, 3703 (1979).
6. D.Enders, P. Fey and H. Kipphardt, Org.Prep.Proc.Int., 17, 1(1985); Org.Synth., 65 (1987), accepted procedure.
7. D.Enders and H. Eichenauer, Angew.Chem., 91, 425 (1979); Angew.Chem.Int. Ed.Engl., 18, 397 (1979); D. Enders, H. Eichenauer, U. Baus, H. Schubert and K.A.M. Kremer, Tetrahedron, 40, 1345 (1984); D. Enders, H. Kipphardt and P. Fey, Org.Synth., 65 (1987), accepted procedure.
8. D.Enders and U. Baus, Liebigs Ann.Chem., 1983, 1439.
9. K. Mori, H. Nomi, T. Chuman, M. Kohno, K. Kato and M. Noguchi, Tetrahedron, 38, 3705 (1982).
10. D. Enders and K.Papadopoulos, Tetrahedron Lett., 24, 4967 (1983).
11. D. Enders and B.E.M. Rendenbach, unpublished results; B.E.M. Rendenbach, dissertation, RWTH Aachen 1987.
12. K. Mori, S. Kuwahara and H. Ueda, Tetrahedron, 39, 2439 (1983).
13. S.E. de Laszlo and P.G. Williard, J.Am.Chem.Soc., 107, 199 (1985).
14. K. Mori and S. Kuwahara, Tetrahedron, 38, 521 (1982).
15. D. Enders, K.Papadopoulos and B.E.M. Rendenbach, Tetrahedron Lett., 27, 3491 (1986).
16. D. Enders and B.E.M. Rendenbach, Tetrahedron, 42, 2235 (1986).
17. M. Bühring, W. Francke and V. Heemann, Z.Naturforsch., 31C, 748 (1976); W. Francke, M. Bühring and K. Horstmann, ibid., 40C, 661 (1985). We thank Prof. Francke for helpful discussions.
18. K. Mori and H. Ueda, Tetrahedron, 38, 1227 (1982).
19. R. Baker , D.C. Billington and N. Ekanayake , J.Chem.Soc.Perkin I, 1983, 1387.

Synthesis of insect antifeedants

Steven V. Ley

Department of Chemistry, Imperial College, London SW7 2AY, U.K.

Abstract - In a short sequence of reactions the syntheses of several sesquiterpene drimane natural products have been achieved including the efficient preparation of the insect antifeedants polygodial and warburganal. Further simple unnatural compounds may also be prepared with equivalent activity to the natural systems. A synthesis of the diterpene clerodane antifeedant ajugarin I has been developed which may be modified to produce structural analogues to define the functional groups responsible for activity. The structure of the limonoid antifeedant azadirachtin obtained from the neem tree has been unambiguously determined by X-ray crystallographic methods. Preparation of several structurally modified azadirachtins are also reported.

The damage caused every year to food and other economic crops by insect pests is enormous. As pest resistance to conventional insecticides increases and as the need to protect our environment in more ecologically acceptable ways becomes more apparent, it is becoming increasingly important to find new methods of insect behavioural control.

Many plants have evolved elaborate chemical defence methods against predatory insect attack. Some of these natural chemicals are especially interesting since they often have low mammalian toxicities, yet can completely disrupt the feeding habits of various insect species. Since the strutures of these so called insect antifeedants are now known we believe the time is right to study these compounds from a synthetic perspective. We therefore developed general synthetic routes to many of these chemicals with several goals in mind. Firstly we wanted to establish synthetic strategies that would be applicable both to total synthesis of the complex natural products as well as providing simple structural analogues. From this work we hoped to be able to study these compounds at a molecular level to define the functional groups responsible for the biological activity. Eventually we hope to design novel compounds which have simple structures yet trigger similar biological responses or, even better, to produce new insecticidal materials. We do not simply want to deter insects from feeding, we wish to have them ingest small quantities of the chemical and subsequently lose all further interest in feeding so that their growth and development is permanently affected. By careful entomological studies we should also be able to design species-selective behaviour modifying chemicals.

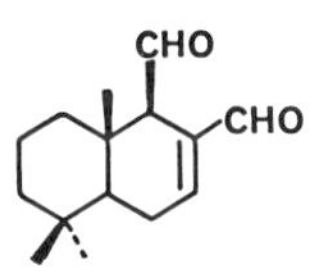

Polygodial

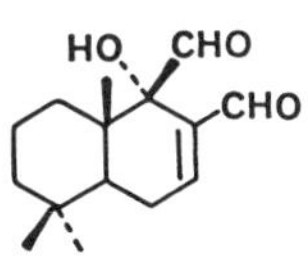

Warburganal

In the first phase of the work concise and highly efficient syntheses of the sesquiterpene drimane insect antifeedants polygodial and warburganal have been achieved.[1] The route to these molecules employs a Diels-Alder reaction between a diene derived from β-cyclocitral and dimethylacetylene dicarboxylate to set up the required decalin ring system. Although other workers[2] had previously used this strategy for drimane systhesis in multistep sequences we showed that the initial Diels-Alder adduct (Scheme 1) could be reductively rearranged using Pd/C and an acid catalyst directly to a trans-fused drimane system in a remarkably efficient fashion. Careful monitoring of the reaction by glc ensured reliable yields of diester (1) of around 80%. Having discovered this method it was possible to prepare large quantitites of material and then reduce the drimane diester with lithium aluminium hydride to afford the key diol (Scheme 1). This diol served as the pivotal starting material for several sesquiterpene natural product syntheses (Scheme 2). Undoubtedly the most important of these syntheses was the extremely efficient oxidation of the diol to polygodial using the Swern oxidation conditions and the rapid conversion to the potent antifeedant warburganal. To date this is still the most efficient synthesis of warburganal from readily available starting materials despite other extensive studies.[3] The diester (1) could also be oxidised by treatment of the resulting enolate with MoO_5 to give an angularly hydroxylated product. This material can be considered as a novel structural analogue of the warburganal system. Reduction of this product with $LiAlH_4$ followed by oxidation with barium manganate gave two further natural products isodrimenin and confertifolin (Scheme 3).

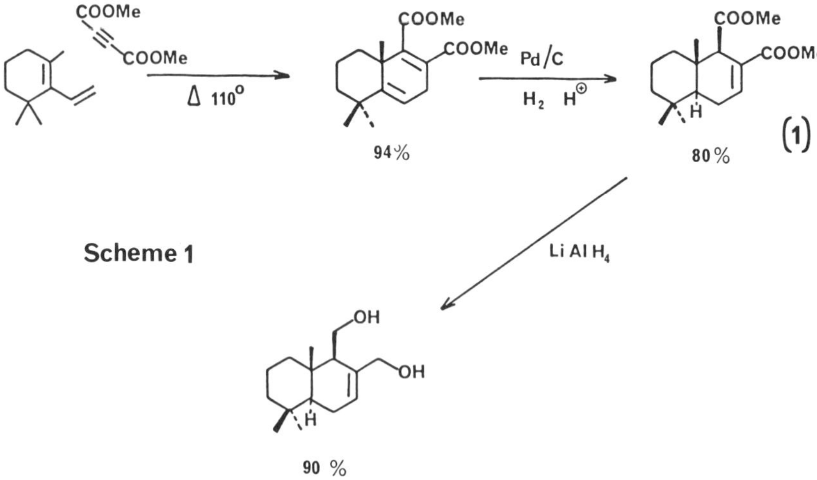

Scheme 1

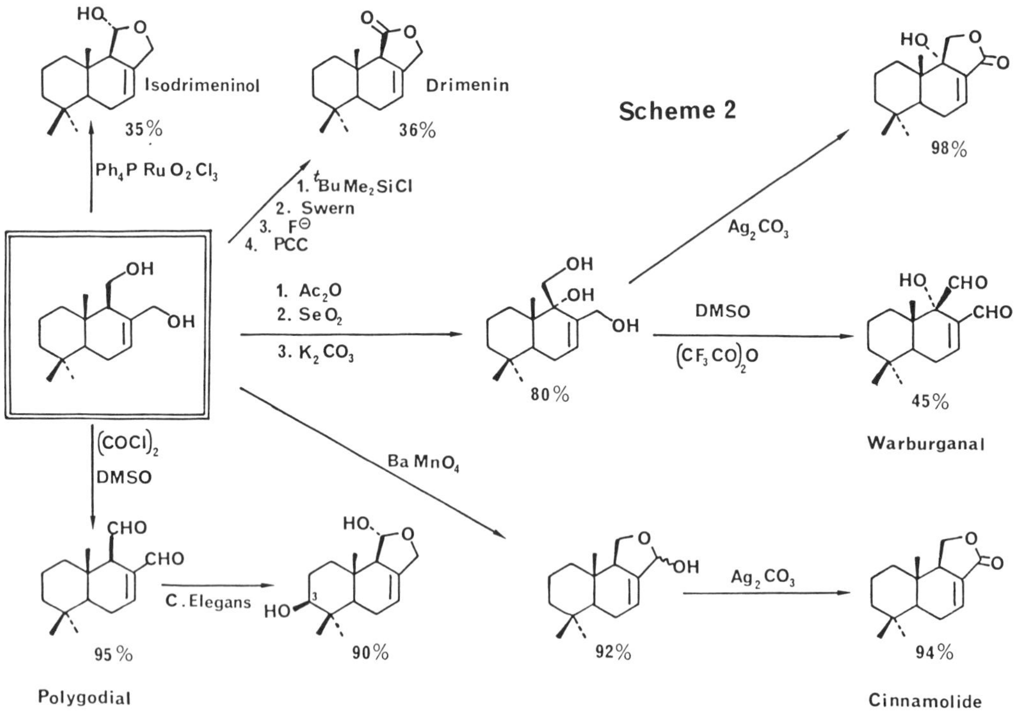

Scheme 2

In an effort to study other structural analogues we have also prepared drimane compounds using the classical biomimetic farnesyl cyclization procedure.[4] (Scheme 4). In this way drimane esters were prepared, the major isomer of which could be oxidised once again using the MoO_5 reagent to give a mixture of angularly hydroxylated materials. The 9-α-hydroxy product from this reaction proved to be a reasonably active insect antifeedant equivalent to polygodial itself in several screens (Table). Further oxidation of these products with SeO_2 gave additional unnatural derivatives for biological evaluation (Scheme 4).

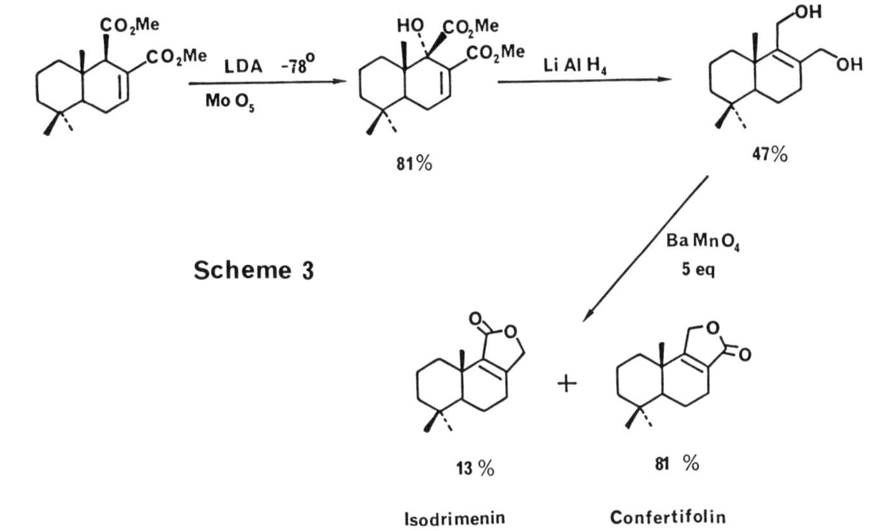

Scheme 3

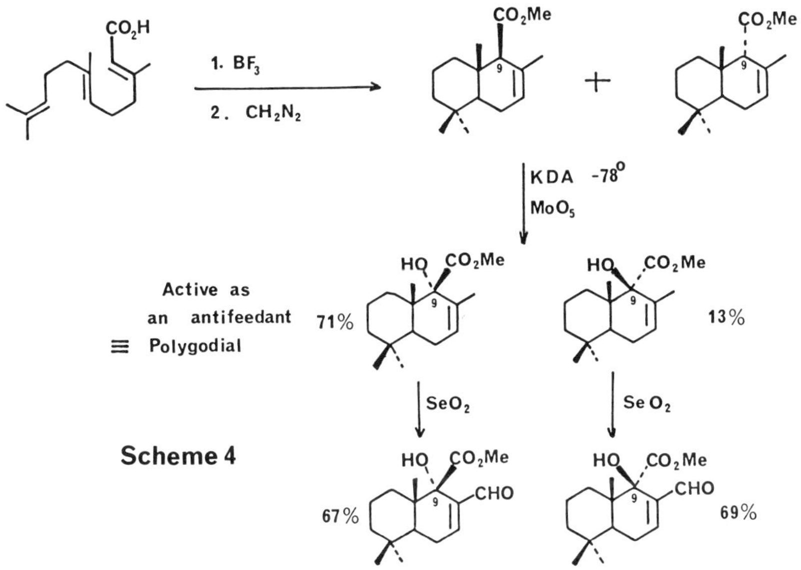

Scheme 4

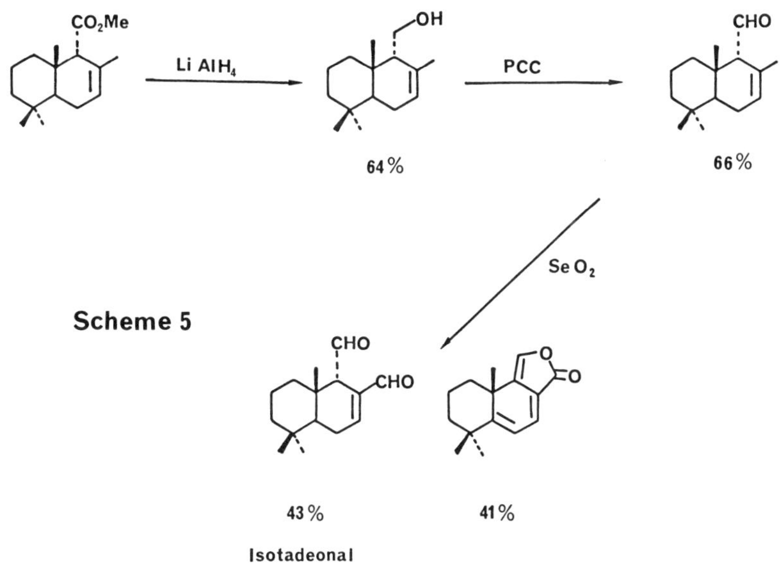

Scheme 5

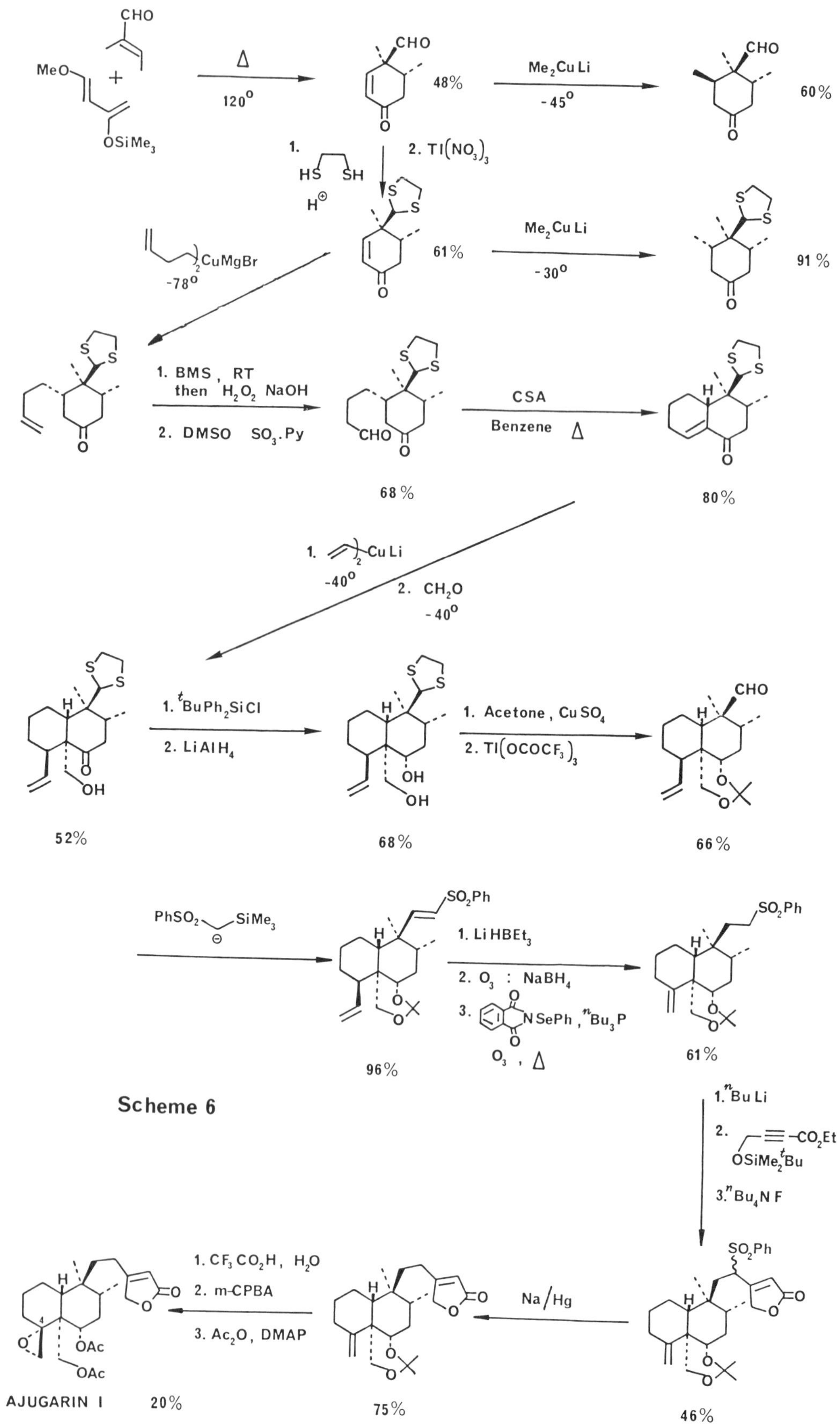
CHO
MeO
OSiMe3
Δ
120°
CHO
48%
Me2CuLi
-45°
60%
1. HS SH H⊕
2. Tl(NO3)3
61%
Me2CuLi
-30°
91 %
CuMgBr
-78°
1. BMS, RT then H2O2 NaOH
2. DMSO SO3.Py
CHO
68%
CSA
Benzene Δ
80%
1. CuLi -40°
2. CH2O -40°
OH
52%
1. tBuPh2SiCl
2. LiAlH4
OH
OH
68%
1. Acetone, CuSO4
2. Tl(OCOCF3)3
66%
PhSO2 SiMe3
SO2Ph
96%
1. LiHBEt3
2. O3 : NaBH4
3. NSePh, nBu3P
O3, Δ
61%
1. nBuLi
2. CO2Et OSiMe2tBu
3. nBu4NF
46%
Na/Hg
75%
1. CF3CO2H, H2O
2. m-CPBA
3. Ac2O, DMAP
OAc
OAc
AJUGARIN I
20%

Scheme 6

Synthesis of the sesquiterpene isotadeonal from the minor isomer of the farnesyl route was also possible (Scheme 5).

Biological testing using electrophysiological, leaf disc and artificial diet methods, of all of the drimane compounds and several others[5], leads us to conclude that it is possible by chemical synthesis to produce simple synthetic analogues with equivalent activity to polygodial, but that the range of functional groups that can be changed is limited.

Table

ANTIFEEDANT EFFECT $\frac{(C-T)}{(C+T)}\%$ OF TEST COMPOUNDS IN DUAL CHOICE TEST

Species: Compound (100 ppm)	Spodoptera exempta	Spodoptera littoralis	Spodoptera frugiperda	Heliothis armigera	Heliothis virescens
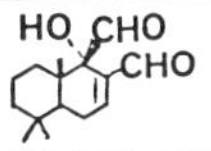	62.3 ± 2.85	63.4 ± 13.21	69.4 ± 4.81	55.7 ±6.92	63.3 ± 15.31
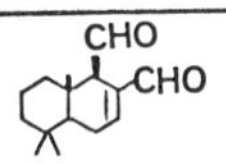	55.0 ± 6.81	56.3 ± 11.41	50.4 ± 8.14	46.3 ± 7.55	33.8 ± 16.50
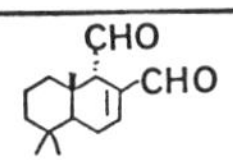	10.9 ± 7.39	-27.3 ± 11.13	9.4 ± 4.11	-25.1 ± 8.99	-5.0 ± 7.47
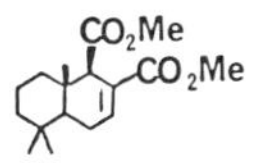	51.3 ± 9.11	57.4 ± 9.96	31.5 ± 10.91	35.4 ± 9.01	30.5 ± 9.53
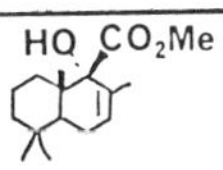	53.9 ± 4.66	52.2 ± 7.94	41.2 ± 9.64	44.2 ± 9.57	27.3 ± 4.88
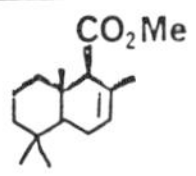	5.3 ± 6.64	7.2 ± 4.97	3.4 ± 6.49	4.2 ± 9.75	3.7 ± 8.48

In the second area of insect antifeedant synthesis we have concentrated on the preparation of diterpene clerodane related compounds. There are now some 400 natural clerodanes known[6] with a wide range of associated biological activities. Total synthesis in this area, to date however, has been limited. We became interested in the deterrent chemicals isolated from the Ajuga remota (Labatiae) plant[7] and especially in the major active component known as ajugarin 1.

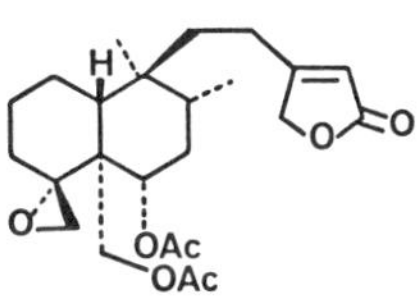

Ajugarin I

We also wished to synthesise compounds related to ajugarin 1 so as to probe the various functional groups responsible for the activity. Synthesis of compounds with the molecular complexity of ajugarin 1 requires very careful planning since they are both base and acid labile and contain a wide range of oxygen functionality. We have completed the synthesis[8] of this diterpene following the route shown in schemes 6. The important features of the synthesis are the use of the 1,3-dithiolane unit to control the stereochemistry of the conjugate addition reaction and the later unmasking to an aldehyde to facilitate introduction of the butenolide side-chain. The homologation of the aldehyde group to the ethylbutenolide side-chain required the development of new synthetic methods which may find application in other systems.

The synthesis of the trans-epoxydiacetate model compound[9] related to ajugarin 1 by a similar synthetic route (Scheme 7) gave a product which showed equivalent biological activity by inhibiting feeding of Locusta migratoria at 100 ppm. Minor structural changes to the epoxide or diacetate groups of this model led to complete loss of activity. Using an alternative route involving a novel selenium-mediated cyclisation reaction a cis-fused decalin epoxy diacetate[10] was also prepared (Scheme 8). This model compound was also active but was some ten times less than the trans-fused analogue. Clearly these results point to the importance, and stereochemical features, of the epoxy diacetate necessary for activity. However, by studying chemical changes in the side-chain of the natural product clerodin hemiacetal we can also shown that this side chain is influential and important for tuning and targeting these antifeedants for particular insects (Scheme 9).

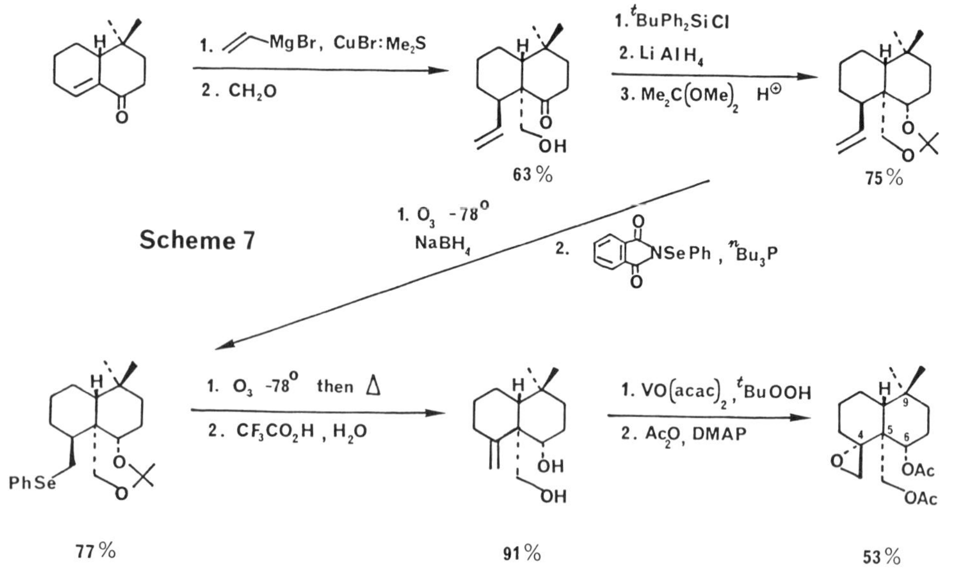

Scheme 7

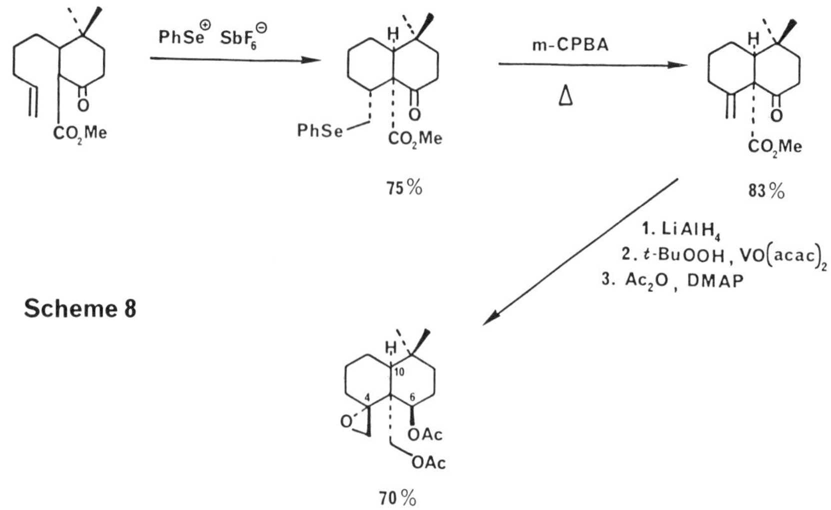

Scheme 8

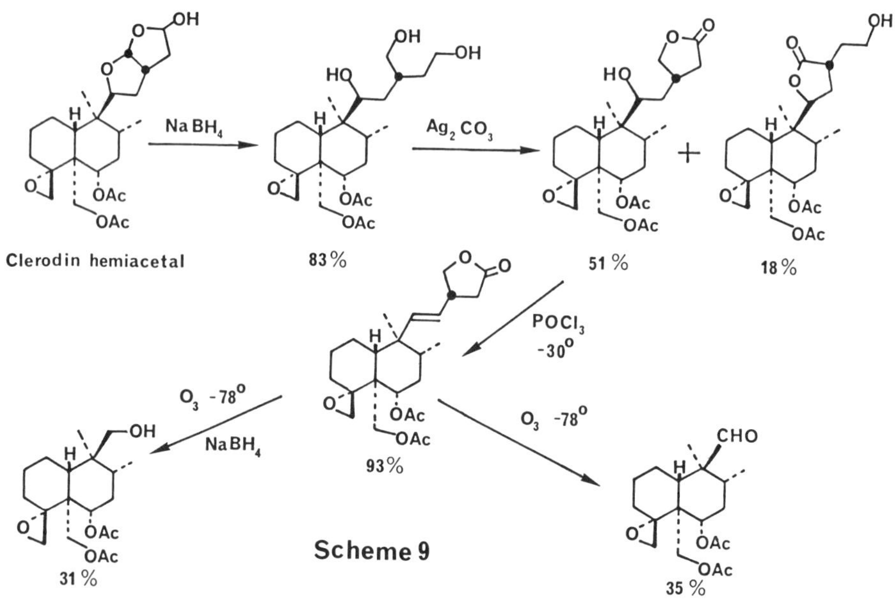

Scheme 9

Finally, a new approach to clerodane natural products has been briefly investigated whereby a Diels-Alder reation was used to set up the furano-spirolactone arrangement common to many clerodanes such as the teupolins (Scheme 10).

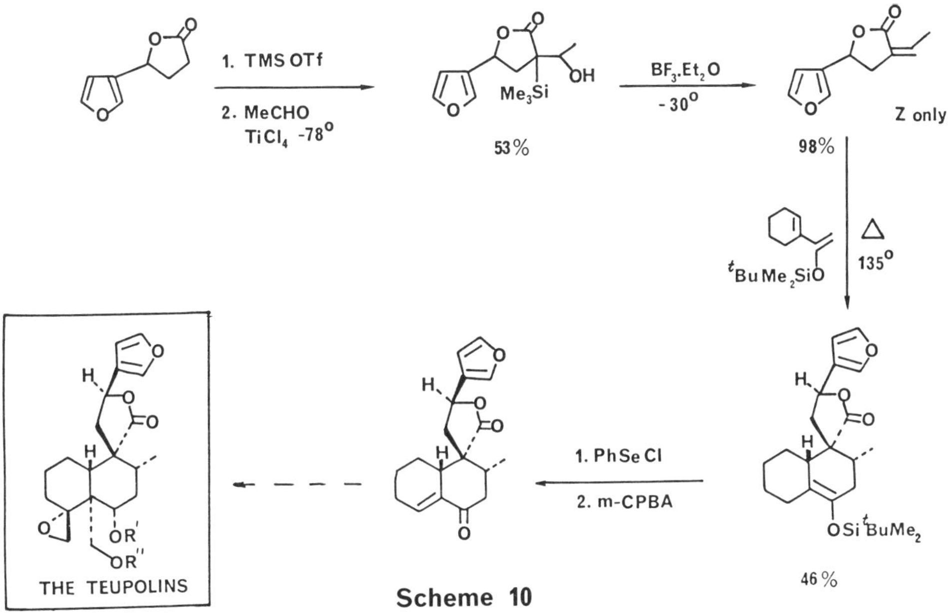

Scheme 10

The last area of study involves the extraordinarily potent insect antifeedant azadirachtin. This compound, isolated from the neem tree <u>Azadirachta indica</u> (A. Juss) has been the centre of interest for a number of years.[11]

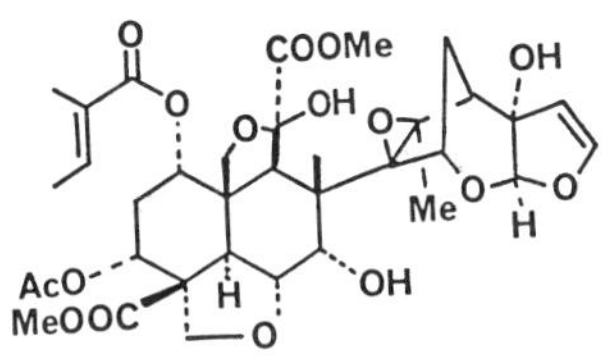

AZADIRACHTIN

The large number of functional groups and the sensitivity of azadirachtin to acids and bases made conventional structural analysis difficult. Although initial structural studies by Morgan[12] revealed key molecular fragments, it was not until 1975 that a complete structural assignment was made.[13] Our recent interests in the synthesis towards these molecules led us to question the original assignment. We felt that the only unambiguous method to solve this structural problem was by X-ray crystallographic methods. Based on these studies we can now confidently assign the structure above to azadirachtin.[14] Elegant n.m.r. studies by Kraus[15] and Nakanishi[16] have independently reached identical conclusions. By chemical studies on azadirachtin itself (Scheme 11) we were able to produce a crystalline sample of a detigloyldihydro derivative (Fig.). These data together with our n.m.r. studies unambiguously allow us to assign structures to azadirachtin and several other synthetic derivatives (Scheme 11). During our work on neem tree extracts we have isolated two further compounds from a Senegal sample of neem seeds. These compounds are 3-acetoxy-7-tigloylvilasinin lactone, whose structure was determined by X-ray crystallography and a second compound which we believe to be 3-desacetyl-11-desoxyazadirachtin. This latter compound, from its rotation, melting point and n.m.r. data suggests it is identical to material previously isolated by Kubo[17] and very similar to a related compound isolated by Kraus[18] where the tigloyl group is located on the three position rather than C-1. Further work is in hand to confirm these results. So far two chemical reactions on this 11-desoxy material have been accomplished (Scheme 12).

We are now very actively engaged in the total synthesis of azadirachtin.

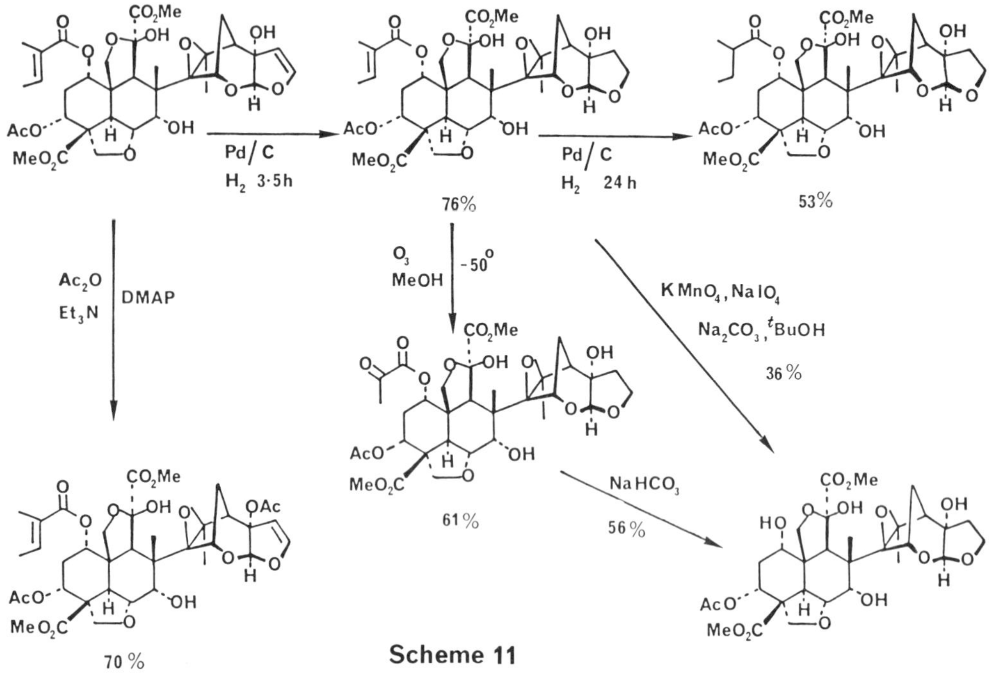

Scheme 11

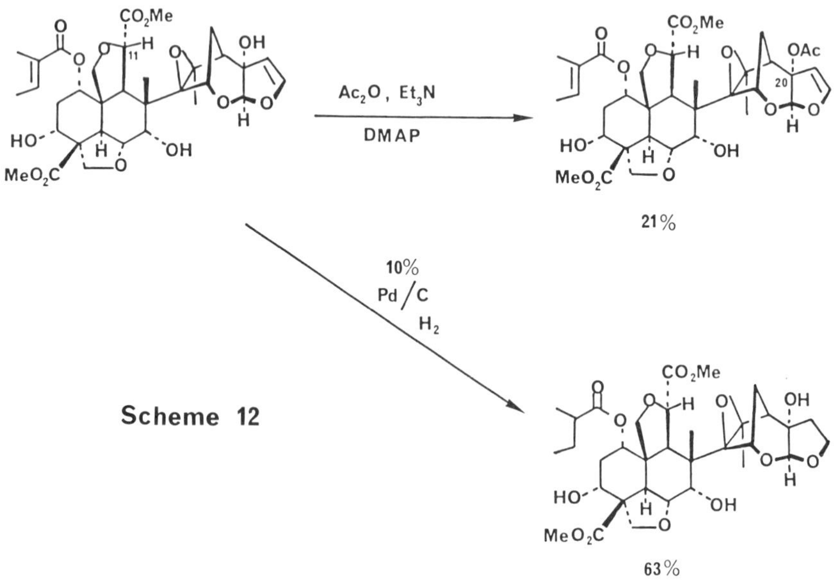

Scheme 12

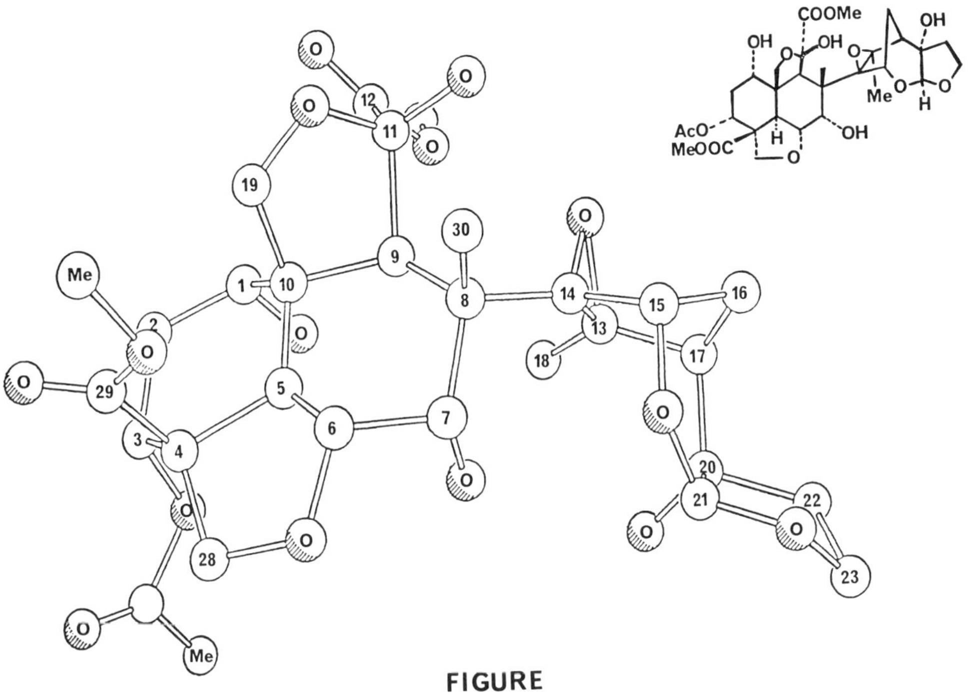

FIGURE

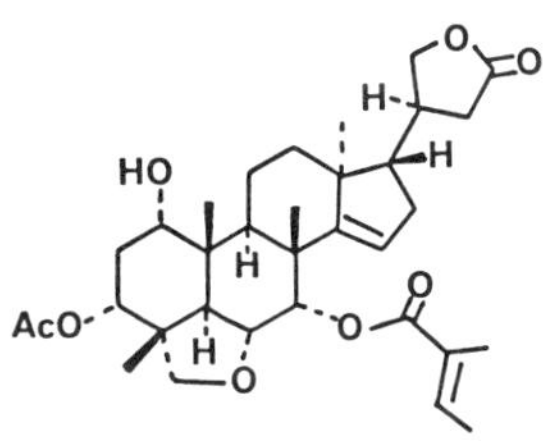

3-Acetoxy-7-tigloylvilasinin lactone

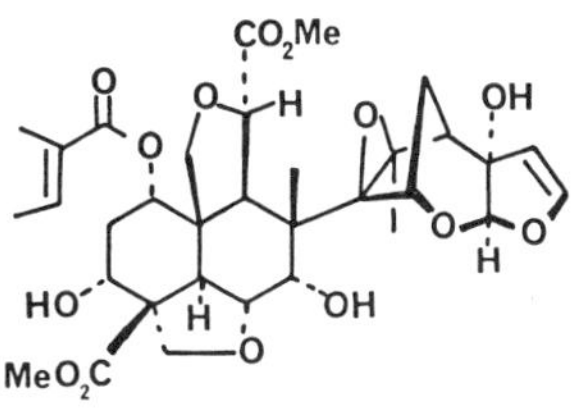

3-Desacetyl-11-desoxyazadirachtin

REFERENCES

1. D.M. Hollinshead, S.C. Howell, S.V. Ley, M. Mahon, N.M. Ratcliffe and P.A. Worthington, J. Chem. Soc., Perkin Trans I, 1579 (1983).
2. a) G. Brieger, Tetrahedron Lett., 4429, (1965); b) J.C. Laperfido, J. Org. Chem., 38, 399, (1973); c) S.P. Tanis and K. Nakanishi, J. Am. Chem. Soc., 101, 4398, (1979).
3. a) T. Nakata, H. Akita, T. Naito and T. Oishi, J. Am. Chem. Soc., 101, 4401, (1979); b) A. Ohsuka and A. Matsukawa, Chem. Lett., 635, (1979); c) A. Kimura and S. Isoe, 22nd 'Symposium on the Chemistry of Natural Products', Fukuoka, p. 198, (1979); d) A.S. Kende, and T.J. Blacklock, Tetraherdon Lett., 3119, (1980); e) D.J. Goldsmith and H.S. Kezar III, Tetrahedron Lett., 3543, (1980); f) T. Nakata, H. Akita, T. Naito, and T. Oishi, Chem. Pharm. Bull., 28, 2172, (1980); g) H. Okawara, H. Nakai, and M. Ohno, Tetrahedron Lett., 23, 1087, (1982); h) P.A. Wender and S.L. Eck, Tetrahedron Lett., 23, 1871, (1982).
4. e.g. G. Stork and A. W. Burgstahler, J. Am. Chem. Soc., 77, 5068, (1955).
5. W.M. Blaney, M.S.J. Simmonds, S.V. Ley, and R.B. Katz, Physiological Entomol., in press.
6. F. Piozzi, Heterocycles, 15, 1489, (1981), J.R. Hanson, Nat. Prod. Rep., 171, 339, 533 (1984).
7. I. Kubo, Y.W. Lee, V. Balogh-Nair, K. Nakanishi, and A. Chapya, J. Chem. Soc., Chem. Commun., 949 (1976); I. Kubo, M. Kido, and Y. Fukuyama, J. Chem. Soc., Chem. Commun., 897 (1980); I. Kubo, J.A. Klocke, I. Miura, Y. Fukuyama, A. Chapya, Chem. Lett., 618 (1982); I. Kubo, Y. Fukuyama and A. Chapya, Chem. Lett., 233, (1983).
8. S.V. Ley, N.S. Simpkins and A.J. Whittle, J. Chem. Soc., Chem. Commun., 503 (1983) and P.S. Jones, S.V. Ley, N.S. Simpkins, and A.J. Whittle, Tetrahedron, in press.
9. S.V. Ley, D. Neuhaus, N.S. Simpkins, and A.J. Whittle, J. Chem. Soc., Perkin Trans I, 2157, (1982).
10. W.P. Jackson and S.V. Ley, J. Chem. Soc., Perkin Trans. I, 1516 (1981).
11. 1st Neem Conference Rottach-Egern Germany 1980, 2nd Neem Conference Giessen Germany 1983, 3rd Neem Conference Nairobi Kenya 1986.
12. J.H. Butterworth, E.D. Morgan, and G.R. Percy, J. Chem. Soc. Perkin Trans I, 2445 (1972)
13. P.R. Zanno, I. Miura, K. Nakanishi, and D.L. Elder, J. Am. Chem. Soc., 97, 1975 (1975).
14. H. B. Broughton, S.V. Ley, A.M.Z. Slawin, D.J. Williams and E.D. Morgan, J. Chem. Soc., Chem. Commun., 46, (1986).
15. W. Kraus, M. Bokel, A. Klenk and H. Pöhml, Tetrahedron Lett., 26, 6435, (1985).
16. C.J. Turner, M.S. Tempester, R.B. Taylor, M.G. Zagorski, J.S. Termini, D.R. Schroeder, and K. Nakanishi, Tetrahedron, in preparation.
17. I. Kubo, A. Matsumoto, T. Matsumoto, J.A. Kloke, Tetrahedron, 42, 489, (1986).
18. A. Klenk, M. Bokel and W. Kraus, J. Chem. Soc., Chem. Commun., 523 (1986).

We acknowledge financial support for this work from the Science and Engineering Research Council, ICI Plant Protection, The British Technology Group and the Rohm and Haas Company. I also acknowledge the efforts and contribution made to the work by my enthusiastic group of collaborators.

Synthesis and biology of the imidazolinone herbicides

Marinus Los

American Cyanamid Company, Agricultural Research Division, P.O. Box 400, Princeton, NJ 08540, U.S.A.

Abstract - The o-(5-oxo-2-imidazolin-2-yl)arylcarboxylates are potent herbicides. Starting from the moderately active phthalimide 1, the project progressed via imidazoisoindoles represented by structure 4 to the first imidazolinone herbicide 5. Analogous derivatives of pyridine and quinoline were prepared. The simplest synthesis of this type of compound is through an aryl-o-dicarboxylic acid anhydride and an α-aminocarboxamide. Among alternative syntheses developed, the use of the imidazolinone group as a directing and stabilizing group for metallations proved to be particularly useful. The imidazolinones inhibit the enzyme acetohydroxyacid synthase. Structural requirements for activity at both the enyzme and whole plant level are deduced. The development of an imidazolinone-resistant line of corn through selection in tissue culture has been achieved.

It is some seventeen years ago that the project which ultimately resulted in the preparation of the herbicidal imidazolinones began. Since 1984, two imidazolinones, ARSENAL® and SCEPTER®, have been brought to the market place while several are in advanced stages of development. The structure of four of these herbicides is shown below:

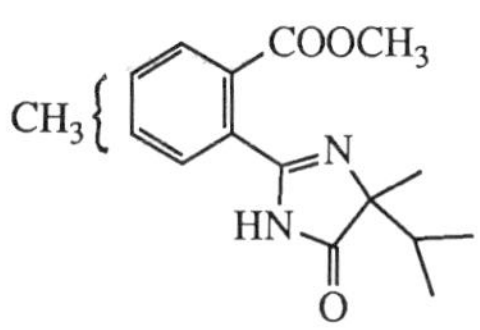

ASSERT® Herbicide

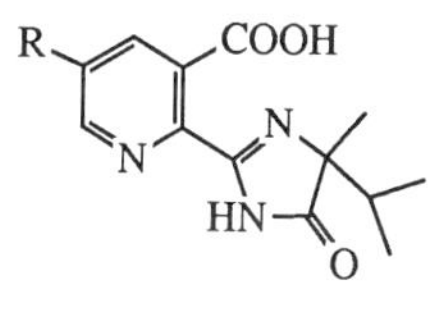

ARSENAL® Herbicide (R = H; isopropylamine salt)

PURSUIT* Herbicide (AC 263,499; R = C_2H_5)

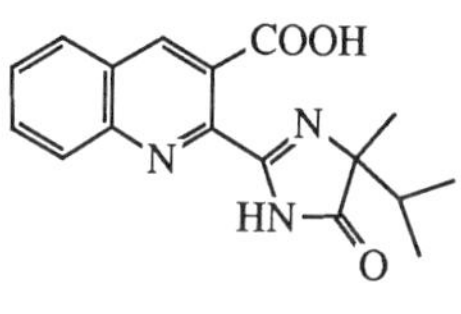

SCEPTER® Herbicide

*Trademark American Cyanamid Co.

Drawn in this manner, the structural requirements for high biological activity appear obvious. An imidazolinone ring, preferably substituted with a methyl and isopropyl group, is attached to an aromatic ring containing a carboxyl group in an ortho position. When the aryl portion of the molecule is not symmetrical, as in the case of pyridine, more than one orientation of these groups is possible. For pyridine, only the structure shown for ARSENAL has high biological activity. Before going into the synthesis and biological activity of the imidazolinones, it is of interest to describe briefly how this class of compounds was discovered.

The account of this achievement is similar to many other structure-activity stories in agricultural and pharmaceutical chemistry involving the discovery of new active classes of compounds in that it starts with a random screening lead. It differs perhaps from many of them in the remarkably strong contribution of organic chemistry to the end result. This is evident from the relatively remote resemblance of the imidazolinone product to the original lead compound. The project started with the finding that the phthalimide, 1, had sufficient herbicidal activity at 4 kg/ha to warrant further synthesis work. This work, particularly that directed to determining the effect of substitution in the phenyl ring, resulted in a series of compounds which demonstrated, not herbicidal, but rather, plant growth regulating activity. One of the most active of these was AC 94,377, 2, and a number of publications on this and related compounds has appeared (ref. 1). These phthalimides behave very much like mimics of gibberellic acid. AC 94,377 gave positive results in many of the tests normally used to demonstrate the presence and activity of GA (ref. 2).

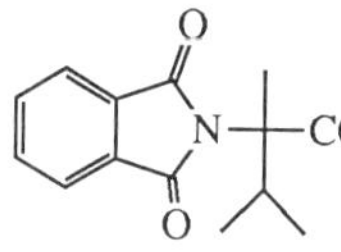

1

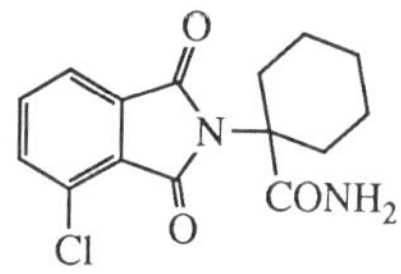

AC 94,377

2

During this synthesis program, a novel cylized product 3 was isolated from a reaction mixture and found to have essentially the same GA-like biological activity as its phthalimide precursor. When the synthesis of this type of structure was pursued, it was found that an analog of 3, namely 4, had a remarkably high level of, not plant growth regulant, but non-selective herbicidal activity. Much effort was expended in this area of chemistry and although no commercial product resulted directly, it did allow the exploration of some new chemistry and, of course, the eventual synthesis of an

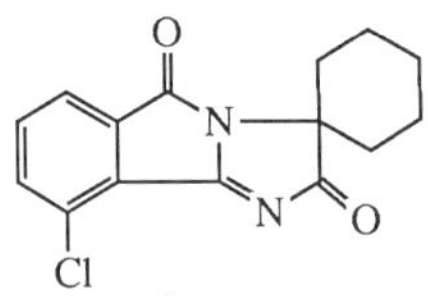

3

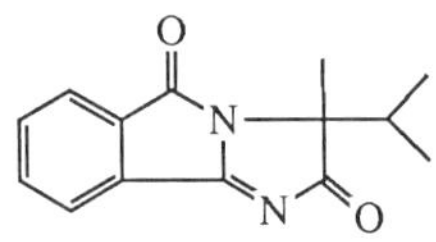

4

imidazolinone. The first imidazolinone 5 was in fact prepared from 4 by the reaction of 4 with methoxide in methanol. This compound too had a high level of herbicidal activity, with a hint of selectivity in rice. This then was the beginning of the synthesis program on imidazolinones proper, which started in 1975 and is still very active.

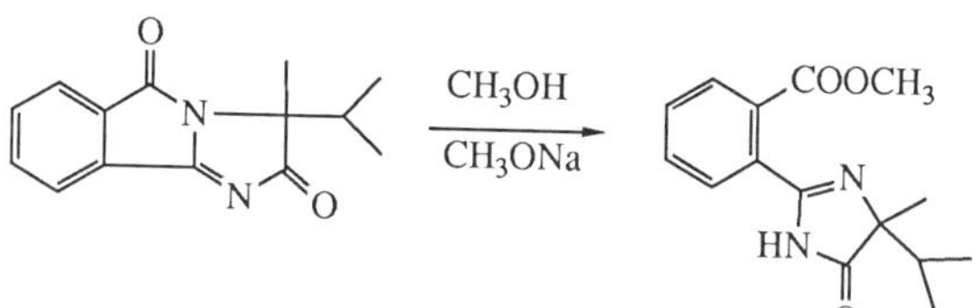

4 5

Although it was stated above that 5 was the first imidazolinone prepared, this is not strictly accurate. In retrospect, an imidazolinone was made earlier without its formation being recognized. During the time the synthesis of the phthalimides as plant growth regulants was in full swing, one of the methods employed for their synthesis involved the following step:

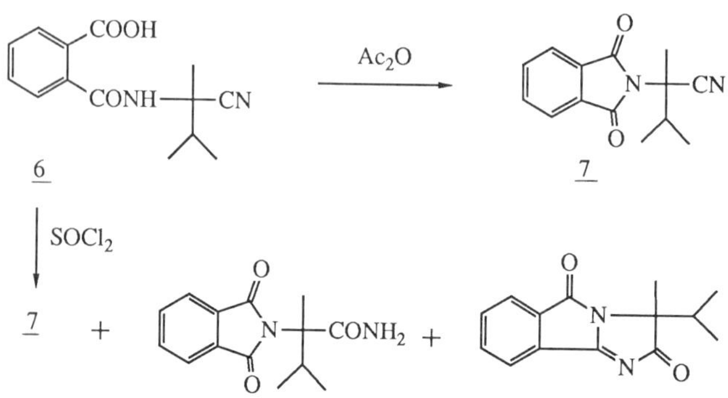

6 7 8 9

Acetic anhydride converted the amide 6 to the imide 7 in excellent yield. Replacement of the acetic anhydride with thionyl chloride resulted in a messy reaction. Thus, although 7 was still the major product, there was also formed the imide 8 as well as small quantities of the tricyclic derivative 9. It was believed that the hydrogen chloride generated in the reaction was probably responsible for this difference. In order to test this, 6 was treated with hydrogen chloride in methylene chloride in a sealed bottle at $60^{\circ}C$. The major products are those shown below although at that time we did not recognize the imidazolinone hydrochloride. Some time later, this reaction was reexamined and it was

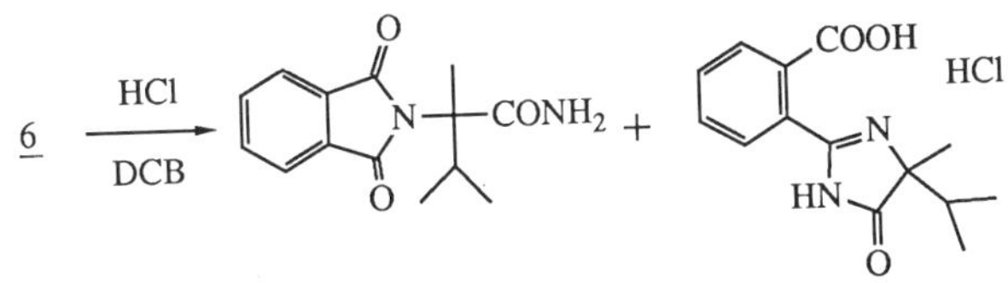

established that this reaction could be carried out conveniently at room temperature in dichlorobenzene to give a good yield of the two products. How does this reaction occur? It is believed that the nitrile is protonated and ring closure occurs to an iminooxazole. Participation of the carboxyl group gives an isoimide which can react in two ways - one involving the amide nitrogen to give the imidazolinone - the other in which the isoimide is isomerized to the corresponding imide. This latter rearrangement has been shown to be both base- and acid-catalyzed and has been well studied by Sauers (ref. 3) and others.

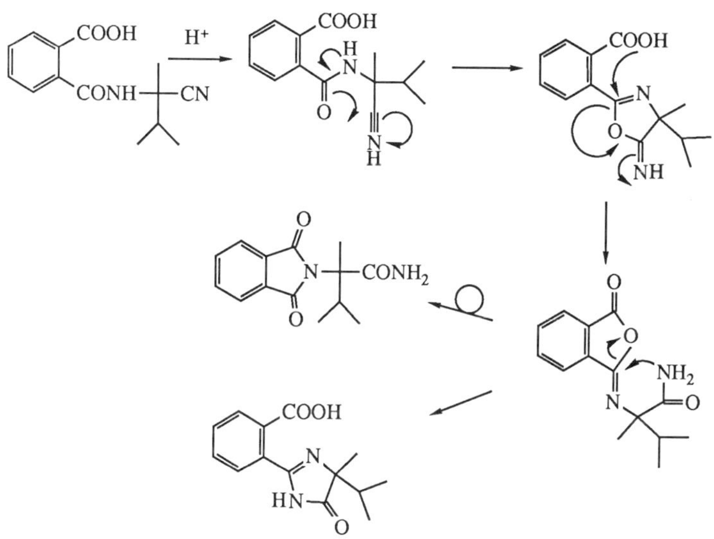

The most direct route to the imidazolinone is via an o-dicarboxylic acid anhydride and an α-aminocarboxamide.

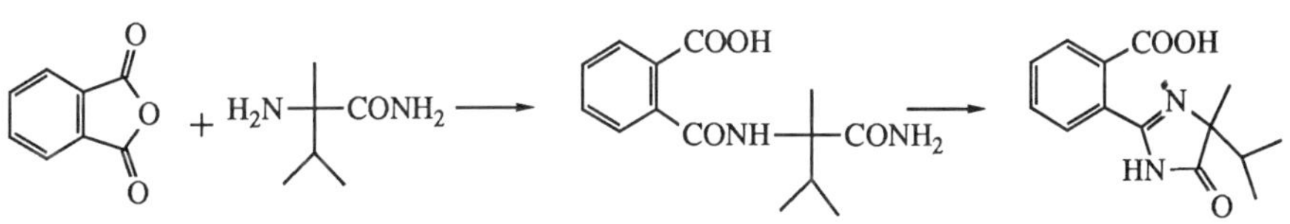

The first step can be conducted in a water-immiscible solvent, the product extracted into 5N sodium hydroxide, and the solution heated at $80^{\circ}C$ to effect the cyclization. This process requires, of course, an anhydride - not always readily accessible particularly for pyridine and quinoline analogs - and leads directly to the acid.

A less direct route to the acids used the tricyclic derivative. This method involved heating the tricyclic compound in a mixture of hydrochloric acid/dioxane, whereby the imidazolinone precipitated from the solution. At first glance this would appear to be a simple hydrolysis but some subsequent experiments suggested it might not be quite that straightforward.

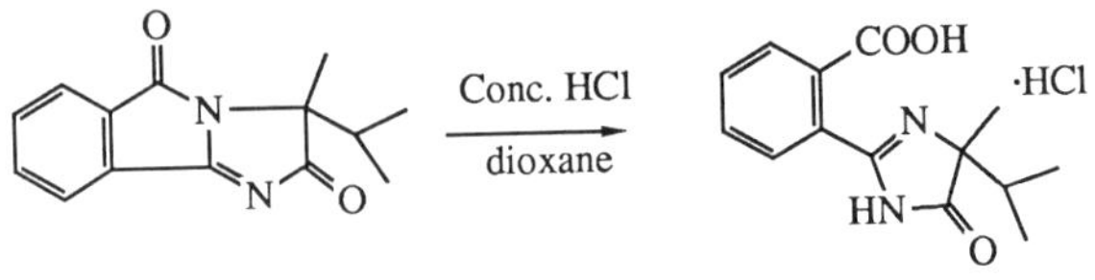

When, for example, a similar reaction was attempted in the pyridine series an unexpected reaction occurred. The tricyclic starting material was a mixture of isomers -85% had the carbonyl group at C3 and 15% at C2 of the pyridine ring.

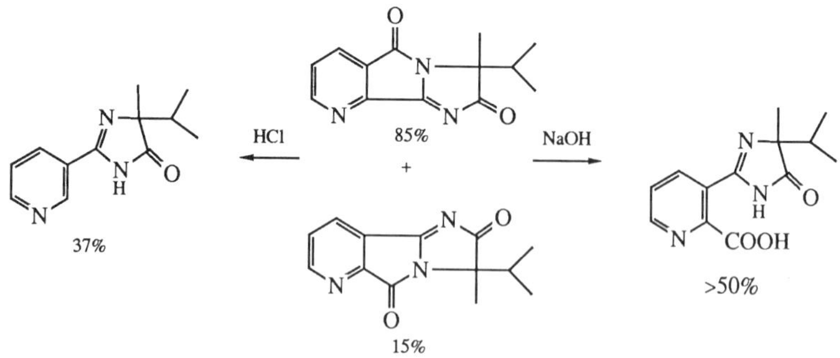

Starting with an unsymmetrical starting material had revealed that something unusual was going on. We interpret this result as follows:

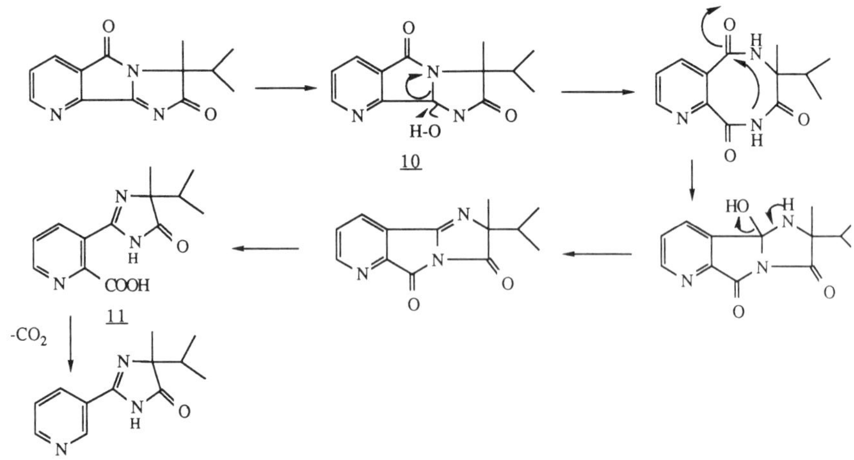

The formation of hydrates such as 10 is usual in this series and this type of product can be isolated. Formation of an 8-membered ring followed by a transanular reaction gives a different tricyclic compound which can now be hydrolyzed to give the picolinic acid 11. Decarboxylation of 11 results in the observed product. The reaction under basic conditions could proceed by the same mechanism but an alternative explanation is available. Ring opening of 10 could result in formation of the imide rather than the 8-membered ring. Hydrolysis at the 2-carbonyl group followed by base-catalyzed cylization would give the product 12.

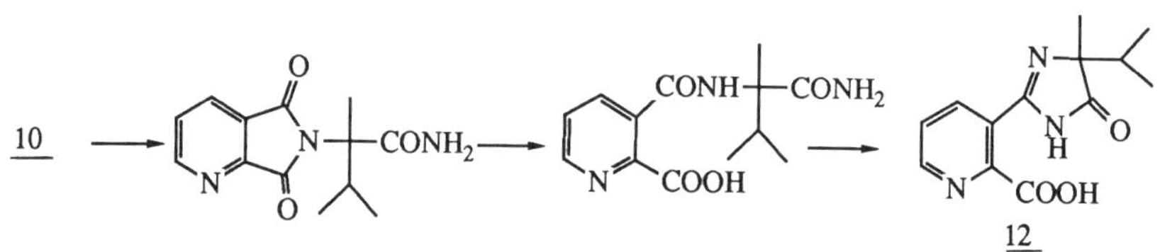

That the formation of an 8-membered ring is not at all difficult is illustrated by work reported by Schmid et al., (ref. 4) who reacted 3-dimethylamino-2,2-dimethyl-2H-azirine with phthalimide and saccharin. The structure of the product from saccharin was deduced by NMR and confirmed by an X-ray structure analysis. When the product from phthalimide was treated with 3N hydrochloric acid, there resulted an imidazolinone hydrochloride.

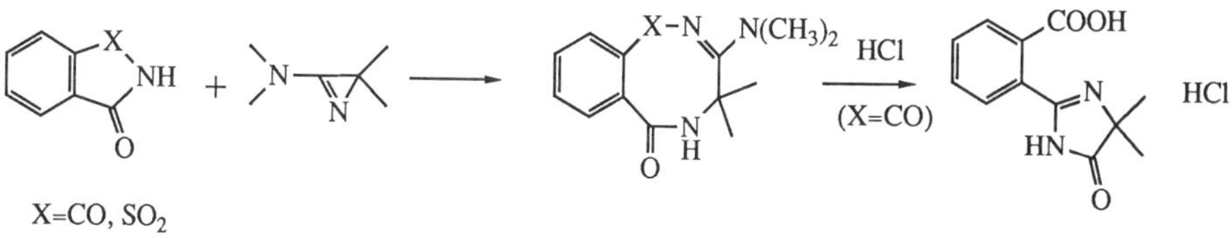

X=CO, SO_2

The objectives for the synthesis program were obvious - prepare a wide variety of arylimidazolinones to study the effect of structure on biological activity. Two obstacles to these goals were apparent. First, not many ortho-dicarboxylic acids were available, particularly in the pyridine and quinoline series and, secondly, there was a regiochemical requirement for optimum activity. Only the product which has the carboxyl group in the 3-position and the imidazolinone ring in the 2-position in the pyridine and quinoline series has high biological activity. The products from the 3,4-dicarboxylic acids have poor activity. The following sections will indicate how some solutions were found for both problems.

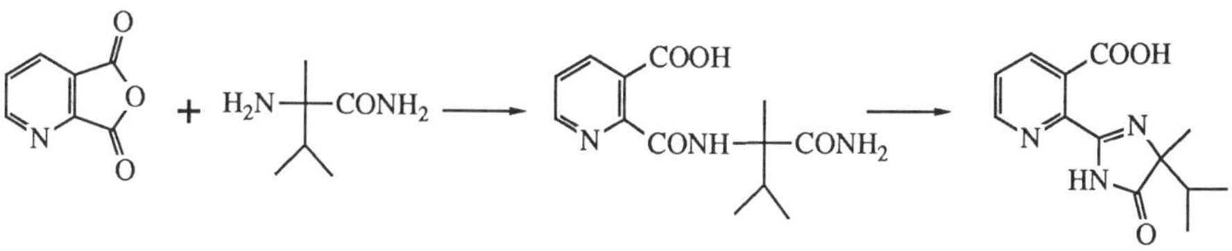

The reaction of the quinolinic anhydride with α-methylvalineamide gives a 3:1 ratio of nicotinic to picolinic acid products. This ratio can be improved considerably if the reaction is carried out in the presence of pyridine. An alternative solution was based on some Russian work (ref. 5). These workers described the reaction of diamide 13 with a mixture of phosphorus oxychloride and penta-

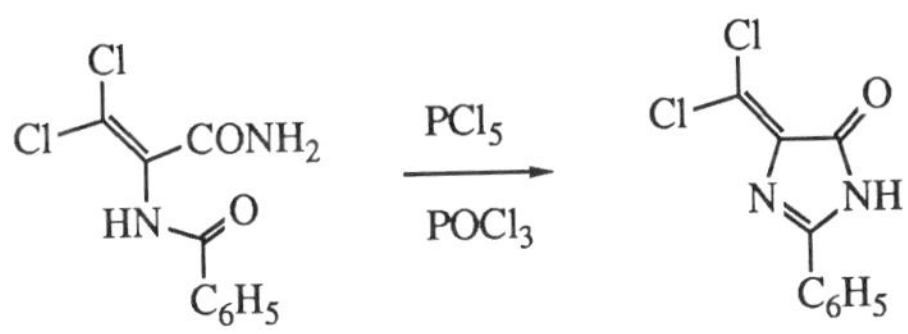

13

chloride. This reaction was the basis for a regioselective synthesis of the desired nicotinate. In this sequence, the imide 14 was reacted with methoxide to give the nicotinate 15 in over 85% yield. This could be the thermodynamic product formed via equilibration through the imide since amines are known to react primarily at the C-2 carbonyl group. Be that as it may, treatment of the ester 15 with phosphorus oxychloride and pentachloride gave a good yield of the desired imidazolinyl nicotinate. A much improved procedure was developed in which the ester 15 was added to a solution of phosphorus pentachloride alone in toluene. In this system the hydrochloride of the product separates from the solution thus facilitating the isolation procedure.

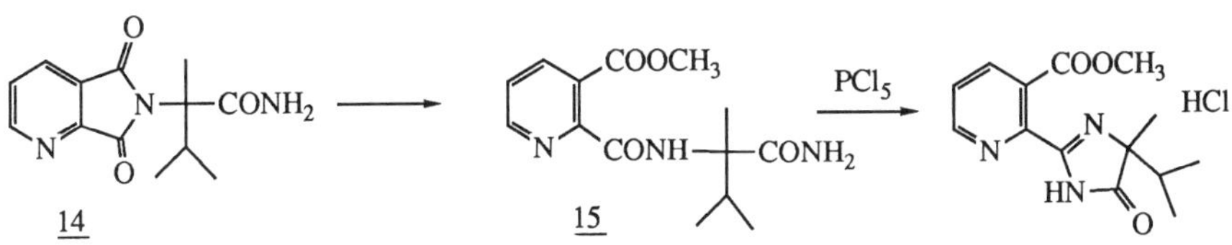

14 15

A procedure which circumvented the necessity for an o-dicarboxylic acid was developed and was invaluable for the synthesis of specific isomers as well as analogs in the pyridine series. Many groups have been developed which can stabilize and direct metallation to the ortho-position in an aromatic system (ref. 6). It was reasoned that the imidazolinone group might also serve this purpose. Picolinic acid was condensed with the aminoamide and the resultant diamide cyclized with base. It required

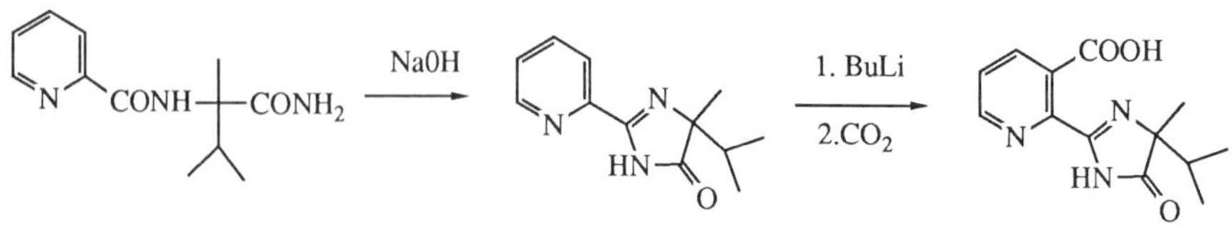

2.2 equivalents of n-butyllithium to generate the dianion which, when quenched with carbon dioxide, gave the desired nicotinic acid in good yield. What was important about this methodology was that it made possible the synthesis of a large variety of otherwise inaccessible analogs. Thus, a number of 4- and 5-substituted pyridine analogs were prepared; substituted benzene analogs could be prepared regiospecifically and functionality other then the carboxyl group could be introduced ortho-to the imidazolinone ring. Of particular interest may be the successful metallation of the alcohol 16 to a trianion and metallation of the 4-chloro pyridine 17. The product from 17 could then be modified by displacement of the chlorine by a variety of nucleophiles.

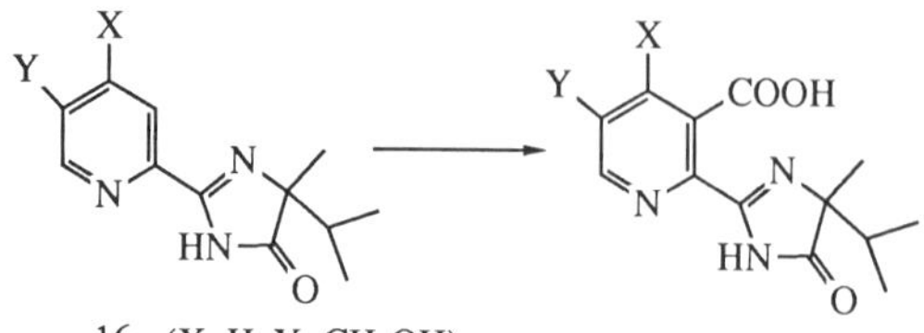

16 (X=H, Y=CH_2OH)
17 (X=Cl, Y=H)

One of the early targets in the synthesis program was the preparation of the 5-ethylpyridine analog. The reason for this was that 5-ethyl-2-methylpyridine is one of the most readily available pyridines. The 5-ethylpyridine analog proved to be an extremely active and selective herbicide and is currently under commercial development.

The initial synthesis was quite traditional. Condensation with benzaldehyde gave the olefin 18 in 75% yield. Ozonolysis of this material with oxidative workup then gave the desired picolinic acid albeit in poor yield. A review of the Willgerodt reaction by E. V. Brown in Synthesis (ref. 7) indicated that the Kindler modification worked quite well on 5-ethyl-2-methylpyridine to give, with aniline, a 65% yield of the thioamide 19. Acid hydrolysis afforded the picolinic acid. Two further steps were then required to give the key intermediate 20 needed for metallation. It was conceivable that the sequence could be shortened if the aniline could be replaced by 2-amino-2,3-dimethylbutyramide in the Willgerodt reaction. When the modification was tried, it was found that not only could this amine indeed replace aniline but that the reaction proceeded further to give the desired imidazolinone 20 in one step.

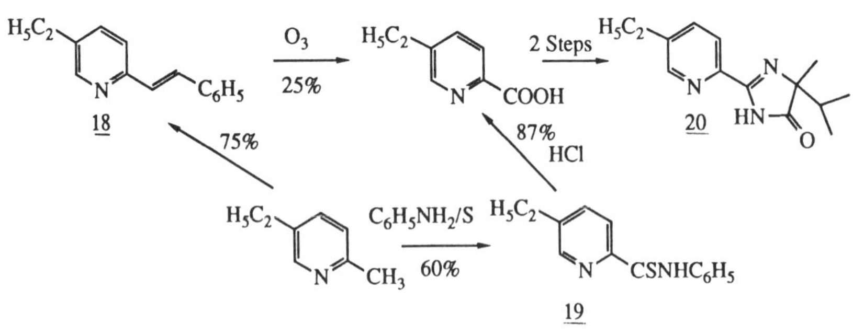

This reaction was examined in some detail. The highest yield of imidazolinone 20 was obtained when the ratio of pyridine to sulfur to amine was 5:3:1. Using this ratio and a temperature of 180-200°C, a yield of 70-75% could be realized. Two minor products were characterized. The first, a thiazolinone 21, presumably arises from the thioamide by expulsion of ammonia rather than hydrogen sulfide. The other product, 22, is the expected Willgerodt product from the pyridine and ammonia.

Conversion of the imidazolinone 20 to the nicotinic acid was achieved by the metallation-carboxylation procedure already described. It should be noted that LDA could not replace an alkyllithium in the metallation step. When LDA was used, metallation occurred at the ethyl group rather than in the ring. Those who follow the patent literature will have seen the number and variety of imidazolinones that we have prepared and tested. It is not possible to cover all these aspects.

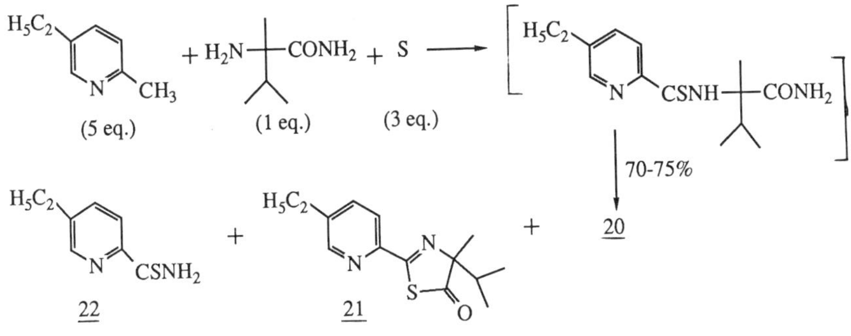

An area of considerable importance to those involved in bioactive compounds is that of prodrugs. This is particularly true in the field of herbicides where the all-important property of crop selectivity is frequently determined by the ability of the crop or the weed to carry out a chemical reaction which the other cannot perform or perform less efficiently. This can be, and often is, as simple as the hydrolysis of an ester. A reasonable prodrug for this series of herbicides might be the corresponding imidazolidinone although further data would be required to prove that it itself was not the active principle. These derivatives could be prepared most conveniently by the Borsh reduction

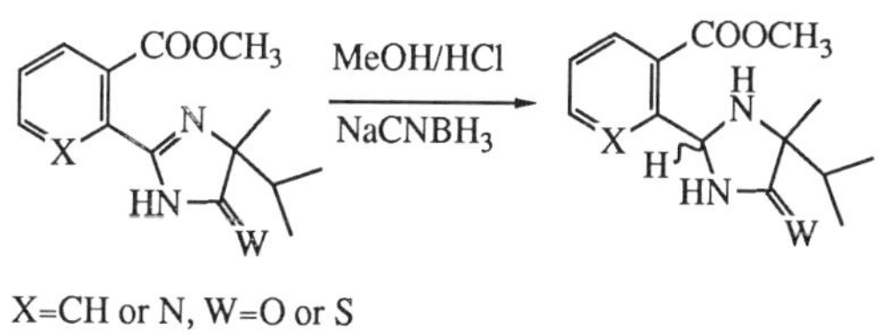

(ref. 8) of the imidazolinone ester. The dihydro products were purified by chromatography to give the cis- and trans-isomers in a ratio of about 3:1. This process did not work in all cases. An alternative procedure is illustrated for the quinoline analog. Once again, a mixture of stereoisomers

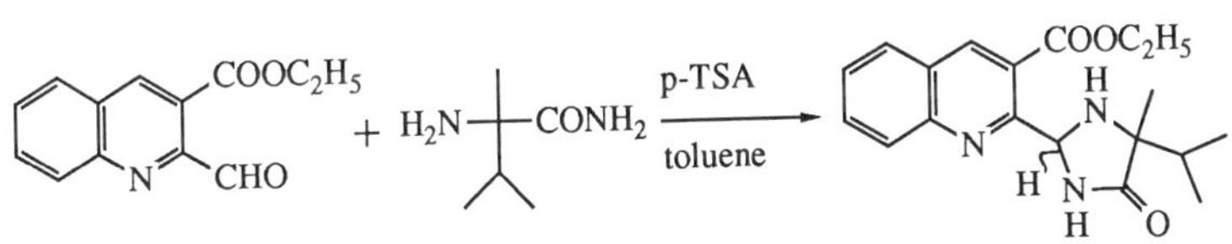

was formed. In other cases, only the Schiff base was formed under those conditions. It was then necessary to treat the Schiff base with acid, for example trifluoroacetic acid in methylene chloride, to effect cyclization. These dihydro esters could be hydrolyzed by base and the acids cyclized to give products of defined sterochemistry. Both the uncyclized and cyclized materials proved to be quite effective herbicides.

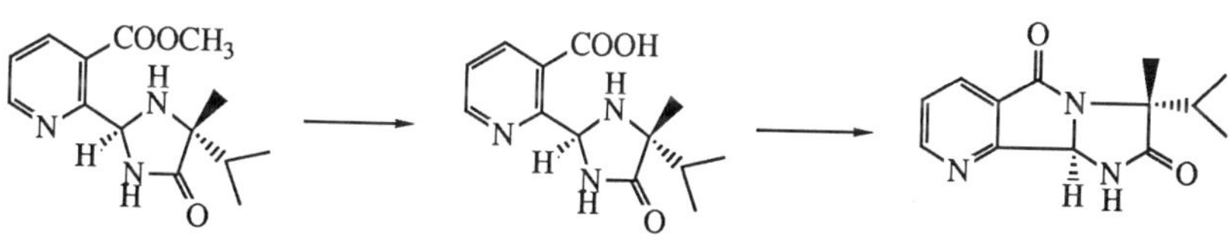

At the same time that the structure-activity work was progressing, the mode of action of these herbicides was investigated both at the whole plant and enzyme level by Shaner and Anderson (ref. 9). In the whole plant, treatment with these herbicides causes cessation of growth, death of the growing points followed by death of mature tissue and of the whole plant. A clue as to what actually happens at the enzyme level came from a study of the levels of individual amino acids within the amino acid pool of imidazolinone-treated corn cells grown in suspension culture. Most increased, with the most dramatic increases occurring in the levels of methionine, cysteine and serine. The levels of three amino acids, valine, leucine and isoleucine decreased. It was surmised that the imidazolinones inhibited the biosynthesis of these amino acids. Indeed, it was found that the herbicidal effects of the imidazolinones could be prevented by supplying these three amino acids to the plant. The first enzyme in the biosynthetic pathway to these amino acids is acetohydroxyacid synthase (AHAS) and the imidazolinones have been found to be potent inhibitors of this enzyme. It should be pointed out that this enzyme is not present in mammals. These imidazolinones are consequently essentially non-toxic to mammals.

With this information available, it now became possible to study the structure requirements for activity at the enzyme level. These studies have shown that the free carboxylic acid group is essential; ester and amides such as 23 are not active on the enzyme. Other examples of structures which do not inhibit the enzyme are the tricyclic derivatives 24 and 25 and the imidazolidinone 26.

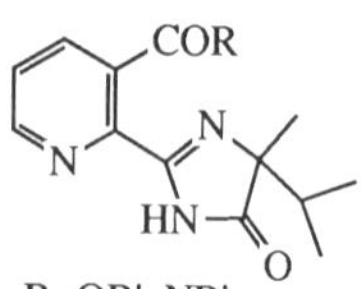

R=OR', NR'$_2$

23

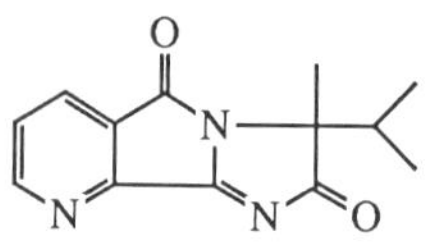

24

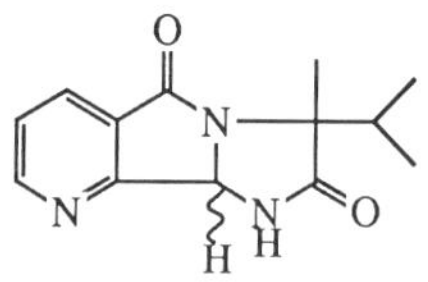

25

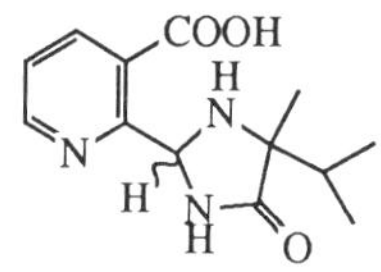

26

As indicated earlier, however, all these materials have considerable herbicidal activity and must therefore be metabolically activated to the inhibitory o-(imidazolin-2-yl)arylcarboxylic acid. An interesting observation has been made that the isomeric acids, 27 and 28 do not differ significantly in their ability to inhibit acetohydroxyacid synthase yet in the whole plant, the nicotinic acid 27 is very much more effective as a herbicide. The reasons for this are obscure at this time.

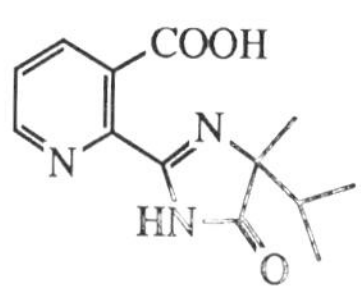

27

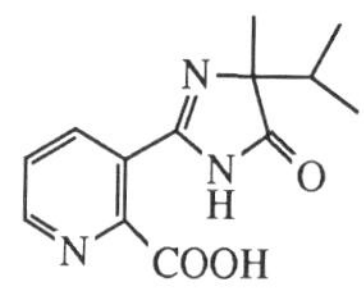

28

One further step has been taken in the development of the imidazolinones. A project sponsored by American Cyanamid at Molecular Genetics Inc. had as its target the development of a line of corn (maize) which was resistant to these herbicides. Dr. Anderson and his colleagues were able to select cells in corn cell culture which were resistant to the herbicide and regenerate a plant from this resistant line (ref. 10). Pollen from this plant was used to cross it with an inbred line and eventually, through backcrossing, a line of corn was generated which was homozygous for the imidazolinone resistant trait. These corn plants as well as the AHAS extracted from them are insensitive to high doses of the imidazolinones.

This project has thus reached a point at which several imidazolinones have been commercialized and others are at late stages of development An enormous body of information has been generated on the chemistry, structure-activity relationships, metabolism and mode of action. This work continues as does an increasing effort to unravel the molecular genetics involved in the biosynthesis of these essential amino acids and the regulation of the process.

ACKNOWLEDGEMENTS

It is a pleasure to acknowledge the many contributions made by members of the two Herbicide Synthesis Groups of the Organic Synthesis Section and the tremendous support given by the Herbicide Discovery Group in the form of testing data and mechanism of action information.

REFERENCES

1. M. Los, C.A. Kust, G. Lamb and R. E. Diehl, HortScience, 15, 22 (1986).
2. J. C. Suttle and D. R. Schreiner, J. Plant Growth Regul., 1, 139 (1982).
3. C. K. Sauers, C. L. Gould and E. S. Iannou, J. Org. Chem., 36, 1941 (1971).
4. S. Chaloupka, P. Vittorelli, H. Heimgartner, H. Schmid, H. Link, K. Bernauer and W. E. Oberhansli, Helv. Chim. Acta, 60, 2476 (1977).
5. B. S. Drach and G. N. Mis'kevich, J. Org. Chem. U.S.S.R., 14, 881 (1978).
6. H. W. Gschwend and H. R. Rodriguez, Organic Reactions, 26, 1 (1979).
7. E. V. Brown, Synthesis, 358 (1975).
8. R. F. Borch, M. D. Bernstein and H. D. Durst, J. Amer. Chem. Soc., 93, 2897 (1971).
9. D. L. Shaner, P. C. Anderson and M. A. Stidham, Plant Physiol., 76, 545 (1984).
10. D. L. Shaner and P. C. Anderson, Biotechnology in Plant Science, p. 287, Academic Press, New York (1985).

Structure and activity of herbicidal inhibitors of phytoene desaturase

Gerhard Sandmann

Lehrstuhl für Physiologie und Biochemie der Pflanzen, Universität Konstanz, 7750 Konstanz, W-Germany

Abstract - Phytoene desaturase, an enzyme in the biosynthetic pathway of carotenoids, is the target of herbicidal inhibitors with different chemical structures. Structure-activity investigations were performed with substituted m-phenoxybenzamides, o-phenoxynicotinamides, and 3'-phenylpyridazinones. The biological activities were assayed by determination of colored carotenoids in algal cells or with a cell-free carotenogenic system which covers the phytoene desaturase reaction. The values obtained were used for comparison of various derivatives and for QSAR analysis. The essential chemical structure established for inhibition of phytoene desaturase is the phenylketo part of the m-phenoxybenzamide molecule. The phenoxy moiety including its substituents increase inhibitory activity by their lipophilicity provided that the phenoxy ring is connected in meta position. The decisive physicochemical parameter of the N-alkyl chains of m-phenoxybenzamides is their chain lengths. The optimum was reached with C_5 in the cellular and C_4 in the cell-free assay. A more important difference between the results obtained with both systems was evident when m-phenoxybenzamides with branched N-alkyl chains were investigated. There was a steric effect of these substituents (described by STERIMOL parameter B_3 in the QSAR equation) in the in-vivo system which could not be observed with the cell-free phytoene desaturase assay. The structural requirements for inhibition of phytoene desaturase by nicotinamides differ in several respects from those of phenoxybenzamides: Nicotinamides are active with a phenoxy ring in ortho position; for high potency as herbicides they have to be N-substituted by a phenyl ring, whereas an additional methyl group on the amide nitrogen counteracts inhibition. In vitro data obtained with 3'-substituted phenylpyridazinones were also suitable for QSAR. The inhibitory activity depends on both the electronic properties and lipophilicity of the substituents in meta position on the phenyl ring.

PHYTOENE DESATURASE AS THE TARGET FOR HERBICIDAL INHIBITORS

Many commercial herbicides are directed against photosynthetic electron transfer reactions through photosystem II. Structure-activity relationships have been formulated for a great number of these herbicides (see refs. 1 and 2 for survey). Furthermore, phytoene desaturase was the first enzyme of a biosynthetic pathway in the chloroplast reported to be a target for herbicides (ref. 3). Since the early seventies numerous herbicidal inhibitors of different chemical structure have been shown to interfere with phytoene desaturase (see ref. 4 for review). As far as it has been determined the type of inhibition is non-competitive or irreversible. Commercially developed herbicides are norflurazon (a phenylpyridazinone), fluridone (a pyridinone), methoxyphenone (a benzophenone), difunone (a phenylfuranone), flurochloridone (a pyrrolidinone), and diflufenican (a nicotinamide). Structure-activity investigations were mainly performed with m-phenoxybenzamides, nicotinamides, and phenylpyridazinones. The immediate goal of these studies was the characterization of the important substituents and essential structures of the inhibitor molecules necessary for inhibitory action on phytoene desaturase. In the long-term information obtained on the structural requirements of inhibitors should help to probe the active and/or inhibition site of phytoene desaturase.

For studies on inhibitors of phytoene desaturase either cellular or cell-free assay systems can be employed. In the first case, herbicides are added to liquid cultures of unicellular green or blue-green algae and after about 2 days decreased formation of colored carotenoids which are a consequence of the phytoene desaturase reaction, can be measured together with the concurrent accumulation of colorless phytoene. The in-vitro assay (ref. 5) converts

^{14}C-labeled geranylgeranyl pyrophosphate into phytoene which is then metabolised further into lycopene and ß-carotene. The ratio of ^{14}C incorporation into phytoene versus ß-carotene is a suitable indicator for inhibition of phytoene desaturase by an added chemical, and can be used for QSAR analysis.

m-PHENOXYBENZAMIDES

From a structural point of view every inhibitor of an enzymic reaction needs an essential molecular arrangement which fits into the inhibition site. Furthermore, the inhibitory potency is modulated by additional substituents. With respect to m-phenoxybenzamides it appears that the phenoxy moiety is not obligatory for inhibition of phytoene desaturase, because this group can be replaced i.e. by CF_3 (ref. 5). Nevertheless, the phenoxy group contributes to inhibitory activity of the molecule provided that it is connected in the meta position. Various substituents on the phenoxy ring show that lipophilicity is the decisive feature.

Fig. 1 shows the chemical modification of the active 3-phenoxy-N-butyl benzamide and the resulting effect on cell-free inhibition of phytoene desaturase. The numbers represent the inhibition ratio IR (= ratio of ^{14}C incorporated into phytoene to ^{14}C-labeled ß-carotene as referred to the same ratio of the untreated control) when the chemicals were added to a final concentration of 50 µM. Some of these compounds can only be assayed in vitro because penetration into plant cells is very poor. Substitution of the phenoxy group by CF_3 increased the inhibitory effect of the resulting compound. However, additional removal of the phenyl ring leaves an inactive acetamide with an IR of 1.2 (a value of 1.0 means no inhibition). In another allied investigation the amide group was replaced either by a ketone or by an ester moiety. Both modifications decreased the inhibitory action to some extent. Nevertheless, substantial inhibition remained. Further cleavage of the ester led to an inactive phenoxybenzoic acid. The sulfonamide in Fig. 1 showed much lower inhibition than the comparable benzamide.

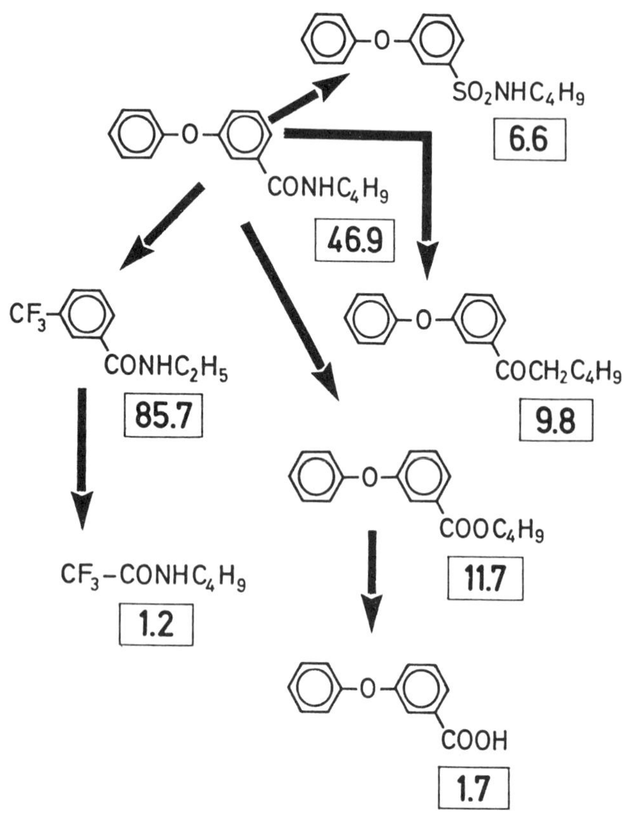

Fig. 1. Inhibition of in-vitro phytoene desaturase by systematically modified m-phenoxybenzamides. The numbers (= IR) indicate the degree of inhibition.

The results of Fig. 1 indicate that the structural part in brackets is essential for inhibition of phytoene desaturase by phenoxybenzamides but it can be replaced to some extent by a corresponding sulfonic acid derivative.

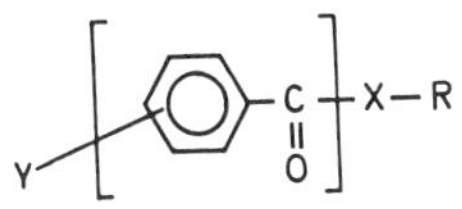

In addition, X resembles NH, O or CH_2. Obviously, electronic properties of these groups determine the inhibitory action. Substituents which lower the positive partial charge on the carbonyl carbon enhanced the inhibitory effect.

R stands for alkyl chains. Increase of the chain lengths up to C_5 intensifies the inhibitory activity in the cellular assay (Fig. 2). Longer alkyl chains, i.e. compounds with up to C_{12} chains that were available, resulted in decreased pI_{50} values. Thus, we obtain a parabolic curve when pI_{50} is plotted against carbon number, Verloop's length parameter L, or lipophilicity π. In order to discriminate between the effects of the latter two physico-chemical parameters, phenoxybenzamides in which a CH_2 group in the alkyl chain was replaced by an ether bridge were investigated. Both compounds possess lower π values but approximately the same length, L, compared to the corresponding alkyl compounds. pI_{50} values for the ethoxy derivatives closely fit on the curve for unbranched alkyl chains in the plot versus L of Fig. 2A. However, in the pI_{50} versus π diagram of Fig. 2B the data points do not match the descending branch of the curve. Therefore, we can conclude that the chain length of the alkyl chain rather than lipophilicity is important for the phenoxybenzamide interaction with phytoene desaturase.

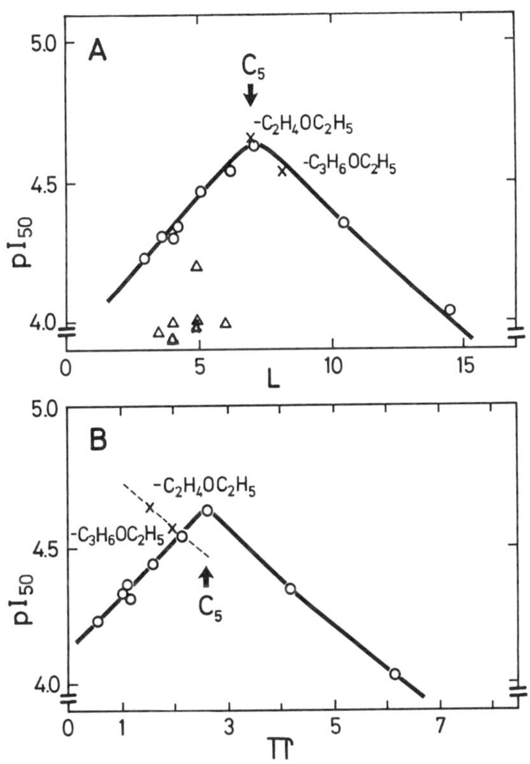

Fig. 2. Plot of pI_{50} for cellular inhibition of carotenoid biosynthesis versus STERIMOL parameter L (A) or lipophilicity π (B) of branched (-Δ-), unbranched (-o-), or unbranched ethoxy (-x-) N-alkyl m-phenoxybenzamides.

Fig. 2A includes *m*-phenoxybenzamides with branched alkyl chains (-Δ-). The pI_{50} values were always lower than for the corresponding n-alkyl derivative. The following QSAR relationship between pI_{50} and alkyl chain length was obtained for n-alkyl *m*-phenoxybenzamides:

$$pI_{50} = 0.196\ (\pm\ 0.041)\ L - 0.013\ (\pm\ 0.002)\ L^2 + 3.759 \qquad (1)$$
$$n = 9,\ r = 0.92,\ s = 0.08,\ F = 16.16$$

When the data with branched alkyl *m*-phenoxybenzamides were included into the QSAR analysis, the STERIMOL parameter B_3 gave the best description of the additional steric effect on inhibition of phytoene desaturase:

$$pI_{50} = 0.187\ (\pm\ 0.049)\ L - 0.012\ (\pm\ 0.003)\ L^2 - 0.364\ (\pm\ 0.047)\ B_3 + 4.479 \qquad (2)$$
$$n = 16,\ r = 0.93,\ s = 0.1,\ F = 25.59$$

QSAR analysis was also performed with the inhibition data for branched and unbranched N-alkyl phenoxybenzamides obtained from in-vitro inhibition of phytoene desaturase. IR for 20 µM concentrations showed correlation with L similar to cellular I_{50} values:

$$\lg IR = 0.553\ (\pm\ 0.079)\ L - 0.034\ (\pm\ 0.003)\ L^2 - 0.839 \qquad (3)$$
$$n = 16,\ r = 0.90,\ s = 0.17,\ F = 27.76$$

In contrast to Eq (2), a significant contribution of STERIMOL parameter B_3 for branched alkyl chains could not be observed. In our opinion, this discrepancy is caused by a swelling of the phytoene desaturase-containing membranes during the preparation procedures leading to an alteration of the environment of the target site.

TABLE 1. Inhibition of carotenoid biosynthesis by selected N-substituted *m*-phenoxybenzamides

		I_{50} value
1a	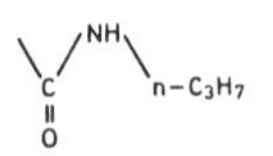	34
1b	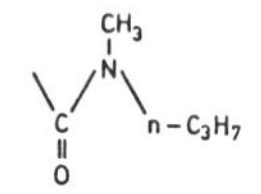	35
2a	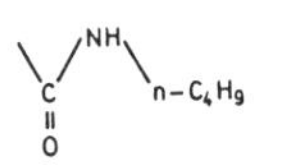	29
2b	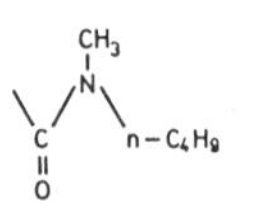	26
3a	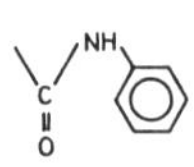	79 (87)*
3b	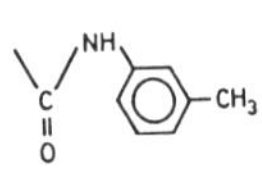	95
3c	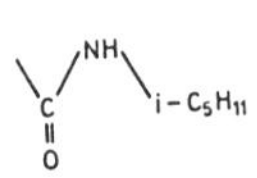	108

*calculated from Eq. (2)

The effect of an additional methyl group on the amide nitrogen was determined (Table 1), and the cellular I_{50} values compared to those of the corresponding amides. For the n-propyl,methyl (1b) as well as for the n-butyl,methyl derivative (2b) the I_{50} values were very close to those of compounds 1a and 2a. The cell-free assay of these compounds gave similar results. Obviously, branching on the amide nitrogen has no negative effect on inhibitory activity of m-phenoxybenzamides. A possible contribution of electron properties of amid substituents was investigated with compounds 3a and b. I_{50} values were determined for N-phenyl and N-tolyl phenoxybenzamides. The former compound (3a) showed better inhibition than the latter (3b) which carries an additional methyl on the phenyl ring in m-position. The I_{50} value calculated from Eq (2) given in parantheses is only slightly higher than the experimental value. From all available aliphatic N-substituted phenoxybenzamides the N-iso-pentyl derivative (3c) has L and B_3 parameters comparable to the N-tolyl or N-benzyl derivatives. Inhibition of all three compounds is in the same range. So, if an electronic contribution of N-substituents is substantial, it is very small and difficult to quantitate either with the cellular or the cell-free assay.

o-PHENOXYNICOTINAMIDES

o-Phenoxynicotinamides inhibit the phytoene desaturase reaction in a non-competitive way similar to the m-phenoxybenzamides. They possess a higher affinity for phytoene desaturase, their I_{50} and k_i values are at least a magnitude of order lower than those for m-phenoxybenzamides. A recently developed herbicide is diflufenican (N-(2,4-difluoro-phenyl)-2-(3-trifluoromethylphenoxy) nicotinamide).

The phenoxynicotinamides show some structural similarities to the phenoxybenzamides. In contrast to the latter, however, the nicotinamides exhibit very high inhibition as o-phenoxy derivatives. In addition, substituent effects on inhibitory activity are not the same as for m-phenoxybenzamides. Table 2 presents some of these differences. The nicotinamide with p-fluorophenyl substitution (compound 1) showed a high IR when added in a 5 µM concentration to the cell-free assay. An additional methyl substituent on the nitrogen strongly decreased inhibition. An even stronger decrease of inhibition of cell-free phytoene desaturation was observed when an ethylene group was inserted between the nitrogen and the phenyl moiety. Since the ortho derivatives of phenoxynicotinamides are active, branching on the amide nitrogen decrease inhibitory properties, and electronic contribution of the N-substituent increases inhibition activity, it is very doubtful whether the essential structure developed for m-phenoxybenzamides from Fig. 1 can also be assigned to the o-phenoxynicotinamides.

2-PHENYLPYRIDAZINONES

Norflurazon (4-chloro-5-methylamino-2-(3'-trifluoromethylphenyl)-pyridazinon-3(2H)one) was the first commercially important herbicide directed against phytoene desaturase. The inhibition properties of norflurazon and other related 3'-phenylpyridazinones are similar to previously described m-phenoxybenzamides and o-phenoxynicotinamides. Comparison of inhibitory activity of phenylpyridazinones substituted at different positions of the heterocycle indicated that substituents at carbon 4 enhance inhibition with increasing σ values (ref. 6). At position 5, however, substituents with low σ values are favorable.

TABLE 2. Inhibition of phytoene desaturase activity by substituted o-phenoxynicotinamides

	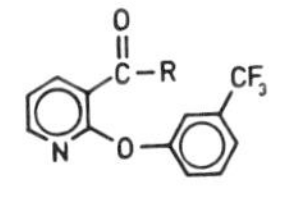	IR (5 µM)
1.	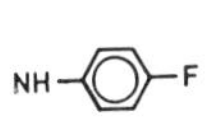	84.2
2.	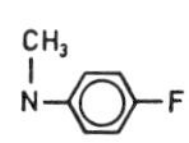	2.7
3.	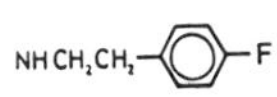	1.4

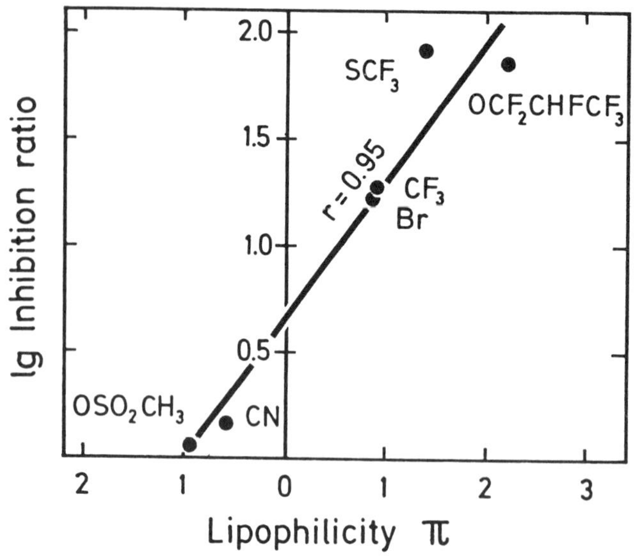

Fig. 3. Correlation of cell-free inhibition of phytoene desaturase by m-substituted phenylpyridazinones with lipophilicity parameter π.

Phenylpyridazinones substituted at the meta instead of the para position show higher inhibition in algal cells. Therefore, several derivatives with different meta substituents of the phenyl ring have been subjected to QSAR analysis (ref. 7) involving π and σ:

$$pI_{50} = 0.90\ (\pm\ 0.39)\pi + 3.57\ (\pm\ 1.19)\sigma_m + 5.25 \qquad (4)$$
$$n = 11,\ r = 0.91,\ s = 0.52,\ F = 19.2$$

Obviously, electronic properties of the substituent exert a very strong effect of inhibitory activity, which can be counteracted by lack of lipophilicity. Steric factors did not show an influence.

Fig. 3 shows that also cell-free inhibition data obtained with m-substituted phenylpyridazinones are suitable for QSAR. The different substituents of the five compounds in Fig. 3 were chosen in a way that their electronic properties were more or less constant; values range from 0.35 to 0.56. Therefore, a highly correlated linear relationship could be established between lg IR and π in accordance with Eq (4). This is another example that IR is a convenient in-vitro parameter and that cellular and cell-free inhibition of phytoene desaturase is comparable.

ACKNOWLEDGEMENTS

The author is grateful to Drs. U. Schirmer, Ludwigshafen, Germany, and P. Böger, Universität Konstanz, for their valuable advice and suggestions during the course of this work. Due thanks are expressed to the chemical companies which synthesized and supplied most of the compounds used in our studies, especially to BASF, Ludwigshafen, Germany; Celamerck, Ingelheim, Germany; and May & Baker Ltd., Daggenham, U.K.

REFERENCES

1. H. Iwamura and T. Fujita, J. Pesticide Sci. 7 , 289-299 (1982).
2. E. Kakkis, V.C. Palmire, C.D. Strong, W. Bertsch, C. Hansch and U. Schirmer, J. Agric. Food Chem. 320 , 133-143 (1984).
3. P.G. Bartels and A. Hyde, Plant Physiol. 45 , 807-810 (1970).
4. G. Sandmann and P. Böger, In: Biochemical Responses Induced by Herbicides (J.B. St. John and F.D. Hess, eds.), ACS Symposium Series No. 181, p. 112-130, Washington DC. (1982).
5. I.E. Clarke, G. Sandmann, P.M. Bramley and P. Böger, Pest. Biochem. Physiol. 23 , 335-340 (1985).
6. G. Sandmann, K.J. Kunert and P. Böger, Pest Biochem. Physiol. 15 , 288-293 (1981).
7. G. Sandmann and P. Böger, Z. Naturforsch. 37c , 1092-1094 (1982).

Asymmetric synthesis of cyclopropanecarboxylic acids. Absolute configuration of 2,3-cis-disubstituted derivatives

T. Aratani*, H. Yoshihara, T. Ageha, Y. Inukai and K. Yanagi

Takatsuki Research Laboratory, Sumitomo Chemical Co., Ltd.
Takatsuki, Osaka 569, Japan

Abstract-- Reaction of alkyl diazoacetate with an olefin in the presence of a chiral copper catalyst gives an optically active cyclopropanecarboxylate. This reaction was successfully applied to the synthesis of cyclopropanecarboxylic acids of practical value: 1R-trans-chrysanthemic acid, 1R-cis-permethrinic acid and 1S-2,2-dimethylcyclopropanecarboxylic acid. An e.e. of more than 90% was achieved when a copper catalyst derived from an optically active α-amino acid was employed.
Other cyclopropanecarboxylic acids of the known absolute configuration were prepared to reveal stereochemistry of the asymmetric synthesis. Cyclic olefins including indene, 1,3-cyclohexadiene, 1,3,5-cycloheptatriene, 1,3-cyclooctadiene and dihydropyran gave the corresponding 2,3-cis-disubstituted cyclopropanecarboxylic acids in optically active forms. The absolute configuration was assigned on the basis of an empirical rule and confirmed by X-ray crystallography.

ASYMMETRIC SYNTHESIS OF CHRYSANTHEMIC ACID AND INVENTION OF CHIRAL COPPER CATALYST

Chrysanthemic acid (**2:** R = H) is an acid component of pyrethroid, an insecticide of low mammalian toxicity (ref. 1). Among the four optical isomers, the most effective isomer is shown to be (+)1R-trans isomer, which is followed by (+)1R-cis isomer. 1S-trans and 1S-cis isomers are almost ineffective (ref. 2). The natural chrysanthemic acid also has the 1R-trans configuration.
In 1966, a novel synthesis of chiral cyclopropane derivatives was reported by Professor Nozaki and his coworkers (ref. 3). They reacted ethyl diazoacetate with styrene in the presence of a chiral copper catalyst to give cis- and trans-2-phenyl-cyclopropanecarboxylate in an optically active form. We applied this reaction to the synthesis of chrysanthemic acid: reaction of alkyl diazoacetate with 2,5-dimethyl-2,4-hexadiene in the presence of a chiral catalyst (Eq 1).

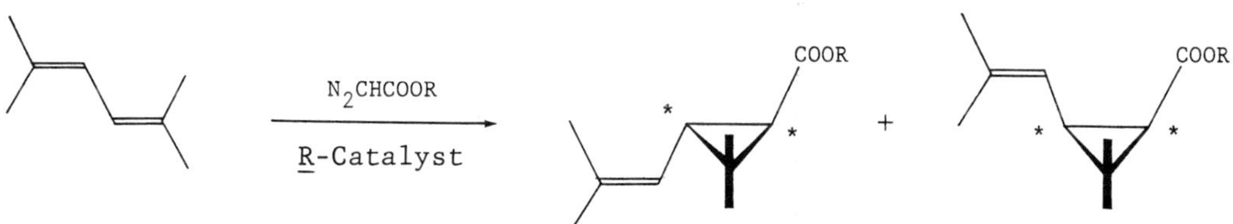

(1)

1 1R-trans-2 1R-cis-2

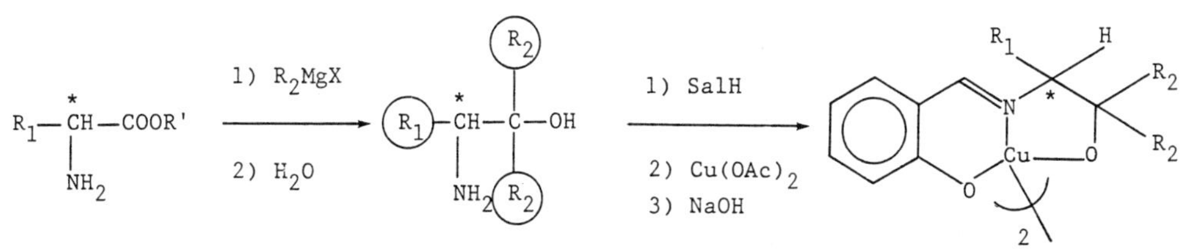

(2)

R-3 R-4 R-5 (Catalyst)

An extensive search for the catalyst was carried out to achieve high e.e. of the product (ref. 4). A series of effective catalysts were prepared as follows (Eq 2). An optically active α-amino ester was reacted with an excess of Grignard reagent to give an amino alcohol, whose Schiff base with salicylaldehyde was treated with cupric acetate to afford a copper complex (**5**). No racemization was detected in these transformations (ref. 5). The binuclear structure of the complex **5**, in which the Schiff base was incorporated as a tridentate ligand, was established by X-ray crystallography (ref. 6).

The complex **5** has two substituents R_1 and R_2. R_1 comes from the α-amino acid and R_2 comes from the Grignard reagent. A variety of copper catalysts were prepared (Eq 2) and examined in the chrysanthemate synthesis (Eq 1). The e.e. of the product was found to increase with the bulkiness of the substituent R_2. An e.e. of 70% was achieved when R_1 was methyl, and R_2 was 5-*t*-butyl-2-octyloxy-phenyl.

R-Amino acid gave the *R*-catalyst, which, in turn, favored the formation of 1*R*-*trans* and 1*R*-*cis* chrysanthemic acid. Ethyl diazoacetate produced nearly equal amount of *cis* and *trans* isomers. Exchange of ethyl group of diazoacetate to bulkier alkyl groups brought about a remarkable improvement in *cis*/*trans* ratio and e.e. (ref. 4). The selected alkyl diazoacetate was *l*-menthyl diazoacetate, when *cis*/*trans* ratio was 7/93 and e.e. of the *trans* isomer was 94%.

ASYMMETRIC SYNTHESIS OF PERMETHRINIC ACID AND 2,2-DIMETHYLCYCLOPROPANECARBOXYLIC ACIDS

Permethrinic acid (**8**) is another acid component producing pyrethroids of higher potency and photostability (ref. 7). The most effective isomer is shown to be 1*R*-*cis* isomer. A stereoselective route to this enantiomer was explored (Eq 3).
Ethyl diazoacetate was reacted with 2-methyl-5,5,5-trichloro-2-pentene in the presence of the catalyst of *S* configuration to give the cyclopropanated product **7**, whose dehydrochlorination and hydrolysis with potassium hydroxide furnished permethrinic acid of the desired configuration. The *cis*/*trans* ratio was 85/15 and e.e. of the *cis* isomer was 91% (ref. 8).

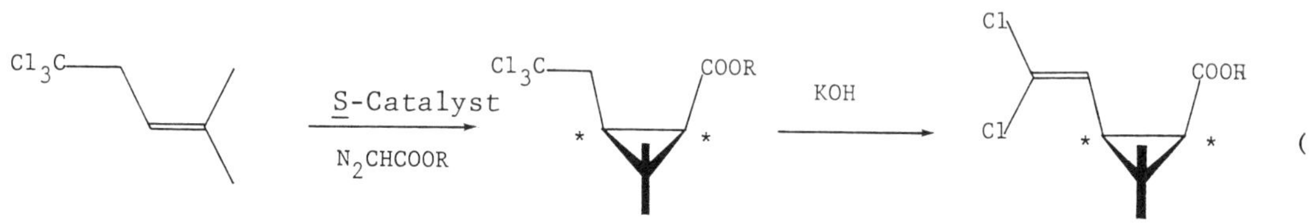

6 7 1*R*-*cis*-8 (3)

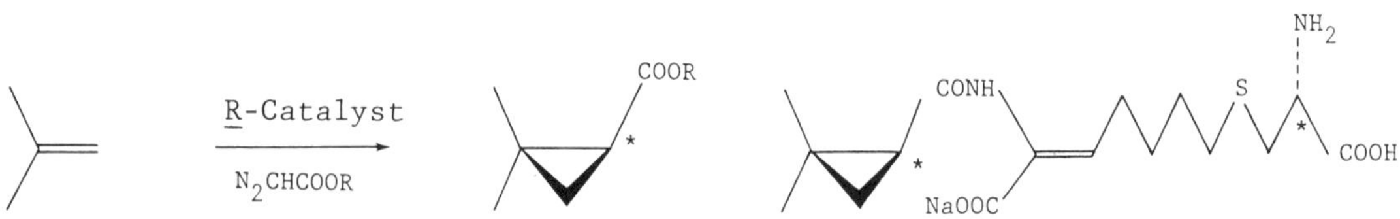

9 1*S*-10 11 (Cilastatin) (4)

The third compound of practical importance came from pharmaceutical field. (+)1*S*-2,2-Dimethyl-cyclopropanecarboxylic acid (**10**: R = H) is a key intermediate in the production of Cilastatin (**11**), a reversible inhibitor for a kidney enzyme, which hydrolyzes a thienamycine derivative, Imipenem. The new antibiotic composite, Imipenem-Cilastatin, has been developed by Merck Sharp and Dohme (ref. 9).

Ethyl diazoacetate was reacted with isobutylene in the presence of the *R*-catalyst to give ethyl ester **10** of 92% e.e. This process works well in a factory scale (ref. 10). Other examples of asymmetric catalysis in industry include *L*-DOPA synthesis by Monsanto (ref. 11) and *l*-menthol production by Takasago Perfumery (ref. 12).

ABSOLUTE CONFIGURATION OF 2,3-cis-DISUBSTITUTED CYCLOPROPANE-1-CARBOXYLIC ACIDS

In the same way, a number of cyclopropanecarboxylic acids of known absolute configuration (ref. 13) were prepared to investigate the mode of chirality-transfer from the catalyst to the products. It was found that chirality of the product is determined by electronic rather than steric factors of the olefins. The empirical rule deduced is shown in Eq 5 (ref. 10).

(1) The prochiral double bond of an olefin is placed on a plane so that the two sp-squared carbon atoms, the carbon, α, capable of production of the more stabilized carbonium ion comes on the left side and the other carbon, β, comes on the right side. There are two possible placements (**12A** and **12B**).
(2) In the presence of the R-catalyst, carbene derived from alkyl diazoacetate prefers to attack the double bond from the front side of the plane with alkoxycarbonyl group downward. The arrangement of **12A** gives the trans product (**13A**) and that of **12B** gives the cis product (**13B**).
(3) Since the arrangements of **12A** and **12B** are mirror images of each other, the preferred products, **13A** and **13B**, have the same absolute configuration at the carbene carbon atom (ref. 14).

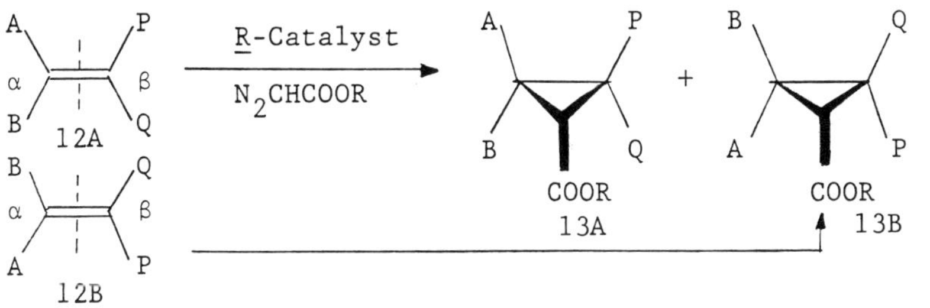

(5)

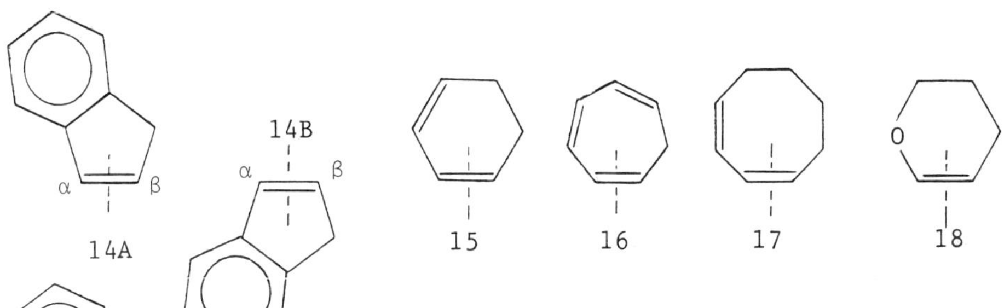

(6)

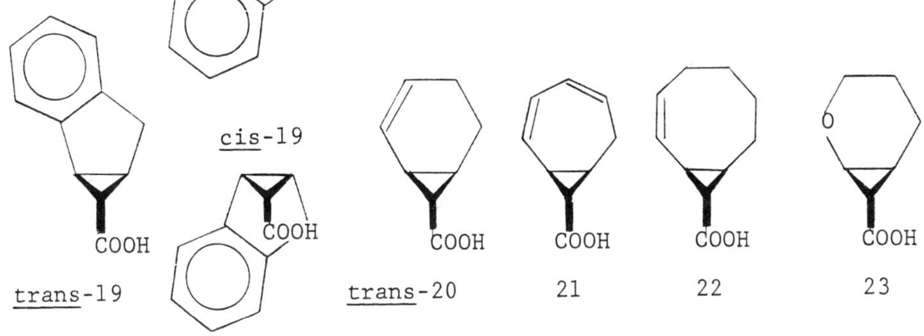

(7)

This rule was applied to the synthesis of 2,3-cis-disubstituted cyclopropane-carboxylic acids. The absolute configuration was predicted as follows.
(1) Cyclic cis-olefins (**14-18**) are placed on a plane as shown in Eq 6. These placements are based on the fact that benzylic, allylic and oxygen-substituted carbocations are more stable than secondary carbonium ions.
(2) In the presence of the R-catalyst, alkoxycarbonylcarbene will attack the double bond from the front side of the plane to give the trans products (**19-23**) of the S configuration at the carbene carbon.
(3) The trans products are accompanied by the cis isomers of the same configuration at the carbene carbon (e.g. 1S-cis-**19** and so on).

Experiments were carried out as follows. Ethyl diazoacetate was reacted with an olefin (**14-18**) in the presence of the R-catalyst to give a mixture of cis- and trans-cyclopropanecarboxylate. Partial hydrolysis of the ester with potassium hydroxide produced trans acid and unreacted cis ester (ref. 15). Optical rotations are summarized in Table 1.

TABLE 1. Asymmetric synthesis of 2,3-cis-disubstituted cyclopropanecarboxylic acid[a]

Olefin	Ester mixture cis/trans	$[\alpha]_D$	cis-Ester $[\alpha]_D$	trans-Acid $[\alpha]_D$
Indene	37/63	(+)48.8[b]	(-)131[c]	(+)178[c]
1,3-Cyclohexadiene	22/78	(+)16.9	(-) 27.9	(+)112
1,3,5-Cycloheptatriene	22/78	(+)53.6	(-)120	(+)188
1,3-Cyclooctadiene	30/70	(-)16.5	(-) 22.4	(-) 4.7
Dihydropyran	6/94	(+) 1.0	(-) 26.3	(+) 5.0

[a]Specific rotations $[\alpha]_D$ were measured in neat for esters and in chloroform solutions for acids.
[b]The e.e. was determined by GC as l-menthyl ester: 89% for the cis isomer and 61% for the trans isomer.
[c]Optically pure acid showed $[\alpha]_D$ (-)293 for the cis isomer and $[\alpha]_D$ (+)323 for the trans isomer (Cf. Ref. 16).

The predictions coupled with the optical data lead the following assignments (ref. 17). Among them, the assignment for trans-**19**, trans-**20** and trans-**22** was confirmed by X-ray crystallography (ref. 18).

(+)1S-trans and (-)1S-cis Tetrahydrocycloprop[a]indene-1-carboxylic acid (**19**)
(+)7S-trans and (-)7S-cis Bicyclo[4.1.0]hept-2-ene-7-carboxylic acid (**20**)
(+)8S-trans and (-)8S-cis Bicyclo[5.1.0]octa-2,4-diene-8-carboxylic acid (**21**)
(-)9S-trans and (-)9S-cis Bicyclo[6.1.0]non-2-ene-9-carboxylic acid (**22**)
(+)7S-trans and (-)7S-cis 2-Oxa-bicyclo[4.1.0]heptane-7-carboxylic acid (**23**)

REFERENCES AND NOTES

1. M. Matsui and I. Yamamoto, Naturally Occurring Insecticides (Edited by M. Jacobson and D.G. Crosby), pp 3-70, Marcel Dekker (1971).
2. For optical resolution and racemization of chrysanthemic acid, see G. Suzukamo, M. Fukao and T. Nagase, Chem. Letters 1799-1802 (1984).
3. H. Nozaki, S. Moriuti, H. Takaya and R. Noyori, Tetrahedron Letters 5239-5242 (1966); Eidem, Tetrahedron 24, 3655-3669 (1968).
4. T. Aratani, Y. Yoneyoshi and T. Nagase, Tetrahedron Letters 1707-1710 (1975); Eidem, Ibid. 2599-2602 (1977).
5. Unpublished results in our laboratory.
6. K. Yanagi and M. Minobe, to be submitted.
7. M. Elliott and N.F. Janes, Chem. Soc. Rev. 7, 473-505 (1978).
8. T. Aratani, Y. Yoneyoshi and T. Nagase, Tetrahedron Letters 23, 685-688 (1982).
9. Chem. Eng. News, December 16, p 18 (1985).
10. T. Aratani, Pure Appl. Chem. 57, 1839-1844 (1985).
11. W.S. Knowles, Acc. Chem. Res. 16, 106-112 (1983).
12. K. Tani, T. Yamagata, S. Akutagawa, H. Kumobayashi, T. Taketomi, H. Takaya, A. Miyashita, R. Noyori and S. Otsuka, J. Am. Chem. Soc. 106, 5208-5217 (1984).
13. W. Klyne and J. Buckingham, Atlas of Stereochemistry, Chapman and Hall (1974).
14. In the presence of the S-catalyst, the placement of olefin is inverted: the α-carbon comes on the right side and the β-carbon on the left side. The same attack of carbene from the front side gives the trans and cis product of the enantiomeric configuration at the carbene carbon.
15. V. Dave and E.W. Warnhoff, Organic Reactions (Edited by W.G. Dauben), Vol. 18, pp 217-401, John-Wiley (1970).
16. Base catalyzed epimerization of (-)-cis-**19** at the C-1 position gave (-)-trans-**19**: S.S. Hixson and J. Borovsky, J. Am. Chem. Soc. 98, 2840-2847 (1976).
17. According to the Brewster's method, molecular rotation $[M]_D$ is calculated to be (+)120 for 1S-trans-**19** and 0 for 1S-cis-**19**.
18. K. Yanagi, M. Minobe and T. Aratani, Acta Cryst. C42, 745-747 (1986); K. Yanagi and T. Aratani, Ibid. to be submitted.

Aldol reactions of triazolyl-acetophenones

D A Griffin

ICI Plant Protection Division, Jealott's Hill Research Station, Bracknell
Berks, RG12 6EY, UK

Abstract - Aldols 1A and 1B are fungicides and plant growth retardants. Both diastereomers display the biological activity, but in general one isomer is more active. The relative stereochemistry of the two isomers has been determined by a combination of chemical derivatisation and observation of nuclear Overhauser effects. The more active isomer was shown to have the 2RS, 3RS configuration. The diastereoselectivity observed in the preparation of these aldols can be rationalised by the so-called Zimmerman-Traxler Model.

INTRODUCTION

Aldols 1A and 1B are fungicides and plant growth retardants (ref 1). Both diastereomers display the biological activity, but in general one isomer is more active. The relative stereochemistry of the two isomers has been determined and the diastereoselectivity observed in their preparation has been rationalised.

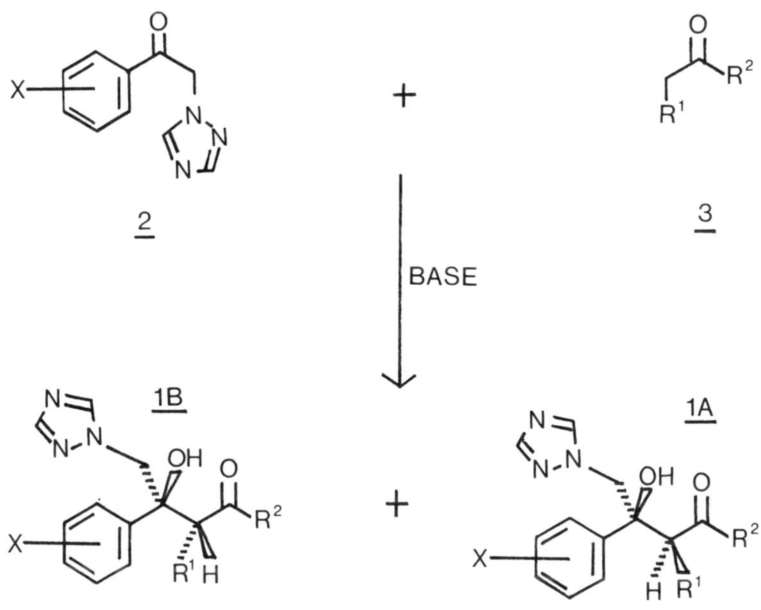

Fig. 1. Aldol Reaction of Triazolyl-Ketones

DETERMINATION OF RELATIVE STEREOCHEMISTRY

The simplest route to such compounds is the base-catalysed aldol reaction between triazolyl-acetophenones 2 and ketones 3. Both diastereomers are generally formed. The 2RS, 3RS diastereomer possesses the higher biological activity. This relative stereochemistry was determined as illustrated by the following example, where R^1 = Me and R^2 = Et.

Pure samples of each diastereomer were readily obtained by HPLC separation. Each was then reduced to the 1,3-diol with sodium borohydride, during which no epimerisation was observed. These diols were converted to cyclic carbonates with carbonyl-di-imidazole. It was then possible to determine the relative orientation of key substituents in these cyclic carbonates by observation of nuclear Overhauser effects (n.O.e), and thus determine the relative stereochemistry of the original aldols. See Fig. 2.

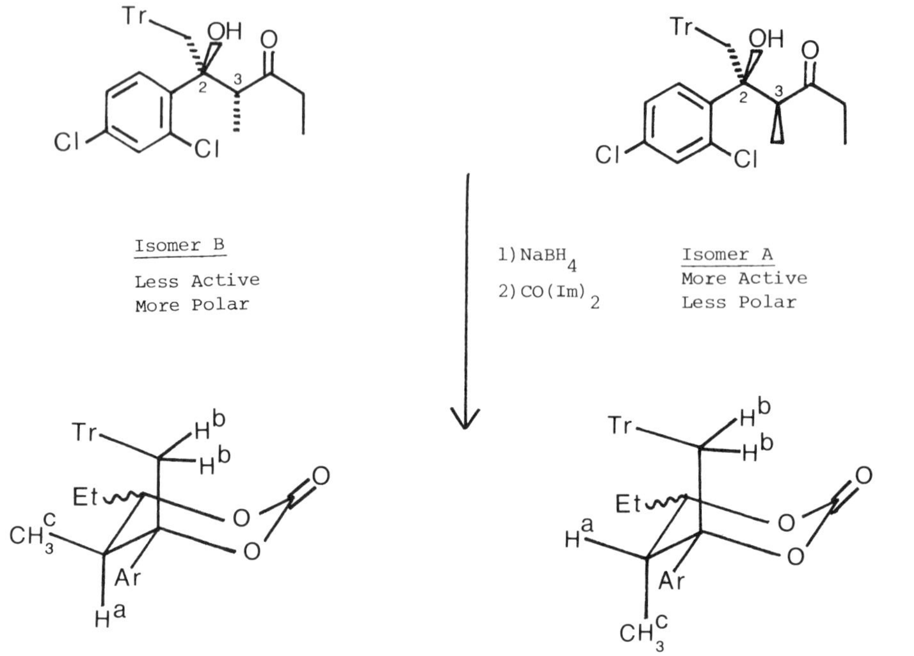

No n.O.e. between H^a and H^b
Small n.O.e. between H^c and H^b

n.O.e. between H^a and H^b
No n.O.e. between H^c and H^b

Thus : isomer A is 2RS, 3RS
isomer B is 2RS, 3SR

(Tr = 1,2,4-triazol-1-yl)

Fig. 2. n.O.e. Experiments

The proportions of isomers formed in the preparation are dependent on the values R^1 and R^2. For example when R^1 = Me and R^2 = Et approximately equal amounts of the diastereomers are formed. However, when R^1 = Me and R^2 = t-Bu exclusively the desired 2RS, 3RS diastereomer results. Similar results are obtained for other cases where R^2 is a bulky group. See Fig. 3.

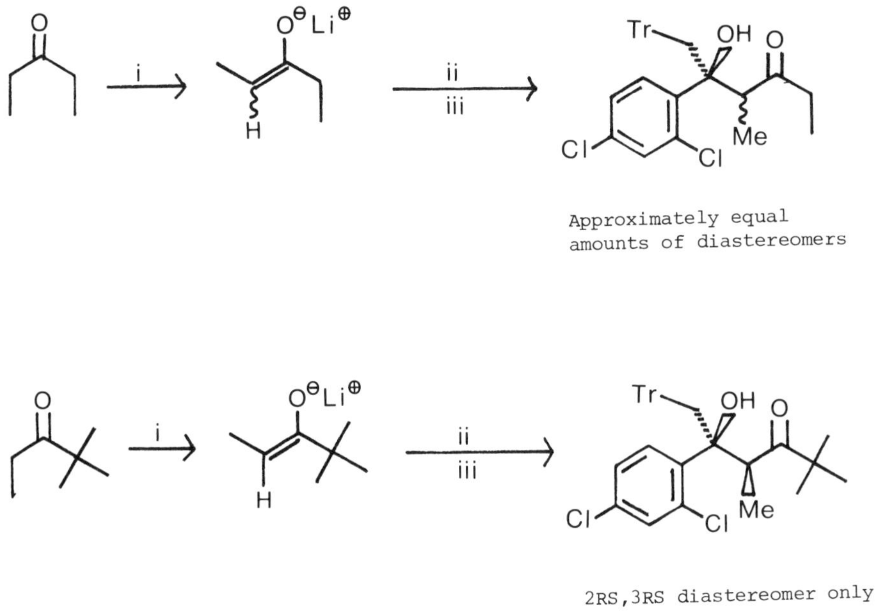

(i) = LDA or $(TMS)_2NLi/-78°$
(ii) = $ArCOCH_2Tr/-78°$
(iii) = $H_3O^+/-78° \rightarrow$ RT

Fig. 3. Variation in Proportion of Diastereomers Formed : Two Examples

RATIONALISATION OF THE STEREOCHEMISTRY

In recent years there has been an abundance of literature on predicting and explaining the stereochemistry of the aldol reaction (ref. 2). Most, if not all of it is restricted to those cases where the electrophile is an aldehyde. The two substituents on the carbonyl are always considerably different in size in an aldehyde, as one is always hydrogen. Where the electrophile is a ketone, as here, the difference in size of the two substituents may be much less marked, making prediction of the stereochemical outcome by models such as that of Zimmerman and Traxler (refs. 2,3) less clear cut. Nevertheless we have found that our results can be rationalised by the Zimmerman-Traxler Model. This postulates the reaction of an enolate of defined stereochemistry in a 6-membered transition state in which the reactants orientate themselves so as to minimise steric interactions.

The example in Fig. 4 illustrates the Zimmerman-Traxler model. Here R^2 is the bulky t-Bu group. In such cases the enolate usually adopts the Z configuration (refs. 4,5). The preference for this single enolate geometry results in a single diastereomer as product. The steric interactions in the transition state determine what the relative stereochemistry of that product will be. The t-Bu group is compelled to adopt a pseudo-axial position. Clearly if the particularly bulky substituted phenyl group adopted a pseudo-axial position there would be a very strong 1,3-diaxial interaction. The phenyl group therefore adopts a pseudo-equatorial position. The triazolyl-methyl group is then pseudo-axial, but this is both smaller and more flexible and thus does not experience such a strong interaction with the t-Bu group.

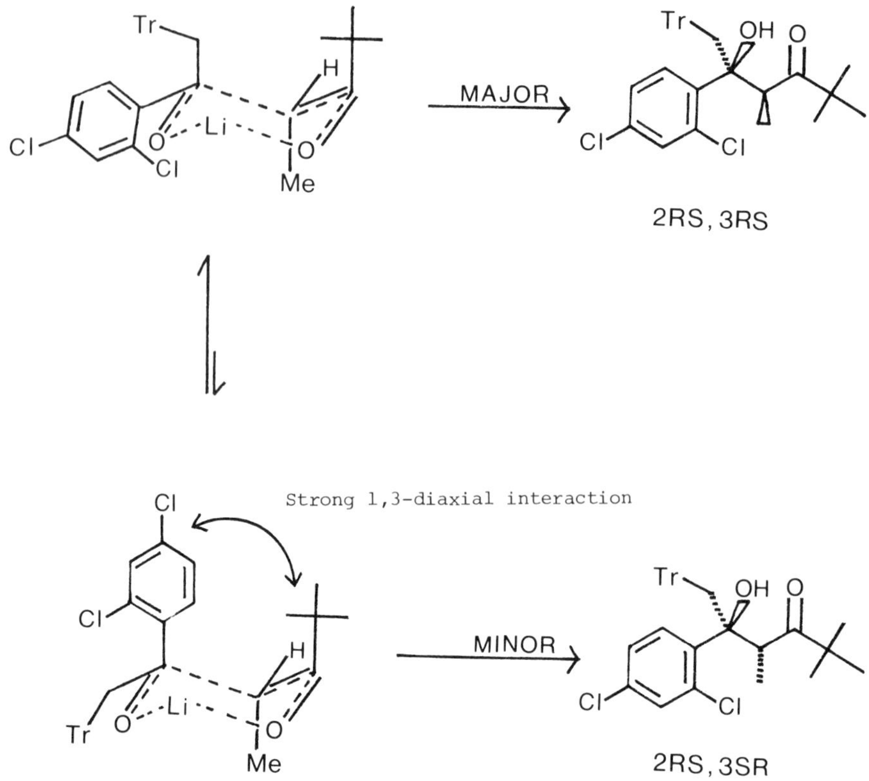

Fig. 4. Zimmerman-Traxler Model

CONCLUSION

The relative stereochemistry of these biologically active triazole-containing aldols has been established. The diastereoselectivity observed in their preparation has been rationalised. This now opens the way to the use of diastereoselective modifications of the aldol reaction for preparing the preferred diastereomer.

REFERENCES

1. ICI Patent : EP86917.
2. See for example : C. H. Heathcock in J. D. Morrison (Ed.) Asymmetric Synthesis Vol. 3 p.111-212, Academic Press, London (1984) and references cited therein.
3. H. Zimmerman and M. Traxler, J.Am.Chem.Soc. 79, 1920-1923 (1957).
4. W. A. Kleschick, C. T. Buse and C. H. Heathcock J.Am.Chem.Soc. 99, 247-248 (1977).
5. C.H. Heathcock et al. J.Org.Chem. 45, 1066-1081 (1980).

Biologically active organosilicon compounds: silicon-containing triazole fungicides

William K. Moberg,* Gregory S. Basarab, John Cuomo, and Paul H. Liang

Agricultural Products Department, Experimental Station, Building 402
E. I. du Pont de Nemours and Co., Inc., Wilmington, DE 19898, USA

Abstract - A program at Du Pont to examine the utility of organosilicon compounds as agrochemicals has produced a new class of sterol-inhibiting fungicides, the silylmethyltriazoles. This paper describes the discovery, synthesis, optimization, and field performance of these compounds. A member of this class, DPX-H6573, is currently under development as a broad spectrum foliar fungicide. This compound establishes the utility of silicon compounds in agriculture, and offers advantages over known agents of the same mode of action.

INTRODUCTION

The discovery of new agrichemicals at Du Pont depends upon a balanced, flexible approach that includes synthesizing novel compounds, enhancing known areas of activity, and studying new biochemical processes for specific inhibitor design. This paper describes our research in organosilicon triazole fungicides, which blended elements of the first two strategies. An interest in studying a largely unexplored area, that of bioactive organosilicon compounds, was combined with a desire to improve upon the activity of a known class of fungicides, triazoles that inhibit ergosterol biosynthesis. The result has been development of DPX-H6573 (ref. 1), an active ingredient in the new Du Pont fungicides Nustar®, Olymp™, and Punch® C.

DISCOVERY

Initially, the two ideas that led to this discovery developed separately. On the one hand, we wanted to investigate organosilicon compounds as agrochemicals, based on several considerations. First, the area was relatively unexplored. Despite extensive research in organosilicon chemistry going back almost 100 years, commercial use of organosilicon compounds has been limited almost exclusively to polymers. To our knowledge the only silicon compounds ever commercialized for use as pharmaceuticals or agrichemicals are chloroethylsilanes (ref. 2), which act as plant growth modifiers by decomposing in the plant to release the hormone ethylene. Second, there was a vast literature on organosilicon chemistry, and academic interest in the area was increasing dramatically. However, in virtually all of this work the interest was not in the silanes themselves. Rather, silicon was used to facilitate chemical transformations, and it disappeared gracefully as the desired product was formed. Third, polymer research had provided several simple but highly functionalized building blocks, available in bulk and at reasonable cost - the monomers for silicone polymers.

On the other hand, we had an independent interest in the emerging class of triazole fungicides. These compounds, which act by interfering with steroid biosynthesis in sensitive fungi, were active against a broad spectrum of economically important plant diseases. However, even the most potent compounds had limitations - for example, in spectrum of diseases controlled for particular crops or in undesired plant growth modifying effects - leaving open the possibility for improvement.

These ideas came together through the observation that many triazole fungicides shared certain structural features (Figure 1). Taking triadimefon (Bayer) and propiconazole (Janssen, licensed to Ciba-Geigy for development) as examples, we noted that each had a triazole ring, a two-atom spacer, and a benzene ring. Representing this common structure schematically with X and Y as the bridging atoms, it occurred to us that silicon might be incorporated in the bridge. We initially ruled out position Y, since N-silylazoles were known to hydrolyze readily, deciding instead to make X silicon. There were negative precedents for this type of substitution - for example, the work of Fukuto on silicon-containing DDT analogs (ref. 3) - but the idea had the great virtue of being easy to test. Thus, we prepared the simplest structure fitting this template, compound I.

FIGURE 1: DISCOVERY STRATEGY

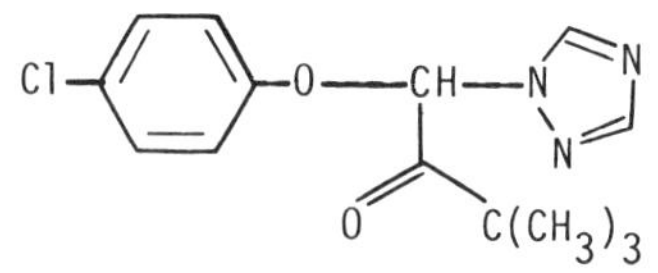

Triadimefon

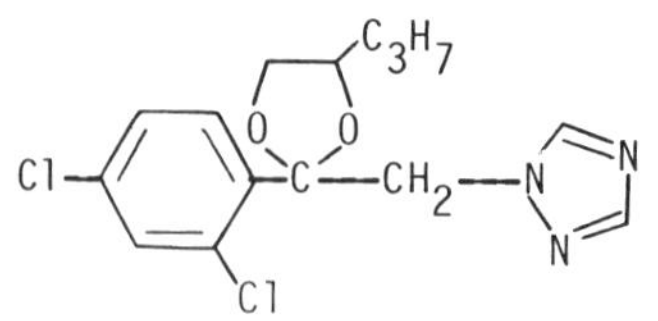

Propiconazole

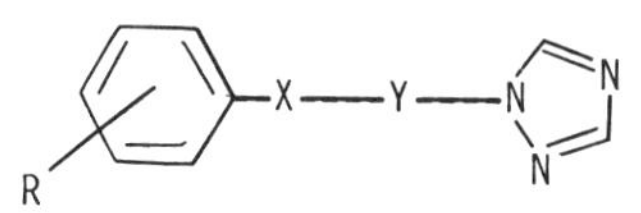

Common structure

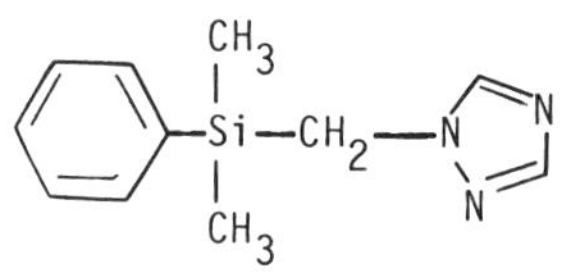

Compound I

Compound I proved to be a classic "chemical lead". Its activity was relatively modest and its spectrum relatively narrow, but it had the kind of activity expected for a sterol inhibitor. Equally important, it was a simple structure that could be modified at several sites. We therefore undertook a synthesis program aimed at optimizing its activity.

SYNTHESIS

A flexible laboratory route to the required compounds was developed, as illustrated below for DPX-H6573. First, a silicone polymer intermediate, in this case dichlorodimethylsilane, is functionalized by light-catalyzed monochlorination. Next, the silicon-bound chlorine atoms are replaced by organometallic reagents. Although two moles of 4-fluorophenyllithium are used in the example, the chlorines can be replaced stepwise by two different organometallics, giving a very versatile synthesis. Finally, the carbon-bound chlorine is replaced by triazole. Starting with other polymer intermediates such as trimethylchlorosilane or methyltrichlorosilane, compounds with a wide variety of substituents about the silicon could be prepared.

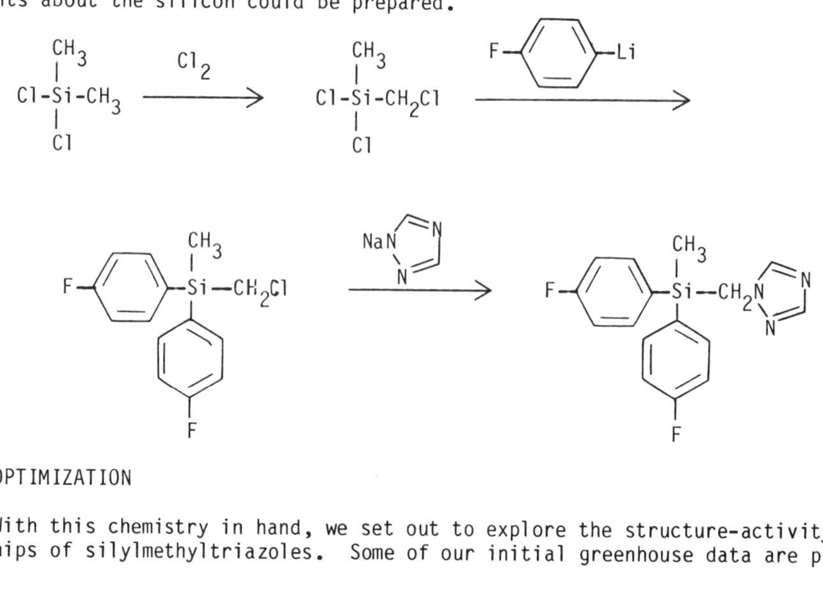

OPTIMIZATION

With this chemistry in hand, we set out to explore the structure-activity relationships of silylmethyltriazoles. Some of our initial greenhouse data are presented in the first section of Table 1. Here the compounds were applied at a concentration of 100 ppm and evaluated for preventive control of cucumber powdery mildew, apple scab, peanut early leafspot, and wheat leaf rust. Starting with our lead compound (first entry), moving a halogen atom around the ring showed that a 4-substituent gave a considerable boost in activity, and in surveying various 4-substituents, we found that phenyl was particularly effective. Extension of the alkyl groups on silicon, or substitution of naphthyl or cyclohexyl for the phenyl moiety, failed to improve activity. Although some compounds from the monoaryl series were active enough to warrant field evaluation, adding a second aryl moiety provided even higher activity. As in the monoaryl series, para-halogenation gave the best potency and spectrum of activity. Extending the series to triarylsilanes caused activity to drop off again.

The most active compounds identified in primary testing were then subjected to thorough greenhouse evaluation. In the second section of Table 1, this process is illustrated for the diaryl series. Here the diseases are the same, but the numbers are now ED90 values, the amount of compound in grams per hectare required to give 90% disease control. These data are derived from dosage response curves obtained under controlled conditions in the greenhouse, and therefore do not translate directly to field rates; however, they provide a good way to compare compounds, and the trends they indicate have been borne out in field testing. One sees that a single para-fluorine boosts activity substantially, and a second fluorine takes the activity to impressive levels across the board. Replacing both fluorines with chlorine gave somewhat reduced activity, as the final entry shows.

TABLE 1: STRUCTURE-ACTIVITY FROM GREENHOUSE TESTING

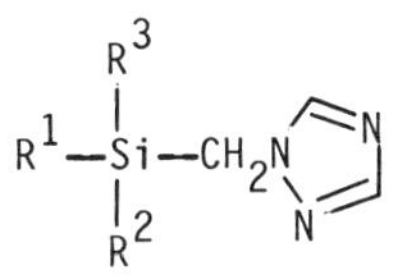

R^1	R^2	R^3	PERCENT CONTROL AT 100 PPM			
			CPM	APS	PCA	WLR
Ph	CH_3	CH_3	100	0	0	90
2-ClPh	CH_3	CH_3	100	90	100	0
3-ClPh	CH_3	CH_3	0	60	0	0
4-ClPh	CH_3	CH_3	100	100	90	100
4-PhPh	CH_3	CH_3	100	100	100	100
4-ClPh	n-C_4H_9	CH_3	100	100	0	80
1-Naphthyl	CH_3	CH_3	100	50	90	0
Cyclohexyl	CH_3	CH_3	100	0	0	0
Ph	Ph	CH_3	100	100	100	0
4-FPh	Ph	CH_3	100	100	100	90
4-FPh	4-FPh	CH_3	100	100	100	100
4-FPh	4-FPh	4-FPh	0	60	90	90
			ED90 (G/HA)			
Ph	Ph	CH3	12	100	70	75
4-FPh	Ph	CH3	4	5	30	120
4-FPh	4-FPh	CH3	0.5	3	10	15
4-ClPh	4-ClPh	CH3	9	7	25	60

Further greenhouse testing, along with extensive field evaluation, led to the selection of DPX-H6573 for development as a foliar fungicide.

FIELD PERFORMANCE

Field testing worldwide has shown that DPX-H6573 can control many important crop pathogens at low application rates. In most instances, multiple pathogens are controlled on a given crop. Some diseases for which good efficacy has been demonstrated are shown in Table 2.

TABLE 2: FIELD EFFICACY OF DPX-H6573

CEREALS:	Powdery Mildew, Rusts, Footrot, Septoria, Helminthosporium, Rhyncosporium
APPLES:	Scab, Powdery Mildew, Rust
GRAPES:	Powdery Mildew, Black Rot
SUGAR BEETS:	Cercospora Leafspot, Powdery Mildew
BANANAS:	Yellow and Black Sigatoka
COFFEE:	Rust
PEANUTS:	Early and Late Leafspots
STONE FRUIT:	Brown Rot

To give a specific example, Table 3 presents trials conducted in France on diseases of wheat and barley. The standards, propiconazole for foliar diseases, and carbendazim and prochloraz for Pseudocercosporella footrot, are taken at recommended field rates. On a weight basis, H6573 is essentially equivalent to propiconazole for the leaf diseases; for footrot, H6573 is comparable to carbendazim and superior to prochloraz. The dashes in this table also make a point - there are currently no compounds effective against both footrot and a broad range of leaf diseases. Table 4 presents further footrot trials, conducted in England. Based on our experience to date, we feel that H6573 offers a uniquely broad spectrum of control for cereal diseases, at rates comparable to or less than those of the best commercial standards. Equally encouraging results have been obtained in field trials on other crops listed in Table 2.

TABLE 3: CEREAL TRIALS (FRANCE, 1983)

		PERCENT CONTROL			
		BARLEY	WHEAT		
	G/HA	MILDEW	RUST	SEPTORIA	FOOTROT
DPX-H6573	80	88	91	78	--
	160	92	98	80	68
	240	--	--	--	76
PROPICONAZOLE	125	91	96	81	--
CARBENDAZIM	200	--	--	--	82
PROCHLORAZ	750	--	--	--	76

TABLE 4: WHEAT FOOTROT TRIALS (ENGLAND, 1983)

	G/HA	PERCENT CONTROL
DPX-H6573	100	51
	200	59
	400	75
CARBENDAZIM	250	30
PROCHLORAZ	400	54

SUMMARY

DPX-H6573 is a new fungicide of great promise. From the chemical point of view, it is to our knowledge the first agrochemical or pharmaceutical to contain silicon in the active form of the molecule. From the biological point of view, it is at least as active as the best commercial standards, sterol inhibitors or otherwise, for many diseases of important crops, and for several of these crops it is either more active than any standard or controls a broader range of diseases.

ACKNOWLEDGEMENTS

We acknowledge with gratitude the many Du Pont biologists who were essential to optimizing and developing this discovery, and Ray Luckenbaugh, the supervisor who provided an atmosphere in which these ideas could grow.

REFERENCES

1. W. K. Moberg, U. S. Patent 4,510, 136 (1984, assigned to Du Pont); T. M. Fort and W. K. Moberg, 1984 British Crop Protection Conference - Pests and Diseases, 3, 413-419 (1984).
2. For example, 2-chloroethyltris(2-methoxyethoxy)silane [etacelasil]: British Patent 1,371,804 (1974, Ciba-Geigy).
3. M. A. H. Fahmy, T. R. Fukuto, R. L. Metcalf, and R. L. Holmstead, J. Agr. Food Chem., 21, 585-592 (1973).

Designed synthesis of pesticides using QSAR procedures: Development of new insect juvenile hormone mimics

Hajime Iwamura and Toshio Fujita*

Department of Agricultural Chemistry, Faculty of Agriculture, Kyoto University, Kyoto 606, Japan

Abstract - Juvenile hormone (JH) mimetic activities of the compounds having terpenoid, 2,4-dodecadienone structures were quantitatively analyzed. The correlations revealed that the steric dimensions of the molecule are among all important for activity, reflecting the accomodation to the receptor. To visualize the results, a "mode-of-action" map was drawn, and as new structures that fit the receptor model, we designed undecen-2-one oxime O-ethers and undecen-2-yl carbamates. They exhibited the activity comparable to natural JH I. To produce higher activity, we transformed the terpenoid structure to non-terpenoid one to develop (4-phenoxyphenoxy)acetaldoxime O-ethers, the activity examined for *Culex pipiens* being about one hundred times higher than that of the parent oximes. Further explorations of more active structures led us to develop propionaldoxime O-ether series of compounds; the most potent member showed the activity about ten thousand times higher than that of the starting, terpenoid oximes.

INTRODUCTION

Recently, successful applications of QSAR analyses to the molecular design have been accumulated in the field of pesticide syntheses (ref. 1,2). In most of these applications, the designed compounds are of the same type of structure as that of compounds included in the primary set for the correlation analysis. More recently, we have shown that the information on the QSAR can be transposed to other types of compounds having the same biological activity in cytokinin and anticytokinin analogs (ref. 3,4,5), synthetic sweetners (ref. 6,7) and JH mimics (ref. 8,9,10). In this article, we review our study for the JH mimics as one of the recent examples of the QSAR application to the pesticide designed synthesis.

We first analyzed QSAR of terpenoid 2,4-dodecadienone type of compounds (ref. 8). Based on the results, we designed undecen-2-one oxime O-ethers and undecen-2-yl carbamates (ref. 9). These compounds have been proved as active as natural JH I against *Culex pipiens* and *Chilo suppressalis* and to be more active than JH I against *Musca domestica*. They were, however, much less potent than methoprene, the most active of the JH mimetic compounds so far known. Thus, to obtain higher activity, we transformed the terpenoid structure to non-terpenoid one, developing (4-phenoxyphenoxy)- and (4-benzylphenoxy)acetaldoxime O-ethers (ref. 9). The activity examined for *C. pipiens* was excellent, a several times more active than the parent, terpenoid compounds. We further explored more active, new skeletal structures, resulting in the development of propionaldoxime O-ethers (ref. 10). One of the most potent members, 3-[4-(3-methylphenoxy)phenoxy]-propionaldoxime O-isopropyl ether, exhibited the activity higher than that of methoprene. The ascent in the potency was about ten thousand times, counting up from the starting, terpenoid oximes.

QSAR OF TERPENOID JH MIMICS

The terpenoid JH mimics, 2,4-dodecadienones and related compounds shown in Fig. 1, are the class that has been intensively investigated by Henrick and his co-workers (ref. 11). Using their data, analyses were performed for activities against *Aedes aegypti* and *Tenebrio molitor*, giving Eqs. 1 and 2.

For *A. aegypti*:

$$pI_{50} = 3.65L_x - 0.35L_x^2 + 1.08D - 0.06D^2 + 1.90 \log P - 0.14(\log P)^2 + 0.57B_x - 0.71I_N + 0.86I_{OR} - 1.39I_{br} - 0.65I_{(-)} - 16.35 \quad (1)$$

$$n = 85, \quad s = 0.53, \quad r = 0.89$$

For *T. molitor*:

$$pI_{50} = 4.63W_x - 0.63W_x^2 + 2.58D - 0.15D^2 + 1.76 \log P - 0.13(\log P)^2 + 0.61\pi_x + 0.87B_x + 2.89I_{NR} - 2.51I_{br} - 23.32 \quad (2)$$

$$n = 84, \quad s = 0.54, \quad r = 0.90$$

I_{50} is the molar concentration for *A. aegypti* larvae and mol/pupa for *T. molitor* to produce 50% inhibition of metamorphosis. In the correlation, the steric parameters indicated in Figs. 2 and 3 were used. L_x is the length along the bond axis of the X end, and W_x is the width across the direction in which the longest chain extends. The D is the sum of the D_x and D_y, the lengths along the axis that passes through the C_1 and C_{11} atoms, and corresponds to the maximum length of the molecule. The B_x manifests the bulkiness toward carbonyl group of α-branch in the

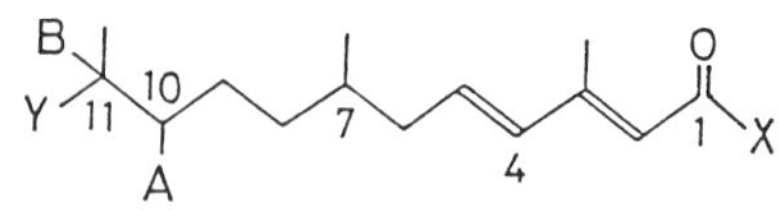

X = OR, SR, NHR, NR_2, Alkyl
Y = OR, SR, OCOR, Me, Et
A = H, OR, SR etc
B = H, Me, Cl

Fig. 1. Structure of 2,4-dodecadienone JH mimics.

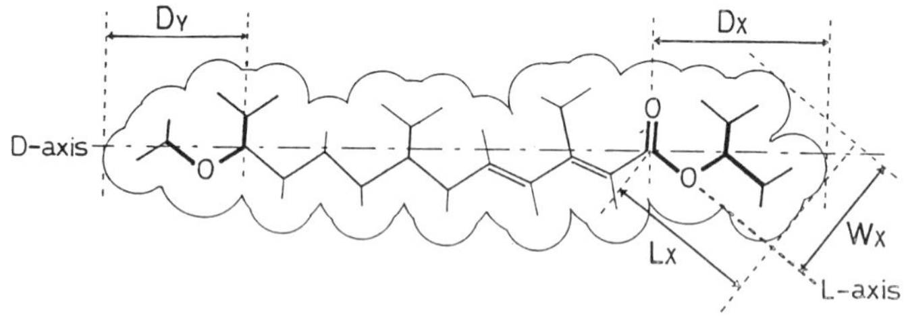

Fig. 2. Schematic representation of the steric parameters.

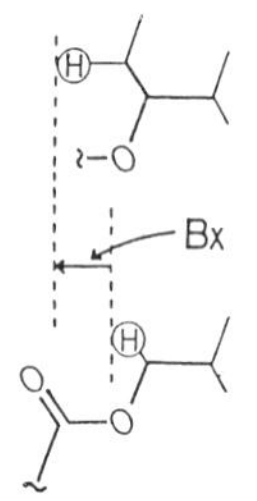

Fig. 3. Steric B_x parameter

alkoxy X. Log P expresses the hydrophobicity of the whole molecule, and π_x is the π value of the C(Me)=CHCO-X moiety.

The optimum log P values calculated from Eqs. 1 and 2 are commonly ca 6.8. The optimum molecular length along the D-axis is calculated to be also common, ca 21 A, suggesting similar dimensions of the receptors in both insect species. The positive sign of the B_x term appears to show that the α-branch obstructs the attack by hydrolytic enzymes to the esters.

For *A. aegypti*, the length L_x is an important factor, whereas it is the width W_x for *T. molitor*. The π_x in Eq. 2 suggests that the region of the *T. molitor* receptor touched by the X end is hydrophobic. The I_{OR} in Eq. 1 is an indicator variable for the compounds whose Y is alkoxy; its positive sign appears to suggest their hydrogen-bonding interaction with an acidic group. For *A. aegypti*, the R-(-)-isomers always exhibited lower activity than the (+)-counterparts, and this is expressed by the $I_{(-)}$ term. The I_{br} in both Eqs. is an indicator variable for branched ketones (X in Fig. 1 = branched alkyl). The I_N in Eq. 1 is that for amides and the I_{NR} in Eq. 2 is for amides and ketones.

Inclusively to show these results, "mode-of-action" maps were drawn and shown in Fig. 4A for *A. aegypti* and Fig. 4B for *T. molitor*. The affixes

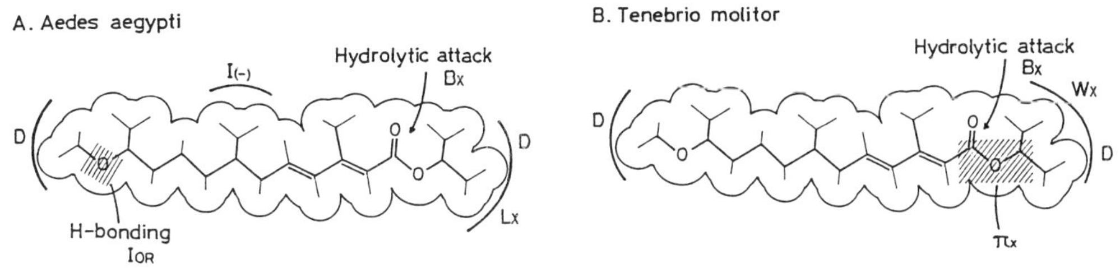

Fig. 4. Receptor models to 2,4-dodecadienones to (A) A. aegypti and (B) T. molitor receptors. The solid lines are the steric interaction sites or spatial walls. The affixes show the parameters incorporated into Eqs. 1 and 2 to suggest their roles.

explain the roles of the parameters incorporated into the correlations. The models help in understanding the overall resemblance as well as species differences in the mode of action or the receptor. In addition, they are very useful in obtaining insight into designing novel structures that confer JH activity.

DEVELOPMENT OF NEW JH MIMICS

Based on the above results, we first designed terpenoid, undecen-2-one oxime O-ethers and undecen-2-yl carbamates shown in Fig. 5. The total length, the steric dimensions at the oxime O-ether and carbamate ends (X) and other structural characteristics were varied so as to conform to the QSAR results.

The activities examined for *Culex pipiens*, *Chilo suppressalis*, and *Musca domestica* of the oxime O-ethers having an epoxy or a methoxy function at the Y end were considerably higher than those of the corresponding 9-ene- and 9,10-unsubstituted compounds. An effect similar to that expressed by the I_{OR} term in Eq. 1 for *A. aegypti* also appears to be at work in the interactions with receptors of the above three insect species. The compounds whose total length corresponds to the optimum value obtained above

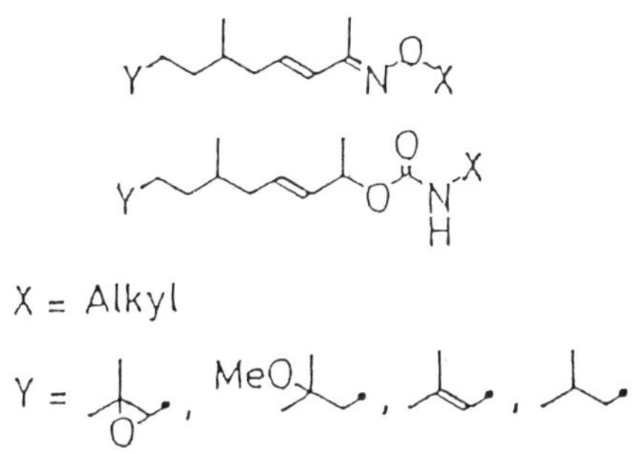

Fig. 5. Structures of the terpenoid oxime O-ethers and carbamates.

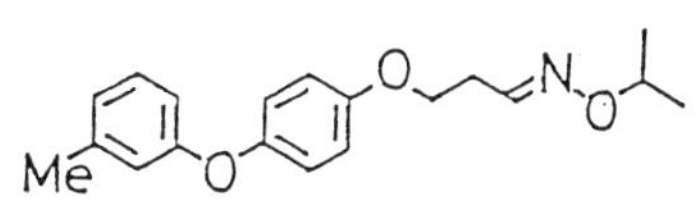

Fig. 7. Structure of 3-[4-(3-methylphenoxy)phenoxy]propionaldoxime O-isopropylether.

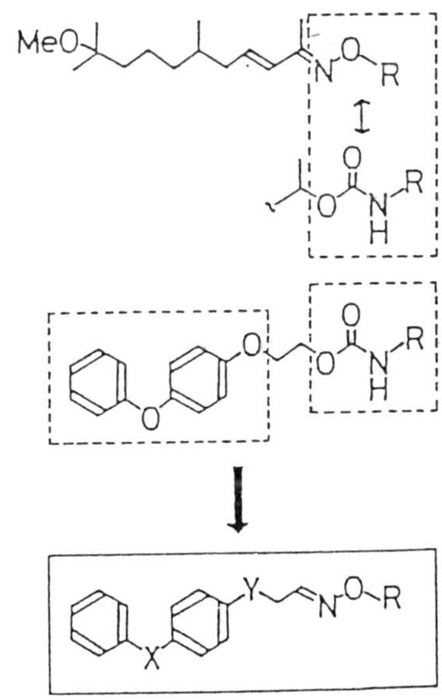

Fig. 6. Development of (4-phenoxyphenoxy)acetaldoxime O-ethers and related compounds.

for *A. aegypti* and *T. molitor*, ca 21 A, showed higher activity than longer and shorter analogs. Accordingly, the most active member of the oxime series was 9,10-epoxy O-propyl ether. The corresponding 9,10-epoxy N-ethyl carbamate was not the most active compound in the series. This apparent discrepancy was, however, investigated to be attributable in part to the hydrophobicity difference. Its log P is by 0.71 lower than that of the corresponding oxime, making the carbamate compound apart from the optimum.

The activity of the most active members was comparable to or higher than that of natural JH I but much lower than that of methoprene. Thus, to obtain higher activity, we transformed the terpenoid structure to non-terpenoid one. Our choice was the 2-(4-phenoxyphenoxy)ethane structure to obtain (4-phenoxyphenoxy)acetaldoxime O-ether type of compounds, since non-terpenoid 2-(4-phenoxyphenoxy)ethyl carbamates have been reported to show high JH activity (ref. 12). The principle of the design was as shown in Fig. 6. The carbamate moiety of the 2-(4-phenoxyphenoxy)ethyl carbamates overlaps that in our terpenoid carbamates, and the latter is convertible to terpenoid oximes. Therefore, the first and last sets of compounds may be interchangeable.

The activity examined for *C. pipiens* of these compounds was excellent, about one hundred times more active than the terpenoid oximes. We prepared the acetaldoxime series of compounds that have systematically varied substituents at both ends. The results of the activity test revealed that an introduction of a methyl group to *meta* position of the terminal benzene and that to the ether end as a branch raise the activity several times.

We further explored more active, new skeletal structures, resulting in the development of propionaldoxime O-ether series of compounds. The activity was dozens of times higher than the corresponding acetaldoxime O-ethers. By optimization of the terminal structures, we finally obtained 3-[4-(3-methyl-phenoxy)phenoxy]propionaldoxime O-isopropyl ether, the most potent member of the class. The activity examined for *C. pipiens* was higher than that of methoprene, and the acent in the potency was about ten thousand times counting up from the starting, terpenoid oximes.

ACKNOWLEDGMENT

Figs. 1-4 and 6 were reproduced from refs. 8 and 9, with permission from the American Chemical Society.

REFERENCES

1. T. Fujita, in Drug Design Fact or Fantasy, G. Jolles and K.R.H. Wooldridge, Eds., pp. 19-33, Academic Press, London (1984)
2. T. Fujita, in QSAR and Strategies in the Design of Bioactive Compounds, J.K. Seydel, Ed., pp. 207-218, VCH Verlagsgesellschaft, Weinheim (1985)
3. H. Iwamura, T. Fujita, S. Koyama, K. Koshimizu and Z. Kumazawa, Phytochemistry, 19, 1309-1319 (1980).
4. H. Iwamura, M. Masuda, K. Koshimizu and S. Matsubara, J. Med. Chem., 26, 838-844 (1983).
5. H. Iwamura, S. Murakami, K. Koshimizu and S. Matsubara, J. Med. Chem., 28, 577-583 (1985).
6. H. Iwamura, J. Med. Chem., 24, 572-583 (1981).
7. H. Iwamura, J. Med. Chem., 23, 308-312 (1980).
8. A. Nakayama, H. Iwamura and T. Fujita, J. Med. Chem., 27, 1493-1502 (1984).
9. A. Nakayama, H. Iwamura, A. Niwa, Y. Nakagawa and T. Fujita, J. Agric. Food Chem., 33, 1034-1041 (1985).
10. A. Niwa, H. Iwamura, Y. Nakagawa and T. Fujita, unpublished results.
11. C.A. Henrick, W.E. Willy and G.B. Staal, J. Agric. Food Chem., 24, 207-218 (1976).
12. F. Karrer and S. Farooq, in Regulation of Insect Development and Behavior, Part 1, F. Sehnal et al., Eds., pp. 289-302, Wroclaw Tech. Univ. Press, Wroclaw (1981).

The application of sequential simplex optimization to pesticide design

E. L. Plummer*, A. A. Liu and K. A. Simmons

Agricultural Chemical Group, FMC Corporation, Box 8, Princeton, NJ 08540

Abstract: The rapid optimization of biological activity is often of primary importance in the process of pesticide design in an industrial environment. Indeed, such optimization may take precedence over a full understanding of structure activity requirements in many lead evaluation decisions. The Sequential Simplex Optimization (SS)) strategy, introduced as a drug design tool in 1974 by Darvas (ref. 1) fits this strategy well, providing an efficient approach to lead optimization as the sole goal. Recently we became involved in evaluation of several analogs of the insect development disrupter (IDD), diflubenzuron. We attempted to apply the SSO i.e. plots in signma-pi space. Such two dimensional plots proved unsatisfactory since the parameters that actually affected activity were not always sigma or pi. We therefore have designed a computer based system for dealing with multidimensional response spaces. The algorithm used is based on the modified SSO strategy suggested by Gillom, et al. (ref. 2). Although our automated process can handle any number of parameters, we have adopted, as a standard approach, the use of pi, F, R and the Verloop STERIMOL parameters L and B1. The choice of the initial SSO set of six compounds is based on cluster analysis of 89 substituents. The system is linked to a database of over t00 additional substituents which are automatically searched for targets that meet the criteria for optimization. The application of the program is illustrated by the optimization of a series of 4-substituted 3-methoxyphenyl benzoyl ureas. Although the search was constrained by the discontinuous nature of the parameter space, the optimum activity was reached in four steps from the unsubstituted compound with a activity increase of 1000 fold.

INTRODUCTION

During our study of biphenyl pyrethroid alcohols (ref. 3) we discovered that the activity of the esters prepared from these alcohols was profoundly affected by the presence of an ortho substituent. When it was reported that the pyrazine analog 2 (ref. 4) of the original Philips-Duphar biphenyl (ref. 5) compounds (1) was similarly enhanced by the addition of an ortho methyl substituent we suspected that a similar phenomena was taking place.

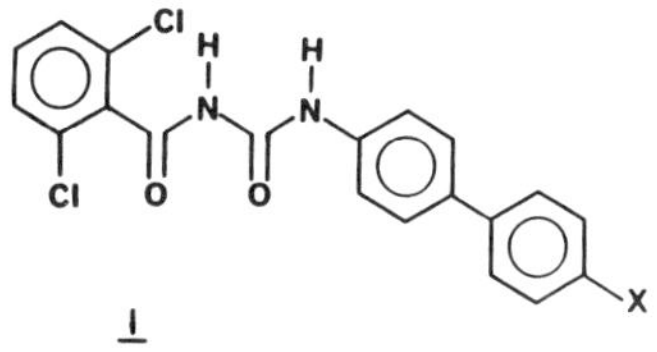

1
X = H,Cl,Br,NO₂
(PHILIPS-DuPHAR)

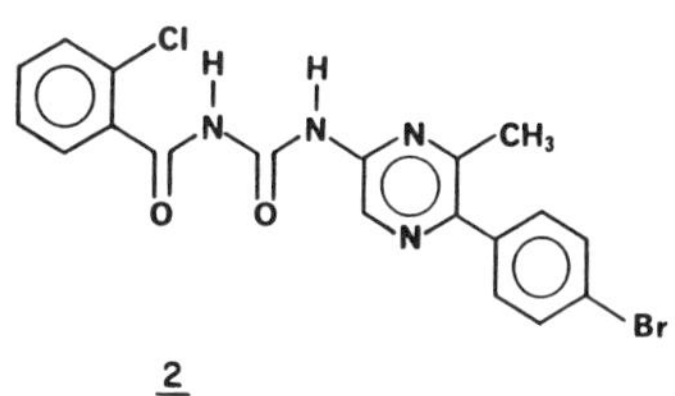

2
(ELI LILLY-EL494)

This seemed like a good place to apply an optimization strategy like SSO, since we were interested in determining the best compound without significant investment of resources in understanding structure activity relationships. In our first attempt at this approach we were seeking the optimal substituent for the *ortho* position in the first aromatic ring. Synthetic ease was considered in the choice of the pyrroles 3 to model the biphenyls.

We started the process by preparing the following compounds:

X	Pi	Sigma(para)	LC_{50}(ppm)
OCH_3	-.02	-.27	2.3
CN	-.57	.66	26.8
CF_3	.88	.54	9.8

The reflection of the simplex was to pi = 1.43 and sigma = -.39. The $OCH(CH_3)_2$ group has pi = 1.05 and sigma = -.45 and was, therefore, prepared. This compound had an LC_{50} greater than 100 ppm. Reapplication of the simplex lead to a reflection back to the original starting point.

Intrigued by the inability of the graphical SSO to improve the biological activity, a set of eight compounds were prepared and subjected to classical regression analysis (Figure 1). The most significant relationship was found for R and B1.

FIGURE 1

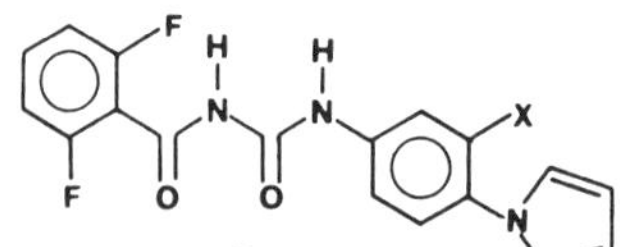

3

X	SAW DIET LC_{50} (PPM)
OCF_2H	11.1
OCH_3	2.3
CL	5.6
CN	26.8
CF_3	9.8
CH_3	7.3
H	58.7
OC_2H_5	11.0

$$\text{Log}\,(1/LC50) = -6.6\,R - 2.7$$

$n = 8 \quad r^2 = 0.38 \quad s = 0.833 \quad F = 3.6$

$$\text{Log}\,(1/LC50) = -9.9\,R + 2.4\,B1 - 6.4$$

$n = 8 \quad r^2 = 0.85 \quad s = 0.444 \quad F = 14.4$

Continuing the optimization of biphenyl diflubenzuron analogs, we turned our attention to structures represented by 4, optimizing the activity of a series of 4'-derivatives holding the (optimal) $3\text{-}OCH_3$ constant. Again we chose to use pi and sigma_p as the two parameters. The first three compounds prepared are listed below:

X	pi	Sigma_p	LC_{50}(ppm)
$C{=}O(CH_3)$	-.64	.50	>200
CF_3	.88	.54	0.15
OCH_3	-.02	-.27	1.70

We were able to prepare the t-butyl analog (pi = 1.98 and sigma_p = -.20) in response to a reflection to pi = 1.50 and sigma = -.04. This compound had an LC_{50} of 0.43 ppm. Reflection from these compounds pointed into a region (pi = 2.88 and sigma_p = .61) where no viable substituents existed.

Since no reasonable substituent existed for the second predicted target, a set of 22 compounds (Figure 2) were prepared and subjected to standard regression analysis. In addition to compounds of structure 4 we included in this analysis a series of 2-Cl analogs 5.

An apparent dependence of biological activity on pi and R was found. Here the graphical SSO in two dimensions (pi and sigma), was responding since $sigma_p$ and R for the set used are heavily intercorrelated, (R^2=0.61). These studies point out that the graphical SSO is subject to chance selection of the proper physiochemical spaces (pi-sigma, F-sigma, MR-pi, etc). Since biological activity can be a function of lipophilic balance (pi), electronics (sigma or F,R), and steric effects (MR or L, B1, B4), we have chosen to run the SSO for all lead areas against parameters representing all three factors. This requires the use of computer facilities since the optimization could be occurring in as many as seven dimensions!

We wrote such a program and included it in our QSAR System. Since the system is designed for use by laboratory chemists, with minimum assistance from a QSAR specialist, one of the primary criteria for modules in our system is user friendliness. We therefore have included an SSO process in which a set of compounds chosen by cluster analysis is suggested to the user:

H, Cl, $C{=}OC_6H_5$, $N(CH_3)_2$, OnC_5H_{11}, $CH(CH_3)_2$

The program will ask for the rank of the compounds by substituent and then after calculating the resultant simplex and reflection point, will search our data set of 500 plus substituents for ones that have the desired parameter values. Instead of searching for a specific value of each parameter, a clearly useless search, a range of + or - 50% of the value is searched (this is adjusted to 75% for very small values). Finally, a list of potential substituents is presented to the chemist for evaluation. The SSO process is a discontinuous one requiring storage of data in a file which is picked up by the program the next time the chemist returns with data.

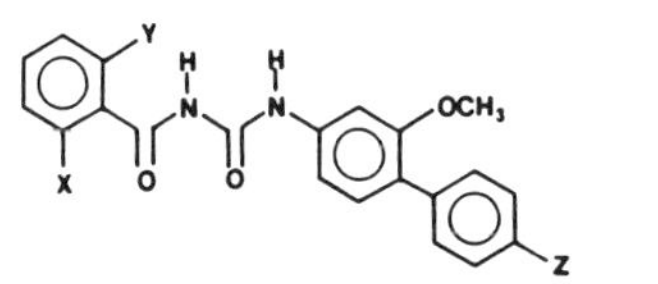

SAW DIET LC_{50} (PPM)

Z	4 X, Y = F	5 X = Cl; Y = H
-H	0.20	3.0
$-OCH_3$	1.70	20.0
-F	1.40	3.2
-Cl	0.15	0.15
$-C(O)CH_3$	>200	110.0
$-CF_3$	0.15	0.20
$-CH_2CH_3$	0.30	0.25
-Br	0.45	0.40
$-OCF_2H$	0.20	0.45
$-C(CH_3)_3$	0.43	11.00
-CN	0.25	2.00
$-OCF_3$	-	0.67

$$\text{Log}(1/LC50) = 3.6\,R + 2.6\,pi - 1.6\,pi^2 + 1.5\,I_{2,6} - .2$$

$n = 22 \quad r^2 = 0.745 \quad s = 0.999 \quad F = 12.4$

optimum pi = 0.8

FIGURE 2

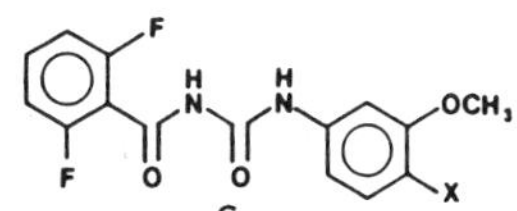

6

X	pi	F	R	L	B1	CL-LC_{50} Diet
-C=OC6H5	1.05	0.30	0.16	4.57	2.35	500.0
-H	0.00	0.00	0.00	2.06	1.00	300.0
-N(CH3) 2	0.18	0.10	- .92	3.53	1.50	100.0
-OCH(CH3) 2	0.85	0.30	- .72	4.59	1.35	20.0
-CL	0.71	0.41	- .15	3.52	1.80	15.0
-OnC5HII	1.55	0.25	- .57	8.11	1.35	8.0
No Subst.	0.27	0.12	-1.1	4.15	0.44	350.0
-SCF3	1.44	0.35	0.18	4.89	1.70	1.0
-CF3	0.88	0.38	0.19	3.30	1.98	0.3
-SO2C2F5*	1.71	0.73	0.26	5.16	1.16	1.5
-SOC3F7*	1.55	0.60	0.14	6.76	2.21	0.5

*F,R VALUES ESTIMATED

pi VALUES CALCULATED USING LEO'S CLogP

L,B1 VALUES CALCULATED USING STERIMOL

FIGURE 3

The program is an implementation of the rules adapted by Gillom et al. (ref. 2) from the work of Morgan and Deming (ref. 6) and Nelder and Mead (ref. 7). As a result the simplex expands, contracts and reverses based on the history of previous moves all of which are stored in a data file. Expansion is suggested by Gillom, et. al. and Deming, et. al., as a means of enhancing the speed of optimization by following in the direction of steepest ascent. However, two-fold expansions, when the original simplex is large, can result in moves outside the bounds of the parameter space. To overcome this difficulty we have added the option to expand from a user selected value between 1.0 and 2.0 (we often use a value of 1.25).

We have now applied this program to several problems. One is a set of diflubenzuron analogs related to the biphenyl benzoylureas that started our program development. Compounds of general structure **6** were prepared. We deviated from the original set by using an $OCH(CH_3)_2$ instead of a $-CH(CH_3)_2$ substituent because of synthesis restrictions (Figure 3).

The decision to use a set of substituents based on cluster analysis was based on the concept that one is unlikely to miss an optimum if one goes to the edges of the parameter space involved for the first substituent set. Unfortunately, the result is likely to be that the first reflection will be outside the parameter space possible. Indeed, this was what we observed. Following the rules of Gillom et al. we assigned a low value to this reflection point and continued. The next compound prepared X = SCF_3 had an LC_{50} = 1 PPM while the ninth compound in the set X = CF_3 was some 1000x more active than the unsubstituted parent for the series. At this point we again ran into the major difficulty of this process, we ran out of substituents that fit the reflected values. In this case we estimated values for F and R based on analogous substituents in our data set and calculated pi using the CLogP program developed by Leo and Hansch (ref. 8). L and B1 values were calculated from a modification of the original STERIMOL program (ref. 9) provided by Hoogenstraaten and Verloop which we have interfaced with the minimization programs of MDL's CHEMLAB (ref. 10). The next two steps lead to no increase in activity suggesting that we had reached an optimum.

We see the Sequential Simplex Optimization strategy as an important tool in the rapid discovery of pesticides. Clearly the method can move design programs to the activity optimum with relatively minor expenditure of resources. A particular strength of the method is that it can be used to optimize multiple dependent variables by application of the proper ranking equations. It also has the advantage of saving testing resources, a critical issue in many industrial testing programs.

We have observed some shortcomings. The discontinuous nature of the process and the sometimes difficult synthesis targets suggested have a tendency to lengthen the time from start to optimization. The major limitation we have found, however, is the lack of substituents with parameters that are indicated on reflection of the simplex. This problem, which is common to all QSAR strategies, can only be overcome by increasing substituents in the data set.

REFERENCES

1. F. Darvas, J. Med. Chem., 17, 799, (1974).
2. R.D. Gillom, W.P. Purcell and T.R. Bosin, Eur. J. Med. Chem., 12, 187, (1977).
3. E.L. Plummer in Pesticide Synthesis Through Rational Approaches, p. 297, P.S. Magee, G.K. Kohn and J.J. Menn Eds., ACS Symposium Series No. 255, American Chemical Society, Washington, D.C. (1984).
4. J.L. Miesel U.S.P. 4,160,834 (July 10, 1979).
5. K. Wellinga, R. Mulder and J.J. van Daalen, J. Agr. Food Chem., 21, 348, 993 (1973).
6. S.N. Deming and S.L. Morgan, Anal. Chem., 45, 278 (1973).
7. J.A. Nelder and R. Mead, Computer J., 7, 308, (1965).
8. A.J. Leo in Pesticide Synthesis Through Rational Approaches, p 213, P.S. Magee, G.K. Kohn and J.J. Menn Eds., ACS Symposium Series No. 255, American Chemical Society, Washington, D.C. (1984).
9. A. Verlopp, W. Hoogenstraaten, J.A. Tipker in Drug Design, Vol. 7, Chapter 4, E.J. Ariens, Ed., Academic Press: New York, N.Y. (1976).
10. A.J. Hopfinger, J. Am. Chem. Soc., 102, 7196, (1980).

Computer assisted molecular modelling (CAMM)

A tool for structure-activity considerations of inhibitors of the enzymatic oxidation of ent-kaurene to ent-kaurenoic acid

U. Müller,* Ph. Huxley, E. Ebert

Agro-Division, CIBA-GEIGY Ltd., Research and Development
CH-4002 Basle, Switzerland

Abstract - A number of structurally different inhibitors of the enzymatic oxidation of ent-kaurene to ent-kaurenoic acid were measured in the cell free preparation from the endosperm of *Cucurbita maxima* Using computer assisted molecular modelling methods (CAMM) we derive by superposition studies basic structural requirements common to all these inhibitors. Strong and weak inhibitors are compared. An explanation for their inactivity based on structure calculations is given. A structural superposition of the most potent inhibitors, with the natural substrate ent-kaurenol is attempted. Molecular orbital calculations of different heterocycles bearing the same lipophilic carbonframe and similar inhibitory strength are discussed.

INTRODUCTION

A number of different inhibitors (triazoles, pyrimidines, diazetins) of the cytochrome P-450 dependent oxidation of ent-kaurene to ent-kaurenoic acid are known. Figure 1 shows the site of action of these "azole-type" inhibitors. The inhibitory activity of these compounds was measured in the cell free endosperm preparation of *Cucurbita maxima* described by Graebe (ref. 1). These in vitro data correlate well with *in vivo* measurements of growth regulation of mono- and dicotyledoneous plants. *In vivo* measurements of growth regulation reflect a combination of factors (uptake, transport, metabolism etc.) while *in vitro* data correspond more closely to intrinsic activity, that is receptor binding affinity. Figure 2 illustrates substructural elements common to the gibberellin inhibitors and inhibitors of the fungal cytochrome P-450 enzyme system catalysing the C-14 demethylation of lanosterol.

The aim of this study is to use *in vitro* data as a basis for Computer Assisted Molecular Modelling methods (CAMM) to:

- Explore the degree of variability of the lipophylic moiety.
- Compare common features of the lipophylic moieties of strong inhibitors with those of the natural substrate.
- Discuss the experimental finding that different heterocycles bearing the same lipophylic moiety are of comparable strength.

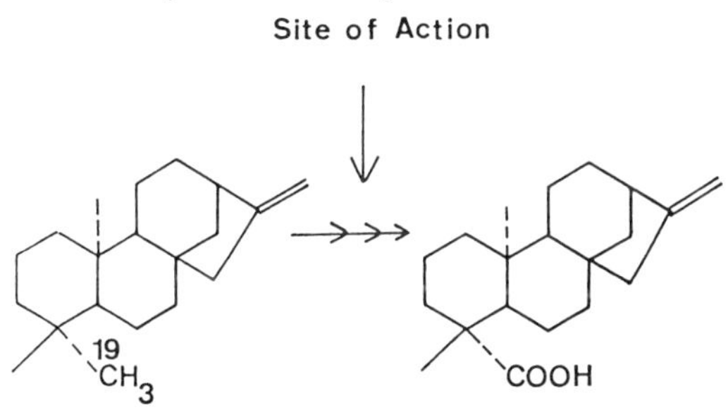

ent-kaurene ent-kaurenoic acid

Fig. 1. Site of action of "azole-type" inhibitors in the biosynthetic pathway of gibberellins (ref. 5)

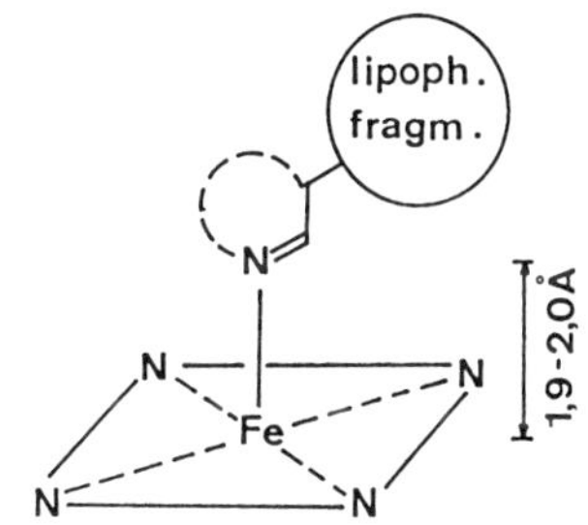

Fig. 2. Proposed mechanism of action of "azole-type inhibitors (ref. 3, 4)

RESULTS

Tables 1 and 3 illustrate the structures included in the study and their *in vitro* activities. The structural differences are striking when compared with the I-50 values which all lie in the range I-50 = 10^{-8} to 10^{-10} mol/l. In attempting to find common features between these structures we have made the following assumptions:
- That the heme bound nitrogen atom in the different inhibitors should be exactly superimposed

Table 1
Enzyme I-50 inhibition values for the racemic mixtures of the inhibitors measured in the cell free preparation of the endosperm of CUCURBITA MAXIMA.

Structure	Common Name (Comp.No.)	I-50 (mol/l)
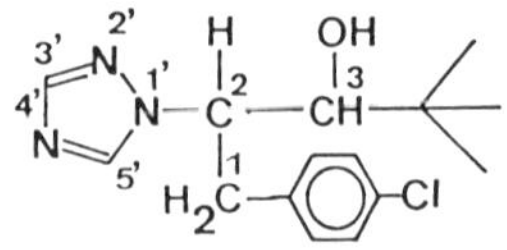	PP-333 Paclobutrazol (1)	$4.5 \pm 1.5 \cdot 10^{-9}$
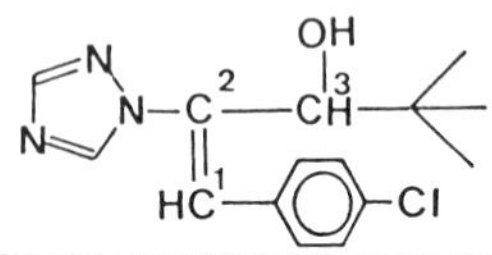	S-3307 Uniconazole (2)	$4.2 \pm 2.4 \cdot 10^{-8}$
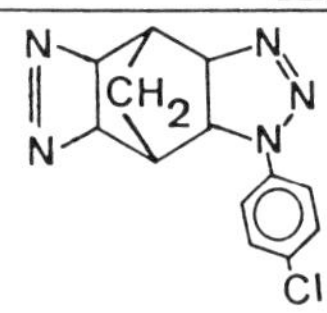	BAS 106W Tetcyclacis (3)	$5.0 \cdot 10^{-10}$
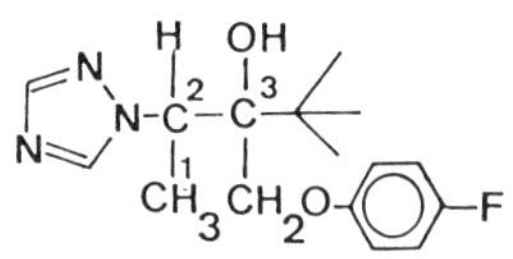	CGA 143 376 (4)	$1.9 \pm 0.6 \cdot 10^{-9}$

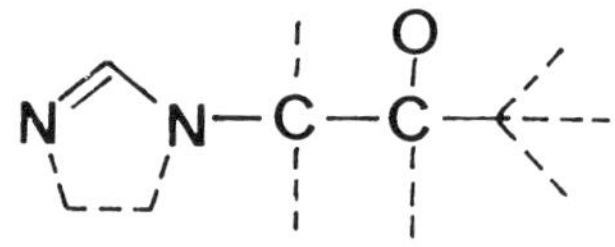

Fig.3. Assigned backbone for PP-333, S-3307, CGA 143376

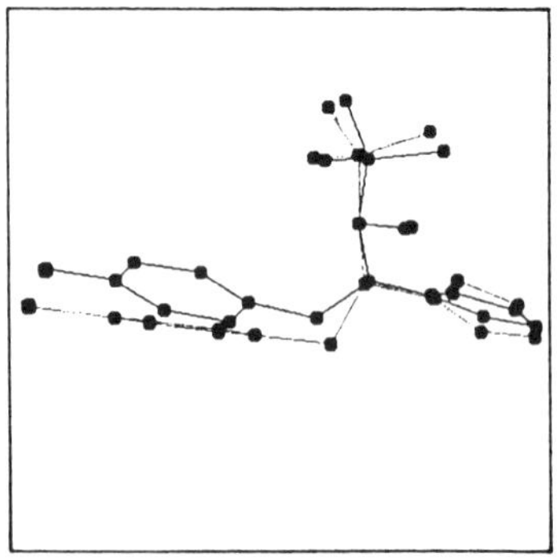

Fig. 4. PP - 333, S - 3307

- That the heterocycles should all be in the same plane
- Since only the (3-S) enantiomer of PP-333 (ref. 2c) and S-3307 (ref. 2b) is active, we assume that compounds 4 and 5 should also have (S)-chirality at C(3).
- PP-333 and CGA 143376 have a second chiral centre at C(2). We have arbitrarily chosen (S) stereochemistry for C(2).

Conformational analysis of compounds 1, 2, 4 and 5

In order to compare compounds 1, 2, 4 and 5 (tables 1 and 3) we calculated the minimum energy geometry of each conformer produced by rotation about the C(2) - C(3) bond. In these calculations the C(2) - C(3) bond was held constant and all remaining rotatable bonds searched for the minimum energy structure using molecular mechanics calculations. In the final step the molecular mechanics geometry was optimised using MNDO calculations. Newman projections and MNDO energies for three conformers of compounds 1, 4 and 5 are shown in table 2. The calculations suggest that compounds 4 and 5 exist in all three conformers and compound 1, PP-333 exists as conformers 1 and 3, but not 2, which is 5.5 Kcal above the lowest energy conformer. Given that PP-333 is active we assume that conformer 2 is not the active conformation. For the remainder of this work we have arbitrarily used conformer 1.

Superpositions of the inhibitors

Using the backbone illustrated in Fig. 3 we have superimposed PP-333, S-3307 and CGA 143376. Our superpositions demonstrate that the phenyl groups on PP-333 and S-3307 superimposed well (Fig. 4), but that the phenyl groups on CGA 143376 and PP-333 cannot be superimposed for an energy cost of less than 5 Kcal (Fig. 5).
CGA 143376 was superimposed on the rigid inhibitor BAS 106W. The C(3)-CH_2-O-Ph moiety on CGA 143376 was twisted in order to maximise the overlap on the two molecules subject to a 2 Kcal energy cost (Fig. 6).

Conclusion

The combined volume for the inhibitors suggests that there are at least two phenyl binding sites (Fig. 6).

Superposition of inhibitors with ent-kaurenol

Assumptions made:
- Substrate and inhibitors were arranged so as to have maximum volume overlap.
- N-4' on the triazoles and one of the nitrogens on the diazetin ring were placed in the proximity of the C(19)-O bond on ent-kaurenol as suggested by Marchington for lanosterol and azole fungicides (ref. 4).

TABLE 2

Energies calculated for conformers of compounds 1, 4 and 5

Conformer 1 | Conformer 2 | Conformer 3

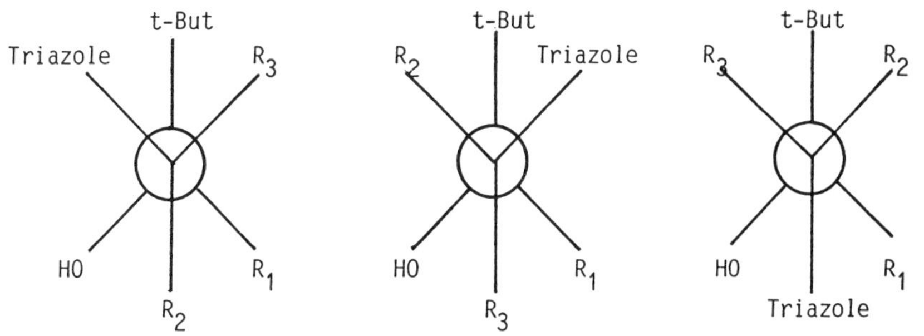

PP-333 R_1 = H, R_2 = CH_2-Phenyl, R_3 = H

CGA 143 376 R_1 = CH_2OPhenyl, R_2 = CH_3, R_3 = H

Compound 5 R_1 = CH_2OPhenyl, R_2 = H, R_3 = H

	Conformer 1	Conformer 2	Conformer 3
PP-333	31 Kcal/mol 1)	36.5 Kcal/mol	31 Kcal/mol
CGA 143 376	-28.6 Kcal/mol	-28.3 Kcal/mol	-29.2 Kcal/mol
Compound 5	-31 Kcal/mol	-31 Kcal/mol	-32.2 Kcal/mol

1) crystal structure: 31.4 Kcal/mol

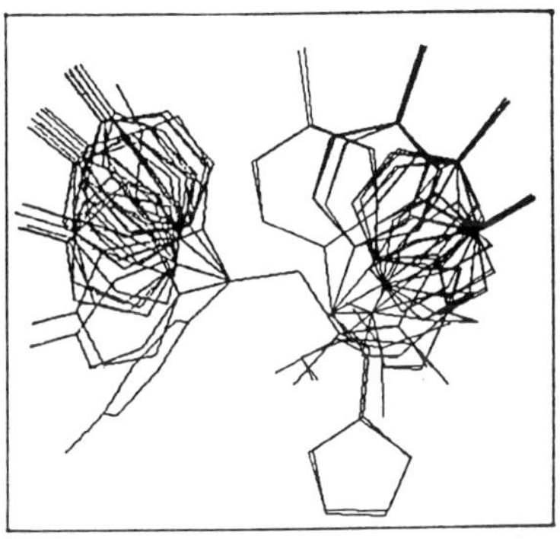

Fig. 5. CGA 143376, PP-333

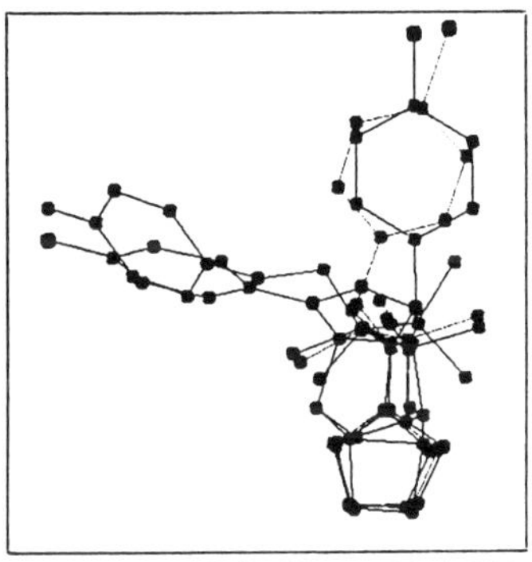

Fig. 6. PP-333, S-3307 BAS 106W, CGA 143376

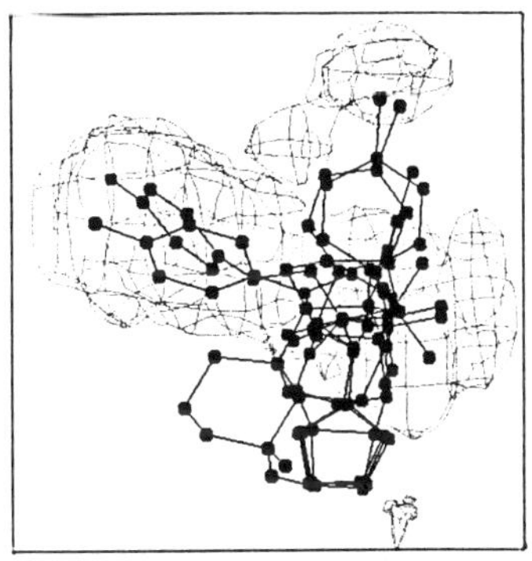

Fig. 7. Difference volume inhibitors - ent-kaurenol

Ent-kaurenol is enveloped by the inhibitors. As can be seen by the difference volume (Fig. 7) the combined inhibitor volume is considerably greater than the natural substrate volume. Detailed speculation about binding sites on the active site is difficult since too little is known about the enzymatic reaction and the enzyme itself.

Comparison of strong and weak inhibitors

Compounds with a substituent on the triazole ring in either the 3'- or the 5'-position are very weak inhibitors (table 3). Figure 8 shows a superposition of the minimum energy conformers of compounds 5, 6, 7 and the difference volume, clearly showing that the inactive compounds have a large volume next to the nitrogen atom complexing to the heme iron atom. Compounds 8, 9 have 5'-substituents on the triazole ring. When compared with compound 5 the five substituted compounds have the triazole ring rotated through 30°, 9, and 24°, 8. The difference volume is shown in Fig. 9. The inactivity of compounds 8, 9 may be due to this additional volume hindering tight binding or it may be due to the triazole rotation. Our analysis of porphyrin - heterocycle crystal structures show that the angle Φ falls within the range 0° - 20° (Fig. 10). Table 4 shows activities and nitrogen lone pair orbital energies for compounds 10 - 14. All the heterocycles are active, except the pyrrazole analogue which lacks a nitrogen atom at the 4'-position of the heterocycle. The remaining heterocycles all have lone pair orbitals lying close to each other, -19.8± 0.7 eV. Theoretical considerations suggest that both the energy of the lone pair orbital and the orientation of the heterocycle with respect to the heme iron atom are important for binding (ref. 6).

Table 3
Enzyme I-50 - inhibition values for substituted 1,2,4-triazole compounds
1) mixture of 3- and 4- isomer

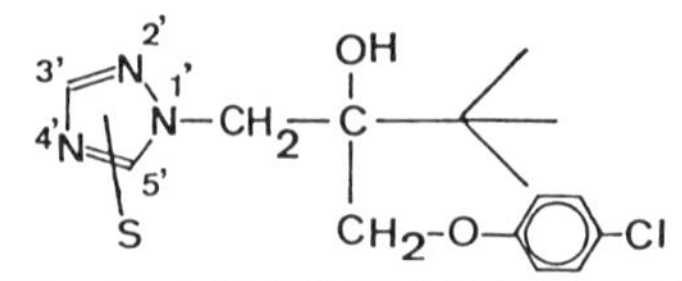

S	Comp. No.	I_{50} (mol/l)
H	(5)	$5.8 \pm 3.6 \cdot 10^{-8}$
3-Br	(6)	$> 10^{-3}$
3-CN	(7)	$> 10^{-3}$
5-Cl	(8)	$> 10^{-3}$
CH_3 1)	(9)	$> 10^{-3}$

Table 4
Comparison of different heterocycles bearing the same lipophilic carbonframe

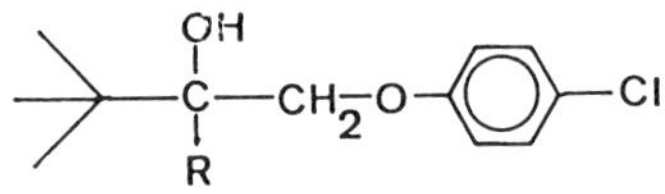

R	Log P(o/w)	eV	I_{50} (mol/l)
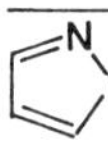 N-CH2- (10)	3.23	-	$> 10^{-3}$
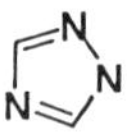 N-CH2- (11)	3.25	-19.9	$5.8 \pm 3.6 \cdot 10^{-8}$
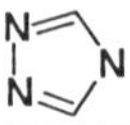 N-CH2- (12)	2.67	-20.6	$7.9 \pm \cdot 10^{-9}$
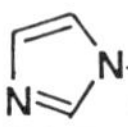 N-CH2- (13)	3.76	-19.3	$4.4 \pm 0.5 \cdot 10^{-9}$
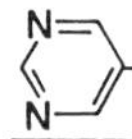 (14)	3.35	-19.9	$3.59 \pm 0.5 \cdot 10^{-8}$

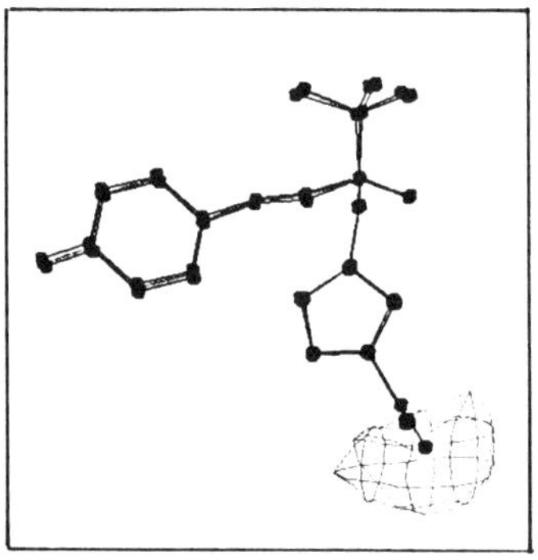

Fig. 8. Difference volume compounds 6,7 - compound 5

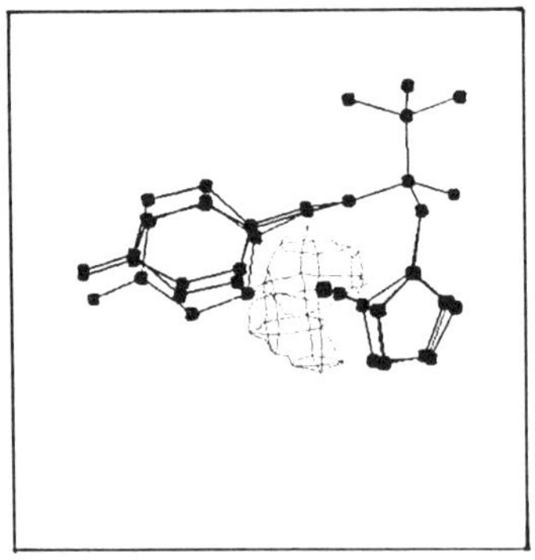

Fig. 9. Difference volume compounds 8,9 - compound 5

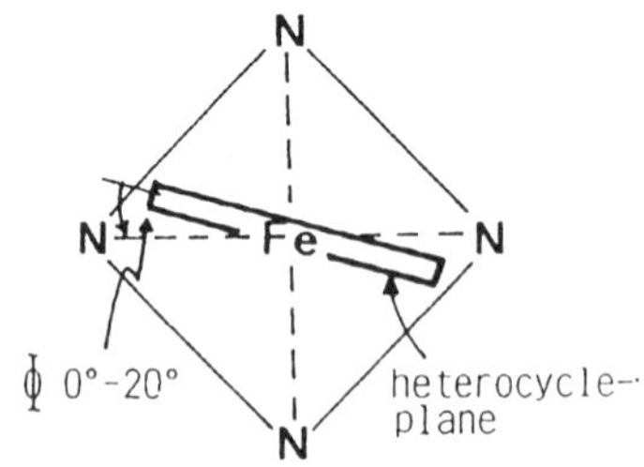

Fig. 10.

REFERENCES

1. J.E. Graebe, Plant Growth Substances, p. 151, Springer, Berlin (1970)
2a. W. Rademacher, J. Jung, J. Agron. & Crop Science, 150 363 (1981)
2b. Izumi Kazuo, Kamiya Yuji, Sakurai Akira, Oshio Hiromichi, Takahashi Nobutaka, Plant Cell Physio., 26, 821 (1985)
2c. Sugavanam Balasubramanyan, Pestic. Sci., 15, 296 (1984)
3. P. Gadher, E.I. Mercer, B.C. Baldwin, T.E. Wiggins, Pesticides Biochemistry and Physiology, 19, 1 (1983)
4. A.F. Marchington, Spec. Publ - R.Soc.Chem., 50, 124 (1984)
5. P. Hedden, The Biochemistry and Physiology of Gibberellins, Vol. 1, Prager (1983)
6. W.R. Scheidt, D.M. Chipman, J.Am.Chem.Soc., 108, 1163 (1986)

ACKNOWLEDGEMENTS

We thank our collegues Dr. A.F. Meyer and Dr. H. Tobler for synthesising some of the compounds, Mr. A. Gaudin for carrying out the in vitro tests and Dr. D. Poppinger for supporting the CAMM work.

All the pictures were produced using Chem-X, developed and distributed by Chemical Design Ltd., Oxford, England.

Chemistry and bioactivity of natural products and prototypes

Insecticides acting as $GABA_A$ receptor antagonists

John E. Casida

Pesticide Chemistry and Toxicology Laboratory
Department of Entomological Sciences, University of California
Berkeley, California 94720, U.S.A.

Abstract - The γ-aminobutyric acid (GABA)-regulated chloride ionophore is disrupted by several types of $GABA_A$ receptor antagonists such as picrotoxinin analogs, trioxabicyclooctanes and polychlorocycloalkanes. These classes of compounds, individually and collectively, are useful models for the development of new insecticides. Structure optimization is facilitated by in vitro radioligand binding assays for the chloride ionophore target. Differences in the GABA receptor-chloride channel complex between insects and mammals may allow the development of selective insecticides.

INTRODUCTION

This review considers recent findings on neurotoxic picrotoxinin (PTX) analogs, 2,6,7-trioxabicyclo[2.2.2]octanes (TBOs) and polychlorocycloalkanes (PCCAs) which establish that the γ-aminobutyric acid (GABA)-regulated chloride ionophore is an important and highly sensitive site at which some insecticides may act.

GABA is the principal inhibitory neurotransmitter of the mammalian and insect central nervous system and the insect neuromuscular junction, where its action is opposed by that

Fig. 1. Diagrammatic representation of $GABA_A$ receptor sites and associated mechanisms in a neuronal membrane (modified from ref. 3). Control of chloride channels is represented by a revolving tap mechanism. GABA activation rotates the 'tap' into the open position whereas the receptor antagonist bicucculine opposes this rotation. Allosteric sites are shown as adjacent wheels which 'rotate' on activation. The cog-toothed wheel represents the binding site of $GABA_A$ receptor antagonists. When they bind the rotation produced prevents the GABA-induced increase in cloride conductance. The pivoted arm illustrates the dependence on external chloride concentration. The "latch mechanism" of barbiturate action keeps the channel in the open position. Benzodiazepines increase the frequency with which the GABA-operated chloride channels open with an opposing effect of antagonists and inverse agonists. For more detailed discussion see ref. 3.

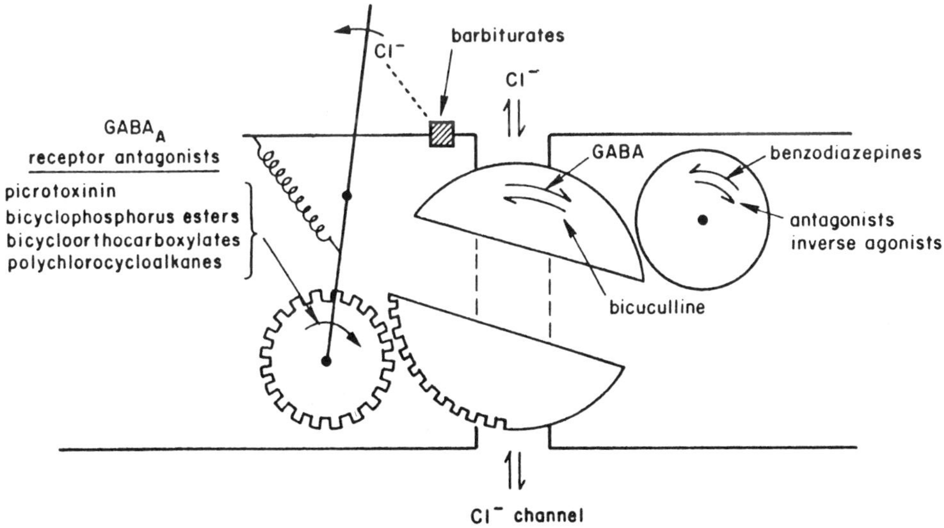

of glutamate (ref. 1 & 2). Compounds acting at the chloride channel or associated $GABA_A$ receptor sites include GABA and GABA-mimetics (e.g., muscimol and 3-aminopropanesulfonic acid), anxiolytics and anticonvulsants (e.g., benzodiazepines and barbiturates), and a variety of convulsants (e.g., PTX, TBOs and PCCAs). Several receptors or binding sites are involved as illustrated for a neuronal membrane in Fig. 1 (ref. 3).

PICROTOXININ ANALOGS (FIG. 2)

PTX, a polycyclic epoxylactone from seeds of Anamirta cocculus L., was the first compound shown to act at the GABA-regulated chloride ionophore (ref. 4) and is the classical $GABA_A$ receptor antagonist (ref. 1-3). [^{3}H]Dihydropicrotoxinin ([^{3}H]DHPTX) is the established radioligand of this type (ref. 5). PTX is moderately toxic to insects with activity values as follows: LD_{50} 0.1 µg injected and 13 µg topical per piperonyl butoxide (PB)-treated housefly; LD_{50} ~0.2 µg injected per PB-treated German cockroach; LD_{95} ~ 6 µg injected/g American cockroach (ref. 6-10). There is strong PB synergism which suggests that PTX undergoes oxidative detoxification but the sites involved remain unidentified. Bridged butyrolactones A and B have moderate toxicity or "convulsant" activity to houseflies (ref. 6 & 7). PTX, coriamyrtin, tutin and many analogs have been examined for convulsant activity in mammals (ref. 11) and the findings used in conjunction with results on alkyl-substituted γ-butyrolactones to propose three-dimensional models of the receptor sites (ref. 12). PTX is more toxic to mice than to cockroaches (ref. 8), indicating that to produce useful pesticides greater insecticidal activity and diminished mammalian toxicity are necessary.

Fig. 2. Structures of picrotoxinin, a relevant radioligand, and some bioactive analogs.

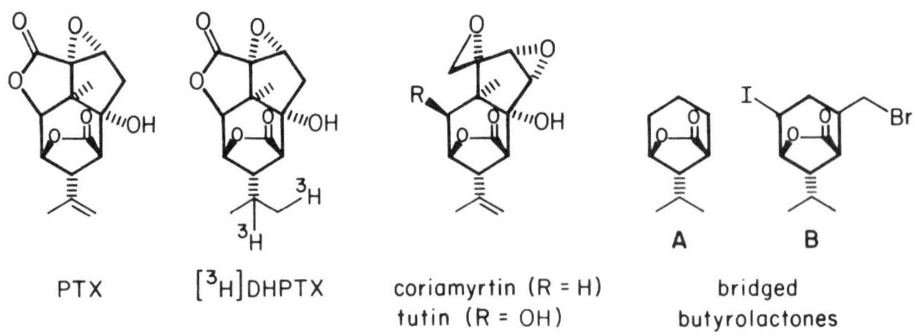

TRIOXABICYCLOOCTANES (FIG. 3)

2,6,7-Trioxabicyclo[2.2.2]octanes of toxicological interest include both bicyclophosphorus esters (BPs) and bicycloorthocarboxylates (BOCs). The BPs were discovered as potent toxicants in routine toxicological tests of industrial chemicals (ref. 13) and in studies on potent organophosphorus convulsants that are not cholinesterase inhibitors (ref. 14). They are hazardous chemical intermediates and products formed on combustion of fire-retarded foams or pyrolysis of synthetic lubricant oils (ref. 14-16). The BPs have high mammalian toxicity and little or no insecticidal activity and therefore a very unfavorable selective toxicity (ref. 9 & 17-19). The preferred BP radioligand for mode of action studies is [^{35}S]t-butylbicyclophosphorothionate ([^{35}S]TBPS) (ref. 21).

Bicycloorthocarboxylates, discovered 30 years ago (ref. 22), have been known for 10 years as toxicants similar to the BPs (ref. 17), but only recently have the BOCs become of interest as candidate insecticides (ref. 19). Substituents conferring high toxicity to houseflies are 4-cyanophenyl, 4-bromophenyl or cyclohexyl in the 1-position and t-butyl or cyclohexyl in the 4-position. The BOCs are synergized 2- to 260-fold by PB. 1,4-Dicyclohexylbicycloorthocarboxylate is selectively toxic to houseflies versus mice. [^{3}H]t-Butylbicycloorthobenzoate ([^{3}H]TBOB) is a suitable radioligand for specific binding site studies (ref. 23).

Fig. 3. Structures of bioactive bicyclophosphorus ester, bicycloorthocarboxylates, and relevant radioligands.

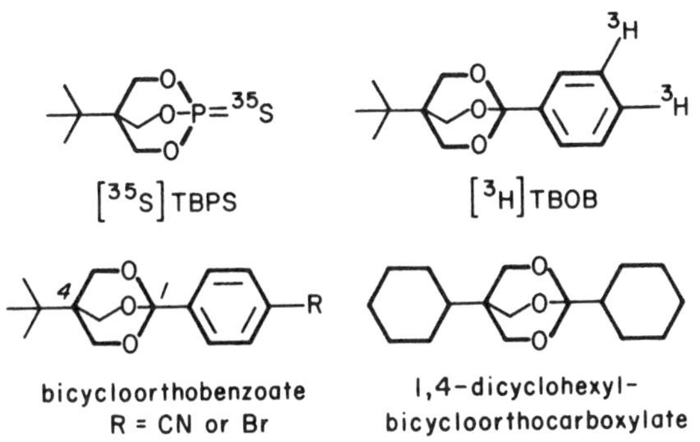

POLYCHLOROCYCLOALKANES (FIG. 4)

The success of DDT prompted the chlorination of many compounds to give candidate insecticides (ref. 24). Chlorination of benzene yields 4 main isomers of hexachlorocyclohexane of which the γ isomer or lindane is the insecticidal component. Chlorination of camphene gives a mixture of polychlorobornanes marketed as toxaphene with two primary active ingredients (heptachlorobornane B and 9-chloro-B) among the >200 components. Reaction of hexachlorocyclopentadiene with various dienes gives the polychlorocyclodiene insecticides with a variety of steric configurations. These 3 classes of chlorinated insecticides have in common the PCCA moiety. They provide economical pest control and ~3 x 10^9 pounds have been used. Many PCCAs are now restricted because of unfavorable persistence, environmental effects and chronic toxicity in mammals. The structure-activity relationships for insecticidal activity and biodegradation are well understood (ref. 24-26). Deschloro-endosulfan is one-tenth as active as endosulfan to PB-treated German cockroaches indicating that the chlorines are important but not essential for insecticidal activity. Potent insecticides of this type have one or two electronegative centers positioned at an appropriate distance from a hydrophobic site (ref. 10).

Fig. 4. Structures of some polychlorocycloalkane and related insecticides.

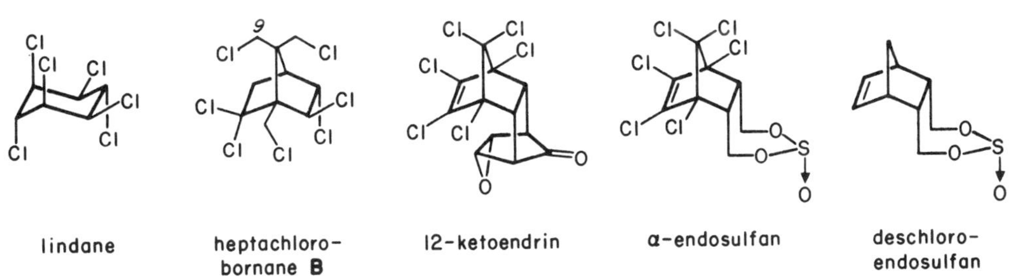

Table 1. Action of Picrotoxinin, Bicyclophosphorus Esters, Bicycloorthocarboxylates, and Polychlorocycloalkanes as $GABA_A$ Receptor Antagonists

Criterion and assay	Type of antagonist and references			
	PTX	BPs	BOCs	PCCAs
Inhibition of GABA-regulated chloride channel				
Neurophysiological	4,27-29	27,28	30	29
^{36}Chloride flux	31-33	32,34	34	31-35
Binding site of vertebrate brain $GABA_A$ receptor				
Preferred radioligand	5	21	23	
Potent inhibitors				
[^{3}H]DHPTX assay	1,5,29	1,5,29		29,35
[^{35}S]TBPS assay	21,32	21,30,32, 36,37	21,30	32,36-38
Competitive inhibition of [^{35}S]TBPS binding	39		23	36-38
Inhibition correlates with toxicity		21,30,36	30	36-38
Stereospecificity of inhibitors				29,36-38
In vivo inhibition	40	40	40	37,40
Coupled mfo/receptor system		41	30	37,41
Localization in brain slices		42	43	
Species specificity	44	44	44	44
Action in insects				
Radioligand assays	29,35	45		
Cross resistance	35,46			35,46
Positive temperature coefficient			19	19
Antidotes	8	17,30	17,30	

NEUROTOXICITY

Table 1 catagorizes the criteria and evidence that PTX, BPs, BOCs and PCCAs act as $GABA_A$ receptor antagonists. The neurophysiological studies were made with GABA-stimulated preparations from crayfish (ref. 4 & 28), cockroach (ref. 29 & 30) and rat (ref. 27). ^{36}Chloride flux studies used cockroach muscle (ref. 31 & 35) or rat or mouse brain vesicles or microsacs (ref. 32-34). The preferred radioligands are [^{35}S]TBPS or [^{3}H]TBOB because of low K_ds and a high percentage of specific binding (80-95%) in filtration assays relative to [^{3}H]DHPTX which has a low percentage of specific binding and requires centrifugation assays (ref. 5, 21 & 23). Early studies on the inhibition of [^{3}H]DHPTX binding (ref. 1, 5 & 35) were verified and greatly extended with [^{35}S]TBPS relative to potent inhibitors and competitive inhibition. The toxicological relevance of the [^{35}S]-TBPS binding site was established for the BPs, BOCs and PCCAs by *in vitro* assay potencies correlating with their mouse ip LD_{50} values (ref. 21, 30 & 36-38) and the same stereospecificity of the PCCAs for inhibition of the [^{35}S]TBPS binding site and for toxicity (ref. 29 & 36-38). Assays of brain membranes from poisoned animals show inhibition with the PCCAs but not with PTX or the BPs and BOCs, indicating a more prolonged inhibition with the PCCAs (ref. 37 & 40). [^{35}S]TBPS also has the advantage that it can be used for coupled brain-specific $GABA_A$ receptor/liver-specific microsomal oxidase (mfo) assays to help define the influence of metabolic activation and detoxification on structure-activity relationships (ref. 30, 37 & 41). For example, heptachlorobornane B is readily detoxified and β-endosulfan and endrin are activated to endosulfan sulfate and 12-ketoendrin, respectively (ref. 37). Both [^{35}S]TBPS and [^{3}H]TBOB have been used to localize the receptors in rat brain slices (ref. 42 & 43) with results similar to filtration assays (ref. 21 & 23). Comparison of brain P_2 membrane preparations from fish, chicken, rat, cow and human with the pharmacological profile of [^{35}S]TBPS binding as the criterion (ref. 44) indicates that the chloride ionophore and coupled components have remained constant during vertebrate evolution (Ref. 44).

To date radioligand studies attempting to examine directly the insect $GABA_A$ receptor are unsatisfactory (ref. 29, 35 & 45). GABA inhibits [^{35}S]TBPS binding in mammalian brain membranes but stimulates binding in housefly preparations (ref. 45). The pharmacological profile in insects also differs greatly from that in mammals with various pharmaceutical agents and insecticides, suggesting either fundamental biochemical differences or an unsuitable radioligand. The latter interpretation is the more likely as the compounds used give relatively high K_ds and have low insecticidal activity, suggesting that a more potent insecticidal compound might be a better candidate.

Cockroaches resistant to polychlorocyclodienes also resist PTX, suggesting that such strains have a modified receptor (ref. 35 & 46). Both the BOCs and PCCAs have a positive temperature coefficient for insecticidal action, in contrast to the negative temperature coefficient for DDT and pyrethroids (ref. 19). The GABA regulated-chloride ionophore may also be involved in an insect antennal attractant receptor since isomers of the chlorocyclohexane attractants for the male medfly displace [^{35}S]TBPS from housefly membranes with the same isomeric specificity for the binding site and attractant actions (ref. 47).

Barbiturates and benzodiazepines protect against the toxic effects of PTX, BPs and BOCs (ref. 8, 17 & 30) (see Fig. 1).

$GABA_A$ RECEPTOR

The rat and cow brain $GABA_A$ receptors have been solubilized (ref. 48-50) and purified to obtain a single protein complex with allosterically-coupled sites for GABA, muscimol, benzodiazepines, and convulsants (PTX and TBPS) (ref. 48 & 49). Ligand binding at one site appears to induce conformational changes in the receptor thereby modulating the binding of ligands at other sites. Thus, TBPS binding is very sensitive to GABA (~0.3 μM for 50% inhibition) which slows radioligand association and speeds up its dissociation (ref. 51). Although TBPS and TBOB appear to have a common binding site, their binding differs in respect to temperature effects, i.e., TBOB but not TBPS has a temperature independent K_d (ref. 23). Inhibition of [^{35}S]TBPS binding sites by photoheptachlor epoxide is insensitive to GABA, suggesting a different type of coupling with the GABA site (ref. 51). The presence of basic neuroproteins similar to GABA-modulin may be a factor in addition to GABA in determining the toxicological properties of TBPS and related convulsants acting on the GABA-regulated chloride ionophore (ref. 52).

$GABA_A$ receptor antagonists acting in vitro below 10 nM and candidate photoaffinity labels (ref. 30) along with improved radioligands will assist in dissecting further the $GABA_A$ receptor complex. They may also indicate how other insecticides, convulsants and drugs interact with the GABA-regulated chloride channel of insects and mammals.

REFERENCES

1. R.W. Olsen, Ann. Rev. Pharmacol. Toxicol. 22, 245-277 (1982).
2. J.M. Clark and F. Matsumura, Eds. Membrane Receptors and Enzymes as Targets of Insecticidal Action, Plenum , New York (1986).
3. N.G. Bowery, G.W. Price, A.L. Hudson, D.R. Hill, G.P. Wilkin and M.J. Turnbull, Neuropharmacology 23, 219-231 (1984).
4. A. Takeuchi and N. Takeuchi, J. Physiol. 205, 377-391 (1969).
5. F. Leeb-Lundberg and R.W. Olsen, pp. 593-606, In Psychopharmacology and Biochemistry of Neurotransmitter Receptors (H.I. Yamamura, R.W. Olsen and E. Usdin, Eds.), Elsevier, New York (1980).
6. T.A. Miller, M. Maynard and J.M. Kennedy, Pestic. Biochem. Physiol. 10, 128-136 (1979).
7. E. Kuwano, K. Ohshima and M. Eto, Agric. Biol. Chem. 44, 383-386 (1980).
8. D.W. Gammon, L.J. Lawrence and J. E. Casida, Toxicol. Appl. Pharmacol. 66, 290-296 (1982).
9. Y. Ozoe, K. Mochida, T. Nakamura, A. Shimizu and M. Eto., J. Pestic. Sci. 8, 601-605 (1982).
10. Y. Ozoe and F. Matsumura, J. Agric. Food Chem. 34, 126-134 (1986).
11. C.H. Jarboe, L.A. Porter and R. T. Buckler, J. Med. Chem. 11, 729-731 (1968).
12. W.E. Klunk, B.L. Kalman, J.A. Ferrendelli and D.F. Covey, Mol. Pharmacol. 23, 511-518 (1983).
13. J.C. Gage, Brit. J. Industr. Med. 27, 1-18 (1970).
14. E.M. Bellet and J.E. Casida, Science 182, 1135-1136 (1973).
15. J.H. Petajan, K.J. Voorhees, S.C. Packham, R.C. Baldwin, I.N. Einhorn, M.L. Grunnet, B.G. Dinger and M.M. Birky, Science 187, 742-744 (1975).
16. D.A. Kalman, K.J. Voorhees, D. Osborne and I.N. Einhorn, J. Fire Sciences 3, 322-329 (1985).
17. J.E. Casida, M. Eto, A.D. Moscioni, J.L. Engel, D.S. Milbrath and J.G. Verkade, Toxicol. Appl. Pharmacol. 36, 261-279 (1976).
18. D.S. Milbrath, J.L. Engel, J.G. Verkade and J. E. Casida, Toxicol. Appl. Pharmacol. 47, 287-293 (1979).
19. C.J. Palmer and J. E. Casida, J. Agric. Food Chem. 33, 976-980 (1985).
20. D.S. Milbrath, M. Eto and J. E. Casida, Toxicol. Appl. Pharmacol. 46, 411-420 (1978).
21. R.F. Squires, J. E. Casida, M. Richardson and E. Saederup, Mol. Pharmacol. 23,

326-336 (1983).
22. W. von E. Doering and L.K. Levy, J. Am. Chem. Soc. 77, 509-513 (1955).
23. L.J. Lawrence, C. J. Palmer, K.W. Gee, X. Wang, H.I. Yamamura and J. E. Casida, J. Neurochem. 45, 798-804 (1985).
24. G.T. Brooks, Chlorinated Insecticides, CRC Press Inc., Cleveland, Ohio, 2 Vol. (1974).
25. S.B. Soloway, Adv. Pest Control Res. 6, 85-126 (1965).
26. G.T. Brooks, World Rev. Pest Control 5, 62-84 (1966).
27. N.G. Bowery, J.F. Collins, and R.G. Hill, Nature 261, 601-603 (1976).
28. D. Gammon and J. E. Casida, Neurosci. Letters 40, 163-168 (1983).
29. K. Tanaka, J.G. Scott and F. Matsumura, Pestic. Biochem. Physiol. 22, 117-127 (1984).
30. J.E. Casida, C.J. Palmer and L.M. Cole, Mol. Pharmacol. 28, 246-253 (1985).
31. S.M. Ghiasuddin and F. Matsumura, Comp. Biochem. Physiol. 73C, 141-144 (1982).
32. J.R. Bloomquist and D.M. Soderlund, Biochem. Biophys. Res. Commun. 133, 37-43 (1985).
33. I.M. Abalis, M. E. Eldefrawi and A.T. Eldefrawi, J. Toxicol. Environ. Health 18, 13-23 (1986).
34. T. Obata, C.J. Palmer, J. E. Casida, H. Laird and H.I. Yamamura, Annual Meeting of Society for Neuroscience, Washington, D.C., Nov. 1986.
35. F. Matsumura and S.M. Ghiasuddin, J. Environ. Sci. Health B18, 1-14 (1983).
36. J.E. Casida and L.J. Lawrence, Environ. Health Perspec. 61, 123-132 (1985).
37. L.J. Lawrence and J. E. Casida, Life Sci. 35, 171-178 (1984).
38. I.M. Abalis, M.E. Eldefrawi, and A.T. Eldefrawi, Pestic. Biochem. Physiol. 24, 95-102 (1985).
39. L.J. Lawrence and J. E. Casida, Science 221, 1399-1401 (1983).
40. L.M. Cole and J. E. Casida, Life Sci., accepted.
41. R.F. Toia and J. E. Casida, Toxicol. Appl. Pharmacol. 81, 50-57 (1985).
42. K.W. Gee, J.K. Wamsley and H.I. Yamamura, European J. Pharmacol. 89, 323-324 (1983).
43. L.H. O'Connor and B.S. McEwen, European J. Pharmacol. 120, 141-142 (1986).
44. L.M. Cole, L.J. Lawrence and J. E. Casida, Life Sci. 35, 1755-1762 (1984).
45. E. Cohen and J. E. Casida, Pestic. Biochem. Physiol. 25, 63-72 (1986).
46. A.A. Kadous, S.M. Ghiasuddin, F. Matsumura, J.G. Scott and K. Tanaka, Pestic. Biochem. Physiol. 19, 157-166 (1983).
47. E. Cohen and J. E. Casida, Life Sci. 36, 1837-1842 (1985).
48. E. Sigel and E. A. Barnard, J. Biol. Chem. 259, 7219-7223 (1984).
49. R.W. Olsen, J.B. Fischer, R.G. King, J.Y. Ransom and G. B. Stauber, Neuropharmacol. 23, 853-854 (1984).
50. J. Seifert and J. E. Casida, J. Neurochem. 44, 110-116 (1985).
51. J. Seifert and J. E. Casida, European J. Pharmacol. 115, 191-198 (1985).
52. J. Seifert and J. E. Casida, European J. Pharmacol. 122, 157-158 (1986).

ACKNOWLEDGEMENTS

A portion of the studies reviewed were supported by United States National Institutes of Health Grant ES00049. Helpful comments were provided by Christopher Palmer and Michael Elliott of this laboratory.

Phytoalexin chemistry and action

Akio Murai

Department of Chemistry, Faculty of Science,
Hokkaido University, Sapporo 060, Japan

Abstract - The present paper describes survey of our chemical studies on the sesquiterpenoid phytoalexins isolated from the Solanaceae with emphasis on the biosynthetic pathways leading to their formation. These pathways have been examined through incorporation experiments using doubly deuterated rishitin, solavetivone, and related compounds. The paper also presents a proposal that the direct and common trigger substance for phytoalexin accumulation might be hydrogen peroxide.

INTRODUCTION

The concept of phytoalexin theory was first proposed by Muller and Borger in 1940 from the phytopathological studies on disease resistance in potato (ref. 1). They defined "phytoalexin" as compounds formed or activated in a living plant cell in contact with fungi, which are antifungal and thereby prevent the full development of the incipient disease. This definition has been interpreted more widely and it has been said that these are synonymous with "abnormal metabolites" or "stress metabolites" (ref. 2), because these are producible even when plants suffer other stresses such as wounding, exposure to chemicals, or to non-pathogens. Thus the term was recently revised as follows; "phytoalexins are low molecular weight, antimicrobial compounds that are both synthesized by and accumulated in plants after exposure to microorganisms" (ref. 3). Phytoalexins have been investigated with increasing attention because of their potential relevance to mechanisms of plant disease as well as some toxicity of diseased food plants. We have studied the chemistry of phytoalexins isolated from the Solanaceae family since our successful first isolation and structure elucidation of rishitin (1) from diseased potatoes in 1968 (ref. 4)

PHYTOALEXINS FROM DISEASED POTATO

Potato cultivar "Rishiri" (Solanum tuberosum × S. demissum) carrying the R_1 gene is known to show a high level of resistance against the attack of Phytophthora infestans. Aged tuber slices of "Rishiri" were inoculated with an incompatible race of P. infestans (race 0) and incubated at 18-20 °C for 2 days. The stress metabolites isolated thus are summarized in Fig. 1a (ref. 5). These are all sesquiterpenes, of which the representatives are rishitin (1), lubimin (2), and oxylubimin (3). Particularly, 1 is the most typical one in respect to accumulation (0.001% yield) (ref. 6) and the first phytoalexin in the Solanaceae family (ref. 4). On the other hand, different results were obtained when we used a compatible race of pathogens (ref. 7). Whole tubers with peel of potato "Rishiri" were wounded with a knife and inoculated with a compatible race of P. infestans (race 1 pathogen) at 18 °C for 18 days. A careful chromatographic separation of the metabolites led to isolation of ten compounds. These metabolites are shown in Fig. 1b. In contrast to the result with the incompatible race, the main phytoalexins were solavetivone (4), phytuberin (5) and 1. It has recently been demonstrated that these, 4 and 5, are produced as the result of interaction of P. infestans (race 1) with Fusarium roseum LK. newly developed during a long inoculation time. Of these compounds, only 1 is considered to play a definite important role in the disease resistance, because of its high local accumulation (ref. 8) and antifungal activity (vide infra)(ref. 9). What role do other metabolites play? Thus we have investigated the biosynthetic correlation of these phytoalexins. Our procedure for biosynthetic

Fig. 1. Phytoalexins isolated from diseased potatoes

a) Phytoalexins isolated from "Rishiri" inoculated with an incompatible race of *Phytophthora infestans* (race 0).

rishitin (1), lubimin (2), oxylubimin (3), epilubimin (6), epioxylubimin (7), lubiminol (8), isolubimin (9), oxyglutinosone (10), rishitinone (11), rishitin M-1 (12), and rishitinol (13)

b) Phytoalexins isolated from "Rishiri" inoculated with a compatible race of *Phytophthora infestans* (race 1).

solavetivone (4), phytuberin (5), rishitin (1), lubimin (2), oxylubimin (3), epioxylubimin (7), rishitinol (13), phytuberol (14), 8-O-acetylrishitinol (15), and 2-epi-10-epilubimin (16).

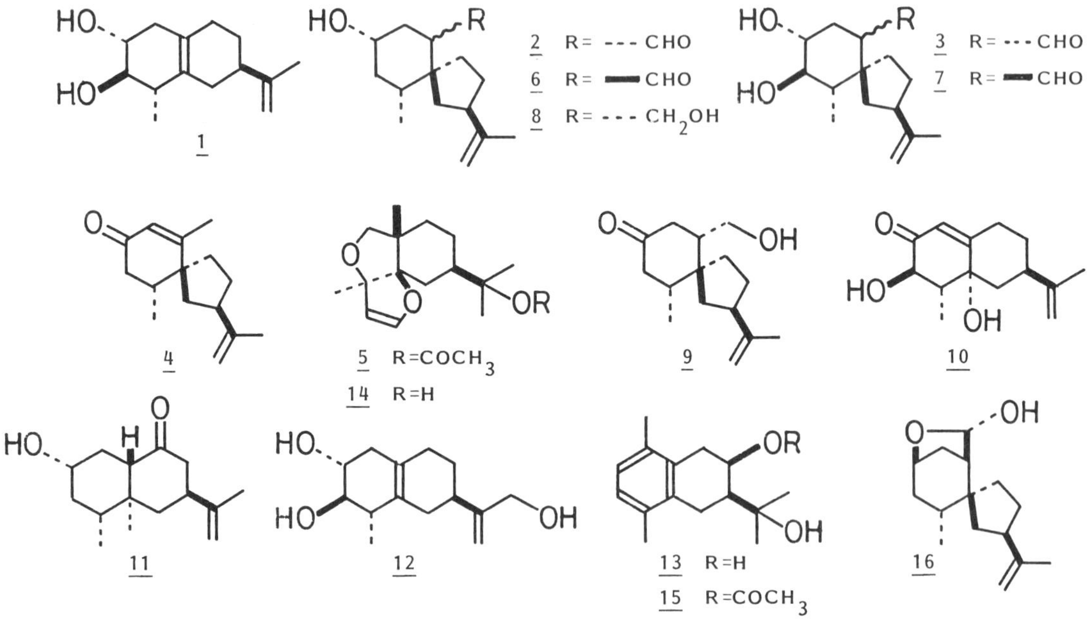

studies is based on the feeding experiments of doubly deuterated metabolic intermediates, which we have synthesized. Total syntheses of representative phytoalexins, 1 (ref. 10), 2 (ref. 11), 3, (refs. 13, 14), 5 (refs. 15, 16), and 13 (ref. 17), have already been carried out by us and other groups.

BIOSYNTHESIS OF RISHITIN

Attempts to elucidate the biogenetic pathways leading to rishitin (1) have been made by Stoessl (ref. 18), Kalan (ref. 19), and our group (ref. 20). The hypothesis proposed by Stoessl *et al.* suggests that 1 and three major spirovetivane phytoalexins, 2, 3, and 4, would be separately biosynthesized from farnesyl pyrophosphate. However, there has been no evidence to support their proposal. On the other hand, Kalan *et al.* and we reported almost the same alternate pathway, in which 4 would be transformed into 1 *via* 2 and 3 successively *in vivo*. While this hypothesis is based on our experiments with ^{14}C-labelled 3, indicating that 3 is an intermediate on the main pathway to 1, the requisite studies with labelled 4 have not been performed yet. In order to confirm this hypothesis, we prepared racemic 4 (4-D) labelled with two deuteriums at C-8 position as shown in Fig. 2. The deuterium contents of 4-D was estimated to be ca. 100% from its EI-mass spectra. Thin slices of aged potato "Rishiri" were incubated with (±)-4-D and related compouuds, and extracted with methanol-chloroform (1:1). The chloroform extracts were fractionated sequentially by column and thin-layer chromatography, giving the deuterated products shown in Table 1. These stress compounds were not detected on a thin-layer chromatogram of the corresponding extracts of the aged potato

Fig. 2. Synthesis of (±)-[8,8-$^2H_2$](4) (4-D)

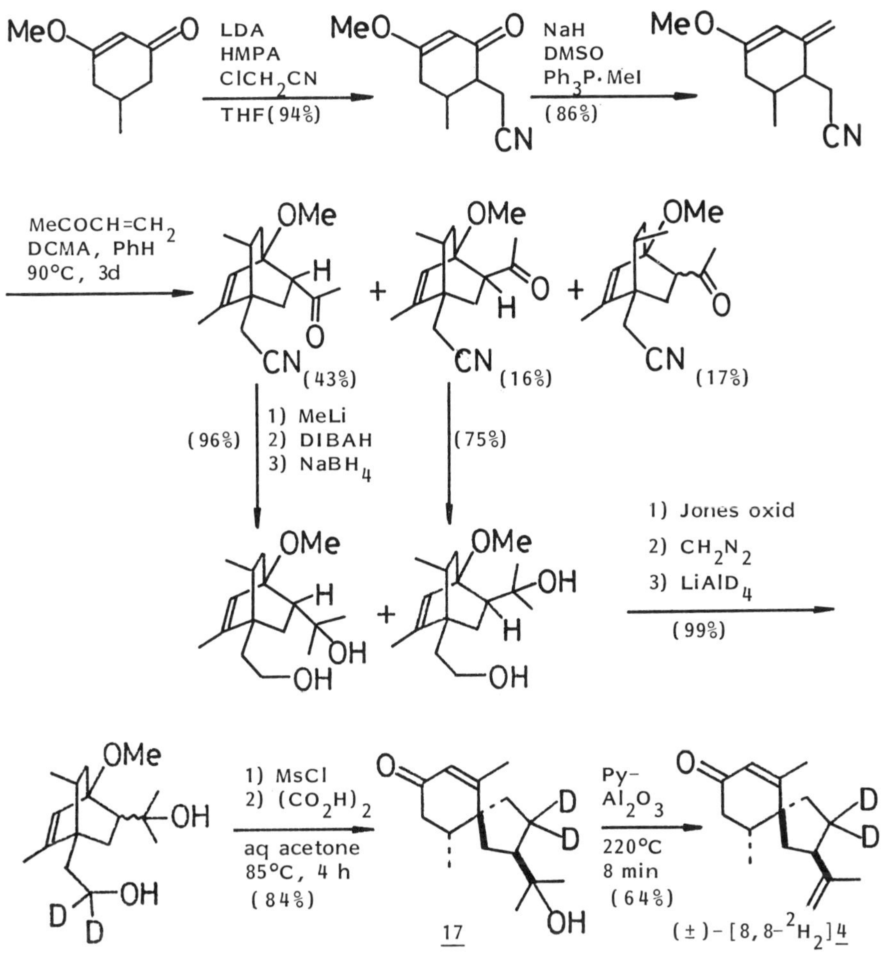

TABLE 1. Transformation of (±)-[8,8-$^2H_2$]4 (4-D) and related compounds in potato tubers.[a]

Run	Precursors	[8,8-$^2H_2$]-Products	Isolated Yield	$[\alpha]_D$[b]	[$^2H_2$]-content[c]
1	(±)-[8,8-$^2H_2$]4, [(±)-4-D], $[\alpha]_D$±0, (ca. 100%)[d]	(+)-4	74%[e]	+82 (-119)	ca. 100%
		(+)-2	29%[f]	+31 (+36)	97.8%
		(+)-3	7%[f]	+17 (+22)	97.7%
		(-)-1	10%[f]	-20 (-35)	93.5%
2	(+)-[8,8-$^2H_2$]4, [(+)-4-D], $[\alpha]_D$+82, (ca. 100%)[d]	(+)-4	71%	+71 (+119)	99.0%
3	(+)-[8,8-$^2H_2$]2, [(+)-2-D], $[\alpha]_D$+31, (97.8%)[d]	(+)-2	9%	+42 (+36)	97.7%
		(+)-3	11%	+12 (+22)	97.7%
		(-)-1	10%	-30 (-35)	97.2%

[a] All reactions were carried out by incubation of the precursors in water-methanol (1:1) to thin slices of aged potato tubers at 23-24 °C for 6 h. [b] Measured in ethanol. The figures in parentheses denote $[\alpha]_D$ of the corresponding unlabelled samples. [c] Determined by comparison of intensities of peaks corresponding to (M^+) and (M^+ -2) in the respective mass spectrum. [d] [8,8-$^2H_2$]-content. [e] Based on the (+)-enantiomer. [f] Based on the (-)-enantiomer.

slices treated only with water-methanol (1:1) in the same manner. The results summarized in Table 1 indicate that (i), (-)-4-D and (+)-2-D are incorporated efficiently into potato slices, (ii), natural (-)-4-D is metabolized into (+)-3-D *via* (+)-2-D. The present result, coupled with our previous one (ref. 20), establishes the main biosynthetic pathway leading to rishitin (1), the sequence (-)-4→(+)-2→(+)-3→(-)-1 (ref. 21). In this pathway, the metabolic mode of 4 to 2 has not been clarified yet. A reduction in time on feeding of 4 in potato led to isolation of a new metabolite (18), $[\alpha]_D$ -91.6°, designated as oxysolavetivone (ref. 22); the deuterated solavetivone, (±)-4-D (D_2-content, ca. 100%), was incorporated into potato slices (Rishiri) at 23 °C for 2 h, giving (-)-[8,8-$^2H_2$]18 (18-D) (D_2-content, 82.1%), $[\alpha]_D$ -5.9° in 17% yield. The corresponding racemic sample of 18-D was prepared from 17 and incorporated into potato slices (Rishiri) at 23 °C for 6 h to afford (+)-2-D (D_2-content, 97.6%), $[\alpha]_D$ +29°, in 7.3% yield. The result clearly indicates that only the natural form of 18 was transformed into 2 *in vivo*. In view of coexistence of various spirovetivane phytoalexins related structurally to 2 in diseased potato (refs. 7, 23), the role of compound (18) would be significant as a biosynthetic key intermediate leading to their production. Thus the main biosynthetic pathway of rishitin (1) has been confirmed as shown in Fig. 3. These results support the biogenetic route to 1 in a successive way from 4 as proposed by Kalan and us. However, it still remains to be elucidated how farnesyl pyrophosphate would be transformed into 4.

Fig. 3. The biosynthetic pathway of rishitin (1).

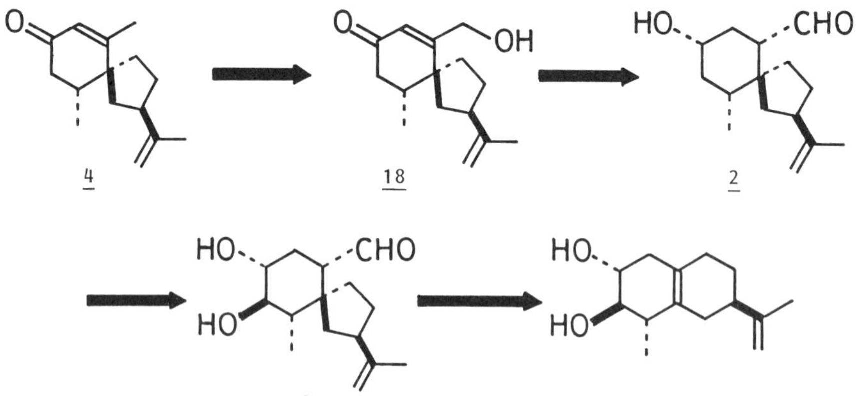

3 1

METABOLISM OF RISHITIN

Rishitin (1), by virtue of its fungitoxic nature (ED_{50} 3.8 × 10^{-4} M against *P. infestans*), constitutes part of the chemical defense mechanism of white potato tubers as described above. On the other hand, healthy tuber tissue is able to metabolize exogenously supplied 1 to the more polar, less toxic compounds, rishitin M-1 (12) and M-2 (19) (ref. 24). These metabolites are nontoxic against the healthy tissues at 1 × 10^{-2} M in water, while 1 is phytotoxic against the healthy tissues at 1 × 10^{-3} M in water. Such metabolism may prevent the accumulation of injurious levels of 1 in healthy tuber tissues. In 1977, we characterized the two metabolites, 12 and 19, and established that the detoxification of 1 involves hydroxylation of the isopropenyl side chain (ref. 25). Recently, Brindle *et al.* demonstrated the origin of the oxygen atoms in 1, 2, 12, and 19, using $^{18}O_2$ (ref. 26). Concerning this study, they proposed that formation of 12 proceeds, almost certainly, by direct attack on C-13 of 1 without any involvement of an epoxide or allylic rearrangement, while 19 would have originated from reduction of 12. Unfortunately, there does not seem to be any evidential experimental data to support this proposal. Their biogenesis has prompted us to investigate the metabolic mode of 1 in details by use of synthetic (-)-[12,12-$^2H_2$]1 (1-D) (ref. 27). Thin slices of aged potato (Rishiri) were incubated with (-)-1-D (D_2-content, >99.4%) at 23 °C for 6 h, and extracted with methanol-chloroform (1:1). The chloroform extracts were separated as usual, resulting in the isolation of deuterated 12 (12-D) and 19 (19-D) in 5.2 and 1.7% yields, respectively, along with an inseparable diastereoisomeric (1.64:1) mixture of [12,12-$^2H_2$]rishitin

Fig. 4. The metabolic pathway of rishitin (1).

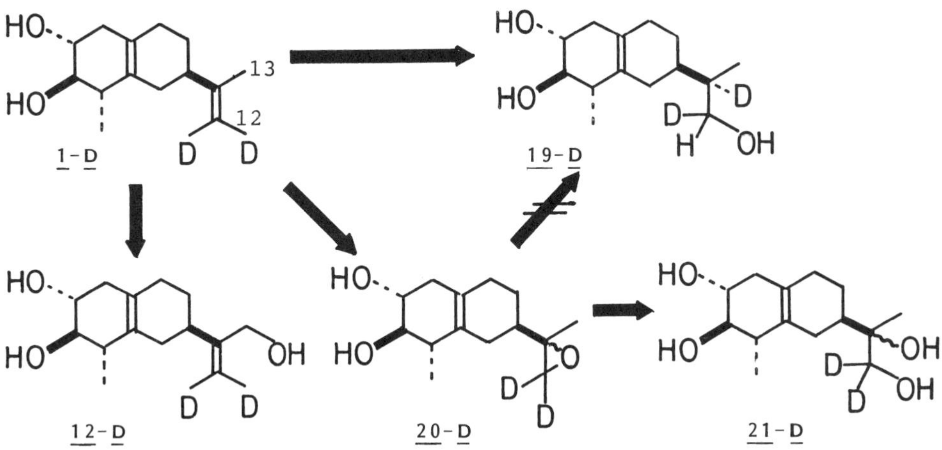

11,12-epoxides (20-D) in 3.8% yield. The ^1H NMR spectra of these metabolites indicated that the D_2-contents in 12-D, 19-D, and 20-D amounted to 91.0, 87.9, and 95.6%, respectively. The ^2H NMR spectra of the respective triacetates (12a-D) and (19a-D) and deuterated rishitin diacetate (1a-D), compared with ^1H n.m.r. spectra of the corresponding cold samples, revealed the respective deuterium distribution in the metabolites (12-D) and (19-D). The result summarized in Fig. 4 implies definitely that 12 originated from direct oxygenation on C-13 of 1, while 19 is produced through a 1,2-migration of one olefinic hydrogen at C-12 of 1 in a stereospecific manner. We know of no precedent for such so-called NIH-like shift of a hydrogen atom on isolated double bond as occurs in this case. The corresponding shift did not occur by incorporation (23 °C, 24 h) of the inseparable diastereomeric (1.44:1) mixture of synthetic 20-D into potato slices, only leading to formation of the corresponding hydrolyzed products (21-D) (ref. 28).

BIOSYNTHESIS OF PHYTUBERIN

Phytuberin (5) is also one of the most representative phytoalexins in the Solanaceae. Stoessl postulated the biogenetic pathway of 5 and phytuberol (14) by using doubly ^{13}C-labelled method (ref. 18). He proposed oxidative cleavage of the C-1 and C-2 bond of 1β,2α,5β,11-tetrahydroxyeudesm-3-ene as a plausible procursor, followed by intramolecular hemiacetal formation and successive generation of the second hydrofuran ring by an S_N2' process leading to formation of two hydrofuran ring moiety in phytuberin skeleton. However, there is no evidence for existence of any biogenetic intermediate *in vivo*. We reported the first total synthesis of (-)-5 and (-)-14 which in its critical stages mimics the proposed biogenetic route (ref. 15), although success of the synthesis does not always support the biogenetic scheme suggested by Stoessl. On the other hand, in view of the co-existence of phytuberin (5) with solavetivone (4) in diseased potato tissues (ref. 7), we postulated that 4 would be a biosynthetic precursor of 5 and 14. Aged tissue plugs of "Rishiri" were inoculated with *Fusarium roseum* Lk. cultivated on Irish Cobbler at 23 °C for 21 h, 5 and 14 being detected on TLC. These were then halved and the cut surface was incubated with (±)-4-D (*vide ante*) at 23 °C for 22 h, and the disks were blended with chloroform-methanol (1:1). The chloroform extracts were fractionated as usual to give pure samples of 5-D and 14-D. Compound 5-D was further saponified to give 14-D for comparison. EI-mass spectra of these phytuberols revealed that they contained at least 0.6-0.7% of D_2 derivatives. The ^2H n.m.r. spectrum of 14-D derivated from 5-D is given in Fig. 5. The broad singlets at 1.95 and 1.73 ppm were provisionally identified as D(8)eq and D(8)ax, respectively. These facts established that phytuberin (5) and phytuberol (14) originate from solavetivone (4) in diseased potato tissues (ref. 29). It has not been completely clarified yet how 4 would be metabolized to 5 and 14. A plausible biogenetic route is illustrated in Fig. 6, which is not contradictory to Stoessl's proposal. Interestingly, 14 seems to be a hydrolyzed product of 5 *in vivo*, because inoculation

Fig. 5. The ^{2}H NMR spectrum of phytuberol-D_2 (14-D) from feeding solavetivone-D_2 (4-D).

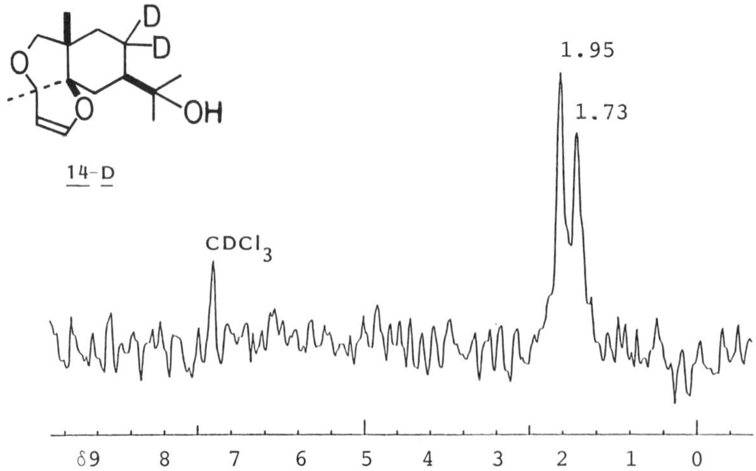

Fig. 6. A plausible biogenetic pathway of phytuberin (5) from solavetivone (4).

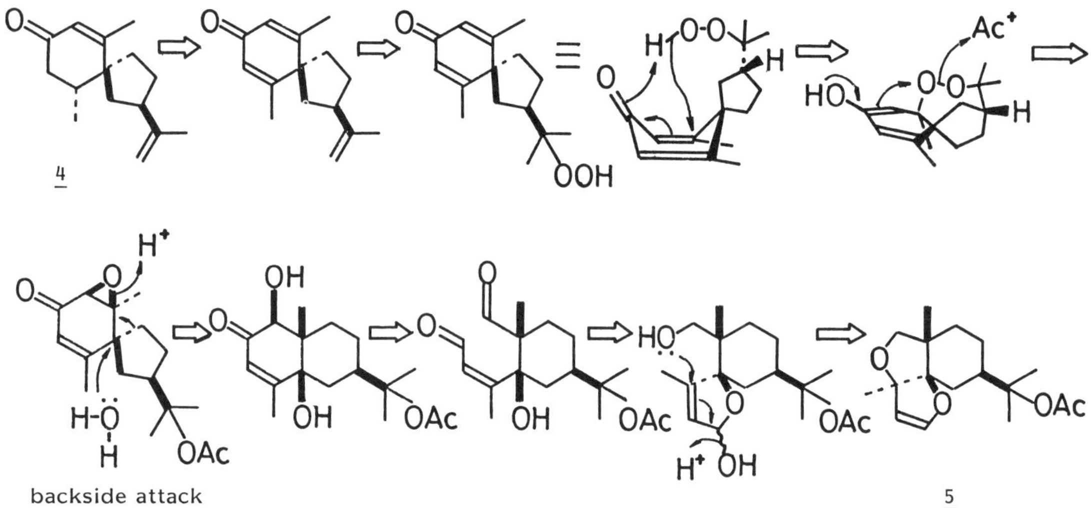

of (-)-5 in healthy thin slices of potato tubers (Rishiri) at 23 °C for 48 h led to isolation of (-)-14 in 13.2% yield (ref. 30). It is to be noted that phytuberin skeleton would be constructed sequentially from eudesmane, spirovetivane, and then eudesmane skeletons.

BIOLOGICAL ACTIVITY OF PHYTOALEXINS

There has been little knowledge concerning the physiological action of the above-mentioned phytoalexins. We checked antifungal activities on the germ-tube growth of the encysted zoospores of *P. infestans* and toxicities to the potato tuber tissue of the sesquiterpenes, solavetivone (4), oxysolavetivone (18), lubimin (2), oxylubimin (3), rishitin (1), rishitin M-1 (12), phytuberin (5), and phytuberol (14) (ref. 9). The results, as shown in Table 2., suggest that increase in hydrophilicity of compounds 1, 2, 4, and 5 by hydroxylation or deacetylation resulted in reduction in the toxicity and antifungal activity. It is noteworthy that each metabolite in the biosynthetic main pathway (from 4 to 12, see Figs. 3 and 4) showed alternately high and low antifungal activities and toxicities. These zig-zag appearance in physiological activities might be of significance in relation to the disease resistance. Although many types of phytoalexins may accumulate

TABLE 2. Inhibition of germ-tube growth of encysted zoospores of Phytophthora infestans by several potato phytoalexins and their toxicity to potato tuber tissue.

Compound	Inhibition of germ-tube growth ED_{50} ($\times 10^{-3}$M)	Toxicity to tuber tissue[a] Conc. of compound ($\times 10^{-3}$M) 8	4	2	1
solavetivone (4)	0.3	+++	++	+	-
oxysolavetivone (18)	1.0	±	-	-	-
lubimin (2)	0.4	++	+	-	-
oxylubimin (3)	1.0	±	-	-	-
rishitin (1)	0.3	++	+	±	-
rishitin M-1 (12)	5.1	-	-	-	-
phytuberin (5)	1.2	++	+	-	-
phytuberol (14)	1.5	-	-	-	-

[a] The levels of injury of the tissues were determined by the colour intensity of the test solution: -, no colour, and +++, reddish brown.

at disease lesions, the total antifungal activity or toxicity of the representative compounds, 1, 2, 4, and 5, will not be largely different from the total antifungal activity or toxicity of all the phytoalexins in the lesions.

HYDROGEN PEROXIDE AS AN ELICITOR

Elicitors are considered to be agents which cause the accumulation of phytoalexins in higher plants (ref. 31). They include substances of biotic and abiotic origin. In 1981, Bostock et al. isolated arachidonic and eicosapentaenoic acids as abiotic elicitors from P. infestans (ref. 32). These fatty acids elicit, as does the fungus, the accumulation of rishitin (1) and lubimin (2) in potato tubers. Their study has led us to propose the following hypothesis: the direct origin of elicitation would not be the fatty acids themselves, but hydrogen peroxide generated probably from the corresponding hydroperoxides, which these fatty acids could produce in contact with lipoxygenase in potato tubers (ref. 33). It has been well known that there are many kinds of specific enzymes to provide hydrogen peroxide from triplet oxygen directly or via superoxide on exposure of higher plants with air. Hydrogen peroxide has the longest life time in active oxygen species. If activities of catalase and peroxidase in plants would be lowered under abnormal conditions, hydrogen peroxide could act as a common trigger substance for eliciting phytoalexin production. On the basis of this hypothesis, we carried out the following experiments (ref. 34). Aged thin slices of potato tubers were inoculated with aq hydrogen peroxide solution at 23.5 °C for 2 days and blended with chloroform-methanol (1:1). The chloroform extracts were separated to afford rishitin (1) and phytuberin (5) as expected. The results are summarized in Table 3. Interestingly, hydrogen peroxide was not responsible for the cultivar-specific differential accumulation of phytoalexins in potato tubers. Attempts to extend the proposal to other plants are now in progress.

TABLE 3. Elicitation of phytoalexins from potato tuber tissue with hydrogen peroxide.

Potato-cultivar	Elicitor	Metabolites (mg/kg, fresh weight) rishitin (1)	Phytuberin (5)
Irish Cobbler	0.3% H_2O_2	1.8	-
	3.0% H_2O_2	4.4	4.2
Rishiri	0.3% H_2O_2	2.3	-
	3.0% H_2O_2	6.8	6.0

REFERENCES

1. K. O. Müller and H. Börger, Arb. Biol. Anst. Reichsant. (Berl.), 23, 189 (1940).

2. A. Stoessl, J. B. Stothers, and E. W. B. Ward, Phytochem., 15, 855 (1976).
3. J. D. Paxton, Phytopath. Z., 101, 106 (1981).
4. K. Tomiyama, T. Sakuma, N. Ishizaka, N. Sato, N. Katsui, M. Takasugi, and T. Masamune, Phytopathol., 58, 115 (1968); N. Katsui, A. Murai, M. Takasugi, K. Imaizumi, and T. Masamune, Chem. Commun., 43 (1968).
5. T. Masamune, A. Murai, and N. Katsui, Kagaku To Seibutsu, 16, 648 (1978).
6. T. Masamune, A. Murai, M. Takasugi, A. Matsunaga, N. Katsui, N. Sato, and K. Tomiyama, Bull. Chem. Soc. Jpn., 50, 1201 (1977).
7. N. Katsui, Y. Takahashi, N. Sato, A. Murai, and T. Masamune, J. Chem. Soc. Jpn., 659 (1981).
8. T. Nakajima, K. Tomiyama, and K. Kinukawa, Ann. Phytopathol. Soc. Jpn., 41, 49 (1975).
9. N. Sato, Y. Yoshizawa, H. Miyazaki, and A. Murai, ibid., 51, 494 (1985).
10. A. Murai, K. Nishizakura, N. Katsui, and T. Masamune, Tetrahedron Lett., 4399 (1975); Bull. Chem. Soc. Jpn., 50, 1206 (1977).
11. A. Murai, S. Sato, and T. Masamune, J. Chem. Soc., Chem. Commun., 511, 513 (1982); Bull. Chem. Soc. Jpn., 57, 2286, 2291 (1984).
12. C. Iwata, Y. Takemoto, H. Kubota, T. Kuroda, and T. Imanishi, Tetrahedron Lett., 26, 3231 (1985).
13. A. Murai, S. Sato, and T. Masamune, ibid., 22, 1033 (1981); J. Chem. Soc., Chem. Commun., 904 (1981); Bull. Chem. Soc. Jpn., 57, 2276, 2282 (1984).
14. K. Yamada, S. Goto, H. Nagase, and A. T. Christensen, J. Chem. Soc., Chem. Commun., 554 (1977); C. Iwata, T. Fusaka, T. Fujiwara, K. Tomita, and M. Yamada, ibid., 463 (1981).
15. A. Murai, M. Ono, A. Abiko, and T. Masamune, J. Am. Chem. Soc., 100, 7751 (1978); Bull. Chem. Soc. Jpn., 55, 1202 (1982).
16. D. Caine and T. L. Smith, Jr., J. Am. Chem. Soc., 102, 7568 (1980); J. A. Findlay, D. N. Desai, G. C. Lonergan, and P. S. White, Can. J. Chem., 58, 2827 (1980); F. Kido, H. Kitahara, and A. Yoshikoshi, J. Chem. Soc., Chem. Commun., 1236 (1981).
17. N. Katsui, A. Matsunaga, K. Imaizumi, T. Masamune, and K. Tomiyama, Tetrahedron Lett., 83 (1971); Bull. Chem. Soc. Jpn., 45, 2871 (1972).
18. A. Stoessl, J. B. Stothers, and E. W. B. Ward, Can. J. Chem., 56, 645 (1978).
19. E. B. Kalan and S. F. Osman, Phytochem., 15, 775 (1976); E. B. Kalan, J. M. Patterson, and D. P. Schwartz, Plant Physiol., 57, (Supplement) Abst., 473 (1976).
20. K. Sato, Y. Ishiguri, N. Doke, K. Tomiyama, F. Yagihashi, A. Murai, N. Katsui, and T. Masamune, Phytochem., 17, 1901 (1978).
21. A. Murai, S. Sato, A. Osada, N. Katsui, and T. Masamune, J. Chem. Soc., Chem. Commun., 32 (1982).
22. A. Murai, Y. Yoshizawa, N. Katsui, S. Sato, and T. Masamune, Chem. Lett., 771 (1986).
23. D. T. Coxon, K. R. Price, J. B. Stothers, and A. Stoessl, J. Chem. Soc., Chem. Commun., 348 (1979); A. Stoessl and J. B. Stothers, Can. J. Chem., 58, 2069 (1980).
24. Y. Ishiguri, K. Tomiyama, N. Doke, A. Murai, N. Katsui, F. Yagihashi, and T. Masamune, Phytopathol., 68, 720 (1978); Y. Ishiguri, K. Tomiyama, A. Murai, N. Katsui, and T. Masamune, Ann. Phytopathol. Soc. Jpn., 44, 52 (1978).
25. A. Murai, N. Katsui, F. Yagihashi, T. Masamune, Y. Ishiguri, and K. Tomiyama, J. Chem. Soc., Chem. Commun., 670 (1977).
26. P. A. Brindle, T. Coolbear, P. J. Kuhn, and D. R. Threlfall, Phytochem., 24, 1219 (1985).
27. A. Murai and T. Masamune, Chem. Lett., 1143 (1984).
28. A. Murai, Y. Yoshizawa, M. Ikura, N. Katsui, and T. Masamune, J. Chem. Soc., Chem. Commun., in press.
29. A. Murai, Y. Yoshizawa, T. Masamune, and N. Sato, unpublished results.
30. A. Murai, Y. Yoshizawa, H. Miyazaki, and T. Masamune, unpublished results.
31. Cf., Phytoalexins, J. A. Bailey and J. W. Mansfield ed., Blackie, Glasgow and London, p.289, (1982).
32. R. M. Bostock, J. A. Kuc, and R. A. Laine, Science, 212, 67 (1981).
33. Cf., E. J. Corey, J. O. Albright, A. E. Barton, and S. Hashimoto, J. Am. Chem. Soc., 102, 1435 (1980).
34. A. Murai, Y. Yoshizawa, T. Hasegawa, and T. Masamune, unpublished results.

Chemistry and biological function of host-specific phytotoxins

Tamio Ueno

Pesticide Research Institute, Faculty of Agriculture, Kyoto University, Kyoto 606, Japan

Abstract - Role of host-specific toxins on host-parasite interaction in some plant diseases was investigated from chemical as well as a plant pathological point of view. As a result of phytopathological studies on host recognition by *Alternaria mali* and *A. kikuchiana*, the causal agents of *Alternaria* leaf spot disease of apple and black spot disease of Japanese pear respectively, AM-toxin I, II and III, and AK-toxin I and II were isolated as the primary factors of the diseases. The chemical structures were elucidated for **1a**, **1b**, **1c**, **2a** and **2b** (Fig. 1) respectively. The isolated AM- and AK-toxins showed not only extremely potent host-specific toxicity, but solutions of these toxins also induced the same necrotic symptoms on the leaves of their host plants as those observed on inoculation with the spores of the causal fungi and caused almost instantaneous electrolyte loss from leaf tissues of susceptible cultivars. Structural features of the toxins that express the host-specific toxicity were also analyzed by a series of experiments which used synthetic analogs of AM and AK-toxins.

INTRODUCTION

Phytotoxins, which are produced by plant pathogens, are known to be significant causal factors in many destructive diseases of plants. However the word phytotoxin also has a different meaning. Vivotoxin has been used to define a substance produced *in vivo*, by a pathogen or host, which functions in disease development but is not the initial inciting agent (ref. 1). At present, the most intriguing problem in the field of plant pathology is "host-specificity", that is, how the pathogens recognize their host plants in the infection process. In order to explain the specific interaction between host plants and pathogens, a number of biochemical mechanisms have been proposed by plant pathologists. In some plant pathological and physiological studies on the disease caused by *Helminthosporium victoriae*, victoria blight of oat, Wheeler and Luke proposed pathotoxin (ref. 2) and later Pringle and Scheffer also proposed host-specific (selective) toxin as descriptive words to explain the metabolite(s) which have host-specific toxicity and were considered to be major determinant(s) of pathogenicity (ref.3). Usually, trivial names of host-specific toxins are derived from abbreviation of mycological names of causal fungi. Since then, eleven more examples of host-specific toxins have been reported in saprophytic fungal diseases, six among them being from the genus *Alternaria* as shown in Table 1, although the number of *Alternaria* host-specific toxins in this category will increase as the studies progress (ref. 4). On the other hand, four examples have been reported on the diseases caused by several species of the genus *Helminthosporium* (see Table 1).

STRUCTURES OF HOST-SPECIFIC TOXINS

Structure elucidation of the host specific taxins have been required for a long time by plant pathologists to promote the investigation of host recognition mechanism and the disease establishment by saprophytic fungi, e.g.*Alternaria* and *Helminthosporium* species, on the molecular level. However, chemical studies on the toxins have encountered some difficulties on account of their unfavorable nature; their content in natural sources is very scant, and they are so unstable as to be easily decomposed during the isolation processes. After the structural elucidation of AM-toxins in 1975 in our laboratory, new state-of-the-art purification and analytical techniques have been applied to overcome the difficulties in the occuring purification and structure determination of host-specific toxins. As a result, eleven examples of their chemical structures have been demonstrated by several groups of natural product chemists (see Fig. 1 and Table 1). In this paper we report on recent progress made in our laboratory, which is related to the structural features of AM-, AK- and AF-toxins that express their host-specific toxicity. Details of the isolation and structural elucidation of AM- and AK-toxins have been previously published in the proceeding of the 5th International Congress of Pesticide Chemistry (Kyoto, Japan, 1982) (ref. 5).
Alternaria mali and *A. kikuchiana* cause leaf spot disease of apple and black spot disease of

TABLE 1. Host-specific toxins reported by the end of 1985

Causal fungus	Disease*	Host-specific toxin**	Susceptible cultivar
Alternaria kikuchiana	Black spot of Japanese pear (1933)	AK-toxin I and II (1983) (ref. 6)	Nijisseiki etc.
A. mali	Leaf spot(Alternaria blotch) of apple (1966)	AM-toxin I(alternariolide), II and III (1975) (ref. 7)	Indo, Red Gold etc.
A. citri	Brown spot of citrus (1966)	ACTG-toxin A, B and C (1986) (ref. 8)	Emperor mandarine, Dancy tangerine
A. alternata f. sp lycopersici	Stem canker of tomato (1976)	AAL-toxin A and B (1981) (ref. 9)	Early park 7
A. fragariae	Alternaria black spot of strawberry (1978)	AF-toxin I, II and III (1985) (ref. 10)	Morioka 16
A. citri	Brown spot of citrus (1979)	ACRL-toxin (1985) (ref. 11)	Rough lemon
A. longipes (tabacco pathotype)	Brown spot of tabacco (1981)	ALT-toxin (unknown)	
Helminthosporium victoriae	Victoria blight of oat (1947)	HV-toxin B, C, D and E (1985) (ref. 12)	Victoria
H. carbonum	Nothern leaf spot of maize (1965)	HC-toxin (1983) (ref. 13)	K-44, K-66 etc.
H. sacchari	Eye spot of sugar cane (1969)	HS-toxin A, B and C (1983) (ref. 14)	51 NG-97
H. maydis race T	Southern leaf bright of maize (1970)	HMT-toxin (1979) (ref. 15)	Texas male sterile cultivar
Periconia circinata	Milo of sorghum (1961)	PC-toxin (unknown)	Giant milo
Phyllosticta maydis	Yellow blight of maize (1973)	PM-toxin (1983) (ref. 16)	T-cell cultivar

* The number in the parenthesis indicates the year of finding the host-specific toxins.
** The number in the parenthesis indicates the year of structure elucidation of the host-specific toxin.

Japanese pear respectively, and produce necrotic blotches specifically on the leaves, shoots and fruits of susceptible cultivars. Various cultivars of apple show a great difference in the susceptibility to A. mali infection, as is the case between cultivars of Japanese pear to A. kikuchiana (see Table 2). As a result of a series of plant pathological experiments on these diseases, Sawamura (ref. 17) and Tanaka (ref. 18) observed that the culture filtrates of these fungi induced the same necrotic symptoms as did their spore inoculations. We isolated the active principles from the culture filtrates and named them AM-toxin and AK-toxin respectively, the structures of which were also elucidated including the absolute configurations (ref. 7 and 6). Recently, a distinct strawberry pathotype of A. alternata (A. fragariae) producing host-specific toxins was found (ref. 19). Interestingly, this pathogen was highly pathogenic not only to the susceptible strawberry cultivar, Morioka 16, but also to certain cultivars of Japanese pear, e.g. Nijisseiki. The toxins were named AF-toxin and these structures were elucidated by Nakatsuka et al. but not the stereochemistry. The chemical bases of the host selection by these three Alternaria species is described.

i) AM-toxin I, II and III
The chemistry of AM-toxins has been extensively developed by two groups in Japan. Both proposed the same structure for AM-toxin I (alternariolide) almost at the same time (ref. 6) which is the first example of structure elucidation of host-spectific toxin (see Fig. 1, 1a). In addition our group reported the isolation and structure elucidation of other host-specific metabolites, AM-toxin II and III (Fig. 1, 1b and 1c) (ref. 7). The proposed structures of which were confirmed by total syntheses (ref. 20).
ii) AK-toxin I and II
For more than fifty years, black spot disease of Japanese pear (Pyrus serotina Rehd.) caused by Alternaria kikuchiana has been one of the most destructive disease in pear orchards in our country. The most popular cultivar, Nijisseiki, is highly susceptible. The report on this disease by Tanaka suggested the presence of host-specific phytotoxic metabolite(s) in the culture filtrate, which is the first description of the host-specific toxin (ref. 18). This enticed many natural product chemists to try the isolation of the active principle(s) from the culture broth of Alternaria kikuchiana, and as a result several phytotoxic metabolites were isolated from it, but none of them showed host-specific toxicity which is characteristic of the culture filtrate. However, isolation of the host-specific active principles was performed by our group in 1983 which were named AM-toxin I and II. The chemical structure of AM-toxin I was characterized as 8-[(2'S,3'S)-2'-acetylamino-3'-methyl-3'-phenyl-propionyl-oxy]-(8R,9S)-9,10-epoxy-9-methyl-deca-(2E,4Z,6E)-trienoic acid from chemical, spectral and X-ray crystallographic data (Fig. 1, 2a). The structure of AK-toxin II was also shown to

TABLE 2. Pathogenicity of A. mali, A. kikuchiana and A. fragariae and host-specific toxicity of AM-, AK- and AF-toxins

pathogenic fungi		apple		Japanese pear		strawberry	
		Indo	Jonathan	Nijisseiki	Chojuro	Morioka-16	Hokowase
name of fungi	assay methods	(S)	(R)	(S)	(R)	(S)	(R)
Alternaria mali	spore inoculation	+++	+	+	-	-	n.t.
	culture filtrate	+++	+	+	-	-	n.t.
	AM-toxin I(10^{-5}M)	+++	+	+	-	-	n.t.
	AM-toxin I(10^{-8}M)	+	-	-	-	-	n.t.
Alternaria kikuchiana	spore inoculation	-	-	+++	-	-	-
	culture filtrate	-	-	+++	-	-	-
	AK-toxin I	-	-	+++	-	-	-
	AK-toxin II	-	-	+++	-	-	-
Alternaria fragariae	spore inoculation	-	-	++	-	+++	-
	culture filtrate	-	-	++	-	+++	-
	AF-toxin I	-	-	++	-	+++	-
	AF-toxin II	-	-	++	-	-	-
	AF-toxin III	-	-	±	-	+++	-

Plus and minus marks indicate positive or negative response. Induced necrosis rating scale is as follows: +++, very severe; ++, severe; +, moderate; and -, nonecrosis.
S and R indicate susceptible and resistant respectively.
n.t. means "not tested".

be 3'-demethyl derivative of AK-toxin I, by comparison the spectral data with those of AK-toxin I (Fig. 1, 2b) (ref. 6).

iii) AF-toxin I, II and III

Another example of Alternaria host-specific toxin in Japan was recently reported by Watanabe (ref. 21). The causal pathogen, A. fragariae, of black spot of strawberry parasitizes not only its susceptible cultivars, e.g. Morioka-16, but also several cultivar of Japanese pear including Nijisseiki as shown in Table 2. This fungus produces three host-specific toxins (named AF-toxin I, II and III, respectively). Isolation and structural elucidation of these toxins were performed quite recently without the steriochemistry by Nakatsuka et al. (ref. 10). Their structures resemble those of AK-toxins and all of them contain 9, 10-epoxy-9-methyl-deca-(2,4,6)-trienoic acid as one of their structural moieties. The differences in the structures are observed in the geometry at the 2, 4 and 6 positions of the trienoic acid as well as another moiety. AF-toxins contain α-hydroxy-β-methyl-pentanoic acid in which a β-hydroxyl group is free in AF-toxin II and esterified with α,β-dihydroxy-β-methyl-butyric acid and α-hydroxy-β-methyl-butyric acid in AF-toxin I and II, respectively (see Fig. 1, 3a-a, 3a-b, 3a-c, 3b-a, 3b-b, 3b-c, 3c-a, 3c-b and 3c-c). AF-toxin Ia (3a-a) and IIa (3b-a) were isolated from the culture filtrate as major components. It should be noted that the geometry at 2E, 4E and 6Z of trienoic parts are different from that of AK-toxins.

iv) other host-specific toxins

Eleven examples of host-specific taxins whose structures were proposed to date indicate that host-specific toxins cover a vast variety of organic compounds; AM-toxins, HC-toxin and HV-toxins relate to cyclic peptide; AK-toxins and AF-toxins to esters composed of amino acids and hydroxy acids with epoxy-hydroxy-unsaturated carboxylic acid, HMT-toxins and PM-toxins to aliphatic polyhydroxy-polyketo compounds, and HS-toxins, ACRL-toxin and ACTG-toxins to terpenes or their related compounds (see Fig. 1). Besides all of them are characterized as new compounds which contain unusual structural moieties. These facts not only attract the interest of natural product chemists but provide a new target for the compounds with selective phytotoxicity.

SYNTHESES AND STRUCTURE-TOXICITY RELATIONSHIP OF HOST-SPECIFIC TOXINS

Total synthesis of host-specific toxins has been reported in the case of AM-toxin I, II and III (ref. 20), HMT-toxins (ref. 22) and AF-toxins (ref. 10). These synthetic efforts represent not only a continuation of a traditional objective of natural product chemistry, but are also very important for proof of proposed structures, especially when X-ray crystallographic procedures can not be applied to the structure elucidation. Further importance of synthetic work should be placed on structurally related compounds of each host-specific toxin in order to investigate structural features to express the host-specific toxicity. In this section, structure -toxicity relationship in AM-toxins and AK-toxins will be discussed on the basis of the experiments developed in our laboratory.

i) AM-toxins

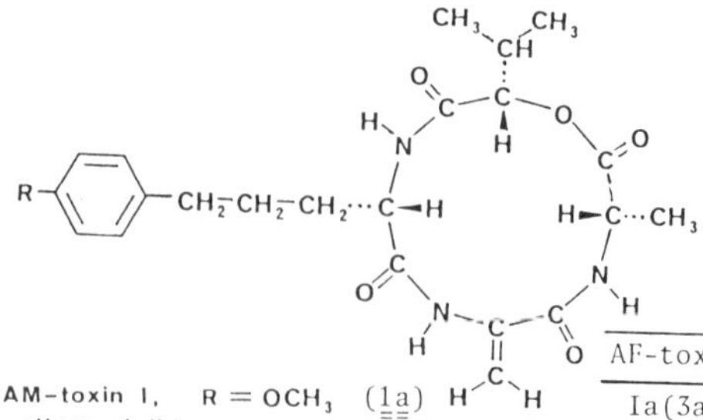

AM-toxin I, R = OCH_3 (1a) alternariolide

AM-toxin II, R = H (1b)

AM-toxin III, R = OH (1c)

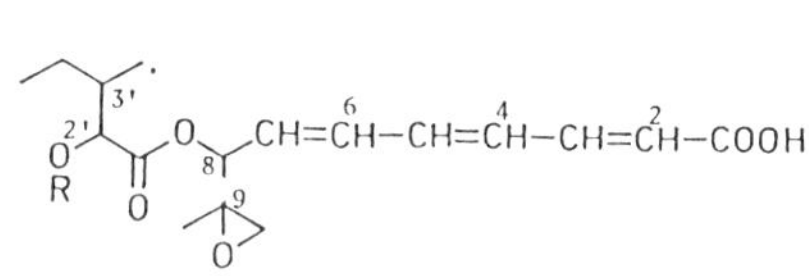

AF-toxin	R	C-2'	C-3'	C-8	C-9	C-2	C-4	C-6
Ia(3a-a)	$COCH(OH)C(CH_3)_2OH$	unknown				*t*	*t*	*c*
Ib(3a-b)	"	"				*t*	*c*	*t*
Ic(3a-c)	"	"				*t*	*t*	*t*
IIa(3b-a)	H	"				*t*	*t*	*c*
IIb(3b-b)	"	"				*t*	*c*	*t*
IIc(3b-c)	"	"				*t*	*t*	*t*
IIIc(3c-a)	$COCH(OH)CH(CH_3)_2$	"				*t*	*t*	*c*
IIIc(3c-b)	"	"				*t*	*c*	*t*
IIIc(3c-c)	"	"				*t*	*t*	*t*

I R=CH_3 (2a)

II R=H (2b)

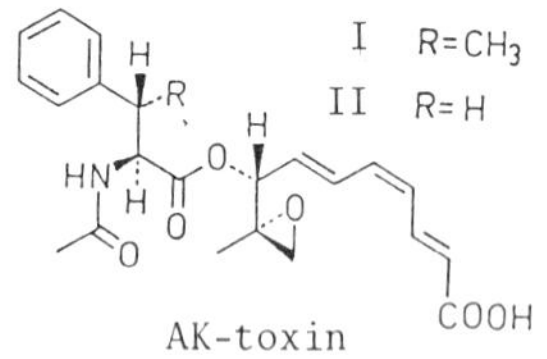

AK-toxin

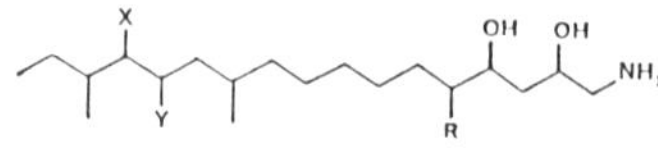

	R	X	Y
AAL-toxin A1.	-OH.	-OH.	$-OOC-CH_2-CH(COO^-)-CH_2-COO^-$
A2.	-OH.	$-OOC-CH_2-CH(COO^-)-CH_2-COO^-$	-OH
AAL-toxin B1.	-H.	-OH.	$-OOC-CH_2-CH(COO^-)-CH_2-COO^-$
B2.	-H.	$-OOC-CH_2-CH(COO^-)-CH_2-COO^-$	-OH

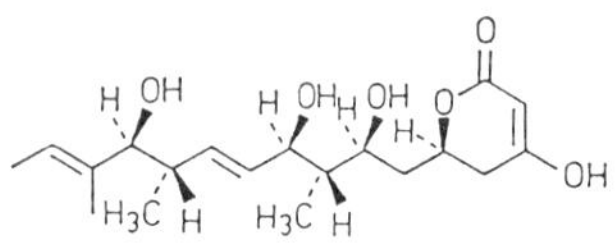

ACRL-toxin

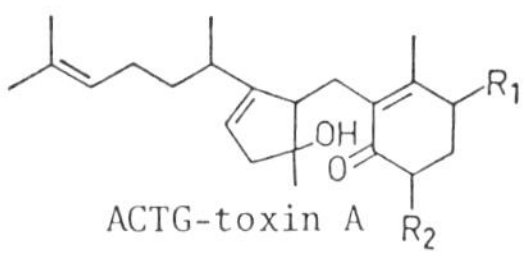

ACTG-toxin A

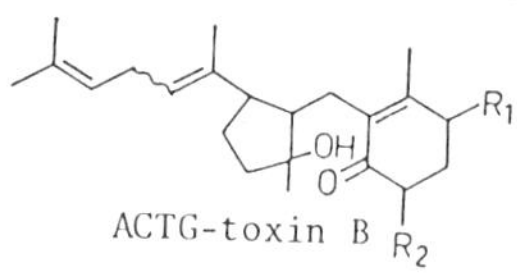

ACTG-toxin B

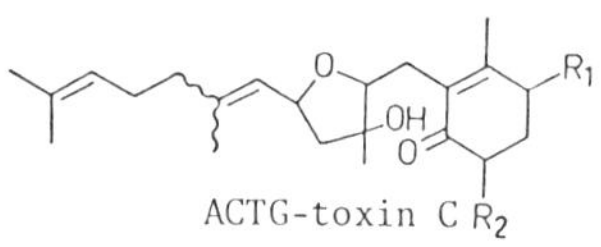

ACTG-toxin C

R_1=OH, R_2=H ⇌ R_1=H, R_2=OH

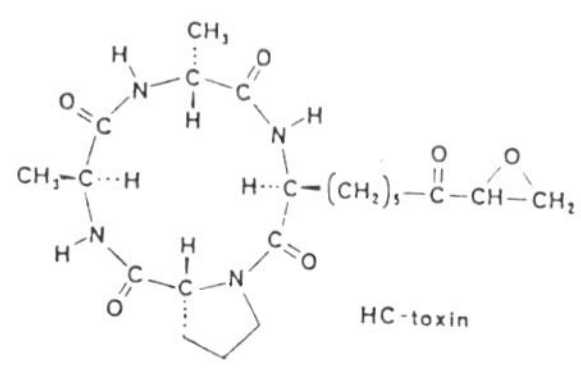

HC-toxin

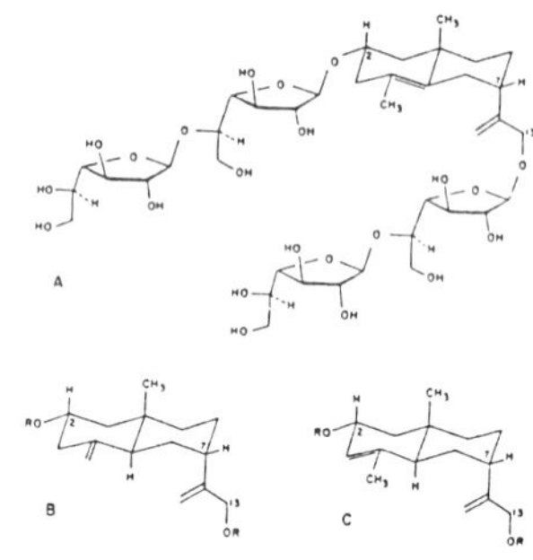

HS-toxin

R=5-O-(β-galactofuranosyl)-β-galactofuranosyl-

Victorin(HV-toxin)

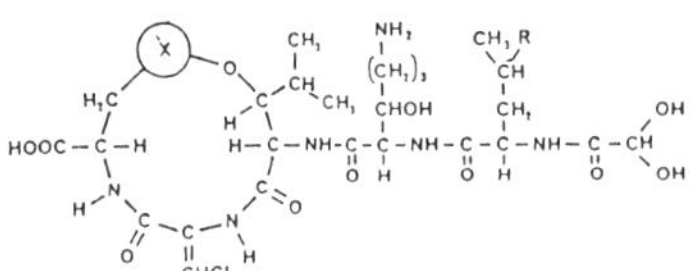

Victorin	X	R
B	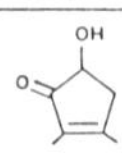	CH_2Cl
C	"	$CHCl_2$
D	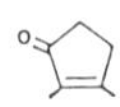	"
E	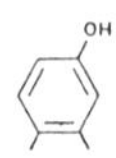	"

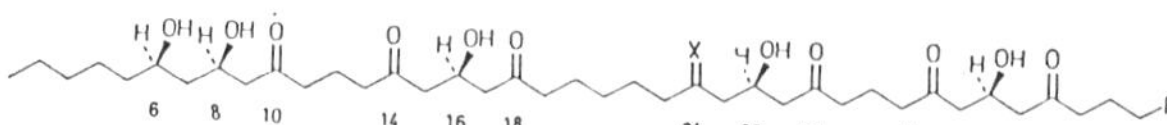

HMT-toxin	Band 1(C_{41})	Band 2(C_{41})	Band 3(C_{39})	Band 1'(C_{39})
R	$COCH_3$	$COCH_3$	H	H
X	O	H OH	O	H OH

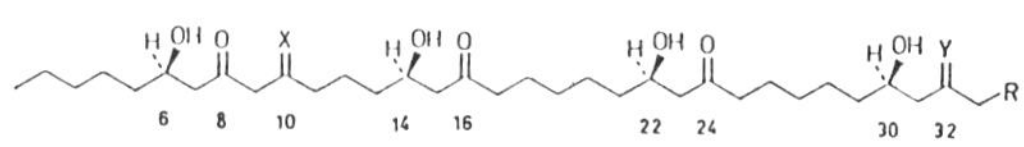

PM-toxin	A(C_{33})	B(C_{33})	C(C_{35})	D(C_{33})
R	H	H	H OH	H
X	H_2	H_2	H_2	H OH
Y	O	H OH	H OH	O

Fig. 1 The structures of host-specific toxins

TABLE 3. Biological activity of AM-toxins and their synthesized analogs inducing necrosis on the leaves of a susceptible cultivar Indo (see ref. 21, 22 and 24)

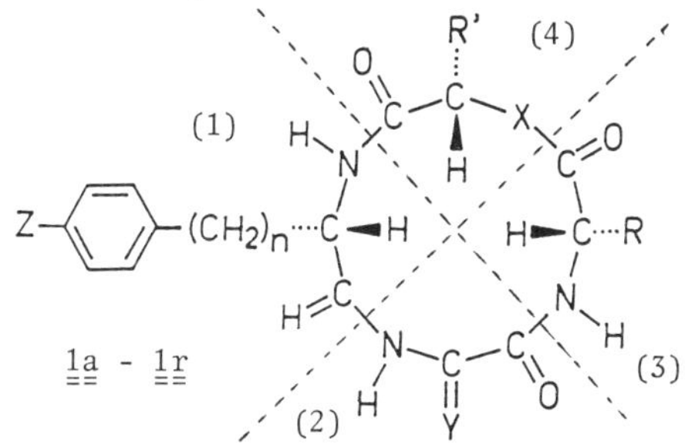

1a - 1r

No.	AM-toxins, synthesized analogs	X	Y	Z	n	R	R'	Activity*
1a	AM-toxin I	0	$=CH_2$	OCH_3	3	CH_3	$CH(CH_3)_2$	10^{-8}M
1b	AM-toxin II	0	$=CH_2$	H	3	CH_3	$CH(CH_3)_2$	10^{-6}M
1c	AM-toxin III	0	$=CH_2$	OH	3	CH_3	$CH(CH_3)_2$	10^{-7}M
1d	[L-Ala^2]AM-toxin I	0	$-CH_3$,•••H	OCH_3	3	CH_3	$CH(CH_3)_2$	>10^{-3}M
1e	[D-Ala^2]AM-toxin I	0	•••CH_3,-H	OCH_3	3	CH_3	$CH(CH_3)_2$	10^{-4}M
1f	[ΔAbu^2]AM-toxin I	0	$=CH-CH_3$	OCH_3	3	CH_3	$CH(CH_3)_2$	>10^{-3}M
1g	[L-Tyr(Me)^1;L-Ala^2]AM-toxin I	0	$-CH_3$,•••H	OCH_3	1	CH_3	$CH(CH_3)_2$	>10^{-3}M
1h	[L-Tyr(Me)^1;D-Ala^2]AM-toxin I	0	•••CH_3,-H	OCH_3	1	CH_3	$CH(CH_3)_2$	>10^{-3}M
1i	[L-Tyr(Me)^1]AM-toxin I	0	$=CH_2$	OCH_3	1	CH_3	$CH(CH_3)_2$	10^{-4}M
1j	[L-Amb^1]AM-toxin I	0	$=CH_2$	OCH_3	2	CH_3	$CH(CH_3)_2$	10^{-6}M
1k	[L-Amh^1]AM-toxin I	0	$=CH_2$	OCH_3	4	CH_3	$CH(CH_3)_2$	10^{-4}M
1l	[L-Val^4]AM-toxin I	NH	$=CH_2$	OCH_3	3	CH_3	$CH(CH_3)_2$	10^{-5}M
1m	[L-Tyr(Me)^1;L-Val^4]AM-toxin I	NH	$=CH_2$	OCH_3	1	CH_3	$CH(CH_3)_2$	10^{-5}M
1n	[L-Abu^3]AM-toxin I	0	$-CH_2$	OCH_3	3	CH_2CH_3	$CH(CH_3)_2$	10^{-5}M
1o	[L-Leu^3]AM-toxin I	0	$=CH_2$	OCH_3	3	$CH_2CH(CH_3)_2$	$CH(CH_3)_2$	10^{-6}M
1p	[L-Phe^3]AM-toxin I	0	$=CH_2$	OCH_3	3	CH_2Ph	$CH(CH_3)_2$	10^{-7}M
1q	[L-Lac^4]AM-toxin I	0	$=CH_2$	OCH_3	3	CH_3	CH_3	10^{-3}M
1r	[L-Hmp^4]AM-toxin I	0	$=CH_2$	OCH_3	3	CH_3	$CH_2CH(CH_3)_2$	10^{-5}M

* The values indicate threshold concentrations inducing necrosis after 18hrs when a droplet of the sample solution was placed on the bottom of a young fresh leaf.

After total syntheses of AM-toxin I, II and III (ref. 20), Izumiya et al. have developed the synthetic work on AM-toxin analogs, 1d - 1r (Table 3), in order to investigate the requisite structural features to express phytotoxicity. The following abbreviations were adopted to indicate the structures of the synthesized analogs: Abu, 2-amino-butyric acid; ΔAbu, 2-amino-2-butenoic acid; Ala, alanine; Amb, 2-amino-4-(p-methoxyphenyl)butyric acid; Amh, 2-amino-6-(p-methoxyphenyl)hexanioc acid; Hmp, 2-hydroxy-4-methyl pentanoic acid; Lac, lactic acid; Leu, leucine; Phe, phenylalanine; Tyr(Me), p-methoxyphenylalanine and Val, valine. The numbers in the brackets indicate the position in the cyclic peptide structure as shown in the figures. The presence of the bulky and hydrophobic side chain in (1) and the cyclotetrapeptide structure are important for exhibiting biological activity, because 1i, 1j and 1k composed of shorter and longer methylene group indicate still high activity. Linear compounds obtained by selective methanolyses of the lactone linkages AM-toxin I, II and III were no longer active. Hydrogenation of the exomethylene part gave [1-Ala^2]- and [D-Ala^2]AM-toxin I, 1d and 1e respectively, the latter of which still retains activity. By replacement of one of the hydrogens of the exomethylene group, 1f loses its activity. This fact suggests that the bulkiness of this part in (2) and the absolute configuration in (2) are closely related to toxicity. The most important feature is observed in the part of (3). Replacement of the methyl group by bulky ethyl, isobutyl and phenylmethyl groups, 1n, 1o and 1p, did not affect the toxicity remarkably. Further interesting features are also observed in the part of (4). A characteristic feature of AM-toxins is the occurence of a lactone bond in their cyclic backbone. Other phytotoxic cyclic terrapeptides such as HC-toxin and tentoxin are exclusively composed of lactam linkages. Therefore, we tried to synthesize AM-toxin I analogs containing L-Val in place of L-2-hydroxy-3-methylbutyric acid, 1l and 1m respectively, which still retain activity. The bulkiness of alkyl group of side chain of the part (4) affacted a slight change in the activity.

The NMR analyses gave an important suggestion. ^{1}H-NMR spectra of 1a, 1b, 1c, 1e and 1i showed that their ring protons appeared well-resolved with quite similar patterns and almost the same chemical shift in the organic solvent used. On the other hand, ^{1}H-NMR spectrum of 1d in the same solvent showed different complex patterns composed of a mixture of more than two conformers. Furthermore, the intensities of the signals which have nearly the same chemical shifts as those of 1e were extremely low. These observations suggest that the main portion of 1d has a different conformation from those of 1a, 1b, 1c, 1e and 1i. This finding implies that the ring conformation of the molecule is strongly related to expression of the biological activity. From the results of conformation analyses using NMR saturation transfer and line shape analyses, it was concluded that the major and minor conformation of AM-toxin I and II, [D-Ala^2] and [L-Ala^2]AM-toxin I corresponded with those shown in Fig. 2 (ref. 23). However several differences were observed between the ^{1}H-NMR spectra of [L-Val^4] and AM-toxin I. The signals of the amide proton of the dehydroalanine residue in the lactam analog shifted down-field from 9ppm to 10.4ppm. Two sets of signals with similar intensity were observed for the aromatic protons and β,γ-protons in the aliphatic side chains at 29° C. These spectra gave only one set of the signals at 100°C in DMSO-d_6. These observations indicate that the lactam analyog exists as a mixture of two conformers with similar populations. It should be noted that AM-toxin I and II are composed of two conformers with very unequal populations (ref. 23,24).

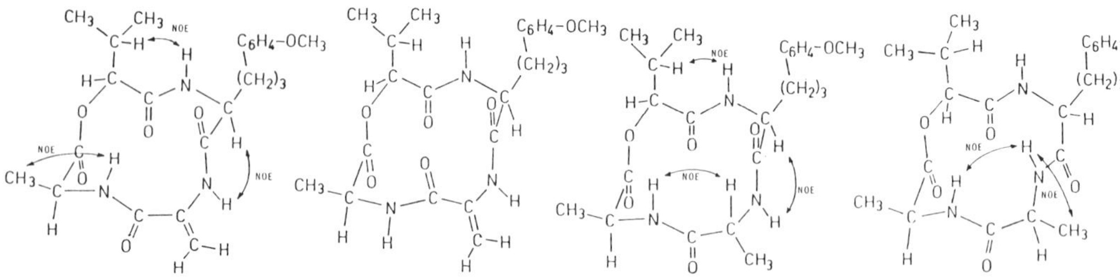

AM-toxin I (major) AM-toxin I (minor) (D-Ala²)AM-toxin I (L-Ala²)AM-toxin I

Fig. 2 Conformations of AM-toxin I and its [Ala^2]-analogs

ii) Syntheses of AK-toxin II, AF-toxin IIc and Their Analogs (see Fig. 3)

The structures of AK-toxin I (2a) and II (2b) were firmly established including the absolute configurations (ref. 6), while the stereochemistry of the AF-toxins remains to be elucidated (ref. 10). Both toxins not only have structural features closely resembling 8-hydroxy-9,10-epoxy-9-methyl-deca-(2,4,6)-trienoic acid, but they express the same host specific toxicity as shown in Table 2, in spite of the different geometries in the trienoic acid. We aimed at the syntheses of these toxins and their congeners to investigate the relationship between their stereostructures and toxicities to the host plant including the elucidation of the absolute configurations in AF-toxins. This report describes an approach to the stereo-selective syntheses of the toxins starting with Vitamin C (4a) (ref. 25).
Oxidation of 5,6-isopropylideneascorbic acid (4b) with potassium permanganate under carbon dioxide gave the potassium salt which was converted to the methyl ester (4c). The Grignard reaction of (4c) with MeMgI gave the diol, which was benzoylated in the usual way to give the mono-benzoate (4d). The olefin compound (4e) was prepared from (4d) by treatment with thionyl chloride in pyridine. Alkaline hydrolysis of the ester group in (4e) and reprotecting the resulting hydroxyl group with *tert*-butyl-diphenylsilyl (TBDPS) group yielded the slefin-silyl-ether (4f). Treatment of the ether (4f) with trifluoroacetic acid in aq. tetrahydrofuran gave the glycol, which was oxidized to the aldehyde (4g) with periodic acid. The Wadsworth-Emmons reaction of (4g) with trimethyl phosphonoacetate and sodium hydride afforded the *trans*-ester (4h). Reduction of (4h) with diisobutylaluminum hydride (DIBAH) gave an alcohol, which was smoothly converted to the aldehyde by manganase dioxide oxydation. Wadsworth-Emmons reaction with methyl dimethylphosphonoacetate and sodium hydride gave the diene-ester (4i) in good yield. The geometry of the double bond formed was confirmed to be *trans*. The same reaction sequence on the diene-ester (4i) gave the trans-triene-ester (4j). Oxidation of each olefinic esters, 4h, 4i and 4j, with *m*-chloroperbenzoic acid gave each pair of two stereoisomeric ipoxides in almost 1:1 ratio in each case (4k, 4k'; 4l, 4l'; and 4m, 4m'). The configuration of the carbon bearing methyl and epoxide oxygen in each oxide was proposed to be S-configur-ation in 4k, 4l and 4m, and to be R-configuration in 4k', 4l' and 4m' by the fact that the former structures showed an AB-type quartet assigned to the methylene protons of the epoxide group similar to the same protons of AK-toxins, while the latter compound showed a singlet due to the same methylene protons. An NMR experiment with the NOE effect using 4k and 4k' provided further proof in support of the stereochemical proposal.
Treatment of all the esters, 4k, 4k', 4l, 4l', 4m and 4m', with tetrabutylammmonium fluoride gave the hydroxy-esters. Acylation of these hydroxy-esters with N-acetyl-(L)-phenylalanine and DCC in the presence of 4-pyrolidinopyridine in methylenechloride afforded two stereoiso-meric esters in each case, thus N-acetyl-(L)-phenylalanine ester and N-acetyl-(D)-phenylala-nine ester in 1:1 ratio. This fact indicates that, in all cases N-acetyl-(L)-phenylalanine was completely racemised during the course of acylation. All compounds which correspond to AK-toxin II analogs are listed in Table 4 together with their biological activity.
On the other hand, acylation of the hydroxy-ester, 4m and 4m', with α-siloxy(S)-β-methyl(S)-pentanoic acid gave esters without racemization. De-silylation of the esters gave the hydroxy-esters, 3d and 3e. Conversely a mild alkaline hydrolyses of the siloxy methyl esters gave the acids, 3f and 3g in good yield. The former (3d) did not show any toxicity but the latter product was toxic to leaves of the host plant, a fact suggesting the structure of AF-toxin IIc to be 3g (see Table 4).

iii) Investigation of Structural Features to Express Host-specific Toxicity in AK-toxins and AF-toxins

After purification of each compound shown in Table 4, the host-specific toxicity inducing necrosis on the leaves of a susceptible cultivar, Nijisseiki, was tested. The results of the bioassay using these compounds resulted in the following conclusions: 1) The absolute configuration at C-2' of N-acetyl phenylalanine in AK-toxin II does not affect expression of the toxicity. 2) The absolute configurations at C-8(R) and C-9(S) in the epoxy-decatrienoic acid are also indispensable to the biological activity in both AK- and AF-toxins. 3) The geometrical arrangements and the free carboxyl group on the conjugated trienoic acid are not indispensable but the length of the conjugated double bond affects expression of the toxicity; monoene>diene>triene. 4) All of the biologically active compounds indicated the host-specific toxicity.

TABLE 4. Biological activity of AK-toxins, AF-toxins and their synthesized analogs inducing necrosis on the leaves of a susceptible cultivar, Nijisseiki.

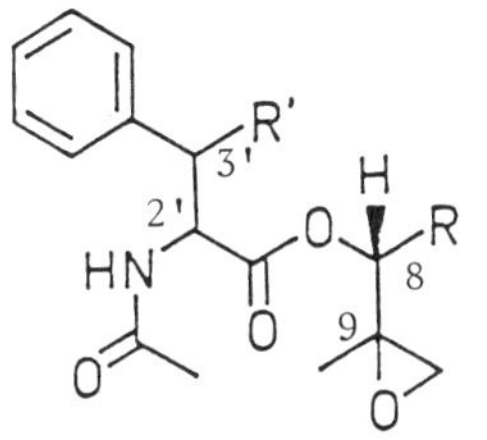

(2a) - (2m)

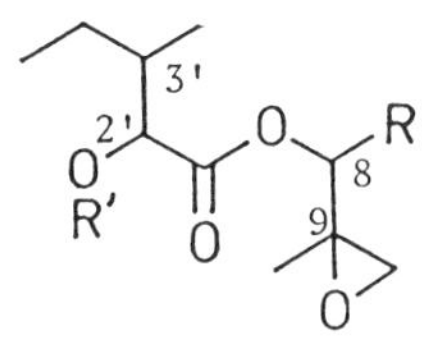

(3a-a) - (3g)

	absolute configurations C-2'	C-3'	C-8	C-9	R'	R (geomeries)	toxicity 10^{-2}M	10^{-3}M	10^{-4}M
(2a) AK-toxin I	*S*	*S*	*R*	*S*	CH_3	CH$\overset{t}{=}$CH-CH$\overset{c}{=}$CH-CH$\overset{t}{=}$CH-COOH	+	+	+
(2b) AK-toxin II	*S*	-	*R*	*S*	H		+	+	+
(2c)	*S*	-	*R*	*S*	H	CH$\overset{t}{=}$CH-$COOCH_3$	+	+	-
(2d)	*R*	-	*R*	*S*	H		+	+	-
(2e)	*S*	-	*R*	*R*	H		-	-	-
(2f)	*R*	-	*R*	*R*	H		-	-	-
(2g)	*S*	-	*R*	*S*	H	CH$\overset{t}{=}$CH-CH$\overset{t}{=}$CH-$COOCH_3$	+	+	±
(2h)	*R*	-	*R*	*S*	H		+	+	±
(2i)	*S*	-	*R*	*R*	H		-	-	-
(2j)	*R*	-	*R*	*R*	H		-	-	-
(2k)	*S*	-	*R*	*S*	H	CH$\overset{t}{=}$CH-CH$\overset{t}{=}$CH-CH$\overset{t}{=}$CH-$COOCH_3$	+	+	+
(2l)	*R*	-	*R*	*S*	H		+	+	+
(2m)	*S*	-	*R*	*R*	H		-	-	-
(2n)	*R*	-	*R*	*R*	H		-	-	-
(3a-a) AF-toxin Ia					$COCH(OH)C(CH_3)_2OH$	CH$\overset{c}{=}$CH-CH$\overset{t}{=}$CH-CH$\overset{t}{=}$CH-COOH	+	+	+
(3b-a) AF-toxin IIa					H	*c* *t* *t*	+	+	+
(3b-b) AF-toxin IIb		unkown			H	*t* *c* *t*	+	+	+
(3b-c) AF-toxin IIc					H	*t* *t* *t*	+	+	+
(3c-a) AF-toxin IIIa					$COCH(OH)CH(CH_3)_2$	*c* *t* *t*	±	-	-
(3d)	*S*	*S*	*R*	*R*	H	CH$\overset{t}{=}$CH-CH$\overset{t}{=}$CH-CH$\overset{t}{=}$CH-$COOCH_3$	-	-	-
(3e)	*S*	*S*	*R*	*S*	H		+	+	+
(3f)	*S*	*S*	*R*	*R*	H	CH$\overset{t}{=}$CH-CH$\overset{t}{=}$CH-CH$\overset{t}{=}$CH-COOH	-	-	-
(3g) = IIc ?	*S*	*S*	*R*	*S*	H		+	+	+

BIOLOGICAL ACTIVITIES OF AK-TOXINS

Biological activities of AM-toxins were discussed in the proceedings of the last congress (ref. 5). Here we report on those AK-toxins from confirmatory plant physiological experiments using the purified samples. The sensitive range of plants to AK-toxin I and II was completely consistent with the host range to the pathogen. The toxins had significant effects on susceptible pear tissues at $5x10^{-9}$M (AK-toxin I) and 10^{-7}M (AK-toxin II), respectively. Resistant pear tissues were not affected even at 10^{-4}M of the toxins. These results indicate that AK-toxin I and II fall under the criteria of host-specific toxin which were reported by Pringle and Scheffer (ref. 3).

We pointed out the importance of AK-toxin released from germinating spores of the pathogen. The toxins are released from germinating spores of virulent isolates, but not from avirulent ones. When avirulent spores were used as inoculum together with dilute solution of AK-toxins, the spores could invade host tissues, indicating that the released toxin was required for initial colonization of the host tissues by the pathogen. Thus, AK-toxins are now considered to be an initial factor for successful pathogenesis.

AK-toxins caused rapid and drastic loss of K^+ from susceptible, but not resistant pear tissues. The first toxin-induced change in susceptible cells detected by electron microscopy was present in the plasma membrane. No ultrastructural changes could be detected in the toxin-treated resistant pear cells. This result is quite different from that of AM-toxins, which affect chloroplast as well as plasma membrane of host plant cells.

The assays described here were done by using two methods: toxin-induced leaf necrosis and K^+ loss from leaves.

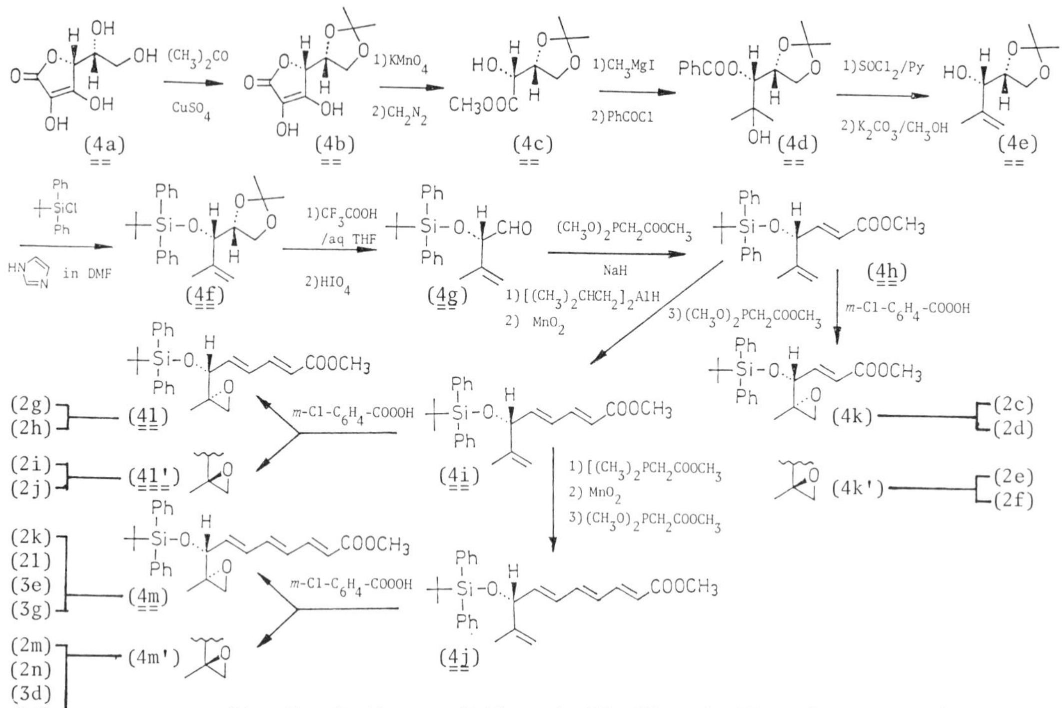

Fig. 3 Syntheses of AK-toxin II, AF-toxin IIc and their analogs

REFERENCES

1. A.E.Dimond and P.E.Waggoner, Phytopathology, 43, 229 (1953).
2. H.Wheeler and H.H.Luke, Annu. Rev. Microbil., 17, 223 (1963).
3. R.B.Pringle andR.P.Scheffer, Annu. Rev. Plantpatol., 2 133 (1964).
4. S.Nishimura and K.Kohmoto, Annu. Rev. Plantpathol.,21, 87 (1983).
5. T.Ueno, T.Nakashima and H.Fukami, IUPAC Pesticide Chemistry(eds, J.Miyamoto et. al) Vol. 2, 75 (1983), Pergamon Press.
6. a) T.Nakashima, T.Ueno and H.Fukami, Tetrahedron Lett., 23, 4469 (1982). b) T.Nakashima, T.Ueno, H.Fukami, T.Taga, H.Masuda, K.Osaki, H.Otani, K.Kohmoto and S.Nishimura, Agric. Biol. Chem., 49, 807 (1985).
7. a) T.Okuno, Y.Ishita, K.Sawai and T.Matsumoto, Chem. Lett.,1974, 635. b) T.Ueno, T.Nakashima, Y.Hayashi and H.Fukami, Agric. Biol. Chem., 39,1115 (1975). c) T.Ueno, T.Nakashima, Y. Hatashi and H.Fukami, ibid., 39, 4469 (1975).
8. Y.Kono, J.M.Gardner and S.Takeuchi, ibid., 50, 801 (1986)
9. A.T.Bottini, J.R.Bowen, D.G.Gilscrist, Tetrahedron Lett.,2723 (1981).
10. S.Nakatsuka, K.Ueda, T.Goto, M.Yamamoto, S.Nishimura and K.Kohmoto, 51st National Meeting of Chemical Society of Japan (Kanazawa), Oct. 1985, 3A-20, p 495.
11. J.M.Gardner, Y.Kono, J.H.Tatum, Y.Suzuki and S.Takeuchi, Agric. Biol. Chem.,49, 1235 (1985).
12. V.Macko, T.J.Wolpert, W.Acklin, B.Jaun, J.Seibl, J.Micili and D.Arigoni, Experientia, 49, 1366 (1985).
13. M.Kawai, D.H.Rich and J.D.Walton, Biochem. Biophys. Res. Commun., 111, 398 (1983).
14. V.Macko, K.Goodfriend, T.Wachs, J.A.A.Renwick, W.Acklin and D.Arigoni, Experientia, 37, 923 (1981).
15. Y.Kono, S.Takeuchi, J.M.Daly and H.W.Knoche, Tetrahedron Lett., 21, 1537 (1980).
16. Y.Kono, Y.Suzuki, S.Takeuchi, H.W.Knoche and J.M.Daly, Agric. Biol. Chem.,49, 559 (1985).
17. K.Sawamura, Bull. Hort. Res. Stn. Japan, C4, 43 (1966).
18. S.Tanaka, Mem. Coll. Agric. Kyoto Imp. Univ., 28, 1 (1933).
19. Y.Watanabe, M.Umekawa, and S.Nishimura, Annu. Phytopathol. Soc. Japan,43, 82 (abstr) (1978).
20. S.Lee, H.Aoyagi, Y.Shimohigashi, N.Izumiya, T.Ueno and H.Fukami, Terahedron Lett.,1977,1141.
21. a) T.Kozono, T.Kanmera, T.Kato, T.Ueno and N.Izumiya, Agric. Biol. Chem., 47, 2631 (1983). b) H.Mihara, H.Aoyagi, T.Kato, T.Ueno and N.Izumiya, Chem. Lett., 1983, 811. c) H.Aoyagi, H.Mihara, S.Lee, T.Kato, T.Ueno and N.Izumiya, Int. J. Peptide Res., 25, 144 (1985).
22. T.Ueno, T.Nakashima and H.Fukami, Mem. Coll. Agric., Kyoto Univ., 122, 93 (1983)
23. T.Higashijima, Y.Shimohigashi, T.Kato, N.Izumiya, T.Ueno and T.Miyazawa, Biopolymers, 22, 1167 (1983).
24. Y.Shimohigashi, S.Lee, T.Kato, N.Izumiya T.Ueno and H.Fukami, Agric. Biol. Chem.,41, 1533, (1977).
25. H.Irie, J.Igarashi, K.Matsumoto, Y.Yamagawa, T.Nakashima, T.Ueno and H.Fukami, Chem. Pharm. Bull., 33, 1313 (1985).

Insect neuropeptides and their physiological degradation: the basis for new insecticide discovery?

David A. Schooley*, Gary B. Quistad, Wayne S. Skinner, and Michael E. Adams

Zoecon Research Institute, Palo Alto, California 94303 USA

Abstract - Proctolin and adipokinetic hormone were the only identified insect neuropeptides until two years ago. We review the considerable recent progress in discovery of additional insect neuropeptides, and discuss how these substances might be utilized for discovery of new methods of insect control. With this in mind, we review what is currently known about the metabolism of two insect peptides, proctolin and periplanetin CC-2. In the cockroach, these peptides are degraded at remarkably different rates, but both are inactivated by cleavage of multiple amide bonds. Design of specific inhibitors of metabolism of the Tyr-Leu bond in proctolin has been accomplished. Until we are aware of the structure of a vital insect neuropeptide which is inactivated predominantly at a single site, other approaches appear more promising for use of insect peptides in pest control. One alternative is to insert genes for peptide hormones into the genomes of insect pathogens or intestinal symbionts, which may soon become practical.

INTRODUCTION

The amino acid sequences of proctolin and adipokinetic hormone (AKH) were determined in 1975 and 1976; until 1984 these were the only chemically identified insect neuropeptides. In the last two years, five additional insect neuropeptides highly homologous to AKH have been sequenced: periplanetin CC-1 and CC-2 from *Periplaneta americana* (ref. 1,2), AKH II-S from *Schistocerca gregaria* (ref. 3), AKH II-L from *Locusta migratoria* (ref. 3), and lepidopteran AKH (with identical structures) from *Manduca sexta* (ref. 4) and *Heliothis virescens* (ref. 5). Recently, partial sequences of three neuropeptides from *Bombyx mori* have been reported, including the 4,000 Dalton prothoracicotropic hormone (4K-PTTH, ref. 6), eclosion hormone (EH, ref. 7), and the melanization and red coloration hormone (MRCH, ref. 8). Furthermore, the full sequence of 4K-PTTH-II has now been established and a report describing it is in press at conference time (ref. 9). It is evident that after a quiescent period, there is now intense research and progress on identification of insect neuropeptides.

How might knowledge of structures of insect neuropeptides be exploited to develop organic chemicals of low molecular weight which interfere with peptide-mediated processes? Can natural insect pathogens, microbial or viral, be genetically engineered so that upon infection of an insect host, a neuropeptide of some type will be expressed in the insect host to accelerate or accentuate the lethal effects of the pathogen? These are but some of the questions being asked in academia, the pesticide industry, and governmental laboratories. A group at Zoecon is investigating the possibility of developing insect control agents that disrupt the neuroendocrine system, an interest evolving from the founding principles of the company: Zoecon synthesized and commercialized two analogs of the nonpeptidal juvenile hormone as pesticides.

In the pharmaceutical industry, analogs of endogenous peptides are being investigated as drugs; this approach seems much less likely for insect control agents because of the very high cost of synthetic peptides. Recently, a group at Merck has used receptor assays for the gut peptide cholecystokinin to screen fermentation broths, and have succeeded in isolating and identifying a natural, nonpeptidal antagonist of CCK having relatively low activity. However, they have synthesized simpler analogs containing a benzodiazepine moiety that have nanomolar activity as CCK antagonists, as well as good oral bioavailability (ref. 10).

A very exciting and successful development has been the design of antihypertensive drugs (e.g. captopril, enalipril) which block angiotensin converting enzyme. The latter provide an interesting precedent for the rational discovery of low molecular weight chemicals which can disrupt peptide-mediated processes. In this paper, we discuss studies on the metabolism of two insect neuropeptides, proctolin and periplanetin CC-2, and the potential utility of inhibitors of their degradation. Inhibition of their inactivation might lead to lethal levels of these chemical messengers, by analogy to the lethal effects of inhibiting acetylcholinesterase, a neurotransmitter-inactivating enzyme.

METABOLISM OF PROCTOLIN

We have published preliminary studies on the metabolism of [3,5-^{3}H-Tyr^2]proctolin (ref. 11) by cockroach tissue homogenates at physiological levels (3.5 nM). This putative neurotransmitter is degraded _in vitro_ with a half-life of less than one minute in buffer-diluted homogenates of proctolin-rich tissues such as the proctodeum and brain. When incubated _in vitro_ with cockroach hemolymph diluted by 20-fold, the half-life of proctolin (3.5 nM) is much longer, ~10 min. From rates of metabolism in these and other tissues, we noticed that proctolin-rich tissues contained much higher levels of proctolin-degrading enzymes, which might be expected in light of the claimed role of proctolin as a neurotransmitter.

In these metabolic studies, we assayed products of incubations by reversed-phase liquid chromatography (LC), so that we could simultaneously determine the amount of unmetabolized proctolin and of several chemically distinct metabolites. We determined the subcellular localization of proctolinase activity; at least 90% of the activity from all tissues studied was in a 100,000 x _g_ supernatant (soluble fraction), with <10% in the pellet (membrane-bound enzymes). We found the predominant labelled metabolite from [3,5-^{3}H-Tyr^2] proctolin to be [^{3}H]Arg-Tyr, accompanied by lesser amounts of [^{3}H]Tyr-Leu-Pro-Thr (Fig. 1).

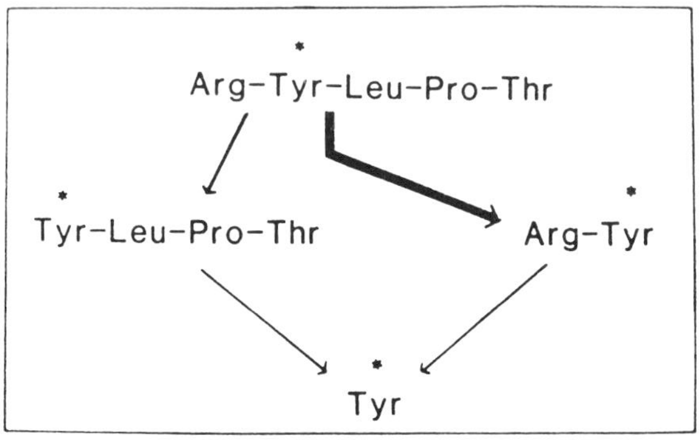

Fig. 1. Metabolism of [3,5-^{3}H-Tyr^2]proctolin by cockroach tissue homogenates yields two primary metabolites, which are ultimately degraded to [^{3}H]tyrosine.

No other cleavage product was detectable at a level of ≥1% at pH 7.2. The level of [^{3}H]Arg-Tyr formed by the aminodipeptidase-type enzyme, as compared to the level of [^{3}H]Tyr-Leu-Pro-Thr from the aminopeptidase-type enzyme, was proportionately higher in proctolin-rich tissues. This suggested that the aminodipeptidase enzyme might have the greater physiological significance for inactivating this presumed neurotransmitter. Both enzymes exhibited high affinity for substrate, with K_m values ~10^{-7}M.

Starratt and Steele (ref. 12) studied the metabolism of [^{14}C-Tyr^2]proctolin with _Periplaneta americana_ treated _in vivo_ with supranormal doses (up to 50 µM, more than 10^4 times the levels used in our _in vitro_ studies). They found the predominant cleavage to be formation of Tyr-Leu-Pro-Thr from action of an aminopeptidase, with a half-life of ~5 min. (At physiological levels, we have found proctolin to be degraded so rapidly _in vivo_ that the half-life was too short for determination [<<1 min]). Steele and Starratt (ref. 13) also studied the inactivation of [^{14}C]proctolin in cockroach hemolymph _in vitro_ at pH 6 and 8. At pH 8, the dominant metabolic pathway was scission of the Tyr-Leu bond, in accordance with our results (ref. 11). At pH 6.0 the main cleavage was a carboxypeptidase-type cleavage to generate Arg-Tyr-Leu-Pro. However, pH 6.0 is not in the normal physiological range of pH 7.3-7.8 in _P. americana_ (ref. 13), and hemolymph contained the lowest level of degradative activity at pH 7.2.

We studied the pH dependence of the rate of proctolin metabolism in proctodeum homogenates at 3.5 nM, simultaneously determining the rate of appearance of various metabolites. As shown in Fig. 2, we observed the carboxypeptidase-type cleavage (ref. 11) to Arg-Tyr-Leu-Pro only at pH 6.2 and below. While the optima for rate of proctolin degradation and Arg-Tyr formation were around pH 6.0-6.5, we conducted all further routine studies at pH 7.2 as this is closer to physiological pH.

In order to characterize the active site of the aminopeptidase and aminodipeptidase enzymes in proctodeum homogenates, we studied the effects of a number of inhibitors. For example, we studied the metabolism of 3.5 nM proctolin in proctodeum homogenates with 1 mM of iodoacetamide or _N_-ethylmaleimide. These sulfhydryl protease inhibitors were without any effect on enzymatic activity. Similar incubations with 1 mM PMSF, a potent serine protease inhibitor, also showed no loss in enzymatic activity. The chymotrypsin inhibitors TPCK and NCDC at 1 mM, and aminopeptidase inhibitors bestatin and pepstatin A (0.1 mM) and bacitracin (50 μM) also failed to inhibit proctolin degradation by either enzyme.

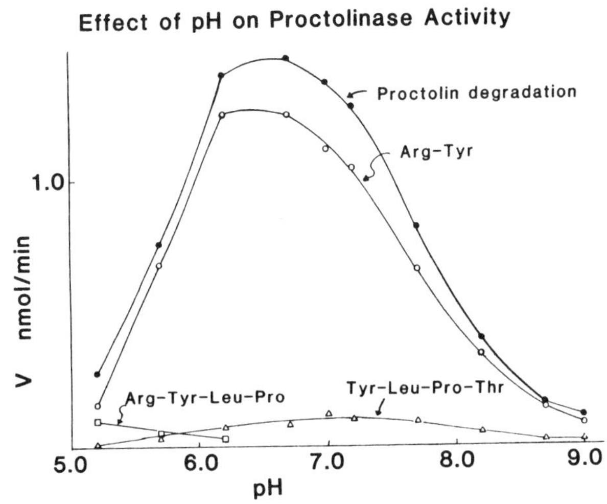

Fig. 2. Rate of degradation of proctolin and rate of appearance of metabolites in cockroach proctodeum homogenates as a function of pH.

On the other hand, inclusion of the chelating agent 1,10-phenanthroline at 0.2 mM inhibited both enzymatic cleavages dramatically (Table 1). Addition of 0.2 mM $ZnBr_2$ together with 1,10-phenanthroline completely relieved the inhibition of Arg-Tyr formation, and partially relieved the inhibition of Tyr-Leu-Pro-Thr formation, and $CoCl_2$ at the same level was even more effective in preventing phenanthroline-induced inhibition of the aminopeptidase (Table 1). Thus, these studies strongly suggest that both enzymes are metalloproteases.

TABLE 1. Effect of 1,10-phenanthroline on proctolin degradation with and without addition of divalent metal ion. All inhibitors were pre-incubated with the enzyme preparation for 10 min before adding substrate.

	conc.	% Inhibition of proctolin degradation	% Inhibition of formation of Arg-Tyr	% Inhibition of formation of Tyr-Leu-Pro-Thr
1,10-phenanthroline	2×10^{-4}M	76	76	94
1,10-phenanthroline	2×10^{-4}M	14	0	87
+ $ZnBr_2$	2×10^{-4}M			
1,10-phenanthroline	2×10^{-4}M	0	0	44
+ $CoCl_2 \cdot 6H_2O$	2×10^{-4}M			

Having established that the proctolinases are metalloproteases, our colleagues in the chemical research department at Zoecon (R. L. Carney, R. M. Scarborough) began the design and synthesis of inhibitors of the major aminodipeptidase enzyme. They began by synthesizing proctolin analogs truncated at the carboxyl terminus. Eventually, they created a modified dipeptide with a hydroxamic acid functionality at the carboxyl terminus, which inhibited the enzyme activity ~50% at 10^{-7}M. This inhibitor is presumed to function by interaction with the aminodipeptidase enzyme at the sites which recognize Arg-Tyr, and by chelating the metal ion (presumably Zn^{++}) at the active site. We were able to obtain a compound with this level of activity after synthesis and assay of ~100 compounds.

Injection of this inhibitor into cockroaches, alone or in combination with proctolin, did not elicit any toxic or other untoward effects. Upon studying the metabolism of proctolin in proctodeum homogenates to which the inhibitor was added, we found that it very effectively blocked cleavage of the Tyr-Leu bond by the aminodipeptidase, but was essentially without effect on the rate of scission of the Arg-Tyr bond by the aminopeptidase. Therefore, to substantially elevate levels of proctolin <u>in vivo</u> by blocking its metabolism, one would have to synthesize as well a potent inhibitor of the aminopeptidase enzyme, and to co-apply this along with the aminodipeptidase inhibitor. From the standpoint of developing an insect control agent, this seems unrealistic. In view of this rapid metabolism of proctolin by multiple pathways, we are also currently unable to assess how lethal the disruption of proctolin-mediated processes might be.

METABOLISM OF THE HYPERGLYCEMIC PEPTIDE, PERIPLANETIN CC-2

We have also studied the metabolism (ref. 14) of the hyperglycemic/myotropic neuropeptide periplanetin CC-2 (pGlu-Leu-Thr-Phe-Thr-Pro-Asn-Trp-NH_2), which has a high degree of sequence homology with glucagon (ref. 1). Together with periplanetin CC-1, this substance exists in large amounts in the corpora cardiaca of the cockroach, P. americana. We synthesized [4-^{3}H-Phe^4]periplanetin CC-2 and studied its metabolism in tissue homogenates (Malphigian tubules and fat body at ~20 mg tissue/ml buffer). At low concentrations (6 nM), its half-life was ~1 hr. Several metabolic fragments were detected, including CC-2(1-4) [pGlu-Leu-Thr-Phe] and CC-2(1-6) [pGlu-Leu-Thr-Phe-Thr-Pro]. Because the half-life was almost 100X that of proctolin at similar concentration, we also studied its metabolism in vivo at very low doses (1-2 pmol/cockroach, or ~1-2 nM). Again, the half-life was found to be ~1 hr. The relatively slow rate of metabolism, combined with the very high content in the corpora cardiaca, suggest that circulating levels of this hormone are more likely controlled by rate of release than by metabolism. Furthermore, injection of large doses of CC-1 or CC-2 into cockroaches does not cause any visually apparent effect. In this case, the lack of toxicity cannot be ascribed to rapid metabolism. In light of the nontoxicity of CC-2 and its metabolism at multiple sites, we feel that attempts to synthesize inhibitors of its metabolism are unlikely to be of practical utility.

CONCLUSIONS

The identification of insect neuropeptides is progressing rapidly, and we should learn the chemical structures of peptides with potent activities affecting various processes. At this time, we are aware of only four publications (excluding abstracts of presentations) on metabolism of insect peptides. Studies on inhibition of peptide-degrading enzymes have demonstrated the potential for rational design of highly specific and potent inhibitors of metabolism of specific bonds in neuropeptides. However, the only two insect neuropeptides whose metabolism has been studied are cleaved at multiple sites; to elevate levels of these neuropeptides by metabolic blockade, it would be necessary to design and administer multiple inhibitors. This situation does not currently appear to offer the potential for creation of insect control agents. Once we are aware of vital insect neuropeptides (or prohormones) which are degraded predominantly at a single site (such as in metabolism of angiotensin I to angiotensin II in mammals), this will be a very exciting approach to design of new insecticides.

In the meantime, the most reasonable prospect for control of insect pests with insect neuropeptides would appear to be the insertion of genes coding for these substances into viral or bacterial pathogens of insects, or even gut symbionts. The success of the latter approach hinges on the identification of peptides which cause irreversible adverse affects to insects upon injection or ingestion.

REFERENCES

1. R.M. Scarborough, G.C. Jamieson, F.K. Kalish, S.J. Kramer, G.A. McEnroe, C.A. Miller and D.A. Schooley, Proc. Natl. Acad. Sci. USA 81, 5575-5579 (1984).
2. J.L. Witten, M.H. Schaffer, M. O'Shea, J.C. Cook, M.E. Hemling and K.L. Rinehart, Jr. Biochem. Biophys. Res. Commun. 124, 350-358 (1984).
3. K. Siegert, P. Morgan and W. Mordue Biol. Chem. Hoppe-Seyler 366, 723-727 (1985).
4. R. Ziegler, K. Eckart, H. Schwarz and R. Keller Biochem. Biophys. Res. Commun. 133, 337-342 (1985).
5. H. Jaffe, A.K. Raina, C.T. Riley, B.A. Fraser, G.M. Holman, R.M. Wagner, R.L. Ridgway and D.K. Hayes Biochem. Biophys. Res. Commun. 135, 622-628 (1986).
6. H. Nagasawa, H. Kataoka, A. Isogai, S. Tamura, A. Suzuki, H. Ishizaki, A. Mizoguchi, Y. Fujiwara and A. Susuki Science 226, 1344-1345 (1984).
7. H. Nagasawa, T. Kamito, S. Takahashi, A. Isogai, H. Fugo and A. Suzuki Insect Biochem. 15, 573-578 (1985).
8. S. Matsumoto, A. Isogai and A. Suzuki FEBS Lett. 189, 115-118 (1985).
9. H. Nagasawa, H. Kataoka, A. Isogai, S. Tamura, A. Suzuki, A. Mizoguchi, Y. Fujiwara, A. Suzuki, S.Y. Takahashi and H. Ishizaki Proc. Natl. Acad. Sci. USA 83, in press.
10. B.E. Evans, M.G. Bock, K.E. Rittle, R.M. DiPardo, W.L. Whitter, D.F. Veber, P.S. Anderson and R.M. Freidinger Proc. Natl. Acad. Sci. USA 83, 4918-4922 (1986); R.S.L. Chang and V.J. Lotti Proc. Natl. Acad. Sci. USA 83, 4923-4926 (1986).
11. G.B. Quistad, M.E. Adams, R.M. Scarborough, R.L. Carney and D.A. Schooley Life Sciences 34, 569-576 (1983).
12. A.N. Starratt and R.W. Steele Insect Biochem. 14, 97-102 (1984).
13. R.W. Steele and A.N. Starratt Insect Biochem. 15, 511-519 (1985).
14. W.S. Skinner, G.B. Quistad, M.E. Adams and D.A. Schooley Insect Biochem. in press, 1986.

Role of intracellular symbiotes in planthoppers

Andreas Fredenhagen, Peter Kenny, Hiroshi Kita,[1*] Hajime Komura, Yoko Naya, Koji Nakanishi,[*] Koushi Nishiyama,[2] Miyoji Sugiura,[3] and Susan Tamura.

Suntory Institute for Bioorganic Research(SUNBOR), Shimamoto-cho, Mishima-gun, Osaka 618 Japan;
[1]IATRON Laboratories Inc., Yakuendai, Funabashi 274, Japan;
[2] National Institute of Agro-Environmental Sciences, Kannondai, Yatabe-cho, Tsukuba-gun, Ibaraki 305 Japan;
[3] National Institute of Agrobiological Resources, address same as NIAES.

Abstract - Certain insects such as planthoppers, leafhoppers and aphides are known to carry intracellular symbiotic microorganisms in their fat body which are transmitted to the offspring through the eggs. The symbiotes can be selectively killed by incubating the eggs(insects) at 35° but the resulting insects are greatly reduced in size and fail to reproduce; occasionally they become covered with mold. However, due to difficulties in isolating and culturing the symbiotes, the chemical basis of these phenomena remained to be clarified. Preliminary studies with several symbiotic cultures have suggested two roles of symbiotes, *i.e.*, to provide sterols to the host insects and to produce antibiotics for protection of hosts. Chemical and biological studies of insect symbiotes offer an intriguing new field of research.

Evidence is presented which suggests that intracellular symbiotes of certain insects exemplified by planthoppers play both nutritional and defensive roles.

1. INTRACELLULAR SYMBIOTES AS A SOURCE OF STEROIDS

Cholesterol is essential for arthropods such as insects and crustaceans. Namely, besides being a component of lipid bilayers to regulate their rigidity, it is also the precursor of 20-hydroxyecdysone, the common molting hormone. However, since arthropods cannot biosynthesize cholesterol, it is incorporated from exogenous steroid sources. Most insect plant feeders metabolize the C-28 and C-29 phytosterols to the C-27 sterols, but some plant sap suckers such as planthoppers can be reared on artificial diets free of steroids and hence do not necessarily rely on dietary steroids. Thus, it has been reported that successive generations of *Laodelphax striatellus* (smaller brown planthopper) could be maintained on a steroid-free artifical diet (ref. 1). In 1965 Nasu discovered the presence of morphologically yeast-like symbiotes in planthoppers (ref. 2). These intracellular symbiotic micro-organisms are found in the vertical section of the host insects' abdomen, and are not recognized as foreign particles since they are contained in myceto-cytes of the host fat body (ref. 3). The symbiotes are transmitted in the eggs to the next generation. Two morphologically and physiologically different yeast-like symbiotes Ls-1 and Ls-2 were isolated from the eggs of *L. striatellus* and by means of immunological techniques were found to live within the insect cell (ref. 4). Similarly, symbiotes Nl-1 and Nl-2 were isolated from eggs of *Nilaparvata lugens* (brown planthopper)(ref. 4). The symbiotes were selectively killed without affecting the eggs upon warming to 37°, the optimal temperature for Ls-1 and Ls-2 being, respectively, 25-31° and 25-35° Upon hatching the heat-treated eggs gave rise to aposymbiotic nymphs which failed to grow properly due to multiple defects (Plate 1). Aposymbiotic nymphs can also be produced upon warming the insects to 35°. Noda *et al.* found that when *L. striatellus* nymphs were kept at 35° for 3 days on a sterol-free diet, the biosynthesis of cuticles was terminated (ref. 3) and that the body cholesterol content was reduced to 1/10 of normal (13-14 days after hatching)(ref. 5). Furthermore, only 6.7% became adults, whereas this increased to 61.6% upon addition of cholesterol and 15.2% upon addition of phytosterol to the diet (ref. 6)(normal insects become adults 22 days after hatching).

Current studies initiated in order to clarify the nutritional role of symbiotes were carried out with the smaller brown planthopper L. striatellus and the brown planthopper N. lugens. After pasteurization of the surface, the eggs were homogenized in saline and cultured on Grace's agar plate. The Ls-1 and Ls-2 thus isolated were cultured on sterol-free Burkholder medium for 4 weeks at 30°. After conventional workup of the culture broth, the TLC area with an Rf value corresponding to free sterols was extracted, silylated, and analysed by gas chromatography(GC)/mass spectrometry(MS). The sterol esters were isolated in a similar manner, hydrolyzed in 10% KOH/MeOH and submitted to GC/MS. In both cases ergosterol was the sole product (Fig. 1). Mixed cultivation of Ls-1 and Ls-2 or addition of mevalonic acid gave similar results.

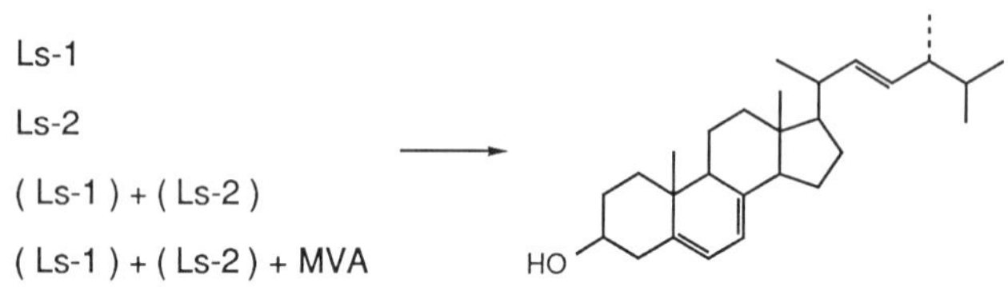

Fig. 1. GC/MS identification of ergosterol from the culture medium of Ls-1 and Ls-2.

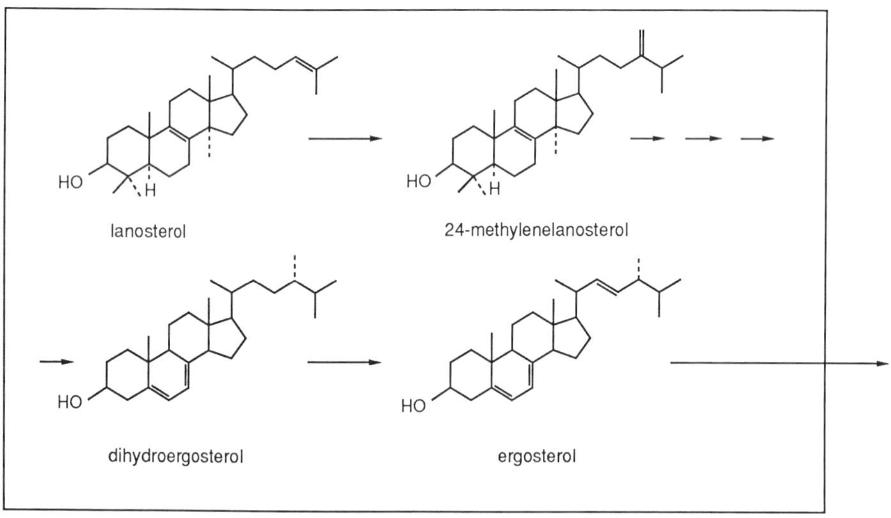

Fig. 2. Sterols detected by GC/MS in the cells of Nl-1 and Nl-2. The compounds are arranged in a conceivable biosynthetic sequence although there are no experimental data to support this at this momemt.

No sterols could be detected, however, from the culture medium of symbiotes Nl-1 and Nl-2 found in the brown planthopper N. lugens. Instead, a series of compounds suggestive of an intracellular biosynthetic buildup of ergosterol from mevalonic acid was detected in the cells (Fig. 2); the cells of Ls-1 and Ls-2 also contained the same set of compounds, although in much smaller quantities. These results demonstrate the contrasting situation between L. striatellus and N. lugens. Namely, in the case of L.s. ergosterol is released from the cells into the culture medium, whereas with N.l. ergosterol and precursors are mostly retained within the cell. Whether this contrasting result is due to lysis of symbiote cells or other factors is as yet unclear. The presence of ergosterol in cells and the culture medium of these symbiotes corroborates their tentative classification as yeasts.

The bodies of the host insect N. lugens, with and without the symbiotes, were next analyzed for the sterols. The crude $CHCl_3$/MeOH extract of normal and aposymbiotic insects was divided into the less polar fraction containing free sterols and their esters and the more polar fraction containing steroidal glycosides (Fig. 3). After base and acid hydrolysis of respective fractions, the free sterols were submitted to GC/MS analysis. In both the rice brown planthopper (BPH) collected in Japan and Leersia BPH collected in Indonesia the sterols consisted of cholesterol, 24-methylenecholesterol and ergosterol (Fig. 4). It is noted that the relative content of 24-methylene-cholsterol is greater in adults than in nymphs. Upon keeping the insects at 35° for 3 days to kill the symbiotes, the total amount of sterols as well as the amounts of ergosterol and 24-methylnecholesterol relative to cholesterol was reduced drastically. Similar results were obtained with the smaller brown planthopper L. striatellus.

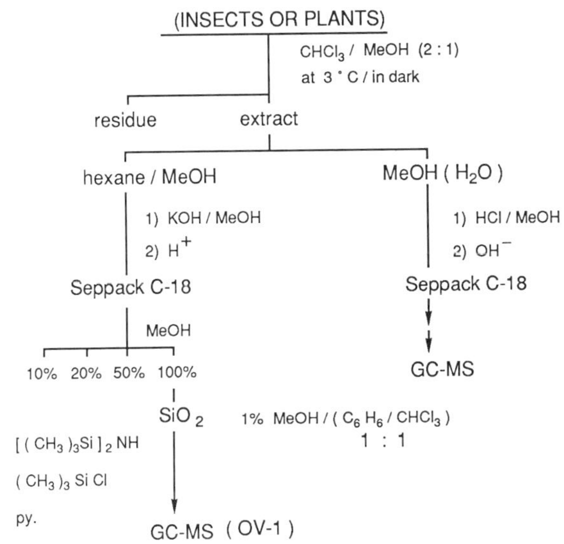

Fig. 3. GC/MS analysis of sterols from N. lugens.

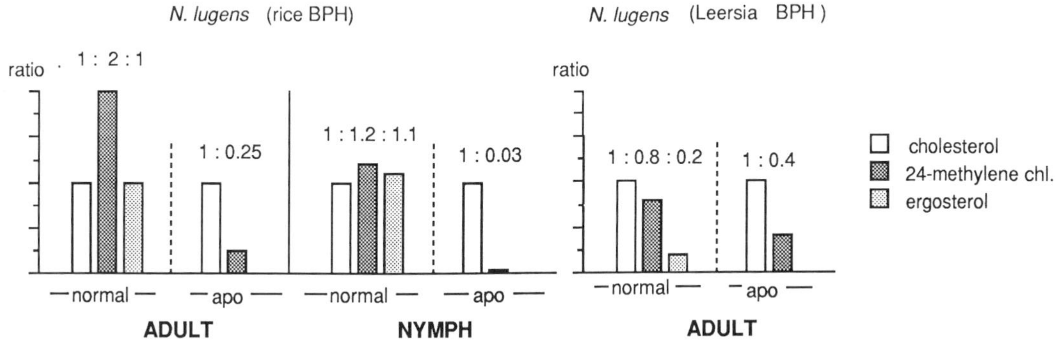

Fig. 4. Sterol constituents in combined sterol fraction from normal and aposymbiotic N. lugens; relative contents of sterols are normalized to cholesterol.

Sterol analysis of the host plants of insects, the rice plant "koshihikari" (host for rice BPH) and Leersia hexandra (host for Leersia BPH)(Fig.3) showed that the former contained β-sitosterol, stigmasterol, campesterol and respective esters, while the latter contained β-sitosterol, stigmasterol, campesterol, cholesterol and the respective esters and glycosides. The presence of cholesterol in L. hexandra may account for the higher ratio of cholesterol in Leersia BPH (Fig. 4). The possibility that ergosterol originates from the food chain can be ruled out; however, the origin of cholesterol shown in Fig. 4 which is unknown at this stage should become clear upon completion of ongoing labeling studies.

In summary, symbiotes cultured on sterol-free media produce lanosterol, 24-methylenelanosterol, dihydroergosterol and ergosterol (Fig. 2). Undoubtedly the host insects must rely on this supply of sterols when kept on sterol-free media since colonies can be maintained for generations. Normal host insects reared on phytosterol-containing plants contain cholesterol, 24-methylene-cholesterol and ergosterol, while aposymbiotic insects contain only a small amount of cholesterol and traces of 24-methylenecholesterol (Fig. 4).

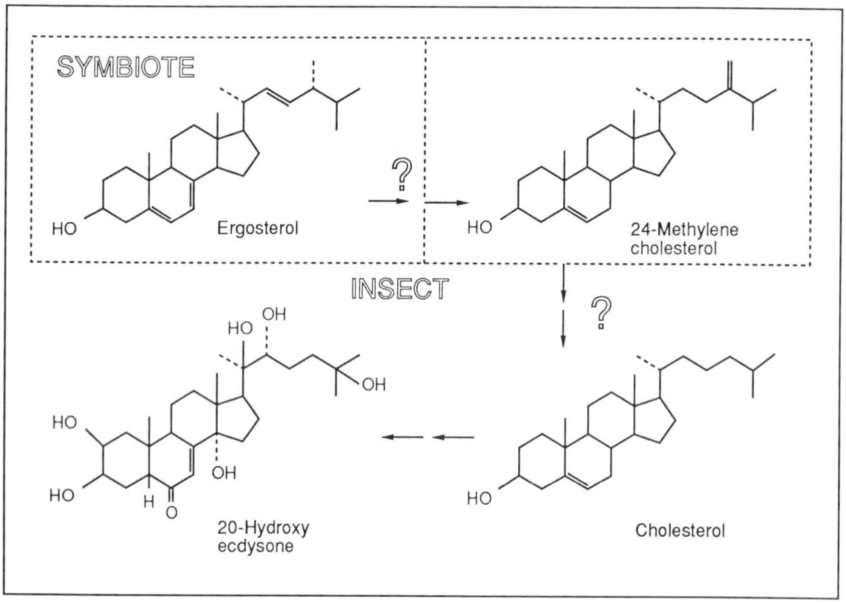

Fig. 5. The role of symbiote as a supplier of sterol.

The crucial question is the extent to which these insects rely on symbiotes for their supply of cholesterol under normal conditions (Fig. 5). Cholesterol could arise from two routes. (i) From ergosterol via 24-methylenecholesterol, a biosynthetic step performed by the symbiote or more likely the insect. The cholesterol found in aposymbiotic insects, Fig. 4, then originates from 24-methylenecholesterol or other cholesterol precursors already present when the symbiotes were killed. (ii) From exogenous plants, the symbiotes providing sterols only under special conditions. The cholesterol in aposymbiotic insects, Fig. 4, then originates from plants.

Judging from the greatly reduced size of aposymbiotic insects (Plate 1), their ability to convert phytosterols into cholesterol is presumably either absent or very weak (ref. 6), the main steroid source being the symbiotes and not the food chain. Tracer experiments currently under study should clarify these points. As mentioned above, the optimal growth temperature of symbiotes Ls-1 and Ls-2 from L. striatellus differed somewhat; on the other hand, since no difference in metabolites has been detected so far the two strains may be present to provide greater chances for survival.

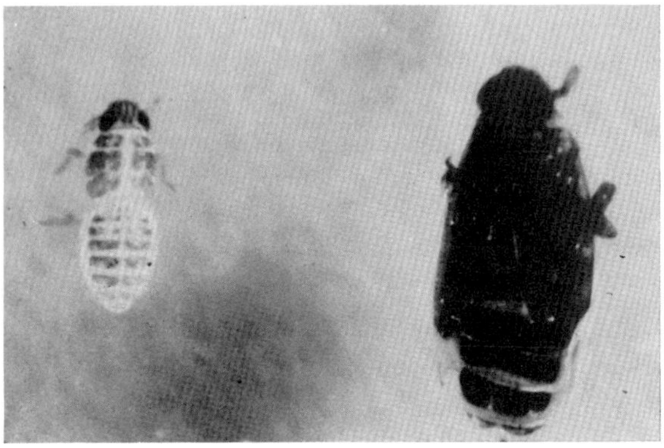

Plate 1. Effect of temperature on the growth of Laodelphax striatellus. Both represent L. striatellus on day 14 after hatching: Normal(right; large) and heat-treated (37°, 6 days)(left; small)

2. INTRACELLULAR SYMBIOTES AS A SOURCE OF ANTIBIOTICS

To compensate for the lack of immune systems against microbial attack, insects protect themselves by physical, chemical and biological means, e.g., chitin formation, terpenoid secretion, rearing of intestinal microorganisms. It is therefore not surprising that intracellular symbiotes produce antibiotic defense substances. Indeed, when symbiotes are killed by heat-treatment, the body of the smaller brown planthopper L. striatellus occasionally becomes covered with mold. Since planthoppers inflict serious damage on the rice plant by sap sucking and transmission of virus disease, a search for antibiotics from various symbiote cultures was initiated a few years ago. The assay was carried out against various viruses and plant pathogenic bacteria including:

Agrobacterium tumefaciens...isolated from crown gall, induces uncontrolled division of plant cells in stone fruits, grape vine, rose.
Corynebacterium michiganense pv. michiganense...responsible for bacterial canker of tomato.
Pseudomonas solanacearum...cause of numerous tropical and subtropical diseases of tomato, potato, tobacco, banana.
Xanthomonas campestris pv. oryzae... cause of bacterial blight of rice, widespread in Asia.

Several antibiotics isolated and characterized to date are summarized in Fig. 6 together with results of the previous section. The culture medium of the symbiote from the planthopper Sogatella furcifera was fractionated using assays against Corynebacterium to yield two known antibiotics 2,4-diacetyl-phloroglucinol and pyoluteorin (Fig. 7), the activity of the latter against Corynebacterium being comparable to gentamycin (MIC 3μg/ml). Since both compounds have been isolated from Pseudomonas it is likely that the present symbiote belongs to the same genus as well. The extracellular symbiote from the ambrosia beetle also gave 2,4-diacetyl-phloroglucinol together with mono-acetylphloroglucinol.

HOST INSECT	SYMBIOTE	MAJOR PRODUCT	FUNCTION
[rice planthopper]			
N. lugens	Nl-1 Nl-2	ergosterol ergosterol	nutrient
L. striatellus	Ls-1 Ls-2	ergosterol ergosterol	
[rice planthopper]			antimicrobial vs.
N. lugens	Nl-T-C1-1W	andrimid	*Xanthomonas*
S. furcifera	Sf-T-TSB-Y3	2,4-diacetyl phloroglucinol pyoluteorin	*Corynebacterium*
[stinkbug]			
P. stali	PS-HB-71412	polymixin E_1	*Salmonella*
[ambrosia beetle]			
S. micado	Sm-S003	monoacetyl phloroglucinol 2,4-diacetyl phloroglucinol	*Corynebacterium*

Fig. 6. The major steroid and antibiotics isolated from intracellular symbiote cultures of planthoppers and other insects.

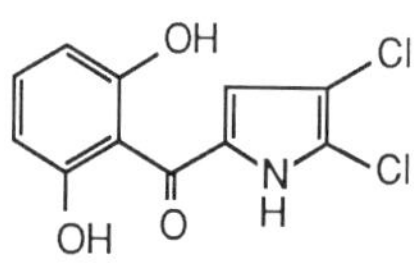

Pyoluteorin

Antibacterial

Antiprotozoan

Protective (Plant Diseases)

Tomato late blight
Cucumber downy mildew
Rice leaf spot
Orange canker
Damping-off of cotton seedlings

Herbicidal

Insecticidal

Isolation from

Pseudomonas aeruginosa

R. Takeda, *Hakko Kogaku Zasshi*, **36**, 281 (1958) *JACS*, **80**, 4749 (1958)
A. J. Birch, *et al*, *JCS*, 2641 (1964)

P. fluorescens

C.R. Howell and R.D. Stippanovic, *Pytopathology*, **70**, 712 (1980)

Sf-T-TSB-Y-3

Synthesis

K. Bailey and A.H. Rees, *Chem. comm.*, 1284 (1969); *Canad. J. Chem.*, **48**, 2257 (1970) *etc.*

Fig. 7. Pyoluteorin produced by a symbiote of Sogatella furcifera (planthopper).

The culture medium (4.5 liters) of the symbiote from the brown-winged green bug (stink bug) gave 2.8 mg of an antibiotic peptide (under study) and 38 mg of polymixin E_1 (colistin, Fig. 8), a potent antibiotic previously isolated from cultures of Bacillus polymyxa(ref. 7).

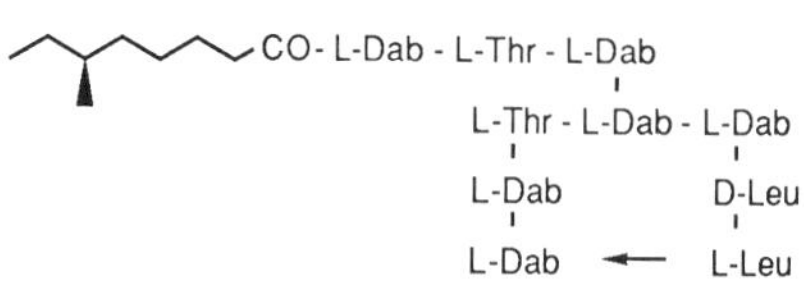

Dab: α,γ-diaminobutyric acid

Fig. 8. Polymixin E_1

The culture broth (25 liters) of symbiotic strain Nl-T-Cl-1W isolated from eggs of N. lugens (brown planthopper collected in Thailand) yielded 36 mg of an antibiotic "andrimid" $C_{27}H_{33}N_3O_5$. It has a unique dipeptidic structure with a polyene moiety and exhibits strong specific activity against the white blight pathogen of rice plants Xanthomonas campestris pv. oryzae at the level of 0.1 µg/ml (ref. 8). Thus a symbiote of the pest insect N. lugens produces a potent antibiotic against the pathogen which is responsible for the disease of the host rice plant. Is it possible that the insects, devoid of immuno-defense systems, are protected by their symbiotes from being destroyed by the pathogens that invade their host plant? Although comparative studies on the resistance of Japanese vs. Thai rice to the white blight pathogen have not been carried out, the disease is not as widespread in Japan as in Thailand. Is this related to the fact that andrimid was isolated only from the symbiote of N. lugens collected in Thailand but not in Japan? Thus the isolation of andrimid from cultures of Thai N. lugens symbiotes, although of practical interest, also raises many questions.

Insects exemplified by planthoppers, and leafhoppers transmit intracellular microorganisms to offspring through their eggs. Despite the fact that symbiotes were known since 1965, the roles could not be clarified on a molecular structural basis due to difficulties in their isolation and culture without extracellular contamination. The success in culturing several symbiotic strains (ref. 4) has made it possible to carry out preliminary studies on the role of intracellular symbiotes. The results suggest that the symbiotes fulfill at least two functions: (i) They provide (or can provide) the host insects with sterols, essential building components of lipid bilayers and the precursor of 20-hydroxyecdysone. (ii) They can produce antibiotic defense substances.

Taking N. lugens as an example, the relative amounts of steroids found in whole insect bodies differ depending on the site of collection (Fig. 4); the production of andrimid is also dependent on the collection site. It therefore appears that the difference in metabolite ratios (Fig. 4) and production of antibiotics are largely governed by necessities arising from the environment of the host insect. The symbiotes open a new field for further studies.

The molting of insects is regulated by a subtle balance betweeen the molting hormone (MH) 20-hydroxyecdysone and juvenile hormones (JH) (ref. 9) (Fig. 9).

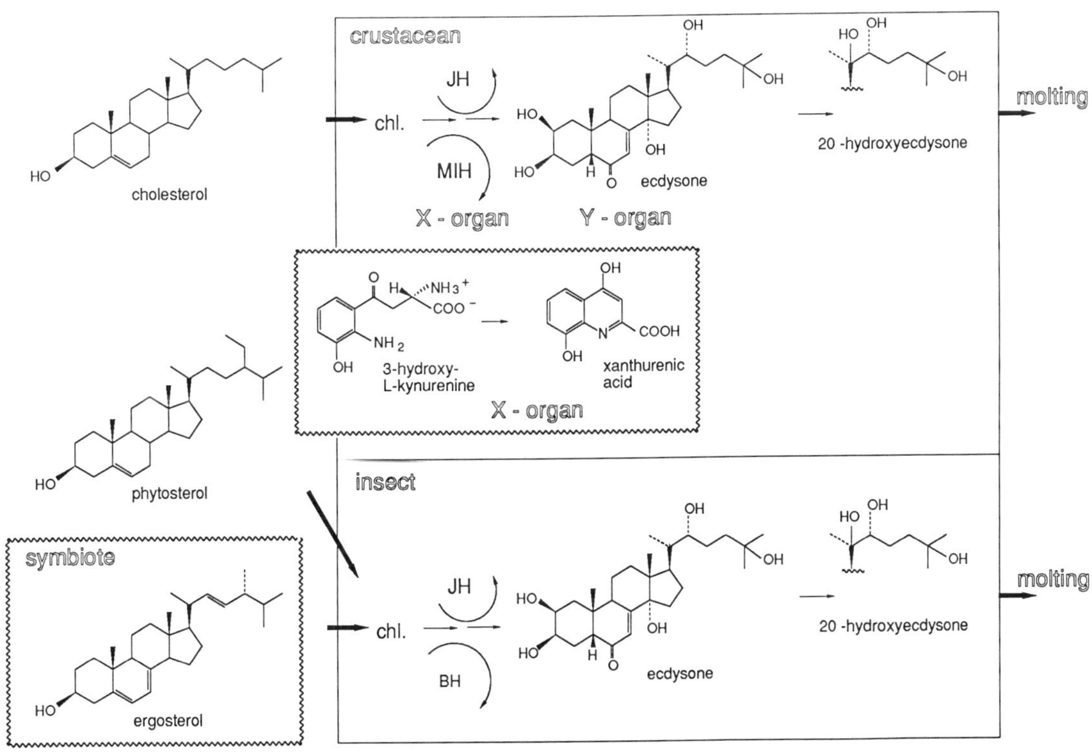

Fig. 9. The molting of crustaceans and insects.

Since insects cannot biosynthesize cholesterol it is either supplied from phytosterols in the food chain or as cholesterol in the case of insects feeding on animals. However, in the insects described above, the symbiotes seem to provide the hosts with the cholesterol precursor, specifically ergosterol. The molting of crustacenas, another class dependent on exogenous steroid, is also induced by ecdysone which is made from cholesterol in the Y-organ; however, in this case, its biosynthesis is controlled by the molt-inhibiting hormone (MIH) produced in the X-organ of the eyestalks. Namely, MIH regulates the P450-dependent hydroxylation(s) of the ecdysone, probably side-chain hydroxylation (it is understood that JH is not involved as directly as in the case of insects but details are not settled). A chemical study of MIH from crustacean eyestalks has recently led to the characterization of 3-hydroxy-L-kynurenine (3-OH-L-K) as one of the factors exhibiting molt-inhibitory *in vitro* activity in crustaceans; it was also found that xanthurenic acid, a metabolite of 3-OH-L-K possibly formed during the incubation for the assay acts directly as "MIH" (ref. 10). The effect of 3-OH-L-K and xanthurenic acid on ecdysone biosynthesis in insects *in vivo* and *in vitro* should further clarify the physiological relation between the two classes, the Insecta and Crustacea.

ACKNOWLEDGEMENTS

Current studies were partially supported by grants from the Science and Technology Agency (Japan), Nos. 278 (1982), 526 (1983), 784 (1984), 645 (1985), and 1049 (1986).

REFERENCES

1. K. Koyama, *Shokubutsu Boeki* (in Japanese), 28, 236 (1974).
2. S. Nasu, *Appl. Ent. Zool.*, 9, 225 (1965).
3. H. Noda and T. Saito, *Appl. Ent. Zool.*, 14, 64 (1979).
4. K. Kusumi, T. Suwa, H. Kita and S. Nasu, *Appl. Ent. Zool.*, 14, 459 (1979); 15, 129 (1980); 16, 88 (1981).
5. H. Noda, K. Wada and T. Saito, *J. Insect. Physiol.*, 25, 443 (1979).
6. H. Noda and T. Saito, *Appl. Ent. Zool.*, 14, 453 (1979).
7. J.S. Glasby, *Encyclopaedia of Antibiotics*, John wiley & Sons, New York (1976).
8. A. Fredenhagen, to be submitted.
9. K. Nakanishi, *Pure & Appl. Chem.*, 25, 167 (1971); *23rd IUPAC Congress Lectures*, Vol. 3, 45 (1971).
10. Y. Naya, K. Kishida, M. Sagane, M. Murata, W. Miki and K. Nakanishi, to be submitted.

Induction of ecdysial stasis with structurally analogous proallatotoxins

Eloi S. Garcia* and Patricia de Azambuja

Department of Biochemistry and Molecular Biology, Fundacao Oswaldo Cruz, Av. Brasil 4365, CEP 21040, C. Postal 926, Rio de Janeiro, RJ, Brasil, and Department of General Biology, Universidade Federal Fluminense, N iteroi, RJ, Brasil

INTRODUCTION

Precocene I and II (PI and PII) are naturally occuring compounds which have been isolated from several species of plants in the genus Ageratum (Bowers, 1976, 1977; Bowers et al., 1976). Many species of insects in contact with precocenes undergo a variety of physiological changes as a result of the absence of juvenile hormone (JH) (see Bowers, 1983). In general, investigations have shown that precocens inhibit the development of the corpus allatum (CA) (Bowers and Martinez-Pardo, 1977; Pender et al., 1978; Unnithan et al., 1979; Schooneveld, 1979) terminating JH production in vivo and in vitro (Bowers et al., 1977; Pratt and Bowers, 1977; Muller et al., 1979; Bowers and Feldlaufer, 1982). Metabolism and mode of action studies suggest that precocenes undergo an oxidative activation by allatal monooxygenase enzymes forming reactive intermediates which condense with macromolecular elements of the cytoplasm causing cell death (Ohta et al., 1977; Brooks et al., 1979; Soderlund et al., 1980). Recently, 3-ethoxy-6-methoxy-2-, 2-dimethylchromene (ethoxyprecocene II, EPII) and 3-ethoxy-4-methoxy-6-iso-pentenylphenol (isopentenylphenol, IPP) were synthesized and shown to produce indistinguishable effects from those of PI and PII (Bowers et al., 1980; Garcia et al., 1982) that causes destruction of CA cells, these compounds are recognized as proallatotoxins. Our knowledge concerning proallatotoxin action is not fully understood. Despite the known effects of these compounds they also induce delay in the molt i.e. many treated nymphs fail to continue ecdysis and remain as nymphs until death (Bowers, 1976, 1977; Bowers et al., 1976; Pener et al., 1978; Tarrant and Cupp, 1978; Masner et al., 1979; Azambuja et al., 1981a, 1981b; Garcia et al., 1984). We have investigated the mechanism of ecdysial stasis induced by proallatotoxins.

PROALLATOTOXINS AND ECDYSIAL STASIS

The PI, PII and EPII induced molting stasis. Although PI has no apparent effect on the induction of precocious metamorphosis as occurs with PII and EPII, the action of PI on molting basis by topical and oral applications is clear. Topical treatment with IPP does not cause effects on precocious metamorphosis and molting. However, IPP by continuous contact treatment is as effective as EPII in inducing these biological effects. The highest production of ecdysial stasis (38%) is obtained by oral treatment with EPII.

ECDYSTEROID TITRES IN PROALLATOTOXIN-TREATED NYMPHS

Since oral treatment with EPII caused 4th-instar nymphs of R. prolixus to molt to either normal 5th instar (18%), to adultoid (44%) or to undergo ecdysial stasis (38%), ecdysteroid titres in haemolymph were determined by radioimmunoassay (RIA) from individual nymphs and correlated to the effects of EPII on molting. In nymphs treated with EPII (20 ug/mL of bloodmeal) three different ecdysteroid-RIA profiles in the haemolymph were observed: (1) with normal molting the ecdysteroid titres were comparable with 4th instar control. However, the ecdysteroid concentration in the peak was half of that observed in the control, and the peak moved from day 6 (control) to day 7 (EPII-treated nymphs). (2) Nymphs molting to adultoids had an ecdysteroid cycle consistent with that of 5th-instar controls. However, the ecdysteroid peak moved from day 12 (5th-instar control) to day 16 (treated-nymphs). The ecdysteroid peak was small, and less sharply defined, increasing and falling off very slowly when compared with the control. (3) Nymphs with ecdysial stasis demonstrated ecdysteroid titres either decreased or abolished.

IN VITRO ASSAY OF PROTHORACIC GLANDS (PG)

The possibility that EPII acts directly and/or indirectly on PG causing alterations in ecdysteroid production was observed by comparing in vitro production of ecdysteroid from groups receiving or not 20 ug EPII/ml in the bloodmeal. A high spontaneous synthesis of ecdysteroid could be observed in the PG control after 5 h incubation at 28^{0}C. However, hormone production was drastically reduced and only a trace of ecdysteroid was observed by RIA that could be detected in incubation of PG derived from nymphs which received EPII. The incubation of PG from control nymphs in a culture medium containing EPII (1 ug/100 uL of medium) also significantly reduced ecdysteroid production.

ECDYSTEROID TITRES IN SURGICAL ALLATECTOMIZED NYMPHS

We have observed that surgical allatectomy could affect the molt cycle of 4th-instar nymphs of R. prolixus. The induced molting inhibition and retardation of the molting could be replaced by application of either JHIII or ecdysone. Ecdysteroid titres in the haemolymph of allatectomized nymphs were significantly reduced. JHIII treatment re-established the hormonal titres in the haemolymph. We also observed that, during the in vitro incubation of PG derived from allatectomized bugs ecdysone synthesis, was reduced or abolished, allatectomized nymphs receiving JHIII treatment had ecdysteroid production as high as the PG control.

DISCUSSION

The present data showed that proallatotoxins are potent agents in inducing molting stasis. Since the effects of proallatotoxins could be reversed by ecdysone therapy (Azambuja et al., 1981a; Garcia et al., 1984), these compounds can limit or abolish with the production of ecdysone by the PG. Proallatotoxin effects therefore could be interpreted as interfering with the neuroendocrine system that controls ecdysone production (JH and prothoracicotropic hormone, PTTH). Histological staining of the neurosecretory cells from the brain revealed a remarkably high accumulation of stainable material in nymphs treated with precocenes (Unnithan et al., 1978; Masner et al., 1979; Azambuja et al., 1981b). It is therefore assumed that the release of PTTH into the haemolymph is reduced or even inhibited through the action of proallatotoxins on the CA, and that it is an important step in causing molting stasis. This interpretation is strongly supported by similar findings observed in allatectomized nymphs of R. prolixus (Garcia, Azambuja and Furtado, in preparation).

Surgical allatectomy may also cause a drastic reduction of ecdysteroid titres which results in ecdysial stasis. Similarly, this molting inhibition could be eliminated by either JH or ecdysone therapy. Since the CA is directly implicated in the storage and/or release of PTTH (Gilbert et al., 1981; Bollenbacher and Granger, 1985) or in other factors to stimulate the PG (Gruetzmacher et al., 1984; Watson et al., 1985), the destruction of the CA by proallatotoxins or its surgical removal could lead to ecdysteroid deficiency. Alternatively, it has been established that precocenes undergoing an oxidative activation by monooxygenase to form highly reactive epoxides (Brooks et al., 1979; Soderlund et al., 1980) that are able to cause destructive alkylation of the allatal tissue causing JH deficiency. Many other tissues, especially gut and fat body, also contain monooxygenases which may be able to undergo similar biotransformation. Since the PG in Robdnius is localized in the inner lobes of the thoracic fat body (Wigglesworth, 1952), it is probable that precocenes could also act directly in PG. This is simply proven by the in vitro incubation of active PG with proallatotoxins; under these conditions the ecdysteroid production was significantly reduced.

In conclusion, JH treatment counteracts the molt inhibitory effect of the proallatotoxins or of the allatectomy. We believe that the ecdysteroid deficiency is an indirect effect caused by JH deficiency in nymphs of Rohodnius prolixus. In addition, however, the possibility of a direct effect of proallatotoxins on the PG can not be excluded, and now is under investigation.

ACKNOWLEDGEMENTS

Original works reported in this chapter was supported by Conselho Nacional de Desenvolvimento Cientifico e Tecnologico (CNPq), Financiadora de Estudos e Projetos (FINEP), and by the UNDP/World Bank/WHO Special Programme for Research and Training in Tropical Diseases. We sincerely appreciate and thank Dr. J.D. O'Connor (UCLA, USA) for supplying with the ecdysone antiserum used in our radioimmunoassay.

REFERENCES

1. P. Azambuja, E..S. Garcia, J.M.C. Ribeiro, Gen. Comp. Endocrinol. 45: 100-104 (1981a).

2. P. Azambuja, E.S. Garcia, A.F. Furtado, C.R. Acad. Sci. Paris 292: 1173-1175 (1981b).

3. W.E. Bollenbacher, N.A. Granger, Endocrinology of the prothoracicotropic hormone. G.A. Kerkut, L.I. Gilbert (eds): Comprehensive Insect Physiology, Biochemistry and Pharmacology, Vol. 7. pp 109-151, Permagon Press, Oxford (1985).

4. W.S. Bowers, Discovery ot insect antiallatotropins. L.I. Gilbert (ed): The Juvenile Hormones. pp. 394-408, Plenum Press, New York (1976).

5. W.S. Bowers, Chemistry and biological activity. Proc. Pontifical Acad Sci, Vatican City, pp 129-142 (1977).

6. W.S. Bowers, The precocenes. R.G.H. Downer, H. Faufer (eds): Endocrinology of Insects. pp. 517-523, Alan R. Liss, Inc., New York (1983).

7. W.S. Bowers, P.H. Evans, P.A. Marsella, D.M. Soderlung, F. Bettarino, Science 217: 647-648 (1982).

8. H.H. Feldlaufer, W.S. Bowers, Gen. Comp. Endocrinol. 47: 120-126 (1982).

9. R. Martinez-Pardo, W.S. Bowers, Science 197: 1369-1371 (1977).

10. W.S. Bowers, T. Ohta, J.S. Cleere, P.A. Marsella, Science 193: 542-547 (1976).

11. G.T. Brooks, G.E. Pratt, R.C. Jennings, Oncopeltus fasciatus, Locusta migratoria. Nature (Lond) 281: 570-574 (1979).

12. E.S. Garcia, P. Azambuja, W.S. Bowers, Arch. Insect Biochem. Physiol. 1: 367-373 (1984).

13. L.I. Gilbert, W.E. Bollenbacher, N. Agui, N.A. Granger, B.J. Sedlak, G. Biggs, C.M. Buys, Am. Zool. 21: 641-653 (1981).

14. M.C. Gruetzmacher, L.I. Gilbert, N.A. Granger, W. Goodman, W.E. Bollenbacher, J. Insect Physiol. 30: 331-340 (1984).

15. P. Masner, W.S. Bowers, M. Kalin, T. Muhle, Gen. Comp. Endocrinol. 37: 156-166 (1979).

16. P.J. Muller, P. Masner, M. Kalin, W.S. Bowers, Experientia 35: 704-705 (1979).

17. T. Ohta, R.J. Kuhr, W.S. Bowers, J. Agr. Food Chem. 25: 478-481 (1977).

18. M.P. Pener, L. Orshan, J. DeWild, Nature 272: 350-354 (1978).

19. G.E. Pratt, W.S. Bowers, Nature 265: 548-550 (1977).

20. H. Schooneveld, Experientia 75: 363-364 (1979).

21. D.M. Soderlung, A. Messegner, W.S. Bowers, J. Agr. Food Chem. 28: 724-731 (1980).

22. C.A. Tarrant, E.W. Cupp, Trans. R. Soc. Trop. med. Hyg. 72: 666-668 (1978).

23. G.C. Unnithan, K.K. Nair, W.S. Bowers, J. Insect. Physiol. 23: 1081-1094 (1977).

24. G.C. Unnithan, K.K. Nair, C.J. Kooman, Experientia 34: 411-412 (1978).

25. R.D. Watson, M. Ciancio, W.P. Gunnar, L.I. Gilbert, W.E. Bollenbacher, J. Insect. Physiol. 31: 487-494 (1985).

26. V.B. Wigglesworth, J. Exp. Biol. 29: 561-569 (1952).

Chemical disruption of insect juvenile hormone biosynthesis

René Feyereisen

Departments of Entomology and Agricultural Chemistry
Oregon State University, Corvallis, Oregon 97331

Abstract - Because insect juvenile hormones (JH) control metamorphosis and reproduction, chemicals that lead to a juvenile hormone deficiency syndrome, for instance by inhibiting JH biosynthesis, should be candidates for evaluation as insect control agents. The discovery of the biological activity on insects of the plant-derived precocenes has "validated" the synthesis of JH as biochemical target, because destruction of the corpora allata is the principal effect of precocenes in sensitive species. In addition to precocene 3,4 epoxide, which is a potent cytotoxin, a few examples of inhibitors of JH biosynthetic enzymes are known. They include inhibitors of isoprenoid biosynthesis (e.g. compactin, mevinolin, fluoromevalonate) and inhibitors of the cytochrome P-450 monooxygenase that epoxidizes the olefin precursor to JH (e.g. methylenedioxyphenyl and 2-propynylether compounds). The selectivity of these inhibitors and the relationship between *in vitro* and *in vivo* activities are still poorly understood. Another way to inhibit JH synthesis is to interfere with the sensitive and selective control of JH synthesis by neuropeptides. Research on agonists and antagonists will start with the structure elucidation of those neuropeptides that inhibit or stimulate JH synthesis by the corpora allata.

INTRODUCTION

Hollingsworth et al. (ref. 1) have listed a number of criteria which should be considered when selecting a target for a biochemically-aided design program. These criteria include the importance of the target in the survival of the pest species as "validated" by lead compounds which disrupt the target. They also include the potential for selectivity/low vertebrate toxicity, and availability of convenient *in vitro* and *in vivo* bioassays. On the basis of such criteria, the biosynthesis of insect juvenile hormones (JH) certainly merits attention as potential target for new insect control agents as noted early on by Matolcsy et al.(ref. 2), and as evidenced by the success of surgical allatectomies in bringing about precocious metamorphosis and/or inhibition of reproduction. *In vivo* bioassays of JH deficiency syndromes, although somewhat more involved than acute toxicity tests are quite easy to perform. Assays of JH biosynthesis by corpora allata incubated *in vitro*, or enzymatic assays of specific steps within the corpora allata have also been developed over the last 10 years (ref. 3,4,5), thus allowing a rational search for inhibitors.

This short paper will review current knowledge on inhibitors of juvenile hormone biosynthesis, which have been discussed in the broader context of anti juvenile hormone agents (ref. 6).

PRECOCENES

The discovery of precocenes (ref. 7) provided the lead needed to establish JH biosynthesis as a "valid target". Precocenes do not specifically inhibit a particular enzyme in the JH synthetic pathway, but instead they destroy the cells of the corpora allata upon epoxidation (ref. 8) by the enzyme methyl farnesoate epoxidase, a cytochrome P-450 monooxygenase (ref. 9, 10). The half-life of the reactive 3,4-ipoxide of precocene I is about 14 s under controlled "physiological" conditions (ref. 11). Diffusion of this reactive opoxide should allow a random attack of cellular nucleophiles by any precocene epoxide molecule which escapes hydration to the dihydrodiols. Precocenes therefore do not qualify as mechanism-based enzyme inhibitors or "suicide substrates". In the moderately sensitive cockroach *Diploptera punctata*, inhibition of JH synthesis by precocene II in short term incubations is achieved at high concentrations (EC50 ca. 0.4 mM). Such concentrations are reached in the hemolymph by precocene when topically applied at an effective dose (ED50 ca. 250 µg/insect) and lead to considerable cellular damage evidenced by electron microscopic examination (ref. 12). The irreversible decline in JH biosynthesis observed *in vitro* is therefore not caused by the selective inhibition of one particular enzyme. The high levels of precocene epoxidation in the corpora allata when compared to epoxidation in other tissues or when compared to other oxidative attacks (e.g. O-demethylation at C6 or C7) may account for the apparently selective toxicity to corpus allatum cells. However, epoxidation of precocenes by cytochrome P-450 monooxygenases in other tissues also has the *potential* of cytotoxicity, as demonstrated in rat liver and kidney (ref. 13,14). There is no single explanation for the selectivity of precocenes for hemimetabolous insects (although high epoxidase / low epoxide hydrolase levels in the corpora allata and peripheral metabolism are probably the two most important factors involved).

INHIBITORS OF BIOSYNTHETIC ENZYMES

Inhibitors of isoprenoid synthesis

Compactin and mevinolin: These sesquiterpene lactones isolated from fungal cultures are extremely potent inhibitors of HMG-CoA reductase, the enzyme which converts 3-hydroxy-3-methylglutaryl coenzyme A to mevalonate. The effects of compactin on insects have been studied (ref. 6). In *D. punctata* corpra allata, mevinolin is slightly more potent than compactin. It is a competitive inhibitor of HMG-CoA reductase with a K_I of 2.4 nM. It also inhibits JH III synthesis *in vitro* with an EC50 of 40 nM (ref. 15). Because the enzyme HMG-CoA reductase also plays important roles in plants, fungi, and animals, the prospects for the design of insect-specific analogs of compactin and mevinolin are not obvious.

Fluoromevalonate: 6-fluoromevalonate is not a specific inhibitor of JH synthesis but like compactin and mevinolin, it also inhibits cholesterol synthesis in vertebrate systems. In rat liver, fluoromevalonate is metabolized by mevalonate kinase to 6-fluoromevalonate-5-pyrophosphate which is a potent competitive inhibitor of pyrophosphomevalonate decarboxylase (K_I =37 nM) (ref. 16). This mechanism of action probably accounts for the anti JH effects of fluoromevalonate in Lepidoptera (ref. 17). Just as for precocenes, there was little success in obtaining active analogs of fluoromevalonate.

Inhibitors of methyl ester formation

Inhibition of the methylation of the carboxylic acid precursor of JH can be achieved by various means *in vitro*: methionine analogs (selenomethionine, cycloleucine) can inhibit the ATP:methionine S-adenosyltransferase thus depressing levels of S-adenosylmethionine (AdoMet). The methylation inhibitor S-adenosylhomocysteine (AdoHcy) can buildup in the corpora allata either by inhibition of AdoHcy hydrolase by 5'-methylthioadenosine or by reversal of the AdoHcy hydrolase reaction (caused by homocysteine and/or adenosine buildup in the glands). Although all these approaches lead to an inhibition of JH synthesis *in vitro* (ref. 15), their practical value has not been tested. However, interesting compounds may eventually spin-off from the closely related anti-viral drug research.

Inhibitors of epoxidation

Following the discovery that methyl farnesoate epoxidase is a cytochrome P-450 monooxygenase (ref. 9,10), P-450 inhibitors were tested either in cell-free methyl farnesoate epoxidase assays (ref. 18), or on intact corpora allata (ref. 19-22). These inhibitors bear moieties which were traditionally used in the design of insecticide synergists such as methylene dioxyphenyl compounds, acetylenics and imidazoles.

Methylenedioxyphenyl (MDP) compounds: The potent JH agonist epoxygeranyl-MDP ether (1) proved to be a relatively good inhibitor of JH synthesis in the *Periplaneta americana* assay developed by Pratt and Finney (ref. 19). The corresponding olefin or 7-ethoxy compounds were 100x less active. When a substrate (methyl farnesoate) analog was designed by maintaining the conjugated methyl ester group instead of the terminal epoxide (2), a considerable gain in inhibitory strength was observed (EC50 = 10 nM, 20). The addition of one isoprenoid group to 2 led to a 300x loss in activity (ref. 20).

Acetylenic compounds: Phenylpropynyl ether- (3) or simpler acetylenic analogs of methyl farnesoate (4) were good inhibitors of JH synthesis *in vitro*, some with an EC50 around 10nM (ref. 20,22). A preliminary study with simple acetylenic fatty acids revealed that 11-undecynoic acid inhibited JH synthesis in *D. punctata* corpora allata (EC50 = 75 μM, 23). The methyl ester was half as active and 12-dodecynoic acid was 5x less active.

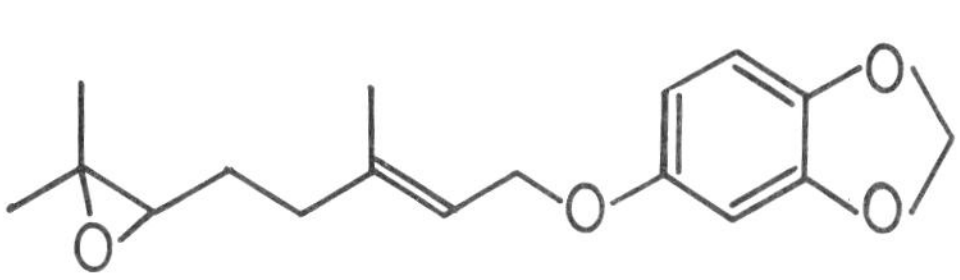

(1)

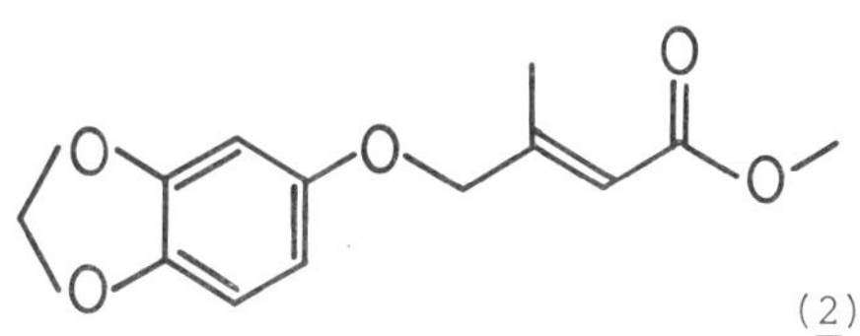

(2)

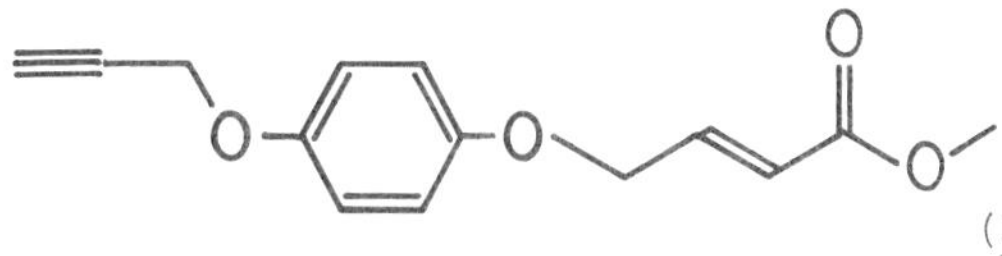

(3)

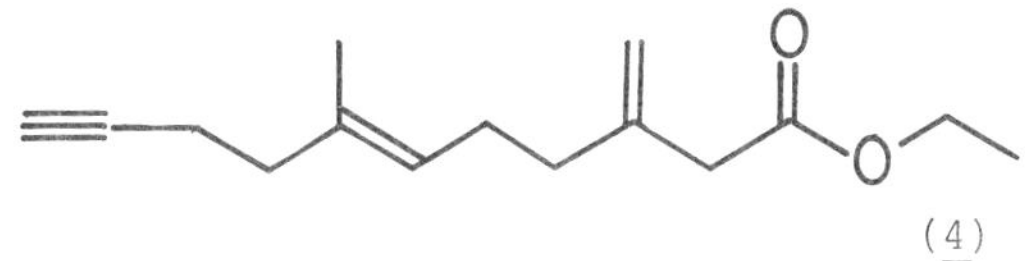

(4)

It would thus appear that inhibition by MDP and acetylenic compounds should be amenable to a QSAR approach for structure optimization. In *P. americana* corpora allata, some epoxidase inhibitors were shown to inhibit the endogenous production of methyl farnesoate at an unknown step situated prior to farnesoic acid methylation (ref. 20,21). Whether this side-effect is caused by the inhibitors themselves or by a feedback mechanism preventing the accumulation of methyl farnesoate in the corpora allata of this cockroach is not yet fully understood (ref.21). The main drawback of these MDP and acetylenic compounds is that the highest inhibitory activity is often linked to a close structural resemblance to the epoxidase substrate (methyl farnesoate). Consequently, such compounds also have a structural resemblance to the product of the epoxidation, JH III, and may act as JH agonists. This explains the low *in vivo* activity of these compounds in anti JH assays.

Imidazoles: Based on the results of Hammock and Mumby, who showed that bromophenoxymethyl imidazoles were potent inhibitors of methyl farnesoate epoxidation in *Blaberus giganteus* corpora allata (ref. 18), Kuwano's group synthesized a number of imidazole compounds some of which insecticidal, some of which with true anti JH activity in *Bombyx mori* (ref. 6). It is not known at this time whether these anti-JH imidazoles such as KK 22 and KK 42 act by inhibition of the epoxidase, but this is a possibility.

A functional group less frequently used in insecticide synergist design, but which is known to cause a mechanism-based inhibition of cytochrome P-450 is the cyclopropylamine group (ref. 24). A cyclopropylbenzamide JH analog (5) inhibited JH synthesis in *D. punctata* (EC50= 42 μM), but was devoid of anti JH activity in this cockroach (ref. 25).

Because precocenes can reach the corpora allata *in vivo*, and can be metabolized there by the epoxidase, it was assumed that precocene analogs might be designed as epoxidase inhibitors. The "MDP-precocene" 6,7-methylene dioxy-2,2-dimethylchromene, and the oxathiole (6) are also good inhibitors of JH synthesis in *P. americana* (ref. 26) but have no anti JH activity. The propynyloxy analog of precocene II was a better inhibitor of JH synthesis than precocene II in *Diploptera*, but its anti JH activity was not improved (ref. 22). In contrast, the simple insecticide synergist (7) was found to have significant anti JH activity in *D. punctata*, but only toxic effect in *Oncopeltus fasciatus* and *Locusta migratoria* (ref. 22,27). I feel that there is considerable room for further synthesis of powerful inhibitors (such as mechanism-based inhibitors) (ref. 28) of the epoxidase which would be devoid of any intrinsic JH agonist activity.

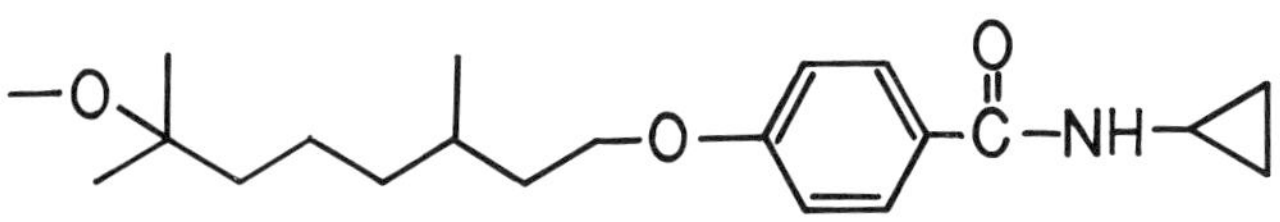

(5)

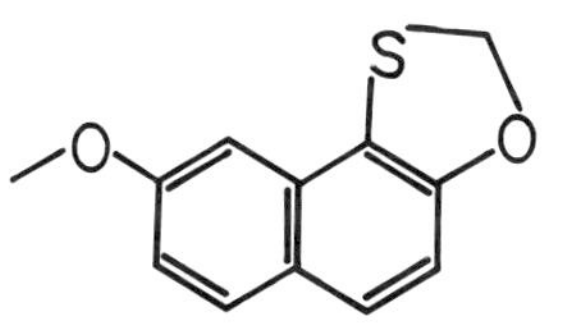

(6)

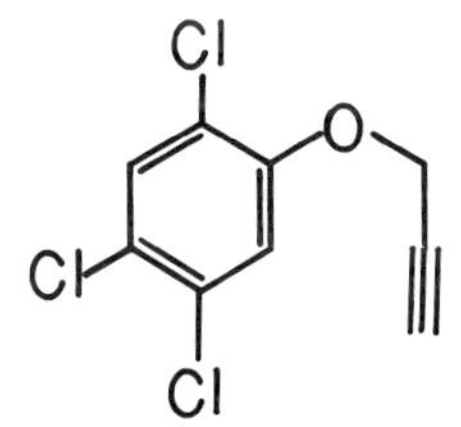

(7)

Inhibitors of JH synthesis with unknown mode of action

Some compounds with anti JH or mixed anti JH / JH agonist activity were shown to inhibit JH synthesis by *Manduca sexta* corpora allata *in vitro*. These compounds include a bisthiocarbamate, dichloropropenylhexanoate and the JH analog ETB (ref. 6) but their mechanism of action is unknown.

SELECTIVITY

The higher homologs of JH III (i.e. JH II, JH I, JH 0 and 4-methyl-JH I) have been identified only in Lepidoptera (ref. 4). These compounds are not produced by methylations of a normal isoprenoid JH III-like structure, but are synthesized *de novo* from 3'-homomevalonate. Homomevalonate is a new homoisoprenoid building block that is produced only in the CA of Lepidoptera from propionyl-CoA via 3-hydroxy-3-ethylglutaryl-CoA. Only few other homoisoprenoids are known in nature (essentially from *Myrmica*, *Acromyrmex*, *Monomorium* and *Solenopsis* ants,ref.29). The major insect predator, parasitoid and pollinator species (chiefly) from the orders Hymenoptera and Diptera) would appear to biosynthesize only JH III, whereas the major agricultural pests from the order Lepidoptera synthesize the higher homologs, JH II, JH I, JH 0, and 4-methyl-JH-I instead or in addition to JH III. Thus, inhibitors of JH synthesis specific for the higher homologs of JH III might have a desirable degree of selectivity build-in. To allow the design of such inhibitors, more needs to be known about the origin of propionyl-CoA in lepidopteran corpora allata. The available evidence indicates that most enzymes of the isoprenoid pathway (with the possible exception of isopentenyl-pyrophosphate isomerase) have a degree of substrate specificity that would tolerate higher homologs of the JH III precursors (ref. 4). However, when combined within a set of enzymes, small differences in specificity would rapidly become limiting. Thus, it is not necessary to look for a single magical enzyme difference in the isoprenoid pathway which would allow Lepidoptera to synthesize homosesquiterpenoid JHs. The supply of propionyl-CoA (possibly from the metabolism of amino acids such as valine, threonine, methionine and isoleucine) remains an important factor allowing the synthesis of JH III homologs. In fact, acetyl-CoA has also been shown to be derived in part from amino acid metabolism in cockroach corpora allata (ref. 15). I feel therefore that selectivity may be achieved by targeting enzymes responsible for propionyl-CoA synthesis not needed for JH synthesis outside Lepidoptera, rather than enzymes of the isoprenoid pathway which are basically of common importance throughout the phylogeny.

CONTROL OF JH SYNTHESIS BY NEUROPEPTIDES

The activity of the corpora allata is under the control of other endocrine centers, in particular the brain (ref. 3). During the last few years, information on the chemical nature of the factors controlling JH synthesis has started to emerge. In *L. migratoria* and *M.sexta* an allatotropin has been shown to stimulate JH synthesis. These factors appear to be neuropeptides. Conversely, inhibitory peptides have been shown to exist in *M. sexta* and *D. punctata*, and certainly more information on the chemistry, distribution and mode of action of these peptides will soon be available. There can be no doubt that chemical design of agonists and antagonists of those peptides will become a intense area of research. Indeed, such neuropeptides usually act at low concentrations and their effect is considerably amplified (e.g. a small change in allatotropin titer may result in a large change in JH titer). Thus, the sensitivity and selectivity of the neuropeptide receptor mechanisms should make them worthwhile targets. The biosynthesis, processing, release, mode of action and degradation of these regulatory neuropeptides should also provide useful new targets for the design of powerful inhibitors of JH biosynthesis.

REFERENCES

1. R.M. Hollingworth, E.M. Johnstone and N. Wright, ACS Symp. Ser. 255, 103-125 (1984).
2. G. Matolcsy, L. Varjas and B. Bordas, Acta Phytopath. Acad. Sci. Hung. 10, 455-463 (1975).
3. R. Feyereisen, Comprehensive Insect Physiology, Biochemistry and Pharmacology, vol. 7, 391-429, Pergamon Press, Oxford (1985).
4. D.A. Schooley and F.C. Baker, ibid., vol 7, 363-389 (1985).
5. R. Feyereisen, Methods Enzymol. 111, 530-539 (1985).
6. G.B. Staal, Ann. Rev. Entomol. 31, 391-429 (1986).
7. W.S. Bowers, T. Ohta, J.S. Cleere and P.A. Marsella, Science 197, 1369-1371 (1976).
8. G.E. Pratt, R.C. Jennings, A.F. Hamnett and G.T. Brooks, Nature 284, 320-323 (1980).
9. B.D. Hammock, Life Sci. 17, 323-328 (1975).
10. R. Feyereisen, A.F. Hamnett and G.E. Pratt, Eur. J. Biochem. 118, 231-238 (1981).
11. A.F. Hamnett, A.P. Ottridge, G.E. Pratt, R.C. Jennings and K.M. Stott, Pestic. Sci. 12, 245-254 (1981).
12. R. Feyereisen, G. Johnson, J. Koener, B. Stay and S.S. Tobe, J. Insect Physiol. 27, 855-868 (1981).
13. M.T.S. Hsia, S. Grossman and K.R. Schrankel, Chem.-Biol. interact. 37, 265-277 (1981).
14. K.R. Schrankel, S.J. Grossman and M.T. Hsia, Toxcol. Lett. 12, 95-100 (1982).
15. R. Feyereisen and D. E. Farnsworth, unpublished results.
16. J.F. Nave, H. d'Orchymont, J.B. Ducep, F. Piriou and M.J. Jung, Biochem. J. 227, 247-254 (1985).
17. G.B. Quistad, D.C. Cerf, D.A. Schooley and G.B. Staal, Nature 289, 176-177 (1981).
18. B.D. Hammock and S.M. Mumby, Pestic. Biochem. Physiol. 9, 39-47 (1978).
19. G.E. Pratt and J.R. Finney, In Crop Protection Agents - Their Biological Evaluation 113-132 Academic Press, London (1977).
20. G.T. Brooks, G.E. Pratt, D.W. Mace and J.A. Cocks, Pestic. Sci. 16, 132-142 (1985).
21. G.E. Pratt, R.C. Jennings and R.J. Weaver, Insect Biochem. 6, 609-614 (1984).
22. R. Feyereisen, D.E. Farnsworth, K.S. Prickett and P.R. Ortiz de Montellano, ACS Symp. Ser. 276, 255-266 (1985)
23. R. Feyereisen, N. Reich and P.R. Ortiz de Montellano, unpublished results.
24. T.L. Macdonald, K. Zirvi, L.T. Burka, P. Peyman and F.P. Guengerich, J. Am. Chem. Soc. 104, 2050-2052 (1982).
25. I. Belai, R. Feyereisen and G. Matolcsy, unpublished results.
26. G.T. Brooks, G.E. Pratt, A.P. Ottridge and J.A. Cocks, J. Pestic. Sci. 9, 211-214 (1984).
27. G.T. Brooks, A.P. Ottridge, R.C. Jennings, D.W. Mace and B.A. Alexander, Pestic. Sci. 16, 571-588 (1985).
28. R. Feyereisen, K.C. Langry and P.R. Ortiz de Montellano, Insect Biochem. 14, 19-26 (1984).
29. A.B. Attygale and E.D. Morgan, J.C.S. Perkin I, 949-951 (1982).

Oregon Agricultural Experiment Station technical paper no 7971.

Myoactive agents from the cockroach *Periplaneta americana*

A.N. Starratt* and R.W. Steele

London Research Centre, Agriculture Canada, University Sub P.O., London, Ontario, N6A 5B7, Canada

Abstract - Studies on three myoactive peptides of defined structure from Periplaneta americana, proctolin, MI and MII, are briefly reviewed together with current work in our laboratory concerned with the isolation of additional substances which are pharmacologically active on the cockroach hindgut.

The presence of peptides in the cockroach Periplaneta americana which act on insect visceral muscles has been recognized for many years but progress towards their identification has been slow despite studies by several groups (ref. 1,2). Early investigations by Brown (ref. 3) from our laboratory concerned constituents of the cockroach corpora cardiaca which were active on the heart and hindgut. Three of these, which appeared to be peptides, accounted for most of the activity in crude extracts. Studies soon shifted, however, to a peptide from foreguts and hindguts of P. americana which Brown recognized as an active principle of major importance (ref. 4). It caused a slow-type, graded contraction of the longitudinal muscles of the hindgut (proctodeum) similar to that induced by electrical stimulation of the nerves innervating these muscles. Brown proposed that this 'gut-factor', later called proctolin (ref. 5), functions as an excitatory neuromuscular transmitter substance in insect visceral muscles (ref. 4, 6).

Because of the potential significance of this substance, the task of isolating pure proctolin in sufficient quantities for structural determination was undertaken. This proved to be difficult due to the low levels present and its association with other substances. Fortunately, a simple and reliable bioassay procedure, which continues to be used for our current studies of cockroach myoactive factors, was available. Samples could be rapidly tested by bath applications to the isolated cockroach hindgut. Eventually 180 µg of pure proctolin were obtained from 125 kg of adult cockroaches by a multistep procedure using separation techniques available at the time (ref. 5). Sequence analysis by the Edman-dansyl method and subsequent synthesis showed that proctolin was H-Arg-Tyr-Leu-Pro-Thr-OH (ref. 7, 8).

Proctolin, the first insect neuropeptide to be identified, has been the subject of many investigations since 1975 concerned with its occurrence, distribution, function and inactivation in insects and other arthropods. A major contributory factor to progress in this area has been the commercial availability of synthetic proctolin and, later, [^{3}H-Tyr^{2}]proctolin of high specific activity.

Evidence has been obtained for proctolin or proctolin-like substances in a variety of insect species in addition to P. americana (ref. 9-14), although Kingan and Titmus (ref. 12) recently reported its absence, based on radioimmunoassay, in a Lepidopteran. In the case of P. americana, several immunological studies have shown that proctolin appears discretely but widely distributed throughout the central nervous system (ref. 15-19) and its presence in individually identified neurons has been demonstrated by HPLC combined with bioassay (ref. 18, 20). This distribution correlates well with bioactivity; as well as acting on the hindgut muscle, proctolin is active on the foregut (ref. 7), heart (ref. 21), hyperneural muscle (ref. 21-23), antennal heart (ref. 24), a coxal depressor muscle (ref. 25), and oviduct (unpublished) of this cockroach.

In vivo studies showed that proctolin was rapidly inactivated in P. americana by an aminopeptidase(s) (ref. 26). Subsequent in vitro metabolism studies (ref. 27, 28) yielded evidence for other enzymes capable of degrading proctolin. The significance of these proteolytic enzymes to proctolinergic action remains an interesting area for speculation.

Recently, the structures of two other myoactive peptides, MI and MII (periplanetins CC-1 and CC-2), from the corpora cardiaca of P. americana have been reported (ref. 29, 30). These belong to a family of invertebrate peptides that includes locust adipokinetic hormone, pGlu-Leu-Asn-Phe-Thr-Pro-Asn-Trp-Gly-Thr-NH_2, whose structure was reported in 1976 (ref. 31). It has since became apparent that a number of factors described earlier in the literature were identical to MI and MII (see Table 1).

Sequence	Reference
(1) pGlu-Val-Asn-Phe-Ser-Pro-Asn-Trp-NH_2	
MI	29
Periplanetin CC-1	30
Neurohormone D	32
Hypertrehalosaemic hormone I	33
HGHI	34
(2) pGlu-Leu-Thr-Phe-Thr-Pro-Asn-Trp-NH_2	
MII	29
Periplanetin CC-2	30
Hypertrehalosaemic hormone II	33
HGHII	34

TABLE 1 Myoactive peptides from P. americana corpora cardiaca.

Using a locust leg assay, Witten et al. (ref. 29) found that these peptides caused an increase in the beating frequency of the myogenic rhythm and an increase in tension at threshold levels of approximately 10^{-7}M for MI and 2 x 10^{-7}M for MII while Scarborough et al. (ref. 30) reported that both were able to accelerate the heartbeat in a semi-isolated P. americana heart preparation at 10^{-8}M. As shown by Fig. 1, they are also active on the cockroach hindgut but can usually be differentiated from proctolin by exhibiting inhibition upon repetitive exposure.

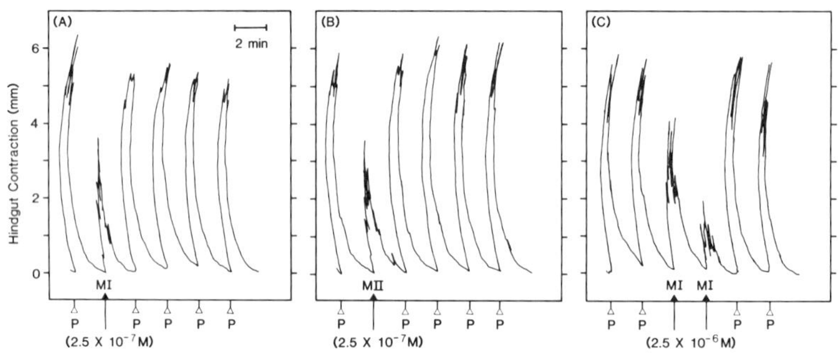

Fig. 1. Effect of MI, MII, and proctolin on a P. americana hindgut. Each peptide was bath applied for 30 sec every 2-min perfusion cycle and induced contractions recorded isotonically against 200 mg tension. (A) Effect of a single exposure to MI at 2.5 x 10^{-7} M. (B) Effect of MII at the same concentration. (c) Effect of consecutive exposures to MI at 2.5 x 10^{-6} M. Proctolin (P) was applied at 5 x 10^{-9}M.

The rapid correlation of a number of factors with MI and MII is a clear demonstration of the importance of the structural determination of pharmacologically active substances. Until such characterization, it remains uncertain whether the observed function(s) is due to a single factor or to two or more active components contained in an impure preparation. In our laboratory, studies have shown additional substances in P. americana which exhibit myoactivity on the cockroach hindgut (ref. 35). On an analytical scale, these may easily be isolated from homogenates of cockroach tissues by Sep-Pak C_{18} extraction followed by reversed-phase HPLC. The addition of [^{3}H-Tyr^2]-proctolin to the homogenates has permitted general assessment of peptide recoveries and enabled proctolin-containing fractions in subsequent chromatography to be distinguished from those containing unknown myotropic factors. Also, by measuring the decrease in contractions induced by proctolin standards introduced to the assay chamber 15 sec after sample application, we have discovered inhibitory factors that we have named neutrolins. Despite variations in relative concentrations, fairly similar patterns of myotropic and inhibitory activities have been obtained for heads, hindguts, foreguts, and various ganglia.

Although Sep-Pak C_{18} cartridges are excellent tools for isolating bioactive peptides from small quantities of insect tissues, their low capacity is a drawback. Consequently, we sought an economical system which could be used for isolation of unidentified myoactive factors from larger extracts.Recent studies have shown that Millipore Norganic resin, a proprietary product marketed for the preparation of HPLC grade water, can serve as a low cost alternative to Sep-Pak C_{18} cartridges for the solid phase extraction and first stage purification of proctolin and other physiologically active constituents (ref. 36). Using [^{3}H-Tyr^2]proctolin as tracer, it was shown that a single Norganic resin cartridge could extract all the proctolin from a homogenate of 4800 cockroach heads or 300 cockroach bodies. Since the resultant extract was still too large to be conveniently chromatographed on available HPLC columns, the bioactive material was further purified using Sep-Pak C_{18} cartridges. This reduced the quantities by approximately 90% with little loss in proctolin and other myoactive substances of interest. The resulting preparation could then be fractionated in a single run on a semi-preparative (15 cm x 19 mm) reversed-phase column. Subsequent runs by reversed-phase and ion-exchange chromatography clearly demonstrate the complexity of the myoactive mixture. These methods offer the potential of yielding sufficient material to expand the inventory of identified insect neuropeptides.

With the exception of proctolin, all the insect neuropeptides identified to date belong to the adipokinetic hormone family. This includes the recently identified myotropic peptide, pGlu-Thr-Ser-Phe-Thr-Pro-Arg-Leu-NH_2, from the cockroach Leucophaea maderae (ref. 37). It remains to be seen whether other myoactive substances from insect sources are structurally related to proctolin or the adipokinetic hormone family or belong to new classes. Recent technological advances such as fast atom bombardment mass spectrometry, used during structural elucidation of myoactive peptides MI and MII (ref. 29) and adipokinetic hormones from several insects (ref. 38-40), should accelerate the progress.

Although insect neuropeptides are not expected to have a practical application in control strategies, much of the recent research in this area has been spurred by the potential of developing new insecticides directed against their actions. Aspects of biosynthesis, storage, release, receptor pharmacoloy, function(s), and inactivation may separately or collectively offer targets for manipulation. The success of such approaches in terms of novel insecticide development may depend on whether the target site(s) can be exploited in a manner such that selective pest control agents can be developed. In this regard, it is noteworthy that proctolin is also active on mammalian muscle (ref. 22, 41) and that proctolin-immunoreactive neurons have been detected in mouse brain (ref. 42).

REFERENCES

1. N. Frontali and H. Gainer, in Peptides in Neurobiology (H. Gainer, ed.), pp. 259-294, Plenum Press, New York (1977).
2. M. Raabe, Insect Neurohormones, pp. 121-140, Plenum Press, New York (1982).

3. B.E. Brown, Gen. Comp. Endocrinol. 5, 387-401 (1965).
4. B.E. Brown, Science 155, 595-597 (1967).
5. B.E. Brown and A.N. Starratt, J. Insect Physiol. 21, 1879-1881 (1975).
6. B.E. Brown, Life Sci. 17, 1241-1252 (1975).
7. A.N. Starratt and B.E. Brown, Life Sci. 17, 1253-1256 (1975).
8. A.N. Starratt and B.E. Brown, Can. J. Chem. 55, 4238-4242 (1977).
9. B.E. Brown, J. Insect Physiol. 23, 861-864 (1977).
10. G.M. Holman and B.J. Cook, Insect Biochem. 9, 149-154 (1979).
11. G.M. Holman and B.J. Cook, Comp. Biochem. Physiol. 62C, 231-235 (1979).
12. T. Kingan and M. Titmus, Comp. Biochem. Physiol. 74C, 75-78 (1983).
13. H. Keshishian and M. O'Shea, J. Neurosci. 5, 992-1004 (1985).
14. J.A. Veenstra, H.M. Romberg-Privee and H. Schooneveld, Cell Tissue Res. 240, 535-540 (1985).
15. C.A. Bishop, M. O'Shea and R.J. Miller, Proc. Natl. Acad. Sci. USA 78, 5899-5902 (1981).
16. M. Eckert, H. Agricola and H. Penzlin, Cell Tissue Res. 217, 633-645 (1981).
17. C.A. Bishop and M. O'Shea, J. Comp. Neurol. 207, 223-238 (1982).
18. M. O'Shea and C.A. Bishop, J. Neurosci. 2, 1242-1251 (1982).
19. H. Agricola, M. Eckert, J. Ude, H. Birkenbeii and H. Penzlin, Cell Tissue Res. 239, 203-209 (1985).
20. M. O'Shea and M.E. Adams, Science 213, 567-569 (1981).
21. T. Miller, Amer. Zool. 19, 77-86 (1979).
22. H. Penzlin, H. Agricola, M. Eckert and T. Kusch, In Advances in Physiological Sciences 22, Neurotransmitters in Invertebrates (K.S.-Rozsa, ed.), pp. 525-535, Akademiai Kiado, Budapest (1981).
23. W. Hertel and H. Penzlin, J. Insect Physiol. 32, 239-248 (1986).
24. W. Hertel, G. Pass and H. Penzlin, J. Insect Physiol. 31, 563-572 (1985).
25. M.E. Adams and M. O'Shea, Science 221, 286-289 (1983).
26. A.N. Starratt and R.W. Steele, Insect Biochem. 14, 97-102 (1984).
27. G.B. Quistad, M.E. Adams, R.M. Scarborough, R.L. Carney and D.A. Schooley, Life Sci. 34, 569-576 (1984).
28. R.W. Steele and A.N. Starratt, Insect Biochem. 15, 511-519 (1985).
29. J.L. Witten, M.H. Schaffer, M. O'Shea, J.C. Cook, M.E. Hemling and K.L. Rinehart, Biochem. Biophys. Res. Commun. 124, 350-358 (1984).
30. R.M. Scarborough, G.C. Jamieson, F. Kalish, S.J. Kramer, G.A. McEnroe, C.A. Miller and D.A. Schooley, Proc. Natl. Acad. Sci. USA 81, 5575-5579 (1984).
31. J.V. Stone, W. Mordue, K.E. Batley and H.R. Morris, Nature 263, 207-211 (1976).
32. E. Baumann and H. Penzlin, Biomed. Biochem. Acta 43, K13-K16 (1984).
33. G. Gäde, Naturwissenschaften 72, 95-97 (1985).
34. K.J. Siegert, P.J. Morgan and W. Mordue, Insect Biochem. 16, 365-371 (1986).
35. R.W. Steele and A.N. Starratt, in Insect Neurochemistry and Neurophysiology (A.B. Borkovec and T.J. Kelly, eds.), pp. 487-489, Plenum Press, New York (1984).
36. A.N. Starratt and R.W. Steele, J. Liq. Chromatogr. In press.
37. R.J. Nachman, G.M. Holman and B.J. Cook, Abstract No. 79, Agrochemicals Division, 192nd American Chemical Society National Meeting, Anaheim, CA, (1986).
38. R. Ziegler, K. Eckart, H. Schwarz and R. Keller, Biochem. Biophys. Res. Commun. 133, 337-342 (1985).
39. G. Gäde, G.J. Goldsworthy, M.H. Schaffer, J.C. Cook and K.L. Rinehart, Biochem. Biophys. Res. Commun. 134, 723-730 (1986).
40. H. Jaffe, A.K. Raina, C.T. Riley, B.A. Fraser, G.M. Holman, R.M. Wagner, R.L. Ridgway and D.K. Hayes, Biochem. Biophys. Res. Commun. 135, 622-628 (1986).
41. H. Schulz, J. Schwarzberg and H. Penzlin, Acta Biol. Med. Ger. 40, K1-K5 (1981).
42. H.-G. Bernstein, M. Eckert, H. Penzlin and A. Dorn, Neurosci. Lett. 45, 229-232 (1984).

Monitoring cutworm populations with sex pheromone-baited traps

J. R. Byers, D. L. Struble* and G. B. Schaalje

Agriculture Canada Research Station, Lethbridge, Alberta, Canada T1J 4B1

Abstract - Reliable estimates of adult cutworm population levels can be obtained and the potential for subsequent larval infestations can be predicted from moth catches in sex pheromone-baited traps. Fluctuations in adult abundances under prairie conditions can be reliably detected by a relatively small number of trapping locations, e.g., 5 per 13 000 km^2, which is a feasible number of traps for long-term surveillance monitoring.

INTRODUCTION

Cutworms are serious pests in many countries and they have caused widespread and major losses on several occasions since the inception of agriculture in North America. Although the use of insecticides has reduced the severity of outbreaks, considerable losses still occur, largely because the initial stages of larval infestations often go undetected. The major cutworm pests on the prairies are native species which at endemic population levels cause little or no damage. Outbreaks are sporadic and occur when a surge in population coincides with conditions that are particularly favorable for survival. Unless the populations are monitored systematically during periods of scarcity as well as abundance, the impending population surge will not be detected. Crucial factors initiating outbreaks may be missed and remain unrecognized because they occur infrequently. Advances in the management of cutworms will be largely dependent on long-term surveillance monitoring.

From 1978 to 1983 the adult population levels of 8 species of noctuids were monitored with sex-pheromone baited traps at 81 fixed locations within a 13 000-km^2 region of southern Alberta. The monitoring program was partially supported by a Farming for the Future grant from the Agricultural Research Council of Alberta. This paper deals with the specificity of the pheromones used, the effect that abundance of the target species has on apparent specificity, the analysis of trap catch variability, precision of population estimates, population fluctuations, and the relationship between trap catches and subsequent larval infestations. A long-term surveillance monitoring system was initiated throughout Alberta based on the results from this intensive monitoring program.

RESULTS AND DISCUSSION

Traps and locations

Cylindrical dry cone-orifice traps (26 cm long, 10.2 cm diameter) were made from fibreglass pipe, and end-cones were made of No. 8 wire mesh with an apical opening of 1 cm. Pheromones were dispensed from rubber septa. Duplicate traps (100 m apart) for each species were placed along cultivated fields at the same 81 locations. Catches were recorded weekly.

Population fluctuations and trends

The species monitored, pheromone blends, specificity of the blends, and total moths captured are summarized in Table 1. The mean moth catches per year for 4 species are shown in Fig. 1. Monitoring for pale western cutworm has been continued since 1983 due to an outbreak in the survey area (Fig. 2A). Usually, year-to-year differences in mean trap catches were significant at the 95% confidence level and it is evident that the monitoring system could detect relatively small changes in population levels between years.

Bertha armyworm and pale western, redbacked and army cutworms reached economic thresholds at some locations during this study. The population of bertha armyworm was very low in 1978 but increased to economic importance in 1981, and then declined. Redbacked and army cutworm populations both peaked in 1979 and there were larval infestations that required control in the spring of 1980. Pale western population was low in 1978 and then it has increased steadily each year to 1985. The large variation in year-to-year population levels of variegated cutworm is indicative of a migratory species.

TABLE 1. Pheromone Monitoring, 1978-1983

Cutworm Species	Pheromone blend	Ratio	Specificity (%)	Total moths (x 1000)
Flight period, April-July				
Bertha	Z9-14Ac, Z11-16Ac	1:19	60	46
Variegated	Z9-14Ac, Z11-16Ac	1:2	73	50
L. commoides	Z9-14Ac, Z11-16Ac, Z11-16OH	5:4:1	97	322
Clover	Z11-16Ac, Z11-16OH	9:1	93	156
Flight period, August-October				
Darksided	Z7-16Ac, Z11-16Ac, Z11-16OH	1:40:0.01	99	240
Redbacked	Z5-10Ac, Z5-12Ac, Z7-12Ac, Z9-12Ac	2:500:2:1	99	556
Pale western	Z5-12Ac, Z7-12Ac	1:2	99	107
Army	Z5-14Ac, Z7-14Ac, Z9-14Ac	100:1:10	99	252

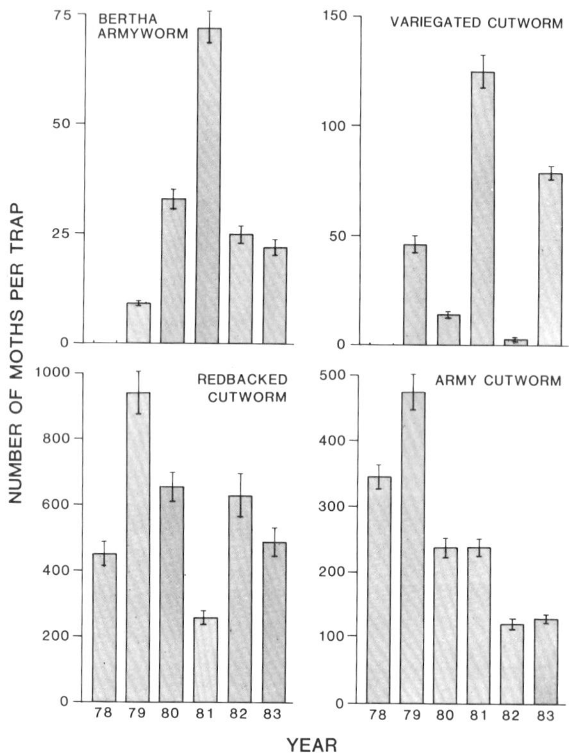

Fig. 1. Mean Moth Catches per Trap at 81 Locations with 95% Confidence Intervals.

Pheromone specificity

The specificities of the pheromone blends were >93% except for variegated cutworm and bertha armyworm (Table 1). These blends involve Z9-14Ac and Z11-16Ac in ratios of 1:2 and 1:19 and their low specificity is due to several factors. Firstly, the variegated cutworm blend became increasingly attractive to bertha armyworm over a 2-month period in the field because the concentration of the more volatile Z9-14Ac declined relative to Z11-16Ac. This is a problem of dispensing chemicals of different molecular weight at a constant relative rate, which cannot be achieved with rubber septa as dispensers. Secondly, the nontarget species attracted have very similar sex pheromones and they probably maintain their reproductive isolation in some other manner, such as specific flight times. Thirdly, the monitor traps were at fixed locations and rarely were they adjacent to fields of canola or mustard, which are the preferred host plants for bertha armyworm. Consequently, catches of bertha armyworm moths were usually low. In a separate experiment, traps in the vicinity of these crops captured about 3 times more moths, and the nontarget species represented only 20% of the catches. It is evident from the data obtained from this study that there is a strong relationship between the mean catch of the target species and the apparent specificity of its attractant. The number of nontarget moths caught was relatively constant and independent of the catches of target species; therefore, the apparent specificity increased with catches of the target species.

The use of high-capacity nonsaturating traps and pheromone blends that are maximally attractive are clearly advantageous in enhancing specificity. High catches also reduce within-site sampling variance and produce more precise estimates of population density. Current studies with different trap designs indicate that the Uni-trap (International Pheromone Systems) is suitable for full-season surveillance monitoring as it is an effective trap for most noctuids and it has a high capacity.

Consistency of trap catches versus locations

The degree of consistency in the relative abundance, or ranking, of moth catches among locations, both within and between years, was determined. The ranking of moth catches among locations within years was highly consistent (Kendall's coefficient of concordance, $W > 0.72$), indicating that the spatial pattern of abundance remained stable during each flight period. The corollary is that differences in population levels among locations were consistently detected by the pheromone traps.

Most of the species exhibited a considerable degree of consistency in pattern of abundance between consecutive years and, to some extent, over all years. The 3 exceptions were army ($W = 0.57$), variegated ($W = 0.51$), and darksided cutworms ($W = 0.49$). The army and variegated cutworms are both vagile species and fluctuations in their patterns of abundance would be expected. Army cutworm moths emerge in June, migrate to the Rocky Mountains for the summer, and return to the prairies in August to mate and lay eggs. Apparently they do not always return to the area of origin. The variegated cutworm can overwinter only in the southern U.S.A. and the moths must migrate northward annually, which explains their variable abundance patterns in western Canada. Darksided cutworm, on the other hand, was the most uniformly distributed within the survey area due to its wide host range, and moderate changes in moth catches among locations resulted in extensive changes in rankings.

Variability of trap catches within and among locations

Estimates of the variability associated with individual traps, between the duplicate traps at locations, and among locations were obtained. The within-location variance was always much less (usually significant at 0.1% level) than the among-location variance, indicating that a useful measure of the relative population level at each location was being obtained. At abundance levels approaching economic thresholds, the likelihood of catches in duplicate traps being within ± 20% of the mean catch for the location was usually greater than 80%. For example, the probabilities (%) over all years were: bertha, 54; variegated, 64; clover, 70; *Leucania commoides*, 76; pale western, 78; redbacked, 86; darksided, 89, and army, 93.

Comparison of the year-to-year changes in population level shown by selected or random subsets of monitoring locations

An important aspect of the project was to determine if a reliable measure of year-to-year population fluctuations could be obtained from a small number of locations such as would be feasible for long-term surveillance monitoring. The most suitable transect of locations for long-term monitoring consisted of 11 locations along Highway No. 3, which runs in an east-west direction through the survey area.

The year-to-year mean catches in the Highway No. 3 transect locations were compared to the yearly mean catches of all 81 locations, and the correlation coefficient was 0.959 which was significant at the 99% level (Fig. 2A and B). Similar results were obtained for random subsets of 5 locations throughout the experimental area (average correlation coefficient of 0.938). These data showed that a useful year-to-year measure of the moth abundances could be obtained with only 5 trapping locations (i.e., 5 per 13 000 km^2).

Because infestations are most liable to occur where the population level is highest, the pattern of fluctuation shown by the 5 locations with the highest moth catch in each year was also examined. The pattern of fluctuation of these locations was similar to that shown by fixed subsets, e.g., the Highway No. 3 transect (Fig. 2B and C).

A long-term surveillance monitoring system was initiated with 14 trap locations throughout Alberta in 1985. This system may be expanded to about 60 sites. Whenever surveillance monitoring indicates that a species is approaching the economic threshold, more intensive monitoring will be required for a detailed forecast of infestations.

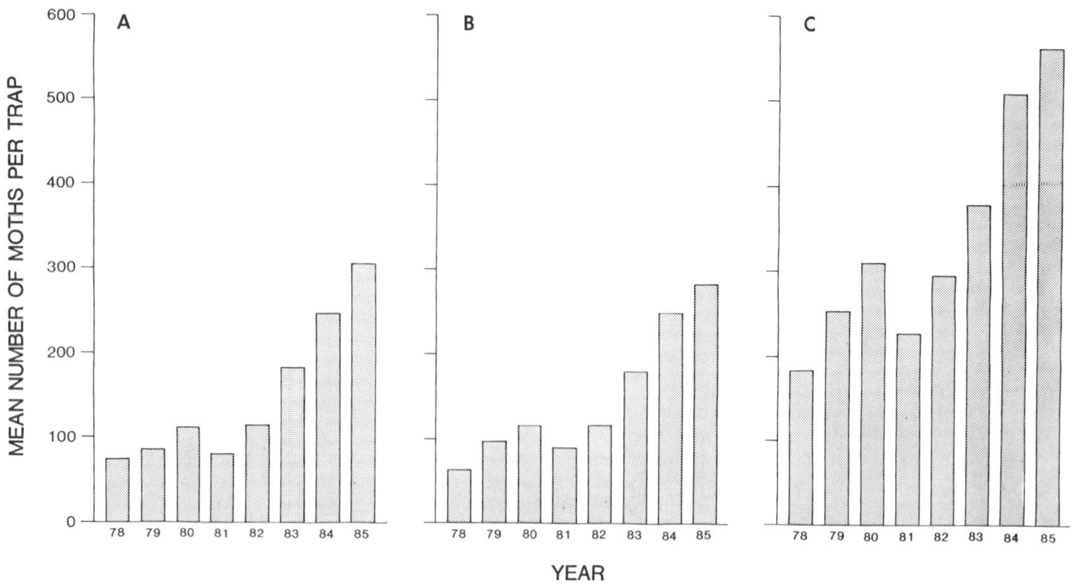

Fig. 2. Mean Catches of Pale Western Cutworm male moths at: A) 81 Locations, B) 11 Locations along Highway No. 3 Transect, and C) 5 Locations with the Highest Catches.

Relationship between moth catches and subsequent larval infestations

The number of moths caught in sex-pheromone traps is influenced by both the abundance of the moths and by factors that affect moth activity. The relationship between catch and absolute population level undoubtedly varies from year-to-year. Catches will be highest in years when conditions are favorable for longevity, flight and mating activity. Therefore, trap catches of male moths are a composite measure of both abundance and the favorableness of conditions likely to affect reproduction, and should be closely linked to the numbers of eggs laid by female moths. The population levels of the next generation are dependent on many factors affecting egg and larval survival. For these reasons, larval control decisions must be based on larval counts and crop conditions, rather than on moth catches.

Economic infestations of bertha armyworm and pale western, redbacked, and army cutworms occurred within the survey area. The corresponding threshold moth catches per trap, before there was economic crop damage by the subsequent larvae were: bertha and pale western, 200, and redbacked and army > 1 200. Redbacked cutworm damage to row crops, e.g., sugarbeet seedlings, can be caused by relatively low numbers of larvae and such infestations are not necessarily preceded by high moth catches. At 200 moths per trap, pale western cutworm infestations occurred in particularly susceptible fields when spring conditions were dry. At progressively higher levels, more fields were at risk. For example, since 1982 catches have exceeded 500 per trap at several locations and larval infestations occurred in many of the adjacent fields. The areas treated for larval control were 5 000, 20 000 and 160 000 ha in 1983 to 1985 respectively. We are predicting that the moth population of pale western cutworm will decrease in 1986 (Aug.-Sept.) due to wet conditions during oviposition in 1985 and the reduced larval infestations experienced in April and May 1986.

The bertha armyworm population peaked in 1981 and larval control was required in a few canola fields. The moth catches at the fixed locations exceeded 200 moths per trap at locations nearest the infested crops. Since bertha armyworm are dependent on the availability of a suitable host crop, particularly canola, reliable assessment of local risk will require placement of monitor traps near the susceptible crop. The economic threshold is about 300 moths per trap for traps adjacent to canola.

Detailed forecast maps of the cutworm population levels were published annually to warn growers of the impending infestations. This prompted them to check their fields for larval damage at the appropriate time. When this was done, the insecticides were applied to the early instar larvae, which improved the efficacy of the control. Agencies can be sure to have sufficient quantities of the registered insecticides on hand and commercial pesticide applicators are also aware of the impending problem. Under these conditions, a major infestation can be controlled in an orderly and efficient manner. A good example is the outbreak of pale western cutworm from 1983-1986, when we were able to predict the areas of heaviest infestations and about 186 000 ha were sprayed for larval control with minimal problems. The estimated cost of insecticides and the crop losses were about $16 M; however, this would have been much greater had producers not been forewarned.

Development of plant-derived antifeedants for crop protection

J.A. Pickett, G.W. Dawson, D.C. Griffiths, A. Hassanali, L.A. Merritt, A. Mudd, M.C. Smith, L.J. Wadhams, C.M. Woodcock and Zhang Zhong-ning

Rothamsted Experimental Station, Harpenden, Herts., AL5 2JQ, U.K.

Abstract - Two plant-derived antifeedants, ajugarin I and (-)-polygodial, are shown to have high activity against feeding by a lepidopteran larva or by aphids, including a highly resistant variant. Large-scale production of (-)-polygodial from cultivated water-pepper plants, isolated by liquid CO_2 extraction, is described and preliminary results of a field trial against aphid-transmitted barley yellow dwarf virus are reported.

INTRODUCTION

Behaviour-controlling chemicals such as pheromones show considerable promise for providing environmentally acceptable methods of pest control, particularly where there are problems of resistance to conventional pesticides, but their narrow target specificity makes commercial development unattractive (ref. 1). However, antifeedants (ref. 2), which are employed by many plants to deter herbivores, may affect a sufficiently broad range of target organisms to warrant commercial interest. Here, two plant-derived antifeedants, ajugarin I and (-)-polygodial, are investigated as potential field-crop protection agents. In these studies, production of the antifeedants is by extraction of the plants, which is the first step in developing new biotechnological approaches for producing these agents for use in agriculture.

AJUGARIN I

Plants in the family Labiatae, particularly those in the genera *Ajuga* and *Teucrium*, produce diterpenoids having the clerodane skeleton, many of which show strong antifeedant activity against lepidopteran larvae (ref. 3). A particularly active example is ajugarin I (Fig. 1.), obtained from *Ajuga remota* (ref. 4). This compound has recently been synthesised by an elegant but necessarily long and expensive route (ref. 5). For the present work, *Ajuga remota* was gathered in the Nairobi region of Kenya and extracted to yield the crystalline ajugarin I (ref. 4).

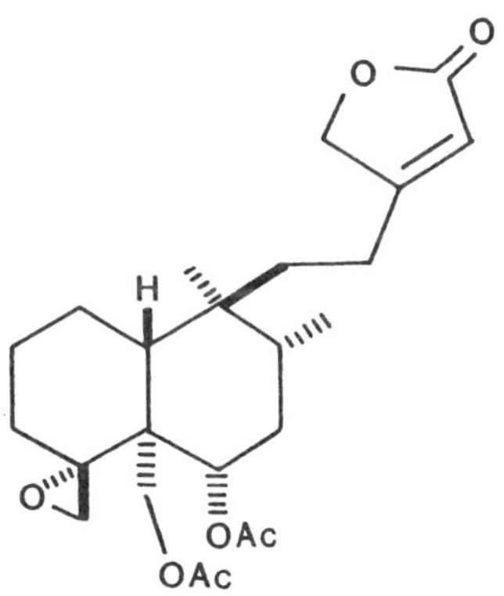

Fig. 1. Ajugarin I

To demonstrate the antifeedant effect against lepidopteran larvae in the laboratory, one half of a chinese cabbage leaf was treated with an ethanolic solution of the ajugarin I. The leaf was then placed in a petri-dish containing larvae of the diamondback moth *Plutella xylostella* (= *maculipennis*) and the amount eaten after 48 h was estimated for 7 replicates (Table 1). Ajugarin I demonstrated high activity against diamondback moth larvae down to 0.1% (*ca* 250 ppm).

TABLE 1. Antifeedant tests with larvae of diamondback moth, *Plutella xylostella*, on leaves treated with ajugarin I.

Concentration	% leaf eaten treated	control	*P*
1%	2.1	29.2	< 0.001
0.1%	4.2	22.5	< 0.05
0.01%	9.1	10.1	N.S.

The activity of ajugarin I against aphids had not been reported. This was tested in a similar manner by treating one half of a chinese cabbage leaf with an ethanolic solution of ajugarin I. Peach-potato aphids, *Myzus persicae*, were then enclosed on the leaf in a petri-dish and the numbers of aphids settled after 2 h were counted for 10 replicates. However, although ajugarin I was highly active against diamondback moth larvae, it was inactive against aphids at 1%. This indicates a different structure/activity relationship for lepidopteran larvae and aphids.

(-)-POLYGODIAL

A number of plants, for example the East African tree *Warburgia ugandensis* and the herbaceous weed water-pepper, *Polygonum hydropiper*, produce dialdehydic sesquiterpenoids having the drimane skeleton, that possess strong antifeedant properties against lepidopteran larvae (ref. 6). The simplest of these molecules is (-)-polygodial (Fig. 2), and material obtained from water-pepper growing wild in England gave good activity against transmission of potato virus Y and beet yellows virus by *Myzus persicae* in the laboratory (ref. 7).

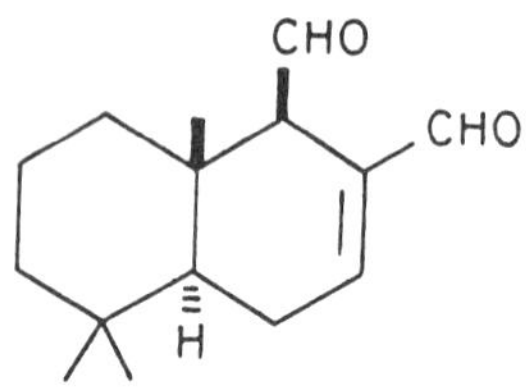

Fig. 2. (-)-Polygodial

(-)-Polygodial, obtained by extraction of water-pepper cultivated under glass, was tested against settling by aphids in the half-leaf test described above, again for 10 replicates (Table 2), using a strain of aphids which was susceptible to insecticides, and a resistant strain (R_2, ref. 8).

TABLE 2. Antifeedant test with aphids, *Myzus persicae*, on leaves treated with 0.05% (-)-polygodial.

Aphid variant	Mean no. aphids settled after 2 h treated	control	*P*
insecticide susceptible	3.4	14.7	< 0.001
highly resistant (R_2)	3.2	14.4	< 0.001

The (-)-polygodial was also active against diamondback moth larvae down to 0.1%. Thus for (-)-polygodial, a similar structure/activity relationship exists for lepidopteran larvae and aphids.

FIELD TRIALS

Because of the broader spectrum of activity for (-)-polygodial, this compound was selected for further investigation and it was necessary to obtain the material on a larger scale. A number of synthetic routes to polygodial are available, e.g. ref. 9. The production by biotechnological means seems promising because of the apparently simple biosynthetic route from farnesyl pyrophosphate, an ubiquitous component of plants (ref. 10). For initial studies reported here, isolation of this material was from large scale production of the plant. Water-pepper seed was produced by cultivation under glass and then grown at a farm in the East Anglian region of England. The aerial parts of the plants were harvested and, without drying, were extracted with liquefied CO_2 under pressure at about 5°C, in the apparatus shown diagrammatically in Fig. 3.

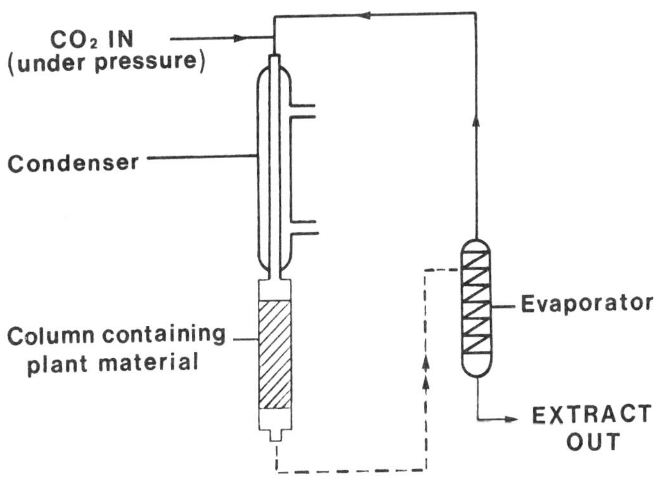

Fig. 3. Extracting plant material with liquid CO_2

Analysis by high performance liquid chromatography (HPLC) on a Spherisorb 5 μ nitrile column (25 x 0.46 cm) using ether-hexane (1:1) and ultra-violet detection at 224 nm showed the extract to contain > 20% of (-)-polygodial, with yields of 0.02 - 0.04% by weight of wet plant (Fig. 4). Extraction using conventional solvents such as ether gave a much less pure extract (Fig. 4).

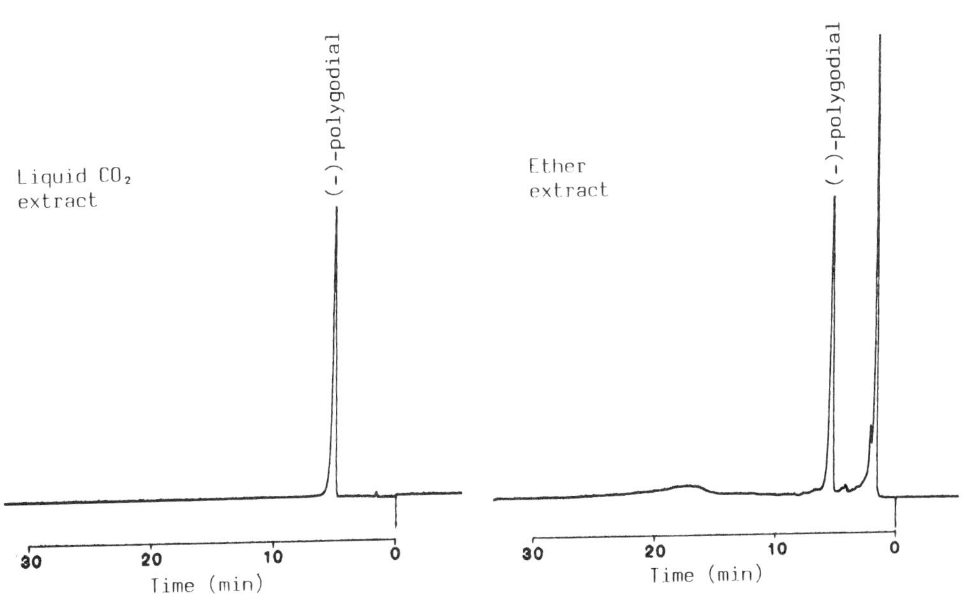

Fig. 4. HPLC of water-pepper extracts.

To test the (-)-polygodial produced by large-scale extraction with CO_2 in the field, transmission of barley yellow dwarf virus by the aphid *Rhopalosiphum padi* was chosen as the target. Winter cereals such as barley and wheat can be seriously affected by barley yellow dwarf virus, which causes stunting of plants, late development and proliferation of weeds (ref. 11). In these trials, winter-sown barley was treated with (-)-polygodial at 50 g/ha on three occasions in early autumn 1985. The (-)-polygodial was sprayed onto the barley at the 2-4 leaf stage. Effective leaf cover was ensured by using an electrostatic spraying system recently developed at Rothamsted (ref. 12). This equipment has the advantage of delivering less than 10 l/ha and the direct electrostatic charging results in minimum soil contamination and spray drift, placing the active material on both upper and lower leaf surfaces (ref. 13). The untreated plots were badly damaged by BYDV but the treated plots showed a good crop.

TABLE 3. Yields of barley from 1985 field trial.

	tonnes/hectare
(-)-polygodial	5.22
control	3.83
S.E.D. 0.781	

This shows that a plant-derived antifeedant can be produced from cultivated plants on a sufficiently large scale for field trials and that this material can be used successfully against an aphid-borne virus disease.

REFERENCES

1. J.A. Pickett. Educ. Chem. 16, 44-47 (1979).
2. D.M. Norris. In: Chemistry of Plant Protection. 1, 97-146. Eds. G. Haug and H. Hoffman (1986).
3. X. Belles, F. Camps, J. Coll and M.D. Piulachs. J. Chem. Ecol. 11, 1439-1445 (1985).
4. I. Kubo, Y-W. Lee, V. Balogh-Nair, K. Nakanishi and A. Chapya. J.C.S. Chem. Comm. 949-950 (1976).
5. S.V. Ley, N.S. Simpkins and A.J. Whittle. J. Chem. Soc., Chem. Commun., 503-505 (1983).
6. K. Nakanishi and I. Kubo. Isr. J. Chem. 16, 28-31 (1977).
7. R.W. Gibson, A.D. Rice, J.A. Pickett, M.C. Smith and R.M. Sawicki. Ann. appl. Biol. 100, 55-59 (1982).
8. R.M. Sawicki and A.D. Rice. Pestic. Sci. 9, 513-516 (1978).
9. S.C. Howell, S.V. Ley and M. Mahon. J.C.S. Chem. Comm. 507-508 (1981).
10. J.A. Pickett. Phil. Trans. R. Soc. Lond. B310, 235-239 (1985).
11. G. Jenkins. Outlook Agri. 14, 61-67 (1985).
12. A.J. Arnold. EPPO Bulletin 13, 451-456 (1983).
13. G.R. Cayley, P. Etheridge, D.C. Griffiths, F.T. Phillips, B.J. Pye and G.C. Scott. Ann. appl. Biol. 105, 379-386 (1984).

Perception of antifeedants by lepidopterous larvae

L.M. Schoonhoven

Department of Entomology, Agricultural University, Wageningen, the Netherlands

Abstract - Antifeedants of plant origin act on receptor cells located in taste sensilla on the mouth parts of caterpillars. Some antifeedants stimulate specialized deterrent receptors, which may vary in their response spectrum between species. Other antifeedants may interact with receptors sensitive to phagostimulants (e.g. sugars). The resulting distortion of normal receptor activity may inhibit food intake. This paper describes the neurophysiological basis of antifeedant perception and the potentials and limitations of the use of electrophysiological methods for screening chemicals for antifeedant activity.

INTRODUCTION

Since feeding behavior is governed by sensory information, studying sensory responses can help to identify chemicals of plant origin which may be used as antifeedants. The number of contact receptors in different insect species varies greatly. Orthopteroid insects, for instance, have large numbers of chemoreceptors, numbering in the thousands. Lepidopterous larvae, on the other hand, select their food plants and optimal feeding sites by means of only a remarkably small number of receptors located on the mouth parts. Therefore lepidopterous larvae are especially amenable to a study of sensory coding of feeding inhibitory compounds. The fact that 40% of the insecticides used world wide is directed to the control of lepidopterans is another motive for investigating the mode of action of substances which modify their feeding behavior.

TASTE PERCEPTION

The sense of taste is located in two hairs on each maxilla and one pair of taste papillae on the inner side of the labrum (i.e. epipharynx). The maxillary sensilla each contain 4 chemosensitive neurones, whereas each epipharyngeal organ is innervated by 3 neurones. The main sensory input informing the insect's brain about food taste is thus accomplished by only 11 (paired) receptor cells. The taste cells of different caterpillar species differ in their stimulus specificity. In many species receptors for sugars, salts and amino acids are present. Stimulation of these receptors induces food intake in a hungry insect. Several species also have receptors tuned to certain chemicals which typically occur only in their host plants, e.g. certain sugar alcohols or glucosinolates. It appears that the association between these insects and their host plant species is based on the perception of such host specific compounds. Stimulation of receptors specialized to detect such compounds usually also enhances feeding intensity.

The natural stimulus, for instance a plant sap, is not perceived as a mere combination of several stimulating compounds, because the different components of the mixture often show considerable interaction with each other at the receptor level. Thus the response of the sugar cell to sugars is modified by salts, and the reverse. The sensory activity pattern evoked by a natural stimulus (e.g. plant sap) cannot be predicted from the responses of individual cells to single compounds (ref. 1).

MODES OF PERCEPTION OF ANTIFEEDANTS

The larval taste system of several lepidopterans possesses, in addition to cells which perceive phagostimulating chemicals, receptors specialized to detect the presence of feeding inhibitory compounds. One of the maxillary hairs of Pieris brassicae caterpillars, for example, contains a 'deterrent receptor' which is stimulated by a variety of alkaloids and terpenoids, and which deters the insect from feeding when stimulated. Strychnine is the strongest stimulus known for this cell, and evokes responses at concentrations above a threshold of 10^{-6} M (ref. 2). The well-known antifeedant isolated from neem tree seeds, azadirachtin, also inhibits feeding in this species by stimulating its deterrent cell (Fig. 1).

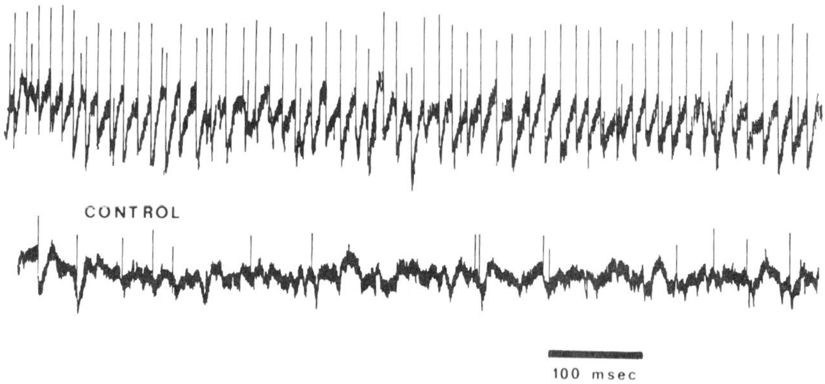

Fig. 1. Nerve action potentials of the deterrent receptor in a maxillary taste hair of a Pieris brassicae larva. Upper trace: responses of predominantly open nerve cell when stimulated with 10^{-3} M azadirachtin + 0.1 M NaCl. Lower trace: 0.1 M NaCl (from ref. 23).

When the responses of deterrent cells to a number of chemicals are compared in different insect species, it appears that, in the few cases studied, there are always interspecific differences with regard to stimulus spectrum (ref. 23). The phenolic compound salicin strongly activates the deterrent cells in Manduca sexta and Mamestra brassicae, but this compound is ineffective on the deterrent cell of Pieris brassicae. Our present knowledge indicates that each species has a unique set of taste receptors, and the deterrent cell of one species shows only partial overlap in its stimulus characteristics with that of another species. Many species possess more than one deterrent receptor in their maxillary taste hairs. Several species are known to have a deterrent cell in each hair, but in all cases they are functionally different, though some overlap is not uncommon.

A remarkable feature of the deterrent receptors is their sensitivity to compounds which do not seem to occur in the insect's environment. The deterrent cell of Pieris brassicae is strongly fired by strychnine and quinine, though it is unlikely that this species has ever been in contact with these compounds in nature. Nor is it probable that the silkworm, which has been cultivated for thousands of years on mulberry leaves, has ever been exposed to caffeine at a sufficient intensity to evolve receptor cells with a marked sensitivity to this compound. The stimulatory properties of these alkaloids cannot be attributed to a general membrane disturbing effect, since the other taste cells function normally.

Several antifeedants are known to have modes of action other than stimulating deterrent sensitive cells. These compounds may interfere with the normal perception of phagostimulating chemicals by distorting normal receptor responsiveness. When a sucrose stimulus is adulterated with azadirachtin the sugar receptor of Spodoptera littoralis is inhibited and therefore misinforms the brain about the real amount of sugar present. When the receptor is subsequently tested with pure sugar solutions, it shows the normal response (ref. 3). Inhibitory interactions by antifeedant compounds have been observed also in other insect species. In Pieris brassicae larvae, strychnine and several other chemicals appear to interfere with normal sugar receptor activity (ref. 4). This phenomenon is not restricted to lepidopteran species, since it has also been observed in beetles (ref. 5) and flies (ref. 6).

Some very potent antifeedants of plant origin have been identified which do not seem to interact instantaneously with sugar reception, but exert a long lasting after-effect. Thus the sugar receptors of Spodoptera exempta larvae show a considerable reduction in sensitivity when they have been exposed to the sesquiterpenoid warburganal. The effect may take up to an hour to develop, depending on the concentration of warburganal used. It has been suggested that receptor inactivation in this case is due to an interaction with the sulfhydryl site of a receptor macromolecule, thereby affecting energy transduction in the chemoreceptory cells in an unspecific way (ref. 7). Toosendanin, a triterpenoid from the chinaberry tree, likewise inhibits the sugar cell as well as the inositol cell in another lepidopterous species. The responses of these cells dropped to 1/3 or less of their original levels and no recovery was observed during at least 40 min (ref. 8).

A number of antifeedants act by unspecific interference with normal receptor activity. Copper sulphate at a concentration of 0.001 M elicits an intense impulse activity in contact chemoreceptors of various caterpillar species (ref. 4). Often the reaction takes some seconds to build up and gradually some, or occasionally possibly all receptor cells increase their firing rate up to very high frequencies. At higher concentrations high impulse frequency starts immediately (ref. 9). Feeding in larvae of Pieris brassicae and Mamestra brassicae is markedly suppressed by copper containing chemicals (ref. 9, 10). Copper ions, warburganal and some other plant-derived compounds appear to disrupt nerve membrane physiology and thereby distort the normal function of several receptor types. This is manifested by increased neural activity and irregular firing patterns. The unspecific responses may develop into bursting patterns, followed by a complete blocking of receptor excitability (ref. 7, 9). Several pyrethroids also have strong antifeedant effects. Larvae of Spodoptera frugiperda show restricted food intake when exposed to one out of 8 synthetic pyrethroids offered in their food at concentrations of 5 ppm (ref. 11). Pyrethroids are known to affect normal nerve cell function by interference with sodium ion pores. A delay in their closing gate mechanism results in prolonged sodium ion influx and consequently trains of action potential, consisting of several to hundreds of impulses, are produced instead of single ones (ref. 12). Antifeedant effects of other pesticides, such as plictran (ref. 13) and rotenone (ref. 14) may be caused by comparable interaction with impulse generating processes.

Still other mechanisms of interference with normal receptor function are conceivable. The basic recognition process in sugar receptors in flies is attributed by several authors to an interaction between the stimulus molecule and glucosidase molecules present in the dendritic membrane (e.g. ref. 15). Some plant alkaloids which are structural analogues of monosaccharides are known to inhibit some glucosidases (ref. 16). When indeed sugar reception in insects is mediated by macromolecules with glucosidase properties, one might expect that these alkaloids show antifeedant potency, which would be due to their binding with receptor molecules. One of them, DMDP (2,5-dihydroxymethyl 3,4-dihydroxypyrrolidine) was tested and appeared to exert an antifeedant effect in a number of insect species. In Spodoptera exempta DMDP caused increased activity in the sugar receptors which after some time developed into bursting activity. Interestingly the latter phenomenon was not observed in Spodoptera littoralis larvae (ref. 17).

A number of alkaloid glucosides occurring in non-host plant species elicit impulse trains in some (though not all) taste cells of tarsal sensilla of the adult Colorado potato beetle. Though only one, or occasionally two cells show bursting activity when stimulated by the alkaloids, there is some evidence that the remaining cells have raised firing thresholds. The alkaloids thus seem to modulate the sensitivity of all cells present (ref. 18). Spilanthol, isolated from Spilanthus mauretania, causes receptor dysfunction in taste hairs of Spodoptera exempta larvae (ref. 19).

In conclusion antifeedants may be perceived by either stimulation of specialized deterrent receptors or by their capacity to distort the normal function of neurons which perceive phagostimulating chemicals. Since sugar in many herbivorous insect species is of paramount importance for sustained feeding, inhibition of sugar receptors is an effective way to reduce feeding activity. Some antifeedants influence feeding activity through a combination of these two modes of action.

STRUCTURE-ACTIVITY RELATIONSHIPS

One of the basic aims of research on chemoreception is to understand the relationship between the chemical structure of stimulants and their efficacy to fire chemoreceptor cells. Although in some cases this goal has been achieved to a limited extent (e.g. ref. 14, 20), there exist many more reports in the literature stressing the point that sensory responses are very difficult to predict on the basis of molecular structure of the stimulus. In insects several taste receptors have been found to react to a wide array of chemicals, though at the same time slight structural changes in the stimulus may decrease its efficacy or even render it ineffective. This feature is more striking in receptors which have a broad specificity and respond to chemicals differing widely in molecular size and shape. Several specialized deterrent receptors seem to fit into this category. The deterrent cell in Pieris brassicae larvae, for instance, is stimulated by many alkaloids, steroids and terpenoids, including the decalin moiety of clerodendrin, a diterpene isolated from the Indian bhat tree. When the two stereoisomers of the decalin unit were tested for their antifeedant activity (using a behavioral test) it appeared that they differed in their ability to reduce food intake by about a factor of 10 (ref. 21). Apparently, small modifications in the spatial arrangement of a molecule may markedly influence its potency to excite a taste receptor, even when this cell is not highly specialized.

Antifeedants of natural origin may sometimes be made more effective by minor changes in their molecular structure. Thus the capacity of clerodendrin to reduce food intake in Spodoptera litura larvae could be raised 15-fold by transforming it into its methanol adduct (ref. 22). This example illustrates the point that antifeedants do not necessarily occur in nature in the forms which maximally inhibit plant feeding insects from attack. There may be several biological or physico-chemical reasons why the plant forgoes producing a more effective variant of a certain antifeedant.

CONCLUSIONS

The search for antifeedants has chemical and biological aspects. After the identification of natural compounds or the synthesis of their analogues, their prospects as antifeedants can only be determined by testing them on key pest insects. The most obvious type of bioassay is based on an analysis of the insect's behavioral responses when it contacts the test chemical. Several types of feeding tests may be used in either a no-choice or a dual-choice situation (ref. 23). A careful choice of experimental conditions is important, since feeding responses may vary with, for instance, the type of foodplant which is used as feeding substrate (ref. 24), and habituation to the presence of antifeedants likewise may depend on food quality (ref. 25). Behavioral experiments are usually labor intensive and often require considerable numbers of experimental insects and fair amounts of the compounds to be tested.

Electrophysiological methods require small quantities of chemicals and limited numbers of insects. The actual tests can be done in a short time. A prerequisite, however, for a reliable electrophysiological test procedure is a thorough knowledge of the insect's chemosensory system. It seems that presently only lepidopterous larvae are amenable to a sufficiently detailed analysis of taste receptors to permit the use of electrophysiological techniques for selecting potential antifeedants.

Different insect species may react differently to a certain antifeedant. In view of this fact the question arises how many insect species should be used in bioassays when searching for potential antifeedants. The answer to this question is determined by practical considerations, like the availability of insects and labor for routine tests. In view of the observation that different species may show considerable differences when the stimulus spectra of their specialized deterrent receptors are compared, it may be expected that striking interspecific variability in behavioral responses to certain chemicals exist. Indeed marked differences can be seen, e.g. in drinking responses to water adulterated with quinine. *Pieris brassicae* larvae and several other caterpillars reject this solution, but some species, e.g. *Lymantria dispar*, imbibe even saturated quinine solutions without any sign of hesitation. Antifeedants which act by distorting the normal function of some or all chemoreceptors interfere probably with general receptor properties and therefore these antifeedants will show less specificity than chemicals stimulating specialized deterrent cells. As a consequence their antifeedants properties will be discovered more easily in a screening program based on only a limited number of species, than compounds perceived by specialized deterrent receptors.

REFERENCES

1. L.M. Schoonhoven, In: E.A. Bernays, R.F. Chapman and J. Stoffolano (Eds.), *Perspectives in Chemoreception and Behavior*, p. 69-97, Springer Verlag, New York (1986).
2. W.-C. Ma, *Meded. Landbouwhogeschool Wageningen*, *72/11*, 1-162 (1972).
3. M.S.J. Simmonds and W.M. Blaney, In: H. Schmutterer and K.R.S. Ascher (Eds.) *Natural Pesticides from the Neem Tree and other Tropical Plants*, p. 163-179, TZ-Verlagsgesellschaft, Rossdorf (1985).
4. L.M. Schoonhoven, unpubl. results.
5. B.K. Mitchell and J.F. Sutcliffe, *Physiol. Entomol.* *9*, 57-64 (1984).
6. K. Hansen and H. Wieczorek, In: R.H. Cagan and M.R. Kare (Eds.) *Biochemistry of Taste and Olfaction*, p. 139-162, Academic Press, New York (1981).
7. W.-C. Ma, *Physiol. Entomol.* *2*, 199-207 (1977).
8. P.-A. Zhou, L.-E. Luo, D.-L. Bai and C.-Y. Liao, *Abstracts 6th Int. Congr. Pesticide Chemistry*, Ottawa, p. 2D/E-16 (1986).
9. L.M. Schoonhoven and T. Jermy, In: N.R. McFarlane (Ed.) *Crop Protection - their Biological Evaluation*, p. 133-146, Academic Press, London (1977).
10. G. Muschinek, *Allattani Kozl.* *63*, 117-122 (1976).
11. G.L. Gist and C.D. Pless, *Florida Entomol.* *68*, 456-461 (1985).
12. J.R. De Weille, *The Modification of Nerve Membrane Sodium Channels by Pyrethroids*, Ph.D. Thesis, Utrecht State University (1986).
13. R.M. Weseloh, *Environm. Entomol.* *13*, 1371-1376 (1984).
14. G.A. Lane, D.R. Biggs, G.B. Russell, O.R.W. Sutherland, E.M. Williams, J.H. Maindonald and D.J. Donnell, *J. Chem. Ecol.* *11*, 1713-1735 (1985).
15. K. Hansen, B. Ritter, U. Welker and H. Wieczorek, *Olfaction and Taste*, *7*, 57-59 (1980).
16. S.V. Evans, L.E. Fellows, T.K.M. Shing, G.W.J. Fleet, *Phytochemistry*, *24*, 1953-1955 (1985).
17. W.M. Blaney, M.S.J. Simmonds, S.V. Evans, L.E. Fellows, *Ent.Exp.Appl.* *36*, 209-216 (1984).
18. B. Stürckow, *Z. Vergl. Physiol.* *42*, 255-302 (1959).
19. J. Clark, Personal communication (1981).
20. H. Wieczorek, *J. Comp. Physiol.* *106A*, 153-176 (1976).
21. R.B.M. Geuskens, J.M. Lutijn and L.M. Schoonhoven, *Experientia* *39*, 403-404 (1983).
22. N. Kato, M. Takahashi, M. Shibayama, K. Munakata, *Agric. Biol. Chem.* *36*, 2579-2585 (1972).
23. L.M. Schoonhoven, *Ent. Exp. Appl.* *31*, 57-69 (1982).
24. M.S.J. Simmonds, W.M. Blaney, F.D. Monacha, M. Marquina-Quhae, G.B. Marini Bettolo, *J. Chem. Ecol.* *12*, 1593-1599 (1985).
25. A. Szentesi and E.A. Bernays, *Physiol. Ent.* *9*, 329-340 (1984).

Synthesis of tricyclic triterpenoid precursors

W. Kraus*, M. Bokel, R. Kröსa, H. Michel, G. Sawitzki, and W. Weber

Department of Chemistry, University of Hohenheim,

7000 Stuttgart 70, W.-Germany

INTRODUCTION

During our work on constituents of Meliaceae which show antifeedant activity towards insects we have isolated a series of tetra-and pentanortriterpenoids related to meliacane (1) (ref. 1). In our attempts to synthesize the parent skeleton (1) we were faced with the problem how to construct a C/D trans fused tetracyclic ring system bearing an α-oriented methyl group at C-13, because this arrangement implies a boat conformation of ring C. Since cyclization reactions of squalene-epoxide and similar compounds lead mainly to chair conformations and β-oriented methyl groups (ref. 2) we tried to solve the problem by fixing the boat form first by bridging, put the methyl group in the final α-orientation, do the cyclization of rings A and B later, and split off the bridge at the end. This led us to investigate the synthesis of tricycloundecanes and -dodecanes (2) in general (ref. 3, 4).

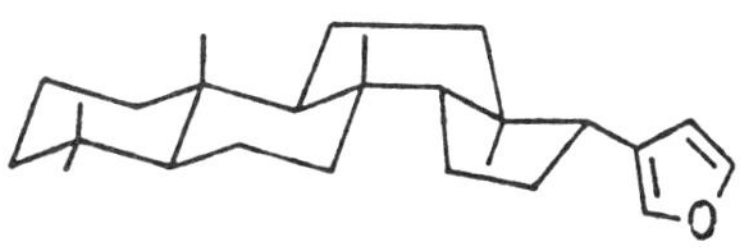

1

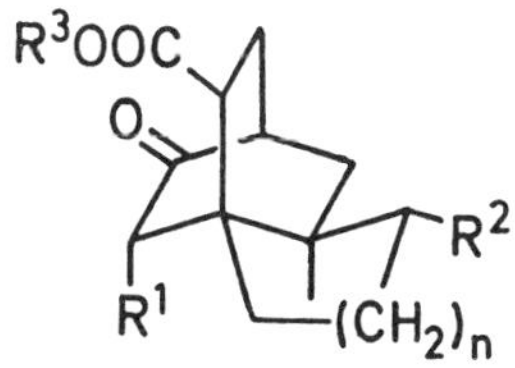

2 n = 1,2

The t.-butyl ether (4) of methyl bicyclo[4.3.0]non-en-ol-one (3), which is easily available (ref. 5) was transformed into the trimethylsilylether (5) in reasonable yields by reaction with lithium diisopropyl amide in tetrahydrofurane and subsequently with trimethylchlorosilane. Diels-Alder reaction of (5) with maleic anhydride gave the tricyclic anhydride (6) stereoselectively in 80% yield. The structure of (6) was established by X-ray diffraction analysis.

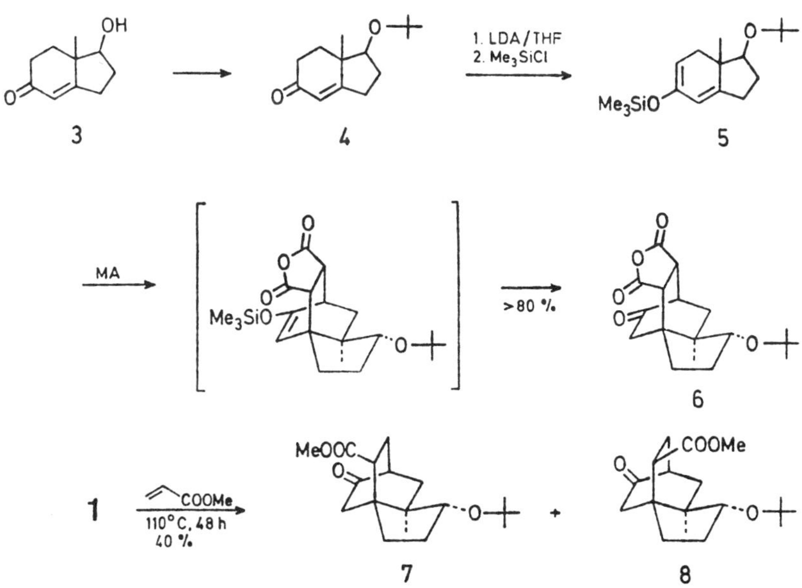

When methyl acrylate was used as the dienophile Diels-Alder adducts (7) and (8) were obtained from (5) in 40% yield (ratio 3:1). The structure determination of (7) and (8) was achieved by high resolution n.m.r. and X-ray diffraction analysis after acidic hydrolysis of the tert. butoxy groups, chromatographic separation of the isomers (silica gel, ether), and conversion into the thioacetales.

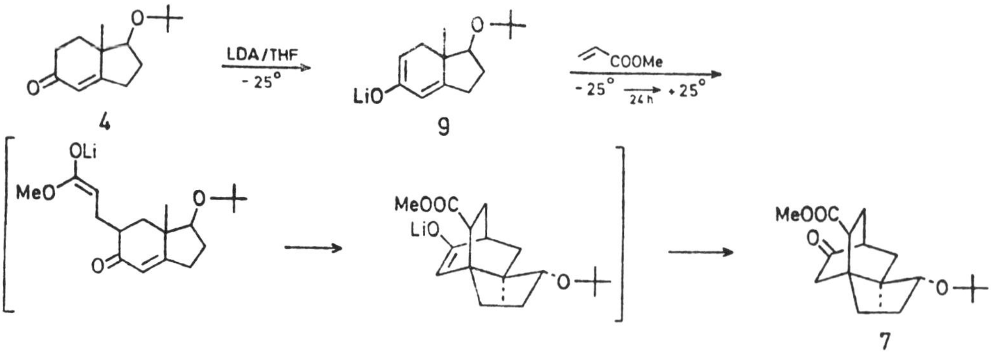

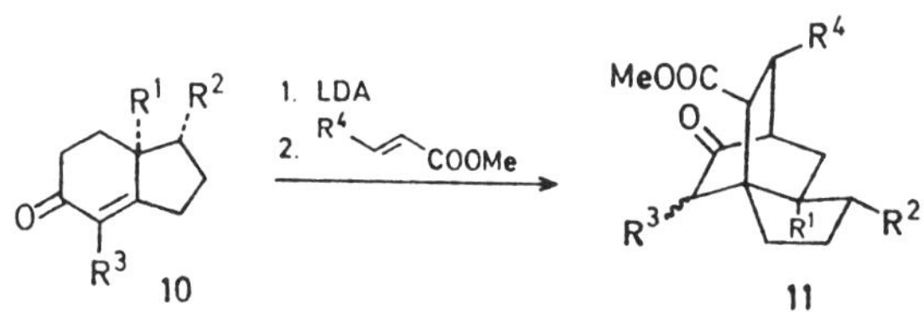

Table 1. Products (11) obtained from (10).

R^1	R^2	R^3	R^4	
H	H	H	H	ca 5%
H	H	H	CH_3	70%
CH_3	H	H	H	28%
CH_3	H	H	CH_3	>90%
CH_3	O-*t*.-C_4H_9	H	H	25%
CH_3	O-*t*.-C_4H_9	H	CH_3	>90%
CH_3	O-*t*.-C_4H_9	CH_3	H	< 5%
CH_3	O-*t*.-C_4H_9	CH_3	CH_3	40%
CH_3	O-*t*.-C_4H_9	OCH_3	H	< 5%
CH_3	O-*t*.-C_4H_9	OCH_3	CH_3	< 5%
CH_3	O-*t*.-C_4H_9	Homo-geranyl	H	30-40%

Compounds of type (7) can be prepared regio- and stereroselectively from lithium dienolates such as (9) via sequential Michael addition (refs. 3, 4, 7). The products (11) obtained in such a way from substituted hydrindenones (10) and methyl acrylate and crotonate, resp., are listed in Table 1. The yields depend strongly on the nature and the position of the substituents in the ketone. In all cases the yields were better if methyl crotonate was used instead of acrylate. The addition occurs with retention of configuration. Substituents in position 4 (R^3) decrease the yield drastically, particularly if R^3 is methoxy. The homogeranyl derivative cannot be worked up and handled easily because of rapid polymerization.

Similar results were obtained starting with octalones (12). The products (13) are listed in Table 2. In all cases the yields are better again with crotonate, probably because of better stabilization of the cationic reaction centre and less polymerization of the unsaturated ester. There is not much influence of the substituents on the formation of the products, except with methoxy in position 3.

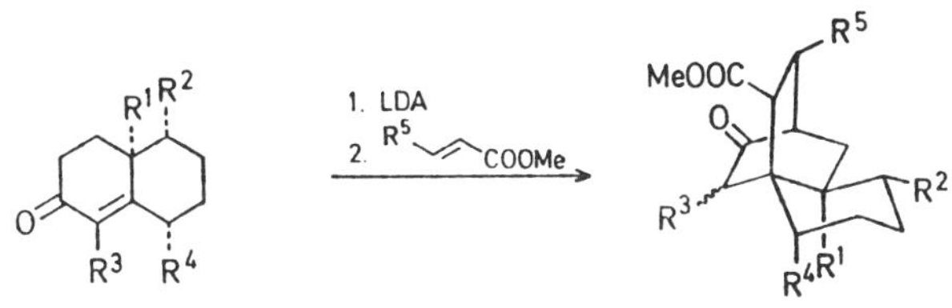

12 13

Table 2. Products (13) obtained from (12).

R^1	R^2	R^3	R^4	R^5	
CH_3	H	H	H	H	90%
CH_3	H	H	H	CH_3	94%
CH_3	CH_3	H	H	H	30%
CH_3	CH_3	H	H	CH_3	68%
CH_3	H	H	CH_3	H	26%
CH_3	H	H	CH_3	CH_3	88%
CH_3	O-*t*.-C_4H_9	H	H	H	46%
CH_3	O-*t*.-C_4H_9	H	H	CH_3	95%
CH_3	O-*t*.-C_4H_9	CH_3	H	H	66%
CH_3	O-*t*.-C_4H_9	CH_3	H	CH_3	78%
CH_3	O-*t*.-C_4H_9	OCH_3	H	H	0%
CH_3	O-*t*.-C_4H_9	OCH_3	H	CH_3	< 5%

The regio- and stereoselectivity of the reaction can be explained on the basis of a mechanism outlined in Chart 1: The orientation of the ester is fixed by the lithium cation being located between the enolate oxygen and the carbonyl oxygen of the ester group, and opposite to the bulky methyl and t.-butoxy group during both reaction steps.

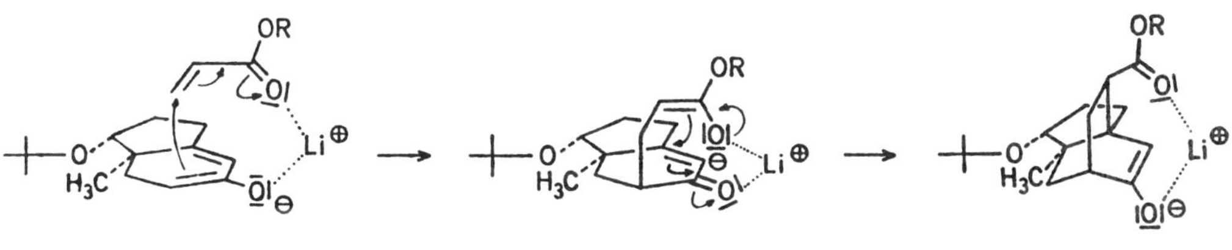

Chart 1

This mechanism is not valid for the octalones if there are no substituents connected to C-10, and/or if there are geminal substituents at C-1 or C-9 as shown in Chart 2 and Table 3. Ketale (14) gives exo and endo products (15) and (16), resp., in low yield in addition to the mono adduct (17) which is in fact the main product. From the homologue enone (18) only the endo adduct (10) was formed with acrylate, whereas both isomers (21) and (22) were obtained in a ratio of 2:1 from (18) with methyl crotonate. Pure endo adducts (24) are found in the reaction of the dienolates obtained from octalones (23). In all cases there is no substituent at C-10 which is obviously responsible for the exclusive formation of the exo products listed in Table 2.

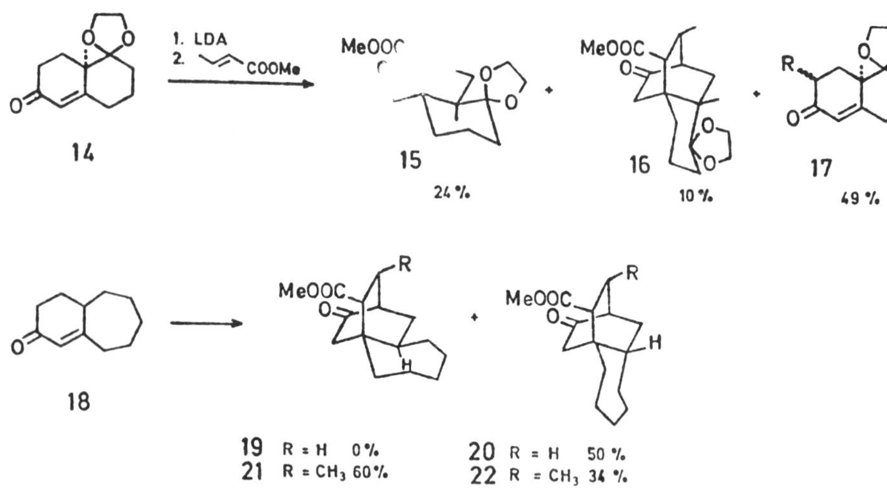

Chart 2

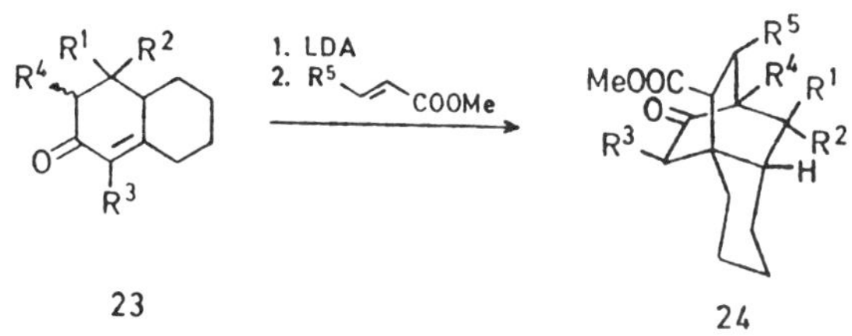

Table 3. Products (24) obtained from (23).

R^1	R^2	R^3	R^4	R^5	
H	H	H	H	H	80%
H	H	H	H	CH_3	92%
H	H	CH_3	H	CH_3	98%
CH_3	CH_3	H	H	CH_3	95% (+1% exo)
H	H	H	CH_3	CH_3	94%

Exclusively monoalkylation occurs with octalone (25) bearing geminal methyl groups at C-7 to give product (26). Inspection of models show that there is strong steric hindrance in this area thus preventing the second Michael addition. The structures of all products were determined using high resolution n.m.r. and, in the case of (22), by X-ray diffraction analysis.

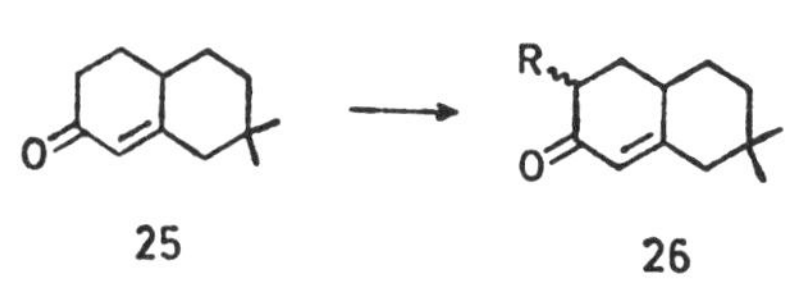

These results show that sequential Michael addition reactions can be used in the stereocontrolled synthesis of tricycloundecane and -dodecane derivatives but that steric hindrance plays a major role in the stereochemistry of the reaction.

ACKNOWLEDGEMENT

Financial support by Deutsche Forschungsgemeinschaft and Fonds der Chemischen Industrie is gratefully acknowledged.

REFERENCES

1. W. Kraus, Stud. Org. Chem., Elsevier Publ. Amsterdam, 17, 331, (1984); W. Kraus, M. Bokel, A. Klenk and H. Pohnl, Tetrahedron Lett., 26, 6435, (1985).
2. W.S. Johnson, Angew. Chem. Internat. Edit., 15: 91, (1976).
3. W. Weber, Dissertation, Universitat Hohenheim (1983).
4. H. Michael, Dissertation, Universitat Tubingen, to be submitted.
5. Z.G. Hajos, R.A. Michaeli, D.R. Parrish and E.P. Oliveto, J. Org. Chem., 32, 3008 (1967).
6. H.O. House, L.J. Czuba, M. Gall and H.D. Olmstead, J. Org. Chem., 34, 2324, (1969).
7. W. Weber, D. Spitzner and W. Kraus, J. Chem. Soc., Chem. Commun., 1213, (1980).

Neem seed oil—a potential antifeedant against insect pests of rice

R. C. Saxena[a]

International Centre of Insect Physiology and Ecology,
P. O. Box 30772, Nairobi, Kenya

Abstract - Feeding by several species of insect pests was reduced on rice plants sprayed with neem (Azadirachta indica A. Juss) seed oil. First instar larvae of chewing insects such as the rice leaffolder, Cnaphalocrocis medinalis (Guenée), the ear-cutting caterpillar, Mythimna separata (Walker), and the armyworm, Spodoptera mauritia acronynctoides (Boisduval), were repelled and their feeding was significantly reduced when offered rice leaves sprayed with 12% crude emulsified neem oil. Likewise, sucking insects such as the brown planthopper, Nilaparvata lugens (Stål), the whitebacked planthopper, Sogatella furcifera (Horváth), and the green leafhopper, Nephotettix virescens (Distant), avoided rice plants sprayed with the oil. When confined on treated plants, the quantity of food ingested per female per 24 h progressively decreased with increasing oil concentration. For N. lugens, feeding duration decreased by 0.93 min per h for every 1% increase in oil concentration, while search and avoidance of feeding sites on treated plants increased correspondingly. Neem oil also disrupted the normal feeding behavior of N. virescens as recorded with an electronic monitoring device. Phloem feeding on oil-treated plants was reduced and was associated with a significant increase in the insect's probing frequency and an increase in durations of salivation and xylem feeding. In both laboratory and field trials, neem oil decreased the transmission of grassy- and ragged stunt virus diseases vectored by N. lugens and the rice tungro virus by N. virescens. Effectiveness against pests, low cost, safety for the environment, and accessibility of neem oil in several rice producing countries of Asia and Africa should make its use attractive for crop protection by the resource-limited farmers.

INTRODUCTION

Insect damage to plants results from direct feeding or from indirect transmission of pathogens during feeding. Therefore, chemicals which reduce pest injury by rendering plants unattractive or unpalatable can be considered as potential substitutes for insecticides (ref. 1). Antifeedants are the chemical substances which when perceived, reduce or prevent insect feeding. Consequently, growth, development, survival, and reproduction are adversely affected. Antifeedants in some plants constitute important defense barriers against insect attack. They are generally pest-specific and therefore harmless to non-target organisms. Because of their biodegradable nature and relative safety to beneficial organisms in the environment, research on the biological activity and chemistry of antifeedants has been emphasized (ref. 2, 3 & 4). Programs have been launched in several countries for developing feeding deterrents known to occur naturally in some plants.

The insect antifeedant properties of the neem tree, Azadirachta indica A. Juss, are well known. Centuries before commercial insecticides were available, farmers in the Indian Subcontinent used neem derivatives to protect agricultural crops from insect attacks. Almost every part of the tree is bitter, but the seed kernel possesses maximum deterrency. The active principals or the "bitters" in neem belong to a group of tetranorterpenoids. Azadirachtin, the best known neem constituent, has shown selective activity against a wide array of insect pests (ref. 4 & 5). In addition to the antifeedant action, azadirachtin and neem seed derivatives have other pronounced behavioral and physiological effects (ref. 6, 7, 8 & 9). This paper reviews the repellent and antifeedant properties of neem oil against selected chewing- and sucking-types of rice insect pests.

Note a: current address: IRRI, P. O. Box 933, Manila, Philippines.

CHEWING INSECTS

In a choice test, significantly less first-instar larvae of the rice leaffolder, Cnaphalocrocis medinalis (Guenée), the ear-cutting caterpillar, Mythimna separata (Walker), and the armyworm, Spodoptera mauritia acronynctoides (Boisduval), settled on leafcuts from rice plants sprayed with neem oil than on leafcuts sprayed with 1.66% 'Teepol' liquid detergent (emulsifier) solution in water and feeding by larvae, based on the weight of dried excreta, was significantly reduced when confined to neem oil-treated leafcuts (Table 1)(ref. 10 & 11). The higher the concentration of neem oil, the greater was the reduction in feeding and excretion. However, the antifeedant effect of neem oil on rice plants got degraded in sunlight in 2 to 4 days. Confinement of fifth instar C. medinalis larvae on leafcuts from rice plants treated with $\geq$ 12% neem oil resulted in pronounced aberrations in behavior and form and enhanced mortality during metamorphosis. Some larvae ceased feeding, shrank in size, became melanized, and died. Others failed to fold the leaves prior to pupation, while a few developed into larval-pupal intermediates which retained larval cuticle patches, head capsule, thoracic legs, etc. Moths emerging from apparently normal looking pupae often had poorly developed or twisted wings, and occasionally failed to emerge from the pupal case. In controls, mortality of C. medinalis larvae during metamorphosis was low and a significantly high percentage of healthy moths emerged. Similar effects were observed when third or fifth instar larvae of M. separata and S. m. acronynctoides were fed neem oil-treated leafcuts of rice plants (ref. 12). Parasitization of C. medinalis larvae and pupae by ichneumonid, encyrtid, and braconid hymenopterans was not affected in rice fields sprayed with neem oil (ref. 10).

TABLE 1. Quantity of food ingested by first instar larvae of C. medinalis, M. separata, and S. m. acronynctoides on leafcuts from rice plants sprayed with neem oil[a]

Concn (%)	Excreta (µg)/10 larvae kept 24 h on treated leafcuts[b]		
	C. medinalis	M. separata	S. m. acronynctoides
1.5	-	310a	169 b
3.0	-	274a	135 b
6.0	-	263a	117 b
12.0	120 b	156 b	28 c
25.0	89 c	86 b	8 c
50.0	48 d	-	-
0.0 (control)[c]	160a	307a	373a

[a]In a column, means followed by a common letter are not significantly different at 5% level by Duncan's multiple range test; average of 25 replications for C. medinalis and 10 replications each for M. separata, and S. m. acronynctoides, respectively.

[b]Weight of excreta oven-dried 24 h.

[c]1.66% Teepol detergent solution.

SUCKING INSECTS

Rice plants sprayed with neem oil using an ULV-spray applicator were not as attractive to the brown planthopper, Nilaparvata lugens (Stål), and the whitebacked planthopper, Sogatella furcifera (Horváth), as compared to the untreated control (ref. 1 & 13). Insect arrival decreased progressively with oil concentration. However, the green leafhopper, Nephotettix virescens (Distant), females were less repelled by neem oil (ref. 13). Feeding by planthoppers and leafhoppers was deterred on rice plants sprayed with neem oil. In a 60-min observation period, starved N. lugens females avoided seedlings sprayed with 50 or 100% neem oil for 15 to 22 min and, even after alighting on them, were restless and took more than 30 min searching for feeding sites (ref. 1). The feeding duration was reduced by 0.93 min/h for every 1% increase in oil concentration while the search for feeding sites was intensified. Food intake was significantly reduced on plants sprayed with 12% neem oil, but the insect fed normally on plants sprayed with a vegetable oil (Fig. 1). Feeding by S. furcifera and N. virescens females was also significantly reduced on neem oil-treated plants (ref. 13, 14).

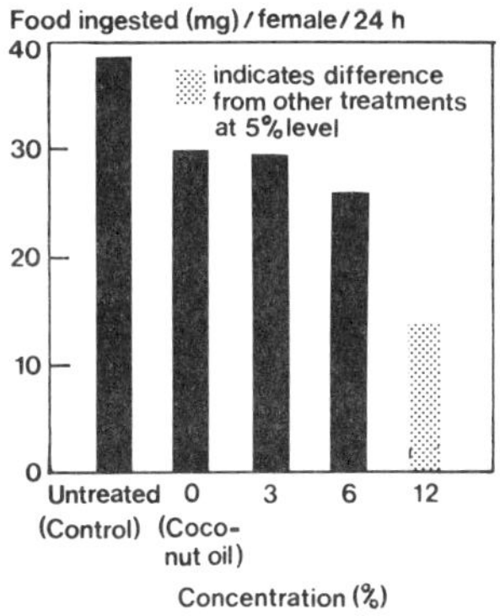

Fig. 1. Quantity of food ingested by N. lugens females on TN1 rice plants sprayed with neem oil. Newly emerged brachypterous females were caged individually in parafilm sachets on plants one day after neem oil application. Average of five replications; means followed by a common letter are not significantly different at 5% level by Duncan's multiple range test. Food ingested = (insect's weight gain due to feeding + weight loss due to wear and tear + water loss by insect + fresh weight of excreta).

Phloem feeding by N. virescens, monitored by an electronic device, on rice plants sprayed with 1.25, 2.5, 5, or 10% neem oil was significantly reduced than on acetone-treated control plants (Fig. 2)(ref. 15). Phloem feeding was erratic on neem oil-treated plants, while probing, salivation, and xylem feeding increased. The increase in xylem feeding served to offset desiccation resulting from repeated probing, profuse salivation, and restlessness. Contact with azadirachtin is known to disrupt food intake and increase insect locomotory activity (ref. 7).

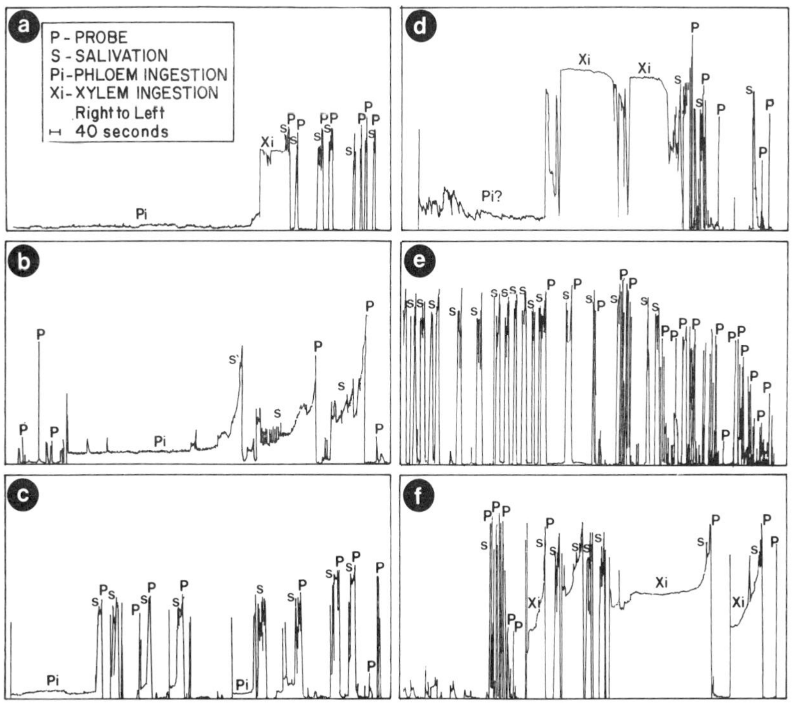

Fig. 2. Electronically recorded waveforms during N. virescens feeding on TN1 rice plants sprayed with (a) acetone (control) and (b) 1.25%, (c) 2.5%, (d) 5%, (e, f) 10% neem oil.

Simons (ref. 16) emphasized the use of oil formulations with antifeedants from plants for control of insect-transmitted virus diseases. Saxena et al. (ref. 1) reported that rice fields sprayed periodically with 12% neem oil emulsion, using an ULV-applicator, had a significantly lower incidence of the N. lugens-transmitted ragged stunt virus disease than unsprayed controls (Table 2). Saxena and Khan (ref. 17) found that the oil reduced the survival of N. lugens and suppressed the transmission of both grassy- and ragged stunt virus diseases.

Neem oil or custard-apple (Annona sp.) oil reduced the survival of N. virescens and its transmission of rice tungro virus (RTV) to rice seedlings (ref. 18). The insect failed to survive and transmit RTV to oil-sprayed seedlings beyond 3 days of exposure, but on control plants the insect lived for more than 20 days and transmitted RTV up to 5 days (Table 3).

TABLE 2. Incidence of N. lugens-transmitted ragged stunt virus disease and grain yield in rice fields sprayed with neem oil[a]

Concn (%)	Ragged stunt (%)[b]	Yield (t/ha)
3	47a	2.52
6	41a	2.96
12	34 b	3.06
Unsprayed (control)	57a	2.26

[a]IR1917-3-17, a susceptible selection, was planted in 11 x 20 m plots; each treatment replicated three times. Neem oil was sprayed with an ULV-applicator at 3ℓ/ha, 20, 40, 60, 80, and 100 days after transplanting.

[b]Means followed by a common letter are not significantly different at P = 0.05 by Duncan's multiple range test.

TABLE 3. RTV infection in TN1 rice seedlings sprayed with custard-apple oil (CAO) or neem oil (NO) after exposure to viruliferous N. virescens[a]

Concn (%)	RTV infection (%) of seedlings[b]			
	1 day		2 days	
	CAO	NO	CAO	NO
5	25 b	27 b	7 b	14 b
10	19 bc	16 bc	4 b	11 b
20	19 bc	30 bc	2 b	10 b
30	20 bc	14 c	11 b	4 b
50	10 c	12 c	0.2 b	10 b
0 (control)[c]	60a		36a	

[a]The experiment was replicated five times using 440 viruliferous insects and 440 seedlings for each replication of 11 treatments. A total of 2200 insects and 4667 treated seedlings, including those of successive feeding were used. RTV symptoms in seedlings were observed 12 days after inoculation.

[b]In a column, means followed by a common letter are not significantly different at 5% level by Duncan's multiple range test.

[c]Control treatment (0.1% Teepol detergent solution) was common to both oils.

Mixtures of custard-apple oil and neem oil in 1:1, 1:2, and 1:4 proportions (vol/vol) at 5, 10, or 20% concentrations were significantly more effective in reducing N. virescens survival and RTV transmission than the spray applications of individual oils in rice seedlings (ref. 19).

Degradation of neem oil by sunlight limits its potential in crop protection. But improved formulations with sunshields can prolong the persistence of its activity. Neem oil and its formulations with 1% carbon or 2% liquid latex were evaluated at the IRRI experimental farm for control of N. lugens. Population of nymphs at 35 and 56 days after transplanting was significantly smaller in plots sprayed with neem oil-carbon formulation than in unsprayed controls (ref. 20). A similar but less marked pattern in nymphal population was observed in plots sprayed with neem oil or neem oil-latex formulation. Natural enemies, such as spiders, mirids, and microveliids were not affected. Yield was significantly higher in plots sprayed with neem oil-carbon formulation than in the control (Table 4).

Table 4. Yield in rice fields sprayed with neem oil formulations[a]

Treatment[b]	Yield (t/ha)[c]
Neem oil + 1% carbon	3.85a
Neem oil + 2% liquid latex	3.65ab
Neem oil	3.70ab
Unsprayed (control)	3.45 b

[a]IR1917-3-17, N. lugens-susceptible selection, was planted at 20 x 20 cm spacing in 11 x 20 m plots; each treatment replicated four times.

[b]Crude neem oil and its formulations were sprayed at 5, 25, 45, 65, and 85 days after transplanting using an ULV-applicator.

[c]Means followed by a common letter are not significantly different at 5% level by Duncan's multiple range test.

For three consecutive croppings in 1984 to 1985, RTV-susceptible IR42 plants treated with custard-apple oil-neem oil mixture alone or in combination with neem cake-urea mixture had lower RTV incidence, and higher yield as compared to the control (Fig. 3)(ref. 21). The insecticide check and neem cake-urea mixture alone were less effective in controlling the vector or the disease.

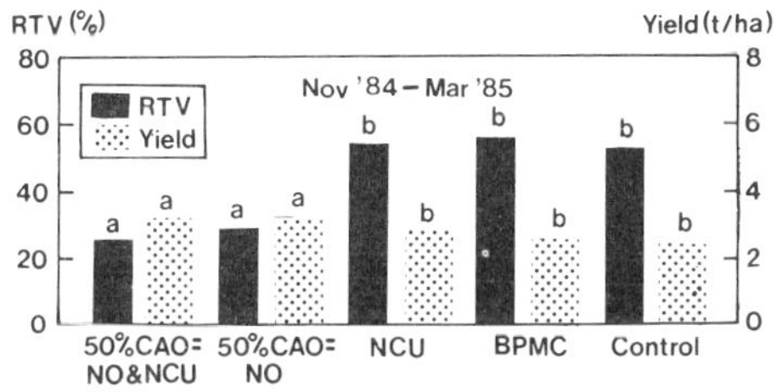

Fig. 3. Comparison of yield and RTV infection in field plots planted to RTV-susceptible IR42 and treated with either custard apple oil and neem oil (CAO:NO), neem cake + urea (NCU), their combination (CAO:NO and NCU), BPMC, or a detergent-water solution (control). Average of 4 replications. Means followed by a common letter are not significantly different at the 5% level by Duncan's multiple range test.

FUTURE CONSIDERATIONS AND CONCLUSION

Neem seed derivatives offer a novel approach in the management of rice insect pests and the virus diseases transmitted by them. Unlike insecticides that kill both pests and predators, neem derivatives are safe to natural enemies, other beneficial organisms, and man. Neem trees are found in several rice growing Asian and African countries. It can withstand climatic and geographic variations and thrives even on poor soils. Neem is being planted in the Philippines, Papua New Guinea, Vietnam, Australia, the Caribbean islands, as well as Central and South America.

India has ca. 14 million trees, but actual seed collection is only 25% (ref. 5). Considering neem's multiple uses, more trees need to be grown. Enough neem derivatives would then become available to rice farmers at low cost for crop protection. Farmers should also grow neem trees on their homestead. A tree becomes fully productive in ca. 10 years. In India, neem flowers from January to May, and fruits mature from May to August, in time for the main rice crop. One tree produces 30-50 kg fruit/year. Thirty kg of seed yields 6 kg of oil and 24 kg cake. Productivity can be increased and stabilized through selection of superior genotypes. The level of variation of azadirachtin content in the neem seed obtained from India and Africa has been studied (ref. 4).

Emulsions made from 8 to 10 ℓ of neem oil are enough to spray 5 to 6 times one ha of rice crop, using an ULV-applicator. Neem cake, the residue after oil expulsion, also possesses pest control potential and can be mixed with urea and applied to rice crop for its nitrification inhibition properties (ref. 22). Rice yield increased by 24% in fields applied with urea mixed with 20% neem cake (ref. 23 & 24).

The use of simple formulations of neem derivatives is ideally suited for crop protection by resource-limited farmers. Neem seed can be processed at the village level using simple methods. Such products may need some definitions of quality control and general biological properties. Efficacy of neem derivatives can be enhanced by using additives, synergists and other botanicals, and improved application techniques. However, pest control effectiveness of neem derivatives should not be equated at par with that of conventional pesticides.

Chances of insect pests developing resistance to neem derivatives are remote (ref. 8). Unlike ordinary insecticides based on a single active ingredient, neem seed derivatives possess an array of chemicals which act concertedly on both behavioral and physiological processes.

An enlightened attitude towards pest control philosophy would be helpful in promoting the use of plant derivatives for pest control. Killing and destruction of pests is not necessary if they can be incapacitated otherwise. The use of neem seed derivatives, which affect insect pests in many ways, offers a harmonious and nonviolent approach to pest management. In the process, pests' natural enemies, other beneficial organisms, and man can be spared from the rigors and hazards of using toxic chemicals. Further research, development, promotion, and use of neem derivatives will pave the way for ecologically sound management of rice insect pests in the future.

REFERENCES

1. R.C. Saxena, N.J. Liquido and H.D. Justo, Jr., Proc. 1st Intern. Neem Conf., GTZ, Eschborn, 171-188 (1981).
2. I. Kubo and K. Nakanishi, Host Plant Resistance to Pests, ACS, Washington, DC, 165-178 (1977).
3. K. Munakata, Host Plant Resistance to Pests, ACS, Washington, DC, 185-196 (1977).
4. M. Jacobson, Natural Resistance of Plants to Pests: Roles of Allelochemicals, ACS, Washington, DC, 220-232 (1986).
5. C.M. Ketkar, Utilization of Neem (Azadirachta indica Juss) and its Bye-products, 234 p., Khadi and Village Ind. Com., Bombay, India (1976).
6. H. Schmutterer, K.R.S. Ascher and H. Rembold, Proc. 1st Intern. Neem Conf., 297 p., GTZ, Eschborn, (1981).
7. L.M. Schoonhoven, Entomol. Exp. Appl. 31, 15-23 (1982).
8. R.C. Saxena, Chemistry and World Food Supplies: The New Frontiers, CHEMRAWN II, Pergamon Press, Oxford, 143-161 (1983).
9. H. Schmutterer and K.R.S. Ascher, Proc. 2nd Intern. Neem Conf., 567 p., GTZ, Eschborn, (1984).
10. R.C. Saxena, G.P. Walbauer, N.J. Liquido and B.C. Puma, Proc. 1st Intern. Neem Conf., GTZ, Eschborn, 189-204 (1981).
11. International Rice Res. Inst. (IRRI), Annual Report for 1983, IRRI, Los Baños, Phil., 191-193 (1984).
12. R.C. Saxena and P.B. Epino, unpublished information.
13. J. v.d. Heyde, R.C. Saxena and H. Schmutterer, Proc. 2nd Intern. Neem Conf., GTZ, Eschborn, 403-412 (1984).
14. R.C. Saxena, P.B. Epino, Tu Cheng-Wen and B.C. Puma, Proc. 2nd Intern. Neem Conf., GTZ, Eschborn, 403-412 (1984).
15. R.C. Saxena and Z.R. Khan, J. Econ. Entomol. 78, 222-226 (1985).
16. J.N. Simons, Vectors of Disease Agents: Interaction with Plants, Prager Publishers, New York, 169-178 (1981).
17. R.C. Saxena and Z.R. Khan, J. Econ. Entomol. 78, 647-651 (1985).
18. V. Mariappan and R.C. Saxena, J. Econ. Entomol. 76, 573-576 (1983).
19. V. Mariappan and R.C. Saxena, J. Econ. Entomol. 77, 519-521 (1984).
20. R.C. Saxena and H.D. Justo, Jr., unpublished information.
21. R.C. Saxena and H.D. Justo, Jr., Intern. Rice Res Newsl. 11(2), 25 (1986).
22. S.S. Bains, R. Prasad and P.C. Bhatia, Fert. News 16, 30-32, 52 (1971).
23. R.N.S. Reddy and R. Prasad, J. Soil Sci., 26, 304-312 (1975).
24. R.N.S. Reddy and R. Prasad, Riso 26, 217-224 (1977).

Herbicidal characteristics of bialaphos

Kunitaka Tachibana

Pharamaceutical Research Laboratories, Meiji Seika, Ltd., Kohoku-ku, yokohama 222, Japan

Abstract - Bialaphos, L-2-[(hydroxy)(methyl)phosphynoyl] butyryl-L-alanyl-L-alanine, controlled a wide range of weeds, including perennials, by foliar treatment. Bialaphos was slower acting than paraquat, but faster than glyphosate. It controlled the regrowth of weeds longer than paraquat, but shorter than glyphosate. ^{14}C-Bialaphos was treated to a leaf of *Rumex obtusifolius* (broadleaf dock) and translocation of radioactivity into the roots and the new leaf was observed with autoradiography. Bialaphos applied before planting did not affect crop emergence nor crop growth. Bialaphos and its metabolite were easily degraded by microorganisms in soil. Adsorption of bialaphos by soils and its rapid degradation in them were considered to be the main factors in loss of herbicidal activity in soils. Ammonia accumulation as the result of glutamine synthetase inhibition is considered to be the main factor of the herbicidal activity.

INTRODUCTION

Bialaphos, a unique tripeptide, is a metabolite of *Streptomyces hygroscopicus*. It controls a wide range of weeds by foliar treatment, and it is the first herbicide produced by fermentation. Meiji Herbiace, 32% liquid formulation of bialaphos, was registered as a herbicide for nonedible crop land use (1984) and for food crop land use (1986) in Japan. Bialaphos, a natural product, is easily metabolised and biodegraded in environment; therefore, it can be used safely. Bialaphos is a fast acting herbicide with long lasting effects. However, it is inactivated rapidly in the soil, and does not affect emergence nor growth of crops through soil. Therefore, it is expected to be used widely for arable land.

TABLE 1 Chemical and physical properties of bialaphos.

Trade name:	Meiji Herbiace
Common name:	bialaphos
Code number:	MW-801, MW-831, SF-1293Na
Chemical name:	Sodium salt of L-2-amino-4-[(hydroxy)(methyl) phosphinoyl]butyryl-L-alanyl-L-alanine
Chemical structure:	

```
          O                        CH3        CH3
          ||                       |          |
  [ H3C-P-CH2-CH2-CH-CONH-CH-CONH-CH-COO-  ] Na+
          |           |
          O-          NH3+
```

Empirical formula:	$C_{11}H_{22}N_3O_6P$ Na
Molecular weight:	345.26
Appearence:	White powder
Melting point:	About 160 C(decomposition)
Optical rotation:	$[\alpha]_D^{25}=-34$ (C=1 water)
Solubility:	Very soluble in water, Soluble in methanol. Insoluble in ethanol, n-buthanol, acetone, ethylether, chloroform, benzene and hexane.

CHEMICAL AND PHYSICAL PROPERTIES

Chemical and physical properties of bialaphos are shown in Table 1.

HERBICIDAL ACTIVITY

(1) Herbicidal spectrum

Table 2 shows the results of a field trial with main annual weeds in Japan. Bialaphos controlled both dicotyledonous and monocotyledonous weeds with a little stronger activity to the former. The dose of bialaphos which gave effective control of annual weeds was between 9.6 to 16 g/a.

TABLE 2 Herbicidal activities of bialaphos on various weeds.

Weed species	Leaf stage	Rate*, g(a.i.)/a	
		9.6	16
Monocotyledonous weeds			
Digitaria adscendens	2-4	9	10
	5-7	4	8
Echinochloa crus-galli	2-4	9	10
	5-7	6	8
Cyperus iria	2-4	10	10
	5-7	10	10
Commelina communis	2-4	10	10
	5-7	10	10
Dicotyledonous weeds			
Polygonum longisetum	2-4	10	10
	5-7	10	10
Chenopodium album	2-4	10	10
	5-7	6	9
Amaranthus viridis	2-4	10	10
	5-7	9	10
Portulaca oleracea	2-4	10	10
	5-7	6	8

* Meiji Herbiace, 32% liquid formulation.
Observation : 3 weeks after treatment. 0: no effect, 10: totoal control.

(2) Selectivity

Phytotoxicity was observed in useful plants to which bialaphos was directly sprayed, and its selectivity has not been found.

(3) Appearence of herbicidal activity and residual effect

Fig. 1 shows the results of time course study of activity of bialaphos, glyphosate and paraquat against *Artemisia princeps*. Symptoms of the herbicidal activity were observed as yellowing or wilting of leaves 2 to 3 days after bialaphos treatment and maximum activity occurred 7 - 10 days after the treatment. Bialaphos was slower acting than paraquat, but faster than glyphosate. It controlled the regrowth of weeds longer than paraquat, but shorter than glyphosate.

(4) Translocation in plants

Residual effect of bialaphos as shown in Fig. 1 suggested that bialaphos acted at the underground part of perennial weeds. ^{14}C-Bialaphos was treated on a leaf of *Rumex obtusifolius* (broadleaf dock). Translocation of radioactivity into the roots and the new leaves was observed with autoradiography as shown in Fig. 2.

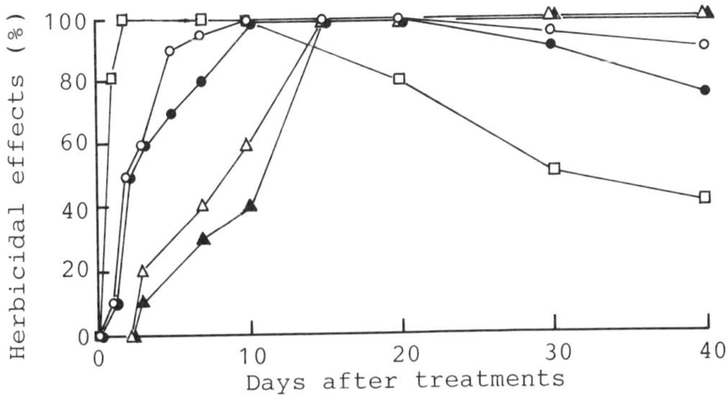

Fig. 1 Herbicidal effects of bialaphos, glyphosate and paraquat on Artemisia princeps.

● : 16 g/a of bialaphos*, ○: 32 g/a of bialaphos*, ▲ : 24 g/a of glyphosate, △ : 48 g/a of glyphosate, □ : 7.2 g/a of paraquat. *Herbiace, 32% liquid formulation. Plant height :20-25 cm.

Fig. 2 Autoradiogram (right) of Rumex obtusifolius at seven days after treatment of C^{14}-bialaphos on the upper surface of middle leaf(↑).

(5) Activity by soil treatment

Table 3 shows at the practical rate (9.6 to 32 g/a) bialaphos did not cause phytotoxicity in pre-emergence treatment. Adsorption of bialaphos to soils and its rapid degradation in them were considered to be the main factors contributing to its loss of herbicidal activity in soils.

TABLE 3 Phytotoxicity of bilalaphos in pre-emergence treatment.

Crops	Rate* g(a.i.)/a			
	10	30	100	300
Corn	0	0	0	3
Soy bean	0	0	1	4
Wheat	0	0	0	4

*Meiji Herbiace, 32% liquid formulation.
0: no effect, 10: all plants dead.

MECHANISM OF ACTION

Glutamine synthetase (L-glutamine : ammonia ligase (ADP), EC 6.3.1.2, GS) assimilates and reassimilates ammonia in higher plants. Decrease of GS activity was observed in plants treated with bialaphos (ref. 1). Fig. 3 shows that ammonia content in the leaf of Amaranthus lividus (livid amaranth) was observed to increase in 4 hours after the bialaphos treatment and reached about 30 times higher than the control in 24 to 48 hours as the results of GS inhibition. High correlation between free ammonia content and herbicidal activity indicates that the toxicity of accumulated ammonia is the primary factor of herbicidal activity of bialaphos (ref. 2). The mechanism of action is summarised in Fig. 4.

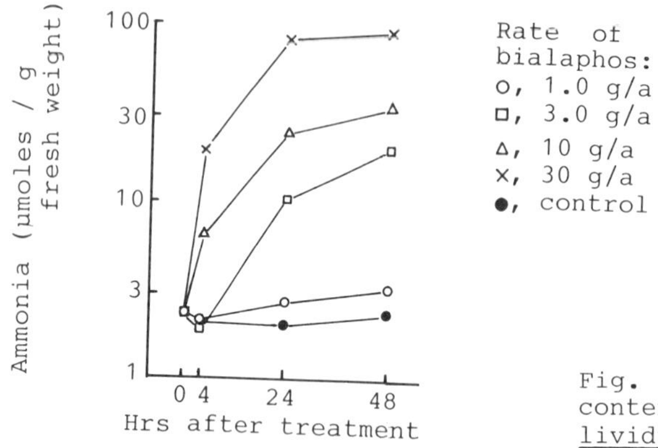

Fig. 3 Time course of free ammonia content in leaves of *Amaranthus lividus* treated with bialaphos.

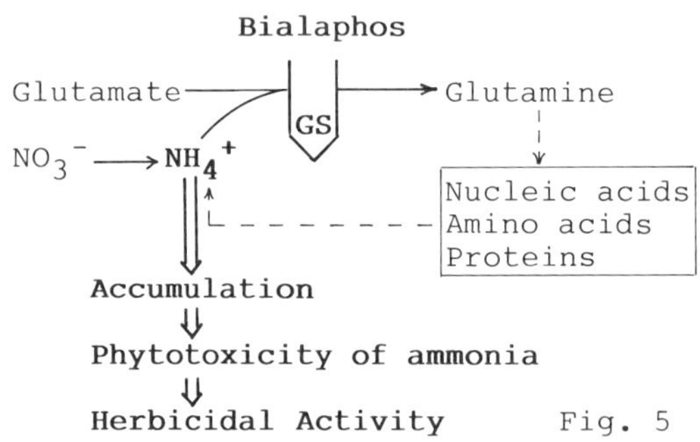

Fig. 5 Action mechanism of bialaphos.

DEGRADATION IN SOILS

Bialaphos and its metabolite were easily degraded in soils. Half lives of bialaphos and its metabolite, L-2-amino-4-[(hydroxy)(methyl)phosphinoyl] butyric acid were about 2 hours and 4 days, respectively.

CONCLUSION

Bialaphos was observed to have potent herbicidal activity by foliar treatment but no activity by soil treatment. Therefore, bialaphos is the herbicide suitable for non-tillage cultivation. In addition, it is also suitable for the control of the bottom weeds in orchards and mulberry fields, and of the weeds between rows of crops, since it it not absorbed through the root system.

REFERENCES

1. K. Tachibana, T. Watanabe, Y. Sekizawa and T. Takematsu, J. Pestcide Sci., **11**, 27-33 (1986)

2. K. Tachibana, T. Watanabe, Y. Sekizawa and T. Takematsu, ibid., **11**, 33-37 (1986)

Chemical and biological basis of pesticide activity and resistance

Advances in understanding insecticide modes of action

Fumio Matsumura

Pesticide Research Center, Michigan State University, East Lansing, Michigan, USA 48824

Abstract - Major biochemical targets recently studied are: (a) the GABA complex, including the GABA recognition site, picrotoxinin receptor and chloride channel, (b) benzodiazepine binding site(s), also part of the GABA complex, (c) receptors for biogenic amines, glutamate receptors, and, (d) neural calcium-sensitive phosphorylation systems. There also has been excellent progress in research on known target areas such as sodium channels and the ACh receptor.

The importance of the GABA complex as a target of insecticides is now receiving due recognition. The picrotoxinin receptor is apparently the major target for all cyclodiene-type insecticides and bicyclophosphates. The accompanying chloride channel is considered to be the main site of action of avermectins in both invertebrate and mammalian species, though in the case of the latter group, other actions such as interference with GABA release and benzodiazepine receptors are involved.

Pyrethroids have been found to inhibit membrane Ca/Mg-ATPase, Na/Ca exchange and calcium-sensitive protein kinases. The importance of such inhibition has been gaining strong support with the discovery that the nerve membranes of the DDT-pyrethroid resistant kdr strains of the German cockroach and the housefly show reduced sensitivities to calcium. Pyrethroids, particularly Type II compounds, have also been found to interfere with some benzodiazepine receptors (e.g., RO-5-4864 sites) in the CNS of both insect mammalian species, indicating that there are common sites of action of these two groups of chemicals.

There has been a surge of new interest in the area of the role of biogenic amines in insects, particularly with regard to their role in triggering release of neuroactive peptides. The specific action of chlordimeform on octopamine receptor in insects is being used as a tool in elucidating these physiological neurosecretion controlling mechanisms.

INTRODUCTION

Studies on modes of action of insecticides are very important from several points of view. First, such knowledge is needed to understand health hazards of those chemicals to man and other non-target organisms. Second, it could enable scientists to design therapeutic measures. Third, it helps chemists to design additional chemicals with similar action mechanisms. Fourth, it could give scientists important clues as to the cause of resistance development in pests, particularly those involving target insensitivity, and thereby helps in designing counter measures to avoid resistance or reverse the development of resistance, and finally the knowledge gained by such studies yields valuable basic information on the nature of target systems (i.e. the weakness of the sensitive insects) in terms of physiological, biochemical and biophysical knowledge of vital biological systems.

Despite the pressing needs for fundamental knowledge in this area, decisive breakthroughs in discovering the basic cause for insecticidal actions are rare. This is because of the difficulties in ascribing the many physiological, biochemical and symptomatological changes evoked in animals by an insecticidal application to a single biochemical lesion, and because of the inexact nature of extrapolating any in vitro study results to in vivo expressions of toxicities.

In the sixties and seventies, the predominant study theme in this area was cholinesterase inhibition by organophosphates/carbamates. The sheer simplicity and the logical nature of the subject matter made it attractive for many scientists to choose this topic. In that time span great strides have been made in research and development on a variety of organophosphates and carbamates. Another area of science in which great progress has been made during this period is in electrophysiological studies. Because of their sensitivity and reproducibility, coupled with great innovations in electronics in the past three decades, such approaches have become very successful, yielding very valuable data.

Despite these spectacular accomplishments, progress in other vital areas has been rather slow. Recently, however, there have been a number of breakthroughs as a result of long term investment of time efforts by a number of scientists in several key areas which makes this symposium a timely one.

MODE OF ACTION OF CYCLODIENE-TYPE INSECTICIDES

The mechanism of action of this type of insecticide has been a mystery for a long time. They have been shown to cause hyperexcitation in insects and higher animals. Their residues have been found in various ecosystems, and because of their potency, there has been a rising concern for their toxicological and ecotoxicological hazards.

During the seventies electrophysiologists have found that these compounds cause stimulation of excitatory neurotransmitter release in the central nervous system (ref. 1, 2). By using such a clue, we began in the late seventies a search for the cause for such a mode of hyperexcitation. Our first attempt was to test many chemical agents which are known to stimulate transmitter release on cyclodiene-resistant and susceptible strains of German cockroaches, *Blattella germanica*. The rationale behind this approach was that these resistant insects are known to possess a nervous system which is insensitive to cyclodienes and to gamma-BHC. This approach uncovered the existence of cross resistance in these insects to picrotoxinin (PTX), a potent naturally occurring excitant. Based on this observation, a working hypothesis was adopted that cyclodiene-like insecticides act in an identical manner to PTX which is known to antagonize the action of the naturally occurring inhibitory transmitter, GABA (gamma-aminobutynic acid), by blocking the GABA-stimulated entry of chloride ion (Fig. 1). By 1982 we obtained preliminary but positive proof that in the coxal muscles of the American cockroach both heptachlor epoxide and gamma-BHC acted as predicted by this hypothesis: i.e. like PTX, they could block GABA-stimulated Cl^- uptake (ref. 3). Subsequent studies using ^{3}H-dihydropicrotoxinin (DHPTX) as a ligand for PTX receptors have yielded equally encouraging data (ref. 4, 5). In three cyclodiene-resistant strains from three different species, the specific binding of ^{3}H-DHPTX to the receptor was much less than that found in their susceptible counterparts (Table 1). Cyclodienes have also been found to interact with the rat brain PTX receptor (ref. 4). The structural similarities between PTX and some cyclodiene insecticides have been pointed out (ref. 4, 5). A comparison of the inhibitory potency of these insecticides on specific ^{3}H-DHPTX binding to their *in vivo* toxicities has shown that in most cases the structure-activity relationships were exactly as predicted (Table 2). Particularly convincing is the case with the BHC isomers where the highly insecticidal gamma-isomer was clearly the most potent inhibitor of ^{3}H-DHPTX binding (ref. 6). Apparently cyclodiene-type compounds owe their toxicities to their molecular similarities to the PTX molecule (ref. 4) (Figure 2). A more recent investigation on the structural requirements for PTX receptor interactions has shown that, among active compounds, there appear to be two electron negative centers and one hydrophobic center situated at proper distances from each other (ref. 7). In the case of bicyclophosphates, there is only one electron negative center in addition to one hydrophobic center indicating that it is not necessary to have all three of these centers. Because of the overall logic and strong supporting evidence, the primary site of attack of cyclodiene-type insecticides is now firmly established to be the PTX receptor in the central and peripheral nervous systems of both insect and mammalian species (ref. 6, 8, 9).

Figure 1. Schematic diagram of drug interactions at a GABA receptor-ionophore complex.

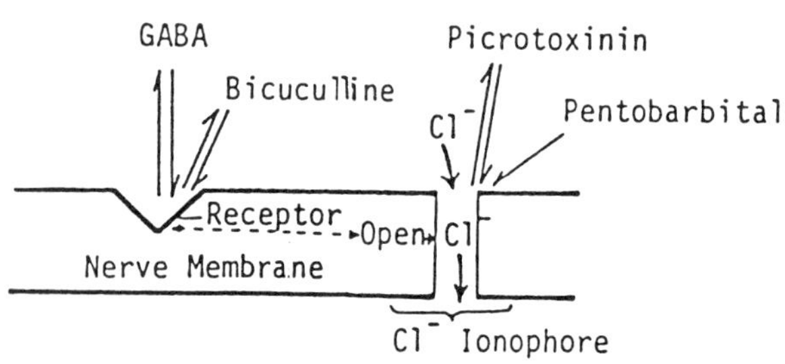

TABLE 1. Strain differences in levels of ^{3}H-DHPTX (dihydropicrotoxinin) specific binding to nerve binding sites and in susceptibility

Strains	Susceptible (S) or Resistant (R)	^{3}H-DHPTX specific binding dpm/mg protein	Susceptibility to dieldrin or picrotoxinin
Aedes aegypti			
Rockefeller	S	2483	16[a]
Trinidad	S	3034	9
Isla Verde	R	1723	48
Musca domestica			
SBO	S	732	13[b]
CLD	R	229	254
Blattella germanica			
CSMA	S	6013	12[c]
LPP	R	1382	182

[a]Against picrotoxinin (30 ppm) LT_{50} hrs. In the presence of 3 ppm piperonyl butoxide. [b]Dieldrin LD_{50} 24 hrs. Expressed in ng/female. [c]Dieldrin LT_{50} hrs. Treated by surface contact method using 3 mg/jar/10 roaches.

Table 2. Effect of Cyclodienes and Other Agents on $[^{3}H]\alpha$-Dihydropicrotoxin Binding[a]

Agents[b]	Percentage inhibition of $[^{3}H]\alpha$-dihydropicrotoxinin specific binding[c]
Dihydropicrotoxinin	100
Aldrin	36.6 ± 0.6
Dieldrin	58.5 ± 2.3
Photodieldrin	79.0 ± 4.3
Heptachlor	36.6 ± 18.6
Heptachlorepoxide	57.3 ± 31.3
Isodrin	35.6 ± 3.8
Endrin	62.3 ± 12.4
γ-Chlordane	56.2 ± 2.0
Oxychlordane	77.8 ± 6.2
Hexachlorocyclo-pentadiene	0.3 ± 2.2
Mirex	14.7 ± 15.7
Kepone	135.5 ± 18.4
Toxaphene	72.3 ± 28.1
DDT	5.4 ± 3.6
α-BHC	21.4 ± 18.5
β-BHC	15.3 ± 15.4
γ-BHC(lindane)	91.5 ± 20.0
δ-BHC	55.0 ± 16.0
Parathion	4.1 ± 26.3
Allethrin	13.2 ± 6.7
Cypermethrin	7.7 ± 3.1
Decamethrin	-0.1 ± 1.0
Fenvalerate	-0.4 ± 5.8
Chlordimeform	7.8 ± 4.7
Pentobarbital	21.1 ± 8.1
Benzodiazepam	45.1 ± 10.8
SQ-65396	21.5 ± 5.7
SQ-20009	-8.3 ± 4.2
Tetramethylene-disulfotetramine	27.1 ± 13.0
t-Butyl bicyclic phosphate	46.6 ± 2.9

[a]$[^{3}H$-$]\alpha$-Dihydropicrotoxinin, 11.1 nM; five American cockroach heads were used for one experiment (one agent), which involves three determinations. [b]Dihydropicrotoxinin, 100 μM; others, 10 μM. [c]Data are expressed as mean ± SE of two or three experiments, each experiment involving three determinations.

Figure 2. Structural comparison of picrotoxinin, heptachlor epoxide and γ-BHC. The parts of picrotoxinin molecule that bear similarities to each pesticide listed are indicated in solid bold lines. (14).

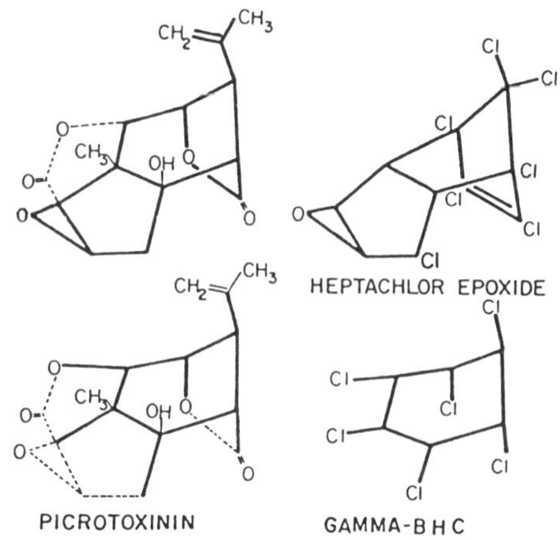

BICYCLOPHOSPHATES

From a completely different point of view Casida's group has been studying the mechanisms of action of bicyclophosphates which are not commercial insecticides per se, but which are known to possess high toxicities to both insects and higher animals. Based on the information from Olsen's laboratory (ref. 10) that some of the bicyclophosphates inhibit the specific binding of ^{3}H-DHPTX and that from Ozoe and Eto that τ-butyl bicyclophosphate is a very potent compound most likely interfering with GABA's action (ref. 11, 12), Casida's group synthesized and successfully tested ^{35}S-τ-butyl bicyclophosphorothioate as an artificial ligand (ref. 13) for studies on the sites of action of bicyclophosphates. This ligand turned out to be very useful, yielding high percentages of specific binding as compared to ^{3}H-DHPTX. Furthermore, the stability of the ^{35}S-TBPS receptor-ligand complex was superior to that of ^{3}H-DHPTX. Indications that ^{35}S-TBPS binding sites overlap with those of some potent insecticides was first obtained by Lawrence and Casida (ref. 14), who found that there is an excellent correlation between the insecticidal activities of type II pyrethroids and their ability to inhibit ^{35}S-TBPS specific binding in a rat brain membrane preparation. In all cases tested, the most insecticidal stereoisomer has been shown to be the most potent one in terms of inhibiting ^{35}S-TBPS binding. Later the same workers (ref. 15) extended their study to cyclodiene-type insecticides to demonstrate that, as in the case of ^{3}H-DHPTX (Ref. 6, 9), there is an excellent correlation between their insecticidal activities and their potencies in inhibition specific ligand binding. A simultaneous publication from Eldefrewi's group (ref. 16) indicates the same result.

One should ask the question of how these three groups of insecticides, bicyclophosphates, cyclodienes, and pyrethroids, affect the same ligand binding. The most plausible answer is that the major site of action for bicyclophosphates and cyclodiene-type insecticides is identical, i.e. the PTX receptor, whereas the inhibition of ^{35}S-TBPS binding by pyrethroids represents a secondary and less important action of type II pyrethroids on the GABA system, occurring only in the mammalian CNS. The supporting evidence for this conclusion is: first, the EC_{50} values for pyrethroids in this regard are rather high; second, it has been shown by Soderlund's group (ref. 17) that type II pyrethroids affect the rat brain GABA system only at relatively high concentrations as compared to those affecting the sodium channel operations, and third, type II pyrethroids have been shown to be totally ineffective in competing against ^{3}H-DHPTX in the American cockroach head preparation (ref. 6). ^{35}S-TBPS is a relatively new ligand compared to the well-studied ^{3}H-DHPTX, and, therefore, it requires further investigation to understand its characteristics. We do know that these two ligands behave differently in one area. ^{3}H-DHPTX binding has been shown to be unaffected by GABA (see ref. 6), which shows the independence of the PTX receptor from the GABA recognition site, whereas that of ^{35}S-TBPS is profoundly affected by GABA.

AVERMECTINS AND MILBEMYCIN

Avermectins and milbemycin are potent insecticidal, acaricidal and antihelminthic microbial products. Their mechanisms of action were reported to be associated with the GABA system by several workers (ref. 18-20) even before these chemicals were fully developed for actual commercial uses. However, there have been some difficulties in assessing their precise site of action, since avermectins have been shown to affect many components of the GABA system in the mammalian CNS. For example, they have been shown to affect GABA release at the presynaptic terminals (ref. 21), to stimulate benzodiazepine and GABA binding, to antagonize ^3H-DHPTX binding, and to increase Cl^- uptake in rat brain preparations (ref. 22). While all of these actions consistently enhance the action of GABA, and therefore the end result of the avermectins' action in vivo is clear, the presence of several different biochemical lesions makes it difficult to pinpoint the basic cause for this series of changes.

To understand the mechanism of action of avermectins in insects, Tanaka and Matsumura (ref. 23) have studied the effects of avermectin B_{1a} on both neuromuscular and CNS preparations from the American cockroach. Our conclusion turned out to be a simple one: i.e. avermectin B_{1a}, one of the most potent avermectin components, acts on the chloride channel, with the end result being excessive Cl^- entry into the postsynaptic region of the GABA-ergic cells. It was surprising that in this insect all other systems which have been shown to be influenced by avermectins in the mammalian CNS were not affected by avermectin B_{1a}. One could interpret the above result to mean that in the mammalian CNS various components of the GABA system are closely interrelated, and that the action of avermectins on the chloride channel probably results in allosteric changes in other GABA components. By contrast, the GABA systems studied by Tanaka and Matsumura (ref. 23) in the American cockroach possess components which do not interact with each other, or at least not as much as those of the mammalian CNS. On the other hand, it must be cautioned that one cannot totally generalize that avermectins affect only the chloride channel in all insects from the results in one species. There are already indications that avermectins also affects some of other GABA components in insects (ref. 24), but the lesson is that in all cases so far the list of affected components always includes the chloride channel.

Also one must note that the action of avermectins on the chloride channel does not always result in increasing Cl^- uptake by the affected cells. The recent report from Soderlunds's laboratory indicates that, in the case of the mouse brain postsynaptic vesicle preparation, the end result was inhibition of Cl^- uptake (ref. 25). Since it has been already shown by Eldefrawi et al. (ref. 26) and subsequently by Soderlund (personal communication 1986) that in the case of the rat synaptic preparations, the effects of avermectins are the stimulation of Cl^- uptake under similar experimental conditions to the mouse brain experiment, the above finding must mean that the avermectins' action on the chloride channel can vary depending on the species involved and perhaps other factors such as tissues, experimental conditions etc. The chloride channels affected by avermectins may be regarded as rigidfied channels occupying either an open or closed state. The positive windfall of such a finding is that it firmly refutes the possibility that avermectins are simple chloride ionophores always inducing increased Cl^- permeability.

PYRETHROIDS

Modern pyrethroids represent one of most potent groups of insecticidal chemicals; consequently there have been intense efforts to understand their mechanisms of action. Even at an early stage of their development, there was a realization that some of the modified pyrethroids have somewhat different insecticidal properties from the original pyrethrin analogs. It was first proposed by Gammon et al. (ref. 27) and later by others (e.g. ref. 28, 29) that there are two types of pyrethroid actions. Generally type I actions are considered to be those caused by DDT and natural pyrethrin, and type II actions are typified by those caused by some newly synthesized pyrethroids, particularly by those containing the α-cyano substituent on the alcohol moiety. Electrophysiological experiments on isolated non-sensory nerve preparations have shown that the former actions are exemplified by the appearance of repetitive discharges and the latter represented by suppression of the spontaneous or evoked generation of action potentials (ref. 28, 30). In fact, both deltamethrin and cypermethrin, two typical type II pyrethroids, suppress repeated firing of axons, whether evoked by repetitive electric stimuli or by spontaneous repetitive discharges (ref. 28, 30, 31).

The basic cause for such neurotoxic actions of pyrethroids is the alteration of the properties of the sodium channel (see ref. 32). Briefly, DDT and all pyrethroids affect the sodium channel in such a way that the channel closes only slowly after each action potential. The only major difference between type I and type II pyrethroids is the time course of their closure. The speed of closure is much slower with type II than type I compounds. The measurement of the tail currents, which are created by the return of the voltage clamped axons to the resting potential, indicates that the time constants for tail current decay are in the order of a few msec for DDT, a few hundred msec for type I compounds (e.g. tetramethrin), and a few min for deltamethrin (ref. 33). The channels so affected become unavailable for the generation of the next action potential during the period of recovery. Therefore, type II pyrethroids are unable to generate action potentials when rapidly stimulated (e.g. ref. 31).

Though the above explanation is logical and the nature of the data are definitive in implicating the important role of the sodium channel in pyrethroid's action, there are indications that other factors are also important in pyrethroid poisoning. Probably the most clear cut support for the above conclusion is that the lethal action of type II pyrethroids on insects is positively temperature correlated, while that of type I pyrethroids and DDT is negatively correlated (e.g. ref. 28). By contrast, the effects of all pyrethroids on the tail current are negatively temperature correlated (ref. 31). One of the most frequently observed effects of type II pyrethroids is their influence on neurotransmitter release and/or changes in neurotransmitter levels in the CNS or neuromuscular junctions. Salgado et al. (ref. 34) have observed in mosquito larvae that a type II pyrethroid increased e.p.s.p. in a mosquito preparation. Cremer (ref. 35) reported very significant increases in the levels of epinephrine and norepinephrine in the rat brain as a result of deltamethrin administration *in vivo*. A part of the cause for such an effect is probably due to the pyrethroids' action on the sodium channel itself (ref. 36, 37). However, there are signs that the sodium channel is not the sole cause for the above physiological changes. First, if such is the case, DDT should also cause the same symptom, which is not the case. Second, recent data from our laboratory (ref. 38) and J.M. Clark's laboratory (ref. 39) indicate that depolarization-induced transmitter release (either with high K^+ or Ca^{2+} added to the external medium) is stimulated by type II pyrethroids *in vitro* in synaptosomal preparations from the squid optic lobe and the rat brain, but that tetrodotoxin (which blocks Na^+ channels) in the μM range could only partially block this effect of pyrethroids. Unfortunately not much can be said about the cause for such actions of type II pyrethroids on transmitter release mechanisms. One promising candidate is the Ca^{2+}-Mg^{2+} ATPase (ref. 40) which is known to play an important role in regulating intrasynaptic concentrations of free Ca^{2+}. Possible involvement of calcium regulatory processes at vital neural sites in the expression of pyrethroid neurotoxicity has been suggested (ref. 41, 42). These areas appear to warrant further investigation.

In this connection it is interesting to note that DDT/pyrethroid resistant insects (i.e. *kdr* insects) show altered nerve sensitivities to Ca^{2+} modulators as compared to their susceptible counterparts (ref. 43, 44, 45). Both resistant *Blatella germanica* and *Musca domestica* show cross resistance to A23187, a well established calcium ionophore *in vivo*. In both cases calcium-dependent nerve protein kinases from the resistant insects have lower Ca^{2+} affinity as compared to those from the susceptible strains. Furthermore, these resistant enzymes exhibit reduced sensitivities to DDT *in vitro*. In the case of the German cockroach such observations were also made on the V_T strain which has been genetically purified through 8 generations of successive backcrossing to the susceptible strain and selection and therefore, any calcium-insensitive factor found in this strain should be regarded as either a genetically integral part of the resistance factor or a very closely linked one. While these observations do not completely prove that the major sites of action of DDT and pyrethroids are calcium-dependent protein kinases which have the function to phosphorylate key neural proteins associated with ion channels, ion-sensitive receptors, and enzymes, they provide additional supporting data to the existing pool of information on Ca^{2+} antagonism in the actions of DDT/pyrethroids in invertebrate nerves (ref. 40, 41). What is clear is that these pesticides affect some of the neural calcium modulatory systems at low concentrations *in vitro* and that *kdr* resistant insects extend their resistance to other calcium modulators such as A23187. However, experimental evidence is lacking to absolutely connect the above observations to the modus operandi of sodium channels, or to other biochemical systems which are known to play a vital role in neural events, or to the disturbances caused by these pesticides.

From a different point of view several groups have been working on methods to assess specific binding sites for pyrethroids *in vitro*. In one of the most complete binding studies, Soderlund's group has studied stereo-specific binding of ^{3}H-NRDC 157 (1*R*, *cis*) in a mouse brain postnuclear membrane preparation (ref. 46). The stereospecific saturable binding site was found to have a half saturation concentration of 40 nM (an apparent K_D). Since the level of NRDC 157 to cause half maximum response in enhancing veratridine-stimulated $^{22}Na^+$ uptake was in the order of 220 nM, Soderlund concludes that the above high affinity binding site most likely represents the sodium channel. Previously it has been shown by Lawrence and Casida (ref. 14) that type II pyrethroids competitively block ^{35}S-TBPS binding in the rat brain preparation at relatively high concentrations (e.g. EC_{50} in 10^{-5}M range) and that such a binding site probably represents a GABA receptor component. Recently Gammon and Sander (ref. 47) and Lawrence et al. (ref. 48) have found that type II pyrethroids competitively inhibit specific binding of ^{3}H-RO5-4864 with an approximate EC_{50} of 10^{-7}M in rat brain membranes. Such a property appears to be confined to insecticidal stereoisomers. RO5-4864 is a peripheral-type diazepin, and its site of action is not limited to the GABA receptor itself (Ref. 49). In view of the relatively high affinity of the site interacting with these pyrethroids it appears worthwhile to investigate the nature of such binding sites.

CHLORDIMEFORM AND RELATED COMPOUNDS

That chlordimeform and related formamidines affect biogenic amine regulatory sites and receptors in many animal species has been well established (ref. 50, 51-53). The best known actions are: first, local anaesthetic action in the mammalian cardiovascular systems (ref. 54) which is likely to be responsible for the acute lethal action of chlordimeform at high doses; second, their inhibitory actions on monoamine oxidases resulting in elevations of monoamines in mammalian species at later stage (several hours) at non-lethal doses (ref. 55), third their action as a pseudo-agonist for the octopamine receptor in the firefly lantern (ref. 56, 57), and the ganglions in the central nerve cord of certain insects, causing excitation (ref. 58, 59), and fourth their nerve membrane destabilization effects to induce DDT-like excitation (ref. 58, 60) in the central nerve cord of the American cockroach. In susceptible insects chlordimeform's action on octopaminergic sites occur at very low concentrations (ref. 59, 61) such as 10^{-7} to 10^{-9} M. Therefore, such interactions are regarded as the major biochemical cause for this pesticide's potent physiological, behavioral effects in vivo. It must be added here that N-desmethyl chlordimeform, a major metabolic product rather than chlordimeform itself is the actual causative agent in such cases (ref. 62). In relatively tolerant species such as the American cockroach the nerve destabilization effect takes place at high concentrations ($>10^{-4}$M) of chlordimeform leading to hyperexcitation and eventually to death.

The octopaminergic properties of chlordimeform and its metabolites are becoming useful tools for insect biochemists and physiologists. For instance, Linn and Roelofs (ref. 63) recently found that chlordimeform, its analogs, and octopamine dramatically increase the effectiveness of pheromones in many lepidopterous insects. The main effect of chlordimeform is to lower the threshold concentrations for male moths to start responding to the species-specific pheromone blend. Orchard (ref. 64) reports that there are two octopaminergic dorsal median neurons in the seventh abdominal ganglion of the desert locust serving the muscles of the oviduct. Both octopamine and chlordimeform caused lowering of the tension of these muscles. According to Orchard there are high titers of octopaminergic neurons throughout the brain and the ganglia of the locust CNS. In addition, elevation of octopamine in the hemolymph apparently signals "orchestration" of specific behavior in this species including stimulation of lipid release from stores to provide energy for sustained flight.

In the case of acarine species, however, the major action mechanism of formamidine pesticides has not been clarified inasmuch as that the roles of octopamine in these organisms are unknown. Thus, despite the excellent acaricidal properties of formamidine pesticides, the critical biochemical effects in these organisms remain unknown. In 1979 Beeman et al (ref. 65) found that the cheese mite has strong monoamine oxidase (MAO) activity and only a weak N-acetyl transferase (NAT) activity, the latter being the main amine degradation enzyme in insects. Recently Scott and Knowles (ref. 66) have found in the bulb mite Rhizoglyphus edinopus a strong MAO using 2-phenylethylamine as a substrate. This enzyme did not metabolize octopamine, but was shown to be highly sensitive to chlordimeform and its metabolites (ref. 67) in vitro. The in vivo data indicate that after 24 hrs. the level of DOPA increases to 331% of the control value in females treated with chlordimeform. Knowing that chlordimeform causes many behavioral changes such as spin-down, walk-off, inhibition of fecundity and egg hatch (e.g. in two-spotted spider mites (ref. 68)) and tick detachment and that biogenic amines play important roles in controlling behavior of various animals, one is inclined to conclude that there is a likely causal relationship between the changes in amine levels and behavioral alteration caused by formamidines. Naturally the meaning of MAO inhibition by formamidines and expected rise in various amine levels must be clarified in the future.

In other areas of mode of action studies, e.g., on chitin synthesis inhibitors, progress has been slow. The major problem has been that there is a notable discrepancy between the results obtained in vivo and in vitro with regard to their actions on chitin synthetase. (See review ref. 69). The major end result of their action in vivo is undoubtedly inhibition of incorporation of N-acetylglucosamine to chitin (i.e. chitin synthesis). However, in the absence of any significant effects of these chemical on the same process in vitro and in the light of the scarcity of fundamental knowledge on chitin synthesis in insect cuticles, this avenue of research has been a frustrating experience to many scientists. One recent finding (70) of great interest is that diflubenzuron, a typical chitin synthesis inhibitor, appears to inhibit transport of UDP-N-acetylglucosamine across the cell membranes of the midgut of the cabbage armyworm. This appears to be a promising area of research as such transporting processes are crucial only in vivo or in intact integuments and not in vitro, providing a reasonable explanation of the above discrepancy. More data would be helpful in elucidating the nature of such transporting systems and their interactions with these chemicals.

In summary, in the last 5 years significant progress has been made in several areas of insecticide toxicology. The most notable gain was made in our understanding of the insect GABA systems as one of the major targets of insecticidal chemicals such as cyclodienes, bicyclophosphates, avermectins and milbemycins. As the nature of this important target site is gradually elucidated a number of new chemicals are being synthesized to affect various components within the GABA system. Significant progress was also recorded in the area of formamidine actions on biogenic amine related systems in insects. Knowledge on the mechanism of action of formamidines has advanced to such an extent that they are now employed as a useful tool in studying the roles of biogenic amines in insect physiology and behavior. In the area of pyrethroids, the importance of their action on the sodium channel has been confirmed. In addition to the past knowledge on their effect on the sodium inactivation mechanism, electrophysiologists have established the importance of the use of "tail current" analysis in uncovering their hidden effects on gate operations.

These results, added to existing knowledge on cholinergic systems, metabolic processes involving xenobiotics, and c-AMP mediated processes, should be regarded as solid contribution to basic science. Toxicologists should be proud of their achievements.

ACKNOWLEDGEMENTS

Supported by Michigan Agricultural Experiment Station and research grant R01-ES01963 from the National Institute of Environmental Health Sciences, Research Triangle Park, North Carolina.

REFERENCES

1. D.L. Shankland and M.E. Schroeder, Pestic. Biochem. Physiol. 3, 77-86 (1973).
2. M. Uchida, T. Fujita, N. Kurihara and M. Nakajima, In Pesticide and Venom Neurotoxicity, D.L. Shankland, R.M. Hollingworth and T. Smyth, Jr., Eds. Plenum Press, New York, pp. 133-151 (1976).
3. S.M. Ghiasuddin and F. Matsumura, Comp. Biochem. Physiol. 73C, 141-144 (1982).
4. F. Matsumura and S.M. Ghiasuddin, J. Environ. Sci. Health B18, 1-14 (1983).
5. A.A. Kadous, S.M. Ghiasuddin, F. Matsumura, J.G. Scott and K. Tanaka, Pestic. Biochem. Physiol. 19, 157-166 (1983).
6. F. Matsumura and K. Tanaka, In Cellular and Molecular Neurotoxicology, T. Narahashi, Ed. Raven Press, New York pp. 225-240.
7. Y. Ozoe and F. Matsumura, J. Agr. Food Chem. 34, 126-134 (1986).
8. F. Matsumura, Neurotoxicology 6, 139-164 (1985).
9. K. Tanaka and F. Matsumura, In Membrane Receptors and Enzymes as Targets of Insecticidal Action, J.M. Clark and F. Matsumura, Eds. Plenum Press, New York pp. 33-50 (1986).
10. F. Leeb-Lundberg and R.W. Olsen, In Psychopharmacology and Biochemistry of Neurotransmitter Receptors, H.I. Yamamura, R.W. Olsen and E. Usdin, Eds., Elsevier, New York, pp. 593-606 (1980).
11. Y. Ozoe and M. Eto, Agr. Biol. Chem. 46, 411-418.
12. Y. Ozoe and M. Eto, In Membrane Receptors and Enzymes as Targets of Insecticidal Action, J.M. Clark and F. Matsumura, Eds. Plenum Press, New York pp. 75-105 (1986).
13. R.F. Squires, J.E. Casida, M. Richardson and E. Saederup, Mol. Pharmacol. 23, 326-336 (1983).
14. L.J. Lawrence and J.E. Casida, Science 221, 1399-1401 (1983).
15. L.J. Lawrence and J.E. Casida, Life Sci. 35, 171-178 (1984).
16. I.M. Abalis, M.E. Eldefrawi and A.T. Eldefrawi, Pestic. Biochem. Physiol. 24, 95-102 (1985).
17. J.R. Bloomquist, P.M. Adams and D.M. Soderlund, Inhibition of γ-amino-butyric acid-stimulated chloride flux in mouse brain vesicles by polychlorocycloalkane and pyrethroid insecticides, Neurotox. 7, In press (1986).
18. L.C. Fritz, C.C. Wang and A. Gorio, Proc. Natl. Acad. Sci. USA 76, 2062-2066 (1979).
19. I.S. Kass, D.A. Larsen, C.C. Wang and A.O.W. Stretton, Exp. Parasitol. 54, 166-174 (1982).
20. S.S. Pong and C.C. Wang, Neuropharmacol. 19, 311-317 (1980).
21. S.S. Pong, C.C. Wang and L.C. Fritz, J. Neurochem. 34, 351-358 (1980).
22. C.C. Wang, In Membrane Receptors and Enzymes as Targets of Insecticidal Action, J.M. Clark and F. Matsumura, Eds. Plenum Press, New York pp. 125-138 (1986).
23. K. Tanaka and F. Matsumura, Pestic. Biochem. Physiol. 24, 124-135 (1985).
24. D.B. Sattelle, GABA receptors of insect CNS. Abstract No. 10, Agrochem. Div. Am. Chem. Soc. Meet., New York City, April 13-18 ACS Symposium Series, R.M. Hollingworth and M.B. Green, Eds. (1986).
25. J.R. Bloomquist and D.M. Soderlund, Neurotoxic insecticides as inhibitors of GABA-dependent chloride uptake by mouse brain vesicles. Abstract No. 19, Agrochemical Division, Am. Chem. Soc. Annual Meeting, New York City, April 13-18 ACS Symposium Series, R.M. Hollingworth and M.B. Green, Eds. (1986).
26. I.M. Abalis, M.E. Eldefrawi and A.T. Eldefrawi, J. Toxcol. Environ. Health 18, In press.
27. D.W. Gammon, M.A. Brown and J.E. Casida, Pestic. Biochem. Physiol. 15, 181-191 (1981).
28. J.G. Scott and F. Matsumura, Pestic. Biochem. Physiol. 19, 141-150 (1982).
29. C.G. Staatz, A.S. Bloom and J.J. Lech, Pestic. Biochem. Physiol. 17, 287-292 (1982).
30. H.P.M. Vijverberg and J. Van den Bercken, Europ. J. Pharmacol. 58, 501-504 (1979).
31. H.P.M. Vijverberg and J.R. DeWille, Neurotoxicology 6, 23-34 (1985).
32. T. Narahashi, Neurotoxicology 6, 3-22 (1985).

33. A.E. Lund and T. Narahashi, Pestic. Biochem. Physiol. 20, 203-216 (1983).
34. V.L. Salgado, S.N. Irving and T.A. Miller, Pestic. Biochem. Physiol. 20, 100-114 (1983).
35. J.E. Cremer, In Developments in the Science and Practice of Toxicology. A.W. Hayes, R.C. Schnell and T.S. Miya, Eds., Elsevier Sci. Publishers, New York pp. 61-72 (1983).
36. R.A. Nicholson, R.G. Wilson, C. Potter and H. Black, In Pesticide Chemistry 3, Mode of action, metabolism, toxicology, J. Miyamoto and P.C. Kearney, Eds., Pergamon Press, New York.
37. J.R. Berlin, T. Akera, T.M. Brody and F. Matsumura, Euro. J. Pharmacol. 98, 313-322 (1984).
38. F. Matsumura, Deltamethrin induced changes in synaptosomal transport of ^{3}H-epinephrine in the squid optic lobes. Comp. Biochem. Physiol. Part C, in press (1986).
39. J.M. Clark, Personal Communication (1986).
40. J. Clark, In Membrane Receptors and Enzymes as Targets of Insecticidal Action. J.M. Clark and F. Matsumura, Eds., Plenum Press, New York, pp. 189-212 (1986).
41. F. Matsumura, In Membrane Receptors and Enzymes as Targets of Insecticidal Action, J.M. Clark and F. Matsumura, Eds., Plenum Press, New York, pp. 173-188 (1986).
42. M.E. Eldefrawi, S.M. Sherby, I.M. Abalis and A.T. Eldefrawi, Neurotoxicology 6, 47-62 (1985).
43. S.M. Ghiasuddin, A.A. Kadous and F. Matsumura, Comp. Biochem. Physiol. 68C, 15-20 (1981).
44. S. Rashatwar and F. Matsumura, Comp. Biochem. Physiol. 81(c), 97-103 (1985).
45. F. Matsumura, In Membrane Receptors and Enzymes as Targets of Insecticidal Action, J.M. Clark and F. Matsumura, Eds., Plenum Press, New York pp. 173-187 (1986).
46. D.M. Soderlund, Neurotoxicology 6, 35-46 (1985).
47. D.W. Gammon and G. Sander, Neurotoxicology 6, 63-86 (1985).
48. L.J. Lawrence, K.W. Gee and H.I. Yamamura, Neurotoxicology 6, 87-98 (1985).
49. R.J. DeLorenzo, S. Burdette and J. Holderness. Science 213, 49-50 (1981).
50. F. Matsumura and R.W. Beeman, Environ. Health Perspec. 14, 71-82 (1976).
51. R.W. Beeman and F. Matsumura, In Pesticide and Venom Neurotoxicity, D.L. Shankland, R.M. Hollingworth and T. Smith, Eds., Plenum Press, New York, pp. 179-188 (1978).
52. R.M. Hollingworth and A.E. Lund, In Insecticidal Mode of Action, J.R. Coats Ed. Academic Press, New York pp. 189-227 (1982).
53. R.M. Hollingworth, E.M. Johnstone and N. Wright, In Pesticide Synthesis Through Rational Approaches, P.S. Magee, G.K. Kohn and J.J. Menn, Eds., American Chemical Society Symposium Series No. 225, Washington, D.C., pp. 103-125.
54. A.E. Lund, D.L. Shankland, C. Chinn and G.K.W. Yim, Toxicol. Appl. Pharmacol. 44, 357-365 (1978).
55. B.A. Bailey, R.J. Martin and R.G.H. Donner, Pestic. Biochem. Physiol. 17, 293-301 (1982).
56. R.M. Hollingworth and L.L. Murdock, Science 208, 74-76 (1980).
57. J.A. Nathanson and E.J. Hunnicutt, Mol. Pharmacol. 20, 68-75 (1981).
58. A.E. Lund, The Mode of Action of Formamidine Pesticides in Mammals and Insects, Ph.D Thesis, Purdue University (1978).
59. P.D. Evans, Nature 287, 60-62 (1980).
60. R.W. Beeman and F. Matsumura, Comp. Biochem. Physiol. 73C, 145-148 (1982).
61. R.M. Hollingworth, H. Hashemzadeh and G. Gad El-Hak, Abstract No. 27, Agrochem. Div., Am. Chem. Soc. Meeting, New York City, April 13-18 ACS Symposium Series, R.M. Hollingworth and M.B. Green, Eds. (1986).
62. H.J. Benezet, K.M. Chang and C.O. Knowles, In Pesticide and Venom Neurotoxicity, D.L. Shankland, R.M. Hollingworth and T. Smyth, Eds., Plenum Press, New York pp. 189-206.
63. C.E. Linn and W.L. Roelofs, Abstract No. 30, Agrochem. Am. Chem. Soc. Meet., New York City, April 16-18 ACS Symposium Series, R.M. Hollingworth and M.B. Green, Eds. (1986).
64. I. Orchard, Abstract No. 26, Agrochem. Div., Am. Chem. Soc. Meet., New York City, April 16-18 ACS Symposium Series, R.M. Hollingworth and M.B. Green, Eds. (1986).
65. R.W. Beeman, F. Matsumura and T. Kikukawa, Comp. Biochem. Physiol. 64C, 149-151 (1979).
66. J.A. Scott and C.O. Knowles, Exp. Appl. Acrol. 1, 227-233 (1985).
67. J.A. Scott, T.L. Johnson and C.O. Knowles, Pestic. Sci. 16, 504-510 (1985).
68. E.J. Franklin and C.O. Knowles, J. Econ. Entomol. 77, 318-323 (1984).
69. E. Cohen and J.E. Casida, In Pesticide Chemistry, J. Miyamoto and P.C. Kearney, Eds. Pergamon Press, Oxford, Vol. 3, pp. 25-32 (1983).
70. T. Mitsui, M. Tada, C. Nobusawa and I. Yamaguchi, J. Pestic. Sci. 10, 55-60 (1985).

Advances in understanding herbicide modes of action

Carl Fedtke* and Achim Trebst

Bayer AG, GB Pflanzenschutz, Biologische Forschung, 5090 Leverkusen, and Biochemie der Pflanze, RUB, 4630 Bochum, West Germany

Abstract - Herbicides take advantage of the peculiarities of plant metabolism in that most structures are directed towards metabolic pathways which are specific for the plant. Many of these specific pathways, e.g. essential amino acid biosynthesis, photosynthesis, plant hormone action, and the regulation of cell division provide important herbicidal targets. This paper describes details for the newly developed herbicidal target acetohydroxyacid synthase as an example of a site of action for potent inhibitors of essential amino acid biosynthesis. We then present an update of information on the photosystem II inhibitor binding niche, integrating sequence data, hydropathy analysis, amino acid changes in resistant mutants, binding studies, X-ray analysis of the bacterial photosynthetic reaction center, and theoretical considerations. Finally we propose an integrated hypothesis on the molecular action of herbicide antidotes which is based on their interference with monooxygenase enzymes.

INTRODUCTION

The herbicide mode of action field has seen important advances in recent years. Not only have new herbicides added new and important herbicidal targets, but also knowledge of the modes of action for established herbicides has been considerably advanced. New herbicides, with targets in the pathways of amino acid syntheses (acetohydroxyacid synthase and glutamine synthetase), and new graminicides, have been a developmental focus. The modes of action of several graminicides are beginning to emerge together with the modes of action of their antidotes. Very important progress has also been achieved in the understanding of the binding site for photosystem II inhibitors. We have chosen to present details on recent advances for the targets acetohydroxyacid synthase and photosystem II. We also present a new hypothesis for the molecular action of antidotes and some herbicides which are affected by antidotes.

BRANCHED CHAIN AMINO ACID SYNTHESIS

Among those herbicides which inhibit the synthesis of different amino acids, the sulfonylurea and imidazolinone compounds constitute a very important addition (Table 1). The sensitive enzyme acetohydroxyacid synthase (AHAS, EC 4.1.3.18.), which initiates the synthesis pathways for the branched chain amino acids valine, leucine, and isoleucine, is particularly sensitive to the sulfonylurea structures, but is also sensitive to feedback inhibition by e.g. L-valine and the L-valine derivative N-phthalyl-L-valine anilide (ref. 4). In the imidazolinone series, the conc. required for 50 % inhibition of the isolated enzyme is 100 to 1000 times higher than that required in cell cultures or the intact plant. However, the inhibition *in vivo* by both sulfonylureas and imidazolinones can be completely reversed by the addition of valine, leucine, and isoleucine (ref. 2,3,5,6). One possible explanation for this discrepancy would be an inhibition at regulatory sites, which have been partially damaged or lost during enzyme isolation.

The enzyme inhibition is of an unusually slow binding type (Table 2) and exhibits biphasic kinetics. The slow tight binding rate is dependent on the presence of pyruvate (ref. 7,8). The inhibition is both reversible and uncompetitive, i.e. it is not antagonized by the participants of the enzymatic reaction pyruvate, TPP, and FAD (ref. 3,7).These characteristics would be in agreement with an allosteric inhibition type, as has already been suggested above for the imidazolinone discrepancy. Much of the enzyme work has been done with the bacterial acetolactate synthase (ALS) enzymes I, II, and III from Salmonella typhimurium (ALS I and II) and Escherichia coli (ALS I and III) (ref. 7,8). Of these enzymes, ALS I, which preferentially uses pyruvate, is completely insensitive. However, only one enzyme exists in plants which efficiently catalyses the conversion of both, pyruvate and α-ketobutyrate, and which is under intricate allosteric control.

TABLE 1. Inhibitors of acetohydroxyacid synthase (ref. 1-4).

Group / Compound		I_{50} (μM)
Sulfonylurea compounds		0.01 - 0.1
Chlorsulfuron (ref. 1,2)	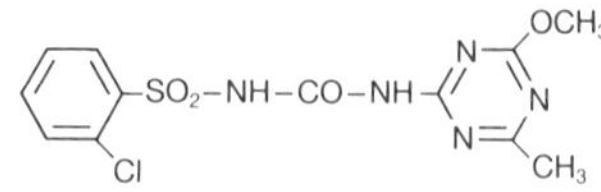	0.02
Imidazolinones		1.0 - 10
Imazaquin (ref. 3)	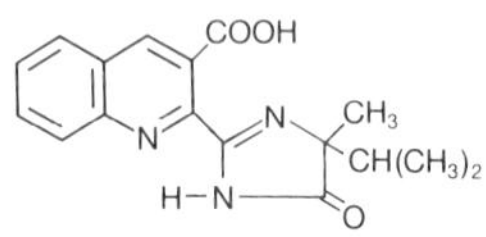	0.8
Other (ref. 4)		-
N-Phthalyl-L-valine anilide	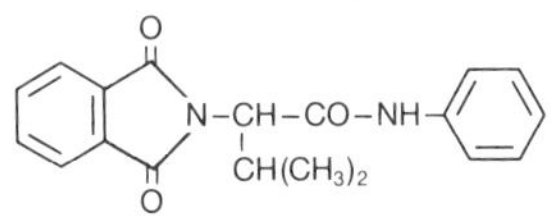	2.3
L-Valine NH_2-CH-COOH / $CH(CH_3)_2$		20,000

TABLE 2. Inhibition and binding characteristics of sulfonylureas and imidazolinones affecting acetohydroxyacid synthase (AHAS).

1. Slow binding type	: 30 - 90 minutes for full inhibition (I_{50} drop from 660 to 65 nM in 90 min.)
2. Tight binding rate	: higher in the presence of pyruvate by a factor of 10
3. Reversible	: slow release during gel filtration
4. Uncompetitive	: allosteric ?
5. Discrepancy in vivo / in vitro	: imidazolinones (I_{50} 3 - 10 nM in vivo / 1-10 μM in vitro)
6. Loss by mutation	: unspecific for the herbicidal group (cross resistance phenomena)
7. Insensitive enzymes	: bacterial acetolactate synthase I

The existance of an insensitive (bacterial) ALS I, together with the other information presented in Table 2, could indicate that the inhibitory action is directed towards allosteric controls, but not towards the catalytic center. Meanwhile we also know a great number of insensitive/resistant mutants which have an altered ALS/AHAS. Table 3 gives an example which has been obtained by selection with an imidazolinone. Very importantly, the changed enzyme obtained was cross resistant with sulfonylureas. In the imidazolinone series, the remote analogy with L-valine and the L-valine derivative (Table 1) could suggest an inhibition at regulatory sites; the cross resistance effects observed with the sulfonylureas again suggest, but do not prove, that all the inhibitors shown in Table 1 interfere with the regulatory capabilities of the enzyme. The very diverse inhibitor structures would then require a rather pleiotropic binding site, a phenomenon which is not uncommon among other herbicidal groups.

The most sensitive physiological processes after chlorsulfuron application in vivo are the cell cycle transitions $G_2 \longrightarrow M$ and secondarily $G_1 \longrightarrow S$ (ref. 9,10). A very sensitive biochemical process in vivo for both herbicidal groups is DNA-synthesis. Interestingly, at the same time, total protein synthesis is unaffected. The inhibition of cell cycle specific RNA-synthesis has been suggested as the primary in vivo consequence of AHAS inhibition (ref. 10).

TABLE 3. Inhibition of acetohydroxyacid synthase from susceptible (WT) and resistant (XA 17) maize (ref. 5).

Herbicidal group	I_{50}		Resistance
	WT	XA 17	factor
Sulfonylureas	0.01 μM	0.3 μM	30
Imidazolinones	1 - 10 μM	3 - 30 mM	3000

PHOTOSYSTEM II HERBICIDES

A large number of herbicides interfere with the acceptor side of photosystem II. In addition to the well known herbicides with a urea, triazine, triazinone, anilide, uracil, or similar structure, a second class of inhibitors of PS II has emerged. To these belong the known phenol compounds (dinoseb, ioxynil) and hydroxyquinoline-N-oxide, but also the recently characterized groups of benzo- and naphthoquinones, of pyrones, chromones, pyridones, and cyanoacrylates. This has often (ref. 11,12) and recently (ref. 13) been summarized. Some chemical structures from the second group are given in Fig. 1.

The comparison of chemical structure and biological activity (here: the inhibition of photosystem II) has already provided detailed knowledge on common essential elements in the different herbicide molecules (ref. 14) and on the dimensions of the binding niche (ref. 15). QSAR showed the size and position of substituents, electronic and steric parameters, and of the charge distribution at essential atoms involved in binding (ref. 16). The charge distributions in the two basically different groups of PS II herbicides, represented by diuron/triazine on one side and phenols, quinones, pyridones, and pyrones on the other side, were respectively positive or negative in the α-position to the carbonyl, suggesting a different compensation charge for this position in the target (ref. 17).

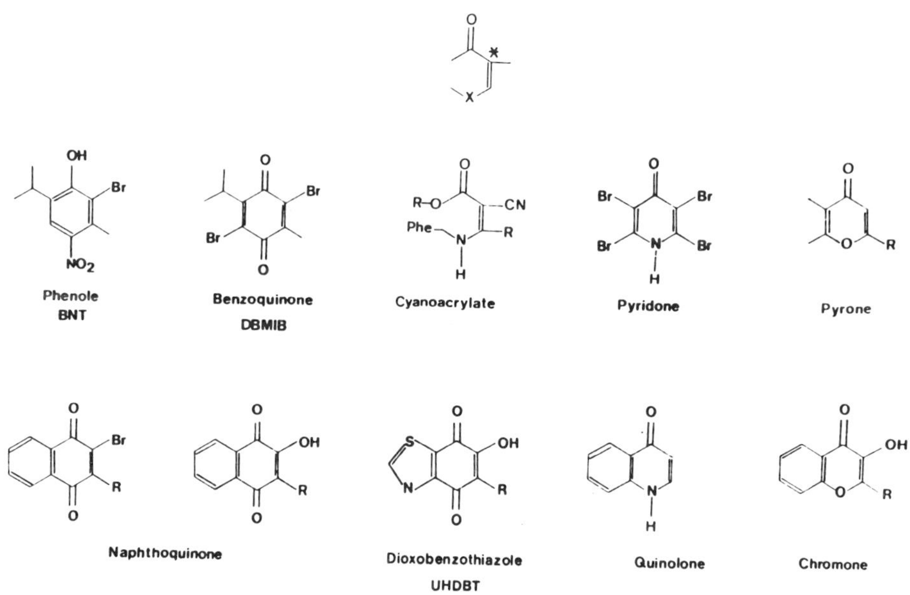

Fig. 1. Basic chemical structures of inhibitors of the second type which effectively block photosynthetic electron transport at photosystem II.

The identification and detailed description of the precise target site has now become possible, because several different and indepedent lines of research (I to VI) have eventually merged (Fig. 2 and 3). (I) The identification of the target site at PS II as a 32 kDa polypeptide subunit was possible by photoaffinity labeling with radioactive herbicides (ref. 18). The photoaffinity label also narrowed down the partial amino acid sequence of the polypeptide which is involved in herbicide binding (ref. 19). (II) Further specific amino acids participating in the binding of herbicides were provided by DNA sequencing of the genes for the 32 kDa polypeptide in herbicide resistant plants (ref. 20-23). In this way, a serine, phenylalanine, valine, and leucine were identified in the binding polypeptide. (III) A particular arginine was postulated to be part of the binding on theoretical grounds (ref. 24). (IV) The mode of action of these herbicides is a displacement of the natural plastoquinone electron acceptor of PS II (ref. 25); information on quinone binding by histidines complexing an iron atom provided by EXAF (ref. 26) is therefore also relevant to herbicide binding. (V) Hydropathy index analysis of the amino acid sequence of the 32 kDa polypeptide suggested the folding of this protein through the membrane in five hydrophobic α-helices (ref. 27) (Fig. 2). (VI) The homology of the 32 kDa subunit of PS II to subunits of the bacterial photosynthetic reaction center was used in support of the folding prediction. The X-ray structure of a crystallized bacterial photosystem finally provided detailed dimensions on amino acids in the binding niche of the quinone (ref. 28,29).

From these data a detailed surrounding of the quinone that is competitively displaced by the herbicide can be constructed (Fig. 3). It indicates hydrogen bridges of the carbonyl groups to particular amino acids (his 215, ser 264), interaction with phenyl (phe 255) and arginine (arg 269) residues below and above the herbicide, and steric constrains by certain hydrophobic side chains (met 215, val 219, leu 275). Although still a model at this stage, is does illustrate already at the angström level the dimensions and chemical reactivities of herbicides in the binding niche, and how these are oriented in the target protein.

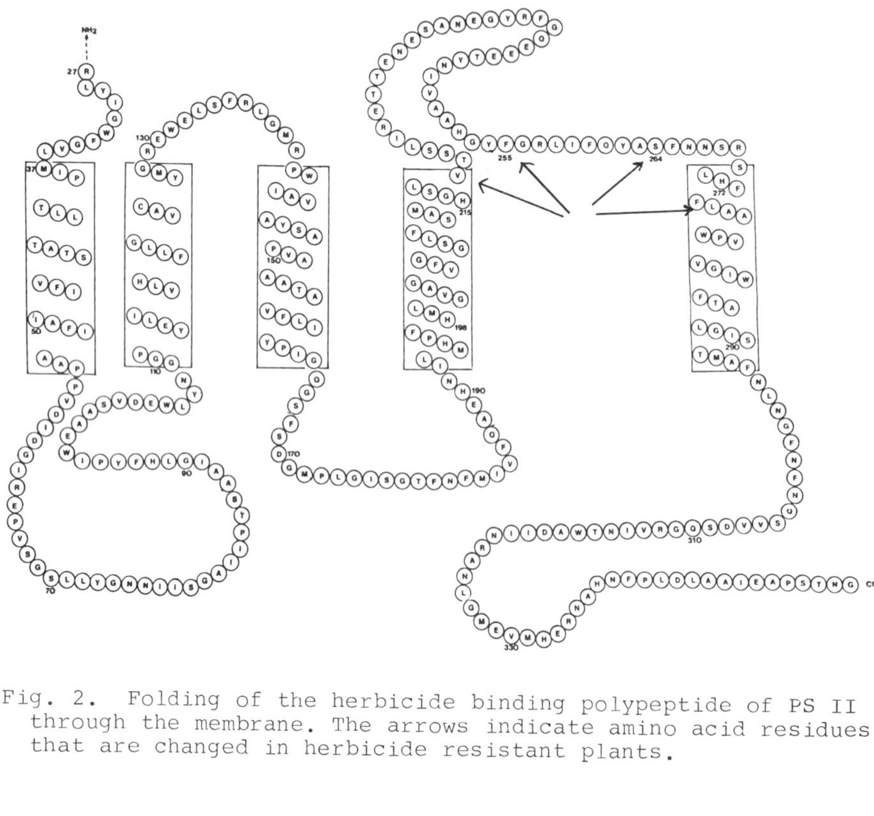

Fig. 2. Folding of the herbicide binding polypeptide of PS II through the membrane. The arrows indicate amino acid residues that are changed in herbicide resistant plants.

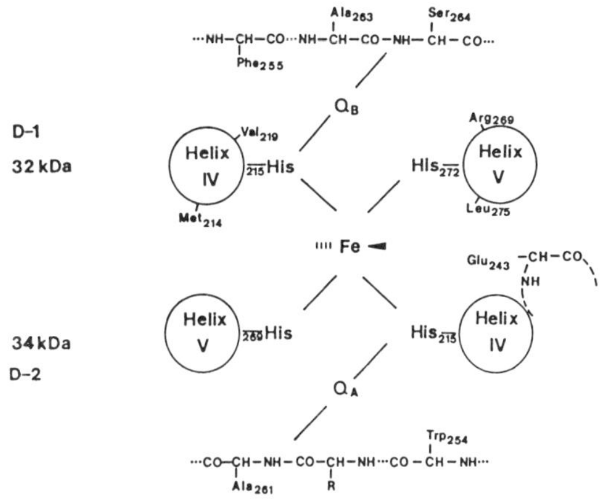

Fig. 3. Orientation of the binding of the plastoquinones A and B (Q_A and Q_B) in two subunits of photosystem II. D-1 is the 32 kDa herbicide binding protein.

ANTIDOTES AND GRAMINICIDES

Very recent reports have shown an involvement of monooxygenase enzymes (formely called mixed function oxygenases) in herbicide antidote action: several different antidotes have been found to increase the activities of monooxygenases (ref. 30); some antidotes and chloroacetamides have been found to act as insecticide synergists, similarly to piperonyl butoxide (PBO), suggesting that they might also act as monooxygenase inhibitors (ref. 31). When comparing and integrating these new results with older information on the action of antidotes and of those herbicides which are most strongly safened by antidotes (chloroacetamides and thiolcarbamates), some interesting similarities can be observed.

TABLE 4. Tabulation of responses to herbicidal antidotes, respective herbicides, and monooxygenase inhibitors. + = positive response in at least one system; - = negative or no response; ? = no data.

Compound / Group	H.	A.	S.	M.	G.	T.	O.
Dichlormid	-	+	?	?	+	+	+
Naphthalic anhydride	-	+	?	?	-	+	+
Cyometrinil	-	+	+	?	-	+	+
Flurazole	-	+	+	?	+	+	?
Chloroacetamides	+	+	+	+	+	+	?
Thiolcarbamates	+	+	?	+	+	+	?
Piperonyl butoxide	-	+	+	+	+	?	?
1-Aminobenzotriazole	(+)	+	?	+	+	?	?

H. = herbicidal activity
A. = herbicidal antidoting
S. = insecticidal synergism
M. = monooxygenase inhibition in plants or insects
G. = glutathione increase in corn, oats, or soybean
T. = glutathione-S-transferase increase in corn
O. = monooxygenase increase in corn and wheat

A survey of those effects which are relevant for the present discussion is tabulated in Table 4 for the herbicide antidotes dichlormid, naphthalic anhydride (NA), cyometrinil, and flurazole, for the herbicidal structure groups chloroacetamides and thiolcarbamates, and for the monooxygenase inhibitors PBO and ABT (1-aminobenzotriazole). ABT, which acts as a suicide substrate for monooxygenase enzymes (ref. 32), has been shown to act as a herbicide synergist by the inhibition of degradation (ref. 33), similar to PBO in insects. It is very interesting to note that not only chloroacetamides and thiolcarbamates (ref. 34), which may act by inhibiting mfo enzymes (ref. 35,36), but also the monooxygenase inhibitors PBO and ABT, can act as herbicide antidotes in a specific test system. This test system is the rooting of oat cuttings in the light or in the dark, as presented in the Figs. 4 and 5, in Table 5, and in a recently submitted paper (ref. 36). In view of the described similarities among herbicides, herbicide antidotes, and the known monooxygenase inhibitors PBO and ABT, the hypothesis is put forward that all the compounds listed in Table 4 act by interfering with monooxygenase enzymes.

In order to further support this hypothesis, the responses of the glutathione and the monooxygenase system after application of the inhibitors are also listed in Table 4. Whereas an increase of the glutathione-S-transferase activity is a very general phenomenon, an increase of the glutathione (GSH) content does not occur with all the tested compounds (ref. 37). Also, the GSH increase has only been found in a few plant species, mainly in corn which is very responsive to antidote action. Interestingly, PBO and ABT both increase the GSH content in oats (Fig. 6), where they simultaneously safen the action of a number of herbicides (Table 5). However, the GSH increase is comparatively small and will not be sufficient for an explanation of the safening effect. The safening effect was also observed with the new rice herbicide mefenacet (ref. 36,39), which has a mode of action similar to the chloroacetamides. A most important finding is the monooxygenase increase observed after antidote application (ref. 30). In the reported case, the monooxygenase increase has been traced by the antidotal action towards sulfonylurea herbicides, which were more rapidly degraded by the presumably increased monooxygenase activities.

Fig. 4. Rooting of oat cuttings after 4 days in the dark (untreated).

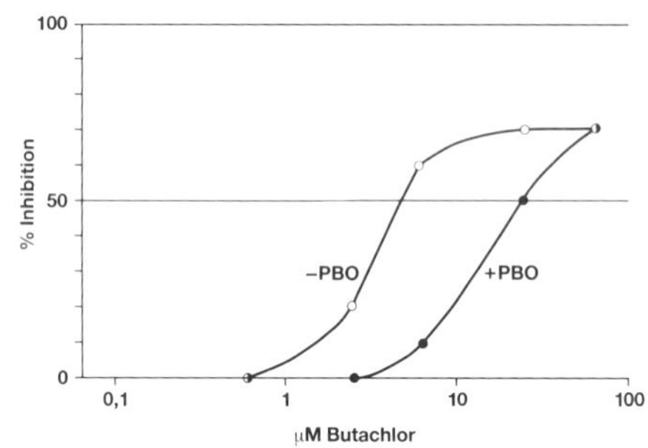

Fig. 5. Rooting of oat cuttings after 4 days in the dark as affected by butachlor in the absence (o——o) or presence (•——•) of 50 mg l^{-1} piperonyl butoxide.

TABLE 5. Reversal of rooting inhibition in oat cuttings by piperonyl butoxide (PBO) and 1-aminobenzotriazole (ABT) at 50 mg l^{-1}. Compare with Figs. 4 and 5.

Herbicide	I_{50} for rooting inhibition (μM)		
	Control	PBO	ABT
Alachlor	4.0	251	16
Butachlor	5.0	25	50
CDAA	0.4	3.4	1.3
Metolachlor	1.3	15.8	2.5
Propachlor	1.6	100	16
Diallate	15.8	200	16
Triallate	12.6	100	13
Mefenacet	0.1	1.0	0.3

The different aspects of the presented hypothesis are summarized in Fig. 7. The different inhibitors of monooxygenase enzymes considered may affect insecticide (PBO) or herbicide (ABT) ring hydroxylation, thereby synergizing insecticide or herbicide action. Other structures may inhibit hydroxylations in the syntheses of complex fatty acids and waxes and/or in the synthesis of complex ring systems, e.g. gibberellic acid (antidotes and herbicides). Apparently, the treated plant, if surviving, can react by increases of the glutathione system and of the monooxygenase system. Monooxygenase induction, following monooxygenase inhibition, is a well known and widespread phenomenon in animals and insects (ref. 40). Plants may respond similarly, thereby leading to increased breakdown and antidotal effects, as e.g. in the sulfonylurea group (ref. 30). However, there is little or no conclusive evidence for an increased herbicide breakdown in the chloroacetamide and thiolcarbamate groups (ref. 37). It is therefore suggested, that in these cases an increase - and possibly also a specificity shift - of monooxygenase herbicide target sites leads to higher herbicide tolerance after antidote application. A comparable situation exists in herbicide tolerant mutants where gene multiplication (e.g. glufosinate), or the introduction of a more effective promoter (e.g. glyphosate), have led to increased target enzyme expression.

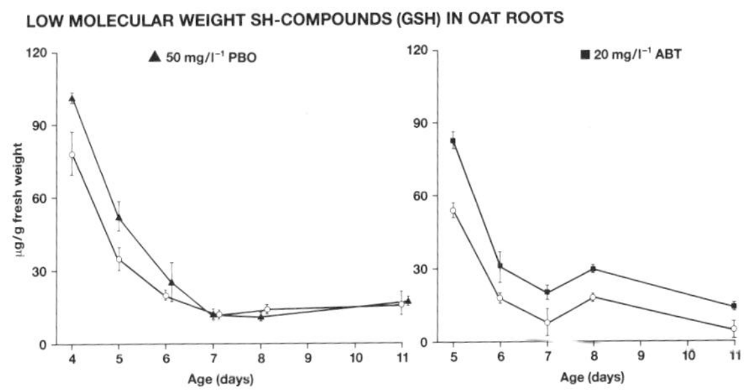

Fig. 6. Low molecular weight SH-compounds (largely GSH) in the roots of control and treated plants (ref. 38). Standard deviations are given by vertical bars (o—o control; ■—■ ABT at 20; ▲—▲ PBO at 50 mg l^{-1}.

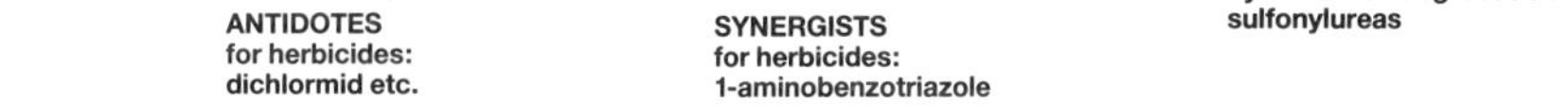

Fig. 7. Suggested inhibitions and responses of monooxygenase enzymes after application of different classes of inhibitors.

It has been suggested that chemical structures which mimic the herbicide molecule without having herbicidal activity would qualify for an antidote for the mimicked herbicide (ref. 41). In view of the presented hypothesis and of the evidence that herbicidal thiolcarbamates (ref. 34) and chloroacetamides (ref. 41) can also be antidotes, this suggestion can be detailed as follows: herbicidal or similar structures can be changed to become herbicidal antidotes, if (a) the obtained structure interferes with monooxygenase enzymes and (b) the safened herbicide is either degraded by a monooxygenase or affects a monooxygenase enzyme.

Although Table 4 lists a number of supporting results for this hypothesis, many of these come from very different systems. Sofar the suggested series of effects and responses has not been demonstrated in the same system, a requirement which will undoubtedly have to be observed in future experimentation.

REFERENCES

1. R.S. Chaleff and C.J. Mauvais, Science 224, 1443-1445 (1984).
2. T.B. Ray, Plant Physiol. 75, 827-831 (1984).
3. D.L. Shaner, P.C. Anderson and M.A. Stidham, Plant Physiol. 76, 545-546 (1984).
4. J.L. Huppatz and J.E. Casida, Z.Naturforsch. 40c, 652-656 (1985).
5. P.C. Anderson and K.A. Hibberd, Eur. Pat. Applic. 0154204 (1985).
6. P.C. Anderson and K.A. Hibberd, Weed Sci. 33, 479-483 (1985).
7. R.A. LaRossa and J.V. Schloss, J. Biol. Chem. 295, 8753-8757 (1984).
8. R.A. LaRossa and D.R. Smulski, J. Bacteriol. 160, 391-394 (1984).
9. T.L. Rost and T. Reynolds, Plant Physiol. 77, 481-482 (1985).
10. T.L. Rost, J. Plant Growth Reg. 3, 51-63 (1984).
11. C. Fedtke, Biochemistry and Physiology of Herbicide Action, p. 24-51, Springer, Berlin (1982).
12. J.R. Corbett K. Wraight and A.C. Baillie, The Biochemical Mode of Action of Pesticides, Second Edition, Acad. Press, London (1984).
13. A. Trebst and W. Draber, Photosynthesis Research, Butler Memorial Volume, Martinus Nijhoff Publisher, in press.
14. A. Trebst and W. Draber, Advances in Pesticide Science, Part 2 p. 223-234, Pergamon Press, Oxford (1979).
15. W. Draber and C. Fedtke, Advances in Pesticide Science, Part 3, p. 475-486, Pergamon Press, Oxford (1979).
16. A. Trebst, W. Draber and W.T. Donner, Human Welfare and the Environmment, p. 85-90, Pergamon Press, Oxford (1983).
17. A. Trebst, W. Donner and w. Draber, Z. Naturforsch. 39c, 405-411 (1984).
18. K. Pfister, K.E. Steinback, G. Gardner and C.J. Arntzen, Proc. Natl. Acad. Sci. USA 78, 981-985 (1981).
19. P.K. Wolber and K.E. Steinback, Z. Naturforsch. 39c, 425-429 (1984).
20. J. Hirschberg and L. McIntosh, Science 222, 1346-1348 (1983).
21. J.M. Erickson, M. Rahire, P. Bennoun, P. Delepelaire, P. Diner and J.D. Rochaix, Proc. Natl. Acad. Sci. USA 81, 3617-3621 (1984).
22. J.M. Erickson, M. Rahire, J.D. Rochaix and L. Mets, Science 228, 204-207 (1985).
23. J.M. Erickson and J.D. Rochaix, First Internatl. Congr. of Plant Molec. Biol., p. 54, the University of Georgia Center for Continuing Education for the Int. Soc. for Plant Mol. Biol., Athens (1985).
24. L.L. Shipman, J. Theor. Biol., 90, 124-148 (1981).
25. B.R. Velthuys, FEBS Lett. 126, 277-281 (1981).
26. G. Bunker, E.A. Stern, R.E. Blankenship and W.W. Parson, Biophys. J. 37, 539-551 (1982).
27. A. Trebst, Z. Naturforsch. 40c, 237-241 (1986).
28. J. Deisenhofer, O. Epp, K. Miki, R. Huber and H. Michel, J. Mol. Biol. 180, 385-398 (1984).
29. J. Deisenhofer, O. Epp, K. Miki, R. Huber and H. Michel, Nature 318, 618-624 (1985).
30. P.B. Sweetser, Brit. Crop Prot. Conf. (Weeds), Vol. 3, p. 1147-1154 (1985).
31. M.L. Ketchersid, F.W. Plapp and M.G. Merkle, Weed Sci. 33, 774-778 (1985).
32. D. Reichhart, A. Simon, F. Durst, J.M. Mathews and P.R. Ortiz de Montellano, Arch. Biochem. Biophys. 216, 552-529 (1982).
33. P. Gaillardon, F. Cabanne, R. Scalla and F. Durst, Weed Res. 25, 397-402 (1985).
34. T. Yuyama, P.B. Sweetser, R.C. Ackerson and S. Takada, Weed Res. Japan 31 Suppl., 101-102 (1986).
35. C. Fedtke, Z. Pflkrankh. Pflschutz. 92, 654-664 (1985).
36. C. Fedtke, Weed Res., submitted.
37. K.K. Hatzios, Adv. Agron. 36, 265-316 (1983).
38. C. Fedtke, Z. Pflkrankh. Pflschutz. Sonderheft IX, 141-146 (1981).
39. R.R. Schmidt, L. Eue, H. Förster, V. Mues and M. Aya, Med. Fac. Landbouww. Rijksuniv. Gent, 49/3b, 1075-1084 (1985).
40. E. Hodgson, Comprehensive Insect Physiology, Biochemistry, and Pharmacology, Vol. 11, p. 226-321, Pergamon Press, Oxford (1985).
41. G.R. Stephenson and G. Ezra, Plant Growth Regulating Chemicals, Vol. II, p. 193-212, CRC-Series, Boca Raton (1985).

Advances in understanding fungicidal modes of action and resistance

Leen C. Davidse

Laboratory of Phytopathology, Agricultural University,
6709 PD Wageningen, The Netherlands

Abstract. Over the past four years considerable progress has been made in understanding the mechanisms of action of the major systemic fungicides and resistance mechanisms in fungi. Vulnerable sites in metabolic pathways and biochemical processes in plant pathogenic fungi have been identified and are explored to find new chemical structures with even higher or broader activity. On the other hand these studies have contributed to our knowledge of the physiology and biochemistry of fungi. The selected examples discussed in this review illustrate how the research area evolved. Early mode of action studies identified targets of fungicides that were empirically found in conventional screening systems. At present the research focuses on the molecular interaction between chemical structures and target sites, aiming at optimization of the activity. In future the target sites would first be selected and characterized before interactive structures are biorationally designed.

INTRODUCTION

Elucidation of the mechanism of action of a fungicide is a long process and requires the dedicated efforts of biologists of various disciplines. The rationale behind starting a research project in this area often is not very obvious and may need some explanation. Getting to know how a compound acts is not a sufficient motive. As yet the mechanism of action of a number of fungicides is still poorly understood, but this does not seem to attract research interest. It clearly needs to be triggered. The rapid development of resistance to benzimidazoles, for instance, definitely provided the incentive for the earlier research on the mode of action of the benzimidazoles and mechanisms of resistance. Initially these compounds were thought to resolve a number of problems in crop protection but the development of resistant strains in target fungi was a serious drawback that obviously needed to be investigated.

One of the long range goals of this type of research is to provide insight in the principles underlying fungicide action and resistance which might lead to the development of better and newer fungicides. Although as yet there are no clear examples of commercial fungicides that have been developed in this way, it cannot be ignored that basic knowledge of the mechanism of action of the inhibitors of sterol biosynthesis already significantly contributes in evaluating candidate fungicides. The research area has gained momentum and will bear fruit in the years to come.

Another, no less important, goal is to increase our knowledge on physiological processes in fungi and their regulation. Specific inhibitors have proved to be important tools in biochemistry. In the long run increased physiological knowledge of plant pathogenic fungi will contribute to a better understanding of pathogenicity. It is difficult to foresee how research in this area will evolve. The perception, however, that mode of action studies ultimately will contribute to the development of better disease control measures either chemically or using the plants own defence mechanism, is highly motivating for continued research on the molecular aspects of fungicide action and resistance.

It is with these ideas in mind that this contribution is written. Advances in understanding modes of action and resistance for a number of fungicides will be discussed in the light of the long range research goals mentioned.

ADVANCES IN UNDERSTANDING MODES OF ACTION AND RESISTANCE

Benzimidazoles

Benzimidazoles act by binding to tubulin, the heterodimeric subunit of microtubules which together with intermediate filaments and actin filaments are the major components of the cytoskeleton, a unique feature of the eukaryotic cell. Binding results in inhibition of microtubule assembly and preexisting microtubules disappear because they are in dynamic equilibrium with tubulin. Since microtubules function in a great number of cellular processes such as mitosis and meiosis, intracellular transport of molecules, particles, and organelles, and in the maintenance of cell shape, inhibition of their assembly leads to a general disturbance of the cell's metabolism. Decreased binding affinity of tubulin to benzimidazoles is an important mechanism of resistance found in *Aspergillus nidulans* and is a proposed mechanism for a number of plant pathogenic fungi. Resistance in a number of *A. nidulans* strains is associated with structural changes of the tubulin molecule. Microtubule poisons play a fundamental role in research on microtubule structure and functioning. In mammalian cells colchicine, the Vinca alkaloids, podophyllotoxins, griseofulvin and recently also nocodazole and taxol are important tools with which to study microtubules and their involvement in cellular processes. Most of the traditional microtubule poisons, however, show only a slight activity, if any, in fungi. The benzimidazole fungicides, therefore, being effective antimicrotubule agents in fungi, have become increasingly important in research on fungal tubulin and microtubules. Mutations conferring benzimidazole resistance have been crucial in the identification of tubulin genes of the fungi *A. nidulans*, *Saccharomyces cerevisiae*, *Schizosaccharomyces pombe* and the slime mould *Physarum polycephalum*.

Much of our recent knowledge about the mechanism of action of the benzimidazoles and resistance mechanisms stems from research in fungal cell biology and molecular genetics. The impact of the benzimidazoles in these areas has recently been reviewed (ref. 1, 2).

The availability of resistant mutants has allowed the isolation and cloning of tubulin genes, which, when they originate from a resistant strain, can be used as a selectable marker in transformation studies (ref. 3). The latter technique, which is rapidly being developed for filamentous fungi, will become increasingly important in the study of gene functioning in fungi. Transforming elements containing part of the tubulin gene sequence have been used for site-directed mutagenesis of tubulin genes in *A. nidulans* and *S. cerevisiae* (ref. 4, 5). Two genes coding for the β-tubulin subunit of the tubulin-dimer and two genes coding for the α-tubulin subunit have been identified in *A. nidulans* (ref. 6). One of the β-tubulin genes functions during conidiation and the other one during vegetative growth. The latter one, however, is also able to function during conidiation when the former is inactivated by site-specific mutagenesis. Whether the α-tubulin genes also have specific functions during development is not known. Differential expression of tubulin genes during development has also been observed in *P. polycephalum* (refs 7, 8). In both *S. cerevisiae* and *S. pombe* one β-tubulin gene and two α-tubulin genes have been identified (refs 3, 9-11) indicating that the number of tubulin genes varies between species.

Benzimidazoles have been directly or indirectly important in these studies. A number of experiments could not have been performed if these compounds or benzimidazole resistant mutants had not been available. The progress made in the molecular genetics of tubulin in model organisms has now provided the probes with which to isolate tubulin genes from plant pathogenic fungi by standard recombinant DNA techniques. Sequencing and comparing the amino acid sequences as predicted from the nucleotide sequences of the isolated genes from benzimidazole sensitive and resistant strains would yield structural information of the benzimidazole binding site on tubulin. Knowledge of the possible changes at this site governing resistance might open the way for biorational design of chemical structures that are able to interact with these altered binding sites. Sequencing tubulin genes from fungi that are naturally resistant to benzimidazoles might even identify amino acid sequences that constitute potential binding sites for yet to be biorationally designed novel fungicides.

In addition to a change in binding affinity of tubulin to benzimidazoles, hyperstability of microtubules also confers resistance to benzimidazoles in *A. nidulans* (refs 12, 13). Hyperstability also causes resistance to the antimicrotubule agents griseofulvin and p-fluorophenylalanine which are structurally unrelated to benzimidazoles, and induces temperature sensitivity for growth. Mutations conferring hyperstability map to the benA gene, one of the two structural β-tubulin genes of this fungus. Other mutations in this gene confer resistance by a change in binding affinity to the benzimidazoles. Because the benzimidazoles release the temperature sensitivity of mutants having hyperstabile microtubules, they apparently still interact with the modified β-tubulin. Otherwise such an effect cannot be explained. Hyperstability can be counteracted by structural alterations in α-tubulin conferred by tubA mutations. In a wild type background the tubA mutations confer decreased stability of microtubules which is expressed as supersensitivity to benzimidazoles, griseofulvin and cold sensitivity. These studies indicate that mutations in the tubulin genes do not only confer benzimidazole resistance based on affinity changes but also confer resistance or supersensitivity to these compounds mediated through altered microtubule stability.

In *A. nidulans*, affinity changes of tubulin usually only confer benzimidazole resistance during vegetative growth. Conidiation is still sensitive to benzimidazoles (ref. 14). Mutants displaying conidiation resistance lack one of the β-tubulin species that is normally present in wild-type strains or strains that are benzimidazole resistant with respect to vegetative growth. Conidiation resistant mutants of *A. nidulans*, therefore, arise by a two-step mechanism, each step involving a distinct β-tubulin gene.

The results obtained with *A. nidulans* raise the question of which mechanism(s) of resistance operate in plant pathogenic fungi. Do mainly affinity changes cause resistance (refs 15, 16) or does hyperstability of microtubules also play a role? And does the two-step mechanism exist? Although it is speculative, altered stability of microtubules might explain the increased sensitivity of benzimidazole-resistant isolates of *Penicillium expansum, Botrytis cinerea, Gerlachia nivalis* and *Monilinia fructicola* to diphenylamine at a low temperature (refs 17, 18). Similarly negatively correlated cross-resistance of benzimidazole-resistant isolates of *B. cinerea, Cercospora beticola, Fusarium nivale, Mycosphaerella melonis, Pseudocercosporella herpotrichoides* and *Venturia nashicola* to N-phenylcarbamates (refs 19-23) might be attributed to altered microtubule stability rather than opposite affinity changes of tubulin to benzimidazoles and N-phenylcarbamates. In view of the potential practical importance of negatively correlated cross-resistance it would be of great value to investigate in more detail the increased sensitivity to diphenylamine and N-phenylcarbamates. The methods have been developed and await application in research on benzimidazole resistance of plant pathogenic fungi.

Inhibitors of sterol biosynthesis

The inhibitors of sterol biosynthesis or SBIs are at the moment undoubtedly the most intensively studied group of fungicides. Their broad spectrum activity, diverse chemical structures and relatively favourable prospects with respect to resistance development have created an extensive research area both in academia and industry. Many publications and review articles have appeared during the past four years. In this contribution only a few advances made in the area can be discussed. Specific data can be found in the papers presented at the Symposium on Ergosterol Biosynthesis Inhibitors (ref. 24) and reviews on the mechanism of action of SBIs (refs 25-27).

The biosynthesis of ergosterol, the major sterol in most fungi, involves a number of reactions and follows pathways that are more or less typical for a particular organism. Inhibition of the formation of ergosterol at a certain step in the pathway changes the sterol composition of the organism by the accumulation of normal sterol intermediates which are then deviated to side pathways to give abnormal sterols. As a result a complex sterol pattern appears which might be unique for a particular organism and can lead to misinterpretation of the site of inhibition.

Interaction with cytochrome-P450 leading to inhibition of the oxidative demethylation of the 14-methylgroups of the lanosterol level substrate has been generally accepted as the mechanism of action of a number of imidazoles, triazoles, pyrimidines, pyridines and piperazines, but uncertainty about the exact site of interference of the morpholines has been solved only recently (ref. 28). These compounds appeared to be potential inhibitors of the sterol Δ^{14}-reductase and the $\Delta^8 \rightarrow \Delta^7$-isomerases, but the sensitivity of each site to a particular compound of this group varies in different organisms. Tridemorph for instance inhibits the $\Delta^8 \rightarrow \Delta^7$-isomerase well in both *S. cerevisiae* and *Ustilago maydis* and inhibits the Δ^{14}-reductase only in yeast. On the other hand fenpropimorph and the related piperidine, fenpropidin, inhibit the Δ^{14}-reductase well in both fungi, the $\Delta^8 \rightarrow \Delta^7$-isomerase in *U. maydis* and probably in yeast. This example illustrates the extreme complexity of elucidation of the exact mechanism of action of an inhibitor of sterol biosynthesis and stresses the need for further development of in vitro assays in which the activity of a potential inhibitor against a particular step in the pathway can be determined (ref. 29, 30).

Elucidation of the specific details of the mode of action of the SBIs would not have been possible without the wealth of information obtained from fundamental research on sterol biosynthesis (ref. 31). Studies on inhibitors in its turn undoubtedly will continue to provide new details.

It is surprising that the present SBIs with agricultural importance fall into only two classes, namely sterol C-14 demethylase inhibitors and inhibitors of both Δ^{14}-reductase and/or $\Delta^8 \rightarrow \Delta^7$-isomerases. What is the reason for this? Is selective inhibition of the sterol biosynthetic pathway only possible at these sites? Is inhibition at other sites less discriminative with respect to target and non-target organisms? Or is it the result of activity optimization of a few lead structures (ref. 32) that happened to interact at these sites? The available data on the mode of action of experimental SBIs and SBIs in human and animal medicine suggest that other vulnerable sites in the biosynthetic pathway indeed do exist (refs 25-27, 33, 34). It is challenging to further explore these sites as potential new targets for agricultural fungicides. Availability of in vitro assays for specific enzymes of the pathway would be very desirable in this respect. Computer aided design of new structures based on knowledge of the interaction between the active center of the enzyme and the natural substrate, and quantitation of the interaction in binding studies will be another approach. These techniques have already been useful in the design of new inhibitors of the C-14 demethylation step (ref. 35-37). Basic knowledge on sterol biosynthesis in fungi will evolve further and may lead to even more sophisticated techniques in the future. Molecular genetic studies on mutants deficient in sterol biosynthesis in particular will be valuable in characterizing in more detail the various steps in the biosynthetic pathway (ref. 38).

The way SBIs actually stop growth of a sensitive fungus is still poorly understood and needs to be clarified. An altered sterol pattern of the membranes obviously has consequences for the metabolism of the cells. The irregular deposition of chitin in SBI-treated cells of a number of fungi (refs 39, 40) and changes of the intermembrane structure (ref. 41) are features of this and may lead to cell death. Toxicity of C-14 demethylation inhibitors is not reversed by adding exogenous ergosterol (ref. 26) which indicates that in addition to ergosterol depletion of membranes additional mechanisms leading to cell death may exist. A careful comparative study with various SBIs would reveal whether this phenomenon is encountered with all types of SBIs. If not, the existence of an additional mechanism of action with some of the SBIs should be considered.

SBIs are readily metabolized by both fungal and plant cells, which influences their performance in disease control. Considerable progress has been made in elucidation the various degradation patterns (refs 42-45). In some cases activation of the parent molecule is necessary for activity, but it cannot be generalized. The fate of each SBI is different for each plant and each fungus, and it would require considerable research efforts to get a complete picture of metabolism-activity relations for the entire SBI group. New SBIs will face this problem in terms of developing costs.

Resistance to SBIs, which can be readily produced with fungal strains in the laboratory, has surfaced now as a reduced sensitivity (particularly in powdery mildews) against a number of C-14 demethylation inhibitors (refs 46-50). The activity of the morpholine fenpropimorph against the strains concerned was not reduced. The mechanism of resistance of barley powdery mildew against triazoles is currently being investigated (ref. 51) but meets great technical problems caused by the biotrophic nature of the organism. This study has already revealed that conidia of this fungus did not contain C-14 methylated sterols even when the conidia had been collected from isolates growing on triadimenol treated plants. Apparently sterol biosynthesis in powdery mildews follows unique pathways. More studies like this are urgently needed to extend our knowledge on resistance mechanisms, which until now has been only investigated with mutants of various target and non-target organisms produced in the laboratory. The energy-dependent efflux mechanism initially described as a mechanism of resistance to fenarimol in *A. nidulans* (ref. 52), also appears to operate in resistant mutants of other fungi including target organisms and against other SBIs (ref. 53). The possibility of influencing the efflux mechanism by inhibitors of energy production may have important practical consequences and its value as a screening system for potential synergists of SBIs has been discussed (refs 54, 55).

In *Penicillium italicum*, which causes rot of citrus fruits, the level of resistance of mutants displaying the efflux mechanism did not lead to loss of disease control by imazalil, when fruits were treated with prescribed dosages (ref. 54). At lower dosages imazalil was clearly less effective against resistant strains than wild-type strains. Imazalil has been used for many years to control this disease and failures of disease control have not been reported. This suggests that in this fungus resistance based on the efflux mechanism cannot increase beyond a level that causes problems or that the fitness of such mutants is so severely reduced that they do not spread in the population. Until now resistance to SBIs has not led to such critical situations as have been encountered with benzimidazoles and phenylamides. The relatively low level of resistance, absence of cross resistance to inhibitors of C-14 demethylation and inhibitors of Δ^{14}-reductase and/or $\Delta^{8}\rightarrow\Delta^{7}$-isomerase, and perhaps reduced fitness of resistant strains have contributed to this relatively favourable picture. But will this continue to be so? Increasing usage of SBIs with a similar mode of action against more diseases in one crop might tip the balance. The decreasing effectiveness of SBIs against powdery mildew of cucurbits (refs 50, 57) after repeated applications is a well documented case and should be a warning.

Dicarboximides

Although a wealth of information is available on the effects of dicarboximides on cellular processes in fungi (ref. 58) as yet their primary mechanism of action has not been elucidated. The various effects observed include genetic effects such as non-disjunction, and effects on cell membrane function and cell wall synthesis which may indicate that in some way the cytoskeleton of the cell is affected. Comparative studies with a number of these compounds and cytochalasin A, a specific inhibitor of actin filament formation, led to the idea that this process may be the primary target (ref. 59). Although speculative the idea is worth further testing, which would not be too difficult in view of our rapidly expanding knowledge of the structure, biochemistry, physiology and molecular genetics of the cytoskeleton and its components.

Resistance to dicarboximides has led to disease control failures in a number of cases (ref. 60) and demands implementation of complicated disease control strategies, mainly because of the absence of other reliable and effective compounds. Resistance in field isolates attains an intermediate level as compared with the high level of resistance that can be induced in the laboratory. Resistance is usually accompanied with slow growth rates and less abundant sporulation. Many field isolates and mutants obtained in the laboratory additionally display a higher sensitivity to the osmotic value of the growth medium. Exceptions do occur, however, and isolates of the non-target organism, *Neurospora crassa*, have been obtained that combine high resistance with low osmotic sensitivity and excellent sporulation (ref. 61).

The fact that in *N. crassa* resistance usually is also combined with osmotic sensitivity opens perspectives for a molecular genetic approach in studying this phenomenon. Isolation and characterization of the osm-gene that is involved in both dicarboximide resistance and osmotic sensitivity may eventually lead to the discovery of the primary target site of this group of fungicides.

Phenylamides

In the past four years considerable progress has been made in elucidation of the mechanism of action of phenylamides. Early effects on nucleic acid synthesis could be related to a specific inhibition of ribosomal RNA synthesis (refs 62-69). In addition an effect on membrane function, expressed as the inhibition of $[^{3}H]$uridine uptake, has been found for benalaxyl and a number of related phenylamide herbicides at higher concentrations (refs 63, 70, 71). The contribution of the latter mechanism to disease control, however, is probably very limited, since benalaxyl did not perform better against phenylamide resistant strains than metalaxyl cyprofuram and oxadixyl, although it equally affected $[^{3}H]$uridine uptake in both resistant and sensitive strains (ref. 70). Studies on isolated nuclei of sensitive and resistant strains revealed that endogenous RNA polymerase activity was differentially inhibited by the phenylamides. Activity of nuclei of resistant strains was either completely insensitive (laboratory mutants of *Phytophthora megasperma* f. sp. *medicaginis*) or only slightly affected (field isolate of *Phytophthora infestans*) in comparison with that of sensitive strains. This supported the idea that resistance is based on a change at the target site. Binding studies, indeed, confirmed that the phenylamide binding site in resistant strains has a very low affinity for metalaxyl, whereas values of the dissociation constants as determined by Sephadex-G25 gel filtration of cell-free mycelial extracts after incubation with $[^{14}C]$- or $[^{3}H]$-metalaxyl were in the order of 0.1 - 0.2 μM for the phenylamide sensitive strains (ref. 72). It would be interesting to know which structural alteration in the phenylamide target site is responsible for the change in affinity. This information would be valuable in interpreting the interaction of the phenylamides with their target site. It may help in redesigning phenylamide structures or even in the design of completely new structures that are able to interact with the altered site.

To achieve this, laborious isolation and characterization of the binding protein will be necessary as a first step. It will be rewarding, however. Many details of the regulation of RNA synthesis are still largely unknown. Knowledge of the molecular interaction between the phenylamides and the target site undoubtedly will contribute to a better understanding of regulation of rRNA synthesis. Once the binding protein has been characterized it would be possible to isolate its structural gene, which will open an extensive research area. The availability of a gene, carrying a mutation governing phenylamide resistance, would provide the molecular geneticist with a selectable marker that could be incorporated in yet to be constructed transformation vectors of fungi belonging to the *Peronosporales*. In future research on host-pathogen relations the latter approach will be instrumental in the identification of fungal genes specifying pathogenicity and/or avirulence. Using the structural gene that specifies the phenylamide binding protein as a probe it would be possible to identify homologous genes in many other fungi. This will yield a picture of how rRNA synthesis evolved in fungi. Proteins homologous to the phenylamide-binding protein but which lack the binding property might be present in fungi that are naturally resistant to phenylamides. Characterizing these might lead to novel chemical structures that are able to interfere with rRNA synthesis in these fungi and that are candidate fungicides. Research along the lines described above is technical feasible now. Whether it will be realized depends on the interest of individual scientists in academia and industry to direct research funds into this area.

Carboxamides

The mechanism of action of carboxamides is relatively well understood and has been thoroughly reviewed recently (ref. 73). Being specific inhibitors of succinate oxidation at the level of the mitochondrial Complex II (succinate-ubiquinone reductase) they have been particular useful in basic research on this process. Characterization of the binding site in beef heart Complex II showed that several components of the complex are involved in carboxamide binding. Single components of the complex do not bind indicating that the site arises as a result of conformation changes induced in the proteins when the complex is formed. Within the complex the polypeptides C_{II-3} and C_{II-4} are principally involved in specific binding of the inhibitors.

It is generally assumed that in sensitive fungi inhibition of succinate oxidation is also mediated by binding at a similar site in the fungal mitochondrion, although data are still lacking. Basic research in the latter area is rather slow because experts on mitochondrial function favour mammalian systems as research objects. For the same reason the genetically and physiologically well-characterized carboxamide resistance in *A. nidulans* and *U. maydis* is still poorly understood at the molecular level. As yet it is not known whether mutations confering decreased sensitivity of the succinate dehydrogenase complex to carboxamides affect the structure of the enzyme molecule or the peptide components of complex II. Recent developments in the genetic manipulation of these fungi, however, may significantly alter the scene within the next few years. Efficient transformation systems will become available allowing the isolation and cloning of the genes that are involved in resistance. The characterization and sequencing of these genes and comparison with genes coding for the various components of the mammalian Complex II would reveal similarities and dissimilarities. Both basic research on mitochondrial functioning in fungi and applied research on carboxamide action and resistance, aimed at the design of novel structures, would benefit from such studies. Many data have already been gathered on structure-activity relationships between carboxamide fungicides and the succinate dehydrogenase complex from wild-type and resistant mutant strains of *A. nidulans* and *U. maydis* (refs 74, 75). These data together with those from studies on the detailed architecture of the binding site will show how the carboxamides fit into the binding site and identify the regions of the inhibitor molecule and the binding site which are involved in the interaction.

Inhibitors of melanin biosynthesis

Inhibitors of melanin biosynthesis (MBIs) which have proved to be effective for practical control of rice blast disease caused by *Pyricularia oryzae*, have attracted considerable research interest. A new mechanism of action, realization of the concept of disease control by interfering with the infection process, and the inability of the compounds involved to inhibit growth on agar media undoubtedly contributed to this popularity. Tricyclazole, one of the first MBIs developed, has been most intensively studied.

MBIs specifically block the capacity of appressoria to penetrate melanized plant epidermal walls by interference with melanization of the appressorial walls (ref. 76). Melanization is prevented by the specific inhibition of the conversion of 1,3,8-trihydroxynaphthalene to vermelone and that of 1,3,6,8-tetrahydroxynaphthalene to scytalone in the polyketide pathway. MBIs have diverse chemical structures. A basic structure required for activity has been proposed (ref. 77).

Basic studies on fungal melanization with melanin-deficient mutants of *Verticillium dahliae* and *P. oryzae* have been very helpful in elucidation of the mechanism of action of tricyclazole. These mutants phenotypically resemble the tricyclazole-treated wild type and are also genetically defective in the melanin biosynthetic pathway at the site blocked by tricyclazole in the wild type. In their turn, studies with tricyclazole helped to elucidate the melanin biosynthetic pathway. The biochemistry of the various reactions is still largely unknown (ref. 78). The enzyme activity inhibited by MBIs is NADPH dependent, but additional data are lacking.

The MBIs have a narrow disease control spectrum. Their practical use at present is limited to the control of rice blast disease, although inhibition of penetration is also observed with other fungi in experimental systems (ref. 79). It is known that both mechanical forces and enzymatic activity are involved in cuticular and cell wall penetration. Since mechanical forces require a rigid appressorial cell wall structure, the selectivity of MBIs against rice blast might be associated with a relatively important role of mechanical forces in the penetration process of *P. oryzae*. With other fungi, having melanized appressoria but the penetration of which is less sensitive to MBIs, enzymatic activity might be more important. The biosynthesis of melanin occurs via a number of steps, two of which involving reduction reactions are inhibited by known MBIs. The enzymatic dehydration reactions are not inhibited as has been shown in in vitro assays using cell homogenates (ref. 78). Screening of antimetabolites using this assay system would be an approach to determine whether the dehydration step can also be inhibited and may yield new lead structures for novel MBIs.

The availability of melanin deficient mutants also allows a molecular genetic approach for further exploration of the melanin biosynthetic pathway as a target for novel fungicides. Complementation of lesions in the pathway by transforming mutants with wild-type derived DNA sequences would identify structural genes for enzymes involved in the conversion of metabolites. Structural analysis of the enzymes would identify the binding site for the natural substrate, allowing a detailed study of the molecular interactions involved in binding of the substrate and potential inhibitors of the conversion reaction.

CONCLUDING REMARKS

This contribution does not pretend to be an extensive review on the mechanism of action of fungicides and resistance mechanisms in fungi. The selected examples serve to illustrate what has been achieved until now and in which direction future research will be going.

The modern specific fungicides owe their activity to high affinity binding to proteins involved in biosynthetic reactions, energy production, and structural organization of the cell.

Molecular interactions leading to binding depend on structural properties of the compound and the architecture of the binding site. The properties required for optimum activity are empirically determined by screening a large number of variants of lead compounds, without knowledge of the structure of the target site. Once the latter is known a more straightforward approach to designing new molecules is possible. New inhibitors of sterol C-14 demethylation are already designed along these lines.

The architecture of the binding site of benzimidazoles, phenylamides and carboxamides is not known yet but the technology has been developed to obtain this information. Resistance to these compounds is based on a change in the binding site and, as has been found in a number of mutants, is associated with an increase in sensitivity to other compounds. When these changes are known, biorational design of novel compounds that either bind to the altered site or perhaps to both the altered and unaltered site would be possible.

Most of the research thus far has been focused on the elucidation of the mechanism of action of existing antifungal compounds that have been found by traditional screening. Vulnerable sites in metabolic pathways and biochemical processes have been identified and techniques have been developed to study the molecular interactions between compounds and binding sites which underly the antifungal activity. The experience thus obtained gives a solid basis to further explore the possibilities of a more biorational approach in the search for new chemicals for plant disease control. The trends towards lower success rates in traditional approaches calls for action (ref. 80). Insight into the biochemistry of plant pathogenic fungi and bacteria and factors involved in pathogenicity will be rapidly increasing in the next decade due to the impact of modern molecular genetic techniques in these areas. A wealth of information will be gathered allowing the identification of new target sites for reliable disease control chemicals. The technology has been developed and awaits application!

REFERENCES

1. N.R. Morris, Exp. Mycology 10, 77-82 (1986).
2. L.C. Davidse, Ann. Rev. Phytopathol. 24, 43-65 (1986).
3. M.J. Orbach, E.B. Porro and C. Yanofsky, Mol. Cell Biol. 6,2452-2461 (1986).
4. N.F. Neff, J.H. Thomas, P. Grisafi and D. Botstein, Cell 33, 211-219 (1983).
5. G.S. May, J. Gambino, J.A. Weatherbee and N.R. Morris, J. Cell Biol. 101, 712-719 (1985).
6. J.A. Weatherbee and N.R. Morris, J. Biol. Chem. 259, 15452-15459 (1984).
7. T.G. Burland, T. Schedl, K. Gull and W.F. Dove, Genetics 108, 123-141 (1984).
8. T. Schedl, J. Owens, W.F. Dove and T.G. Burland, Genetics 108, 143-164 (1984).
9. P.J. Schatz and D. Botstein, J. Cell Biochem. Suppl. 9C, 111 (1985).
10. T. Toda, Y. Adachi, Y. Hiraoka and M. Yanagida, Cell 37, 233-242 (1984).
11. Y. Hiraoka, T. Toda and M. Yanagida, Cell 39, 349-358 (1984).
12. B.R. Oakley and N.R. Morris, Cell 24, 837-845 (1981).
13. J. Gambino, L.C. Bergen and N.R. Morris, J. Cell Biol. 99, 830-838 (1984).
14. J.A. Weatherbee, G.S. May, J. Gambino and N.R. Morris, J. Cell Biol. 101, 706-711 (1985).
15. L.C. Davidse and W. Flach, J. Cell Biol. 72, 174-193 (1977).
16. C. Gessler, D. Sozzi and H. Kern, Ber. Schweiz. Bot. Ges. 90, 45-54 (1980).
17. D.A. Rosenberger and F.W. Meyer, Phytopathology 75, 74-79 (1985).
18. M.S. Sharom and L.V. Edgington, Can. J. Plant Pathol. 7, 389-394 (1985).
19. P. Leroux and M. Gredt, BCPC Monograph 31, 443-446 (1985).
20. T. Kato, K. Suzuki, J. Tokahashi and K. Kamoshita, J. Pesticide Sci. 9, 489-495 (1984).
21. G.L. Bateman, C. Smith, M.F. Creighton, K.Y. Li and D.W. Hollomon, Trans Br. Myc. Soc. 85, 335-338 (1985).
22. N.Q.R. Nathaniels, K. Wilson and J.T. Fletcher, Ann. appl. Biol. 107, 151-154 (1985).
23. H. Ishii, Y. Yamase and J. Dekker, BCPC Monograph 31, 323-326 (1985).
24. Contributors to the Symposium on Ergosterol Biosynthesis Inhibitors, Pestic. Sci. 15, 40-96, 133-220, 268-323 (1984).
25. P. Leroux and P. Bienveniste, BCPC Monograph 31, 67-78 (1985).
26. H.D. Sisler and N.N. Ragsdale, in: Mode of action of antifungal agents (eds A.P.J. Trinci and J.F. Ryley), pp 258-282, Cambridge University Press, Cambridge (1984).
27. D. Berg, ACS Symposium Series 304, 25-51 (1986).
28. R.I. Baloch, E.I. Mercer, T.E. Wiggins and B.C. Baldwin, Phytochemistry 23, 2219-2226 (1984).
29. P. Gadher, E.I. Mercer, B.C. Waldwin and T.E. Wiggins, Pestic. Biochem. Phys. 19, 1-10 (1983).
30. L.G. Copping, R.J. Birchmore, K. Wright and D.H. Godson, Pestic. Sci. 15, 280-284 (1984).
31. E.I. Mercer, Pestic. Sci. 15, 133-155 (1984).
32. K.H. Büchel, ACS Symposium Series 304, 1-23 (1986).
33. D. Berg, W. Kramer, E. Regel, K.-H. Buchel, G. Hollmwood, M. Plempsel and H. Scheinpflug, Br. Crop Prot. Conf. 1984, 887-892 (1984).
34. A.C. Oehlslager, R.H. Angus, A.M. Pierce, H.D. Pierce and R. Shrinivasan, Biochemistry 23, 3582-3589 (1984).
35. A.F. Marchington, 10th Int. Congr. Pl. Prot. 1983, 201-208 (1983).
36. T.E. Wiggins and B.C. Baldwin, Pestic. Sci. 15, 206-209 (1984).

37. H. Van den Bossche, W. Lauwers, G. Willemsens, P. Marichal, F. Cornelissen and W. Cools, Pestic. Sci. 15, 188-189 (1984).
38. D.J. King, A. Wiseman, D.E. Kelly and S.L. Kelly, Curr. Genet. 10, 261-267 (1985).
39. A. Kerkenaar and D. Barug, Pestic. Sci. 15, 199-205 (1984).
40. A. Kerkenaar, J.M. van Rossum, G.G. Versluis and J.W. Marsman, Pestic. Sci. 15, 177-187 (1984).
41. S. Hippe, Pestic. Sci. 15, 210-214 (1984).
42. M. Gasztonyi and G. Josepovits, Pestic. Sci. 15, 48-55 (1984).
43. A.H.B. Deas, T. Clark and G.A. Carter, Pestic. Sci. 15, 63-70 (1984).
44. A.H.B. Deas, T. Clark and G.A. Carter, Pestic. Sci. 15, 71-77 (1984).
45. A.H.B. Deas, Pestic. Sci. 17, 69-70 (1986).
46. M.S. Wolfe, Bull. OEPP/EPPO Bull. 15, 451-457 (1985).
47. H. Buchenauer and K.H. Hellwald, Bull. OEPP/EPPO Bull. 15, 459-466 (1985).
48. D.W. Hollomon, T. Locke and M. Proven, Bull. OEPP/EPPO Bull. 15, 467-471 (1985).
49. M.A. de Waard, E.M.C. Kipp, N.M. Horn and J.G.M. Van Nistelrooy, Neth. J. Pl. Path. 92, 21-32 (1986).
50. H.T.A.M. Schepers, Neth. J. Pl. Path. 91, 105-118 (1985).
51. D.W. Hollomon and R.T.S. Loeffler, 8th Reinhardsbrunn Symp. Systemic Fungicides and Antifungal Compounds, in press (1986).
52. M.A. de Waard and J.G.M. Van Nistelrooy, Pestic. Biochem. Physiol. 13, 255-266 (1980).
53. M.A. de Waard and J.G.M. Van Nistelrooy, Neth. J. Pl. Path. 90, 143-153 (1984).
54. M.A. de Waard and J.G.M. Van Nistelrooy, Pestic. Sci. 13, 279-286 (1982).
55. M.A. de Waard and J.G.M. Van Nistelrooy, Pestic. Sci. 15, 56-62 (1984).
56. M.A. de Waard, H. Groeneweg and J.G.M. van Nistelrooy, Neth. J. Pl. Path. 88, 99-112 (1982).
57. F. Huggenberger, M.A. Collins and G. Skylakakis, Crop Protection 3, 137-149 (1984).
58. P. Leroux and R. Fritz, in: Mode of action of antifungal agents (eds A.P.J. Trinci and J.F. Ryley), pp. 207-237, Cambridge University Press, Cambridge (1984).
59. T. Kato, in: Pesticide chemistry, human welfare and the environment (eds J. Miyamoto, P.C. Kearny, P. Doyle and T. Fujita), vol. 3, pp. 153-157. Pergamon, Oxford (1983).
60. P. Leroux and M. Clerjeaux, Crop Protection 4, 137-160 (1985).
61. M. Grindle and W. Temple, Tr. Br. Myc. Soc. 84, 369-372 (1985).
62. L.C. Davidse, in: Mode of action of antifungal agents (eds A.P.J. Trinci and J.F. Ryley), pp. 239-255, Cambridge University Press, Cambridge (1984).
63. L.C. Davidse, O.C.M. Gerritsma and G.C.M. Velthuis, Pestic. Biochem. Physiol. 21, 301-308 (1984).
64. L.C. Davidse, A.E. Hofman and G.C.M. Velthuis, Exp. Mycology 7, 344-361 (1983).
65. D.J. Fisher and A.L. Hayes, Pestic. Sci. 13, 330-339 (1982).
66. A. Kerkenaar, Pestic. Biochem. Physiol. 16, 1-13 (1981).
67. R. Wollgiehn, E. Bräutigam, B. Schuhmann and D. Erge, Zeitschr. allgem. Mikrobiol. 24, 269-279 (1984).
68. D.J. Fischer and A.L. Hayes, Crop Protection 3, 177-185 (1984).
69. K.Y. Kang and J.W. Eckert, 8th Reinhardsbrunn Symp. Systemic Fungicides and Antifungal Compounds, in press (1986).
70. L.C. Davidse, 8th Reinhardsbrunn Symp. Systemic Fungicides and Antifungal Compounds, in press (1986).
71. D.J. Fischer and A.L. Hayes, Crop Protection 4, 501-510 (1985).
72. L.C. Davidse and G.C.M. Van den Berg-Velthuis, to be published.
73. P.J. Kuhn, in: Mode of action of antifungal agents (eds A.P.J. Trinci and J.F. Ryley), pp. 155-183, Cambridge University Press, Cambridge (1984).
74. G.A. White, J.M. Phillips, J.L. Huppitz, B. Witrzens and S.J. Grant, Pestic. Biochem. Physiol. 25, 163-168 (1986).
75. G.A. White and S.G. Georgopoulos, Pestic. Biochem. Physiol. 25, 188-204 (1986).
76. H.D. Sisler and N.N. Ragsdale, Agricultural Chemicals of the Future (BARC Symp. 8), Rowman and Allanheld, Totowa, pp. 175-188 (1985).
77. S. Inoue, T. Uematsu, T. Kato and K. Ueda, Pestic. Sci. 16, 589-598 (1985).
78. M.H. Wheeler, Exp. Mycol. 6, 171-179 (1982).
79. Y. Kubo, K. Suzuki, I. Furusawa and M. Yamamoto, Pestic. Biochem. Physiol. 23, 47-55 (1985).
80. F. Schwinn and H. Geissbühler, Crop Protection 5, 33-40 (1986).

Ryania insecticide: chemistry, biochemistry and toxicology

John E. Casida, Isaac N. Pessah, Josef Seifert* and Andrew L. Waterhouse*

Pesticide Chemistry and Toxicology Laboratory, Department of Entomological Sciences, University of California, Berkeley, California 94720

Abstract - Ryania, the powdered stem wood of *Ryania speciosa*, controls important lepidopteran larval pests at 3 to 16 grams alkaloid equivalent per acre, making it one of the most potent natural product insecticides. The principal active ingredients are ryanodine and 9,21-didehydroryanodine. Ryanodine is a muscle poison binding at the Ca^{2+}-activated open state of the channel involved in the release of contractile Ca^{2+} from the sarcoplasmic reticulum (SR), thereby uncoupling the electrical signal of the transverse tubule from the Ca^{2+}-release mechanism of the SR. The Ca^{2+}-ryanodine receptor complex of skeletal and cardiac muscle is highly specific, both activated and inactivated by Ca^{2+}, biochemically relevant and pharmacologically unique. The alkaloid binding site of skeletal muscle is a high molecular weight oligomer consisting primarily of Ca^{2+}-ATPase and a 228 kDa protein. The potency of 15 ryanoids at the skeletal muscle SR specific binding site generally parallels their toxicity to mice suggesting the toxicological relevance of the Ca^{2+}-ryanodine receptor. The hydrolysis products ryanodol and 9,21-didehydroryanodol have low toxicity to mice and little activity at the mammalian SR receptor yet they are potent knockdown agents for insects, establishing selective toxicity within the ryanoids and suggesting a possible difference in the target sites of mammals and insects.

INTRODUCTION

The powdered stem wood of *Ryania speciosa* Vahl (family *Flacourtiaceae*) (Fig. 1) has been used for almost 40 years as Ryania insecticide (ref. 1 & 2). Ryania, with ~0.22% active alkaloids, was extensively used to control European corn borer larvae at ~16 grams alkaloid equivalent per acre. An improved preparation is effective in controlling codling moth larvae at 3 to 4 grams alkaloid per acre. It protects against feeding attack without high mortality due in part to inducing "flaccid paralysis". Except in specialty markets, Ryania is not competitive with currently available synthetic organic insecticides. The active ingredient ryanodine is a muscle poison which specifically uncouples the electrical signal of the tranverse (t)-tubule from the Ca^{2+}-release mechanism of the sarcoplasmic reticulum (SR) (ref. 3 & 4). Renewed interest in ryanoids was prompted by identification of a new toxic component, using this chemical to prepare [^{3}H]ryanodine, and in turn using the labeled compound as a radioligand to define the mode of action.

CHEMISTRY

Ryanodine was first isolated from the roots of *R. speciosa* by solvent extraction and recrystallization with monitoring by assaying for the protection against clothes moth larval damage (ref.5). Elegant studies by Wiesner and associates (ref. 6) led to the structural identification of ryanodine (Fig. 2) based ultimately on X-ray crystallographic analysis of its hydrolysis product ryanodol as the *p*-bromobenzyl derivative (ref. 7). Ryanodine and a "closely related compound" in commercial ryanodine (ref. 8), isolated by HPLC, were subjected to total ^{1}H and ^{13}C NMR assignments using a new fully-coupled two-dimensional technique (ref. 9 & 10). This new ryanoid differs from ryanodine only in having a 9,21-double bond, *i.e.* 9,21-didehydroryanodine (ref. 9). This didehydro compound was not recognized in the initial isolation of ryanodine (ref. 5) probably because it would have been removed on the extensive recrystallization used to obtain pure ryanodine. Didehydroryanodine is equitoxic with or more toxic than ryanodine and a modified

*Current addresses: JS at Department of Agricultural Biochemistry, University of Hawaii, Honolulu, Hawaii 96822 and ALW at Department of Chemistry, Tulane University, New Orleans, Louisiana 70118.

analytical method for ryanoid components revealed a didehydroryanodine:ryanodine ratio of 3:1 for a sample of stem wood from Trinidad; on this basis, the principal toxic ingredient is the didehydro compound (ref. 9-11). Five other minor naturally-occurring ryanodine derivatives are: 18-hydroxy; 9,10-dehydro-8-keto; 9-hydroxy-10-epi; 8-hydroxy-4-methoxy-10-epi; 9,21-didehydro-8-hydroxy-4-methoxy-10-epi (ref. 11 & 12).

Fig. 1. Ryania insecticide is the ground stem wood of Ryania speciosa. Stages in the grinding process are illustrated with commercial material from Trinidad graciously provided by Thomas B. Harding, Jr. of AgriSystems International, Wind Gap, Pennsylvania 18091.

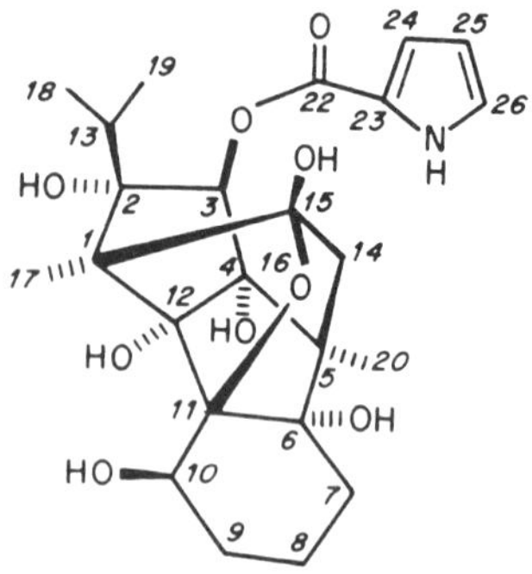

Fig. 2. Structure of ryanodine.

Two procedures for preparing $[^3H]$ryanodine start from tribromoryanodine and didehydroryanodine (Fig. 3). Tribromoryanodine (ref. 13) in the presence of palladium on carbon with tritium gas provides access to $[24, 25, 26\text{-}^3H_3$-pyrrole]ryanodine (2.9 Ci/mmol) (ref. 13). Addition of an acid scavenger (K_2CO_3) (ref. 11) to the reaction mixture (ref. 13) improved the yield of the reductive debromination reaction. Reduction of 9,21-didehydroryanodine with tritium gas yields high specific activity (60 Ci/mmol) $[9,21\text{-}^3H_2]$ryanodine mixed with $[9,21\text{-}^3H_2]$9-epiryanodine, readily separable by HPLC (ref. 9). The ratio of ryanodine, 9-epiryanodine and impurities is dependent on the solvent (ref. 9 & 14).

Ryanodine and didehydroryanodine were converted to a series of ryanoids of widely varying biological activity (ref. 11) discussed below. Hydrolysis of the ester group yields pyrrolecarboxylic acid and ryanodol or didehydroryanodol from ryanodine and didehydroryanodine, respectively (ref. 6 & 11). Alkylation requires a base to deprotonate the pyrrole NH and the alcohol groups. Alkylation occurs first at the pyrrole nitrogen and second at the 15-hemiketal. Methylation then proceeds to the 4, 6 and 10 positions. Acetylation with acetic acid-dicyclohexylcarbodiimide occurs preferentially at the 10-position. Didehydroryanodine yields the 9,10-epoxide with peracid and the 9,10-diol with osmium tetroxide. Swern oxidation (ref. 15) converts ryanodine to 10-ketoryanodine and didehydroryanodine to 8,9-didehydro-10-ketoryanodine. Reaction with ozone results in opening of the pyrrole ring.

BIOCHEMISTRY

Ryanodine has extremely potent myogenic activity (ref. 16) which varies with the type of muscle, inducing progressive contracture leading to complete and irreversible tetanus of mammalian skeletal muscle while decreasing contractile tension in mammalian cardiac and insect skeletal muscles (ref. 17-20). Ryanodine directly interferes with the mechanical properties of muscle fibers without significantly altering the electrical features of the surface membrane (sarcolemma) or interfering in crossbridge formation of the contractile elements (ref. 3). Attention was therefore focused on the "uncoupling" action of ryanodine at the terminal cisternae (TC) of the SR, i.e. the specialized region of the SR which makes contact with the t-tubule membrane and the site at which excitation-contraction (EC) coupling takes place. The mechanism of EC coupling or how depolarization of the t-tubule triggers the release of Ca^{2+} from the TC SR is one of the major unsolved problems of muscle biology (ref. 21). Ryanodine is the most specific chemical probe to clarify this process. Although ryanodine clearly alters the permeability of the TC SR in striated muscle to Ca^{2+}, its positive or negative inotropic effects (i.e., increased or decreased permeability to Ca^{2+} of SR membranes) depends on the type of muscle (cardiac vs skeletal), the preparation (whole muscle fibers, skinned fibers, or SR membrane vesicles), the assay conditions, and the alkaloid concentration (ref. 8 & 22-28). These studies suggest that ryanodine acts at the TC SR and specifically interferes with EC coupling.

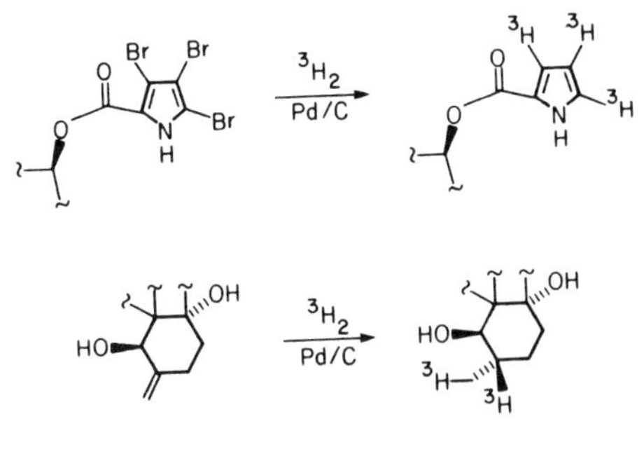

Fig. 3. Reactions for preparing [^{3}H]ryanodine labeled in the pyrrolecarboxylate and ryanodyl moieties by reduction with tritium gas.

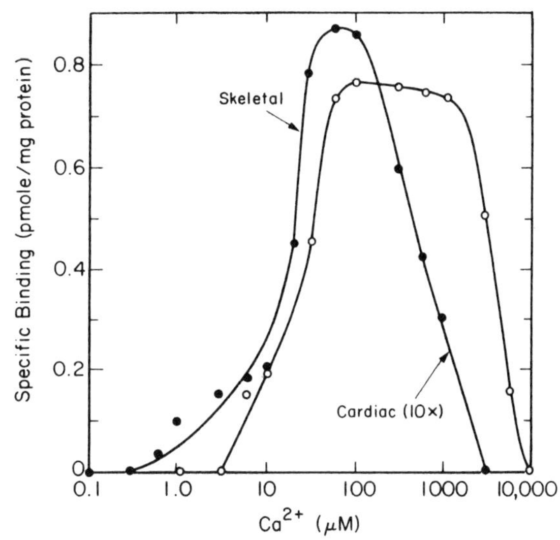

Fig. 4. Ca^{2+} regulation of [^{3}H]ryanodine binding to "heavy" SR membrane preparations from rabbit skeletal and cardiac muscle (adapted from ref. 32).

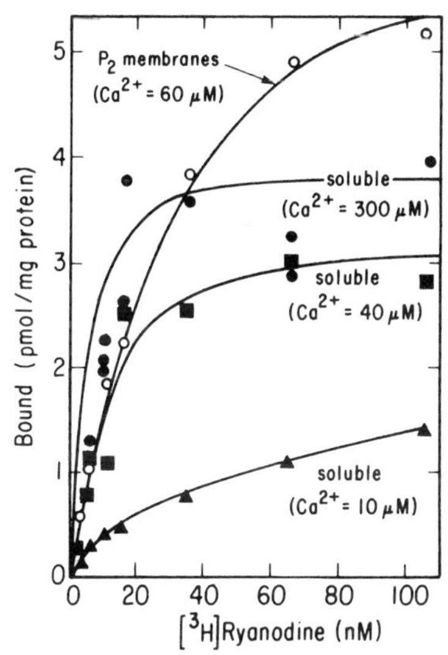

Fig. 5. Influence of Ca^{2+} concentration on specific binding of [^{3}H]ryanodine to soluble receptor shown as binding isotherms. Under optimal conditions of Ca^{2+} the soluble preparation has 60% of the receptor binding site density (on a protein basis) and a 4-fold higher affinity when compared to the respective membrane preparation.

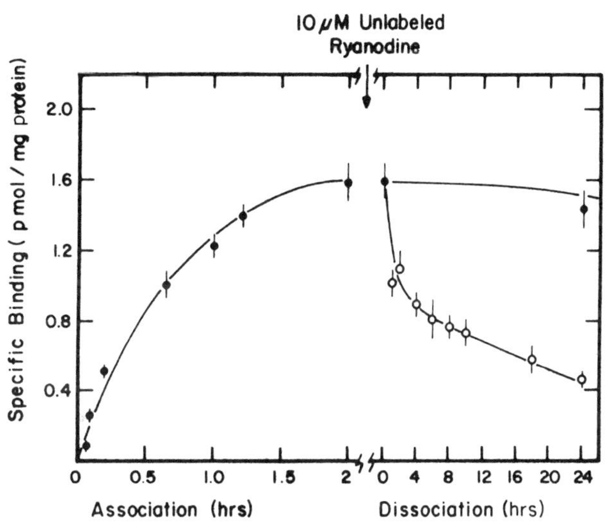

Fig. 6. Kinetics of association and dissociation with soluble receptor at optimal Ca^{2+} level. Half time to equilibrium is 24 min while dissociation of the equilibrium complex is much slower (half time = 15 hr).

Initial studies with [^{3}H]ryanodine and "light" or "heavy" SR membrane preparations (primarily longitudinal and TC SR vesicles, respectively) of rabbit fast skeletal muscles failed to detect specific binding sites for the alkaloid (ref. 29). Success was dependent on preserving the integrity of the TC junctional "feet" processes and activation by Ca^{2+}. Modification of a procedure for preparing morphologically-intact triads from fast skeletal muscle (ref. 30), utilizing 20 mM Na pyrophosphate instead of 0.6 M KCl (which disrupts the junctional feet processes) to dissociate the contractile elements from SR membranes, resulted in SR vesicles having morphological (exhibiting intact junctional feet) and biochemical (enriched in calsequestrin) markers of the TC (ref. 31). These membrane preparations with appropriate levels of Ca^{2+} allowed the recognition of highly specific, saturable binding sites for [^{3}H]ryanodine in skeletal and cardiac muscle (ref. 32). The highest density of [^{3}H]ryanodine binding sites is associated with the SR membrane fraction which is enriched in elements of the TC (ref. 31).

The [^{3}H]ryanodine binding domain is completely regulated by Ca^{2+} including both activation and inactivation. Skeletal and cardiac membrane preparations exhibit thresholds of approximately 1 µM Ca^{2+} for [^{3}H]ryanodine binding and although binding in both cases increases as a direct function of the free Ca^{2+} level the skeletal preparation is inactivated at 100 µM or greater Ca^{2+} while the cardiac receptor requires Ca^{2+} above 1 mM for inactivation (Fig. 4). Regulation by Ca^{2+} involves an uncompetitive mechanism where both the number of alkaloid binding sites and their affinity are decreased with suboptimal levels of Ca^{2+} suggesting that formation of a Ca^{2+}-receptor complex is prerequisite to ryanodine binding (Fig. 5) (ref. 31 & 32). The term "Ca^{2+}-ryanodine receptor complex" was coined to emphasize the Ca^{2+}-regulated nature of the ryanodine binding domain (ref. 32). The irreversible action of ryanodine on muscle tissue is explained by its receptor binding kinetics (Fig. 6). Dissociation of the receptor-alkaloid complex is extremely slow having a half-time of approximately 15 hours. The calculated Kd from kinetic rate constants is two orders of magnitude lower than that estimated from equilibrium binding experiments providing strong evidence that the Ca^{2+}-ryanodine receptor complex undergoes a rapid conformational change which occludes the alkaloid binding domain (ref. 31).

The Ca^{2+}-ryanodine receptor complexes of cardiac and skeletal origin are both inhibited by ruthenium red and seven ryanoid derivatives with the same ranking in potency but are unaffected by voltage-dependent Ca^{2+} channel blockers (ref. 32 & 33). Skeletal muscle preparations reveal a single class of high affinity binding sites while cardiac preparations contain two (or more) classes of binding sites of greatly differing affinities (ref. 31-33). These tissues also differ in sensitivity to inhibition by physiological levels of Mg^{2+} and relatively lower sensitivity of the cardiac receptors to inhibition by Cd^{2+}, Ba^{2+} and La^{3+} (ref. 32 & 33).

The physiological action of ryanodine requires extravesicular Ca^{2+} to disrupt Ca^{2+} permeability (ref. 34-37). More specifically, nanomolar levels of ryanodine block $^{45}Ca^{2+}$ reuptake into actively or passively loaded SR vesicles of skeletal and cardiac origin suggesting that ryanodine interferes with the inactivation of the Ca^{2+} release channels localized at the TC SR (ref. 34-36), while above 10 µM ryanodine blocks Ca^{2+} release in either tissue source (ref. 35). Further, ryanodine interferes with the inactivation of Ca^{2+}-induced release of Ca^{2+} from actively-loaded skeletal vesicles (ref. 36). In fact, regulation of [^{3}H]ryanodine binding is similar in several respects to regulation of the Ca^{2+}-triggered Ca^{2+} release channels of the TC SR including: 1) activation by Ca^{2+} with an apparent K_d of 20-40 µM, 2) inactivation by higher Ca^{2+} concentrations and physiological levels of Mg^{2+} (less pronounced with cardiac tissue), 3) inhibition by micromolar levels of ruthenium red, 4) pronounced increase in the Ca^{2+} activator site's affinity for Ca^{2+} in the presence of caffeine (ref. 38), 5) stimulation by adenine nucleotides (ref. 31, 35 & 38), and 6) the presence of critical sulfhydryl groups at the Ca^{2+} activator site (ref. 39 & 40). Ryanodine therefore acts at the Ca^{2+}-activated Ca^{2+} release channels of the TC SR, binding only to the Ca^{2+}-activated open state of these channels (ref. 31) and possibly preventing their complete closing (ref. 35 & 36).

The functional Ca^{2+}-ryanodine receptor complex is solubilized from both skeletal (ref. 31) and cardiac (ref. 33) muscle on treatment with 3-[(3-cholamidopropyl)-dimethylammonio]-1-propanesulfonate in the presence of high ionic strength. The soluble skeletal receptor maintains about 60% of the binding site density of the membrane-associated receptor (Fig. 5) and undergoes uncompetitive regulation by Ca^{2+}, but the soluble skeletal receptor has a higher affinity for the alkaloid and a decreased sensitivity to inactivation by Ca^{2+} and Mg^{2+} (ref. 31). Proteins which comprise the [^{3}H]ryanodine binding sites elute from a Sepharose column as a high molecular weight oligomer made up primarily of Ca^{2+}-ATPase (116 kDa) and a 228 kDa protein (Fig. 7). Other smaller aggregates lack [^{3}H]ryanodine binding sites and are primarily comprised of calsequestrin (63 kDa) and a 55 kDa protein (Fig. 7).

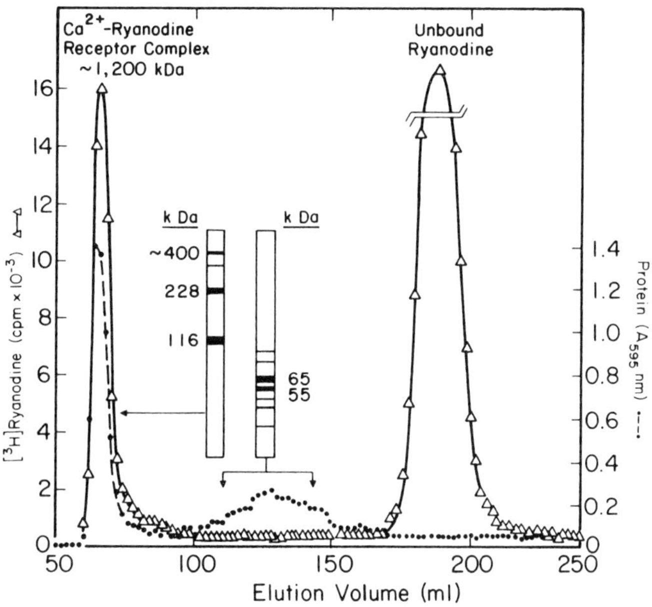

Fig. 7. Elution of Ca^{2+}-[^3H]ryanodine receptor complex and unbound [^3H]ryanodine from Sepharose 6B column and sodium dodecyl sulfate polyacrylamide gels of the respective pooled fractions.

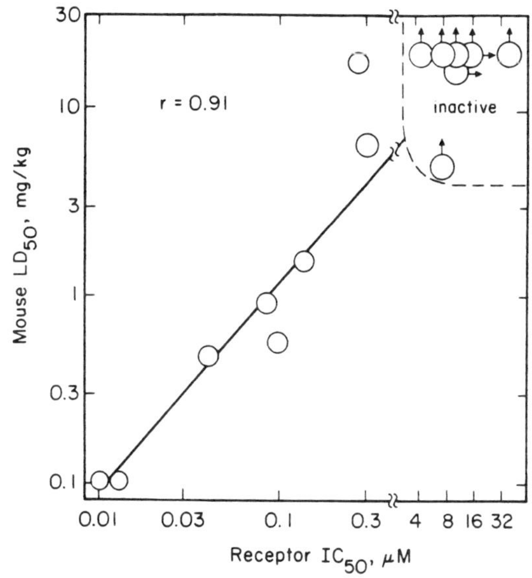

Fig. 8. Relation of potency for decreasing [^3H]ryanodine binding to the Ca^{2+}-ryanodine receptor of rabbit skeletal muscle SR and for intraperitoneal toxicity to mice. Compounds include 8 active ryanoids and 7 inactive ryanoids.

TOXICOLOGY

The natural ryanoids ryanodine and 9,21-didehydroryanodine are much more potent inhibitors of the Ca^{2+}-ryanodine receptor complex and/or toxicants than 18 other natural ryanoids, ryanoid derivatives or degradation products examined (ref. 11). The 18-hydroxy analog is less active by ~8-fold. Formation of the N-benzyl and N-butyl derivatives and progressive methylation indicates that both the pyrrole and hydroxyl groups at each of C_4, C_6, C_{10} and C_{15} play some role in binding. Anhydroryanodine, with a different overall shape due to opening of one bridging ring, has negligible activity. The conformation and presence of the pyrrole are important since tribromoryanodine, the pyrrole-ring-opened product and ryanodol are essentially inactive in mammalian systems. The potency of 15 ryanoids at the skeletal muscle SR specific binding site generally parallels their toxicity to mice establishing the toxicological relevance of the Ca^{2+}-ryanodine receptor (ref. 11 & 32) (Fig. 8).

Ryanodine and didehydroryanodine are nonselective as injected toxicants on comparing the LD_{50} to mice and KD_{50} to houseflies pretreated with piperonyl butoxide (an inhibitor of oxidative metabolism) (ref. 11). 10-Ketoryanodine is selectively toxic to mice and epoxyryanodine to houseflies. Most interestingly, both ryanodol and didehydroryanodol show high selective toxicity towards houseflies and/or cockroaches compared with mice. These findings have two important implications. They suggest a possible difference in the target sites of mammals and insects. They also indicate the possibility of selective toxicants acting at the Ca^{2+}-induced Ca^{2+}-release channel (ref. 11).

ACKNOWLEDGMENT

These studies were supported in part by National Institutes of Health grant 5 P01 ES00049.

REFERENCES

1. R.E. Heal, Agr. Chem. 4(5), 37 (1949).
2. D.J. Crosby, In "Naturally Occurring Insecticides", eds. M. Jacobson and D.G. Crosby, Dekker, New York, p. 198, 1971.
3. D.J. Jenden and A.S. Fairhurst, Pharmacol. Rev. 21, 1 (1969).
4. J.L. Sutko and J.L. Kenyon, J. Gen. Physiol. 82, 385 (1983).
5. E.F. Rogers, F.R. Koniuszy, J. Shavel Jr. and K. Folkers, J. Am. Chem Soc. 70, 3086 (1948).
6. K. Wiesner, Adv. Org. Chem. 8, 295 (1972).
7. S.N. Srivastava and M. Przybylska, Can J. Chem. 46, 795 (1968).
8. J.L. Sutko and J.T. Willerson, Circ. Res. 46, 332 (1980).
9. A.L. Waterhouse, I. Holden and J.E. Casida, J. Chem. Soc., Chem. Commun., 1265 (1984).
10. A.L. Waterhouse, I. Holden and J.E. Casida, J. Chem. Soc. Perkin Trans. II, 1011 (1985).
11. A.L. Waterhouse, I.N. Pessah, A.O. Francini and J.E. Casida, submitted.
12. L. Ruest, D.R. Taylor and P. Deslongchamps, Can. J. Chem. 63, 2840 (1985).
13. A.S. Fairhurst, J. Labelled Compd. 7, 133 (1971).
14. J.L. Sutko et al., J. Labelled Compd. Radiopharm. 23, 215 (1986).
15. K. Omura and D. Swern, Tetrahedron 34, 1651 (1978).
16. L. Procita, Federation Proc. 13, 395 (1954).
17. L. Procita, J. Pharmacol. Exp. Ther. 117, 363 (1956).
18. I.W. Hillyard and L. Procita, J. Pharmacol. Exp. Ther. 127, 22 (1959).
19. W.L. Haslett and D.J. Jenden, J. Cell Comp. Physiol. 57, 123 (1961).
20. P.N.R. Usherwood, Comp. Biochem. Physiol. 6, 181 (1962).
21. A.N. Martonosi, Physiol. Rev. 64, 1240 (1984).
22. A.S. Fairhurst and W. Hasselbach, Eur. J. Biochem. 13, 504 (1970).
23. W.G. Nayler, P. Daile, D. Chipperfield, and K. Gan, Am. J. Physiol. 219, 1620 (1970).
24. A.S. Fairhurst. Biochem. Pharmacol. 22, 2815 (1973).
25. M. Frank and W.W. Sleator, Res. Commun. Chem. Pathol. Pharmacol. 11, 65 (1975).
26. L.R. Jones and S.E. Cala, J. Biol. Chem. 256, 11809 (1981).
27. D.W. Hilgemann, M.J. Delay and G.A. Langer, Circ. Res. 53, 779 (1983).
28. A. Fabiato, Federation Proc. 44, 2970 (1985).
29. I.N. Pessah, unpublished results.
30. R.D. Mitchell, P. Palade, and S. Fleischer, J. Cell Biol. 96, 1008 (1983).
31. I.N. Pessah, A.O. Francini, D.J. Scales, A.L. Waterhouse and J.E. Casida, J. Biol. Chem. 261, 8643 (1986).
32. I.N. Pessah, A.L. Waterhouse and J. E. Casida, Biochem. Biophys. Res. Commun. 128, 449 (1985).
33. J. Seifert and J.E. Casida, Biochim. Biophys. Acta, in press.
34. S. Fleischer. E.M. Ogunbunmi, M.C. Dixon and E.A.M. Fleer, Proc. Natl. Acad. Sci. USA 82, 7256 (1985).
35. G. Meissner, J. Biol. Chem. 261, 6300 (1986)
36. I. N. Pessah, K.W. Anderson and J.E. Casida, Biochem. Biophys. Res. Commun. 139, 235 (1986).
37. M.D. Stern, D.J. Pelto,. M.C.Capogrossi and E.G. Lakatta, Biophys. J. 49, 15a (1986).
38. I.N. Pessah, R.A. Stambuk, A.O. Francini and J. E. Casida, submitted.
39. G. Salama and J. Abramson, J. Biol. Chem. 259, 13363 (1984).
40. I.N. Pessah, R.A. Stambuk and J.E. Casida, unpublished results.

Biochemical analysis of paraquat resistance in Conyza leads to pinpointing synergists for oxidant generating herbicides

Y. Shaaltiel and J. Gressel*

Plant Genetics, Weizmann Institute of Science, Rehovot, Israel

Abstract - A paraquat resistant biotype of Conyza bonariensis had higher levels than the wild type of three chloroplast enzymes: superoxide dismutase, ascorbate peroxidase and glutathione reductase. These enzymes detoxify the superoxide generated by the interaction of electron transport systems with paraquat and other oxidant generating herbicides. Superoxide dismutase contains both copper and zinc. Chelators which complex copper and/or zinc inactivated this enzyme synergizing paraquat, atrazine and acifluorfen in the cellular and plant systems tested.

INTRODUCTION

A paraquat resistant biotype of Conyza bonariensis evolved in the Tahrir irrigation district in Egypt following monthly treatments of paraquat for many years (ref. 1). Various data have been used to explain this resistance: (a) an elevated level of total cellular superoxide dismutase (ref. 2), as was found in other species (refs. 4,5), and (b) a compartmentation of paraquat near veins (ref. 3). Neither we (ref. 6) nor others (ref. 3) could find elevated total superoxide dismutase in Conyza. Elevated superoxide dismutase would result in the superoxide generated by the interaction of photosystem I and paraquat being transformed into hydrogen peroxide (Fig. 1). Hydrogen peroxide itself is toxic and in the presence of Fe^{+2} can be transformed by the Fenton reaction into hydroxyl radicals (ref. 7) which are even more toxic than peroxide and superoxide. The data that support the compartmentation hypothesis were inconclusive because they lack kinetics, used stem-fed radioactive paraquat, and did not show that the radioactivity was associated with the parent paraquat (ref. 3) and was not a catabolite or a conjugation product. We thus studied the enzymology of paraquat resistance and the properties of those enzymes. The suppression of these enzymes could lead to synergists for paraquat as well as other oxygen radical generating herbicides, if: (a) the same enzymes detoxify these radicals, with other oxidant generating herbicides, and (b) the variability in herbicide rate requirement between and within species is due to different levels of these enzymes. Such a synergist might not be as active with weeds that require low levels of herbicide to kill, but should be effective in decreasing the level of herbicides needed to kill the more recalcitrant species.

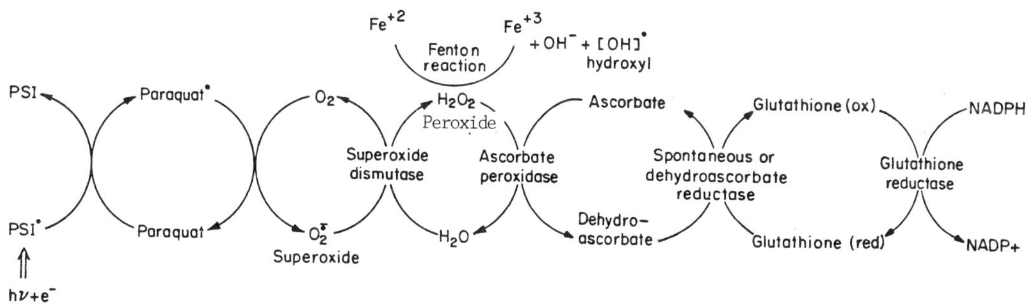

Fig. 1. The formation of toxic superoxide and hydroxyl radicals in the presence of paraquat, light, superoxide dismutase and ferrous ions and the Fenton reaction (ref.7). Superoxide can be detoxified by a related pathway using three or four enzymes as outlined by Halliwell (ref. 8) and Asada (ref. 9). This pathway presumably competes with the Fenton reaction.

DETOXIFICATION OF OXYGEN RADICALS

Whereas we were unable to find any consistent enrichment of superoxide dismutase activity in whole leaf extracts of the resistant Conyza biotype, we found higher levels of one electrophoretically separated isozyme in Class I chloroplasts (Table 1). Two other enzymes of the pathway (Fig. 1), ascorbate peroxidase and glutathione reductase were elevated to much higher levels than superoxide dismutase (Table 1). Although dehydroascorbate reductase is reported to be part of the pathway and has been found in chloroplasts (ref. 10), we did not assay for this enzyme, because the reaction can take place spontaneously at a sufficient rate for detoxification at cellular pHs. Elevated levels of all three enzymes were found in the F_1 of reciprocal hybrids of resistant and sensitive plants.

TABLE 1. Increases in the enzymes of the oxygen detoxifying pathway in chloroplasts of *Conyza bonariensis* biotypes.

Enzyme	paraquat sensitive	paraquat resistant	ratio
	(enzyme units per mg protein)		
Superoxide dismutase	1.04	1.67	1.60
Ascorbate peroxidase	0.74	1.85	2.50
Glutathione reductase	0.64	1.87	2.92

Enzymes were measured as outlined in (ref. 6).

PREVENTION OF TOXICITY

We have previously tested various compounds to see if they could act as protectants against paraquat damage. We found that benzylviologen could protect in an *in vitro* system using duckweed (ref. 5) and recently found that it also protected peas (ref. 11). Peroxide is the least toxic of the oxygen radicals. We hypothesized that if the Fenton reaction (ref. 7) could be inhibited, then there might be a modicum of protection, especially in a cell suspension system. Hydrogen peroxide is diffusable, and is found in the media of algae cultured in high light (ref.8). We added iron chelators (hydroxyquinoline and ethylene diamine) to the medium, and obtained protection against paraquat damage (Table 2). This protection would probably be only short term, because the chelator should cause an iron deficiency in plants if used for long periods.

TABLE 2 - Protection against hydroxyl ion toxicity by blocking the Fenton reaction with chelators

	μM paraquat				
	35	45	55	65	75
chlorophyll content (% of untreated control)					
no chelator	100	83	30	8	0
0.8mM ethylene diamine	100	100	100	100	100
3mM hydroxy-quinoline	100	100	100	100	100

Asparagus cells were isolated according to ref. 13 and incubated with treatments for 14 h under 0.45 mE m^{-2} s^{-1} (PAR) light prior to extraction of chlorophyll.

SYNERGIZING OXIDANT ACTION

We have previously found that the paraquat resistant Conyza has a modicum of tolerance to other oxygen radical stresses, such as those elicited by atrazine and sulfur dioxide (ref. 14). Thus it seemed that if we could stop the detoxification of the oxygen radicals, we might be able to synergize the action of all oxygen radical generating herbicides. If superoxide dismutase could be blocked, the plant might die due to lipoxidation by the superoxide. If ascorbate peroxidase, dehydroascorbate reductase or glutathione reductase could be blocked, the plant might die due to action of superoxide and hydroxyl radicals.

The latter two enzymes contain sulfhydryl groups (Table 3) that can be blocked by mercurials and other compounds with high mammalian toxicity. Thus sulfhydryl inhibitors were not tested. Chelators should afford the best synergism as the superoxide dismutase in the Conyza plastids is a copper and zinc containing enzyme (ref. 6) and ascorbate peroxidase contains iron and copper. Iron-containing chelators have a protectant effect (Table 2) and thus chelators of copper and zinc only should synergize. Indeed, sodium diethyl dithiocarbamate (DDC) synergized both paraquat and atrazine in isolated cells (Table 4). We similarly found that an oxime of 4-octyl-2-acetyl phenol, another copper chelator (ref. 19) and di-2-ethyl-hexyl-phosphoric acid, a zinc chelator (ref. 20) were synergistic, as expected.

TABLE 3. Properties of the oxygen radical detoxifying enzymes

Enzyme	Contains	inhibitors
Superoxide dismutase	Cu/Zn[15]	CN^-DDC N_3
Ascorbate peroxidase	Cu/Fe thiol[16]	CN^-DDC and thiol reagents
Dehydroascorbate reductase	thiol[17]	IAC PCMB NEthMal
Glutathione reductase	thiol[18],	UAC PcmB NEthMal Znson

TABLE 4. Synergism of diethydithiocarbamate (DDC) with herbicides using paraquat resistant and sensitive Conyza bonariensis biotypes:

	paraquat				acifluorfen	
	0	1	30	300	30	100
		Visual rating[a]				
Sensitive						
control	0	0	1.8	3	2	3
DDC	1	2	2.0	2.8	3	3
Resistant						
control	0	0	0	0	1	2.5
DDC	0	0.8	0.6	2.4	2.5	3.0

0.75% DDC was used with paraquat and a 0.5% DDC with acifluorfen. Herbicide concentrations are in (μM) and spraying was to runoff.
[a] visual ratings (scored 1 day after treatment):
0-no damage to slight burn; 1-moderate burn (may regenerate); 2-severe burn (will die); 3 - death.

TABLE 5. Synergism between herbicides and diethyldithiocarbamate (DDC) at the cellular level using isolated Asparagus sprengeri cells

Concentration DDC (%)	No herbicide	10μM Paraquat	30μM Atrazine	100μM Atrazine
	% of control chlorophyll			
0	100	107	95	92
0.02	118	11	40	8

Paraquat toxicity was measured 24 h and atrazine toxicity was measured 40h after treatment and incubation at 0.3mEin.$m^{-2}s^{-1}$ PAR.

This synergism was also apparent with Conyza plants (Table 4). Whereas sodium diethyldithiocarbamate had little or no effect by itself, it considerably lowered the levels of paraquat and acifluorfen required to severely damage both the resistant and sensitive biotypes. The manganese (maneb) or the zinc (zineb) salts of this chelator were not synergistic, as expected, because the ability to chelate copper was reduced. We found that this chelator also synergized paraquat with bean.

CONCLUDING REMARKS

The use of copper and zinc chelators as synergists for oxidant generating herbicides can considerably lower the level of herbicide required. This will decrease the environmental load of such herbicides by substantially decreasing the amounts used. It is envisaged that the first use will be in no-till situations or wherever a general herbicide is needed. Such compounds may well broaden the spectrum of activity of many herbicides.

ACKNOWLEDGEMENTS

The technical assistance of Peninit Birkenfeld is gratefully acknowledged, as are useful discussions with and synthesis of some chelators by Dr. A. Warshawsky of the Chemistry Department. This research was supported in part by the Israel Academy of Sciences and Arts fund for basic research and the Yeda Fund for Applied Research. J.G. holds the Gilbert de Botton Chair of Plant Sciences.

REFERENCES

1. J. Gressel, H.U. Ammon, H. Fogelfors, J. Gasquez, Q.O.N. Kay and H. Kees in Herbicide Resistance in Plants (H. LeBaron and J. Gressel, eds.), pp 31-55, Wiley, New York (1982).
2. R.J. Youngman and A.D.Dodge, Proc. 5th Intl. Cong. Photosynthesis. pp 537-544, Balaban, Philadelphia (1981).
3. E.P. Fuerst, H.Y. Nakatani, A.D. Dodge, D. Penner, and C.J. Arntzen, Plant Physiol. 77, 984-989 (1985).
4. B.M.R. Harvey and D.B. Harper, in Herbicide Resistance in Plants (H. LeBaron and J. Gressel, eds.) pp 215-233, Wiley, New York (1982).
5. E. Lewinsohn and J. Gressel, Plant Physiol. 76, 125-130 (1984).
6. Y. Shaaltiel and J. Gressel, Pestic. Biochem. Physiol. 26, 22-28 (1986).
7. B. Halliwell and J. Gutteridge (eds) Free Radicals in Biology and Medicine. Oxford Science Publications (1985).
8. C.H. Foyer and B. Halliwell, Planta 133, 21-25 (1976).
9. K. Asada, Y. Nakano and M.A. Hossain in Oxidative Damage and Related Enzymes (G. Rotilio and J.V. Bannister, eds.) pp. 342-347, Harwood, London (1984).
10. M.A. Hossain and K. Asada, Plant Cell Physiol. 25, 85-92, (1984).
11. Y. Shaaltiel, D.J. Gillham, A.D. Dodge and J. Gressel (unpublished),(1986).
12. E. Tel-Or, M. Hufleijt and L. Packer, Biochem. Biophys. Res. Comm. 132, 533-539 (1985).
13. B. Colman, B.T. Mawson and G.S. Espie Can. J. Bot. 57, 1505-1510 (1979).
14. Y. Shaaltiel, A. Glazer and J. Gressel (unpublished), (1984).
15. Y. Sawada, T. Ohyama and T. Yamazaki, Biochem Biophys. Acta 268, 305-312 (1972)
16. K. Asada, Y. Nakano and M.A. Hossain, In "Oxidative Damage and Related Enzymes" G. Rotilio and J.V. Bannister (Eds) Harwood ACademic Publisher London pp. 342-347 (1984)
17. M. Anwar Hossain and K. Asada, Plant and Cell Physiol. 25, 85-92 (1984)
18. B. Halliwell and C.H. Foyer, Planta 139, 9-17 (1978)
19. A. Warshawsky, A. Deshe and Y. Shay. Israel J. Chem. 26. 48-56 (1985).
20. D.S. Flett Chem. and Ind. 17, 706-712 (1977).

Effects of sethoxydim on the metabolism of excised root tips of corn

Kanji Ishihara, Hideo Hosaka, Masae Kubota, Hideo Kamimura, Nobuo Takakusa and Yasushi Yasuda

Odawara Research Center, Nippon Soda Co. Ltd., 345 Takada, Odawara 250-02, Japan

Abstract - The effects of a herbicide sethoxydim on the selected metabolic processes of root tips of corn (*Zea mays* L.) were studied. Sethoxydim was rapidly absorbed into root tips, and inhibited its elongation (IC_{50}=0.1μM, IC_{90}=1μM; 48hr after treatment). The incorporation of ^{14}C-uridine into RNA or ^{14}C-leucine into protein was not affected by sethoxydim even at 1μM, while ^{14}C-thymidine into DNA and ^{14}C-glucose into cell wall fraction was inhibited at 120min. Furthermore, the incorporation of ^{14}C-acetic acid into lipids was inhibited at 30min. Analyzing ^{14}C-lipids from the treated root tips, polar lipids (phospholipids) were alreary decreased within 15min. It is suggested that a possible mechanism of the herbicidal action of sethoxydim is the interference with a delicate lipid metabolism at an early stage after treatment.

INTRODUCTION

Sethoxydim, (±)-2-(1-ethoxyiminobutyl)-5-[2-(ethylthio)propyl]-3-hydroxycyclohex-2-enone, is a selective herbicide to control grasses in broadleaf crops, and affects meristematic regions of gramineous plants (ref. 1). There are some studies of the herbicide on the mode of action including on lipid metabolism (ref. 2). In this report, the effects of sethoxydim on DNA, RNA, protein, cell wall and lipid syntheses were studied by using excised root tips of corn (*Zea mays* L.), which were most sensitive tissues to the herbicide.

MATERIALS AND METHODS

Plant materials

Corn (*Zea mays* L. "Goldencrossbantam") seeds were surface sterilized, and the following procedures were carried out under aseptic conditions. Three days after germination, each terminal 1-cm root tip was removed into White's medium (pH6) with 2% sucrose and cultured with constant shaking (100cycles/min) in the dark at 25°C. After 24-hr cultivation, root tips were used for following experiments.

Effect on elongation

Root tips were incubated in culture medium containing sethoxydim (0.01 - 10μM) for 48hr, then each length was measured.

Uptake of ^{14}C-sethoxydim

Root tips were incubated in culture medium containing 1μM of ^{14}C-sethoxydim (s.a. 10.3mCi/mmol) for 15, 30, 60 and 120min. Each terminal 2-mm root tip was radioassayed with a liquid scintillation counter (Aloka 903; LSC) after combusting in a sample oxidizer (Packard Tri-Carb 306).

Effects on the metabolisms

Root tips were incubated in culture medium containing sethoxydim (0, 0.1 or 1μM) with 0.1 to 1μCi per ml of ^{14}C-labeled thymidine (s.a. 56mCi/mmol), uridine (s.a. 53mCi/mmol), L-leucine (s.a. 66mCi/mmol), glucose (s.a. 270mCi/mmol) or sodium acetate (s.a. 55mCi/mmol) to measure incorporation into DNA, RNA, protein, cell wall or lipid fraction. After incubation (15, 30, 60 and 120min), uptake of each ^{14}C-precursor into 2-mm terminal was determined as described above. For DNA, RNA and protain synthesis assay, the samples were ground in cold 5% trichloroacetic acid, and then precipitates were radioassayed (ref. 3). To assay cell wall synthesis, the samples were ground in cold medium and precipitates were radioassayed.

The samples for lipid synthesis assay were extracted with boiling chloroform-methanol and then partitioned with 1% sodium chloride (ref. 4). Radioactivity measurements were carried out on both layers (chloroform and water) and on residues. The chloroform layer was concentrated and analyzed by thin-layer chlomatography (TLC) on silica gel, autoradiography with X-ray film and LSC.

RESULTS AND DISCUSSION

Properties of excised root tip sistem for study of sethoxydim

Sethoxydim inhibited elongation of the excised root tips. IC_{50} and IC_{90} were 0.1 and 1μM, respectively, which were comparable to the effective dose in a hydroponic study reported previously (ref. 1). ^{14}C-sethoxydim was rapidly absorbed into corn root tips (data not shown), and various ^{14}C-labeled precursors used were continuously incorporated into cell constituents (Fig. 1). Therefore, this system was considered to be suitable for estimating effects of sethoxydim on the metabolic processes. This is the first finding obtained by using excised root tip system for study of herbicides such as growth and/or cell division inhibitors.

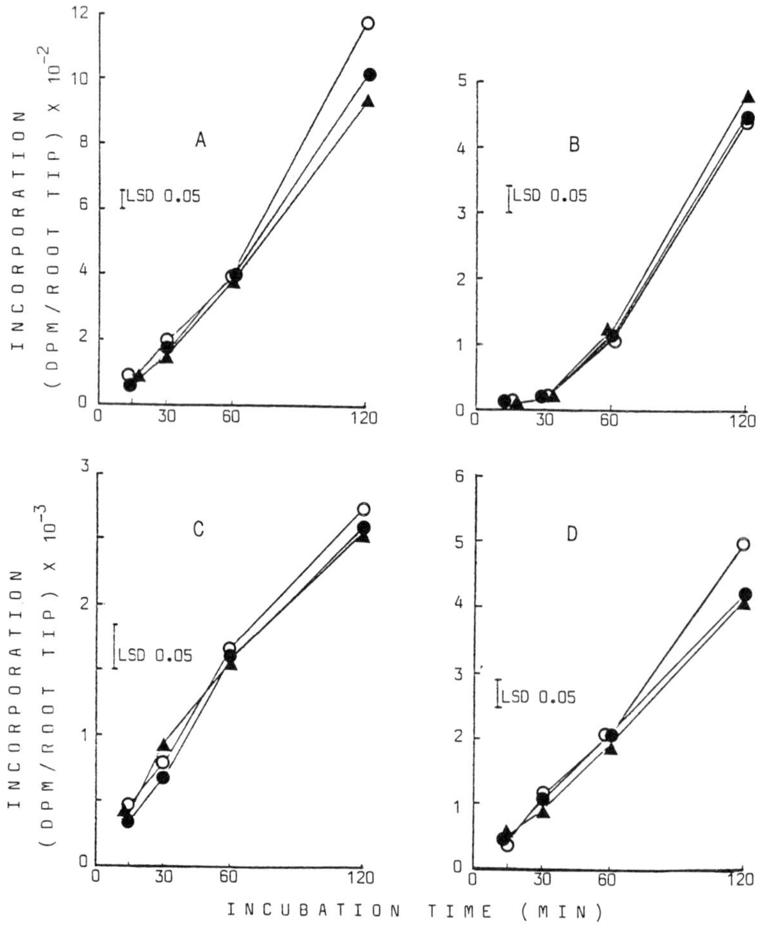

Fig. 1. Effects of sethoxydim on the incorporation of ^{14}C-precursors into cell constituents of excised root tips of corn.
Sethoxydim concentrations are 0μM(○), 0.1μM(●) and 1μM(▲).

A: ^{14}C-thymidine, B: ^{14}C-uridine, C: ^{14}C-leucine, D: ^{14}C-glucose

Effects of sethoxydim on the metabolic processes

The uptake of ^{14}C-precursors were not remarkably affected by sethoxydim (data not shown). The incorporation of ^{14}C-uridine into RNA or ^{14}C-leucine into protein fraction was not affected (Fig. 1 B, C), but that of ^{14}C-thymidine into DNA and ^{14}C-glucose into cell wall fraction was inhibited at 120min (Fig. 1 A, D). The incorporation of ^{14}C-acetic acid into lipids was significantly inhibited by sethoxydim treatment at 30min (Fig. 2 A).

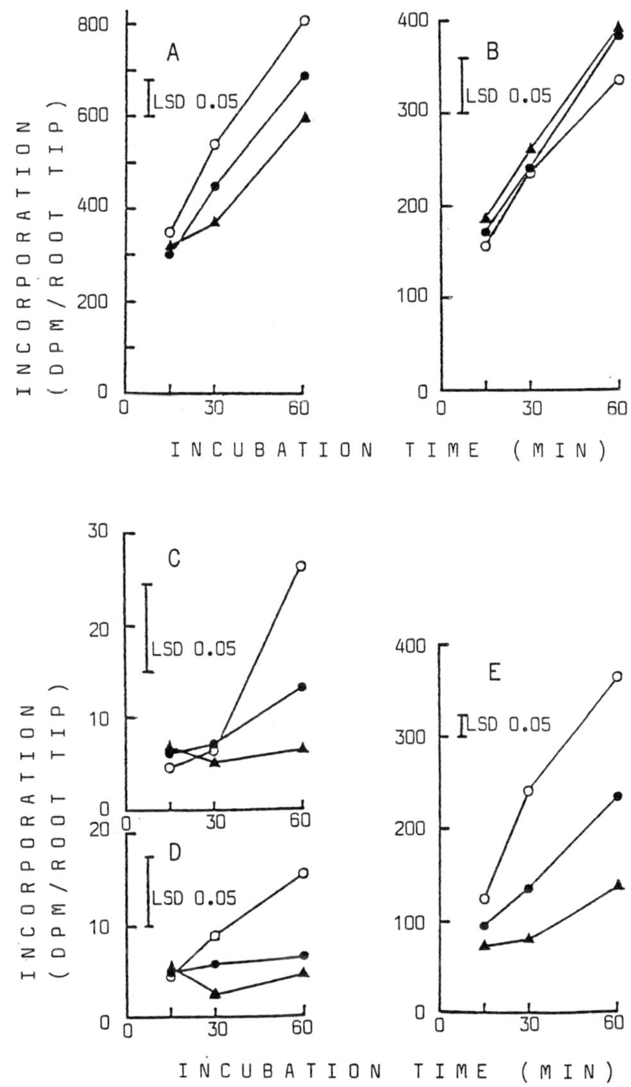

Fig. 2. Effects of sethoxydim on the incorporation of ^{14}C-acetic acid into lipid fractions.
Sethoxydim concentrations are 0μM(○), 0.1μM(●) and 1μM(▲).

A:total lipids, B:sterols, C:diglycerides, D:triglycerides, E:polar lipids

Analyzing the ^{14}C-lipids by TLC, polar lipids (phospholipids) were already decreased at 15min (Fig. 2 E, Table 1). Diglycerides and triglycerides were slightly decreased at 60min, but sterols were not (Fig. 2 B, C, D). Burgstahler and Lichtenthaler have already reported that glyco- and phospholipid accumulation in seedling of corn were inhibited by sethoxydim over 3days (ref. 2). In this report, the effects of sethoxydim on lipid metabolism were observed in a much shorter time by using ^{14}C-acetic acid and the simple culture system. It is suggested that a possible mechanism of the herbicidal action of sethoxydim is the interference with a delicate lipid metabolism.

TABLE 1. Effects of sethoxydim on the incorporation of ^{14}C-acetic acid into polar lipids.

Incubation	Sethoxydim	Inhibition (%)					
Time (min)	(μM)	PA	PI	PS	PC	PE	PG*
15	0.1	53	44	29	62	65	25
	1	82	79	47	86	72	79
30	0.1	47	49	52	70	65	64
	1	79	88	62	88	90	88
60	0.1	52	37	10	4	9	15
	1	82	85	22	54	66	77

*) PA:phosphatidic acid, PI:phosphatidylinositol, PS:phosphatidylserine, PC:phosphatidylcholine, PE:phosphatidylethanolamine, PG:phosphatidylglycerol

REFERENCES

1. H. Hosaka, H. Inaba, A. Satoh and H. Ishikawa, Weed Sci. 32. 711-721 (1984).
2. R.J. Burgstahler and H.K. Lichtenthaler, Structure, Function and Metabolism of Plant Lipids, p.619, Elsevier Sci. Publishers, B.V. (1984).
3. K.K. Hatzios, Plant Cell Reports 1. 87-90 (1982).
4. H.H. Hoppe, Pestic. Biochem. Physiol. 23, 297-308 (1985).

The mode of action of cymoxanil in *Botrytis cinerea*

P. LEROUX*, R. FRITZ and D. DESPREAUX

Institut National de la Recherche Agronomique
Station de Phytopharmacie, 78000 Versailles, France

Abstract - Using radiolabelled precursors, inhibition of several biosyntheses has been studied in mycelial cells of a sensitive strain of *Botrytis cinerea*. Metabolism of [14C]-uridine and its incorporation into RNA appear to be strongly inhibited by cymoxanil. The other biosynthetic processes and respiration are less affected and the inhibitory effects disappear with time. The fact that cysteine, glycine and serine have antagonistic effects, suggests that cymoxanil disturb the synthesis of this group of amino acids.

When [14C]-cymoxanil is added to liquid cultures of insensitive strains of *B. cinerea*, it is slowly metabolized (about 80 % of cymoxanil is recovered after an incubation time of 5 hours). On the other hand, a sensitive isolate is able to completely decompose this fungicide after an incubation of 6 hours. The insecticide carbaryl inhibits the transformation of cymoxanil in the sensitive strains of *B. cinerea* and presents an antagonistic action. These observations indicate that an hydrolase is probably responsible for the activation of cymoxanil. The localization of this enzyme, its specificity and the nature of the fungitoxic metabolite(s) are discussed.

INTRODUCTION

Cymoxanil [1-(2-cyano-2-methoxyiminoacetyl)-3-ethylurea] is toxic to numerous Perenosporales ; however its activity varies between species. For example in the genus *Phytophthora*, this fungicide is highly effective (in-vitro tests on mycelial growth) against *P. infestans* (ED 50 < 1 µg/ml) whereas *P. cactorum*, *P. cinnamomi* or *P. palmivora* are not sensitive (ED 50 > 10 µg/ml).

In France, cymoxanil is used mixed with various multi-site fungicides (copper, captafol, folpet, mancozeb, metiram-zinc, propineb, zineb) against downy mildew of vine (*Plasmopara viticola*), potato or tomato blight (*P. infestans*) and downy mildew of lettuce (*Bremia lactucae*). Some combinations containing oxadixyl (phenylamide group) or fosethyl-Al have recently been developed (ref. 1).

On tomatoes, under experimental conditions with hydroponic cultures, cymoxanil is translocated from the roots to the leaves where it prevents the development of *P. infestans* (ref. 2). In practice this translocation is limited ; however cymoxanil has some curative effects against various Peronosporales on the treated parts of the plants. This phenomenon has been demonstrated with downy mildew on vine (ref. 3).

At the present time, resistance to cymoxanil has never been reported in practice. However, in laboratory tests, various responses to this fungicide have been observed in *B. cinerea*. Most strains are highly resistant to cymoxanil (ED 50 mycelium > 100 µg/ml), some ones are inhibited between 10 and 50 µg/ml and only the isolate B (from a Bordeaux vineyard) is very sensitive (ED 50 mycelium = 0.7 µg/ml) (ref. 4).

This report summarizes the work in our laboratory on the biochemical mode of action of cymoxanil. Most of the data which have never been published, concern the relation existing between cymoxanil metabolism and the resistance to this fungicide in *B. cinerea*.

BIOCHEMICAL MODE OF ACTION OF CYMOXANIL

The experiments conducted on mycelial suspensions of *B. cinerea* (strain B, sensitive to cymoxanil) produced in shaken medium (ref. 5,6). The incubation of mycelium (30 min to 6 h) with cymoxanil (1 to 100 µg/ml) indicates that the respiration, the protein biosynthesis and the lipid synthesis of *B. cinerea* are hardly affected (ref. 5, 6). On the other hand, the uptake of [14C]-uridine by the fungal cells, and its incorporation into nucleic acids (mainly RNA) are inhibited by cymoxanil (table 1).

TABLE 1. Effect of cymoxanil on uptake (ethanolic extracts of mycelium) and incorporation into nucleic acids of [14C]-uridine in *B. cinerea*-strain B (percent of the control)

cymoxanil (µg/ml)	incubation time 1 h		4 h	
	uptake	incorporation	uptake	incorporation
1	160	157	42	17
10	98	106	32	14
100	66	45	7	2

The various metabolites of [14C]-uridine in the ethanolic extracts of mycelium are subjected to thin layer chromatography on polyethyleneimine (PEI)-cellulose plates (elution by water and then lithium chloride 0.15 to 1.5 M). The bands of radioactive compounds are removed from the plates and oxidized for quantitative determinations of the radioactivity (Fritz, unpublished data). From the results (Table 2), it appears that cymoxanil inhibits the formation of UDP-glucose and UDP-N-acetylglucosamine ; this effect disappears with time.

TABLE 2. Effect of cymoxanil on the metabolism of [14C]-uridine in the mycelium of *B. cinerea*-strain B (percent of the various metabolites in the ethanolic extracts).

radioactive metabolites of uridine	incubation time 1 h cymoxanil (µg/ml)				4 h cymoxanil (µg/ml)			
	0	1	10	100	0	1	10	100
uridine	21	19	50	71	18	23	16	21
uracile	2	5	4	6	5	6	8	5
UDP-N-acetyl-glucosamine	11	27	10	0	27	23	21	22
UDP-glucose	58	40	29	0	38	37	44	36
UMP	8	9	7	23	7	9	9	14
UDP + UTP	0	0	0	0	5	2	2	2

The studies conducted with *P. infestans* by L. Davidse (unpublished data) indicated that cymoxanil, like the phenylamide fungicides (benalaxyl, metalaxyl, ofurace, oxadixyl) inhibited RNA biosynthesis. The target of the phenylamides is a ribosomal RNA polymerase whereas that of cymoxanil remains unknown. The absence of a positive cross resistance between cymoxanil and the phenylamides indicates that they do not have the same biochemical mode of action (ref. 4).

ABSORPTION AND METABOLISM OF CYMOXANIL BY MYCELIUM OF *BOTRYTIS CINEREA*

- Materials and methods

Radiolabelled cymoxanil, 1-(2-cyano-2-methoximino-[2-14C] acetyl)-3-ethylurea (351 k Bq/mg), was purchased from Du Pont de Nemours, Wilmington, USA (ref. 7). The mycelial suspensions (dry weight of 2 mg/ml) of *B. cinerea* are treated by 20 µg/ml of [14C]-cymoxanil. At the end of the incubation time (15 min to 6 h) samples of *B. cinerea* cultures (5 mL) are collected and the mycelium separated from the liquid medium by filtration on a 155 mesh cheese-cloth. The mycelium is extracted with a mixture of ethanol + acetone (1+1, by volume) ; and then the extract is separated from the mycelium by filtration on Whatman GF/A glass-fibre filter. The residual mycelium (unextractable radioactivity) is submitted to combustion in an Oxymat apparatus. The radioactivity of the various fractions is counted in a liquid scintillation spectrometer.

To study the metabolism of cymoxanil in the culture medium, each sample (5 mL) is lyophilized and the residue solubilized in 0.5 mL of the mixture, acetone + ethanol + water (5+5+1, by volume). Analyses are conducted by thin layer chromatography (Silical gel 60 F plates - elution by ethyl acetate + hexane + acetic acid : 60 + 40 +1, by volume) and the radioactive compounds are identified by autoradiography. To estimate the proportion of the various radio-active compounds, their bands are removed from the plates and combusted for quantitative determinations of radioactivity. Similar studies have also been conducted with mycelial extracts.

Some complementary experiments have been performed with non-radioactive cymoxanil. The estimation of the amount of intact cymoxanil in the culture medium is done by using a bioautographic method. Cymoxanil is extracted from the aquous medium by ethyl acetate. After concentration of this organic extract, the samples are analyzed by thin layer chromatography (silica gel 60 F plates - elution by ethyl acetate + hexane : 6 + 4, by volume). The detection is done by spraying (at 50°C) an agar medium containing spores of *B. cinerea*-strain B. After an incubation time of 30 h the only fungitoxic compound extracted from the culture medium was cymoxanil (Rf value : between 0.4 and 0.5).

- Results

The amount of extractable radioactivity from mycelium of *B. cinerea* is higher in the sensitive strain B than in the highly resistant one, VI. The difference is greater during the first 2 hours of incubation and then it is reduced (Fig. 1). The radioactivity bound to the mycelium presents the same pattern, but in both strains it represents less than 2% of the radioactivity detected inside the mycelium (Fig. 1). Such results can indicate that the resistance to cymoxanil in *B. cinerea* is due to a reduced uptake of this fungicide by the fungal cells. However analysis of the radioactivity from the culture media and mycelial extracts indicates that radioactive cymoxanil is rapidly metabolized by the strain B whereas such a phenomenon is limited for the strain VI (Table 3). This result has been confirmed with non-radioactive cymoxanil and the reduced metabolism of cymoxanil has been observed in several resistant strains, including VI (Table 4).

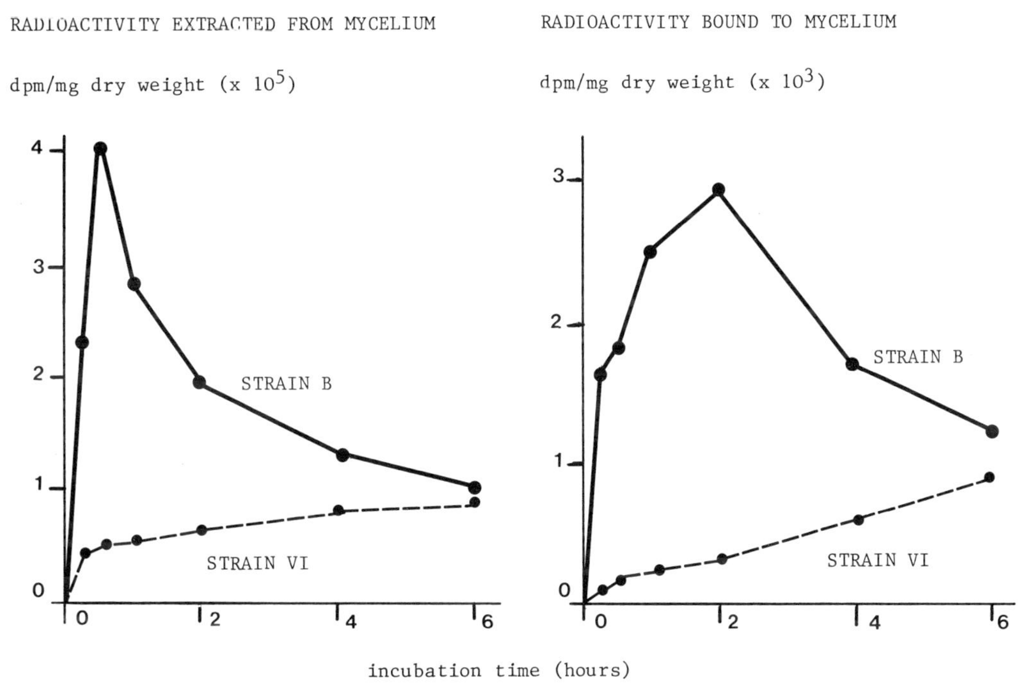

Fig. 1. Uptake of [14C]-cymoxanil by the mycelium of a sensitive strain (B) and a resistant strain (VI) of *B. cinerea*.

As observed in plants and animals (ref. 7, 8), a part of the radioactivity detected in the culture medium and in the mycelium of strain B is found in gylcine (Fig. 2). This amino acid and several other polar compounds (RF values below 0.2) stained by the ninhydrin reagent represent about 50 % of the radioactivity after 1 to 6 h of incubation. In the mycelial extracts, during the first 2 h, we detect a compound (10 % to 20 % of radioactivity) which is stained by the ninhydrin reagent ; its Rf value is about 0.85 and it is not observed in the culture media. Several radioactive compounds with Rf values greater or smaller than that of cymoxanil (Rf value about 0.5) have been detected in the culture medium and in the mycelium of the strain B. Among them is one with a Rf value of 0.65; whose concentration increases with time and after 6 h represents between 20 and 30 % of the radioactivity in both culture medium and mycelium.

TABLE 3. Degradation of [14C]-cymoxanil by a sensitive strain (B) and a resistant strain (VI) of *B. cinerea* (radioactive cymoxanil as percent of the whole radioactivity in the culture media or in the mycelia)

incubation time	culture media		mycelial extracts	
	strain B	strain VI	strain B	strain VI
15 min	74	83	15	82
60 min	44	84	5	74
120 min	6	77	3	79

TABLE 4. Degradation of non-radioactive cymoxanil by *B. cinerea* (intact cymoxanil in the culture media as percent of the initial amount)

experimental conditions [a]	incubation time 1 h	5 h
strain B, 3 mg dry weight/ml	45	7
strain B, 1,5 mg dry weight/ml	70	10
strain VI, 3 mg dry weight/ml	100	80
strain B, 1,5 mg dry weight/ml + carbaryl, 25 µg/ml	100	50
culture medium from strain B	100	100

a : the initial concentrations of cymoxanil are always 25 µg/ml.

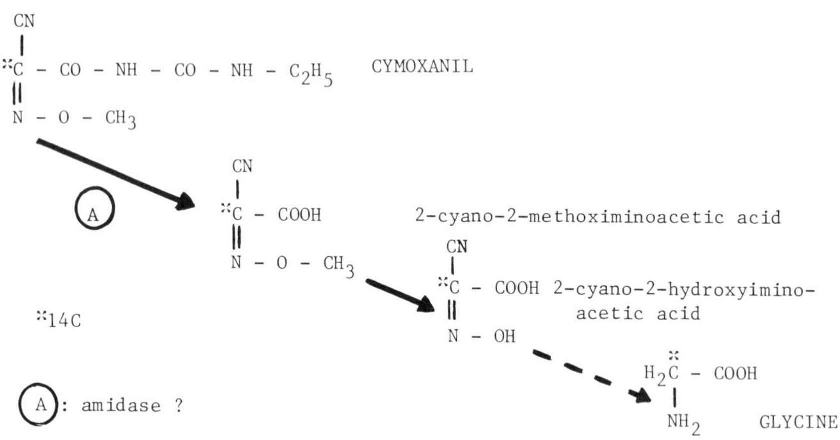

Fig. 2. Metabolism of cymoxanil in animals and plants (from ref. 7, 8).

When carbaryl, a known inhibitor of hydrolases (especially amidases), (ref. 9, 10) is added to liquid cultures of *B. cinerea*-strain B, the metabolism of cymoxanil is reduced (Table 4). This result indicates that probably the first steps of cymoxanil transformation are controlled by hydrolases. If the degradation pathways are similar in fungi, plants and animals (ref. 7, 8), amidases could be involved (Fig. 2). The characterization of the enzyme from *B. cinerea* is under investigations and at present we know that it is not excreted in the culture medium (Table 4) and it does not hydrolyze acetamide on propanil (an arylamide) (Leroux, unpublished data). As strain B does not produce cymoxanil it is of interest to seek for natural substrate(s) of the enzyme.

In the culture media and in the mycelial extracts of *B. cinerea*-strain B, the only fungitoxic compound is cymoxanil. This observation can indicate that the active metabolites produced inside the fungal cells (and which are probably unstable) are not toxic when they are given outside the mycelium. The characterization of such active metabolites is under investigations.

INTERACTIONS BETWEEN CYMOXANIL AND CARBARYL OR AMINO ACIDS

We have studied the effect of carbaryl and various amino acids on the fungitoxicity of cymoxanil towards the strain B of *B. cinerea*. The interaction has been investigated using a crossed paper strip bioassay (ref. 11, 12). Filter paper strips (width 1 cm) are dipped in ethanolic solutions of cymoxanil (300 to 3000 μg/ml), carbaryl (500 to 2000 μg/ml) or amino acids (20 to 100 mM). The strips are dried and transferred on an agar medium seeded with spores of *B. cinerea*-strain B. The inhibition of the fungal growth is assessed after 2 or 3 days of incubation. Several experiments have also been done according to the method of Leroux and Gredt (ref. 13). The mixtures of cymoxanil with other compounds are incorporated in an agar medium; the inoculations are conducted with mycelial plugs of *B. cinerea*-strain B.

TABLE 5. Interaction between cymoxanil and carbaryl or amino acid derivatives on *B. cinerea*-strain B (the mycelial growth rates are expressed as percentages of those observed in the presence of only carbaryl or amino acid derivatives)[a]

compounds in mixture with cymoxanil	cymoxanil in μg/ml 1	3	10
none [b]	40	15	10
carbaryl, 20 μg/ml	80	45	20
carbaryl, 40 μg/ml	90	65	35
L-alanine, 10 mM	50	15	10
L-cysteine, 10 mM	100	100	95
L-glycine, 10 mM	80	55	35
L-methionine, 10 mM	50	20	10
L-serine, 10 mM	100	80	70
D-serine, 10 mM	45	20	10
O-phospho-L-serine, 5 mM	50	15	15
glycyl-L-serine, 10 mM	100	100	90
L-glutathione (reduced form), 5 mM	50	20	10

[a] The tests are done according to the method of Leroux and Gredt (ref. 13)

[b] The comparisons are done with the fungus cultivated in the absence of any compounds.

In both tests carbaryl exhibits an antagonistic effect towards cymoxanil (table 5). This result confirms the hypothesis of the activation of cymoxanil by an hydrolytic enzyme sensitive to carbaryl. Within the amino acids tested, antagonistic effects are observed only with L-glycine, L-serine and L-cysteine which do not reduced the metabolism of cymoxanil. Such observations suggest that cymoxanil or its actives metabolites, interfere with the biosynthesis of this group of amino acids. Biochemical studies are required to verify this hypothesis.

CONCLUSIONS

The investigations conducted on *B. cinerea* indicate that the fungitoxicity of cymoxanil is correlated to its metabolism ; so this compound can be considered as a pro-fungicide. Some of the metabolities like glycine are not toxic, but others, probably unstable, are highly active inside the fungal cells. The activation of cymoxanil is due to an enzyme (or several enzymes) which is inhibited by carbaryl, a known inhibitor of hydrolases (including amidases).

The fungitoxic metabolites have not been characterized, however it is possible that some of them are also produced in animals and in plants (ref. 7, 8). As several metabolites in these organisms have structures close to those of amino acids, the fungitoxicity of cymoxanil may be due to inhibitions of amino acid biosyntheses. The fact that glycine, serine and cysteine are antagonists of cymoxanil suggests that active metabolites of cymoxanil inhibit a step of the biosynthesis of this group of amino acids. Even though this biochemical mode of action is not proven (an inhibition of synthesis of RNA has also been observed), it seems that the biosynthesis of amino acids is a good target for pesticides ; at present the main applications concern herbicides (ref. 14).

REFERENCES

1 . G. Dubois, Index phytosanitaire, p 223-312, ACTA, Paris (1986).
2 . H.L. Klopping and C.J. Delp, J. Agric. Food, Chem., 28, 467-468 (1980).
3 . M. Clerjeau, R. Lafon and Y. Bugaret, Phytiatr. Phytopharm., 30, 215-234 (1981).

4 . P. Leroux and M. Gredt, Phytiatr. Phytopharm., 30, 273-282 (1981).
5 . D. Despreaux, R. Fritz and P. Leroux, Phytiatr. Phytopharm., 30, 245-256 (1981).
6 . R. Fritz, D. Despreaux and P. Leroux, Tag. Ber. Akad. Landwirtsch. - DDR, Berlin, 222, 65-69 (1984).
7 . I.J. Belasco and F.J. Baude, Pestic. Sci., 12, 27-36 (1981).
8 . I.J. Belasco, J.C.Y. Han, R.L. Chrzanowski and F.J. Baude, Pestic. Sci., 12, 355-364 (1981).
9 . R.P. Lanzilotta and D. Pramer, Appl. Microbiol. 19, 301-306 (1970).
10. R.P. Lanzilotta and D. Pramer, Appl. Microbiol. 19, 307-313 (1970).
11. M.A. De Waard and J.G.M. Van Nistelrooy, Pestic. Sci., 15, 56-62 (1983).
12. Y. Uesugi, M. Katagi and O. Nada, Agric. Biol. Chem., 38, 907-912 (1974).
13. P. Leroux and M. Gredt, Phytopathol. Z., 91, 177-181 (1978).
14. J.R. Corbett, K. Wright and A.C. Baillie, The biochemical mode of action of pesticides, 2nd ed., p. 234-290, Academic Press, London (1984).

Biological activity and mode of action of diniconazole

H. Takano* and T. Kato

Takarazuka Research Center, Sumitomo Chemical Co. Ltd.,
Takarazuka, Hyogo 665, Japan

Abstract - Diniconazole ((E)-1-(2,4-dichlorophenyl)-4,4-dimethyl-2-(1,2,4-triazol-1-yl)-1-penten-3-ol) is a new triazole fungicide. The R(-) isomer of diniconazole was more fungitoxic than the S(+) isomer whereas the S(+) isomer showed plant-growth regulating activity at much lower dosages than did the R(-) isomer. Both isomers inhibited ergosterol biosynthesis in Ustilago maydis at the steps of 14α-demethylation and Δ^{22}-desaturation. The fungitoxicity of both isomers coincided with the inhibitory potency against sterol biosynthesis both in intact cells of U. maydis and in a cell-free system of Saccharomyces cerevisiae. The R(-) isomer of diniconazole bound to the cytochrome P-450 in yeast microsomes and produced a type II difference spectrum.

INTRODUCTION

Diniconazole ((E)-1-(2,4-dichlorophenyl)-4,4- dimethyl-2-(1,2,4-triazol-1-yl)-1-penten-3-ol) is a newly developed triazole fungicide showing a broad antifungal spectrum. It is effective by preventive or curative treatment in controlling powdery mildews, rusts, and Venturia and Cercospora diseases. The target of diniconazole in fungi has been shown to be 14α-demethylation in ergosterol biosynthesis (ref. 1), as demonstrated in many azole fungicides, and thus it causes the accumulation of 14α-methyl sterols in fungi. The 14α-methyl sterols are considered to be unsuitable to maintain the stability of fungal membranes (ref. 2) and therefore the membrane function may be disturbed. Most of the azole fungicides possess some isomers due to the existence of one or more asymmetric carbon atoms in their structures. In some cases, a difference in biological activity has been recognized among isomers (ref. 3, 4). Diniconazole also has one asymmetric carbon atom and hence it is possible for the structure to exist as two optical isomers. This study was undertaken to compare the biological activity and mode of action of the R(-) and S(+) isomers of diniconazole.

ANTIFUNGAL ACTIVITY

The antifungal activity of the R(-) isomer on an agar nutrient medium was more than sixty times higher than that of the S(+) isomer in Monilinia fructicola, Venturia inaequalis, Sclerotinia sclerotiorum, Rhizoctonia solani and Penicillium digitatum (Table 1). Against Phytophthora infestans, however, both isomers were similarly toxic.

In a liquid culture, the R(-) isomer inhibited sporidial multiplication and dry weight increase of Ustilago maydis at 10^{-8} and 10^{-7}M respectively (Fig. 1A), whereas 10^{-4}M S(+) isomer was required to suppress fungal growth (Fig. 1B). It is interesting to note that the R(-) isomer was more inhibitory to sporidial multiplication than to dry weight increase, while this difference was not apparent with the S(+) isomer.

PLANT-GROWTH REGULATING ACTIVITY

In contrast, the R(-) isomer showed much weaker inhibition of plant growth than the S(+) isomer which suppressed stem growth of cucumber seedlings by 73% even at a concentration of 3.1ppm (Fig. 2). On the other hand, the R(-) isomer caused no significant effect on growth at less than 50ppm.

TABLE 1. Antifungal activity of diniconazole.

isomer	ED_{50}^{a} (μM)					
	Pi[b]	Mf	Vi	Ss	Rs	Pd
R(-)	12	0.025	0.003	0.003	0.52	0.26
S(+)	16	2.7	0.49	0.2	58	55

[a]Fungicide concentration required to inhibit mycelial growth by 50% on PDA medium.
[b]Fungal species. Pi; Phytophthora infestans , Mf; Monilinia fructicola , Vi; Venturia inaequalis , Ss; Sclerotinia sclerotiorum , Rs; Rhizoctonia solani , Pd; Penicillium digitatum.

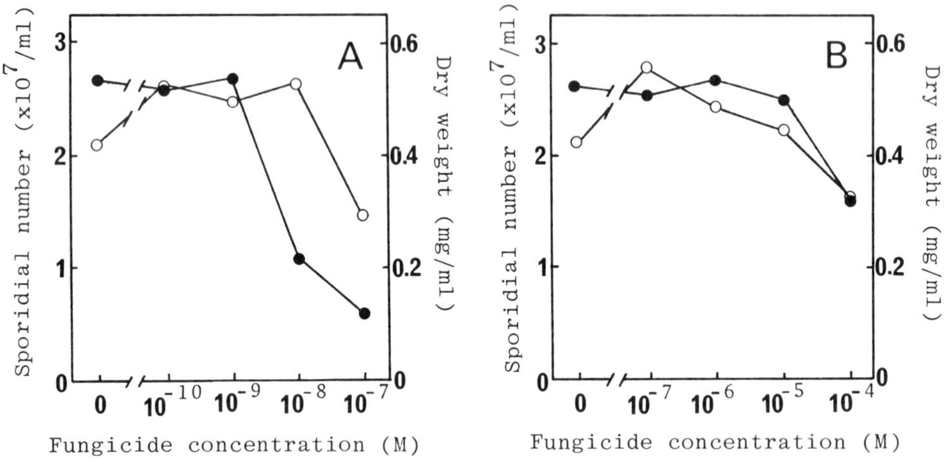

Fig. 1. Effects of diniconazole on sporidial multiplication (●) and dry weight increase (o) of U. maydis.
(A), the R(-) isomer (B), the S(+) isomer

EFFECT ON STEROL BIOSYNTHESIS

The R(-) isomer inhibited ergosterol biosynthesis in U. maydis at concentrations higher than 10^{-9}M with the concomitant accumulation of 24-methylenedihydrolanosterol (Table 2). Furthermore, obtusifoliol also accumulated at fungicide concentration greater than 10^{-8}M. The results showed that the R(-) isomer inhibited 14α-demethylation in ergosterol biosynthesis. At concentrations of 10^{-9} and 10^{-8}M, $\Delta^{5,7}$-ergostadienol also accumulated in the R(-) treated cells. This result indicated that Δ^{22}-desaturation was also inhibited by the fungicide. However, $\Delta^{5,7}$-ergostadienol was not detected at fungicide concentrations higher than 10^{-7}M. This suggested that the inhibition of 14α-demethylation, which is known to occur prior to Δ^{22}-desaturation (ref. 5), was completely interrupted.

On the other hand, the S(+) isomer moderately inhibited ergosterol biosynthesis both at the steps of 14α-demethylation and Δ^{22}-desaturation at more than 10^{-6}M. Its inhibitory potency against the former reaction seemed to be weaker than that against the latter and this is probably the reason why the accumulation of obtusifoliol was not found in the S(+)-treated culture.

In a cell-free system of Saccharomyces cerevisiae, the R(-) isomer greatly inhibited incorporation of radioactivity from ^{14}C-labeled mevalonate into 4-desmethyl sterol fraction, whereas the S(+) isomer moderately inhibited the conversion of the radiolabeled substrate into 4-desmethyl sterols (Fig. 3). Radioactivity accumulated in the fraction of 4,4-dimethyl sterols as a result of the treatment by either isomer. It was apparent that the R(-) isomer was more inhibitory to 14α-demethylation than the S(+) isomer also in the cell-free system.

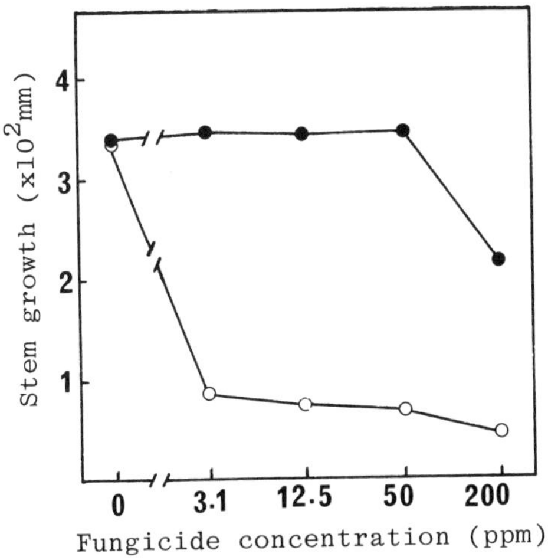

Fig. 2. Effect of diniconazole on stem growth of cucumber seedlings.
(●), the R(-) isomer
(○), the S(+) isomer
Fungicide was applied 14 days before observation.

TABLE 2. Sterol composition in U. maydis treated by diniconazole.

Treatment	Sterol[a]				
	A	B	C	D	E
None	76[b]	-[c]	21	-	3
R(-) 10^{-9} M	68	17	-	-	15
R(-) 10^{-8} M	35	13	-	12	39
R(-) 10^{-7} M	19	-	-	22	58
R(-) 10^{-6} M	17	-	-	21	62
S(+) 10^{-7} M	74	-	20	-	5
S(+) 10^{-6} M	64	26	-	-	10
S(+) 10^{-5} M	46	37	-	-	17
S(+) 10^{-4} M	46	36	-	-	18

[a]Sterols were identified by relative retention times on gas-liquid chromatography and by fragmentation patterns using gas-liquid chromatography-mass spectrometry.
A, ergosterol ; B, $\Delta^{5,7}$-ergostadienol ; C, Δ^{7}-ergostenol ;
D, obtusifoliol ; E, 24-methylenedihydrolanosterol
[b]% of total sterols.
[c]- means not detected.

BINDING OF THE R(-) ISOMER OF DINICONAZOLE TO CYTOCHROME P-450 IN YEAST MICROSOMES

14α-Demethylation in the ergosterol biosynthetic pathway has been thought to be catalyzed by microsomal monooxygenase system containing cytochrome P-450 (ref. 6). The addition of the R(-) isomer to microsomes containing cytochrome P-450 prepared from S. cerevisiae gave rise to a type II difference spectrum with the absorbance maximum at 425nm and minimum at 390nm (Fig. 4). Similar spectral change was also caused by other azole fungicides, such as diclobutrazol and prochloraz (ref. 7,8). This result indicated that a nitrogen atom of the triazole ring in diniconazole, perhaps in position 4, binds to the cytochrome P-450 iron (ref. 9).

DISCUSSION

It is interesting that both isomers of diniconazole showed moderate antifungal activity against P. infestans which belongs to the Oomycete fungi. Similar results were also obtained with other sterol-inhibiting fungicides (ref. 10). These results indicate that the azole fungicides including diniconazole may possess a secondary site of action apart from the inhibition of ergosterol biosynthesis, since the Oomycete fungi are generally considered to be incapable of synthesizing sterols (ref. 11).
The R(-) isomer was more potent in fungitoxicity than the S(+) isomer, although the difference of activity between isomers varied among fungal species. However, the S(+) isomer was more than sixty times stronger in plant-growth regulating activity than the R(-) isomer. The reversal in the order of

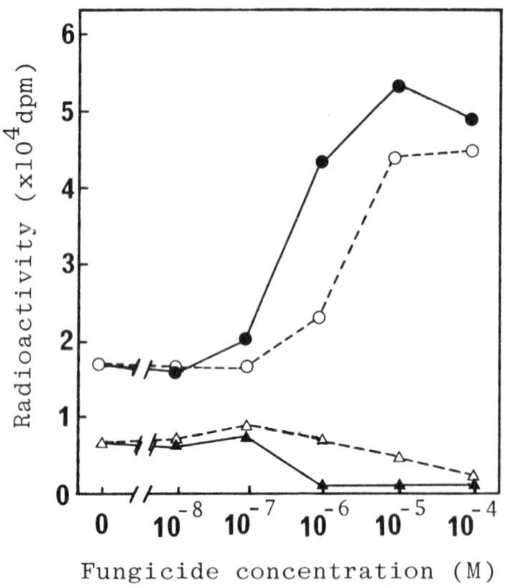

Fig. 3. Incorporation of radioactivity from ^{14}C-mevalonic acid into sterol fractions in a cell-free system of S. cerevisiae.
▲,△ ; 4-desmethyl sterols
●,○ ; 4,4-dimethyl sterols
(—); the R(-) isomer
(---); the S(+) isomer

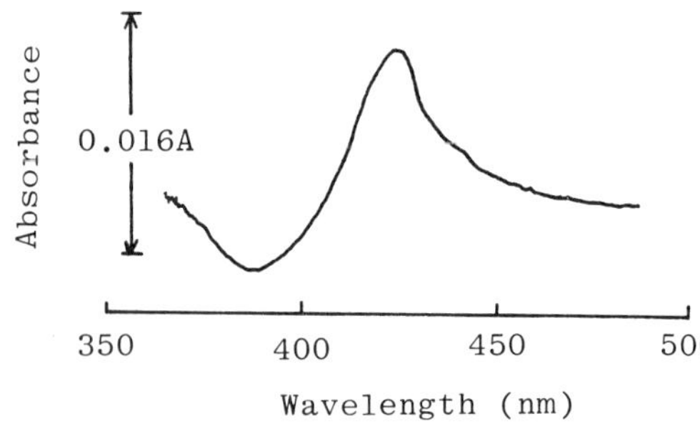

Fig. 4. The difference spectrum of yeast microsomes with the R(-) isomer of diniconazole. One ml of microsomal preparation (2mg protein/ml) was added to both sample and reference cuvettes. Then 10μl of the fungicide solution in methanol was added to the sample cuvette (final concentration of the fungicide was 10^{-5}M). The same amount of solvent was added to the reference cuvet.

activity against fungi and plants of enantiomers has also been reported with paclobutrazol which has a chemical structure similar to diniconazole (ref. 4). The plant-growth regulating activity of diniconazole may be due to the interference with gibberellin biosynthesis as reported with a sister compound, uniconazole ((E)-1-(4-chlorophenyl)-4,4-dimethyl-2-(1,2,4-triazol-1-yl)-1-penten-3-ol) (ref. 12).

The binding of the R(-) isomer to microsomal cytochrome P-450 is strong evidence of direct interaction between the fungicide and the enzyme catalyzing 14α-demethylation in ergosterol biosynthesis. Both isomers inhibited Δ^{22}-desaturation in ergosterol biosynthesis of U. maydis. Triarimol, a pyrimidine-type sterol inhibitor, is also known to inhibit Δ^{22}-desaturation in addition to inhibiting 14α-demethylation and reduction of the C-24(28) double bond (ref. 13). Hata *et al*. reported the involvement of cytochrome P-450 in Δ^{22}-desaturation in ergosterol biosynthesis (ref. 14). Therefore, it is suggested that diniconazole also binds to the cytochrome P-450 catalyzing Δ^{22}-desaturation.

REFFERENCES

1. H. Takano, Y. Oguri and T. Kato, J. Pesticide Sci. 8, 575-582 (1983).
2. K. Bloch, Curr. Topics Cell. Regul. 18, 289-299 (1981).
3. B. C. Baldwin and T. E. Wiggins, Pestic. Sci. 15, 156-166 (1984).
4. B. Sugavanam, Pestic. Sci. 15, 296-302 (1984).
5. T. Nishino, S. Hata, T. Osumi and H. Katsuki, J. Biochem. 88, 247-254 (1980).
6. M. Ohba, R. Sato, Y. Yoshida, T. Nishino and H. Katsuki, Biochem. Biophys. Res. Commun. 85, 21-27 (1978).
7. T. E. Wiggins and B. C. Baldwin, Pestic. Sci. 15, 206-209 (1984).
8. J. L. Riviere, Pestic. Biochem. Physiol. 19, 44-52 (1983).
9. C. R. Jefcoate, Methods Enzymol. 52, 258-279 (1978).
10. H. Buchenauer, Z. Pflanzenkr Pflanzenshutz 86, 341-354 (1979).
11. J. W. Hendrix, Mycologia 58, 307-312 (1966).
12. K. Izumi, Y. Kamiya, A. Sakurai, H. Oshio and N. Takahashi, Plant Cell Physiol. 26, 821-827 (1985).
13. N. N. Ragsdale, Biochim. Biophys. Acta 380, 81-96 (1975).
14. S. Hata, T. Nishino, M. Komori and H. Katsuki, Biochem. Biophys. Res. Commun. 103, 272-277 (1981).

MG-191: a new selective herbicide antidote

F. Dutka* and T. Kőmives

Central Research Institute for Chemistry, Hungarian Academy of Sciences, H-1525 Budapest, P.O.Box 17, Hungary

Abstract. - The concept is described that led to the discovery of a new, highly active antidote (MG-191: 2-dichloromethyl-2-methyl-1,3-dioxolane) against phytotoxic injury to corn by thiocarbamate herbicides. The chemical acts by increasing herbicide detoxication potency of the crop plant.

INTRODUCTION

Chemical antidotes are compounds reducing or eliminating the injury by herbicides to crops without decreasing the herbicidal potential. 1,8-Naphthalic anhydride was the first commercial antidote to protect corn against injury from the herbicide EPTC (S-ethyl-N,N-dipropylthiocarbamate) (ref.1). R-25788 (2,2-dichloroacetyl-N,N-diallylamine) was introduced by Stauffer Chemical Co. as a combined formulation with EPTC under the trade name "Eradicane", a highly selective preparation to corn (ref.2). Later several other antidotes were developed and patented primarily as sorghum protectants against chloroacetanilide herbicide injury (ref. 1, 3, 4).

Research for new herbicide antidotes is prompted by biological, environmental and economic reasons. Discovery of new antidotes would possibly permit the widespread use of herbicides having little or no crop selectivity and of older herbicides that are highly cost effective and have been shown to be safe to the environment. Moreover, a major breakthrough in this area could have a more significant impact on the future course of chemical weed control than the successful development of several new herbicides (ref. 5).

CONCEPT

Structure-activity studies led to the conclusion that the most active antidotes of EPTC to corn were N,N-disubstituted-2,2-dichloroacetamides; generally, mono- and trichloroacetamides proved to be less active. Variations in nitrogen substituents resulted in different degrees of protective activity. Usually, dichloroacetamides having two substituents on the nitrogen atom are more effective than those with one substituent (ref. 2, 6, 7).

These results seemed to suggest that the antidotal activity can primarily be attributed to the dichloroacetamido group

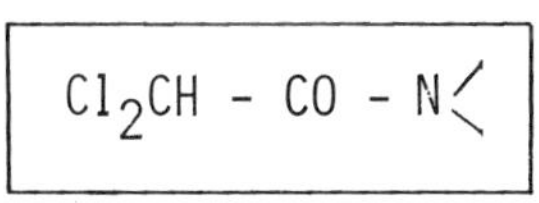

in the molecule. Probably, this is the reason why research was focussed on this family of compounds and antidotes for use in corn developed after the discovery of R-25788 are also dichloroacetamides:

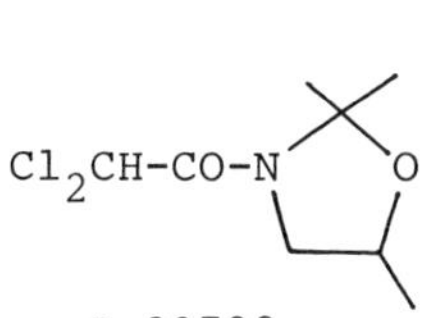

R-29788

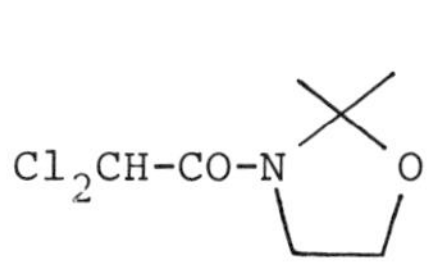

R-28725

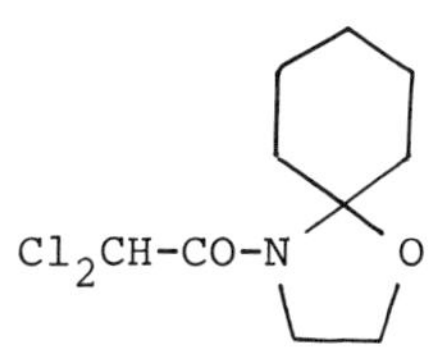

AD-67

The method by which antidotes were recognized appears to be random screening rather than rational deduction. Even, the degree of success suggested that other methods are not vital and the possibility of finding new species behaving like R-25788 in specific protecting ability was thought to be remote (ref. 3).

Our earlier chemical reactivity-biological activity studies on amides, esters, and thiolesters led to the conclusion that compounds having chemically activated acyl groups (dichloroacetyl, substituted benzoyl, etc.) may generally have antidotal activity, i.e., in dichloroacetamides not the dichloroacetamido group but the

$$\boxed{Cl_2CH\text{-}CO\text{-}}$$

function is primarily responsible for the antidotal activity (ref. 7, 8).

Now, it was interesting to speculate that this was also true for chemicals containing preformed (potential) chemically activated acyl group(s). Such compounds are acetals, ketals, etc., in which the central carbon atom is a "pro--acyl carbon" (ref. 9). To examine the reliability of this hypothesis, the antidotal activity of the compound

$$Cl_2CH\text{-}CH(OCH_2CH_3)_2$$

was investigated. Since it showed high biological activity in laboratory and field experiments, a large number of derivatives characterized by the general formula

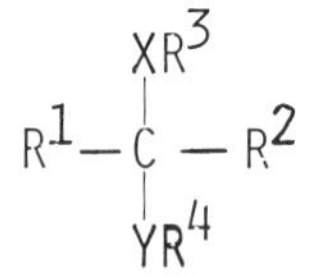

X = O, S, NR^5
Y = O, S, NR^5
R^1-R^5 = alkyl, aryl

were synthesized and analyzed for antidotal potency in laboratory, green house and field experiments. Compound MG-191 showed optimum biological activity.

MATERIALS AND METHODS

MG-191 was synthesized according to ref. 10. Technical grade MG-191 for field studies as well as thiocarbamate herbicides (EPTC; vernolate: S,N,N-tripropylthiocarbamate; butylate: S-ethyl-N,N-diisobutylthiocarbamate) were gifts from Nitrokemia Chemical Works (Fűzfőgyártelep, Hungary). Enzyme substrates, solvents, and other chemicals not indicated above were from commercial sources and were used without further purification. Seeds of maize (Pioneer 3950 hybrid) were grown in sand culture for bioassay (ref. 8) or were germinated in petri dishes for assays of levels of glutathione (GSH, ref. 11), cytosolic and microsomal glutathione S-transferase (GST, EC. 2.5.1.18, ref. 12) and glutathione reductase (GR, EC. 1.6.4.2, ref. 13). Field experiments were made in a randomized complete block design with four replications at two locations in 1982-1985 (ref. 14).

CHEMICAL STRUCTURE

MG-191 is a colourless liquid (bp. 85°C/2 kPa) and has the formula

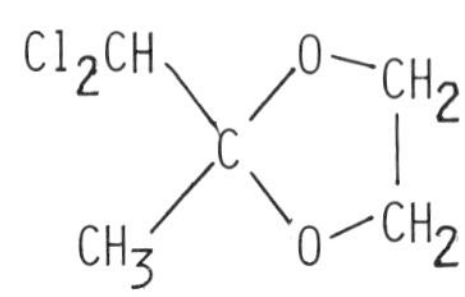

ANTIDOTAL PROPERTIES

Results of laboratory and field experiments are in good agreement and indicate superior protective activity of MG-191 against EPTC and other thiocarbamate herbicides (Tables 1,2) even in sandy soil, in which these herbicides are most toxic to corn (ref. 14).

TABLE 1. Effectiveness of MG-191 to counteract thiocarbamate herbicide injury (expressed in plant height relative to untreated control) to corn (Pioneer 3950 hybrid) in sand culture after 21 days of treatment

Herbicide	Dose (μM)	MG-191 (μM)				
		0	0.1	1	3	10
EPTC	100	35[a]	50[a]	91	101	103
Vernolate	100	20[a]	35[a]	93	100	101
Butylate	200	33[a]	58[a]	95	102	100

[a]Significantly different from controls at the 5% level.

TABLE 2. Plant injury (expressed in per cent of plants showing visual herbicide damage after 21 days of treatment) and corn yield[a] (per cent of hand weeded control) in response to herbicide + MG-191 applications in 1985 (Pioneer 3950 hybrid; calcaraeous sandy soil; humus: 0.6%; clay: negligible)

Herbicide	Dose (kg/ha)	MG-191 (wt% of herbicide a.i.)			
		0		10	
		yield	injury	yield	injury
EPTC	6.4	98	52[b]	103	0
	12.8	59[b]	85[b]	92	4.5
	19.2	31[b]	93[b]	94	6.0
Vernolate	6.4	75[b]	52[b]	118	0
	12.8	58[b]	89[b]	98	6.9
	19.2	49[b]	85[b]	102	15.0[b]

[a]Yield for hand weeded control was: 6.5 t/ha;
[b]Significantly different from control at the 5% level.

No protection of weeds was observed when mixtures of EPTC and MG-191 were applied under laboratory and field conditions (data not shown). When applied alone, MG-191 showed no phytotoxic effects up to 100-fold of its normal application rate (10 μM). The antidotal activity of MG-191 is concentration dependent. Significant activity was found at levels higher than 0.1 μM, and complete protection against EPTC (100 μM), vernolate (100 μM) and butylate (200 μM) was achieved by applying MG-191 concentrations higher than 3 μM (Table 1, 2).

MODE OF ACTION

In corn seedlings treated with MG-191 increased levels of glutathione and glutathione-related enzymes were found (Table 3). Since thiocarbamate herbicides are metabolized via sulfoxide formation (ref. 15, 16) and this derivative is detoxified by corn plants with glutathione conjugation, our data are consistent with the possibility that MG-191 protects corn by increasing the plant's ability to detoxify the herbicide.

TABLE 3. Effect of 24-h exposure of 2.5-day old corn seedlings (Pioneer 3950 hybrid) to different concentrations of MG-191 on the levels of root GSH, shoot cytosolic and microsomal GST, and shoot GR

MG-191 (μM)	GSH (μM/g fwt)	Cytosolic GST (μM/min.mg protein)	Microsomal GST	GR
			(nM/min.mg protein)	
0	0.49±0.05	0.96±0.11	69± 9	0.17±0.02
0.1	0.48±0.07	1.09±0.17	67±10	0.23±0.03
1	0.77±0.10	1.63±0.08	89±18	0.25±0.07
10	1.03±0.10	2.31±0.34	101±21	0.33±0.06

CONCLUSIONS

The acylation hypothesis based on results of chemical reactivity-bioactivity studies proved to be a fruitful concept for discovering an original, highly active herbicide antidote. Studies to find experimental evidence for the (bio)transformation of the pro-acyl carbon to a genuine acyl carbonyl carbon are in progress.

ACKNOWLEDGEMENT

This work was supported by a grant from Nitrokemia Chemical Works (Fűzfőgyártelep, Hungary).

REFERENCES

1. K.K. Hatzios, Adv. in Agronomy 36, 265-316 (1984).
2. F.M. Pallos, R.A. Gray, D.R. Arneklev and M.E. Brokke, J. Agric. Food Chem. 23, 821-822 (1975).
3. C. Parker, Pestic. Sci. 14, 40-48 (1983).
4. I. Paradies, E. Ebert and E.F. Filstner, Pestic. Biochem. Physiol. 15, 209-212 (1981).
5. C.G. McWhorter, Weed Sci. 32, 850-855 (1984).
6. G.R. Stephenson, N.J. Bunce, R.I. Makowski and J.C. Curry, J. Agric. Food Chem. 26, 137-141 (1978).
7. F. Dutka and T. Kőmives, IUPAC Pesticide Chemistry: Human Walfare and Environment (Eds. J. Miyamoto et al.) pp. 213-218, Pergamon Press, Oxford (1982).
8. F. Dutka, T. Kőmives, A.F. Márton, Á. Hulesch, K. Fodor-Csorba and M. Kárpáti, Proc. 19th Hung. Ann. Meet. Biochem. 253-256 (1979).
9. Y. Chiang, A.J. Kresge, P. Saloma and C.I. Young, J. Amer. Chem. Soc. 96, 4494-4499 (1974).
10. D.C. Mueller and R. Seyferth, J. Amer. Chem. Soc. 91, 1754-1758 (1969).
11. C. Fedtke, Z. Pflanzenkrankheiten u. Pflanzenschutz, IX. 141-146 (1981).
12. V.A. Kőmives, T. Kőmives and F. Dutka, Cereal Res. Comm. 13, 253-257 (1985).
13. T. Kőmives, V.A. Kőmives, M. Balázs and F. Dutka, Proc. Brit. Crop. Prot. Conf. Vol. 3, 1155-1162 (1985).
14. E. Széll, I. Csala, F. Fodor, T. Kőmives and F. Dutka, Cereal Res. Comm. 13, 55-61 (1985).
15. M.M. Lay and J.E. Casida, Pestic. Biochem. Physiol. 6, 442-456 (1976).
16. T. Kőmives, M. Balázs, V. Kőmives and F. Dutka, Cytochrome P-450, Biochemistry, Biophysics and Induction. Proc. 5th Intl. Conf. Cytochrome P-450, (Eds. L. Vereczkey, K. Magyar), pp. 451-454, Akadémiai Kiadó, Budapest (1985).

Dose-related neurophysiological effects and penetration studies of a congeneric series of synthetic pyrethroids

D.S. Buckley, C.J. Brealey, M.G. Ford and L.D. Leake

Department of Biological Sciences, Portsmouth Polytechnic, Hants. PO1 2DY U.K.

Wellcome Research Laboratories, Ravens Lane, Berkhamsted, Herts. HP4 2DY U.K.

Abstract - The pharmacokinetics and pharmacodynamics of a congeneric series of substituted benzyl cyclopropane-1-carboxylate pyrethroids and permethrin have been studied in houseflies. Target site data from a haltere preparation show that the molar threshold concentration (MTC) for neurotoxicity is positively correlated with both knockdown and kill activities. MTC is well defined by physicochemical characteristics of the pyrethroids (a partition term from ΔRm; R, Swain-Lupton resonance parameter describing π-orbital electron density; and 3B4, a Sterimol parameter describing size of a meta substituent). Pharmacokinetic parameters vary widely between compounds but could not be correlated with activity using linear regression methods. Although these pyrethroids can be classified as Type I, some of them showed Type II activity at high concentrations.

INTRODUCTION

The use of compartmental models to study the pharmacokinetics of pyrethroid insecticides has resulted in a set of parameters which describe measured data and correlate with observed toxicities e.g. for the mustard beetle (Phaedon cochlaeariae Fab. (ref. 1,2). However, until recently, measures of pharmacodynamic potency have been less well correlated with toxic action. Use of the housefly has enabled us to measure both reliable pharmacokinetic values and neurophysiologically-derived pharmacodynamic data on a single insect species.

In this study, the pharmacokinetics and pharmacodynamics have been measured for a congeneric series of synthetic (±)trans-pyrethroids (QSAR 2,4,6,12,14 and 18) synthesised at Portsmouth (>96% by GC), together with those for (±) 40/60 cis, trans-permethrin (purified technical material, >96% by GC) (toxicology and pharmacodynamics) and (±)trans-permethrin (98% pure by HPLC) (pharmacokinetics). The presence of the (±)cis-isomer of permethrin present in toxicological and pharmacodynamic experiments may complicate the comparisons with the other compounds (all pure (±) trans isomers). However, the ranges of the observed insecticidal and pharmacodynamic activites (500-fold) are much greater than differences attributable to the activities of cis and trans isomers (approximately 2-fold). To examine the pharmacokinetics, the penetration of the compounds after topical application was followed by measuring the amounts of insecticide in acetone surface washes up to 24 hours. For pharmacodynamics, several parameters of the neurophysiological response were measured after exposing isolated predominantly sensory haltere nerve preparations (ref. 3,4) to a range of concentrations of the compounds for various periods of time.

Finally, as the physico-chemical properties of these pyrethroids have been well described quantitative structure-activity relationships could be readily analysed.

THE QSAR PYRETHROIDS

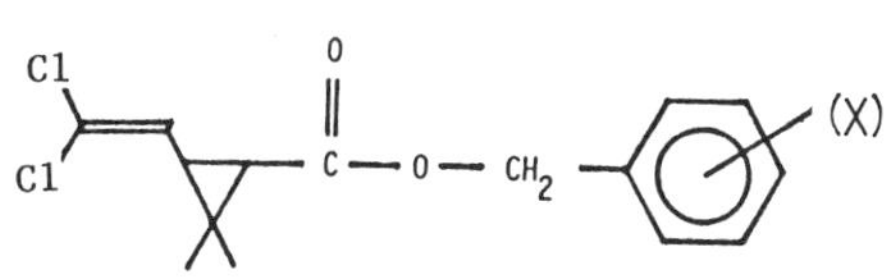

Compound	Substituted benzyl alcohol (X)
QSAR 2	2,3,4,5,6-pentafluorobenzyl
4	2,3,4,5,6-pentamethylbenzyl
6	3-iodobenzyl
12	3,5,-dinitrobenzyl
14	2,6-dichloro-3-nitrobenzyl
18	3-nitrobenzyl

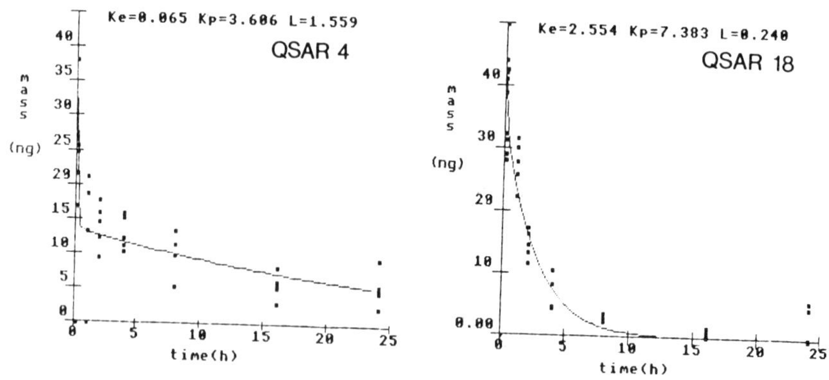

Figure 1. Two-compartment model fitted to data for penetration of QSAR 4 and 18.

TOXICOLOGY, PHARMACOKINETIC AND PHARMACODYNAMIC PROPERTIES OF THE PYRETHROIDS

This group of QSAR pyrethroids was chosen from a larger set of more than 20 compounds synthesised to provide a series with a large range of relatively uncorrelated physico-chemical properties. These features have resulted in a wide spectrum of potency both in vivo and at the target site (Table 1). Thus killing activities range over 500-fold; knockdown activities over 5000fold; and neurotoxic activity (MTC's) over 10^6-fold. As observed previously (ref. 5,6) there is some correlation between kill and knockdown (correlation coefficient for these data = 0.72). As an insecticide the pentafluoro-analogue, QSAR 2, approaches permethrin in killing activity and bioallethrin in knockdown activity

Pharmacokinetics - the two compartment model describing pharmacokinetics (ref. 1) fitted the observed data well for the loss of material from the surface of the flies for up to 24 hours, allowing estimates to be made of kp, ke and λ (Table 1 and Figure 1). The estimated M_0 values (dose at zero time) were very close to the nominal dose (41ng) in all cases. Penetration curves for various compounds are clearly different. For example, QSAR 6 (a compound which penetrates and is eliminated rapidly) shows total loss of material from the surface in 24 hours. Controls for mechanical loss and evaporation

TABLE 1 Toxicological, pharmacokinetic and pharmacodynamic properties of the pyrethroids

	TOXICOLOGY		PHARMACOKINETICS				PHARMACODYNAMICS		
Compound	Killing[a] Activity	Knockdown[b] Activity	M_0^c	kp[d]	ke[e]	f	MTC[g]	Slope[h]	Intercept[j]
Permethrin	1.39	0.50	39	5.01	0.15	0.76	-13	-0.095	-0.12
QSAR 2	1.08	0.82	41	2.10	0.17	2.25	-12	-0.24	-0.98
QSAR 4	-0.29	-2.26	35	3.61	0.07	1.56	- 9	-0.28	-1.25
QSAR 6	-0.04	-0.88	42	7.38	2.55	0.24	- 7	-0.24	0.46
QSAR 12	-1.31	-1.59	41	3.22	0.31	0.43	- 7	-0.28	-0.54
QSAR 14	-1.00	-0.18	44	5.98	0.21	1.23	- 9	-0.060	0.97
QSAR 18	-0.73	-0.38	37	0.07	0.08	12.39	- 7	-0.20	0.10

Note

a log ($1/LD_{50}$) μg/fly♀ at 48h after topical application of 0.3 μl Cellosolve solution of insecticide to the dorsal thorax.
b log (1/dose) μg/fly♀ to produce knockdown in individual flies in 10 min.
c dose (ng) estimated at zero time
d first order penetration rate constant (h^{-1})
e first order elimination rate constant (h^{-1})
f ratio of the average capacities of the two compartments (ref. 1)
g Molar threshold concentration for pyrethroid-induced activity (logM)
h Slope of log plot of time to onset of activity vs. log molarity of irrigating solution
j Intercept of log plot of time to onset of activity vs log molarity of irrigating solution

showed this was not an artifact. By contrast, QSAR 4 (moderate penetration, but very slow elimination) and QSAR 18 (slow penetration and elimination) retained more than 60% of the dose at 24 hours. In general, the penetration process is faster in houseflies than mustard beetles, with kp values ranging from 0.08 - 7.38 h^{-1} in the fly but only 0.008 - 0.44 h^{-1} for the beetle (ref. 7). Furthermore, evidence suggests that penetration will not be saturated at the dose applied to these experiments (ref. 2).

Elimination rates were also faster in the housefly (ke 0.02 - 2.6h^{-1}) compared with the mustard beetle (ke 0.008 - 0.44h^{-1}) (ref. 7) presumably reflecting the much greater metabolic capacity of the fly.

Distribution of the insecticide between compartments however, was very similar in both species (values of λ for flies were 0.24-9.8 and for beetles 0.17-11.0 - unpublished data).

Pharmacodynamics - neurotoxicities (MTC) covered a 10^6-fold range (Table 1), with the most toxic compound (QSAR 2) approaching permethrin in activity. The slopes of the dose-response curves (Table 1), fell into two groups of parallel lines. The four with steep slopes were the least potent, whereas the three with the shallow slopes included the two most potent compounds, permethrin and QSAR 2. Finally, the time to onset of response measured under standard conditions by the intercept of the plots (Table 1) showed that there is no simple correlation between speed of action at the target site and knockdown (correlation coefficient 0.33). For example, QSAR 14 acts rapidly *in vitro* and gives fast knockdown, whereas QSAR 4 which also has rapid action *in vitro*, is the slowest knockdown compound of the group.

In addition, compounds were monitored for Type I and Type II effects over a range of concentrations (10^{-14} - 10^{-3} M). Some compounds in this set (previously classified as Type I, ref. 8) produced a spectrum of neurophysiological effects. At high doses both QSAR 4 and permethrin produced a pattern of response lacking repetitive bursts of action potentials. This pattern is more often associated with Type II compounds (usually containing the α-cyano group). It seems likely that classification of neurophysiological response in terms of Type I and II compounds is not a clear one. Several effects may be observed which depend on applied dose, elapsed time, and type of nerve preparation (ref.9)

TABLE 2 Physico-chemical properties of the pyrethroids

Compound	Partition[1] Term	Swain-Lupton[2] Resonance, R	Sterimol[3] 3B4
Permethrin	-4.28	-0.26	5.89
QSAR 2	-6.06	-1.16	1.35
QSAR 4	-5.30	-0.48	2.04
QSAR 6	-4.84	-0.07	2.15
QSAR 12	-5.04	0.10	2.44
QSAR 14	-5.62	-0.23	2.44
QSAR 18	-4.24	0.05	2.44

Note

1 Slope of a plot of ΔRm vs. proportions (acetone/water) of the TLC-eluting solvent, indicating the sensitivity of the partition process to solvent composition (ref. 10).
2 Describes π-orbital electron density of the benzyl alcohol moiety (ref. 11).
3 A Verloop Sterimol parameter defining the maximum breadth of substituents at the 3-position for this series of pyrethroids (ref. 12).

A RATIONAL BASIS FOR THE BIOLOGICAL ACTIVITY OF THESE PYRETHROIDS

In attempting to separate the various factors contributing to and limiting the toxicity of pyrethroids, we have examined pharmacokinetic and target site effects in the same organism. Both knockdown and killing activity are correlated with neurotoxicity measured as the minimum concentration to produce a neurotoxic effect (expressed as MTC), equations 1 and 2:

$$\text{Kill activity} = -3.38 - 0.36\,(\text{MTC}) \qquad n = 7,\ \sigma = 0.58,\ r = 0.86 \qquad (1)$$

$$\text{Knockdown activity} = -3.23 - 0.29\,(\text{MTC}) \qquad n = 7,\ \sigma = 0.91,\ r = 0.66 \qquad (2)$$

The target site activity (measured by MTC) is very well described by two of the physico-chemical parameters listed in Table 2 which define binding at the receptor in terms of electron-distribution (R) and shape (3B4) of the substituents in the benzyl ring:

$$\text{MTC} = -4.24 + 5.16\,(R) - 1.27\,(3B4) \qquad n = 7,\ \sigma = 0.34,\ r = 0.98 \qquad (3)$$

It is appropriate to use multiple linear regression to define this relationship since the correlation between R and 3B4 is very weak ($r = 0.28$). Thus high neurotoxicity (ie low MTC value) is associated with electron-donating substituents and, within the narrow size range of the substituents included in this study, a large maximum breadth at C_3 of the aromatic ring of the alcohol moiety. The partition term (Table 2) makes no significant contribution to the definition of MTC for this limited group of analogues where the values for the partition term span a relatively narrow range. However, when the range of lipophilicities of analogues is wider, this term becomes important. This suggests that the lethal effect on flies depends primarily on the potency of a compound at the target site. However, the difference in the strengths of these correlations (knockdown is correlated more weakly with MTC) suggests that other, unidentified variables affect the knockdown process. This is unlikely to arise from errors in the assessment of knockdown and kill since the variances for the two assays are almost identical ($\sigma^2 = 1.2096$ for knockdown and 1.05929 for kill). Although compounds of different potencies show markedly differing pharmacokinetic properties, it has not been possible to determine a general relationship between insecticidal activities and these attributes using linear regression methods on a data set of this size. The pharmacokinetic parameters are themselves highly correlated in these data for houseflies, as shown in the matrix for values of r:

	kp	ke
ke	0.76	-
λ	-0.85	-0.90

The relationships between pharmacokinetics, action and structure of these pyrethroids are under investigation using multivariate techniques.

These studies demonstrate that for these pyrethroids high target site potency is essential for good insecticidal activity. However, other factors (as yet unidentified) may modify this neurotoxicity to result in compounds with varying speeds of action.

REFERENCES

1. M.G. Ford, R. Greenwood and P.J. Thomas, Pestic. Sci., 12, 175-198 (1981).
2. M.G. Ford, R. Greenwood and P.J. Thomas, Pestic. Sci., 12, 265-284 (1981).
3. D.C. Sandeman and H. Mark , J. exp. Biol., 85, 43-60 (1980).
4. P.E. Burt and R.E. Goodchild, Rothamsted Experimental Station Rep., pp 127-128 (1981).
5. G.G. Briggs, M. Elliott, A.W. Farnham and N.F. James, Pestic. Sci., 5, 643 (1974).
6. M.G. Ford, Pestic Sci., 10, 39-49 (1979).
7. R.M. Szydlo, M.G. Ford, R. Greenwood and D.W. Salt, QSAR and Strategies in the Design of Bioactive Compounds (J.K. Seydel (Ed.) pp. 219-228, VCH, USA, (1984).
8. D.W. Gammon, M.A. Brown and J.E. Casida, Pestic. Biochem. Physiol., 15, 181-191 (1981).
9. S. Buckley, M.G. Ford, L.D. Leake, D.W. Salt, P.E. Burt, M.D.V. Moss, C.J. Brealey and D.J. Livingstone, Proc. 6th European Symp. on Chemical Structure - Biological Activity: Quantitative Approaches (in press).
10. M.G. Ford, R. Greenwood, L.D. Leake, R.M. Szydlo and C.H. Turner, Pestic. Sci., 16, 673-683 (1985).
11. C.G. Swain and E.C. Lupton, J. AM. Chem. Soc., 90, 4328 (1968).
12. A. Verloop, W. Hoogenstraaten and J. Tipker, Drug Design, 7, (E.J. Ariens, Ed.) pp. 165-207. Academic Press, New York (1970).

ACKNOWLEDGEMENTS: D.S.B. was supported by a SERC (CASE) award and Wellcome Foundation. P.E. Burt, R.J. Hart and M.D.V. Moss are thanked for assistance to D.S.B.

The physical form of flamprop-ethyl herbicide on sprayed leaves influences absorption and subsequent efficacy

F.D. Hess[a], J. R. Goss, D.L. Bucholtz and R.H. Falk

Department of Botany and Plant Pathology, Purdue University, West Lafayette, IN 47907; Shell Development Co., Modesto, CA 95352; and Department of Botany, University of California, Davis, CA 95616

Abstract - The ethyl ester of flamprop (N-benzoyl-N-(3-chloro-4-fluorophenyl)DL-alanine) has been evaluated in the U.S. for control of wild oat (*Avena fatua* L.) in wheat (*Triticum aestium* (L.) Beauv.) and barley (*Hordeum vulgare* L.). The solubility of flamprop-ethyl in solvents commonly used in the preparation of emulsifiable concentrates is very low. During the development process several formulation mixtures were evaluated for their efficacy in control of wild oat. Formulation variables were the amount of active ingredient and the amount and type of solvent, oil and emulsifier. Two formulations that differed significantly in their efficacy were evaluated to determine the reason for the varying activities. In wild oat plants, a high degree of correlation was obtained between the presence of crystalline deposits of flamprop-ethyl on the leaf surface and reduced effectiveness of wild oat growth inhibition. The formulation yielding crystalline deposits on the leaf surface absorbed into leaf tissue less than the formulation where crystals were not apparent. The research reported here suggests that in order to assure maximum efficiency of post-emergence herbicide applications, the physical form of the active ingredient, in the end use formulation, should be determined on the leaf surface.

INTRODUCTION

The degree of weed control achieved by applying a herbicide directly to weeds present in a field (postemergence application) is ultimately controlled by the amount of herbicide absorbed by the plant. For a given amount of herbicide which adheres to the tissue after postemergence application, the degree of absorption is controlled by formulation composition, growth habit of the target weed species, and environmental conditions before and at application (1-5). Whereas the influence of plant species and environmental conditions on herbicide absorption and subsequent efficacy are well studied (6-9), the influence of formulation composition and resulting physical form of the active ingredient after application is not characterised.

A parameter that should impact herbicide efficacy is the physical form of the active ingredient on the leaf surface after application. The physical form of many purified herbicides is crystalline. To achieve efficient application, crystalline herbicides are often solubilized in organic solvents and formulated as emulsifiable concentrates. After postemergence application these herbicides may return to their crystalline form on the leaf surface (10). Because of the stable nature of crystalline substances, absorption into plant tissue and subsequent herbicidal efficacy may be adversely affected.

Flamprop-ethyl, a crystalline compound, was a development candidate from Shell Chemical Company primarily for the control of wild oat in wheat and barley production. A preliminary evaluation of weed control achieved by several formulations revealed a large range of activity for wild oat control with the same dosage of flamprop-ethyl. The purpose of this research was to determine if observed differences in herbicidal activity could be explained

---

a. Current address: Zoecon Research Institute, a division of Sandoz Crop Protection Corporation, 975 California Avenue, Palo Alto, CA 94304, USA.

by differences in formulation performance (e.g. recrystallization on the leaf surface). Two formulations of flamprop-ethyl were selected and evaluated in field, greenhouse, and laboratory studies.

FORMULATION COMPOSITION

The composition of the two formulations used in the experiments are shown in Table 1. These two were chosen for study because of a difference in herbicidal activity. Formulation 1 (the least active) contained more flamprop-ethyl and less emulsifier than Formulation 2 (the most active). Formulation 1 contained naphthenic oil and Formulation 2 paraffinic oil. Formulation 2 contained nearly twice as much emulsifier as Formulation 1. The solvent composition and concentration in the two formulations were similar.

TABLE 1. Formulation components.

Formulation Component	% Ingredient (w/w)	
	1	2
flamprop-ethyl	13.0	8.0
emulsifier	6.5	12.0
solvent	75.3	77.0
naphthenic oil	5.2	0
paraffinic oil	0	3.0

FIELD AND GREENHOUSE STUDIES

To confirm the above two formulations would yield different herbicidal performance field and greenhouse efficacy studies were conducted. The field study took place at Modesto, California, during the winter of 1981. The application volume of each formulation was equivalent to 75 L/ha and the treatments were replicated four times. At the time of application the wild oat plants where in the 1 to 2 tiller stage of growth. At all concentrations tested, Formulation 2 performed better than Formulation 1 (Fig. 1). The greatest difference in herbicidal activity between the two formulations occurred between 0.22 and 0.28 kg/ha. At these concentrations the differences in activity were 52 and 54%, respectively.

Because planned analyses would utilize greenhouse grown plants, the field efficacy study was repeated in the greenhouse. In this assay, the application volume was 75 L/ha and the treatments were replicated three times. The wild oat seedlings had two to four true leaves at the time of application. At all concentrations tested (0.06 kg/ha to 0.28 kg/ha) Formulation 1 yielded less than 20% reduction in wild oat growth (Fig. 2). With Formulation 2, reduction in growth increased as the concentration of flamprop-ethyl was increased from 0.08 kg/ha to 0.28 kg/ha. At 0.28 kg/ha Formulation 2 reduced the growth of wild oat 88%. Maximum differences between formulation performance (55 to 70%) occurred between a concentration range of 0.17 kg/ha and 0.28 kg/ha.

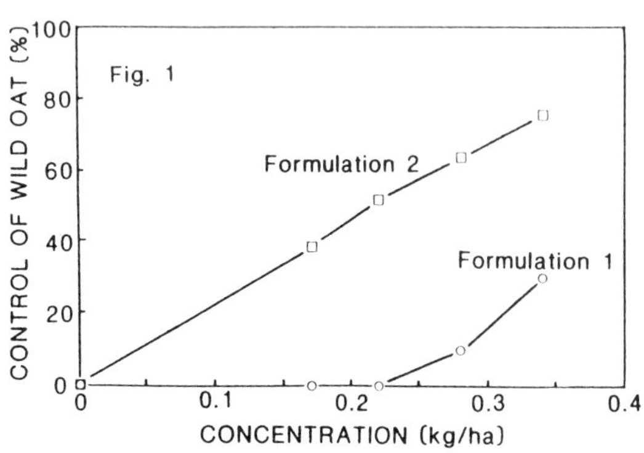

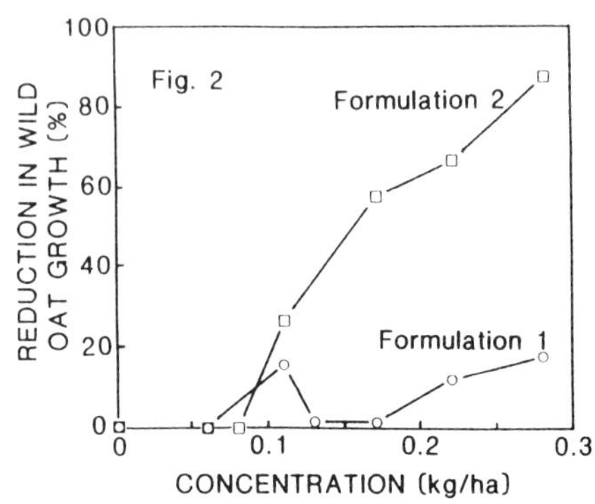

Fig. 1 and 2 The influence of flamprop-ethyl formulations on wild oats in a field (Fig. 1) and greenhouse (Fig. 2) application.

ANALYSIS OF THE PHYSICAL FORM OF FORMULATED FLAMPROP-ETHYL

Flamprop-ethyl has a low solubility in organic solvents commonly used in the preparation of emulsifiable concentrate formulations. Because of this characteristic, the physical forms of Formulations 1 and 2 were evaluated by scanning electron microscopy (SEM). In a preliminary experiment Formulation 1 (Fig. 3A and 3B) and Formulation 2 (Fig. 3C and 3D) were diluted with distilled water and drops of solution applied directly to specimen holders. The concentration of formulation in water was 5% v/v (Fig. 3A and 3C) or 10% v/v (Fig. 3B and 3D). The samples were observed in the SEM immediately after the water evaporated. Both the 5 and 10% concentrations of Formulation 1 showed large crystal development. At 5% Formulation 2 a few crystals were present, whereas at the 10% concentration no crystals were observed.

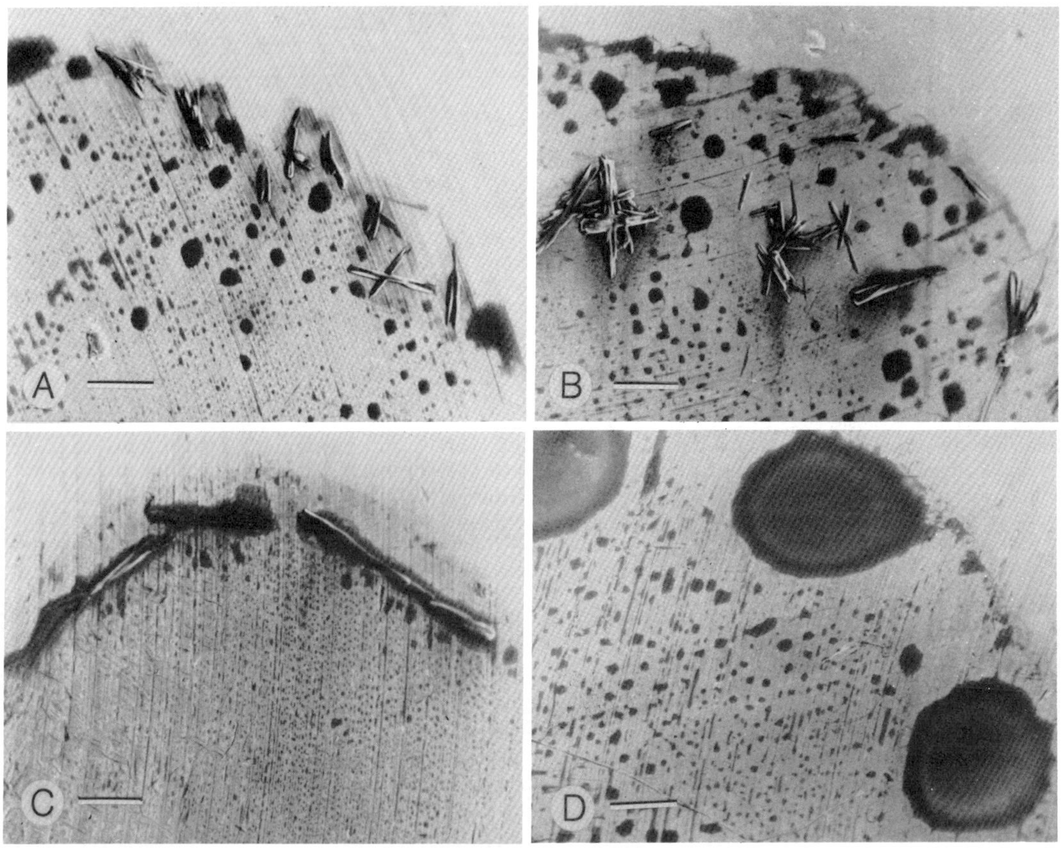

Fig. 3 A and B are Formulation 1; C and D are Formulation 2; A and C are from a 5% v/v solution in water; B and D are from a 10% v/v solution in water. Scale marker = 0.1 mm.

To characterize the physical forms of Formulations 1 and 2 on leaf surfaces, leaves from sprayed wild oat plants were evaluated with an analytical SEM. Plants at the two leaf stage of development were sprayed with an amount of Formulation 1 or 2 equivalent to 0.22 kg/ha, and at an application volume equivalent to 75 L/ha. This concentration was chosen because of the significant difference in herbicidal performance in the field (Fig. 1) and greenhouse (Fig. 2) studies. Observations were begun within 2 hours after treatment. One-cm pieces of leaf were excised from the plants and adhered, adaxial side up, to specimen holders with silver conductive paint. Samples were immediately placed in the specimen chamber of the SEM. The SEM was operated at an accelerating potential of 20 KV and the detection systems used were secondary electron and cathodoluminescence (CL). At least four leaf samples were evaluated for each parameter studied.

Leaves treated with complete formulations not containing the active ingredient, flamprop-ethyl, did not emit a CL image, whereas formulations containing active ingredient emitted a strong CL image. Therefore, CL analysis of sprayed leaves could be used to determine the location of the active ingredient. CL images on leaves sprayed with Formulation 1 were conspicuous (Fig. 4). The periclinal surface of epidermal cells were covered with large numbers of small crystals. The components of Formulation 1 remaining after the spray drops dried on the leaf surface did not keep flamprop-ethyl dissolved. Formulation 2 resulted in amorphous spray deposits on the cuticle surface (Fig. 5). There was no evidence of crystallization, thus the flamprop-ethyl remained dissolved in the formulation components after the aqueous spray drops dried on the leaf surface.

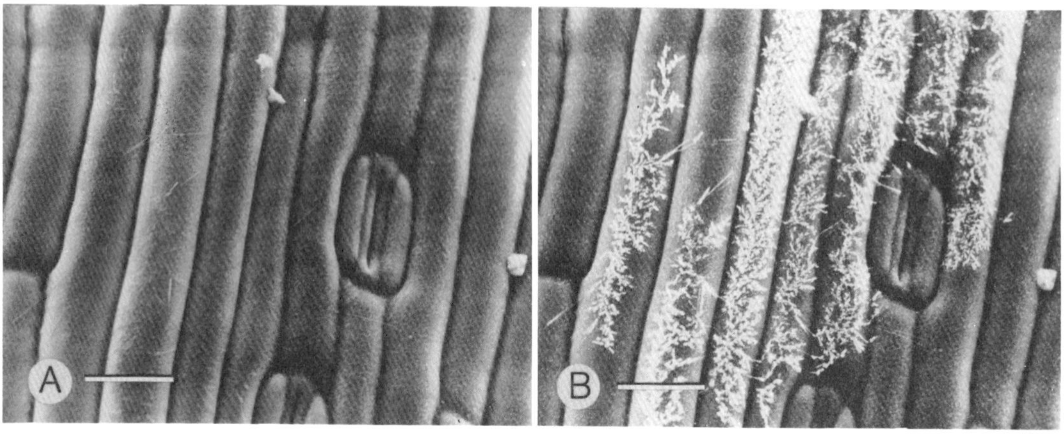

Fig. 4 A. Secondary electron SEM image of a wild oat leaf sprayed (75 L/ha) with Formulation 1. B. Composite secondary electron - CL image. Scale Marker = 25 μm.

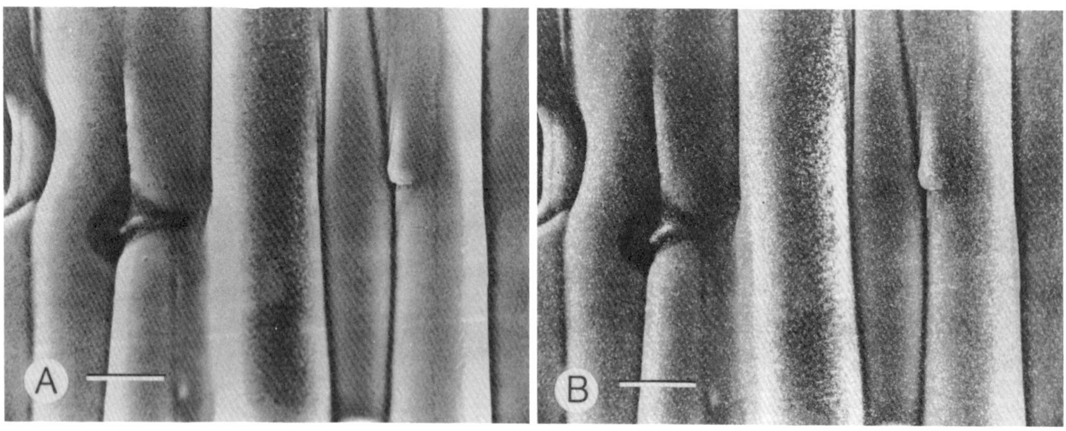

Fig. 5 A. Secondary electron SEM image of a wild oat leaf sprayed (75 L/ha) with Formulation 2. B. Composite secondary electron - CL image. Scale marker = 25 μm.

Because the amount of herbicide and formulation components in an emulsified mixture influenced recrystallization (Fig. 3), the previous application of herbicide (0.22 kg/ha) was repeated using an application volume of 374 L/ha (Fig. 6 and 7). The plant size, spray conditions, sample preparation and SEM operation were the same as in the previous study. On leaves sprayed with Formulation 1, an abundance of large needle-like crystals were present over the entire leaf surface (Fig. 6). The crystal shape, size and distribution was significantly different than the crystals observed with an application volume of 75 L/ha (Fig. 4). Only a few crystals were found on plants treated with Formulation 2 (Fig. 7). To locate any crystals several fields of view had to be analyzed with the CL detector.

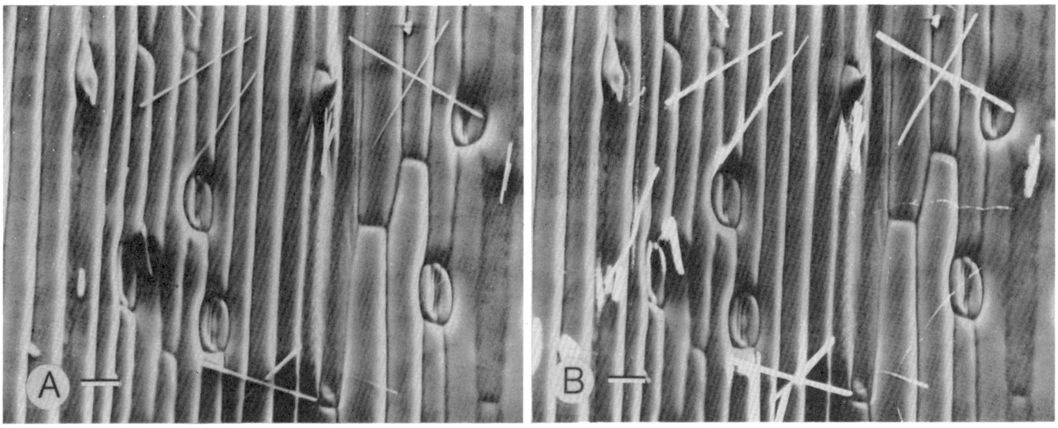

Fig. 6 A. Secondary electron SEM image of a wild oat leaf sprayed (374 L/ha) with Formulation 1. B. Composite secondary electron - CL image. Scale marker = 25 μm.

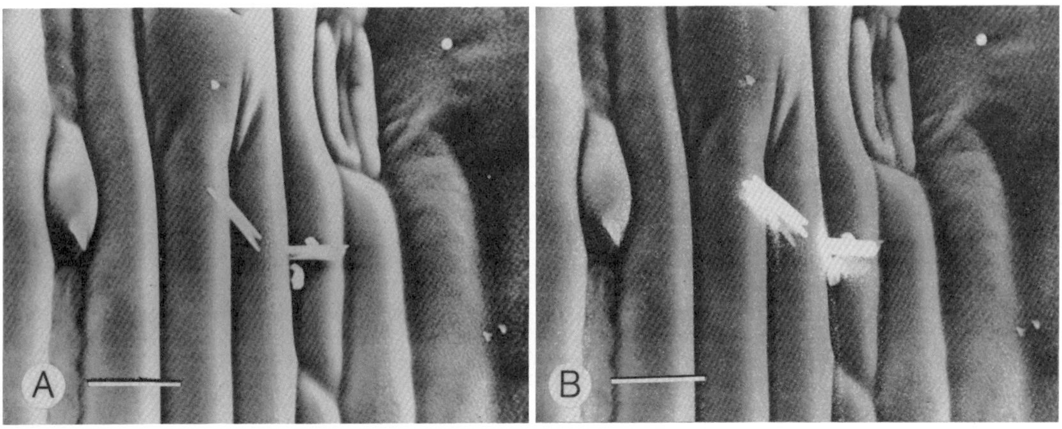

Fig. 7 A. Secondary electron SEM image of a wild oat leaf sprayed (374 L/ha) with Formulation 2. B. Composite secondary electron - CL image. Scale marker = 25 μm.

ABSORPTION OF FLAMPROP-ETHYL

In a final series of experiments, absorption of flamprop-ethyl into leaf tissue when formulated in system 1 or 2 (Table 1) was evaluated. Preliminary experiments identified growth chamber conditions that would yield herbicidal performance differences similar to those observed in the field and greenhouse studies. Suitable conditions were: 200 $\mu Em^{-2}s^{-1}$ light (PPFD), 14 hour day length, $20^{o}C$ day and $13^{o}C$ night temperatures, and ambient relative humidity. To test flamprop-ethyl absorption, the distal ends of the second true leaf of wild oat plants (two leaf stage of growth) were treated with a 2 μL drop of treatment solution. Each 2 μL drop contained the appropriate amounts of formulation components, active ingredient (including 27,000 dpm of ^{14}C-flamprop-ethyl; 23.9 mCi/mmol) and water to be equivalent to the 75 L/ha spray solution. At various times after treatment, leaves were removed from plants and the surfaces washed for 1 minute with 2 ml methanol and then for 1 minute with 2 ml of chloroform. After surface washing the treated leaf, the tissue was combusted (11) and the collected radioactivity assayed by liquid scintillation spectrometry. The experiment was replicated three times, and repeated three times. The tissue absorption of flamprop-ethyl from Formulation 2 was significantly greater than from Formulation 1, 6 hours and longer after application (Fig. 8). Very little absorption of Formulation 1 occurred over time, whereas absorption of Formulation 2 reached a level 175% greater than Formulation 1, 24 hours after application.

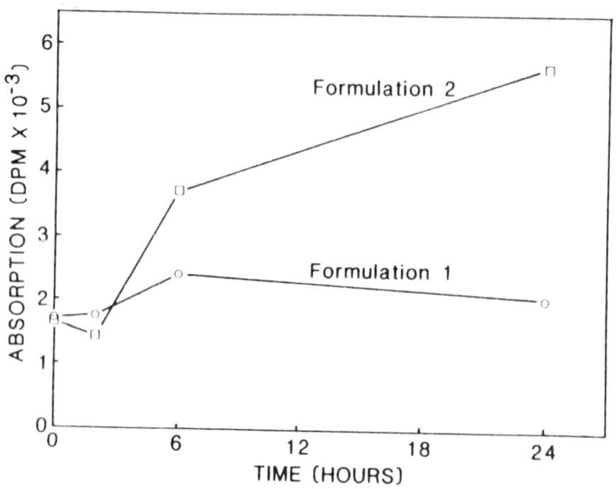

Fig. 8 The influence of formulation components on absorption of flamprop-ethyl into wild oat leaves.

DISCUSSION

This research has shown formulation components of emulsifiable concentrates do influence the physical form of herbicides after spraying onto plant surfaces (compare Fig. 4 and 5). In addition, the concentration of the emulsifiable concentrate formulation in an emulsified mixture influences the physical form of herbicides on sprayed plant surfaces (compare Fig. 4 with Fig. 6). No crystal substances were seen in light microscopic observations of the spray solutions prior to application, suggesting the crystallization occurred while the spray drops were drying on the leaf surface. Based on the formulation components shown in Table 1, the differences in the physical form of flamprop-ethyl on the leaf surface was the result of the concentration of active ingredient, concentration of emulsifiers, or the type of oil used in the formulations. The most effective formulation in preventing crystallization contained less active ingredient (8.0% vs. 13.0%) and a greater proportion of emulsifier (12.0% vs. 6.5%). Whereas the type of oil (paraffinic vs. naphthenic) may have influenced crystallization, the concentration of oil in the formulation could not have been a factor. Apparently adequate formulation components in Formulation 2 remained on the leaf surface after spray application, and water and solvent evaporation, to keep the flamprop-ethyl dissolved on the leaf surface. For postemergence application perhaps achieving a formulation with a high active ingredient concentration may not yield the best formulation with respect to performance.

Formulation 1, which yielded crystalline deposits on the leaf surface (Fig. 4), did not absorb into wild oat leaves as efficiently as Formulation 2 (Fig. 8), which did not crystallize on the leaf surface. This is not surprising considering the stable state of molecules in crystalline form. The decreased absorption of Formulation 1 was translated to decreased herbicide effects on wild oat plants (Figs. 1 and 2). This research shows the physical form of the herbicide on sprayed plant surfaces after postemergence application can influence performance. To assure crystallization is not limiting herbicidal efficacy, the physical form of herbicides on sprayed plant surfaces should be evaluated by SEM or other suitable analytical procedure. Because crystallization can occur during spray drop drying, light microscopic evaluation of the spray solution or sieving the spray solution is not adequate.

REFERENCES

1. F.D. Hess, Weed Physiology, VII Herbicide Physiology, S.O. Duke, Ed., CRC Press, Boca Raton, Florida, 191-214 (1985).
2. M.J. Bukovac, Herbicides: Physiology, Biochemistry, Ecology, 2nd ed., L.J. Audus, Ed., Academic Press, New York, 335-364 (1976).
3. J.L. Hammerton, Weeds 15, 330-336 (1967).
4. H.M. Hull, Residue Reviews 31, 1-155 (1970).
5. H M. Hull, H.L. Morton and J.R. Wharrie, Bot. Rev. 41, 421-452 (1975).
6. C.E. Price, The Plant Cuticle, D.F. Cutler, K.L. Alvin and C.E. Price, Eds., Academic Press, New York, 237-252 (1982).
7. E.A. Baker, Pestic. Sci. 11, 367-370 (1980).
8. T.J. Muzik, Herbicides: Physiology, Biochemistry, Ecology, 2nd ed., L.J. Audus Ed., Academic Press, New York, 203-247 (1976).
9. J.D. Skoss, Bot. Gaz. 117, 55-72 (1955).
10. F.D. Hess, D.E. Bayer and R.H. Falk, Weed Sci. 29, 224-229 (1981).
11. D.L. Bucholtz and F.D. Hess, Weed Sci. 31, 159-162 (1983).

New benzimidazole fungicides: their efficacy against benomyl-sensitive and benomyl-resistant isolates of *Botrytis cinerea*

Mikio Chiba* and John Northover

Research Station, Agriculture Canada, Vineland Station, Ontario L0R 2E0 Canada

Abstract - Four alkyl isocyanate homologues of benomyl were compared for their toxicity towards benomyl sensitive (S) and benomyl resistant (R) isolates of *Botrytis cinerea* using a laboratory spore germination test. For the inhibition of germ tube length, MBC-MIC, MBC-EIC and MBC-PIC (methyl, ethyl and propyl homologues of benomyl, respectively), were as effective against the S isolate as MBC-BIC (benomyl), but they were more fungitoxic than MBC-BIC towards the R isolate. MBC-MIC was equally effective in inhibiting spore germination of both the S and R isolates, and was more effective than the other homologues against S isolates. However, MBC-EIC and MBC-PIC were more fungitoxic to the R than to the S isolate and thereby showed negative cross resistance to benomyl. For the protection of wounded apples, MBC-EIC was much more effective than benomyl against the R isolate and was only slightly less effective than benomyl against the S isolate. Accordingly MBC-EIC and possibly MBC-MIC are benzimidazole compounds with potential for use against fungal isolates both sensitive and resistant to benomyl.

INTRODUCTION

Benomyl is a widely used fungicide which has been effective worldwide against many diseases of commercially important crops. However, the repeated use of benomyl and other methyl 2-benzimidazole carbamate (MBC or carbendazim) generating fungicides has resulted in the selection of benomyl resistant fungal isolates against which benomyl has very much reduced activity.

The objective of the present study was to find new benzimidazole fungicides which would provide control of both benomyl resistant and benomyl sensitive isolates of the important fungal pathogen *Botrytis cinerea*.

Four alkyl isocyanate homologues of benomyl were synthesized and compared with benomyl and MBC for their toxicity towards germination and germ tube growth of *B. cinerea* spores. The efficacy of three of these compounds for the protection of wounded apples against benomyl sensitive and benomyl resistant *B. cinerea*, was also examined.

RESULTS

Synthesis

Four benzimidazole compounds, homologues of benomyl were synthesized by the reaction between MBC, and methyl, ethyl, propyl and hexyl isocyanates, and the authenticity of synthesized products (expressed as MBC-MIC, MBC-EIC, MBC-PIC and MBC-HIC, respectively) was confirmed by elemental analysis, H^1-NMR and mass spectrometry.

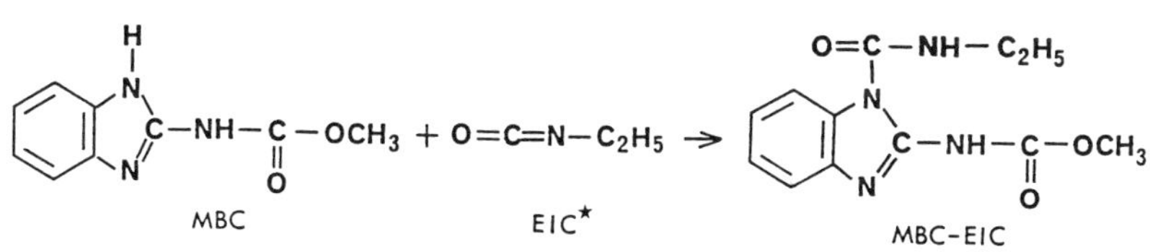

*MIC, PIC and HIC were treated same way.

Fungitoxicity

Using spore suspensions of two monoconidial isolates of *Botrytis cinerea*, one sensitive (S) and another resistant (R) to benomyl, a germination test was conducted at 21°C and was terminated after 20 h with Na azide. The percentage of germination and individual germ tube length of germinated spores were measured. Results were plotted on probit log graph paper and EC_{50} values were calculated.

A: Inhibition of germ tube growth

For the inhibition of germ tube length, MBC-MIC, MBC-EIC, and MBC-PIC were as effective as MBC-BIC (benomyl) against benomyl sensitive (S) *Botrytis cinerea*, (EC_{50} values 0.18-0.23 uM). Interestingly, against benomyl resistant (R) *B. cinerea*, these three compounds were markedly more effective than MBC-BIC and MBC, although there were resistant factors ($MEC_{50}(R)/EC_{50}(S)$) of 3.1, 4.3 and 11.3 with MBC-MIC, MBC-EIC and MBC-PIC, respectively.

B: Inhibition of germination

In the germination inhibition studies, only MBC-MIC was more toxic than MBC-BIC towards the S isolate (EC_{50} 18 uM versus 161 uM, respectively), whereas towards the R isolates, MBC-MIC, MBC-EIC and MBC-PIC were more fungitoxic (EC_{50} 7-55 uM) than MBC-BIC and MBC (EC_{50} $>$317 and $>$564 uM, respectively). MBC-EIC and MBC-PIC were negatively cross resistant and showed lower EC_{50} values for germination inhibition for the R isolate than for the S isolate. MBC-HIC showed a similar fungitoxicity to that of MBC-BIC.

Efficacy: Protection of wounded apples

MBC-EIC provided protection at 317 uM comparable to that of MBC-BIC and MBC at 100 uM, against the S isolate. Against the R isolate, MBC-EIC gave complete protection at 1000 uM in contrast to the total ineffectiveness of MBC-BIC and MBC at 3170 uM.

DISCUSSION

Benzimidazole compounds generally show less effect on spore germinaton than on the subsequent growth of the germ tube, and this was confirmed in this experiment.

The results of the apple protection experiment can not be directly extrapolated to practical application because the conditions used in this study were different from those used in the storage of apples at 1°C for extended period of storage of 1-8 months. However, the results indicate the potential for protection within the practical concentration range for commercial fungicides.

It is interesting to note that although these compounds generate MBC, MBC-MIC and MBC-EIC showed low resistance factors in the germ tube length study, and little or negative cross resistance in the germination study, and thereby indicated their potential as useful fungicides against both benomyl sensitive and benomyl resistant *B. cinerea*.

Fungicide resistance in *Penicillium* attacking harvested fruits

Joseph W. Eckert

Department of Plant Pathology, University of California, Riverside, CA 92521, USA

Abstract - Synthetic fungicides are the primary means for controlling Penicillium decay of citrus and pome fruits after harvest. In recent years, the effectiveness of most registered compounds has been diminished by the buildup of fungicide-resistant biotypes in the pathogen populations. Strategies to retain the effectiveness of registered fungicides have emphasized the: 1) use of multi-site fungicides while reserving selective compounds for unique situations; 2) alternation of fungicides with different biochemical targets; 3) development of compounds and treatments that suppress *Penicillium* spore production and dissemination. Trials involving these techniques with citrus and pome fruits have given promising results. The frequency of benzimidazole-resistant biotypes in the *Penicillium* population decreased when these fungicides were discontinued, but the effectiveness of these compounds was not restored proportionately.

INTRODUCTION

Citrus and pome fruits are often stored for weeks to months after harvest to maintain and improve the quality of the fruit for marketing. Synthetic organic fungicides are applied to these fruits after harvest to prevent infection by several species of *Penicillium* during the storage period. The intensive and continuous use of these treatments throughout the world has diminished their effectiveness due to the emergence of fungicide resistant biotypes of *Penicillium digitatum* and *P. italicum* on citrus fruits and *P. expansum* on apples and pears (refs.1,2,3,4,5,6).

The selection pressure of a postharvest fungicide treatment is considerably greater than that exerted by a field application of the same compound. The postharvest treatment results in almost complete coverage of the fruits and the fungicide residue decreases only slightly during the storage period. The treated fruit are stored or "cured" under environmental conditions that permit growth and sporulation of the fungicide-resistant biotypes of the pathogen. Under optimum environmental conditions, the number of *Penicillium* spores can multiply about 10^8 fold in two to three weeks. *Penicillium* biotypes that are resistant to one fungicide are usually cross-resistant to structurally related compounds (eg, biphenyl and 2-phenylphenol; thiabendazole and benomyl). Treatment of a fruit lot with one fungicide before storage usually results in the buildup of resistant biotypes so that a second treatment with the same fungicide after storage is not effective. The impact of the fungicide resistance problem upon effective decay control depends upon the virulence of the resistant biotypes and their dissemination. Strategies to minimize the problem are directed to delaying the buildup of resistant biotypes in the population and controlling disease in situations where the resistant biotypes are already a major component of the pathogen population.

INTERACTION OF RESISTANT AND SENSITIVE BIOTYPES IN THE PENICILLIUM POPULATION

The relative fitness of a pathogen biotype in the absence of the selecting fungicide is important in strategies aimed at reducing its frequency in the population. Isolates of *P. digitatum* and *P. expansum* that are resistant to the benzimidazole fungicides (benomyl, thiabendazole, carbendazim, thiophanate-methyl) appear to be less fit ecologically than the wild type and their frequency in the population decreases when the benzimidazole treatment is discontinued (refs.2,5,7). Fenarimol-resistant isolates of *P. italicum* behave in a similar manner (ref.8). In contrast to *P. digitatum*, a benzimidazole-resistant isolate of *P. italicum* persisted at a constant frequency in the population of that pathogen during several infection cycles in fungicide-free oranges (ref.3). A reasonable number of isolates should be evaluated to establish the relative fitness of the resistant and sensitive biotypes of the pathogen species, rather than the unique behavior of a few isolates. A study of 27 combinations of nine isolates of P. digitatum resistant to butylamine (sec-butylamine) and six wild-type isolates, revealed that the frequency of the butylamine resistant spores increased in 70% of the combinations, decreased in 19%, and remained constant in 11% (ref.9).

In natural populations, benzimidazole-resistant isolates of P. digitatum may not behave in a manner predicted by their generally lower fitness compared to wild-type isolates. The observation that the frequency of benzimidazole-resistant isolates does not decrease rapidly in packinghouses where the benzimidazole fungicide has been discontinued may be explained by the existence of Penicillium biotypes with resistance to the benzimidazoles as well as 2-phenylphenol and butylamine (refs.7,10). Application of any one of these fungicides resulted in the buildup of the benzimidazole-resistant phenotype in the population (refs.2,7). Another anomalous observation is the poor performance of benzimidazole fungicides in controlling a population of P. digitatum or P. expansum that has only a small component of resistant biotypes (refs.2,5,7). This phenomenon may be due to a synergistic influence of the benzimidazole-sensitive isolates upon the infection of fruit by the resistant isolates (ref.11).

STRATEGIES FOR MANAGEMENT OF FUNGICIDE-RESISTANT BIOTYPES

Several strategies have been utilized to control decay of citrus fruits and pome fruits caused by fungicide-resistant biotypes of Penicillium: 1) suppress production and dispersal of Penicillium spores; 2) mix or rotate fungicides with different biochemical mechanisms, emphasizing multi-site inhibitors when possible; 3) develop new fungicides and treatments that do not encourage the buildup of pathogenic fungicide-resistant biotypes of Penicillium.

Citrus fruits

Fungicide-resistant biotypes of Penicillium spp. infect fruit when their spores accumulate, in sufficient numbers, at a suceptable site (eg, an injury) on the fruit surface (ref.11). Treatments and practices that interfere with spore production or dissemination will reduce the probability of fruit infection. The principal methods for reducing spore inoculum levels in packinghouses are: 1) treat fruit with an antisporulant compound, such as imazalil; 2) disinfect surface-contaminated fruit with a contact sporicide such as hypochlorite or 2-phenylphenol; 3) isolate and eliminate decayed fruit and exhaust the spore-contaminated atmosphere from the packinghouse. A study of the benefits of an inoculum-reduction program was carried out in three California lemon packinghouses (ref.12). Packinghouse layout and sanitation significantly influenced the size of the population of thiabendazole-resistant spores and the level of decay in lemons shipped to distant markets. Effective strategies included the isolation of sources of fungicide-resistant biotypes, disinfestation of contaminated equipment and containers and a program for monitoring the level of fungicide-resistance in the packinghouse.

Application of several fungicides with different biochemical targets, in an appropriate sequence, can delay the buildup of fungicide-resistant biotypes of Penicillium (ref.2). This strategy is especially valuable in situations where citrus fruit are treated with fungicides both before and after storage. The fungicides applied before storage are chosen on the basis that they will not encourage the proliferation of Penicillium biotypes that are resistant to the fungicides applied after storage. The post-storage treatments are of major importance since they protect the fruit from decay during shipment and marketing. For example, lemons may be treated with fungicides chosen from the group, sodium carbonate, butylamine, potassium sorbate, borax, and imazalil before storage and, after storage, with 2-phenylphenol, thiabendazole (or benomyl), and biphenyl. The pre-storage treatments suppress biotypes of Penicillium that are resistant to 2-phenylphenol, thiabendazole, and biphenyl. The success of this strategy depends upon an efficient sanitation program, because the buildup of a large population of spores during storage would increase the frequency of isolates with resistance to several fungicides (refs.10,13,14) and enhance the possibility of a synergistic interaction between fungicide-sensitive and fungicide-resistant biotypes (ref.11). The simultaneous application of two unrelated fungicides was not effective in delaying the buildup of benzimidazole-resistance in Penicillium digitatum on oranges (ref.7) or P. expansum on pears (ref.5). Mixtures of thiabendazole with butylamine (ref.15) or 2-phenylphenol (ref.16) have been recommended for improved control of Penicillium decay of citrus fruits, but without any claims regarding the control of benzimidazole-resistant biotypes.

Over the past decade, several new fungicides that control benzimidazole-resistant isolates of Penicillium have been tested (refs.17,18). Imazalil, prochloraz, and etaconazole belong to a class of compounds that inhibit the biosynthesis of ergosterol, an essential component of the fungus cell membrane. Generally, mutants "resistant" to these compounds show only a low level of tolerance and are often less virulent than wild-type isolates of the pathogen. For these reasons, it is generally believed that biotypes of Penicillium possessing tolerance to these fungicides will buildup very slowly, if at all. Imazalil has been extensively used on a commercial scale in most citrus production areas for over five years without any report of field resistance. Laville (ref.19) transferred fungicide-sensitive isolates of P. digitatum and P. italicum weekly for three years on growth medium containing benomyl or imazalil. Mutagenized spores of these fungi were subjected to the same treatment. Many benomyl-resistant mutants appeared, but no imazalil-resistant mutants were found. DeWaard et al. (ref.8) isolated P. italicum strains that possessed elevated levels of resistance to imazalil. Treatment of spores of a "resistant" strain with mutagenic agents, followed by further selection on imazalil-containing medium, produced mutants of greater resistance that were pathogenic to oranges. The sensitive (wild) isolates were controlled by treatment of inoculated fruit with a solution of imazalil (220 mgl^{-1}), whereas the resistant mutants were not well controlled at this dosage.

Pome fruits

Apples and pears are harvested during a relatively short season and are stored for several months to one year. Usually, pome fruits are treated immediately after harvest with an anti-scald agent (diphenylamine or ethoxyquin) and a fungicide, and stored at 0-3° C. Blue mold, caused by P. expansum is the principal postharvest rot of pome fruits in many production areas (refs.4,5,6,20). During the 1960s, captan was frequently applied to pome fruits after harvest to control blue mold, but this treatment was replaced by benomyl and thiabendazole in the 1970s because the latter fungicides were more effective. Years of continuous use of thiabendazole and benomyl have resulted in the emergence of resistant strains of P. expansum in most apple production areas, and a simultaneous deterioration in the efficacy of the postharvest benzimidazole treatments (refs.4,5,20,21).

Several investigators have demonstrated that benzimidazole-resistant biotypes of P. expansum can be controlled on pome fruits by postharvest application of imazalil, etaconazole, prochloraz, iprodione, and vinclozoline (refs.4,5,6,13). However, isolates of P. expansum resistant to both iprodione and vinclozoline have been identified and these are not well controlled by postharvest treatments of these fungicides (ref.13). Although imazalil has provided good control of benzimidazole-resistant blue mold on pome fruits, Prusky and Ben-Arie (ref.20) reported the isolation of an imazalil-resistant Penicillium spp., P. crustosum, from a treated decayed pear. Gullino and DeWaard (ref.22) found that imazalil-resistant biotypes of P. expansum were easily isolated from natural spore populations and these isolates were not controlled on apples by treatment with ergosterol-inhibitor fungicides. Since imazalil, etaconazole, and prochloraz are not currently registered for postharvest treatment of pome fruits, Rosenberger et al. (ref.4) recommended treatment of apples before storage with a mixture of captan and benomyl. The mixture provided excellent control of benzimidazole-sensitive P. expansum and about 80% reduction in decay of apples that were inoculated with a benzimidazole-resistant isolate. The mixture was effective when the treated apples were stored at 0° C, but not at 20° C. Prusky et al. (ref.5) observed that benzimidazole-resistant isolates of P. expansum characteristically showed a lower parasitic fitness than wild-type isolates and investigated fungicide strategies that would reduce the frequency of the resistant biotypes in the population. The most rapid decline in the frequency of the resistant isolates resulted from a fungicide program consisting of a combination of two fungicides (eg. thiabendazole plus captan), alternated with imazalil, etaconazole, or iprodione. Spotts and Cervantes (ref.21) also observed that benomyl-resistant isolates of P. expansum from apple and pear packinghouses were less virulent than benomyl-sensitive isolates and did not increase in the population over a five year period.

Clearly, the EBI fungicides are a more formidable barrier than the benzimidazole fungicides to the emergence of highly resistant biotypes of Penicillium (ref.23). However, imazalil-resistance in P. expansum is inherited by a polygenic mechanism and the level of resistance is positively related to the number of resistance genes accumulated in a biotype (ref.24). Imazalil-resistant biotypes of P. italicum and P. expansum are pathogenic to oranges (ref.8) and apples (ref.22), respectively. Whether or not these fungi can acquire several EBI resistance genes without sacrificing pathogenicity is a moot question at present.

Fungicide-resistant biotypes of Penicillium may exhibit negatively-correlated cross-resistance to a second compound, ie., they are more sensitive to the latter than wild-type isolates. Leroux and Gredt (ref.25) described the in vitro sensitivity of benzimidazole-resistant biotypes of P. expansum and Botrytis cinerea to several N-phenylcarbamate herbicides. Following this lead, methyl 3,5-dichlorophenylcarbamate and isopropyl 3,4-diethoxyphenylcarbamate were reported effective for control of several plant diseases caused by benzimidazole-resistant biotypes of plant pathogens, especially gray mold caused by Botrytis cinerea, without damaging the host plant (refs.26,27). Negatively-correlated cross-resistance in this case appears to depend upon the enhanced affinity of tubulin in the benzimidazole-resistant isolates for the N-phenylcarbamate compounds (refs.27,28). Rosenberger and Meyer (ref.29) observed that benzimidazole-resistant biotypes of P. expansum were often sensitive to diphenylamine whereas wild-type isolates were not. Combinations of diphenylamine with benomyl, thiabendazole, and thiophanate-methyl provided better control of benomyl-resistant P. expansum isolates in inoculated apples stored at 2-4° than either diphenylamine or the fungicides alone.

DeWaard et al (ref.8) evaluated cross-resistance of fenarimol-resistant mutants of P. italicum to other EBI fungicides. All fenarimol-resistant mutants were cross-resistant to other EBI fungicides tested except fenpropimorph, a morpholine derivative (refs.8,30). Biotypes with a high degree of resistance to fenarimol, and cross-resistant to other EBI fungicides, exhibited negatively correlated cross-resistance to fenpropimorph and dodine (refs.30,31). Fenpropimorph was more effective against decay of oranges inoculated with fenarimol-resistant isolates of P. italicum than against decay caused by the wild-type isolates.

The application of negatively correlated cross-resistance as a strategy against practical problems of fungicide-resistance is attractive in theory and may be useful in practice in certain disease situations. Many reported examples of negatively correlated cross-resistance indicated that the phenomenon is unique to certain species and biotypes of pathogens (refs.25,29,30). Many benzimidazole-resistant isolates of Penicillium spp. are not sensitive to phenylcarbamates and diphenylamine. However, most benzimidazole-resistant isolates of Botrytis cinerea do exhibit negatively correlated cross-resistance to phenylcarbamate compounds. Furthermore, fenarimol(EBI)-resistant isolates of Penicillium spp. are invariably more sensitive to fenpropimorph and dodine than wild-type isolates (refs.30,31). This strategy for control of fungicide-resistant Penicillium spp. should be evaluated further since it offers a rare opportunity to selectively reduce the frequency of fungicide-resistant biotypes in the pathogen population.

REFERENCES

1. R.E. McDonald, L.A. Risse and B.M. Hillebrand, J. Am. Soc. Hort. Sci. 104, 333-335 (1979).
2. J.W. Eckert and B.L. Wild, in: Pest resistance to pesticides (eds. G.P. Georghiou and T. Saito), pp. 525-556, Plenum Publishing Corp., New York (1984).
3. Y. Gutter, A. Shachnai, M. Schiffmann-Nadel and A. Dinoor, Phytopathology 71, 482-487.
4. D.A. Rosenberger, F.W. Meyer and C.V. Cecilia, Plant Dis. 63, 1033-1037 (1979).
5. D. Prusky, M. Bazak and R. Ben-Arie, Phytopathology 75, 877-882 (1985).
6. W. Koffmann, L.J. Penrose, A.R. Menzies, K.C. Davis and J. Kaldor, Sci. Hortic. 9,31-39 (1978).
7. B.L. Wild, Doctoral Thesis, Univ. Calif., Riverside, 89 pp. (1980).
8. M.A. DeWaard, H. Groeneweg and J.G.M. Van Nistelrooy, Neth. J. Pl. Path 88, 99-112 (1982).
9. J.L. Smilanick and J.W. Eckert, Phytopathology 76, 805-808 (1986).
10. B.A. Dave, H.J. Kaplan and J.F. Petrie, Proc. Fla. State Hort. Soc. 93, 344-347 (1980).
11. B.L. Wild and J.W. Eckert, Phytopathology 72, 1329-1332 (1982).
12. M.N. Bancroft, P.D. Gardner, J.W. Eckert and J.L. Baritelle, Plant Dis. 68, 24-28 (1984).
13. D.A. Rosenberger and F.W. Meyer, Plant Dis. 65, 1010-1013 (1981).
14. B.L. Wild, Ann. Appl. Biol. 103, 237-241 (1983).
15. Y. Gutter, Crop Protection 4, 346-350 (1985).
16. D.J. Hall, J.R. Bice and P.J. Lewis, Proc. Fla. State Hort. Soc. 91, 159-161 (1978).
17. J.W. Eckert and G.E. Brown, in: Fresh Citrus Fruits (eds. W.F. Wardowski, S. Nagy, W. Grierson), pp. 315-360, Avi Publ. Co., Westport, CT (1986).
18. Y. Gutter, A. Shachnai, M. Schiffmann-Nadel and A. Dinoor, Phytopath. Z. 102, 127-138 (1981).
19. E.Y. Laville, Proc. Int. Soc. Citricult. 2, 783 (1981).
20. D. Prusky and R. Ben-Arie, Plant Dis. 69, 416-418 (1985).
21. R.A. Spotts and L.A. Cervantes, Plant Dis. 70, 106-108 (1986).
22. M.L. Gullino and M.A. DeWaard, Neth. J. Pl. Path. 90, 177-179 (1984).
23. S.G. Georgopoulos, EPPO Bull. 15, 513-517 (1985).
24. J.M. Van Tuyl, Doctoral Thesis, Agricultural University Wageningen, Netherlands 137 pp. (1977).
25. P. Leroux and M. Gredt, 6th Reinhardsbrunn Symp. Systemic Fungicides and Antifungal Compounds pp. 297-301 (1982).
26. T. Kato, K. Suzuki, J. Takahashi and K. Kamoshita, J. Pestic. Sci. 9, 489-495 (1984).
27. S. Nakamura, T. Kato, H. Noguchi, J. Takahashi and K. Kamoshita, IUPAC 6th Internatl. Congress Pestic. Chem., Ottawa, Canada, Abstr. No. 3E-01 (1986).
28. K. Suzuki, T. Kato, J. Tahahashi and K. Kamoshita, J. Pestic. Sci. 9, 497-501 (1984).
29. D.A. Rosenberger and F.W. Meyer, Phytopathology 75, 74-79 (1985).
30. M.A. DeWaard and J.G.M. Van Nistelrooy, Neth. J. Pl. Path. 88 231-236 (1982).
31. M.A. DeWaard and J.G.M. Van Nistelrooy, Neth. J. Pl. Path. 89, 67-73 (1983).

Progress and development in formulation chemistry and technology

Controlled release formulation

Kozo Tsuji

Pesticides Research Laboratory, Takarazuka Research Center,
Sumitomo Chemica Co., Ltd. Konohana-Ku, Osaka 554 Japan

Abstract-Controlled release technology can be defined as a technique or a method in which active ingredients (AI) are made available to a specified target at concentration and duration designed to accomplish an intended effect. This technology has received increasing attention in pesticide formulation. Research has been done extensively on this subject, mainly on slow or sustained release formulation, which releases AI from the formulation at a slow and continuous effective rate into the environment. Controlled release formulations have various advantages over conventional ones as follows; 1) satisfactory effect for longer periods of time, 2) smaller dosage, 3) longer application interval, 4) decrease in mammalian toxicity, 5) decrease in phytotoxicity, 6) decrease in environmental degradation, 7) decrease in environmental pollution. Controlled release formulations have many advantages which are in good accordance with recent needs. The development of the ideal controlled release formulation would be comparable to the invention of ideal pesticides. Therefore much more work should be done to develop true controlled release formulations which show proper functions.

INTRODUCTION

Recent requirements for the pesticide industry have become more severe, and pesticides are required to satisfy the following conditions.

1)Higher safety; pesticides should be safe for workers and give no harmful effect nor impact on the nontarget organisms and the environment.
2)Higher efficacy; Pesticides should show good initial and residual efficacy at lower dosage.
3)Cheaper price; Manufacturing cost should be low and cost/performance should be good.

When new pesticidal compounds are sought, it is not easy for candidate compounds by themselves to satisfy all these requirements. Therefore screening probability of new pesticides has decreased and it has become significantly difficult to develop new pesticides. It becomes very important to improve formulation and application technology in order to satisfy the above requirements. Labor-saving on spraying is also desired. In order to respond to such recent demands, various kinds of research work has been done on the pesticide formulations. One of the most important among them is controlled release formulation.

Controlled release technology can be defined as a technique or a method in which AI is made available to a specified target at concentration and duration designed to accomplish an intended effect. This technology has received increasing attention in the various industrial fields including pesticide formulations (ref.1-15). The research work on the controlled release technology in pesticide formulation is mainly on slow, sustained or prolonged release formulation, which releases AI from the formulation at a slow and continuous effective rate into the environment for a long time. Therefore the term "Controlled release formulation" is often used instead of "Slow release formulation, "Sustained release formulation" or "Prolonged release formulation", although, strictly speaking, they are not the same.

SIGNIFICANCE AND ADVANTAGE OF PESTICIDE CONTROLLED RELEASE FORMULATION

From a viewpoint of biological efficacy alone, the persistent pesticide is preferred. But such persistent pesticide is slow to decompose, remains for a prolonged time in the environment and is apt to be taken into the food chain. Therefore the use of these persistent pesticides has become severely restricted recently. When a nonpersistent pesticide is applied by a conventional formulation, excess initial concentration or repeated application is needed to assure not only initial but also long-lasting effect. Such overdosing wastes much of pesticide and often causes toxicity problems for nontarget organisms. Repeated application is contrary to the current trend for labor-saving. When a nonpersistent pesticide is applied by a controlled release formulation, a specific concentration can be maintained at an action site for long time, which is a little higher than the minimum effective concentration and lower than the maximum tolerated one. Therefore the controlled release formulation can maintain the concentration above the effective one with less side effects.

In order to understand the advantage of the controlled release formulation, mathematical equations are shown in the following (ref.4). When the conventional formulation is applied, pesticides dissipate from their application sites due to leaching, evaporation and degradation. The rate of removal follows the first-order kinetics, which is given by Equation (1).

$$dM/dt = -k_r M \tag{1}$$

where dM/dt is the rate of removal, k_r is the rate constant and M is the amount of pesticide at time t. By integrating Equation (1), the following Equation (2) is obtained.

$$\ln(M_\infty/M_e) = k_r t_e \tag{2}$$

Where M_∞ is the amount of pesticide applied initially, M_e is the minimum effective level of pesticide and t_e is the effective duration time. Equation (2) shows that to prolong the effective duration time t_e, exponentially greater quantities of the pesticides must be applied. On the other hand, when the controlled release formulation is considered, if the pesticide could be maintained at the minimum effective level M_e by a continuous supply to compensate for dissipated fraction from the controlled release formulation, the following Equations (3) and (4) are obtained.

$$dM/dt = -k_r M_e + k_d M_e = 0 \tag{3}$$

$$t_e = (M_\infty - M_e)/K_d M_e \tag{4}$$

where k_d is the rate constant for pesticide delivery. Equation (4) shows that to prolong the effective duration time t_e in this case, linearly greater quantities of the pesticides must be applied. Therefore when we want to prolong the effective duration time it is clear that much excess amount of the pesticides is needed for the conventional formulation, compared with the controlled release one, and also a large fraction of the pesticides is wasted for the conventional formulation.

According to the argument described above, it is easily understood that the controlled release formulation has mainly the following advantages.

1) satisfactory effect for longer period of time (residual effect)
2) smaller dosage (resource-saving, safety)
3) longer application interval (labor-saving)

Moreover the following advantages may also be expected.

4) decrease in mammalian toxicity
5) decrease in phytotoxicity
6) decrease in environmental degradation
7) decrease in environmental pollution
8) decrease in leaching and evaporation
9) decrease in reactivity
10) changing liquid material into solid
11) masking of odor
12) easier handling

Some other advantages might be obtained for some controlled release formulations, which will be explained later.

KIND OF CONTROLLED RELEASE TECHNOLOGY

Controlled release technology is mainly attained by interaction of pesticides with polymers. Several methods have been developed and they are classified according to the mechanism or materials used as follows.

(1) Diffusion of pesticides in polymeric medium

1) Monolithic system. Pesticides are dissolved (solid solution) or dispersed (dispersion) uniformly in polymeric or elastomeric matrices.
2) Reservoir system. Pesticides are covered by membranes. Laminate, capsule, coating, liposome etc. are examples.

The Hercon® controlled release system of dispensing AI consists of several layers of laminated polymeric materials. The specially formulated inner layer serves as a reservoir for AI. AI migrate continuously, due to the concentration gradient from the reservoir layer through outer layers to the exposed surface. Reservoir layer is mainly made of flexible PVC. Release rate is affected by the following factors, (1) reservoir concentration, (2) membrane thickness, (3) polymer stiffness, (4) co-diffusant, (5) molecular weight of the diffusant, and (6) chemical functionality. By combining these factors properly, desired release rate can be attained. This Hercon® dispenser is used mainly for insecticides and pheromones, and good biological results have been obtained (ref. 6).

(2) Polymeric pesticide

1) Pesticide is bound via covalent or ionic bonds to polymers. Polymeric pesticide is also obtained by binding pesticide first to monomers, followed by polymerization. Pesticide is released from the polymeric pesticide by hydrolysis of the polymer-pesticide bond in the environment. Pesticides that contain very reactive functional groups such as carboxyl, hydroxyl, sulfhydryl and amino groups can react with monomers or polymers directly. If pesticides don't have any functional groups, some bridging agents are needed in order to bind pesticides to monomers or polymers. By these methods, various kinds of pesticides are bound to natural polymers such as kraft lignin, bark and sawdust (ref. 6). Polymerizable monomers containing pesticides are also synthesized using vinyl derivatives of carboxylic acids and alcohols. The homopolymer obtained by homopolymerization of monomers containing pesticides, however, is hard to hydrolyze. Introduction of hydrophilic monomers into the polymer chain accelerates hydrolysis.

2) Difunctional monomeric pesticide is homo- and co-polymerized with other monomers, and pesticide is incorporated into the polymer main chain. Most of the pesticides in the polymer main chain, however, are very hard to hydrolyze.

(3) Erosion of polymers; Pesticide is released by chemical or biological erosion of polymers from monolithic or reservoir system.

(4) Porous materials; Pesticide is retained in capillary channels such as hollow fibers, porous plastic, foam and so on. This technique is used mainly for insect pheromones, which vaporize slowly from the dispenser.

(5) Inclusion compound; Cyclodextrin and urea form inclusion compounds with various pesticides. Inclusion of high-volatile pesticide into cyclodextrin causes reduction of vaporization and gives longer residual activity. Inclusion compounds of pyrethroids in cyclodextrin become stable against light and usable even for agriculture.

(6) Adsorptive carriers; Montmorillonite clays and attapulgite clays have adsorptive ability for various compounds including pesticides. The formulation thus obtained is released under controlled conditions and longer residual activity is expected. Activated charcoal is also used for this purpose.

(7) Propesticide; A modified molecule releases active pesticide under certain conditions. One of the examples of propesticide for controlled release is pyrazolate. Pyrazolate is hydrolyzed in the paddy field into destosylpyrazolate (DTP), which shows herbicidal activity. It was found that pyrazolate granule gives better residual activity than DTP granule. This may be caused by the difference of water solubility of these two compounds; 0.05 ppm for pyrazolate and 4500 ppm for DTP(ref.16).

(8) Osmotic pump; Pesticide is covered by semipermeable membrane with small holes.

MATERIALS USED FOR CONTROLLED RELEASE FORMULATION

Various kinds of polymers are used for controlled release formulations. They should satisfy the following conditions in general (ref.6).

(1) The polymer should have proper molecular weight, glass transition temperature and molecular structure in order to get proper diffusion and release rates of the pesticide.

(2) The polymer should not be reactive with pesticides.

(3) The polymer and its degradation products must not cause any environmental impact.

(4) The polymer should be stable during storage and usage generally.

(5) The polymer should be easily manufactured and fabricated into the desired product.

(6) The polymer should not be expensive.

For agricultural application, they should be biodegradable to avoid environmental pollution. For this purpose, several derivatives from natural polymers such as cellulose, starch, sawdust and bark have been developed for controlled release formulations (ref. 6). There are several reports on starch xanthate, starch-alkali earthmetal adduct and starch borate for microcapsules (ref.4). Polymeric pesticides derived from sawdust, bark and kraft lignin are also reported. Polylactic acid, polyglycolic acid and their copolymers are used for controlled release formulations as biodegradable and erodible polymers (ref.6). Recently inorganic materials were used for wall materials of microcapsules of pesticides. Silica microcapsules are prepared by interfacial reaction as follows;

$$Na_2O \cdot xSiO_2 + 2NH_4X \dashrightarrow xSiO_2 \cdot hydrate + 2NaX + 2NH_3$$

Various kinds of pesticides are encapsulated by this method, but stability of the pesticide depends largely on the nature of the pesticide (ref. 17).

EXAMPLES OF CONTROLLED RELEASE FORMULATION

Several controlled release pesticide formulations have been developed and commercialized. Some examples are shown in Table 1 and microcapsules are explained below in more detail.

Microencapsulation of pesticides is carried out by several chemical, physical and mechanical methods. Microcapsules are a typical controlled release formulation type but they are not always useful for all pesticides. Microencapsulation is used predominantly for insecticides. Pesticide microcapsules are often formulated as a slurry but they can also be used as dry powders. In the case of slurry formulation, pesticide is released to the aqueous phase until the solution is saturated. Once it is sprayed and there is no water left around the microcapsule, pesticide is released continuously.

One of the typical examples of commercially available pesticide microcapsules is Penncap M® (ref.6). This is produced by microencapsulation of toxic methylparathion by polyamide-polyurea membrane, which is formed by the interfacial reaction of difunctional acid chloride with diamine followed by crosslinking with polyisocyanate. This formulation is approved by EPA. Particles are about 25-30 µm in diameter and dispersed in a slurry. The release rate of AI from the microcapsule is controlled by composition, degree of crosslinking of the wall material, thickness of the wall, particle size and additives, and the release is thought at present to be by a diffusion-controlled mechanism, rather than release by leakage of the contents through numerous holes in the capsule wall. Penncap M® has been reported to control some insects at 1/4 lb AI per acre up to 5 to 7 days in the field. Under the same conditions, an EC formulation of methylparathion gave control for only 1 to 2 days. Therefore the use of Penncap M® causes fewer applications to result in less insecticide used per acre. When Penncap M® is used on apples, it is highly effective in controlling codling moth and has caused no russetting. Toxicity decreases several times by microencapsulation. The most serious problems of Penncap M® was its toxicity to bees (ref. 18). When pollen is contaminated with microcapsules which are nearly the same size and are carried back to the hive, bees and larvae may be killed by eating the pseudo-pollens. According to the recent report, however, harmful effect of the microcapsule is less than that of EC (ref. 15).

Knox Out® 2FM is a microencapsulated formulation of diazinon with a polyamide-polyurea wall (ref. 19). This microcapsule has relatively low oral and dermal mammalian toxicity, outstanding cockroach control properties and reduced volatility of AI. It is also reported that laboratory and field tests have confirmed that Knox Out® provides excellent control of diazinon-resistant cockroaches as well. Cockroaches rub their legs together and comb the legs and body with their antennae, to remove foreign substances from their bodies. In turn, they clean their antennae by passing them through their mandibles. This indicates that the enhanced activity of Knox Out® is due to pick up of the microcapsules and subsequent ingestion followed by stomach poison activity. Diazinon usually controls cockroaches by acting as contact poison. The enhanced activity against resistant cockroaches is another big benefit of the microencapsulateld pesticide.

Sumithrin® is microencapsulated by using polyurethane wall, and it is reported that biological efficacy against cockroaches does not change depending on the kind of spray surfaces (plywood panel and overlaid plywood panel), while those of wettable powder and emulsifiable concentrate change very much (ref. 20). Another benefit of microencapsulated pesticides is that they always give consistent biological efficacy irrespective of spray surfaces.

It is reported that the edifenphos release from the microcapsule increases by addition of a suitable surfactant at a proper concentration either to the water medium or the core, and stops by addition of another suitable surfactant above a given temperature (ref. 21).

In-flight encapsulation is also reported (ref. 22).

It is well known that pyrethroid insecticides give strong fish toxicity, but it was found that microencapsulation reduces this toxicity significantly. This is a big improvement of drawbacks of pyrethroids.

It was recently reported (ref. 23) that fenitrothion microcapsule as a residual spray formulation against cockroaches shows very interesting behavior. The release of AI from the microcapsule is not by a diffusion-controlled mechanism, and the main way of action is to be trampled by cockroaches, which causes burst release of fenitrothion resulting in their death, just like a land mine destroys a tank. This is a novel mechanism of release, that is to say,

TABLE 1. Examples of controlled Release Formulations

Name	Pesticide	Material	Structure
Hercon Insectape®	Propoxur diazinon chlorpyrifos	PVC/polyester	laminate
Luretape™	disparlur	vinyl/acrylic polymer	"
Lur NKill™ Flytape	resmethrin vanillin	PVC/acrylic polymer	"(double)
Staph-Chek®	organotin	nylon-reinforced fabric	"
Penncap M®	methylparathion	polyamide-polyurea	microcapsule
Penncap E®	parathion	"	"
Knox Out® 2FM	diazinon	"	"
Penncapthrin® 200	permethrin	"	"
Sectrol™	pyrethrin	polyurea	"
Tossits®	DDT/oil	gelatin	"
Altosid® SR-10	methoprene	polyamide	"
SumiCue-lure microcapsule sol	fenitrothion Cue-lure	gelatin/gum arabic	"
Capsolane®	EPTC	polyurea	"
Fonofos Seed Treatment®	fonofos	"	"
Envirocap-C	copper sulphate	polyethylene/wax	"
Cap-Cyc	chlormequat	--------	"
Tox-Hid®	warfarin	--------	"(erodible)
Dursban ME	chlorpyrifos	--------	"
Bug-X™	chlorpyrifos	--------	"/acrylic emulsion
Lasso® micro-Tech®	alachlor	--------	"
No-Pest® strip	DDVP	PVC	plate
Vaporette™	"	"	"
Star-bar™	"	"	"
Zodiac®	propoxur	"	collar
No Foul®	TBTO	neoprene rubber	sheet
Dursban® 10CR	chlorpyrifos	chlorinated PE	pellet
Bio Met® SRM	TBTO	natural rubber	"
CBL-9B	TBTF	"	"
14ACE-B	2,4-D	"	"
Incracide® E-51	copper sulphate	ethylene-propylene elastomer	"
" EC-100	copper sulphate monohydrate	ethylene-propylene copolymer	"
" EC-13	TBTF	"	"
" 10A	copper sulphate	styrene-butadiene copolymer	"
Ecopro™ 1700	temephos	ethylene-propylene copolymer	"
" 1330	TBTF	"	"
" 1043	copper sulphate	"	"
suSCon 140G	chlorpyrifos	thermoplastic resin	"
Conrel® microfiber	gossyplure	polyamide	hollow fiber
Conrel® hollow fiber	endo- and exo-brevicomin	polyester	"
No Mate™ Chokegard	Z-11-hexadecenal	black cellucon	"
Killmaster™	chlorpyrifos	--------	paint
Pyrethrin CD wettable powder	pyrethrin	cyclodextrin	powder
Ectrin™ Insecticide cattle ear tag	fenvalerate	PVC	plate
Flectron fly tag	cypermethrin	"	"
Auriplak® ear tag	permethrin	"	"

release of AI by destruction of the microcapsule with mechanical power, although biting of the microcapsule in the case of rodenticide might also be assumed to be another example of mechanical destruction of the microcapsule. According to this mechanism, AI is released only when the cockroaches trample the microcapsules. Therefore AI remains in the microcapsule for a long time, if no trampling takes place. This behavior is illustrated in Figure 1 and 2. This mechanism is very desirable because AI is released only when it is necessary, and AI in the microcapsule remains for a long time with a minimum loss. Therefore residual activity is expected to continue until all microcapsules are destroyed. In this case the trampling effect depends on the strength of the microcapsule. D/T ratio (the ratio of mass median diameter (D) and wall thickness (T)) was proposed as a parameter of the strength of the microcapsule. The higher the D/T ratio was, the greater the amount of microcapsules that were trampled. The most optimal value of the D/T ratio to get microcapsules having both good initial and residual efficacy was found to be around 150. One example of change of residual activity of three kinds of microcapsules is shown in Figure 3.

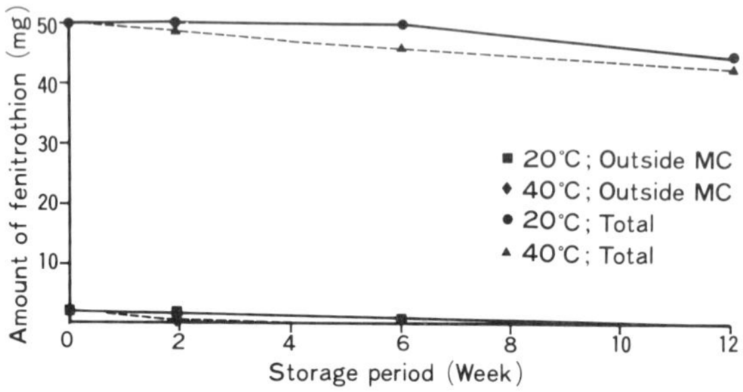

Fig. 1. Release of fenitrothion from the microcapsule *vs.* storage period of the microcapsule-treated Petri dish.
MC : Microcapsule
Outside MC : Amount of fenitrothion outside the microcapsule
Total : Total amount of fenitrothion on the dish

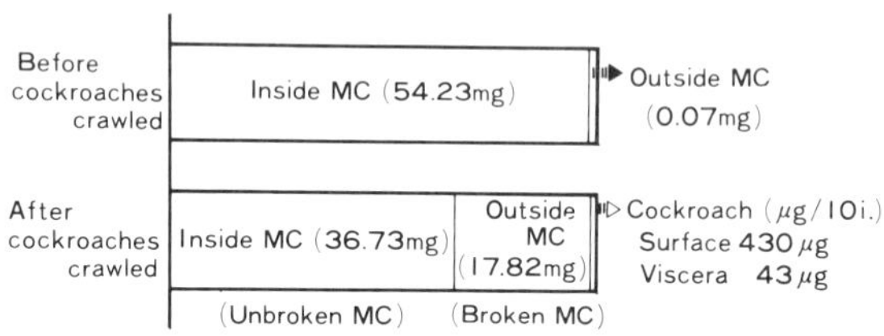

Fig. 2. Trampling effect; location of fenitrothion(inside and outside the microcapsule) before and after German cockroaches crawled on the microcapsule-treated Petri dish.
Outside MC : Amount of fenitrothion released by diffusion and/or trampling effect
Inside MC : Amount of fenitrothion inside the microcapsule

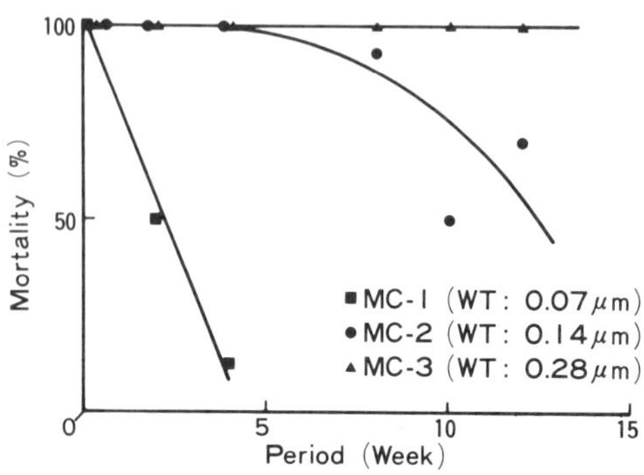

Fig. 3. Residual contact efficacy of fenitrothion microcapsules whose mass median diameter is 50 µm against German cockroaches.
WT : Wall thickness
Method : Two hours confined contact on the microcapsule-treated plywood panel(500 mg AI/m^2)

FUTURE LINES OF RESEARCH

The controlled release formulations have various advantages such as labor-saving, safety for workers, smaller dosage, fewer applications and less environmental impact which are in good accordance with recent needs. Therefore much more research work will be done.

The development of the ideal controlled release formulation would be comparable to the invention of the ideal pesticide. For example the observation of reduction in phytotoxicity on some crops by microencapsulation suggests re-examination of old pesticides, which could not be developed due to phytotoxicity, in microencapsulated form. Also currently used pesticides should be reformulated as microcapsules where the conventional formulation can not be used due to crop injury. Reduction in fish toxicity of pyrethroid by microencapsulation will broaden the application field very much. If some candidate compound cannot be developed due to high cost with the conventional formulation, the controlled release formulation will help considerably, where smaller dosage and fewer application can be achieved, and by this method the candidate compound can be developed.

As the cost of development and registration of new AI continues to increase year by year, it is attractive to find new applications and uses of currently registered pesticides by means of controlled release formulation.

Therefore in the future economics of new pesticide development will favor evaluation of the new formulations including controlled release formulation of the existing pesticides as an economical alternative to the development of new pesticides.

There are many preparation methods for controlled release formulations, but generally speaking the manufacturing cost of the controlled release formulation may be more expensive than that of the conventional ones. However manufacturing cost alone should not be the only consideration because the controlled release formulations provide various advantages as described above. These advantages should be taken into consideration overall.

Controlled release formulations in their truest sense should be developed as much as possible. At first "a specified target" should be considered. It would be good to introduce the concept of targeting in the pharmaceutical field to the pesticide formulation. Of course the concept of targeting is not the same in both fields, but some analogies exist. Secondly, release timing is important. The release of AI should start at the time when the controlled release formulation encounters the target organisms. In that case what will be the driving force for the release? Mechanical power is one of them, and this mechanism is useful as in the case of fenitrothion microcapsules trampled by cockroaches. Some other possibilities are pH, temperature, moisture, water, light, enzyme and so on. For example, the microcapsules whose release rate changes depending on pH, temperature, light, ultrasonic, and calcium ion have been investigated (ref. 24). The microcapsule which is destroyed after evaporation of aqueous medium around it is made by in situ polymerization (ref. 25). Controlled release formulation which can change the release rate by temperature change is also reported (ref. 26). It might be possible to control the release rate from the microcapsule by the influence of some kinds of enzymes which are released from the target organisms. Fruitful results are expected in the future from controlled release formulations.

REFERENCES

1. A. C. Tanquary, R. E. Lacey, Controlled Release of Biologically Active Agents, Plenum Press, New York (1974).
2. D. R. Paul, F. H. Harris, Controlled Release Polymeric Formulations, ACS Symp. Ser., No. 33 (1976).
3. N. F. Cardarelli, Controlled Release Pesticides Formulations, CRC, Florida (1976).
4. H. B. Scher, Controlled Release Pesticides, ACS Symp. Ser., No. 53 (1977).
5. R. Baker, Controlled Release of Bioactive Materials, Academic, New York (1980).
6. A. F. Kydonious, Controlled Release Technology; Methods, Theory and Applications, Vol. 1 and 2, CRC, Florida (1981).
7. D. H. Lewis, Controlled Release of Pesticides and Pharmaceuticals, Plenum Press, New York (1981).
8. T. J. Roseman, S. Z. Mansdorf, Controlled Release Delivery Systems, Marcel Dekker, New York (1983).
9. R. S. Baker, H. K. Lonsdale, Chemtech, 5, 668 (1975).
10. G. G. Allan, J. F. Friedhoff, W. J. McConnel, J. C. Powell, J. Macromol. Sci., Chem., A 10, 223 (1976).
11. K. Tsuji, Kaigai Kobunshi Kenkyu, 24, 232 (1978), 25, 239 (1979), 27, 218 (1981), 29, 211 (1983).
12. K. Tsuji, J. Pesticide Sci., 7, 539 (1982).
13. K. Tsuji, High polymers, Japan, 32, 790 (1983).
14. K. Tsuji, Chemical Engineering, 48, No. 5, 345 (1984).
15. H. B. Scher, Adv. Pesticide Formulation Technology. ACS Sypm. Ser., No. 254 (1984).

16. K. Ishida, T. Matsui, T. Yanai, K. Kawakubo, T. Honma, K. Tanizawa, M. Nakagawa, H. Okudaira, Ann. Rept. Sankyo Research Lab. 36, 44, (1984).
17. Y. Nakahara, Solid Handling and Processing Industry, 16, No. 7, 23 (1984).
18. R. J. Barker, Y. Lehner, M. R. Kunzman, Z. Naturforsch., 34 C, 153 (1979).
19. J. R. Lowell, Jr., C. B. DeSavigny, G. D. Curl, Polym. Prepr. 21, No. 1, 113 (1980).
20. H. Fuyama, G. Shinjo, K. Tsuji, ref. 8, p. 315.
21. Y. Wada, T. Nakahara, T. Orii, Y. Okano, M. Aya, K. Yasui, A. Kamochi, Y. Yamada, O. Katsumata, S. Sakawa, Y. Kurahashi, Pesticide Chemistry, 4, 257 (1983).
22. T. P. Anderson, D. R. McGuffog, 10th Internat. Symp. Controlled Release Bioactive Materials, Preprint. p 116, San Francisco (1983).
23. S. Tsuda, T. Ohtsubo, H. Kawada, Y. Manabe, N. Kishibuchi, G. Shinjo, K. Tsuji, 11th Ann. Meeting, Pesticide Soc. Japan, B202, 203, March (1986).
24. K. Okahata, Res. Rept. Ind. Technol. Bounty, Asahi Glass Co., 46, 159 (1985).
25. Japanese Patent, Publication (laid open) No. 59-16802 (1984), Toa Gosei Chem.
26. Japanese Patent, Publication (examined) No. 60-50762 (1985), Uniroyal Inc.

Optimization of biological efficacy through formulation

J. M. Deming, R. W. Magin*

Monsanto Company, 800 N. Lindbergh, St. Louis, MO, 63167

Abstract - Use of Controlled Release formulations by the pesticide industry is expected to increase significantly as costs become competitive with conventional formulations. Cost improvement coupled with increasing environmental and marketing demands should sustain a compounded 12% yearly increase in commercial exploitation of pesticide Controlled Release technology over the next five years.

While there probably is no misunderstanding about its real meaning, the title of this presentation is somewhat pretentious because it implies that we would recognize optimum biological performance if we saw it. Improvement in Biological Efficacy Through Formulation is a more realistic title; and for simplicity, we will refer to this endeavor as bio-targeting. Bio-targeting, for the purposes of this presentation, includes physical, chemical or biological manipulation employed by a formulator to improve the quantity and timing of targeting a pesticide to its active biological site.

Controlled release is usually the area that comes to mind when we think of biological targeting and will be a major topic. However, there are several other programs in our industry which meet the bio-targeting definition; and brief mention of some of them will be made prior to addressing the main topic.

Seed safeners - Crop selectivity, accomplished by the use of safeners, falls under the bio-targeting definition. This is an area which has made considerable progress this past decade. There are commercial products on the market, as well as a variety of candidates undergoing field evaluation. Early employment of this concept involved inclusion of the safener in the same package as the herbicide. Therefore, as is the case in most bio-targeting, biologists and formulators worked together to bring this concept to fruition. Single-package mixtures of herbicide and safener remain the preferred commercial form. In many instances, the formulator is asked not only to provide physical compatibility of the components in the package but also some means of improving the effectiveness of the safener, which can usually be utilized more efficiently as a direct seed treatment. To date, placement of the chemical mixture in the soil and selection of safeners with higher vapor pressures and soil mobilities than the herbicide component have done more to enable co-formulation than have any manipulations by the bio-formulator. However, microencapsulation of either or both components offers intriguing possibilities.

Microbial repressants - Microbial repressants have been used not only to preserve the chemical integrity of pesticide concentrates but, on occasion, to extend the life of soil-applied chemicals. As might be expected, the most effective soil microbial repressants have been those with long half-lives and with soil mobilities similar to those of the chemicals they are protecting. Adsorption, coating, matrixing and encapsulation are all possible means of matching the mobilities of the active and its preservant.

Contact chemicals plant entry and translocation - Bio-targeting manipulation with contact chemicals may involve anything from simple leaf wetting to the use of adjuvants to enhance leaf penetration and plant transport. Workshop III-D, of this Congress, addresses this topic; and only two points will be made here from a formulation standpoint.

a. Surfactants, organic solvents, humectants, salts and acid buffers have all been successfully used to improve efficacy of contact herbicides. Attempts to find universal penetration or translocation aids effective for all contact chemicals have not been successful. Adjuvant levels of surfactants, well above those required for good leaf wetting, have come the closest to providing consistent enhancement of contact chemicals.

b. It is somewhere between a mistake and impossible for a formulator to take on a plant entry and transport enhancement problem without strong biological support. As an example, it is not unusual for a formulation mechanism which might improve rate of leaf entry to be detrimental to the ultimate goal of plant transport because flooded entry caused the chemical to shut down the plant's transport mechanism.

Wind drift consideration - Wind drift consideration of both solid and liquid applications has an obvious influence on efficacy. Formulation manipulations to reduce drift are in common practice and are strongly application equipment dependent.

Granule properties - Particle count, dispersion characteristics and adsorption properties all influence efficacy of direct-applied solids. The need to control these three parameters has become so well understood and commonplace that currently these factors are seldom associated with bio-targeting. There are other bio-targeting aspects of granules that will be covered in the controlled release section.

Physical handling properties - Pesticide formulators spend a major portion of their time improving or developing desirable physical handling, compatibility or delivery characteristics to their formulations. The obvious need for these manipulations for efficacy maintenance should not go unacknowledged just because of their obvious and essential nature.

Controlled release - The technologies available in the area of controlled release consist of a variety of entrapment mechanisms, encompassing macro and micro encapsulation, matrixing, chemical tie-up by pendant linkage, physical tie-up on a strongly absorbing surface, impregnation into preformed hollow shells such as fibers or spheres with constricted egress, or any combination thereof. There are a few limited applications of our industries' actives that respond economically and, in some cases, require these entrapment mechanisms. Household insecticide platelets, flea and tick collars, placement and longevity enhancement of herbicides and insecticides in waterways, and anti-foulants are included along with volatile attractant/ repellant insecticides which are providing a growing market. These topics have dominated the Pesticide presentations, which started out accounting for more than 80% of the agenda and now only account for about 20%. The pesticide industry has, to date, not made significant inroads in exploiting the extensive controlled release technologies available to it. This is particularly true if we consider application to the larger volume of pesticides used for agronomic and horticultural crops.

These entrapment bio-targeting mechanisms have been relatively expensive, often increasing the cost of formulated product 100% or more. The tremendous attrition of our candidates in the pesticide screening programs essentially required the new survivors to have considerable bio-targeting built into their molecules. It is understandable that formulation bio-targeting endeavors involving these actives seldom provide efficacy responses of a magnitude able to justify the incurred high cost.

Recently, a survey by Chemical Week and High Technology Associates provided an optimistic view of the present status and projected growth of the controlled release industry including pesticide industry participation.

Since costs have frequently been a deterrent in the development of controlled release formulation, I would like to call attention to several areas where economically-feasible utilization of controlled release formulations have been achieved.

Microencapsulation by interfacial polymerization currently accounts for more than 80% of the Pesticide Industries' commercial exploitation of Controlled Release formulations. It was first employed on a large-scale pesticide use by Pennwalt, under U.S. Patent 3,577,515 on methyl parathion, where both safety in handling and protection against U.V. light breakdown of the active were bio-targeting benefits. Pennwalt has, on a smaller scale, extended their practice to other actives. Stauffer's microencapsulation of thiocarbamates, under U.S. Patent 4,046,741, was well noted by our industry, as they dealt with the EPA on capsule clearance about the same time that Pennwalt, the EPA and honey bees were making the news.

U.S. Patent 4,280,833 describes one-step, high-concentration encapsulation capable of providing concentrates economically competitive with conventional emulsifiable concentrates or water-based flowable formulations. The encapsulation mechanism and components in this patent were all established in the industry prior to its issue and its significance centers entirely on the fact that it accomplishes entrapment without exceeding the cost of unencapsulated formulations.

In recent developments, Insulated Technologies announced last fall that they expected to be able to employ their encapsulation technology at costs in the 20-30¢/lb range. Moreover, Pennwalt has developed 4 lb/gal experimentals which are certain to provide cost advantages over their 2 lb/gal systems.

An important characteristic of interfacial polymerization of organic pesticides is that it provides a built-in release mechanism for the active ingredient. The mechanics of interfacial polymerization require that one of the components of the shell wall polymer be soluble in the organic pesticide being encapsulated. Meeting this requirement essentially assures sufficient miscibility of pesticides in the shell wall. Film composition and thickness can be used to modify diffusion rate, but rate is also strongly influenced by the sink in contact with the capsule to which the diffusing pesticide migrates. If water is the suspending medium of the microcapsule concentrate and the encapsulated pesticide has low water solubility, there is very little tendency for the active to escape into the aqueous environment in the concentrate. However, if the capsule encounters the strong and extensive adsorption sites provided by the soil, diffusion is rapid.

The lack of adsorption or sink properties of surface plant residue inherent in minimum-till cultural practices presents a unique targeting opportunity for microencapsulation to improve pre-emergent herbicide efficacy. Various levels of limited tillage are practiced both as a soil conservation measure and to reduce energy, equipment and labor costs. Adoption of these practices has increased in direct proportion to the availability of herbicides to eliminate weed competition without tillage. Surface residue has limited the effectiveness of pre-emergent herbicides by intercepting some of the chemical spray before it can reach its target, the soil. Microencapsulation can minimize this problem.

The water-based microcapsules have less tendency to absorb into the trash than do emulsifiable concentrates of the same active. This provides greater penetration of the applied apray, as well as greater susceptibility to wash down from rain and other mechanisms. Moreover, that portion of encapsulated herbicide which is intercepted remains primarily within the shell wall due to the low adsorption characteristics of un-decomposed organic matter. It is, therefore, protected against volatility while awaiting transport to the soil.

Water-dispersible granules - Another means of minimizing trash interception of applied herbicides is to use granules which tend to fall through surface trash. On the premise that bio-targeting of minimum-till herbicides would be enhanced through increased use of granules, a brief discussion of granule use trends is warranted here.

Utilization of low-cost encapsulation technology can facilitate economic conversion of liquid or low-melting, solid actives into water-dispersible granules. This essentially permits development of high-concentration dispersing granules. Pesticide mobility limitations accentuated by high-concentration compared to low-concentration granules can usually be compensated for by the ability of the granule to disperse to micro particles from contact with soil moisture. The use of smaller granules (40/80 mesh) could be employed, if necessary, to further compensate for lack of mobility. Direct application of agglomerated micro granules would require special attention to air curtains and equipment skirting to minimize wind drift. Complimentary to these formulation trends are the evolving granular applicators capable of accurately metering and distributing as little as one pound of granules per acre. These developments indicate, in the not far distant future, there should be a sufficient number of high-concentration solid formulations that were originally developed for spray applications to justify farm purchase of low-volume, direct-application equipment. Such equipment could treat 100 or more acres from one hopper loading. At the same time, bio-targeting benefit from penetration of surface trash would be realized.

We have spoken of microencapsulation's possible competitive cost extension of granule use. If this extended use does take place, it should also extend the use of other controlled release mechanisms that are difficult, if not impossible, to attain through the use of any other formulation mode. Entrapment through matrixing, macro-coating or adsorption are all associated with solid systems which might be dust or wettable powders but usually take the form of granules.

Future considerations - There are developing needs in the pesticide and agricultural industries that portend increasing use of bio-targeting in the future. It has already been utilized to reduce toxic exposure to the farmer and, with environmental concerns proliferating, will be further utilized, whether by direct envelopment or as an aid to development of formulation modes which simplify handling and package disposal. Some applications to minimum-till, cultural practices have already been described.

New chemicals on the pesticide horizon have, for the most part, high unit activity and low-use rates. These lower rates more easily justify the cost of bio-targeting formulations. Moreover, expanded synthesis around this new, highly-active chemistry is very likely to provide molecules with even stronger recognized needs for bio-targeting.

In the past, the greatest obstacle to effective bio-targeting has been insufficient target definition. One has only to look at some of the topics under discussion here at this Congress and couple them with technologies like genetic engineering and molecular biology to illustrate the major strides currently being made in understanding the biological systems with which we deal. Much of this knowledge can, and is, being applied to better define bio-targeting needs to the formulator, to enable him to provide solutions to these needs.

Those of you who work in, or close enough to, molecular biology or genetic engineering programs to have discussed the future isolation and delivery needs with these people would agree that bio-targeting will have to make tremendous strides from its present state if it is going to properly service those needs.

One last trend which seems to be gaining a foothold in our industry is the positioning of the bio-targeting formulator at as early a stage as possible in the screening/secondary testing program. This provides, essentially, a mutual benefit. Some of the active chemicals will be saved from extinction by the bio-targeter; and for the most part, those chemicals will be more responsive to formulation and more easily justify bio-targeting formulation costs than was true for the few chemicals that survived past screening programs.

In light of the above trends, a projection of $500M yearly worth of Controlled Release Pesticides by 1990 seems realistic and probably on the conservative side.

Physical test methods for formulation development and quality standards

H. J. Niessen

Bayer AG, Pflanzenschutz-Produktion FE, Institut für Formulierung, D-5090 Leverkusen, Federal Republic of Germany

Abstract - The dependence of the biological properties of a crop protection product upon its physical properties is so extremely complex that despite considerable research work it is still not comprehensible in its entirety. Nonetheless, the formulation chemist is able to achieve specific improvements of certain product properties by systematic research. For this purpose, he has at his disposal a host of methods some of which are very elaborate. Although they do not always provide comparable values, they do however permit trends and interrelationships to be recognized. Test methods for quality control, on the other hand, must be easy and inexpensive to perform, and feature adequate reproducibility of a level established by collaborative tests in different laboratories. Many of the methods published by international organizations are in need of improvement with respect to reproducibility. New methods must be devised for new formulation types or for changing application conditions. They must be fully orientated to product behaviour in field conditions if they are to be meaningful. Finally, automation of tests is a desirable objective.

INTRODUCTION

A high quality crop protection product is expected by users to be easy and safe to handle and easy to apply with the equipment available, that it will provide acceptable control of the target pest complex, and at the same time display good crop compatibility. Users also expect that the product will have a satisfactory shelf life, in other words that its properties will not undergo any significant changes during a certain period of storage.

The quality of a crop protection product as defined in this way can only be measured in the laboratory by determination of certain individual parameters. Parameters of major importance are of course the nature and the amount of the active ingredient contained in the formulated product. Given identical formulation, there is a clear dose-effect relation for a certain active ingredient, characterized by an S-shaped curve slope. "Given identical formulation" already implies that the type of formulation and hence the physical properties thereby endowed on the product, play a decisive role. However, the correlations between different measurable physical parameters and that what is defined as quality are frequently of a complex nature and difficult to fathom.

In this paper, it is distinguished clearly between two categories of physical test methods. The first category comprises those methods employed by formulation chemists to elucidate correlations between different parameters in order to achieve wherever necessary specific improvements of certain properties of the compound that is to be formulated. The second category embraces those methods by which products can be tested for conformity with certain quality requirements and which permit declaration of limit values in specifications.

METHODS FOR FORMULATION DEVELOPMENT

The methods of the first category that are accessible to formulation chemists virtually cover the entire range of present-day physical testing technology, from light or electron microscopic methods through techniques for measurement of rheological characteristics, interfacial tension, adsorption isotherms and particle sizes to combined laser-MS. Limitation to these possibilities are not less by the imagination and creativity of the formulation chemist than by his budgetary problems because more highly sophisticated and more highly sensitive equipment is also more costly. Therefore, the formulation chemist must always make a renewed estimate of what data are obtainable at what price. This sounds very unscientific but it fully reflects the actual situation in daily practice.

Of course the formulation chemist could be told to concentrate firstly on the most important properties of his formulation and try to optimize them. But when the problem is tackled this way, it will soon be found that the "most important" property is a different one in almost every formulation, in fact the very one that is hardest to manipulate.

Parameters mainly related to biological properties

The most important property is biological activity. A compound will only display optimal biological activity when it is present at the site of action at the right time and at the right concentration. There are many examples to illustrate that different formulation types satisfy this requirement in a very different manner. Frequently, it is emulsifiable concentrates that display better biological activity although sometimes also poorer crop compatibility than water dispersible powders or suspension concentrates. Addition of adjuvants to the spray mixture may enhance biological activity in many cases but in others such additives are useless or perhaps even cause a loss of activity (ref. 1). Consequently, there has been no lack of studies to investigate retention of sprays on leaf surfaces and leaf penetration of pesticidal compounds through wax layers and cuticles. The physical test methods used for such studies are of a most varied nature, ranging from measurement of contact angles and interfacial tensions through light and electron microscopy to radiotracer techniques or employment of a scanning electron microscope fitted with an X-ray analyzer system (ref. 2). It is commonly revealed by such intricate studies that spray retention and compound uptake are determined not only by the formulation but also by the plant species, spray chemical dose and droplet size, in fact even by the age of the respective leaf and the growth environment (ref. 3). Additional influential factors are the unpredictable weather conditions. What influence is exerted on a spray deposit by dew, rain and wind? Will resistance to rain perhaps be reduced when the wetting property of a formulation is improved? But even the rain resistance parameter is very difficult to measure because it is affected on the one hand by the nature of the substrate and on the other hand by the highly diversified nature of rainfall ranging from slight drizzle to torrential downpours.

These examples have been given with the objective of demonstrating what we are still far remote from being able to blueprint at the office desk the ideal formulation for a certain pesticide on the grounds of its chemical and physical data and the knowledge of its mechanism of action. Too many complexly interrelating factors are involved, and therefore it is necessary to take account of the results of biological field trials which inlcude a host of natural marginal conditions. Nonetheless, such systematic studies are extremely important in order to deepen our knowledge of the mode of action of pesticides and, in combination with field trials, to improve certain properties of formulations. From these studies, interesting results are expected to emerge in future, which will be attainable through the use of increasingly sensitive methods for the detection of active ingredients.

Parameters mainly related to application properties

What goes for biological properties similarly holds also for the more application-orientated physical properties. More profound physical tests disclose relations in partial areas and permit specific alteration of certain properties.

To give two examples. In many WP's, suspensibility is barely affected by water hardness although in some cases there is pronounced sensitivity to hard water (Fig. 1) (ref. 4). It is noted that formulation A is still acceptable in soft to normally hard water whereas in extremely hard water its suspensibility is not adequate. In order to improve formulation A in the direction of the curve for B and C, the formulation chemist cannot be content with one-point measurements as contained in specifications but instead he must plot whole curves so that he can monitor the influence exerted by the type and amount of different dispersing agents which must now be tested. Plotting of time-related sedimentation curves fitted to the individual points provides additional information. If this also fails, measurements of the zeta potentials may prove helpful since they are capable of providing interesting results even for emulsions. Lee and Tadros (ref. 5) found, as illustrated in Fig. 2, that surface tension has hardly any direct influence on droplet stability but that it increases substantially parallel to zeta potential.

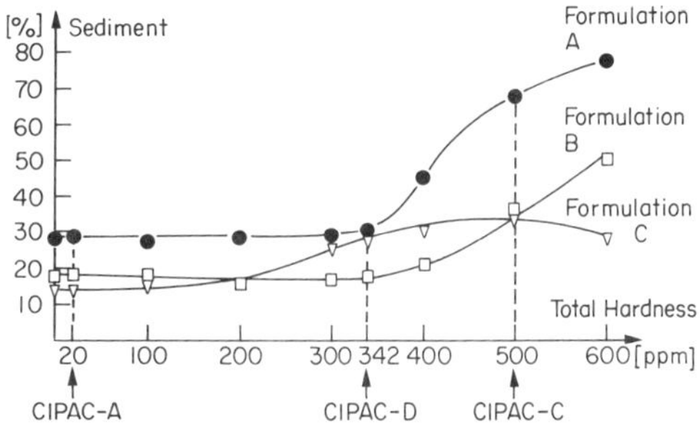

Fig. 1. Correlation between suspensibility and water hardness of different WPs

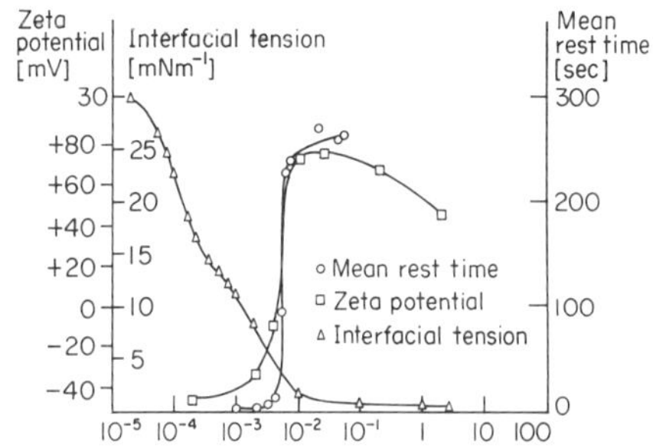

Fig. 2. Correlation between mean rest time, interfacial tension, and zeta potentials of emulsion droplets

The second example: In another product, a DS, problems are posed by the flowability of the powder. Table 1 shows the limited possibilities open to the formulation chemist in this case.

TABLE 1. Interpendence of Properties

Premises	Variables	Dependent Properties
A.I. properties	(A.I. concentration)	biological activity
formulation type: here: DS	inerts	plant compatibility
	particle size	adhesion to seeds
A.I. ratio	oil concentration	flowability
		dustability
		storage stability

Given are the physical and chemical properties of the active ingredients, the formulation type and the composition of the active ingredients to be used, whilst the concentration rate usually can still be varied within limits. Only few variables are available. A reduction of particle size results in improved sticking property but in a reduction of flowability and an increase in dustiness. By in- creasing the oil content, dustiness is improved but flowability is again reduced. The optimum formulation sought by a formulation chemist is therefore a compromise between different properties that often undergo change in opposite directions.

An extremely important property of most solid formulations and suspension concentrates is particle size which is of interest especially in the range between 0.1 and 10 µm and can be measured only with considerable outlay. As measurements are possible only in diluted suspensions, sample preparation attaches decisive importance, in particular careful preparation of saturated active ingredient solutions for dilution and the required energy for disintegration of loose agglomerates. Evidence that different instruments produce different results despite identical sample preparation has been furnished by a test in which 5 different instruments were compared (Fig. 2) (ref. 6).

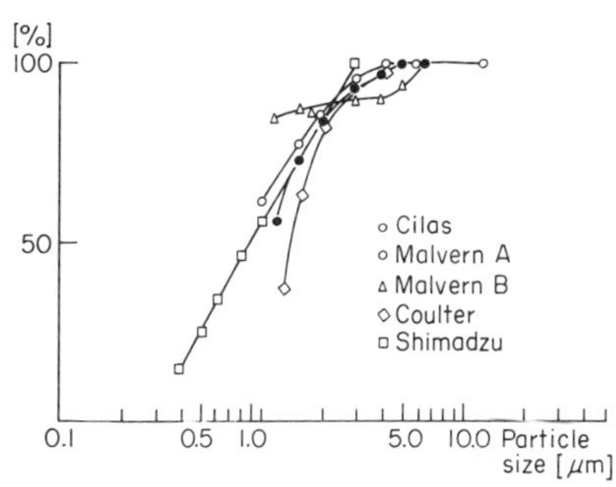

Fig. 3. Particle size distribution of Baytan[R] FS measured by different equipment

Three instruments operate on the laser diffraction principle (1 Cilas, 2 Malverns of different age), one instrument measures the particle volume on passing through a capillary (Coulter Counter), and one measures sedimentation in a centrifuge (Shimadzu). It is seen that in this case the centrifuge is the best instrument as regards measuring principle because it is still capable of sizing particles in the range between 0.4 and 1 µm. Extrapolations in the lower range are extremely problematic and are at the most still permissible by curve fitting when there are still values well below the maximum in the frequency distribution, as seen in the left part of Fig. 4. It easily leads to wrong results when the maximum is not determined as in the right part, where a computer still may be able to make a mathematical evaluation.

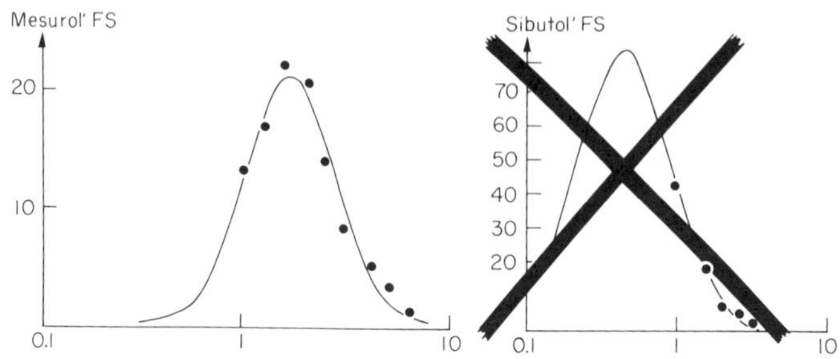

Fig. 4. Extrapolation of particle distribution curves with sufficient data (left) and unsufficient data (right)

The divergencies of the measured results in Fig. 3 are substantial. Nevertheless, all instruments are capable of correctly performing relative comparisons between samples of different fineness if grinding. It is, however, extremely difficult to confirm the correctness of measured results expecially when particle shape differs greatly from spherical shape. Therefore, one must be wary of comparing values that have not been measured by instruments of the same design and same type!

METHODS FOR QUALITY CONTROL

Such methods that are described in specifications must satisfy two main requirements. They must be easy and inexpensive to perform, and their results must enable the inspector to sort out inferior batches of a product. They must also have been tested in collaborative studies and have been widely approved. As they are used in different laboratories for quality control, they should display good reproducibility. But it is just in this respect that many of the methods employed today are still in need of improvement.

Wettable powders

The parameters that require to be tested in a specification should be orientated to the application conditions. Therefore, in the case of a WP, instead of determining a mean particle size, a procedure that involves considerable outlay, the quality of grinding is tested by measuring suspensibility and by sieving usually at a mesh size of 75 µm. Determination of the wetting time will furnish information on the desired easy mixing of a spray.

The last two methods can be taken as examples for discussing the problem of poor reproducibility.

Fig. 5 shows a titration curve. It is at once seen that an end point can be determined with good reproducibility because a small addition of alkali induces a considerable alteration of the pH close to the point of equivalence. The curve for the wetting time (Fig. 6) reveals quite a different situation. The end point is in the flat tail part of the curve, and it is left more or less to chance when the last lump of powder will become wetted and sink. Reproducible data would permit only an evaluation of the steep part of the curve, say the expression of a W_{50} value or extrapolation to the abscissa in the indicated manner, but experimentally the curve cannot be measured in detail by simple means.

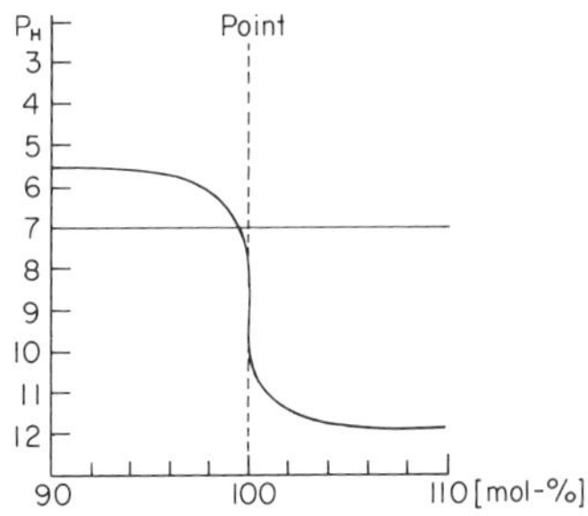

Fig. 5. Titration curve of acidic acid

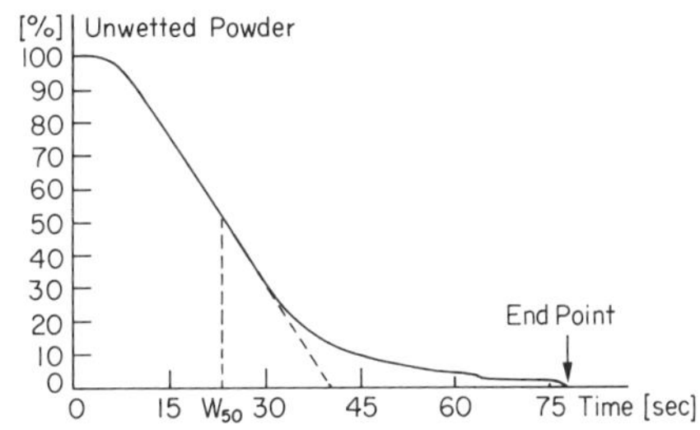

Fig. 6. Wetting time of WPs

Comprehensive collaborative analyses in the fertilizer industry (ref. 7) have revealed that in <u>sieve analyses</u>, repeatability is good but reproducibility is very poor. This is due mainly to the sieve fabrics that are used. Fig. 7 shows that only small changes of mesh size and that are still within the tolerances, can lead to marked alterations of the results. Additionally, the tolerances specified in the different standards are far exceeded in many test sieves particularly after they have been in use for a lengthy period.

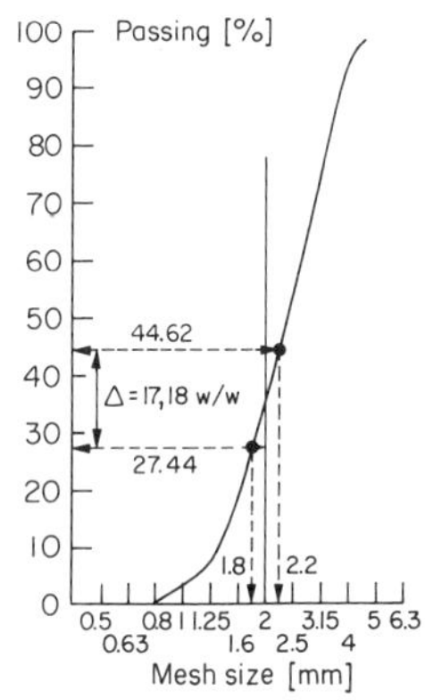

Fig. 7. Influence of tolerated variations of mesh size on the results of sieving

These uncertainties involved in the generation of data must be borne in mind and taken into account when it comes to deciding whether to accept or to reject a certain batch on the grounds of what is revealed by the check on its specification. In view of such inaccuracies of the methods, some latitude of decision is in fact quite justified, particularly considering that many numerical values declared in specifications tend to represent "typical" data generated for a good product rather than actual tolerances in the sense that every slight transgression would provide an indicator of a "poor" product. If, for example, a wetting time of 100 or 110 seconds is measured instead of a specified maximum time of 90 seconds, this alone certainly does not make a product unfit. On the other hand, however, one knows that shorter wetting times are considered to be "normal".

Suspension concentrates

A most difficult proposition is the specification of suspension concentrates. A good product should be easy to homogenize after transportation and storage, and when diluted with water it should form a homogeneous spray mixture that is easy to apply. Factors that essentially co-determine quality are viscosity and particle size. During the development of a product, they must be controlled very carefully by the formulation chemist. However, these factors are excluded from the specification of quality for the following reasons, and must be described by other parameters:

- Most SC's display non-Newtonian flow characteristics, that is to say viscosity changes with shearing stress and cannot be expressed as a simple numerical value. For users, it is after all only of interest to know whether the containers can be easily emptied and whether the product readily disperses in water. This can be determined by tests for pourability and dispersibility.

- As already pointed out, there are no internationally accepted simple methods for determination of particle size range. Also in this case application-orientated parameters must be quantified that are influenced by particle size. These parameters are wet sieve analysis, suspensibility, and pourability.

Another critical property of suspension concentrates that poses still greater difficulty in quantifying is claying, that is to say the formation of a non-dispersible sediment after storage. The onus of responsibility for establishing whether claying has occured is on the sampler who must inspect the commercial containers very carefully before taking samples to verify product quality. The container must be shaken gently, and after this has been done the product must appear homogeneous, and the pack must not contain a nondispersed sticky layer at the bottom, which can be checked easily by probing with a glass rod. Viscosity can of course be quantified in a "poor" product by measuring it in different layers (ref. 8) (Fig. 8), but for this purpose samples aged in suitable test vessels are needed, which are not available to the quality inspector. Certain indications of storage stability are of course provided also by the heat test embodied in the specification. Notwithstanding, the fact remains that particularly in the case of SC's and of course also in the case of FS's, the point concerning product description and the work of the sampler attaches very special importance.

However, the formulation chemist will carry out very intensive measurements of viscosity during the development of a product. These measurements will enable him, among other things, to state data on storage stability. Fig. 9 shows, for example, the change in viscosity at a certain shearing stress in relation to storage time and temperature.

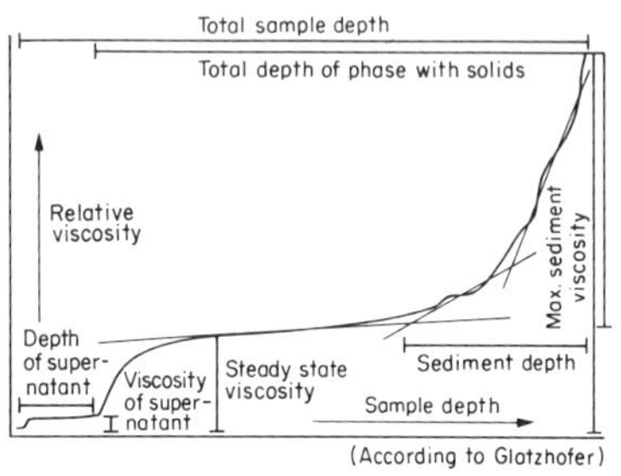

Fig. 8. Typical graph of relative viscosity vs. sample depth

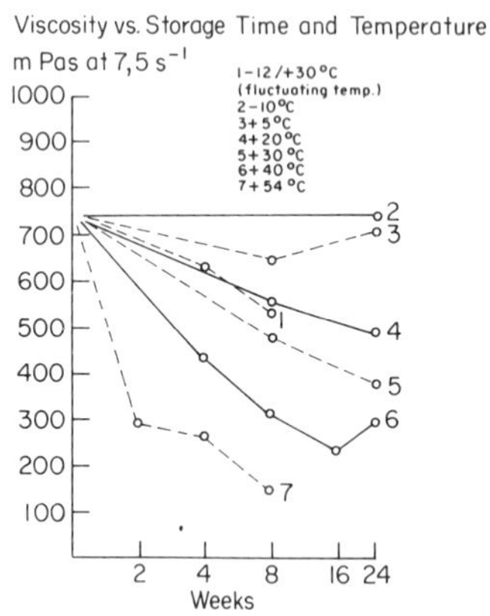

Fig. 9. Viscosity readings after storage at different temperatures

Emulsifiable concentrates

Sometimes established methods must be modified or new ones must be devised whenever application conditions change or when new formulation types are developed or marketed. For example, the official emulsion tests that require 5 % emulsions will no longer be optimal for new compounds that are to be applied in substantially greater dilutions (ref. 9). They are, however, completely unsuitable for such EC's containing solid active ingredients that may crystallize in the aqueous spray mixtures. For these formulations, for which the new term "Dispersible Concentrates" and the code DC has been suggested, new test methods must be sought that are very closely matched to field application conditions, in other words to the behaviour of the products in field sprayers (ref. 10). A procedure that keeps spray mixtures in motion for a lengthy period by defined stirring, followed by wet sieving, might perhaps be contemplated for these formulations.

Water dispersible granules

In Germany, an international working group is currently engaged in the development of test methods for the relatively new WG formulation type. But also in this case it is not an easy matter to find test methods that display good reproducibility and relevance to field conditions, especially with regard to quantification of dispersion spontaneity, abrasion and dust content, parameters that are all of importance and bear relevance to quality. The difficulties encountered in discovering appropriate test methods are of such considerable magnitude particularly because the products are manufactured by greatly differing methods thereby potentially subjecting the properties to strong influences. Fig. 10 shows a granular product manufatured by the fluidized bed technique, and Fig. 11 shows a spray-dried one. Both products differ not only in size but also in their surface and grain structure.

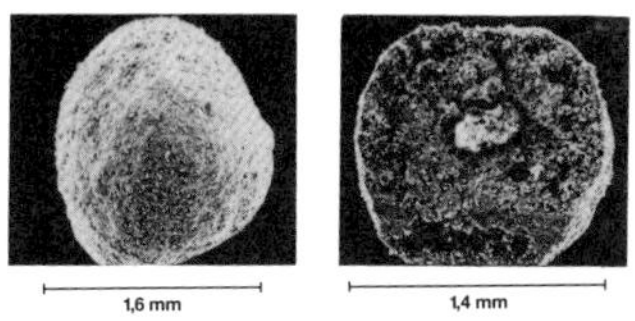

Fig. 10. Surface and cross section of a fluidized bed granulated WG (Sencor R WG 70)

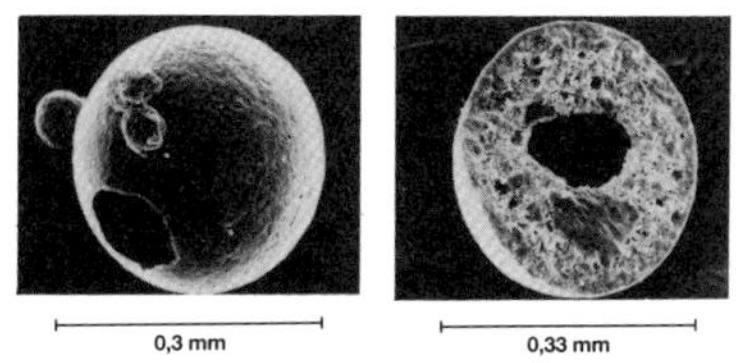

Fig. 11. Surface and cross section of a spray-dried WG (Goltix R WG 70)

We expressed the demand that test methods for quality control should be easy and inexpensive to perform. But low-cost implementation also implies greater reduction of work time. This in turn calls for automation. In our laboratories we have installed a system designed to automate the suspensibility test. A central unit fitted with a pump supplies 12 sedimentation vessels which are controlled by a microprocessor and are capable of separating the bottom 10 percent from the rest of the suspension within the preset time. In this area of automation, there is still very wide scope for the apparatus industry.

OUTLOOK

Many test methods for determination of quality have been published by CIPAC and WHO as international organizations, most of them after completion of collaborative studies. Many of these methods have proved to be excellent and are constantly in use. Others, as demonstrated by the example of the wetting time of WP's, are in need of improvement due to the wide range of analytical error. Properties that are difficult to quantify are particularly those that are partially dependent upon sample pretreatment, like the flowability of powders or abrasion of granules.

I came across the following sentences in an article published by Herman F. Beckmann in 1963 (ref. 11):

"Presently, evaluations of quality are generally on an empirical and personal basis. Research needs to be done to put this evaluation on a scientific basis. Agreement on what constitutes "quality" and how to make evaluations will be difficult to obtain."

It could be shown that many of the questions still open in 1963 have been answered during the past 23 years by the development of appropriate methods and by the preparation of specification guidelines and product specifications. But undoubtedly there is still cause to reiterate Beckmann's call for more research to be done in both categories of physical test methods, because the paper presents a catalogue of problems still awaiting a satisfactory solution.

REFERENCES

1. G. Maas and H. Niessen, DLG-Mitteilungen 1984, 252-253
2. E.A. Baker, G.M. Hunt and P.J.G. Stevens, Pestic. Sci. 14, 645-658 (1983)
3. E.A. Baker and G.M. Hunt, Ann. appl. Biol. 106, 579-590 (1985)

4. H. Tengler and E. Radlmann, Correlation between suspensibility of water dispersible powders and water hardness. Paper given at the CIPAC-Symposium, Vienna 1986
5. G.W.J. Lee and Th.F. Tadros, Colloids and Surfaces 5, 129-135 (1982)
6. A. Bürkholz, R. Böck and K. Wangermann, unpublished (1986)
7. S. Vielhauer, Kali und Steinsalz 8, 117-125 (1981)
8. J.P. Glatzhofer, In: Scher, H.B. (Ed.): Advances in pesticide formulation technology, ACS Symposium Series 254, Washington 1984, 77-88
9. E. Neuenschwander and E. Hofmann, Stability of aqueous emulsions. Paper given at the CIPAC-Symposium, Vaerloese, Denmark, 1985
10. H. Niessen: Emulsion testing of emulsifiable concentrates comprising solid active ingredients. Paper given at the CIPAC Symposium, Vienna 1986
11. H.F. Beckman, In: Zweig, G. (Ed.): Analytical methods for pesticides, plant growth regulators, and food additives, Vol. I, New York and London, 1963, 7-23

Pesticide application technologies

John Maybank

Saskatchewan Research Council, Saskatoon, Saskatchewan, S7N 2X8, Canada

Abstract - While the past two decades have seen major research advances in chemicals and formulations, improvement in application techniques has been relatively modest. Most pesticides are applied in a broadcast manner as an aqueous solution, using very simple nozzle devices to effect liquid breakup.

Efforts have been made to modify flat fan nozzles so as to improve the emission pattern and/or to reduce the volume present in the small, drift-prone droplet fraction. The primary problem with most of these attempts to improve spray technology lies in the interactive nature of the controlling parameters, viz. nozzle geometry, boom orientation, hydraulic pressure, emission height and solution properties. Attempts to improve the pattern frequently worsen off-target droplet drift, and vice versa. A second major problem lies in the broadness of the droplet size range emitted by almost all nozzles in use today. Increasing, but still fragmentary, evidence indicates that certain portions of the total droplet spectrum may provide more efficient weed or insect control per unit volume than other droplet sizes.

New application technologies are clearly needed, and several promising developments are in testing and demonstration phases. Controlled droplet applicators (CDA) should provide, in principle, a droplet spectrum concentrated in the most useful range. CDA devices should also allow for the de-coupling of pattern and emission rate control parameters. More specific approaches towards doing this are potentially available with controlled solution applicators (CSA), wherein the chemical is monitored into the solvent only as it is needed, either along the spray boom pipe or even right at the nozzle. Finally, practical drift control measures may be incorporated into any of these new systems.

INTRODUCTION

The intent of pesticide use in agriculture is not actually to control or reduce the prevalence of weeds and insects within the farm field, but to increase the yield return to the farmer, to an extent that provides an acceptable benefit - cost ratio. All proposed improvements to application technology, as indeed in the introduction of new chemicals, must be tested against this criterion.

In almost all cases pesticides are broadcast; that is, they are spread everywhere within the target area in the expectation that some fraction of the active ingredient will meet up with and kill the weed or insect pest that it is designed to control. Most commonly, the broadcasting is done as a spray of fine droplets, but for some pesticides granulated powders may be disseminated in the dry form. In what follows emphasis will be placed on the former, liquid application techniques.

Once released it is naturally intended that all such droplets or dry particles will move in the air from the point of emission to the target area soil or foliage, and that following deposition the active chemical will begin its control action. The three physical stages in this process are identified as dissemination, transport and deposition.

Inherent to the broadcast scheme for pesticide application are three significant aspects of the operation, for each of which it has been customarily assumed that satisfactory conditions or limits pertain.

(i) Droplet heterogeneity. It is generally accepted that the volume (mass) of pesticide chemical needed to affect the target is independent of the droplet size in which it is applied; thus, eight 100 μ droplets per unit area will be as effective as one 200 μ droplet of the same solution strength. While this may be true for both herbicides and insecticides over a fairly wide droplet spectrum, concrete evidence is frequently deficient. As nearly every spray device in common use produces a wide range of droplet sizes an alternative interpretation is widely held: some fraction of the droplet spectrum may be the most effective, so that one need only ensure that it is adequately represented. This may lead to the situation of possible over-application (Fig. 1).

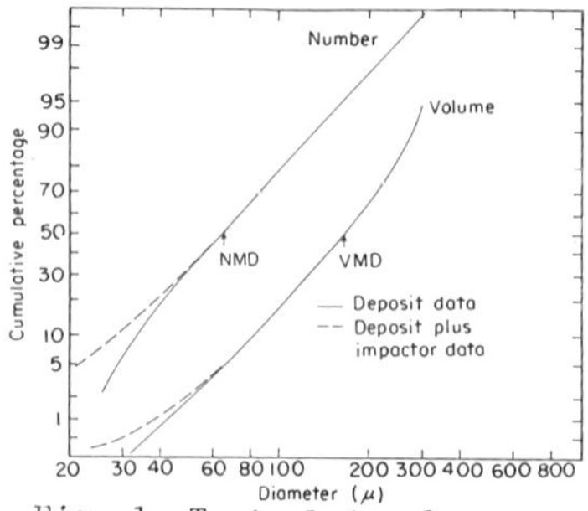

Fig. 1. Typical droplet spectrum for flat-fan nozzle.

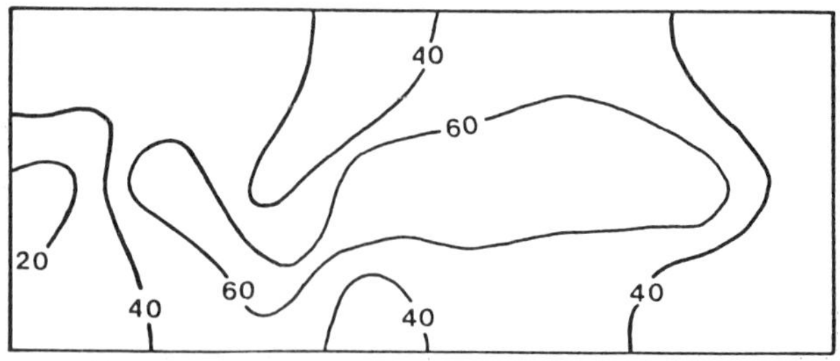

Fig. 2. Typical groundrig deposit patter.. (mg/m^2).

(ii) Pattern uniformity. The mass of pesticide, or the number of droplets, per unit area should be reasonably constant. If not, certain areas of the target field or foliage will receive too little pesticide for effective control. The easiest way of avoiding this is to overdose so that even the lowest receipt per unit area exceeds the minimum desired control density. Apart from cost this approach can, however, result in crop damage (with herbicides) on those areas receiving significantly greater deposit densities than the mean (Fig. 2).

(iii) Off-target drift. It is known that a fraction of the droplet spray will be caught up in horizontal air motions and so be carried beyond the target area. Insofar as this loss reduces the target deposit, it may diminish the desired control, but as it customarily varies downward from around 20 to less than 1%, it can in practice be easily accounted for, by a commensurate increase in the application rate. Unwanted pesticide effects outside the target area are another matter of course (Figs. 3 and 4).

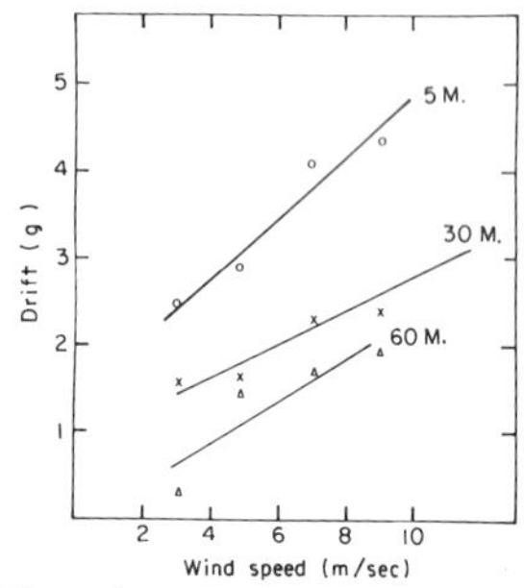

Fig. 3. Off-target drift, per 100 g groundrig spray.

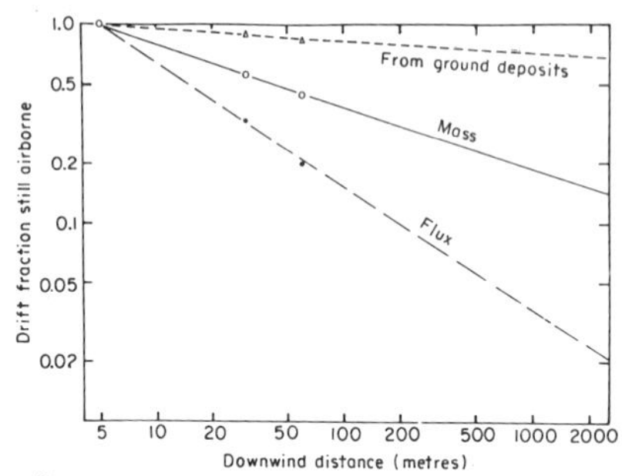

Fig. 4. Decrease in off-target drift with distance.

As long as the chemicals, farm fuel and farmer's time remained cheap, there was a requirement that the application system remain cheap as well. To a great extent that has been why research effort has been low. The complexity of the entire spray process is another reason, with "ad hoc" tests on minor improvements taking precedence over coherent multivariate approaches. However, rising costs for chemical application, the multiplicity of available pesticides, and recognition of the hazards of off-target drift have in recent years led to major investigations of the spray process and to attempts to develop truly new and innovative sprayer systems.

PRINCIPLES OF SPRAY PRODUCTION

There is a complex mix of factors that govern droplet spray production, starting from the dissemination stage at the nozzles and leading, after transport and diffusion in the atmosphere to deposition on the target soil or foliage. Research to date has addressed various combinations of these stages and factors, but seldom have all aspects been investigated in a single experiment or theoretical analysis. The entire process is just too complicated for a comprehensive treatment, so restrictions and simplifying assumptions have to be made.

The two primary factors may be considered to be the droplet and air motions relative to each other, with the former being a function of droplet size. For each diameter, it is possible to calculate a terminal velocity, which in still air, and for a droplet starting from rest, permits a determination of settling time from any release height. If the droplets were all of the same size and the ambient windspeed constant in both space and time, the downwind resting place of the emitted spray could be easily found. There would be, in such circumstances, no distinction possible between "target" and "off-target" deposits: the entire spray would come out as a single deposit band. Its width would be relatively narrow, being set by the length of the spray boom nozzle array, and by whatever variations might occur in release height or initial droplet trajectory.

Even for heterogeneous droplet spray, the problem would not be very difficult to handle in a uniform air flow, if only the droplet size spectrum were known and truly controllable. Then, each size interval (which could be set as acceptably narrow as desired) could be assigned the appropriate terminal velocity and the position on the ground of each size of droplets calculated. Unfortunately, of course, the variations in windspeeds, together with naturally-occurring air turbulence variations, make it impossible to obtain realistic deposit data from this simplistic approach.

The problem of predicting successfully a deposit pattern actually arises even before air motions come into play, for the droplet spectrum is itself a complex and frequently varying product of several factors right at the spray vehicle. The most common type of nozzle presently used in both groundrig and aircraft application consists of a metal tip or plug having a small orifice hole machined into it. Through this opening the spray solution is forced under pressure. It then spreads out as an unstable sheet or cone of fluid, followed in short order (of a few milliseconds or so) by breakup into filaments and then into droplets. This dissemination process is affected by such properties of the pesticide solution as viscosity and surface tension, together with the emission pressure on the fluid, and the dimensions of the orifice as well as its configuration. Increased values for the first two will tend towards retarding the liquid break-up, thereby increasing the droplet sizes. Also, a larger orifice opening will generally result in a coarser spray at the same hydraulic pressure than would a smaller one. On the other hand, the breakup is increased and a finer drop spectrum ensues when the pressure is increased. However, the relationship is extremely complex, and in practice it is impossible to predict the drop spectrum to be expected from any combination of these four factors (viscosity, surface tension, pressure and orifice diameter).

There are several other nozzle designs presently in use besides the flat fan and cone ones, but which may also be characterized as "passive" or static types, with the fluid flow being forced through them by some impressed pressure (or even just by gravity). Their flow rates and droplet size patterns will therefore depend on the same factors as the flat-fan ones do, though perhaps to a different degree. On the other hand, several new approaches utilize an "active" or dynamic nozzle design in which the nozzle interacts with the fluid flow to enhance breakup and droplet formation. These commonly take the form of a rapidly spinning disc or cone, onto the surface of which the solution is fed at relatively low pressure. Centrifugal forces effect the breakup of the fluid sheet as it reaches (or extends beyond) the periphery of the rotor. In principle, this type of atomizer should produce homogeneous droplets, or at least a relatively narrow droplet spectrum, with the size dependent, in a predictable manner, on surface tension of the fluid, rotor diameter and rotational speed. In practice, the equation holds only for very low fluid flow rates. At higher values, breakup occurs not at the rotor edge but at some fluid periphery of varying diameter, resulting in a heterogeneous droplet spray (but one with distinctly narrower size range nevertheless, than those produced by static nozzles). The size spectrum is no longer independent of solution viscosity, but remains independent of hydraulic pressure and fluid feed rate, over quite a wide range at least.

It is important to note that neither the droplet pattern nor the initial size spectrum is independent of the forward motion of the vehicle from which the spray is being emitted. At aircraft speeds breakup of the fluid sheets and filaments is significantly altered towards smaller sizes by the air flow past the nozzles. The emission pattern too is affected, a situation that commonly varies nozzle by nozzle according to the changing turbulence along the wings. The much lower speed of groundrig spray vehicles does not produce much effect on droplet pattern or spectrum by itself, but boom and wheel bounce due to rough terrain affect both features in a most erratic manner.

Following dissemination, the emitted droplets spread out from the nozzle and commence their descent towards the ground. Their initial trajectories and emission velocities are quickly superseded by the airflow, which initially is a vector combination of the ambient wind speed and the spray vehicle velocity. The effective release height is commonly determined by where this flow change occurs. For a groundrig, the point lies close to the nozzle outlet, and so boom height becomes the useful measure of this parameter. For fixed-wing aircraft, wing tip vortex is a major input into the initial condition of the spray droplet cloud, and sets an effective release height that may be several times the nozzle-to-ground distance. The same is true for helicopters where it is the rotor vortex. In both cases, while a mean release height may be observed, and in some cases predicted, on the basis of aircraft weight, speed and wing loading, variation about it for individual droplets can be considerable. This feature plays a major role in spreading the initial spray and in causing large variations in off-target deposition close in to the sprayed area (i.e. at relatively short downwind distances).

A major factor affecting droplet transport is the rapid evaporation of the water carrier of the spray solution (or indeed the volatility of any major component). In typical operations, between 90 and 99% of the spray volume will be water, and this will evaporate very quickly under the air temperature and humidity conditions commonly present during spraying activity (e.g. 15-30°C and 30-70% relative humidity). Depending on actual values for these two controlling parameters, a 100μ diameter solution droplet initially containing 98% water, say, will diminish in size to around 30μ in a very few seconds. For a 500μ droplet from the same spray solution, the time is considerably longer, typically several minutes. They will therefore reach the ground at sizes near to their initial values, while the smaller droplets may reach their final sizes before deposition occurs.

Once the airborne droplet reaches close to the foliage or ground surface, the deposition stage may be considered to have been reached. It is particularly difficult to identify and assess the various factors here, as leaf and stem diameters, leaf orientation to the wind, foliage density, etc., all become potentially important to the final motion and deposit point for the droplets. Turbulent eddies do not end once the canopy is encountered but their time and space scales will change abruptly. Fortunately for many modelling approaches, the droplet exposure to canopy-controlled turbulence is relatively short and its effects can therefore be ignored. This is a generally satisfactory approach for deposition onto bare soil or water surfaces, or even onto low crops and pasture grasses, but is distinctly less so for tall brush, shelter belt, etc., or where the pesticide in question may have different activity levels for different portions of susceptible plants (stems vs leaves for instance), or for different insect stages and resident sites.

EFFORTS TO IMPROVE SPRAY TECHNOLOGY

Research on nozzles, spray boom design, pressure settings and solution properties has been generally directed towards either improving the deposit pattern or reducing drift. Only infrequently has it involved attempts to simultaneously assess and optimize both, or to develop realistic "trade-offs". Hence new systems or combinations of parameters have often been proposed to reduce drift which are subsequently found to have worsened the swath deposit uniformity (and vice versa). This apparent lack of research intensity or coordination in sprayer design may be contrasted with the wealth of comprehensive information on the chemistry and biological effects of various pesticides. The difference may be ascribed partly to the relative size of the companies engaged in manufacturing the equipment components (generally small and/or into other lines as well), and those making the chemicals (generally large). A second reason is very likely the absence of regulations governing equipment performance.

Nevertheless, a great many different styles of nozzle have been designed and are manufactured. Variations between them commonly relate to orifice size for different output rates. Shape of the actual opening has been subjected to careful attention to improve the uniformity of emission, and hence of the deposit pattern on the ground. Overlap behavior of adjacent nozzles has been investigated in conjunction with boom height and spacing to reduce gap or overdose occurrences. Most of this research has been carried out "in-house" by the various nozzle manufacturers and has led to the wide range of types presently on the market. Only rarely has the most important property of all, the initial droplet size spectrum, been provided in the brochure information available to users.

More intensive work has involved the influence of hydraulic pressure, coupled with release height, in attempts to decrease the number of smaller droplets produced or their likelihood of drifting beyond the swath. It was clearly recognized that such efforts must run counter to achieving optimal deposit pattern, inasmuch as high pressures and boom height should improve this aspect, while changes in the opposite direction were indicated for drift control. Unless the two were measured for each proposed change (not usually done) effective compromises were missed. This was especially true when thickening agents, emulsifiers, etc., were added to the solution in an effort to reduce production of the smaller droplets. Once again the coarser spray, and hence lower drift, was accompanied by a major reduction in uniformity of target coverage. On the other hand, a few of these "ad hoc" approaches to the problem of reducing the small droplet fraction were promising, notably the development of the so-called "low-pressure" type of flat-fan nozzle (ref.1). Surveys of the interactions of the factors affecting droplet size spectra at the nozzle and initial spray drift characteristics may be found in refs. 2, 3 and 4.

In addition to the difficulty of independently improving drift control and deposit pattern, much of the effort directed towards coarsening the spray suffered from a property common to droplet-producing devices that emit a wide range of sizes. Simply shifting the entire spectrum upward, as measured by its most commonly used characteristic, the Volume

Median Diameter (VMD - that diameter dividing the entire spectrum such that 50% of the emitted volume is in smaller droplets and 50% in larger), might result in an even more heterogeneous spray. That is, such an upward shift could occur without reducing the fraction of the volume being emitted as fine droplets. More seriously, it could reduce the volume of spray solution being emitted in droplet sizes providing the most effective biological control.

This aspect of effective drop size has been addressed for a variety of insect-insecticide combinations, but has been all but ignored until recently for herbicides. For the former the optimum for most species appears to be 80 - 150 but might well be lower (ref. 5). It is obviously going to depend on insect size, mobility, and location within a crop canopy. There is less specificity (but less research) on the range of droplet sizes that might be most effective for weed control in an agricultural setting. Most evidence to date indicates that 200 - 300 droplets are likely to provide the most efficient control per unit mass of herbicide, but the limits of this range are not well-identified and might vary with weed size and mode of herbicide action (refs. 6 and 7). Nevertheless, this range has been identified as acceptable in the development of controlled droplet applicators (CDA) and of ultra low volume (ULV, commonly considered to be around 1-10 L/ha). For both, the emphasis has been on clearly identifying the initial droplet spectrum and in attempting to narrow it to the most appropriate size for biological effectiveness while minimizing the production of smaller, drift-prone droplets (ref. 8).

The spinning disc or cone type atomizers being introduced for CDA spraying are potentially usable on both ground-rig and aircraft, while the nozzles and formulations involved in the ULV approach are primarily suited to aerial application. Inherent in ULV spraying is the replacement of aqueous solutions by oil-based carriers, to eliminate droplet spectrum change through evaporation of the solvent component.

Considerable research has been undertaken on aircraft spray operations, including efforts to reduce drift. Much of this effort has gone into boom nozzle spacing and arrangement, in an effort to avoid unwanted effects imposed by the aircraft vortex, specifically non-uniform patterns and/or excessive spray drift. For aircraft, as for helicopters, it is potentially possible to produce nozzle type, pressure and formulation combinations designed to reduce the production of small droplets, without necessarily affecting the swath deposit pattern thereby. The reason is that the latter is much more controlled by wing tip vortex effects (or rotor vortex for helicopters) than by the initial droplet emission pattern at the nozzle. Modelling of vortices has been attempted by Morris (ref. 9) and leads to the establishment of "effective" release heights in place of nominal ones as set by the boom and nozzle location above ground. Helicopter research has been much less extensive, and tends to be limited to specific application situations (ref. 10).

NEW DIRECTIONS TO APPLICATION IMPROVEMENT

One of the major difficulties with conventional sprayers is the inter-active nature of the parameters controlling droplet production. Emission rate, droplet size spectrum and initial pattern at the nozzle are not easily separable. For conventional groundrig sprayers with flat fan nozzles, tractor speed may frequently have to be reduced in the course of a spray operation, and this can result in unacceptably high deposit densities. The only readily available recourse is to reduce the hydraulic pressure. Some recently - introduced controlled emission applicators (CEA) do this automatically by linking the pressure in the solution tank to the wheel rotation rate. This approach avoids the time-delay problem but fails to address the fact that the hydraulic pressure also controls the droplet size spectrum and the emission pattern at the nozzle.

Controlled droplet applicators (CDA), in principle, allow for the number of droplets emitted per unit time to be varied without affecting either pattern or spectrum. However, their deposit pattern uniformity has been found in various field trials to be less than satisfactory.

Another new approach presently finding its way into the market place involves what might be appropriately called controlled solution applicators (CSA). These emit solvent at a fixed rate per unit time, with the active ingredient being monitored into the carrier at a rate set by the forward speed of the sprayer. Improved deposit uniformity is the primary purpose of CSA sprayers, but two other significant advantages may accrue: (i) whenever two pesticide chemicals are to be applied in a field, they may be introduced separately, with the more expensive one (presumably) being emitted only as needed, rather than being sprayed over the entire field; and (ii) on completion of the application the operator would not have the disposal problem of a mixed tank solution.

There are two drawbacks to CSA sprayers at the present time, one actual and one potential. The former concerns the difficulty of obtaining a good solution mix within the line, especially if injection takes place at or near the nozzle. The second possible problem concerns the fact that the solution strength varies with sprayer speed: at slow speeds the pesticide is applied as a more dilute solution. It is not known in general how this might affect weed or insect control, with the variety of pesticides that could be used in such sprayers.

In an endeavor to utilize smaller droplets so that the greater droplet number per unit area might provide improved coverage and potentially greater efficiency, fine nozzle arrays have been used in conjunction with groundrig boom covers. These shields or shrouds are designed to reduce the otherwise higher off-target droplet drift that might be expected from the smaller size spectrum. While the shrouded-boom sprayer is promising as a drift reduction device, it has not yet been established that the smaller droplets do provide for increased weed control efficacy, say, when herbicides are applied.

Other new approaches involve utilizing airfoils and the aircraft speed to produce the droplet spray in aerial applications, rather than relying on hydraulic pressure within the spray tank system, incorporating air blast jets into gravity-fed groundrig nozzles, and "swirl chamber" devices to mix air and pesticide solution just prior to dissemination. Several novel variations on the spinning-disc or cone are being developed for market introduction. Finally, electrostatic charging is being tried, both to break up the liquid at the nozzle and to impress on the resultant droplets an electric charge that is intended to improve their impaction efficiency onto foliage.

Although each of these new nozzle and spray boon designs is obviously being tested by the innovator as the development proceeds, most of the research is "ad hoc" and proprietary, so that valid comparisons between sprayers is not possible. This is perhaps understandable, given the high cost of comprehensive field trials undertaken to evaluate pattern uniformity and drift fraction, together with the absence of licensing and regulatory controls. Such tests are, however, essential and should lead to establishing whether any of these new devices truly provides significant improvement in pest control efficacy. Some may.

A more coherent approach to pesticide improvement is needed. A list of research activities that could lead to major breakthroughs would include ways to achieve the following:

(i) the determination of optimal droplet size ranges for particular chemical-pest combinations;

(ii) the effective setting of droplet number and total pesticide volume for maximum effect in various circumstances;

(iii) the ability to vary application rate without altering the desired droplet spectrum;

(iv) the ability to separate control of volume being applied per unit area from the speed setting of the sprayer;

(v) the safe utilization of low volatile carriers to replace water;

(vi) the determination, for various chemicals and prevalent weed or insect species, exactly where the net return value begins to diminish significantly as the amount being applied is decreased.

This last is clearly the most important, yet the most difficult. It should, however, be the general goal towards which the other investigations also lead, that is, the making of pesticide application a much more exact and skillful operation than it is now.

REFERENCES

1. J. Maybank, K. Yoshida and R. Grover. J.Air Poll. Control Assoc. 28, 1009-1014 (1978).
2. W.E. Yates, N.B. Akesson and R.E. Cowden. Trans. Amer. Soc. Agric. Eng. 17, 627-632, (1974).
3. J. Maybank, K. Yoshida and R. Grover. Can. J. Plant Sci. 54, 541-546 (1974).
4. A. Nordby, and R. Skuterud. Weed Res. 14, 385-395 (1979).
5. T.J. Lawson, and S. Uk. Atmos. Environ. 13, 711-715 (1979).
6. J.R. Lake. Pestic. Sci. 8, 515-520 (1977).
7. P. Ayres and C.R. Merritt. Weed Res. 18, 209-214 (1978).
8. C.E. Goering, L.E. Bode and D.B. Smith. Int. Conf. Liquid Spray Systems, Tokyo. 51-57 (1978).
9. D.J. Morris. ASAE Conf., Chicago: paper no 78-1506 (1978).
10. W.E. Yates, J.M. Ogawa and N.B. Akesson. Trans. Amer. Soc. Agric. Eng. 17, 633-639 (1974).

Analysis of the herbicide flamprop *m*-isopropyl (single enantiomer) using capillary gas chromatography

A. Gray, B.L. Mathews* and C. Self

Shell Research Limited, Sittingbourne Research Centre, Sittingbourne, Kent, ME9 8AG, UK.

Abstract - The introduction of chiral chromatographic stationary phases for both HPLC and capillary GLC has enabled the rapid, specific determination of the active enantiomers of flamprop isopropyl and flamprop methyl. Analytical methods have been developed using S5 CHIRAL (HPLC) and Chirasil-Val (capillary GLC) for both compounds for use in research and development laboratories.

For quality control analysis purposes the use of packed column gas chromatography combined with optical rotation measurements will remain the favoured approach for the immediate future.

INTRODUCTION

Flamprop M-isopropyl (trade mark SUFFIX BW) is a post emergence herbicide giving very good control of wild oats (Avena spp.) in cereals. This is an example of a pesticide which can exist in two optically active (chiral) forms and where only one of the enantiomers is biologically active. The R(-)isomer replaced the racemate which was less active and less selective. The structure is shown below:

Flamprop-M-isopropyl

R(-)enantiomer, active WL43424 S(+)enantiomer, inactive

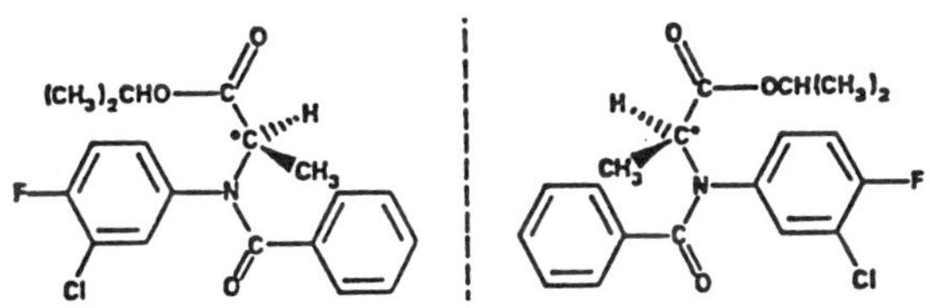

isopropyl n-benzoyl-N-(3-chloro-4-fluorophenyl)-D-alaninate
* denotes asymmetric carbon atom.

Possible approaches to enantiomer separation

1) Combined optical rotation/chromatography
2) Derivatisation/chromatography
3) NMR with chiral shift reagent
4) Liquid chromatography with chiral stationary phase or eluant containing chiral additive
5) Gas chromatography with chiral stationary phase

RESULTS

Analytical liquid chromatography employing chiral stationary phase S5 CHIRAL. Chiral stationary phase containing N-formyl-(L)-isoleucine covalently bound via aminopropyl groups to a microparticulate silica

$$-Si.CH_2.CH_2.CH_2.NH.\overset{O}{\overset{\|}{C}}.\overset{NH.CHO}{\overset{|*}{C}H}.\overset{*}{C}H(CH_3).C_2H_5$$

S5 CHIRAL (UMIST Batch) stationary phase

Chromatographic conditions

Column	:	0.25m x 4.6 mm ID stainless steel packed with S5 CHIRAL (UMIST Batch) 5 μm ex Phase Separations Ltd. Packed using the solvent arklone P.
Mobile phase	:	Propane-2-ol plus hexane premixed (0.35 + 99.65 v/v) at a flow rate of 0.5 ml/min
Temperature	:	Ambient (approx. 18°C)
Detection	:	UV absorbance at 226 nm
Sample	:	20 μl of nominal 0.1 mg/ml solutions made up in the mobile phase.
Retention times	:	WL43424 (S(+)enantiomer) 26.5 min. flamprop-M-isopropyl (R(-)enantiomer) 28 min.

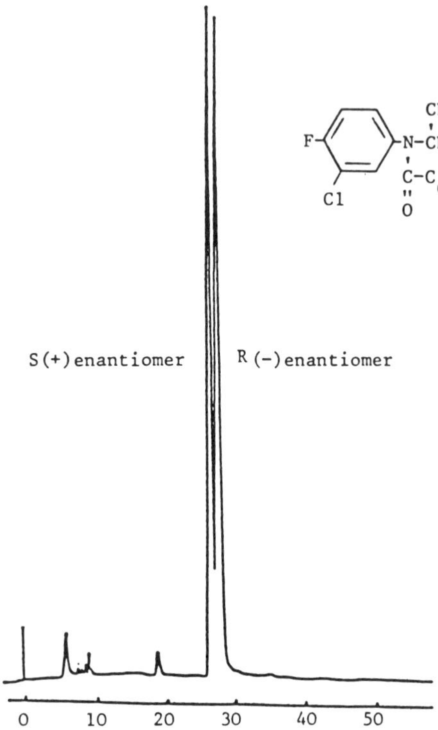

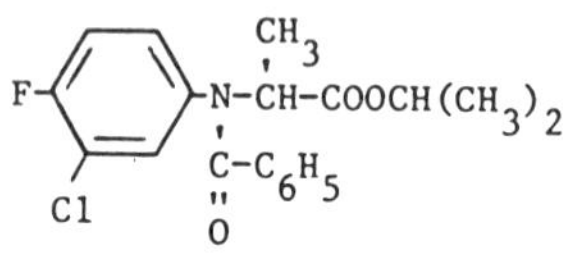

Liquid chromatographic resolution of flamprop-isopropyl enantiomers

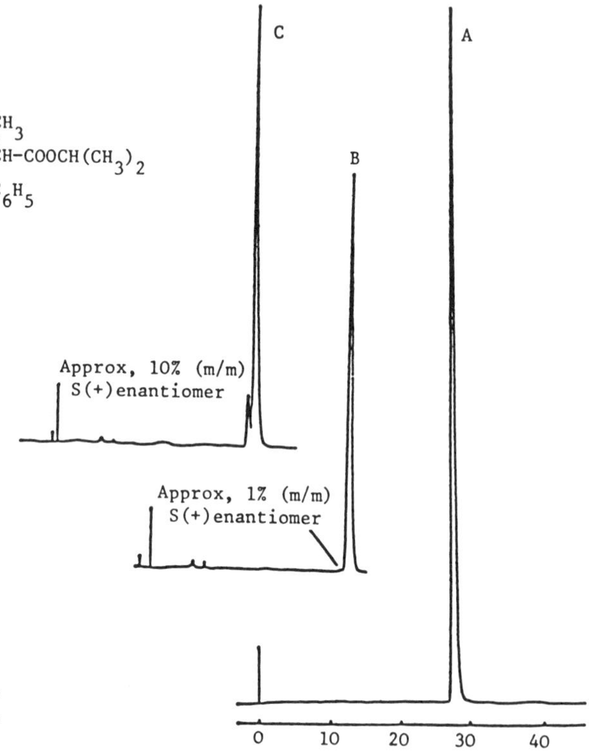

Liquid chromatograms of flamprop-M-isopropyl

(A) analytical standard (B) standard plus approx. 1% (m/m) of S(+)enantiomer

(C) standard plus approx, 10% (m/m) of S(+)enantiomer

The closely related wild oat herbicide MATAVEN is poorly resolved when applying the HPLC conditions used for SUFFIX BW. Modification of the mobile phase from propan-2-ol to methanol improved the enantiomer resolution. Mobile phase methanol plus hexane (0.4 + 99.6 v/v) at a flow rate of 0.5 ml/min and column temperature 22°C (see chromatogram overleaf.

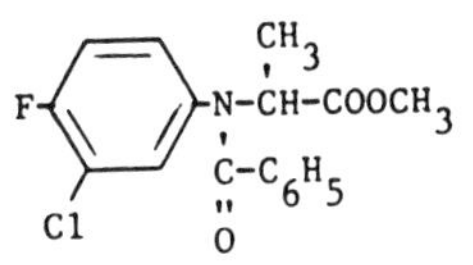

MATAVEN

Separation of flamprop methyl enantiomers by HPLC

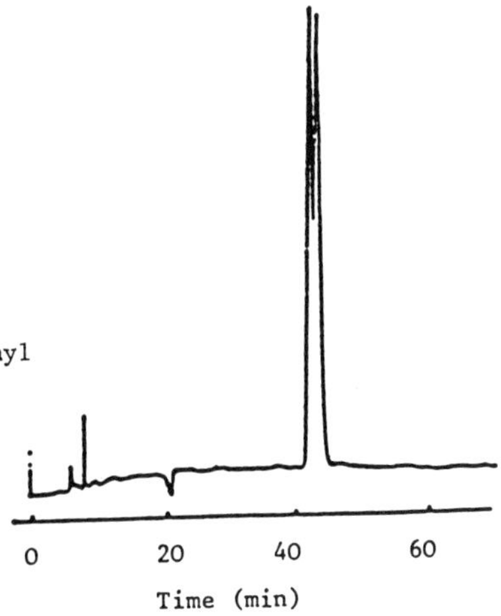

CAPILLARY GC USING CHIRAL STATIONARY PHASE, CHIRASIL-VAL

Chiral stationary phase containing N-propionyl-L-valine-t-butylamide polysiloxane groups A 7:1 ratio is required for solute penetration/interaction and hence separation.

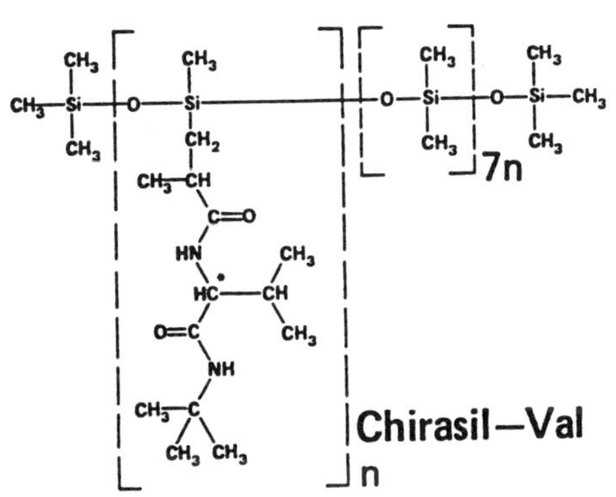

General chromatographic conditions:

Column	:	18m x 0.3 mm ID glass WCOT capillary coated with Chirasil-Val
Injection	:	Split, ratio 17:1
Carrier gas	:	Hydrogen, μ = 0.38 m/sec
Detection	:	Flame ionisation

BARNON (racemate)

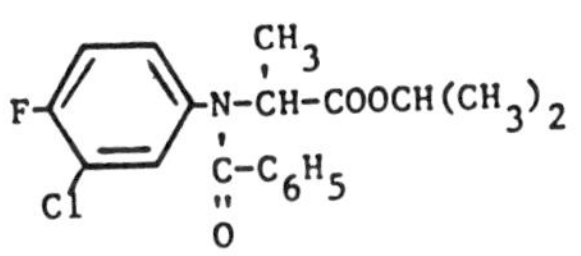

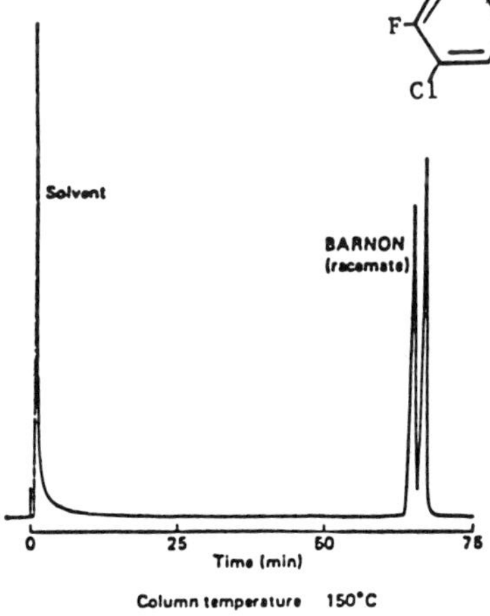

SUFFIX BW

R(–) isomer of BARNON

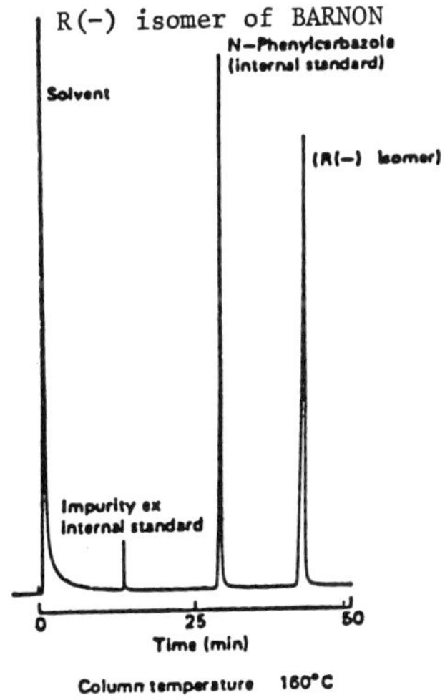

DISCUSSION

The analysis of technical material and formulations of flamprop M-isopropyl (and flamprop M-methyl) requires methodology whereby the active enantiomer can be accurately measured in the presence of the inactive enantiomer and other impurities/formulation adjuvants, etc.

High Performance Liquid Chromatography (HPLC)

This paper presents results obtained using HPLC with a chiral stationary phase S5 CHIRAL (UMIST Batch). This phase contains N-formyl-(L)-isoleucine covalently bound via aminopropyl groups to microparticulate silica. The phase produces quite different enantiomer resolution from the apparently similar Pirkle phases (which employ N-(3,5-dinitrobenzoyl)-(D)-phenylglycine or (L)-leucine bonded either ionicially or covalently to microparticulate silica). S5 CHIRAL has proved very useful for resolution of enantiomers of 'wild oat' herbicides. In the case of flamprop isopropyl enantiomers good separation is achieved using a mobile phase of propan-2-ol plus hexane (0.35 + 99.65 v/v). Surprisingly, under the above chromatographic conditions the closely related compound flamprop-methyl was relatively poorly resolved. Modification of the mobile phase to methanol plus hexane (0.4 + 99.6 v/v) improved the enantiomer resolution. This may be caused by flamprop-methyl binding more strongly to the stationary phase (via hydrogen bonds and π-π interactions) so necessitating a more polar modifier.

Increased enantiomer resolution was achieved at reduced (5°C) column temperature. This may result from the specific interaction mechanism between the solute and stationary phase required to produce chiral separation. This is thought to involve a conbination of π-π bond attraction, transient hydrogen bonding and steric hindrance interactions. As the temperature is reduced, the structures would be expected to become more rigid and so enhance the steric hindrance interaction.

Capillary GLC

The stationary phase, Chirasil-Val is synthesised by coupling the N-propyl-L-valine-t-butylamide moiety onto a 7:1 copolymer of dimethyl and carboxyalkylmethyl-silicones to form the structure shown above. The 7:1 ratio is necessary to allow solute enantiomers sufficient space for penetration between and interaction with the L-valine-t-butylamide groups. The reaction between the chiral sorbent and solute enantiomers is thought to involve the formation of transient diastereoisomeric complexes without co-valent bonding.

Stereochemically significant interactions between chiral sorbent and solute enantiomers required for resolution.

BARNON/SUFFIX-BW

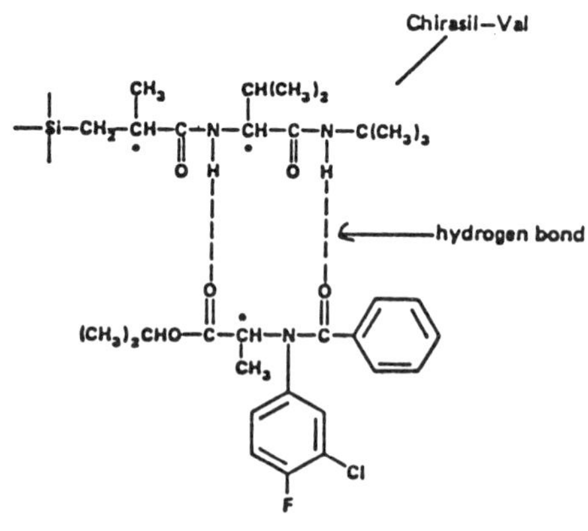

Enantiomer resolution effected by steric hindrance between $-CH(CH_3)_2$ and $-CH_3$ groups. Thus, the Chirasil-Val method is a direct gas chromatographic procedure for the separation of the enantiomers of flamprop isopropyl and flamprop methyl.

Rheological investigations of some antisettling systems used in suspension concentrates

K. Parkhouse[1] and Th.F.Tadros[2]

ICI Plant Protection Division, Jealott's Hill Research Station, Bracknell
Berks, RG12 6EY

Abstract - Rheological investigations (steady state shear stress-shear rate, oscillating and shear wave measurements) are described for two antisettling systems, namely silica + Kelzan and bentonite + Kelzan. For the silica + Kelzan system, both the yield value τ_β and shear modulus G_∞ increased rapidly with increase in Kelzan concentration, particularly at pH 5. This was attributed to the interaction between the silica particles (which are populated with silanol groups) and the Kelzan chains, by hydrogen bonding. Viscoelastic measurements on a 4% bentonite suspension showed increases in the complex modulus G*, the storage modulus G' and loss modulus G" with reduction in the pH of the suspension from 9 to 4. This was attributed to face-to-edge interaction which was enhanced at low pH values as a result of the positive edges and the negative faces of the clay particles. Addition of Kelzan caused an increase in all rheological parameters, but the interaction between bentonite particles and polymer chains was not as strong as with the silica system. Moreover the rheology was less sensitive to pH changes, making the bentonite + Kelzan system more suitable for application in pesticidal suspension concentrates. Addition of a nonionic surfactant at low concentrations (<2%) caused extensive flocculation of the clay suspension and this caused a rapid increase in G* and G_∞. However, above 2% surfactant, restabilisation of the clay suspension occurred and the moduli decreased to low values. These results illustrate the need to use rheological techniques to study the interactions that take place between the various components in suspension concentrates.

INTRODUCTION

In the formulation of pesticidal suspension concentrates, it is essential to prevent settling and formation of dilatant sediments. The sediments form due to the larger density of the particles relative to that of the continuous medium. Since the particle size is usually outside the colloid range (an average particle radius of 1-2 μm is usually the case), then the gravitational force (that is equal to $(4/3)\pi a^3 \Delta\rho\, gL$, where a is the particle radius, $\Delta\rho$ the density difference between the particles and the medium, g the acceleration due to gravity and L the height of the container) exceeds the Brownian diffusion force (which is in the region of kT, where k is the Boltzmann constant and T the absolute temperature). When the suspension is colloidally stable, the repulsive force necessary to ensure stability enables them to move past each other to form a compact layer, which is technically referred to as "cake" or "clay" (1). These cakes or clays are difficult to redisperse and hence they should be prevented from forming on standing.

Several methods are applied in practice to control settling and formation of dilatant clays (2). The most commonly used method is to add to the continuous medium a high molecular weight material (e.g. Xanthan gum), a finely divided inert solid (e.g. silica or bentonite) or a combination of the two. These materials, which are referred to as antisettling systems, form a "gel" structure in the continuous phase with a very high viscosity at low shear, thus preventing particle settling. Moreover this "gel" structure is usually elastic thus preventing extensive separation.

In this paper, we shall describe some preliminary results on the rheology of two commonly used antisettling systems, namely silica + Xanthan gum (Kelzan supplied by the Kelco company U.S.A.) and bentonite + Kelzan. Three main rheological techniques were applied; namely steady state shear stress-shear rate, oscillatory and shear wave propagation measurements.

---

(1) Industrial Trainee from Bath University

(2) Author to whom correspondence should be addressed

Measurements were carried out at constant silica or bentonite concentration, whilst varying the concentration of Kelzan. The effect of pH was investigated. Moreover, the effect of addition of a nonionic surfactant (namely $R-(CH_2-CH_2-O)_X-OH$ with $R=C_{13}/C_{15}$ and X=7) on the rheology of bentonite and bentonite + Kelzan suspensions was also investigated.

EXPERIMENTAL

Materials

Silica (Aerosil 200, Degussa, W. Germany) with a surface area of 200 $m^2\ g^{-1}$ was used as received. A stock dispersion of 12.8% (w/v) was prepared by dispersing the powder for approximately 6 hrs using a Silverson mixer (an average particle size of 635 nm, as measured by the Nanosizer was produced). The suspension was left stirring using a paddle stirrer and before use in rheological measurements it was further sheared for 3 hrs until a constant viscosity was produced. The pH of the dispersion was ~5.

Bentonite (sodium montmorillonite) was a commercial material (pharmaceutical grade Bentopharm B20, ex Bromhead and Denison, Ltd., U.K.) which was also used as received. A stock 10% (w/v) was prepared by dispersing the powder into water using a Silverson mixer for approximately five hours.

Xantham gum (Kelzan, ex Kelco Co. U.S.A) was also used as received. A stock 2% (w/v) solution was prepared by dissolving the powder in water and a small amount of a bactericide (Proxel AB, ex ICI Organics Division) was added to prevent bacterial attack of the polymer.

Dispersions of silica + Kelzan and bentonite were prepared from the above stock solutions. The concentration of silica was kept constant at 8% (w/v) and that of bentonite at 4% (w/v), while that of Kelzan was varied from 0 to 0.5%. The pH of the dispersions was adjusted by addition of HCl or NaOH solutions.

A nonionic surfactant, namely Synperonic A7 (ex ICI Plastics and Petrochemicals Division) was used as received. The surfactant contains an alkyl chain length of 13-15 and 7 moles of ethylene oxide.

Methods

Steady state shear stress-shear rate curves were carried out using Haake-Rotovisco rheometer (models RV100 and RV2) fitted with measuring heads MK150 and MK50 as appropriate. Concentric cylinder platens were used for the measurements. All samples were pseudoplastic and the shear stress-shear rate curves were fitted to the Bingham or Casson's models to obtain the yield vale τ_β and τ_C respectively, i.e.

$$\tau = \tau_\beta + \eta_{app}\,\dot{\gamma}_{app} \qquad (1)$$

and $$\tau^{1/2} = \tau_C{}^{1/2} + (\eta_{app}\,\dot{\gamma}_{app})^{1/2} \qquad (2)$$

where τ is the applied shear stress, $\dot{\gamma}_{app}$ is the shear rate and η_{app} is the apparent viscosity. τ_β was obtained by extrapolation of τ to $\dot{\gamma}_{app} = 0$, whereas τ_C is the intercept of the linear plots of $\tau^{1/2}$ versus $\dot{\gamma}_{app}{}^{1/2}$.

Oscillating measurements were carried out using a Bohlin rheometer (Bohlin Rheologie, Lund, Sweden). In these dynamic measurements, a sinusoidally varying strain with an angular frequency, ω (cycles per sec or Hz) is applied to the material and the stress response is monitored as a function of time. The stress response also varies sinusoidally with time, and its amplitude is proportional to the strain amplitude, oscillating at the same frequency but out of phase with the strain. From the shift in phase angle, δ, one is able to split the complex modulus, G*, into its storage component, G' and its loss component, G", i.e.

$$G^* = G' + i\,G'' \qquad (3)$$

where $$G' = G^* \cos\delta \qquad (4)$$

$$G'' = G^* \sin\delta \qquad (5)$$

and i is a constant, equal to $(-1)^{1/2}$.

G' is a measure of the energy stored elastically in a cycle, i.e. it gives the elastic component of the complex modulus, whereas G" is a measure of the energy dissipated as heat during a cycle of deformation, i.e. it gives the viscous component of the complex modulus.

Shear modulus measurements were carried out using a pulse shearometer (Rank Bros., Bottisham, Cambridge, U.K.). Basically, the dispersion is placed between two parallel perspex (or metal) plates each connected to a piezo-electric crystal. Using a pulse generator, a shear wave is generated at the bottom plate (at high frequency of ~200 Hz and low amplitude, less than 10^{-4} rad s^{-1}), which propagates through the dispersion and is detected by the upper plate crystal. The shear wave velocity through the dispersion, u, is determined by measuring the time, t_s, for a shear wave to propagate from a maximum to a minimum amplitude as a function of plate separation d (u is the slope of the d versus t_s plot). The instrument is provided with a microcomputer for processing the data. From u, the shear modulus G (which is the high frequency modulus, i.e. a measure of the elasticity of the dispersion) is given by,

$$G = u^2 \rho \qquad (6)$$

where ρ is the density of the suspension.

RESULTS AND DISCUSSION

Fig. 1 shows the variation of τ_β and G_∞ with % Kelzan for an 8% (w/v) silica dispersion at three pH values, namely 5, 7 and 9. All rheological data show the same trend, namely an increase in τ_β and G_∞ with increase in Kelzan concentration.

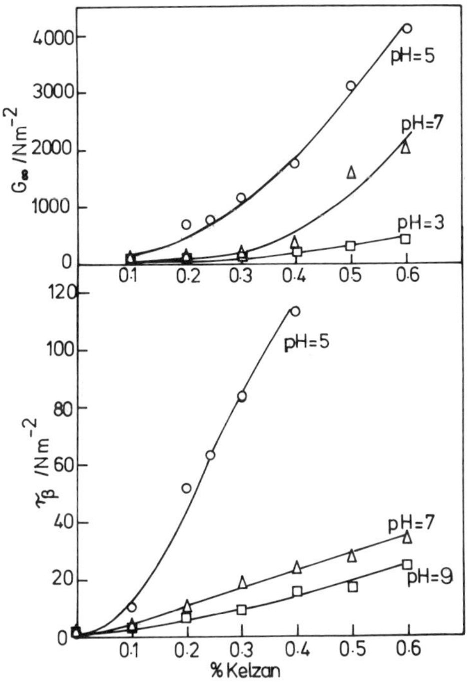

Fig. 1 - Variation of τ_β and G_∞ with Kelzan concentration for 8% (w/v) silica dispersion at three pH values

This is as expected since addition of Kelzan, which is a high molecular weight poly saccharide (in excess of 10^6) contributes significantly to the rheology of the silica dispersion. The polysaccharide molecule, which forms long chains in aqueous solution (with a high axial ratio) produces a viscoelastic gel in the continuous phase (3). However, the high yield values and the high values of the modulus G_∞, indicate interaction between silica and the polysaccharide chains. This is particularly the case at low pH values. Indeed the results of Fig. 1 show that at any given Kelzan concentration, τ_β and G_∞ increase with decrease in pH value of the suspension. This can be readily explained in terms of the interaction between silica and Kelzan, which depends on the pH. The pH determines the number of undissociated silanol groups (4) which increases with decrease in pH. Assuming that the interaction between silica and Kelzan occurs through hydrogen bonding between the silanol groups and the polysaccharide chains, then such interactions should increase with decrease of pH. A similar interpretation was put forward for the interaction between silica and poly(vinyl alcohol) (5,6). Thus, one would expect the present silica + Kelzan system to be suitable as an antisettling system for pesticidal suspension concentrates, particularly at low pH values. It should be mentioned, however, that when applying such an antisettling system, one should be careful since the "gel" structure changes with time, depending both on the procedure used for dispersing the silica and further adjustment of other parameters, e.g. pH, electrolyte. With time the silica particles undergo flocculation particularly at low pH values and this could lead to an increase in viscosity of the suspension on storage.

Fig. 2 shows plots of G*, G' and G" for a 4% (w/v) bentonite suspension at various pH values.

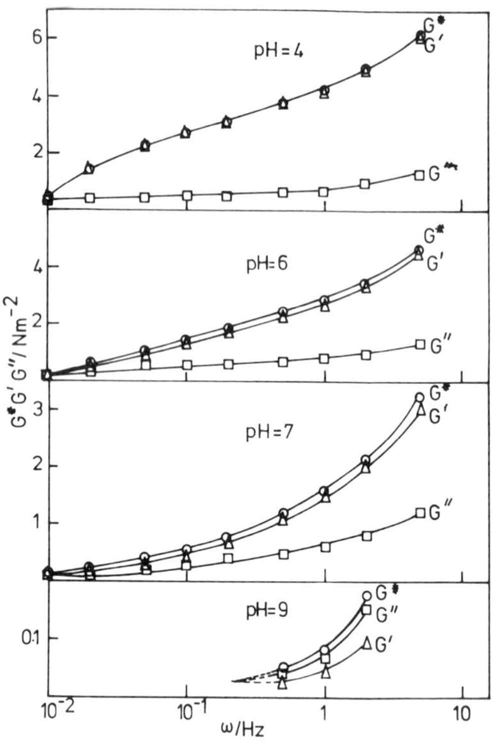

Fig. 2 - Variation of G*, G' and G" with frequency for a 4% bentonite suspension at different pH values

At pH = 9, all moduli were very low (note that the scale is expanded for clarity) and the loss modulus (the viscous component) is higher than the storage modulus (the elastic component). This is due to the weak particle interactions at such high pH values. At pH9, the edges of the clay platelets are negatively charged and hence there will be no edge-to-face interaction, i.e. "the house of card structure" is broken down (7). Any interaction at pH = 9, is due to double layer repulsion (8), which in the presence of electrolyte (from impurities in the bentonite) is relatively weak.

With decrease of pH of the bentonite, the edges acquire a positive charge and edge-to-face interaction is now possible. This results in an increase in all modulus values and now the storage modulus G' becomes higher than the loss modulus G". This is clearly shown in Fig. 2 at pH = 7,6 and 4. It is also clear that with reduction in pH from 7 to 4, G' becomes very close to G*, indicating the strong face-to-edge interaction at pH = 4. This reduction in interaction with increase of pH is also reflected in the yield value which shows a rapid decrease with increase of the pH of the suspension.

Fig. 3 shows plots of G*, G' and G" for a 4% (w/v) bentonite dispersion at pH = 9 and at various concentrations of Kelzan. It is clear from this figure that addition of Kelzan caused an increase in all modulus values, as expected.

Similar trends were obtained at the other pH values. Figure 4 shows plots of the Bingham yield value, τ_β, complex modulus G* (at 0.2 Hz) and shear modulus G_∞, as a function of Kelzan concentration.

This figure illustrates the increase in rheological parameters with increase in Kelzan concentration. However, the antisettling system of bentonite + Kelzan differs from that of silica + Kelzan in two respects: firstly, the much smaller dependence of rheological parameters on pH and secondly, the much less rapid increase in these parameters with increase in Kelzan concentration. Both these results imply that the interaction between bentonite and Kelzan is not as strong as that between silica and Kelzan particularly at low

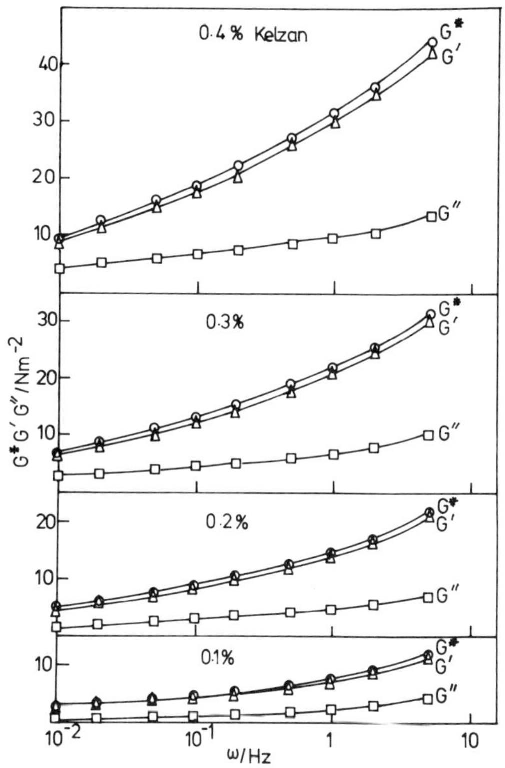

Fig. 3 - Variation of G*, G' and G" with frequency for a 4% bentonite suspension at pH = 9 and various Kelzan concentrations

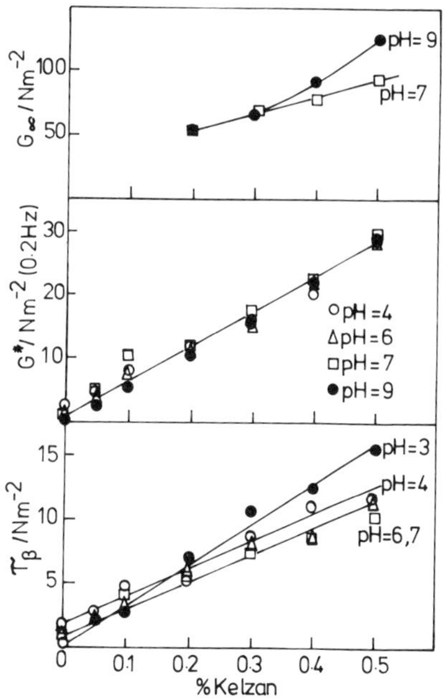

Fig. 4 - Variation of τ_β, G* and G_∞ with Kelzan concentration for a 4% bentonite suspension at various pH values

pH values. This can be explained in terms of the difference between the two particles surface characteristics. With silica, at low pH, the surface is populated with silanol groups which interact strongly with Kelzan by hydrogen bonding. The faces of the bentonite particles are negatively charged and this charge is constant and independent of pH (7). Thus, it seems likely that the addition of Kelzan to bentonite suspensions cause an increase in the viscoelastic properties as a result of the presence of the polymer molecules which interact with each other (3). This does not of course rule out interaction between the polymer chains and the bentonite particles. Such interaction is still possible, although its mechanism is far from being clear.

The above results with bentonite + Kelzan show that such an antisettling system is suitable for prevention of settling and prevention of caking and claying in pesticidal suspension concentrates. This is for two reasons. The first is the much smaller dependence on pH (provided such pH does not fall much below 4 since at very low pH the bentonite particles become chemically unstable) and therefore any drift in the pH of the suspension has a smaller effect on rheology. Secondly, due to the lack of strong interaction between bentonite particles and Kelzan, aging effects (which usually result in increase in viscosity on storage) are minimal. It should be mentioned, however, that care should be taken in applying this antisettling system since swelling of the bentonite suspension (which takes a long time depending on pH and electrolyte concentration) can be a problem in practice.

Fig. 5 shows the variation of the Cassonn yield value ${}_{C}$, complex modulus G* (at 0.2 Hz) and shear modulus G with Synperonic A7 concentration, for a 4% bentonite suspension at pH = 9 and also in the presence of 0.1% Kelzan. These results show a striking behaviour, namely a rapid increase in rheological parameters on addition of Synperonic A7, reaching a maximum at 1-2% surfactant.

Above 2% Synperonic A7, the yield value drops and then gradually increases with increase in surfactant concentration. On the other hand, both G* and G show a rapid decrease above 1-2% Synperonic A7, followed by a slow increase in these rheological parameters. Moreover, below 2% Synperonic, the suspensions are more elastic than viscous (G'>G"), whereas above 2% Synperonic A7, G">G'. These results point towards strong interaction between the bentonite particles at low surfactant addition (<2%). The most likely mechanism is adsorption of the surfactant molecules with the hydrophobic ethylene oxide units pointing towards the particle

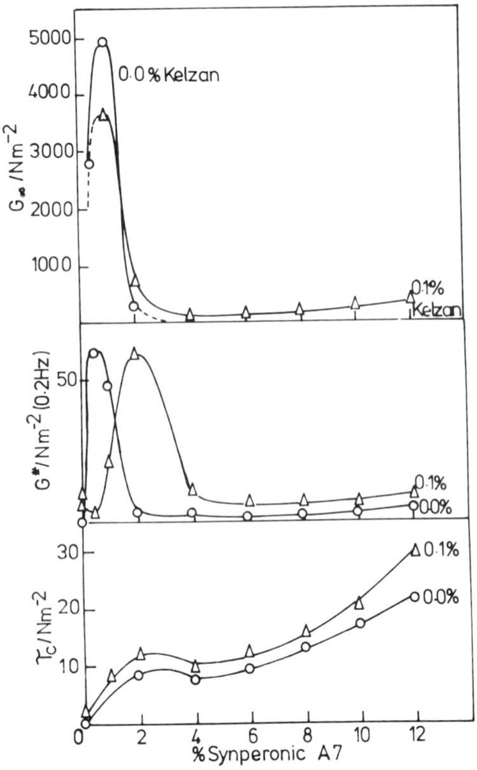

Fig. 5 - Variation of τ_C, G* and G_∞ with Synperonic A7 concentration

surface, leaving the hydrophobic alkyl group pointing towards the solution. This results in hydropholic interaction between the bentonite particles (which now become hydrophobised by the surfactant molecules) leading to a rapid increase in G* and G_∞. With further increase in surfactant concentration (above 2%), a second layer of surfactant molecules could be adsorbed with the hydrophobic alkyl group now adsorbing on the hydrophobised bentonite particles, leaving the ethylene oxide chain dangling in solution, i.e. the particles now become hydrophilic. This leads to stabilisation of the bentonite particles against flocculation, resulting in a reduction in G* and G_∞.

The above results clearly show that addition of nonionic surfactants to pesticidal suspension concentrates, which are sometimes necessary for enhancing the biological efficacy, can cause a dramatic effect on the rheology of the suspension. This is reflected in some cases in extensive separation of the suspension and formation of compact layers which are difficult to redisperse. Thus, one should consider all possible interactions when designing an antisettling system for pesticidal suspension concentrates. Rheological techniques are very useful in optimising these systems.

ACKNOWLEDGEMENTS

Th.F.Tadros is indebted to Miss R. Cheney and Mr R. Sohm for helping Miss K. Parkhouse in carrying out the experiments described in this paper. These experiments were part of a six month industrial training programme for Miss Parkhouse who graduated from Bath University.

REFERENCES

1. Th.F. Tadros, Advances Colloid Interface Sci. 12, 141 (1980).
2. Th.F. Tadros, in "Solid-Liquid Dispersions", Editor Th.F. Tadros, Academic Press, in preparation.
3. F. Durmon, M. Gover, R. Sohm and Th.F. Tadros, to be published.
4. Th.F. Tadros and J. Lyklema, J.Electronol Interfacial Electrochem. 17, 267 (1968).
5. Th.F. Tadros, J.Colloid and Interface Sci. 64, 36 (1978).
6. D. Heath and Th.F. Tadros, J.Colloid Interface Sci. 93, 320 (1983).
7. H. van Olphen An Introduction to Clay Colloid Chemistry, Interscience, London, 1977.
8. K. Norrish, Disc. Faraday Soc. 18, 120 (1954).

Parameters governing dose transfer

Franklin R. Hall

Laboratory for Pest Control Application Technology
The Ohio State University, Wooster, Ohio U.S.A. 44691

Abstract - The biological effect of a toxicant depends upon its toxicological properties and the quantity available at the point of action. Spray application is a complex and dynamic process involving many interdependent components. In order to optimize formulation/application effects on dose transfer, there first must be a definition of the droplet pattern in space and time required for optimum biological effect. The processes involved in spray retention involve the physical/chemical nature of the spray, dynamics of droplet delivery, and the morphology and chemistry of the plant surface. Pests can respond to discrete droplet patterns as well as differences in formulations and concentrations of AI. Predictive responses in spatial disruption and age structure of specific pests are essential to the design of useful formulation/application protocols of each AI. Monosize droplet atomization techniques are thus essential in order to allow a precise study of droplet/plant/pest relationships. Formulation specialists have to be more closely aligned with biologists, in order to achieve significant advances in dose transfer.

INTRODUCTION

Spray application is a complex and dynamic process involving many interdependent components. As summarized by Young (ref. 1), the factors of atomization, transport to target, impaction, deposition formation, movement in/on plant, and biological effect are influenced by each other, the external factors of environment and operating conditions, target properties, as well as the formulation, and the active ingredient (AI). Improvements in pesticide use efficiency will require a more intensive study of the physical and physiochemical parameters controlling droplet dynamics, deposition on defined targets and biological response. The biological effect depends on the toxicological properties of the AI, its formulation, the concentration and pattern of the dose at the point of action. Target coverage and ultimate response of insect and mite pests are frequently not well correlated. This suggests that there is a lack of basic understanding of the dose transfer processes.

The search for more active agents has continued although as Graham-Bryce (ref. 2) succinctly observed at the last session of this conference, there is a great disparity between field and laboratory rates. Consequently, as rates of application decrease, weathering, inactivating and redistribution forces are likely to be more influential on the amount of toxicant reaching the target organism. The delivery of a very active agent in smaller droplets also suggest that gains in activity may be offset by inefficient mechanisms of delivery. Consequently, Graham-Bryce concludes that it may be more rewarding to concentrate on properties which affect biological availability, reliability or costs of production and ease of application. Clearly, availability to the target, physiochemical properties, and a fundamental understanding of the biological characteristics of a population (resistance) and some explanation of substrate specificity of the target process and resistance mechanisms will be fruitful areas for needed research.

To optimize dose transfer, the distribution of a toxicant in space and time should be matched to the target characteristics (ref. 1). In practice, less than 1-3% of an agrochemical may reach this site and contribute to the biological effect (ref. 3). Atomization of a particular formulation, and its subsequent fate in the environment will depend in part upon the formulation properties of solubility, volatility, partition characteristics and stability. Considerations of these properties are reviewed and discussed by Hartley and Graham-Bryce (ref. 4), and more recently by McCann and Whitehouse (ref. 5). Past studies of pesticide formulations have tended to focus on reliability, ease of handling, and safety. Clearly these are important, but enhancement of biological activity via formulation and application research deserves more attention then it has thus far received. While applications made at different dosages may give similar areas of deposit, the rates of volatilization in a given environment may be the same. However, the loss represents a greater proportion of the toxicant applied at the lower dose. Therefore, losses may become more serious for more active compounds applied at lower rates (ref. 5). In order to pursue a fundamental understanding for optimizing formulation/application effects on dose transfer, there first must be a definition of the droplet pattern in space and time required for optimum biological effect. Thereafter, application and formulation methodology designed for a specific distribution and release rate can be devised. Heretofore, the comprehension of volume, droplet size, and formulation interactions on biological effects have been partly hidden because of the wide range of droplet sizes produced from most conventional nozzles. The development of CDA nozzles has lessened the range of droplet sizes and some success has been obtained with the use of laboratory monosized droplet devices. It is these devices which offer the unique opportunity to study the effects of formulation on the biological effect as it is influenced by droplet size, pattern of distribution, and concentration (ref. 6,7).

HERBICIDES

Upon reaching a leaf surface, a pesticide can be lost by photodecomposition, volatilization, rainfall, and foliar uptake. Clearly, the uptake of a herbicide is important, and the extent to which each factor may contribute to the loss of an effective dose may, in part, depend on the nature of the AI. McCann and Whitehouse (ref. 5,6) review the magnitude of such factors. The problem of rain fastness, in general, appears to be related more to water-soluble products although mechanical dislodging can also occur. Foliar uptake can be affected by environmental features of temperatures and relative humidity. Two formulation approaches are proposed by McCann and Whitehouse (ref. 6): (1) promotion of foliar uptake, and (2) direct protection of deposits. Reduction of volatile losses may be achieved with use of salts vs. esters, mixtures with involatile carriers, or with surfactants to reduce rates of diffusion, but loss in activity may also be a result. Attempts to improve rain fastness in many cases includes the use of adjuvants. Surfactant combination with oils and with salts or humectants have shown to be successful with some herbicides. Film formers, ie. water soluble polymers, emulsions, spray gels, etc. appear to have widely varying results to date. The form of the deposit is of some importance, but the conditions under which penetration is increased, may be quite specific. Some surfactants can promote the action of herbicides with increasing doses of surfactant, even though differing considerably in their influence on surface tension and wetting ability (ref. 7). There are many adjuvants registered for use with herbicides and the efficacy with some products is well documented. At the same time, individual studies on certain formulation/additive combinations illustrate the potential for a vast myriad of biological responses depending upon the specific AI, method of application, and target species (ref. 8). Consequently, the rationale for their general use is not well established, and the results of the vegetable oil-pesticide combinations should alert us to the fact that caution in their selection and detailed study on their use is needed for each type and use pattern.

GROWTH REGULATORS

A frequent limitation to the general use of plant growth regulators (PGR) is a high degree of inconsistency in performance. The basis for this inconsistency is not always clear, but studies have focused on the contribution of foliar absorption, and the relationship between the active ingredient, formulation and the spray droplet:plant surface interaction (ref. 9). The spray retention value (spray retained per unit area of surface) establishes the dose of PGR available for penetration. Where low volume conditions are utilized, this relationship takes on greater significance and may influence the bounce potential of the droplets.

The processes involved in spray retention are many, and complex and involve the physical/chemical nature of the spray, dynamics of droplet delivery, and the morphology and chemistry of the plant surface (ref. 1,10). Surface tension values have been utilized to explain some of the phenomena, but dynamic surface tension is thought to be of more value. In any case, there is likely to be a critical value for a specific chemical formulation, droplet size and plant surface interaction. Recent data on conditions for droplet bounce (ref. 11) differ from data reported earlier and suggest that velocity, size, and chemical and physical properties of a formulation need to be more closely studied.

Macroscopic features and other morphological and chemical aspects of plant surfaces play a major role in wettability, hence droplet retention. The amount of surface wax, and its chemistry, (functional groups and degree of packing), and the fact that leaf age changes wettability as do the features of upper and lower surfaces, all contribute to the numerous questions about penetration now being addressed by researchers. Recent studies by Bukovac and co-workers (ref. 9) showed the contribution of veins to ethephon impingement on sour cherry leaves, and copper sulfate on pear leaves. The chemistry of the surfactant plays a significant role in covering leaf wax fine structure. Tween -20 deposits appear as a polymer-like film, while Triton X-100 deposits appear to soften and deform secondary branches of wax projections. More recently, studies have shown that the degree of polymer-like deposit was related to the moles of oxyethylene present on the octylphenol hydrophobe (ref. 12). Bukovac (ref. 9) concludes that while there has been some greater understanding of deposit formation, there is still no clear association between the nature of the deposit, penetration, and performance of a PGR.

INSECTICIDES

The concept of dose transfer involves the delivery of a toxicant (atomization to deposit) to a point where there is contact with an organism sufficient to result in a measurable, predictable biological response. It is clear that this is a complex, dynamic process which is far from being efficient. With the arrival of CDA and ULV, and use of low rates of carrier fluids, i.e., water, oils, etc., changes can occur in resistivity, volatility, drift potentials, as well as redistribution phenomena. The delivery and subsequent impingement of specific droplet sizes can be enhanced by adjustments in formulation parameters, as well as by the addition of adjuvants.

Pest responses to distribution of discrete droplet patterns are a unique feature of the pest control function. Hall (ref. 13) showed changes in twospotted spider mite (TSSM) Tetranychus urticae (Koch) response to droplet density (Table 1). More recent data showed significant differences in response of TSSM to two different formulations of permethrin at 0.118 gm AI/l concentration (Table 2). Recent data from Adams and co-workers (unpublished) shows lethal number (LN) values can be halved with certain droplet sizes, surfactant/permethrin combinations.

TABLE 1. TSSM responses to formulation and droplet density (ref. 10)[a/].

	1 Droplet/Disk		4 Droplets/Disk	
Treatment	Walk	Feed	Walk	Feed
Permethrin 3 ED[b/]	45.7	254.3	251.8	48.1
Permethrin 2 EC[c/]	--	--	63.8	236.2
Water	--	--	59.9	240.1

a/ Data in seconds.

b/ ED = Electrodyn®-generated permethrin at 0.1 ml/s flow rate.

c/ EC used at concentration of 0.06 g AI/l

TABLE 2. Effect of permethrin formulations on behavior of TSSM at 48 hr[a/].

Treatment	Avg. % Alive	Avg. % Eggs/Disc	Avg. # Feeding Scars
Permethrin 2 E	75.0 a	10.5 a	26.9 a
Encapsulated Permethrin	100.0 b	28.5 b	165.0 b
Check	100.0 b	29.9 b	141.0 b

a/ Means in each column followed by same letter are not significantly different at .05 level DNMRT.

Biological effects of pyrethroids on mites, for example, include changes in age structure, dispersal (hyperactivity), alarm response, non-feeding, fecundity and resurgence (ref. 15). Certain formulations/toxicant combinations can disrupt populations in randomness and behavior. Predictive responses in spatial disruption and age structure of specific pests are essential to the design of useful formulation/application protocols of each AI. Where CDA situations are simulated, lowering the volume and concentrating the AI also concentrates the formation additives. Considering the retention and redistribution parameters, this only complicates the "whim" of adding a sticker.

DISCUSSION

The problem area remains the failure to recognize that these highly active AI's present dynamic and complex interactions with the application/formulation process from the point of atomization to the biological effect. Should we continue to utilize constant pesticide rates over a wide carrier volume assuming that there is no volume effect and that the biological response is dose dependent? Concentrating surfactants in ULV applications, droplet size (atomization and delivery), carrier solvent, redistribution phenomena, and target definition with subsequent changes in application protocols, all remain as opportunities to optimize the biological effect of the pest control agent.

Weathering has been studied fairly intensively by researchers, but laboratory rain may be quite different from natural rain both in quantity and quality. Prolonged drizzle can have significant effects on effective biological deposition patterns. Still, we have numerous unpublished examples of adding adjuvants to increase rainfastness with a resulting decrease in biological availability. In many cases, enhancement of biological activity was not the result of increased spreading as shown by SEM methodology and needs to be explored further. Differences in residues and resulting deposits within various canopy structure and leaf morphology could have interesting potential for formulation expertise if adopted to various needs and/or regional climatic regions.

Munthali and Wyatt (ref. 16) conclude that it is vital to choose the correct droplet size and concentration, otherwise, 50 or more times the AI may be required to achieve the same results. In some cases, increasing the concentration may reduce effectiveness, which has far-reaching implications for formulation of ULV sprays (ref. 15,16,17). The mechanisms remain unclear as to whether it is a physical constraint on the toxicant, or whether it restricts movement within the leaf ultrastructure. However, the situation presents some unique opportunities for studies on the physicochemical properties of an AI at an early stage in product development.

High speed cinematography can be used to study atomization, droplet formulation and transport, deposition and deposit formation phenomena with various pesticide/additive combinations. Surfactants can affect deposit formation by directly affecting epicuticular wax fine structure as viewed by SEM and X-ray dispersive analysis (ref. 7,8). However, the dynamics of toxicant transfer to various pests remain unclear where formulation/additive combinations change pest perception of the AI and subsequent behavior.

Monosize droplet atomization techniques (Fig. 1) are thus essential in order to allow a precise study of droplet/plant/pest relationships. As noted by Seaman (ref. 18), the scatter

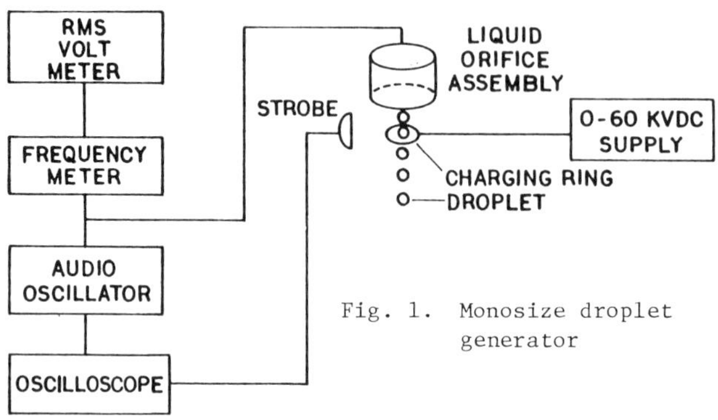

Fig. 1. Monosize droplet generator

in data for adhesion of drops from a surfactant solution may be a result of the wide range in droplet sizes, as well as possible secondary impactions. Almost anything that is done to the AI can affect its biological properties. Droplet spread is particularly important for immobile pests, and is affected by non-volatile, highly viscous and hydrophobic organic solvents. Before pick up, the organism must encounter the AI, which is affected by loss mechanisms such as penetration, evaporation, weathering, leaching, mechanical dislodging, and biotransformation (ref. 19). The probability of the encounter can be increased by diffusion and droplet spread. Transfer to insects and a target site is enhanced by a favorable adhesion force (to insect cuticle). All the modes of transfer, including that of behavior are dependent upon the amount and distribution reaching a target surface. In view of the problems in application efficiency, and poorly calibrated, inefficient delivery by users, and combinations with other pesticides, especially in low volume situations, many of the changes in biological response resulting from formulation changes can remain hidden in field tests. Nevertheless, an increase in the understanding of particle size, and formulate that change chemical and physical properties of the AI is vital. The case of insect growth regulators (IGR) is an excellent example of the need for a well detailed examination of this dynamic interface.

Baker and Hunt (ref. 17) observed increased uptake of a growth regulator (PGR) with increased surfactant concentration in the presence of increased temperature and drop size, and exposure to dew. More significantly, they showed that combinations with various fungicides (in place of Agral 90) can <u>increase</u> droplet reflection and <u>decrease</u> PGR uptake. Once again, there is the conclusion that the variability observed in retention and uptake associated with the delivery of a wide range of droplet sizes, identifies a requirement for careful selection <u>and</u> calibration of spray equipment for improvements in formulants in order to achieve a high and uniform coverage of the target surface (ref. 17).

The concept of prescription dosages and application parameters for certain targets and uses (ref. 20), brings with it the assumption that there are optimal pesticidal rates. One might ask what is the target, i.e.: a mobile or sessile pest, or is it foliage with its associated canopy structure? The British Crop Protection Council (BCPC) approach is an attempt to narrow the spectrum of atomizing devices to a recognized and predictable range of application scenarios for a particular chemical and its intended use. Where a specific droplet size range is denoted, the formulation/concentration relationships thus become important in achieving that objective, and hence the desired optimal biological response. The escalation of resistance phenomena should show us that there is an opportunity to "optimize" the biological response with a more defined and innovative delivery/formulation combination. Ford and Salt (ref. 19) logically suggest that we determine (prior to the development of application/use patterns), the pattern and properties of the spray deposit necessary for optimum biological effects. These properties and conditions can thus determine the formulation and spray characteristics needed for a particular product and target.

Pesticide-induced changes in plant physiology are recognized as contributing to alterations in overall crop productivity. Future research on quantitative structure/activity relationships should also include studies of pesticide effects on crop yields. Consequently, we could optimize pesticide formulations to have <u>maximum</u> efforts on pest species and <u>minimal</u> effects on plant physiology (ref. 21). A review of Graham-Bryce's discussion of pesticide research presented at the 5th International Pesticide Chemistry Congress, reveals a persistent plea for more basic research on dose transfer, physicochemical properties, and a greater attention to the assessment of subtle, but significant changes in delivery and transfer processes. In our haste to produce evermore active agents, have we ignored the significant advances which can be made by these types of cooperative multidisciplinary research investigations? Indeed, full advantage of these possibilities have still yet to be taken.

CONCLUSIONS

Improvements in dose transfer will require active multidisciplinary teams and cooperative industry/academia programs. The delivery of a practical optimal dose transfer system still lacks expertise on decision rules, on-farm records management, accurate weather predictions, and most importantly, basic information on placement rules. Formulation and pesticide application technology specialists have to be more closely aligned with biologists, if significant advances in resistance management, delivery efficiency and dose transfer are to be attained as part of a rational pesticide strategy.

REFERENCES

1. B. W. Young, Outlook on Agriculture, 15, 80-87 (1986).
2. I. J. Graham-Bryce, Proc. 5th Int. Pesticide Chemistry, 1, 21-42 (1983).
3. F. R. Hall, Proc. ARI Mtg., Improving Agrochemical and Fertilizer Application Technology, Feb-Mar, 148 pp (1985).
4. G. S. Hartley and I. J. Graham-Bryce, Physical Principles of Pesticide Behavior, Academic Press, London (1980).
5. A. W. McCann and P. Whitehouse, Asp. Appl. Biol., 4, 329 (1983).
6. A. W. McCann and P. Whitehouse, Span, 28, 98-99 (1985).
7. P. Ayres and D. J. Turner, Symp. Application and Biology, BCPC Mono 28, 155-160 (1985).
8. J. C. Streibig and K. E. Thonke, Symp. Application and Biology, BCPC Mono 28, 147-154 (1985).
9. M. J. Bukovac, In Agr. Chemicals of the Future, Rowman and Allanheld, Totowa, NJ, 85-90, (1985).
10. M. J. Bukovac, In Proc. ARI Mtg., Improving Agrochemical and Fertilizer Application Technology, 25-38 (1985).
11. D. L. Reichard, R. D. Brazee, M. J. Bukovac, and R. D. Fox, Paper No. AA84-006 Amer. Soc. Agr. Eng., St. Joseph, MI (1984).
12. P. J. Q. Stevens and M. J. Bukovac, British Crop Protection Conference - Weeds, 3C-7, 309-315 (1985).
13. F. R. Hall, In 4th Symp., Pesticide Formulations and Application Systems, ASTM 875, 122-130 (1985).
14. F. R. Hall, In 5th Symp., Pesticide Formulations and Application Systems, In press (1986).
15. F. R. Hall, Ohio State Univ. Res. Cir. 290, 49-51 (1986).
16. D. C. Munthali and I. J. Wyatt, Pestic. Sci. 17, 155-164 (1986).
17. E. A. Baker and G. M. Hunt, Ann. Appl. Biol. 106, 579-590 (1985).
18. D. Seaman, Chemistry and Industry (London), 5, 128-130 (1979).
19. M. Ford and D. Salt, In Pesticides on Plant Surfaces, series of Critical Reports in Applied Chemistry, ed. H. Cottrell, Blackwell Sci. Pubs., In press (1986).
20. I. Rutherford, In Application and Biology, BCPC Mono 28, 5-9 (1985).
21. V. Jones, N. Toscano, M. Johnson, S. Welter, and R. Youngman, Bull. Ento. Soc. Amer., 32, 103-109 (1986).

Formulation and application of controlled release pesticides

Müfit Bahadir and Friedhelm Korte

Institute of Chemistry, TU Munich, D-8050 Freising, and
Institute of Ecological Chemistry of GSF, D-8042 Neuherberg, F.R.G.

Abstract - Granule, strip and film shaped controlled release formulations of selected pesticides in LDPE and EVA were prepared by extrusion at 150-170 °C. Pesticides were chlorinated hydrocarbons, organophosphates, carbamates, derivatives of triazines and of further chemical classes. Migration of active ingredients (a.i.) from the matrix into water and into vapour phase was studied, values of diffusion coefficients of a.i. to $D_o = 10^{-8} - 10^{-12}$ (cm^2/s) were obtained. Kinetic models for different pathways of release were evaluated. Several formulations were tested under laboratory and field conditions. Desmetryn containing covering films showed a significant growth suppression of weeds in field trials for cultivation of vegetables. Herbicidal granules inhibited growth of duckweed in a simulated running water at steady state concentrations of approx. 20 ppb terbutryn, and endosulfan containing blue colored film formulations had a high effectiveness in controlling tsetse and mediterranean fruit flies.

INTRODUCTION

The use of plastic films for cultivation of early growing vegetables requires a prophylactic application of herbicides before covering the fields to suppress the growth of undesired weeds (ref. 1). To achieve a sufficient duration of herbicidal activity, application of higher than usual amounts of active ingredients (a.i.) might be necessary. By means of incorporation of bioactive substances into the polymer matrix of films and their diffusion controlled release, which is defined as controlled release formulations (CRF) (ref. 2), avoidable residues in food crops and/or in the environment could be diminished.

Monolithic controlled release formulations were prepared by extrusion of mixtures of selected pesticides with LDPE and EVA copolymers (ref. 3). Parameters of preparation, physicochemical behavior of a.i. in the polymer matrix under laboratory and outdoor conditions including biological test results will be described.

PREPARATION AND CHARACTERIZATION OF CRF

Mixtures of selected pesticides with polymer ratios of granules/powder (9/1) were extruded at 150-170 °C to granule, strip or film (70 μm) shaped CRF. As polymer matrices LDPE and EVA copolymers (10 - 20 % VAc) were used. Pesticides applied were carbofuran (1), chlorfenvinfos (2), chlorpyrifos (3), dichlorvos (6), (α+ß) endosulfan (7), lindane (9), parathion (10), propoxur (11), atrazine (12), chloridazon (13), desmetryn (17), metoxuron (21), monolinuron (22), terbutryn (26), dichlofluanid (28) and pentachlorophenol (29). Except for 11, which decomposed to 2-isopropoxyphenol and methyl isocyanate, all a.i. were thermally stable under the conditions of extrusion. With analytical and IR-spectroscopical methods neither degradation products nor any chemical interactions of a.i. with the matrix could be observed. In comparison with KBr spectra some absorption bands were shifted (νNH: + 100 - 200 cm^{-1}, δNH: - 20 cm^{-1}, νCO: + 10 cm^{-1}).

Solubilities of a.i. in polymer matrices depended on polarities of a.i. and of the matrix (VAc content) as well. Furthermore, the solubilities are linearly correlated with the Rf-value of a.i. on RP-TLC as shown in Fig. 1. Unsoluble surplus of a.i. sweats out of the matrix during or after extrusion and may contaminate machines, working rooms and staff. Therefore, an estimation of presumable solubilities of a.i. according to this method prior to extrusion is of great importance.

During weathering of selected CRF such as EVA-films under outdoor conditions different loss rates of a.i. were detected (Fig. 2). Photochemical isomerizations were found for endosulfan (ß $\rightarrow$ α) and chlorfenvinfos (trans $\rightarrow$ cis). Photodegradation and -transformation reac-

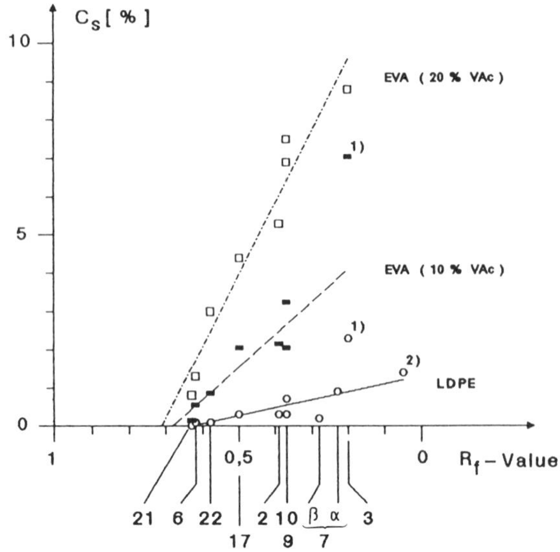

Fig. 1. Dependence of saturated concentrations of a.i. in polymer films (70 μm) from Rf-values on RP-TLC (C_{18}, 75 vol.-% acetonitrile in water) 1) oversaturation

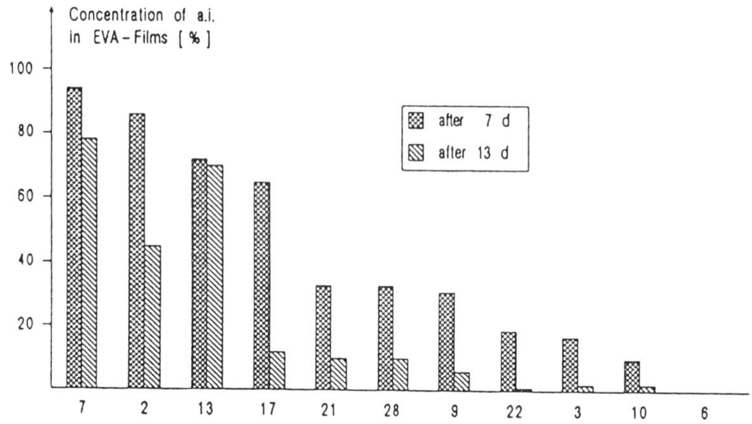

Fig. 2. Loss of pesticides from EVA-films (70 μm) during weathering

tions dominated also in the case of 3, 10 and 22. 3 was mono-dechlorinated (2 isomers), 10 showed an O-S phenyl rearrangement, and photodecomposition of 22 led to N-(p-chlorophenyl)-N-methyl-O-methyl carbamate, p-chlorophenyl isocyanate and phenyl isocyanate. Because of high vapour pressure of 6 (10^{-2} hPa) and steam volatility of 9, both compounds showed a rapid loss into the atmosphere. Migration and photochemical reactions interfered in the case of 17, 21 and 28. Experiments to stabilize a.i. in the polymer matrix photochemically with the help of 0.5 - 1 % light protectants (quencher and UV-absorber) were without success.

During migration experiments with 1, 17, 22 and 26 as wettable powder, EVA granules and calcium alginate gel formulations, of which the preparation and release kinetics are described in ref. 4, the most effective retardation of a.i. was found in monolithic EVA formulations (Fig.3).

Diffusion coefficients obtained from equation (2) which is derived from the 2. Law of Fick, amounted to $D_o = 10^{-8} - 10^{-12}$ (cm^2/s) for most of the a.i. investigated:

$$m_t/m_g = 4 \, [\, D_o \cdot t \, / \, \pi \cdot a^2 \,]^{1/2} \quad \text{for } 0 \le m_t/m_g \le 0.6 \qquad (1)$$

$$D_o = [\pi \cdot a^2 \, / \, 16 \cdot m_g^2 \,] \cdot [\, \Delta m_t^2 \, / \, \Delta t \,] = \text{const} \cdot [\, \Delta m_t^2 \, / \, \Delta t \,] \qquad (2)$$

where a (cm) is diameter of CRF, m_t (mg) migrated a.i., m_g (mg) incorporated a.i. and t (s) time (ref. 5, 6).

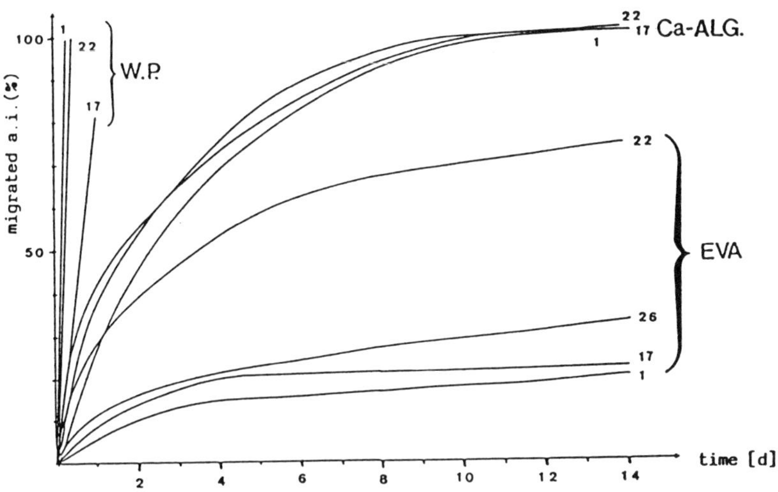

Fig. 3. Release rates of pesticides from different types of formulations into water (Ca-alginate and EVA as granules Ø 2 mm)

During application of pesticidal film formulations to cover the plant/soil system, the a.i. (e.g. herbicides) can be transported downwards to the soil by condensed water from the bottom surface of films, or the a.i. can be lost to the atmosphere by evaporation from the top surface. However, in laboratory experiments with desmetryn (herbicide) it was found that migration of a.i. from 70 µm EVA-films into water was complete within 6 hours (Fig. 4), whereas evaporation into vapour phase lasted more than 60 days (Fig. 5). Temperature dependence of evaporation betweeen 0 and 75 °C was negligible (factor 1.5).

In the field test, a.i. transport occurs by dripping of condensed water from bottom surface of covering films, which leads to a reduced concentration of a.i. in comparison with the top surface. Because of this, main diffusion of a.i. in the matrix is directed downwards, that is, towards the soil/plant system. Thus, losses to the atmosphere are negligible. Desmetryn release in actual field experiments with EVA-films was finished within 3-4 weeks according to the slower re-formation of condensed water compared with the water movement in migration experiments at laboratory scale (Fig. 4).

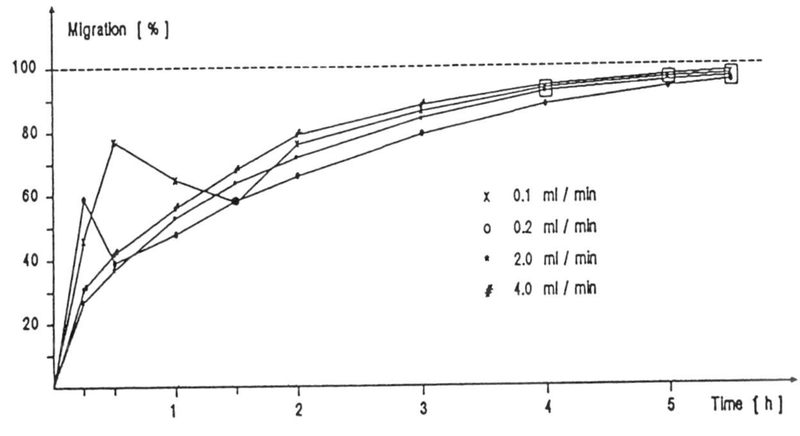

Fig. 4. Migration of desmetryn from EVA-films into water at different flow rates (0.2 % a.i.; 70 µm film thickness)

BIOLOGICAL TESTS

Herbicidal polymer films containing 0.1 - 1.0 % desmetryn were applied in field trials for early cultivation of cabbage. The field plots had a size of 1.5 m x 3 m, 4 replications per each, and were additionally sown with weeds as indicator plants. The formulations significantly suppressed growth of weeds and, therefore, had mostly positive effects on the growth

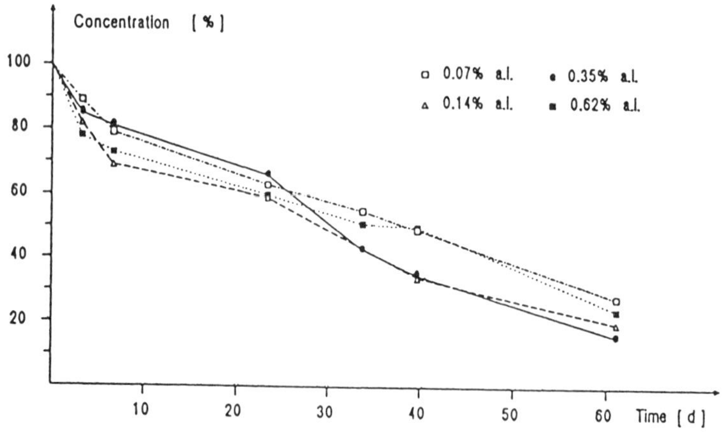

Fig. 5. Migration of desmetryn from EVA-films (70 μm) into the vapour phase (22 °C, 60 % R.H., 600 m^3 air/h)

of food crops. Results of these experiments are shown in Table 1. Best results were obtained with 0.2 and 0.1 % a.i. in the films, while conventional spraying under the films at normal field rate did not show any suppression of growth of weeds and led to low yield of cabbage. Except for 1.0 % a.i. in EVA-films (concentration of a.i. 25 times higher than recommended) no residues could be found in plants or soil.

Table 1. Cultivation of white cabbage under EVA-films containing desmetryn; a) kg/m^2; b) g/m^2

test plots desmetryn in films	marketable cabbage plants number of	yield of[a]	weeds (mean weight)[b]
without film	53	2.749	954
0 %	72	3.028	999
0.1 %	79	4.813	415
0.2 %	75	5.101	31
0.5 %	78	4.723	0
1.0 %	44	2.349	0
conventional spraying (37.5 mg/m^2)	68	2.525	840

In further bioassays the biological activities of 28 or 29 containing EVA-granules (2 % a.i.) against Trichoderma viride could be proved. Blue colored film formulations of 7 had a high effectiveness in controlling tsetse flies (Glossina morsitans) and mediterranean fruit flies (Ceratitis capitata). Herbicidal effectiveness of 26 as EVA-granules was investigated in a laboratory scale simulated flow system device with duckweed plants (Lemna minor) as test organism. Growth inhibition of duckweed could be achieved at approx. 20 ppb 26 in water at steady state (ref. 7).

REFERENCES

1. W. Kampe, Z. Pfl.krankh. Pfl.schutz, 81, 321-336 (1974)
2. A.F. Kydonieus, Controlled Release Technologies: Methods, Theory and Applications, pp. 3-18, CRC Press, Boca Raton (1980)
3. M. Bahadir, G. Pfister, R. Herrmann, P. Moza, Angew. Makromol. Chem., 116, 139-148 (1983)
4. G. Pfister, M. Bahadir, F. Korte, J. Controlled Release, 3, 229-233 (1986)
5. R. Langer, N. Peppas, JMS - Rev. Macromol. Chem. Phys., C23, 61-126 (1983)
6. K. Rombusch, G. Maahs, Angew. Makromol. Chem., 35, 55-70 (1973)
7. M. Bahadir, G. Pfister, Ecotox. Environ. Safety, 10, 197-201 (1985)

Preparation and properties of concentrated emulsions

H. Röchling,* K. Albrecht and R. Heinrich

Hoechst AG, Frankfurt/M. West Germany

INTRODUCTION

Concentrated emulsions in water are a new type of formulation for plant protection agents. They can replace the classical emulsifiable concentrates which contain high quantities of solvents.
Concentrated emulsions contain less or (in some cases) no solvent; they are based on water and therefore have a number of important advantages over emulsifiable concentrates:

- Ecologically beneficial and less offensive smelling.
- The formulations have higher flash points and are therefore safer in transport and storage .
- Solvents can be harmful. In particular they can enhance skin penetration of (sometimes toxic) active ingredients. An extensive substitution of solvent by water therefore normally means reduction of the user-toxicity.
- In some cases an improved biological efficacy and reduced phytotoxicity was obtained by using a concentrated emulsion instead of an emulsifiable concentrate.

The preparation of concentrated emulsions became possible through the application of new types of phosphorylated emulsifiers.
Methods for development of this new formulation-type are shown.
Concentrated emulsions can be obtained as milky-white macroemulsions or as transparent microemulsions.
Different emulsifiers for the preparation of both emulsion-types are described as well as particle-size and properties of macro- and microemulsions during storage.

RESULTS

1. Preparation

In the development of a concentrated emulsion one has first to find a suitable solvent in which the active ingredient is soluble in a high concentration. Especially useful are: ketones like cyclohexanone and acetophenone, alcohols like cyclohexanol, esters like oxo-hexylacetate and diethylphthalate, alkylated aromatics like xylene and methyl-naphthalene.

Only in some cases can an emulsion be obtained without solvent:

> HOE 36275 has a water-solubility of 4%. By application of a combination of phosphorylated and non phosphorylated emulsifiers we obtained a 20% HOE 36275-microemulsion in 61% water.

As emulsifiers for macro- and microemulsions we have applied phosphorylated tensides; we have been particularly successful with two groups:

- styryl-substituted phenols, ethoxylated, phosphorylated and neutralized with triethanolamine.
- salts of phosphorylated ethylenoxide-propylenoxide-ethylenoxide-bloc-copolymers.

For the preparation of microemulsions a special combination of phosphorylated and non phosphorylated emulsifiers proved to be useful:

- The main component is tristyrylphenol, ethoxylated with 18-20 mole EO, phosphorylated and neutralized with triethanolamine.
- The combination-partner is preferably alkyldiglycolethersulfate-sodium salt where alkyl means 70-75% C_{12} and 25-30% C_{14}.

With this combination of emulsifiers several active ingredients (insecticides, nematicides, fungicides and herbicides) could be formulated as microemulsions with the sometimes without application of solvents.

An example in which solvents have to be used is the 300 g/L-microemulsion of binapacryl:

28.96 wt.-% binapacryl 97.8% (≙ 300 g/L pure A.I.) are dissolved in
25.00 wt.-% cyclohexanone and
10.00 wt.-% cyclohexanol

In a second flask a solution of emulsifiers in water is prepared:

15.00 wt.-% tristyrylphenol + 20 EO, phosphorylated and neutralized with triethanolamine
3.02 wt.-% alkyl-diglycolether-sodium-sulfonate
18.02 wt.-% distilled water

The solution in water is added with thorough stirring to the solution of binapacryl. Stirring is contined for about 2 hours. A transparent, homogeneous emulsion is obtained.

As EC-formulations of the nematicide HOE 36275 have high dermal toxicities, a concentrated emulsion was developed without use of a solvent. Again the combination of ethoxylated and phosphorylated tristyrylphenol with alkyl-diglycolether-sodium-sulfonate could be applied:

16.5 wt.-% tristyrylphenol + 20 EO, phosphorylated and neutralized with triethanolamine
2.5 wt.-% alkyl-diglycolether-sodium-sulfonate
61.07 wt.-% distilled water

The mixture is stirred until all components are dissolved. 19.93 wt.-% O-ethyl-S-(N-phenyl-N-methyl-carbamoyl-methyl)-isopropyl-amido-thiophosphate (HOE 36275) ≙ 200 g A.I./L are added and stirring is continued until a transparent, homogeneous emulsion is obtained.

Acute dermal toxicity was substantially reduced in the concentrated emulsion in comparison to EC-formulations of HOE 36275:

	LD_{50} female rat
EC-40 in solvesso 200	57.5 mg/kg
EC-40 in methyldiglycole	282.2 mg/kg
200 g/L-emulsion in water	2000.0 mg/kg

For the preparation of macroemulsions we used the salts of phosphorylated ethylenoxide-propylenoxide-ethylenoxide-bloc-copolymers and obtained very stable formulations.

A typical composition is the 360 g/L-macroemulsion of diclofopmethyl:

36.6 wt.-% diclofopmethyl 93 % (≙ 360 g/L pure A.I.) and
6.0 wt.-% ricin-oxethylate (40 EO) are dissolved in
18.0 wt.-% xylene

In a second flask a solution of emulsifier and antifreezing agent in water is prepared:

2.0 wt.-% potassium salt of phosphorylated ethylenoxide-propylenoxide-ethylenoxide-bloc-copolymer
10.0 wt.-% ethylenglycol
27.4 wt.-% distilled water

With thorough stirring the organic phase is slowly added to the solution in water. The mixture is stirred for a further 30 min. and a white emulsion is obtained.

If the receipe of the formulation has been found, the technical preparation of concentrated emulsions is not problematic and one can proceed in one of the following ways:

- Normally the A.I. is dissolved in a solvent or solvent-mixture; the solution of emulsifiers in water is added to the organic phase with thorough stirring.
- It is also possible to dissolve the emulsifiers together with the A.I. in the solvent and to add water with vigorous stirring.
- Or (in the invert way) A.I. or solution of A.I. is added to the solution of emulsifiers in water.

2. Droplet-Size

As can be expected the droplet-size of transparent and milky-white emulsions differ considerably. They are smaller than the wavelength of visible light (400 nm) in transparent and larger in white emulsions. By measurement with the Malvern Autosizer® we found that liquid particles of microemulsions are in the range of 3 to 10 nm.

On heating the droplets of some of our microemulsions coagulate and become larger: from an average size of 2 nm at 20° C to an average size of 17 nm at 55° C:

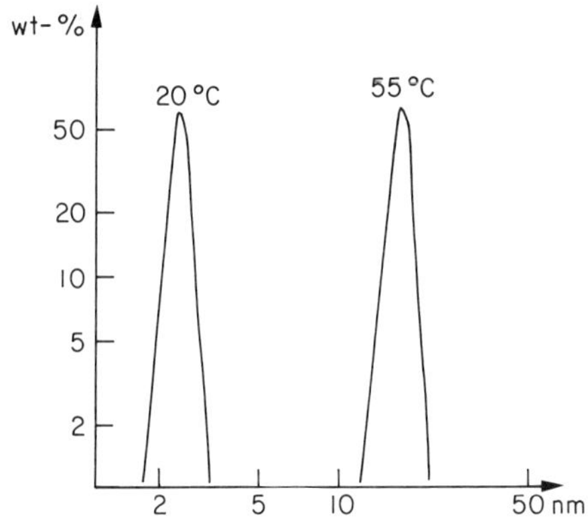

Fig. 1 Droplet-size distribution of transparent HOE 36275 200 g/l-emulsion at 20° C and 55° C (Malvern Autosizer II c®)

On further heating liquid particles enlarge over 400 nm and the emulsion becomes milky-white as well. In a well-composed formulation either no coagulation occurs or the described process of droplet-enlargement is reversible.

Droplets of macroemulsions range from 500 to 1000 nm, e.g.:

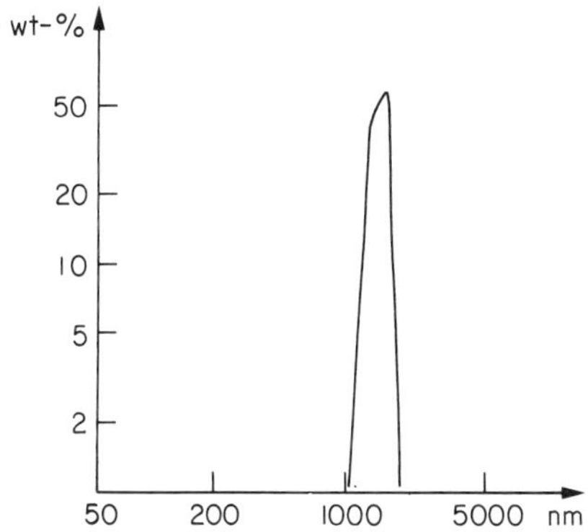

Fig. 2 Droplet-size distribution of milky-white HOE 23408 360 g/l-emulsion (Malvern Autosizer II c®)

A narrow distribution curve indicates uniformity of droplet-size which is favourable for the stability of a concentrated emulsion: An emulsion in which droplet-size varies considerably (=broad distribution cruve) is less stable as large droplets attract smaller ones. Coagulation and finally separation into two layers can occur.

3. Conductivity

Water content of our concentrated emulsions varied between 18 and more than 60%. It seemed possible that in those cases in which the water content is low the formulations are not oil-in-water- but water-in-oil-emulsions.
We determined the conductivity and compared it with that of water and with conventional emulsifiable concentrates and ULV-formulations (Table 1).

Conductivities of all concentrated emulsions were in the range of water or higher. EC- and ULV-formulations in which only solvents are used have considerably lower conductivity-values. These results suggest that the coherent phase in the concentrated emulsions is water and the emulsion type is "oil in water".

TABLE 1 - Conductivity of Concentrated Emulsions

Product	Emulsion-type	% a.i.	% water	Conductivity (µs)
HOE 36275 200 g/l-emulsion	transparent	19	61	2470
triazophos 400 g/l-emulsion	transparent	36	20	430
binapacryl 300 g/l-emulsion	transparent	28	18	310
diclofopmethyl 360 g/l-emulsion	milky-white	34	27.4	250
fenoxaprop-ethyl-60 g/l-emulsion	milky-white	5.6	27.7	285
water				590
conventional EC- or ULV-formulation				< 10

4. General Requirements

Concentrated emulsions of plant protection agents have to meet the following requirements:

Low Temperature Stability. During storage at low temperature (0° C to -10° C) no irreversible separations or crystallizations should form by freezing of the water or too low solvent-content. If necessary antifreezing agents like glycerol, isopropanol, mono-, di- or tripropylenglycol-monomethylether, cyclohexanol or urea have to be incorporated.

Heat Stability. Above 45° C two layers can form by coagulation of the emulsified droplets. This has to be prevented by selection of suitable emulsifiers and the right proportions of active ingredient, solvent and emulsifier. Using the defined combination of ethoxylated and phosphorylated tristyrylphenol and alkyldiglycolether-sodium-sulfonate or the phosphorylated ethylenoxide-propylenoxide-bloc-copolymers we could obtain concentrated emulsions with excellent stability.

Stability of Active Ingredient. Stability of some active ingredients during storage at elevated temperatures in the presence of water is not sufficient. By maintaining a pH-value above or below 7 through suitable buffer-solutions stabilization of our products could be achieved in most cases.

SUMMARY

The advantages and methods for the preparation of concentrated emulsions are discussed. Microemulsions can be prepared with a combination of ethoxylated and phosphorylated tristyrylphenol and alkyl-diglycolether-sodium-sulfonate and macroemulsions with phosphorylated ethylenoxide-propylenoxide-bloc-copolymers.
The use of droplet-size of concentrated emulsions is described together with their behaviour during heating. Conductivity measurements give evidence that our concentrated emulsions exist in form of oil-in-water emulsions. It was found that the dermal toxicity of a nematicide could be reduced by changing the formulation from an emulsifiable concentrate to a concentrated emulsion.

Suspension emulsions—a new look at tank mix technology in one pack

Patrick J Mulqueen, Graham Banks, David L Press

Chemistry Development Group, Dow Chemical Company,
King's Lynn, Norfolk, England.

Abstract - Suspension-Emulsion formulations containing a mixture of three phases; an organic (discontinuous phase) dispersed in an aqueous (the continuous phase), i.e., the oil-in-water emulsion, and a solid phase dispersed throughout the continuous phase can offer a convenient means of combining the tank-mix components of an emulsifiable concentrate with an aqueous suspension concentrate in a single-pack product. These formulations have their own particular problems to be overcome in their development and manufacture. The development of a typical suspension emulsion is described with reference to fluroxypyr-1-methylheptyl ester, a new broad-leaf herbicide developed by The Dow Chemical Company, in combination with other herbicide components. Identification of suitable test methods to characterise these formulations is discussed. Manufacturing implications are discussed and an outline development scheme proposed.

INTRODUCTION

Herbert B Scher commented in his overview at a symposium in Advances in Pesticide Formulation Technology in August 1983 that emulsion type flowables (oil-in-water flowables) had only just begun to receive widespread attention and were still very much in the experimental stage. That also summed up The Dow Chemical Company's position in 1983.

However, with the development of a new herbicide, we found ourselves with the need to attempt to develop such formulations. Starane is the trademark of The Dow Chemical Company for herbicidal products containing 4-amino-3,5-dichloro-6-fluoro-2-pyridyl oxyacetic acid (common name: fluroxypyr) as an active ingredient. Fluroxypyr is a readily translocatable non-phenoxy herbicide exhibiting a high degree of activity in post-emergence foliar applications to a wide spectrum of broad-leaved plants. It is extremely effective for the control of several major economically important broad-leaved weeds in cereals (including Galium aparine and Stellaria media) as well as showing considerable promise for the control of deep-rooted perennials such as Convolvulus arvensis, Calystegia sepium and Rumex species, in orchards, vineyards and pastures.

Based on biological evaluation of a range of derivatives, fluroxypyr is manufactured and used in all formulations as the 1-methylheptyl ester (Formula 1).

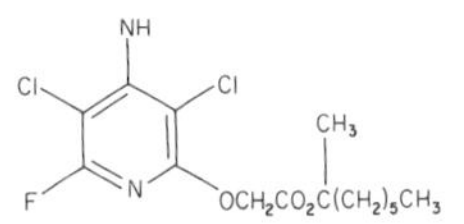

The properties of fluroxypyr 1-methylheptyl (fluroxypyr-1MHE) ester are shown in Table 1.

TABLE 1. Properties of fluroxypyr 1-methylheptyl ester

Empirical Formula:	C15 H21 C12 F N2O3
Molecular Weight:	367.3
Melting Point:	56 - 57°C
Octanol/Water Coefficient:	6140 : 1
Solubility in Water:	0.9ppm at 20°C
Solubility in Xylene:	40% w/w at 20°C

Based on these physical properties, fluroxypyr 1-methylheptyl ester has been formulated as emulsifiable concentrates, where possible, in aromatic solvents, especially in combination with hydroxybenzonitrile esters and phenoxy esters, the formulations containing typically up to 250g/L fluroxypyr acid equivalent (a.e.).

In order to develop new markets, products containing fluroxypyr with hydroxybenzonitriles and urea-herbicides were required to broaden the spectrum of activity in cereals, especially against Alopecurus myosuroides, Apera spica-venti and Avena fatua.

Such urea herbicides have typically low (1% w/v) solubility in organic solvents and low (approximately 100mg/Kg) solubility in water. With relatively high melting points (about 150°C) these products are conveniently formulated as aqueous suspension concentrates at up to 500g/L, as well as high concentrate wettable powders.

Laboratory studies with fluroxypyr 1-methylheptyl ester had previously shown problems in the production of aqueous suspension concentrates and wettable powders, as well as reduction in activity against Galium aparine. An evaluation of the possibility of producing useful emulsifiable concentrates of ureas soon ruled out the possibility of a mixed emulsifiable concentrate. Since the products were individually available either as an emulsifiable concentrate (fluroxypyr plus hydroxybenzonitrile) or aqueous suspension concentrate (urea herbicide) and the tank mix of their components was biologically effective under field conditions, it was decided to attempt to produce this tank-mix as a concentrate. This suspension-emulsion formulation was called a 'Suspo' within the Dow Chemical Company.

Since this was such an obvious solution to a formulation problem, why was there so little patent literature available and so few 'Suspo' formulations being produced? Three explanations could be; other formulation alternatives available, packaging non-availability or formulation difficulty. Each point is dealt with below.

Other formulation alternatives available
With established phenoxy acid products or hydrobenzonitriles there is a choice of aqueous soluble salt or oil-soluble ester giving the formulator flexibility as to how he can formulate such a product. Should the need arise to add a water-insoluble component, then the aqueous soluble option rather than the oil-soluble option can conveniently be chosen without too much formulation difficulty.

Packaging non-availability
Whilst the many plastic materials now available in large quantity (and variety) have been accessible for the past 5 - 10 years, it has only been in the last two years that a formulator has had an economic choice of package. Today he can use any metal container (with a host of lacquers), and a wide range of plastic materials from HDPE through co-extruded bottles to polyethylene terephthalate (PET). With this wide choice, at approximate parity in cost, he now has a much better chance of finding a pack with suitable barrier properties for the mixture he has put together in a glass bottle in the laboratory.

Formulation difficulty
From our experience, these formultions are difficult to put together, requiring great attention to detail. They also require careful choice of pack and manufacturing route.

The rest of this review will deal with some of our experiences in developing this area of technology.

The following points will be considered with respect to a model system:- (i) Requirements for a 'Suspo', (ii) Tests and methods of testing for performance, (iii) Development of a 'Suspo' and (iv) Manufacturing routes.

Requirements for a 'Suspo'
The prime requirements for the components of a 'Suspo' formulation are:-
(i) An oil-phase with an active content as high as possible, ideally a liquid active, but also such oil-phase having no water solubility (i.e. less than 10ppm).
(ii) A dispersed solid phase whose active is not only insoluble in water (as per a normal aqueous suspension concentrate) but which should also be insoluble in the dispersed oil-phase.

Other components such as water soluble salts of actives can also be considered for addition to the aqueous phase.

Tests and Methods of Testing For Performance
Such a product has to perform on dilution both as an aqueous suspension concentrate and as an emulsifiable concentrate. So, as well as evaluating characteristics of the concentrated product, several aspects of the dilution behaviour need to be evaluated. These can be listed as:-

Concentrate Characteristics
Active Content (and chemical stability) Particle Size - emulsion phase - dispersed particulate phase - sieve retention Rheology Freeze-thaw Stability pH

Dilution Characteristics
Spontaneity Suspensibility Particle size

Development of a 'Suspo'

A model system for study was chosen as the following:-

3-(3-chloro-p-tolyl)-1,1-dimethyl urea(chlortoluron)	2400g/Ha
3,5-dibromo-4-hydroxy benzonitrile(bromoxynil)	500g/Ha
fluroxypyr (a.e.)	100g/Ha

It was envisaged that fluroxypyr (as 1-methylheptyl ester) and bromoxynil (as octanoate) would be combined in the oil phase and dispersed into a chlortoluron aqueous suspension concentrate.

This model system was broken down into a 6L/Ha product concept containing:-

	% w/v
chlortoluron	40.0
bromoxynil phenol	8.33
(bromoxynil octanoate)	(12.12)
fluroxypyr a.e.	1.67
(fluroxypyr 1-methylheptyl ester)	(2.41)

Development was broken down into 3 stages, namely (i) Identification of a suitable chlortoluron aqueous suspension concentrate, (ii) Identification of a suitable efficient solvent system for fluroxypyr 1-methylheptyl ester and bromoxynil octanoate and (iii) Combination and stabilisation of (i) and (ii). By careful selection, a ternary blend of surfactants was identified (at 2,3 % w/w) which allowed the preparation of a 65% w/w (80% w/v) chlortoluron aqueous flowable feedstock. This low level of surfactant was selected in an attempt to reduce any active transfer between phases, as it was felt that high levels of surfactants could cause such effects. This low level subsequently caused problems with the stability of the diluted dispersion and had to be increased. The particle size of this aqueous flowable feedstock will be discussed later in connection with methods of determining particle size. No rheological additives were incorporated into this flowable feedstock.

The emulsion phase was investigated by (i) Choice of Solvent (ii) Choice of colloid stabilizer.

A particularly advantageous colloid stabiliser for the emulsion was found to be polyvinyl alcohol (PVA). A range of solvents were evaluated and four types chosen for further testing.

(i) An aromatic C_9 blend (solvent 1).
(ii) An aliphatic ester (solvent 2).
(iii) A chlorinated aromatic hydrocarbon (solvent 3).
(iv) A chlorinated aliphatic hydrocarbon (solvent 4).

PVA stabilized emulsions of fluroxypyr and bromoxynil were made up as in Table 2, using a high speed emulsifier-head mixer. After 9 days and after 3 months storage at 55°C, the products were physically unchanged. The chemical stability was as in Table 3.

TABLE 2. Compositions of emulsion phases containing fluroxypyr 1-MHE and bromoxynil octanoate. All figures expressed as % w/w

SAMPLE	A	B	C	D
fluroxypyr 1-MHE	5.0	5.0	5.0	5.0
bromoxynil octanoate	24.7	24.7	24.7	24.0
Solvent 1	15.0	-	-	-
Solvent 2	-	15.0	-	-
Solvent 3	-	-	15.0	-
Solvent 4	-	-	-	15.0
PVA and Water	bal	bal	bal	bal

TABLE 3. Chemical stability of fluroxypyr 1-MHE and bromoxynil octanoate mixed emulsions. All figures expressed as % w/w

SAMPLE	A	B	C	D
9 days storage				
bromoxynil octanoate	24.9	24.0	25.0	25.7
fluroxypyr 1-MHE	5.1	5.0	5.1	5.2
3 months storage				
bromoxynil octanoate	24.6	24.2	25.3	24.9
fluroxypyr 1-MHE	5.1	5.0	5.2	5.0

It was with confidence in the chemical stability of the emulsion that an emulsion of fluroxypyr and bromoxynil in a C_9 aromatic solvent blend was prepared and blended with the chlortoluron suspension concentrate feedstock. Combination of these two components tended to cause thickening, dependent upon the intensity of mixing and the type of colloid stabiliser employed. However by addition of an appropriate surfactant a suitable formulation was initially prepared containing the following, chlortoluron 35.3% w/w, fluroxypyr ester 2.17% w/w and bromoxynil octanoate 10.89 % w/w, which was considered acceptable for initial field trials. This was formulation P. No physical changes were apparent after 8 days at 55°C and on 12 hour 0 - 30°C cycles.

A comparative study was carried out in the glasshouse between Formulation P and a laboratory tank mix containing the same chemical products for action against Galium aparine, Matricaria indora and Alopecurus myosuroides and selectivity to wheat and barley. All plants were hardened off in an outdoor growth area before application of chemicals and remained in this outdoor area for the duration of the study after application of chemicals. The results are shown in Table 4.

TABLE 4. Efficacy/selectivity study using formulation P

CODE	TREATMENTS	RATE	CLEAVERS			BLACKGRASS			MAYWEED			BARLEY			WHEAT		
			DAY 5	DAY 9	DAY 12	DAY 5	DAY 9	DAY 12	DAY 5	DAY 9	DAY 12	DAY 5	DAY 9	DAY 12	DAY 5	DAY 9	DAY 12
1A	Formulation P	6L/ha	3	4	4	2	3	4	5	5	5	2	2	2	1	1	1
B	"	3L/ha	2	3	3	2	2	3	4	5	5	2	2	1	0	0	0
C	"	1.5L/ha	2	2	3	0	0	2	3	5	5	2	0	0	0	0	0
D	"	0.75L/ha	2	2	2	0	0	0	3	5	5	1	0	0	0	0	0
2A	Tankmix(fluroxypyr/	6L/ha	4	4	4	3	3	3	5	5	5	4	4	4	2	2	2
B	chlortoluron/	3L/ha	3	4	4	2	1	2	5	5	5	2	2	3	2	1	1
C	bromoxynil	1.5L/ha	3	3	3	2	0	2	5	5	5	0	0	1	0	0	0
D	"	0.75L/ha	2	2	2	0	0	0	4	5	5	0	0	0	0	0	0
3A	Fluroxypyr e.c.	102gai/ha	3	3	3	0	0	0	4	4	4	2	0	0	0	0	0
B	"	51gai/ha	3	3	3	0	0	0	4	4	4	1	0	0	0	0	0
C	"	25.5gai/ha	3	3	3	0	0	0	4	4	4	0	0	0	0	0	0
4A	Chlortoluron s.c.	2400gai/ha	1	1	0	0	0	0	1	5	5	0	0	0	0	0	0
B	"	1200gai/ha	1	0	0	0	0	0	0	5	5	0	0	0	0	0	0
C	"	600gai/ha	1	0	0	0	0	0	0	5	5	0	0	0	0	0	0
5A	Bromoxynil e.c.	510gai/ha	3	3	3	2	2	3	5	5	5	1	0	1	0	0	0
B	"	255gai/ha	3	3	3	2	1	2	5	5	5	0	0	0	0	0	0
C	"	127gai/ha	2	2	2	1	0	1	5	5	5	0	0	0	0	0	0
6	Control			0						0			0			0	

KEY: Scale 0 to 5 where 0 = Normal Healthy Plants 5 = dead

Formulation P can be considered as effective as the tank mix and at least as safe as the tank mix to crops. This was subsequently borne out in field trials.

However, on longer term testing, formulation P showed two problems:-

(i) The formulation was found to be non freeze-thaw stable.

(ii) An initial trend of bromoxynil octanoate loss continued such, that at 10 weeks the following losses were recorded.

Bromoxynil octanoate 15% relative loss at 40°C
28% relative loss at 55°C

Fluroxypyr and chlortoluron levels were constant (within method error).

There was also a slight upwards movement in particle size.

Studies of the freeze-thaw stability indicated the colliod stabilizer at fault. It appeared to dehydrate on freezing and did not reconstitute quickly enough on thawing causing failure of the system. It was necessary to establish the cause of loss of bromoxynil octanoate. The early emulsion phase chemical stability indicated that bromoxynil octanoate and fluroxypyr 1-MHE were chemically compatible, so the problem appeared to stem from the addition of the chlortoluron aqueous flowable feedstock.

Two series of tests were conducted with the emulsion phase adding a variety of potential stabilizers such as oxiranes, antioxidants as well as buffering the pH at 4, 5 and 6. One series contained the emulsion without added chlortoluron, the other contained added chlortoluron. All mixtures containing added chlortoluron showed loss of bromoxynil octanoate. No loss of bromoxynil octanoate was observed in the absence of chlortoluron. Detailed assays on Formulation P showed that although loss of bromoxynil octanoate was excessive, the total phenol content, based on a summation of octanoate and free phenol remained constant as shown in Table 5.

This situation was not considered satisfactory for a product. From further consideration of and work on the problems, it was decided that the only way to overcome the chemical incompatibility between bromoxynil octanoate and chlortoluron would be to remove bromoxynil octanoate from the oil-phase and put it into the aqueous phase as a salt. The triethanolamine salt was chosen as a convenient starting point.

TABLE 5. Bromoxynil content of formulation P after storage at various temperatures

Temp of Storage	Broxoxynil Octanoate %w/w	(% loss)	Broxoxynil %w/w	Total phenol* %w/w
-5°C	10.5	(0)	ND	7.21
ambient	10.8	(0)	ND	7.42
40°C	9.72	(7.4)	0.423	7.10
55°C	7.15	(31.9)	2.39	7.30

* Expressed by calculation as a summation of bromoxynil octanoate and bromoxynil.

The formulation was therefore re-prepared by producing a fluroxypyr emulsion in an aqueous solution of the triethanolamine salt of bromoxynil and then mixing with the previously prepared chlortoluron aqueous flowable feedstock. The product was stabilized by xanthan gum addition. This was found to be satisfactory, being both chemically and physically stable. A development of this work was the use of a micro-emulsion of the oil-phase. Due to the low oil-phase obtained by using an aqueous salt solution of bromoxynil, it was now relatively easy to prepare a micro-emulsion of fluroxypyr and incorporate as with a normal emulsion. Recent patent publications by Hoechst have indicated the suitability of phthalate esters as solvents for this type of formulation. These have been evaluated and indeed offer potential for use in these formulations. Similarly, a range of formulations has now been prepared employing fluroxypyr-1MHE as an emulsion phase in these suspo-type formulations.

Having developed physically and chemically stable formulations, 3 important test parameters became obvious namely (i) Particle size determination, (ii) Suspensibility of the formulation and (iii) Rheology of the formulation.

In order to be able to produce a formulation commercially it was considered that these measurements are essential in determining the quality of the product. The suspensibility of the formulation was chosen to be determined by an active method, since it was considered probably the best way of determining an overall performance figure for the formulation. Both the WHO and the AID method could be used.

The rheology of a formulation is determined in order to characterise a product, to indicate potential problems and to detect changes in physical properties at an early stage in a products lifetime. Subsequent transfer of this knowledge allows viscosity calculation from a rheogram produced by an instrument such as the Haake RV2 rheometer, calculated at two points on the down curve. A typical product would have a sheared-down viscosity of 50-200 mPa sec. Such a curve is shown in Figure 1.

FIGURE 1. (Haake RV2 Rotovisco flow curve of formulations with MV2 measuring head

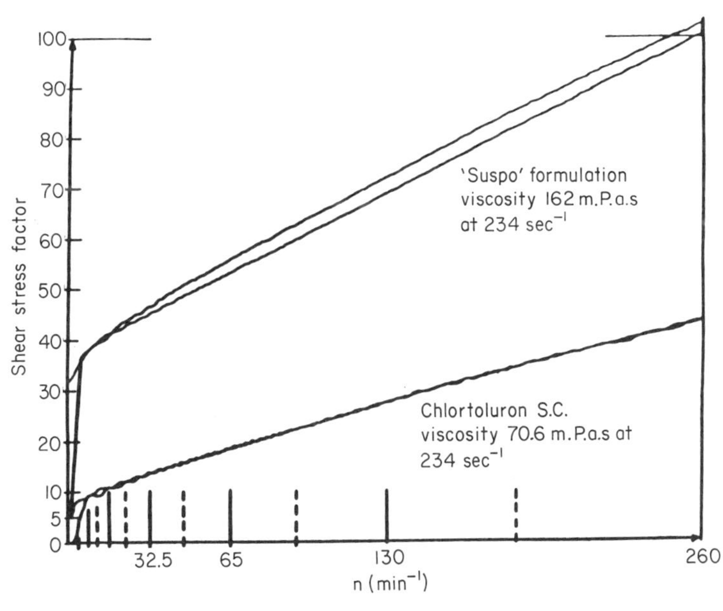

Commercial suspension concentrates have been used as feedstocks for some of our formulations but owing to the very high viscosity of these products, the resultant 'Suspo's' have been even thicker. This aspect must be bourne in mind in developing suitable products from commercially available suspension concentrates.

The measurement which has caused most problems has been the determination of the particle size of the finished product. Typically one would expect an aqueous suspension concentrate to be between 1 and 10µm (VMD) and an emulsion could be as fine as 0.1µm (VMD) - or smaller in the case of a micro-emulsion. Typical equipment used to determine the particle size of suspension concentrates such as the Coulter Counter TAII or the Malvern 3600E were found to be unsuitable for evaluating the particle size of the emulsion phase. However, the Leeds and Northrop Microtrac small particle size analyser was found to give interesting results and in some work was very useful. Typical data is shown in Figure 2.

FIGURE 2. Leeds and Northrop Microtrac small particle size analysis data for suspension emulsions

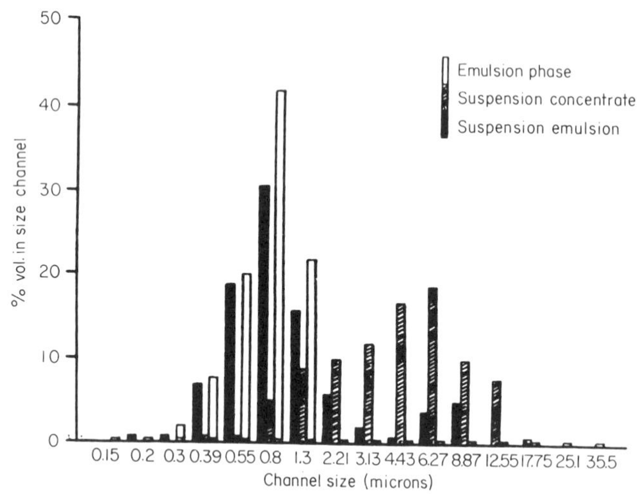

These results are far from ideal since whilst the emulsion and suspension phases can be seen when tested separately, the mixture does not give an accurate size determination of both components. Different oil/solids volumes give vastly different effects. Coupled with problems of flocculation at the high dilutions employed in this test equipment there remains a need for equipment (other than an optical microscope) for determining the particle size in such mixtures. Dow's current approach is to control independently the particle size of the emulsion phase and the suspension concentrate and measure change/quality of particle size by other physico-chemical tests.

Manufacture of 'Suspo's'

Having prepared suitable formulations on a laboratory scale, to manufacture these the following procedure can be recommended:-

(i) Keep things simple.
(ii) Use pre-prepared manufacturing concentrates such as commercial aqueous flowables, or manufacture a high quality mill base.
(iii) Prepare a stable emulsion.
(iv) Blend together.

A rule of thumb here is that if all components can be mixed together under high shear (e.g. Silverson mixing) the formulation will probably be sufficiently robust. If it cannot be prepared in this way, it will probably not be storage stable.

From our studies, the following formulation development scheme for a 'Suspo' formulation can be outlined.

(i) Minimise solubility problems in the oil-phase by appropriate selection of solvents, crystallisation inhibitors. The oil-phase should be completely liquid at all temperatures likely to be encountered.
(ii) The emulsion phase, when prepared, should be stable.
(iii) Prepare a suspension concentrate with as high a solids loading as possible.
(iv) Chemical incompatibilities must be minimised by choice of physical form of each active.
(v) Optimise pack choice by careful screening of candidate formulations in potential packs.

The influence of molecular weight and degree of sulfonation of kraft lignin on the rheological properties of concentrated suspensions of agrichemicals

H. T. Delli Colli

Polychemicals Department
Westvaco Corporation
P. O. Box 70848
Charleston Heights, South Carolina 29415-0848
U.S.A.

Abstract - Synthetically accessible changes in the molecular weight and degree of sulfonation of kraft lignin based dispersants have been shown to have significant effects on the flow properties of concentrated suspensions of agricultural chemicals in which the dispersants are used.

The changes in rheology based upon changes in the chemistry of dispersants can in turn influence storage stability, transport stability, and handleability of finished formulations in manufacturing plants. Additionally, the suitability of a concentrated suspension for further processing such as spray drying can be altered by changing the chemistry of the dispersing agent used in the suspension.

In the present work, a series of dispersants based on technical unsulfonated kraft lignin as a parent compound were prepared and incorporated into concentrated aqueous suspensions of atrazine, diuron, and copper oxychloride. The results suggest that higher molecular weight dispersants regardless of degree of sulfonation can be used to develop pseudoplastic suspensions with yield points. On the other hand, a shift in molecular weight distribution towards lower values by chemical reaction or ultrafiltration results in a general lowering of viscosities. In this instance, increases in the degree of sulfonation of the dispersant can be used to control the rheology of the suspensions.

INTRODUCTION

As agricultural chemical formulations have become more complex and varied, so has the job of the formulation chemist. What was at one time often a random selection of several dispersants, wetting agents, and emulsifiers followed by a period of mixing and stirring, with hopefully crossed fingers, has now become a calculated and well planned exercise in the physical chemistry of dispersions.

This type of formulation development activity requires that the chemist have more basic information about those surface and interfacially active substances he intends to utililze. Statements printed on product labels which give the approximate concentration of active ingredients contained in the product or which make qualitative or semi-quantitative statements on use of the product contained therein are no longer sufficient. Information on molecular weight/molecular weight distributions, ionic charges, charge to mass ratios, and solubility parameters are just some of the properties of surfactants which today's formulation chemists are now trying to understand and base their formulation selections on.

The scope of the present work is to relate two of these properties, molecular weight/molecular weight distribution and degree of sulfonation of a series of dispersants, sodium and ammonium salts of sulfonated kraft lignin, to the rheological properties of dispersions of three agricultural chemicals, atrazine, diuron, and copper oxychloride. The dispersant properties selected are those which are synthetically accessible. That is, molecular weight and degree of sulfonation can be deliberately changed in order to change the properties of the dispersants themselves as well as those of the concentrated suspensions in which the dispersants are used.

The present work is based upon a series of kraft lignin derivatives prepared for the purpose of quantifying the impact of molecular weight and charge to mass ratio changes of the dispersants on suspension viscosity.

BACKGROUND

Experience with dispersions of dyestuffs, coal + water slurries and cement as well as agricultural chemicals has shown that the primary manifestation of molecular weight and charge of sulfonated kraft lignin in concentrated suspensions is the low shear rheological behavior of the suspension.

For example, as molecular weight of a lignin based dispersant is increased, the molecular volume usually increases. This results in an increase in the interaction between lignin molecules and between lignin and the continuous phase of the dispersion. This results in viscosity increases (1).

On the other hand, as the degree of sulfonation of a lignin derivative increases, the coulombic repulsive forces between lignin molecules and the particles upon which they are adsorbed also increases. This phenomenon results in viscosity decreases at low shear.

Since these two properties appear to produce opposing results in concentrated suspensions, the formulation chemist is faced with the requirement of carefully identifying the specific properties of his formulations based upon their intended end use. For example, is the formulation intended to be shipped, sold, and on-farm manipulated in bulk or will it reach its end user in small packages of a few hundred milliliters and require very careful measurements of small volumes of formulation for spray tank admixture? Each of these uses would favor different rheological properties. For example, bulk formulations or even 2 1/2 and 5 gallon containers would favor extreme fluidity to effect complete transfer from container to spray tank. Small measuring errors would not be likely to effect biological activity or exceed label limitations. On the other hand some of the new very low dosage pesticides do require extremely precise measurements of dosages because of their biological activity and/or label requirements. Slightly higher viscosities would favor this use.

Differences in transport and storage conditions and anticipated shelf lifetimes can and often do benefit from different rheological characteristics. The implications of formulation rheology are obvious here.

To these ends, what properties of dispersants will be most important and for what reason do they exert these specific effects on formulations? These are questions which formulation chemists find themselves called upon to answer almost daily.

EXPERIMENTAL

Technical pesticides were used as received from their respective manufacturers, atrazine (IPiCi), diuron (E. I. DuPont De Nemours Co., Inc.) and copper oxychloride (Sandoz - Brazil). The sodium ammonium salts of sulfonated kraft lignin were received as solutions containing 20 - 25% solids (w/w) and used as such.

Formulations were prepared by slurrying the pesticides in the continuous phase in which all of the water soluble components such as dispersants, wetting agents and antifreeze agents have been dissolved. They were then ground in an attritor until the mean particle size of the formulations was 4 ±1 microns.

Particle size measurements were made on a Microtrack 7995-10 particle size analyzer.

Rheological measurements were made with a Haake RV-12 Rotational Viscometer. Rotors and samples were thermostated at T = 25 ±0.5°C during measurements. After the sample cups were filled with the required volume of suspension concentrate, the rotors were accelerated from zero revolutions per minute to 512 rpm over a period of three minutes, held at maximum shear for 1 minute and decelerated to zero rpm over a 3 minute period. Values of shear stress and viscosity were taken from this initial acceleration leg of the curve rather than the deceleration or down speed leg. This was preferred since it more closely reflects the rheological behavior of a system which has been setting undisturbed for a given period of time rather than being exposed to a shear rate of almost 2800 sec^{-1}. Measurements of molecular weights distribution were made by passing solutions of the lignin derivatives through columns packed with Sephadex G-100 and measuring the UV absorbance at 290 nanometers with a Beckman Model DU-8 spectrophotometer.

RESULTS AND DISCUSSION

Molecular weight

The first property to be studied was that of molecular weight. In the present case, three experimental ammonium lignosulfonates coded 5125-61, 5347-72D, and 5347-61 were utilized. Each dispersant had a nominal degree of sulfonation of 2.7 moles of sulfonic acid groups per 1000 unit molecular weight of lignin.

Differences in molecular weight among the three dispersants are described on Fig. 1. The low elution volume end of the chromotograms indicate that the 5125-61 contains the highest relative amount of high molecular weight material of the series and 5347-61 contains the smallest amount. 5347-61 also contains the greatest relative percentage of lower molecular weight material.

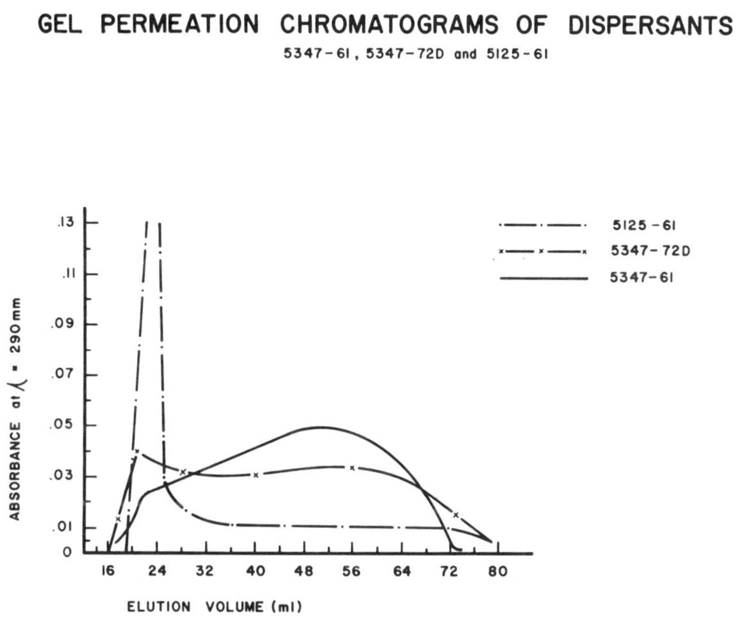

Fig. 1 Gel Permeation Chromatograms of Dispersants 5347-61, 5347-72D and 5125-61.

Each of the above lignin derivatives was then incorporated into an atrazine suspension containing 50% by weight technical atrazine and 3% dispersant. Viscosity measurements were made 72 hours after grinding.

Fig. 2 shows the viscosity curves of the three suspensions. Curve 1 representing dispersant 5125-61 indicates the highest viscosity at shear rates between 0.47 sec^{-1} and 1000 sec^{-1}. Curves 2 and 3 representing the two remaining dispersants show lower viscosities. The vertical displacement of these viscosity curves follow the same order as the relative amounts of high molecular weight material and reverse order as the relative amounts of low molecular weight lignin. This influence of the net molecular weight of one dispersant relative to another, all other factors being equal, is one of direct proportion, i.e., the dispersant showing the greatest average molecular weight (weight average) or as in the present case whose gel permeation chromatogram shows the greatest relative amount of high molecular weight material will as a rule produce the higher viscosities at low shear and presumably at rest.

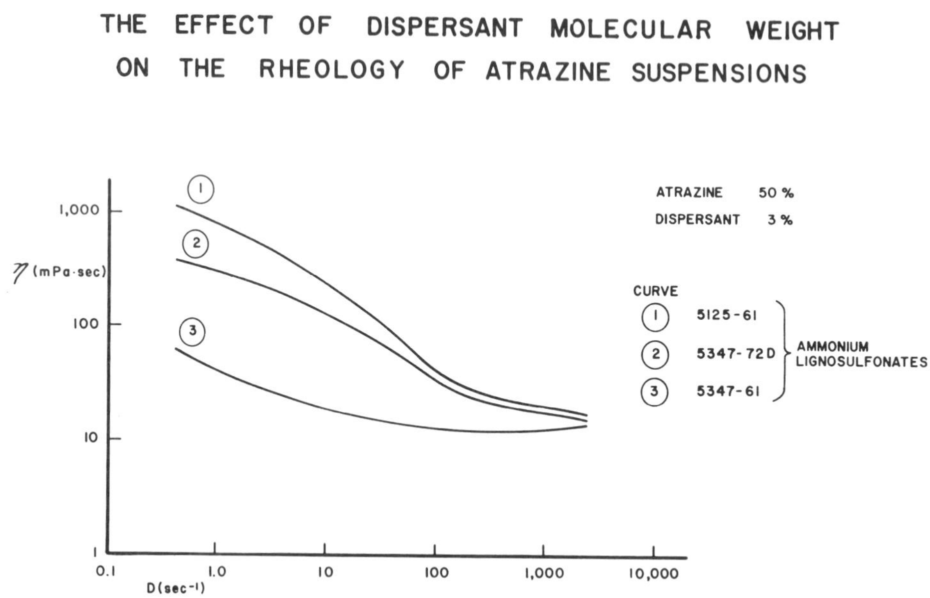

Fig. 2 The Effect of Dispersant Molecular Weight on the Rheology of Atrazine Suspensions.

Each of the formulations described on Fig. 1 was placed on a vibration platform oscillating at 60 cycles per second. After 2 1/2 hours, the dispersion based upon 5347-61 showed an appreciable amount of hard packed clay-like sediment. Analysis of the supernatant after vibration showed a suspended atrazine content of approximately 22.7%. The suspension using 5347-72D showed a slight trace of sediment with a 47.3% suspended atrazine level in the supernatant. Dispersant 5125-61 resulted in a suspension which showed no visible evidence of sediment and had a 49,8% suspended atrazine level relative to an initial pre-vibration level of 50.2%. The implications for transport stability are obvious. The apparent structure within the continuous phase of the suspension resulting from the higher molecular weight status of 5125-61 was more able to resist vibration induced destabilization of the formulation than that resulting from 5347-72D or 5347-61.

In another related experiment, a sample of 5347-72 was fractionated by ultrafiltration using a membrane with a molecular weight cutoff of approximately 30,000 Daltons.

A series of copper oxychloride suspension concentrates was prepared using 100% high molecular weight fraction, 100% low molecular weight fraction, blends containing 9/1, 5/5, and 1/9 mixtures, and the original unfractionated product. The viscosity curves describing the behavior of the resulting suspension are presented on Fig. 3.

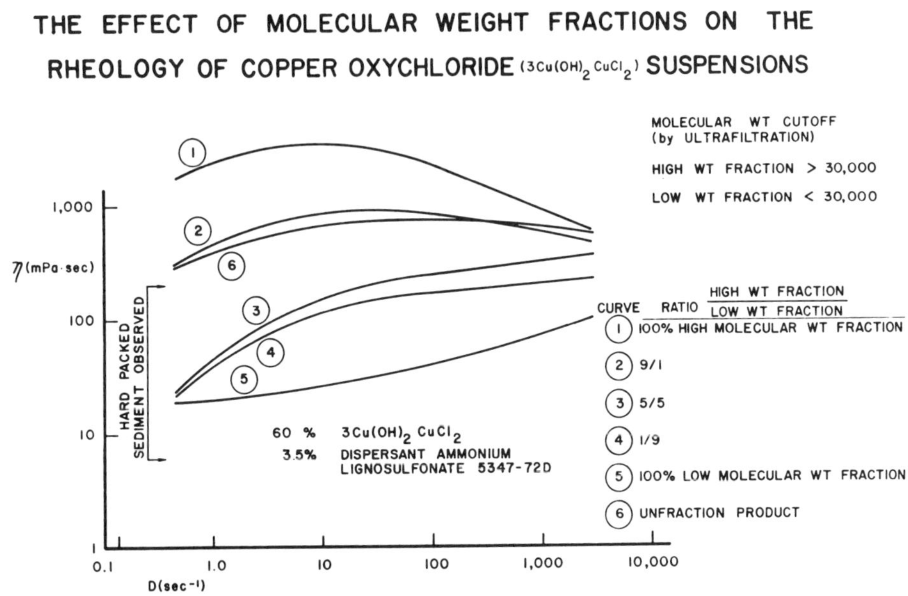

Fig. 3 The Effect of Molecular Weight Fractions on the Rheology of Copper Oxychloride Suspensions.

As these data indicate, the higher viscosities were observed with the suspension containing 100% high molecular weight dispersant (Curve 1) while the lowest were noted when the dispersant used was 100% low molecular weight fraction. As the ratio of high molecular weight to low molecular weight fractions was varied from 9/1 - 1/9, the viscosity curves were displaced downward.

The formulations described by Curves 3, 4, and 5 showed large amounts of hard packed sediment after being placed on a vibrating table for 2 - 2 1/2 hours while a trace of sediment was noted for systems represented by Curves 2 and 6. None was observed with the suspension based on 100% high molecular weight material.

Degree of Sulfonation and Charge to Mass Ratio

The degree of sulfonation of sulfonated kraft lignin based dispersants is, as stated before, an important synthetically accessible property which influences the rheology of concentrated suspensions. Historically, it has been the major reason for selecting amongst sulfonated kraft lignins, the REAXES and POLYFONS.

Four commercially available products were selected. They are INDULIN AT®, REAX 85A®, REAX 83A®, and REAX 88B® with nominal degrees of sulfonation of zero, 1.0, 2.0, and 3.8 moles of $-SO_3^-$ per 1000 grams of lignin, respectively.

Again, a series of atrazine suspensions was prepared. Viscosity curves are described on Fig. 4. As expected, the viscosity curves are positioned in ascending order of nominal degree of sulfonation. Caution must be exercised here, since some differences in molecular weight do exist among the four products. REAX 88B is considered lower in molecular weight than the parent starting material, REAX 85A has a higher molecular weight, and INDULIN AT and REAX 83A have molecular weights similar to that of the starting material. However, the impact of the degree of sulfonation is clearly in evidence when one compares Curve 1 and 3 which represent formulations based upon the unsulfonated INDULIN AT with REAX 83A (2 moles of sulfonation). What is significant, however, is the charge to mass ratio of each product. They can only be compared qualitatively here since accurate quantitative measurements of either the weight average or number average molecular weights of the products are not available. If H represents a higher molecular weight than the starting material, S - a similar value and L a lower value, one can represent the quantity as O/S, 1/H, 2/S, and 3.8/L for the three sulfonated lignins. As can be seen, variations in direct proportion to a property common to all four lignin products is evident. As the charge to mass ratio increases, the viscosity at all shear rates from 0.47 to 2800 sec^{-1} continues to decrease. Although viscosity differences at the lower shear rates (below 100 sec^{-1}) are considered the more important under most storage, transport, and on-farm use conditions encountered by suspension concentrates.

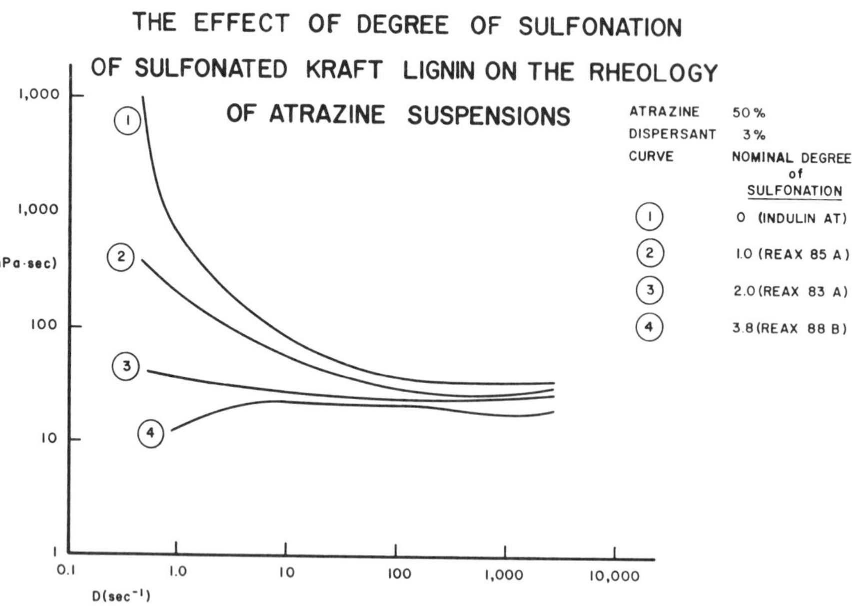

Fig. 4 The Effect of Degree of Sulfonation of Sulfonated Kraft Lignin on the Rheology of Atrazine Suspensions.

The last segment of the present work to be considered is the effect of the charge to mass ratio on short term storage stability of concentrated suspensions. Here a series of 45% diuron suspensions was prepared. The dispersant package consisted of 3% sulfonated kraft lignin derivatives and 1% condensed naphthalene sulfonate, Morwet® 425 (DeSoto Inc.). The lignins used were POLYFON O®, REAX 83A®, and REAX 88B® with nominal degrees of sulfonation of 1.2, 2.0, and 3.8 moles of SO_3 per 1000 grams of lignin. Relative to starting material, molecular weights can be considered lower (L), similar (S), and lower (L) respectively. Of the two materials with molecular weight lower than the starting material, REAX 88B and POLYFON O, REAX 88B is considered the lower molecular weight product of the two. Here, refer to its charge to mass ratio as 3.8/S' and that of POLYFON O as 1.2/S. Obviously, 3.8/S' is greater than 1.2/S since S'<S.

The suspensions studied here, were prepared as described and stored at 0°C, 25°C, and 50°C for 3 months prior to viscosity measurements. The data on Fig. 5 indicate that for this series of formulations, viscosities are lowest at a given shear rate with REAX 88B and higher with POLYFON O for 0° and 25°C storage conditions. What was unexpected was the situation with materials stored at 50°C. The smallest viscosity increases occurred with REAX 88B, the dispersant with the highest charge to mass ratio. This is contrary to the previously held belief that for conditions under which agricultural chemical formulations will be stored or used the more highly sulfonated lignin derivatives would desorb from the substrate, and suspension viscosities would increase to a greater extent than would be expected for the lignin with the lower degree of sulfonation (2).

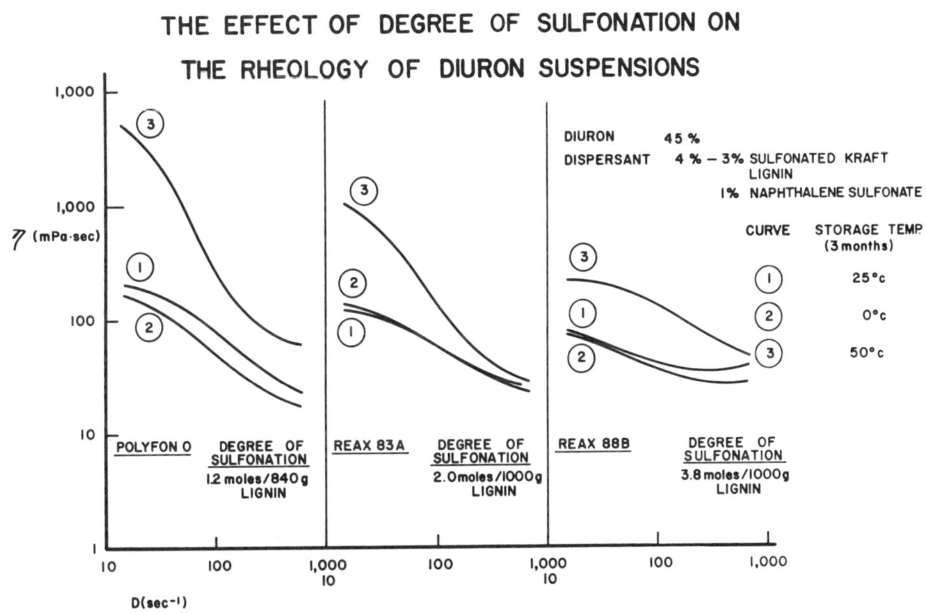

Fig. 5 The Effect of Degree of Sulfonation on the Rheology of Diuron Suspensions.

CONCLUSION

While the work described here is only the initial phase of a long term effort to quantify the impact of properties of sulfonated kraft lignin which can be controlled by industrial scale synthetic processes on formulations, some preliminary conclusions can be drawn.

All other things being equal with sulfonated kraft lignin the rheology of concentrated suspensions varies in direct proportion with molecular weight. That is, higher molecular weight products, as a rule, result in higher suspension viscosities at low shear (under 100 sec^{-1}) than those kraft lignins with low molecular weights. In like fashion, all other things being equal, those sulfonated kraft lignins with higher degrees of sulfonation result in lower viscosity suspensions than those with lower degrees of sulfonation. (3)

In the real world, however, "all other things" are not necessarily equal. Consequently, some particular composite property of a lignin based dispersant which encompasses more than one discrete physico-chemical property may have to be substituted. Presently, this composite property appears to be the charge to mass ratio. Based upon the work presented here, it does appear that as the charge to mass ratio of sulfonated kraft lignin based dispersants increases, the viscosity of suspensions utilizing these dispersants decreases provided factors such as dispersant loading, total dissolved and dispersed solids contents and particle size distributions are held similar.

This work is continuing and will be the subject of subsequent publications.

ACKNOWLEDGEMENT

The author wishes to acknowledge the contributions of the following individuals.

Peter Dilling	Westvaco, Charleston Research Center - synthesis of dispersants.
William H. Gillette James E. Davis Catherine Sweat	Westvaco, Polychemicals Department - Rheological measurements.
Gerald Prazak Howard H. Moorer	Westvaco, Charleston Research Center - Discussion of results.

REFERENCES

1) S.A. Herb and S. Ross, Colloids and Surfaces, 1, 57-77 (1980).

2) Th F. Tadros, in Advances in Pesticide Science, Geissbuhler et al. Editors, Pergamman Press, New York, pp. 820-826, 1979.

3) H.T. Delli Colli, The Use of Kraft Lignin and Its Derivatives in Agricultural Chemical Formulations. Westvaco Corporation Internal Report, April 1979.

Factors affecting the phytotoxicity of solvents used in pesticide formulations

M.R. Krenek,* K.G. Reed and D.N. King

Exxon Chemical Company
5200 Bayway, Baytown, Texas 77522 U.S.A.

Abstract - Phytotoxicity induced by the direct application of neat solvents to growing crops has been observed and rated based upon visual assessment of leaf damage in studies in Europe and North America. A relative phytotoxicity rank order list was generated for all solvents tested. Direct relationships of phytotoxicity to solvent composition, solvency power, and aromatics content of hydrocarbons were demonstrated, with the stronger solvents and more aromatic hydrocarbons causing the most phytotoxic damage. The North American work also suggested a relationship with polarity (Hansen parameter), but this was not borne out by the European results. Surface tension was seen to be a contributing factor. No general correlation between phytotoxicity and volatility was found. Molecular structure (linear versus branched) appeared to be a factor in one case.

INTRODUCTION

The phytotoxicity of a material is its capacity to damage plants, and the term is usually used to refer to the undesirable effects caused to the crops or plants being protected through the use of pesticides. Most materials exhibit some degree of phytotoxicity, and the solvents are no exception. There has been little reported in the open literature on solvent phytotoxicity, however, and there is a definite pesticide industry need for these data.

Two research projects are discussed here. One was conducted at Texas A&M University, College Station, Texas, while the other was carried out at Imperial College, London, U.K. The test methods used in each study were different, but the work was coordinated between the two, and the research goals were essentially the same.

EXPERIMENTAL PROCEDURES

North American Research In this work, 20 products were tested with four crops: corn, soybeans, wheat, and cotton. Neat, undiluted solvents were applied "over the top" in a one-time application to plants grown in a greenhouse. The solvents were applied using a spray chamber at a controlled nominal rate of 32.7 liters/hectare (3.5 gals/acre) with a fan spray nozzle. The crops were grown side-by-side in flats and were approximately 10-12 days post emergent at the time of application. Two replications were run for each treatment. After application, observations were made at varied intervals over a 56 hour period.

In field ec and ulv applications, actual solvent concentration in the final application is often only in the range of 1-4 L/ha, and little or no phytotoxic effects are noticeable. The much higher spray rate of 32.7 L/ha was chosen to "force" a range of phytotoxic effects that could be easily seen, rated and compared.

European Research Eight crops (bean, citrus, cotton, maize, rape, soybeans, tomato, and wheat) and 22 solvents were tested. The solvents were applied neat using a hand-held spinning disc applicator at rates of 10, 20, 40, and 80 L/ha. As with the U.S. work, these unrealistically high rates were used to force a broad range of phytotoxic responses. At the time of application, the crops were at the three to five leaf stage, one to two months post emergent. Table 1 is a list and description of the solvents tested in either the North American or European programs.

TABLE 1. Descriptions and properties of solvents and oils tested

	Product Name/ General Description	Wt % Aromatics	Boiling Rng, deg C	Polarity δ_p	Pendamethalin Solubility[a]
ISOPAR® L	Iso Paraffinic Solvent	0.06	188-207	0.0	14
ISOPAR M	Iso Paraffinic Solvent	0.2	206-253	0.0	13
ISOPAR V	Iso Paraffinic Solvent	0.65	255-295	0.0	11
NORPAR® 12	Normal Paraffinic Solvent	0.6	188-219	0.0	13

NORPAR 13	Normal Paraffinic Solvent	0.2	226-243	0.0	12
NORPAR 15	Normal Paraffinic Solvent	0.01	252-277	0.0	10
EXXSOL® D 60	Narrow Cut Aliphatic Solvent		186-217	0.0	-
EXXSOL D 80	Narrow Cut Aliphatic Solvent	0.8	202-232	0.0	17
EXXSOL D 100	Narrow Cut Aliphatic Solvent	0.9	233-252	0.0	12
Hydrog. SOLVESSO® 150	Naphthenic Solvent		167-196	-	-
Decalin	Decahydronaphthalene		186-194	-	-
LOPS®	Deodorized Kerosene	3.0	197-252	-	16
VARSOL® 18	Mineral Spirits	7.7	159-203	-	28
VARSOL 1	Mineral Spirits	16.0	159-204	0.1	33
VARSOL 60	Mineral Spirits	20	187-216	-	-
HAN® 857	Highly Aromatic Solvent	86	182-277	0.4	55
Xylene		99.9	138-140	0.5	54
Aromatic/SOLVESSO 100	Xylene Range Aromatic Solvent	98	155-173	0.5	49
Aromatic/SOLVESSO 150	Alkyl Benzene Aromatic Solvent	98	183-210	0.3	47
Aromatic/SOLVESSO 200	Alkyl Naphthalene Aromatic Solvent	98	226-279	-	53
THN	Tetrahydranaphthalene		205-209	-	-
ACTREL® 400	Synthetic Heavy Aromatic		332-355	-	-
Isophorone	Cyclic Ketone	-	210-218	4.0	55
EXXATE® 600	Oxo-Hexyl Acetate	-	162-176	3.4	55
EXXATE 700	Oxo-Heptyl Acetate	-	176-200	3.3	55
N-C_7 Acetate	Straight Chain Acetate	-	192	-	-
ORCHEX® 796	70 sec. premium spray oil	b	337-415	-	-
N-Octanol	Straight Chain Alcohol	-	195	-	-
Iso-Octanol	Branched Chain Alcohol	-	184	-	-
Soybean Oil	Straight Chain Vegetable Oil	-	-	-	-
DINP	Di-isononyl phthalate (ester)	-	-	-	-

[a] Grams pendimethalin per 100 ml solvents @ 25 deg C.
[b] Minimum unsulfonated residue is 92.0 volume percent.
ISOPAR, NORPAR, EXXSOL, LOPS, VARSOL, HAN, SOLVESSO, EXXATE, and ACTREL are registered trademarks of Exxon Corporation.

RESULTS

<u>North American Research</u> None of the plants died or failed to resume what appeared to be normal growth at the end of the 56 hour observation period. Ratings were made using a scale of 0 to 100 percent leaf damage based upon visual assessment only, with the untreated controls rated 0. These numerical ratings were then converted to a simplified, sliding "A to D" scale independently calculated for each crop, where "A" represented the minimum and "D" the maximum phytotoxic effects observed for that crop. These ratings are in Table 2.

TABLE 2. Relative phytotoxicity ratings[a]

Solvent/Oil	Corn	Wheat	Cotton	Soybeans
ISOPAR/NORPAR/EXXSOL/LOPS (all grades)	A	A	A	A
VARSOL 18	B	A	A	B
VARSOL 1	B	B	C	B
HAN 857	C	C	B	C
Xylene	B	C	C	B
Aromatic 100	C	D	C	C
Aromatic 150	C	D	C	C
Aromatic 200	D	D	C	C
Isophorone	B	B	B	D
EXXATE 600	C	D	C	C
EXXATE 700	C	D	D	C

[a]Key: A-No Observed Effect B-Slight Effect C-Moderate Effect D-Maximum Effect

<u>European Research</u> Three assessors were used, and they visually estimated damage at various time intervals up to two months after application. Effects were reported as <u>less than</u> or <u>more than</u> 10% damage on the assumption that any visual damage greater than 10% would be unacceptable to the farmer or consumer. Where assessors failed to agree whether damage was above or below 10%, a "borderline" rating was assigned.

Based upon the detailed results, two phytotoxicity indices were calculated for each solvent, and they are reported in Table 3. In Method 1, the solvents were rank ordered and numbered 1, 2, 3, etc. for phytotoxic effect on each crop (from least to most). The rankings were totalled for each solvent and crop, averaged, and multiplied by 100 to come up with the final index. In Method 2, rating points were assigned to each result for a given solvent/crop combination: a positive number where no phytotoxic effects were observed ("1" for 10 L/ha, "2" for 20 L/ha, "4" for 40 L/ha and "8" for 80 L/ha) and a negative number for an unacceptable response at a given rate ("-1" for 80 L/ha, "-2" at 40 L/ha and so forth).

"Borderline" ratings for any crop and all data for wheat (it did not grow well) were not taken into account in these calculations, and the accuracy of the indices is therefore diminished. There were also "gaps" in the data where crops were not tested at every rate. In total, however, the indices are still a good relative measure of phytotoxicity.

TABLE 3. European relative phytotoxicity indices

Product[a]	Index 1[b]	Index 2[c]	Product[a]	Index 1[b]	Index 2[c]
NORPAR 12	100	650	Isophorone	450	67
ISOPAR L	112	600	SOLVESSO 200	488	- 88
EXXSOL D 100	133	667	Hydrg. SOLVESSO 150	525	-162
Soybean Oil	167	583	THN	550	-175
EXXSOL D 60	212	412	Iso-octanol[d]	550	-212
DINP	250	400	SOLVESSO 100	575	-300
NORPAR 15	283	483	ACTREL 400	600	-250
VARSOL 60	375	88	Iso-octanol[d]	638	-450
EXXATE 700	375	62	N-C_7 Acetate	638	-325
EXXSOL D 80	383	300	SOLVESSO 150	650	-375
DHN	400	12	N-Octanol	688	-633

a Listed from least phytotoxic to most phytotoxic according to Index 1 values.
b Smallest values correspond to the lowest phytotoxicity.
c Largest positive values correspond to the lowest phytotoxicity.
d Two different samples from two different sources were tested.

CONCLUSIONS

Relative Solvents Phytotoxicity Rank Order Since the North American and European studies used different test methods, it is difficult to compile a meaningful, calculated consolidated index. Instead, the authors have subjectively assessed the combined results and generated a relative phytotoxicity rank order list of all the solvent types tested. Also, a Relative Scale of 1-10 is included to help demonstrate the degree of difference in phytotoxicity among the product groups. This rank order list is in Table 4.

TABLE 4. Relative solvents phytotoxicity rank order

Products	Relative Scale	
ISOPAR Solvents	1.0 - 2.0	Lowest
NORPAR Solvents	1.0 - 2.0	
DINP (Phthalates)	2.0	
EXXSOL Solvents	1.0 - 3.0	
Mineral Spirits	5.0 - 6.0	
Naphthenic Solvents	5.0 - 6.0	
Isophorone	6.0	
EXXATE Solvents	6.0 - 8.0	
THN (Naphthalenes)	6.0 - 8.0	
Iso-alcohols	8.0 - 9.0	
N-acetates	8.0 - 9.0	
Aromatics	9.0	
N-alcohols	10.0	Highest

Composition Results from both studies clearly showed that relative solvent phytotoxicity is directly related to solvent composition, and this fact formed the basis for the rank order list in Table 4. Several other conclusions were also apparent, however. For hydrocarbon solvents, there was a definite direct relationship between aromatics content and phytotoxicity. The very low, almost non aromatic paraffinic solvents exhibited a very low relative phytotoxicity and compared favorably with soybean oil and ORCHEX 796, oils considered to be "phytobland". The aromatics, on the other hand, were among the most phytotoxic solvents tested. However, none of the aromatics were significantly more phytotoxic than the "xylene range aromatic solvent" (such as SOLVESSO 100 and Aromatic 100) that is considered an industry standard for ec's. Products that contained aromatics in amounts less than 95-100% exhibited phytotoxicity in proportion to their aromatic content, e.g., an "intermediate" phytotoxicity for VARSOL 1 with about 16% aromatics. The purely naphthenic solvents exhibited mid-range phytotoxicity between the paraffins and the aromatics.

Among the oxygenated solvents, the alcohols appeared to be the most phytotoxic while the phthalates were the least. Isophorone fell in the middle. The EXXATE solvents, however, were harder to classify. In the North American work, both EXXATE 600 and EXXATE 700 exhibited phytotoxicity essentially equal to aromatics. In Europe, though, the EXXATE 700 caused noticeably less response than aromatics. As a compromise, the authors have elected to rate the EXXATE solvents, on the whole, as less phytotoxic than the aromatics.

Solvency Power The results show a definite relationship between solvency power and phytotoxicity: in general, the stronger the solvent, the greater the phytotoxicity. This relationship appears to be linear when phytotoxicity is plotted versus the solubility data (Table 1) calculated for pendamethalin by Exxon (plot not shown).

Polarity The North American results demonstrated an asymptotic relationship between polarity (when expressed as the Hansen parameter for polarity) and phytotoxicity (plot not shown). The European work, however, with its broader range of solvents, did not appear to support this relationship. As a result, polarity would not seem to be a highly accurate predictor of phytotoxicity. Interestingly, though, even the European data showed a very rough trend line of polarity and phytotoxicity.

Surface Tension The North American results suggested that surface tension differences can impact the phytotoxicity of products of similar solvency power: lower surface tension directionally increases the effect. In the pendamethalin solubility tests, Isophorone and the two EXXATE solvents demonstrated the same ability to dissolve the active ingredient, 55 grams per 100 ml of solvent. Yet, the Isophorone caused less average damage than the two acetates. One reasonable possibility is that the much lower surface tensions of the EXXATE solvents led to better leaf wetting and increased the area of damage versus the Isophorone. The Isophorone was seen to "bead up" on the leaves after spraying and not spread out across the leaf. Another possible explanation is that the the lower surface tension lead to greater stomatal entry by the EXXATE solvents into the leaves and thus caused more damage. The European work is at odds with this conclusion because the EXXATE 700 tested on average less phytotoxic than the Isophorone.

Volatility The data from both tests did not show a strong relationship between volatility and phytotoxicity. The results for a number of the different product groups (the ISOPAR, NORPAR, EXXSOL and Aromatic or SOLVESSO solvents) show similar phytotoxic effects for each grade within the class despite often large differences in volatility. There were cases where the heaviest, least volatile grade had the highest effect, but this was more the exception than the rule. Xylene was also an exception since in the North American work it was clearly less phytotoxic than the other aromatic grades. This suggests there may be volatility "thresholds" or break points (rather than close step-by-step correlations) that come into play across a broad range of volatilities or boiling ranges for solvents.

Molecular Structure Differences in molecular structure and their relationship to phytotoxicity have been discussed to some degree under "Composition". Some additional observations, however, can be made based upon the European work. NORPAR 12/ISOPAR L, n-heptyl acetate/EXXATE 700, and n-octanol/iso-octanol were all product pairs consisting of a linear chain and a branched chain solvent, respectively, of essentially equal carbon number. The results show little difference in phytotoxic response between the alcohol and the NORPAR/ISOPAR pairs, but there was a noticeably lower response for the mixed, branched EXXATE 700 versus the linear n-heptyl acetate. The North American work included few such product pairs and offered little additional insight.

Note: Oils are routinely added to many ulv sprays as a means of adjusting the formulation viscosity or volatility. In such cases, applicators often choose vegetable oils over petroleum oils or solvents because the "conventional wisdom" is that vegetable oil is more "non-phytotoxic" than the petroleum oils. This is often attributed to the linear hydrocarbon chain make-up of the vegetable oils. The results here do not support this, and there is no evidence to support the belief that the linear chains of the vegetable oils make them more non-phytotoxic than a corresponding petroleum oil or solvent.

SUMMARY

This work has shown that the phytotoxic response of a given crop can vary greatly with the type of solvent applied when relatively high application rates are used. Although the solvents were applied neat in this research, it is believed the relative rank order of the solvents' phytotoxicity would be carried through to pesticide formulations. That is, in general, the "less" phytotoxic solvents applied neat would also be less phytotoxic in a formulated pesticide versus the "more" phytotoxic solvents. The other components in the final formulation could alter this outcome, however, and the authors do not believe phytotoxic effects are uniformly additive. Instead, they may be synergistic, antagonistic, or additive depending upon the formulation. Finally, all of the solvents tested here are generally suitable for use in ec's from a phytotoxicity standpoint since none of the products were observed to be significantly more phytotoxic than the xylene range aromatic solvent. There are, however, a number of solvent types (the ISOPAR, NORPAR, EXXSOL solvents and phthalates) that cause much lower phytotoxic effects and thus the pesticide formulator also has a choice of solvents for use in very low phytotoxic applications (e.g., ulv).

Pesticide residue methodology

Trends in pesticide residue methodology

Helmut Frehse

Bayer AG, Pflanzenschutz-Forschung, Inst. für Rückstandsanalytik, Monheim
D-5090 Leverkusen, Fed. Rep. of Germany

Abstract - Methodology, for the purpose of this paper, is interpreted also as approach or performance. Examples are discussed showing how residue analysis, in an era of diminishing resources, could be made more economical, through simplifications in both approach and method. It is essential to consider establishing a new relation between demand and rational analytical possibilities in order to satisfy demand. Account must be taken, in this context, of the fact that residue analysts are overtaxed when called upon to satisfy constantly changing or very greatly differing requirements, which is accompanied also by loss of the comparability of results. Besides these aspects, the paper highlights a number of analytical techniques with regard to their future potential and importance, from the viewpoint of the author.

Analysis for organic pesticide residues had its inception about 40 years ago. I have been allotted only 2 ppm of that period for the presentation of my opening paper. Nonetheless, I shall try to draw some conclusions and to outline a few perspectives from my own personal viewpoint. In doing so, I shall confine my comments to aspects of quantitative analysis. Moreover, I propose to address more general problems that have repercussions on quantitation and thereby to interpret methodology also in the broader sense of approaches and needs. More specialized aspects and technical details will be discussed in the other contributed papers at this Symposium and in the Poster/Workshop Sessions of this Main Topic.

Information we could well afford to do without (or to replace)

In our profession, there are members who would appear to be blind to the signs of the time. There are those who are still preoccupied with the development of methods for very old pesticides. To demonstrate this, I have chosen three compounds at random (Fig. 1).

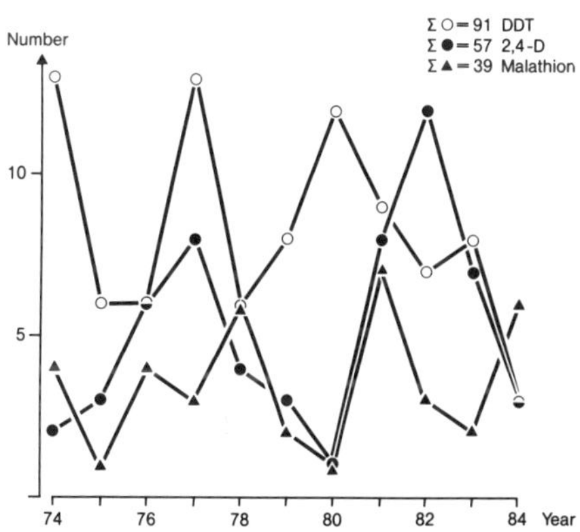

Fig. 1: Number of papers on residue analytical methods for three "old" pesticides. Source: Analytical Abstracts 1974-1984.

What do we want with 90 "new" methods for DDT? By simply using proven methods, those analysts could make themselves useful in other ways, for example by providing us with more information

- on the intercomparability of residues levels measured in field trials for different crops (later only crop A need be analyzed when it is known that higher residue levels are not to be expected in crop B) or
- on losses of residues incurred during household-scale and industrial processing (this would help to relieve consumers of many fears and reduce to absurdity the principle still employed by many registration authorities of predicting actual dietary exposure of consumers to pesticide residues by using the concept of potential or theoretical daily intake that is calculated from the MRLs (maximum residue limits) for the pesticides).

Other residue analysts of the non-aligned category devote their attention to an avenue of research I would call Pure Compound Methodology, for example research into the applicability of separation or determinative procedures to pesticides in the absence of a matrix. This line of research can well assume the nature of a quiz, it being left to others to find out whether the results obtained by those colleagues have any practical significance. For residue analysts to preoccupy themselves with work directed towards expanding or sophisticating instrumentation is quite acceptable provided the work has prospects, in an era of diminishing resources, of speeding up or simplifying residue analysis, in short to make it more economical and yet more exact. But such work obviously will be less acceptable if its results do not serve this purpose and/or cannot be utilized by other analysts.

Does it make sense or is it worth it?

However, residue analysts often are compelled also to make exceptional efforts. There is a growing tendency among registration authorities and/or legislators to widen the scale of their requirements still more on the grounds that this will enhance user, consumer and environmental safety. Requirements like the conduct of degradation studies in bean straw come to mind. But I would prefer to draw attention to two examples of a yet more aggravating nature, namely

- the trend towards inclusion of highly polar or even conjugated metabolites in the definition of the residue of a certain pesticide or, in Europe,
- the EC Directive relating to Drinking Water (ref. 1), which prescribes an MRL of 0.1 ppb for every single pesticide, a level of 0.5 ppb as the total residue for simultaneously occurring pesticides, and a margin of error not exceeding ± 10 %. Needless to say, in the foreseeable future compliance with this requirement surely will be feasible for only a small percentage of the pesticides in usage.

Regarding the first example, one must ask how such compounds can at all be incorporated into a multiresidue method suitable for regulatory or monitoring purposes. The second example concerning the EC directive means that for the majority of the pesticides, the residue analysts are confronted with the problem of firstly having to create the prerequisites necessary for enforcing a legislative measure. This is an astonishing turn but it is symptomatic of the rift between politics, public opinion and science, and in which residue analysis has also become caught. Most distressing of all in this respect is that the requirements of registration authorities and/or legislators are often so poorly harmonized that the data laboriously generated by the analysts cannot be used internationally.

Merger a long way off

There are two divisions in residue analysis, industrial and public. Prospects of them merging technically in the foreseeable future would appear extremely remote. One is responsible for supplying data for the registration of pesticides, and the other for the enforcement of MRLs and for monitoring residues in foods, in feeds and in the environment. Both, however, must adapt themselves to the emergence of new pesticides many of which now are applied at dose levels several orders of magnitude lower than those needed for older pesticides. And also common to both is the compelling need for rationalization and for the adoption of more cost-effective approaches. Both divisions would therefore be well served by establishing a system of more intensive communication between their analysts because after all, the methods and data generated by the industrial analysts form the cornerstone of the legislation which the public analysts must enforce with the aid of their methods. I consider it most important that metabolism researchers and residue chemists also cooperate very closely.

The pesticide manufacturer, the registrant, must submit to the registration authority residue data from supervised field trials. These data constitute the basis for the establishment of MRLs and safety intervals (however, it would also be of interest to learn from the legislators what criteria they apply for establishing MRLs on the basis of a set of data supplied by the analysts). Guidelines have been issued for the conduct of such supervised field trials and for reporting the results (ref. 2). The methods of analysis that have been developed for them mostly are matched optimally to one particular pesticide and its significant metabolites, and often they must have a limit of determination extending down to 0.01 mg/kg or even less. These are the compound-specific or single methods. They also can be used for the confirmation of results obtained from regulatory analyses as well as by public analysts in instances where the respective residue cannot be determined by a multi-method.

Multi-residue procedures (MRPs) are the main tools of the public analysts (ref. 3). Their targets are predominantly parent compounds. However, there is no MRP which alone embraces the wide range of compound/matrix combinations that ought to be analyzed for regulatory or monitoring purposes. What is more, there are many pesticides and metabolites that cannot be determined at all by MRPs. This is a well-known problem.

Most of these methods consist of only a few recurrent steps such as extraction with a limited number of solvents, liquid-liquid partitioning, adsorbent column chromatography or gel permeation chromatography, and determination mostly by GLC, HPLC or TLC. An experienced analyst can modify these procedures by interchanging or modifying the individual steps in order to obtain the best result for the given purpose. Unfortunately too little use is made of this possibility.

A comparison of the different procedures reveals, however, that those used in analytical practice are not necessarily restricted to the pesticides specified by the authors but have the potential to cover many more compounds. Hence, it would be to the benefit of analysts if extensions of standardized procedures were to be published regularly. Information obtained on compounds that are not extractable, not recoverable from purification steps, or not detectable is equally important. A detailed list of those compounds that were sought should be included in every survey based on MRPs. More efforts than those made in the past should be undertaken to develop methods for the analysis of residues of such chemically related compounds that have a certain moiety in common. An interesting line of research would also be to develop broadly applicable procedures for quantitative hydrolysis of conjugates potentially present in a matrix. However, this would necessitate reaching agreement as to which compound types are to be regarded as significant residues.

Time or money

The laboratories of public analysts could exploit their resources more economically if they were to concentrate on residues that are of real toxicological significance and at the same time on such foods that are constantly consumed in significant quantities. This subject will be discussed in Workshop 5 E. The problem is illustrated schematically in Fig. 2. The shaded section in the top diagram symbolizes those pesticide/commodity combinations that should command the special interest of legislators and analysts. By establishing the proportion that can be embraced by proven multi-residue methods (bottom diagram), the easier it will be for analysts to concentrate on the development of suitable methods for the remainder and thus contribute to an even more effective protection of consumers. Such a system lends itself to application both on a broad international scale and within defined geographical or cultural or ethnical regions.

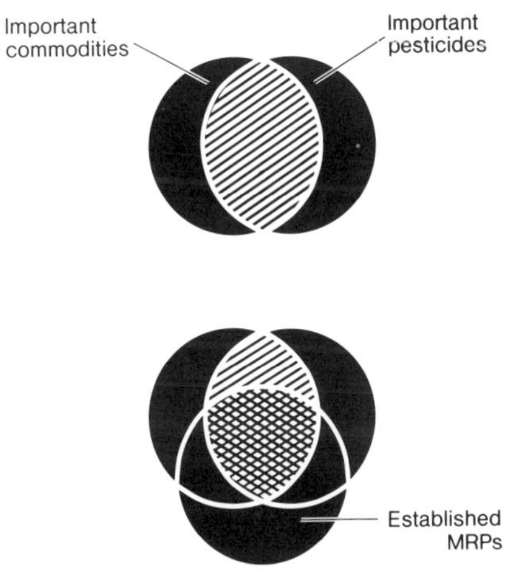

Fig. 2: See body of text for explanation.

More profound consideration also should be given to an approach in which an indicator compound is selected to represent the residue of a certain pesticide and its metabolites. By this approach, the indicator compound alone is analyzed, and permits conclusions to be drawn on the residue level of the respective pesticide. For the selection of a single compound to represent the residue, the following criteria have been suggested (ref. 4):

- its concentration should bear a known relationship to the concentration of the toxicologically significant residue;

- it should be a) sufficiently stable to allow reproducible analysis, b) available as an analytical standard, and c) recoverable in MRPs.

The selection of indicator compounds could be made by experts on the basis of international agreement; for example, in line with the deserving efforts of FAO and WHO to bring about international harmonization of MRLs through the agency of the Codex Alimentarius Committee on Pesticide Residues (CCPR), an undertaking which because of national egoism and the introduction of their own measures by supranational blocs, is making slow headway. Notwithstanding, the CCPR meanwhile has established MRLs for about 150 compounds and more than 2500 pesticide/commodity combinations, based on twenty years of scientific evaluation by the FAO/WHO Joint Meeting on Pesticide Residues (JMPR). All these efforts relate to such pesticides whose residues may be contained in foods moving in international trade. In this context, I would draw attention also to the recommendations of the CCPR regarding proven analytical methods (ref. 5a), as well as with respect to sampling (ref. 5b), the portions of commodities to which MRLs apply and which should be analyzed (ref. 5c), and the metabolites to be included in the analyses (ref. 5d), because these are factors that are handled differently, consequently making it difficult (and sometimes impossible) to compare results obtained worldwide.

Analysts engaged in the monitoring of foods can or should confine their work to finding out whether established MRLs have been exceeded. They must analyze the foods for a very large number of pesticides at largely differing levels accompanied by correspondingly great differences in terms of time and cost incurment. Herein lies scope for simplification of methods, which evidently has not yet been systematically explored. Twenty years ago, the use of a three-step screening system was advocated by Francis A. Gunther who passed away last year. I wish to take this opportunity of paying special tribute to him in appreciation of his outstanding pioneering achievements in the field of pesticide residue analysis and in expression of my own personal gratitude to him.

The steps of the system he described, and which can be used in combinations, are 1) constituent screening (detection of sought compounds; with previously established limits of minimum detectability); 2) segregative screening (separating above-tolerance from below-tolerance samples); 3) quantitative screening (determination of sought compounds) (ref. 6). In the 1984 Survey of Pesticide Residues in Food, published by the British Association of Public Analysts (ref. 7), the laboratories were explicitly requested to work to reporting limits set at specific levels for different classes of pesticides.

There is considerable merit in using specified levels of presence as "reporting limits" which are readily achieved by all laboratories, provided the set level is below the MRL by a given factor. In this context, a proposal of the GIFAP Residue Committee (ref. 8) also deserves mention. The proposal is that it should be acceptable to define a minimal concentration to be determined (MCD) in a sample. The MCD is specified as a fraction of the (existing or estimated or proposed) MRL (Fig. 3). Analogous concepts could be applied also to analyses of environmental samples. Such approaches allow safety to be assessed in a cost-effective manner and are useful in enabling the analyst to devise suitable methods.

MRL (mg/kg)	MCD (mg/kg)
≥ 5	0.5
$0.5 \rightarrow 5$	$0.1 \rightarrow 0.5$
$0.05 \rightarrow 0.5$	$0.02 \rightarrow 0.1$
< 0.05	MRL/2

Fig. 3: Proposed minimal concentrations to be determined in relation to MRLs.

Decision-making

A problem that has not yet been solved satisfactorily is that concerning what criteria an analyst should apply in reaching decisions on whether MRLs have been exceeded or not. For this purpose, use could be made of the latitude concept by which proof of an infringement is furnished only when the whole of the confidence interval of the analytical value is above the MRL. It is simpler to follow an approach recommended by the CCPR (ref. 9), by which the result of the analysis is rounded off to one significant figure and this rounded-off value is compared with the MRL. A precondition for the use of this approach is that the MRL also has only one significant figure.

Speaking of significant figures recalls to mind that residue analysts still have not agreed on a definition of limits of detection and/or determination. A final statement by an analyst that a sample does not contain any detectable or quantitatively determinable residues is therefore a reflection not only of his skill but also of the construction he places on these limits. Once again we find ourselves confronted with effusions of Pure Compound Methodology whenever analysts characterize the smallest measurable signal obtained from standard solutions as the limit of determination or detection. The definition of the limit of determination in residue analysis embraces three criteria: it cannot be lower than the limit of detection; the recovery may not be less than 70 to 80 %; the coefficient of variation should not exceed 20 % (ref. 10).

Where do we stand, and where do we go from here?

In their 1985 review on the analysis of pesticides (ref. 11), Sherma und Zweig reported, inter alia, that analytical techniques which appear to have made significant advances are capillary gas chromatography (GC); GC coupled with mass spectrometry for the identification of pesticide residues, metabolites, and contaminants; high-performance liquid chromatography (HPLC) with reversed and other bonded phases and electrochemical detectors; high-performance thin-layer chromatography (HPTLC) and quantitation by automated densitometry. These authors report further that the "sensitivity" of most methods has not gone beyond the picogram level, although capillary GC-electron capture is capable of detecting femtogram quantities. In their view, the rule seems to persist that a clean sample gives the best results!

It must be the aim of extraction to achieve near complete retention of matrix components, at the same time enriching the residues by several powers of ten. This is not achievable with liquid/liquid extraction, and for this reason it is being increasingly replaced by liquid/solid extractions that also can be directly coupled to HPLC.

Great hopes are placed, and in my view quite justifiably, on miniaturization. Analytical procedures can be miniaturized already in the very first step, namely extraction. However, we still do not have enough supporting data to show by how much the sample size can be reduced without destroying the representative nature of the field sample. It would nonetheless appear that the weight of an analytical sample can be reduced to between 2 and 5 g without appreciable loss of accuracy provided that the analytical sample is taken from a properly homogenized field sample and that the residue will not be destroyed by such homogenization. Miniaturization will often be concentrated on aliquots from a conventionally produced extract. But even then astonishing savings in terms of equipment, reagent use, time und laboratory space can be realized through reduction in the size of cleanup procedures, for example by using mini-columns, and by scaling down the determinative step (ref. 12). Miniaturized adsorption columns, however, will not be very promising for efficient removal of co-extractives, particularly in MRPs, because many of the residues in question and the co-extractives to be removed do not differ greatly in their polarity. For true miniaturization of pesticide residue analytical methods special equipment is required and tools like those used in the field of organic microchemistry have proved suitable (ref. 12).

Analytical precision in miniaturized residue methods is improved by the use of an internal standard (IS) (ref. 12). The use of internal standards, however, has not been commonly accepted in residue analysis. The main argument is that co-extractives or derivatizations could change the size of the standard peak and cause false results. However, by careful choice of a suitable IS, or by using two standards, this can be avoided. Alternatively, isotope-labelled forms of the sought compounds can be used as standards. The advantages to be gained from the use of IS methods are substantial. They are independent of absolute recovery, i.e. lower efficiencies for individual steps can be tolerated; they are not so critically dependent upon the skills of the residue analyst; each sample acts as its own recovery check thus precluding the need to run further recovery samples; in multiresidue analyses, pairs of ISs can be chosen which behave in such way that they bracket a certain class of pesticides; and most particularly there will be no need to measure the volume of the final solution that can be concentrated to as little as 1 or 2 drops, although it may prove to be a handicap that as yet not all autosampler systems are designed to handle such minimal volumes.

In the area of chromatographic methodology there are no epoch-making innovations in sight, although many promising developments are reported. Continuation of the great advances achieved in liquid chromatography is hampered by the non-availability of atom-specific detectors. The use of electrochemical detectors (with their problem of likely electrode contamination) and photoconductivity detectors will remain confined to a few small groups of pesticides. A new trifunctional detector simultaneously monitors ultraviolet absorbance, fluorescence, and conductivity in a single flow cell (ref. 13). It remains to be seen what impact this detector and others (e.g. flame ionization, electron capture) will have. However, the performance of HPLC is so convincing that there is no longer any question mark against its usefulness in residue analysis particularly as the development of new column materials for this technique is continuing to advance. These materials include reversed phases based on stabilized polymers for the separation of polar compounds, an example being Macrophase MP-1 developed by Interaction Chemicals Inc. (ref. 14). Further, quantitations are now also possible in combined LC/MS.

Most HPLC is currently conducted with conventional large-bore columns of 4 to 5 mm internal diameter, but there is a continuing interest in the utilization of columns with reduced internal diameters (ref. 15). Three different types of column have emerged, namely narrow-bore or small-bore packed columns, packed microcapillaries, and open-tubular capillary columns. All of them are often referred to as "microbore" columns. At present, only narrow-bore packed columns appear to be suitable for routine work, and these are essentially conventional HPLC columns with reduced internal diameters.

More experience from routine analysis is required by which the various claims and counter claims concerning the advantages of narrow-bore columns can be tested. Conventional HPLC equipment is not generally suitable for operating them and often requires extensive modifications by the analyst. On the other hand, large savings on consumables are certainly significant. It is also claimed that the reduction in peak volumes facilitates lower detection limits (increase in "mass sensitivity"), that such columns can be coupled together in order to solve difficult separation problems, that they provide an effective method of achieving rapid separations, and that the reduced diameters could make temperature programming feasible. Narrow-bore columns have been shown at least to be advantageous with several types of coupling techniques.

Although GC can be considered a mature technique, it continues to evolve to provide higher levels of performance. Recent improvements in column and detector technology have contributed to this evolution (ref. 16). Wide-bore capillary columns with internal diameters ranging from 0.32 to 1 mm although mostly 0.53 mm have become the systems of choice as replacements for the packed column (ref. 17, 18). Speed, reproducibility, inertness of surface, chemical and thermal stability, lower bleed, and ease of instrument interchangeability - especially for fused silica columns with chemically bonded stationary phases - are reported to be their most prominent advantages. The ability to manipulate the film thickness (up to 5 µm) reproducibly has led to columns that can be tailored for analysis of very volatile or very high-boiling mixtures and for various column capacities. The combination of a conventional packed column with a wide-bore capillary column and N/P/S detection also has promising potential (ref. 19).

Another new technique I wish to mention is supercritical fluid chromatography (SFC) in packed and capillary columns (ref. 20, 21). Although supercritical fluid chromatography itself is not new, recently there has been a resurgence of interest in it, and it would seem well worthwhile for residue analysts to follow this trend. SFC does not suffer from volatility limitations as GC does. It can analyze thermally labile and underivatized compounds and solutes of much higher molecular weight than can be analyzed by GC. Capillary column SFC is compatible with many types of GC and HPLC detectors, and there are indications that SFC has resolving powers of five times that of HPLC, and analysis times five to ten times faster than HPLC. Mention should also be made of the possibility for using supercritical fluids for the extraction of bound residues from soil and plants (ref. 22).

TLC has mostly replaced paper chromatography, retaining its status as a valid and simple method for qualitative and even quantitative analysis (ref. 23, 24) although the difficulties in interpreting "spots" should never be underestimated! Many papers on this subject originate from researchers in developing countries. This is plausible considering the lack not only of sophisticated instrumentation but also of service, spares and consumables in these regions. It also underlines how important it is to continue expanding this technique and never to neglect it.

A highly promising new technique of TLC is Automatic Multiple Development (AMD; ref. 25, 26), an improvement of Perry's Programmed Multiple Development technique devised in 1973 (see ref. 25). In the AMD method, the TLC plate is subjected to a multiple chromatographic development. A chromatogram generally involves 10 to 25 individual runs, each run being longer than the previous one by a constant increment of about 3 to 5 mm. Between the individual runs, the mobile phase is removed from the chamber, and the plate is then dried by applying vacuum to the chamber, followed by recharging of the chamber with fresh mobile phase. Normally, eluants of gradually decreasing polarity are used. This procedure results in a focusing effect which yields extremely narrow compound bands on the chromatogram. In a 25-step programme, the total time taken to complete the chromatographic process is about 4 hours.

AMD permits gradient elution on HPTLC silica gel, covering a wide range of polarity. The fully automated system is commercially available (CAMAG). The compounds or extracts are applied as 4-mm long streaks by means of an automatic appliance so that 18 chromatograms can be developed on one plate. The chromatograms are evaluated by using a TLC scanner combined with a multicolour plotter and a computer. Quantitative evaluation of each consecutive chromatogram can be achieved by measurement of remission in the visible spectral region or in the UV region at optional wavelengths or by measurement of fluorescence. In this way, a compound is characterized not only by its distance of travel but also by its absorbance at the different wavelengths. The evaluation of a plate having 18 chromatograms by one of these procedures takes a total of 35 minutes per wavelength.

This technique has been used successfully in our company's laboratories for the determination of pesticide residues in groundwater and drinking water in the sub-ppb range (ref. 27). AMD also can be used for analyses of residues in plant material, by coupling it with a narrow-bore HPLC-column, which at a flow rate of about 30 µl/min permits direct application on the TLC plate (ref. 27).

Ease and expectations

Regarding the much discussed simplified methods, I consider that a few terms ought to be clarified. If it really were a simple matter to make residue analyses simpler, then residue analysis would have been revolutionized accordingly a long time ago because there can be hardly any residue analysts who have dedicated their lives to self-castigation.

Therefore, the gap between complexity and simplicity can be bridged only by reducing the demand. This automatically implies that simplified methods can never be a substitute for sophisticated methods.

Unlike cost-effective approaches, what is required here is to develop complete methods that are easy to handle, that is without the use of complicated equipment. A distinction should be made between simplified methods that will work satisfactorily only in the hands of experienced analysts and those that require less skill. Methods to be used for analyzing samples from field trials must satisfy other requirements than those to be used for regulatory purposes, that is to say simplified methods would be intended mainly for screening purposes rather than as a basis for legal action. General requirements for methods to be classified as "simplified" were proposed by the CCPR in 1985 (ref. 5a; Fig. 4). Of the some 290 listed methods recommended by the CCPR, so far 22 have been selected according to these criteria. It is obvious that most of these criteria are satisfied only by few techniques, for example TLC and colorimetry.

Criteria for classifying a method as "simplified":

- the use of TLC, spectrophotometry or basic GLC or HPLC for the determinative step
- the use of small volumes of solvents
- no, or only elementary, clean-up
- no expensive or rare reagents
- a procedure robust enough to withstand less than ideal laboratory conditions

Fig. 4.

TLC already has become very widely established in this area, largely through its use by analysts for whom there was no alternative. In consequence, colorimetric methods have suffered what I consider to be too big a loss of favour. The lack of specificity they often have could now be compensated for in many cases, at least partly, by the quasi-selective pre-cleanup procedures that are available today. This is an area where further development work would be worthwhile. More attention should also be given to neglected techniques like steam or sweep co-distillation. Little progress unfortunately seems to have been made in the development of simple field tests that require virtually no analytical experience for their application. We must wait and see what potential applications lie in store for the re-emerging test paper technique (e.g. enzyme tickets) or for immunoassays adapted for outdoor use. (I will not discuss immunoassays in this paper because they will be dealt with in detail by other contributors).

What areas are there where we have still failed to achieve anything real? We have not yet made a decisive breakthrough in the automation of extraction and/or clean up of extracts, which would enable us to simplify the steps that dictate speed in residue analysis. In this respect, it will be interesting to follow the development of laboratory robotics (ref. 28, 29) should it prove feasible for essential steps prior to the determinative one to be schematized in such way that they can be performed by one robot alone or by several operating in series, and provided the robot can be instructed to communicate with analytical instruments.

What should we strive for? Residue analysts should not seek to demonstrate at all costs what traces they are capable of analyzing. They should regard it as their more important duty to explain the relevance of their findings and to interpret the data. They should generate more willingness to adopt existing methods (it is said that a residue analyst would rather use a colleague's toothbrush than his method). This is all associated with the need to raise the analytical capacity of the laboratory. Statistical criteria should be generated for the verification and the acceptance of results. The recovery technique is still in need of improvement, and in connection with the development of analytical methods I wish to urge authors in future to give concrete and clearly written guidance explaining how in cases of doubt, confirmation of their results can be obtained by another technique. This should be made an obligatory component of every fundamental description of analytical methods, especially single compound methods.

But perhaps all these endeavours are unnecessary. At the 1985 Pittsburgh Conference, about 1200 lectures were presented to a community of over 20 000 analysts; 12 of these papers dealt with pesticide analysis (ref. 30). From this, it could be concluded that either we already know everything there is to be known or that we still know nothing.

ACKNOWLEDGEMENT

I wish to thank my colleagues A. Brauner, R. Brennecke, S. Gorbach, W.D. Hörmann, W. Maasfeld, H.P. Thier, Gabriele Timme and U. Werthmann for the good suggestions they contributed to the compilation of this paper and also for dissuading me from the inclusion of less good ones. I am grateful also to J. Edwards for translating the original German manuscript.

REFERENCES

These references are given as recent examples only, and no attempt has been made to achieve a fuller literature coverage of the topics treated in this paper.

1. Official Journal of the European Communities, L 229, Vol. 23, 11-29, Aug. 30, 1980.
2. Bates, J.A.R. and S. Gorbach, Pure & Appl. Chem. 54, 1361-1450 (1982).
3. A. Ambrus and H.-P. Thier, Pure & Appl. Chem. 58, 1035-1062 (1986).
4. FAO/WHO, FAO Plant Production and Protection Paper 56, p. 4-5, Rome, 1984.
5. CCPR, Guide to Codex Recommendations Concerning Pesticide Residues, FAO/WHO, Rome a) Part 8, CAC/PR 8 (1985); b) Part 5, CAC/PR 5 (1984); c) Part 6, CAC/PR 6 (1984), d) Part 2, CAC/PR 2 (1985).
6. W.E. Westlake and F.A. Gunther, Residue Rev. 18, 175-217 (1967).
7. R.S. Nicolson, J. Assoc. Publ. Analysts 24, 27-39 (1986).
8. GIFAP, Residue Committee, Doc. C.14533, October 3, 1985.
9. FAO/WHO, CCPR Meeting 1986, Room Document 9.
10. DFG, Rückstandsanalytik von Pflanzenschutzmitteln, Verl. Chemie, Weinheim, Abschnitt XI (1982).
11. J. Sherma and G. Zweig, Anal. Chem. 57, 1R-15R (1985).
12. R.J. Hemingway et al., Pure & Appl. Chem. 56, 1131-1152 (1984).
13. J.R. Gant and P.A. Perronne, Intern. Clinical Prod. Rev. 5, No. 3, 40-47 (1986).
14. D. McBlane and J.R. Benson, Eine Reversed-Phase-Säule mit einem C18-derivatisierten Polymer, ict Handelsgesellschaft, Antoniterstr. 27, D-6230 Frankfurt 80.
15. R. Gill and B. Law, J. Chromatogr. 354, 185-202 (1986).
16. H.M. McNair, M.W. Ogden and J.L. Hensley, Intern. Lab. 16, No. 1, 14-21 (1986).
17. O.L. Duffy, Intern. Lab. 16, No. 3, 78-87 (1986).
18. R.T. Wiedemer, S.L. McKinley and T.W. Rendl, Intern. Lab. 16, No.4, 68-77 (1986).
19. W. Blaß, private communication (1986).
20. C.M. White and R.K. Houck, J. High Resolut. Chromatogr. Chromatogr. Commun. 9, 3-17 (1986).
21. B.W. Wright and R.D. Smith, J. High Resolut. Chromatogr. Chromatogr. Commun. 9, 73-77 (1986).
22. P. Capriel, A. Haisch and Sh. U. Khan, J. Agric. Food Chem. 34, 70-73 (1986).
23. T.E. Beesley, J. Chromatogr. Sci. 24, 525-531 (1985).
24. G. Becker, D. Eichler, H.-G. Nolting and H.-P. Thier, Die Dünnschichtchromatographie in der Rückstandsanalytik von Pflanzenschutzmitteln und Metaboliten, DFG-Forschungsbericht, in press.
25. D.E. Jaenchen, in R.E. Kaiser (ed.), Proc. 3rd Intern. Symp. Instrumental High-Performance Thin-Layer Chromatography, Würzburg, West Germany, April 17-19, 1985, p. 71 - 82, Inst. for Chromatography, D-6702 Bad Dürkheim, W. Germany (1985).
26. K.D. Burger and H. Tengler, in R.E. Kaiser (Ed.), Planar Chromatography, Vol. 1, p. 193-205, Huethig Verlag, Heidelberg, Basel, New York (1986).
27. K.D. Burger, private communication (1986).
28. J.R. Strimaitis and G.L. Hawk (Ed.), Advances in Laboratory Automation Robotics 1985, Zymark Corporation, Hopkinton, MA, USA.
29. C.H. Lochmüller, K.R. Lung and M.R. Cushman, J. Chromatogr. Sci. 23, 429-436 (1985).
30. K.W.C. Burton, W.O. George and B.C. Thomas, Analyt. Proc. 22, 164-168 (1985).

DFG = Deutsche Forschungsgemeinschaft, Postfach 20 50 04, D-5300 Bonn 2.

GIFAP = Groupement International des Associations Nationales de Fabricants de Produits Agrochimiques, Avenue Hamoir 12, B-1180 Brussels.

Pesticide analysis by mass spectrometry

Joseph D. Rosen

Department of Food Science, Cook College, Rutgers University, New Brunswick, NJ 08903

Abstract - The analysis of pesticides and their metabolites at the parts per million level and below in complex matrices such as food, soil and animal tissue presents a serious challenge to the analytical chemist. The key to successful analyses is to separate the materials of interest from the matrix by chemical and/or physical methods and to use specific, sensitive detectors for the final determination. The mass spectrometer is an excellent detector of chromatographic effluents as it is very sensitive and very selective. Gas chromatography and mass spectrometry are ideal partners. Recently, advances have been made in the coupling of liquid chromatographs to mass spectrometers so that polar, non-volatile and thermally labile materials can also be identified and/or quantified at trace levels. Tandem mass spectrometry (MS/MS) also confers a high degree of selectivity. Here, the first mass spectrometer screens out all ions except those of interest for analysis by the second mass spectrometer. Theoretically, such a system should allow analysis without clean-up. In reality, sample clean-up is still necessary, but need be less extensive than for conventional methods of analysis. Another new technique, fast atom bombardment (FAB), allows for the mass spectrometric identification of polar materials such as glucuronides and gluthatione conjugates.

INTRODUCTION

Not too long ago, the use of mass spectrometry (MS) for the analysis of pesticides or pesticide metabolites involved separation of the pesticide or its metabolites in a mixture by column or preparative thin layer chromatography or by repeated injection/collection on a gas chromatograph with subsequent direct insertion into the mass spectrometer. The author still has unpleasant memories of sitting in front of a gas chromatograph for several days trying to collect enough pentachloro-aldrin (a photoproduct of aldrin) in order to have enough material for identification by mass spectrometry (ref. 1). This situation changed dramatically in the late sixties with the commercialization of interfaces which allowed for the coupling of a gas chromatograph to a mass spectrometer (GC/MS). Now, materials in complex mixtures could be separated by GC and identified by MS "on the fly". The mass spectrometrist would watch for the appearance of a peak on the total ion monitor trace and scan the magnet. If there were two closely-eluting peaks, the mass spectrometrist would obtain spectra at the upside of the first peak and downside of the second peak in the hope of obtaining reasonably pure spectra of each. This situation became more difficult when there were more than two poorly-resolved materials. Introduction of dedicated data systems (GC/MS/DC) alleviated these problems considerably. All information could be stored in memory and recalled in various forms. For example, instructing the computer to plot out a chromatogram containing only the molecular weight of DDT, for example, helped considerably in the rapid identification of this pesticide in a complex mixture. The data system also allowed one to automatically compare the mass spectrum of an unknown with that of a standard spectrum stored in the computer library, thus facilitating identification. Of particular importance, was the data system's ability to control accelerating voltage thereby allowing for selected ion monitoring (SIM) analysis. SIM is a highly sensitive and highly selective technique for quantifying materials in a mixture. When used in conjunction with a labeled internal standard, SIM gives analytical results corrected for losses during sample clean-up as well as results of high precision. One problem with early mass spectrometry was that the only method of sample ionization was electron impact (EI), which imparted an energy of 70 eV to the material. For many compounds, this resulted in the absence of a molecular ion in the spectrum and one could sometime not be certain of the molecular weight of the compound. Positive ion chemical ionization mass spectrometry (PICI), a gentler method of ionization (10-15 eV) allowed for determination of the molecular weights of materials which did not produce strong EI molecular ions. A variation of PICI, negative ion chemical ionization (NICI) also gives strong molecular ions and is highly sensitive to compounds that contain electron-capturing atoms. In general, materials that are sensitive to detection by electron capture GC detectors are also sensitive to detection by NICI. CI, however, did not eliminate the problem of analyzing non-volatile or thermally-sensitive compounds. Thus, identification of pesticide metabolites such as glucuronides, glycosides, sulfates, glutathion conjugates and mercapturic acids could not be performed by mass spectrometry. The discovery of sputtering phenomna by Barber et al. (ref. 2) lead to the development of the fast atom bombardment (FAB) technique. In this technique, a sample is dissolved in a matrix such as glycerin and the solution is bombarded with a Xenon or Argon beam. Polar and ionic materials are sputtered proton affinities 20-30 kcal/mole less than that of ammonia exhibit no ions in the "filament off" mode. For these analytes, e.g. chlorinated carboxylic acids, CI-type spectra are obtained in the "filament on" mode (ref. 20). Cairns et al. (ref. 21) also performed LC/MS on some carbamate and urea pesticides using the moving belt interface and methane CI. The moving belt was designed to remove mobile phase by flash evaporization, thus minimizing the thermal decomposition problems experienced with earlier moving belt models. Fig. 8 shows the SIM chromatograms for lettuce spiked at the 1 ppm level with monuron, diuron and linuron. Shalaby (ref. 22) evaluated DLI/LC/MS for the separation analysis of the herbicides chlorsulfuron and sulfometuron methyl. In both cases, she was able to obtain a mass spectrum with complete fragmentation pattern and an abundant protonated molecular ion. The detection limit was estimated at 2 µg injected on column as DLI requires a 100:1 split before introduction of the sample into the mass spectrometer source. This detection limit could be reduced in the SIM mode. Better sensitivity for chlorosulfuron (50 ng on-column) was obtained with TSP/LC/MS and this technique proved more adaptable to reversed phase liquid chromatography. The DLI mass spectrum gave more structural information. Rosen and Dziedzic (ref. 23) constructed an inexpensive DLI interface by connecting (after a 36:1 split) a 1.5 m x 25 µ i.d. fused silica line between the exit of an analytical HPLC column and the mass spectrometer ion source through the line-of-sight inlet normally used for capillary GC columns. The advantages of this interface are low cost, ability to use a magnetic instrument (the non-conducting fused silica line is compatible with instruments having accelerating voltages in the kilovolt range), and the ability to obtain true EI mass spectra. In addition, the clogging problems associated with commerical DLI interfaces are solved by simply replacing the clogged line with another 1.5 m x 25 µ i.d. piece of capillary column. With the use of this interface, Rosen and Dziedzic obtained the reversed phase HPLC/MS chromatograms of technical COMMAND® herbicide, 2-[(2-chlorophenyl)methyl]-4,4-dimethyl-3-isoxazolidinone, as shown in Fig. 9. Interpretable mass spectra were obtained from impurities present at levels as low as 0.025%. Normal phase HPLC/MS permitted separation and analysis of the cis and trans isomers of both permethrin and cypermethrin (Fig. 10). Replacement of the 25 µ i.d. line with a 75 µ i.d. line allowed for the use of mobile phases containing 0.1 M ammonium acetate and thermospray type mass spectra. LC/MS is currently undergoing rapid development and evaluation. Each of the interfaces, as one would expect, have their own advantages and disadvantages and it is not yet clear which one, if any, will become most commonly used. What is clear, however, is that LC/MS will eventually play an important role in pesticide residue analysis. HPLC, of course, permits separation of pesticides too polar or thermally unstable for GC, while the mass spectrometer confers much greater selectivity and sensitivity than UV detectors. The current multi-residue procedure for polar pesticides and for pesticides that do not elute from Florisil (ref. 24) omits clean-up on Florisil columns and relies on specific GC detectors for specificity. Such a procedure leads to rapid deterioration of the GC column as well as an increased rate of false positives. The development of LC/MS procedures holds promise for improved analyses of these pesticides.

Tandem mass spectrometry (MS/MS)

Johnson and Yost (ref. 25) have published a very short and very understandable explanation of MS/MS. This technique uses the first mass spectrometer to isolate an ion of interest (parent ion) and a second mass spectrometer to record the mass spectrum of this ion (daughter ions), i.e., the first mass spectrometer is used to separate an analyte from the other mixture components. The technique, therefore, greatly reduces chemical noise, and confers a high degree of specificity and sensitivity to an analysis. Greater sensitivity is achieved by monitoring a limited number of parent-daughter pairs instead of scanning. This technique is called selected reaction monitoring (SRM) and is formally analagous to SIM. Hunt et al. (ref. 26) applied MS/MS to the analysis of US Environmental Protection Agency priority pollutants with little or no clean-up. Detection limits were at the 10-100 ppb level and total analysis time was typically 25-30 minutes per sample. Trehy et al. (ref. 27) analyzed aldicarb, aldicarb sulfoxide and aldicarb sulfone in citrus juice using short, open tubular fused silica GC columns and PICI/MS/MS. In Fig. 11, the SIM chromatograms for m/z 191 $(M + H)^+$, m/z 116 and m/z 89 are compared to the SRM chromatogram for the $116^+ \rightarrow 89^+$ reaction for 36 ppb aldicarb in citrus juice. Aldicarb could not be detected by SIM because of high chemical background but the elimination of chemical noise by MS/MS permitted detection. Repeated injections of citrus juice extracts (which contain high concentrations of nonvolatile components) quickly contaminated the injection port and the column. Sample clean-up was thus necessary. While MS/MS appears to work well without sample clean-up for relatively clean matrices such as water (ref. 26), samples in food cannot be analyzed routinely without some sort of clean-up. The advantage of MS/MS, however, is that because of its selectivity, less time-consuming clean-up procedures are needed and the GC separation does not need to be as good as with other techniques. Thus, the analyst can save time during clean-up by eliminating several steps and by using a much shorter column for the GC separation. For example, Yost et al. (ref. 28) analyzed hexachlorobenzene (HCB) and 2,4,5-trichlorophenol (TCP) in human blood serum and urine much more quickly by GC/SRM than by capillary column GC/MS/SIM. Clean-up for SRM consisted of a simple 1:1 extraction of serum or urine with either hexane (for HCB) or ether (for TCP). For twenty-five samples, a preparation time of 36 hours for GC/MS/SIM was reduced to 1 hour for GC/SRM. In addition, a 50 cm GC column (on which 100 injections per hour could be made) was substituted for an 18 m column (on which only 2 injections per hour could be made). Overall, the MS/MS method reduced the analysis time for 25 samples from 186 to 27 hours. An application involving desorption chemical ionization and SRM was published by Tway et al. (ref. 29). Ivermectin, a mixture of dihydroavermectin B_{1a} (molecular weight, 874) and dihydroavermectin B_{1b} (molecular weight, 860), is highly non-volatile and thermally labile. A mass spectrum of ivermectin could be obtained by ammonia desorption chemical ionization. Monitoring the ammonium adduct ions at m/z 892 and 874 gave unacceptable interferences even after extensive clean-up of a liver sample. Selective reaction off the surface of the glycerin matrix and are then mass analyzed. Very little fragmentation occurs and ions corresponding to molecular weight are observed. FAB has been used for the analysis of peptides, oligosaccharides, ionic surfactants, etc. The limitations of FAB are that it often gives very little structural information and it cannot be used for complex mixtures unless used in conjunction with tandem mass spectrometry. FAB spectra also exhibit a high level of chemical noise. In the past few years, a major thrust in instrument development has been the design of interfaces between liquid chromatographs and mass spectrometers (LC/MS) so that mixture analysis could be performed on materials (e.g. some carbamates, triazines) difficult to separate by gas chromatography. Three different interfaces, a moving belt, a thermospray and a direct liquid inlet have now been commercialized. Although many problems have been encountered with this technique, some successes (discussed below) have been achieved. Again, as in the case of FAB, mainly molecular data with very little structural information is obtained by LC/MS when used with the thermospray interface. The absence of structural information in CI, FAB and TSP/LC/MS can be alleviated, at least in part, by focusing the molecular ion (parent ion) of interest into a collision activation region where the ion is fragmented by collision with a gas at an applied potential. The so-formed daughter ions are then mass analyzed by a second mass spectrometer operating in tandem with the first mass spectrometer. This process is called tandem mass spectrometry, collisionally activated dissociation or mass spectrometry/mass spectrometry (MS/MS). This technique, at least theoretically, also allows one to obtain qualitative as well as quantitative information on pesticide residues in environmental samples without sample clean-up or chromatographic separation. Essentially, the specifity of the first mass spectrometer separates the molecular ion of the analyte from all of the other ions produced in the ion source. Matrix effects, however, lead to the production of artifacts in many applications. On a practical basis, samples that have not been cleaned up very quickly contaminate the mass spectrometer ion source and this necessitates shutting down and cleaning the instrument. However, GC/MS/MS (using a short, packed column) with some prior clean-up is a promising technique for the analysis of pesticide residues with greatly reduced sample preparation time as well as greatly reduced GC retention time. Mass spectrometry is the method of choice for qualitative analysis with confirmation of identity of parts per million or even parts per billion levels of pesticides in environmental matrices. For quantitative analysis, mass spectrometry, because of its excellent selectivity (SIM, GC/MS/MS/SIM) and sensitivity is superior to other methods. The major problem with mass spectrometry is that its high cost is beyond the reach of most residue laboratories. The past few years, however, have witnessed the commercial introduction of Finnigan's Ion Trap Detector and Hewlett Packard's Mass Selective Detector. Both of these instruments sell for about $35,000 US and can be directly interfaced to a capillary gas chromatograph.

SELECTED APPLICATIONS OF MASS SPECTROMETRY IN PESTICIDE ANALYSIS

The literature concerning applications of mass spectrometry to pesticide residue determination is growing rapidly with references too numerous to discuss in this short review. Instead, some selected examples of applications involving GC/MS, GC/MS/SIM, FAB, LC/MS and MS/MS will be presented. Although clean-up procedures will not be thoroughly discussed, the reader must be aware that in most cases good results depend on good clean-up procedures as well as good chromatography in spite of the highly sensitive and selective powers of mass spectrometry.

Gas chromatography/mass spectrometry (GC/MS)

The major application of mass spectrometry in pesticide residue methodology is the use of GC/MS to confirm the identity of materials found in environmental samples. If the pesticide residue analyst has no access to a mass spectrometer, confirmation of identity must be obtained by determination of retention time on two different-polarity GC columns. While this works well most of the time, errors are undoubtedly made. An extreme example was reported several years ago (ref. 3) when a pesticide residue in imported onions was tentatively identified as bromophos on the basis of co-elution with a reference sample on both 10% OV-101 and 4% SE-30/6% SP2401 columns using electron capture, microcoulometric, flame photometric and nitrogen-phosphorous detectors. GC/MS analysis of the cleaned-up onion extract, however, demonstrated that the pesticide residue was trichlorinate rather than bromophos. Because so many pesticides contain electron-capturing atoms, the use of GC/NICI/MS has become important. Stout and Steller (ref. 4) discussed both the advantages and limitations of this technique. Pendimethalin, the active ingredient in PROWL® herbicide, contains two nitro groups and exhibits a very large NICI response when methane is used as the reagent gas. Fig. 1 shows the mass chromatogram (m/z 281) for 5 pg pendimethalin standard as compared to the solvent blank. At the 100 pg level (corresponding to 0.05 ppm pendimethalin in an analyzed commodity), the mass spectrometer can be run in the full scan mode to routinely confirm electron capture analyses. Fosthietan, the active ingredient in NEM-A-TAK® nematicide-insecticide gives the methane NICI spectrum shown in Fig. 2 for a 12.5 ng on-column injection. The sensitivity achieved was approximately fifty times better with NICI than with PICI. At the 0.05 ppm level in corn, good correlation between analysis by flame photometric detection (FPD) and NICI was obtained. Because of a greater number of electron-capturing than flame photometric sensitive materials in the corn extract, FPD gas chromatography was more selective than NICI. As a result, the selective power of SIM (monitoring of the m/z 241, 166 and 58 ions) was used to obtain the results shown in Fig. 3 for corn grain fortified at the 0.05 ppm level with fosthietan.

Selected ion monitoring (SIM)

As stated above, the SIM technique permits one to confirm the identity of low levels of material in an environmental sample and also to quantify the amount present. Greater sensitivity is obtained in the SIM mode than in the full scan mode. For example, assume a GC component takes four seconds to elute and the mass spectrometer scans between m/z 44 and m/z 444 at 4 sec/decade. Thus, the mass spectrometer spends only 0.01 second in recording the output of each ion in the spectrum. If instead of scanning the entire mass range, the mass spectrometer is set to record the current from only one ion (pre-selected on the basis of the known mass spectrum of the analyte), sensitivity is greatly enhanced because all four seconds of elution time are used for detection of that particular ion. Monitoring of a single ion at this level of sensitivity may, however, lead to misconfirmation. Better selectivity at moderate loss in sensitivity is obtained if three ions, instead of one, are monitored. Millard (ref. 5) has calculated an analytical identity confidence ratio of 10^6:1 if three ions above m/z 350 are monitored. Further selectivity is obtained from GC retention time as well as type of fragmentation (EI, PICI, NICI) and extraction/clean-up. Some examples of SIM for pesticides include the analysis of atrazine, lindane, pentachlorophenol and diazinon in soil and water samples with detection limits of 0.1-1.0 ppb (ref. 6); ethafluralin and trifluralin in soil at 5 ppb (ref. 7); parathion and methyl parathion in marine sediment at the parts per trillion level (ref. 8). An advantage of selected ion monitoring over all other methods of detection is that a deuterated analogue of the analyte may be used as an internal standard. Thus, adjustments for losses of analyte during clean-up may be made. Furthermore, constructing a calibration curve of the ratio of areas obtained from analyte and constant amount of internal standard vs. increasing levels of analyte gives analytical data of high precision by eliminating the problem of mass spectrometer detector variability. For example, Nielsen (ref. 8) used [$^2H_4$] parathion (molecular weight, 295) and [$^2H_4$]-methyl parathion (molecular weight 267) as internal standards for the analysis of parathion (molecular weight, 291) and methyl parathion (molecular weight, 263), respectively. Using a constant amount of the deuterated standard (50 ng) and varying amounts of parathion or methyl parathion (1-50 ng) he obtained the calibration curves shown in Fig. 4. He then used (Fig. 5) selected ion monitoring (m/z 291 and m/z 154) for quantification of parathion using [$^2H_4$]-parathion (m/z 295) as the internal standard. Methyl parathion (m/z 263 and 154) was quantified with comparison to the molecular ion (m/z 267) of the [$^2H_4$]-methy parathion.

For some applications, specificity is achieved by the use of high resolution selected ion monitoring instead of multiple ion selected ion monitoring as was done recently for the analysis of 2,3,7,8 -tetrachlorodibenzo-p-dioxin (TCDD) and its isomers in human adipose tissue (ref. 9). Selected ion monitoring was conducted at 10,000 resolution and the limit of detection was approximately 1 part per trillion. My colleagues and I (ref. 10) recently demonstrated the power of medium resolution SIM for the analysis of a mycotoxin, aflatoxin B_1, in peanuts. At 1,000 resolution, the monitored molecular ion (m/z 312) was completely obscured by a $C_{23}H_{36}^+$ (m/z 312) fragment which probably arose from peanut oil. By increasing the resolution of the mass spectrometer to 3,000 we were able to separate the molecular ion of aflatoxin B_1 (m/z 312.063) from the interfering peak (m/z 312.218) and achieve a limit of detection of approximately 0.1 ppb.

Fast atom bombardment (FAB)

The fast atom bombardment technique is best suited for the detection of polar pesticides and metabolites. Molecular ions are deduced from $(M + H)^+$, $(M + Na)^+$ and/or $(M + K)^+$ ions. Larsen and Ryhage (ref. 11) used FAB to obtain the mass spectra of the glutathione, cystein (Fig. 6) and mercapturic acid (Fig. 7) conjugates of propachlor, the active ingredient of RAMROD® herbicide. These spectra were routinely obtained from 3-30 µg samples. Wickramanayke et al. (ref. 12) obtained both positive and negative ion liquid secondary ion mass spectra of several chlorophenyl- and nitrophenylglucuronides. Fujiwara and Wratten (ref. 13) analyzed the polar pesticide glyphosate (N-phosphonomethyl glycine), the active ingredient in ROUNDUP® herbicide, by FAB/MS using aminoalcohols instead of glycerin as the matrix because of the poor solubility of glyphosate in glycerin. Identification of rat metabolites of the experimental rice fungicide, 1,2,5,6-tetrahydro-4H-pyrrolo-[3,2,1-i,j]quinolin-4-one, (pyroquilon), was greatly facilitated by FAB (ref. 14). Positive and negative FAB spectra of a glucuronide metabolite and of a sulfate conjugate were obtained.

Liquid chromatography/mass spectrometry (LC/MS)

The development of an interface for combined LC/MS has been more difficult than development of a GC/MS interface because it is much more difficult to remove a liquid mobile phase than a gas mobile phase. In spite of severe difficulties, three systems, direct liquid introduction (DLI), moving belt and thermospray (TSP) have been developed. A review of these techniques is beyond the scope of this paper and the reader is referred to several review articles (refs. 15, 16, 17). Voyksner and Haney (ref. 18) studied the TSP/HPLC/MS of several trizaine and organophosphate pesticides and found that the sensitivity was superior to HPLC/UV because of the greater specificity of the MS detector. The mobile phases used for the HPLC separations contained 0.1M ammonium acetate and the mass spectra of the materials resembled those obtained by ammonia CI. In the "filament off" mode, ammonium acetate is necessary for the ionization process and gives $(M + NH_4)^+$ or combinations of $(M + NH_4)^+$ and $(M + H)^+$ ions, depending on the proton affinity of the analyte. Ammonium acetate is an ideal buffer for LC/MS applications because it is volatile and does not contaminate the source. Many HPLC separations, however, do not include ammonium acetate as a buffer, but TSP/HPLC/MS in the "filament off" mode can not be used unless it is present. Voyksner et al. (ref. 19) overcame this problem by adding ammonium acetate postcolumn for the TSP/HPLC/MS analysis of a mixture containing fourteen carbamate and urea pesticides. Detection limits of 20 ng were achieved for most compounds when mass chromatograms of the base peak were recorded. Detection limits were reduced in the SIM mode; 200 pg being sufficient for the detection of carbofuran and 2-sec-butylphenyl-N-methylcarbamate (BPMC). Analytes with monitoring of the $892^+ \rightarrow 567^+$ transition, however, permitted the analysis of ivermectin in liver at the 5-10 ppb level. MS/MS can also overcome a major problem associated with TSP/LC/MS, i.e. lack of structural identification. For example, the thermospray mass spectra of propazine and terbutylazine are essentially identical because they both have molecular weights of 229. Using TSP/LC/MS/MS, Voykener et al. (ref. 30) obtained spectra which permitted identification of these herbicides (Fig. 12).

CONCLUSIONS

It is quite apparent that mass spectrometry has had, and will continue to have, an important role in the analysis of pesticide residues as well as in the identification of pesticide metabolites and environmental degradation products. For applications with stringent sensitivity requirements, such as the analysis of dioxin in environmental samples, no instrumental method compares with mass spectrometry. For non-volatile or thermally unstable pesticides, the use of soft ionization techniques or LC/MS is gaining importance. Tandem mass spectrometry techniques offer the advantage of reducing sample analysis time considerably. Whether this time and labor saving is enough to offset the costs of an MS/MS system depends on the mission of the laboratory as well as the nature of the samples analyzed.

REFERENCES

1. J.D. Rosen, J. Chem. Soc., Chem. Commun., 189-190 (1967).
2. M. Barber, R.S. Bordoli, R.D. Sedgwick and A.N. Tyler, J. Chem. Soc., Chem. Commun., 325-327 (1981).
3. T. Cairns, E.G. Siegmund, R.A. Jacobson, T. Barry, G. Petzinger, W. Morris and D. Heikes, Biomed. Mass Spectrom., 10, 301-315 (1983).
4. S.S. Stout and W.A. Steller, Biomed. Mass Spectrom., 11, 207-210 (1984).
5. B.J. Millard, Quantitative Mass Spectrometry, pp. 127-131, Heyden, London (1978).
6. V. Lopez-Avila, P. Hirata, S. Kraska, M. Flanagan and J.H. Taylor, Anal. Chem., 57, 2797-2801 (1985).
7. J.R. Moyer and J.L. Elder, J. Agric. Food Chem., 32, 866-868 (1984).
8. P.G. Nielsen, Biomed. Mass Spectrom. 12, 695-698 (1985).
9. D.G. Patterson, J.S. Holler, C.R. Lapeza, L.R. Alexander, D.F. Groce, R.C. O'Conner, J. Smith, J.A. Liddle and L.L. Needham, Anal. Chem., 58, 706-713 (1986).
10. R.T. Rosen, J.D Rosen and V.P. DiProssimo, J. Agric. Food Chem., 32, 276-278 (1984).
11. G.L. Larsen and R. Ryhage, Xenobiotica, 12, 855-860 (1982).
12. P.P. Wickramanayake, M.L. Deinzer and A.L. Burlingame, Biomed. Mass Spectrom., 12, 127-133 (1985).
13. H. Fujiwara and S.J. Wratten, in Applications of New Mass Spectrometric Techniques in Pesticide Chemistry, J.D. Rosen, Ed., Chemical Analysis Series, Vol. 91, Wiley, New York, in press.
14. P. Kriemler and W. Muecke, ibid.
15. A.P. Bruins, J. Chrom., 322, 99-111 (1985).
16. D.M. Desiderio and G.H. Fridland, J. Liq. Chrom., 7(S-2), 317-351 (1984).
17. Vestal, M.L., in Mass Spectrometry in the Health and Life Sciences, A.L. Burlingame and N. Castagnoli, Eds., p. 99-118, Elsevier, Amsterdam (1985).
18. R.D. Voyksner and C.A. Haney, Anal. Chem., 57, 991-996 (1985).
19. R.D. Voyksner, J.T. Bursey and E.D. Pellizzari, Anal. Chem., 56, 1507-1514 (1984).
20. R.D. Voyksner, as in ref. 13.
21. T. Cairns, E.G. Siegmund and G.M. Doose, Biomed. Mass Spectrom., 10, 24-29 (1983).
22. L.M. Shalaby, Biomed. Mass Spectrom., 12, 261-268 (1985).
23. R.T. Rosen and J.E. Dziedzic, as in ref. 13.
24. M.A. Luke, J.E. Froberg, G.M. Doose and H.T. Masumoto, J. Assoc. Offic. Anal. Chem., 58, 1020-1026 (1975).
25. J.V. Johnson and R.A. Yost, Anal. Chem., 57, 758A-764A (1985).
26. D.F. Hunt, J. Shabanowitz, T.M. Harvey and M. Coates, Anal. Chem., 57, 525-537 (1985).
27. M.L. Trehy, R.A. Yost and J.G. Dorsey, Anal. Chem., 58, 14-19 (1986).
28. R.A. Yost, D.D. Fetterolf, J.R. Hass, D.J. Harvan, A.F. Weston, P.A. Skotnicki and N.M. Simon, Anal. Chem., 56, 2223-2228 (1984).
29. P.C. Tway, G.V. Downing, J.R.B. Slayback, G.S. Rahn and R.K. Isensee, Biomed. Mass Spectrom., 11, 172-176 (1984).
30. R.D. Voyksner, W.H. McFadden and S.A. Lammert, as in ref. 13.

ACKNOWLEDGMENTS

The author gratefully acknowledges the assitance of Trudy Kozlosky in the typing of this manuscript.

This is New Jersey Agricultural Experiment Station Publication #F-10201-1-86.

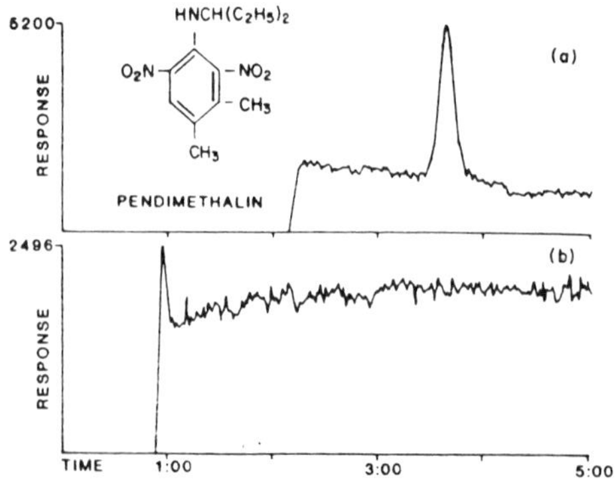

Figure 1. Mass chromatograms of m/z 281$^-$ from gas chromatographic NICI (CH_4) analyses of (a) 5 pg pendimethalin and (b) solvent blank.

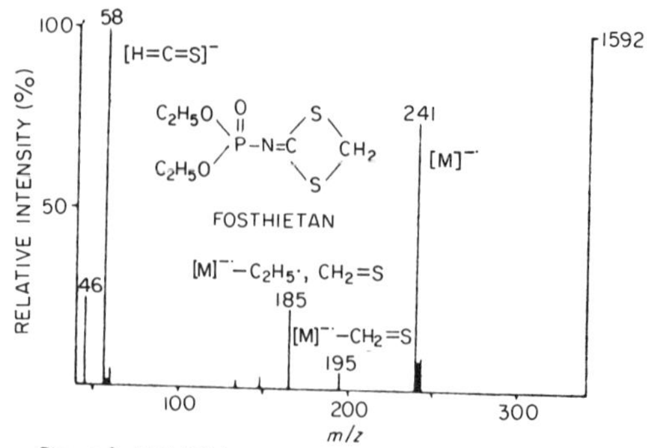

Figure 2. NICI (CH_4) mass spectrum of 12.5 ng of fosthietan.

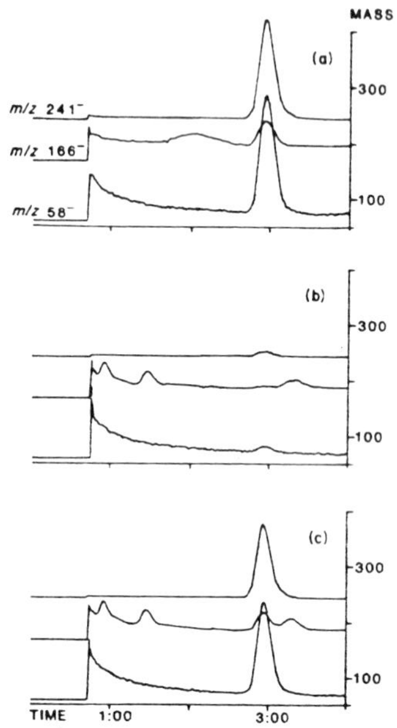

Figure 3. Ion maps from SIM analyses in gas chromatography NICl (CH_4) of (a) 1.25 ng standard, (b) control corn grain, and (c) corn grain fortified at 0.05 ppm.

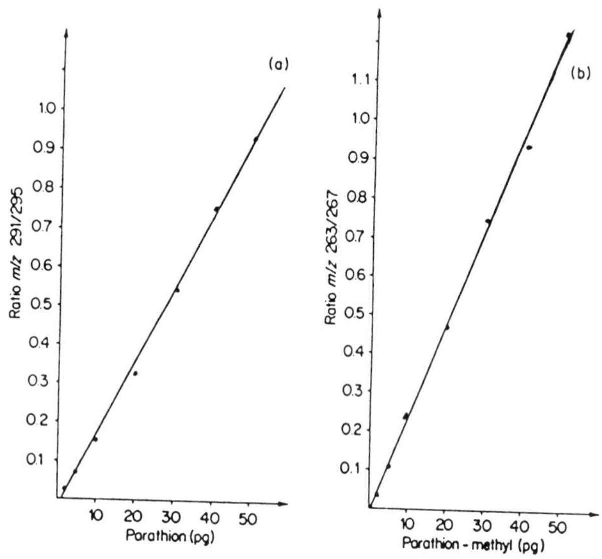

Figure 4. Lower part of calibration curves for (a) parathion and (b) parathion-methyl with constant amount (50pg) of the deuterated analogues as internal standards. Selected ion monitoring mode of the molecular ions (m/z=263,267 for parathion-methyl and m/z=291,295 for parathion).

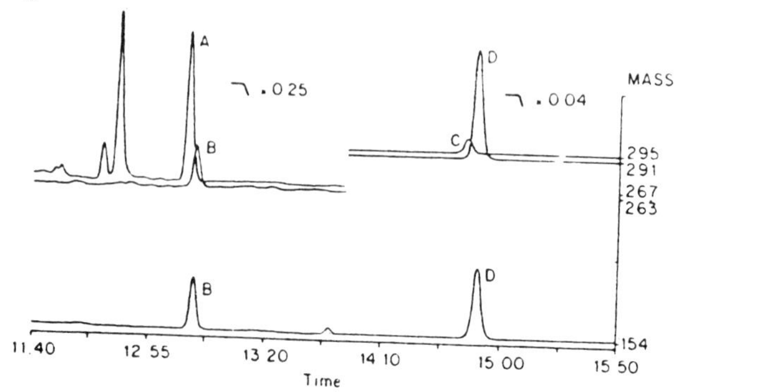

Figure 5. Selected ion monitoring traces of sediment sample spiked with 50 ng of [$^2H_4$] parathion-methyl (A, m|z = 267) and [$^2H_4$] parathion (C, m/z = 295). 1.7 ppb of parathion-methyl (B) was observed from the selected ion monitoring traces of m/z = 154 and m/z = 263 and 70 ppb of parathion (D) from m/z = 154 and m/z = 291. The selected ion monitoring traces are reduced with the noted magnification factor.

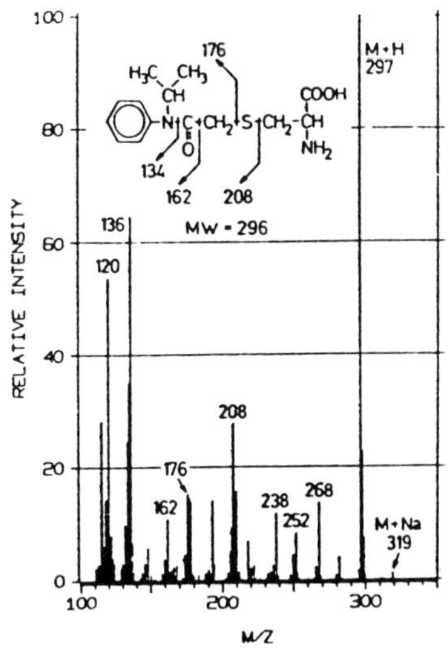

Figure 6. FAB mass spectrum of the cysteine conjugate of propachlor.

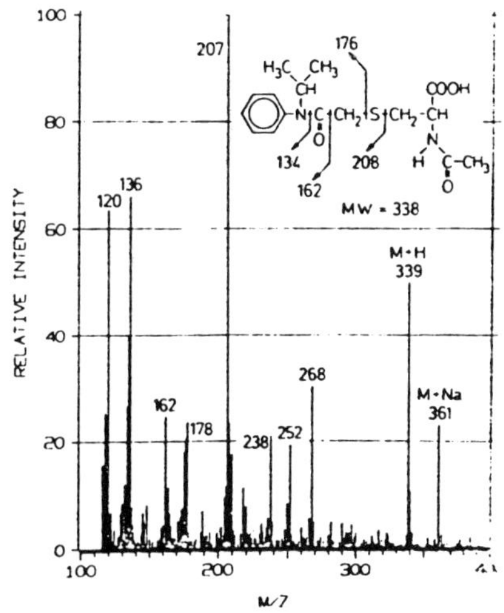

Figure 7. FAB mass spectrum of the mercapturic acid conjugate of propachlor.

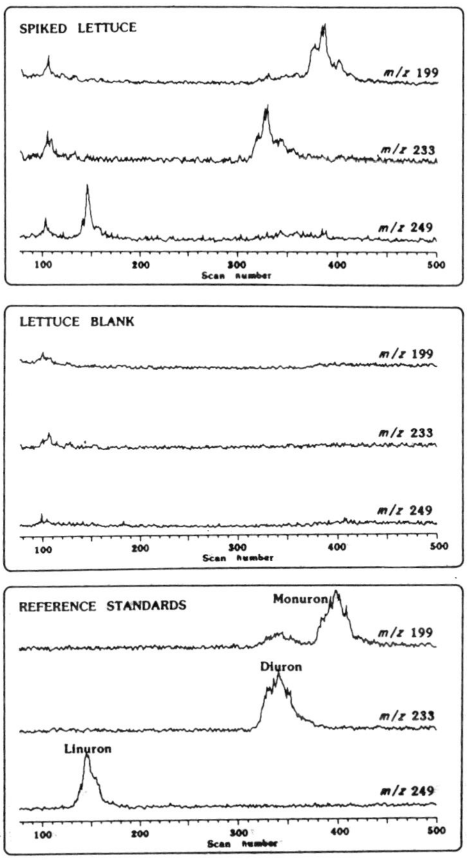

Figure 8. Selected ion monitoring chromatograms (LC/MS CI-CH_4) of lettuce spiked with linuron, diuron and monuron at the 1 ppm level, sample blank and reference standards using *m/z* 249, 233 and 199 for detection and quantification of linuron, diuron and monuron respectively.

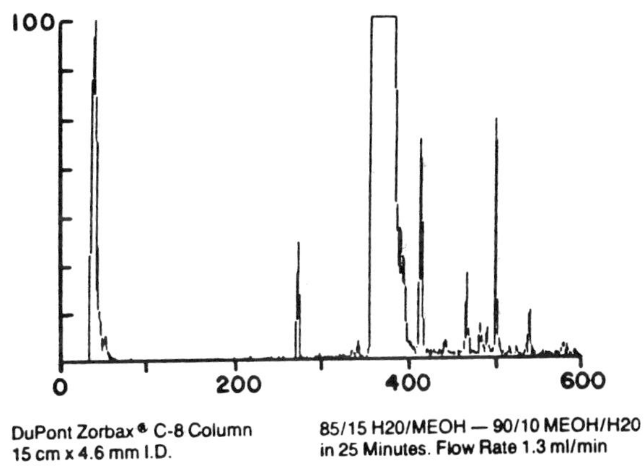

Figure 9. Reconstructed HPLC/MS chromatogram of a technical sample of Command® herbicide.

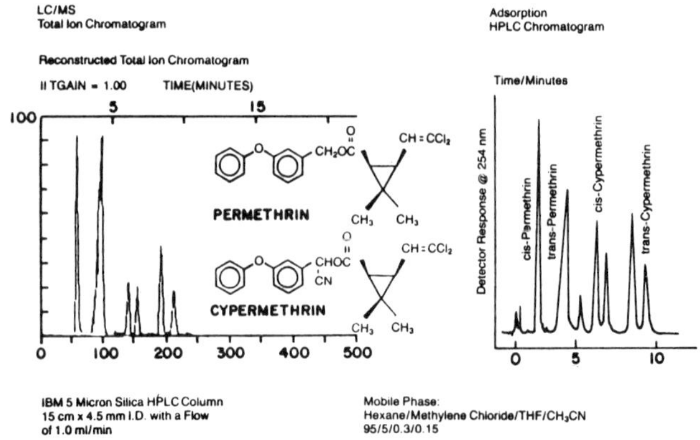

Figure 10. Normal phase separation of permethrin and cypermethrin isomers.

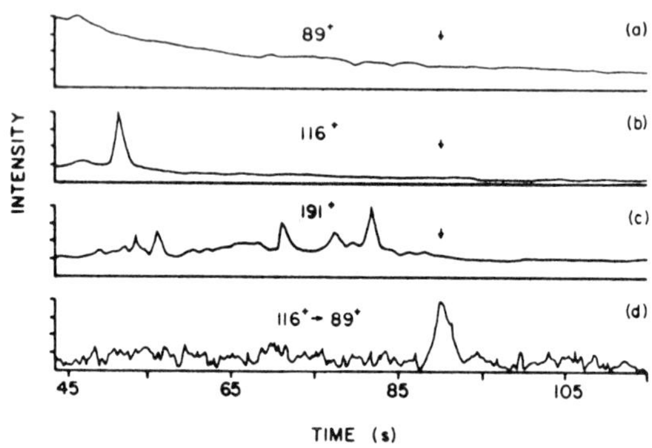

Figure 11. Analysis of an aldicarb-spiked sample of citrus juice by chemical ionization MS and MS/MS. The mass chromatograms for the m/z 89, 116, and 191 ions in a, b and c, respectively, do not indicate the presence of aldicarb, while the MS/MS technique, d, shows the presence of aldicarb. The arrows indicate the retention time of aldicarb on the 1-m DB5 open tubular column.

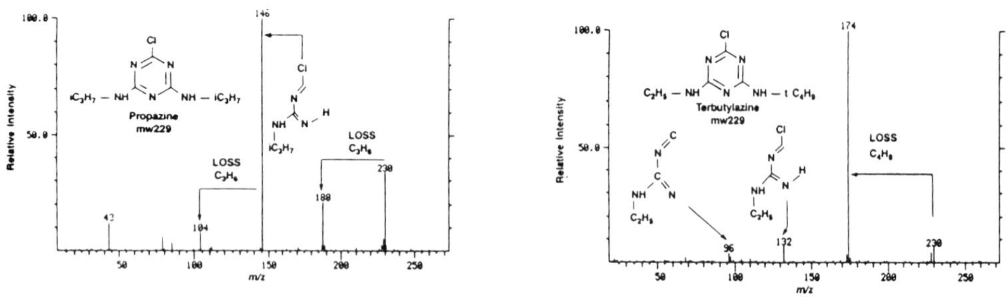

Figure 12. TSP/HPLC/MS/MS of propazine and terbutyazine.

Acknowledgement is given to the following publishers for permission to reproduce figures.
Figs 1,2,3,4,5 and 8, reprinted by permission of John Wiley & Sons Ltd, © 1983, 1984 and 1985 from articles by Stout & Steller 1984, Nielsen 1985 and Cairns et al. 1983 in Biomedical Mass Spectrometry.
Figs 9, 10 and 12 from J.D. Rosen (ed.) Applications of New Mass Spectrometry Techniques in Pesticide Chemistry, 1987 John Wiley & Sons Ltd.
Figs 6 and 7 reprinted with permission from the publishers of Xenobiotica © 1982 Taylor & Francis Ltd.
Fig. 11 reprinted with permission from Analytical Chemistry, **58**, (1), © 1986, American Chemical Society.

Utility of immunoassay in pesticide trace analysis

B.D. Hammock, S.J. Gee, P.Y.K. Cheung, T. Miyamoto, M.H. Goodrow,
J. Van Emon, and J.N. Seiber

Departments of Entomology and Environmental Toxicology, University of
California, Davis, California 95616, U.S.A.

Abstract - Immunoassays are one of several powerful biotechnologies immediately applicable to pesticide chemistry. The power of immunochemical technology to solve analytical problems in the pesticide field is becoming widely appreciated. This chapter reviews some of the applications of immunochemical technology to problems in the pesticide field.

INTRODUCTION

Immunochemical methods of analysis are widely used in biochemistry, endocrinology, and clinical chemistry, but they rarely have been used in environmental chemistry. The past failure of environmental chemists to apply this powerful technology largely is due to historical precedent rather than to inadequacies in the technology. In 1980 Hammock and Mumma (ref. 1) reviewed the potential of immunochemistry in pesticide analysis and urged its acceptance. Six years later the interest of many academic, governmental, and industrial laboratories indicates that the potential of the technology finally is beginning to be realized (see Mumma, this volume). During the last decade the application of immunochemical technology to problems in environmental chemistry has been explored in this and other laboratories. Our conclusion is that the technology is not a panacea, but that it complements classical analytical procedures in many applications, it offers dramatic advantages over more classical techniques for some problems, and may be the only viable technology for still other areas.

In this chapter no attempt will be made to describe how one develops an immunoassay since this is well documented in other reviews. However, the status of immunochemistry in pesticide analysis will be detailed in light of the relative advantages and disadvantages of this technology. Some of the past and current research from this laboratory will be reviewed on pesticide analysis by immunoassay to illustrate the advantages and limitations of immunochemical approaches as well as some of the special applications of immunoassay in pesticide analysis. Analysis of products of research in biotechnology represents a major challenge for environmental chemists, and the application of immunochemical technology to this problem will be discussed. Immunoassay often is thought to be simply an inexpensive alternative to classical technology. As outlined below, the availability of inexpensive analytical data is likely to impact the pesticide field in a variety of novel ways far outweighing the impact of the technology in classical uses. Finally, recommendations for the implementation of this technology in the pesticide field will be presented.

ADVANTAGES OF IMMUNOCHEMICAL TECHNOLOGY

Immunoassays are generally applicable to pesticide chemistry. In addition they are highly sensitive, specific, and precise which leads to rapid, cost effective assays. In addition the technology is very adaptable to a variety of analytical problems.

Applicability to pesticide chemicals

Immunochemical assays represent an immediately practical application of biotechnology. However, these assays are physical assays based on the law of mass action and are not bioassays. They owe their great sensitivity and specificity to biological systems which can produce the reagent antibodies which will bind with high affinity to compounds of interest. Immunoassays probably can be applied to more diverse structures than any other analytical technology. Since the affinity of the antibody for the compound to be analyzed depends upon the summation of a variety of noncovalent interactions, it is difficult to prepare antibodies to exceptionally small molecules. Fortunately, there appears to be no upper limit on the size of the compounds that can be analyzed. As our field moves toward the utilization of increasingly complex synthetic molecules (ie diflubenzuron and chlorsulfuron), (refs. 2-5), fermentation products such as avermectin, and even proteins such as the toxins of *Bacillus thuringiensis* (ref.6), it will become increasingly important to have accepted techniques for the analysis of such large molecules.

Immunoassays are usually performed in aqueous solution, thus the molecule to be analyzed should be at least moderately stable in water. Surprisingly, solubility in water is only rarely a problem since even very lipophilic compounds often are soluble at the pico or femtomolar concentrations used in an immunoassay. Even highly insoluble compounds often can be presented to the antibody in micelles or with water soluble cosolvents. Thus, problems associated with the analysis of highly lipophilic compounds are more commonly associated with removing them from an oily matrix than with absolute solubility (ref. 7). As a general rule immunoassays are most applicable to molecules difficult to analyze by gas liquid chromatography, which makes immunoassays a very attractive complementary technology. Although it is easiest to devise assays for water soluble compounds, with sufficient imagination, immunochemical technology can be applied to the analysis of almost any compound. With little difficulty, the technology certainly can be applied successfully to most current pesticides, and it is likely to be even more appropriate for the next generation of compounds.

Balancing the advantages of immunoassay

In 1974 we initiated studies to explore the possibility of using immunochemical technology for the analysis of pesticides and other chemicals. Much of our research has been designed to evaluate the advantages and limitations of immunochemical technology in general to the field of environmental chemistry. Some of these advantages and limitations are indicated in Table 1. These criteria are not firm rules, for one can

TABLE 1. Advantages and disadvantages of immunochemical technology

Advantages	Disadvantages
Generally applicable	New technology in environmental labs
Highly sensitive	Too sensitive
Highly specific	Hard to apply to multianalyte problems
Highly precise	Cross reactivity and interference
Very rapid	Reagents not available
Cost effective	Confusing terminology
Highly adaptable	Large sample load required

overcome many of the limitations of immunoassays with innovative applications of this very adaptable technology. However, these criteria can be used to indicate the likelihood of a successful project. It also is critical to remember that immunoassays are physical assays sadly lacking in magic. With a working knowledge of the technology, one often can balance the advantages and limitations to obtain an assay of the desired properties. However, one cannot have all of the advantages of an immunoassay simultaneously. For instance it is possible to design very inexpensive immunoassays of moderate sensitivity that can be performed in the field with no specialized equipment or highly trained personnel. One cannot expect these assays to be highly sensitive and precise as well. Although immunoassays excel as single analyte methods, it is possible to design an immunoassay capable of detecting a class of compounds (acyl urea insecticides), (refs. 2-4) or a mixture of compounds (nonionic surfactants), (ref. 8). However it is not reasonable to expect these same assays to be highly specific.

Many years of experience in clinical chemistry and increasing experience in environmental chemistry serve to verify the tremendous advantages of immunoassay. The high sensitivity of immunoassays is based upon the reversible but tight binding of an antibody to the compound to be analyzed. Since this binding is based on the sum of many noncovalent interactions (especially weak molecular interactions that are highly dependent upon the proximity of interacting groups), the interaction is highly specific. Specificity and sensitivity often permit one to carry out an analysis directly in a crude biological matrix as shown in Fig. 1A for paraquat in serum. This advantage often allows one to avoid some extraction and/or clean up steps leading to rapid, inexpensive assays of increased precision. This was demonstrated with an acyl urea insecticide where immunoassay allowed 100 times as many milk samples to be processed per worker day with a reduction in cost and an increase in precision and sensitivity (ref. 3).

As more effective compounds are developed such as the pyrethroid and acyl urea insecticides or sulfonyl urea herbicides, it will be important to develop more sensitive assays to detect toxicologically relevant residues. The tremendous sensitivity possible with some immunoassays with minimal clean up is shown for the acyl urea diflubenzuron in Fig. 1B.

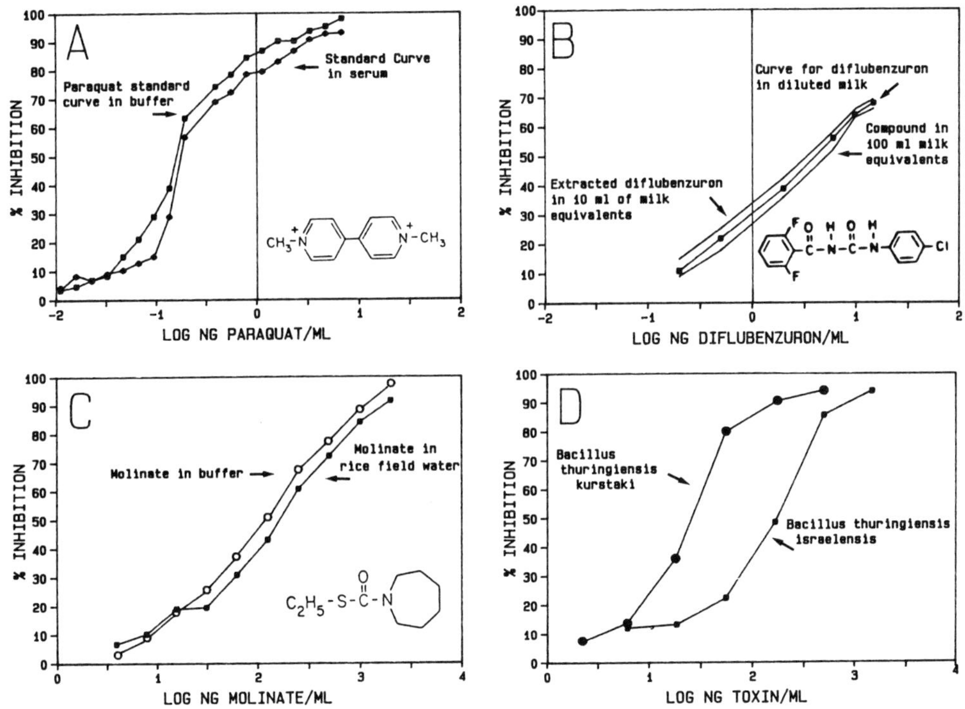

Figure 1. Some examples of applications for immunoassay in pesticide analysis. Percent inhibition in a competitive ELISA was plotted against log ng analyte/ml. Panels A-C show a standard curve of the analyte run in buffer solution vs one or more curves of either compound run directly in the environmental matrix or extracts of that matrix. A. Analysis of paraquat (1,1'-dimethyl-4,4'-bipyridinium) in sheep serum. These data demonstrate that up to 33 percent serum (♦,or lymph) can be analyzed for paraquat without shifting the standard curve. Larger amounts of serum can be analyzed by running a standard curve in serum, or paraquat can be extracted from the serum. The coating antigen concentration was 1 μg/ml and the antibody dilution was 1/4000. B. Analysis of diflubenzuron (Dimilin, N-[[(4-chlorophenyl)amino]-carbonyl]-2,6-difluorobenzamide) in whole milk (■) and extracts of whole milk (solid lines). Whole milk was extracted with ethyl acetate using the AOAC approved method. These data indicate that the antibody is able to extract diflubenzuron from micelles in the milk by mass action. Simple extraction procedures followed by adequate presentation to the antibody can result in highly sensitive assays. Up to 50% milk was added with no statistical difference in the standard curves. Within run variation was 3%, between run variation was 5% over several months. Antibody dilution was 1/1200. C. Analysis of molinate in rice field water samples. A field water sample (♦) was buffered with a concentrated PBS solution and spiked with molinate. No significant effect on the standard curve was observed when the final concentration of field water in the assay was 50%. Increased sensitivity may be obtained by extracting water samples into ethyl acetate or toluene. Extracts are concentrated, the molinate exchanged into propylene glycol, and added to antibody. Up to 5% propylene glycol may be used in the ELISA. The coating antigen concentration was 2 μg/ml and the antibody dilution was 1/4000. D. Standard curves for the purified crystal toxins of *Bacillus thuringiensis israelensis* (■) and *Bacillus thuringiensis kurstaki* (♦) run in formulated material. These data indicate that immunoassays can be used for the quantitative analysis of the proteins which are active ingredients of biological pesticides and genetically engineered material. These assays can be used to supplement bioassay for the monitoring of production and for quality control of formulations. Coating antigen concentration was 1 μg/ml. Antibody dilutions were 1/4000 for *Bacillus thuringiensis israelensis* and 1/800 for *Bacillus thuringiensis kurstaki*.

However, of the many advantages of immunoassay, the one most feared is sensitivity. There is the possibility that immunoassays can be employed to reduce further the levels of a compound that can be detected. Certainly this possibility is true, but the same fear applies to all physical assays. Fantastically sensitive immunoassays can only be developed with the loss of speed, precision, and economy. The true advantage of the method is in offering toxicologically relevant levels of sensitivity at a tremendous savings. In fact, the developer of an immunoassay has more control over the ultimate sensitivity of that assay than the developer of any other physical assay.

All immunoassays are based on the law of mass action and depend upon measuring the antibody bound to a compound and/or the free antibody and compound. However, they offer the advantage of being highly adaptable to the problem at hand. Using the same antibody one can develop a rapid field assay, a highly precise laboratory assay, or an assay to monitor production in real time. Many physical methods are used for the analysis of bound vs free material including scintillation counting, turbidity, polarization of light, visible, ultraviolet and electron spin resonance spectrophotometry, and a host of other methods. Each of these methods has numerous variations. Even if we address one major approach, that of the enzyme linked immunosorbant assay (ELISA), there are many subclasses of ELISA and the same basic assay is renamed (Western blot, dot blot) when it is applied to different matricies or when it is packaged for sale. Thus, the great advantage of phenomenal adaptability has led to excessively confusing terminology for a basically simple technology. Once one is familiar with one immunoassay format, the mysticism falls from the other formats as well. However, the terminology still serves to intimidate many analytical chemists. All of the variations in immunoassay format rely on a reversible antibody-compound interaction. Given an excellent antibody, almost any of the formats can be used to design a qualitative or quantitative immunoassay. In approaching this technology, the analytical chemist already has mastered the difficult concepts of sampling, sample handling, sample presentation, and the basic concepts of analysis. With the three working definitions of the terms below, immunochemistry easily can be added to an analysts' repertoire.

ANTIBODY-One of a class of serum proteins which binds to an antigen.

ANTIGEN-A molecule (usually a protein) which will elicit the production of antibodies and bind to them.

HAPTEN-A compound (usually a small molecule) which will bind with antibodies but will not alone cause the production of antibodies

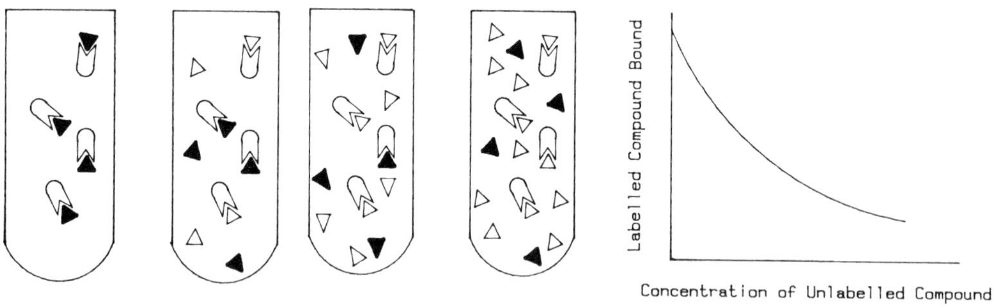

Figure 2. All immunoassays depend upon the reversible interaction of an antibody (⊂) with a compound (△). All assay formats determine the amount of compound free vs bound. A common way to carry out this procedure is to label small, constant amounts of the compound (▲) as shown above for a competition radioimmunoassay. Following separation, the antibody bound to radiolabeled compound is related to concentration to yield a standard curve useful for the determination of compound in unknown samples.

APPLICATIONS TO PESTICIDE CHEMISTRY

The advantages and limitations of immunochemical technology listed in Table 1 translate into some immediate applications to pesticide chemistry listed below. Some of the work from this and other laboratories is reviewed below in the context of these applications.

Analysis of compounds difficult to analyze by classical methods

The most obvious application of immunochemistry is for compounds difficult to analyze by classical methods. Gas liquid chromatography is most applicable to compounds that are volatile (this usually means hydrophobic and small), heat stable, and which have functionalities easily detected by available selective detectors. High performance liquid chromatography is less restrictive, but still depends upon specific functionalities for detection. The above criteria are not necessary for a successful immunoassay. Since there is a trend indicating that immunoassays are most easily developed to large hydrophilic compounds, immunoassays are complementary to existing technology and can even be used as HPLC detectors. Since it is human nature to only try a new technology when an existing technology fails, immunoassays unfortunately will first be applied to apparently intractable analytical problems. In most cases immunoassay will excel in this role as savior. However, if immunoassays are introduced into a laboratory by untrained personnel for the analysis of certain "nightmare" compounds (highly reactive, very small, very lipophilic, present in very low levels, immunosupressive), the entire technology may fail to be accepted.

We have applied immunochemical technology to a variety of compounds difficult to analyze by classical methods. The pyrethroid allethrin lacked a functionality easily recognized by GLC or HPLC; the acyl urea insecticides and paraquat both required a multistep work up procedure before GLC could be performed and HPLC procedures were too insensitive, while Triton X was nonvolatile, hard to extract, and yielded multiple peaks on HPLC (refs. 2-4 and 9-11). The structure and high activity of the sulfonylureas indicates that immunoassays also will be very applicable to those structures (ref. 5).

Discrimination of chirality and closely related compounds

Due to the increasing cost of starting materials and environmental concerns one can anticipate that an increasing number of compounds will be sold as optically rich or pure materials. Other compounds are racemates which may be degraded differently in the environment. Thus, the ability to distinguish among optical isomers at a residue level becomes important. The insecticide allethrin consists of 8 optical and geometrical isomers with most of the biological activity in the 1R, 3R, 4'S isomer. We developed an immunoassay which was very selective for this single isomer (refs. 9, 10) illustrating the potential of this technology to deal with chirality.

In the very competitive area of pesticide chemistry, compounds are often developed which have almost identical physical properties. This is the case for Dimilin and a number of other active compounds such as BAY SIR 8514. A very high efficiency HPLC column is needed to separate these compounds while we developed immunoassays which were able to detect the class of compounds and others that could distinguish clearly between Dimilin and a host of related compounds including some active compounds whose structures were released after the development of the assay (refs. 3, 4).

Development of immunoassays is most cost effective shortly after a compound class is discovered. Such class specific immunoassays could be very useful in the early development and screening of a new class of insecticides.

Analysis of human body fluids and as biomarkers

To evaluate human exposure it often is necessary to analyze human body fluids such as urine and blood. As we move toward relating exposure to toxicity, analysis of parent compounds, metabolites, and indicators of toxicity in occupationally and environmentally exposed individuals is of increasing importance. Immunoassays have been optimized in the clinical area just for this task, and highly sensitive immunoassays often require only trivial amounts of biological material. For instance immunoassays for paraquat are used for clinical diagnosis of poisoning (refs. 12-14), and we have found that crude blood, urine and lymph can be analyzed directly in an immunoassay for paraquat with greater sensitivity than other published techniques (Fig. 1A). Such rapid assays make it possible to carry out the pharmacokinetic studies needed to evaluate the significance of occupational exposure to pesticides.

Analysis of large numbers of samples

Immunoassay is so easy and inexpensive to automate that it is ideal for analyzing very large numbers of repetitive samples. This attribute makes immunoassays perfect for registration procedures, quality control work, and residue procedures where one major compound is suspect. Even if classical assays are used for confirmation, immunoassays are useful for eliminating negatives and ranking samples in terms of concentration for subsequent analysis. In this regard we have developed an immunoassay for the thiocarbamate herbicide molinate. The compound is very valuable in rice culture, but if released prematurely into drains it can result in fish kills. An assay has been developed which can handle the tremendous numbers of samples generated in an attempt to evaluate the dynamics of molinate in rice culture. Fig. 1C indicates that water from rice fields can be added directly to an immunoassay while retaining adequate sensitivity for the determination of release times for the water. These data also indicate that it is possible to develop assays for small, volatile, water unstable molecules. The assay can be adapted following extraction to a highly sensitive analytical tool for trace analysis. In order to integrate pesticide usage with social concerns it will be increasingly important to develop such markers for realistic levels of environmental contamination. Assays for relatively water soluble herbicides such as the thiocarbamates and 2,4-D and 2,4,5-T (ref. 15) are likely to be valuable in a surface water monitoring program while rapid assays for triazines and acetanalides (ref. 16) could be very valuable in monitoring of ground water.

Rapid and/or field analysis

In the above applications immunoassay simply replaced classical methods when it was more appropriate. However, there are many applications where immunoassay represents the only viable technique. For instance the molinate assay is being configured to allow data to be collected very rapidly in the field using little or no equipment. This assay will allow farmers and county agricultural personnel to test a field before draining water. Similarly assays can be used to check a field for toxic chemicals before reentry or to monitor drift. Such applications will be very important with highly toxic compounds such as paraquat (ref. 11) and parathion (refs. 17-20). Such assays also can be used to monitor a field before planting for phytotoxic herbicide residues. This has been important in the use of some triazine herbicides and will likely be important in the use of some sulfonylureas. The same basic technology could be used by a farmer to monitor veterinary drugs and pathogens, plant pathogens, and pesticides. This represents an entirely new concept in residue analysis by placing the technology in the field where the contamination occurs and by making the analytical technology available to the individual who needs it. This technology will be critical to wholesalers interested in insuring that residues are low in key crops and to pesticide companies in stewardship programs.

Analysis of products from research in biotechnology

Until recently most compounds introduced into the environment were, for economic reasons, relatively small molecules of simple structure. Biotechnology will certainly change this situation. For instance fermentation technology has already given us avermectin and ivermectin while a complex exotoxin from *Bacillus thuringiensis* is not allowed in current formulations and has been considered for development itself. These materials are far more complex than the compounds normally analyzed by GLC and HPLC. Certainly their analysis can be approached by classical means, but immunoassay seems particularly advantageous with these molecules. We can anticipate the development of more such complex molecules, and society needs the analytical technology to deal with them.

The major public concern in the immediate future will not be over fermentation products but over genetically engineered materials. It is certain that the vast majority of products from this type of research will be peptides and proteins. For such materials immunochemical methods are unsurpassed. In the last year the protein toxin of *Bacillus thuringiensis kurstaki* has been cloned into and expressed in plants, and Monsanto is still trying to test the same toxin in the field following expression in an alternate species of bacteria. In this regard we have developed immunoassays for the crystal toxin of the *kurstaki* and *israeliensis* strains (Fig. 1D, refs. 21-24). These assays are valuable for monitoring production, quality control, and may be adapted to actual residue procedures. In situations when foreign genes are released into the environment, immunochemical procedures are necessary to monitor expression as well as the products of expression in the environment. Thus, immunoassays become complementary to the genetic probes which will be necessary to monitor the actual gene. Fortunately, most companies will have developed both immunoassays and genetic probes in the research leading to a successful genetically engineered product. Thus, the analytical tools will be available, but regulatory agencies must have the foresight how to use them in environmental chemistry and how to obtain administrative and legal acceptance of the procedures. Thus, this technology has come the full circle. Not only do immunoassays represent a biotechnology themselves, but they may well be the best way to analyze for products of biotechnology research.

RECOMMENDATIONS

Although immunochemical technology is not a panacea, it is very unfortunate that it has not been used extensively in environmental chemistry for the last 20 years. It is important that the technology be implemented in both the public and private sector. Numerous small companies are capitalizing on the technology and will serve a major role in introducing and perfecting the technology in the pesticide area. However, if it is to be used effectively, both the primary producers of pesticides and regulatory agencies must have expertise on the technology in-house. Hopefully the trend toward increased use of immunoassay in analytical chemistry will continue (refs. 25, 26).

Validation of immunoassays

For immunoassays to gain acceptance as an analytical technique they should first be treated as any other analytical technique and compared against an established method. For such work it is important that an assay format be used which will give quantitative rather than qualitative data. The extensive comparisons needed for such validation have not yet been reported for any compound in the pesticide field. However, Wie and Hammock (ref. 2) have reported an extensive study on within, and between run variation for an ELISA for diflubenzuron (3 and 5% respectively) and the influence of the biological matrix on the immunoassay. Comparisons of immunochemical and classical procedures are becoming more prevalent in the literature. For instance, in the fungicide area, such data are available for benomyl, metalaxyl, and triadimefon (refs. 27-29). Van Emon et al. (ref. 11) found that GLC and colorimetric procedures for paraquat compared favorably with immunoassay in the analysis of samples generated in a worker exposure study, while Schwalbe et al. reported a good correlation between an enzyme immunoassay and fluoroimmunoassay for the herbicide diclofop-methyl (ref. 30).

For initial studies, immunoassays should be compared with classical techniques as if they were simply a new GLC procedure. However, immunoassays do have their own unique properties and regulatory agencies should move toward the establishment of validation and quality control protocols to address these properties. Such protocols will be of even greater importance as immunoassays are applied to the analysis of biological and genetically engineered pesticides for which no physical assay exists.

Introduction of immunoassays into the analytical laboratory

Both regulatory agencies and primary producers of pesticide chemicals must have in-house expertise on immunochemistry if they are to evaluate critically, analytical procedures developed in academia and the private sector. In developing in-house expertise it is important that a team approach be followed. It often is ignored that the most critical members of the team already are present. These individuals are the ones who appreciate the analytical and regulatory problems faced with a compound. Another very important member of the team often overlooked in the press for biotechnology is the chemist. Some toxins are easily coupled to proteins for analysis, while others will require innovative and difficult chemistry for attachment. There also is the temptation to mask key recognition sites for antibody specificity by using them for attachment. Very expensive approaches involving the screening of numerous antisera or clones then are needed to develop a useful assay when inexpensive chemistry should have solved the entire problem. There may be a long delay in the development of immunoassays due to the design and synthesis of an appropriate hapten. If this development is initiated in the early stages of compound development, appropriate haptens are likely to exist as synthetic intermediates, structural analogs or metabolite standards.

The antibody is the critical component of the immunoassay regardless of the format. Surprisingly, in the early phases of this program the production of antibodies is the task most easily contracted. If hapten and antigen preparation is careful, the production of antibodies is straightforward and can be done by numerous immunochemical companies with no loss of quality. As discussed earlier by Hammock and Mumma (ref. 1), polyclonal antibodies will be adequate or even preferable to monoclonal antibodies for many projects (ref. 31). In the past, rabbit polyclonal antibodies often have yielded assays more sensitive than the corresponding murine monoclonal antibodies. There also is no guarantee that monoclonal antibodies will yield a more specific assay. The success of an initial project should not be sacrificed simply to use monoclonal antibodies for the sake of apparent timeliness.

There is no doubt that hybridoma technology will dominate the immunoassay field in a few years for both commercial and scientific reasons. There are problems such as the analysis of multiple peptides responsible for the toxicity of *Bacillus thuringiensis israelensis* (ref. 32) where hybridoma technology is indispensible. The cost of production of monoclonal antibodies is dropping rapidly, especially in laboratories able to take advantage of economies of scale. The use of multiple, genetically diverse strains of mice for initial immunization followed by screening of crude serum before fusion is resulting, in some cases, in monoclonal antibodies of greater sensitivity than polyclonal antibodies.

Before undertaking such a project, the researcher should determine that a monoclonal antibody is needed and should develop a rapid screening procedure for isolating exactly the clones desired and eliminating those not desired. The researcher also should realize that the true strength of hybridoma to technology is in having a library of well characterized antisera. Thus, at some stage it will be important to acquire expertise in hybridoma technology.

Haptens, ligands, and antibodies are the reagents for an immunoassay, but they must be combined into an analytical method. There are numerous recipe procedures, but forging an assay is a craft, and the expertise to do this must be in-house. The assay must then be applied to real analytical problems. In both the development and application one can draw on expertise in the clinical area, but the matrices and ultimate analytical goals are much different. Thus, one needs an environmental chemist trained in immunoassay or a close collaboration between two individuals.

The sampling strategies, clean up procedures, data handling and a myriad of other tasks in analytical chemistry are identical regardless of whether one is relying on GLC-MS or immunoassay. The major expertise needed for immunoassays to be successful is already in-house in the form of good analytical chemists.

Selection of initial compounds for analysis

In introducing any new technology, it is important that the initial in-house application of the technology is successful. Thus, selection of the first problem to be addressed with immunochemical technology becomes critical. This selection process will vary with the company or agency but several general criteria can be considered.

The compound should be at least of moderate water solubility and should not be notorious for adhering to surfaces or the matrix for analysis. The compound should be nonvolatile and stable in water. The chemistry and physical properties should be familiar to the personnel involved, and at least one matrix of interest for analysis should be easy to handle.

Certainly the compound should be of interest to the research group, but it need not be the worst problem the group faces. The number of analyses performed on the compound should be high. If the analytical data can first be used in-house, frustrations with delays in acceptance by outside agencies can be avoided. There should be good classical procedures for the compound for validation studies. Thus, the decision making process regarding which compound to target is not complex, but it is essential that a good decision be made if the technology is to be accepted quickly. Fortunately, such sound decisions now are being made in a variety of industrial and regulatory laboratories.

ACKNOWLEDGMENTS

This work was supported by NIEHS Grant 2 R01 ES02710-06 and a grant from the California Department of Food and Agriculture. The assistance of K.D. Wing and S.I. Wie in early phases of this work is gratefully acknowledged.

REFERENCES

1. B.D Hammock and R.O. Mumma, Recent Advances in Pesticide Analytical Methodology, p. 321, American Chemical Society Publications, Washington D.C. (1980).
2. S.I. Wie, A.P. Sylwester, K.D. Wing and B.D. Hammock, J. Agric. Food Chem. 30, 943-948 (1982).
3. S.I. Wie and B.D. Hammock, J. Agric. Food Chem. 30, 949-957 (1982).
4. S.I. Wie and B.D. Hammock, J. Agric. Food Chem. 32, 1294-1301 (1984).
5. M.M. Kelley, E.W. Zahnow, W.C. Petersen and S.T. Toy, J. Agric. Food Chem. 33, 962-965 (1985).
6. A.I. Aronson, W. Beckman and P. Dunn, Microbiol. Rev. 50, 1-24 (1986).
7. J.J. Langone and H. Van Vunakis, Res. Commun. Chem. Pathol. Pharmacol. 10, 163-171 (1975).
8. S.I. Wie and B.D. Hammock, Anal. Biochem. 125, 168-176 (1982).
9. K.D. Wing, B.D. Hammock and D.A. Wustner, J. Agric. Food Chem. 26, 1328-1333 (1978).
10. K.D. Wing and B.D. Hammock, Experientia 35, 1619-1620 (1979).
11. J. Van Emon, B.D. Hammock and J.N. Seiber, Anal. Chem., 58:1866-1873 (1986).
12. D. Fatori and W.M. Hunter, Clin. Chim. Acta 100, 81-90 (1980).
13. T. Levitt, Lancet 8033, 358 (1977).
14. Z. Niewola, S.T. Walsh and G.E. Davies, Int. J. Immunopharmac. 5, 211-218 (1983).
15. D.F. Rinder and J.R. Fleeker, Bull. Environ. Contam. Toxicol. 26, 375-380 (1981).
16. S.J. Huber and B. Hock, Z. Pflanzenkrankh. Pflanzenschutz 92, 147-156 (1985).
17. C.D. Ercegovich, R.P. Vallejo, R.R. Gettig, L. Woods, E.R. Bogus and R.O. Mumma, J. Agric. Food Chem. 29, 559-563 (1981).
18. R.P. Vallejo, E.R. Bogus and R.O. Mumma, J. Agric. Food Chem. 30, 572-580 (1982).
19. K.W. Hunter and D.E. Lenz, Life Sci. 30, 355-361 (1982).
20. A.A. Brimfield, D.E. Lenz, C. Graham and K.W. Hunter, Jr., J. Agric. Food Chem. 33, 1237-1242 (1985).
21. S.I. Wie, R.E. Andrews, Jr., B.D. Hammock, R.M. Faust and L.A. Bulla, Jr., Appl. Environ. Microbiol. 43, 891-894 (1982).
22. S.I. Wie, B.D. Hammock, S.S. Gill, E. Grate, R.E. Andrews, Jr., R.M. Faust, L.A. Bulla, Jr. and C.H. Schaefer, J. Appl. Bacteriol. 57, 447-454 (1984).
23. P.Y.K. Cheung and B.D. Hammock, Current Microbiol. 12, 121-126 (1985).
24. P.Y.K. Cheung and B.D. Hammock, Appl. Environ. Microbiol., in press.
25. T.R. Roberts, Trends Anal. Chem. 4, 3-7 (1985).
26. J.M. Van Emon, J.N. Seiber and B.D. Hammock, Bioregulators for Pest Control, p. 307, American Chemical Society Publications, Washington D.C. (1985).
27. W.H. Newsome and J.B. Shields, J. Agric. Food Chem. 29, 220-222 (1981).
28. W.H. Newsome, J. Agric. Food Chem. 33, 528-530 (1985).
29. W.H. Newsome, Bull. Environ. Contam. Toxicol. 36, 9-14 (1986).
30. M. Schwalbe, E. Dorn and K. Beyermann, J. Agric. Food Chem. 32, 734-741 (1984).
31. J. Vinas, Pure & Appl. Chem. 57, 577-582 (1985).
32. R.M. Roe, P.Y.K. Cheung, B.D. Hammock, D. Buster and A.R. Alford, Bioregulators for Pest Control, p. 279, American Chemical Society Publications, Washington D. C. (1985).

Automation and data handling in the residue analysis laboratory

William J. McKinney

Analytical Chemistry Department, Shell Agricultural Chemical Co., California, USA

Abstract - Residue analysis has many opportunities to improve efficiencies using the emerging technologies of information handling and automation of instruments and procedures (robotics). In the future robotics will incorporate other technologies such as flow injection analysis and new approaches to old problems will allow these systems to duplicate or mimic most, if not all, standard laboratory operations. Recently, the capabilities of robots to link together several procedures into methods has been demonstrated. The use of sensors and rudimentary artificial intelligence will combine to make robots much more flexible. Laboratory Information Management Systems (LIMS) are available off the shelf and are capable of sample management, reporting and accounting functions. Data generated will either be passed on to LIMS or custom software which can easily be tailored to retrieve and report information into a word processing file. The future will find all of these areas integrated in sophisticated systems.

INTRODUCTION

We are in a period where the rate of technical change is greater than anything we have been accustomed to in our lives. In fact we cannot expect this rate to remain static, it is but an indication of the future. John Naisbitt in his book, Megatrends talks about the coming information society (ref. 1). Others refer to it as the post-industrial period (ref. 2).

As in many cases the scientific community is at the forefront in incorporating these new technologies into their daily work practices. All analytical groups, including residue analytical groups, tend to be among the first impacted in these areas, primarily because of the interdisciplinary nature of their work, but also because of continuing challenges to improve efficiency for the generation of information. There are four areas of research and development which are currently converging, which will produce a great impact on the ability of residue analytical groups to generate information in an efficient manner. These areas are: (1) Artificial Intelligence (AI)/Software; (2) Semi-conductor/Hardware Developments; (3) Robotics; and (4) Sensors. As these technologies converge, and their capabilities integrated, the result will be something that is much greater than the sum of its parts. An analogy can be found in nuclear energy. When a single uranium atom undergoes fission, a small amount of energy is generated. Adding subsequent atoms increases the energy generated per unit time proportionately. However, when critical mass is reached, we have a thermal nuclear chain reaction which can be used to generate much larger quantities of energy in a given time period. The growth of capability for generating residue analytical information as the above technologies are integrated into systems, will be dramatic.

In the cinema, robots are humanoid in appearance and capability. They function in familiar surroundings. Tomas Hirschfeld has more aptly described this view versus the actual capabilities of a robot (ref. 28).

> "We think of it as a tireless graduate student doing perfect work 24 hours a day at triple speed. This is nice, but not real. More truthfully, our new worker has no legs, one drastically crippled arm, two fingers, is deaf and blind (unless you have a lot of spare money and are willing to wait five years for good software), and exceedingly dumb to boot. He also needs huge amounts of your time for training, and can make real trouble if left unsupervised in a less than absolutely-controlled environment."

In fact, robots will not be human-like in familiar settings. Nor will the artificial intelligence that they use be capable of executing the variety of complex decisions made by a human in the foreseeable future. At least initially, AI will be used in contrived situations to make simple structured decisions.

It is interesting to note that manufacturing technologies are proceeding on parallel courses with automation in the laboratory. In fact there are many analogies that can be made between sample preparation and manufacturing a product. The difference is that the manufacturing group makes a product that is tangible, whereas the residue analysis group provides information. The importance of these parallel trends is that they spread the cost of developing the necessary concepts and capabilities over a larger group of people. Due to the size and importance of the manufacturing effort it brings more resources to bear on the problems.

In looking at automation of the residue analysis laboratory we shall examine the forces driving automation, analyze the current capability, look at future trends and finally, look at the impact of this activity on the way residue information is generated. Implementation strategies will also be discussed.

FORCES DRIVING AUTOMATION

The need to efficiently generate valid data has increased substantially over the past 30 years. The number and sophistication of regulatory/environmental questions has continued to grow with each revision of FIFRA (Federal Insecticide Fungicide Rodenticide Act). Early concerns with the chlorinated hydrocarbons led to many questions on the environmental fate of pesticides and their ability to concentrate in food chains. The projected toxicological impact in humans results in the setting of an ADI (Allowable Daily Intake) for any compound registered by the EPA. This concept assures that human exposure is below the no-effect level by applying significant safety factors to the ADI. The protection of workers has led to a large number of analyses to define the type of exposure they will be subjected to in different jobs, or in the laboratory. With increasing recognition of the ecological system in which foods are grown there have been many studies mandated on the impact of pesticides on non-target organisms. Recently there has been an increasing emphasis on the contamination of ground and surface waters. These data are now all generated under GLP (Good Laboratory Practices), a set of rules designed to guarantee to the extent possible that the data are valid for making the intended decisions.

Against this backdrop of desire to obtain sufficient knowledge of the impact of pesticides in the environment is the fact that industrial, governmental and academic researchers are faced with limited resources. The staff required to generate these data are highly trained. They view themselves as decision makers and are in a period of rapid technical change in which they have to spend a growing proportion of their time incorporating new capabilities and concepts into their knowledge base. This is, in fact, desirable for the knowledge worker who has spent a great deal of his life preparing for decision-making work, and is commensurate with the salary this person commands.

In addition, the cost of capital equipment has escalated continuously. This includes the sophisticated instrumentation required by the modern residue analyst, as well as the space required to house both this person and the instrumentation. Unfortunately, this space and instrumentation are used less than 24 hours a day. Starting up the instrument and obtaining equilibrium can take a significant portion of this work day.

Many of the advantages to automation are quite obvious. There is the potential for significant capital avoidance or improved capital utilization because equipment can function on a 24-hour day, 7-days a week basis. With automated equipment, worker exposure to chemicals can be minimized. Automation should produce faster sample turnaround time, and one would hope to obtain at least equal, if not better, precision from automated equipment (ref. 4). Utilization of automated equipment must permit automatic data handling of the information that is generated (ref. 5). If one looks at the required functions to obtain information from the sample there is log in, sample preparation, introduction to an instrument, analysis by the instrument, recording of data produced by the instrument, interpretation of the data and feedback, documentation and archiving (ref. 5). Increases in the efficiency or capability in any area can put severe strains on other aspects of the total process. Laboratory automation must address all aspects of the process in an integrated system with the flexibility to accommodate hardware and instrumentation produced by reputable vendors.

ANALYSIS OF CURRENT CAPABILITY

Currently there are a number of systems for automating individual LUOs (Laboratory Unit Operations) as well as entire analytical procedures. Depending on the type and volume of sample, various types of autosamplers, robotics, FIA (Flow Injection Analysis) and/or segmented flow systems have been utilized. Data management systems are being developed on an analogous track, but in many cases their use is being frustrated by communication problems between systems manufactured by different vendors. First attempts to integrate these capabilities into flexible, integrated systems for moderate to low sample volumes are being made, and the future is being actively discussed by virtually all instrument companies (ref. 6).

Sample Preparation and Analysis

A number of laboratory unit operations have been effectively automated. We are all familiar with the auto samplers for chromatographic systems and there are auto samplers for most analytical instruments. Automated valve stations for directing fluids to different parts of chromatographic systems (ref. 7) or to run multiple samples through small, disposable, absorbent tubes for clean-up are commercially available (ref. 8). The use of valve and syringe devices to dilute, preconcentrate, and derivitize samples has also been demonstrated (ref. 9). The use of an automated valve-switching device to provide cleanup of oily samples on a gel permeation chromatograph has also been successfully demonstrated (ref. 10), and recently, the ability to automatically remove solvent from the collected fractions has been added.

New detection methods are having a tremendous impact on the ability to automate and improve the efficiency of residue analysis. For example, the use of mass selective detectors (MSDs) can in many cases greatly simplify sample clean-up. The use of immunoassay procedures to quantitate analytes at low levels in complex matrices has been demonstrated in several instances. The ability of post-column reaction detection to provide sensitive and selective detection for pesticides at low levels has also been demonstrated. The promise of the use of super-critical fluids for both extracting analytes from samples, or as a mobile phase in a chromatographic system is also being pursued by many instrument companies. Significant decreases in the price of MS/MS instruments could also make them an effective analytical tool for trace analyses. Although these new detection methods are not strictly related to automation, they in fact do have an enormous impact on sample preparation prior to final separation and detection.

Laboratory robotics have made significant advances recently (ref. 11). As can be surmised from Table 1, they are currently being used to address specific analytical problems with significant sample volumes, or a group of similar analytical problems. The ability to handle homogeneous liquid samples is well advanced. Successful work has also been reported where liquid and solids are mixed and adaptation of conventional laboratory methodology is relatively straight forward.

Substantial effort is required to design and install robotics applications. An application described recently (ref. 17) indicates research groups are beginning to address the problem of more versatile robotics to handle situations where multiple analyses for moderate sample loads are needed. The design utilized put up to ten procedures in a single work area for a robot and further enhanced the flexibility of the system by making the robot mobile so that it could be "plugged" into different work areas with different LUO's using pegboards mounted on the wall. Analytical procedures using headspace gas chromatography, colorimetry, potentiometry and standard preparation utilizing a single robot were described. Whereas the first method developed on this system took approximately four months to implement, the last took about one hour since it was essentially a matter of linking together previously automated procedures. Even though much effort is required to intially get involved in robotics (ref. 16), a base is built which facilitates future efforts. This direction is certainly being enhanced by vendors' efforts to provide preprogrammed routines for many common laboratory LUO's (ref. 6).

Other approaches to improving the flexibility of robotics have also been reported. One method has been to gang a large variety of LUO stations around a single robot for multi-product sample preparation for LC and GC (ref. 18). Another approach has been to place two robot arms such that they form a figure eight and have a common working area (ref. 19). This configuration raises the question of potential robotic conflicts in which the primary loss (capital and information) is incurred by the analyst. The use of sensors to mitigate this problem has already been incorporated in a commercial system (ref. 6). Safety aspects of this feature should also be noted. In addition, tactile sensing capability has been added which allows greater speed, since it provides error checking.

Robotics have to date demonstrated enormous capability. Problems remain to be solved. Innovative approaches and imagination as well as common sense (Table 2) will be required to effectively utilize robots. As has been demonstrated to date, it is not an impossible task.

Data/Information Handling

Instrument data systems are well advanced and frequently are integrated with the instrument control function. Those that are not integrated with this function are capable of controlling or being controlled by external events. Early attempts at using AI concepts to design chromatographic separations are being made and extensions of AI technology to other areas is being attempted and discussed. Communications between instruments, data systems and robotics systems are a continuing source of problems.

Laboratory Information Management Systems (LIMS) are proliferating at a rapid rate. The central role of the data base management system has been recognized. LIMS Systems have been designed with considerable flexibility to adapt to different laboratories and instruments. Communications between various computers are still, however, a core problem. Many groups have been putting their information into data bases for a number of years. How these data bases will interact with the new LIMS system and with the individual instruments that make up the LIMS system is frequently a difficult problem to resolve. The cost of integrating an old data base with a new LIMS system can be staggering. The basic problem is how each computer will interpret the string of 0's and 1's being transmitted to it. If the data is merely stored in an open text field, it will not be easy to search. If it is sent in a highly-structured manner, then it requires a very highly-structured file to receive it. Otherwise spread sheets and graphs will be merely images of the data they represent. The central theme is to process data into information and not vice-versa. As a result of the communications difficulties and the probability that much programming would have to be duplicated, raw data will not generally be resident in a central data system. Rather, if questions arise which require a re-examination of these data, the person will have to go back to the analyst and the instrument involved where these data will be archived and the appropriate programming will be available.

As the LIMS systems are being developed, barcodes are being used extensively to reduce data entry/transcription errors. Barcodes are being used to identify samples and blanks, instruments, calibration standards, analysts, methods to be applied to the sample and even to instruct instruments on what method they should be running. Barcodes are a major advance for sample control and identification and will be key to future systems.

Custom reports are generally desired on all of the off-the-shelf systems to meet special information needs and work practices of the purchaser. Generation of new searches, reports, etc. requires knowledge of the database structure and query language. In some cases this can be done by laboratory support staff, but will only occur after extensive training. Limited EPA formats are available on some of the systems but these are generally of the "fill-in-the-blank" type.

A residue analysis information handling system has requirements generally similar to the LIMS which are currently being offered. As stated previously, custom programming would be required to devise an integrated system for a particular laboratory. A general description of the requirements for a LIMS for residue analysis would include the following: Data would be entered into the data base on sample history, sample processing, laboratory work up and analytical results. The latter should be reviewed prior to actual entry into the data base to avoid erroneous data being propagated. The data would come from multiple sources. This could include analytical instruments, keyboard entry, bar code entry, etc. Upon loading all these data into a data base it would be desirable to perform the following operations: Inventory and control of the samples, comparison of the data received with the samples with the protocols for the experiment, output of the data in an EPA-acceptable format and retrieval of sample data for control or information purposes (see Figure). Such a system has been in operation at Shell Agricultural Chemical Company laboratory at Modesto, California for six years.

Design and implementation of MRIS (Modesto Residue Information System) and other LIMS have demonstrated several principles which were important to their success. The most basic and essential concept is; if you as the customer do not understand the job you wish performed in minute detail, the programmer will not either. Reducing any function to code requires that all sample and information flow be completely understood. All decision points and the criteria for their execution need to be reduced to terms comparing fields such as equal to, not equal to, greater or less than, etc. Another concept is to allow variable complex input, e.g. titles, abstracts, to be input as opposed to designing a program capable of generating these aspects of a report. To this end having report files which are compatible with word processing systems may save a great deal of effort. In addition, other small changes for non-routine reports may greatly improve the flexibility of the system. Finally, involvement of the affected staff is not only essential for success but offers a perfect vehicle for a team building exercise which can pay off handsomely in terms of people's improved understanding of the interrelationship of their jobs. This may in fact lead to more effective ways of conducting business.

FUTURE TRENDS

There will be a continued increase in the capabilities of all elements of the systems previously mentioned: LIMS, robots, AI, sensors, instrumentation and communication; all augmenting the primary thrust--INTEGRATION. In looking into the future, it is anticipated the following will have been demonstrated technically in the next five years and widely implemented in the next ten years.

Robotics

In this time frame, robots will be capable of performing or mimicking all standard laboratory operations. Reliability will be one of yesterday's questions and high-quality consumables will be available for most, if not all operations that the robot performs.

Sensors will be in wide use to provide inputs in feedback loops for controlling the robots operations as well as for analytical sensors. Some major technologies that will impact on sensors are microelectronics and micromechanics. Use of microelectronics to provide analytical sensors is well known and has been the subject of several reviews (refs. 21, 22, 23). Applications of the microelectronic sensors can expect to be greatly facilitated by immunological approaches due to the potential selectivity and sensitivity advantages (refs. 22, 23). Micromechanics in which small mechanical structures are machined onto silicon chips to sensors capable of measuring gas flow, temperature, strain, acceleration, vibrations, etc., can be expected to provide the inputs for control of robotic functions (refs. 24). In addition, micromechanics will be used to construct instruments on chips. Gas chromatography has been demonstrated and other devices ranging from liquid chromatographs to low resolution mass spectrometers are deemed feasible (ref. 23). Devices such as these are capable of high information content feedback for robotics control, if not final quantitation.

Samples will be processed in a serial as opposed to a batch mode for the following reasons:[4)] (1) In a serialized procedure the robot will tend to utilize laboratory equipment on a continuous basis as opposed to a batch operation where much of the equipment tends to be sitting idle while another LUO is being performed; (2) In well-designed serial procedures it is possible to assure uniform sample histories improving precision for many operations. Indeed sources of error may be more apparent in a serialized procedure (ref. 12); (3) Serialized procedures will provide results on the initial samples faster than would be the case in a batch operation. Programs to integrate the robots movements and optimize use of laboratory equipment so that multiple serialized procedures can be run concurrently will be part of the robots' operating systems.

LIMS

Most of the concepts required for an effective LIMS have already been demonstrated and were previously discussed. Improvements will come in terms of ease of daily use and flexibility of design using a basic system or tool kit. These changes will occur as a result of continued increases in computational power per dollar as well as significant decreases in the cost of data storage. Relational databases with natural language interfaces backed by embedded AI programs will be in common use. Input to these systems via keyboard will have peaked with mouse, barcode and limited voice entry being applied to this area.

Utilization of databases will require less and less systems knowledge. Data dictionaries combined with relational data bases and networking will facilitate access to data residing on separate computers. It will be reasonable to make data, local to specific instruments, appear globally to all nodes on the network. The dictionary and network software will handle the details of how to retrieve the data. It will be essential for vendors to conform to a commonly-accepted standard such as the ISO (International Standards Organization) open system networking model. Examples would be the standard Ethernet protocols or the MARS protocol being used by General Motors to network their manufacturing facilities. It will become appropriate to exclude vendors who do not provide this capability with their lab equipment or instrumentation.

Integration

Harry S. Hertz, director of the Center for Analytical Chemistry at the National Bureau of Standards, has said the last remaining piece in automating the laboratory of the future is the incorporation of feedback loops and expert systems. Feedback loops have already been demonstrated with robotics (refs. 12, 13, 14). Reports of the use of programs such as Personal Consultant, a tool kit for building expert systems, by two engineers with little programming experience to build three expert systems in one week including one that selects materials for steam generators to minimize corrosion are being published in popular business magazines (ref. 25). Tremendous strides are being made in this area as well.

Use of expert systems to provide the decision process for feedback loops does not require the same degree of sophistication as required by DENDRAL to deduce the structure of an organic compound from spectroscopic data. Rather the majority of applications could be handled by systems of similar complexity to the expert advisor program for chromatography being developed by Varian, instrument control of a triple-quadrupole mass spectrometer or determining the mineral constituents of a rock sample from a powder diffraction pattern (ref. 26). In this article J. Karnicky of Varian predicts that in the near term, expert systems will:

* "Routinely perform consultative and educational functions (e.g., automated methods development and management of structured data bases); and

* Implement intelligent response optimization strategies that use rule-based methods to control starting conditions, process parameters, and termination conditions."

Management of structured data bases integrates both the needs for handling data from robots but also provides the LIMS link.

The final hurdle to be overcome is communications between various instruments and data systems. Some progress has already been made in providing standard communication ports and protocols. Certainly standards to solve the problem are technically possible. The argument can be made that as automation gathers momentum, thought out or defacto standards will be adopted and vendors who do not support them will see their markets reduced. The use of expert system software may also impact the problem.

CONCLUSION

Laboratory automation has arrived for Residue Analysis. The basic robotics systems for many of the LUO's required for residue procedures have been demonstrated. In addition feedback loops to robotics are being utilized to optimize analytical procedures. Flexible LIMS systems are being offered by a number of vendors and the ability to custom program a system to handle information management for a Residue Analytical Laboratory has been demonstrated. The basic concepts and capabilities for integrated automation are in place. Future trends will allow utilization of this technology without the detailed systems knowledge currently required.

Carl Djerassi has addressed human fears of being replaced by expert systems. He feels these attitudes must be changed since expert systems, properly used, complement human abilities rather than substitute for them (ref. 27).

ACKNOWLEDGEMENT

I would like to acknowledge the combined efforts of many who, when called upon, gave of their time and effort in the preparation of this paper. For their ideas, comments and suggestions, I am truly grateful.

Table 1. SELECTED APPLICATIONS OF ROBOTICS TO TRACE ORGANIC ANALYSIS

Sample Matrix	Analyte	Conc. Range	LUO's	Comments	Ref.
Water	Triallate	ng/ml	R,P,B,A	Comparable precision to manual	11
Rodent Chow	Chlorpyrifos	ug/g	E,C,CC,W,A	Method optimization (extraction time)	12
Animal Chow	Methyleugenol, disulfiram & DEHP	--	E,C,F,LC/GC	Method optimization, decision tree to chose extraction solvent by comparing results spiked feeds at two temperatures	13
Blood/Urine	Tolazimide	ug/ml	CC,R,C,LC	Method optimization of column cleanup	14
Blood Plasma	Prostaglandin E, analogue	pg/ml	V,CC,R,LC	LC fractions analyzed by RIA	15
Rodent Chow	Tridipphane	ug/g	CC,E,F,W,R,A	Sequential sample processing	16

R = Reagent addition
P = Partition
B = Break emulsion
A = Sample into vial
E = Extract
C = Centrifuge
CC = Column cleanup
W = Weigh
F = Filter
LC = HPLC
GC = GC
V = Evaporation

Table 2. THINKING LIKE A ROBOT

Classical Problem	Human Approach	Robotics Problem	Robotic Approaches
Paritioning	Shake phases in separartory funnel	Emulsions	- Rotating Test Tube - FIA - Disposable Column Separations
Extraction of Analytes from Solids	Grind and Extract	Adaption of Grinding Operation	- Vortex mixer - Dissolve Sample and Flash Off Solvent - Imbibe Super Critical Fluid and Explosively Decompress
Sample Routing and Identification	Labels	Limited Vision, Pattern Recognition Capability	- Barcodes

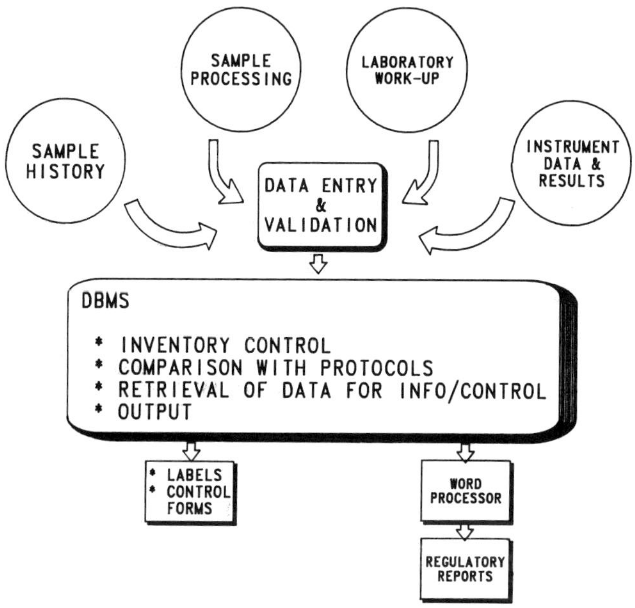

Fig. Sample/Information Flow Chart.

REFERENCES

1. J. Naisbitt, Megatrends, p. 1, Warner Books, New York (1984).
2. J.W. Wilson and J.H. Dobrzynski "And Now, The Post Industrial Corporation", Business Week, p. 64, McGraw-Hill (March 3, 1986).
3. R. Brandt, "How Automation Could Save the Day", ibid., pp. 72-74 (March 3, 1986).
4. F.H. Zenie, Advances In Laboratory Automation - Robotics, pp. 1-16, Zymark Corporation, Maine (1984).
5. F.H. Zenie, Advances In Laboratory Automation - Robotics, pp. 43-59, Zymark Corporation, Maine (1985).
6. Chemical and Engineering News, American Chemical Society, Washington, DC, pp. 21-32, (March 24, 1986).
7. E. Hillier, R. Cotter and M. Andrews, Am. Lab., p. 67 (August, 1986).
8. L. Yago, Am. Lab., p. 118 (October, 1985).
9. F. Verillon and R. Glandian, ibid., 142.
10. D.L. Stalling, R.C. Tindle and J.L. Johnson, J. Assoc. of Anal. Chem., 55, 32 (1972); and R.C. Tindle and D.L. Stalling, Anal. Chem., 44 1768 (1972).
11. D.M. Haile, and R.D. Brown, Advances In Laboratory Automation - Robotics 1985, p. 123, Zymark Corporation, Maine.
12. B.E. Kropscott, et al., ibid, p. 131.
13. J.J. Rollheiser, K.M. Stelting and W.A. Schmidt, ibid., p. 149.
14. S.F. Kramer, M.J. Levitt and M.M. Passarello, ibid., p. 465.
15. G. Schoenhard, et al., Advances In Laboratory Automation - Robotics 1984, p. 61, Zymark Corporation, Maine.
16. M. Dulitzky, Advances In Laboratory Automation - Robotics 1985, p. 163, Zymark Corporation, Maine.
17. M. Markelov, M. Antloga and S.A. Schmidt, ibid., p. 209.
18. C. Hatfield, et al., ibid., p. 599.
19. B.P. Hatton, P. Abley and T.J. Lux, ibid., p. 621.
20. C.R. Knipe, C.R., ibid., p. 61.
21. W.R. Seitz, Anal. Chem. 56, 16A (1984).
22. J.D. Czaban, Anal. Chem., 57, 345A (1985).
23. H. Wohltjen, Anal. Chem., 56, 87A (1984).
24. R. Brandt, "Micromechanics: The Eyes and Ears of Tomorrow's Computers", p. 88, Business Week, McGraw-Hill, (March 17, 1986).
25. "Expert Systems That Can Be Written In A Jiffy", p. 90E, Business Week, McGraw-Hill, (March 3, 1986).
26. J. Karnicky, H.B. Woodruff, C.M. Wong, and S. Ennis, Anal. Chem., 56, 1312A (1984).
27. R.E. Dessy, ed., Anal. Chem., 56, 1200A (1984).
28. Tomas Hirschfeld, Trends In Analytical Chemistry, 4, 195 (1985).

A two-dimensional reaction gas chromatographic system for congener specific determination of polychlorinated organics

S. Kapila, D. Duebelbeis, R. Malhotra, A.F. Yanders and S.E. Manahan

Environmental Trace Substances Research Center and Department of Chemistry, University of Missouri, Columbia, MO 65201

Abstract - Polychlorinated compounds have been characterized by two dimensional gas thromatogrphy using an intermediate electron capture cell. The pattern of the neutral products obtained from electron capture reactions are in many cases quite distinct and can be used in confirmatory analysis of chlorinated aromatics compounds including PCB's down to picogram levels.

INTRODUCTION

The identification and quantitation of constituents of complex environmental and biological samples is a challenging task. However, the importance of such characterization for toxicological, environmental impact and medical diagnostic studies can hardly be overstated. The magnitude of the task increases exponentially when analysis of individual isomers at trace and ultra trace quantities is desired. A number of analytical techniques ranging from radioimmunoassay to gas chromatography interfaced to tandem mass spectrometry (GC-MS/MS) have been employed for the purpose. However, even with high resolution gas chromatography - high resolution mass spectrometry the confirmatory determination of positional isomers is quite tedious, since the determination is largely dependent on separation capabilities of the chromatographic system. The mass spectra for such positional isomers are generally quite similar and thus offer limited information in identifying the individual isomer. The gas chromatography fourier transform infrared spectroscopy (GC-FTIR) systems with cryogenic trap has shown promise in this area,but the sensitivity of technique does not compare favorably with sensitivity routinely obtained with such systems as GC-ECD and GC-MS. Laborious sample clean-up, fractionation and/or derivatization steps are often required and incorporated into the analytical schemes. Multidimensional chromatogrpahy and derivatization reactions have been gaining popularity to eliminate the shortcomings and enhance the capabilities of the routine analytical schemes (refs. 1,2).

The combination of two dimensional gas chromatogrpahy with a gas phase reactor offers unique solutions to overcome the limitation of gas chromatographic systems. In this approach, the preliminary separation of the sample is carried out by one column and selected components are transferred to a reaction chamber, the products and the parent molecules are then separated by a second column, and under suitable circumstances distinct "product spectrum" can be obtained and used for confirmatory analysis.

A number of polychlorinated molecules, in the presence of suitable reaction moderators have been shown to give distinct neutral products (ref. 3-5). The data obtained so far seem to suggest that these products are produced by random loss of chlorine. The "product spectrum" thus obtained is quite specific for various polychlorinated molecules. Thus an electron capture detector can be used as a reactor for polychlorinated compounds. A simple system showing the feasibility of the concept was reported by Aue and Kapila (refs. 6-8). The present study deals with a refined system and shows application in the confirmatory analysis of polychlorinated molecules.

EXPERIMENTAL

The flow schematic of the chromatographic arrangement is shown in figure 1. The chromatographic system consisted of two independently operable gas chromatographic ovens. Column 1 was housed in one oven and columns 2 and 3 were housed in the second oven. Column 1 consisted of a 30 mm bonded methylsiloxane fused silica capillary while columns 2 and 3 consisted of 60 m capillary with methyl (+5% phenyl) siloxane and methyl-phenyl siloxane, stationary phases, respectively. The heart of the system is the reactor, which consisted of a low dead volume electron capture cell. The selected peaks eluting from column 1 were introduced into the reactor with the aid of a switching valve. The

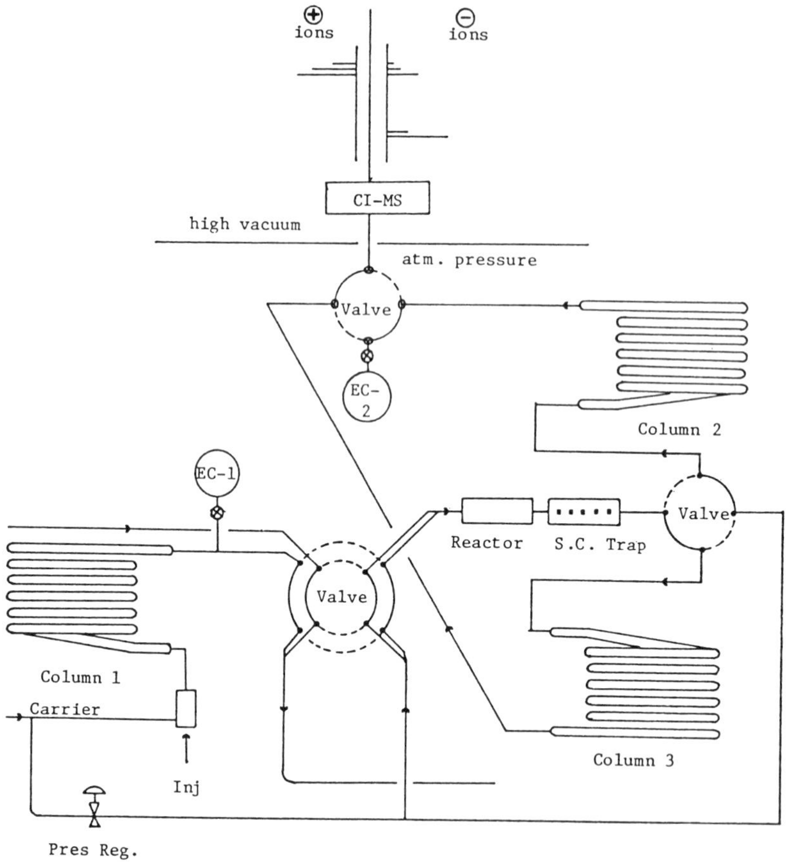

Figure 1: Schematic of two-dimensional reaction chromatographic system.

parent molecule and/or products were trapped with serial cryogenic traps consisting of independently controlled cooling and heating zones. The trapped components were introduced into the second column as a plug by rapid heating of the traps. The components eluting from the second column were detected by a second electron capture detector or identified by a quadrupole mass spectrometer operating in electron impact (EI) or chemical ionization mode (CI).

RESULTS AND DISCUSSION

The use of chemical modification (prior to chromatographic separation) for confirmatory analysis of pesticide residues has been reported (refs 8,9). The chromatographic arrangement described in the present report can be thought of as an in-line derivatization system. The system allows the reactions to be carried out on selected components of a mixture as opposed to classical derivatization approach which generally involves the reaction of the total sample rather than the reaction of a single constituent, even though the latter would be preferable since it eliminates the possibility of interference from other compounds present in the sample. The two dimensional chromatography offers a workable solution for carrying out reactions on individual components of complex mixtures.

A number of polychlorinated molecules and nitroaromatics including many chlorinated pesticides yield neutral products from an electron capture detector. These neutral products of the dissociation resonance capture are sometimes capable of reacting with thermal or near thermal electrons. The extent of the reactions is strongly dependent on electron energy and is affected by the presence of moderating agents. The electron energy and extent of reaction can be easily controlled by varying the applied potential.

In the case of polyhalogenated and nitro aromatics, the data so far obtained in our studies seem to suggest that the products are formed by random loss of halogens and nitro groups. A number of halogenated pesticides in the presence of iso-octane as a moderating molecule give distinct products. Figure 2 shows the chromatograms of products and residual parent compounds for tetrachloronitrobenzene and tetrachlorbenzene. The predominant products were found to be tetra and trichlorobenzenes. The characteristic products were also obtained from isomeric compounds such as from hexachlorocyclohexanes (ref. 5). The observation that the products are formed by random loss of chlorine is supported by the product patterns obtained form tetrachlorobenzene, figure 3. As would be expected 1,2,4,5-tetrachlorobenzene gave just one trichlorobenzene, 1,2.3.4 tetrachlorbenzene gave two trichlorobenzenes and 1,2,3,4 tetrachlorobenzene resulted in the formation of three trichlorbenzenes. Furthermore, the proportion of trichlorobenzene products was found to agree with the predicted values, figure 4. The distinctive product patterns hold promise in conformatory screening of a wide variety of halogenated organics at extremely low

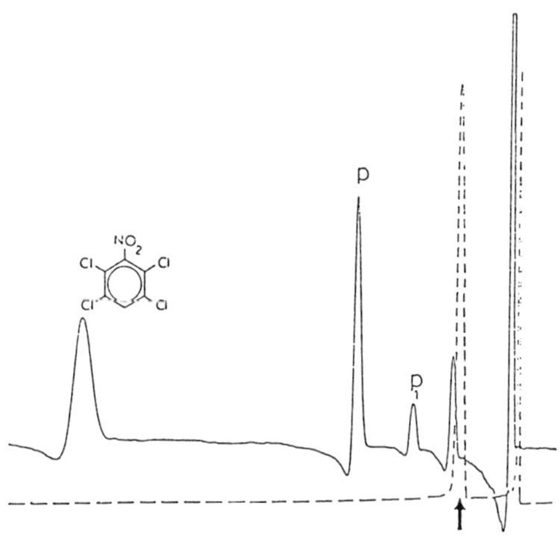

Figure 2: Chromatography of TCNB (200pg) and its electron capture reaction products.

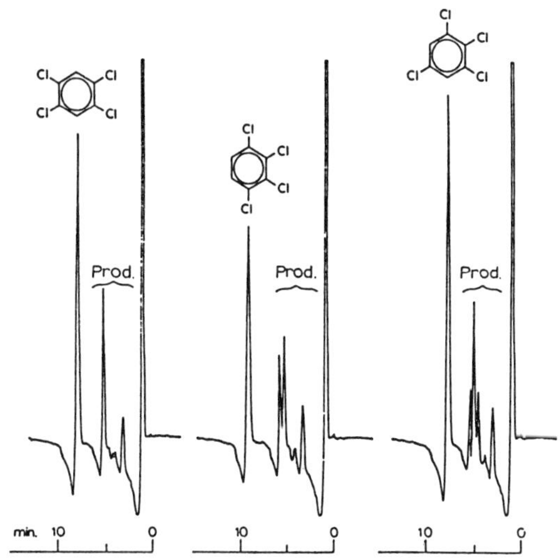

Figure 3: Chromatography of tetrachlorobenzene isomers (100pg each) and their electron capture reaction products.

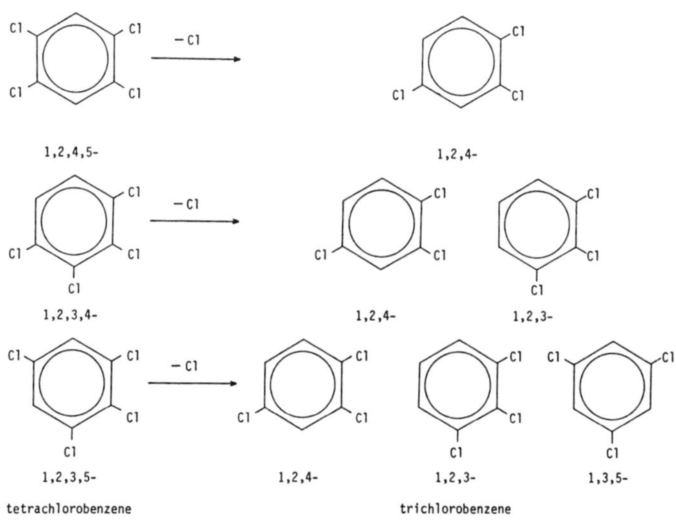

Figure 4: Trichlorobenzene isomers expected as a result of random loss of chlorine from tetrachlorobenzenes.

levels, such as congener specific determination of polychlorinated biphenyls in human blood, figure 5, this is especially true since resolution of all components of these samples is very difficult which is further complicated by lack of ready availability of individual standards. Furthermore, the system can also be useful in establishing the identity of isomeric standards of polyhalogenated molecules.

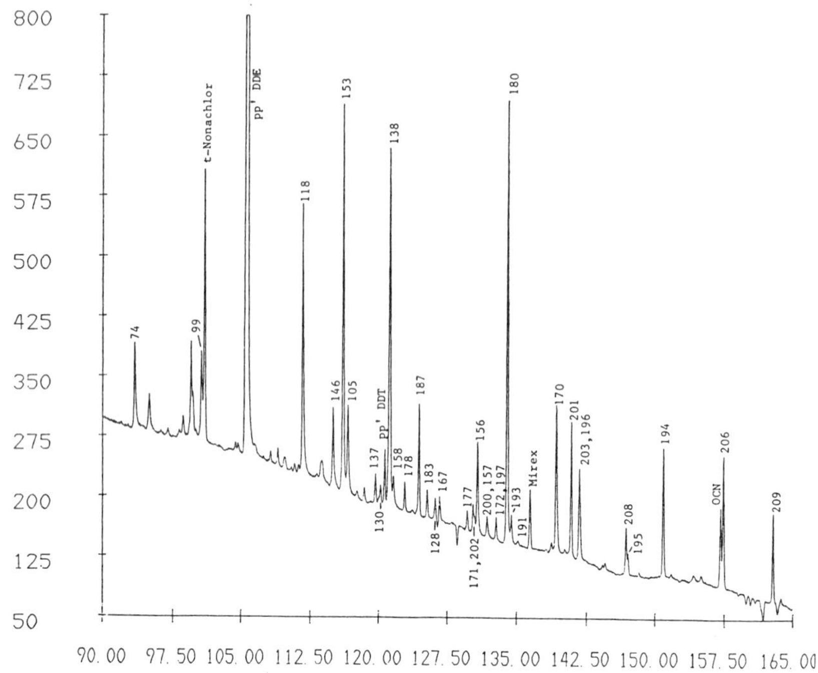

Figure 5: Chromatography of polychlorinated biphenyls and chlorinated pesticides found in human blood serum.

REFERENCES

1. G. Schomburg, H. Koetter, D. Stoffels and W. Ressig, Chromatographia, 19, 382-390 (1984).
2. G. Schomburg, F. Weeks, F. Mueller and M. Oreans, Chromatographia, 16, 87-91 (1982).
3. C.R. Hastings, T.R. Ryan and W.A. Aue, Anal. Chem., 47, 1169-1172 (1975).
4. S. Kapila and W.A. Aue, J. Chromatogr., 108, 13-17 (1975).
5. S. Kapila and W.A. Aue, J. Chromatogr., 148, 343-315 (1978).
6. W.A. Aue and S. Kapila, Anal. Chem., 50, 536-538 (1978).
7. W.A. Aue and R. Aigner-Held, J. Chromatogr., 189, 119-125 (1980).
8. S. Kapila and W.A. Aue, J. Chromatogr. Sci., 15, 569-572 (1977).
9. J. Lawerence, Pesticide Analysis. Confirmatory Tests in Pesticide Analysis, K.G. Das (ed.), Dekker: New York, NY, 1981.
10. J.L. Sericano and A.E. Pucci, Bull. Environ. Contam. Toxicol., 33, 138-143 (1984).

Chiral phase HPLC columns

P G Baker

Laboratory of the Government Chemist, Cornwall House, Waterloo Road, London SE1 8XY

Abstract The recent development of chiral phase columns means that the separation of optical isomers of individual pesticides can now be achieved. The paper examines some of the chiral phase columns which have been used with pesticides to date.

INTRODUCTION

The timely introduction of chiral phase columns has greatly facilitated the analysis of the widely used synthetic pyrethroids, including the newly developed single isomer or single diastereoisomer compounds. These chiral phase columns can also be used for the resolution of other optically active pesticides. Work is now being carried out on the development of numerous chiral phases, although only a limited number are available commercially at the present time.

RESULTS

1. Pirkle columns.

The most widely used chiral phase column for pyrethroids to date has been the Pirkle 1-A column. This is an aminopropyl packing which has been modified with the N-3,5-dinitrobenzoyl derivative of D-phenyl glycine.

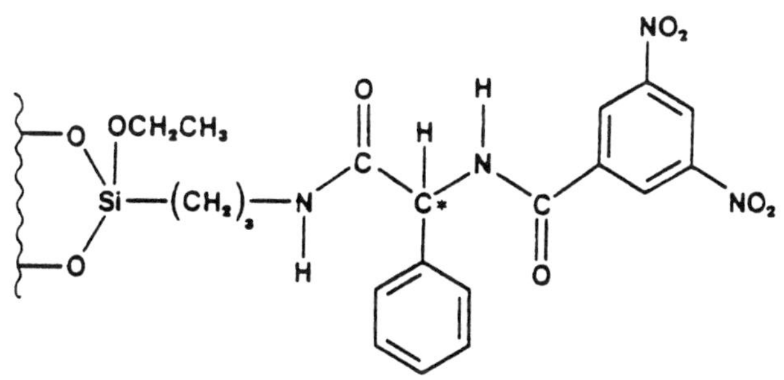

Figure 1 Structure of Pirkle covalent 1-A packing.

This column with a mobile phase of hexane and small amounts of a polar modifier has been used to separate many of the major pyrethroids. For example cypermethrin can be separated into seven (or possibly eight) components and fenvalerate into four. The mixture of 16 isomers obtained by adding bromine to cypermethrin (tralomethrin) can be resolved into 11 components.

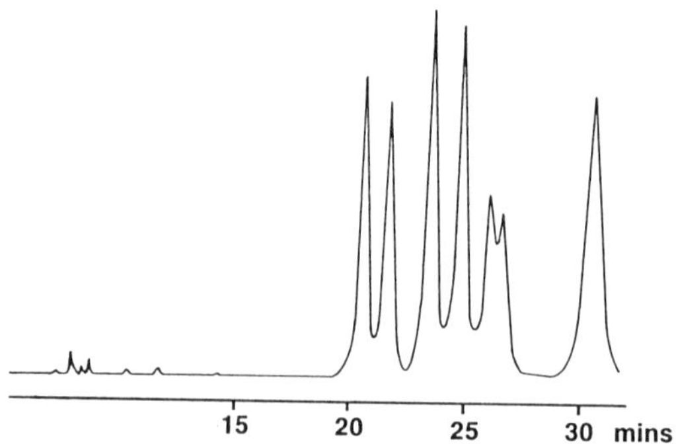

Figure 2 Separation of cypermethrin.
Mobile phase - 0.2% propan-2-ol in hexane, flow rate 1.0 ml/min., oven temperature 40°C, wavelength 220nm.

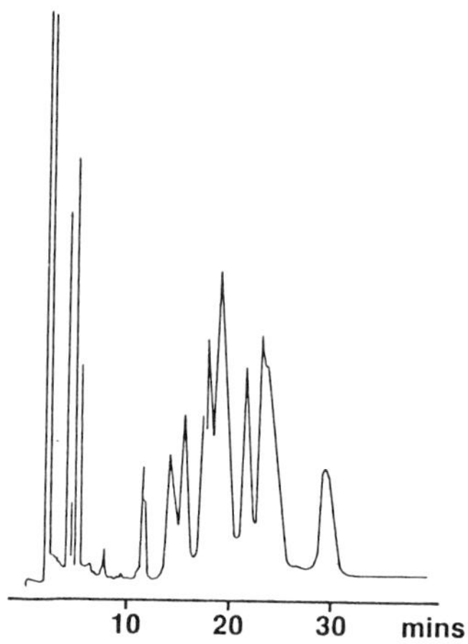

Figure 3 Separation of the isomers of tralomethrin [a-cyano-3-phenoxybenzyl-3-(1,2-dibromo-2, 2-dichloroethyl)-2,2-dimethylcyclopropane-carboxylate.
Mobile phase - 0.05% propan-2-ol in hexane, flow rate 2 ml/min (ref. 1).

Silica bonded polymethacrylate columns can be used for the separation of the isomers of pyrethroids such as phenothrin.

2. U IP column

This is an ion-pair column, which when run under normal phase conditions can be used for the separation of the optical isomers of fungicides such as fenarimol, triadimefon and triadimenol.

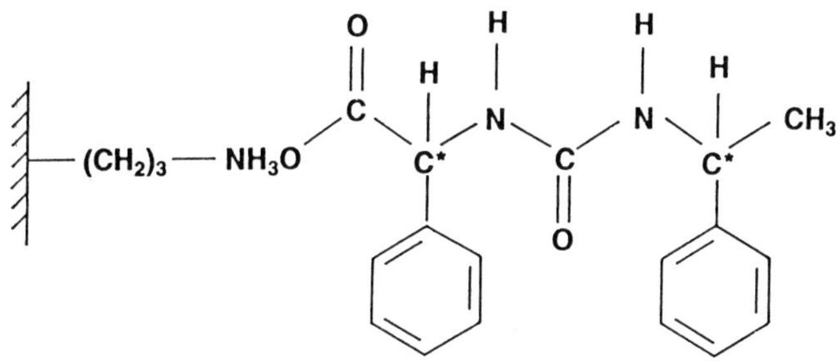

Figure 4 Structure of U IP packing

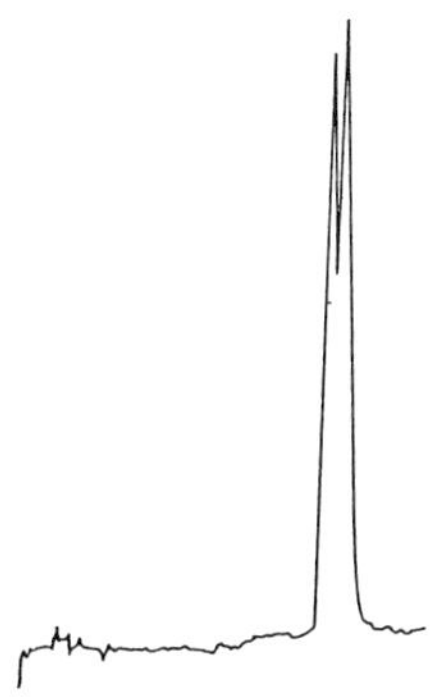

Figure 5 Separation of fenarimol

1.0% propan-2-ol in hexane at 0.5 ml/min. Wavelength 236 nm.
G. R. Cayley and B. W. Simpson - unpublished.

3. Cyclodextrin columns

The isomers of these fungicides can also be separated on cyclodextrin columns. These consist of α, β or γ cyclodextrin molecules (six to eight glucose units respectively) covalently bonded to silica. The glucose units, each of which contains five chiral atoms, form a cavity where separation takes place via the formation of an inclusion complex. This type of interaction is spatial in nature and isomers can form complexes of considerably differing stability. The columns are usually run in the reverse phase mode, with methanol/water as the mobile phase.

4. Albumin columns.

Bovine serum albumin covalently bound to 300 A pore silica can be used to resolve (±) warfarin. The column is usually operated with a phosphate buffer and propan-1-ol as a modifier.

TABLE 1 - Applications of Chiral Phase columns in Pesticide Analysis.

Applications of Chiral Phase Columns in Pesticide Analysis.

Column	Pesticide
Pirkle 1-A	Cypermethrin Permethrin Fenvalerate Cyhalothrin Allethrin Resmethrin Fenpropathrin Fluazifop-Butyl
β Cyclodextrin	Diclobutrazol Fenarimol
β Cyclodextrin (acet.)	Tripenthanol Flutriafol
U IP	Triadimefon Triadimenol Fenarimol
Polymethacrylates	Phenothrin
Polysaccharide Cellulose	Fluazifop-Butyl Flutriafol Paclobutrazol
Bovine Serum Albumin	Warfarin

DISCUSSION

The ability to separate individual isomers is becoming increasingly important as new single isomer or single diastereoisomer pesticides are introduced on to the market. These new compounds can be applied at much lower dose rates than the parent mixtures and therefore are more economical, less toxic to non-target species and result in lower residues. For the residue analyst, chiral phase columns provide the possibility of determining those isomers which are environmentally the most persistent, and for the formulation analyst they allow isomer specific analysis of products. The technique has the advantages that it employs conventional HPLC apparatus and can be scaled up for preparative work.

A great deal of work is being carried out on chiral phase columns at the present time. However, the majority of published work has so far been done on standard solutions and it remains to be seen how robust these columns are when actual samples are run on them.

REFERENCE

G. R. Cayley and B. W. Simpson, J Chromatog. 356, 123-134 (1986).

Application of miniaturization for clean-up in pesticide residue analysis

Hans-Peter Thier

Institute of Food Chemistry, University of Münster, Münster, Federal Republic of Germany

Abstract - Miniaturized clean-up combined with optimum use of sensitivity and selectivity of gas-chromatographic detectors available offer a highly valuable tool for improving analytical methods with regard to effectiveness and savings of cost and time. Miniaturization will be greatly promoted by introducing gel permeation chromatography and RP-18 small disposable cartridges into current residue methodology. The advantages are demonstrated by means of two clean-up procedures of broad applicability.

INTRODUCTION: ESSENTIALS OF MINIATURIZATION

In conventional residue analytical procedures it is customary to use large amounts of solvents and other chemicals for extraction and clean-up.

This high input is rather disproportionate, for the amount of compound required for gas-liquid chromatographic (or HPLC) determination is much smaller than that contained in the solution to be analyzed.

The best way to improve this unfavourable ratio is reduction of scale (ref. 1). The size of the analytical sample can be reduced only to a limited extent for the residues may be unevenly distributed. Therefore it is the clean-up which should be miniaturized, e.g. by starting from a much smaller aliquot of the sample extract than is usually employed. This will result in

- decrease of the amount of co-extractives to be removed,
- reduction of scale of laboratory equipment,
- less chemicals and solvents being required,
- considerable savings in cost and time.

For compensating for the reduced quantity of compound present in the final solution, the following measures should be taken:

- Evaporate the final solution to a very small volume of 1 or 2 drops which need not to be exactly defined.
- Add two internal standards not interfering with pesticides or co-extractives for quantitative measurement.
- Replace packed GLC columns by capillary columns for attaining best separation and peak shape to enhance detectability.
- Use splitless or on-column injection techniques in order to apply the total volume injected for analysis.
- Select most appropriate detector operating conditions for achieving optimum sensitivity and selectivity.

Effective miniaturization requires a careful coordination of the clean-up and determination steps. Its possible extent is only limited by

- the detector response to pesticides to be analyzed,
- the minimum levels of residues which must be detected in the analysis,
- the portion of co-extractives in the final solution interfering with determination.

RESULTS

Miniaturization will be greatly promoted by introducing gel permeation chromatography and RP-18 small disposable cartridges into current multi-residue methodology. They can both be used either for an advance removal of the bulk of co-extractives, followed by any miniaturized clean-up step more specific for the compounds analyzed, or as the sole clean-up step for some groups of pesticides, when followed by a highly selective GLC detection (e.g. flame photometric detector).

Gel permeation chromatography (GPC)

GPC using Bio-Beads S-X3 and cyclohexane - ethyl acetate 1 + 1 as eluting mixture is well known for its capability of removing more than 95 % of high-molecular co-extractives and lipids from crude sample extracts. It can be applied

- in an automated mode with Autoprep™ or other equipment commercially available,
- in a miniaturized mode using reduced column diameter.

In both cases, GPC is best followed by a small (1 g or less) silica gel column. Figure 1 demonstrates analytical details of a well proven clean-up sequence (ref. 2,3).

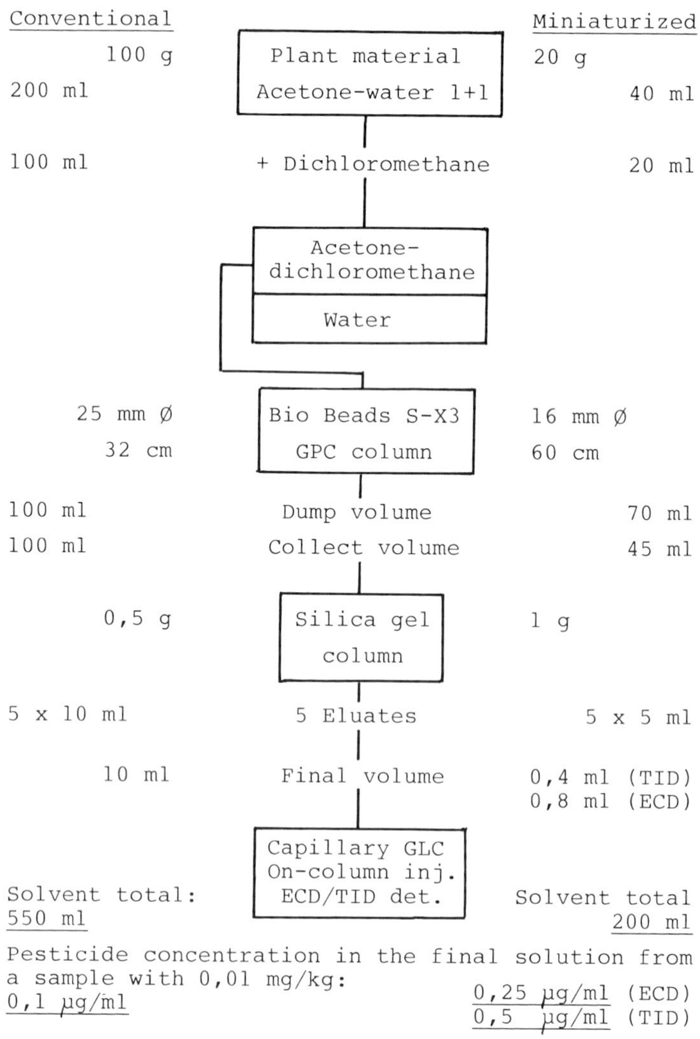

Fig. 1. Miniaturization of a clean-up procedure for organochlorines and organophosphates using GPC and mini-silica gel column chromatography

Sep-Pak RP-18 disposable cartridges

They offer an effective way for small scale clean-up of plant material in the analysis of some groups of more polar pesticides. Figure 2 presents the analytical sequence of a multiresidue procedure for the determination of methylcarbamate insecticide residues in fruits and vegetables (ref. 4).

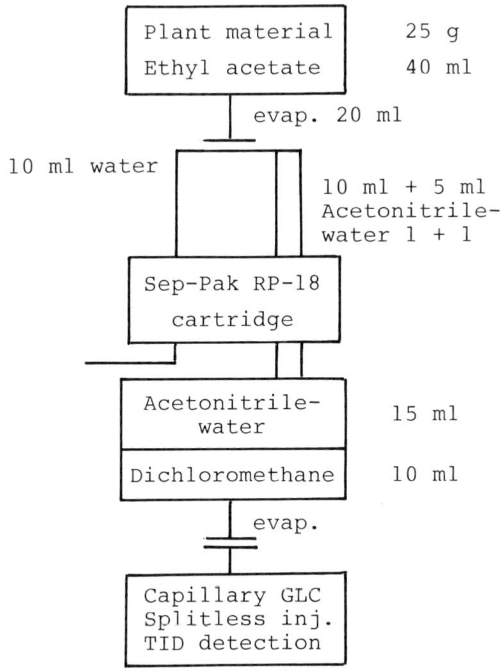

Fig. 2. Small scale multiresidue clean-up for determination of methylcarbamate insecticide residues using Sep-Pak cartridge

DISCUSSION

Cost effectiveness

In the first example (Fig. 1), the conventional clean-up (ref. 2) requires a total of 550 ml of solvents as compared with 200 ml in the miniaturized mode (ref. 3). This affords a saving of 350 ml in each analysis, hence resulting above all in a substantial reduction of cost.

In the second example (Fig. 2), the saving of time for analysis is predominant. Clean-up is accelerated in such a way that 4 - 6 samples can be processed to be ready for determination within 2 hours.

Detection limit

When the determinative step is optimized according to the recommendations given in the introduction, miniaturization will even result in an increased sensitivity. The limits of detection were found to be

- in example 1: 0,01 µg/kg for organochlorines, 0,06 - 0,1 µg/kg for some organophosphates (ECD) and about 1 µg/kg for organophosphates (TID P-mode), respectively, in raw crops and total diet samples,
- in example 2: 1 - 5 µg/kg for methylcarbamates (TID N-mode) in fruits and vegetables.

Internal standards

Internal standardisation is required to control the final volume of the solution to be analyzed.For use in multiresidue procedures, it is favourable to have pairs of internal standards which bracket the pesticides in the chromatogram, e.g. 2,2',6,6'-tetrachlorobiphenyl and mirex (Fig. 1) or n-dodecanonitrile and n-octadecanonitrile (Fig. 2).

Use of capillary columns for GLC
They are particularly useful in multiresidue analysis and for separation of complex mixtures such as polychlorinated biphenyls. Because the retention time can be determined much more exactly than on packed columns, they result also in a higher reliability for identification.

Since many co-extractives are separated from pesticides on the capillary column, the clean-up can often be largely reduced. In this case, however, an increased risk of misinterpretations must be taken into account.

Capillary column injection
No prediction can be made for the best alternative. It depends on the stability of the pesticides in question, but also on the extent of discrimination effects and the deposition of non-volatile co-extractives which will change the adsorptivity of the column surface. In example 1, on-column injection yielded optimum results, whereas for the methylcarbamates in example 2, splitless injection proved to be superior.

REFERENCES

1. R.J. Hemingway, N. Aharonson, P.A. Greve, T.R. Roberts and H.P. Thier, Pure Appl. Chem. 56, 1131-1152 (1984).
2. W. Specht and M. Tillkes, Fresenius Z. Anal. Chem. 301, 300-307 (1980); 322, 443-455 (1985).
3. J. Vogelgesang and H.P. Thier, Z. Lebensm. Unters. Forsch. 182, 400-406 (1986).
4. S. Brauckhoff, Thesis Univ. Münster (1986).

Progress in clean-up and derivatization techniques for pesticide residue analysis

Mikio Chiba

Research Station, Agriculture Canada, Vineland Station, Ontario LOR 2EO Canada

Abstract - A rapid clean-up method for analysis of oxamyl residues in potato tubers was developed with the use of SEP-PAK Florisil cartridges. Another clean-up method was developed for GC determination of residues of aldicarb and its toxic oxidation products in chrysanthemum leaves with the use of a disposable pipet column which was packed with 0.2 g of Nuchar-Attaclay. A gas chromatographic method was developed to determine oxamyl residues after derivatizing to dinitrophenylmethylamine. A unique method without clean-up was developed for individual analyses of residues of benomyl and its degradation compound carbendazim in apple leaves by normal-phase HPLC. In another method benomyl in aqueous media was derivatized to STB while co-existing carbendazim remained unchanged, and then both were individually determined by reversed-phase HPLC. Residues of benomyl and carbendazim in apples were derivatized to BBU and 2-AB, respectively and determined by reversed-phase HPLC. On-column clean-up was accomplished for analyses of aldicarb and two of its oxidation compounds in soil by HPLC with the use of a size-exclusion column.

INTRODUCTION

Clean-up is an essential step for accurate determination of pesticide residues, but many conventional methods are costly and time consuming. It is rather impractical, however, to expect to find a universally available and simple clean-up technique because characteristics of analytes are markendly different and the nature of the impurities to be removed is extremely intricate. Cleanup techniques currently in use include liquid-liquid partitioning, open column chromatography, thin-layer chromatography, distillation and low temperature precipitation (ref. 1). Of several new approaches successfully used in modern methodologies use of small columns may be the most notable (ref. 1-6).

Derivatization is used frequently in residue analysis to alter characteristics of analytes so that derivatized products are more suitable for specific analytical systems, mainly GC and HPLC (ref. 1); the major characteristics to be altered are volatility, sensitivity and selectivity. To improve sensitivity and selectivity, the most notable advances have been made in post-column derivatization on HPLC systems (ref. 7-11). One of the greatest advantages of this approach is the less stringent requirements for clean-up .

There is a clear trend in the realm of clean-up and derivatization; that is the simplification of clean-up procedures by utilizing both material and intellectual resources wisely. Reported in this paper are several examples of new methods which have been successfully developed in our laboratory for specific pesticide residue analyses. These include simple clean-up methods, derivatization methods and novel approaches in which the need for clean-up was eliminated. Concentrated effort has been made throughout these studies to develop methods which will determine parent compounds per se by clearly separating them from their natural degradation compounds.

SIMPLIFIED CLEAN-UP AND LIQUID CHROMATOGRAPHIC DETERMINATION OF OXAMYL RESIDUES IN POTATO TUBERS

Sample potatoes were extracted with methanol, which was diluted with 10% Na_2SO_4 (aq) and partitioned against dichloromethane. A 4-mL aliquot of dichloromethane extract after concentration (representing 2 g potato) was passed through a SEP-PAK Florisil cartridge. Elution was made with 4 mL of methanol + dichloromethane (1+9). After evaporating the eluate, the residue was redissolved in water (2 g potato/mL). Liquid chromatographic (LC) determination was performed using a Zorbax PSM60 size-exclusion column with an acetonitrile+ water (1 + 9) mobile phase and UV detection at 254 um. There was essentially no interference and the minimum detectable concentration of oxamyl was 0.01 ug/g (ref. 12).

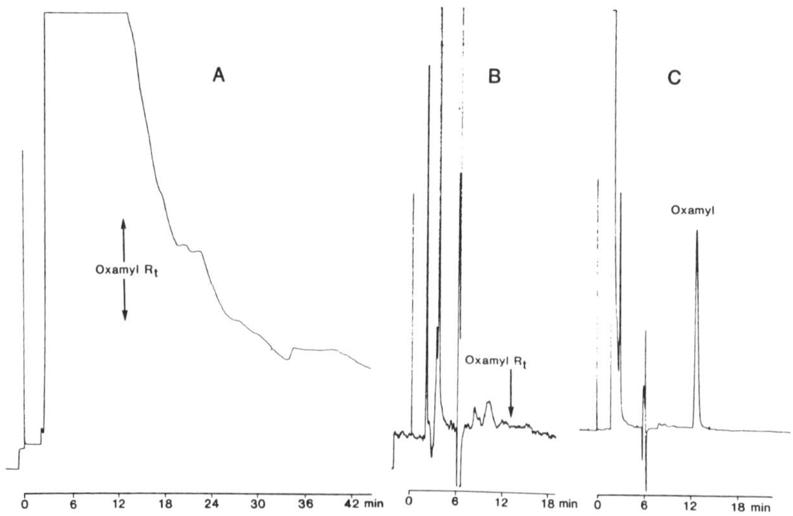

Figure 1. LC chromatograms of potato tuber extracts (100 uL injections, representing 200 mg of tuber); A, untreated sample without prior clean-up, detector range 0.16; B, same sample as above, after clean-up, detector range 0.02; C, treated sample, after clean-up, detector range 0.16 (oxamyl concentration 0.90 ug/g).

A SIMPLE AND EFFICIENT METHOD FOR THE CLEAN-UP AND CAPILLARY GAS CHROMATOGRAPHIC ANALYSIS OF RESIDUES OF ALDICARB AND ITS OXIDATION PRODUCTS IN CHRYSANTHEMUM LEAVES

Sample leaves were extracted with methanol and the extracts were cleaned up using columns consisting of 0.2 g Nuchar-Attaclay on top of a silanized glass wool plug in disposable pipets. A 2-mL aliquot of the methanol extract (representing 0.24 g leaves) was passed through the column. The analytes were eluted with an additional 1 mL methanol. After evaporating the methanol, the residue was dissolved in ethyl acetate. The solution thus prepared was ready for quantitation by capillary gas chromatography with splitless injection and thermionic specific detection. Aldicarb, aldicarb sulfoxide and aldicarb sulfone were clearly resolved and individually determined with a 1 uL injection (ref. 13).

DETERMINATION OF OXAMYL BY GAS CHROMATOGRAPHY AFTER DERIVATIZING TO DINITROPHENYLMETHYLAMINE

Oxamyl, after alkaline hydrolysis, was derivatized to dinitrophenylmethylamine (DNPMA) and determined by gas chromatography with electron capture detection. Hydrolysis was performed at pH 12 at 80°C for 10 min and the liberated methylamine was reacted with dinitrofluorobenzene at 80°C for 10 min to yield DNPMA. The column used was a 1.8 m x 2.5 mm i.d. glass column packed with 3% XE-60 on Gas Chrom Q 80/100 mesh. With this method oxamyl can be determined without interference from its corresponding oxime, which is a common non-toxic degradation product of oxamyl and present in essentially all oxamyl treated samples. Actual residues of oxamyl in tobacco leaves and soils were analysed at a level as low as 0.1 ppm (Chiba and Lee, unpublished data).

A NOVEL ANALYTICAL APPROACH FOR STABILIZATION, EXTRACTION AND DETERMINATION OF RESIDUES OF BENOMYL AND CARBENDAZIM IN APPLE FOLIAGE WITHOUT CLEAN-UP

Sample leaves in a Mason jar were freeze-dried and tumbled for extraction in chloroform containing 5000 ug n-propylisocyanate (PIC)/mL at 1°C. n-Butylisocyanate (BIC) at 5000 ug/mL was added to the extract, and 20 uL of the mixture was injected into the HPLC system. Residues of carbendazim, which is difficult to dissove in chloroform, were converted to 1-(n-propylcarbamoyl)-2-benzimidazole carbamate (MBC-PIC) during the extraction. MBC-PIC along with benomyl, was easily extracted with chloroform at 1°C, at which temperature the degradation of benomyl was undetectable. However, addition of BIC was necessary after extraction for HPLC analysis to prevent degradation of benomyl at room temperature. Normal-phase HPLC with a mixed mobile phase of chloroform - hexane (4 + 1) saturated with water was employed with UV detection at 280 nm. No clean-up was necessary because only a very small amount of impurities was extracted with chloroform at 1°C (ref. 14). (Note: carbendazim is expressed as MBC.)

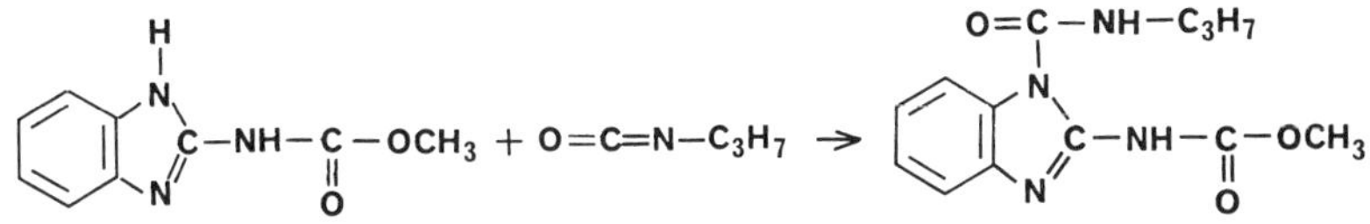

MBC PIC MBC-PIC

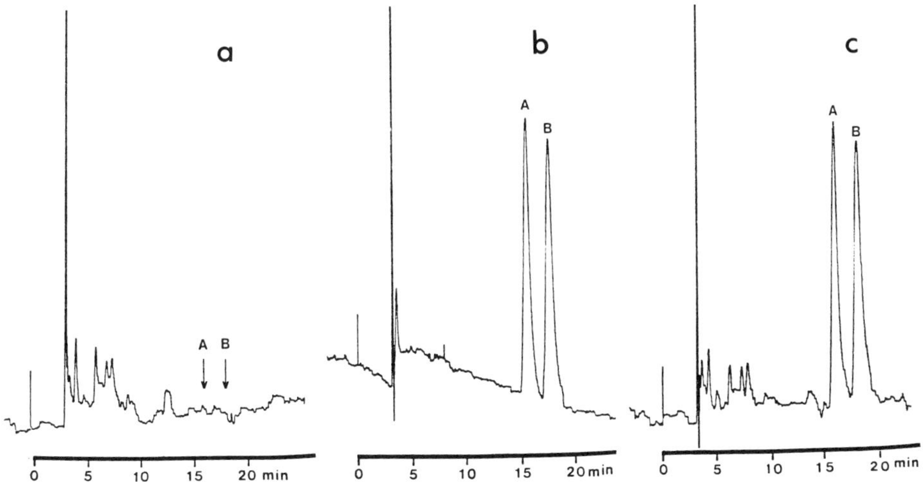

Figure 2. HPLC chromatograms of 20 uL sample solutions; peak A is benomyl and peak B is MBC-PIC: a, background of apple leaf extract; b, standard solution containing 2 ug/mL each of benomyl and MBC-PIC; c, standard solution with apple leaf extract (combination of a and b).

HIGH PERFORMANCE LIQUID CHROMATOGRAPHIC METHOD FOR SIMULTANEOUS DETERMINATION OF BENOMYL AND CARBENDAZIM IN AQUEOUS MEDIA

It is essentially impossible to accurately and individually determine concentrations of intact benomyl and carbendazim in solutions because benomyl decomposes to carbendazim. In addition, isocratic and individual analysis of both compounds by HPLC is impractical because of a wide difference in the polarity of the two. In this method benomyl in aqueous media is quantitatively converted in 20 min to 3-butyl-2,4,dioxo-s-triazino [1,2-a] benzimidazole (STB) with NaOH at pH 13 and determined as STB. In contrast carbendazim, present in essentially all samples treated with benomyl, is unaffected by this alkaline treatment and determined as carbendazim. Both STB and carbendazim are individually determined by isocratic reversed-phase HPLC. STB thus formed is stable and its resolution from carbendazim is easily optimized with mobile phase mixtures consisting of acetonitrile and pH 7 aqueous buffer in different proportions. This method is applicable to essentially all water samples including spray mixtures and environmental waters. The minimum detectable limit is 0.05 ug/mL with a 50 uL injection (ref. 15).

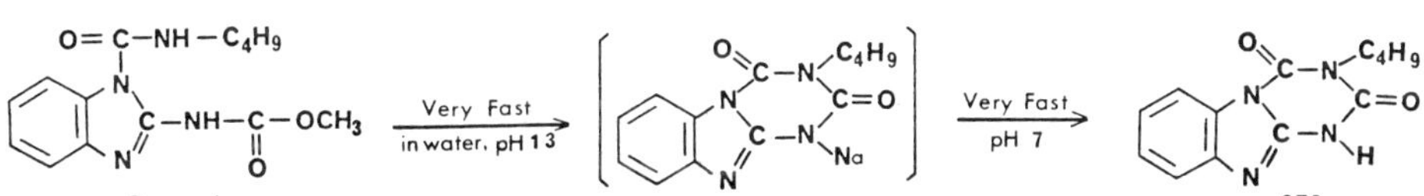

Benomyl STB

SIMULTANEOUS DETERMINATION OF RESIDUES OF BENOMYL AND CARBENDAZIM IN APPLES BY HIGH PERFORMANCE LIQUID CHROMATOGRAPHY

Sample apples were macerated by using a Polytron homogenizer without using any solvent. The macerated apple was mixed with NaOH solution and the mixture (pH 13.2) was refluxed for 2 hrs at 100°C. Benomyl was quantitatively converted to 1-(2-benzimidazolyl)-3-n-butylurea (BBU) while carbendazim was quantitatively converted to 2-aminobenzimidazole (2-AB). The derivatives were extracted in ether + 2-butanol (1 + 1). After evaporation of the solvent, The residues were dissolved in 30% CH_3OH in water and determined by the following HPLC system with UV detection at 280 nm. A short size-exclusion column was connected to a CN column via a switching valve. After removing interfering coextractives through the size-exclusion column the effluent was passed through the CN column where resolution of BBU and 2-AB was achieved satisfactorily. The mobile phase used was a mixture consisting of CH_3OH and an aqueous buffer solution (Chiba, unpublished data).

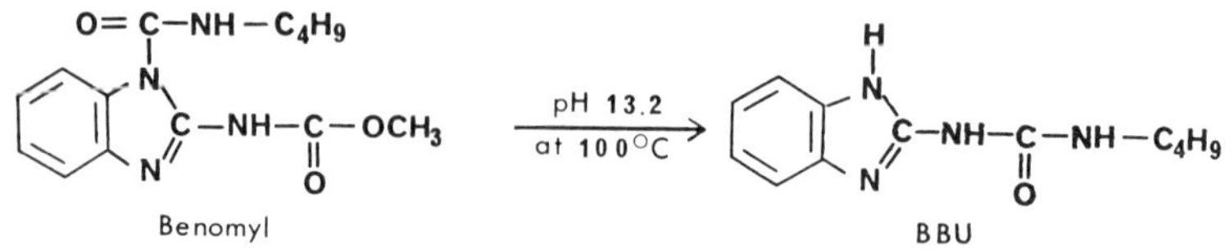

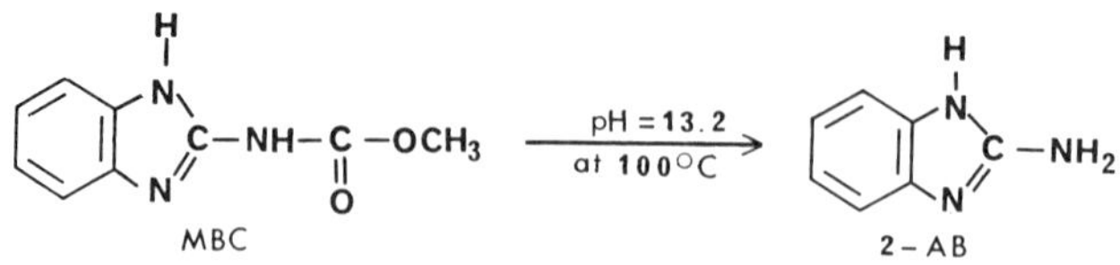

DETERMINATION OF ALDICARB, ALDICARB SULFOXIDE AND ALDICARB SULFONE IN WATER AND SOILS BY SIZE-EXCLUSION HPLC

Residues of aldicarb and its two oxidation products in soil were extracted in acetone; the solvent was evaporated and the residue was dissolved in a solvent consisting of acetone + water (1 + 9), and analyzed by HPLC. A Zorbax PSM60 size-exclusion column was used with an isocratic mobile phase consisting of acetonitrile and water (without buffer) and UV detection at 210 nm. A syringe filtration was the only step required before injection in this method. Impurities were eluted early and the three compounds were then clearly resolved with a total run time of 10-16 min. The significance of this method is the analysis of all three compounds with an isocratic mobile phase, without any clean-up. Reversed-phase HPLC, which is commonly used for this analysis, requires a gradient mobile phase or multiple injections using different isocratic mobile phases in order to determine all three compounds. It is important to prepare both standards and samples in the same solvent, which consists of acetone + water (1 + 9); peak distortion due to differences in acetone concentration is substantial (Chiba, Zimmerman, McGarvey, unpublished data).

REFERENCES

1. J.F. Lawrence, Organic Trace Analysis by Liquid Chromatography, Academic Press, New York (1981).
2. H. Agemian and A.S.Y.Chau, J. Assoc. Off. Anal. Chem., 60, 1070-1076 (1977).
3. W. Specht and M. Tillkes, Fresenius Z. Anal. Chem., 322(5), 443-455 (1985).
4. S. Forbes and A.J. Dutton, Pest. Sci., 16, 404-408 (1985).
5. C. Bicchi, A. D'Amato and I. Tonutti, Chromatographia, 20 (4), 219-222 (1985).
6. H. Nijhuis, W. Heeschen and P. Meibohm, Milchwissenschaft, 40(2), 78-80 (1985).
7. R.T. Krause, J. Chromatogr., 255, 497-510 (1983).
8. R.E. Majors, H.G. Barth and C.H. Lochmuller, Anal. Chem., 56, 300R-349R (1984).
9. R.W. Frei and U.A.Th. Brinkman, Anal. Chem., 57, 1529A-1539A (1985).
10. S.R. Priebe and J.A. Howell, J. Chromatogr., 325, 53-63 (1985).
11. R.L. Roseff, Food Technol., 39(9), 87-90 (1985).
12. B.D. McGarvey, M. Chiba and Th.H.A. Olthof, J. Assoc. Off. Anal. Chem., 68, 753-756 (1985).
13. B.D. McGarvey, M. Chiba and A.B. Broadbent, J. Assoc. Off. Anal. Chem., 69, in press (1986).
14. M. Chiba and D.F. Veres, J. Assoc. Off. Anal. Chem., 63, 1291-1295 (1980).
15. M. Chiba and R.P. Singh, J. Agric. Food Chem., 34, 108-112 (1986).

(Note: Figures 1 and 2 have been reproduced from ref. 12 and ref. 14, respectively, with permission from the Association of Official Analytical Chemists.)

Immunological assays for agrochemicals

Ralph O. Mumma* and James F. Brady

Pesticide Research Laboratory and Graduate Study Center, Department of Entomology, The Pennsylvania State University, University Park, PA 16802, USA

Abstract - Immunological assays offer considerable potential for rapid, inexpensive, and sensitive methods for analysis of agrochemicals and their metabolites. Radioimmunological (RIA) and enzyme-linked immunosorbent assays (ELISA) have been developed for numerous pesticides, antibiotics and plant pathogens. Reported procedures range in time of analysis from seconds to hours and have sensitivities of detecting nanograms and, in some cases, picograms of chemicals. Methods may involve heterogeneous or homogeneous procedures or a combination of both. Horseradish peroxidase, alkaline phosphatase and glucose-6-phosphate dehydrogenase are enzymes commonly used with the ELISA procedure. Monoclonal antibodies have been developed but most reported methods use polyclonal antibodies. Techniques employ liquid scintillation counting or spectrophotometric methods for detection; however, a quantitative system comparing colored spots formed during the analysis with standardized colored spots has also been developed.

INTRODUCTION

There is a need for simple, reliable and inexpensive procedures for analysis of agrochemicals and other xenobiotics, since costs of analysis and environmental concerns are increasing. Immunological techniques offer these advantages. Immunological assays have been routinely used in clinical laboratories for analyses of small molecules such as hormones and drugs. The procedures are becoming so simple that these analyses are now being conducted in medical doctors' offices and even in private homes. The general principles of immunoassays have been adequately reviewed (refs. 1, 58); this article will briefly introduce immunoassays, review what has been accomplished relative to analysis of agrochemicals (herbicides, insecticides, antibiotics, plant pathogens and xenobiotics), describe products currently available commercially, and discuss areas which we consider will be the future thrusts.

Immunological reactions are no different from any other chemical reaction. Reagent A and protein B react to produce a product A-B. That B is a biosynthesized reagent (antibody) with a certain specificity of binding to A, should not discourage us from considering the process like any other chemical reaction. Antibody development is a two-phase process. First, the small molecule to be analyzed (hapten) is covalently bound to a much larger immunogenic molecule, usually a protein. The resulting conjugate is then injected into an animal host which mounts a defensive response culminating in synthesis of antibodies of varying specificities (polyclonal), some of which are directed against the hapten. Alternatively, monospecific (monoclonal) antibody preparations can be developed using *in vitro* cell culture and fusion techniques. Antibodies are durable reagents that retain their activity even after prolonged storage. One of the first immunoassays developed for an agrochemical was that for dieldrin by Langone and Van Vunakis in 1975 (ref. 2). The anti-dieldrin antibody was recently tested in an homogeneous assay and it exhibited strong binding to dieldrin after 11 years of storage (ref. 3).

Although analytical methods vary, each technique monitors antibody-hapten complex formation. This may be accomplished by measuring unreacted species or products depending on assay design. Thus, immunoassays usually involve two steps: a prior incubation of antibody with the chemical under test followed by a second reaction to determine the extent of initial product formation. Radioimmunoassay (RIA) involves incorporation of a radiolabel (^{3}H, ^{14}C, ^{125}I) into a reagent, typically the hapten, facilitating its quantification. Enzymatic techniques may employ enzyme-labeled antibodies which bind to antibodies directed against the hapten. Horseradish peroxidase or alkaline phosphatase are the enzymes most frequently used and are covalently bound to the second antibody. Anti-hapten antibodies yet unbound after incubation with sample are permitted to complex with hapten conjugated to a protein carrier adsorbed to a solid phase (nonhomogeneous assay). The extent of such binding is determined by sequential addition of enzyme-labeled anti-globulin and chromogenic substrate. The overall reaction is then evaluated spectrophotometrically. These kinds of analyses are referred to as enzyme-linked immunosorbent assays (ELISA) and often require 24 hours to give quantitative data. Qualitative nonhomogeneous assays, however, may take only several minutes to complete. Recently, homogeneous assays for agrochemicals have been developed which require only a few seconds for analysis (ref. 3); this will be addressed in more detail later.

ACCOMPLISHMENTS

We will briefly summarize the published literature. Readers should be aware that many accomplishments exist within the agrochemical industry, but these successes have not been and may never be published. Table 1 presents a list of agrochemicals for which immunoassays have been developed. The lower limit of detection, type of antibody, method of analysis, and reference source is given. We have included in this table several plant hormones and the xenobiotic dioxin, since all of these molecules are of agricultural interest.

There has been considerable concern for the level of antibiotics that may exist in meat and food products. Antibiotics are difficult to detect by classical means and immunoassays have been developed for a number of products (Table 2). Some companies have developed immunoassays for their manufactured antibiotics and use these assays for in-house operation; several diagnostic kits are now commercially available. Plant pathogens cause millions of dollars of damage to agronomic crops each year and early detection and identification of these organisms would assist pest management decisions. Pathogens possess characteristic proteins and/or antigenic surface structures which can readily be detected through immunological procedures. Table 3 lists some of the plant pathogens for which immunoassays have been developed.

COMMERCIAL PRODUCTS

Many companies have expressed interest in immunodiagnostics, but only a few have developed commercial products. Apparently, industries prefer to have immunoassays for their products available for company use, but are not particularly interested in marketing diagnostic products. Presently, two companies are now marketing test kits, while others are developing them.

Idetek, Inc. (San Bruno, CA, USA) markets monoclonal antibodies for quantification of several plant hormones: indole-3-acetic acid methyl ester, abscisic acid, zeatin riboside, and dihydrozeatin riboside (ref. 7). These ELISA assays measure in the range of 0.02 to 5.0 pmole/0.1 ml for the last three compounds. Indole-3-acetic acid methyl ester is determined in the range of 0.5 to 100 pmole/0.1 ml. In this case, the antibody does not bind to the free acid, but only binds to the ester. This particular monoclonal antibody does not have a high degree of specificity for the indole ring; it seems to have more specificity for the chain and bridge coupling the carrier protein to the hapten.

Granite Division - Environmental Diagnostics, Inc. (Burlington, NC, USA) markets E-Z Screen test kits for agrochemicals. The kits consist of QUIK-CARDs and reagents (antibody and substrates for enzymes). The cards contain an indicator which changes color if the chemical to be tested is present. This technique only requires a few minutes, is semiquantitative, and indicates whether the sample contains more than a certain level of test chemical. Because of the simplicity and stability of the QUIK-CARD technique, it may have wide applicability for rapidly screening large numbers of samples. Currently, Environmental Diagnostics markets kits for two pesticides (paraquat and parathion), six antibiotics (chloramphenicol, gentamicin, neomycin, sulfamethazin, sulfadimethoxine, and tylosin), triton-x, aflatoxin and soluble soy (ref. 8).

While not yet marketing its product, Immunosystems, Inc. (Biddeford, ME, USA) has developed homogeneous immunological assays for several agrochemicals (ref. 3). These assays are based upon the EMIT® procedure, which measures the effect of antibody binding on the activity of an enzyme (glucose-6-phosphate dehydrogenase) covalently bound to hapten. When antibody binds to enzyme-bound hapten, the enzyme's rate of reaction decreases proportionately to the amount of binding. If the antibody preparation is first incubated with a sample containing the test chemical, some antibody would be bound by free hapten and, thus, not available to bind to enzyme-coupled hapten. In this manner, the concentration of the agrochemical can be measured. The analysis is extremely fast (15 to 45 seconds), reproducible, and seems to eliminate some of the troublesome problems, such as wall binding and washings, associated with nonhomogeneous assays. Since the EMIT® assay requires a spectrophotometer to determine the rate of reaction, nearly all of the procedures can be automated.

Table 1. Agrochemicals For Which Immunoassays Have Been Developed.

Chemical	Detection Limit	Polyclonal (P) or Monoclonal (M) Antisera	Method	Reference
Abscisic acid	66 pg	P	RIA	4
	13 pg	P	ELISA	5
	4 pg	M	RIA	6
	5 pg	M	ELISA	7
Aldicarb	5 ng	P	EMIT	3
Aldrin	700 pg	P	RIA	2
alfatoxin	0.5 ng	P	ELISA	8
2-aminobenzimidazole	100 pg	P	Fluorometric	9
Atrazine	215 pg	P	ELISA	10
BAY SIR 8514	9.6 ng	P	ELISA	11
Benomyl	0.1 ng	P	Fluorometric	9
	1.25 ng	P	RIA	12
S-bioallethrin	1 ug	P	Immunodiffusion	13,14
	0.5 ng	P	ELISA	55
Bacillus thuringiensis				
subs. kurstaki	105 ng[a]		ELISA	15
subs. berliner	220 ng[a]	P	ELISA	15
2-(Chlorophenyl)-isovaleric acid	2 ng	P	RIA	19
Chlorsulfuron	0.1 ng	P	ELISA	20
Cypermethrin	---	---	---	21
2,4-D	13 ng[a]	P	RIA	22
	2.5 ng	P	EMIT	3
Diclofop-methyl	20 ng	P	ELISA	23
	10 ng	P	Fluorometric	23
Dieldrin	150 pg	P	RIA	2
	10 ng	P	EMIT	3
Diflubenzuron	3.9 ng	P	ELISA	11
Dihydrozeatin riboside	7 pg	M	ELISA	7
Dimethylallyl-adenineriboside	4 ng	P	RIA	24
Dioxin	1 ng	P	RIA	24
	25 pg	P	RIA	25
	1 ng	M	ELISA	56
Gibberellic acid	2 pg	P	RIA	26
Gypsy moth nuclear polyhedrosis virus	0.01 ug	P	ELISA	27

Table 1. continued

Chemical	Detection Limit	Polyclonal (P) or Monoclonal (M) Antisera	Method	Reference
IAA methyl ester	200 pg	P	RIA	28
	94 pg	P	RIA	29
	3 pg	P	ELISA	30
	940 pg	M	RIA	31
	94 pg	M	ELISA	31
	200 pg	M	ELISA	7
Iprodione	0.2 ng	P	ELISA	32
Maleic hydrazide	6-100 ppb	M	ELISA	33
	6 ppb	M	ELISA	34
Metalaxyl	63 pg	P	ELISA	35
Methyl-2-benzimidazole carbamate	125 ng	P	RIA	12
Paraoxon	275 ng	P	ELISA	36
	1 ug	M	ELISA	37
Paraquat	10 ng, 0.1 ng	P	RIA	38
	0.5 ng	P	RIA	39
	0.1 ng	P	ELISA	40
Parathion	4 ng	P	RIA	41
	7.8 ng	P	ELISA	43
	20 ng	P	RIA	44
	20 ng	P	ELISA	44
	2.5 ng	P	ELISA	8
Penfluron	6.8 ng	P	ELISA	11
Permethrin	---	---	---	21
Soluble soy	10 ng	P	ELISA	8
Surflan	---	M	ELISA	45
Terbutryn	4.8 ng	P	ELISA	46
Triadimefon	1 ng	P	ELISA	47
2,4,5-T	3.3 ng[a]	P	RIA	22
Zeatin riboside	15 pg	P	RIA	48
	7 pg	M		7
Zeranol	100 pg	P	RIA	49

[a] Sensitivity expressed as I_{50}.

Table 2. A Partial List Of Antibiotics For Which Immunoassays Have Been Developed.

Antibiotic	Detection Limit	Polyclonal (P) or Monoclonal (M) Antisera	Method	Reference
Chloamphenicol	10 pg	P	RIA	50
	5 ng	P	ELISA	8
Gentamicin	10 ng	P	ELISA	8
Hygromycin B	0.02 units	P	RIA	51
Neomycin	10 ng	P	ELISA	8
Penicillin G	---	---	---	52
Sulfadimethoxine	10 ng	P	ELISA	8
Sulfamethazin	10 ng	P	ELISA	8
Tylosin	10 ng	P	ELISA	8

Table 3. A Partial List of Plant Pathogens For Which Immunoassays Have Been Developed.

Plant Disease	Pathogen	Reference
Brown Patch	Rhizoctonia	53
Dollar Spot	Sclerotinia homoeocarpa	53
Pythium Blight	Phythium spp.	53
Tobacco Ringspot Virus		54

Immunodiagnostic kits for a number of plant pathogens are under development. These products are not designed to be a research tool but are aimed at the horticultural, turf grass, and fruit industries. AGRI-DIAGNOSTICS Assoc. (Cinnaminson, NJ, USA) claims it is currently field testing three turf disease kits to detect pythium blight, brown patch, and dollar spot, utilizing ELISA procedures and monoclonal antibodies (ref. 53). If the activities and claims of companies result in marketable products, we should soon see a new and growing industry directed at detecting plant pathogens.

PROBLEM AREAS

The development of hapten-specific antibodies requires two major steps: first, the coupling of the antigenic carrier molecule to the hapten via a bridge, and second, the protocol of injection of the animal in raising the antibodies. Both of these steps may vary for optimization of antibody production against a particular molecular configuration. Certain immunogenic carrier molecules (i.e., bovine serum albumin, hemocyanin and egg albumin) are more desirable than others when associated with a particular hapten. The use of a specific bridge group (its chemical type, length and geometry) coupling the hapten to the larger carrier molecule determines the spacial orientation of the hapten to the carrier and, consequently, affects the recognition of the antibody for the hapten. The anti-indole acetic acid antibody (refs. 7, 29-31), for example, seems to have greater specificity for the methyl ester than for the free acid, indicating recognition of the bridging structure is important. In some cases, such as during the development of the immunoassay for dioxin (ref. 25), antibodies specific to the bridge may be selectively removed without affecting specificity for the hapten. The protocol of injection, number of booster shots, and time of sera collection after injection give rise to antibodies of varying sensitivity and specificity.

Nonhomogeneous immunological assays are complex, time consuming, and differ with the individual performing the task. In these assays, molecules must be adsorbed to a solid phase, usually polystyrene beads or microtiter plates, must be rinsed with detergents, used at optimum concentrations, and allowed to react for prescribed times. Because of the many manual manipulations and nonuniformity in laboratory supplies, results may vary. Until we can increase the reproducibility and shorten the time of analysis, we can not expect immunological methods to be adopted.

Use of in vitro cell lines as sources of antibody production should be explored. A consistent, well-characterized antibody preparation is required for commercialization of diagnostic kits. Experimental animals produce variable concentrations and specificities of antibody over the course of immunization. In contrast, once a desirable cell line is developed and isolated, practically unlimited amounts of antibody are available for as long as needed. However, use of polyclonal antisera should not be off-handedly rejected. Instead, the task at hand should be evaluated to determine which source of antibodies provides the best results. Monoclonal antibodies, for instance, exhibit notoriously low affinity constants and may, therefore, be undesirable for certain applications.

Perhaps the greatest difficulty affecting potential commercialization of agrochemical immunoassays may be convincing industry that there is a significant market for immunodiagnostics. It is expensive to develop test kits, but immunoassay will never become a routine procedure until such kits are marketed. Unfortunately, most assays were developed in academic laboratories and have not yet seen commercial application.

FUTURE THRUSTS

In this section we present our opinion of the future of agrochemical immunoassays based upon our own prejudicial perspectives. We expect to see a great expansion of both academic and industrial involvement in immunoassays for environmentally sensitive chemicals and pathogens. All concerned parties need to apply immunoassay to practical environmental problems to demonstrate the versatility of this analytical tool. There will be an increase in use of enzyme-linked assays and a much greater use of homogeneous assays due to their high reproducibility and short analytical times. Immunological techniques can readily complement existing analytical procedures. Antibodies bound to a solid support may serve as affinity columns that could be used to concentrate pesticides from large volumes of water. The pesticide may then be eluted and analyzed by chromatographic means. Large columns could even be used to purify water of undesirable environmental pollutants. Perhaps the best way to have immunological assays accepted for analyses of pesticides is to demonstrate to the classically trained analytical chemist that immunochemical techniques complement normal analytical procedures, and are not intended to replace them.

We expect to see greater application of monoclonal antibodies and the use of dedicated instruments. The need for immunological techniques employing simple instruments or color charts is particularly important to lesser developed countries, because these nations lack the trained personnel and analytical capabilities of more developed countries. Consequently, they may readily accept immunoassay due to its low cost and ease of operation.

Immunoassays have been developed for bacterial toxins isolated from varieties of Bacillus thuringiensis used to control lepidopterous and dipterous insects (ref. 15). Other large molecular pesticides of biological origin are under development. Such molecules are difficult to quantitate by classical procedures, such as gas-liquid chromatography, but are ideally suited for immunochemical detection. Biotechnological methods now permit incorporation of these toxins into plants and other organisms; immunoassay will be the method of choice to monitor these new products and assist risk assessment of these new developments.

Although analytical laboratories are busy determining the level of residues of agrochemicals and xenobiotics, the public is demanding even greater control of environmental quality. Unfortunately, classical analytical procedures are expensive, time consuming, limited by the number of capable laboratories, and require highly trained personnel and sophisticated equipment. When agrochemical immunoassays become routine, rapid, and low cost procedures, we will see an even greater demand for analysis of environmental samples. This will not occur, however, until regulatory agencies cast a more favorable outlook upon immunotechnology applied to environmental problems.

ACKNOWLEDGMENTS

Supported in part by Northeastern Regional Research Project (NE-115) and Paper No. 7465 in the Journal Series of the Pennsylvania Agricultural Experiment Station.

REFERENCES

1. B.D. Hammock and R.O. Mumma, ACS Symposium Series 136, 321-352 (1980).
2. J.J. Langone and H. VanVunakis, Res. Commun. Chem. Pathol. Pharmacol. 10, 163-171 (1975).
3. B.S. Ferguson, Sixth Int. Congress of Pesticide Chemistry, Poster Session (1986).
4. E.W. Weiler, Planta 144, 255-263 (1979).
5. E.W. Weiler, Physiol. Plant. 54, 510-514 (1982).
6. R. Mertens, B. Deus-Neumann and E.W. Weiler, FEBS Lett., 160, 269-272 (1983).
7. Idetek, Inc., San Bruno, CA 94066 USA
8. Granite Division - Environmental Diagnostics, Inc., Burlington, NC, USA.
9. H.R. Lukens, C.B. Williams, S.A. Levinson, W.B. Dandliker and D. Murayama, Environ. Sci. Technol. 11, 292 (1977).
10. S.J. Huber, Chemosphere 14, 1795-1803 (1985).
11. S.I. Wie and B.D. Hammock, J. Agric. Food Chem. 30, 949-957 (1982), 32, 1294-1301 (1984).
12. W.H. Newsome and J.B. Shields, J. Agric. Food Chem. 29, 220-222 (1981).
13. K.D. Wing, B.D. Hammock and D.A. Wustner, J. Agric. Food Chem. 26, 1328-1333 (1978).
14. K.D. Wing and B.D. Hammock, Experientia 35, 1619-1620 (1979).
15. S.I. Wie, B.D. Hammock, S.S. Gill, E. Grate, R.E. Andrews, Jr., R.M. Faust, L.A. Bulla, Jr. and C.H. Schaefer, J. Appl. Bacteriol. 57, 447-454 (1984).
16. S.I. Wie, R.E. Andrews, Jr., B.D. Hammock, R.M. Faust and L.A. Bulla, Jr., Appl. Environ. Microbiol. 43, 891-894 (1982).
17. P.Y.K. Cheung and B.D. Hammock, Current Microbiol. 12, 121-126 (1985).
18. P.Y.K. Cheung and B.D. Hammock, Appl. Environ. Microbiol., in press.
19. M.M. Kelley, E.W. Zahnow, W.C. Peterson and S.T. Toy, J. Agric. Food Chem. 33, 962-965 (1985).
20. J. Ohnishi, Y. Nishikawa, Y. Suzuki, T. Matsuda and J. Miyamoto, Sixth Int. Congress of Pesticide Chemistry, Poster Session (1986).
21. M.J. Wraith, E.J. Hitchings, A.P. Woodbridge, E.R. Cole and T.R. Roberts, Sixth Int. Congress of Pesticide Chemistry, Poster Session (1986).
22. D.F. Rinder and J.R. Fleeker, Bull. Environ. Contam. Toxicol. 26, 375-380 (1981).
23. M. Schwalbe, E. Dorn and K. Beyerman, J. Agric. Food Chem. 32, 734-741 (1984).
24. W. DeGreef, M. Dekegel and R. Hamers, Arch. Internat. de Physiologie et Biochimie. 88, B134-B135 (1980).
25. P.W. Albro, M.I. Luster, K. Chae, S.K. Chaudhary, G. Clark, L.D. Lawson, J.T. Corbett and J.D. McKinney. Toxicol. Appl. Pharmacol. 50, 137-146 (1979).
26. E.W. Weiler and U. Wieczorek, Planta 152, 159-167 (1981).
27. M. Ma, J.K. Burkholder, R.E. Webb and H.T. Hsu, J. Econ. Entomol. 77, 537-540 (1984).
28. W. Pengelly and F. Meins, Jr., Planta 136, 173-180 (1977).
29. E.W. Weiler, Planta 153, 319-325 (1981).
30. E.W. Weiler, P.S. Jourdan and W. Conrad, Planta 153, 561-571 (1981).
31. R. Mertens, J. Eberle, A. Arnscheidt and E.W. Weiler, Planta 166, 389-393 (1985).
32. W.H. Newsome and P.G. Collins, Sixth Int. Congress of Pesticide Chemistry Poster Session (1986).
33. Maryland Agric. Exp. Sta. Annual Report (1985) Regional Project NE-115.
34. R.O. Harrison, A.A. Brimfield, K.W. Hunter, Jr. and J.O. Nelson, Sixth Int. Congress of Pesticide Chemistry, Poster Session (1986).
35. W.H. Newsome, J. Agric. Food Chem. 33, 528-530 (1985).
36. K.W. Hunter and D.E. Lenz, Life Sci. 30, 355-361 (1982).
37. A.A. Brimfield, D.E. Lenz, C. Graham and K.W. Hunter, J. Agric. Food Chem. 33, 1237-1242 (1985).
38. D. Fatori and W.M. Hunter, Clin. Chim. Acta 100, 81 (1980).
39. T. Levitt, Proc. Analyt. Div. Chem. Soc. 16, 72-76 (1979).
40. J. Van Emon, B.D. Hammock and J. N. Sieber, Anal. Chem. 58, 1866-1873 (1986).
41. C.D. Ercegovich, R.P. Vallejo, R.R. Gettig, L. Woods, E.R. Bogus and R.O. Mumma, J. Agric. Food Chem. 29, 559-563 (1981).
42. R.P. Vallejo, E.R. Bogus and R.O. Mumma, J. Agric. Food Chem. 30, 572-580 (1982).
43. A.Y. Al-Rubae, Ph.D. Thesis, The Pennsylvania State University (1978).
44. J.F. Brady, M.S. Thesis, The Pennsylvania State University (1984).
45. A.H. Kuniyuki and S. McCarthy, Sixth Int. Congress of Pesticide Chemistry, Poster Session (1986).
46. A.U. Huber and S.J. Hock, J. Plant Dis. Prot. 92, 147-156 (1985).
47. W.H. Newsome, Bull. Environ. Contam. Toxicol. 36, 9-14 (1986).
48. E.W. Weiler, Planta 149, 152-162 (1980).
49. J.P. Duchakel, Ann. Rech. Veterinaires 16, 93-97 (1985).
50. D. Arnold and A. Somogyi, J. Assoc. Off. Anal. Chem. 68, 984-990 (1985).

51. M.A. Foglesong and D.S. LeFeber, J. Assoc. Off. Anal. Chem. 65, 48-51 (1982).
52. P. Rohner and J. Nicolet, J. Food Prot. 48, 59-62 (1985).
53. AGRI-DIAGNOSTICS Assoc., Cinnaminson, NJ, USA.
54. C.P. Romaine, S.R. Newhart and D. Anzola, Phytopathol. 71, 308-312 (1981).
55. J. Van Emon, J.N. Seiber and B.D. Hammock, Sixth Int. Congress of Pesticide Chemistry Poster Session (1986).
56. M. Landerlaan, B. Watkins, R. Devivar and L. Stanker, Sixth Int. Congress of Pesticide Chemistry, Poster Session (1986).
57. P.W. Albro, K. Chae, M. Luster, J.D. McKinney, S. Chaudhary, G. Clark, J. Fawkes and J. Corbett, Environ. Health Persp. 20, 247 (1977).
58. B.D. Hammock, S.J. Gee, P.Y.K. Cheung, T. Miyamoto, M.H. Goodrow and J.N. Seiber, Sixth Int. Congress of Pesticide Chemistry, Poster Session (1986).

Determination of iprodione in foods by ELISA

W. Harvey Newsome

Food Research Division, Bureau of Chemical Safety, Health and Welfare Canada, Ottawa, Ontario K1A 0L2, Canada

Abstract - An enzyme-linked immunosorbant assay (ELISA) was developed capable of determining the fungicide iprodione in an ethyl acetate extract of foods without prior clean-up. When tested on 4 commodities to which 0.1 - 1.0 mg/kg iprodione was added, at least 86% recovery was obtained. Recoveries correlated well with a GLC method employing N/P detection. Iprodione hydrolysis/rearrangement products reacted similarly to iprodione in the assay, while related fungicides vinclozolin and procymidone were 3.5 and 10 times more active than iprodione.

INTRODUCTION

Iprodione [1] is a contact fungicide registered for use in Canada on a variety of crops. Different methods have been published for residue determination, including quantitation by GLC after clean-up on Florisil (ref. 1,2) and reversed phase HPLC (ref. 3).

As an alternative to conventional analyses, immunochemical assays have been developed for a number of pesticide residues in foods, including diflubenzuron in milk (ref. 4), metalaxyl in fruit (ref. 5), and dichlofop-methyl in sugarbeet, wheat, and soybean (ref. 6). Since these procedures offer the possibility of a specific determination employing a minimum of sample preparation, the technique was investigated as a means of providing a mechanism for analysing a large number of samples in a cost-effective manner.

METHODS

ELISA

Immunogen. 3-(3,5-Dichlorophenyl)-2,4-dioxo-1-imidazoline propionic acid [5] was coupled to human serum albumin with 1-[3-(dimethyl-aminopropyl]-3-ethylcarbodiimide hydrochloride at pH 6.6 in 0.2 M phosphate.
Antiserum. New Zealand White rabbits were injected with 0.5 mg of immunogen in 0.5 mL an emulsion consisting of Freund's Complete Adjuvant and saline (1:1).
Plate sensitization. Sensitizing protein was prepared by condensing [3] with ovalbumin using a mixed anhydride reaction as described by Wie and Hammock (ref. 7). A solution of 4 ug|mL of the conjugate in bicarbonate buffer, pH 9.5 (200 uL) was added to each well of a microtiter plate. The plate was incubated at 4 deg C for 16 h, then washed 3x with 0.1% Tween 20, twice with distilled water and stored below 0 deg. C.
Sample preparation. Sample (10g) was extracted by homogenizing with 40 mL of ethyl acetate, filtered, and made to 50 mL. An aliquot of 50 uL was taken to dryness in a 12x75 mm glass tube, then dissolved in 25 uL of DMSO.
Immunoassay protocol. A 1:3000 dilution of antiserum in 0.1% gelatin (1.0 mL) was added to the samples and 0, 0.2, 0.4, 0.8, 1.6, 3.2, 6.4, and 12.8 ng of iprodione standard in 25 uL of DMSO. Tubes were incubated at 4 deg C for 30 min. Aliquots of 200 uL were added in triplicate to the wells of the previously sensitized microtiter plates. After incubation at 4 deg C for 1 h, wells were emptied and washed 3x with 0.1% Tween 20. A 1:1000 dilution of horseradish peroxidase labelled anti-rabbit IgG (Sigma Chemical Co.) (225 uL) was added, and after 30 min at room temp. the plate was washed and 225 uL of subtrate (o-phenylene-diamine/hydrogen peroxide) added. The reaction was stopped after 15 min by the addition of 0.5 M sulfuric acid (50 uL), and the OD read on a Titertek MCC colorimeter.

GLC

Sample preparation. The ethyl acetate extract prepared for ELISA was evaporated to dryness and partitioned between 25 mL of water and 25 mL of dichloromethane. The aqueous layer was extracted a second time with 25 mL of dichloromethane and the extracts combined, dried, and taken to dryness.

Column chromatography. Dried extracts were taken up in dichloro methane (5mL) and chromato-graphed on a 180x6 mm column packed with 3 g of Florisil in dichloromethane. The column was eluted first with 20 mL of dichloromethane which was discarded, then 25 mL of 2% ethyl acetate in dichloromethane which was taken to dryness and the residue dissolved in 2 mL of toluene for GLC determination.

Gas liquid chromatography. A Varian 1400 instrument was fitted with an N/P detector maintained at 260 deg C and a 20 m x 0.25mm id fused silica column coated with 0.25 u of DB 5. Samples (1.0 ul) were injected with a J&W on-column injector using a cold-trapping technique. Gas flows were as follows:

Helium carrier	33 cm/sec
Air	175 mL/min
Hydrogen	4.5 mL/min
Nitrogen makeup	30 mL/min

The sample was injected at 110 deg C and after a 45 s delay rapidly programmed to 245 deg C. A 1.8 ug/mL standard gave a 50% fsd response at an attenuation of 2. Retention time was 5.3 min.

RESULTS AND DISCUSSION

A typical standard curve is shown in Fig. 1. A linear relationship was found between log concentration of iprodione and the optical density of the wells after the enzymic reaction had occurred. Mean CV's of triplicate determinations over the concentration range studied were 3.7%. The slope of the standard curve was increased from 0.17 to 0.20 when gelatin was used instead of bovine serum albumin in the diluent, indicating a more sensitive assay. The minimum detection limit, based on 3x the mean standard deviation of the points within the standard curve was 0.03 mg/Kg. The minimum quantitation limit, based on 10x the mean standard deviation was 0.1 mg/Kg.

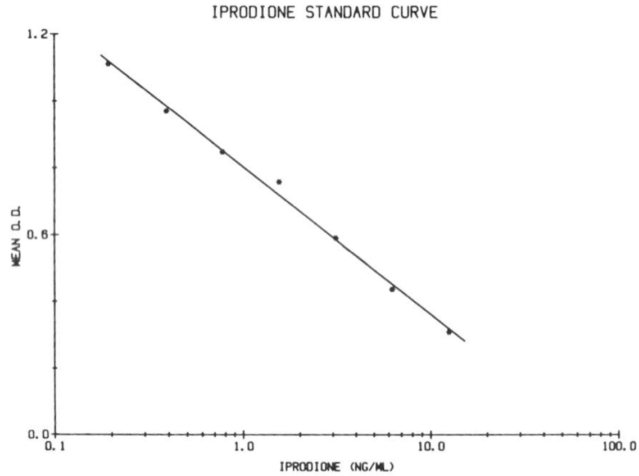

Fig. 1. Standard curve for ELISA of iprodione

The recovery of iprodione from several commodities as determined by ELISA and by GLC is given in Table 1. A strong correlation between the two methods is evident from the magnitude of the coefficient. In all cases, with the exception of GLC determinations on strawberry, recoveries exceeded 85%.

TABLE 1. Recoveries of iprodione from various commodities

Iprodione Added (mg/kg)	Iprodione Found (mg/kg) Tomato ELISA	Tomato GLC	Cucumber ELISA	Cucumber GLC	Grape ELISA	Grape GLC	Strawberry ELISA	Strawberry GLC
0	0.008	0.006	0.008	0.008	0.019	0.010	0.010	0.002
0.10	0.082	0.080	0.102	0.102	0.095	0.102	0.084	0.078
0.26	0.211	0.225	0.216	0.222	0.205	0.239	0.222	0.208
0.52	0.467	0.475	0.429	0.464	0.488	0.481	0.526	0.400
1.00	0.968	1.00	1.04	0.824	1.04	0.957	0.865	0.840
Mean Recovery (%)	86	92	91	88	91	94	88	78
R^2	0.997		0.974		0.998		0.975	

Within-run repeatability experiments with tomato spiked at 0.5 mg/Kg gave a CV of 7.8% by ELISA compared to 3.8% by GLC. The between-run CV for experiments conducted on 3 different days of ELISA was 9.9%.

Figure 2 shows the relative response of several compounds related to iprodione [1]. The response was determined by plotting the OD obtained from various concentrations of compound in question against the log of the concentration and determining the point at which the value was 50% of that obtained in the absence of inhibitor.

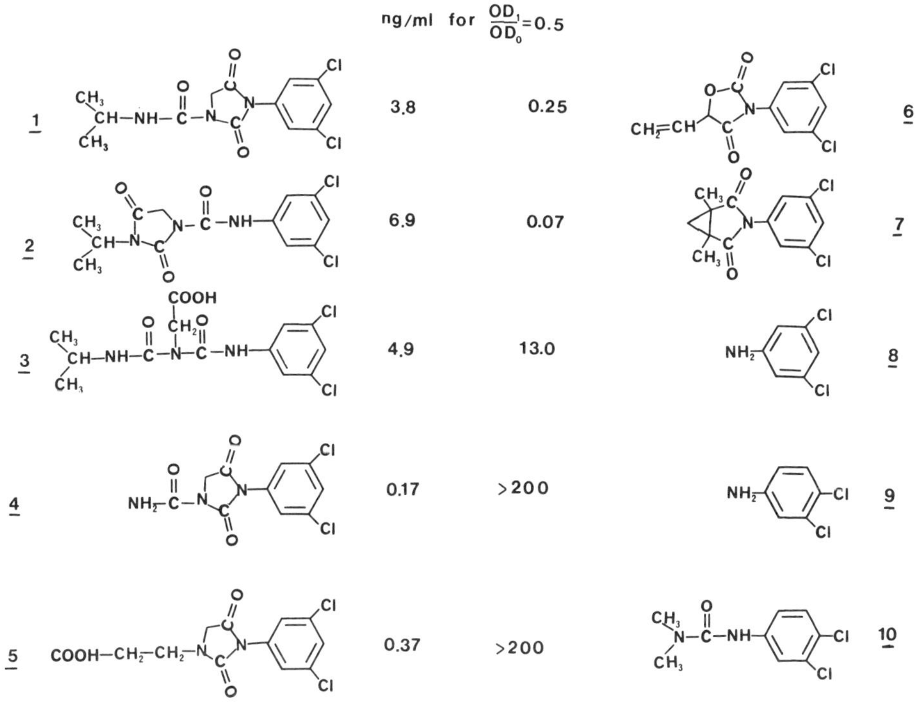

Fig. 2 Binding of iprodione and related compounds to antiserum directed toward [5].

The metabolite [2] and hydrolysis product [3] gave similar responses to iprodione, while the 3,5-dichloroaniline [8] is less inhibitory to antibody binding. 3,4-dichloroaniline [9] is readily distinguished from 3,5-dichloroaniline, as is the herbicide diuron [10]. Vinclosolin [6] and procymidone [7] are not registered for use in Canada and are not expected to occur on domestic crops. The use of these fungicides in foreign countries, particularly on grapes can result in residues on imported commodities which would give erroneously high values for iprodione as measured by ELISA, due to their greater binding.

Approximately 40 samples may be processed by one technician using ELISA compared to 10 by GLC. A screening role is suggested for the method, in view of the significant cross reactivity of the assay with related fungicides.

REFERENCES

1. L. Lacroix, M. Laurent, M. Buys, In Analytical Methods for Pesticides and Plant Growth Regulators, Vol. XI pp. 247, G. Zweig and J. Sherma Eds. Academic Press. New York (1980).
2. E. Lemperle, N. Emmanouilidis, N., E. Kerner, Deut. Lebensm. Rundschau. 78, 6 (1982).
3. P. Cabras, P. Diana, M. Meloni, F. Pirisi, R. Pirisi, J. Chromatography 256, 176 (1983).
4. S.I. Wie. B.D. Hammock, J. Agric. Food Chem. 32, 1294 (1984).
5. W.H. Newsome, J. Agric. Food Chem. 33, 528 (1985).
6. M. Schwalbe, E. Dorn, K. Beyerman, J. Agric. Food Chem. 32, 734 (1984).
7. S.I. Wie, B.D. Hammock, J. Agric. Food Chem. 30, 949 (1982).

Optimization of determination and clean-up parameters for sensitive multi-residue analysis of pesticides

L. Gyorfi[1], A. Ambrus[1] and E. Bolygo[2]

[1]Plant Protection and Agrochemistry Centre, Ministry of Agriculture and Food, Budapest, XI., Budaorsi ut 141-145, H-1502 Budapest, P.O. Box 127. HUNGARY

[2]Plant Protection and Agrochemistry Station
Vas County, 9762. Tanakajd, Ambrozi sêtany 2., HUNGARY

Abstract To facilitate the correct use of multi-residue procedures, the behaviour of a number of pesticides was studied using gas chromatography (GC) and compounds susceptible to the GC conditions were identified. The loadability of both GC columns and TLC plates was tested with apple, carrot and spinach extracts, which were applied after clean-up. General conclusions are drawn regarding the combined use of some clean-up and detection methods.

INTRODUCTION

The high selectivity of the flame photometric (FPD), thermionic (P-TID, in the phosphorus mode), and electrolytic conductivity (HCD) detectors as well as for some TLC detection methods enables the determination of some pesticide residues present in extracts, without the need of any clean-up. The direct injection of concentrated extracts both speeds up and simplifies the analyses and is used in many laboratories. However, the direct application of crude extracts may adversely affect the resolution of the chromatographic separation, change the interaction of some sensitive compounds with the chromatographic column and alter the relative retention time. Consequently the relative response factors (RRF) and the minimum detectable quantities (MDQ) may vary from time to time depending on the load and condition of the chromatographic columns.

For the proper use of multi-residue procedures (MRP), several parameters important (ref. 1). The detection conditions have to be continuously monitored to ensure that the sensitivity of labile compound is adequate for example. Therefore, the behaviour of several pesticides selected from the different chemical classes was studied using packed columns loaded with plant extracts in order to:

- identify the labile compounds which require special attention, especially when MRPs are used for samples of unknown origin;
- determine the loadability of short, low capacity GC columns.

Samples of apple, carrot and spinach were extracted with acetone, partitioned into methylene chloride and purified on a mixed adsorbent consisting of 1 g active carbon, 2 g magnesia and 4 g diatomaceous earth, as described in ref. 2. The effect of coextractives on TLC analysis was also tested using the same extracts applied to precoated plates.

SELECTION OF GC CONDITIONS

Inertness of GC columns

Labile molecules may decompose in the injector port and also on the chromatographic column. In both cases the catalytic effects from the injector port and column and the temperature govern the rate of decomposition. By using acid washed, silane-treated pyrex glass columns, together with the inert packings and silanized glass wool, the decomposition of most of the labile compounds can be eliminated or kept at a constant low level. Repeated injection of concentrated plant extracts or larger amount of the labile compounds (ref. 3) and the post silylation of the column (refs. 4, 5) further improves the inertness of the chromatographic system. By contrast a contaminated injector can also partly or completely decompose labile materials.

Comparing the response of the selected pesticides; carbaryl, monocrotophos, propiconazole and p-p'DDT under a variety of column conditions it was found that they could readily be detected at the 10 ng and 10 pg level with FPD or NPTID and ECD, respectively. The chromatographic conditions used to obtain this response were also suitable for the determination of the other pesticide residues which were recovered by the MRPs most widely used (ref. 1).

Among the organophosphorus pesticides analysed, the most problematic compounds were the polar ones which contained a side chain containing:
a) double bond, (mevinphos, phosphamidon, monocrotophos),
b) amide moiety (methamidophos, fenamiphos, phosfolan),
c) carbamoyl moiety (monocrotophos, omethoate, phosphamidon, dimethoate),
d) ester group (malathion, phenthoate, mevinphos),
e) or such a structure which is capable of forming hydrogen bond.

In addition, special attention is required for the analysis of the carbamate (carbaryl, dioxacarb, carbofuran, pirimicarb), uracils (lenacil bromacil, terbacil) and oth er compounds including endrin, DDT, metalaxyl, imazalil, dinosebacetate, chloridazion, propiconazole, chloropropylate, bromopropylate, dicofol, fenarimol and thiabendazol.

Column size

Short (50-90 cm) and narrow (1.75-2 mm i.d.) packed columns proved to be very useful for both screening and the confirmation of pesticide residues (refs. 2,6,7). In comparison with the traditional 180 cm x 3 mm columns, both the short and narrow columns provided 2 to 5 times faster analyses, with 5 to 10 times lower MDQs which enables the use of lower column temperatures factors which faciliates the analysis of labile compounds.

Loading columns with plant extracts

Short column (45 x 0.3 cm) packed with 3% OV-101 Chromosorb WAW 60-80 Mesh treated with Carbowax 20M were loaded with repeated injections of concentrated extracts equivalent to 50 to 100 mg of apple, carrot and spinach. Using an FPD and a temperature programme of (140°/2 min program at 10°C/min, 220°C/10 min) no change was observed in RRF values, relative retention times or the resolution when a maximum load of 2000 mg was applied. While isothermally at 180°C, both the RRF for monocrotophos (A) to parathion-methyl (B), and the resolution decreased continuously with increased loading as shown in Fig. 1.

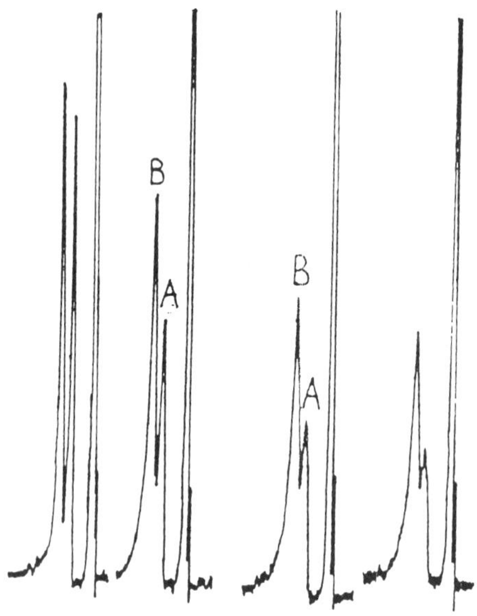

Fig. 1 Chromatograms of Monocrotophos (A) and Parathion-methyl (B) after loading the column with left to right, 0, 200, 300 and 500 mg spinach extracts.

The deterioration of the chromatographic column caused by loading it with 200 mg spinach was approximately equal to that resulting from injection 1400 mg apple or 600 mg carrot extracts. Loading the columns with extracts which had been purified on the mixed adsorbent did not change the column performance even when the total load was 3000 mg.

These results indicate that the loadability primarily depends on the type of extract and column temperature, and is influenced to a lesser extent by the capacity of the column. The approximate loadability of columns can be quickly determined by checking the RRF values after repeated injection of extracts representing 50 to 200 mg of a sample, which contains a suitable pair of test compounds e.g. monocrotophos/parathion-methyl or carbaryl/propham.

The injections are continued until the RRF of labile/stable compounds decreases to 80% of the starting value, which is defined as the maximum total load (L_M). A single injection can contain 0.03 L_M sample providing that the silanized glass wool is changed and the column is reconditioned at maximum operating temperature plus 30°C for 8 hours after every 20 sample injections.

When determining the maximum permissible load, the selectivity of the detector also has to be considered. In the case of the ECD, practically every extract needs some clean-up. Similarly, in the case of the NPTID, purified extracts are required. The phosphorus selective TID usually allows the injection of 2-10 mg plant extracts, while in case of FPD and HCD the loadability of the columns is the limiting factor. To ensure the best performance of a GC system, the columns should be loaded with the minimum amount of sample needed to achieve the limit of determination (mg/kg) as required for the sample-pesticide matrices.

LOADABILITY AND TLC

The loadability of silicagel and aluminium oxide TLC plates was tested with various plant extracts, eluents and pesticides. The coextractives usually gave diffuse, tailing spots and influenced the distribution of pesticides migrating with lower linear velocity as well as their levels of detectability. Consequently the loadability of the layer varied with the solvent strength of the mobile phase, retention of the pesticides and was also dependant on the mode of detection. The maximum load which enables reproducible separation and detection together with the limits of determination (LD) are shown for some cases in Table 1, the detection methods used were:

a) carbaryl and dioxacarb with p-nitrobenzene/diazonium/fluoroborate after hydrolysis;
b) atrazine with o-tolidine/potassium iodide after chlorination;
c) carbendazim with bioassay using spore suspension of Aspergillus niger L;
The extracts were purified with mixed adsorbent.

TABLE 1. Change of the limit of determination (mg/kg) of some pesticides depending on the TLC conditions

Compound	Eluent	R_f	MDQ /ug	Carrot: Purified extr. Sample mg	Carrot: Purified extr. LD	Carrot: Crude extr. Sample mg	Carrot: Crude extr. LD	Spinach: Purified extr. Sample mg	Spinach: Purified extr. LD	Spinach: Crude extr. Sample	Spinach: Crude extr. LD
Carbaryl	EtAc	0.58		500	0.2	100	1	1000	0.1	100	1
	PE+E=1+2	0.23	0.05	2000	0.05	250	0.4	1000	0.05	50	2
	CH_2Cl_2	0.11		500	0.1	200	0.5	1000	0.05	200	0.1
Dioxacarb	EtAc	0.41		500	0.4	100	2	1000	0.1	100	2
	PE+E=1+2	0.06	0.1	2000	0.05	250	0.8	1000	0.1	50	4
	CH_2Cl_2	0		500	0.4	200	1	1000	0.2	200	1
Atrazine	EtAc	0.59		250	0.2	100	0.5	1000	0.1	100	1
	PE+E=1+2	0.27	0.05	500	0.2	200	0.5	2000	0.05	100	1
	CH_2Cl_2	0		400	0.25	100	1	1000	0.1	250	0.4
Carbendazim	EtAc	0.22		500	0.2	500	0.2	2000	0.04	250	0.3
	PE+E=1+2	0.02	0.01	400	0.2	100	0.8	2000	0.02	250	0.2
	CH_2Cl_2	0		500	0.2	50	1.6	1000	0.30	500	0.3

PE+E = petroleum ether + ethyl ether

Limits of determination below levels of 0.5 to 1 mg/kg can rarely be achieved when the plant extracts are applied directly to the TLC plate even if bioassay methods (enzyme or fungi spores) were used for detection. The experiments carried out revealed that a bioassay detection methods, being specific for certain type of pesticides, are not sensitive in the presence of large amount of coextractives. The clean-up of crude extracts on the mixed adsorbent reduced the LD approximately three to five fold. If the limits of determination below 0.1 mg/kg are required then an efficient clean-up must be used.

CONCLUSION

The thermal stability of pesticides to gas chromatographic conditions varies to a great extent and it is affected by the type of sample extract. To ensure that sensitive detection of pesticides recovered by the MRPs is obtained, proper GC conditions must be maintained and checked regularly during the analyses of the samples. Monitoring of determinative parameters can be conveniently carried-out using test mixtures containing both labile and stable compounds.

Thus, the type of sample extract, the chromatographic operating conditions and the mode of detection all influence the loadability of chromatographic columns and TLC plates. The direct injection of some concentrated extracts is possible when the P-TID, FPD or HCD gas chromatographic detectors are employed. When TLC detection is used for multi-residue analysis only purified extracts should be applied.

REFERENCES

1. A. Ambrus, H.P. Thier, Pure and Applied Chem. 58. 1035-1062, (1986)
2. A. Ambrus, et al., J. Assoc. Off. Anal. Chem. 64. 733-768, (1981)
3. D.L. Lewis, D.F. Paris, J. Agric. Food. Chem. 22. 148-149, (1974)
4. M. Riva, A. Carisano, J. Chromatogr. 42. 464-469, (1969)
5. P.R. Laski, P.R. Watts, J. Assoc. Off. Anal. Chem. 56. 328-332, (1973)
6. E. Bolygo and F. Zakar, J. Assoc. Off. Anal. Chem. 66. 1013-1017, (1983)
7. M.A. Luke et al., J. Assoc. Off. Anal. Chem. 64. 1187-1195, (1981)

Some thoughts on multiresidue methods

Brian D. Ripley

Provincial Pesticide Residue Testing Laboratory, Agricultural Laboratory Services Branch, Ontario Ministry of Agriculture & Food, c/o University of Guelph, Guelph, Ontario Canada N1G 2W1

Abstract - The pesticide residue analyst receives several types of requests for the analysis of a particular substrate for pesticides. How should this analysis be conducted? What method should be used? This paper discusses some thoughts on multiresidue methods.

Increased public awareness and concern about the use of pesticides (and chemicals in general), their residues in foods and the environment, and their toxic effects has increased the demands to detect and measure these residues at lower and lower concentrations. While the toxicological significance of these trace concentrations may be unknown, it is incumbent on the residue analyst to report residue data derived from a method that has reliability, consistency, and reproducibility.

In general, two types of requests are received by the pesticide residue analyst. In the easiest case, the request is for a single pesticide in a specific substrate, often after a known application. A specific, sensitive method may be developed to handle the separation, qualitation and quantitation of the specific pesticide. Frequently this method is extended to include specific metabolites of the designated compound. This method may or may not be applicable to other pesticides or substrates. Alternatively, the request may be for a large number of different compounds representing several classes of pesticides, often with no background information on which, if any, pesticide residue is likely to be present.

The present state of multiresidue methodology has been discussed at previous Congresses (ref. 1) and elsewhere at this Congress. Several options are available to the residue analyst. The paramount question in deciding which method to use is - What is the purpose of the analysis?

In regulatory or compliance laboratories, as well as those laboratories conducting environmental monitoring and surveillance programs, the goal of the analysis is often to obtain a rapid screening of many samples to identify and quantitate as many different pesticides as possible. The potential residues in the commodity may be known or suspected as with domestic commodities or unknown as with some imported food products or environmental substrates. In some cases, sensitivity only to the maximum residue limit is required while in other cases the compound must be determined to the detection limit of the method. Invariably, a wide array of matrices must also be analyzed. To fulfill these needs, multiresidue methods are used.

The FDA Pesticide Analytical Manual (ref. 2) describes "methods which detect multiple residues." More correctly, these multiresidue methods compile those pesticides that are recovered from various substrates under specified conditions during extraction and partitioning, those pesticides that are separated and recovered from clean-up columns, and those chemicals that are detected with various detectors under specific chromatographic conditions. One must realize that a "multiresidue" method is not capable of detecting or measuring <u>all</u> pesticides or even all pesticides in a single class of compounds. Specifically, multiresidue methods allow the measurement of more than one residue under certain conditions.

The three pillars of analytical chemistry are : (1) separation, (2) identification, and (3) measurement (ref. 3). Within these objectives, the analyst has two general goals in the development of a residue method - specificity and sensitivity.

Analytical methodologies evolve from a series of steps. The three main steps involved in a pesticide residue method are extraction, clean-up, and determination. Each step in a procedure may influence the next; for example, a relatively clean water sample may not require clean-up whereas plant or animal extracts may require two or more clean-up steps. Usually the method of final determination or detection influences the degree to which co-extractives or interferences need to be removed. Moreover, the extraction procedure may have to be modified depending on the goal of the analysis, i.e., whether metabolites, particularly conjugates, are to be analyzed.

Each step in the analytical procedure is devoted to one or more of the above goals. For example, extraction is designed to separate often minute amounts of pesticide(s) from the bulk of the sample matrix; adsorption column chromatographic clean-up can further separate the pesticide(s) from unwanted co-extractives and provides a tentative identification of the residue based on a known elution pattern from the column; and gas chromatography can separate the pesticide from other pesticides and co-extractives, provide tentative identification based on retention time and specific detector response, and allows measurement or quantitation of the amount of pesticide eliciting the response. One must bear in mind, however, that GC (or HPLC) is only a tool that serves the analyst as a remarkable clean-up, separation and detection device when the proper selection of column, detector and chromatographic conditions are used. These techniques must not be considered as an absolute "identifying" device but rather an aid to suggest possible identification. Various confirmatory techniques such as alternate column or detector examination, chemical derivatization, or mass spectrometry should be employed.

Dao et al. (ref. 4) noted that the primary criteria for choosing an extracting solvent have been utilitarian ones such as (1) absolute extraction efficiency, (2) minimal amounts of co-extractives and subsequent interferences, (3) minimal clean-up steps of the extract, and (4) reproducibility of residue recovery. Many conventional residue methods, especially multiresidue methods, do not extract bound residues, that is, those residues remaining in soil, plant material or animal tissue after exhaustive solvent extraction (ref. 5). For example, Wheeler et al. (ref. 6) compared the extraction efficiency of several solvent systems to extract field-weathered residues. The significance of these residues is still to be determined in many cases. Consideration must also be given to the ionic nature of some residues and an appropriate pH must be chosen for extraction or partitioning.

Satisfactory clean-up techniques are limited only by the imagination of the analyst in terms of the chemical and physical characteristics of the analyte or substrate (ref. 7). However chosen, they must be reproducible and efficient. When no clean-up is necessary or efficient, this step should be eliminated. For example, accepted multiresidue methods extract most organochlorine, organophosphorus and carbamate insecticides. However, Florisil clean-up is necessary and efficient only for the electron capture examination of OCs. Therefore, an aliquot of the extract can be removed for gas chromatographic separation and specific detection of the OPs with a flame photometric detector and the carbamates with a nitrogen-phosphorus detector for example (ref. 8). One must remember, however, that no detector is wholly "specific".

Often several clean-up or fractionation steps may be required to achieve optimum separation of co-extractives from the analyte(s). Much ingenuity has been used to separate organochlorine insecticides from PCBs or dioxins. Once the latter analytes are isolated in a relatively "clean" extract, isomer-specific analysis may be applied. Again, the ingenuity of the analyst may have created data that requires critical toxicological interpretation.

With HPLC and non-specific ultraviolet detection, extracts often have to be cleaned up to a high degree often surpassing the requirement for electron capture detectors. On the other hand, use of multiple-ion-monitoring mass spectrometry allows for specific detection of several selected pesticides in a relatively dirty extract. Frequently, the clean-up step produces the most significant loss of analyte. In many cases, this step can be eliminated through a judicious choice of extraction procedure and GC-detector system.

Horwitz (ref. 9) noted that analytical methods have two types of characteristics - scientific and practical. Scientific qualities determine the reliability of a method (accuracy, precision, specificity, detection limit) while the practical aspects describe the utility of a method (time, cost, availability of equipment). Therefore, some compromise must be made between these two characteristics in deciding what method to apply to a specific analytical problem. A multiresidue method should be designed to achieve the most information on a sample, with the fewest analyses, and in the fastest and cheapest manner that is realistic.

Often the use of a modified "official" method of analysis is acceptable and necessary for certain circumstances or resources. Each analyst has his favorite procedure usually determined in part by cost, availability of equipment and facilities, ease of manipulation, the analyte and the substrate. These favorite procedures are most evident in the determinant step, especially when derivatization is used. Despite some "all"-encompassing multiresidue methods, our laboratory still prefers to use smaller multiresidue (or multiclass) methods, as well as some specific methods, to generate residue data on many pesticides.

Pesticides are often divided into groups according to their chemical structure. Methods are similarly applied depending on the chemical properties of the analyte or the detection system used. In selecting a method, consideration must also include the wide range of polarities of pesticides, fat and moisture content of the substrate, and the expected concentration of the analyte. Whenever possible, it is desirable to incorporate newly developed or registered pesticides into existing multiresidue methods so that simultaneous screening of numerous chemicals is possible. Some minor changes in extraction, clean-up, or chromatographic conditions may be necessary for optimum determination of all compounds (ref. 10, 11).

The prime consideration for an acceptable method is that the residue data generated by it are valid. Collaborative studies are the most widely used criterion and several multiresidue methods have been validated through interlaboratory assessment and are widely applied in different laboratories. All analysts should have confirmed that the methods they use are reproducible and accurate within their laboratory. Using their own method in a check sample program will demonstrate the utility or deficiency of the method, and also the performance of the analyst.

A pesticide analyst must have many and varied skills and, above all, he must be a good analytical chemist. Often the analyst must be a "Quincy" to solve a particular analytical problem or investigation into a suspected pesticide incidence or occurrence. How much information is "thrown out with the dishwater"? Even though requested to make a specific analysis, the analyst must have an awareness of what may have happen in a poisoning incident, what possible residues could be present in a particular sample or what other residues could be detected with a particular methodology. For example, our laboratory was asked to specifically analyze well waters to determine the extent of alachlor contamination. Fortunately, a multiresidue method was used and many triazine and chloroacetamide herbicides were found in the wells. Had the method used only examined for alachlor, additional information could have been lost since there were only "a few extraneous, non-interfering peaks" present. Furthermore, the additional residue information provides a "complete" and unbiased picture of the situation (or does it?).

NEW APPROACHES TO MULTIRESIDUE METHODS

Several approaches have been made to reduce costs yet increase productivity. Automatic liquid-samplers are commonly used on gas or liquid chromatographs while reporting integrators have simplified some chromatographic interpretation steps. Automation has also been applied to the partitioning (ref. 12) and cleanup (ref. 13) steps of the AOAC method; sweep co-distillation (ref. 14,15), and gel permeation (ref. 16) have shown some cost saving and increased precision. Full automation from sample entry to data reporting is probably impractical except for some substrates such as water. However, any gain in cost effectiveness and reproducibility is desirable provided there is no sacrifice in the integrity of the analytical data.

Sep-Paks (ref. 17, 18) and other adsorbent cartridge-type systems, such as XAD resins (ref. 19), have been employed for one-step sampling, extraction and clean-up. Such techniques are known as "trace enrichment" since a large volume of water is passed through the cartridge with the pesticide(s) being adsorbed. Subsequently, a small volume of organic solvent is used to desorb the pesticide for GC or LC determination. Such devices are useful for field sampling of water since they allow a time-integrated and large volume of sample to be examined and eliminate the need for glass bottles. Desorption of pesticides or pollutants can be carried out in the laboratory. Cartridges filled with Florisil, alumina or XAD resin can also be used as a clean-up step for extracts, and since small volumes of eluant are required there is considerable savings in glassware and solvents. Often a solid sample can be "turned into" an aqueous-based sample and trace enriched thus achieving both concentration and clean-up. Air and pesticide vapours have also been successfully sampled using similar techniques.

SUMMARY

Ideally, a multiresidue procedure entails a simple and efficient extraction and/or clean-up procedure that recovers a wide array of pesticides of different polarity and representing diverse chemical classes. Thereafter, the total time and number of pesticides detected is dependent on the extent of the determinative steps used to examine the extract (ref. 20). Perhaps this is why an examination of the literature on pesticide analytical methods reveals a high plurality of determinative procedures and relatively few extraction or clean-up methods. From a practical point of view, the choice of the method may depend on availability of certain instrumentation, specialized glassware and reagents, facilities, monetary and, most of all, time constraints.

And finally, a thought on reporting residue data. Horwitz (ref. 9), Gunther (ref. 7), and Freshe and Timme (ref. 21) have elegantly discussed the role of the analyst in the interpretation and reporting their own residue data; these articles should be mandatory reading for all pesticide residue analysts. How low and how accurately can we determine trace residues? Often we delude ourselves by thinking that we can precisely measure low concentrations of contaminants. Both accuracy and precision of measurement decrease exponentially as the concentration of analyte decreases. Typically, at 1 mg/kg levels residue determinations should have a latitude of ±10-25% while at 1 ug/kg the latitude is ±50-200% in real samples. Residue data should only be reported to, at most, two significant figures. Detection limits should be reported for the lowest concentration at which the analytical method can be reproduced and not at the sensitivity of the measuring device. The "vanishing zero" (ref. 22) is truly disappearing and the analyst must be certain of the quality and significance of the data they report.

REFERENCES

1. Pesticide Chemistry, Human Welfare and the Environment, J. Miyamoto and P.C. Kearney, eds., Vol. 4, pp 117-140, Pergamon Press, Oxford (1983).
2. Pesticide Analytical Manual, Vol. 1, Food and Drug Administration, Washington, DC, (1985).
3. I.A. Selby, Chem. in Britian 14, 606-613 (1978).
4. T.H. Dao, T.L. Lavy and J. Dragun, Residue Rev. 87, 91-104 (1983).
5. S.U. Khan, Residue Rev. 84, 1-25 (1982).
6. W.B. Wheeler, N.P. Thompson, P. Andrade and R.T. Krause, J. Agric. Food Chem. 26, 1333-1337 (1978).
7. F.A. Gunther, Residue Rev. 76, 155-171 (1980).
8. B.D. Ripley and H.E. Braun, J. Assoc. Off. Anal. Chem. 66, 1084-1095 (1983).
9. W. Horwitz, in The Pesticide Chemist and Modern Toxicology, S.K. Bandal, G.J. Marco, L. Golberg and M.L. Leng, eds., ACS Symp. Ser. 160, American Chemical Society, Washington, DC, chapt. 24 (1981).
10. H.E. Braun and J. Stanek, J. Assoc. Off. Anal. Chem. 65, 685-689 (1982).
11. R.T. Krause and E.M. August, J. Assoc. Off. Anal. Chem. 66, 234-240 (1983).
12. F.M. Gretch and J.D. Rosen, J. Assoc. Off. Anal. Chem. 67, 108-111 (1984).
13. F.M. Gretch and J.D. Rosen, J. Assoc. Off. Anal. Chem. 67, 783-789 (1984).
14. B.G. Luke, J.C. Richards and E.F. Dawes, J. Assoc. Off. Anal. Chem. 67, 295-298 (1984).
15. B.G. Luke and J.C. Richards, J. Assoc. Off. Anal. Chem. 67, 902-904 (1984).
16. J.A. Ault and T.E. Spurgeon, J. Assoc. Off. Anal. Chem. 67, 284-289 (1984).
17. Application Notes, Waters Assoc., Milford Mass.
18. B.S. Clegg, D.H. Schellenberg, B.D. Ripley and N.D. Chapman, Proc. 6th Int. Cong. Pest. Chem., Ottawa, Abstr. (1986).
19. J.E. Woodrow, M.S. Majewski and J.N. Seiber, J. Environ. Sci. Health B21, 143-164 (1986).
20. M.A. Luke, J.E. Froberg, G.M. Doose and H.T. Masumoto, J. Assoc. Off. Anal. Chem. 64, 1187-1195 (1981).
21. H. Freshe and G. Timme, Residue Rev. 73, 27-47 (1980).
22. G. Zwieg, J. Assoc. Off. Anal. Chem. 61, 229-248 (1978).

Methods of assessing consumer exposure to pesticide residues

Leslie G. Ladomery

Joint FAO/WHO Food Standards Programme, FAO, Rome, Italy

Abstract - Consumer protection regarding pesticide residues in food is achieved through setting maximum residue limits (MRL) at the lowest possible level consistent with "good agricultural practices" and through a comparision of the "acceptable daily intake" (ADI) with estimates of daily residue intake. It is suggested that an accurate knowledge of actual pesticide residue intake is needed where it is necessary to demonstrate that the ADI has been exceeded. To show the opposite, it is possible to use reliably exaggerated intake estimates (which do not require detailed information on food consumption data, pattern of use etc.) resulting in significant savings in laboratory resources.

INTRODUCTION

As long as there are living organisms which cause disease in plants or feed on crops grown for human consumption, there will be a need to use pesticides. The development of new pesticides is also unlikely to stop, since resistance to chemicals used in agriculture, just like the pests themselves, is likely to continue to plague those who work the land in order to produce food. The fundamental similarity between the biochemical mechanisms of living cells, whether of man or of pests is not likely to permit the development of pesticides with completely selective toxicity to the unwanted living organisms which damage crops and stored food. For these reasons, there will always be a need for vigilance concerning the proper use of pesticides and their effects on man.

The use of pesticides, whether pre or post harvest, inevitably leads to residues in the food sold to the consumer. Damage to health due to these residues depends on the intrinsic toxic nature of the residue and the amount ingested daily as well as the length of exposure. Unfortunately, this time/quantity expression of consumer exposure is normally not practical to estimate, since it would be necessary to assess the length of exposure to a given quantity of pesticide residue by individual consumers or groups of consumers. For this reason, any estimate of exposure to pesticide residues takes into account only daily intakes and it is assumed that this amount is ingested each day for a lifetime.

Acceptable daily intake of pesticide residues

Risk assessment calls for a determination of that amount of pesticide residue which may be safely ingested for a lifetime. This amount is determined in test animals on the basis of internationally accepted protocols and is referred to as the "acceptable daily intake" or simply ADI, expressed as mg/kg body-weight. It need hardly be pointed out that the ADI, like any end point based on informed judgement involving observed or measured parameters, is subject to errors. Paynter and Schmitt (ref. 1) have described the various errors which are inherent to the ADI. However, these errors are seldom known. Furthermore, the ADI incorporates an arbitrary "safety factor" which is determined by toxicologists in the light of all relevant facts known at the time. None-the-less, the ADI will be treated in this paper as a maximum exposure level above which action is required in the interest of consumer protection, but below which consumers do not run any appreciable risks to their health. It is not the purpose of this paper to discuss what action should be taken when the total daily intake of a pesticide residue approaches or exceeds the ADI. Rather, this paper is intended to discuss methods to ascertain whether or not the ADI has been exceeded. Clearly, where there is strong evidence that daily exposure to a pesticide residue is above that which is acceptable, appropriate steps have to be taken involving informed persons in various fields, including toxicology.

Maximum limit for pesticide residues in food

It is wrong to assume, as some do, that consumer protection can only be secured through an assessment of pesticide residue intakes and their comparison with the applicable ADIs. In fact, the "maximum residue limits" (MRLs) set for individual food products also protect health, since they enforce the principle that no more of a pesticide should be used than is necessary in accordance with "good agricultural practice" (GAP). MRLs are based on residues data from supervised field trials representing GAP and not on animal toxicity tests and, therefore, do not represent upper limits based on possible health effects. The numerical value of an MRL, therefore, depends on parameters relating to good agricultural practices, nature of the pest, climatic conditions and other such considerations (ref 2).

MRLs cover variation in the distribution of residues in food products, errors in sampling and analysis and variation in what is "good agricultural practice". MRLs, therefore, do not represent actual amounts of residues which are always present in a food product. In fact, residues in food seldom approach MRLs for the reasons mentioned above and also in view of the degradability of most pesticides. Thus, MRLs cannot be used to calculate daily intakes of pesticide residues accurately. They can be used, however, to demonstrate that the ADI is not likely to be exceeded as will be shown later. In view of the role of MRLs in protecting health, it is essential that they be set at the lowest possible level so that they will be effective in keeping the use of pesticides to a minimum. Pieters (ref. 2) has described the role of MRLs from the point of view of consumer protection.

ASSESSMENT OF PESTICIDE RESIDUE INTAKES AS AN INDICATION OF CONSUMER RISK

As has been mentioned earlier, the ADI represents the daily exposure to a pesticide residue for a lifetime which appears to be without appreciable risk to health, based on all facts known at the time. This is explained by the Joint FAO/WHO Meeting on Pesticide Residues as the "practical certainty that injury will not result even after a lifetime of exposure"(ref. 1). This means that if the daily intake of the pesticide residue is equal to or less than the ADI no further attention need be given to chronic toxicological considerations relating to the daily amounts of pesticide residues ingested by the general population. On the other hand, other aspects such as, allergenicity, sensitivity, idiosycrasy and synergistic, additive and pontentiating effects, vulnerable groups of the population such as infants and children, state of health, pregnancy, age, prematurity at birth, high selective consumption of given foods and other such considerations must be taken into account in addition to the comparison of intake estimates with the ADI.

Notwithstanding the above remarks, estimates of pesticide residue intake are an essential component in the assessment of risk to health from pesticide residues in food. This means that attempts must be made to estimate, from time to time, the daily intake of pesticide residues by man.

There are two ways of approaching to the problem. One approach is to try to assess (eg. in mg/day/man) actual pesticide residue intakes by 50th, 90th or some other percentile of the population (ref. 12). The other approach is to determine, with acceptable confidence, that the daily pesticide residue intake will not exceed the ADI without trying to quantify residue intake. As a matter of fact, both approaches are needed, as will be seen later in this paper. While the first approach is more informative since it tends to quantify intake, the second approach is by far the more simple one and the one which requires less data. From a consumer protection point of view, both approaches should be acceptable, since the ADI is based on a maximum "no observed effect level" (NOEL) of the residue in animal tests which means that neither exposure to amounts of residue equal to the ADI nor, obviously, to amounts less than the ADI will cause injury during an entire lifetime.

Calculation of highest intake limit (HIL)

The expression "intake limit" rather than "intake" is proposed to suggest that calculations involving theoretical and grossly exaggerated assumptions do not result in estimates of intake but in exaggerated limiting values which can be used to demonstrate that the ADI will not be exceeded, but which cannot be used to indicate the opposite.

One such "intake limit" expression could be derived as follows:

$$\text{HIL/ADI RATIO} = \frac{\text{highest authorized MRL (mg/kg) x 3(kg/person)}}{\text{ADI x 60 (mg/person)}}$$

where:

HIL = highest intake limit = residue in 3 kg total food resulting from presence of residue at highest authorized MRL (mg/person/day)

Highest authorized MRL = the highest MRL in a list of MRLs, established for a given pesticide in various foods, excluding animal feed or non edible commodity (mg/kg)

ADI = acceptable daily intake estimated for the pesticide and its residue (mg/kg body-weight)

60 = assumed average human body weight (kg)

3 = total amount of food consumed, excluding water and beverages (kg)

The Highest Intake Limit (HIL) assumes that all food contains residue at a level permitted by the highest MRL in a series of MRLs for a given pesticide. This means that we can ignore variation from one country or one season to another in the pattern of use of the pesticide. The value of the HIL may change, of course, if an additional and even higher MRL is introduced through the national authorization process or through the Codex mechanism (ref. 13). The HIL is also based on a high total daily food intake and a relatively low body-weight, both of which work in favour of the consumer.

The HIL/ADI ratio was calculated using Codex MRLs (Ref. 13,14) and the following results were obtained:

TABLE 1 Frequency distribution of HIL/ADI ratios

HIL/ADI Ratio	Frequency (Number of Pesticides)	Frequency (%)
0.01- 1.00	7	5.88
1.01- 1.25	2	1.68
1.26- 2.00	7	5.88
2.01- 4.00	8	6.72
4.01- 6.00	8	6.72
6.01- 8.00	3	2.52
8.01- 10.00	13	10.92
10.01- 20.00	18	15.13
20.01- 30.00	8	6.72
30.01- 40.00	10	8.40
40.01- 60.00	10	8.40
60.01- 80.00	4	3.36
80.01-100.00	6	5.04
100.01-150.00	1	0.84
150.01-200.00	8	6.72
>200	6	5.04
TOTAL	119	99.97

As can be seen, 9 out of the 119 pesticides examined (7.56%) can be cleared on the basis of the HIL/ADI ratio or 16 out of 119 (13.44%) if we are prepared to go to a HIL/ADI ratio of 2 without having access to national food consumption data or residues information from monitoring or disappearance studies. It also means that the respective Codex MRLs should be acceptable on grounds that the daily intake of pesticide residues is unlikely to exceed the ADI.

Calculation of adjusted intake limit (AIL)

Following examination of about 26,000 food samples analyzed during 1974-1976, Frawley and Duggan (ref. 6) calculated the frequency rates of occurrence of various classes of pesticides. Examination of these data in relation to 70 total diet studies was used by these authors in the calculation of "concentration ratios" reflecting the reduction of the residue from time of harvest to time of ingestion. Frawley and Duggan applied these reduction factors to "theoretical daily intakes" (TDI) in order to arrive at the more realistic "estimated daily intakes" (EDI). TDI is calculated from legal limits (MRLs) and daily food consumption data, assuming that the residue is always present at the legal limit at the time of consumption, for a life time. Various food intake data can be used, eg. 50th or 90th percentile (ref. 11).

Even a highly theoretical and exaggerated estimate such as the TDI requires the use of food intake data which are not always available. It was thought to be useful, therefore, to apply the factors of Frawley and Duggan (see Table 2) to the HIL values derived as described above, in order to see what proportion of pesticides could be cleared on this basis. The HIL to which these factors were applied are referred to as "adjusted intake limits" (AIL). It was also thought to be useful to compare all available TDIs estimated by WHO (Ref. 7, 11) and all available intake estimates from total diet studies (Ref. 3,4,5,8,9,10) with the HIL and AIL. (see Table 3)

TABLE 2. Factors used to calculate AIL from HIL (Ref. 11)

Pesticide group	Frequency of occurrence in foods	Disappearance	Combined
carbamates	0.08	0.14	0.0112
organophosphates	0.16	0.84	0.1344
chlorinated hydrocarbons	0.44	0.37	0.1628
other	0.14	-	0.14

TABLE 3. Proportion of intake limits and daily intakes $\leqslant$ ADI

Pesticide Group	No. in Group	HIL	AIL	TID	EDI	DDI
ORGANOPHOSPHATES	45	4/45 (8.89%)	8/45 (17.78%)	19/36 (52.78%)	34/36 (94.44%)	20/20 (100%)
CARBAMATES	11	2/11 (18.18%)	10/11 (90.91%)	5/6 (83.33%)	6/6 (100%)	4/4 (100%)
CHLORINATED HYDROCARBONS	8	0/8 (0%)	1/8 (12.50%)	2/8 (25%)	6/8 (75%)	8/8 (100%)
MISCELLANEOUS	55	4/55 (7.27%)	24/55 (43.64%)	23/31 (74.19%)	29/31 (93.55%)	14/14 (100%)
TOTAL	119	10/119 (8.40%)	43/119 (36.13%)	49/81 (60.49%)	74/81 (91.36%)	46/46 (100%)

Results of the calculations of HIL and AIL are given in Table 3 together with available TDI and EDI estimates and intake data from total diet studies. The data show that roughly 36% of pesticides can be cleared on the basis of the AIL, ie. without any knowledge of detailed food consumption data. This approach can be criticized on the basis that the factors used are only from US data from 1974, 1975 and 1976. However, the point to note is that the factors can be changed on the basis of more recent and more extensive information, leaving the principle of the approach unaffected. Another point to note is that the AIL is still a gross overestimate and is only seldom exceeded by the TDI which, in turn, is also a gross overestimate. Even the EDI is an overestimate when compared with actual data derived from direct estimates of intake (DDI).

It is interesting to note that 91% of the pesticides evaluated by FAO/WHO (ref. 14) can be cleared with sufficient confidence on the basis of theoretical claculations involving only food intake data and available extensive monitoring and food surveillance data.

DISCUSSION

Calculations of pesticide residue intakes referred to as "theoretical daily intake", "potential daily intake". "estimated daily intake" etc, are often criticized on the basis that they are gross overestimates. Such calculated intakes are, therefore, categorically rejected by many on the basis that they are not accurate estimates of actual intake. The author believes that deliberate and reliable overestimates of pesticide residue intakes have the same value in demonstrating that the ADI has not been exceeded as the more accurate actual daily intakes. It is to demonstrate the opposite that such "actual" estimates of pesticide residue intake need to be made. This is illustrated in Table 4 using some selected examples from the 119 pesticides examined.

As can be seen, with the exception of acephate, aldrin and chlorothalonil from among the examples chosen, a more accurate knowledge of intake (ie. DDI) is not necessary to demonstrate that the ADI is not exceeded. With parathion-methyl, cartap, bendiocarb and 2, 4-D a simple calculation of HIL gives the same information as the more accurate "actual daily intake" about the ADI not being exceeded. On the other hand, none of the calculations, eg. HIL, AIL, TDI and EDI, could be used to show the opposite, as evidenced by the results of total diet studies (DDI). It is also clear from Table 4 that, depending on the pesticide, it is necessary to refine the estimation of residue intake by moving towards the more accurate DDI. This is borne out in Tables 3 and 4, which show that the degree of overestimation decreases as follows: HIL > AIL > TDI > EDI > DDI. This is why Schmitt (ref. 11) proposes a scheme (Limit of Exposure Scheme) in which estimates of intake are made starting with the most exaggerated estimates and working towards estimates which are closer to actual intakes.

TABLE 4 Examples of HIL, AIL, TDI, EDI and DDI for some selected pesticides. Expressed as mg/person/day.

	ADI X 60	HIL	AIL	TDI	EDI	DDI
ACEPHATE	0.03	15	2	1.2	0.16	0.00016
DIAZINON	0.12	6	0.8	0.33	0.044	0.00082
FENITROTHION	0.18	30	4	0.15	0.020	0.00005
PARATHION	0.3	3	0.4	0.27	0.10	0.00066
PARATHION-METHYL	1.2	0.6	0.08	0.31	0.042	0.000052
ALDRIN	0.006	0.3	0.049	0.067	0.01	0.0069
DDT	1.2	15	2.44	2.54	0.41	0.08
HEPTACHLOR	0.03	0.6	0.098	0.039	0.0052	0.0004
CHLORDANE	0.06	0.3	0.049	0.067	0.01	0.00039
LINDANE	0.6	9	1.5	1.34	0.22	0.034
CARBARYL	0.6	30	0.34	2.1	0.023	0.007
CARBOFURAN	0.6	6	0.067	0.21	0.0024	0.000018
CARTAP	6	6	0.067	0.071	0.00079	ND
BENDIOCARB	0.24	0.3	0.0034	ND	ND	ND
PIRIMICARB	1.2	6	0.067	0.21	0.0024	0.000012
METHIOCARB	0.06	15	0.17	ND	ND	ND
CHLOROTHALONIL	0.03	75	10.5	4.22	0.59	0.011
DICOFOL	1.5	15	2.1	2.19	0.30	0.0018
CAPTAN	6	150	21	3.49	0.49	0.0042
THIABENDAZOLE	18	30	4.2	0.66	0.092	0.0023
2,4-D	18	6	0.84	0.17	0.024	0.0006

ND = no data available

The scheme also allows pesticides to be ranked in order of priority for monitoring purposes and serves to liberate valuable analytical time for the investigation of pesticide residues and other substances of health or environmental concern. It cannot be stressed enough that the exercise of comparing estimates of intake or intake limits with ADIs is only one component in the overall task of assessing the safety of food containing pesticide residues, as has been described earlier in the paper. Conversely, it should be realized and stressed that calculations such as given in Table 4 should be used with caution in rejecting internationally recommended MRLs on the ground that the ADI is exceeded. As can be seen, some pesticides in Table 4 have TDIs or even EDIs which "exceed" the ADI. As pointed out earlier, such estimates do not prove that the ADI is actually exceeded, and this is in fact shown by the results of total diet studies (DDI).

As a final remark the author wishes to point out that the various estimates included in this paper are not necessarily fully comparable. They were used to illustrate various points and to demonstrate various approaches to assessing the safety of pesticide residues in food. The data have to be refined and adapted to existing situations both at the national and international level.

REFERENCES

1. O.E. Paynter and R. Schmitt, Proc. 4 IUPAC Congress of Pesticide Chemistry (Zurich), pp 60-65, Pergamon Press, London (1978)
2. A.J. Pieters, Proc. 4 IUPAC Congress of Pesticide Chemistry (Zurich), pp 66-73, Pergamon, (1978)
3. H. Galal Gorchev and Charles F. Jelinek, Bull. WHO, 65, 945-969
4. C. Wuthrich, F. Muller, O. Blaser and B. Marek, Mitt. Gebiete Lebensm. Hyg. 76, 260-276 (1985)
5. Results of USA Total Diet Studies for Pesticide Residues, 1982-1984 (Room Document 8 for 18th Session of the Codex Committee on Pesticide Residues).
6. J.P. Frawley and R.E. Duggan, Proc. 4th Int. Cong. Pesticide Chemistry (Zurich), pp 82-86, Pergamon Press, London (1978)
7. G. Vettorazzi, Eighth Session of Codex Committee on Pesticide Residues, CX/PR/75/5.
8. Pesticide Residue Levels in Foods in the United States from July 1, 1969 to June 30, 1976, FDA/AOAC publication.

9. Pesticide Residue Intake in the Netherlands (Working Paper submitted to the Codex Committee on Pesticide Residues, Room Document 8, May 1982). (Total diet study).
10. Pesticide Residues and their Intake in Food in Finland 1977-1980 (Working Paper submitted to the Codex Committee on Pesticide Residues, Room Document 10).
11. R.D. Schmitt, WHO/EHE/FOS/86.8 (1986)
12. Guidelines for the Study of Dietary Intakes of Chemical Contaminants, WHO, Geneva (WHO Offset Pub. 87).
13. Codex Committee on Pesticide Residues of the Codex Alimentarius Commission (see Guide to Codex Maximum Limits for Pesticide Residues, Part 1).
14. Volume XIV of the Codex Alimentarius (Ref. CAC/VOL XIII-Ed. 2 in preparation).

Organochlorine insecticide residues in predatory birds—long-term trends and bioaccumulation

C.H. Walker* and P.I. Stanley[1]

Department of Physiology & Biochemistry, University of Reading, P.O. Box 228 Reading, Berks. RG6 2AJ
[1]Slough Laboratory, Ministry of Agriculture, Fisheries and Food, London Road, Slough

Abstract - Predatory birds are efficient accumulators of certain lipophilic organochlorine compounds. In North America and Britain the analysis of their tissues and eggs has provided a useful means of monitoring the levels of persistent compounds such as pp'DDE and dieldrin in ecosystems. Although various restrictions and bans have brought a general reduction of these residues in predators, the decline has been slow, and there have been some notable exceptions; in particular pp'DDE levels in British sparrowhawk have remained relatively constant over 20 years.
The tendency of predators to efficiently bioaccumulate pp'DDE, dieldrin and other persistent organochlorine pollutants is related to very low activities of the hepatic microsomal enzymes that are responsible for the effective elimination of these compounds. A kinetic model is discussed for the prediction of bioaccumulation risks from metabolic studies with liver microsomes and other tissue preparations *in vitro*.

INTRODUCTION

Since the early 1960s there has been much public concern about the presence in the living environment of persistent residues of certain organochlorine insecticides including dieldrin, pp'DDT and its metabolite pp'DDE, and heptachlor epoxide. Some of the highest concentrations of such compounds have been found in the tissues of predatory birds, and have been associated with population declines of species such as the sparrowhawk (*Accipiter nisus*) and peregrine (*Falco peregrinus*) in Britain, and of the bald eagle (*Haliaetus leucocephalus*), peregrine and osprey (*Pandion haliaetus*) in North America.

Restrictions and bans were placed on the use of aldrin, dieldrin, heptachlor, DDT and other organochlorine insecticides in many countries. Monitoring schemes were set up to follow the long term trends in environmental levels of organochlorine residues following the implementation of restrictions and bans. Some species of predatory birds were included in monitoring schemes because of their tendency to accumulate relatively high levels of organochlorine compounds. Thus, in Britain, the Monks Wood Research Station (National Environment Research Council) has monitored dieldrin and pp'DDE in tissues and/or eggs of sparrowhawks, peregrine falcon, kestrel (*Falco tinnunculus*), barn owl (*Tyto alba*), golden eagle (*Aquila chrysaetus*) and heron (*Ardea cinerea*), and in the USA, organochlorine residues were monitored in the bald eagle and the brown pelican (*Pelecanus occidentalis*).

This paper will be concerned with two related issues - (1) The major trends found in levels of pp'DDE and dieldrin and (2) The question of the bioaccumulation of organochlorine insecticides and other lipophilic pollutants by predatory birds.

LEVELS OF pp'DDE AND DIELDRIN IN PREDATORY BIRDS

Since 1963, a countrywide monitoring scheme has operated from Monks Wood Experimental Station, Huntingdon, U.K., to investigate the levels of organochlorine pollutants in predatory birds. Results up to and including 1977 have been reviewed by Cooke *et al.* (1982) (ref. 1). Ten species were collected and analysed from the inception of the programme, namely - kestrel, sparrowhawk, barn owl, heron, golden eagle, peregrine, great-crested grebe (*Podiceps cristatus*), kingfisher (*Alcedo atthis*), rough-legged buzzard (*Buteo lagopus*) and long-eared owl (*Asio otus*). In the case of each of the first four species, more than 100 individuals were analysed. Collections were also made of eggs of the following species - heron, golden eagle, sparrowhawk and peregrine. Recently Newton and Haas (ref. 2) have produced data on pp'DDE and dieldrin residues in livers of sparrowhawk, kestrel and heron between 1963 and 1982/3 and these results are featured in this presentation.

There were seasonal variations in liver concentrations of pp'DDE and dieldrin in kestrel, sparrowhawk, barn owl and heron. Residue levels reached a peak during the period March-June and then fell away during the subsequent autumn and winter months (Cooke et al. 1982).

In the following discussion, 3 yearly mean values are given, the cited years being the second of each group of three (see Table 1). Regarding dieldrin levels, the highest mean concentrations recorded in the sparrowhawk were in the range 1-2 ppm between 1964 and 1966. The highest values for heron were 3.8-3.9 ppm between 1967 and 1968. In both cases these levels had fallen to 0.04-0.40 by the years 1981 and 1982 (ref. 2). In the kestrel, levels were between 0.9 and 2 ppm over the period 1963-1974 but had fallen to 0.40-0.48 ppm during the period 1979-1982 (ref. 2). A downward trend was followed in the barn owl between 1963 and 1977 (ref. 1). Dieldrin levels in the eggs of golden eagles fell steadily between 1963 and 1974, and in sparrowhawks between 1963 and 1969, although, in the latter species, there appeared to be an upward trend between 1970 and 1972.

Mean pp'DDE concentrations in liver showed a downward trend in kestrels and herons between 1968-1970 and 1980-1981, but there was an indication of a temporary upward trend during 1978-1979 in both species (ref. 2); a downward trend was found in the barn owl between 1969 and 1977 (ref. 1). Contrasting with this, pp'DDE concentrations in sparrowhawk livers have shown little change over the period 1963-1982, the mean values falling within the range 3.4-6.9 ppm over the entire period (ref. 2). pp'DDE concentrations in sparrowhawk eggs were 12.3 ppm in 1963-1965 and 8.41 ppm in 1966-1971.

In summary, in kestrel, sparrowhawk and heron, dieldrin levels in liver samples taken in the late 1970s and early 1980s had fallen substantially below the peak values recorded in earlier years. pp'DDE levels also tended to decrease, but to a lesser extent, and levels in sparrowhawk livers have shown little change over the period 1963-1983.

In the USA residues of pp'DDE and dieldrin have been monitored in the brown pelican and the bald eagle. Brown pelican eggs were taken from two locations in South Carolina during the period 1969-1975 (ref. 3). Throughout this period there was a fall in mean pp'DDE concentration from 5.45 ppm in 1969 to 1.40 ppm in 1975, with surprisingly little interindividual variation in the later years. Mean dieldrin concentrations fell from 1.16 ppm to 0.46 ppm between 1969 and 1971, but subsequently remained reasonably constant. A similar decline in pp'DDE occurred between 1969/70 and 1974 in pelican eggs from three study areas in Florida where dieldrin levels fell in two areas, but showed a slight increase in one area.

The monitoring of residues of organochlorine insecticides in the tissues of bald eagles has been carried out at the Patuxent Wildlife Research Centre, Laurel Md. (ref. 4). Samples were collected from many areas of the USA. A systematic sampling scheme was not possible, so considerable variability was inevitable. Throughout the period 1964-1977 residues were measured in carcase and brain (ref. 5). For both dieldrin and pp'DDE no clear trends were evident between 1964 and 1971, but in the period 1972-1977 there was a steady fall in the concentrations of both compounds in carcases. Thus, in 1970 mean concentrations were 18.0 ppm (pp'DDE) and 0.74 ppm (dieldrin) whereas the corresponding figures for 1977 were 4.0 ppm and 0.22 ppm respectively (ref. 5).

The general picture in Britain and North America is one of slowly declining residue levels during the 1970s and early 1980s, as might be expected with the restrictions and bans that have been placed upon DDT, aldrin and dieldrin. [In Britain restrictions on the use of aldrin and dieldrin came into force in 1962, 1964 and 1969 with the final; withdrawal from the agricultural market in 1975]. The slowness of disappearance of these compounds is attributable mainly to their marked chemical and biochemical stability. The maintenance of relatively high levels of pp'DDE in British sparrowhawks is a striking example of this. pp'DDE has a very long half life in soils (over 10 years in British soils), and an important route of transfer to this species is from worms to thrushes to sparrowhawks. A study of migrant thrushes carried out during 1979-1981 showed that they sometimes contain substantial residues of pp'DDE (ref. 7).

BIOACCUMULATION OF ORGANOCHLORINE INSECTICIDES RESIDUES BY PREDATORY BIRDS

The accumulation by predatory birds of substantial levels of organochlorine compounds such as pp'DDE and dieldrin requires explanation. Are there relatively high concentrations of these compounds in their food, are predators highly efficient bioaccumulators, or is it a combination of these two things The basic difficulty in answering these questions is the uncertainty about levels of organochlorine compounds in the prey taken in the field. It cannot be assumed that a predator will take a 'representative' sample of prey that contains the average concentration of pp'DDE and dieldrin for the prey population. Where levels of contamination are high, similar to that of the prey population as a whole, there is a distinct probability that the predator will take the most contaminated individuals because they show symptoms of poisoning.

In a study of the marine ecosystem of the Farne Islands, England, during the period 1962-1964, (ref. 8) carefully monitored residues of pp'DDE and dieldrin in organisms from different trophic levels. Some of the highest levels were found in the cormorant (*Phalacrocorax carbo*) and shag (*Phalacrocorax aristotelis*), fish-eating birds from the highest trophic levels of the ecosystem. The shags contained levels of dieldrin and pp'DDE at least 50-fold higher than those of their principal prey. This apparently high bioaccumulation factor was not explicable in terms of selective predation because residue levels in the prey population did not vary greatly between samples taken at different times. Other studies also suggest that, in the longer term, predatory birds can bioaccumulate persistent organochlorine compounds to levels well above those in their prey. Investigating deaths of owls at London Zoo, Jones *et al*.(ref. 9) found that that the birds contained considerably higher levels of dieldrin in their tissues than were present in the mice upon which they fed.

In conclusion, there is evidence that efficient bioaccumulation is an important factor in determining the high levels of some organochlorine pollutants in predatory birds.

PREDICTION OF BIOACCUMULATION RISKS

Unless they are efficiently metabolised, highly fat soluble pollutants have very long half lives in terrestrial species (including fish-eating birds), because the rates of excretion of the unchanged compounds are very low (refs. 10,11). Thus, pp'DDE has a half-life of 240 days in the feral pigeon (*Columba livia*) (ref. 12), and dieldrin has a half life of 270 days in man. On the other hand analogous lipophilic compounds that are rapidly metabolised have much shorter half lives (< 4 days) (ref. 10). The metabolism of dieldrin is principally by microsomal monooxygenase attack, and the same appears to be true of pp'DDE since the main metabolites found *in vivo* are phenols, which are typical products of this type of enzymic degradation. It is, therefore, noteworthy that the activities of hepatic microsomal monooxygenase towards organochlorine substrates are particularly low in predatory birds such as the shag, cormorant, razorbill (*Alca torda*), guillemot (*Uria aalge*), Manx shearwater (*Puffinus puffinus*) and sparrowhawk (refs. 13,14). The activities per unit body weight for the shag, cormorant, guillemot, Manx shearwater and sparrowhawk were between 4% and 9% of the corresponding activities for the male rat. This and other evidence strongly suggests that the marked bioaccumulation of compounds like pp'DDE and dieldrin by these predators is due, at least in part, to a deficiency of the hepatic microsomal monooxygenase system (ref. 15).

The risks of bioaccumulation of lipophilic pollutants may, therefore be predictable from their rates of metabolism by appropriate enzyme systems *in vitro*. In vertebrates many lipophilic pollutants are metabolised largely by hepatic microsomal enzymes (especially monooxygenases). A new approach to the prediction of bioaccumulation risks would be to study metabolism by hepatic microsomal enzyme systems over a wide range of substrate concentrations, down to environmentally realistic levels (ref. 11). From the data obtained, Lineweaver-Burk plots could be constructed expressing reaction velocity as mg pollutant metabolised per kg body weight per unit time, and substrate concentration as ppm per mg microsomal protein or ppm per g liver. From the plots an estimation could be made of the concentration of pollutant present in the liver at the steady state, for any defined concentration of pollutant in food. The rate of loss of the pollutant should equal the rate of absorption under these conditions; if nearly all of the loss is due to liver metabolism, then the concentration needed in the liver to give a metabolic rate that will balance the rate of absorption can be deduced from the Lineweaver-Burk plot, and from this the bioaccumulation factor (BF) for the liver calculated (BF = conc. in liver/conc in food). The bioaccumulation factor for the whole body can be estimated from this, if the ratio conc. in liver/conc. in whole body is known at the steady state; (this ratio will depend upon the lipid content of the body, which can be corrected for). Lineweaver-Burk plots define the maximum metabolic rate that is possible, and could be used to account for the consequences of induction, and of interactions between different pollutants at the metabolic level.

Where significant metabolism occurs in tissues other than in the liver, the model could be adapted to take account of this. If such a model is successfully developed it will be possible to make predictions of bioaccumulation risks of lipophilic pollutants in a very wide range of species that could not otherwise be investigated.

REFERENCES

1. A.S. Cooke, A.A. Bell and M.B. Haas, *Predatory birds, pesticides and pollution*. pub. Institute of Terrestrial Ecology (NERC), Huntingdon. (1982)

2. I. Newton and M.B. Haas, Data from Pesticide Monitoring Scheme, held at Institute for Terrestrial Ecology (NERC), Monks Wood Experimental Station, Huntingdon, U.K. (1986)
3. L.J. Blus, T.G. Lamont and B.S. Neely, Pesticides Monitoring Journal, 12, 172-184 (1979).
4. W.L. Reichel, E. Cromartie, T.G. Lamont, B.M. Mulhern and R.M. Prouty, Pesticides Monitoring Journal, 3, 142-144 (1969).
5. T.E. Kaiser, W.L. Reichel, L.H. Locke, E. Cromartie, A.J. Krynitsky, T.G. Lamont, B.M. Mulhern, R.M. Prouty, C.J. Stafford and D.M. Swineford, Pesticides Monitoring Journal, 13, 145-149 (1980).
6. A.R. Hardy, M.R. Fletcher and P.I. Stanley, Veterinary Journal. In press (1986).
7. M.R. Fletcher, A.R. Hardy and R.A. Hoodless, 6th I.U.P.A.C. Meeting, Abstracts (1986).
8. J. Robinson, A. Richardson, A.N. Crabtree, J.C. Coulson and G.R. Potts, Nature, 214, 1307-1311 (1967)
9. D.M. Jones, D. Bennett and K.E. Elgar, Nature, 272, 52 (1978).
10. C.H. Walker, Progress in Pesticide Biochemistry, I, 247-285 (1981).
11. C.H. Walker, Marine Env. Research, 17, 297-300 (1985).
12. S. Bailey, P.J. Bunyan, B.D. Rennison and A. Taylor, Toxicol. Appl. Pharmacol., 14, 23-32 (1969).
13. G.C. Knight and C.H. Walker, Comp. Biochem. Physiol, 73C, 211-221 (1982).
14. C.H. Walker, I. Newton, S. Hallam and M.J.J. Ronis, Submitted ms. (1986).
15. C.H. Walker, G.C. Knight, J.K. Chipman and M.J.J. Ronis, Mar. Env. Res., 14, 416-419 (1984).

TABLE 1 Mean residue levels in selected years

Samples	1966	1970	1973	1975	1977	1979	1981
	pp'DDE						
Sparrowhawk, U.K. liver	5.34	4.29	4.14	4.83	6.93	4.63	3.39
Kestrel, U.K. liver	1.38	2.87	1.54	1.48	2.05	1.06	0.61
Heron, U.K. liver	5.20	8.35	3.34	2.32	2.57	3.80	0.91
Brown pelican, S. Carolina, eggs		3.58*	2.09*	1.40*			
Bald eagle, N. America, carcase		18*	12*	4.5*	4.0*		
	Dieldrin						
Sparrowhawk, U.K. liver	1.32	0.256	0.700	0.852	0.465	0.368	0.309
Kestrel, U.K. liver	0.95	1.24	1.73	1.09	0.671	0.483	0.408
Heron, U.K. liver	2.33	0.61	0.97	0.46	0.33	0.095	0.041
Brown pelican S. Carolina, eggs		0.82*	0.45*	0.40*			
Bald eagle, N. America, carcase		0.74*	0.74*	0.60*	0.22*		

Data taken from (ref.1-4). Residue concentrations are in parts per million based on fresh weights. Three yearly geometric means are given, unless otherwise indicated, the cited year being the middle one of three. Values marked with an asterisk (*) are means for one year's samples only. In the case of each species, at least 300 individual samples have been analysed.

Environmental fates and effects

Fate modelling of pesticides in groundwater

Michael Matthies

Gesellschaft für Strahlen- und Umweltforschung mbH München, Projektgruppe Umweltgefährdungspotentiale von Chemikalien, 8042 Neuherberg, Federal Republic of Germany

Abstract - An overview is given on the mathematical methods (analytical and numerical) developed for the transport and fate of chemicals in soil/groundwater systems. The interrelationships between solute movement, molecular diffusion, and hydrodynamic dispersion in saturated and unsaturated porous media, the retardation by sorption processes, and the elimination by chemical and biological processes are discussed. Crucial field conditions and pesticide characteristics, which pose in combination, a potential threat to the quality of groundwater resources, are identified.

INTRODUCTION

Concerns about groundwater contamination with agrochemicals have resulted in calling for an effective pollution control to ensure proper drinking water quality. Groundwater is the main source for drinking water in many countries (USA: 45 %, FRG: 70 %) of which only a small percentage undergoes treatment. Standards for regulation have been issued by various authorities (e.g. CEC: 0.1 ug/l for each pesticide, but 0.5 ug/l total) (ref. 1). About 30,000 tons of pesticides are applied annually in the Federal Republic of Germany, i.e. about 2.5 kg/ha (ref. 2). In 1983 1,800 pesticides were registered in the FRG, similar data are available for other countries. Pesticides with higher water solubility are increasingly used thus raising the potential for leaching through the upper soil zone.

Monitoring studies and recent findings from various countries, e.g. USA, Canada, FRG, show that pesticides of different chemical classes have been observed in groundwater samples and in drinking water wells (ref. 3). Various authors reported concentrations of pesticides in groundwater in the ppb - range resulting from agricultural uses. Cohen et al. (ref. 4, 5) compiled a list of 17 pesticides which have been found in the United States. At several symposia (ref. 5, 6, 7) the assessment and evaluation of groundwater contamination from pesticide use have been discussed. In the FRG positive results for the herbicide atrazine and the nematocide 1,3-dichloropropene have been reported for various sites (ref. 2, 6). However, for most pesticides applied to agricultural land only fragmentary knowledge exists about their potential threat of groundwater contamination.

In order to predict the transport of pesticides in the surface and subsurface mathematical models can be used which integrate the physical, chemical and biological processes that are known or thought to affect pesticides (or their residues) in the unsaturated and saturated soil zones. The ability to describe the mechanisms of transport which are not directly observable and the ease with which models can be used have led to a tremendous interest in developing computerized contaminant transport models. Transport and mobility of pesticide in the upper soil zone have intensively been studied and models have been tested with laboratory and field studies. Most of the model approaches for groundwater transport (deeper soil zones) were developed for radionuclides, heavy metals, or organic trace contaminants (ref. 8). Applications for pesticides are just at the beginning.

To deal with the problem systematically two problem areas have to be considered:

1. What are the governing processes which determine the distribution and fate of pesticides in the unsaturated and saturated soil zone?
2. How can we assess the potential of groundwater contamination with pesticides and their transport and fate in aquifers?

TABLE 1. Physical, chemical and biological processes in unsaturated and saturated zones.

Physical processes
- advection
- hydrodynamic dispersion
- liquid diffusion
- gaseous diffusion/volatility
- filtration

Chemical processes
- solubility in soil solution and groundwater
- equilibrium partitioning between phases
- adsorption/desorption and ion exchange
- ionization/dissociation (acid-base)
- complexation
- oxidation/reduction
- hydrolysis

Biological processes
- microbial degradation/transformation
- bioavailibility and bioconcentration
- microbial cell synthesis

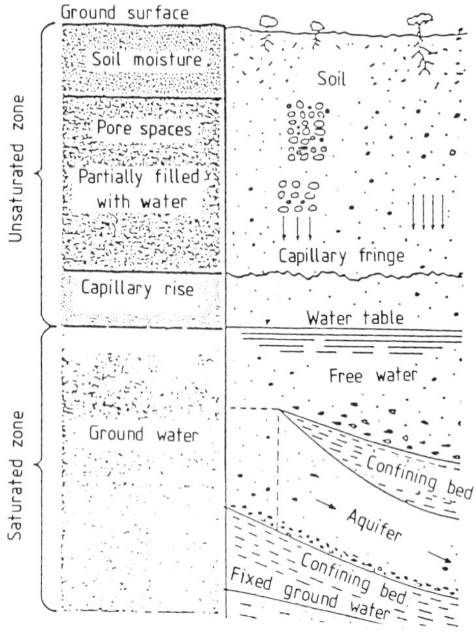

Fig. 1. Groundwater profile. Relationship between unsaturated and saturated zones (ref. 9).

PHYSICAL, CHEMICAL AND BIOLOGICAL PROCESSES

Pesticides move in soil groundwater by a complex interaction of physical, chemical, and biological processes (Table 1). They can reach groundwater flow systems by vertical leaching or diffusion through the unsaturated soil zone (Fig. 1). Within an aquifer the horizontal advection and hydrodynamic dispersion mainly govern the transport of contaminants. Models for simulating fluid flow will not be discussed here (see for example (ref. 10)).

Advection and dispersion

With advective transport, contaminants dissolved in water are moved from one point to another by the water motion alone. In the unsaturated soil zone, water moves vertically within soil pores depending on precipitation, evapotranspiration, water binding, runoff and water uptake by plants. Downward infiltration into initially unsaturated soil generally occurs under the combined influence of matrix potential and gravity gradients. Dispersion is a term that is used to describe a phenomenon that is primarily mixing within the water-saturated, porous medium. The overall effect of dispersion is to spread or dilute an advective contaminant from which moves through a porous medium. In simplified approaches dispersion is split up in a longitudinal (along the flow path) and a transverse dispersion coefficient (perpendicular to the flow path) thus resulting in a tracer plume.

Sorption and retardation

Pesticides can be adsorbed by organic constituents, by ion exchange with minerals, and by polar interactions. Usually, the adsorption/desorption processes are described by the retardation factor R_d which is the ratio of the flow velocities of the water v_a and the contaminant v. Assuming only reversible sorption processes within an aquifer the retardation factor can be expressed by

$$R_d = v_a/v = 1 + \rho_b/n_e\ K_d \qquad (1)$$

where ρ_b = bulk density [g cm^{-3}], n_e = effective porosity [-], K_d = distribution coefficient [cm^3 g^{-1}]

This equation is an oversimplification since adsorption and desorption have to be determined seperately. Freundlich or Langmuir adsorption isotherms are normally used to describe the kinetics of adsorption. For nonpolar neutral organics the K_d-value is related to the organic carbon partition coefficient K_{oc}

$$K_d = f\ K_{oc}/100 \qquad (2)$$

with f = organic carbon content of the considered soil layer [%]

The organic carbon content decreases with deeper soil layers. However, for some soil types remarkable amounts of organic carbon have been found in rather deep soil layers. Sorption on minerals strongly depends on ion exchange properties of the materials. Some pesticides are in the ionized form depending on the pH-value of the soil, e.g. the phenoxy acids. For those compounds the acid dissociation constant can be used to determine the ionization state. Cations are much stronger sorbed than anions because of the cation exchange capacity of the soil materials. Sorption of pesticides was almost solely studied with materials from soil root zones for which the hydrophobic interactions with humic and fulvic acids dominate. Data for sorption to geological materials are usually not available. Sorption to subsurface materials and hence retardation in groundwater flow systems have therefore to be based on K_{oc} values. Often not even K_{oc} values are available, therefore water solubilities have to be used to estimate K_{oc} and K_d values. Various regression equations are reported in the literature (ref. 11). Such regression equations can be misleading, for example for simazine and atrazine (ref. 6).

Transformation and degradation

Removal and/or transformation of pesticides from the surface and subsurface environment can occur through the action of microorganisms or by chemical kinetics, e.g. hydrolysis. Photolysis is also observed for pesticides after application and before entering deeper soil layers. Abiotic processes can be simulated in laboratory experiments taking into account hydrolysis, oxidation/reduction processes, dissolution/precipitation and complexation. Microbial degradation depends on the aerobic/anaerobic conditions and the cell population. The number of cells decreases from about 10^8 - 10^9 cells/m^3 in the surface layer to 10^6 - 10^7 cells/m^3 in deeper layers. Microbial interactions in the surface environment have been studied much more extensively than those in the subsurface. Most of the models assume degradation processes of first or pseudo-first order:

$$\frac{dc}{dt} = \sum_i \lambda_i \quad c \qquad (3)$$

where λ_i = degradation rate constants for hydrolysis microbial degradation or oxidation/reduction

Reported half-lives can be transformed by $\lambda_i = \ln 2/T_{1/2}$

Other approaches assume a second order kinetic which can be adapted to different cell numbers or a Monod-kinetic which is derived from bacteria growth studies.

Diffusion and volatilization

Molecular diffusion can occur in the liquid and gaseous phase. Liquid diffusion is only important when the velocity of water flow is very small, otherwise the hydrodynamic dispersion dominates. Under most conditions, the contributions of liquid diffusion is neglegible. Gaseous diffusion can play an important role for pesticides with remarkable partitioning into the soil air phase. The partition between liquid and gaseous phase is described by Henry's law. For most pesticides Henry's law constant has not been measured, but can be estimated from the vapor pressure P_v and the water solubility S_w:

$$H = P_v/S_w \qquad [Pa\ m^3\ mol^{-1}] \qquad (4)$$

or if the ideal gas law is assumed: $H' = H/RT = P_v/(RT\ S_w)$. Effective liquid and gaseous diffusion coefficient (Fick's law) can be estimated from the binary diffusion coefficients corrected by the effective pore volumes. An example is the tortuosity model of Millington and Quirk (ref. 12):

$$D_{g,eff} = \frac{(n - \theta)^{10/3}}{n^2} D_g \quad \text{and} \quad D_{w,eff} = \frac{\theta^{10/3}}{n^2} D_w \qquad (5)$$

where n = porosity of the unsaturated zone, θ = volumetric water content, D_g, D_w = binary diffusion coefficient in air and water, resp.

MODEL APPROACHES

Three different approaches will briefly be discussed
(i) benchmark ranking
(ii) analytical methods
(iii) numerical methods

Benchmark ranking

Data interpretation by benchmark comparison is accomplished by comparing a series of fundamental properties of pesticides whose environmental behavior is unknown with the standards whose environmental behavior is known (ref. 13). Benchmark comparison is essentially no model approach since no attempt is made to calculate environmental concentration levels. However, the method is useful to give a relative ranking according to leaching, volatilzation and retardation. Laskoswski et al.derived (ref. 13) a series of mathematical equations that express the behavior of pesticides on a relative basis:

Leaching potential = $S_w/(P_v\ K_{oc})$
Leaching index = $S_w\ T_{1/2}/(P_v\ K_{oc})$
Volatility potential = $P_v/(S_w\ K_{oc})$
Volatility index = $P_v\ T_{1/2}/(S_w\ K_{oc})$

This approach is based on the assumption that partitioning between the three phases in unsaturated soil zone is the governing pesticide characteristic for volatility and leaching. Cohen (ref. 5) suggested a list of 8 pesticide characteristics and 4 field conditions as a first decision support for considering computer modeling and field testing. Generalized pesticide characteristics are: - water solubility greater than 30 ppm; - K_d-value less than 5, and usually less than 1 or 2; - K_{oc} less than 300 - 500; - Henry's law constant less than 10^{-2} atm m^3/mol; - speciation, negatively charged at ambient pH; - hydrolysis half-life greater than about 25 weeks; - photolysis half-life greater than about 1 week; - soil half-life greater than about 2 - 3 weeks. Crucial field and weather conditions are: - total recharge greater than 250 mm/yr; high levels of nitrate in groundwater as indicator; - unconfined aquifer; - soil pH providing high stability to the pesticide residues.

Another benchmark approach is the comparison of retardation factors R_d for the saturated soil zone. Retardation by sorption is one of the main pesticide characteristic properties. R_d values can be calculated from hydrogeological parameters (bulk density, soil porosity, organic carbon content) and the K_{oc}-values. Assuming typical values for sand and gravel aquifers a relative ranking and grouping of pesticides can be derived according to their mobility.

Transport in water-saturated zone

In an isotropic, homogeneous porous medium the three-dimensional transport of a contaminant can be expressed by the general advection/dispersion equation

$$\frac{dC}{dt} = \text{div } (D \text{ grad } C) - \text{div } v\, C + \sum_i S_i \qquad (6)$$

where C = concentration of the contaminant, D = dispersion tensor, v = contaminant flow velocity, S_i = source and sink terms

The main difficulty in solving the above equation is estimating the value of D. Some simplifications can be made in order to obtain analytical solution. Numerical solutions, e.g. by finite difference method, are better suited for more complex hydrogeological conditions.

Analytical models. Assuming line injection in the z-direction (vertical) and injection quantities which do not affect density of groundwater, the dispersion tensor D can be reduced to a longitudinal D_L and a transverse coefficient D_T. Furthermore, it is possible to extend the transport equation to include effects of retardation of contaminant transport through sorption, chemical reaction, biological transformation and degradation. The transport equation can then be written in a two-dimensional simplified form

$$\frac{C}{t} = \frac{D_L}{R_d}\frac{\partial^2 C}{\partial x^2} + \frac{D_T}{R_d}\frac{\partial^2 C}{\partial y^2} - \frac{v_a}{R_d}\frac{\partial C}{\partial x} - \sum_i \lambda_i \; C \qquad (7)$$

where D_L, D_T = longitudinal and transverse dispersion coefficients [$m^2\ d^{-1}$],
R_d = retardation factor [-], v_a = pore water velocity [$m\ d^{-1}$],
$_i$ = degradation rate constants [d^{-1}]

The longitudinal dispersion coefficient is related to the flow velocity v_a through the dispersivity D_L

$$D_L = \alpha_L\, v_a^b \qquad \text{with } 0.9 < b < 1.2 \qquad (8)$$

The dispersivity D_L is characteristic for each type of subsurface material and is related to the flow distance s

$$D_L = c\ s^d \qquad (9)$$

Seiler (ref. 14) reported values of c = 0.29 and d = 0.72 for gravels with distances between 10 m and 1,200 m.

The transverse dispersion coefficient D_T is related to the longitudinal dispersion coefficient by

$$D_T = \begin{cases} D_L/10 & \text{for homogeneous material} \\ D_L/100 & \text{for stratified aquifer in sand and gravel} \end{cases}$$

The simplified dispersion equation can be solved analytically assuming specific initial and boundary conditions. One-dimensional and two-dimensional analytical solutions are available and can be found in various publications (ref. 15, 16). Computer codes are distributed by the International Ground Water Modeling Center (ref. 17).

The governing pesticide characteristics are the retardation by sorption and the elimination by abiotic and biotic degradation. Some results with a two-dimensional analytical solution (ref. 16) for atrazine are shown in Fig. 2 and Fig. 3. Typical values for sand/gravel aquifer have been used and the influence of the retardation has been studied. Elimination by biotic degradation is rather low and has no significant influence.

Figure 2 shows that the concentration in the main flow path increases with the pore water velocity and decreases with increasing sorption. The resulting iso-concentration lines for an estimated R_d-value of 1.7 and a medium water velocity can give indications for expected pesticide concentrations within an aquifer (Fig. 3).

Numerical models. Numerical methods are generally required to solve complex equations describing coupled or uncoupled processes under various initial and boundary conditions. Various methods have been developed (e.g. finite-difference, finite-element, statistical models) for which extensive surveys are available (ref. 16, 17). An international intercomparison study of various numerical models is now being conducted in order to elucidate the advantages and disadvantages of the different solutions (ref. 18).

The basic method is to break up continuous space into cells, approximate the governing partial differential equations by differences between the values of the parameters over the network and then compute the discrete parcel. This is accomplished by finite-element or finite-difference schemes. Table 2 gives a short compilation of models for saturated zones.

Transport in water-unsaturated soil zone

In addition to the transport with the water flow gaseous diffusion has to be incorporated into the general advective/dispersion equation. Furthermore, the soil pore volume is split into a gas-filled and a water-filled fraction. Moisture content as well as temperature can influence the movement and degradation of pesticides (ref. 19).

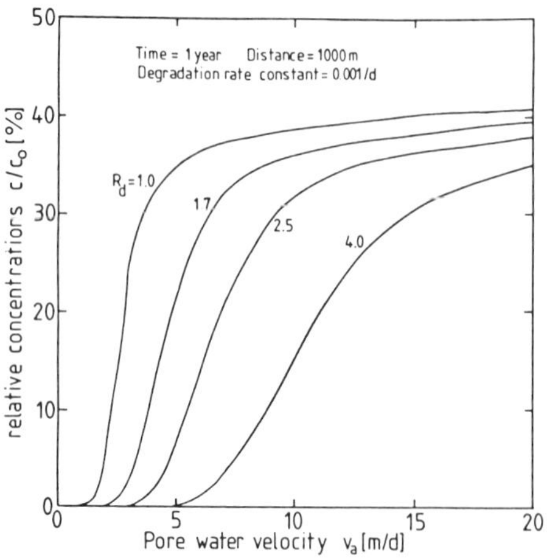

Fig. 2. Relative concentrations of a pesticide with variable R_d-values in the main flow path for various pore water velocities.

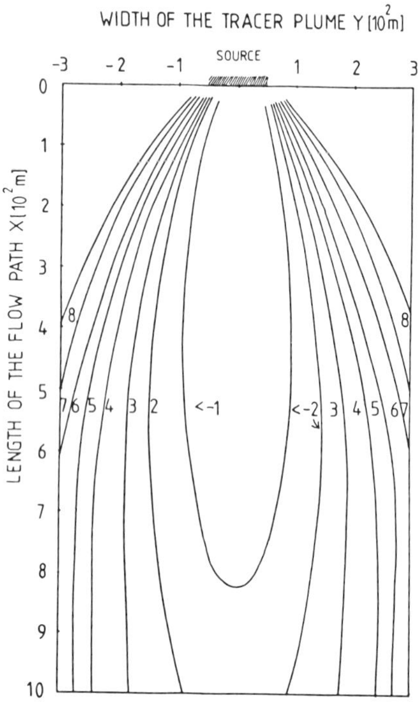

Fig. 3. Iso-concentration lines log c/c_o of a pesticide in an aquifer, v_a = 3 m/d, R_d = 1.7, λ= 0.001/d, continuous source with length = 100 m, time = 1 year

As has been shown in the preceding paragraphs the leaching potential is highest for pesticides with low K_{oc} values and low Henry's law constants. However, some nematocides e.g. 1,3 dichloropropene and EDB are found in groundwater samples and drinking water wells. This can be explained by diffusion in the soil pores. A second reason for introducing gas phase transport in soil results from considering volatility as a removal process from soil.

Since climate and weather conditions influence the behavior in the unsaturated soil zone much more than in the saturated soil zone, mathematical models should be combined with long-term weather statistics (ref. 20).

Analytical models. Analytical solutions of the advective/dispersion equation are restricted to a few specific cases. Mostly one-dimensional vertical transport through a homogeneous soil layer is assumed. Various model approaches have been reported, which mostly neglect gaseous transport or hydrodynamic dispersion (ref. 13). In the field, K_d-values, rate constants, ion exchange capacity, organic carbon content and other soil key parameters vary with depth as the soil profile changes from surface to subsurface. Therefore, most recent model approaches are numerical solutions of the vertical advection/dispersion equation taking into account the heterogeneous soil layer structure.

Numerical models. Various transport models have been developed for the unsaturated soil zone in the last few years, (ref. 21, 22, 23, 24, 25, 26). Most of them require a numerical solution because of complex dynamic processes of solute movement in soil. Table 2 gives a short description of some recent developed models.

Some of them have been evaluated with data from field experiment, e.g. PESTANS (aldicarb) (ref. 21), PRZM (metalaxyl) (ref. 27). Because of its complexity computer facilities are required to solve the differential equation systems numerically.

TABLE 2. Principles of selected transport models for unsaturated and saturated zones.

Model	Characteristics	Fluid Flow Processes	Solute Transport Processes	Ref.
PESTANS I	analytical one-dimensional steady-state unsaturated	average recharge rate from annual data	advection, sorption, degradation 1. order	21
PESTANS II	numerical two-dimensional dynamic unsaturated	transient hydraulic model	advection, sorption, degradation	22
PRZM	numerical one-dimensional finite-difference unsaturated	transient hydraulic model; SCS curve number technique for estimating infiltration and runoff	advection, ad/desorption; dispersion; degradation 1 order; plant uptake	24
Jury, Spencer, Farmer	analytical steady-state unsaturated	average recharge rate; boundary layer model for evaporation	partitioning, diffusion, degradation 1. order, volatilization, leaching	12
SESOIL	numerical one-dimensional dynamic unsaturated	statistical and seasonal data for the hydrological cycle	advection, ad/desorption, diffusion, degradation 1. order, volatilization	25
EXSOL	numerical one-dimensional dynamic unsaturated	average or statistical recharge rate from annual data	advection, partitioning, dispersion, diffusion, degradation 1. order, plant uptake, volatilization	26
SWIFT	numerical three-dimensional finite-difference saturated	transient, well-bore, non isothermal convection	advection, dispersion, diffusion, sorption, decay	28
Javandel, Doughty, Tsang	analytical semi-analytical numerical saturated	Handbook of mathematical models for groundwater transport		16
IGWMC	analytical numerical saturated	Collection of program codes and documentation for groundwater models (1983: 550 groundwater model annotations)		17

CONCLUSIONS

Mathematical modelling of pesticides fate in surface and subsurface can give better insight in the spatial and temporal behavior in water-saturated and unsaturated zones. Combination of dominant pesticide properties and field parameters are suited for a first relative ranking of groundwater contamination potential. Analytical solutions of the advection/dispersion equation can be used for screening purposes and for checking against numerical results. Predicted pesticide concentrations have to be compared with field levels in order to validate model results. Model validation and field testing of pesticide fate in groundwater are required to achieve confidence in models as practical tools for solving a variety of questions involving groundwater systems. Models can then be useful for: - preliminary studies prior to field investigations; - interpretative studies after the field studies; - predictive studies to estimate the potential groundwater situation; - assessment of the potential for groundwater contamination within the pre-market licensing and registration process.

REFERENCES

1. Commission of the European Communities, Directive No. 80/778/EEC, Qualität von Wasser für den menschlichen Gebrauch, Amtsblatt der EG No. L229 vom 30.8.1980, pp. 11-29 (1980)
2. Rat von Sachverständigen für Umweltfragen, Umweltprobleme der Landwirtschaft, Kohlhammer-Verlag, Stuttgart (1985)
3. C. Baier, K. Hurle and J. Kirchhoff, Datensammlung zur Abschätzung des Gefährdungspotentials von Pflanzenschutzmittel - Wirkstoffen für Gewässer, DVWK-Schriften no. 74, Verlag P. Parey, Hamburg (1985)
4. S.Z. Cohen, Pesticide Contamination of Ground Water Resulting from Agricultural Practices in the USA, Grundwasserbeeinflussung durch Pflanzenbehandlungsmittel, Fischer, Stuttgart (1986), in press.
5. S.Z. Cohen, C. Eiden and M.N. Lorber, Monitoring Ground Water for Pesticides in the USA, Pesticides in Ground Water, ACS Symposium Series (1986), in press.
6. S.Z. Cohen, S.M. Creeger, R.F. Carsel and C.G. Enfield, Potential Pesticide Contamination of Groundwater from Agricultural Uses, Treatment and Disposal of Pesticide Wastes, ACS Symposium Series 259, Washington (1984)
7. G. Milde and P. Friesel (Eds.), Gewässer- und Pflanzenbehandlungsmittel V. Grundwasserbeeinflussung durch Pflanzenbehandlungsmittel, Fischer, Stuttgart (1986)
8. D.M. Mackay, P.V. Roberts and J.A. Cherry, Transport of Organic Contaminants in Groundwater, Environ. Sci. Technol. 19 (5), 384-392 (1985)
9. P.N. Cheremisinoff, K.A. Gigliello and T.K. O'Neill, Groundwater-Leachate: Modeling/Monitoring/Sampling, Technomic Publ., Lancaster Pean. (1984)
10. R.A. Freeze and J.A. Cherry, Groundwater, Prentice-Hall, Englewood Cliffs (1979)
11. W.J. Lyman, Adsorption Coefficients for Soil and Sediments, Handbook of Chemical Property Estimation Methods, Ch. 4, McGraw-Hill, New York (1982)
12. W.A. Jury, W.F. Spencer and W.J. Farmer, Behavior Assessment Model for Trace Organics in Soil: I. Model Description, J. Environ. Qual. 12 (4), 558-564 (1983)
13. D.A. Laskowski, C.A.I. Goring, P.J. McCall and R.L. Swann, Terrestrial Environment, Environmental Risk Analysis for Chemicals, pp. 198-240, Nostrand Reinhold, New York (1982)
14. K.P. Seiler, Results of Field Experiments on Hydrodynamic Dispersion in Quarternary Gravels of Southern Germany, Scientific Basis for Water Resources Mangagement (IAHS Publ. no. 153, 1985)
15. M.T. Van Genuchten, J.M. Davidson and P.J. Wierenga, An Evaluation of Kinetic and Equilibrium Equations for the Prediction of Pesticide Movement through Porous Media, Soil Sci. Soc. Am. J. 38, 29-34 (1974)
16. I. Javandel, C. Doughty and C.F. Tsang, Groundwater Transport: Handbook of Mathematical Models, American Geophysical Union, Washington D.C. (1984)
17. P. van der Heijde, Y. Bachmat, J. Bredehoeft, B. Andrews, D. Holtz and S. Sebastian, Groundwater Management: The Use of Numerical Models, 2nd edition, Amer. Geophys. Union (Water Resources Management, Monograph 5), Washington (1985)
18. Swedish Nuclear Power Inspectorate (Project secretariate), INTRACOIN International Nuclide Transprot Code Intercomparison Studies, Progress Reports, Stockholm (1981-1983)
19. L.W. Parker and K.G. Doxtader, Kinetics of the Microbial Degradation of 2,4-D in Soil: Effects of Temperature and Moisture, J. Environ. Qual. 12, 553-558 (1983)
20. G.R. Oliver and D.A. Laskowski, Development of Environmental Scenarios for Modeling the Fate of Agricultural Chemicals in Soil, Environ. Toxicol. Chem. 5, 225-231 (1986)
21. C.G. Enfield, R.F. Carsel, S.Z. Cohen, T. Phan and D.M. Walters, Approximating Pollutant Transport to Groundwater, Ground Water 20, 711-721 (1982)
22. C.G. Enfield, R.F. Carsel and T. Phan, Comparison of a One-Dimensional, Steady-State Hydraulic Model with a Two-Dimensional, Transient Hydraulic Model for Aldicarb Transport through Soil, Quality of Groundwater, Elsevier Sci. Publ. (1981)
23. M.J . Frissel and P. Reiniger, Simulation of Accumulation and Leaching in Soils, Wageningen (1974)
24. R.F. Carsel, C.N. Smith, L.A. Mulkey, D. Dean and P. Jowise, User's Manual for the Pesticide Root Zone Model (PRZM). Release I., EPA, Athens, Georgia
25. M. Bonazountas and J. Wagner, SESOIL, A Seasonal Soil Compartment Model, Office of Toxic Substances, U.S. Environmental Protection Agency, Washington D.C. (1981)
26. M. Matthies, R. Brüggemann and R. Trenkle, Multimedia Modelling Approach for Comparing the Environmental Fate of Chemicals, Environmental Modelling for Priority Setting among Existing Chemicals, Ecomed, Landsberg/Lech (1986)
27. R.F. Carsel, W.B. Nixon and L.G. Ballantine, Comparison of Pesticide Root Zone Model Predicitions with Observed Concentrations for the Tobacco Pesticide Metalaxyl in Unsaturated Zone Soils, Environm. Tox. Chem. 5, 345-353 (1986)
28. R.T. Dillon, R.B. Lautz and S.B. Pahwa, The Sandia Waste Isolation Flow and Transport (SWIFT) Model, Report NUREG/CR-0424 (1978)

Non-extractable pesticide residues in soil

Fritz Führ

Institute of Radioagronomy, Nuclear Research Centre Jülich, GmbH, D 5170 Jülich, Federal Republic of Germany

Abstract - Studies of the fate of ^{14}C-labelled pesticides in soil have shown that radiocarbon from active ingredient molecules is integrated into the stable carbon fraction of soil. By varying and combining extraction procedures, information has been obtained about the sorption and bonding mechanisms responsible for this integration. Improved methods are required for isolating, characterizing, and determining the bound residues.
Standardized biotests with plants or other test organisms have provided data concerning the biological availability of the bound residues. Lysimeter studies with 14C-labelled pesticides have shown that aging of the active ingredients in the soil leads to a considerable reduction in their biological availability to plant roots. If the extractable residue fractions are removed from the soil, the biological availability of the bound residues decreases by a factor between 5 and 500. So far no ecobiological effects of bound residues have been observed.

INTRODUCTION

Currently approximately 10^6t of active ingredients are applied annually worldwide in various pesticides. Most of the pesticides applied reaches the soil by preemergence spraying with herbicides, by seed dressing and spraying with fungicides and insecticides, by soil fumigation with nematocides, by precipitation, and by working-in treated plants. The pesticides are subjected to degradation processes in soil which, depending on the physico-chemical behaviour of the active ingredient and the prevailing soil and climatic conditions, lead to different mineralization and fixation rates in the soil (ref. 1). Chemical and biochemical degradation processes compete with sorption processes in which the active ingredients or the metabolites are sorptively attached, complexed or chemically incorporated at binding sites of the soil organic matter or clay minerals (ref. 2). These bound pesticide residues partly defy analysis because they can no longer be extracted from soil by conventional extraction methods (ref. 3).

DEFINITION OF BOUND RESIDUES

In 1975 at a symposium of the American Chemical Society (ref. 4) the following interim definition was discussed on the basis of the state of knowledge at that time: A soil bound residue is "that unextractable and chemically unidentifiable pesticide residue remaining in fulvic acid, humic acid, and humin fractions after exhaustive sequential extraction with non polar organic and polar solvents." Since that time attempts have been made to elucidate the bonding and fixation of numerous 14C-labelled pesticides in soil. Attempts have also been made to assess the bioavailability of bound residues in order to be able to evaluate the biological significance of these residues. On the basis of wide-ranging studies, the Commission on Pesticide Chemistry of the International Union of Pure and Applied Chemistry (IUPAC) defined non-extractable pesticide residues in plants and soil as follows (ref. 5,6): "Non-extractable residues (sometimes referred to as "bound" or "non-extracted" residues) in plants and soils are defined as chemical species originating from pesticides, used according to good agricultural practice, that are unextracted by methods which do not significantly change the chemical nature of these residues. These non-extractable residues are considered to exclude fragments recycled through metabolic pathways leading to natural products".

Especially in the past ten years, extensive efforts have been made to characterize the chemically and analytically non-detectable active ingredient residues in the soil. This requires the application of tracer labelling, especially labelling with radiocarbon ^{14}C (ref. 1). Only in this way is it possible to show that carbon atoms from active ingredient molecules still persist in the soil after intensive extractions with polar and non-polar solvents. Numerous studies concerning the degradation of active ingredients in soil have confirmed the general situation described above (ref. 7 and 8). Depending on the physico-chemical behaviour of an active ingredient, the degradation conditions prevailing in the soil, and especially the quality and quantity of the soil organic matter, varying fractions of the carbon skeleton are mineralized to the end product CO_2 and active ingredient fractions and metabolites are simultaneously adsorbed, fixed or incorporated in the soil. As a consequence, the fraction of biologically available pesticides or their metabolites in the soil solution is drastically reduced with increasing residence time (Fig. 1, ref. 9).

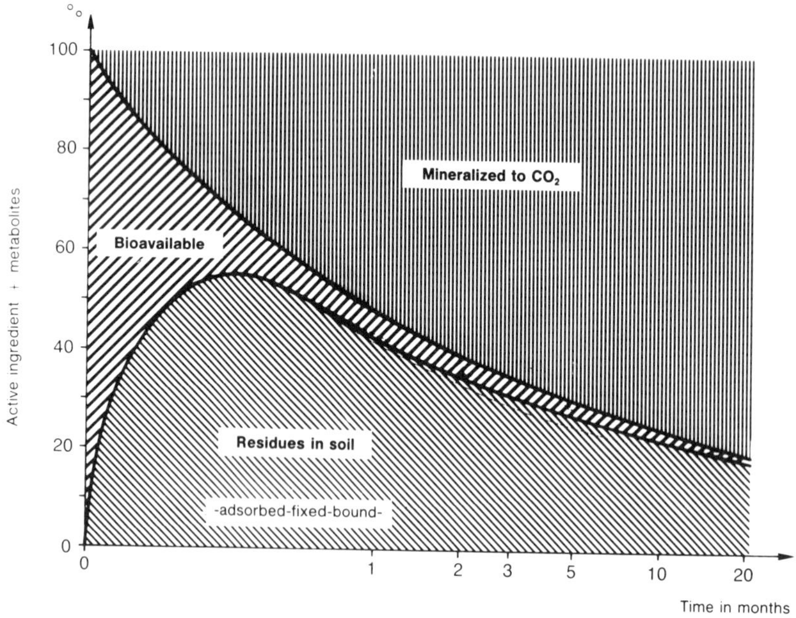

Fig. 1
The general fate of organic chemicals in soil

The influence exerted by the physico-chemical behaviour of the active ingredient on these processes is illustrated by studies showing that paraquat is fixed very strongly to soil clay minerals (ref. 10) and thus largely eludes degradation, whereas 2,4-D is not fixed strongly and is very rapidly mineralized (ref. 11). These studies demonstrate the dominant role of clay minerals and other soil constituents in fixing pesticides (ref. 12) and the decisive influence of microbial biomass development in soil with respect to conversion, mineralization and incorporation of pesticides and their metabolites into soil organic matter constituents (ref. 13).

The key results of studies of the bonding of pesticides and their metabolites in soils have been discussed in several review papers, and an interpretation and evaluation of the results has been attempted with respect to relevance in practice (ref. 6,7,8,14,15). The International Atomic Energy Agency (Vienna) has initiated and conducted a coordinated research programme to elucidate the binding and bioavalability of soil-bound pesticide residues (ref. 16). On the initiative of the Bayer AG, Leverkusen, a workshop has been recently held on "Bound Residues in the Soil", at which published and unpublished studies of the formation, characterization, and long-term behaviour of bound residues, as well as their potential effects, were discussed intensively (ref. 17).

INCORPORATION OF PESTICIDE RESIDUES INTO SOIL ORGANIC MATTER FRACTIONS

During the past ten years, extensive efforts have been made to characterize the chemically and analytically non-detectable active ingredient residues in soil and to gain information on their bioavailability to plant roots and soil organisms. By applying numerous 14C-labelled compounds, it became apparent that radiocarbon from active ingredient molecules is integrated predominantly into the stable carbon fractions of soil organic matter - the humic acids and the humin (ref. 18,19). The question therefore arises as to the extent to which active ingredient residues or molecular fractions which can no longer be characterized are converted into stable carbon fractions. The mineralization rate of radiocarbon from a defined position in the active ingredient molecule from aged, no longer extractable residues could provide an indication of the stability of binding. It is known from many and varied studies of the formation of soil organic matter using uniformly 14C-labelled plant residues or their major components (polysaccharides and lignins) that there is not a clear separation between the labile and stable carbon pools in soil. There are apparently transitions between the two carbon pools which can be rapid under certain conditions. An equilibrium between the formation and further degradation of the bound residues, probably closely linked to that of the humic substances, is also established for pesticides and their metabolites in these carbon pools with increasing residence time in the soil (ref. 19). Long-term experiments with 14C-labelled herbicides are being carried out on this topic (ref. 20).

Many studies have shown that the physically and chemically stabilized fractions account for 80 - 90 % of the total carbon fraction of a soil (ref. 21,22). This stable fraction is further differentiated into fulvic acids, humic acids and humins on the basis of its solubility, and these fractions differ significantly with respect to their stability and turnover rate. Recent ^{13}C NMR spectra studies have indicated that fulvic acids have a low aromaticity, whereas the aromaticity of humic acids and humin ranges from less than 20 % to 70 % (ref. 23). Generally the three soil organic matter fractions consist of polymerized aromatic components with aliphatic side chains, and especially of condensed polyphenols (ref. 12). Humic acids from arable soils are generally richer in phenolic fragments and fragments derived from polysaccharides with hexoses and pentoses as constituent units (ref. 24).

Foreign molecules of similar structure can react with these soil organic fractions and many of these reactions lead to relatively stably bound pesticide residues which behave in part like the organic soil substance itself and are released only when the C-skeleton is degraded. Dragun and Helling (ref. 25) found in a systematic study that a great number of synthetic aromatic chemicals can be oxidized in soil to radicals which react with the high molecular weight matrix of humin components. Haider (ref. 26) points out that only a relatively small fraction of the humic materials are rapidly built up or degraded. This fraction is possibly especially active in bonding foreign molecules (ref. 12). However, currently available data do not permit the unambiguous conclusion that active ingredients and metabolites are exclusively incorporated via the path of this labile carbon fraction. Indeed, recent studies with relatively stable triazine compounds have indicated that there is a rapid and preferred bonding or fixation of these compounds into the humin fraction of the soil (ref. 27,28).

The physico-chemical behaviour of active ingredients and metabolites, especially their reactivity, is decisive for their incorporation into the stable humus form. In studying various substances occurring in the soil, Kassim et al. (ref. 29,30) established that reactive, readily oxidizable compounds are autoxidatively or enzymatically bonded to the stable polyphenol bodies of the humus. Among such compounds are hydroxylized cinnamic acids, pyrocatechol, phenolcarboxylic acids and lignin dervatives such as coniferyl and alcohol, sinapyl alcohol, and p-coumaryl alcohol. Other authors have confirmed these results (ref. 26,31) and have pointed out that these reactive compounds usually have several free hydroxyl groups on their molecules, whereas organic substances that are largely mineralized to CO_2 in soil are mainly of an aliphatic nature or have on or more methylated hydroxyl groups on their aromatic nuclei (ref. 32).

Active ingredients and metabolites from organic pesticides are also subjected to the processes described, whereby the physico-chemical character of the humic material colloids as well as the sorption properties of the soil itself are of significance for the type and intensity of the reaction. Stevenson (ref. 2) suggests hydrogen bridge bonds with functional groups, reactions of alkyl residues, copolymerization and radically induced bondings as possible reaction mechanisms for bonding of pesticides and their metabolites. Thus pesticides and their metabolites do not behave differently in the soil from the organic components of plant residues (e.g. straw or roots), which are partly degraded to CO_2, but which also contribute to polymerized, stable compounds in the humus fraction (ref. 31,33,34,35). The degradation rate of the humus fraction, about 2 % per annum, is therefore decisive for the further persistence of the carbon from pesticides and their metabolites incorporated in the humus fraction (ref. 22,36). From field lysimeter studies with ^{14}C-labelled herbicides (ref. 20,37), it appears that in the second and third year after spraying, losses of radiocarbon from the soil considerably exceeded this rate. On the whole, therefore, pesticides display a higher mineralization and conversion rate before they are apparently increasingly stabilized and incorporated into the soil. The soil has too long been regarded as a uniform system. Haider (ref. 38) could demonstrate that soil contains microcompartments that differ markedly with respect to the microbial conversion of organic compounds. It is also important to keep in mind that the stable carbon fraction in the soil is in the order of 40,000 kg C ha-1 and that the labile, hydrolyzable carbon fraction amounts to approx. 5 - 10 % of this (ref. 21). According to Haworth (ref 39), humic acids have a highly aromatic core to which polysaccharides, proteins, phenols and metals are attached by chemical or physical bonds. Bound residues from different pesticides applied to crops at rates between 100 g and a few kg ha-1 a-1 are transferred into these carbon fractions of the soil during the continuous soil organic matter formation processes. Besides making analysis more difficult, this greatly complicates the evaluation of findings from intensive studies of the degradation and characterization of these residues.

TABLE 1. Carbon transformation in an agricultural soil, estimated production of carbon dioxide. Degraded loess soil (Parabraunerde) with 1 % soil organic matter carbon = 40 t · ha^{-1}
Cultivated crop: winter barley
Preceding crop: winter wheat

Carbon source	Mineralization rate %	Total mineralized carbon kg · ha^{-1} · a^{-1}
Soil organic matter[1)]	2	800
Wheat straw and roots[2)]	60	1500
Barley roots and root exudates[3)]	30 - 60	1000
Soil microbial biomass[4)]		200
Total mineralized carbon		3500
Pesticides applied to winter barley	50	4

1) ref. 22 3) ref. 40 and 41
2) ref. 34 and 35 4) ref. 42 - 44

As can be seen from Table 1, the total mineralization of an arable soil is in the order of 3500 kg C ha-1 a-1. This mineralization involves a large number of chemical and in particular microbial conversion processes involving highly stable organic structures from the long-lived carbon pool of the soil, various plant constituents, and the continually renewed fauna and flora of the soil. In contrast, the addition of carbon in intensive agriculture in the form of pesticides amounts to 3 - 10 kg of C ha-1 a-1. All studies of the degradation of these substances have shown that, with the exception of a few compounds (ref. 10), the pesticides and their metabolites are subjected to the same intensive conversion processes in the soil, especially in the active rhizosphere soil (ref. 11,40). According to Martin and Packridge (ref. 41), four compartments are involved in the intensive C-transfer in rhizosphere soil: root tissue, soil biomass, soil organic matter, and rhizosphere. Agronomic practices, including inorganic and organic amendments, can greatly influence the turnover rates and the flow of C between these compartments by leading to a more rapid mineralization of soil bound pesticides (ref. 46) and to an increased stabilization of pesticide residues in the soil humin fraction (ref. 47).

In addition to this direct incorporation into the large organic molecules of the soil, there is also the possibility of a fixation reaction with humus constituents, that is not necessarily a consequence of incorporation but is rather a special case of particularly stable sorption (ref. 48). Together with the incorporated residues, this type of fixed residue forms the bound residue fraction (Fig. 1). However, the magnitude of the fraction of bound residues is dependent on a series of processes in the soil and on the physico-chemical behaviour of the active ingredient and/or its metabolites (Table 2). The analytically detectable quantity of these residues is then a result of the extraction processes applied. It cannot be ruled out that additional bonding processes are created by the various extraction procedures used to detect bound residues.

TABLE 2. Factors influencing the degradation, fixation and bioavailability of non-extractable bound pesticide residues in the soil

1. Physico-chemical behaviour of the active ingredient
2. Quantity, type and time of application (e.g. spraying on the soil surface, granular application in the seedbed, seed dressing) and concentration in soil
3. Physical, chemical and biochemical reactions in the soil depending on soil type and properties: clay minerals, organic matter, pH, Eh, oxygen content and water content
4. Climatic factors: precipitation, temperature, relative atmospheric humidity, soil moisture and insolation
5. Plant species
   a) root development
   b) transpiration
   c) nutrient status
6. Argonomic practices
   a) cultivation
   b) fertilization
   c) organic amendments, plant residues
   d) irrigation

Although detailed studies on the degradation of organic chemicals by microorganisms are undoubtedly valuable the conclusions from such studies which can be applied to natural soil conditions are limited. According to Alexander (ref. 49), a conclusion relevant for practice requires "environmentally relevant investigations of the physiology, biochemistry and ecology of the microorganisms as well as field assessment of the transformation of synthetic chemicals". Pesticides are synthetic chemicals whose concentration in soil is extremely low, or becomes extremely low after application due to soil cultivation, degradation and translocation processes. As shown by extensive lysimeter experiments with metamitron (ref. 20), the concentration of this pesticide in soil is very low at the time of harvesting (0.3 mg kg-1 soil of metamitron equivalents) due to translocation and loss via mineralization. Thus even in the case of multiple application during the course of crop rotation, only a very low concentration of an active substance can be built up, except in the case of certain persistent pesticides.

Soils contain a large number of microorganisms that have the ability to degrade pesticides. Novik and Alexander (ref. 50) studied the decomposition of low concentrations of pesticides in soil, sewage sludge, and water and found that in each case a cometabolism of certain pesticides such as propachlor or alachlor took place, the degradation of propachlor by cometabolism being increased by glucose. Alexander (ref. 49) concluded from systematic studies on the degradation of foreign chemicals by microorganisms that a certain minimum concentration is required to establish the acclimated population needed for biodegradation. Mineralization is therefore also a function of concentration (Table 2), as has been established for the degradation of glucose in soil (ref. 31). Studies of the decomposition and stabilization of ring-^{14}C-labelled catechol in soil (ref. 51) showed that mineralization of this compound was retarded when it was applied at a low concentration (1 mg kg^{-1} loamy soil). They also indicated that the apparent high stability of catechol in soil is caused by a rapid polymerization of this compound into existing soil polymers. This may explain why pesticides with phenolic structures are bound rapidly and intensively in the soil (ref. 19,28). Recent unpublished data from experiments with simazine and bentazone indicate that more stable non-extractable residues are formed in soil when the pesticides reach the soil after plant uptake and are decomposed with the plant material. The nutrient status of a soil thus still has a considerable influence on the conversion processes (Table 2). With a high nutrient supply and a high supply of energy, the diversity and population of microorganisms are greater (ref. 52), so that although the mineralization rate increases a more intensive incorporation of the non-extractable residue into the humin fraction is also observed.

In conclusion, it can be stated that the basic principles for the formation of soil-bound pesticide residues are similar to those for the formation of stable humic substances from a wide range of natural plant constituents (ref. 17):

- Both charged and neutral molecules are bound.
- The nature of the bonding ranges from entrapment to covalent bonding of the pesticide
- Microbial and chemical processes result in reactive compounds leading to bound residues.

- Soil organisms incorporate active substances or their metabolites.

Further research on physical and chemical characterization by effective chromatographic and spectroscopic methods may yield additional information.

LONG-TIME BEHAVIOUR AND BIOAVAILABILITY OF NON-EXTRACTABLE RESIDUES

The question arises as to the extent to which bound, sorbed or fixed residues and/or metabolites become biologically available again in the formation and degradation of organic carbon in the soil as a result of desorption processes (Figure 1). Essential information was obtained in lysimeter experiments with ^{14}C-labelled pesticides under field conditions (ref. 1,3,37,53). As an example, the results from the experiment with methabenzthiazuron showed that with increasing duration of the experiment, the extractability of the compounds and/or its metabolites remaining in the soil is greatly reduced. At the same time, however, radioactivity from the active ingredient labelling is still taken up by subsequent plants in the second and third year of crop rotation (ref. 1,3,53,54,55). The aged active ingredient residues remaining in the soil are thus still biologically available for subsequent crops (Table 3), but their availability is reduced by a factor up to 18 in comparison with the uptake by treated crops (ref. 37,53). The biological availability apparently greatly decreases with increasing residence time of the pesticides in soil. An equilibrium is established between fixed, adsorbed and incorporated pesticides and/or their metabolites and radiocarbon derived from the active ingredient and metabolites, that are freely available in the soil solution (Figure 1).

TABLE 3. The total uptake of active ingredient equivalents after spray application of [carbonyl-^{14}C] methabenzthiazuron to winter wheat at the 3–4 leaf-stage (ref. 50). Methabenzthiazuron applied at a rate of 2 kg Tribunil · ha^{-1}
Applied ^{14}C activity = 100 %
Average values of three 0.5 m^2 lysimeters.

	Hectare yield	Methabenzthiazuron equivalents[1)]	
		in dry matter	in the hectare yield
	t · ha^{-1}	mg · kg^{-1}	g · ha^{-1}
1984 **Winter wheat**			
straw	9.4	2.05	19.30
chaff	1.8	0.40	0.70
grain	6.2	0.05	0.31
Total	17.4		20.31
1985 **Winter barley**			
straw	8.9	0.12	1.12
chaff	0.6	0.05	0.03
grain	8.0	0.02	0.15
Total	17.5		1.30

[1)] Calculated on the basis of the specific ^{14}C activity of the [carbonyl-^{14}C] methabenzthiazuron applied.

In field studies it is not possible to distinguish between aged residues still analytically detectable and characterizable and non-extractable residue fractions with respect to their biological availability. Führ (ref. 56) therefore suggested studying the relative biological availability of bound residue fractions with the aid of a biotest in special microecosystem which would simultaneously permit determination of the mineralization rate. Similar experiments have since been implemented by various research units (ref. 16,19, 57-64). However, the results of these experiments cannot be compared with each other because the quantities of active ingredients used, the soils, the conditions to produce non-extractable residues, the plants, the duration of the experiments, and the experimental conditions with respect to illumination and temperature have differed greatly.

Nevertheless, these experiments, which can only be regarded as preliminary tests, indicate that mere aging of the active ingredients in soil leads to a considerable reduction in their bioavailability to plant roots. If the extractable residue fractions are then removed from the soil, the biological availability of the bound residues decreases even further. In work to compare active ingredients freshly added to soil, the relative biological availability of the bound radiocarbon was reduced by a factor of approximately 4 to 10 (Table 4) and in experiments with metamitron and simazine this factor was as high as 600 (ref. 19). Additional experiments indicated that the findings in the simazine experiment are related to the drastic treatment of the soil during extraction. Microecosystem experiments therefore only provide data on the relative biological availability of non-extractable residues and must be considered separately with respect to the prevalent experimental conditions.

In some studies soils containing non-extractable residues were also given to earthworms in small-scale experiments (ref. 58,59). In these experiments the uptake of radiocarbon from non-extractable residues of methylparathion or hydroxymonolinuron was approximately three times less than that oberserved with active ingredients freshly added to soil. No damage to earthworms was observed.

TABLE 4. The comparison of ^{14}C uptake from soil containing freshly added pesticides, aged or non-extractable bound residues
Standardized experiments with maize or sugar beet plants (ref. 19,64).
Radioactivity in soil at the beginning of the experiment = 100 %.

Pesticide	Total ^{14}C uptake by plants from			Relative uptake ratios		
	freshly added %	aged %	"bound" residues %	freshly added	aged	"bound" residues
[carbonyl-^{14}C] methabenzthiazuron	4.80	2.76	0.99	5	3	1
[3-^{14}C] metamitron	16.80	0.01	0.03	559	1	1
[ring-^{14}C] simazine	20.28	10.17	0.03	675	338	1
[benzene ring ^{14}C] anilazine	3.14	0.60	0.78	4	1	1
[benzene ring ^{14}C] bentazon	36.02	12.47	0.50	72	25	1

The disadvantage of all these laboratory experiments is that they do not reflect the real situation. Under practical agricultural conditions numerous factors influence root uptake and the degradation of active substances in soil (Table 2), so that quite different conditions arise there. It has not been taken into consideration that in laboratory experiments a) the soil has been treated with chemical solvents and b) in small-scale experiments the ratio of root to soil or test object to soil is much more restricted than in the field soil so that in microecosystems of this type uptake rates can be up to 50 times higher than those overserved in the field (ref. 65). Such experiments can therefore provide only indications and clues of the way in which the biological availability of the non-extractable residue (usually measured as radiocarbon) behaves in comparison to the aged residue in the soil not treated with solvents with respect to uptake by plants or effects on different soil organisms.

Experiments have not yet been conducted to discover what has been changed in the soil simply by the various extraction steps. In addition to extracting the aged residues with organic and inorganic solvents, it is therefore suggested that 0.01M $CaCl_2$ solution should be used because it can be regarded as a simulated soil solution. The use of this solution for desorbing and washing out soluble active ingredients and metabolites does not seriously affect the soil structure, and it reflects the situation in the soil during wetting, drying and rewetting. It must, however, be emphasized once again that these microecosystem experiments can only provide data on the relative biological availability of non-extractable residues and that they must be combined with realistic crop rotation experiments involving application of ^{14}C-labelled pesticides in field lysimeters (ref. 37). As can be seen from the data in Table 3, even as much as 40 % of the ^{14}C applied as [carbonyl-^{14}C]methabenzthiazuron at a Tribunil rate of 2 kg ha^{-1} was still in the plough layer at harvest time of the succeeding crops (ref. 20) and the ^{14}C found in the total hectare yield of 8 tonnes of barley represented 150 mg methabenzthiazuron-equivalent.
In another lysimeter experiment in which the herbicide [3-^{14}C]metamitron (ref. 55) was applied at a concentration of 4.9 kg ha-1 (corresponding to 7 kg Goltix ha-1 in the preemergence spraying of sugar beets) at the time of harvesting (188 days after metamitron application), the beets contained active ingredient equivalents of 0.1 mg kg^{-1} fresh weight computed via the specific radioactivity of the [3-^{14}C]metamitron applied. Since sugar beets develop their storage organs during the second half of the vegetation period, these labelled residues must be considered to represent compounds from the fraction of aged residues. The ^{14}C activity in the beets represented approx. 3 g of metamitron equivalents in the yield per hectare of 30 tonnes of sugar beets, of which 800 mg was identified as incorporated into sucrose (ref. 66). This is therefore no longer a residue in the sense of the accepted definition. In another metamitron experiment (ref. 53), in the subsequently cultivated crops (winter wheat and winter barley) metamitron equivalents of approx. 150 mg were calculated in the yields per hectare of 7 to 8 tonnes, although approx. 50-60 % of the applied ^{14}C activity was still measured in the top soil. It can be

assumed from the results of these sugar beet experiments with metamitron (ref. 53,55,66) that a considerable fraction was assimilated as $^{14}CO_2$ produced by mineralization of the labelled triazine ring.
With respect to the long-term behaviour of non-extractable bound residues, the conclusion can be drawn from the data available that, with some exceptions, the pesticides lose their bioactivity and in most cases even their identity in the degradation and bonding processes. This must lead to different mean residence times in the soil depending on the soil carbon pool and the bonding in which the residues are present. It is known from experiments with highly degradable organic compounds that, for example, as much as 10-15 % of the carbon in glucose can enter the stable carbon fraction and that some of this carbon is converted to non-hydrolyzable forms through secondary metabolic reactions. During the continuously progressing changes in the stable carbon fractions of the soil, molecular constitutents from the non-extractable pesticide residues are proportionately released. This leads to the availability of residues which are then, however further degraded. This long-term behaviour should be studied in more detail particularly with respect to possible translocation processes in the soil. Depending on soil type and sorption conditions (Table 2), soluble organic compounds are translocated out of the top soil into deeper soil zones. During this processes a translocation of pesticide residues must also be expected. Realistic data can be predicted from lysimeter experiments employing ^{14}C-labelled pesticides (ref. 1,20,37). Apart from detecting the labelled residue compounds, however, the total content of translocated native organic compounds in the soil must also always be determined in order to evaluate the results.

ACKNOWLEDGEMENTS

Special thanks are due to all participants of the Bayer Workshop on "Bound Residues" and especially to Dr. K. Haider, Braunschweig, and Dr. J.M. Bremner, Ames, Iowa, USA for many valuable comments.

REFERENCES

1. F. Führ, In "Isotopes and Radiation in Agricultural Sciences", Vol 2, M. L'Annunziata and J.O. Legg (ed.), 239-270 (1984).
2. F. J. Stevenson, ACS Symposium Series, 29 180-207 (1976).
3. W. Klein and I. Scheunert, In "Agrochemicals: Fate in Food and the Environment using Isotope Techniques", IAEA Vienna, 177-205 (1982).
4. D. D. Kaufman, ACS Symposium Series, 29, 1-9 (1976).
5. P. C. Kearney, J. Assoc. Off.Anal. Chem., 65, 1030-1032 (1982).
6. T. R. Roberts, Pure & Appl. Chem., 56, 945-956 (1984).
7. E. P. Lichtenstein, Residue Rev., 76 147-153 (1980).
8. S. V. Khan, Residue Rev., 84 1-25 (1982).
9. F. Führ, Rheinisch-Westfälische Akademie der Wissenschaften, Vorträge N 326 Westdeutscher Verlag, Opladen, 5-35 (1984).
10. D. Riley, W. Wilkinson and B. V. Tucker, ACS Symposium Series, 29, 301-351 (1976).
11. K. Seibert, F. Führ and W. Mittelstaedt, Landw. Forsch., 35, 5-13 (1982).
12. F. J. Stevenson, Humus Chemistry, John Wiley & Sons, New York, 403-419 (1982)
13. J. P. E. Anderson and K.H. Domsch, Can. J. Microbiol., 21 314-322 (1975).
14. J. C. B. Ottow, Landw. Forsch., 35, 238-256 (1982).
15. F. Führ, R. Kloskowski and P. Burauel, In "Pflanzenschutzmittel und Boden", Berichte über Landwirtschaft, SH 198, 106-116 (1985).
16. Anonymous, Radiotracer Studies of Bound Pesticide Residues in Soil, Plants and Food. IAEA, Vienna, IAEA-TECDOC-306 (1984).
17. Anonymous, Workshop on Bound Residues in Soil - Summary and Conclusions. 22-24 Jan., 1986, Bayer Pflanzenschutzzentraum Monheim, F.R. Germany (1986).
18. R. Bartha, ASM-News, 46, 356-360 (1980).
19. R. Kloskowski, F. Führ and W. Mittelstaedt, 6th Int. Congr. of Pesticide Chemistry, Proceedings, Blackwell's Scientific Publishers, (in press).
20. R. Kubiak, Diss. Univ. Bonn, Jül-Ber., 2055 (1986).

21. C. A. Campbell, In "Soil Organic Matter", M. Schnitzer and S.U. Khan (ed), Elsevier, Amsterdam, 173-271 (1978).
22. D. S. Jenkinson and J. H. Rayner, Soil Sci., 123, 298-305 (1977).
23. M. Schnitzer and C.M. Preston, Soil Sci. Soc. Am. J. 50 326-331 (1986).
24. K. Haider and H.R. Schulten, J. Anal. Appl. Pyrol., 8, 317-331 (1985).
25. J. Dragun and C. S. Helling, Soil Sci., 139 100-111 (1985).
26. K. Haider, ACS Symposium Series, 29, 244-257 (1976).
27. R. Kloskowski and F. Führ, Chemosphere, 12, 1557-1574 (1983).
28. W. Mittelstaedt, F. Führ and R. Kloskowski, J. Environ. Sci. Health, B (in press).
29. G. Kassim, J. P. Martin and K. Haider, Soil Sci. Soc. Am. J., 45, 1106-1112 (1981).
30. G. Kassim, D. E. Stott, J. P. Martin and K. Haider, Soil Sci. Soc. Am. J., 46 305-309 (1981).
31. J. P. Martin and K. Haider, Soil Sci. Soc. Americ. J., 43, 917-920 (1979).
32. K. Haider and J. P. Martin, Z. Pflanzenern. Bodenkde. 142, 456-475 (1979).
33. D. Sauerbeck, Landw. Forsch., 21, 91-102 (1968).
34. D. Sauerbeck and F. Führ, Mitt. Dsch. Bodenkde Ges., 10, 174-177 (1970).
35. H. E. Oberländer and K. Roth, Landw. Forsch., 33, 179-188 (1980).
36. D. S. Jenkinson, Soil Sci., 111, 64-70 (1971).
37. F. Führ, Weed Sci.,33, Suppl. 2, 11-17 (1985).
38. K. Haider, IUPAC-Pesticide Chemistry, Human Welfare and the Environment, J. Miyamoto et al. (ed.), Pergamon Press, Oxford, 351-356 (1983).
39. R.D. Haworth, Soil Sci., 111 71-79 (1971).
40. K. Seibert, F. Führ and H.H. Cheng, Proc. EWRS-Symp. on Theory and Practice of the Use of Soil Applied Herbicides, Paris, 137-146 (1981).
41. J. K. Martin and D. W. Packridge, In: The Cycling of Carbon, Nitrogen, Sulphur and Phosphorus in Terrestrial and Aquatic Ecosystems, I.E. Galbally and J. R. Freney (ed.), Australian Acad. Sci., Canberra, 77-82 (1982).
42. D. Sauerbeck and G. B. Johnen, In: Soil Organic Matter Studies, Proc. Symp. 1976 IAEA, Vienna, 1, 141-148 (1977).
43. J. M. Lynch and L. M. Panting, Soil Biol. Biochem., 12, 29-33 (1980).
44. J. M. Lynch and L. M Panting, Soil Biol. Biochem., 12, 547-550 (1980).
45. J.P.E. Anderson and K.H. Domsch, Soil Sci., 130, 211-216 (1980).
46. Z. Gerstl and C. S. Helling, Soil Biol. Biochem., 17, 667-673 (1985).
47. F. Azam, F. Führ, W. Mittelstaedt and R. Kubiak, Z. Pflanzenern. Bodenkde. (in preparation).
48. M. B. H. Hayes, Residue Rev., 32, 131-174 (1970).
49. M. Alexander, In "Environmental Biogeochemistry", R. C. Hallberg (ed.), Ecol. Bull. (Stockholm), 35, 503-510 (1983).
50. N. J. Novick and M. Alexander, Appl. Environ. Microbiol., 49, 737-743 (1985).
51. J. P. Martin, K. Haider and L. F. Linhares, Soil Sci. Soc. Am. J., 43, 100-104 (1979).
52. M. Alexander, Environ. Sci. Technol., 18, 106-111 (1985).
53. M. Hansper, Diss. Univ. Bonn, Jül-Bericht 2067 (1986).
54. F. Führ and W. Mittelstaedt, Landw. Forsch., SH 32, 286-294 (1976).
55. W. Mittelstaedt and F. Führ, Landw. Forsch., SH 37, 666-676 (1981).
56. F. Führ, ACS Symposium Series, 29, 356-357 (1976).
57. F. Führ and W. Mittelstaedt, J. Agric. Food Chem., 28, 122-125 (1980).
58. T. W. Fuhremann and E. P. Lichtenstein, J. Agric. Food Chem., 26, 605-610 (1978).
59. A. Haque, J. Schuphan and W. Ebing, Pestic. Sci., 13, 219-228 (1982).
60. C. S. Helling and A. E. Krinovak, J. Agric. Food Chem., 26, 1164-1172 (1978).
61. S. U. Khan, J. Agric. Food Chem., 28, 1096-1098 (1980).
62. T. R. Roberts and M. E. Standen, Pestic. Sci., 12, 285-296 (1981).
63. K. Sato, S. Maki, Y. Kato, O. Matano and S. Goto, J. Pestic. Sci., 9, 49-59 (1984).
64. K. Lee, F. Führ and W. Mittelstaedt, Chemosphere (in press).
65. F. Führ and W. Mittelstaedt, 5th Int. Congr. of Pesticide Chemistry, Human Welfare and the Environment, Pergamon Press, Oxford, 183-188 (1983).
66. L. Müller, W. Mittelstaedt, J. Pfitzner, F. Führ and H. J. Jarczyk, Pestic. Biochem. Physiol., 19, 254-261 (1983).

An environmental screening model for agrochemicals

W. Brock Neely

EnviroSoft, 4302 Cruz Dr., Midland, MI 48640, USA

Abstract - A computer model has been developed for the purpose of establishing an environmental profile for agrochemicals. The model is based on established algorithms that have been modified to work in the Personal Computer. The model, given the acronymn PEEC (Predicting the Environmental Exposure of Chemicals), is illustrated with the pesticides DDT, DCP,Alaclor and Atrazine.

INTRODUCTION

With the ban on certain pesticides such as DDT, increased attention has been focused on the environmental behavior of the replacement chemicals. Decisions to proceed or cancel the development process must be made prior to the introduction of the chemical into the environment. To accomplish this objective it is necessary to have reliable, economical methods to generate the required information. In view of the large number of candidate chemicals, simulation modeling of environmental systems becomes a necessary adjunct to the screening operation. Such models are not new. For example, Knisel with the aid of many investigators (ref.1), assembled a model (CREAMS) for evaluating the amount of pesticide in surface runoff. In addition Donigian et al (ref.2) developed HSPF (Hydrological Simulation Program Fortran) as a tool for evaluating the fate of chemicals in a large watershed. These models are site specific requiring a great deal of input and computer time to make them operate. Consequently, they are unsuitable for the present purpose which is to rank a series of chemicals from an environmental perspective. What is required is a model that uses a minimum amount of data, is easy to operate, has a short turn around time and is a reasonable reflection of the environmental processes that the chemical will be subjected to in the real world. Jury, Spencer and Farmer (refs.3-6) addressed the same problem and were successful in developing a screening model to describe the environmental profile of a chemical added to the soil. The model to be presented in this paper will be a further simplification, allowing the investigator to more quickly answer key environmental questions.

The starting point for the present study will be the approach introduced by Mackay (ref.7). The model and the associated transfer constants are illustrated in Figure 1 and have been referred to as the "Unit World" (ref.8). The name was derived from the observation that the dimensions and properties (Table 1) simulated important conditions on earth. Previous use has dealt almost exclusively with steady state equilibrium (refs.8-11). However, the movement of pesticides in a terrestrial environment is far from equilibrium, accordingly a different approach is required. Early attempts at using a steady state non equilibrium model have been presented (refs.12-14). In these studies the estimation of the transfer coefficient, D_{13} between soil and air (see Fig.1) was not too reliable. A procedure for generating D_{13} based on the studies of Jury et al (refs.3-6) will be described and incorporated into a steady state non equilibrium model. Finally, the model will be tested with a number of well known environmental contaminants.

MODEL

For the terrestrial portion of the model the rate constants in Figure 1 will be discussed. As chemicals are deposited on the ground either intentionally or inadvertently, there are three main driving forces controlling the movement. The first is absorption to the soil particles, which are then subjected to runoff processes and may end up in the water. The second is leaching causing the chemical to move into the groundwater. Finally, there is loss due to volatility. Once the chemical reaches surface water via runoff there are two

TABLE 1 Environmental Parameters for Screening Model *

Compartment	Volume(m^3)	Depth(m)	Area(m^2)
Air	6×10^9	6000	
Water	7×10^6	10	7×10^5
Ground	4.5×10^4	0.75	3×10^5

Properties of the Ground

Organic Fraction	0.02
Porosity	0.4
Volumetric air content	0.2
Volumetric water content	0.2
Bulk Density	1.5×10^6 (g/m^3)

Other Properties

Average rainfall	0.7 m/year
Average Temperature	25 C
Exit to Stratosphere k_s	4.1×10^{-5}/day (7)
Runoff Constant k_{32}	7.7×10^{-6} m/day

$K_{12} = P \times 16.04/TS$ (20)
$K_d = oc \times (0.6K_{ow})$ (19)

* See Reference 7 for rational of the dimensions.

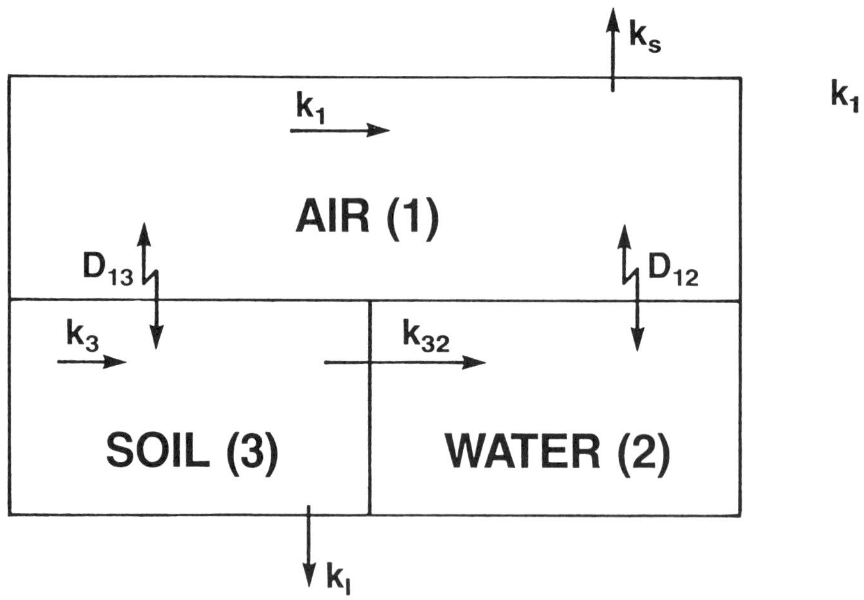

Fig.1 A diagram indicating the compartments and the rate constants that are in PEEC

additional diffusive mechanisms operating. One controls the loss to the atmosphere while the second is concerned with the water/sediment layer. These latter two processes have been discussed by Mackay et al (ref.13) and Neely (ref.14) and will not be included in this model.

Prior to describing transport across media borders the concept of fugacity (f) will be reviewed. This term has units of pressure and may be visualized as a measure of the "escaping tendency" of a molecule from one phase to another (7). At equilibrium fugacities are equal. Consequently, on the approach to equilibrium, diffusion must be driven by the Δ f between two phases. Therefore, in a non equilibrium situation it is necessary to determine the fugacity for each compartment. Once this is accomplished, the concentration (C) is estimated using equation (1).

$$C_i = Z_i \times f_i \qquad (1)$$

Mackay and Paterson (ref.9) have reviewed the procedures for calculating Z in different compartments (The symbols have been collected in a Glossary at the end of this paper.)

In addition to transport being controlled by concentration gradients, there is a second mechanism of interphase transport movement. This occurs when a chemical moves in association with another material like sediment or water. Diffusive transport is reversible while material transport is essentially one way. There are many transport mechanisms operating such as transfer by either wet or dry deposition from the air, settling of sediments to the bottom of the water compartment and runoff. Only volatility, runoff and leaching will be discussed in this article.

Volatility is a diffusion process where the transfer occurs in both directions and is controlled by the fugacity difference that exists between the air and ground compartments. Thus the transfer in moles/day between ground and air is given by equation 2. The direction of transfer depends

$$N_{13} = D_{13}(f_1 - f_3) \qquad (2)$$

on the sign associated with the fugacity difference. For most chemicals applied to the ground $f_3 > f_1$ and transfer will occur from soil to air.

The reciprocal of diffusion is an easier concept to visualize. Thus, if the resistance is large the diffusion will be small and little material will be transferred. The total resistance is the sum of the resistances that occur in the boundary layers between the soil and air. Resistance in air has been described previously (refs.6,7) and is given by equation 3.

$$R_1 = 1/(G_g\ A_{13}\ Z_1) \qquad (3)$$

Equation 4 (ref. 13) shows the resistance in the soil compartment.

$$R_3 = d/(M^s\ A_{13}\ Z_i) \qquad (4)$$

The fugacity capacity Z_i will depend on which compartment (soil water or soil air) dominates as far as soil movement is concerned. In moist soil, chemicals at the interface are maintained by diffusion of the compound through the soil water from its absorption sites. Since the vapor diffusion coefficient is approximately 10,000 times larger than the solution phase coefficient, a Henry's coefficient for the distribution between soil air and soil water of 1×10^{-4} (g/cm^3 in soil air / g/cm^3 in soil water) may be considered as a transition for determining when vapor or nonvapor becomes dominant (eg most pesticides which have a Henry's coefficient much larger than 1×10^{-4}) will diffuse mainly in the soil air while those with coefficients less than 1×10^{-4} (water soluble materials like ethylene glycol) will diffuse mainly in the solution phase. The influence of vapor phase diffusion becomes less significant as bulk density increases and air filled porosity decreases, leaving fewer pathways for which vapor can diffuse.

For the present model the soil vapor diffusion coefficient is equated to the air diffusion coefficient multiplied by a tortuosity factor to account for the reduced flow area and increased path length. Various models have been described for estimating this factor and the one chosen by Jury et al (ref.3) will be used.

$$M^s_1 = (\alpha^{3.33}/\varphi^2)\ M_1 \qquad (5)$$

In a similar manner the soil water coefficient M^s equals the water coefficient multiplied by a tortuosity factor (ref.3).

$$M^s_2 = (\Theta^{3.33}/\varphi^2)\ M_2 \qquad (6)$$

The diffusion coefficients in water and vapor were estimated by the methods outlined by Tucker and Nelker (ref.18). These two coefficients were than combined into a single equation 7 (ref.3).

$$M^s = (K_{12}\ M^s + M^s)\ /\ K \qquad (7)$$

In this equation K is the ratio of the total concentration to the concentratin in the liquid phase as defined by Jury et al (ref.3)

$$K = (\beta K_d) + \Theta + (K_{12}\alpha) \qquad (8)$$

The diffusion coefficient as defined in equation 7 is used in equation 4.

The total resistance to mass transfer is the summation of Equations 3 and 4. The reciprocal of the sum becomes the diffusion transfer coefficient D_{13}. R_3 is usually greater than R_1 and diffusion will be controlled in the soil layer.

$$D_{13} = 1/(R_1 + R_3) \quad (9)$$

For most pesticides Henry's constant $K_{12} > 1 \times 10^{-4}$ consequently, the net diffusion coefficient may be approximated by equation 10

$$M^s = K_{12} M_1^s /K \quad (10)$$

and the fugacity capacity factor for air will be used in equation 4.

Runoff- In previous reports (refs.13-15) a rational was presented for choosing 7.7×10^{-6} m/day for k_{32} the sediment runoff constant. While this is about 5 times larger than the value estimated for Lake Erie (ref.15) it does generate a reasonable sediment burial rate (ref.13). The present model will use this value to simulate conditions of pesticide applications to soils that are susceptible to sediment erosion. Hence the results will exaggerate the impact of runoff as a route of loss from the soil

In addition to the movement of chemical from the ground to the stream *via* sediment the movement of chemical in the actual runoff water must also be considered. Once a chemical has been added to the soil the fraction of chemical in the water phase will be given by Equation 11.

$$fr_2 = 1/(K_d \times \text{concentration of sediments} + 1) \quad (11)$$

For the soil compartment the concentration of soil in the field is approximately equal to the bulk density. Thus it is apparent that even for materials with very low absorption coefficients the fraction of chemical sorbed to the soil is greater than the amount in the water phase. It will only be during an intense storm immediately following application that much chemical will be found in the surface water. Consequently, for screening purposes the amount of chemical transported in the water phase will be ignored.

Leaching McCall *et al* (ref.16) demonstrated that the leaching or movement of chemicals through soils followed equation 12.

$$\text{cm moved} = \frac{1}{K_d}(\text{cm of water entering the soil}/((1-Q^{2/3})\beta) \quad (12)$$

This equation was developed from column studies using disturbed soils and a constant head of water. Such a situation is atypical of what is found in the field, however, it was felt that the equation would correctly predict the relative movement of a series of chemicals. Thus for an average rainfall of 0.7m/yr where 90% of the rain enters the soil (ref.17) equation 12 may be converted to 13.

$$k_1(\text{m/day}) = (\text{Rain(m/yr)} \times 0.36)/K_d \quad (13)$$

The porosity (φ) and density in equation 12 have been assigned values of 0.4 and 1.5 g/ml respectively. Leaching will have occurred when the chemical moves below the 2 meter depth.

Mass Balance Equations. The mass balance for each compartment may be expressed in words as follows:

Input = [Loss from a compartment] - [Gain from an adjacent compartment] (14)

where Loss is from : reaction, diffusion
advective and material transport.

Gain is from : diffusion and material
transport

Input is from: annual environmental release and advective flows from adjacent compartments

For this screening model advective flows will not be considered. Expressing equation (14) in mathematical terms generates 15

$$I_i = [V_i Z_i (k_i + k_{ij}) + \Sigma D_{ij}] f_i - [V_j Z_j k_{ji} + \Sigma D_{ji}] f_j \quad (15)$$

where the summations are carried out for all j adjacent to compartment i

Assuming that the input can only be added to either ground, air or water a mass balance equation for each of the three compartments may be written as follows:

$$I_1 = f_1 X_1 - f_2 D_{12} - f_3 D_{13} \quad (16)$$

$$I_2 = -f_2 D_{12} + f_2 X_2 - f_3 R \quad (17)$$

$$I_3 = -f_3 D_{13} + f_3 X_3 \quad (18)$$

where $X_1 = V_1 Z_1 (k_1 + k_s) + D_{12} + D_{13}$
k_s = exit to the stratosphere
$X_2 = V_2 Z_2 k_2 + D_{1,2}$
$X_3 = V_3 Z_3 (k_3 + k_{32}(1 - fr_3) + k_1) + D_{13}$
fr_3= fraction of chemical sorbed to soil
$R = V_3 Z_3 k_{32} (1 - fr_3)$

Using matrix notation equations 16-18 can be illustrated in equation 19

$$I_i = f_i \, [A] \quad (19)$$

where

$$A = \begin{bmatrix} X_1 & -D_{12} & -D_{13} \\ -D_{12} & X_2 & -R \\ -D_{13} & & X_3 \end{bmatrix}$$

By inverting the matrix, the fugacities are given by:

$$f_i = [A]^{-1} \; I_i \quad (20)$$

At steady state the moles of chemical in each compartment are:
$V_i Z_i f_i$
where the loss is represented by:

a)volatility = $D_{13} (f_3 - f_1)$
b)runoff = $V_3 Z_3 k_{32} f_3 fr_3$
c) leaching below 2 meter = $V_3 z_3 k_L f_3 fr_3$ and
d) degradation = $V_3 Z_3 k_3 f_3$

MODEL EVALUATION

The equations described above along with the standard environmental conditions listed in Table I were programmed for the IBM-PC. The output was formatted as a pie chart where the percent lost by volatility, runoff, leaching and degradation could be illustrated (Figures 2-3).

The chemicals examined in PEEC (Predicting Environmental Exposure of Chemicals) are given in Table 2 along with the key properties. The properties were taken from several different references and represent typical values for these pesticides.

TABLE 2. Properties of Chemicals to be Evaluated in PEEC [1]

Chemical	Molecular Weight	Water Solubility ppm	Vapor Pressure Pascals	log K_{ow}	Degradation (1/day)
Atrazine	216	70	3.9×10^{-5}	2.82	1.98×10^{-3}
DCP	110	3600	5526	1.5	Stable
Alaclor	270	242	2.8×10^{-3}	3.83	1.38×10^{-2}
DDT	354	.003	2.6×10^{-5}	6	Stable

[1] PEEC is the name that was assigned to the Computer Program for evaluating the set of equations and drawing the pie charts.

Alaclor The diagrams for this herbicide are shown in Figure 2a. The profile indicates that the the amount below 2 meters increases as the soil type goes from a clay with a low porosity and a high organic content to a sandy soil where the porosity and bulk density increase. In a monitoring study performed in Nebraska, Spalding et al (ref.22) found Alaclor in 2 out of 14 wells that were sampled, both of which were in sandy soils.

Atrazine- Figure 2b demonstrates the potential for atrazine to move into groundwater. This conclusion has been substantiated by monitoring analysis (refs.22,23). Figure 2b demonstrates that if the lipophilicity of the molecule could be increased the mobility would be reduced and the extent of degradation increased. Thus in the design phase of a synthesis program it might prove worthwhile to add either two more methyls or an ethyl group to one of the nitrogens in atrazine.

DCP (Dichloropropane) -DCP is a well known soil fumigant and the profile in the standard environment is shown in Fig. 3a. In spite of the high vapor pressure the fumigant, once in the soil volatilizes to the atmosphere at reduced rate. Consequently, the water soluble chemical migrates downward and is found as a contaminant in the groundwater. This observation has been confirmed in many studies ranging from Maryland to

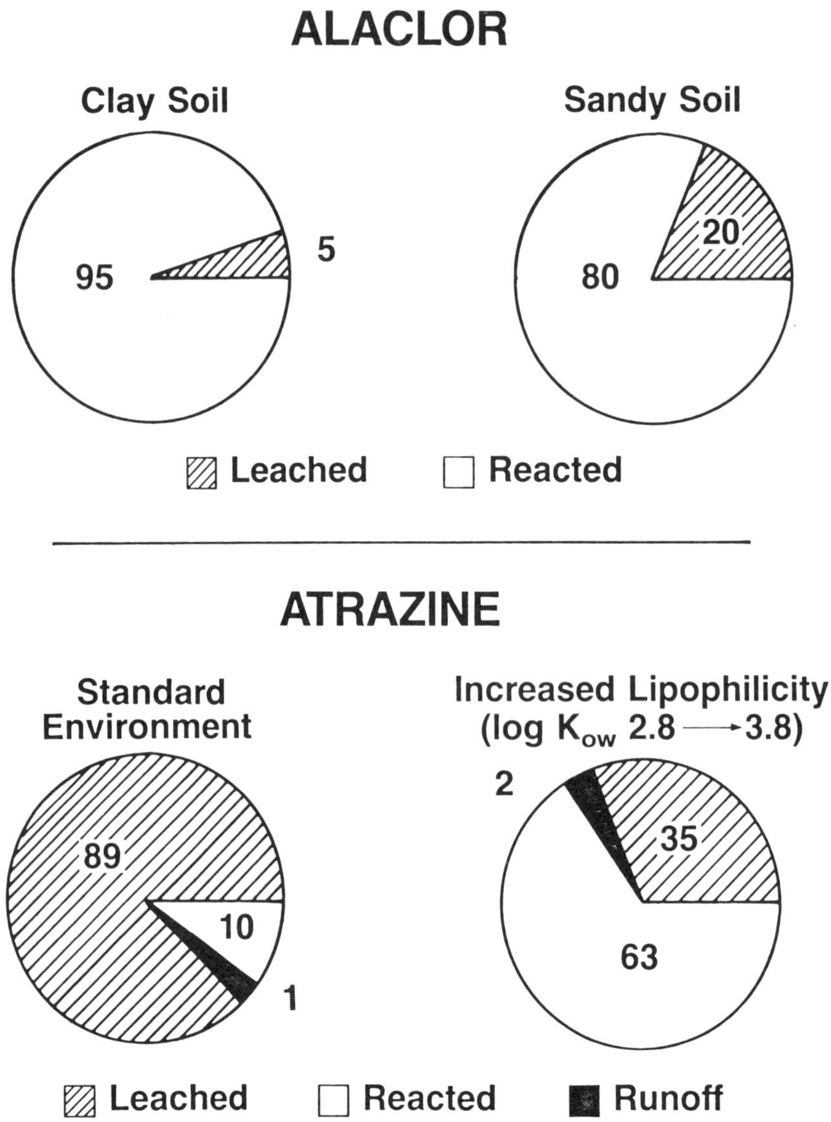

Figure 2. Piechart showing the results of using PEEC on the two chemicals: a.ALACLOR and b.ATRAZINE.

New York and California (ref.23). From a development point of view if it were possible to synthesize a molecule with a lower solubility (1500 ppm compared to 2600 ppm) an agent with a greater opportunity to be lost to the atmosphere (Fig. 3a) would be possible. It is interesting to note that a chemical possessing a soil half life of at least one month would be necessary before degradation becomes a significant process (Fig. 3a).

DDT The profile of this well known pesticide is illustrated in Figure 3b. The diagram illustrates a chemical that has a high probability of being lost via runoff. There is no question that the strong affinity of DDT for soil will cause the molecule to move into the surface water. Once in the water the known volatility of DDT will hasten the global distribution.

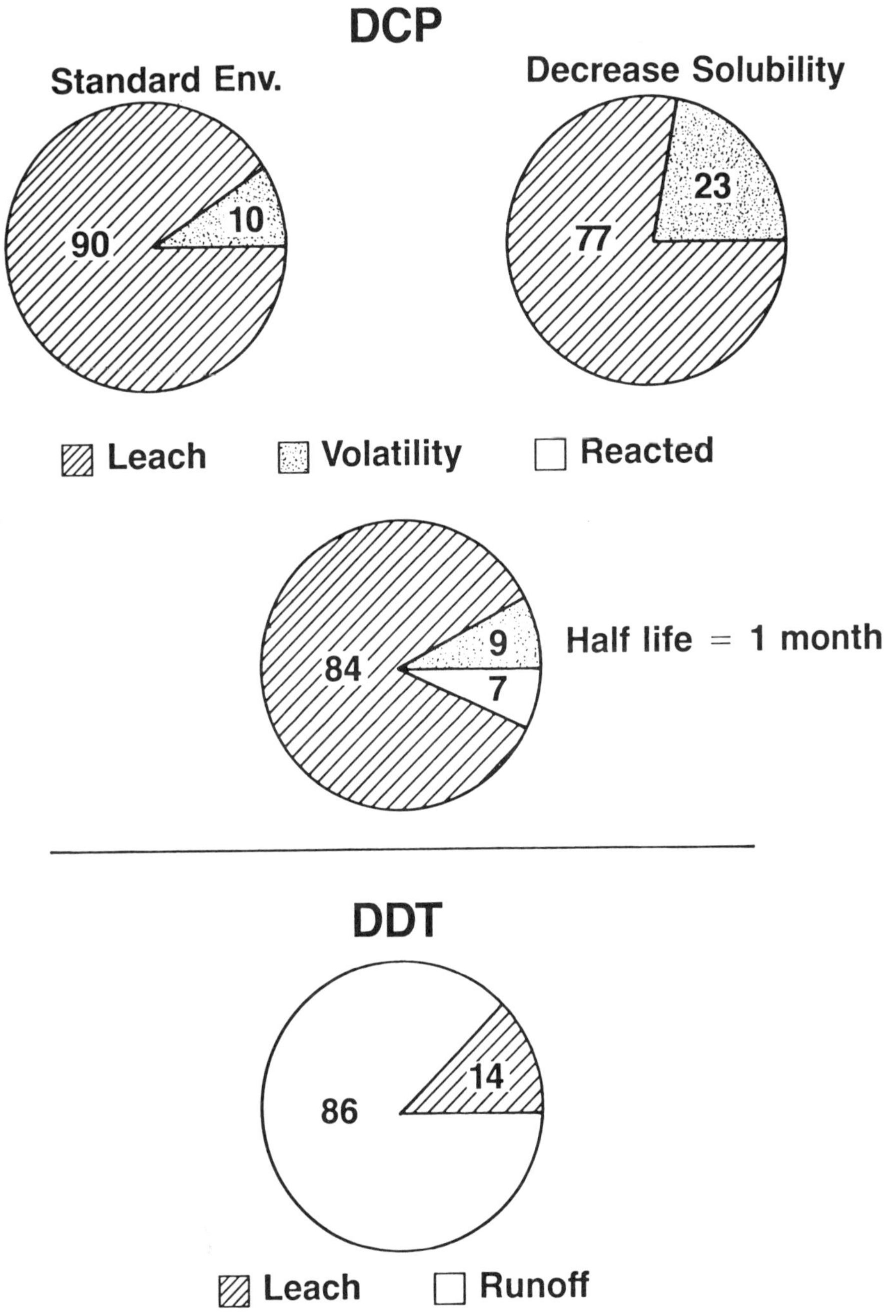

Figure 3. Piechart showing the results of using PEEC on the two chemicals: a.DCP and b.DDT.

SYMBOLS USED IN THE EQUATIONS

Subscripts i and j refer to the compartments and may take on values ranging from 1 to 3 where 1 = air, 2 = water and 3 = ground.

A_{ij} = Cross sectional area between compartments (m^2).
C_s = adsorbed chemical concentration (g/kg of soil)
C_w = chemical concentration in soil water (g/m^3 of soil water)
C_T = Total chemical concentration (g/m^3 of bulk soil)
d = average depth of soil (1/2 vertical depth).
D_{ij} = diffusion transfer coefficient between compartments (moles/Pa/day).
fr_2 = fraction of chemical in water - dimensionless
fr_3 = fraction of chemical in soil - dimensionless
f_i = fugacity (Pa) in the compartments
G_g = gas exchange coefficient assigned a value of 750 m/day (7).
I_i = input into the compartment (moles/day).
K_d = soil partition coefficient (g/ml)
K_{12} = Henry's constant (conc. in air/conc. in water)
K = overall partitioning coefficient between soil and the soil water.
K_{oc} = soil organic carbon partition coefficient (m^3 /kg)
k_{32} = mass transfer coefficient (m/day).
k_i = rate constant within the compartment (1/day).
k_s = exit to the stratosphere (1/day)
k_l = leaching rate constant (1/day).
K_{ow} = octanol water partition coefficient.
M_1 = diffusion coefficient in air (m^2/day)
M_1 = diffusion coefficient in soil air (m^2/day)
M_2 = diffusion coefficient in soil water (m^2/day).
M_2 = diffusion coefficient in water (m^2/day).
M^s = net diffusion coefficient in soil (m^2/day).
N_{ij} = Transfer constant in moles/day.
oc = fraction of organic carbon in the soil.
P = vapor pressure (Pa)
S = water solubility (moles/m^3)
R = gas constant 8.314 Pa m^3/mol K
Re_{ij} = Resistance between the compartments ($1/D_{ij}$).
T = absolute temperature 298 K
V_i = volume of compartment (m^3).
Z_i = fugacity capacity factor (moles/m^3/Pa)
Z_1 = fugacity capacity factor in air (1/RT)
Z_2 = fugacity capacity factor in water (S/P)
Z_3 = fugacity capacity factor in soil (K_d b Z_2)
φ = soil porosity
Θ = volumetric water content in soil (1 - α).
β = soil bulk density (g/m^3).
α = volumetric air content in soil.

REFERENCES

1. W.G. Knisel, CREAMS: A field-scale model for chemicals, runoff, and erosion from agricultural management systems. USDA Conservation Research Rep 26. US Government Printing Office, Washington DC (1980).
2. A.S. Donigian, J.C. Imhoff, B.R. Bicknell, and J.O.L. Kittle, Guide to the Application of the Hydrological Simulation Program - Fortran (HSPF). EPA Contract No. 68-01-6207 USEPA Athens, GA (1982).
3. W.A. Jury, W.J. Farmer, and W.F. Spencer, J. Environ. Qual., 12, 558 (1983).
4. W.A. Jury, W.J. Farmer, and W.F. Spencer, J. Environ. Qual., 13, 567 (1984).
5. W.A. Jury, W.F. Spencer, and W.J. Farmer, J. Environ. Qual., 13, 573 (1984).
6. W.A. Jury, W.F. Spencer, and W.J. Farmer, J. Environ. Qual., 13, 580 (1984).
7. D. Mackay, Environ. Sci. and Technol., 13, 1218 (1979).
8. W.B. Neely, and D. Mackay, Evaluation Model for Estimating Fate of Chemicals in "Modelling the Fate of Chemicals in the Aquatic Environment" eds. Dickson, K.L., Maki, A.W., and Cairns, J., Ann Arbor Science, Ann Arbor Mich. p. 127 (1982).
9. D. Mackay, and S. Paterson, Environ. Sci. and Technol., 15, 1006 (1982).
10. W.B. Neely, Environ. Toxicol. and Chem., 1, 259 (1982).
11. P.J. McCall, D.A. Laskowski, R.L. Swann, and H.J. Dishburger, Residue Reviews 85, 231 (1983).
12. W. Klopffer, G. Ripper, and R. Frische, Ecotoxicol. Environ. Safety, 6, 294 (1982).
13. D. Mackay, S. Paterson, B. Cheung and W.B. Neely, Chemosphere 14, 335 (1985).

14. W.B. Neely, Workshop on Multi Media Modelling, Los Angeles, CA Jan 22-24, 1986. Proceedings to be published by Plenun Press New York NY.
15. W.B. Neely, Abstracts American Chemical Soc., Division of Environmental Chem. 25, 175 (1985).
16. D. Mackay and S. Paterson, Environ. Sci. and Technol., 16, 654 (1982).
17. P.J. McCall, R.L. Swann, D.A. Laskowski, S.A. Vrona, S.M. Unger, and H.J. Dishburger, Prediction of Chemical Mobility in Soil form Sorption Coefficients in "Aquatic Toxicology and Hazard Assessment" ed. by Branson, D.R. and Dickson, K.L., ASTM p. 49 (1981).
18. V.T. Chow, "Handbook of Applied Hydrology" McGraw Hill Book Co., New York NY (1964).
19. W.A. Tucker and L.H. Nelker, Diffusion Coefficients in Air and Water in "Chemical Property Estimation Methods" eds. Lyman, W.J., Reehl, W.F., and Rosenblatt, D.H., McGraw Hill Book Co. New York NY Chap. 17 (1982).
20. S.W. Karickhoff, Pollutant Sorption in Environmental Systems in "Environmental Exposure From Chemicals", eds. Neely, W.B. and Blau, G.E., CRC Press, Boca Raton FL Volume I, p. 49 (1985).
21. W.L. Dilling, Environ. Sci. and Technol., 11, 405 (1977).
22. R.F. Spalding, G.A. Junk, J.J. Richard, Pesticide Monitoring J., 14, 70 (1980).
23. J.J. Richard, G.A. Junk, M.J. Avery, N.L. Nehring, J.S. Fritz, H.J. Svec, Pesticide Monitoring J., 9, 117 (1975).
24. S.Z. Cohen, S.M. Creeger, R.F. Carsel, and C.G. Enfield, Treatment and Disposal of Pesticide Wastes, to be published.

An aerosol smog chamber for testing abiotic degradation of compounds with low volatility

Wolfgang Behnke, Friedrich Nolting and Cornelius Zetzsch*

Fraunhofer-Institut für Toxikologie und Aerosolforschung
Nikolai-Fuchs-Str. 1, D 3000 Hannover 61, FRG

Abstract - An aerosol smog chamber facility is described, which is used to simulate the atmospheric, abiotic degradation of gaseous chemicals in the presence and absence of airborne aerosols,and also to simulate the degradation of condensed chemicals adsorbed on the aerosols. The aerosol materials SiO_2, TiO_2, Fe_2O_3, coal fly ash and NaCl are investigated in a photosmog of sun-irradiated NO_x/hydrocarbon mixtures at the ppb level. The production of OH radicals (and Cl atoms in the presence of NaCl) in the gas-phase are computed from the degradation pattern of the hydrocarbons, and a heterogeneous degradation of hydrocarbons on the surfaces of irradiated TiO_2 and Fe_2O_3 is detected. The kinetics for the consumption of a monolayer of the organic compound di(2-ethylhexyl)-phthalate on SiO_2 and Fe_2O_3 by reaction with gas-phase OH is studied in detail. The rate constant of this second order process, $1.4 \times 10^{-11} cm^3 s^{-1}$, yields a half-life for di(2-ethylhexyl)phthalate of about 1 day at the global average concentration of OH in the troposphere.

INTRODUCTION

Smog chamber studies are a generally accepted method to study the photochemical decay of organic compounds under simulated tropospheric conditions (ref. 1). Previous investigations have been limited to gases and vapours of rather low boiling compounds in the homogeneous gas-phase. Less volatile compounds have only been investigated in the bulk adsorbed on sand or on silica gel in static or rotating drum photoreactors (ref.2-6), where the active species of the gas phase can neither be monitored nor controlled. Hence there is a lack of data describing the photochemical degradation of less volatile compounds (e.g. plasticisers, pesticides or PCBs) under tropospheric conditions in the airborne state. In this work experiments in a smog chamber are performed in the presence of inorganic airborne aerosols, and results are reported using these materials as carrier for the plasticiser di(2-ethylhexyl)phthalate (DEHP).

EXPERIMENTAL

The experiments are performed in a cylindrical reactor made of pyrex glass (1 m in diameter, 3 m high) placed in an air conditioned laboratory. The lid is kept at room temperature, and a small temperature gradient of - 3 K ensures mixing of the gases and optimizes the residence time of the airborne aerosols. In the dark residence times up to two days are obtained at optimized particle diameter (0.5 µm), under illumination up to one day (ref. 7). The irradiation spectrum is adjusted to match the spectral distribution of the sun in the photoactinic region. It corresponds to the threefold intensity of a European summer day. The solar spectrum of a hazy day is simulated by filtering the light by acryl glass (ref. 8).

The aerosol is generated by atomizing a suspension of water containing the aerosols (or a solution in the case of NaCl). The droplets are dried in a flow of purified air, yielding particles of 0.5 µm mean diameter. The aerosol mass concentration is around 3 mg/m^3, corresponding to a number density of 10^5 cm^{-3}. The following materials are used for the production of aerosols:

1) SiO_2 (Aerosil 200, Degussa)
2) TiO_2 (Anatase, Bayer)
3) Fe_2O_3 (Sicrotram Orange, BASF)
4) Coal fly ash
5) NaCl

In some experiments the aerosol materials are coated with DEHP prior to atomizing: Either a solution of DEHP in Freon 113, containing the aerosol material, is dried and then suspended in water or the dry aerosol (coal fly ash and NaCl) is coated by vapour phase deposition. A photosmog is produced in the presence of the aerosol and the hydrocarbons at initial concentrations of 100 ppb NO and 200 ppb NO_2 by irradiation with the simulated sunlight. Up to 25 compounds are injected simultaneously at 50 ppb each and are monitored every 20 minutes, using capillary gas chromatography from 20 ml samples down to a level of 1 ppb. Two hours after the beginning of the experiment NO is injected at 60 ppb h^{-1} continuously in order to keep the level of photochemically produced O_3 below 1 ppm. The OH radical concentration in the gas-phase is computed from the decay rates of hydrocarbons, ranging from propane to nonane, using known rate constants.

DECAY OF SUBSTANCES WITH HIGH AND MEDIUM VAPOUR PRESSURE

In the presence of SiO_2 aerosol no significant differences of the photochemical reactions are observed in comparison to pure gas phase. During these experiments OH-levels of about 10^7 cm^{-3} are observed. This inert behaviour of SiO_2 enables us to expose almost non-volatile compounds (adsorbed on the aerosol) to the active constituents of a photosmog, containing OH and O_3. By this method the abiotic degradability of such chemical compounds in the atmosphere can be tested (see below), similar to the common smog chamber technique for gaseous compounds.

In the presence of TiO_2 aerosol under similar conditions we observe a strong increase of OH up to 10^8 cm^{-3} during the experiment. This behaviour of TiO_2 may be typical of aerosols with semiconductor properties, and the impact of photoactivated aerosols on tropospheric chemistry (ref. 8) should be investigated further with various materials.

In addition, in the presence of TiO_2 aerosol we observe rapid consumption of hydrocarbons (ref. 9), especially for those with a low vapour pressure. To suppress the gas-phase degradation of the hydrocarbons by reaction with OH, we made experiments in the absence of NO_x and with a UV cut-off filter at 360 nm (acryl glass). The additional loss process can then be observed separately. Its strength is proportional to the inverse of the vapour pressure of various compounds. Branched chain alkanes, cycloalkanes, and aromatic hydrocarbons fit into the same scheme as n-alkanes (ref. 8), indicating a heterogeneous process.

The heterogeneous photodegradation in the presence of TiO_2 can be explained from its UV absorbing and semiconducting properties, forming radicals on the surface according to the following mechanism:

$$TiO_2 + h\nu\ (\lambda < 400\ nm) \rightarrow TiO_2\ (h^+ + e^-)$$
$$O_2\ (g) + e^- \rightarrow O_2^-\ (ads)$$
$$O_2^-\ (ads) + H^+ \rightarrow HO_2\ (ads)$$
$$h^+ + OH^- \rightarrow OH\ (ads)$$

The production of gas-phase OH in the presence of NO_x might be explained by the reaction:

$$HO_2\ (ads) + NO\ (g) \rightarrow NO_2\ (ads) + OH\ (g)$$

In a similar way a slow heterogeneous degradation occurs in the presence of Fe_2O_3 aerosols. It can only be detected for the aromatics ethylbenzene, p-xylene and o-xylene. Here the heterogeneous decay of o-xylene, the substance with the smallest vapour pressure of our series, is observed to correspond to the consumption by an OH-concentration of 5 x 10^5 cm^{-3}, whereas in the presence of TiO_2 the heterogenous decay of the more volatile n-octane would already correspond to the consumption by OH at 10^8 cm^{-3}.

In the presence of fly ash aerosol from a coal power plant no hint of a heterogeneous photodegradation could be detected.

In the presence of NaCl aerosol the decay pattern of the hydrocarbons can be explained by the simultaneous presence of OH radicals and Cl atoms (ref. 10). The production of Cl atoms may be an important process during the well-known conversion of sea spray (of chloride to nitrate) in the presence of NO_x in the atmosphere (ref. 11). The high reactivity of Cl atoms may facilitate a significant contribution of Cl to tropospheric chemistry, especially for alkanes.

DECAY OF SUBSTANCES WITH LOW VAPOUR PRESSURE

A new area is the investigation of chemical substances with low vapour pressure, since these substances can only be degraded in a heterogeneous way. We chose di(2-ethylhexyl)-phthalate (DEHP) as a model for other substances with high boiling point. DEHP was adsorbed as a monolayer on the aerosol. As to be expected from the degradation of substances with medium vapour pressure in the presence of TiO_2, DEHP adsorbed on TiO_2 was observed to have a very short half-life (less than ten minutes).

Considering the dominating presence of SiO_2- and Fe_2O_3-aerosols in the environment, the decay processes on these aerosols may be of higher relevance to natural conditions. Figure 1 shows the decay of DEHP (adsorbed as a monolayer on airborne SiO_2 aerosol) as a function of time in a semilogarithmic diagram. In the presence of gas-phase OH the portion of DEHP on the aerosol filter samples decreases much faster than the total aerosol mass concentration. Both decays can be described by single exponentials, thus defining decay rates, τ. The decay rate of DEHP, generated in the photosmog, is observed to be independent of the concentration of O_3 (checked at up to 1 ppm).

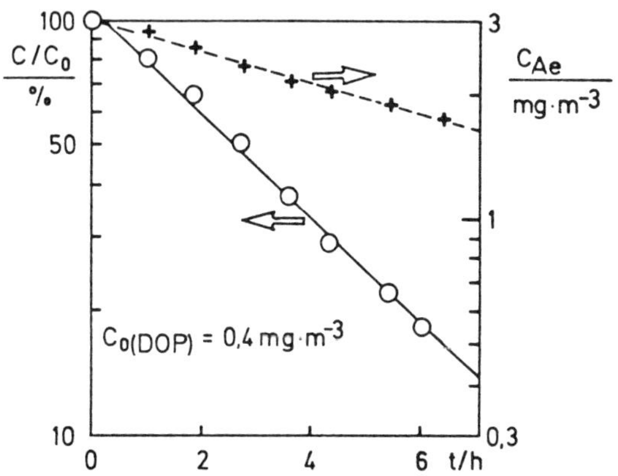

Fig.1 Decay of DEHP (initially adsorbed as an approximate monolayer on SiO_2 aerosol) in the presence of gas-phase OH at $5x10^6 cm^{-3}$ (left hand scale) and decay of the total aerosol mass density (right hand scale).

Figure 2 shows the dependence of the decay rate, τ^{-1}, of DEHP on the concentration of OH radicals from various experiments. The decay rate of DEHP on Fe_2O_3 was measured with a UV cut-off filter at 360 nm. The decay rates show a good correlation with the concentration of gas-phase-OH on SiO_2 aerosol as well as on Fe_2O_3 aerosol. From the slopes of the lines in Fig. 2 a rate constant of $1.4 \times 10^{-11} cm^3 s^{-1}$ can be deduced for the heterogeneous reaction of OH with DEHP. Considering the concentration of OH in the troposphere ($5 \times 10^5 cm^3$ as a global average), the half-life of DEHP in the atmosphere is determined to be less than 2 days.

On the other hand, it is necessary to interprete the (almost imperceptible) decay in the absence of OH radicals, considering an OH concentration of only $5x10^5 cm^{-3}$ in the environment. Hence various experiments were performed with a UV cut-off filter and in the dark with up to 20 layers of DEHP adsorbed on the aerosol. Since the decay rate of DEHP increases with the thickness of the adsorbed layer, the decay of DEHP observed on SiO_2 in the absence of OH ($\tau^{-1} = 2x10^{-5} s^{-1}$) is interpreted to be due to desorption. The decay of DEHP, adsorbed on Fe_2O_3, shows an additional, slight contribution of hydrolysis.

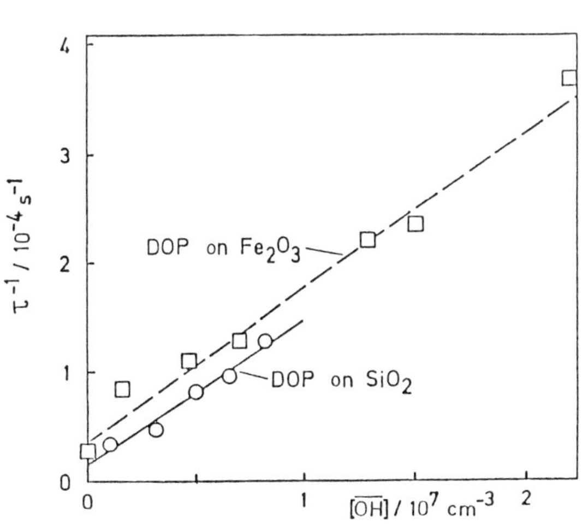

Fig. 2. Decay rate of DEHP (a well-known plasticiser, adsorbed as a monolayer on the carrier-aerosols SiO_2 and Fe_2O_3) as a function of the concentration of OH in the smog chamber. The slope yields a rate constant of $1.4x10^{-11} cm^3 s^{-1}$ for the reaction of OH with DEHP.

CONCLUSIONS

The long residence times of aerosols achieved in an optimized smog chamber enable us to study atmospheric chemistry in the presence of airborne aerosols. The influence of aerosols on the degradation of gaseous compounds and the degradation of particle-borne organic compounds by reaction with gaseous active species can be investigated in detail.

An impact of aerosols on the chemistry of gaseous compounds in the atmosphere is observed with a semiconductor aerosol, TiO_2, when photoactivated by the sunlight. An enhanced concentration of OH and a heterogeneous photodegradation of hydrocarbons are observed, which may significantly contribute to the degradation of substances with low and medium vapour pressure in the troposphere. An almost imperceptible heterogeneous degradation is detected for aromatics in the presence of Fe_2O_3 aerosol.

Sodium chloride aerosol is photochemically decomposed in a photosmog containing NO_x, and the formation of Cl atoms is observed. The contribution of Cl atoms (from sea spray) to the chemistry of the troposphere may be significant and competing with other active species like OH, O_3, NO_3, HO_2 and others.

Silicon dioxide aerosol can be used as an inert carrier, to study the photodegradation of almost unvolatile compounds by reaction with gas-phase OH. This abiotic degradation process (with a well-known, simple reaction mechanism) can easily be converted to environmental tropospheric half-life .

The results obtained for di(2-ethylhexyl)phthalate, a plasticiser, open a perspective on the testing of pesticides for abiotic degradation in the airborne state.

ACKNOWLEDGEMENT

Support of this study by the Bundesminister für Forschung und Technologie and by the Verband der Chemischen Industrie (companies BASF, Bayer, Henkel, Hoechst and Hüls) is gratefully acknowledged.

REFERENCES

1. R. Atkinson, Chem. Rev. 85, 69 - 201 (1985).
2. S. Gäb, H. Parlar, S. Nitz and F. Korte, Chemosphere 4, 251-256 (1975).
3. S. Gäb, J. Schmitzer, H.-W. Thamm, H. Parlar and F. Korte, Nature (London), 270, 331-333 (1977).
4. L. Blau and H. Güsten, Polynuclear Aromatic Hydrocarbons: Physical and Biochemical Chemistry, M. Cooke, A.J. Dennis and G.L. Fischer, eds.,pp. 133 - 144, Battelle Press, Columbus (Ohio) and Springer Verlag, Berlin, (1982)
5. H. Güsten, Chemistry of Multiphase Atmospheric Systems, W. Jaeschke, ed., pp. 567 - 592, Springer, Springer, Heidelberg, (1986).
6. H. Güsten, Z. Bozicevic and L. Klasinc, Staub Reinhalt. Luft 44, 448 - 451 (1984).
7. W. Holländer, W. Behnke, W. Koch and G. Pohlmann, Proc. Third European Sympos. on Physico-Chemical Behaviour of Atmospheric of Pollutants, B. Versino and G. Angeletti, eds.,pp. 309 - 319, Reidel Publ. Co., Dordrecht (1984).
8. W. Behnke, F. Nolting and C. Zetzsch, J. Aerosol Sci. (in press).
9. W. Behnke, W. Koch, F. Nolting and C. Zetzsch, Air Pollution and Plants, C. Troyanowsky, ed., pp. 97-103, VCH-Verlagsgesellschaft, Weinheim (1985).
10. W. Behnke, F. Nolting and C. Zetzsch, Abstr. XVIth Informal Conf. on Photochemistry, J.G. Anderson, ed., paper T-21, Harvard University (1984).
11. C.S. Martens, J.Rech. Atmos. 8, 988-991 (1974).

The uptake of non-extractable soil-bound pesticide residues by roots—standardized experiments with four pesticides

R. Kloskowski*, F. Führ and W. Mittelstaedt

Institute of Radioagronomy, Nuclear Research Centre Jülich
D 5170 Jülich, Federal Republic of Germany

Abstract - [carbonyl-^{14}C] methabenzthiazuron, [3-^{14}C] metamitron, [ring-14C] simazine, [benzene ring-U-14C] and [triazine ring-U-14C] anilazine were exhaustively extracted with various solvents after aging for 100-190 days in a degraded loess soil, either at 22°C and 65 % of the maximum water-holding capacity of the soil or under field conditions in a lysimeter with vegetation. Between 12 and 83 % of the applied radioactivity remained in the soil as non-extractable residues. The soils were either used directly or thoroughly mixed with fresh untreated soil in the ratio of 1:1 and planted with maize or sugar beets in special pots. During the four-week plant experiment, the plants took up between 0.03 % (simazine) and 0.78 % (anilazine) of the applied radioactivity. In a parallel experiment, the radioactive pesticides were applied to the plants immediately before starting the experiment. The uptake rates indicated that the biological availability of the non-extractable residues was reduced by a factor of 4-8 (anilazine, methabenzthiazuron) up to 675 (simazine) in comparison to freshly applied active substance. Mere aging of the active substances in the soil led to a four-fold decrease in their bioavailability for the plant roots, and the uptake of metamitron residues aged under field conditions was greatly reduced (by a factor of more than 1000).
Between 60 and 80 % of the non-extractable radioactivity was found in the stable carbon (humin) fraction of the soil, whereas the fulvic acids contained 10-20 % and the humic acids contained 8-17 % of this radioactivity.

INTRODUCTION

Pesticide active substances form bound residues in the soil which can range from 7 % (ref. 1) to 90 % of the quantity applied (ref. 2). The structure of the bound compounds still remains largely unexplained since there has been a lack of appropriate methods to study the incorporation of the foreign chemicals into the soil's own organic structures and to extract the artefact-free residues from the soil for analytical determination. The results published to date on characterizing the non-extractable residues in soil (ref. 3,4,5,6) provide indications that part of these residues may represent the unchanged initial molecule.

In order to evaluate the significance of the non-extractable bound residues, pragmatic experiments on the bioavailability of these residues for plants and microorganisms were carried out under laboratory conditions with 14C-labelled compounds. The results have been compiled in several review articles (ref. 7, 8,9,10). The uptake of radioactivity from bound residue fractions has ranged from 0.03 to 2.6 %, depending on the active substance and soil used and the experimental system. Due to the differences in the experimental approaches it is possible to make a comparative evaluation of the results obtained only to a limited extent since different active substances and quantities thereof, soil mixtures, plant species, experimental duration and conditions were selected. Führ (ref. 11) proposed a standardized experimental approach specifically for the study of the bioavailability of non-extractable pesticide residues for plant roots. In this experimental approach, the bioavailability of the freshly added, aged and non-extractable residues of four 14C-labelled active substances was studied in parallel for plant roots and microorganisms.

MATERIAL AND METHODS

Active substances: The four active substances studied are shown in Table 1.

TABLE 1. Chemical designation, structural formula and labelling position of the active substances used

Chemical designation	Structural formula and labelling position (*)	Application
Methabenzthiazuron 1,3-di-methyl-3-(benz-thiazol-2yl)-urea	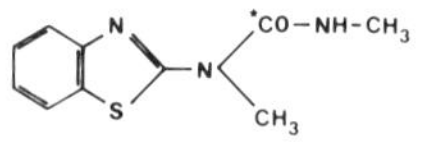	herbicide (cereal cultivation)
Metamitron 3-methyl-4-amino-6-phenyl-1,2,4-triazine-H-on	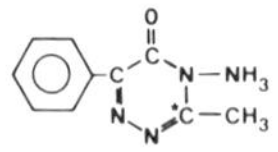	herbicide (sugar beet cultivation)
Simazine 2-chloro-4,6-bis-ethylamino-s-triazine	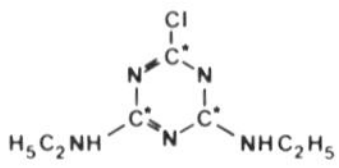	herbicide (maize cultivation)
Anilazine 4,6-dichloro-N-(2-chlorphenyl) 1,3,5-triazine-2-amine	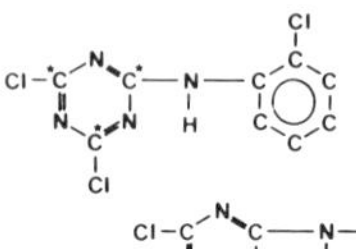 [T] 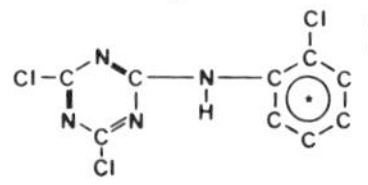 [B]	leaf fungicide (cereal cultivation)

Soil: The degraded loess soil (alfisol) from Merzenhausen near Jülich represents the most widespread, largely agriculturally used type of soil ("Parabraunerde") in the Federal Republic of Germany (ref. 12). The essential soil data are: 1.1 % organic C; 0.11 % N; 9.9 % clay; 84.6 % silt; 5.5 % sand; pH(KCl) 6.8; exchange capacity 8.5 mval/100 g.

Microecosystem: The plant experiments were carried out in specially constructed experimental pots that contained 1.5 kg of soil and permitted both separation of the shoot or leaf atmosphere from the root atmosphere and continuous aeration of the soil (ref. 13,14). The CO2 from the soil and root respiration, and the $^{14}CO_2$ from the degradation of labelled active substances were sorbed in 1N NaOH, and volatile organic compounds were trapped in further absorber systems. In each case, four test plants (maize or sugar beet) were inserted at the 2-leaf stage into openings in the pot lids which were sealed with permanently plastic putty (Prestik).

Production of aged and non-extractable bound residues: In order to obtain both aged and bound residues, an aging experiment was carried out before each plant experiment in accordance with the regulations of the Federal Biological Institute (ref. 15). The active substances (simazine and anilazine) were applied in quantities usual in practice and were aged for 100 days in the laboratory at 22°C and 65 % maximum water-holding capacity of the soil. In the case of the active substances methabenzthiazuron and metamitron, residues aged in a field lysimeter experiment were already available (ref. 16). The 0-5 cm soil layer was removed in each case after harvesting the treated crops. Winter wheat was harvested 140 days after spraying with methabenzthiazuron, sugar beet was harvested 190 days after spraying with metamitron. The bound residues in the harvested soils were exhaustively extracted with various solvents (Table 2).

Experimental approach: The organization of the experimental approach was adapted to the respective active substance and somewhat modified in the course of time taking new findings into consideration. However, the modifications were not fundamental so that in general three different variants were selected for each of the four active substances:

1. Soil with freshly added active substance
2. Soil with aged residues
3. Soil with non-extractable, bound residues

In the case of the anilazine experiment, the extracted soil of variant 3 was mixed with fresh untreated soil in the ratio of 1:1. The root uptake from the fraction of bound residues was compared with the corresponding uptake from the fraction of aged and freshly added active substances. This standardized experiment accordingly allowed assessment of the relative bioavailability of the non-extractable bound residues. The experiments lasted 3 or 4 weeks and were conducted in an isotope greenhouse at approx. 25°C with additional illumination for 14 hours (ref. 17,18).

RESULTS AND DISCUSSION

Depending on the active substance, duration and conditions of incubation, between 0.1 and 50 % of the applied radioactivity was mineralized to CO_2 or recorded as a loss in the field lysimeter experiment (Table 2). From 12 to 83 % of the radioactivity applied remained in the soil after exhaustive extraction (Table 2).

TABLE 2. Formation of bound residues in the soil
Applied radioactivity = 100 %

Active substance	Conditions	Residence time in the soil d	Degradation ($^{14}CO_2$) or losses in the field %	Solvent used for extraction	Non-extractable (bound) residues %
[carbonyl-^{14}C] methabenzthiazuron	lysimeter, winter wheat, 0–2.5 cm layer	140	50	H_2O/acetone ethyl acetate chloroform	20.8
[3-^{14}C] metamitron	lysimeter, sugar beet, 0–2.5 cm layer	190	50	H_2O/acetone ethyl acetate chloroform	12.2
[ring-^{14}C] simazine	laboratory 22°C, 65 % of the maximum water holding capacity	108	0.95	0.2 M ascorbic acid/ 0.2 M citric acid/ acetone (1:1:2)	35.1
[benzene ring-U-^{14}C] anilazine	laboratory 22°C, 65 % of the maximum water holding capacity	100	0.1		82.6
[triazine ring-U-^{14}C] anilazine		100	0.1	acetone/ 0.01 M $CaCl_2$/ methanol	77.9

UPTAKE OF RADIOACTIVITY

In spite of the different quantities of bound residues formed in the soil during aging, their total uptake by the test plants was always less than 1 % of the radioactivity applied (Table 3). The radioactivity taken up by the roots was, however, distributed differently in the plants depending on the active substance. In the case of methabenzthiazuron residues, there was increased transport into the leaves. The anilazine residues taken up were distributed uniformly over the shoot and roots. The different transport and distribution patterns also influence the significance of the residues taken up because, depending on the fate of the plant parts after harvesting, a further intensive conversion of the incorporated residues can take place

TABLE 3. Total uptake of radioactivity from the fraction of non-extractable bound residues
Radioactivity originally applied to the soil = 100 %

Active substance	Test plant	Experimental duration	Mineralization ($^{14}CO_2$)	Uptake by the plants leaves	roots	total uptake	Desorption ($CaCl_2$) after the experiment
		weeks	%	%	%	%	%
[carbonyl-^{14}C] methabenzthiazuron	maize	4	0.2	0.24	0.12	0.36	2.2
[3-^{14}C] metamitron	sugar beet	3	0.6	0.02	0.01	0.03	1.1
[ring-^{14}C] simazine	maize	4	0.01	0.01	0.02	0.03	0.5
[benzene ring-U-^{14}C] anilazine	maize	4	0.17	0.34	0.44	0.78	6.4
[triazine-U-^{14}C] anilazine	maize	4	0.08	0.09	0.34	0.45	6.3

(ref. 19). It must be emphasized once again that these microecosystem experiments can only provide data on the relative biological availability of non-extractable, bound residues. It must be remembered that in microecosystems the ratio of roots to soil is much closer than in field soil so that uptake rates up to 50 times higher can result in such microecosystems (ref. 20). The uptake from the fraction of bound metamitron residues was reduced by a factor of approximately 560 in comparison to the uptake of freshly applied active substance (Table 4). However, mere aging of metamitron residues in the soil reduced the bioavailability to an even greater extent (ref. 18). This indicates that the bonding of this active substance in the soil can be altered by extraction with various solvents.

In the case of simazine the root uptake of non-extractable radiocarbon was reduced by 675 times; this extremely low uptake can in part be attributed to inhibited growth of the test plants on the exhaustively extracted soil (ref. 21). As in the case of metamitron, aging of simazine residues in the soil also reduced the uptake considerably (by a factor of about 340) (Table 4). The radioactivity taken up by plants from the soil with aged or bound methabenzthiazuron and anilazine residues was less by factors of 4 to 9 than uptake from the fraction of freshly added active substances.

TABLE 4. Comparison of the root uptake from the fractions of freshly added, aged and non-extractable residues in the soil
Radioactivity originally applied to the soil = 100 %

Active substance	Uptake of radioactivity: from the fraction:			Relative uptake ratio
	fresh %	aged %	bound %	fresh/aged/bound
[carbonyl-^{14}C] methabenzthiazuron	1)	1.44	0.36	–/4/1
[3-^{14}C] metamitron	16.8	0.01	0.03	559/0.3/1
[ring-^{14}C] simazine	20.28	10.17	0.03	675/338/1
[benzene ring-U-^{14}C] anilazine	3.14	0.60	0.78	4/0.8/1
[triazine-U-^{14}C] anilazine	3.96	0.53	0.45	8.8/1.2/1

1) Not investigated

Relatively small, but nevertheless significant, quantities of ^{14}C-carbon were converted to $^{14}CO_2$ from the fraction of bound residues during the plant experiments. Depending on the active substance, the mineralization amounted to from less than 0.1 % to 0.6 % of the radioactivity applied (Table 3).

At the end of the plant experiments the quantity of water-soluble ^{14}C-labelled material in the experimental soil was determined by a multiple-stage desorption with a simulated soil solution (0.01M $CaCl_2$). It became apparent that part of the previously non-extractable residues can be retransferred into the soil solution due to renewed cultivation of the soil. However, these residues were not available to the root and microorganisms in the same order of magnitude (Table 3), so that in this experimental approach water solubility does not provide an index of possible bioavailability. It is to be expected that extraction of a soil with organic solvents (Table 2) will affect its structure and reduce its biological activity (ref. 21). Based on the fact that only the soil solution and rain water can be taken into consideration as solvent, extraction and transport media in the field, it has been suggested (ref. 10) that the soil should be extracted only with water or a simulated soil solution (0.01M $CaCl_2$) for experiments on the bioavailability of bound pesticide residues and that the fraction not so extracted should be defined as "bound". This experimental variant was also included in the experiments with methabenzthiazuron and metamitron. After the extraction of the methabenzthiazuron residues aged in a lysimeter under field conditions using organic solvents, 20.8 % of the applied radioactivity remained in the soil (Table 5). This fraction increased to 30.7 % after extraction with 0.01M $CaCl_2$ solution had been repeated five times. Mineralization and root uptake by the plants during the 28-day experiment were increased two-fold. The total uptake of radioactivity rose from 0.4 to 0.9 % in comparison to the variant in which extraction was carried out with various solvents. In the case of the metamitron, approx. 12.5 % of the applied radioactivity remained in the 0-2.5 cm soil layer in the form of non-extractable residues, irrespective of the extraction method used (Table 5). The bioavailability of these residues for plants and microorganisms was slightly reduced after $CaCl_2$ extraction so that the total uptake was less than 0.01 %.

TABLE 5. Non-extractable, bound [carbonyl-^{14}C] methabenzthiazuron and [3-^{14}C] metamitron residues in the soil: comparison of various extraction methods as well as mineralization and uptake by plants
Radioactivity sprayed to winter wheat or sugar beet = 100 %

Active substance	Soil layer in the lysimeter	Non-extractable residues in the soil after extraction with:		Plant experiment	
				mineralization ($^{14}CO_2$)	uptake by plants
	cm		%	%	%
[carbonyl-^{14}C] methabenzthiazuron (winter wheat)	0–2.5	water/acetone ethyl acetate chloroform	20.8	0.2	0.4
		0.01 M $CaCl_2$	30.7	0.6	0.9
[3-^{14}C] metamitron (sugar beet)	0+2.5	water/acetone ethyl acetate chloroform	12.3	0.6	0.03
		0.01 M $CaCl_2$	12.7	0.4	0.01
	2.5–5.0	water/acetone ethyl acetate chloroform	7.6	0.6	0.01
		0.01 M $CaCl_2$	7.5	0.4	0.00(3)

INCORPORATION OF NON-EXTRACTABLE BOUND RESIDUES INTO SOIL ORGANIC MATTER FRACTIONS

The soils with the bound pesticide residues were extracted once again after the plant experiment with the organic solvents listed in Table 2. The soil was subsequently extracted with 0.1M $Na_4P_2O_7$ and the extract was separated into fulvic and humic acids by acidification. With all four active substances studied, the fulvic acid fraction contained less than 10 % of the applied radioactivity (Table 6). Only in the case of methabenzthiazuron could the unaltered active substance be determined in the organic phase of the aqueous fulvic acid solution (ref. 6). With the other compounds, the bound radioactivity thus extracted could not be characterized.

TABLE 6. Distribution of the bound ^{14}C residues in the soil after the plant experiment and repeated extraction with organic solvents
1) Applied radioactivity = 100 % 2) Bound radioactivity = 100 %

Active substance	Fulvic acids % 1)	Fulvic acids % 2)	Humic acids % 1)	Humic acids % 2)	Soil residues (humin) % 1)	Soil residues (humin) % 2)
[carbonyl-^{14}C] methabenzthiazuron	2.8	19.7	2.4	16.9	9.0	63.4
[3-^{14}C] metamitron	2.3	21.9	1.1	10.5	7.1	67.6
[ring-^{14}C] simazine	6.3	24.7	4.5	17.6	14.7	57.7
[benzene-ring-U-^{14}C] anilazine	9.7	11.9	6.6	8.1	65.4	80.0
[triazine-U-^{14}C] anilazine	9.6	11.7	6.6	8.0	65.9	80.3

Depending on the active substance, the humic acid fractions isolated contained between 1.1 and 6.6 % of the applied radioactivity, whereas 7-66 % of this radioactivity was incorporated into the humin fraction. With all four active substances examined, the humin fraction is involved to the greatest extent in binding the residues. This fraction is only subjected to slight changes (ref. 22,23) so it is to be expected that the bound residues will be included only to a slight extent in further conversion processes of the organic soil phase. The availability of bound residues for plants and for further microbial degradation is thus also reduced as shown by the example of the active substances studied. Further research approaches should aim at the fate of the bound pesticide residues in this fractions of the soil humus in order to thus obtain, if possible, more detailed information on the long-term behaviour of the non-extractable bound residues.

ACKNOWLEDGEMENTS

This study was supported by the Deutsche Forschungsgemeinschaft.

REFERENCES

1. Ch.S. Helling and A.E. Krivonak, J. Agric. Food Chem., 26, 1156-1163 (1978).
2. R. Viswanathan, J. Scheunert, J. Kohli, W. Klein and F. Korte, J. Environ. Sci. Health, 13, 243-259 (1978).
3. S.U. Khan, J. Agric. Food Chem., 28, 1096-1098 (1980).
4. B.L. Worobey and G.R.B. Webster, J.Agric. Food Chem., 30, 164-169 (1982).
5. P. Capriel, A. Haisch and S.U.Khan, J. Agric. Food Chem., 33, 567-569 (1985).
6. F. Führ and W. Mittelstaedt, J. Agric. Food Chem. ,28, 122-125 (1980).
7. E.P. Lichtenstein, Residue Rev., 76, 147-153 (1980).
8. S.U. Khan, Residue Rev.,84, 1-25 (1982).
9. T.R. Roberts, Pure and Appl. Chem.,56, 945-956 (1984).
10. F. Führ, R. Kloskowski and P. Burauel, Berichte über Landwirtschaft, SH 198, 106-116 (1985).
11. F. Führ, ACS Symp. Ser., 29, 356-357 (1976).
12. E. Mückenhausen, Entstehung, Eigenschaften und Systematik der Böden der Bundesrepublik Deutschland, DLG-Verlag Frankfurt/M (1977).
13. W. Mittelstaedt and F. Führ, Pflanzenschutz-Nachr. Bayer, 28, 353-358 (1975).
14. K. Seibert and F. Führ, Z. Pflanzenernähr u. Bodenkde, 145, 505-509 (1982).
15. W. Weinmann and K. Schinkel, BBA-Merkblatt Nr. 36, Braunschweig (1976).
16. R. Kubiak, Diss. Univ. Bonn, Jül-Ber. Nr. 2055 (1986).
17. R. Kloskowski, F. Führ and W. Mittelstaedt, J. Env. Sci. and Health, Part B (in press).
18. R. Kloskowski and F. Führ, J. Env. Sci. and Health, Part B (in press).
19. H.H. Cheng, F. Führ and W. Mittelstaedt, Env. Qual. and Safety, Suppl. Vol. 3, 271-276 (1974).
20. F. Führ and W. Mittelstaedt, In: Pesticide Chemistry: Human Welfare and the Environment. J. Miyamoto and P.C. Kearney (Eds.), Pergamon Press, Oxford, 183-188 (1983).
21. R. Kloskowski and F. Führ, Chemosphere, 12, 1557-1574 (1983).
22. D.S. Jenkinson, Soil Sci., 111, 64-70 (1971).
23. J.P. Martin, H. Zunino, P. Peirano, M. Caiozzi and K. Haider, Soil Biol. Biochem., 14, 289-293 (1982).

Fate of pesticides in soil

Winfried K. Ebing

Department for Ecological Chemistry, Federal Biological Research Center, Koenigin-Luise-Str. 19, D-1000 Berlin 33, Fed. Rep. Germany

Abstract- Suggestions have been made for improving our knowledge about long-term fate of pesticides in soil. Quantitative data are needed to fulfil the small pool of informations about pesticides residence in soil including bound residues, metabolism and translocation. The most suitable ways and methods are cited from the literature as well as the possible conclusions from data obtained by laboratory tests and experimental ecosystem models are discussed.

INTRODUCTION

Pesticides are designed to be applied mainly on terrestrial areas. Thus, they enter soil either by direct application or by washing off plant surfaces during rainfall. Apart from volatilization, this is the main transport route for the major part of formulations. Depending on the phenotype and the density of the plant culture, generally 35 - 50 % of the plant protection material is deposited on soil, immediately after spraying. Sooner or later, in addition to the original active ingredient, there are also metabolites present in the soil.

DEFINITIONS

I should like to define e c o c h e m i c a l s t u d i e s as those which lead to information about the fate of the environmental chemical itself. That means its distribution, its transportation and its conversion to other chemical structures. E c o t o x i c o l o g i c a l s t u - d i e s differ in that they are focussed on the chemical's effects on biota.

RESIDUE BEHAVIOR

The ecochemical behaviour of pesticides in soil involves residence, movement and metabolism. These aspects contribute to the general subject of residue behaviour. The formation of residues is mainly restricted by
(a) Microbial metabolism
(b) Metabolism by other organo-chemical systems, such as higher soil organisms and plants, including the formation of soluble conjugates
(c) Degradation by photolysis
(d) Other abiotic processes such as hydrolysis and oxidation
(e) Volatilization and co-distillation characteristics.

The formation of residues is promoted by
(a) Low water solubility
(b) Binding capacity of inorganic soil constituents, that means mainly sorption by clay minerals
(c) Binding capacity of organic soil constituents such as fulvic acids, humins etc.

BOUND RESIDUES

These latter processes (b) and (c) can lead to formation of non soluble, so-called bound residues where pesticide molecular structures are linked to insoluble, high-molecular material of natural origin. Only small portions of these bound residues may liberate active substances in the course of time, but the percentages of bound residues formed in soil are high, frequently near 50 %.

Until now, only about 40 pesticides have been examined with respect to the quantification of their bound residue in soil. In the future, several activities should contribute to enlarging our knowledge to produce an information base of a reasonable number of pesticides. Radiotracer techniques are suitable for such measurements. Biological, enzymatic (ref. 1) and physical (ref. 2) disintegration methods should be developed further in order to improve our understanding of the importance of bound residues for the agricultural practice. Further research work on humin molecular structure, as well as on interactions and bond types between pesticide molecules and humins or minerals may be helpful in this context. The main criterion remains - according to the IUPAC Pesticide Commission report - "the bioavailability to plants and to soil organisms, the persistence and the mobility in soil" (ref. 3).

BIOAVAILABILITY OF RESIDUES

Basic knowledge about several of these topics has been summarized by EWRS (ref. 4). Today, the bioavailability of residues in soil may be examined by a set of test methods (ref. 5, as well as other workers). Within the same group of the European Weed Research Society, a prognosis system has been developed (ref. 6) to predict bioavailability of residues in soil to rotational crops. This involves multiple residue degradation experiments in the laboratory and glasshouse, which are then mathematically related to soil characteristic features and climate data of the outdoor area under consideration. Calculations are made using the mathematical simulation model of Walker & Barnes (ref. 7). The theoretical plant availability of a pesticide residue in the soil is based on its distribution between the portion adsorbed and the portion diluted in soil solution. Predicted bioavailability results are obtained by hydroponic trials with the culture plant of interest exposed to concentrations corresponding with the theoretical plant availability (ref. 8). It should be a rewarding task to examine the range of applicability of these procedures which are intended for practical use on farms. Here like elsewhere, one should have in mind that this method involves several simplifications.

Tests used for decisions in governmental regulation systems make more simplifying assumptions.

BASIC LABORATORY TESTS AND FIELD RELEVANCE

A lot of laboratory tests have been developed to estimate the effectiveness of selected factors on the substance to be investigated under standardized laboratory conditions. In general, results of such basic laboratory tests are interpreted with respect to the behaviour of the substance and are one important basis for regulatory decisions.

However, several workers have reported that the results from these laboratory standard methods do not reflect the real fate of the pesticide in the field. For example, results of the recent studies by Führ (ref. 9) with methabenzthiazuron. Using the official German standard residue tests with glass vials and soil columns, he obtained six times higher residue levels than from field containing the same soil type. On the other hand, Scheunert und coworkers ref. 10) showed, that somewhat more sophisticated laboratory tests can be correlated with field experiments. The results from the laboratory volatilization test (Fig. 1), the volatility/mineralization/conversion test (Fig. 2) and the biodegradation test (Fig. 3) could be mathematically related to the respective outdoor data. (Fig. 4-6).

LYSIMETERS

At present time, several workers are turning to lysimeter trials as a realistic compromise. A core of undisturbed soil from outdoor area (about 60 cm in diameter and about 50 cm deep) is transferred to an heat-insulated cylinder and cultivated with the plant culture in question (Fig. 7). Biocoenosis similar to that in the respective field can be reestablished. Because of outdoor localization, climate conditions are the same as in field experiments. In this manner, pesticide studies may be conducted similar to field experiments over the whole vegetation period within a much

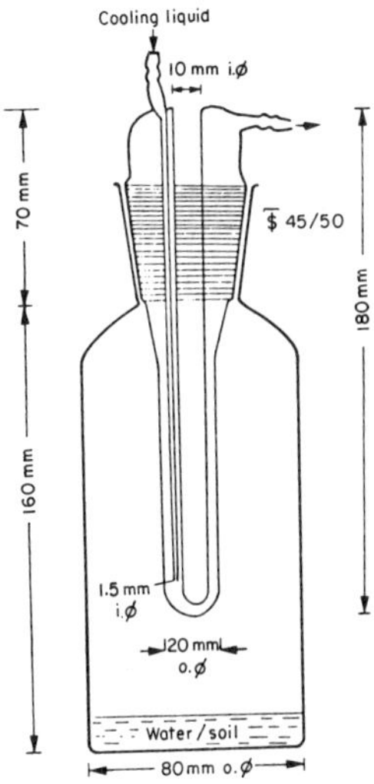

Fig. 1. Apparatus for screening the volatility of chemicals from the soil in laboratory (ref. 10)

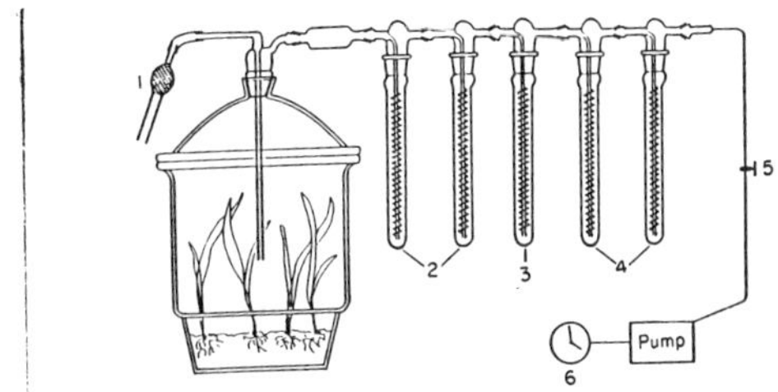

Fig. 2. Apparatus for screening volatility, mineralization, and conversation of chemicals in the plant-soil system in the laboratory (ref. 10)

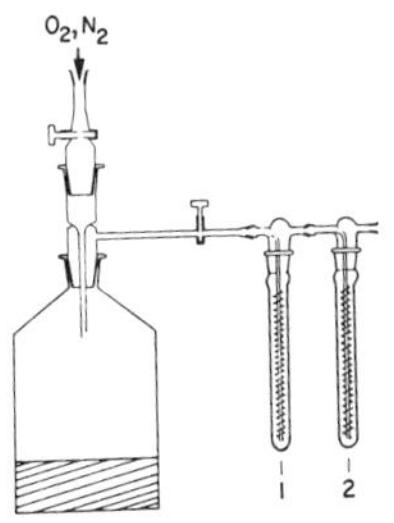

Fig. 3. Apparatus for screening the biodegradation of chemicals in soil-water in the laboratory (ref. 10)

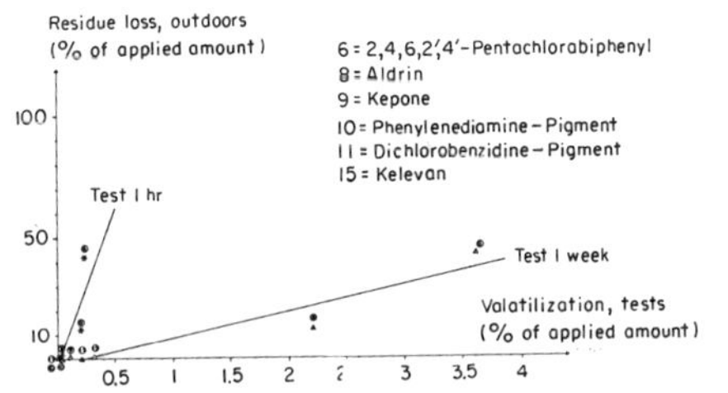

Fig. 4. Correlation of volatilization rates of chemicals from laboratory tests with outdoor residue losses from soil and/or the soil-plant system (ref. 10)

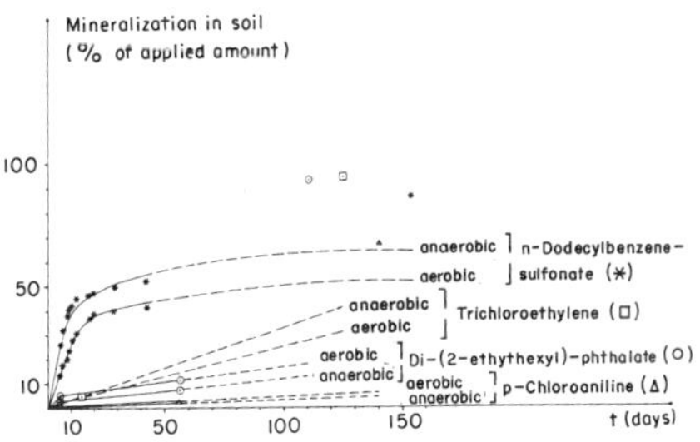

Fig. 5. Correlation of mineralization rates of chemicals from soil-plant tests in the laboratory with outdoor mineralization rates (ref. 10)

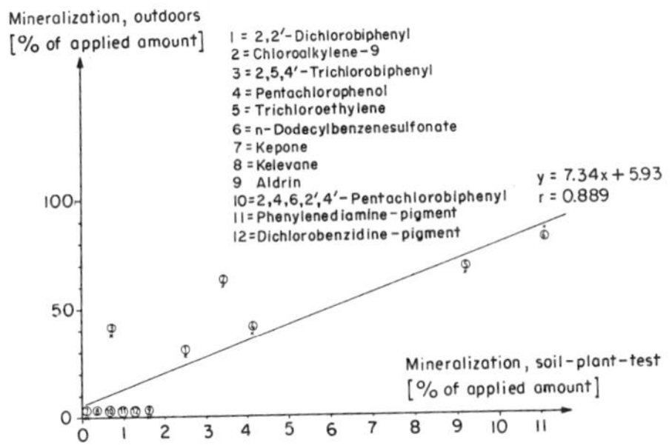

Fig. 6. Correlation of conversion rates of chemicals in soil-plant tests in the laboratory with outdoor conversion rates (ref. 10)

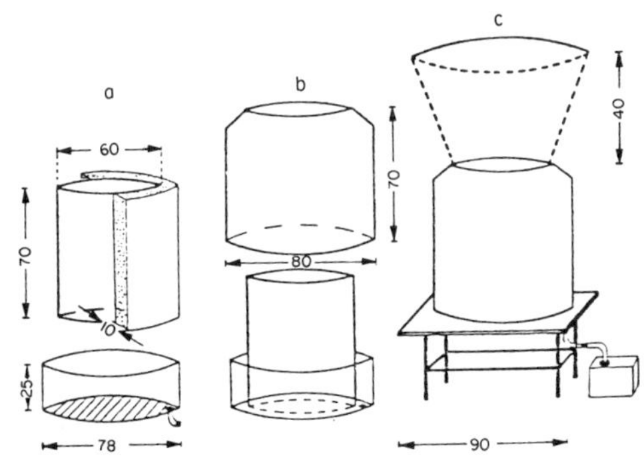

Fig. 7. Construction of an outdoor lysimeter

smaller area and using less materials. Owners of radioactivity licences are allowed to apply radio-labelled pesticide formulations onto such a lysimeter. So, they are able to measure the pesticide and its possible transformation products in all compartments of the biocoenosis: the crop plant, weeds, soil organisms, soil and leaching (ground) water even in minimum quantities. In addition, studies can be designed to define metabolites; translocation and nutrient chain conclusions may be deduced. Several questions from the ecological point of view may be answered. Today, the most modern lysimeter system seems to be that built at the Institute for Radioagronomy in KFA Jülich, FRG (Fig. 8).

Fig. 8. Outdoor lysimeter system of KFA Jülich GmbH

But even these systems have some disadvantages. Besides the certainty, that only a few institutions will receive a licence for operations with radioactivity outdoors, pesticide losses into the local and atmospheric environment cannot be controlled and each experiment depends on the local climate which applies only to the one experimental year. Repeating the experiment three times for obtaining average data will be much too expensive and time consuming. Thus the results do not represent an average situation nor can they be compared with others.

MODEL ECOSYSTEMS

To attempt to overcome these problems, typical field segments or (agricultural) ecosystem sections have been put into experimental laboratory chambers to observe the chemical fate of a (radio-labelled) pesticide applied to a crop. In best cases, the typical and most important constituents of a defined outdoor ecosystem are incorporated, following an intelligent program. It is evident, that such systems are more or less artificial but one tries to keep cultivation conditions as close to realistic average conditions as it is possible. This type of experiment may be repeated relatively quickly yielding reproducible results. Series of experiments with different chemicals may be conducted and lead to comparable results.

The earliest (purely agricultural) terrestrial chambers have been described by Metcalf's group (e.g. ref. 11). 50 corn seeds are grown in a 19 L glass jar on vermiculite plus nutrition solution plus some invertebrates. The apparatus allows volatile compounds and leaching to be monitored. Schu-

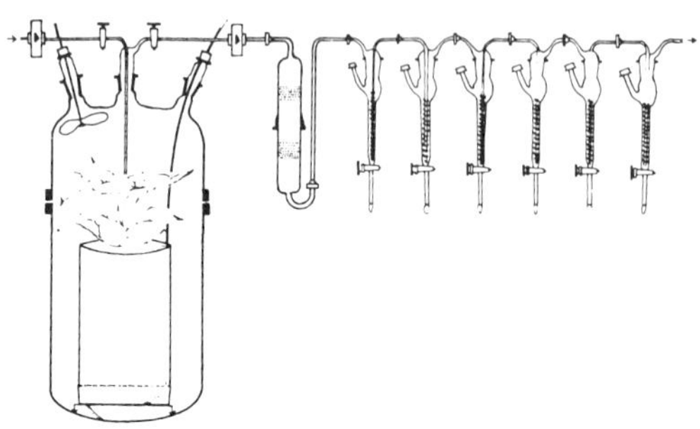

Fig. 9. Closed cultivation system (BBA Berlin) for laboratory metabolism studies

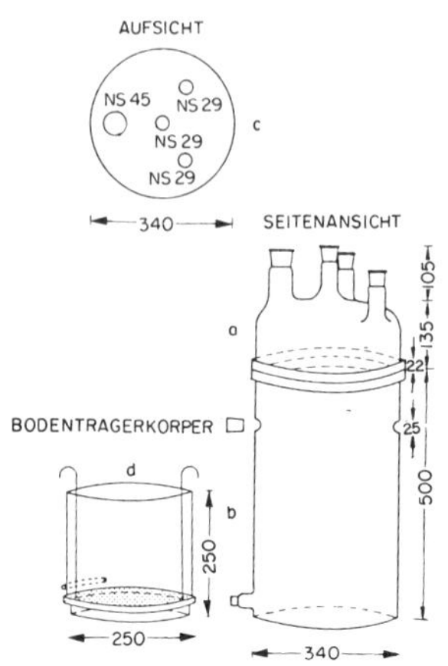

phan (ref. 12) developed this apparatus further leading to his so-called closed cultivation system (Fig. 9). This provides glass containers of different size not only adaptable to the type of plants, but also allowing the measurement of $^{14}CO_2$ evolved in case of mineralization of the labelled pesticide. By these means, establishment of total balances including original compound, transformation and degradation products in plants, soil animals, soil, leaching-water and atmosphere has become possible. About 20 pesticides and several of their metabolites had been studied with these types of chambers.

Gillet & Gile (ref. 13) used an aerated terrarium of the dimensions 1 x 0.75 x 0.61 m filled with 'artificial soil' (sand) for their so-called terrestrial microcosm chamber (Fig. 10). With this chamber similar experiments - sometimes enriched by vertebrates - involving about a dozen pesticides have been studied with respect to balance and translocation.

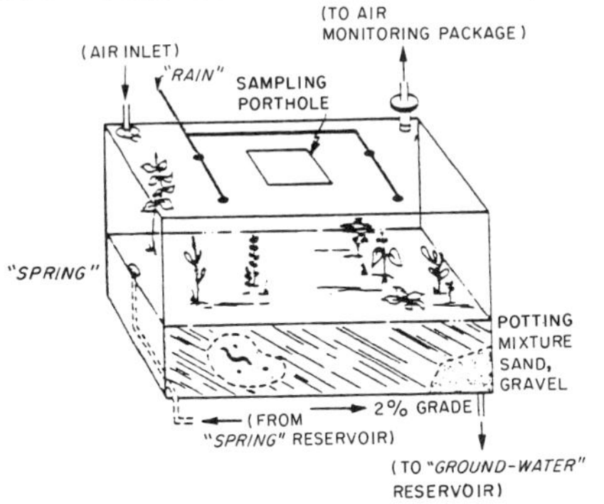

Fig. 10. Terrestrial microcosm chamber from Gillet & Gile (ref. 13)

For studying a more realistic and broader ecosystem section Nash and coworkers (ref. 14) built an enlarged agroecosystem chamber (Fig. 11) of the dimensions 1.5 x 0.5 x 1.25 m. In this case problems with technical support increased so much that the authors operated it without trapping $^{14}CO_2$. This disadvantage has been eliminated by the vegetation chamber described in the thesis of Schärer (ref. 15) and is presented by my colleague Dr. Schuphan (ref. 16) at his congress (Fig. 12). Our chamber is of similar size and construction as that of Nash but may be used with undisturbed soil blocks. Several years before, Figge (ref. 17) introduced his so-called 'metabolism box' (Fig. 13). Its area is 0.7 m in diameter covered by a cupola made from quartz. It is supported by highly sophisticated control systems in order to adjust standard climate conditions as well as climate settings in parallel to actual field climates.

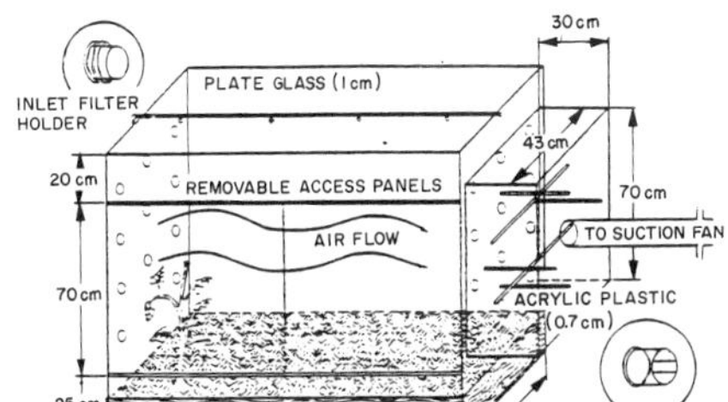

Fig. 11. Agroecosystem chamber from Nash et al. (ref. 14)

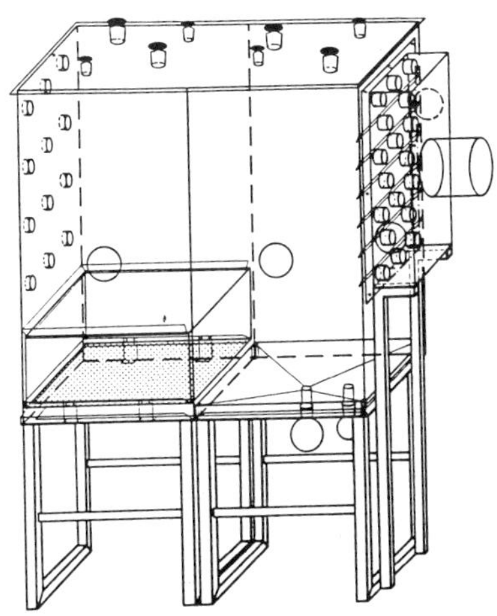

Fig. 12. Vegetation chamber of BBA Berlin

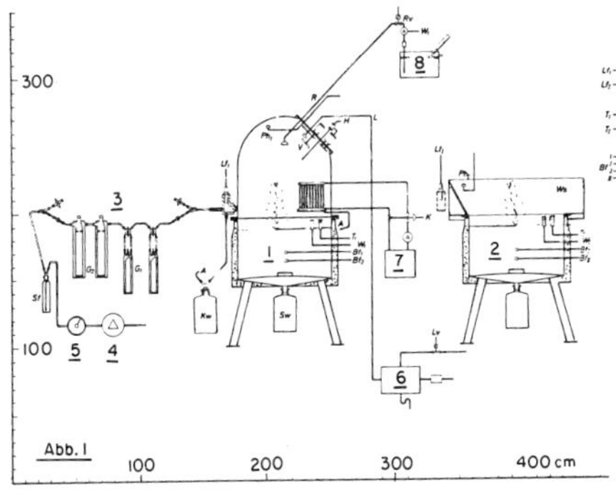

Fig. 13. Metabolism box from Figge (ref. 17)

Besides the purely terrestrial laboratory ecosystem models, combined terrestrial/aquatic systems are described, such as the aquarium of Metcalf or the multicompartment system of Lichtenstein. These models, constructed rather for research with a pronounced ecological emphasis are not further discussed in this context.

CONCLUSION

In my opinion, these techniques are well suited to obtain sufficient ecochemical data to evaluate risks for soil and groundwater loading by pesticides. Quantitatively complete balance data are needed for each pesticide. So far, such complete data sets are available for less than 1 % of the pesticides known today. Because of the immense work which must be done in future, laboratory model research using radiolabelled test substances is urgent. There is no need to characterize all metabolites which can be separated from the extracts by TLC, HPLC, gel permeation chromatography etc., but extracts or several fractions obtained during the separation procedure should be subjected to ecotoxicological tests with vertebrates, invertebrates, soil organisms etc. If a result indicates a possible danger, the metabolite fraction should be further characterized, the relevant compound identified and its content in the sample or compartment quantified.

At present, we are far from having a comprehensive knowledge of pesticides in the soil. For example, in case of insecticides and fungicides registered in 1985 for marketing in the FRG, for 95 of them, no quantitative data for residue dynamics in soil have been reported in the literature (ref. 18). Much still remains to be done to obtain sufficient information about the long term fate of all pesticides in soils.

REFERENCES

1. H. Løkke, Bull. Environ. Contam. Toxicol., 13, 730-736 (1975).
2. S.U. Khan and H.A. Hamilton, J. Agric. Food Chem., 28, 126-132 (1980).
3. P.C. Kearney, J. Assoc. Offic. Anal. Chem., 65, 1030-1032 (1982).
4. R.J. Hance, 'Interactions between herbicides and the soil', Academic Press, London, New York (1980).
5. W. Pestemer, L. Stalder and B. Eckert, Weed Res., 20, 349-353 (1980).
6. L. Stalder and W. Pestemer, Proc. EWRS Symposium 'Theory and Practice of the Use of Soil-Applied Herbicides', 216-225 (1981).
7. A. Walker and A. Barnes, Pesticide Sci., 12, 123-133 (1984).
8. W. Pestemer, 'Untersuchungen zum Verhalten und zu Nebenwirkungen von Herbiziden im Boden'. Habilitationsschrift. University of Hannover, Fachbereich Gartenbau (1984).
9. F. Führ, IRA Ergebnisbericht 1985. Aus: Forschungs- und Entwicklungsarbeiten 1985. Wissenschaftlicher Ergebnisbericht der KFA Jülich. Kernforschungsanlage Jülich GmbH (1985).
10. I. Scheunert, D. Vockel, J. Schmitzer, R. Viswanathan, W. Klein and F. Korte, Ecotox. Environm. Safety, 7, 390-399 (1983).
11. L.K. Cole, R.L. Metcalf and J.R. Sanborn, Internat. J. Environm. Studies, 10, 7-14 (1976).
12. I. Schuphan, Chemosphere, 6, 5-10 (1977).
13. J.W. Gillett and J.D. Gile, Internat. J. Environm. Studies, 10, 15-22 (1976).
14. R.G. Nash, M.L. Beall, jr., and W.G. Harris, J. Agric. Food Chem., 25, 336-341 (1977).
15. E. Schärer, Mitt. Biol. Bundesanst. Land- u. Forstwirtsch., No. 215 (1983).
16. I. Schuphan, E. Schärer, M. Heise and W. Ebing, Sixth International Congress of Pesticide Chemistry (IUPAC) 10-15 Aug. 1986, Ottawa. Proceedings.
17. K. Figge and W. Rehm, Z. Pflanzenkrankh. Pflanzensch., 84, 385-409 (1977).
18. W. Ebing, Ber. Landwirtsch. 198. Sonderh., 35-69 (1985).

Bound pesticide residues and their bioavailability

Shahamat U. Khan* and Stéphane Dupont

Land Resource Research Centre, Research Branch, Agriculture Canada, Ottawa, Ontario, Canada K1A 0C6

Abstract - The use of ^{14}C-labelled pesticides has resulted in an increased awareness of the presence of bound residues in soil and plants. A considerable portion of bound residues in soil may also comprise the parent pesticide molecule. The organic fraction of a soil appears to have the potential for forming bound residues with pesticides or products arising from their degradation. In addition to chemical binding, the pesticides or their metabolites are also firmly retained by organic fractions (humic materials) by a process that more likely involves adsorption on external surfaces and entrapment in the internal voids of a molecular sieve-type structural arrangement. Evidence has been presented to demonstrate the bioavailability of soil-bound pesticide residues to plants. However, the plant availability of ^{14}C from soil containing only ^{14}C-bound residues, as compared to the freshly ^{14}C-pesticide-treated soil, is considerably lower. Soil microbes can potentially release bound residues from soil. A major portion of bound pesticide residues in plant tissues may be associated with lignin. Feeding of plant materials containing ^{14}C-bound residues to monogastric animals indicates that these residues may be considered to be of little toxicological concern and may not present any problem as long as such residues do not accumulate in significant amounts.

INTRODUCTION

In recent years 'bound' or 'unextracted' pesticide residues which are frequently encountered in metabolism studies ;when radiolabelled compounds are used have been the subject of numerous studies. These residues are difficult to extract from the soil or plant samples, are difficult to identify and generally escape detection in the analytical procedures conventionally used in residue analysis. Therefore, it is possible that the soil or plant burden of total pesticide residues may be underestimated.

The work reported in a number of papers during the past few years has led to a growing awareness of the environmental significance of such pesticide residues. This paper describes the trends in the study of bound pesticide residues, and attempts to summarize current knowledge on the bound pesticide residues in soil and plants, and their bioavailability.

BOUND RESIDUES IN SOIL

Table I shows a summary on bound residues in different soils for different classes of pesticides. In most of these studies, radiolabelled pesticides were used and the bound radioactivity was determined by combusting the extracted soil to produce $^{14}CO_2$, which was then quantitated by scintillation counting. The bound residues expressed as % of applied pesticide ranged from 7 to 90%. A significant quantity of bound residues is being 'any amount of unextractable residue, greater than 10% expressed as the parent compound, remaining one year after a single treatment' (Ref. 1). Therefore, in situations where significant concentrations of bound pesticide residues occur in soil, some relevant properties of these residues, such as bioavailability, persistence, and mobility in soil should also be investigated.

In laboratory experiments the formation of bound residues have been shown to increase with incubation time. Utilizing ^{14}C-ring labelled prometryn, the amount of bound ^{14}C residues in an organic soil increased steadily over one year incubation period (Ref. 2). It is possible that in addition to the parent pesticide, the degradation products formed may also become bound in the production of soil-bound residues. In a recent study Capriel and Haisch (Ref. 3) demonstrated that under outdoor conditions nine years after atrazine application the soil still contained about 83% ^{14}C of the initially applied radioactivity. It was observed that a considerable portion of the ^{14}C residues (50%) in soil remained in the bound form even after exhaustive solvent extraction. In a latter study it was found in addition to the parent herbicide a considerable proportion of these residues was comprised of the hydroxy analogues of atrazine and their dealkylated products (Ref. 4).

It has been suggested that there is a relationship between soil-binding tendency of pesticides to their chemical structure (Ref. 5). Thus, chlorinated compounds form only small

TABLE 1. BOUND RESIDUE LEVELS OF SOME PESTICIDES IN SOILS

PESTICIDE	BOUND ^{14}C-RESIDUES (% OF APPLIED)
Herbicides	7-90
Insecticides	
OP	18-80
OC	5-25
Carbamates	32-49
Pyrethroids	23

portions of bound residues in soil, whereas phenols and anilines, and their derivatives have a high binding tendency. Carbamates, triazines and organophosphates also form considerable amounts of bound residues. Persistent pesticides (dieldrin, DDT) differ from nonpersistent pesticides (methyl parathion, fonofos) in their relative low binding properties and their high extractability from soils (Ref. 6-8).

The localization of binding sites in soil fractions have been investigated by several workers. Bound pesticide residues have been detected in the three classical organic matter fractions of soil, i.e. humic acid (HA), fulvic acid (FA) and humin. Khan (Ref. 6) determined the distribution of bound ^{14}C residues among various humic fraction of an organic soil treated with ^{14}C-prometryn. Proportions of total bound ^{14}C in humin, HA, and FA fractions were 57, 11, and 26% respectively. The incorporation of ^{14}C-bound residues into humic materials observed was consistent with those reported for other pesticides described earlier. Recently Capriel et al. (Ref. 4) observed that a considerable proportion of ^{14}C residues was present in the bound form in humic materials 9 years after the application of ^{14}C-atrazine under outdoor conditions. Thus, in terms of the initial ^{14}C applied in the field, the distribution of bound ^{14}C in HA, FA and humin was 7, 17 and 22%, respectively. Capriel et al. (Ref. 4) also observed that the parent herbicides was still present in humic materials in the bound form.

The general concensus appears to be in the formation of bound residues, the pesticide or its metabolites are chemically bound into the soil organic matter. Thermoanalytical methods were applied to soil and humic materials containing bound residues to demonstrate that in addition to the chemical binding, physical binding of the parent molecule and/or metabolites also plays an important role in the formation of soil bound residues (Ref. 9,11).

RELEASE AND PLANT UPTAKE OF BOUND RESIDUES FROM SOIL

The question of the potential release and biological availability of soil-bound residues has been addressed in several studies. Small amounts of soil bound ^{14}C-dichloroaniline residues were mineralized to $^{14}CO_2$ upon incubation of soil containing bound residues with fresh soil (Ref. 12). This mineralization was decreased with sterilization or anerobic incubation. Roberts and Standen (Ref. 13) reported that between 25 and 40% of soil bound ^{14}C-cypermethrin residues were mineralized to $^{14}CO_2$ after a 26 week incubation of soil containing bound residues to which fresh soil had been added. Khan and Ivarson (Ref. 14) investigated the microbiological release of bound residues from an organic soil treated with ^{14}C-prometryn. Extracted soil samples were incubated in a Warburg apparatus with a liquid inoculum obtained from an untreated and unextracted control soil. It was observed that the loss of radiocarbon due to evolution of $^{14}CO_2$ during incubation period of 22 days was negligible from the soil containing ^{14}C-bound residues. The extractable and unextractable (bound) radioactivity in the soil after incubation with an inoculum for 22 days amounted to 27.0 and 71.6%, respectively, of the total ^{14}CO-bound residues. Examination of the extractable residues indicated the presence of prometryn, hydroxypropazine, and a small amount of the partially N-dealkylated compound. Their results suggest that the microbes initially released the bound prometryn, which was then degraded in the incubated soil to hydrolyzed and dealkylated products. In a subsequent study Khan and Ivarson (Ref. 13) observed that soil bound ^{14}C residues were released by different physiological groups of microorganisms. Recently, Racke and Lichtenstein (Ref. 16) investigated the importance and significance of soil microorganisms relative to the release of bound ^{14}C residues from soils, which previously had been treated with ^{14}C-parathion. They observed that with an increase in soil microorganism populations a concomitant increase in the release of soil bound ^{14}C residues occurred. The release of soil bound residues may also be affected by the various soil treatment (Ref. 17-19).

BOUND RESIDUES IN PLANTS

Many studies using radiolabelled compounds have revealed that a considerable portion of the pesticide residues may become bound in the plant (Ref. 23-36). Since bound residues in plants cannot be removed by conventional solvent extraction techniques designed to extract and detect the parent compound and certain metabolites, more attention should be directed towards evaluating their residual nature in plants.

Chin et al. (Ref. 32) observed that in rice straw, bound residues of the herbicide SWEP were in the form of a lignin-SWEP complex. Wargo et al. (Ref. 25) have shown that ^{14}C from nitrofen ^{14}C can be incorporated into starch in rice and wheat grain. Rice and wheat can metabolize nitrofen-^{14}C in such a way as to produce lignin fraction containing radioactivity (Ref. 26,30). It was thought that this association is either in the form of a lignin-nitrofen conjugate or as an integral part of one or more of the molecules which comprise the lignin. Khan (Ref. 18) investigated the nature of bound ^{14}C residues in oat plants grown in ^{14}C-prometryn treated solvent extracted soil containing bound residues. It was observed that plant-bound ^{14}C residues were smaller in roots and greater in shoots amounting to 20 and 40% respectively, of the total ^{14}C recovered from each tissue. The bound residues in the plant tissues comprised mainly of a mono-N-dealkylated and traces of di-N-dealkylated compounds. The total lignin (free + bound) from roots and shoots contained 10.5 and 28.5%, respectively, of the total ^{14}C-bound residues in the plant tissue (Ref. 24). The remaining solid tissue of roots and shoots, which presumably consists mainly of carbohydrate and protein contained relatively smaller amounts of ^{14}C residues.

Several studies have been reported indicating that lignin is the major bound residue fraction of aniline and chloroaniline compounds in plants (Ref. 31-40). Still et al. (Ref. 28) investigated the molecular nature and identity of bound 3-chloroaniline and 3,4-dichloroaniline residues in rice plants grown hydroponically. More than 40% of the ^{14}C was present in the isolated lignin fraction from the roots. These workers suggested that chloroanilines may be bonded covalently to lignin via 1,6-addition to a quinone methide intermediate during the lignin synthesis. The benzylic-carbon in the lignin side chain was considered to be the most likely chloroaniline nitrogen binding site (Ref. 28). Other studies have shown that major binding sites in plants in the formation of bound pesticide residues include hemicellulose (Ref. 40-43) and pectic polysaccharide (Ref. 27).

BIOAVAILABILITY TO ANIMALS

It is known that urinary and/or biliary excretion of a compound or its metabolites signifies that the material is bioavailable. However, quantitative fecal elimination without biliary excretion indicates that a material is not bioavailable. Only a few studies have been reported which have examined the bioavailability of plant bound pesticide residues. Paulson et al. (Ref. 44) observed that alfalfa plants containing profam-phenyl-^{14}C bound residues were not bioavailable when fed to rats, since 86% of the dose was eliminated in the feces. Sutherland (Ref. 39) investigated the bioavailability of propanil-phenyl-^{14}C bound residues from rice plants when fed to white albino rats. It was observed that 89% of the dose was eliminated in the feces and 11.5% in the urine. He concluded that because of the low absorption by monogastric animals, the bound propanil residues should be considered to be of little toxicological concern. Khan et al. (Ref. 45) investigated the potential bioavailability in rats of bound residues from the aerial portion of corn plants grown to the silage stage and treated with ^{14}C-atrazine. The chemical identity of bound ^{14}C residues in plant material before feeding to rat, as well as in the excreted and feces was also determined. The results demonstrated a low degree of bioavailability in rats. However, unlike bound residues, freshly added ^{14}C-atrazine to plant material when fed to rats was readily bioavailable. In a recent study Khan et al. (Ref. 46) demonstrated that bound residues in radishes treated with dieldrin and carbofuran had low degree of bioavailability in rats. It was suggested that if we are to properly assess the bioavailability and/or toxicological significance of bound residues, information on their chemical identity must be obtained. In studies reported earlier we do not know for certain whether these ^{14}C bound residues in crop materials fed to animals, as well as the ^{14}C residues excreted, were present as the parent compound or in the form of their breakdown products.

BOUND RESIDUES DETERMINATION

Determination of the nature and quantities of bound pesticide residues in soil and plants has occupied many research workers during the past few years and still remains a challenging problem. The use of radiolabelled pesticides in this field of study has been indispensible. It has made us aware that a portion of the ^{14}C residues of these compounds in soil and plants becomes bound, since combustion of the extracted materials yields $^{14}CO_2$.

In most of the studies reported in the literature, quantification of ^{14}C-bound residues in soil or plants has been achieved by total combustion. This technique is limited to the quantitative determination of ^{14}C-bound residues and cannot be used to determine the chemical form of the bound residues. Attempts have been made to extract bound herbicide residues by the milder to harsher methods. Drastic extractive procedures that destroy the structure of soils or plants by solubilizing the materials, and strong acid or base hydrolysis often result in the destruction of the identity of bound residues.

A noval technique for identification of the bound ^{14}C herbicide residues has been reported by Balba et al. (Ref. 40). For the chloroaniline compounds studied pyrolysis in the absence of oxygen released more than 80% of the radioactivity as volatile products. Gas chromatography-mass spectrometry for the pyrolysate indicated that the major radioactive products were the chloroaniline compounds. A similar technique has been used by Khan and Hamilton (Ref. 2). The technique involved high-temperature distillation (HTD) of the extracted soil or plant sample to release bound ^{14}C residues. However, this technique has a drawback that during the distillation the bound residues are partially or completely decomposed to $^{14}CO_2$. Recently Capriel et al. (Ref. 47) used a new solvent extraction procedure at supercritical conditions for the isolation and identification of bound ^{14}C residues in soil and plants. The fractions were subjected to column clean up and then analysed by GC and GC-MS. In general, the results were similar to those obtained by HTD on the same samples.

REFERENCES

1. Anonymous: "Registration of Pesticides in the United States-Proposed Guidelines", Fed. Reg. 40, 123 (1975).
2. S.U. Khan and H.A. Hamilton. J. Agr. Food Chem., 28, 126, 1980.
3. P. Capriel and A. Haisch. Z. Pflanzenernaehr. Bodenk, 146, 474, 1983.
4. P. Capriel, A. Haisch and S.U. Khan. J. Agric. Food Chem., 33, 567 (1985).
5. W. Klein and I. Scheunert. Bound pesticide residues in soil, plants and food with particular emphasis on the application of nuclear techniques, in "Agrochemicals: Fate in Food and the Environment", International Atomic Energy Agency, Vienna, SM-263/38, 177 (1982).
6. Katan, J. and E.P. Lichtenstein. J. Agr. Food Chem., 25, 1404 (1977).
7. E.P. Lichtenstein. Residue Reviews. 76, 147 (1980).
8. E.P. Lichtenstein, J. Katan and B.N. Anderegg. J. Agr. Food Chem., 25, 43 (197).
9. S.U. Khan. J. Agr. Food Chem., 30, 175, 1982.
10. M. Schnitzer and S.U. Khan. Humic substances in the environment, New York, Marcel Dekker, (1972).
11. C.S. Helling and A.E. Krivonak. J. Agr. Food Chem., 26, 1156, 1978a.
12. T. Hsu and T. Barthan. Soil Sci. 118, 213 (1974).
13. T.R. Roberts and M.E. Standen. Pestic. Sci. 12, 285 (1981).
14. S.U. Khan and K.C. Ivarson. J. Agr. Food Chem., 29, 1301 (1981).
15. S.U. Khan and K.C. Ivarson. J. Environ. Sci. Health. B17, 737 (1982).
16. K.D. Racke and E.P. Lichtenstein. J. Agr. Food Chem., 33, 938 (1985).
17. Z. Gerstl and C.S. Helling. Soil Biol. Biochem., 17, 667 (1985).
18. I.C.M. Rae. Soil Biol. Biochem., 18, 221 (1986).
19. D. Yee, P. Weinberger and S.U. Khan. Weed Sci., 33, 882, (1985).
20. C.S. Helling and A.E. Krivonak. J. Agr. Food Chem., 26, 1164 (1978b).
21. T.W. Fuhremann and E.P. Lichtenstein. J. Agr. Food Chem., 26, 605 (1978).
22. A. Haque, I. Schuphan and W. Ebing. Pestic. Sci. 13, 219 (1982).
23. F. Fuhr and W. Mittelstaedt. J. Agr. Food Chem., 28, 122 (1980).
24. S.U. Khan. J. Agr. Food Chem., 28, 1096 (1980).
25. J.P. Wargo, R.C. Honeycutt and I.L. Adler. J. Agr. Food Chem., 23, 1095 (1975).
26. R.C. Honeycutt and I.L. Adler. J. Agr. Food Chem., 23, 1097 (1975).
27. A. Haque, I. Weisgerber and W. Klein. Chemosphere. 5, 167 (1976).
28. G.G. Still, H.M. Balba and E.R. Mansager. J. Agr. Food Chem., 29, 739 (1981).
29. J. Wieneke. Amer. Chem. Soc. Symp. Ser., 29, 166 (1975).
30. R.C. Honeycutt, J.P. Wargo and I.L. Adler. Amer. Chem. Soc. Symp. Ser., 29, 170, 1975.
31. G.G. Still, F.A. Norris and J. Iwan. Amer. Chem. Soc. Symp. Ser., 29, 156, 1976.
32. W.T. Chin, N. Kucharczyk and A.E. Smith. J. Agr. Food Chem., 21, 506, 1973.
33. S.U. Khan, L.Z. Zhang and M.H. Akhtar. J. Agr. Food Chem., 32, 1141 (1984).
34. W.T. Chin, R.P. Stanovick, T.E. Cullin and G.C. Holsing. Weeds, 12, 201, 1964.
35. S.U. Khan, G.D. Stratton and W.B. Wheeler. J. Agr. Food Chem., 32, 1189 (1984).
36. W.T. Chin, G.M. Stone and A.B. Smith. J. Agr. Food Chem., 18, 709, 1970.
37. N.M. Mallipuch and T.R. Fukuto. Arch. Environ. Contam. Toxicol. 10, 505 (1981).
38. R.Y. Yih, D.A. McRae and H.F. Wilson. Science, 161, 376, 1968.
39. M.L. Sutherland. Amer. Chem. Soc. Symp. Ser., 29, 153, 1976.
40. H.M. Balba, G.G. Still and E.R. Mansager. J. Assoc. Official Anal. Chemists, 62, 237, 1979.
41. J. Rouchaud, J.R. Dacallonne and J.A. Meyer. Pest. Sci., 9, 74, 1978.
42. J. Rouchaud, C. Moons and J.A. Meyer. Pest. Sci., 10, 509, 1979.
43. J.B. Pillmoor, J.K. Gaunt and T.R. Roberts. Pestic. Sci., 15, 375, 1984.
44. G.D. Paulson, A.M. Jacobsen and G.G. Still. Pest. Biochem. Physiol., 5, 523, 1975.
45. S.U. Khan, S. Kacew and S.J. Molnar. J. Agr. Food Chem., 33, 712 (1985).
46. S.U. Khan, S. Kacew, G. Straton and W.B. Wheeler. J. Agr. Food Chem. (in press).
47. P. Capriel, A. Haisch and S.U. Khan. J. Agr. Food Chem., 34, 70 (1986).

Degradation of kelthane in soil by *P. aeruginosa* BS827 carrying plasmid pBS3

Ludmila A. Golovleva*, Raisa N. Pertsova, Vasily M. Travkin

Institute of Biochemistry and Physiology of Microorganisms
USSR Academy of Sciences, 142292 Pushchino, Moscow region,
USSR

Abstract - The strain P. aeruginosa BS827 with the plasmid pBS3 encoding the ability of cells to grow on naphthalene (Nah^+) and salicyclate (Sal^+) as carbon sources, as well as to degrade kelthane was introduced into soil containing the acaricide kelthane (DDT analogue).
The study of the dynamics of the BS827 strain and the stability of inheriting Nah^+ and Sal^+ features by cells showed that the strain preserves its number of cells on the level of 10^5-10^3 cells per g of soil for three months.
The number of cells and their stability depended on the presence of kelthane in soil. They were the highest during the active kelthane degradation which occurred only in the presence of the strain BS827.

INTRODUCTION

The removal from the environment of residues of persistent pesticides is an acute problem. One of the basic trends of ecological biotechnology aimed at the fulfilment of this task is screening and molecular breeding of new active strains capable of degrading persistent toxicants. There are already definite achievements in the field and some laboratories possess strains able to degrade chemicals earlier considered recalcitrant (ref. 4, 6). Usually these strains contain plasmids which carry out the genetic control of biodegradation. At the same time many questions connected with the release of these strains to natural ecosystems still remain unanswered. It is unclear whether the strains are stable and whether the plasmids they carry are also stable and their fate after they have fulfilled their main goal, i.e. degradation of xenobiotic, is unknown.

It was the purpose of the present work to study the ability of strain P. aeruginosa BS827 to degrade the acaricide kelthane under natural conditions, to study the dynamics of its cell number in soil and the stability of the inheritance by this strain of naphthalene degradative plasmid pBS3.

MATERIALS AND METHODS

The experiments were carried out for three months in natural soil ecosystems namely 1.5 x 1.5 m areas of garden soil loosened to a depth of 10 cm, as well as earlier reported model soil columns (ref. 4).

Microorganisms: Strain Pseudomonas aeruginosa BS827 resulted from introduction to plasmid-free strain P. aeruginosa 640x of naphthalene and salicylate degradative plasmid pBS3 (ref. 2) which favours the development of cells on naphthalene (Nah^+ sign) and salicylate (Sal^+ sign). To genetically mark this strain, we isolated a rifampicine (Rif^r) resistant variant with mutation in a chromosomal gene. The cell numbers of inoculated strain and natural microflora were equal.

The total abundance of soil microflora was determined by inoculation on meat-peptone agar, of strain BS827 by inoculation of the same agar supplied with rifampicin. The availability of Nah^+ and Sal^+ signs was revealed on minimal agar media containing salicylate or naphthalene as sole carbon sources, salicylate added at a concentration of 2.0 g/l, naphthalene applied to a petri plate cover.

In a natural ecosystem, kelthane was introduced at a quantity of 100 mg/kg of soil. In the laboratory, kelthane solution (100 mg/l) was passed through soil columns at a flow rate 60 to 70 ml/day.

The kelthane quantity was determined in soil samples and eluates, respectively. Soil suspension resulting from soil dilution by equal amounts of water and eluate was acidified to pH 2 and extracted thrice with ether followed by removal of ether on a rotor evaporator. The residue was dissolved in heptane.
Qualitative analysis was performed by thin-layer chromatography (TLC) on Silufol UV-254 plates. Plates were developed in a mixture of n-heptane-ethylacetate-acetic acid (30:10:0.1). Chlorine-containing and phenolic compounds were revealed using corresponding reagents (ref. 3, 8).

Gaschromatographic/mass-spectrometry analysis was performed on LKB-2091 (Sweden). Separation conditions: glass column 180 x 0.2 cm, carrier "Chromosorb W/HP" 100-120 mesh with phase 3% OV-101. Temperature of the injector was 250°C, column 90-280°C, heating rate 10 degree/min, flow-rate of carrier 20 ml/min. Mass-spectra were registered at ionizing voltage 70 eV. Compounds were identified by comparing mass-spectra obtained with that of a standard.

Quantitative analysis was performed by gas chromatography on a Pye-Unicam equipped with an electron-capture detector on a glass column 120 x 0.2 cm, phase OV-101 (1.5%), carrier Superlcoport W/HP 800-1000 mesh. Carrier gas, nitrogen 40 ml/min. Temperature of injector and column 210°C.

RESULTS

Fate of strain P. aeruginosa BS827 in soil.

During the first two weeks the cell number of strain BS827 in soil without xenobiotics dropped sharply (by a factor of 100), then stabilized and remained constant (5.10^3 cells/g) throughout the 3 months of the experiment (Fig. 1A).

The analysis of cells for the availability of plasmid genes showed the instable inheritance of Nah^+ and Sal^+ by the strain: after two weeks 43% of the cells acquired phenotype $Nah^- Sal^-$, while 21% of the cells contained either $Nah^+ Sal^-$ or $Nah^- Sal^+$. Nine weeks later phenotype $Nah^- Sal^-$ was observed in 92% of the cells (Table 1).

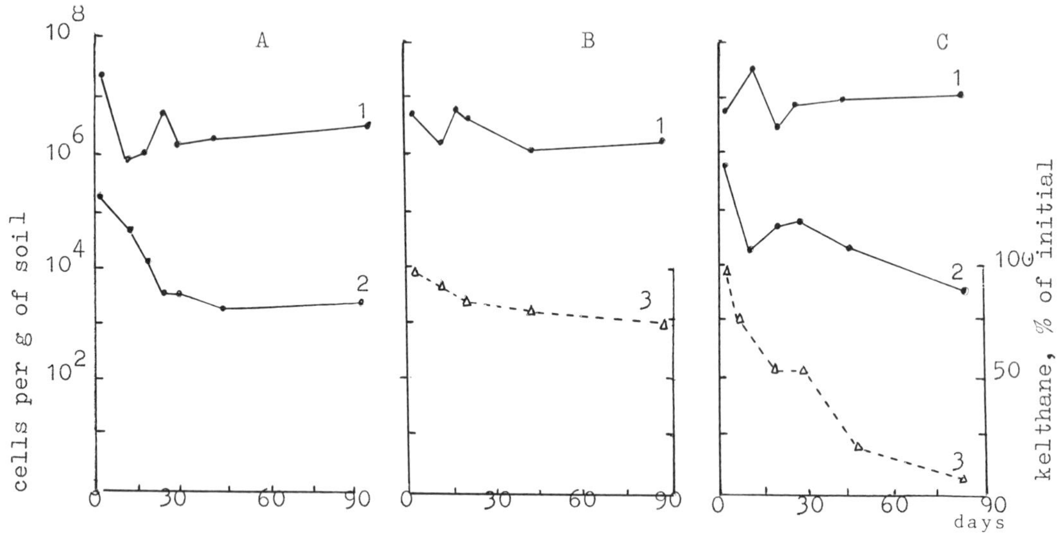

Fig. 1 Number of cells of P. aeruginosa BS827 (2) and fate of kelthane (3) in natural soil ecosystem
I - total abundance of microorganisms; A - without kelthane;
B - without strain BS827; C - strain BS827 in the presence of kelthane.

The presence of kelthane in soil somewhat changed the picture. Firstly, the decrease in the cell number was not so sharp (10-fold only) and not long (a week, compare with two weeks of the control). Secondly, during the period preceding the kelthane active degradation, the cell number was 20-fold higher than in the control (Table 1). Thirdly, the percentage of cells with phenotype Nah$^+$ Sal$^+$, i.e. carrying plasmid genes, was also significantly higher than in the control: 82%, 72% and 100% after 2, 2.5 and 6 weeks respectively (compare with 39%, 19.6% and 0.1% respectively in the control) (Table 1). One should mention however, that in the control, the cells with the partially lost phenotype, i.e. Nah$^+$ Sal$^-$ or Nah$^-$ Sal$^+$ were more abundant: 21, 32 and 8% after 2, 2.5 and 6 weeks, respectively (compare with 17, 11 and 0.1% in kelthane treated soil).

TABLE 1. Stability of P. aeruginosa BS827 carrying plasmid pBS3 in soil.

Duratin of experiment, days	Phenotype of strain	% cells of population	
		with kelthane	without kelthane
1	Nah$^+$ Sal$^+$	100,0	100,0
13	Nah$^+$ Sal$^+$	83,0	39,0
	Nah$^-$ Sal$^+$	13,0	19,0
	Nah$^+$ Sal$^-$	4,0	1,8
	Nah$^-$ Sal$^-$	0	43,0
17	Nah$^+$ Sal$^+$	72,0	19,6
	Nah$^-$ Sal$^+$	0	18,7
	Nah$^+$ Sal$^-$	11,0	13,1
	Nah$^-$ Sal$^-$	17,0	48,6
41	Nah$^+$ Sal$^+$	100,0	0
	Nah$^-$ Sal$^+$	0	0
	Nah$^+$ Sal$^-$	0	8
	Nah$^-$ Sal$^-$	0	92

Kelthane degradation in soil by strain BS827.

Dynamics of kelthane degradation in soil treated with strain BS827 and in the control is shown in Fig. 1. In the latter variant, the residual kelthane quantity comprised 99% after a week and 79% three weeks later, then it remained practically constant (Fig. 1). Quite another pattern of kelthane bioconversion was observed in the presence of strain BS827: already in a week the pesticide residue comprised; 65%; during the following two weeks its quantity remained almost unchanged and varied within the narrow limits from 57 to 54%. This was followed by a sharp decrease resulting in only 18% residual kelthane 6 weeks later. By the end of the experiment (3 months), the amount of xenobiotic in soil was no higher than 8% versus 72% in the control.

The pathways of kelthane metabolism were studied in soil columns through which the solution of xenobiotic was passed at a low rate. The eluate was collected and used to analyse the metabolites. For short time intervals, the fates of strain BS827 in the column and in the natural ecosystem were alike, the only difference being that in the column degradation of kelthane occurred at a significantly higher rate than in nature and thus facilitated the analysis of metabolites.

The latter showed that in soil an aliphatic fragment of the kelthane molecule was dechlorinated then oxidized yield 4,4-dichlorobenzophenone and 4,4-dichlorobenzyhydrol. One of the pesticide rings was split yield chlorbenzoic acid. In addition there were trace quantities of the products of kelthane resulting from dehydrochlorination and dehydration.

DISCUSSION

The DDT analogue kelthane is an extremely stable compound hence it may accumulate in soil in appreciable amounts (ref. 5). We showed earlier that introduction of naphthalene and salicylate degradative plasmid pBS3 to strain P. aeruginosa 640x able to degrade DDT under metabolic conditions (ref. 1) results in the development of new properties by strain BS827 (ref. 2). For instance, in addition to the enhancement of its activity against DDT, the strain acquired the ability to actively degrade its analogue kelthane which was transformed by the pure culture of this strain in two weeks under cometabolic conditions (ref. 2).

All the results presented above show that kelthane is degradable by strain BS826 both in model soil columns and in nature. In the former, the kelthane residue was 2% already two weeks later; in the latter it was 18% after six weeks and 8% by the end of the experiment (3 months). Therefore, the pesticide bioconversion was much slower in nature (one should take into account however that during the second and third weeks i.e. one third of the 6-week period kelthane remained essentially unchanged therefore the time of its active degradation accounts for about one month). It is noteworthy, that precisely during "retardation" of kelthane degradation, the soil humidity diminished two-fold and was 11% versus 20-22% and 18-19% during the first week as well as 4 and 5 weeks respectively. This probably exerted a negative effect on the intensity of microbial kelthane degradation, which may be confirmed by the data on the 95% microbial degradation of 2,4,5-T, at 25% humidity of soil (compared with 78% degradation at 15% humidity) (ref. 4).

The analysis of kelthane degradation products demonstrated the similar microbial metabolism of this compound with the DDT bioconversion described earlier (ref. 1).

The abundance of strain BS827 in soil is not strongly variable since a sharp drop in the cell number (10- to 100-fold) was followed by its stabilization at about 5.10^3 cells/g soil throughout the 3 months experiment, and this level was independent of the availability of xenobiotic though the latter still affected the dynamics of the cell number (e.g. during the intensive kelthane degradation, the number of cells of the strain was 20-fold higher than in the control). A similar picture was observed with another plasmid-carrying strain P. putida 87 released in soil simultaneously with 3-chlorbenzoic acid which is consumed as a carbon source.

More striking was the "presence" of a xenobiotic in studies of the stability of phenotype Nah^+ Sal^+ in strain BS827. As shown earlier, under non-selective conditions cells appeared (at 1% frequency) with the phenotype Nah^- Sal^- resulting from the elimination of corresponding plasmid pBS3. At a lesser frequency cells were formed with the partially lost phenotype either Nah^-Sal^+ or Nah^-Sal^- due to the deletion of a part of plasmid DNA (data not presented). In untreated soil, about 43% cells acquired phenotype Nah^-Sal^- after two weeks reaching 92% by the end of the sixth week. The presence of xenobiotic stabilized the population with respect to these signs. This was most clear-cut during periods of the active kelthane degradation. The results in Table 1 and Fig. 1 show that in soil with kelthane, Nah^- Sal^- variants were not found during its active transformation, e.g. by 13th and 14th days, while during "retardation" of kelthane degradation (17th day), Nah^-Sal^- variants appeared both in the soil treated and untreated with xenobiotic at a rate of 17% versus 48.6% respectively.

Therefore, kelthane is degraded in soil treated with strain P. aeruginosa BS827. The availability of kelthane in soil and its degradation are selective factors for maintaining the population of strain P. aeruginosa BS827 carrying plasmid pBS3.

REFERENCES

1. L.A. Golovleva, A.M. Zyakun, B.P. Baskunov, R.N. Pertsova and G.K. Skryabin, Izvestiya Akad. Nauk SSSR, biol., No. 2, 243-253 (1980).
2. L.A. Golovleva, R.N. Pertsova, A.M. Boronin, B.V. Grishchenkov, B.P. Baskunov and A.V. Polyakova, Mikrobiologiya 51, 973-978, (1982).
3. J. Kirchner, Mir Publishers, 1, 229, (1981).
4. D.K. Chatterjee, J.J. Kilbane and A.M. Chakrabarty, Appl. Environ. Microbiol., 44, 514-516 (1982).
5. P.J. Van Dijck and H. Van de Voorde, Eur. J. Appl. Microbiol., 2, 277-281 (1976).
6. J.J. Kilbane, D.K. Chatterjee and A.M. Chakrabarty, Appl. Environ. Microbiol., 44, 1697-1700, (1983).
7. R.N. Pertsova, F. Kunc and L.A. Golovleva, Folia Microbiol., 29, 242-247, (1984).
8. A. Szokolay and A. Madaric, J. Cromatogr., 42 509-519, (1969).

Abiotic processes in the degradation of pesticides in natural waters

William M. Draper[1] and N. Lee Wolfe[2]

[1]California Public Health Foundation
2151 Berkeley Way, Berkeley, California 94704

[2]U. S. Environmental Protection Agency
Environmental Research Laboratory, Athens, Georgia 30613

Abstract - Pesticides and other trace micropollutants released to the aquatic environment are subject to hydrolysis, oxidation-reduction, and photodecomposition. Kinetic treatments of hydrolysis and direct photolysis processes have been successful in allowing accurate extrapolation of laboratory studies to field conditions. The most recent hydrolysis models account for not only specific acid-base catalysis, but also the impact of suspended sediments. Nonbiological oxidations and reductions, although demonstrated in numerous recent studies, are not well enough understood at this time to be described by mathematical models based on chemical kinetics. Indirect photolysis processes involve a variety of excited state and/or reactive chemical species. Recent research in this area has been directed at measurement or estimation of steady-state reactant concentrations in natural waters. Further basic research into aquatic chemistry and photochemistry will provide the basis for accurate modelling of other abiotic, pesticide degradation processes.

INTRODUCTION

The focus of this workshop session has been to assess the relative importance of biological and abiotic processes in pesticide degradation. To this end it is useful to define biologic processes as those that occur within living cells -- abiotic processes then are strictly extracellular. This distinction is arbitrary, of course. Biologists may argue that extracellular reactions are indeed metabolic because cell-free enzymes may be involved. On the other hand, chemical and photochemical reactions may produce primary transformation products that are preferred microbial substrates. Certain intracellular processes may be uncatalyzed. Hydrolytic reactions, for example, may be mediated by cytosolic water and not a specific enzyme system. By their very existence microbial populations change the chemical and photochemical characteristics of the environmental compartments.

For atmospheric transformations the question is more easily addressed because the biosphere is largely limited to the aquatic environment and the upper layers of the lithosphere. Many chemicals including pesticides are rapidly degraded in the atmosphere by sunlight and chemical reaction with hydroxyl free radicals and ozone. Recently a third atmospheric species, the nitrate radical, has been identified that appears to oxidize micropollutants in the nighttime sky (ref. 1). Biological and nonbiological processes are inextricable in the soil and aquatic compartments.

Because it is futile to attempt to address this question, at least with our present knowledge, we will turn our efforts to a more manageable task. This brief report will discuss the chemical and photochemical processes that are known to degrade pesticides in the environment. Kinetic treatment of these processes is emphasized, not only as a conceptual tool, but also because chemical and photochemical kinetics are essential in weighting the importance of competing natural processes. Kinetics also forms the basis of mathematical modelling which allows tremendous flexibility in predicting chemical behavior. Here we focus on direct and indirect photolysis, hydrolysis reactions and oxidation-reduction.

KINETICS OF ABIOTIC PROCESSES IN AQUATIC ENVIRONMENTS

To model the fate and transport of pollutants in natural waters, soils and sediments and to predict exposure concentrations in environmental compartments, it is necessary to have quantitative descriptions of the reaction processes (ref. 2,3). These descriptions are usually kinetic expressions that often include the concentration of the compound, a reaction rate constant that incorporates the reactivity of the molecule for the reaction under consideration and a measurement of activity of the system reactants such as sunlight intensity, pH or Eh. This level of understanding is necessary to extrapolate results based on laboratory studies or a given set of field conditions to another ecosystem with different physical, chemical and biological characteristics. The kinetics of direct and indirect photolysis, hydrolysis and redox reactions in the water column and sediments are addressed below.

Kinetics of photolysis

Environmental photochemists generally divide photochemistry into two types of processes, direct and indirect. Both of these pathways can occur in natural waters, in the presence of suspended sediments, dissolved humic substances and biota. Direct photolysis has received the most attention and is the most straight forward to model.

Direct photolysis. In the case of direct photolysis, the compound absorbs light and the electronically-excited species, P*, undergoes bond cleavage and bond formation.

$$P \xrightarrow{h\nu} P^* \longrightarrow \text{products} \qquad (1)$$

Kinetics of direct photolysis are determined by two factors, light absorption and an efficiency term, the quantum yield (ref. 4).

$$k_p = Ia\,\Phi \qquad (2)$$

where Ia is the system specific rate of light absorption and Φ is the quantum yield, a measure of the efficiency with which absorbed light energy results in a chemical reaction. Under most environmental conditions, the rate of photolysis of a pesticide (P) is given by:

$$-d[P]/dt = k_p[P] \qquad (3)$$

where k_p is the photolysis rate constant and [P] is the concentration of the compound. Because this is a first-order reaction, the half-life is given by:

$$t_{1/2} = 0.693 / k_p \qquad (4)$$

Indirect photolysis. Indirect photolysis can be generalized by two pathways shown in equation 5. Indirect photolysis requires a sensitizer, S. In the first case, the sensitizer absorbs light and transfers energy to the pesticide which then undergoes bond cleavage and formation (ref. 5).

$$S \xrightarrow{h\nu} S^* \qquad (5)$$

$$S^* + P\ (O_2) \longrightarrow \text{products}$$

$$R\text{-}X \xrightarrow{h\nu} R\cdot + X\cdot$$

$$R\cdot\ (X\cdot) + P\ (O_2) \longrightarrow \text{products}$$

In the alternative mechanism, light absorption results in homolytic cleavage of a chemical bond (R-X) with formation of a radical R· and subsequent reaction of the radical with the pesticide (ref. 6). Presently, equations are not available that describe the kinetics of the indirect photochemical processes. One study, however, that offers some quantitative insight has been reported for the photolytic transformation of 2,4,5-T in natural water samples and in fulvic acid solutions (ref. 7). This study gives a linear relationship between the UV absorbance of the water at 312 nm and the photolysis rate constant at the surface. The mechanism in these solutions is not known but it may well involve singlet oxygen or hydroxy radicals.

Although indirect processes are not well defined, many recent studies demonstrate their importance in the transformation of pesticides. Photochemically generated reactive species will be addressed later in greater detail.

Kinetics of hydrolysis

Hydrolysis is receiving increased attention as a degradation pathway in groundwaters. This process takes on increased importance as photolytic processes are curtailed as a result of lack of sunlight penetration. Also, microbial processes are likely to be retarded by reduced organism activity.

Hydrolysis used in the usual context refers to reaction with water to cleave an existing bond of a compound and form a new bond with oxygen. When these reactions are mediated by acid (H^+) or base (OH^-) respectively, they are referred to as specific acid-base catalyzed hydrolysis (ref. 8). These processes have been extensively studied for many classes of pesticides.

Several other types of catalyzed processes have been suggested to enhance hydrolysis in natural waters. General acid-base catalysis in which hydrolysis is promoted by inorganic or organic species such as phosphate, carbonate, bicarbonate or acetate has been addressed (ref. 9). Metal catalyzed hydrolysis by copper and manganese species has been investigated (ref. 10). Cell free enzymes and humic substances have also been proposed to contribute to hydrolysis of pesticides (ref. 11). Although enhanced hydrolysis has been demonstrated by these catalyzed processes under laboratory reaction conditions, under field conditions where very low activities of the catalysts are found, these processes are generally not likely to be important.

Hydrolysis in solution. Under most reaction conditions found in natural waters, specific acid-base and neutral hydrolysis reactions are all that is required to adequately model the hydrolysis pathways. The equations that describe specific acid-base hydrolysis are:

$$-d[P]/dt = k_H[P][H^+] + k_{HOH}[P] + k_{OH}[P][OH^-] \quad (6)$$

where hydrogen ion and hydroxide ion concentration (actually activity) are specified as are acid, neutral and alkaline rate constants. The activities of H^+ and OH^- are readily obtained from the measurement of the pH of the water. If it is assumed that the H^+ and OH^- activities remain constant as a result of buffering of the water, then the observed rate constant is given by:

$$k_{obs} = k_H[H^+] + k_{HOH} + k_{OH}[OH^-] \quad (7)$$

Hydrolysis rate constants (k_{obs}) have been measured for a large number of pesticides in natural water samples. The rate constants compare favorably with the calculated rate constants obtained from data determined in distilled water. Thus to forecast the hydrolysis half-life of a pesticide in the water column, it is necessary only to know the pH of the water and the hydrolysis rate constants at the temperature of the water.

Hydrolysis in sediment suspensions. Recent studies have extended hydrolysis in the water column to hydrolysis in the presence of suspended sediments (ref. 12). The model considers sorption-desorption kinetics between the bulk water phase and organic carbon associated with the solid phase. It also assumes that reaction can occur in the bulk water phase and in the sediment sorbed phase.

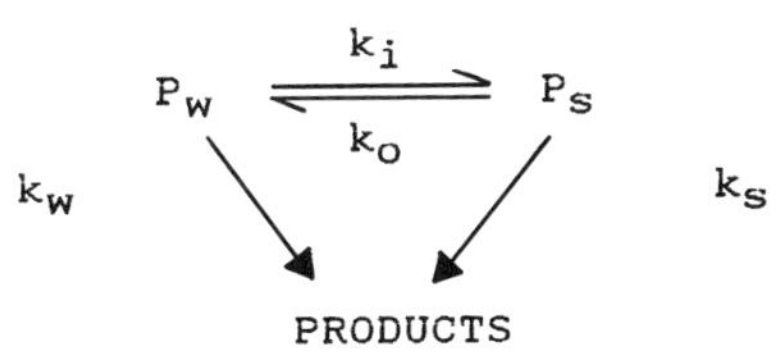

(8)

$$d[P]_w/dt = -(k_w + k_i)[P]_w + k_o[P]_s \qquad (9)$$

$$d[P]_s/dt = -(k_s + k_o)[P]_s + k_i[P]_w$$

$$d[P]_t/dt = k_s[P]_s + k_w[P]_w$$

$$[P]_t = [P]_w + [P]_s$$

$$d[P]_t/dt = -k_{obs}[P]_t$$

where $[P]_w$ and $[P]_s$ are concentrations of the the compound in the water and sediment sorbed phases respectively. Also, k_i and k_o are the respective sorption and desorption rate constants. The constants, k_w and k_s, are the respective hydrolysis rate constants for the water and sediment phases. $[P]_t$ is the total amount of compound in the sediment-water system at time t.

In this kinetic model, the reaction rate constant in the aqueous phase, k_w, is given by the hydrolysis rate constant in distilled water. Thus, there is no heterogenous catalysis by the solid particles. The compound in the sediment sorbed phase reacts with a rate constant, k_s, and must be determined experimentally.

The studies carried out to date indicate that water phase hydrolysis is accurately described by the specific acid-base and neutral kinetic expressions. On sediment surfaces, however, alkaline hydrolysis is retarded, neutral hydrolysis is unaltered and acid hydrolysis is unaltered or accelerated relative to the water phase processes.

It is not possible at this time to predict the magnitude of k_s for alkaline or acid mediated hydrolysis reactions. Based on kinetic data for the compounds studied so far, it appears that alkaline hydrolysis is retarded by at least two orders of magnitude. Acid catalyzed hydrolysis is accelerated by as much as an order of magnitude. The rate constant for neutral hydrolysis reaction is the same as that for distilled water. Additional work is needed to expand the data base and to determine the factors that dictate the magnitude of the rate effects in the sediment sorbed phase.

Redox processes

Redox transformations have been observed under a variety of environmental reaction conditions. The literature contains many qualitative descriptions of pesticides undergoing redox reactions. Reduction has been observed in anaerobic water columns and sediments while oxidation has been observed in surface waters on soil surfaces (ref. 13).

The types of reduction reactions that have been observed are numerous and include: dehalogenation (ref. 14); nitroreduction to amines (ref. 15); azo reduction to hydrazo and amine derivatives (ref. 16); quinone conversion to semiquinones and hydroquinones (ref. 17); and reduction of sulfones to sulfoxides and sulfides (ref. 18). Both abiotic and biotic pathways have been proposed to account for these processes. Chemical reducing agents include inorganics such as ferrous and manganese species. Bioorganics, such as flavins, porphyrins and extracelluar enzymes, have been studied in the laboratory as model reducing systems. Microbial pathways including metabolism and cometabolism have received the most attention. Unfortunately, these studies are of a qualitative nature and generally do not distinguish whether the reactions proceed by chemical or microbial pathways.

Oxidation pathways in the water column have received renewed attention. Proposed chemical pathways include direct oxidations by ozone or peroxides and autooxidations initiated by metals in the presence of peroxides. Biological pathways that have been suggested include monoxygenases, dioxygenases, and oxidative couples such as lactones, peroxidases and tryptophanases.

REACTIVE CHEMICAL SPECIES IN THE ABIOTIC DEGRADATION OF PESTICIDES IN NATURAL WATER

The preceding section has outlined in some detail the kinetic treatment of abiotic processes which simultaneously compete to degrade pesticides in the environment. A particular facet of this topic that has gained increasing attention over the past 10 years is that of indirect photolysis (ref.

19,20). At this time indirect photolysis is not well enough understood to be adequately described by rate laws and mathematical modelling. Indirect photolysis is given expanded coverage here, however, because there have been major advancements since the fifth IUPAC Congress of Pesticide Chemistry in elucidating mechanisms and identifying and quantitating reactive chemical species in natural waters. This attention is not meant to imply that indirect photolysis is important for all pesticides in all "types" of natural water.

Photochemical formation of reactive chemical species

Dissolved substances in natural waters generate reactive species by a variety of mechanisms outlined in equation sequence 10. In this scheme, natural aquatic solutes (so-called humus) absorb sunlight energy generating electronically-excited species. The first excited singlet state ($^1S^*$) may be quenched by interaction with the solvent or undergo intersystem crossing (ISC) to the relatively long-lived triplet state ($^3S^*$). Humus triplets may promote pesticide reactions in various ways including: 1) energy transfer; 2) abstracting H-atoms or electrons; and 3) generating electronically-excited molecular oxygen, again by energy transfer (ref. 21).

Alternatively, the humus chromophore may photoionize on absorbing solar radiation. The free electrons expelled are rapidly solvated and scavenged by molecular oxygen under aerobic conditions. Superoxide radical anion ($O_2^{-}\cdot$) rapidly disproportionates (dismutates) to hydrogen peroxide. The photoionization pathway, then, generates a number of highly reactive chemical species.

Hydrogen peroxide can be further activated to hydroxyl free radicals (OH·) on absorbing additional solar energy or on reduction by dissolved metal ions as in "Fenton-like" reagents. Nitrate ions also produce OH· on photolysis and are believed to represent the principal source of OH· in natural waters (see below). Finally, organic hydroperoxy compounds (SOOH), possibly formed on interaction of dissolved organic matter with $^1O_2^*$ or free radicals, may be cleaved photolytically to alkoxy radicals (RO·) and OH·. Reduction of SOOH by dissolved metal ions yields alkoxy radicals and hydroxide ions.

$$S \xrightarrow{h\nu} {}^1S^* \qquad (10)$$

$$^1S^* \xrightarrow{ISC} {}^3S^*$$

$$^3S^* + O_2 \longrightarrow {}^1O_2^*$$

$$S \xrightarrow{h\nu} S^{+}\cdot + e^{-}_{aq}$$

$$e^{-}_{aq} + O_2 \longrightarrow O_2^{-}\cdot \longrightarrow HOOH$$

$$HOOH \xrightarrow{h\nu} HO\cdot \qquad NO_3^{-} \xrightarrow{h\nu} HO\cdot$$

$$HOOH + M^{+X} \longrightarrow HO\cdot + OH^{-} + M^{+X+1}$$

$$SOOH \xrightarrow{h\nu} SO\cdot + HO\cdot$$

where S represents a naturally-occurring solute (aquatic humus).

Steady-state concentrations of these reactive chemical species in natural water ($[RS]_{ss}$), when combined with approximate second-order reaction rate constants, allow us to identify significant chemical processes. For this purpose, the pseudo-first-order half-life (equation 11) is a useful parameter.

$$-d[P]/dt = k\,[P]\,[RS]_{ss} \qquad (11)$$

$$t_{1/2} = 0.693 / k\,[RS]_{ss}$$

Singlet molecular oxygen

The occurrence of singlet molecular oxygen in natural waters was first reported by Zepp and coworkers in 1977 (ref. 22). The capacity of a natural water to generate $^1O_2^*$ appears to correlate with the dissolved organic matter content and is highest in highly colored, humic-containing waters.

Near surface $[^1O_2^*]_{ss}$ are in the range of 3 X 10^{-15} to 3 X 10^{-14} M per mg of DOC/L in midday summer sunlight (ref. 23,24). Accordingly, $[^1O_2^*]_{ss}$ as high as 5 X 10^{-13} M can be expected in moderately high DOC waters. $^1O_2^*$ also is formed on soil surfaces (ref. 25) and in elevated levels in surface oil films (ref. 26).

While $^1O_2^*$ appears ubiquitous and abundant in natural water, it is a selective oxidant reacting rapidly with relatively few organic structures. Aromatic molecules vary widely in their reactivity with second-order reaction rate constants of 10^4 - 10^9 M^{-1} s^{-1} -- products are transannular peroxides resulting from charactistic cycloaddition reactions. Other singlet oxygen acceptors include sulfides that are converted to sulfoxides ($k = 10^5 - 10^7$ M^{-1} s^{-1}) and allylic olefins. Members of the latter group are often used as $^1O_2^*$ probes, i.e., dimethylfuran (ref. 27) and furfuryl alcohol (ref. 28), since they are oxygenated to distinctive hydroperoxide products via an "ene" reaction.

Disulfoton (ref. 29), captan (ref. 20) and the synthetic pyrethroids, resmethrin (ref. 30), phenothrin and tetramethrin (ref. 31), are examples of pesticides that are efficient $^1O_2^*$ acceptors.

While dissolved humic materials can promote indirect photolysis processes, they also limit direct photodecomposition of pesticides by markedly reducing penetration of sunlight into the water column (ref. 4).

Hydroxyl free radicals

Much is known about the chemistry of OH·, both in water and the gas phase because this species has been extensively investigated by radiation and atmospheric chemists, respectively. OH·, unlike $^1O_2^*$, is an indiscriminate oxidant consuming most substrates in rapid reactions (i.e., most substrates react with second-order reaction rate constants of 10^9 - 10^{10} M^{-1} s^{-1}). OH· reacts by radical addition to aromatic pesticides to give hydroxycyclohexadienyl radicals that autooxidize to phenolic products (ref. 29,32). OH· also adds to olefins and, in the case of aldrin the epoxide derivative, dieldrin, is a product. In addition, OH· initiates other free radical reactions by H· and electron transfer.

Earliest reports of OH· in natural waters suggested concentrations of <10^{-17} M, too low to play a significant role in indirect photolysis (ref. 33). Since the role of nitrate ion (and not HOOH) as the principal photochemical OH· precursor has been clarified (ref. 34,35,36,37), it now appears that $[OH\cdot]_{ss}$ may be as high as 5 X 10^{-15} M in near surface waters. Direct measurement of OH· using a probe molecule is complicated by its high reactivity and tendency to produce a multitude of products.

$[OH\cdot]_{ss}$ is very sensitive to natural water chemistry because OH· is scavenged by dissolved organic matter, bicarbonate ion, and, in the marine environment, bromide ion. Accordingly, [OH·] is highest in natural waters high in nitrate and low in dissolved organic carbon. Certain agricultural practices, e.g., use of high nitrate groundwater for irrigation or distribution of ammonium nitrate fertilizer in irrigation water may promote OH·-mediated indirect photolysis of pesticides under field use conditions.

In scavenging OH·, dissolved organic matter and bicarbonate ion generate daughter radicals of greater selectivity. Additional research should be directed at characterization of these secondary radicals and defining their impact on pesticides and other micropollutants.

Hydrated electrons, superoxide radical anion and hydrogen peroxide

Hydrogen peroxide has been detected in agricultural irrigation water (ref. 38), surface water (ref. 39,40) and ground water (ref. 40) exposed to sunlight. In natural waters peroxide concentrations are generally below 10^{-5} M. The mechanism of peroxide formation, outlined here in equation sequence 10, is substantiated by detection of both e^-_{aq} (ref. 41,42) and ($O_2^{-}\cdot$) (ref. 40,43) in irradiated natural water. Steady-state, aquatic peroxide levels appear too low to consume most pesticides in the absence of sunlight or metal ion activation.

Little is known about the organic chemistry of e^-_{aq} and $O_2^{-}\cdot$ in aqueous solution, e.g. reaction rate constants, products and so forth. Breugem and coworkers (ref. 42) suggest that $[e^-_{aq}]_{ss}$ in natural water may reach 10^{-16} M and is proportional to dissolved organic matter. The restricted lifetime of

e^-_{aq} due to scavenging by O_2 suggests that this species may be confined to the surface of aquatic humus particles and possibly responsible for reduction of sorbed organics and complexed metal ions.

Superoxide is both an oxidant and reductant and an effective nucleophile (ref. 44) in aprotic solvents. Halogenated pesticides studied in organic solvents with potassium superoxide are subject to a variety of selective elimination reactions (ref. 45). In aqueous media nucleophilic displacement of chlorines is observed (ref. 46). Superoxide, however, is not extremely reactive with organic molecules. Typical second-order reaction rate constants with chlorinated alkanes and DDT are in the range of 10^1 - 10^3 M^{-1} s^{-1} (ref. 44).

CONCLUSION

Tremendous strides have been made in the past 10 years in elucidating mechanisms responsible for abiotic degradation of pesticides. Advancements in understanding pesticide fate processes have largely been due to basic research in aquatic chemistry and photochemistry. Similar progress can be expected if studies are focused in the following areas:

- Measurement of physical/chemical- and reaction rate constants for pesticides
- Characterization of reactive species in natural water and determination of the impact of water chemistry on their steady-state levels
- Application and refinement of mathematical models

ACKNOWLEDGMENT

We would like to thank R. G. Zepp for his time and useful discussions on the kinetics of photochemical reactions. D . M. Mar provided expert technical assistance with manuscript preparation. The National Institutes of Health (grant R01 OH02119-01 from the National Institute for Occupational Safety and Health of the Centers for Disease Control) provided support for W.M.D. during preparation of this manuscript. We thank the Program Committee for their invitation to participate and generous support.

REFERENCES

1. A. M. Winer, R. Atkinson and J. N. Pitts, Jr., Science (Washington, DC) 224, 156-159 (1984).
2. L. A. Burns and D. M. Cline, U. S. EPA Report, EPA-600/3-85/038, 1-83 (1985).
3. L. A. Burns, D. M. Cline and R. R. Lassiter, U. S. EPA Report, EPA-600/3-82/023, 1-443 (1982).
4. R. G. Zepp and D. L. Cline, Environ. Sci. Technol. 11, 359-366 (1977).
5. R. G. Zepp, P. F. Schlotzhauer, M. S. Simmons, G. C. Miller, G. L. Baughman and N. L. Wolfe, Fresenius Z Anal. Chem. 319 (1984).
6. R. G. Zepp N. L. Wolfe, J. A. Gordon and G. L. Baughman, Environ. Sci. Technol. 9, 1144-1150 (1975).
7. U. I. Skurlatov, R. G. Zepp and G. L. Baughman, J. Agric. Food Chem. 31, 74-81 (1985).
8. N. L. Wolfe, R. G. Zepp, J. A. Gordon, G. L. Baughman and D. M. Cline, Environ. Sci Technol. 11, 88-93 (1977).
9. E. M. Perdue and N. L. Wolfe, Environ. Sci. Technol. 17, 635-642 (1983).
10. J. A. A. Ketelaar, H. R. Gersman and M. M. Beck, Nature 177, 392-393 (1956).
11. E. M. Perdue and N. L. Wolfe, Eviron. Sci. Technol. 16, 847-852 (1982).
12. D. L. Macalady and N. L. Wolfe, J. Agric. Food Chem. 33, 167-173 (1985).
13. D. L. Macalady, P. G. Tratnyek and T. J. Grundl, J. Contam. Hydrol. 1, 1-28 (1986).
14. G. M. Klecka and S. J. Gonsior, Chemosphere 13, 391-402 (1985).
15. P. A. Wahid, C Ramakrishana and N. Sethunathan, J. Environ. Qual. 9, 127-130 (1980).
16. P. Dubin and K. L. Wright, Xenobiotica 5, 53-571 (1975).
17. D. Liu and W. M. J. Strachan, Water Res. 15, 353-359 (1981).
18. C. J. Miles and J. J. Delfino, J Agric. Food Chem. 33, 455-460 (1985)

19. D. G. Crosby, Pest. Soil: Ecol., Degradation Mov., Int. Symp., 1970, 1970, 86.
20. N. L. Wolfe, R. G. Zepp, G. L. Baughman, R. C. Fincher and J. A. Gordon, U. S. EPA Report, EPA-600/3-76-067 (1976).
21. R. G. Zepp, G. L. Baughman and P. F. Schlotzhauer, Chemosphere 10, 109-117 (1981).
22. R. G. Zepp, N. L. Wolfe, G. L. Baughman and R. C. Hollis, Nature 267, 421-423 (1977).
23. W. R. Haag and J. Hoigne', Environ. Sci. Technol. 20, 341-348 (1986).
24. R. G. Zepp, P. F. Schlotzhauer and R. M. Sink, Environ. Sci. Technol. 19, 74-81 (1985).
25. K. Gohre and G. C. Miller, J. Agric. Food Chem. 31, 1104-1108 (1983).
26. L. K. Sydnes, S. H. Hansen and I. C. Burkow, Chemosphere 14, 1043-1055 (1985).
27. C. S. Foote, M. T. Wuesthoff, S. Wexler, I. G. Burstain, R. Denny, G. O. Schenck and K- H. Schulte-Elte, Tetrahedron 23, 2583-2599 (1967).
28. W. R. Haag, J. Hoigne', E. Gassman and A. M. Braun, Chemosphere 13, 631-640 (1984).
29. W. M. Draper and D. G. Crosby, J. Agric. Food Chem. 29, 699-702 (1981).
30. K. Ueda, L. C. Gaughan and J. E. Casida, J. Agric. Food Chem. 22, 212-220 (1974).
31. L. O. Ruzo, Ian H. Smith and J. E. Casida, J. Agric. Food Chem. 30, 110-115 (1982).
32. W. M. Draper and D. G. Crosby, J. Agric. Food Chem. 32, 231-237 (1984).
33. T. Mill, D. G. Hendry and H. Richardson, Science (Washington, DC), 207, 886-887 (1980).
34. D. Kotzias, H. Parlar and F. Korte, Naturwissenschaften 69, 444 (1982).
35. H. Russi, D. Kotzias and F. Korte, Chemosphere 11, 1041-1048 (1982).
36. W. R. Haag and J. Hoigne', Chemosphere 14, 1659-1671 (1985).
37. R. G. Zepp, J. Hoigne' and H. Bader, Environ. Sci. Technol. (in press).
38. W. M. Draper, D. G. Crosby and J. B. Bowers, "Measurement of photochemical oxidants in agricultural irrigation water", Proceedings of the 172nd National Meeting of the American Chemical Society, San Francisco, CA (1976).
39. W. M. Draper and D. G. Crosby, Arch. Environ. Contam. Toxicol. 12, 121-126 (1983).
40. W. J. Cooper and R. G. Zika, Science (Washington, DC) 220, 711-712 (1983).
41. A. M. Fischer, D. S. Kliger, J. S. Winterle and T. Mill, Chemosphere 14, 1299-1306 (1985).
42. P. S. Breugem, P. van Noort, S. Velber, E. Wondergem and J. Zijlstra, Chemosphere 15, 717-724 (1986).
43. R. M. Baxter and J. H. Carey, Nature (London) 306, 575-576 (1983).
44. J. L. Roberts and D. T. Sawyer, J. Am. Chem. Soc. 103, 712-714 (1981).
45. P. Dureja, J. E. Casida and L. O. Ruzo, Tetrahedon Lett. 23, 5003-5004 (1982).
46. W. M. Draper and D. G. Crosby J. Agric. Food Chem. 31, 734-737 (1983).

Predicting, measuring and minimizing the effects of pesticides on farmland wildlife on intensively managed arable land in Britain

Nicolas W. Sotherton* and Michael R.W. Rands

The Cereals and Gamebirds Research Project, the Game Conservancy, Fordingbridge, Hampshire, SP6 1EF, United Kingdom

Abstract - The decline of the grey partridge (*Perdix perdix* L.) in Britain over the past thirty years is used to illustrate how it is thought that pesticides may have affected some species of farmland wildlife on intensively farmed areas of lowland Britain. Long-term monitoring studies are described that have been used to predict annual rates of chick mortality and how well these explain the observed declines of grey partridges.
Two methods are investigated to minimize the effects of pesticides on farmland wildlife.
1. The screening of compounds to assess the extent of the spectrum of activity against non-pest, non-target insects. Combined laboratory and field trials are described in a screening process and two examples are cited.
2. The modification of pesticide use on small areas of cereal fields for the benefit of farmland wildlife. This involves the exclusion of certain pesticides at certain times of the year on limited areas of cereal fields.

INTRODUCTION

In lowland Britain, the grey or Hungarian partridge (*Perdix perdix* L.) has suffered a severe decline in numbers over the past thirty years. Breeding stock densities have been surveyed by the Game Conservancy since 1933 and the national decline since 1952 has been from approximately twenty-five pairs per km^{-2} to approximately five pairs in the early 1980's; a decline of about 80%. Similar declines have been observed in North America and Eastern Europe (ref. 1).

The reason for this decline in Britain is considered to be the increased levels of chick mortality rates observed during this decline phase (ref. 1). For the first two to three weeks of life, grey partridge chicks rely almost entirely on insects in the diet for their sources of food before switching to a plant diet at about three weeks of age (ref. 2). Decreased chick survival rates are considered to be caused by the lack of sufficient insect food in the diet of chicks during these crucial early weeks (ref. 1) (Fig. 1).

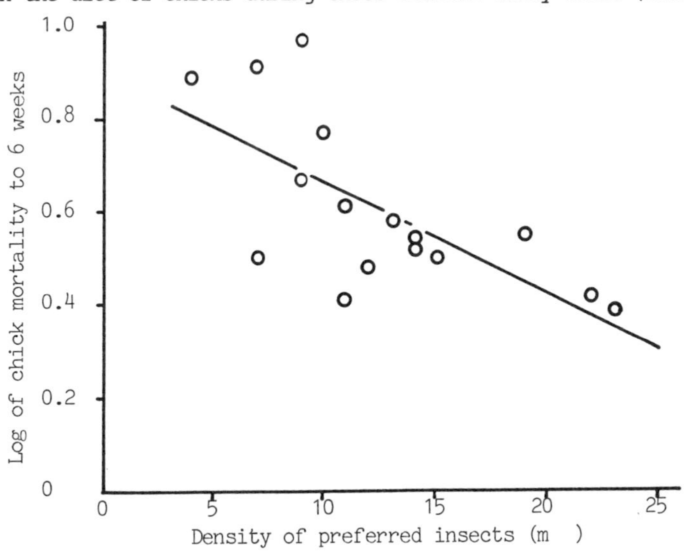

Fig. 1. Annual chick mortality in relation to the density of preferred insects, third week of June, Sussex U.K., 1967-1985.

Broods of grey partridges forage in lowland Britain on arable land that has become more and more intensively farmed since the 1950's. Intensification and the associated increased use of pesticides during this time has been shown to be responsible for some of these observed increases in chick mortality (Table 1). Indeed, many groups of non-target insects especially those known to be preferred food items of partridge chicks (ref. 2) have shown to have declined since the late 1960's from data accumulated in the Game Conservancy's Sussex Monitoring Data Bank (ref. 1). Beginning in 1969, in each year up to 1986, insects on the Sussex study area in southern England (ref. 3) were sampled from between 100-150 cereal fields (ref. 4). The pesticides currently used on cereals are not directly toxic to grey partridge chicks. Their effects on partridge populations are via a removal of preferred insect food items. Many of these chick food items have been shown to be reduced directly by the use of broad-spectrum compounds, either insecticides (ref. 5, ref. 6) but also fungicides (ref. 7, ref. 8, ref. 9) and herbicides (ref. 10). Many of these preferred cereal dwelling arthropods are weed feeders. Therefore, their numbers can also be reduced by the removal of their host plants by herbicide use (ref. 11, ref. 12, ref. 10).

TABLE 1. Estimates of partridge chick mortality rates grouped according to time and pesticide use (ref. 1).

	Chick survival rate (%)**		
	Up to 1952 (no herbicide)	1953 - 1961 (some herbicide)	1962 - 1985 (herbicide and some insecticide)
U.K. National*	46.7	35.5	30.9
Hampshire U.K.	47.8	31.6	25.7
Sussex U.K.	--	35.9	22.4
Mainland Europe	51.3	42.7	35.5
North America	56.2	52.5	43.7

* based on The Game Conservancy's National Game Census
** confidence limits given in original text

Pesticide use on the Sussex study area increased dramatically over the monitoring period. In 1969, 10% of cereal fields received no pesticides and no fields received either insecticide or fungicide. In 1984, only one field of eighty-seven sampled received no pesticides. All of the fields received at least one application of a fungicide and 37% of fields received an application of insecticide (ref. 1).

This long-term monitoring exercise has enabled the declines in insects to be detected and measured, and to explain the observed decline of a species of bird dependent on farmland. It also enables us to measure and predict the effects of pesticides on this indicator species. This will be illustrated in more detail later.

Multiple regression analyses of the effects of the weather and insect abundance on chick mortality rates, using data from this monitoring work allows us to predict chick mortality annually. A total of 58% of the variation in chick mortality was explained by this method (ref. 1):

$$\text{Chick mortality rate} = 1.532 - 0.016x_i, - 0.048x_{ii}$$

where x_i = density of preferred insects (m^{-2}) in cereal fields, third week in June.
x_{ii} = mean daily temperature 10 June - 10 July.

Calculations of the effects of pesticides on the density of preferred insects can then be incorporated into this equation and used to predict the effects of pesticide use on chick mortality.

If, as we suggest, the effects of pesticides have greatly contributed to the observed declines in the number of grey partridges in Britain then, the future for this species appears to be uncertain, especially as trends in pesticide use on cereals in Britain do not appear to be changing. Indeed, methods of alleviating the potential adverse effects of pesticides must be found in the British cereal growing situation where today's increased quantity and quality of grain have been brought about largely because of the use of pesticides and in a situation where standards for the acceptance of grain into intervention are becoming increasingly more stringent in times of surplus production.

The Game Conservancy's farmer funded "Cereals and Gamebirds Research Project" has been set up specifically to devise methods of alleviating adverse effects of modern farming. The Project has put forward two approaches to achieve a balance between the needs of farmland game and wildlife and the present intensive nature of cereal production.

Firstly, advocating the use by farmers of those pesticides without a broad-spectrum of activity against non-target arthropods and secondly, the exclusion or modification of pesticides on small areas of cereal fields.

Assessing the spectrum of activity of a compound has involved laboratory and field screening. The Cereals and Gamebirds Cereal Project has begun work by screening the foliar fungicides approved for use in United Kingdom cereal fields. So far, of twenty-seven single active ingredient compounds screened in the laboratory only one (pyrazophos) has been shown to have significant insecticidal properties (Table 2). Use of pyrazophos in cereal crops showed the extent of the insecticidal action under normal field conditions (Table 3). Mortalities of up to 69% were recorded amongst the chick food insects where pyrazophos had been used compared to only 10% where two other fungicides had been applied. Reductions of preferred food insect densities by 60% by the use of pyrazophos were predicted to have reduced chick survival rates from 31% to 12% in 1984, and from 20% to 16% in 1985.

TABLE 2. Laboratory screening to assess the mortality of three foliar fungicides to species of non-target arthropods compared to mortality caused by a toxic standard and a control (distilled water). Pesticides applied at field dose rate and dilution. Mortality estimated after 24 hrs (after Sotherton, unpubl. data).

	Carabidae (Ground beetles)	Staphylinidae (Rove beetles)	Miridae (Plant bugs)	Chrysomelidae (Leaf beetles)	Tenthredinidae (Sawfly larvae)	Syrphidae (Hoverfly larvae)
pyrazophos	100.0	100.0	75.0	100.0	90.0	72.0
propiconazole	0.0	0.0	0.0	6.0	5.0	4.0
triadimefon	0.0	0.0	20.0	1.5	5.0	10.0
toxic standard	100.0	100.0	100.0	93.3	100.0	96.0
control	0.0	0.0	0.0	0.0	13.0	4.0

TABLE 3. The effects of three foliar fungicides on non-target, beneficial arthropods in United Kingdom cereal fields (after Sotherton, unpubl. data).

	% reduction due to pesticide compared to untreated areas				
	Propiconazole*	Triadimefon*	Pyrazophos*	Pyrazophos**	Pyrazophos***
Total chick food insects	9.6	3.3	44.6	69.4	31.0
Total predatory arthropods	NIL	NIL	24.9	25.1	47.6
Total aphid-specific predators	NIL	16.9	83.8	80.4	34.6
Parasitoids	NIL	NIL	55.2	36.0	34.0

* = 1984, winter wheat, sprayed at GS 50, sweep samples
** = 1985, winter barley, sprayed at GS 37, suction samples
*** = 1984, spring barley, sprayed at GS 60, suction samples

The activity-spectrum of insecticides to control cereal aphids in summer in Britain have also been studied so that farmers can be given advice concerning their potential effects on preferred insects. Preliminary work based on field scale experiments has shown a likely reduction in insects of 50% following the use of pirimicarb, 58% using demeton-s-methyl and 80% using dimethoate (ref. 1).

Exclusion of certain pesticides at certain times of the season has also been proven as a technique for the alleviation of the effects of agrochemicals on wild game and farmland wildlife. The technique which has become known as "unsprayed headlands" involved the exclusion of all pesticides (herbicides, insecticides and fungicides) in the spring and summer from the outermost six metres around the crop edge. From 1983 to 1986, large blocks of cereal fields were either sprayed normally or had the headlands left unsprayed on many farms throughout lowland Britain. Production of wild grey partridges greatly increased in all experiments (ref. 13, ref. 14). The dependence of wild pheasant chicks on insect food in cereal crops (ref. 15) has also led to an observed increase in pheasant numbers on areas containing unsprayed headlands (ref. 14). The effects of unsprayed headlands on wild game production are summarised in Table 4. The use of unsprayed headlands is clearly of major benefit to the production of gamebird chicks on intensively farmed cereals in the United Kingdom. Moreover, where other game conservation techniques are followed (eg. nest predator control and the provision of adequate nesting cover), long term changes in grey partridge population densities can be obtained. On our major study area in Hampshire where these techniques have been continually practised, spring pair densities (per km^2) were approximately fourteen pairs before 1950. Throughout the five years prior to our experiments, the density had fallen to around four to five pairs. After three years of experimenting with unsprayed headlands to alleviate the adverse effects of pesticides, on only a proportion of the total cereal acreage, the 1986 spring pair count was up to twelve pairs per km^2 (Rands, unpubl. data).

TABLE 4. The effects of unsprayed headlands in cereal fields on wild gamebird chick production in United Kingdom cereal fields, 1983 - 1985 (ref. 13, ref. 14).

	Sprayed headlands			Unsprayed headlands		
	Mean brood size	No broods counted	Chick survival rate (%)	Mean brood size	No broods counted	Chick survival rate (%)
			Grey partridge			
a) Hampshire						
1983	4.7	18	19	8.4	21	47
1984	7.4	34	37	10.0	34	67
1985	3.3	9	12	5.7	14	31
b) East Anglia						
1984	4.7	71	20	7.8	57	40
1985	2.7	19	9	4.0	19	15
			Pheasant			
a) Hampshire						
1984	3.2	18	—	6.9	29	—
1985	3.0	3	—	4.6	8	—
b) East Anglia						
1985	2.6	30	—	3.7	35	—

Project work has now expanded to cover the benefits to other forms of farmland wildlife other than these indicator species. Work at present is underway to include farmland butterflies (ref. 16) songbirds, wild flowers and insectivorous mammals (ref. 17). In close conjunction with this work, the Project is also quantifying the agronomic implication of the technique of using "unsprayed headlands" and developing a management advice based on a much more rigorous definition of "unsprayed" and "headland" (ref. 18). This will enable us to give advice to United Kingdom cereal growers on wildlife conservation on intensively farmed land.

REFERENCES

1. G.R. Potts, *The Partridge: pesticides, predation and conservation*. Collins, London (1986).
2. R.E. Green, *J. Appl. Ecol.* 21, 817-830 (1984).
3. G.R. Potts and G.P. Vickerman, *Adv. Ecol. Res.* 8, 107-197 (1974).
4. G.R. Potts, *Proc. NERC ITE Monks Woods Exp. Sta.* , 128-134 (1984).
5. G.P. Vickerman and K.D. Sunderland, *J. Appl. Ecol.* 14, 767-777 (1977).
6. K.C. Brown, S.W. Shires and J.H. Lawton, *Environ. Entom.* (in press).
7. G.P. Vickerman, *Proc. Br. Crop Prot. Conf.* 1, 121-128 (1977).
8. G.P. Vickerman and N.W. Sotherton, *Pestic. Sci.* 14, 405-411 (1983).
9. N.W. Sotherton and S.J. Moreby, *Ann. Appl. Biol.* 104, 16-17 (1984).
10. N.W. Sotherton, *Z. Angew. Ent.* 94, 466-451 (1982).
11. T.R.E. Southwood and D.J. Cross, *J. Anim. Ecol.* 38, 497-509 (1969).
12. G.P. Vickerman, *Proc. Br. Weed Cont. Conf.* 3, 929-939 (1974).
13. M.R.W. Rands, *J. Appl. Ecol.* 22, 49-54 (1985).
14. M.R.W. Rands, *Ibis* 128, 57-64 (1986).
15. D.A. Hill, *J. Appl. Ecol.* 22, 645-654 (1985).
16. M.R.W. Rands & N.W. Sotherton, *Biol. Cons.* 36, 71-82 (1986).
17. N.W. Sotherton, *Ann. Rev. The Game Conservancy 1985* 17, 54-55 (1986).
18. N.D. Boatman, *Ann. Rev. The Game Conservancy 1985* 17, 61-65 (1986).

Use of laboratory model ecosystems to evaluate quantitatively the behaviour of chemicals

Ingolf Schuphan*, Evelyn Schärer, Margarete Heise and Winfried Ebing

Abteilung für ökologische Chemie, Biologische Bundesanstalt, D-1000 Berlin 33, Federal Republic of Germany

Abstract - A concept is presented for the quantitative evaluation of xenobiotics in soil-plant systems and laboratory model terrestrial ecosystems. Characteristic features of these systems are closed cultivation containers with controlled ventilation conditions which comprise monoliths containing the natural plant and animal organisms as well as artificially composed model ecosystems. These "ecochambers" are constructed of all Tempax^R or Duran^R glass. Quantitative determination of volatile parts of the organic ^{14}C-labelled chemical applied and its terminal breakdown product $^{14}CO_2$ is feasible out of the airstream of up to 60 airchanges per hour. The different chambers can be constructed easily and they are relatively low in price. Small glass containers (10 up to 100 L) are mainly suitable for testing chemicals in single foodchains or a plant-soil system under defined reproducible conditions. Ecochambers up to 450 L serve for studies of chemicals as influenced by sections of ecosystems and partially deliver data of the effects of chemicals on the organisms. A model was designed to represent the agroecosystem "savoy cabbage field". The main components of the model-biocenosis were savoy cabbage *(Brassica oleracea L.)* as the cultural plant, larvae of the large white butterfly *(Pieris brassicae L.)*, and their most important endoparasite, the braconide wasp *Apanteles glomeratus L.*. The fate and impact of ^{14}C-labelled contaminants were balanced in this system in two studies using ^{14}C-HCB and ^{14}C-monolinuron. An urban grassland (non fertilized meadow) was included into the studies as a characteristic non-agroecosystem. The urban grassland model-section consists of four graminaceous plant species (two main species are *Poa pratensis L.* and *Lolium perenne L.)* and nine broadleaved plant species. The original fauna is enriched by Carabidae (four genera, mainly two species *Calathus fuscipes Goez.* and *Amara aenea Degeer)*, earthworms *(Allolobophora caliginosa Sav.)* and pillbugs *(Porcellio scaber Latr.* and *Oniscus asellus L.)*. The fate of ^{14}C-pentachlorophenol had been balanced in this urban modelecosystem.

INTRODUCTION

Field tests conducted with chemicals yield only data on residues of the agents, i.e. in plants and soil, and even at best they can only indicate the decrease in concentration of the agents in the course of time. Such tests trace the time- dependent disappearance of the initial chemical, as it is brought about by atmospheric influences, e.g. leaching by rain, volatilization due to certain temperature and wind conditions, conversion by the action of UV light and by physiological events. In fact, however, all that has probably happened is that the active ingredient or its metabolites have merely been relocated by such events, and generally speaking only a very small quantity has been broken down into the final products of mineralization such as carbon dioxide and water; this can be proven by means of quantitative balances, but not by field tests carried out under open growth conditions (ref. 1).

CLOSED SYSTEMS

However, such quantitative data can be obtained in closed systems in which studies are carried out under conditions of controlled ventilation. By designing special vegetation containers and using, for example, ^{14}C- labelled chemicals the volatile organic components and also the ultimate breakdown product carbon dioxide can be specifically and quantitatively determined and thus approximately complete (100 %) balances can be calculated (ref. 2)

Working with closed systems, e.g. to trace the quantitative conversion and breakdown characteristics of ^{14}C-environmental chemicals in soils <u>without</u> plants, while simultaneously detecting ^{14}C-carbon dioxide, is not new and is a method that is already being employed (ref. 3). The studies become more difficult when plants are present in the system. In this case, it is not possible first to remove the carbon dioxide from the inflowing air because it is vitally needed for the assimilation processes taking place in the plants. In addition, a particularly high rate of airflow must be provided to prevent ^{14}C-carbon dioxide formed by the chemical under test will to a large extent be reassimilated by the plant and will not be available for detection (ref 4). Therefore, one should try to achieve high ventilation rates which approximate the air movements encountered in the field. Achieving a high airflow in closed vegetation chambers creates technical problems, involves a high consumption of chemicals, and necessitates a great deal of maintenance work. The tests described below have shown that an approximately 60-fold replacement of the air in the chamber per hour is technically possible and just about feasible in practical terms (ref 5). The replacement rate of 60 exchanges of the chamber air per hour which is proposed here, and which has occasionally been achieved, corresponds to a wind speed of no more than 1.25 m/min. In the field (Berlin) the average values at a point 0.5 m above the ground are in the order of 49.8 m/min. The high rates of exchange of the chamber air which are called for in closed vegetation chambers with controlled ventilation systems create some major technical problems for the experimenter as far as the quantitative absorption of $^{14}CO_2$ is concerned. Consequently, the closed cultivation systems described in the literature are generally only fitted with devices to adsorb the organic components without detecting the $^{14}CO_2$ (ref. 6-8).

In contrast, a very expensive "plant metabolism box" has been developed with devices for simulating various climatic parameters prevailing in the field as well as for determining the mineralization of chemicals into $^{14}CO_2$ at low rates of chamber airflow namely between 2.5 and 10 replacements per hour (refs 9-10). This complicated and expensive "plant metabolism box" can be used to conduct studies commissioned by third parties.

A description will be given of two cultivation chamber systems; each of them is simple to set up in any laboratory and they are cheap and technically easy to operate for the following purposes:

- Tracing the fate of organic chemicals quantitatively in soil-plant systems. These containers are used to obtain <u>ecochemical</u> data in a small number of plants and in the soil after the application of the active ingredient to the soil or to the plants [volatilization characteristics, uptake in the plant, determination of the non-metabolized residue, quantitative determination of metabolites, including the ultimate breakdown product $^{14}CO_2$, formed from the chemical under test (ref. 11].

- Tracing the fate of environmental chemicals in laboratory model ecosystems in <u>complex</u>, closed vegetation chambers. These systems are used to obtain <u>ecochemical</u> and <u>ecotoxicological</u> data. Further data can be gathered on enrichment factors between organisms at different trophic levels. Included host-parasite (harmful organism-beneficial organism) systems deliver additional ecotoxicological data (ref. 5, 12).

STRUCTURE AND FUNCTION OF CLOSED CULTIVATION CONTAINERS FOR TESTS INVOLVING SOIL-PLANT SYSTEMS

The closed cultivation container system, with a controlled air supply, consists of the specially constructed cultivation container and adsorption devices for the volatile organic components of the active ingredients or of the metabolites, or for the $^{14}CO_2$ formed by the mineralization of the chemical. The design principle of the system is shown in diagrammatic form in Fig. 1 (ref. 13).

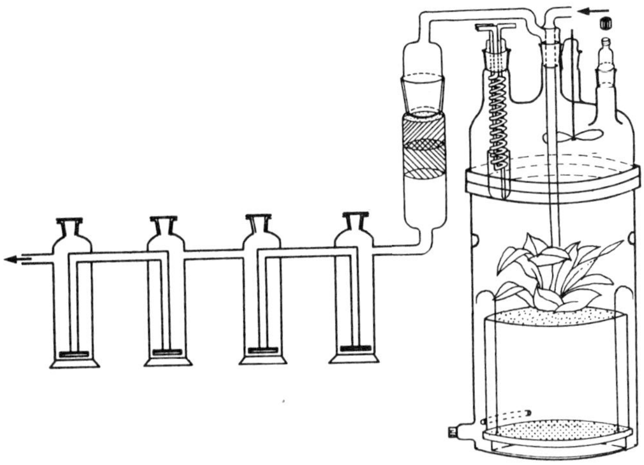

Fig. 1. Closed cultivation chamber system for balance studies using radioactively labelled chemicals.

The plants are treated with a ^{14}C-labelled chemical and continue growing in the glass container. A current of air is drawn through the apparatus to supply the plants with the CO_2 needed for photosynthesis and at the same time to remove the $^{14}CO_2$ formed by metabolic breakdown of the ^{14}C-labelled chemical. Fig. 2 presents the details of one type of cultivation container.

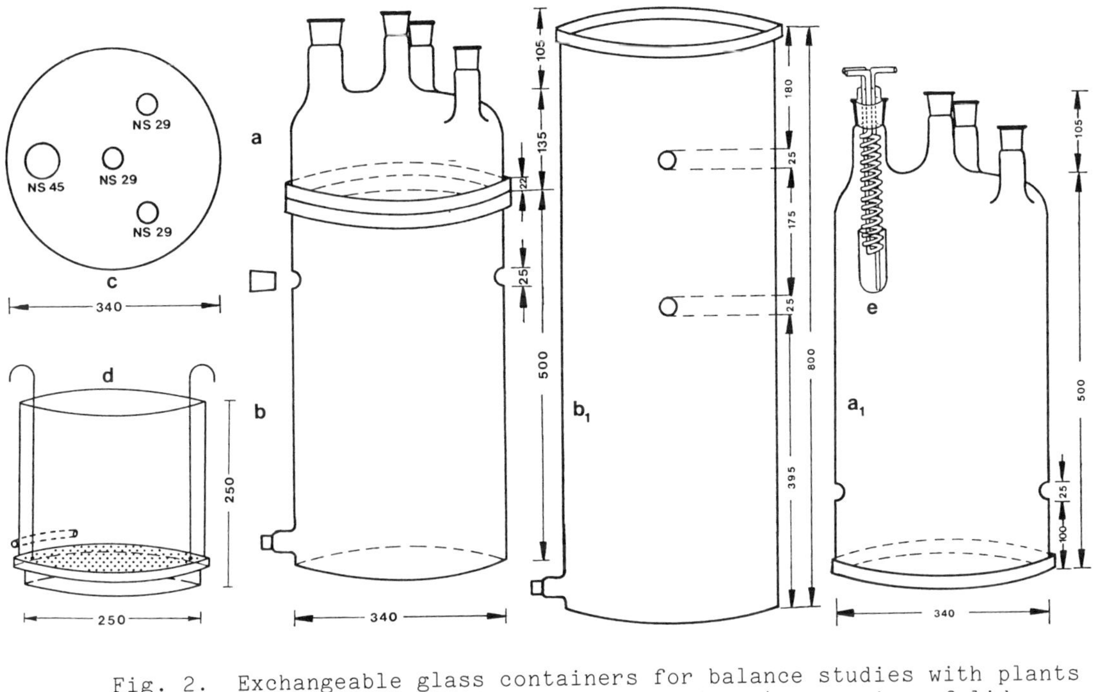

Fig. 2. Exchangeable glass containers for balance studies with plants (dimensions in mm). Lids (a, a_1); Bodies (b, b_1); Top view of lid (c); Soil container (d); Self-depleting condensing water separator (e).

BALANCE DATA OBTAINED FROM THE APPLICATION OF SODIUM [^{14}C] PENTACHLOROPHENOL IN A SOIL-GOLDENROD SYSTEM

Out of studies concerning the fate of xenobiotica in urban terrestrial ecosystems Table 1 presents results obtained with sodium [^{14}C]pentachlorophenol (Na-PCP) after soil application in a soil-goldenrod *(Solidago gigantea)* system.

Detection of 7.5 % as the mineralization product carbon dioxide in only 19 days is the most important result. Volatilization of pentachlorophenol is of minor interest in the evaluation of alkaline Na-PCP. Minerali-

TABLE 1. Radioactivity balance following application of sodium [^{14}C] pentachlorophenol to soil in a soil-goldenrod system taken as a core into the closed glass container (duration of cultivation: 19 days)

System components	Percent of applied radioactivity		
	Extractable	Non-extractable	Sum
Solidago gigantea			
plant (shoot)	0.37	0.18	0.55
roots	0.13	0.07	0.2
Weeds	0.18	0.13	0.31
Soil			
0-1 cm			84.4
1-3 cm			1.8
3-8 cm			0.6
CO_2			7.52
Volatile organics			0.64
Washings			0.27
Total[^{14}C]balance			96.3

zation of PCP as the free phenol is much slower in comparable studies (results not shown).

CLOSED VEGETATION CHAMBERS WITH CONTROLLED VENTILATION SYSTEMS FOR ECOCHEMICAL AND ECOTOXICOLOGICAL STUDIES OF THE BEHAVIOUR OF ENVIRONMENTAL CHEMICALS IN LABORATORY MODEL ECOSYSTEMS

Consistent further development of the concept for deriving quantitative ecochemical data on organic compounds progressed from soil-plant systems to laboratory model agroecosystems. The aim was to trace the pathway and fate of a chemical through to the individual components of a laboratory ecosystem. Fig. 3 shows in diagrammatic form the design of the all-glass vegetation chamber.

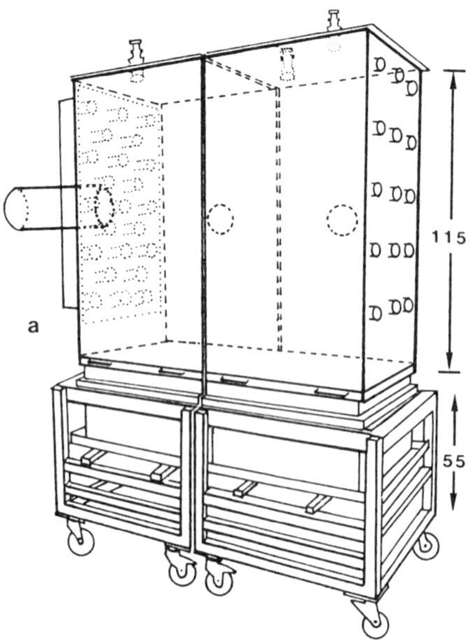

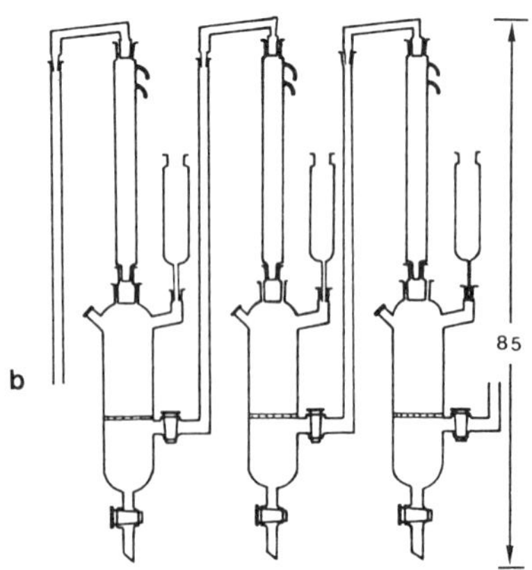

Fig. 3. All glass vegetation chamber (a); $^{14}CO_2$-absorption system (b), dimensions in cm

The chamber consists of UV-permeable Tempax glass; it has a base area of 120 x 60 cm and is 115 cm high. It accepts monoliths of the soil as well as the plants and animals and it is built in two separate, joinable halves, each of which is mounted on movable stands to make it easier to load and also to carry out the analyses at a later date - it also permits the transition of chemicals from one half of the chamber system to the other to be studied. The climate in the chamber is regulated by devices for ventilation, lighting and moisten the air as well as to simulate rainfall and to control the irrigation of deeper soil layers. Volatile components of the test substance and also of organic metabolites are adsorbed from the exhaust air on polyurethane foam plugs mounted in the air outlet holes in the chamber as well as in the main exhaust air pipe which conducts away all the air leaving the chamber. The quantitative determination of the metabolite carbon dioxide in the exhaust air, which was a central problem to be solved when designing the well ventilated chamber system, is achieved by means of separating out part of the airflow from the overall airflow in a controlled and reproducible manner for the purpose of analysis, after the air has passed through the polyurethane foam plugs. Large, cooled absorption traps are used to quantitatively bind the carbon dioxide (inactive and radioactive) from the analytical airflow by forming carbonate-carbamate compounds. The precisely defined and controlled split ratio permits the total amount of $^{14}CO_2$ leaving the system to be calculated. The total balance is given by volatile ^{14}C-components together with the residues of ecosystem elements obtained from extraction or combustion processes.

COMPOSITION OF THE LABORATORY MODEL "CABBAGE FIELD" AGROECOSYSTEM

The model system described here is intended to represent a field of savoy cabbage reduced to its essential components (ref 5) The central elements in this agrarian biocenosis are the "crop plant" savoy cabbage with a phytophagous "pest" (harmful organism), namely the larva of the large cabbage white butterfly *(Pieris brassicae)*, as well as the natural antagonist ("beneficial organism") of the cabbage white, namely the endoparasitic braconid *Apanteles glomeratus*. This "harmful-beneficial" relationship is of special interest because through it the transfer of the test chemical via the trophic level can be simultaneously combined with possible influences on the agriculturally useful organism. The model system also contains other primary producers of biomass, i.e. frenchweed *(Galinsoga parviflora)* and common grounsel *(Senecio vulgaris)*. An additional phytophagous organism is present in the form of the common garden slug *(Deroceras reticulatium)* while the saprophagous organisms are represented by two species of earthworm *(Lumbricus terrestris, Allolobophora* spec.), woodlice *(Armadillidium vulgare, Trachelipes rathkei)*, millipedes *(Cylindroiulus londinensis, Brachydesmus* spec., *Polydesmus* spec.) and the zoophagous organisms are represented by centipedes *(Lithobius forficatus, Geophilus longicornis)*.

MODEL APPLICATION OF MONOLINURON; A PHENYLUREA HERBICIDE

The use of monolinuron as a model chemical to test the performance of the vegetation chamber, including the model "cabbage field" agroecosystem, was justified by the experience gained with the metabolism and mineralization of this substance, namely its breakdown into CO_2. In addition to determining the volatile organic components of the active ingredient and metabolites, the characteristic feature of the vegetation chamber that was developed was to quantitatively determine the ultimate breakdown of environmental chemicals into CO_2.

The non-parasitized L_2 *Pieris* larvae living on the undersides of the cabbage leaves while the herbicide was being applied pupated after 14 days, and in the case of the parasitized *Pieris* larvae the *Apanteles* larvae began to bore their way out of their hosts, starting on day 16. Over the entire period of the test 2-3 % of the amount of ^{14}C-monolinuron applied was mineralized relatively uniformly at 3-day intervals. After 21 days of testing 15 % of the ^{14}C carbonyl group had been oxidized to $^{14}CO_2$ and nearly 20 % of the active ingredient plus organic metabolites volatilized. These results underscore the importance of the closed method of operation (Table 2).

Converting the detected radioactivity into residues gives the values presented in Table 3 for the cabbage- *Pieris-Apanteles* food chain which is of central interest here.

TABLE 2. Radioactivity balance after the application of 0.4 kg/ha [^{14}C]carbonyl-monolinuron in the "model cabbage field" system, after a test duration of 21 days.

System components	Percent of applied radioactivity			
	Extractable		Non-extractable	Total
	Monolinuron	Metabolites		
Soil	16.0	6.9	27.10	50.0
Savoy cabbage	1.6	2.7	0.59	4.9
Pieris - non-parasitized				1.23
Pieris - *Apanteles* complex				0.07
Deroceras				0.06
Armadillidium				0.07
Brachydesmus, Polydesmus, *Cylindroiulus*				0.02
Lithobius				< 0.01
Allolobophora, Lumbricus				0.94
Volatilized monolinuron				18.65
Breakdown product $^{14}CO_2$				15.1
Acetone rinse + water				0.05
Total [^{14}C]-balance				91.0

TABLE 3. Residue values calculated for monolinuron from the radioactivity measured in the various links of the food chain ("pest - beneficial organism" relationship) at the end of the 21-day test.

System components	Monolinuron residues mg/kg	
	Fresh weight	Dry weight
Savoy cabbage leaves	3.51	27.65
Pieris host larvae	2.87	15.96
Apanteles parasites (pupae incl. cocoons)	5.24	14.31
Apanteles imagines	1.38	4.86
Pieris -non-parasitized (L_5)	4.79	30.09
Pieris -non-parasitized (pupae)	3.45	17.34
Pieris imagines	3.21	8.76

COMPOSITION OF THE LABORATORY MODEL "URBAN GRASSLAND" ECOSYSTEM

Urban grassland was chosen as a model representing characteristic biotops in and nearby of cities. Grassland sections were taken between November and June from an unfertilized park area being cut only once a year. Preincubation of the grassland monoliths prior to an experiment in a climatisized greenhouse under reproducible conditions guaranteed comparable developing stages at the beginning of each experiment. Using several of the monolith carrier frames it is possible to preincubate such model sections parallel to running experiments in the vegetation chamber.

The urban grassland model section consists of four graminaceous plant species *Poa pratensis* and *Lolium perenne* as two dominant and *Agrostis tenuis, Bromus mollis* as minor dominant species. Nine broadleaved plant species are present of which *Cerastium fontanum, Taraxacum officinale, Potentilla argenta* and *Ranunculus acris* are the most important. The original fauna is enriched by Carabidae of four genera, mainly two species, *Calathus fuscipes* and *Amara aenea* and earthworms *(Allolobophora caliginosa)* and pillbugs *(Porcellio scaber* and *Oniscus asellus).*

These standardized model sections were developed for tests of chemicals under reproducible conditions. The evaluation of the results may lead to a decision, whether this system may serve as a model giving results precisely enough to predict the behaviour of new chemicals in an urban environment.

The results of two balance experiments in succession with sodium [^{14}C] pentachlorophenol are listed in Table 4.

TABLE 4. Radioactivity balances of two experiments in succession after the application of 0.4 kg/ha sodium [^{14}C] pentachlorophenol in the model "urban grassland" ecosystem, after a test duration of 23 days.

System components	Percent of applied radioactivity	
	I	II
CO_2 + volatile parts	12.9	10.6
Soil	60.2	65.8
Plants	19.0	22.1
Animals	0.16	0.06
Washings	0.04	0.01
Total [^{14}C] balance	92.3	98.6

The radioactivity distribution of experiment II is shown as an example in detail in Table 5.

TABLE 5. Distribution of radioactivity in the system components of experiment II.

System components	Percent of applied radioactivity	
Air		10.6
CO_2	5.0	
Volatile organics	5.6	
Soil		
1. Layer 0-1 cm		42.7
Root part	11.8	
Soil part	30.9	
2. Layer 1-3 cm		19.2
Root part	8.5	
Soil part	10.7	
3. Layer 3-5.5 cm		2.1
Root part	0.3	
Soil part	1.8	
4. Layer 5.5-8 cm		1.8
Root part	0.2	
Soil part	1.6	

System components	Percent of applied radioactivity	
Plants		
Monocotyle		17.3
Stalk + leaves	15.8	
Inflorescence	1.5	
Dicotyle		4.8
Cerastium fontanum	2.4	
Taraxacum officinale	1.0	
Potentilla argentea	0.6	
Ranunculus acris	0.5	
Veronica arvensis	0.2	
Polygonum aviculare	0.1	
Animals		0.06
Earthworms	0.05	
Carabidae	0.004	
Pillbugs +	0.003	
Spiders +	0.001	
Grasshoppers	0.0005	

DISCUSSION AND CONCLUSIONS

The closed test procedure described, using smaller cultivation containers as well as greater vegetation chambers enable

- to obtain quantitative balance data of chemicals in organism systems
- to include data concerning volatile parts of compounds and chemical degradation products normally not obtained
- to operate ecosystem sections using mainly defined and reproducible conditions
- incomplicated use of this method in all laboratories because of low price replica and easy maintenance.

Regarding to other systems the following advantages involve

- obtaining of even the terminal degradation product $^{14}CO_2$ of organic chemicals in a simple, easy to handle experimental arrangement
- independance of climatic variations as they appear e.g. in outdoor lysimeter experiments using ^{14}C-labelled chemicals
- possibility to compare data of chemicals between each other because of standardized experiment conditions
- keeping or breeding mobile organisms in the system.

The use of model ecosystem sections enable

- to trace the fate of a chemical in a collective of organisms
- to observe distinct foodchains e.g. host-parasite (pest-beneficial organism) systems
- to calculate enrichment-derichment factors
- to view more completely the behaviour of chemicals as influenced by many ecological parameters.

ACKNOWLEDGMENTS

These studies were supported in part by the German Research Foundation (DFG), Bonn-Bad Godesberg, as well as the Environmental Agency (Umweltbundesamt) of the FRG, Berlin.

REFERENCES

1. I. Schuphan, Mittlg. Biol. Bundesanst. Berlin, 203, 12-28 (1981).
2. I. Schuphan, Chemosphere 6, 5-10 (1977).
3. P.C. Kearney, A. Kontson, J. Agric. Food Chem. 24, 424-426 (1976).
4. F. Führ, Isotopes and Radiation in Agricultural Sciences. II. M.L'Annussziata, Ed., Academic Press, London, 239-270 (1984).
5. E. Schärer, Mittlg. Biol. Bundesanst. Berlin, 215, 1-169 (1983).
6. M. L. Beall Jr., R.G. Nash, P.C. Kearney, Proceed. of EPA Conf. on Environmental Modeling and Simulations, April 20-22, Cincinnati, Oh. 790-793 (1976).
7. J.W. Gillett, J.D. Gile, Intern. J. Environmental Studies 10, 15-22 (1976).
8. R. G. Nash, M.L. Beall, W.G. Harris, J. Agric. Food Chem. 25, 336-341 (1977).
9. K. Figge, J. Klahn, GIT Fachz. Lab. 26, 680-685 (1982).
10. K. Figge, J. Klahn, J. Koch, Regulatory Toxicol. Pharmacol. 3, 199-215 (1983).
11. I. Schuphan, W. Ebing, Pesticide Biochem, Physiol. 9, 107-118 (1978).
12. E. Hoffmann, I. Schuphan, Z. Angew. Entomol., in press.
13. I. Schuphan, Habilitation, Johannes Gutenberg-Universität, Mainz (1985).

A physical-chemical microcosm for ricefield environmental fate studies

Richard M. Higashi* and Donald G. Crosby

Department of Environmental Toxicology, University of California Davis, Davis, CA 95616 U.S.A.

Abstract – Physical-chemical microcosms can serve as a basis for prediction of persistence. The California flooded rice field environment was modeled because of its well-defined history, relatively few biological interactions, and the large amounts of agrochemicals introduced into it yearly. We have constructed a computer-controlled aquatic microcosm to model volatilization and photolysis of the herbicides molinate and bentazon, respectively. Modeling was conducted with the dissipation forces (temperature and UV light) maintained both steady-state (field average) and profiled (diurnally varied as in a corresponding field study).

Molinate dissipation in the steady-state microcosm (28°C, $t_{1/2}$=1.60 days) was in discordance with experiments under profiled temperatures (18-32°C, $t_{1/2}$= 7.22 days), but the latter was similar to the field dissipation. This difference was attributed to an interaction of the profiled temperature with the formulation. For bentazon, the dissipation under constant UV and profiled temperatures (16-28°C, $t_{1/2}$=3.00 days) matched the field dissipation rate ($t_{1/2}$= 2.90 days) well, but was faster than the steady-state average temperature experiment (23.5°C, $t_{1/2}$=4.12 days). In addition, cycling the UV light diurnally resulted in the persistance of a major photochemical product, although it did not accumulate under constant UV.

These results indicate that accurate representation of variable field forces is essential for the development of a realistic microcosm system.

INTRODUCTION

Microcosms are laboratory systems that seek to mimic the environment. Although most designs have been used to determine the equilibrium compartmentation of a chemical [ref.1], in reality the environment is dynamic. As a consequence, processes of dissipation are in continual nonequilibrium, so that the rate of processes such as hydrolysis are constantly changing. If it were possible to control the forces that drive processes so that they mimic the actual field profiles, we should be able to identify and prioritize them, and thereby account for the variable dissipation rates caused by diurnal or irregular cycling of forces.

Logically, the development of a dynamic microcosm should entail four steps: (a) definition of the environmental system and chemicals to be modeled; (b) identification of the physical and chemical forces that drive the processes of dissipation; (c) control of these forces sufficient to mimic field profiles; and (d) comparison of the experimental results against that of a field study, repeating steps b–d to add to the system's capabilities. Eventually the microcosm could also accomodate biological interactions. The completion of these steps may qualify the microcosm for the estimation of persistance and fate of other chemicals in the same environment, or for the testing of computational models.

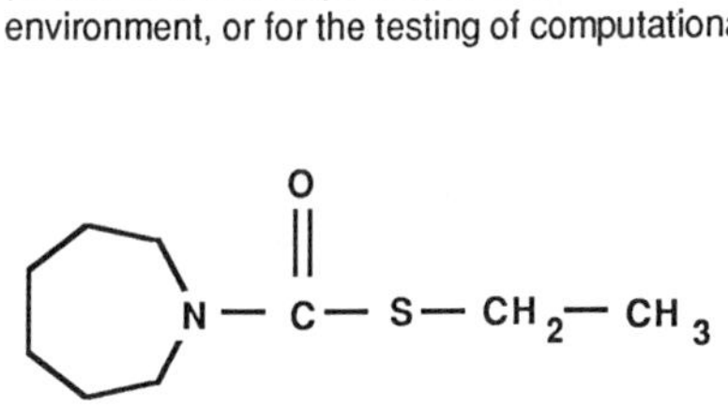

Molinate

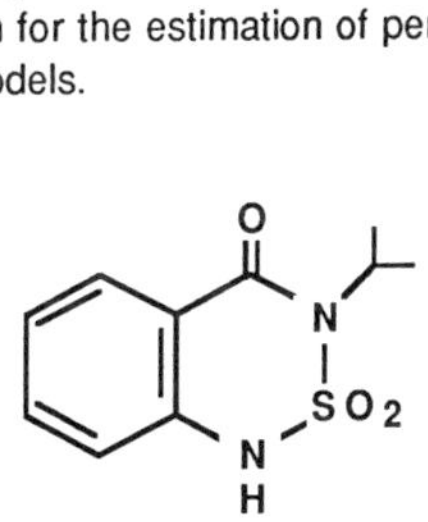

Bentazon

Fig. 1. Structures of molinate and bentazon.

THE CALIFORNIA RICE FIELD MICROCOSM

In order to test the feasibility of the above approach, we have constructed a microcosm to model the physical and chemical dissipative processes of newly-flooded California rice fields. This environment was chosen mainly because of its well-defined physical and chemical characteristics. For instance, the water is of uniform depth and held stagnant for 2-8 days with minimal leaching following application of a pesticide. In addition, the newly-flooded field has relatively few biological interactions, and large amounts of agrochemicals that can serve as model chemicals are introduced into it yearly [ref.2]. The selection of this particular environment reduces the major non-biological dissipative processes to soil adsorption, volatilization, and chemical breakdown. In this paper, we describe the modeling of volatilization and photochemical processes using the herbicides molinate and bentazon, respectively (Fig. 1).

The modeling of these processes in the rice field environment required an aquatic microcosm that has good control over the pertinent forces, including airflow, temperature, and UV intensity. Figure 2 describes this microcosm, which is a 75.7 l aquarium fitted with airflow manifolds, an integrated heating and cooling system, and UV lamps. The airflow, temperature, and lighting systems are controlled by the computer, which can be programmed to follow the diurnal profiles of a particular field study (profiled), or even that of a hypothetical day. To begin the experiment, either the analytical grade or formulated chemical was added, at field application rates, to the microcosm water. Although only the experiments that used air and water are described here, others using air-water-soil gave similar results. As a guide for microcosm experiments, commerical agricultural applications were monitored for the dissipation behavior of molinate and bentazon.

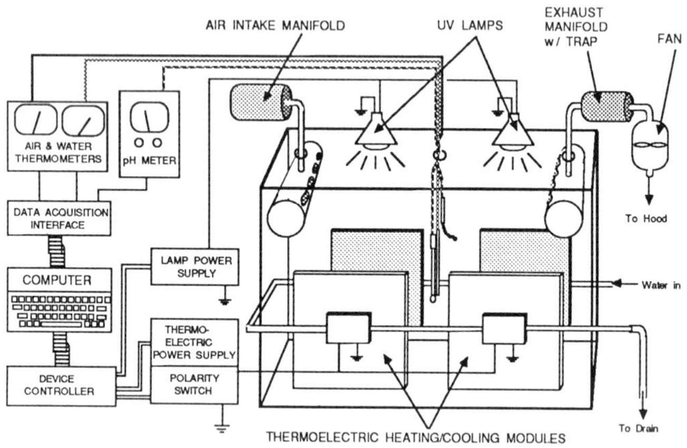

Fig. 2 Schematic of the microcosm system

The analytical methods were as follows. Analysis of molinate was by extraction of a 20 ml water sample with a C_{18} bonded-phase resin cartridge, elution with 1 ml of ethyl acetate, and injection of the eluate into the GLC (180 °C isothermal program on 0.53 mm i.d. X 10 m RSL-300 open tubular column, N-P detection). Bentazon and its products were analyzed by C_8 resin cartridge extraction of an acidified 20 or 200 ml water sample, followed by elution with 1 ml of methanol and analysis of the eluate by HPLC (gradient: 10-90% MeCN:H_2O +0.5% HOAc over 30 min, 1.0 ml/min flow, 4.6 X 250 mm 5µ C_{18} column, 300 nm detection).

VOLATILIZATION MODELING

The grass herbicide molinate was chosen for modeling volatilization because of its heavy use on rice and its principal dissipation route, which is volatilization [ref.3]. It exhibits no direct photochemical breakdown at sunlight wavelengths (>290 nm) and is commercially applied as Ordram® 10G, a granular formulation. In the rice fields, the concentration of molinate in the water typically reaches a maximum in the first 12–16 h following application, during which desorption from the granules is completed. Thereafter, the concentration decline is due 85–90% to volatilization [ref.3] before the held water is released at about the 8th day. The particular field study we were modeling was unusual in that the herbicide was applied in the evening instead of the usual morning application.

Table 1 lists the experiments performed, together with the resulting half-lives (in days) of the parent compound. The results of the volatilization modeling were reasonable, and dissipation rates were reproducible to within 5%. Compared with the field study (I), the microcosm results under profiled temperature (II) showed a 20% longer $t_{1/2}$. This difference is consistent with the report that 10-15% of molinate in the field dissipates by means other than volatilization [ref.3].

However, an experiment at constant 28 °C (III) resulted in a much shorter $t_{1/2}$, similar to the volatilization of the pure chemical from water (IV, V) [ref.3; Higashi and Crosby, unpublished data]. Experiments IV and V demonstrated that the diurnal variation of temperature *per se* was not responsible for this discrepancy. It was still possible that the formulation and the profiled temperature were interacting in some way, since these factors were the differences between experiments II–V. Studies of desorption of molinate under profiled temperatures (data not shown) established that the normal 12–16 h period for complete desorption coincided with the steadily dropping temperature of the evening application modeled in II. This retarded the completion of desorption for up to 30 h after application, which interfered with the volatilization measurement (14–48 h after application).

Experiment VI verified this scenario dramatically. This experiment was identical to II except that the temperature profile was shifted by 12 h (i.e., to simulate early morning as opposed to evening application). In this case, rapid desorption was encouraged instead of retarded (as in II), and the results were in keeping with the average-temperature experiments (III and V). Thus, even in a simple air-water-formulation system, unexpected interactions can occur, and the field situation must be reproduced carefully.

TABLE 1. Field and microcosm experiments

	Profiles		Chemical	$t_{1/2}$
Experiment[a]	Temperature[b]	UV	Formulation	(days)[c]
Molinate experiments [d]				
I Field study	Field, 18–32°C	Field	Ordram 10G	6.02
II Microcosm	Field, 18–32°C	none	Ordram 10G	7.22
III Microcosm	Constant, 28°C	none	Ordram 10G	1.60
IV Microcosm	Field, 18–32°C	none	molinate	1.70
V Microcosm	Constant, 28°C	none	molinate	1.80
VI Microcosm	Field, 12h shifted	none	Ordram 10G	1.60
Bentazon experiments [e]				
VII Field study	Field, 16–28°C	Field	Basagran	2.90
VIII Microcosm	Field, 16–28°C	Constant	bentazon	3.00
IX Microcosm	Constant, 23.5°C	Constant	bentazon	4.12
X Microcosm	Constant, 23.5°C	Field	bentazon	6.39

[a] All microcosm experiments were air-water systems, and airflows were constant at 5 l/min.
[b] Constant temperatures are the time-weighted average temperatures from the corresponding field study.
[c] Half-lives for Ordram formulations were measured starting at 12 h after application, by which time desorption of molinate from the granules is normally complete (see text).
[d] Molinate experiments ran for ~48 h, and a no-airflow experiment gave <3% dissipation over 120 h.
[e] Bentazon experiments were ~50 h long, and there was negligible dissipation in the dark control.

PHOTOCHEMICAL MODELING

The broadleaf herbicide bentazon was chosen due to its increasing usage on rice, and its major dissipation route (photochemical) [ref.4,5; Higashi and Crosby, unpublished data]. In contrast to molinate, its volatility is negligible, and its commercial form is an emulsifiable concentrate known as Basagran®. Typically, the rice field water is held for 24–48 h following application, after which the water may be released. Investigation of bentazon dissipation in rice fields has not been reported.

Results from the field and microcosm studies of bentazon also are shown in Table 1. As with the volatilization modeling, using the field temperature profile (under constant UV, VIII) gave results similar to that of the field (VII). By substituting constant temperature for the field profile, the $t_{1/2}$ was 37% longer (IX), suggesting that rates of the prevailing photochemical processes were nonlinear with temperature.

Note, however, that the microcosm required constant illumination to match the field dissipation rate. This indicated that either naturally-occuring oxidants [ref.6, 7] or the sunlight intensity and wavelengths were not adequately modeled. Experiment X, which substituted diurnal light for constant, had a $t_{1/2}$ twice that of VIII, indicating that the UV intensity in the microcosm was insufficient by a factor of about 2. Experiments using higher intensity lamps are planned.

In addition to dissipation rates, the breakdown product distribution over time can be modeled. Bentazon has over 16 products that can be simultaneously monitored by HPLC, and the time courses for one major photoproduct under constant and diurnal light is shown in Fig. 3. The identity of this product is currently under investigation. Under constant UV, the concentration of this product reached a peak after 12 hours of illumination, but fell to low levels by 38 h. On the other hand, under diurnal UV, the product accumulated and reached a steady-state concentration. It is likely that the photochemically formed product itself is photolabile, and that diurnal UV maximizes its 12-hour formation while limiting its photodegradation. As Fig. 3 shows, this leads to a significant difference in the predicted concentration of this major product at the end of the second day, when the held water may be released into the waterways.

As with volatilization modeling, the photochemical modeling demonstrates that the use of profiled instead of steady-state experimental conditions can lead to drastically different predictions of chemical persistence.

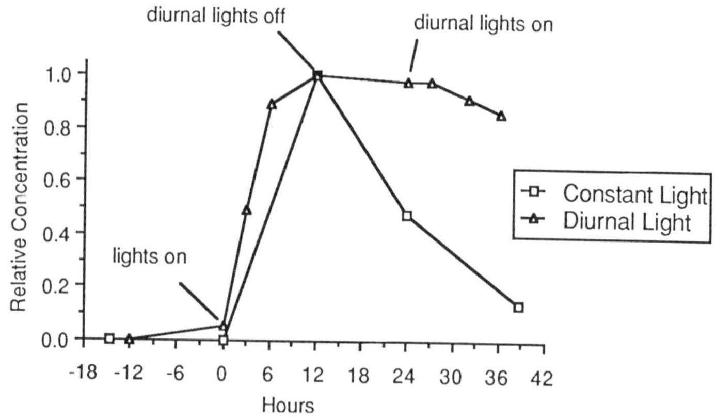

Fig. 3. Persistence of a photochemcial product of bentazon under constant and diurnal light.

CONCLUSION

The experiments described here demonstrated that the dynamics of environmental forces can profoundly affect the dissipation processes of chemicals. For molinate, the timing of the diurnal temperature variation (not the variation *per se*) reduced molinate dissipation in the field by four-fold, preventing direct interpretation of the field data based on volatilization theory and steady-state experiments. Similarly for bentazon, diurnal cycling of temperature and light affected overall field dissipation rates and product distribution, impairing interpretation by simple chemical kinetics. In all these cases, modeling the forces in a dynamic fashion enabled interpretation of the field dissipation by normalizing the variable process rates.

Prediction of persistence is presently an area of uncertainty in environmental chemistry [ref.8]. The importance of accounting for effects of the dynamics of the environment cannot be overstated. Even in a simple air-water microcosm system, we have demonstrated that a photoproduct of bentazon may be predicted either to be insignificant, or to build up to significant levels at the time of water discharge into public waterways. When we consider the much greater complexity of the real world, it becomes clear that accurate modeling of ever-changing forces are essential for an understanding and prediction of environmental chemical fate.

REFERENCES

1. J.W. Gillet IN Dynamics, Exposure and Hazard Assessment of Toxic Chemicals, R. Haque (ed.), p.231–249, Ann Arbor Science Publishers, Ann Arbor (1980).
2. CDFA, Pesticide Use Report 1983, Calif. Dept. of Food and Agriculture, Sacramento, CA (1984).
3. C. J. Soderquist, J.B. Bowers, and D.G. Crosby, J. Agr. Food Chem. 25, 940–946 (1977).
4. G.P. Nilles and M.J. Zabik, J. Agr. Food Chem. 23, 410–415 (1975).
5. S. Otto, P. Beutel, N. Drescher, and R. Huber, Adv. Pest. Sci. 3, 551–556 (1978).
6. W.M. Draper and D.G. Crosby, J. Agr. Food Chem. 32, 231–237 (1984).
7. R.D. Ross and D.G. Crosby, Environ. Toxicol. Chem. 4, 773–778 (1985).
8. W.B. Neely, Environ. Toxicol. Chem. 1, 259–266 (1982).

ACKNOWLEDGEMENTS

This research was supported, in part, by the California Rice Research Foundation. The authors wish to thank R. Barton Draper, Amy Pettibone, and Teresa Fan for stimulating discussions and suggestions.

Short-term biochemical indicators of chronic insecticide toxicity to *Daphnia magna*

Charles O. Knowles* and Michael J. McKee

Department of Entomology, University of Missouri, Columbia, Missouri 65211
USA

Abstract - *Daphnia magna* were exposed to chlordecone, fenvalerate or di-2-ethyl-hexyl phthalate for 21 days. On days 7 and 21 of exposure, levels of total protein, ribonucleic acid (RNA), and deoxyribonucleic acid (DNA) were measured, and the results were related to protein growth, survival and reproduction during the 21-day test period. The static parameters of growth at day 7, protein, RNA and DNA, had significant positive correlation coefficients with protein growth, survival and reproduction at day 21. The coefficients for protein and RNA were greater than 0.80 which may be sufficiently high to support a predictive model. Significant positive correlation was observed between the dynamic bioindicator protein:RNA:DNA ratios were negatively correlated. No correlation was observed using the protein: DNA ratio. The protein:RNA:DNA ratio yielded the highest correlation with chronic toxicant effects on protein growth, survival and reproduction for the dynamic variables tested.

INTRODUCTION

Insecticide impact on aquatic receiving systems is managed mainly on information collected from a battery of tests based on guidelines promulgated by regulatory agencies. These tests provide data on the survival, growth and reproduction of organisms and form the basis of the water quality criteria for that chemical. Although these criteria may provide protection of aquatic resources, they have been criticized as being non-responsive to site specific variations in toxicity. Moreover, some chemicals do not have sufficient toxicity to warrant the extensive testing scheme. The responsiveness and cost effectiveness may be enhanced by the development of bioassays which are more sensitive in the laboratory and which have potential application in field monitoring programs associated with compliance monitoring and hazard assessment (Refs 1, 2).

A considerable amount of research has been devoted to studying biochemical indicators of stress in aquatic organisms. One goal of this research has been to develop biochemical indicators that aid in understanding and predicting chronic toxicant effects on survival, growth and reproduction of aquatic organisms in the laboratory. Another goal has been to use biochemical indicators to assess the "health" of aquatic organisms *in situ*. The laboratory components of biochemical indicator research may provide short-term alternatives to chronic toxicity tests currently required by regulatory agencies and, in addition, provide basic information that will be helpful in assessing their utility in field applications.

Although many laboratory studies of the biochemical effects of toxicants in aquatic organisms have been conducted, the relationship between these effects and whole organism responses such as survival, growth and reproduction need further elucidation (Refs 3, 4). Recent investigations have demonstrated that chronic effects of chlordecone, fenvalerate and di-2-ethylhexyl phthalate, (DEHP) on survival and reproduction of *Daphnia magna* Strauss were preceded by biochemical changes after 7 days of exposure (Refs 5-7). These biochemical changes were classified as static if they involved a change in biomolecule content per individual or as dynamic if they involved a change in the ratio of biomolecules. The objective of this paper is to compare the results of these studies with regard to the effects of chlordecone, fenvalerate and DEHP on static and dynamic changes involving protein, ribonucleic acid (RNA) and deoxyribonucleic acid (DNA) and their relationship to survival and reproduction of *D. magna*.

MATERIALS AND METHODS

Organisms and Compounds

D. magna were cultured in well water at $22^{\circ}C \pm 2^{\circ}$, at a photoperiod of 16:8 light:dark, lighting of about 80 ft-c, and at a density of 20 adults/L. Chemical characteristics of the well water follow: hardness (as $CaCO_3$), 300 mg/L; alkalinity (as $CaCO_3$), 250 mg/L; pH, 7.9; conductivity, 473 μmhos/cm; dissolved oxygen, saturation. Daphnids were fed about 1.2 x 10^7 cells of *Selenastrum capricornutum* and a mixture of 10 mg of yeast and PR 11 trout chow per liter of culture each day. The culture was cleaned three times each week.

Technical grade samples of chlordecone, fenvalerate, and DEHP were used.

Test System

The test system, which was patterned after that recommended by the American Society for Testing and Materials (Ref. 8), consisted of five toxicant concentrations and one solvent control. Tests were conducted under conditions similar to those for culture except that flow-through rather than static conditions were used. Each test concentration comprised four replicate 1-L beakers that received 3.6 turnovers of solution per day. The study was initiated by adding 20 first instar daphnids (<24 h old) to each beaker. The system was cleaned and reproduction assessed three times each week. Daphnids were fed as mentioned above. Adult daphnids were sampled for biochemical analysis on days 7 and 21. Since all animals were used in the biochemical assays, total protein, specified in this paper as protein growth, was used as the index of growth.

Isolation, Analysis, and Recovery of Biomolecules

On day 7, samples were collected by pipetting 10 adult daphnids from each of four replicates per concentration making a total of 24 samples (control + five treatments) for each compound. Samples on day 21 were composed of the remaining adult daphnia. Procedures for isolation and analysis of biomolecules have been described in detail (Ref. 5). Briefly, daphnids from each replicate were homogenized in 0.2 N perchloric acid, and an aliquot was analyzed for total protein as described by Lowry *et al.* (Ref. 9) and modified by Miller (Ref. 10). To isolate RNA and DNA, the Schmidt-Thannhauser method described by Munroe and Fleck (Ref. 11) and modified by Shibko *et al.* (Ref. 12) and Vytasek (Ref. 13) was used. RNA was quantified by UV detection at 260 nm (Ref. 14), and DNA was determined by the fluorometric method of Kissane and Robins (Ref. 15) as modified by Vytasek (Ref. 13). Recovery and analytical variation as the coefficient of variation were 99.1 ± 9.0% and 3.6 ± 1.0% for protein, 88.7 ± 3.0% and 3.6 ± 1.0% for RNA, and 86.3 ± 6.5% and 6.6 ± 1.5% for DNA.

RESULTS AND DISCUSSION

Table 1 shows the effects of the three chemicals on survival, protein growth and reproduction of *D. magna* after 21 days of exposure. The maximum allowable toxicant concentration (MATC) based on survival, protein growth and reproduction was between 15 and 30 μg/L for chlordecone, between 0.13 and 0.25 μg/L for fenvalerate, and between 600 and 1200 μg/L for DEHP.

Toxicant induced reductions in protein growth at day 7, protein growth at day 21, reproduction at day 21, and survival at day 21 were associated with reduced levels of static bioindicators (Table 2). Positive correlation was highest for protein growth at day 7 and reproduction at day 21. The static bioindicators protein and RNA yielded correlation coefficients with reproduction that were greater than 0.80 which confirmed that biochemical indicators at day 7 were significantly related to chronic toxicant effects. The static bioindicators had relatively low correlations with day 21 survival (Table 2).

Protein:RNA ratios were positively correlated to toxicant related reductions in survival, protein growth and reproduction; however, the correlation coefficients were low, and this would make prediction of chronic effects difficult (Table 3). Barron and Adelman (Ref. 16) found that decreased growth in fathead minnows as a result of toxicant exposure was associated with an increased protein:RNA ratio (reported as decreased RNA:protein ratio). Few investigations of toxicant effects on protein:RNA ratios have been reported making definitive relationships difficult to establish.

RNA:DNA ratios have been shown to be positively correlated with the growth rate in many organisms (Refs 17, 18). Barron and Adelman (Ref. 16) found that reduced growth of the fathead minnow exposed to toxicants was highly correlated with decreases in the RNA:DNA ratio. We found a negative correlation between RNA:DNA ratios and protein growth at day 7 and reproduction at day 21 (Table 3). This negative correlation was not expected since the literature indicated that a positive correlation is more general. However, a definitive relationship between RNA:DNA ratios in zooplankton remains to be demonstrated (Ref. 18). We are currently investigating the relationship between growth rate and RNA:DNA ratio in *Daphnia magna* under non-stressed conditions.

TABLE 1. Effects of chlordecone, fenvalerate and DEHP on Daphnia magna survival, protein growth and reproduction[a]

Concentration µg/L	% of Control: Survival	Protein Growth	Reproduction
Chlordecone			
3.8	95.0 (16.8)	102.8 (7.5)	104.1 (10.6)
7.5	100.0 (4.1)	94.6 (6.9)	107.3 (5.0)
15.0	95.3 (8.8)	87.1 (4.2)	109.2 (3.6)
30.0	0.0 (0.0)		43.9 (5.3)
60.0	0.0 (0.0)		17.8 (3.3)
Fenvalerate			
0.03	98.8 (2.5)	114.6 (5.4)	104.5 (2.0)
0.06	93.3 (4.5)	114.0 (5.9)	100.3 (3.5)
0.13	94.3 (8.5)	109.7 (4.3)	104.2 (6.9)
0.25	93.5 (13.0)	99.8 (6.3)	78.5 (5.6)
0.50	80.0 (21.6)	49.7 (10.4)	27.6 (6.2)
DEHP			
75	97.5 (5.0)	110.1 (6.7)	103.8 (1.1)
150	100.0 (0.0)	96.4 (14.0)	99.9 (5.3)
300	94.5 (6.4)	114.7 (12.4)	108.6 (4.5)
600	97.5 (5.0)	100.5 (13.1)	98.4 (4.3)
1200	45.0 (53.0)	100.5 (4.3)	89.3 (3.1)

[a] Values are means with standard deviations in parentheses, n=4.

The relationship between the protein:RNA:DNA ratios and growth and reproduction has not been previously reported; however, this ratio provided the highest correlation of the dynamic bioindicators investigated (Table 3). The correlations were potentially high enough that linear models may have utility in predicting chronic effects on growth and reproduction. The correlation with survival, however, was too low for use in a predictive capacity. Although little is known about this ratio, it is encouraging that agreement exists between increases in this parameter and decreases in growth and reproduction.

TABLE 2. Correlation matrix for the relationships among static biochemical indicators at day 7 and protein growth (day 7), protein growth (day 21), reproduction and survival[a]

	Protein	RNA	DNA
Protein Growth (Day 7)	0.91069 P<0.01 60	0.89817 P<0.01 60	0.82715 P<0.01 57
Protein Growth (Day 21)	0.72259 P<0.01 50	0.63428 P<0.01 50	0.42199 P<0.01 47
Reproduction (Day 21)	0.83047 P<0.01 58	0.83636 P<0.01 58	0.77960 P<0.01 55
Survival (Day 21)	0.52548 P<0.01 58	0.51642 P<0.01 58	0.52119 P<0.01 55

[a] Values are Pearson product-moment correlation coefficient, probability and n, respectively.

TABLE 3. Correlation matrix for the relationships among dynamic biochemical indicators at day 7 and protein growth (day 7), protein growth (day 21), reproduction and survival[a]

	Protein/RNA	Protein/DNA	RNA/DNA	Protein/RNA/DNA
Protein Growth (Day 7)	0.36271 P<0.01 60	-0.22952 P=0.09 57	-0.44250 P<0.01 57	-0.81371 P<0.01 57
Protein Growth (Day 21)	0.48871 P<0.01 50	0.05642 P=0.71 47	-0.22070 P=0.14 47	-0.67258 P<0.01 47
Reproduction (Day 21)	0.25535 P=0.05 58	-0.26749 P=0.05 55	-0.43696 P<0.01 55	-0.78559 P<0.01 55
Survival (Day 21)	0.14390 P=0.28 58	-0.18599 P=0.17 55	-0.27248 P<0.05 55	-0.45859 P<0.01 55

[a] Values are Pearson product-moment correlation coefficient, probability and n, respectively.

In general, these data indicated that static bioindicators of growth at day 7 were highly correlated with chronic (21 day) toxicant effects on growth and reproduction but not survival. Therefore, the static parameters appear to have utility in predicting chronic effects using short-term tests. The protein:RNA:DNA ratio was the only dynamic indicator that has potential predictive capabilities.

ACKNOWLEDGEMENT

This research is a contribution from the Missouri Agricultural Experiment Station, Columbia, Journal Series no. 10139. It was a cooperative endeavor between the Department of Entomology of the University of Missouri and the Columbia National Fisheries Research Laboratory and was supported in part by the Missouri Research Assistance Act and Monsanto Company. The assistance of Mr. D.U. Palawski and Ms. J. Quade is gratefully acknowledged.

REFERENCES

1. E.E. Herricks and D.J. Schaeffer, in Freshwater Biological Monitoring, Pergamon Press, New York, 153-166 (1984).
2. R.A. Kimerle, Environ. Toxicol. Chem. 5, 113-115 (1986).
3. P.A. Mehrle and F.L. Mayer, Environ. Health Perspec. 34, 139-143 (1980).
4. P.A. Mehrle and F.L. Mayer, in Fundamentals of Aquatic Toxicology: Methods and Applications, Hemisphere Pub. Corp., Washington,D.C., 264-282 (1985).
5. M.J. McKee and C.O. Knowles, Environ. Poll. Ser. A, In press (1986).
6. M.J. McKee and C.O. Knowles, Ecotoxicol. Environ. Safety, In press (1986).
7. C.O. Knowles, M.J. McKee and D.U. Palawski, Environ. Tox. Chem. submitted (1986).
8. American Society for Testing and Materials, ASTM E-35.21 Draft #5 (1985).
9. O.H. Lowry, N.J. Rosenbrough, A.L. Farr and R.J. Randall, J. Biol. Chem. 93, 265-275 (1951).
10. G.L. Miller, Anal. Chem. 31, 964 (1959).
11. H.N. Munroe and A. Fleck, Analyst 91, 78-88 (1966).
12. S. Shibko, P. Koivistoinen, C.A. Tratnyek, A.R. Newall and F. Friedman, Anal. Biochem. 19, 514-528 (1967).
13. R. Vytasek, Anal. Biochem. 120, 243-248 (1982).
14. M.J. Dagg and J.L. Littlepage, Mar. Biol. 17, 162-170 (1972).
15. J.M. Kissane and E. Robins, J. Biol. Chem. 233, 184-188 (1958).
16. M.G. Barron and I.R. Adelman, Can. J. Fish. Aquat. Sci. 41, 141-150 (1984).
17. L.J. Buckley, T.A. Halavik, G.C. Laurence, S.U. Hamilton and P. Yevich, Trans. Am. Fish. Soc. 114, 114-124 (1985).
18. Q. Dortch, T.L. Roberts, J.R. Clayton, Jr. and S.I. Ahmed, Mar. Ecol. Prog. Ser. 13, 61-71 (1983).

Metabolism in plants and animals

Current problems, trends, and developments in pesticide metabolism in plants

Gerald L. Lamoureux* and D. Stuart Frear

USDA, ARS, Metabolism and Radiation Research Laboratory, State University Station, Fargo, ND 58105 USA

Abstract - The use of fast atom bombardment mass spectrometry for identification of plant pesticide metabolites is illustrated with metabolites of propachlor (2-chloro-N-isopropylacetanilide), atrazine [2-chloro-4-(ethylamino)-6-(isopropylamino)-s-triazine], metribuzin (4-amino-6-tert-butyl-4,5-dihydro-3-methylthio-1,2,4-triazin-5-one), fluorodifen (2,4'-dinitro-4-trifluoromethyl diphenylether), EPTC (S-ethyl-N,N-dipropylthiocarbamate), chlorsulfuron {2-chloro-N-[(4-methoxy-6-methyl-1,3,5-triazin-2-yl)aminocarbonyl]benzenesulfonamide}, metazachlor [2-chloro-N-(2,6-dimethylphenyl)-N-(1-pyrazolylmethyl)-acetamide], and tridiphane [2-(3,5-dichlorophenyl)-2-(2,2,2-trichloroethyl)oxirane]. Metabolites identified include glutathione-, homoglutathione-, malonylcysteine-, glucoside-, malonylglucoside-, malonamide-, malonylthiolactic acid-, thioacetic acid sulfoxide-, and sulfonic acid conjugates. Several problems associated with the use of FAB are also discussed. Recent evidence that some herbicide antidotes function by increasing the rate of herbicide metabolism to glutathione conjugates in corn and sorghum is reviewed and the possibility that antidotes may also produce a protective effect by altering the activity of enzyme systems unrelated to herbicide metabolism is considered. Herbicide synergism by inhibition of herbicide metabolism is discussed in relation to tridiphane, a synergist of atrazine. Both tridiphane and two metabolites of tridiphane inhibited glutathione S-transferase activity; however, the metabolites were more potent inhibitors than tridiphane. One of these metabolites also inhibited glutathione reductase activity. Genetic engineering for herbicide resistance is briefly considered.

INTRODUCTION

Several areas of plant metabolism research are particularly exciting at this time because they can either increase the speed and ease with which metabolism studies can be conducted or lead to the direct and beneficial use of pesticide metabolism knowledge. These areas of research include: (1) application of fast atom bombardment mass spectrometry to pesticide metabolite identification, (2) antidotes or synergists that increase or decrease the rate of herbicide metabolism, (3) biological activity of metabolites, and (4) genetic engineering for herbicide resistance by increasing the rate of herbicide metabolism.

FAST ATOM BOMBARDMENT MASS SPECTROMETRY (FAB)

From 1965 to 1975, plant pesticide metabolism studies were focused primarily on metabolite purification and identification. During the 1970's, high performance liquid chromatography (HPLC) greatly simplified metabolite purification, but metabolite identification remained a difficult problem. Mass spectrometry was a widely utilized tool for metabolite identification, but most plant metabolites were polar conjugates that could not be readily analyzed by conventional mass spectrometry unless derivatized to increase their volatility. Since these conjugates were frequently labile and the nature of these conjugates was not always known, derivatization frequently produced artifacts. Although a soft ionization technique that could be used with non-volatile compounds, field desorption mass spectrometry, had been available for ca 18 years, it was not widely used because it was an experimentally difficult technique (ref. 1). In 1981, a new soft ionization technique was described that could be used with non-volatile compounds, fast atom bombardment mass spectrometry (FAB). Several reviews have reported that FAB is applicable to compounds similar to pesticide conjugates found in plants (refs. 1,2,3) and it is apparent that FAB is the most important advancement applicable to pesticide metabolism technology since the introduction of HPLC. FAB usually provides information on the molecular weight of metabolites and frequently provides fragmentation data that is useful in structural assignments. Although other soft ionization techniques show considerable promise (ref. 2), at present, FAB appears to be the most useful general tool for the identification of pesticide conjugates isolated from plants. The technique is simple and can be adapted to many existing mass spectrometers.

These authors have used FAB in the identification of approximately 25 different conjugates from 10 different pesticides isolated from plant systems without a failure. However; as with any technique, FAB does not work equally well with all classes of compounds. Some metabolites derived from glutathione (GSH) conjugates, such as malonylcysteine conjugates, do not produce good spectra when analyzed in pure glycerol; however, when these metabolites are dissolved in glycerol that contains oxalic acid, they yield very intense spectra. The FAB spectra of the malonylcysteine conjugate of fluorodifen in the absence and presence of oxalic acid is shown in Fig. 1. This metabolite was isolated from a peanut cell suspension culture. Similar results were observed when the malonylcysteine conjugate of EPTC, isolated from corn, was analyzed in the absence or presence of oxalic acid. A FAB spectrum of the γ-glutamylcysteine conjugate of atrazine was obtained in a pure glycerol matrix, but the quality and intensity of the spectrum was greatly improved by the addition of oxalic acid. In some laboratories, thioglycerol is used instead of glycerol to obtain FAB spectra of GSH-related conjugates (ref. 3,4).

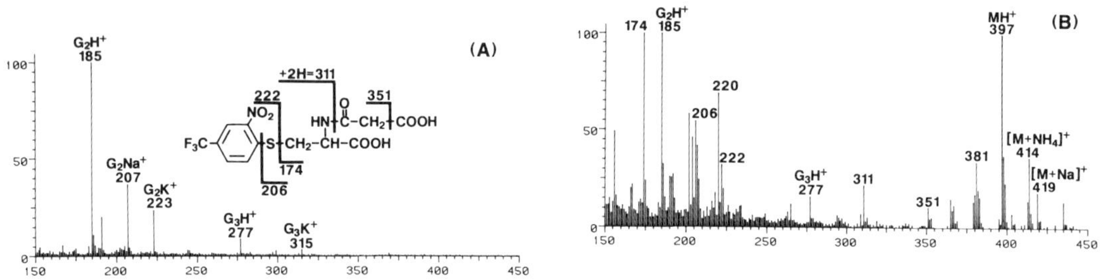

Fig. 1. FAB spectra of 2 µg of a malonylcysteine conjugate of fluorodifen in 0.4 µl glycerol: (A) in the absence of oxalic acid and (B) in the presence of 10 µg oxalic acid. In these and all other spectra, GNa^+, G_2H^+, G_2K^+, etc. indicate ions due to glycerol (G) associated with protons, sodium, or potassium.

The success of FAB is often dependent on the purity of the metabolite. With conventional mass spectrometry, glucosides, GSH conjugates, etc., are usually derivatized. The derivatives usually have very different properties than the original metabolites and the derivatives can be further purified prior to mass spectrometry. Samples analyzed by FAB are usually not derivatized; therefore, the initial purification should be more rigorous when FAB is used. Alkali salts are among the common contaminates of conjugates that frequently have a deleterious effect on FAB. The γ-glutamylcysteine conjugate of atrazine, as originally isolated from sorghum by ion exchange chromatography (ref. 5), produced a weak FAB spectrum characterized by ions associated with sodium and potassium. A much improved spectrum resulted when this metabolite was further purified by reverse phase HPLC, Fig. 2.

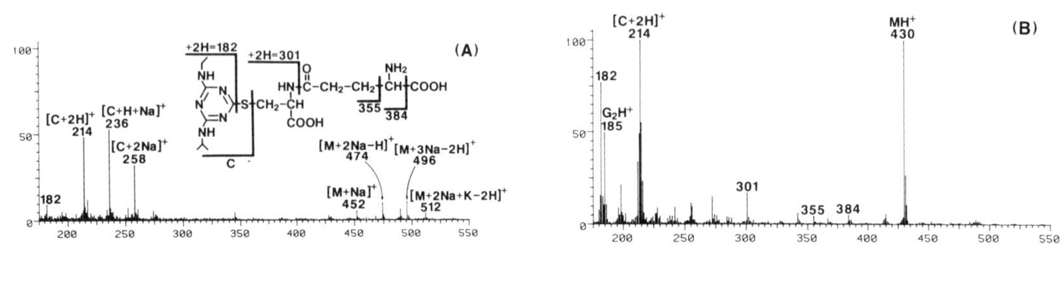

Fig. 2. FAB spectra of the γ-glutamylcysteine conjugate of atrazine in 0.3 µl glycerol and 10 µg oxalic acid: (A) 6 µg as originally isolated by ion exchange and (B) 2 µg after further HPLC.

The addition of 2 μg of sodium carbonate to 2.5 μg of pure γ-glutamylcysteine conjugate of atrazine decreased the intensity of the MH^+ from 100% to 60% and increased the $[M+Na]^+$ from 9% to 33% and the $[M+2Na-H]^+$ from 2% to 8%. The ion fragment at m/z 236, associated with Na, increased from 8% to 52% with a resulting decrease in the ion at m/z 214. Corresponding changes were observed when 2 μg of potassium carbonate was added to this metabolite. When 7 μg of sodium carbonate was added, the spectrum was very similar to that of the sample as originally isolated by ion exchange methods, but no MH^+ was observed. Similar effects were observed when sodium or potassium carbonate or sodium chloride was added to the malonylcysteine conjugate of fluorodifen. A glucoside of chlorsulfuron, isolated from corn, yielded a weak MH^+ (1%); when 10 μg of sodium chloride was added to 3 μg of the glucoside, the MH^+ was lost and an $[M+Na]^+$ (1.6%), $[M+2Na-H]^+$ (8%) and an $[M+3Na-2H]^+$ (1%) appeared. Because of the very weak MH^+ ion and the intensity of ions due to the glycerol matrix, it was difficult to determine which ions were due to the matrix and which were due to the glucoside of chlorsulfuron. By analyzing the sample with glycerol-D_5 as the matrix, it was possible to distinguish between ions related to the matrix and those related to chlorsulfuron (ref. 6). It is apparent that FAB spectra can become either very weak or very complex if the sample being analyzed is contaminated with inorganic salts or both organic impurities and inorganic salts.

Atrazine is metabolized to a GSH conjugate in sorghum. This conjugate was originally isolated by ion exchange and TLC and identified by degradative reactions and amino acid analyses (ref. 5). More than one year was required to isolate and identify this metabolite. Metabolites such as these can now be easily purified by HPLC and TLC and the FAB spectra of these conjugates usually produce an intense MH^+ and ion fragments that are diagnostic of a GSH conjugate, Fig. 3A. This GSH conjugate was reported to be metabolized to an unusual lanthionine conjugate in sorghum (ref. 5). The identification of the lanthionine conjugate isolated from sorghum has now been confirmed by FAB, Fig. 3B. Tridiphane is a herbicide that synergizes atrazine toxicity. Enzymes from corn metabolized tridiphane to a metabolite characterized as a GSH conjugate by positive and negative ion FAB (ref. 7), Fig. 3C.

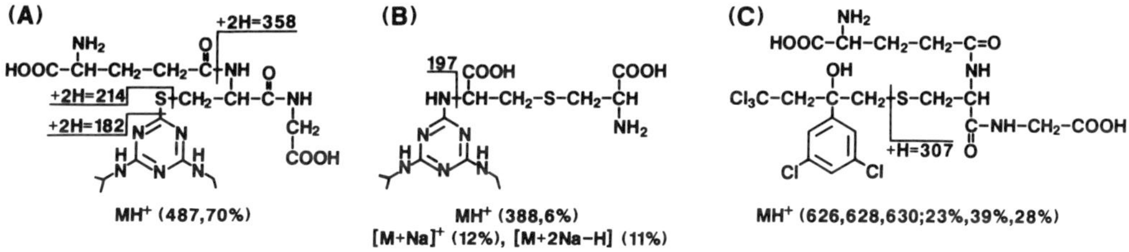

Fig. 3. Major FAB ions from: (A) the GSH conjugate of atrazine, (B) the lanthionine conjugate of atrazine, and (C) the GSH conjugate of tridiphane.

In corn, EPTC was metabolized to a malonylcysteine conjugate via the GSH pathway and also to a second unidentified metabolite (ref. 8). Attempts to identify the second metabolite were hampered by the lability of the metabolite during purification and derivatization. Both the unidentified metabolite and its breakdown product yielded the same compound when esterified or esterified and acetylated. This compound appeared to be the 3-thiolactic acid conjugate methyl ester of EPTC [S-(N,N-dipropylcarbamoyl)-3-thiolactic acid]. By careful control of the conditions during purification, the intact metabolite was isolated in pure form and analysis by FAB clearly showed that the metabolite was S-(N,N-dipropylcarbamoyl)-O-(malonyl)-3-thiolactic acid, Fig. 4. This appears to be the first report of an acylated thiolactic acid conjugate of a xenobiotic. Identification would have been much more difficult without FAB. This metabolite was also produced by a corn cell suspension culture and by cotton (ref. 9).

The major metabolite of metribuzin in soybean was identified as a homoglutathione conjugate by a variety of techniques, including FAB. In addition, a minor, highly acidic metabolite was also detected (ref. 4). When the minor metabolite was esterified and analyzed by conventional mass spectral methods, it appeared to be an N-acetylated derivative of metribuzin; however, when this metabolite was analyzed directly by FAB, it was found to be a malonamide, Fig. 5.

Propachlor is initially metabolized to a GSH conjugate in several plant species (refs. 8, 10). However, it was recently shown that acifluorfen and metribuzin were metabolized to homoglutathione (γ-glutamylcysteinyl-β-alanine) rather than GSH conjugates in soybean (ref.

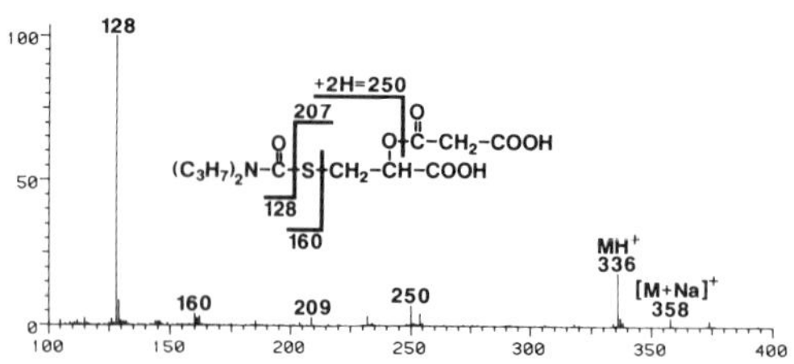

Fig. 4. FAB spectrum of an EPTC metabolite from corn. Glycerol/oxalic acid matrix.

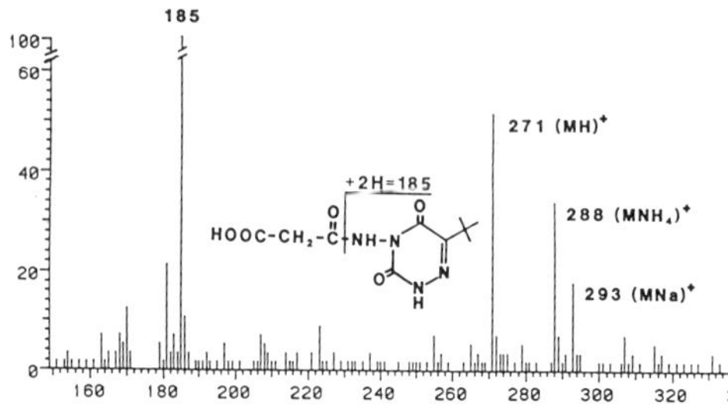

Fig. 5. FAB spectrum of a malonamide conjugate of metribuzin from soybean. Thioglycerol/ammonium acetate matrix.

5). Subsequently, when excised soybean leaves were treated with propachlor, the primary metabolite detected after 4 hr was shown by FAB to be the homoglutathione conjugate, Fig. 6A. Total time required for treatment, isolation and identification was three days. The malonylcysteine conjugate of propachlor was identified by conventional mass spectral methods as one of the major terminal metabolites of propachlor in soybean plants treated and grown to maturity in hydroponic cultures (refs. 8,10). This metabolite has now been shown to produce an excellent FAB spectrum with an intense MH^+ ion. The malonylcysteine sulfoxide conjugate of propachlor was also a major metabolite and it was originally characterized by mass spectral analysis of the methyl ester (ref. 8). However, the methyl ester was prepared in low yield and apparently underwent a Pummerer rearrangement. Identification by EI and CI mass spectrometry was ambiguous. Direct FAB analysis yielded an intense MH^+ and unambiguous fragment ions that verified the metabolite to be a simple sulfoxide. Sulfoxides of the homoglutathione, cysteine, and thiolactic acid conjugates were also metabolites and they produced unambiguous FAB spectra as well. The FAB spectrum of the thioacetic acid sulfoxide conjugate of propachlor is shown in Fig. 6B. The sulfoxide of the GSH conjugate of another herbicide, tridiphane, also gave an excellent FAB spectrum (unpublished data). It appears that sulfoxide conjugates can be readily identified by FAB.

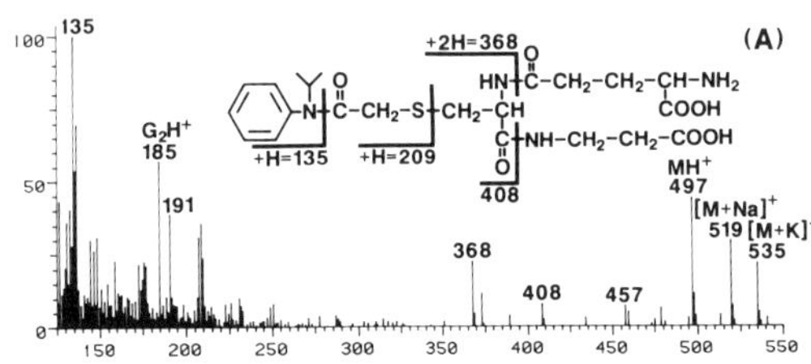

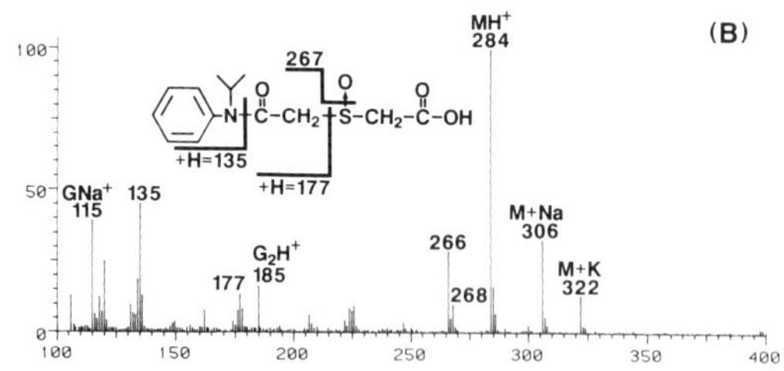

Fig. 6. FAB spectra of the (A) homoglutathione and (B) thioacetic acid sulfoxide conjugates of propachlor run in glycerol/oxalic acid.

Propachlor-treated soybeans also contained a polar metabolite that was degraded to a compound with a different retention time after purification by HPLC in a system that contained 1% acetic acid. After esterification and acetylation, this metabolite produced a mass spectrum consistent with C_6H_5N(isopropyl)-OC-CH_2O-OC-CH_3. When this labile metabolite was carefully purified by HPLC and analyzed by FAB, it was apparent that the metabolite was a malonylglucoside, Fig. 7A. The artifact produced during chromatography or concentration of solvents was shown to be the simple glucoside by FAB, Fig. 7B.

Metabolites in soybean plants grown in propachlor-treated soil were different from those in plants grown and treated in hydroponic culture (ref. 8). Some of these differences were thought to be due to the metabolism of propachlor in the soil and uptake of the soil metabolites by the soybean plants. Because of difficulties in identification, the nature of these differences was not completely defined. This problem was greatly simplified when FAB was

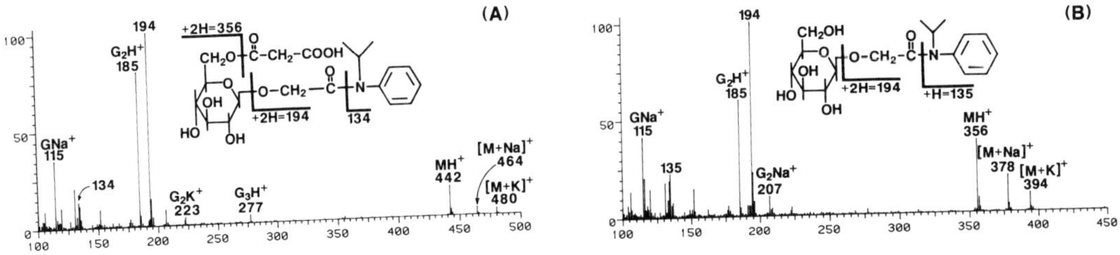

Fig. 7. FAB spectra of propachlor metabolites from soybean: (A) malonylglucoside (B) glucoside. Approximately 4 µg sample in 0.3 µg glycerol containing 10 µg oxalic acid.

used for metabolite identification. One month after treatment of sandy loam soil with [^{14}C]propachlor, three soluble metabolites were detected in the soil (unpublished data). The most abundant metabolites, C_6H_5-N(isopropyl)-CO-COOH (oxamic acid derivative) and C_6H_5-N(isopropyl)-CO-CH_2-SO_3H (sulfonic acid derivative), accounted for 28% and 27%, respectively, of the soluble ^{14}C in the soil. The oxamic acid derivative was identified as the methyl ester by EI and DCI mass spectrometry, but was also identified without derivatization by FAB. The structure of the sulfonic acid metabolite was confirmed by comparing it to a standard by FAB. The third metabolite accounted for 12% of the soluble ^{14}C in the soil. This metabolite produced an excellent FAB spectrum and its structure was confirmed by synthesis to be the thioacetic acid sulfoxide conjugate, Fig. 6B. The most abundant metabolites present in soybean plants grown to maturity in propachlor treated soil were the oxamic acid and sulfonic acid derivatives of propachlor. These metabolites were not detected in plants treated and grown in hydroponic culture. Some of the differences observed in the metabolism of propachlor in soybeans grown in hydroponic culture versus soybeans grown in soil are similar to those reported for 2-chloro-N-(2,3-dimethylphenyl)-N-(1-methyl)acetamide metabolism in soil and soybean, (ref. 11).

FAB and various other methods are continuing to facilitate the speed and ease with which pesticide metabolism studies can be conducted. The knowledge gained from these studies will lead to a better understanding of how herbicides are metabolized and also aid in the understanding and use of herbicide selectivity and toxicity in plants.

ALTERATION OF THE RATE OF PESTICIDE METABOLISM

Herbicide antidotes.

A number of chemicals, frequently referred to as antidotes, are known to decrease the toxicity of certain herbicides to selected plant species. Two of the most well known antidotes are dichlormid (N,N-diallyl-2,2-dichloroacetamide) and naphthalic anhydride. A number of theories have been proposed as to the modes of action of these compounds and they have been the subject of several reviews (refs. 12,13,14). Lay and Casida proposed that dichlormid protects corn from EPTC injury by increasing the rate of metabolism of EPTC to a GSH conjugate by induction of increased levels of GSH and glutathione S-transferase (GST) activity (ref. 15). The mode of action of EPTC and most other herbicides protected by dichlormid or related antidotes is not known. This complicates interpretation of data on the mode of action of the antidotes because the antidotes may cause effects related to the mechanism of action of the herbicide in addition to or instead of altering the rate of herbicide detoxification. However, studies with dichlormid and several other antidotes used with thiocarbamate or chloroacetamide herbicides provide additional documentation that certain of these antidotes do increase the rate of herbicide metabolism by GSH conjugation.

Mozer et al. showed that flurazole [2-chloro-4-(trifluoromethyl)-5-thiazolecarboxylic acid benzyl ester] induced a new GST in both the root and shoot tissues of corn (ref. 16). The new GST was active with both alachlor [2-chloro-2',6'-diethyl-N-(methoxymethyl)-acetanilide] and 1-chloro-2,4-dinitrobenzene (CDNB). GST activity assayed with CDNB was 2.3 X and 2.7 X higher in induced root and shoot tissues, respectively. Several other antidotes, including dichlormid and naphthalic anhydride also induced GST levels. Lay and Niland studied EPTC and dichlormid uptake and distribution, GSH and GST levels, and EPTC injury in various hybrid and inbred corn lines treated with one or both of these compounds (ref. 17). Both corn lines that were protected and those that were not protected by dichlormid were included in this study. GSH and GST levels were measured in the roots and EPTC related injury was measured by leaf malformation 18 days after planting. Dichlormid generally caused an increase in GST and GSH levels in the roots, but this did not appear to be related to protection from leaf malformation measured 18 days after planting. GST levels were measured in the coleoptiles of only one corn line (De Kalb XL-72B) and no increase in GST was observed after a 24-hr dichlormid treatment. In contrast, Komives et al., showed that dichlormid caused a signficant elevation in the cytosolic and microsomal GST levels in corn (Pioneer 3950) shoots as well as an increase in glutathione reductase activity. After a 48 hr treatment with dichlormid, all of these activities were from 2.6 X to 2.8 X those of the control values (ref. 18). Fuerst and Gronwald examined the effect of five herbicide antidotes on metolachlor injury, uptake, and metabolism in sorghum (ref. 19). Metolachlor inhibited both root and shoot growth, but was much more effective in inhibiting shoot growth. All five of the following antidotes gave some degree of protection to inhibition of shoot growth: CGA-92194 {α-[(1,3-dioxolan-2-yl-methoxy)imino]benzeneacetonitrile}, cyometrinil {(Z)-α[(cyanomethoxy)-imino]benzeneacetonitrile}, fluorazole, naphthalic anhydride, and dichlormid. CGA-92194 increased tolerance to metolachlor by 10-fold, but fluorazole was the most effective antidote. All of the antidotes caused an increased rate of metolachlor metabolism, presumably by GSH conjugation. An excellent correlation was observed between protection and the amount of metolachlor metabolized to water-soluble products in the sorghum shoots (Fig. 8). Ezra et al. obtained similar results in studies with the chloroacetamide herbicide, metazachlor, in corn (ref. 20). Both dichlormid and an experimental antidote, 145-138 [1-dichloroacetylhexahydro-3,3-8a-methylimidazo(1,2-a)pyrimidin-6-(2H)-one], caused an increased rate of metazachlor metabolism in the roots and shoots of corn plants pretreated for 2-7 days with 5 µM 145-138 or dichlormid. An increased level of GST activity was also observed. In corn shoots, metazachlor metabolism via GSH conjugation was confirmed by FAB and synthesis. The antidote, 145-138, was effective in protecting corn against metazachlor injury, but dichlormid was only marginally effective.

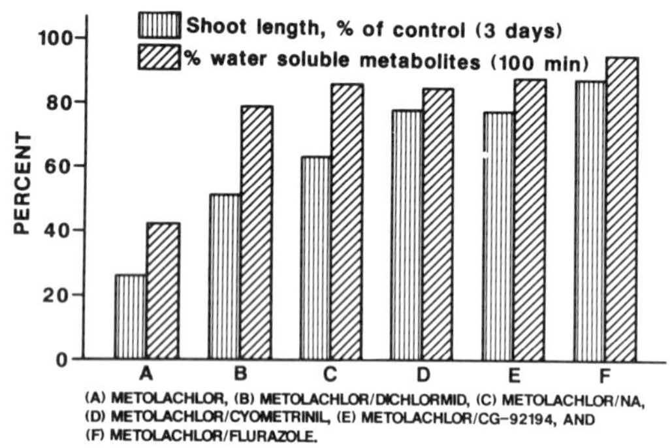

Fig. 8. Effect of several antidotes on shoot growth and rate of metolachlor metabolism in sorghum treated with metolachlor. Data adapted from Fuerst and Gronwald, Weed Sci., 34, 354 (1986).

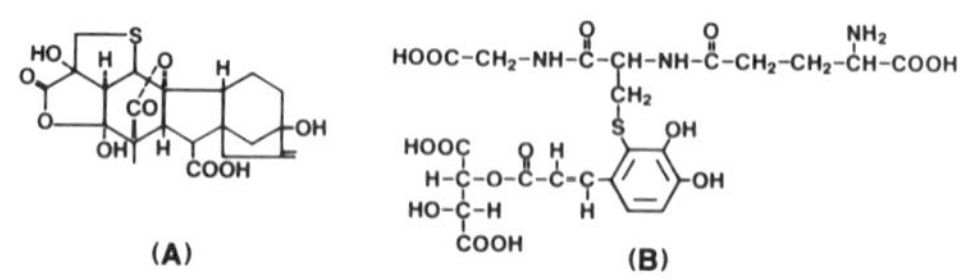

Fig. 9. Structures of gibberethione (A) and (B) 2-S-glutathioneyl caftaric acid.

These studies do not prove that the primary mode of action of any of the antidotes is by increasing the rate of herbicide GSH-conjugation; however, it is very likely that this is one factor involved in the protection of corn and/or sorghum against certain chloroacetamide and thiocarbamate herbicides by antidotes such as dichlormid and fluorazole. In some cases, the increased rate of herbicide metabolism associated with the induction of GSH and GST levels may be incidental to other more important changes that occur within the plant. GSH conjugation may be involved in normal biological processes and alteration of GST enzymes could effect these processes. A biologically inactive sulfur containing derivative of gibberellin GA_3 (gibberthione) was isolated from Japanese morning glory (Pharbitis nil) (ref. 21, Fig. 9A). This derivative may have been formed by addition of GSH to 3-keto GA_3 and subsequent metabolism to a thiolactic acid intermediate that underwent ring closure. Thus, GSH conjugation could play a role in hormone metabolism and/or regulation. The recent isolation of GSH conjugates of caffeic acid derivatives in grape juice also suggests that GSH conjugation of natural products may be important in plants (ref. 22, Fig. 9B). Antidotes could also induce other enzymes and thereby alter herbicide toxicity. The induction of GSH reductase activity by dichlormid (ref. 18) could have a signficant protective effect against certain types of herbicide activity. Oxidative enzymes might also be induced. Although signficant induction of P-450 enzymes by dichlormid was not demonstrated by Komives et al. (ref. 23), many oxidative plant enzymes are highly substrate specific and induction of a key oxidative enzyme might be difficult to detect. Also, oxidative enzymes other than P-450 may be involved.

Herbicide synergists.
Herbicide toxicity can be synergized by inhibition of herbicide metabolism. An excellent example of this is the increased toxicity of propanil to rice that results when propanil (3',4'-dichloropropionanilide) is used in combination with organophosphorus or carbamate insecticides (ref. 5). These insecticides inhibit amidase enzymes involved in the detoxification of propanil. Although herbicide synergism is not a commonly used practice in weed control, recent studies with tridiphane, a synergist of atrazine, indicates that this concept may be useful. Tridiphane synergizes atrazine toxicity in young giant foxtail without causing a similar effect in corn. In young giant foxtail, tridiphane appears to inhibit atrazine metabolism to a GSH conjugate, but this same effect is not observed in corn (ref. 24).

BIOLOGICAL ACTIVITY OF METABOLITES

GST enzymes from corn and giant foxtail catalyze the conversion of tridiphane to a GSH conjugate. Both tridiphane and its GSH conjugate appeared to inhibit atrazine GSH conjugation (ref. 7). The GSH conjugate of tridiphane (Fig. 3C), but not parent tridiphane, inhibited GSH conjugation reactions with several other substrates. Tridiphane was converted to a GSH conjugate more rapidly than atrazine by GST preparations from both corn and giant foxtail. It was not proven whether tridiphane or its GSH conjugate was primarily responsible for inhibition of atrazine GSH conjugation. Giant foxtail GST was 4 X more sensitive to inhibition by the tridiphane GSH conjugate than GST from corn. Ten compounds were compared as inhibitors of GST activity from corn and giant foxtail. Only triphenyl tin chloride (I_{50} 2 µM) was more effective than the tridiphane GSH conjugate (I_{50} 3.5 µM) as an inhibitor of GST from giant foxtail. The GSH conjugate of tridiphane also inhibited GSH conjugation of PCNB, propachlor and fluorodifen when GST from pea epicotyl was used. I_{50} values ranged from 1 to 9 µM. The tridiphane GSH conjugate was more stable in crude enzyme preparations from giant foxtail than in similar enzyme preparations from corn. The sulfoxide of the tridiphane GSH conjugate was also a potent inhibitor of GST activity (unpublished data). Since tridiphane is a herbicide, it was difficult to determine whether inhibition of atrazine metabolism in giant foxtail was a result of inhibition of GST or some indirect herbicidal effect of tridiphane. However, tridiphane was not insecticidal to the house fly, but it synergized diazinon toxicity. This synergism was related to inhibition of diazinon metabolism. Both tridiphane and diazinon are metabolized by GSH conjugation in the house fly. *In vitro*, the GSH conjugate of tridiphane was a potent inhibitor of house fly GST with diazinon and several other substrates (unpublished data). The data indicate that it may be possible to inhibit GSH conjugation *in vivo* by certain GSH conjugates.

A high ratio of GSH to GSSG may be important to plants for a variety of reasons, including optimal protein synthesis (ref. 25). Tridiphane GSH conjugate, but not tridiphane, was an inhibitor of spinach glutathione reductase (I_{50}=30 µM). Thus, the possiblity that pesticide metabolites might alter GSH:GSSG ratios should not be overlooked. The induction of GSH reductase in corn by dichlormid (ref. 18) may also point to an alternate mechanism for the antidotal effect of this and related compounds. Antidotes should be assayed, not only for their effect on levels of total glutathione, but also for their effect on GSH:GSSG ratios and levels of GSH reductase. In mammals, the synthesis of GSH is a feed-back regulated reaction. Some GSH conjugates interfere with the feed-back regulation mechanism and cause an increased synthesis of GSH (ref. 26). Compounds that form GSH conjugates may actually cause an increase in the levels of GSH. Rubin *et al*. (ref. 27) recently studied a series of N-phenylmaleimides and related compounds that were expected to conjugate readily with GSH. Some of these compounds were antidotes of alachlor and also increased GSH levels. Perhaps some of the antidotes, such as dichlormid, are metabolized to GSH-conjugated intermediates as proposed by Mialus *et al*. (ref. 12) and this is one mechanism by which these compounds increase GSH levels. Some chloro-acetamide herbicides are known to cause an elevation in GSH levels and these herbicides are generally metabolized by GSH conjugation (ref. 14).

Certain amino acid conjugates of 2,4-D and related compounds have biological activity. Although these compounds may be hydrolyzed back to the parent herbicide, currently available evidence does not justify this conclusion (ref. 5). The fungicide, vitavax, was metabolized to malonanilide in peanut and it was reported that malonanilide has potent growth regulating capacity (ref. 5). It is apparent that conjugated pesticide metabolites should be tested for biological activity; however, decisions as to which metabolites should be examined for biological activity and by what methods will be difficult.

FUTURE TRENDS

Although recent advancements in HPLC and mass spectrometry have facilitated the rapid isolation and identification of pesticide metabolites, similar progress in the area of pesticide biochemistry and enzymology has been slow to develop. This is especially true in higher plants where sites of action and enzyme detoxication/activation systems responsible for the bioregulation of pesticide activity are difficult to isolate and characterize. Fortunately, recent interest in the biorational design of herbicides and the genetic engineering of herbicide tolerance has focused attention on the need for increased research on the biochemistry of pesticide action and metabolism in higher plants. Indeed, current approaches to genetic engineering and target-directed herbicide synthesis have been advanced primarily through basic biochemical investigations of the primary sites of herbicide action (ref. 28,29) and key enzymatic detoxication reactions (ref. 16,30,31) in higher plants. Continued interest on the biochemistry of pesticide metabolites, synergists and antagonists should also provide new insights and approaches to the enzymatic bioregulation of pesticide activity in plants.

REFERENCES

1. G.D. Paulson, G.L. Lamoureux and V.J. Feil, J. Toxicol.-Clin. Toxicol., 19 (6 and 7), 571-608 (1982-83).
2. C. Fenselau and L. Yellet, in Xenobiotic Conjugation Chemistry, ACS Symposium Series Vol. 299 (G.D. Paulson, J. Caldwell, D.H. Hutson and J.J. Menn, Eds.) ACS, Washington, pp 159-176 (1986).
3. C. Fenselau, J. Natural Products 47, 215-225 (1984).
4. D.S. Frear, H.R. Swanson and E.R. Mansager, Pestic. Biochem. Physiol., 23, 56- . (1985).
5. G. Lamoureux and D. Rusness, in Xenobiotic Conjugation Chemistry, ACS Symposium Series Vol. 299 (G.D. Paulson, J. Caldwell, D.H. Hutson and J.J. Menn, Eds.) ACS, Washington, pp 62-105 (1986).
6. D.S. Frear, H.R. Swanson and E.R. Mansager, Proc. 6th Int. Cong. Pesticide Chemistry (IUPAC) Ottawa (1986): this volume.
7. G.L. Lamoureux and D.G. Rusness, Pestic. Biochem. Physiol. (accepted for publication Feb. 12, 1986).
8. G.L. Lamoureux and D.G. Rusness, in Pesticide Chemistry Human Welfare and the Environment, Vol. 3 (J. Miyamoto and P.C. Kearney, Eds.) Pergamon Press, New York, pp 295-300 (1983).
9. G.L. Lamoureux and D.G. Rusness, J. Agric. Food Chem. (submitted for publication 5/1/86).
10. G.L. Lamoureux and D.G. Rusness, in Sulfur in Pesticide Action and Metabolism, ACS Symposium Series 158 (J.D. Rosen, P.S. Magee, and J.E. Casida, Eds.) ACS, Washington pp 133-164 (1981).
11. M. Hussain, I.P. Kapoor, C.C. Ku and S. Stout, J. Agric. Food Chem. 31, 232-236 (1983).
12. F. M. Pallos and J.E. Casida (Eds.), Chemistry and Action of Herbicide Antidotes Academic Press, New York (1978).
13. C. Parker, Pestic. Sci. 14, 40-48 (1983).
14. G.R. Stephenson and G. Ezra, in Plant Growth Regulating Chemicals, Vol II (L.G. Nickell, Ed) C.R.C. Press, Boca Rotan, Florida pp 193-212 (1983).
15. M.-M. Lay and J.E. Casida, Pestic. Biochem. Physiol. 6, 442-456 (1976).
16. T.J. Mozer, D.C. Tiemeier and E.G. Jaworski, Biochemistry 22, 1068-72 (1983).
17. M.-M. Lay and A.M. Niland, Pestic. Biochem. Physiol. 23, 131-140 (1985).
18. T. Komives, V.A. Komives, M. Balazs and F. Dutka, Proceedings of the British Crop Protection Conference, Vol. 3, 1155-1160 (1985).
19. E.P. Fuerst and J.W. Gronwald, Weed Sci. 34, 354-361 (1986).
20. G. Ezra, G. Stephenson and G. Lamoureux, National Meeting of Weed Sci. Soc. America, Houston, Texas (1986): Poster.
21. T. Yokota, H. Yamane and N. Takahashi, Agr. Biol. Chem. 40, 2507-8 (1976).
22. V.F. Cheynier, E.K. Trousdale, V.L. Singleton, M.J. Salgues, and R. Wylde, J. Agric. Food Chem. 34, 217-221 (1986).
23. T. Komives, M. Balazs, V. Komives, F. Dutka, Proceedings of the 5th International Conference on Cytochrome P-450 (L. Vereczkey and K. Magyar, Eds.) Akademiai Kiado, Budapest, pp 30-33 (1985).
24. P.S. Zorner and G.L. Olson, Proceedings of the North Central Weed Control Conference 36, 115 (1981) (Abstract).
25. H. Rennenberg, Phytochem. 21, 2771-2781 (1982).
26. T. Kondo, N. Taniguchi and Y. Kawakami, Eur. J. Biochem. 145,131-136 (1984).
27. B. Rubin, O. Kirino and J.E. Casida, J. Agric. Food Chem. 33, 489-494 (1985).
28. C.J. Arntzen and J.H. Duesing, in "Advances in Gene Technology: Molecular Genetics of Plants and Animals," pp. 273-294, Academic Press (1983).
29. R.A. LaRossa and S.C. Falco, Trends in Biotechnology, 2, 158-161 (1984).
30. D.S. Frear, H.R. Swanson and F.S. Tanaka, in "Recent Advances in Phytochemistry: Structural and Functional Aspects of Phytochemistry," pp. 225-246, Academic Press (1972).
31. D.M. Shah, C.M. Hironaka, R.C. Wiegand, E.I. Harding, G.G. Krivi and D.C. Tiemeier, Plant Molec. Biol., 6, 203-211 (1986).

Current trends in animal metabolism: the importance of interorgan relationships and intermediary metabolism in the disposition of xenobiotics

G.D. Paulson

USDA, ARS, Metabolism and Radiation Research Laboratory, State University Station, Fargo, ND 58105

Abstract - Knowledge of intermediary xenobiotic metabolism in animals is important because some intermediary metabolites have more biological activity than both the parent xenobiotic and the end products of metabolism observed in the excreta. Qualitative and/or quantitative differences in intermediary metabolism are often responsible for the varying toxicological responses to a xenobiotic by animals of different species. In addition, information concerning intermediary biotransformations of a xenobiotic often provides the basis for understanding how factors such as age, sex, health and nutritional status of an animal may affect the disposition of the xenobiotic. Special techniques and procedures, in addition to the classic whole animal and in vitro techniques, are used to determine the intermediary metabolism and interorgan relationships involved in the disposition of a xenobiotic in animals. This review discusses selected methods that have been used to study the sites and routes of xenobiotic metabolism in animals and the types of information that have been obtained by the use of these methods.

INTRODUCTION

There is a growing body of information which indicates that intermediary metabolites of some xenobiotics are more biologically active than the parent compound and the end products of metabolism observed in excreta. Therefore, information concerning the intermediary metabolism of a xenobiotic is important when evaluating toxicological data. A comparison of the intermediary metabolism of a xenobiotic in the species used for toxicology testing with the intermediary metabolism in target or nontarget species that may be exposed to the compound may be especially important. Factors such as age, sex, nutritional status and health status may drastically alter xenobiotic metabolism - sometimes with adverse consequences for the animal (ref. 1, 2). The concept of genetic polymorphism as it applies to xenobiotic metabolism and the adverse consequences that sometimes occur when an individual, deficient in an enzyme(s) normally involved in the metabolism of a xenobiotic, is exposed to the xenobiotic was the subject of a recent volume (ref. 3). Knowledge of the intermediary metabolism of a xenobiotic in "normal animals" is helpful in recognizing, and predicting how inherited enzyme deficiencies may alter the disposition and thereby the biological activity of the xenobiotic.

Many of the early studies on xenobiotic metabolism in animals were restricted to dosing animals with the xenobiotic and examination of excreta for the parent compound and/or metabolites (ref. 2, 4, 5). These types of experiments are valuable for estimating the rates of absorption and urinary excretion of the xenobiotic related residues; however, they are of limited value in determining the site(s) of metabolism and the nature of a intermediate biotransformation product(s). For example, if all of the ^{14}C-activity given orally to an animal as a ^{14}C-labeled xenobiotic is excreted in the feces, this does not necessarily mean that the xenobiotic is not absorbed from the gastrointestinal tract (GIT). The ^{14}C-xenobiotic may be absorbed from the GIT, metabolized to a ^{14}C-labeled metabolite(s) by the liver or another organ(s), excreted in the bile, and then eliminated in the feces. Further metabolism of the biliary metabolites by bacteria in the GIT and enterohepatic circulation of the ^{14}C-xenobiotic related residues may further complicate the disposition of the xenobiotic.

A variety of in vitro techniques can provide useful information such as the nature of enzymes and cofactors involved in specific biotransformations of xenobiotics (ref. 6). However, attempting to describe the overall metabolism of a xenobiotic in an animal on the basis of information obtained with only in vitro studies is usually not justified. For example, information obtained using a liver homogenate may have little or no relevance for predicting the *in vivo* metabolism of a xenobiotic that, when ingested, is extensively metabolized by bacteria or mucosal cells in the gastrointestinal tract.

Studies designed to identify intermediate xenobiotic metabolites, the site(s) of metabolism and interorgan transport mechanisms for the xenobiotic and its metabolites, in addition to the *in vitro* and whole animal studies discussed above, are useful in describing the overall disposition of a xenobiotic in an animal. The sites and route of xenobiotic metabolism can sometimes be determined by collection of an intermediate(s) from a strategic location in a donor animal. The intermediate(s) may then be administered at the same location in a recipient animal and the disposition followed. In some instances, evidence for intermediates can be obtained by comparative studies with specially prepared animals (germfree or antibiotic treated animals, animals fed special diets, animals deficient in key enzyme systems, etc.). Evidence for short lived metabolites can sometimes be obtained with trapping agents or by other indirect methods. Examples from recent publications which illustrate how studies of these types have been conducted and the information that has been obtained are the subject of this brief review.

BILIARY EXCRETION AND ENTEROHEPATIC CIRCULATION

Xenobiotic related residues in the vascular system can enter the GIT by several routes other than biliary excretion; these include pancreatic and gastric secretions, secretion by the salivary glands or lungs followed by swallowing, and passive or active transport from the vascular system directly into the intestine (ref. 7). A study by Davison et al. (ref. 8) illustrates this point. Twenty three hours after ^{14}C-monensin was intravenously administered to a bile duct cannulated horse 13% of the ^{14}C-activity was detected in the feces and contents of the GIT. These investigators did not determine the mechanism involved; however their results clearly indicated that the monensin and/or its metabolites did not enter the GIT via the bile. Dayton et al. (ref. 7) reviewed evidence that compounds such as quinine, dieldrin, emperonium, phenytoin, mefloquine, chlordecone, chlorophenol red, and certain antibiotics are transferred from the vascular system into the GIT by a mechanism(s) other than biliary excretion. Thus, xenobiotic-related residues in the feces from an orally dosed animal are not always the sum of the unabsorbed xenobiotic and the xenobiotic and/or its metabolites which are excreted in the bile and escape enterohepatic circulation. Nevertheless, biliary excretion is probably the most common mechanism for transfer of xenobiotic-related residues from the vascular system into the GIT and collection and examination of the bile is a basic element of a thorough animal metabolism study. If a xenobiotic related compound(s) is detected in the bile, it is instructive to perfuse bile collected from a donor animal into the small intestine of a recipient animal to determine if there is bacterial metabolism and/or enterohepatic circulation of the compound(s) in the bile.

Nussbaum et al. (ref. 9) reported that the formation of tumors in the bladder and small intestine of animals exposed to 3,2′-dimethyl-4-aminobiphenyl (DMAB) was related to the rate of biliary excretion of a conjugated metabolite. Both the rat and the hamster metabolized DMAB to the N-glucuronic acid conjugate of N-hydroxy-DMAB, and then to the ultimate carcinogen, nitroso-DMAB. In the hamster, the glucuronic acid conjugate of N-hydroxy-DMAB was excreted in both the bile and urine thereby leading to exposure of both the bladder and small intestine to the carcinogen nitroso-DMAB. Tumors were formed in both tissues. In contrast, the rat very efficiently excreted the N-glucuronic acid conjugate of N-hydroxy-DMAB in the bile (resulting in tumor formation in the small intestine) and thereby sparing the bladder from nitroso-DMAB exposure and tumor formation.

Bakke (ref. 10) observed that rats orally dosed with ^{14}C-labeled bis-methylthiotetrachlorobenzene (bis-MTTB) excreted 97% of the ^{14}C activity in the feces and that at least 83% of the ^{14}C-activity in the feces was present as bis-MTTB. However, when ^{14}C-labeled bis-MTTB was given to bile duct cannulated rats approximately half of the ^{14}C-activity was excreted in the bile and approximately half of the ^{14}C-activity in the bile was present as mercapturic acid pathway metabolites; no ^{14}C-labeled bis-MTTB was detected in the bile. When the ^{14}C-labeled bile from these animals was perfused into the small intestine of recipient rats, the major ^{14}C-labeled product in the feces of the recipient rats was identified as bis-MMTB. These and subsequent studies showed that at least 50% of the bis-MTTB was metabolized by replacing one of the methylthio groups with glutathione, and catabolism of the glutathione conjugate through a series of intermediates to a thiol and finally reformation of bis-MTTB which was excreted in the feces (ref. 10-12). Thus what appeared to be a case of "nonmetabolism" of bis-MTTB (on the basis of studies with intact animals) was shown to be a complex sequence of metabolic transformations ultimately regenerating the starting compound.

Naphthalene was oxidized to naphthalene-1,2-oxide and this intermediate was partially converted to 1-naphthol and 2-naphthol in certain *in vitro* systems (ref. 13). This and related information led to the conclusion that the same mechanism was responsible for the formation of the naphthols and conjugated naphthols observed in the urine of animals dosed with naphthalene. However, more recent studies by Bakke et al. (ref. 14), who compared the urinary metabolites from intact control rats and bile duct cannulated rats orally dosed with naphthalene, indicated that this conclusion should be reconsidered. They observed 1,2-dihydro-1-hydroxy 2-S-(N-acetyl)cysteinyl naphthalene and 1,2-dihydro-1,2-dihydroxynaphthalene glucuronide in the urine of intact control rats and bile duct cannulated rats and 1,2-dihydro-1-hydroxy-2-S-cysteinyl naphthalene in the bile from rats orally dosed with naphthalene. The isolation and identification of these three metabolites clearly indicated that their precursor, naphthalene-1,2-oxide, was formed in vivo. The intact control rats also excreted ^{14}C-labeled naphthols, conjugated naphthols and 1,2-dihydro-1-hydroxy-2-methylthio naphthalene glucuronide in the urine after dosing with ^{14}C-naphthalene. In contrast, these metabolites were not observed (limit of detection less than 0.5% of dose) in the urine of bile duct cannulated rats. However, large amounts of naphthols and 1,2-dihydro-1-hydroxy-2-methylthionaphthalene (free and conjugated) were observed in the urine of rats given 1,2-dihydro-1-hydroxy-2-S-cysteinylnaphthalene or 1,2-dihydro-1-hydroxy-2-S-(N-acetyl)cysteinyl-naphthalene orally or by intracecal injection. These results indicated that the naphthols and the methylthio metabolites were not derived directly from naphthalene 1,2-oxide. Rather the evidence suggested that naphthalene 1,2-oxide was conjugated with glutathione and that the glutathione metabolite was catabolized to the corresponding cysteine and mercapturic acid conjugates which were subsequently excreted in the bile and metabolized to naphthols and the methylthio metabolite.

When rats were dosed with 3-phenoxybenzoic acid (3-PBA) the major metabolite in the urine was identified as the sulfate ester of 4′-hydroxy-3-PBA which suggested that 3-PBA was merely hydroxylated, conjugated with sulfate and eliminated (ref. 15). However further studies showed that most of the 3-PBA was initially metabolized to glucuronic acid conjugates of 3-PBA and 4′-hydroxy-3-PBA; the latter metabolites were excreted in the bile, hydrolyzed in the GIT and the aglycones were metabolized to the sulfate ester of 4′-hydroxy-3-PBA. Therefore if bile had not been examined, the formation of the intermediate glucuronic acid conjugates would have gone undetected. Similarly, the sulfate ester and glucuronic acid conjugates of 3′-o-methylquercetin (a weak mutagen) was excreted in the bile of rats dosed with quercetin. However 3′-o-methylquercetin was not excreted in the urine or feces of these rats and therefore would have gone undetected if the bile had not been examined for metabolites (ref. 16).

Struble et al. (ref. 17) observed substantial amounts of ^{14}C-activity in both the urine and feces of rats after administration of 9-hydroxy-9,10-dihydro-10-cysteinyl[^{14}C]phenanthrene (^{14}C-HDCP) by intravenous (IV), intraduodenal (ID) and intracecal (IC) injection. The major ^{14}C-metabolite in the urine of rats dosed by IV and ID injection was identified as N-acetyl-S-(9-hydroxy-9,10-dihydro-10-phenanthryl)cysteine; however this metabolite was not detected in the urine of the rats dosed by IC injection. In contrast ^{14}C-labeled 9-hydroxy-9-10-dihydro-10-methylthiophenanthrene, 9,10-dihydro-9,10-dihydroxyphenanthrene and phenanthrol were excreted in the urine of all three groups of rats dosed with ^{14}C-HDCP by IV, ID and IC injection. These studies indicated that ^{14}C-HDCP was partially absorbed from the small intestine, acetylated and excreted in the urine as the mercapturate. They also indicated that much of the ^{14}C-HDCP that was not absorbed from the small intestine was cleaved (C-S lyase) when it entered the cecum and the resulting thiophenol was absorbed from the cecum, methylated and partially excreted in the urine as 9-hydroxy-9-10-dihydro-10-methylthio-phenanthrene. Some of the ^{14}C-HDCP, when administered IV, ID or IC, was converted to 9,10-dihydro-9,10-dihydroxyphenanthrene (dihydrodiol) and a phenol (OH at 1, 2, 3 or 4 position and protons at the 9 and 10 positions). This provided persuasive evidence that some of the ^{14}C-HDCP was converted to phenanthrene and the latter compound was metabolized to the dihydrodiol and the phenol that were observed in the urine (ref. 17).

There has been considerable interest in the importance of biliary excretion and enterohepatic circulation of benzo[a]pyrene and related metabolites. Weyland and Bevan (ref. 18) found that the percentage of the ^{3}H-activity excreted in the bile after intratracheal administration of ^{3}H-benzo[a]pyrene to rats, hamsters, and guinea pigs was a function of animal species and the size of the dose. Most of the metabolites in the bile were glucuronic acid and thioether conjugates; however as the size of the dose was increased, the fraction present as the glucuronic acid conjugates increased. Bowes and Renwick (ref. 19) observed that approximately one-third of ^{14}C-activity given to guinea pigs as ^{14}C-benzo[a]pyrene by intravenous injection was excreted in the bile. The biliary excretion of ^{14}C-activity was not altered by high-fat or high-cholesterol diets; however both of these dietary factors increased the proportion of dihydrodiol glucuronides in the bile. This effect was attributed to inhibition of glutathione S-transferase by the larger quantity of bile acids produced in response to the high-fat and high-cholesterol diets - thereby resulting in a larger proportion of the benzo[a]pyrene metabolites available for conjugation with glucuronic acid. When ^{3}H-benzo(a)pyrene-4,5 epoxide (^{3}H-BP-epoxide) was perfused into the mesenteric vein of bile duct cannulated rats, 76% of the radiolabel was excreted in the bile within 3 hours; the major metabolites in the bile were the glucuronic acid and the glutathione conjugates of ^{3}H-BP-epoxide (ref. 20). When the glucuronic acid conjugate of ^{3}H-BP-epoxide was perfused into the duodenum of bile duct cannulated rats, 40% of the radiolabel was excreted in the bile as the administered compound. In contrast, when the glutathione conjugate of ^{3}H-BP-epoxide was perfused into the duodenum, 26% of the radiolabel was excreted in the bile as a single unidentified metabolite (ref. 20). The metabolite(s) of [^{3}H]benzo[a]pyrene in the bile of rats and rabbits, when incubated with β-glucuronidase, yielded a product(s) which was mutagenic and which irreversibly bound to protein and DNA *in vitro* (ref. 21). However, when bile containing the metabolite(s) of [^{13}H]benzo[a]pyrene was perfused into the duodenum, ^{3}H-activity was absorbed from the GIT but no ^{3}H-labeled adducts were observed in the liver, lung or intestinal wall - indicating that a protective mechanism(s) was operating *in vivo* (ref. 22).

Watari et al. (ref. 23) administered the glucuronic acid conjugate of acetoaminophen (AG) by intravenous, intraduodenal, intraileal, and intracecal injection and determined the kinetics of absorption of this and related metabolites from the GIT and excretion in the bile and urine. These studies provided evidence that almost all of the AG was subject to enterohepatic circulation. Some of the AG was directly absorbed from the proximal intestinal lumen and part of it was cleaved in the distal region of the GIT. The free acetoaminophen was reabsorbed from the GIT and much of it was converted to the sulfate ester and excreted in the urine. However, some of the AG given to bile duct cannulated rats by intravenous injection was excreted in the urine as the sulfate ester, indicating that the AG was hydrolyzed at another site(s) in addition to the GIT.

There is evidence that some acyl glucuronic acid conjugates rearrange *in vivo* by migration of the aglycone moiety away from the C^1-position to other positions on the glucuronic acid moiety. These isomeric glucuronic acid conjugates are not hydrolyzed by β-glucuronidase enzymes. Therefore, some investigators have questioned how the formation of these isomeric "glucuronidase resistant" conjugates may affect the overall disposition of certain xenobiotics. Ruelius et al. (ref. 24) found that there was extensive enterohepatic circulation, and therefore slow excretion in the feces, of residues related to drug WY-41,770. This was attributed to fact that the glucuronic acid conjugate of this compound (excreted in the bile but resistant to rearrangement) was hydrolyzed by β-glucuronidase in the GIT and the aglycone was rapidly reabsorbed. In contrast, drug WY-18,251 was excreted in the bile as an acyl glucuronide which rapidly rearranged to a β-glucuronidase resistant isomer(s) that was therefore rapidly excreted in the feces.

PERFUSION OF OTHER ORGANS *IN VIVO*

A few investigators have perfused xenobiotics, or their metabolites, at strategic locations in the vascular system and then examined the metabolites at other key sites in the animal to determine the ability of certain organ or organ systems to metabolize the compound(s) in question. The results of some studies of this type have resulted in modification of earlier theories that were based upon more circumstantial evidence. For example, earlier studies which involved intravenous perfusion of glutathione and related conjugates and quantitation of metabolic products in the liver, kidney and urine suggested that conversion of glutathione conjugates to mercapturic acid conjugates was accomplished by the following sequence of events: (1) hydrolysis of the glutathione conjugates by the brush border membranes in the kidney to form the cysteine conjugate; (2) transport of the cysteine conjugate to the liver; (3) acetylation of the cysteine conjugate in the liver; (4) transport of the resulting mercapturic acid conjugate to the kidney and; (5) excretion of the mercapturic acid conjugate in the urine (ref. 25). However, studies by Davison et al. (ref. 26) involving perfusion of chicken kidneys *in situ* (ref. 27, 28) suggest that this proposed sequence of events should be reconsidered. Results obtained by perfusion of ^{14}C-labeled propachlor and glutathione and related conjugates of propachlor into the isolated portal system of one kidney and comparing the ^{14}C-products in the urine from that kidney with the ^{14}C-products in the urine produced by the other kidney (which received blood only from the general vascular system) indicated the chicken kidney had all of the enzymes required to convert propachlor to the mercapturic acid conjugate of propachlor (ref. 26).

When ^{14}C-labeled 9-hydroxy-9,10-dihydro-10-cysteinylphenanthrene (^{14}C-HDCP) was injected into the mesenteric vein of rats, one of the ^{14}C-labeled metabolites in the urine was identified as S-(9-hydroxy-9,10-dihydro-10-phenanthryl) mercaptoacetic acid; however when ^{14}C-HDCP was injected into the duodenum of rats the mercaptoacetic acid metabolite was not detected in the urine. This suggested that either ^{14}C-HDCP was not absorbed from the GIT or that the GIT (or microorganisms in the GIT) converted ^{14}C-HDCP to a metabolite(s) that was not a precursor to the mercaptoacetic acid metabolite (ref. 17).

METABOLISM BY MICROORGANISMS IN THE GASTROINTESTINAL TRACT

There is much evidence that bacteria in the GIT metabolize certain orally administered xenobiotics and/or xenobiotic metabolites that enter the GIT via the bile or other routes. If bacteria in the GIT do metabolize a xenobiotic, or xenobiotic metabolites formed by the animal, then it is apparent that factors which alter the microbial population in the GIT (antibiotics, diseases and nutritional factors, etc.) may alter the disposition of that xenobiotic. Thus the question of whether transformation of a xenobiotic is due to animal metabolism per se or due to microorganisms in the GIT is relevant. Comparing the disposition of a xenobiotic in conventional animals with the disposition in germfree or antibiotic treated animals is a useful method to determine if bacteria in the GIT are involved in the metabolism of the xenobiotic. For instance, both conventional and germfree rats orally dosed with pentachloromethylthio[^{14}C]benzene excreted most of the ^{14}C-activity in the feces (ref. 29). However, the major ^{14}C-labeled products in the feces from conventional rats were bis(methylthio)tetrachlorobenzene and insoluble residues and the major ^{14}C-product in the feces of the germfree rats was N-acetyl-S-(methylthiotetrachlorophenyl)cysteine. This and related information provided evidence that the formation of the cysteine conjugate was due to animal metabolism and that the cleavage of the C-S bond in the cysteine conjugate was due to bacterial metabolism.

When conventional rats were orally dosed with ^{14}C-propachlor, urinary excretion of the mercapturic acid conjugate accounted for 17% of the dose and 2-methylsulfonylacetanilides accounted for 23% of the dose; 19% was present in the feces as nonextractable residues. In contrast, germfree rats metabolized orally administered ^{14}C-propachlor exclusively to cysteine and mercapturic acid conjugates which were eliminated in the feces and urine (ref. 30) providing evidence that the bacteria in the GIT were directly involved in the formation of the metabolites which contained the methylsulfonyl moieties.

Based on studies with ^{35}S-Prontosil, it appeared that the intestinal mucosa was the major site of glucuronic acid conjugation (ref. 31). When ^{35}S-Prontosil was injected into the duodenum of rats there was much more ^{35}S-Prontosil-N-glucuronide in the bile than when the ^{35}S-Prontosil was given intravenously. When the ^{35}S-Prontosil was injected into the cecum of rats most of the ^{35}S-activity was excreted in the urine as azo-reduction products (free and acetylated sulfonilamide). In contrast, these azo-reduction products were only minor metabolites in the urine of rats given ^{35}S-Prontosil by intraduodenal or intravenous injection. This indicated that cecal bacteria, and not the liver, was the primary site of azo-reduction of Prontosil. These observations were consistent with the conclusion that the quantitatively important features of Prontosil metabolism in rats were as follows: (1) absorption and glucuronic acid conjugation by the small intestine; (2) biliary excretion of Prontosil-N-glucuronide; (3) conversion of Prontosil-N-glucuronide to sulfonilamide by bacterial action in the cecum; (4) absorption of sulfonilamide from the GIT; (5) acetylation of some of the sulfonilamide and; (6) excretion of sulfonilamide and N^{4}-acetylsulfonilamide in the urine. Partial hydrolysis of Prontosil-N-glucuronide in the small intestine and enterohepatic circulation (thereby delaying urinary excretions) may have played an important part in the disposition of this compound (ref. 31).

Isomeric glucuronic acid conjugates of valproic acid that were formed by migration of the aglycone moiety from the C^{1}-position to other positions on the glucuronic acid moiety were resistant to glucuronidase enzymes. When the isomeric glucuronic acid conjugates were perfused into the duodenum of rats they were disposed of by three major routes (ref. 32). Some were excreted in the feces, presumably without further metabolism; some were absorbed from the GIT intact and excreted in the bile and urine without further metabolism; and some were hydrolyzed by nonspecific esterases (presumably microbial) in the GIT and the aglycone was reabsorbed. Ruelius et al. (ref. 24) reported that the cleavage of resistant isomers of other acyl glucuronides was due to bacterial action in the cecum.

Larsen (ref. 33) compared homogenates of bacteria isolated from the GIT of animals to homogenates of animal tissues with respect to: GSH-transferase activity; (2) C-S lyase activity and; (3) S-methyltransferase activity. GSH-transferase was widely distributed in animal tissues but little or no GSH-transferase activity was detected in the bacteria. This suggests that electrophilic xenobiotics are not conjugated with GSH until they reach the mucosa of the gastrointestinal tract. C-S lyase for aryl cysteine conjugates was detected in both bacteria and animal tissue but only the bacteria showed C-S lyase activity toward alkyl cysteine conjugates. S-methyltransferase was not detected in the bacteria but was high in kidney and liver homogenates. Thus, these studies provided evidence that alkyl cysteine conjugates or their precursors were excreted in the bile, cleaved to form the thiol in the GIT and then the thiol was absorbed from the GIT and methylated in the liver, kidney or other tissues (ref. 33).

Larsen et al. (ref. 34) conducted substrate specificity studies with purified bacterial enzymes to determine which of the mercapturic acid pathway metabolites are cleaved by bacterial C-S lyase in the gut. They found that the cysteine conjugate of propachlor was rapidly cleaved by a C-S lyase isolated from a gastrointestinal bacterium, *Fusobacterium necrophorum*; however this enzyme had little or no activity when the glutathione, mercapturic acid, and the S-oxide of mercapturic acid conjugates were provided as substrates (ref. 34). These data indicated that the cysteine conjugate of propachlor was the immediate precursor to the thiol metabolite and the latter compound was converted to the methylthio and methylsulfoxides observed in animals systems.

Studies by Rowland et al. (ref. 35) indicate that caution should be exercised in attempting to predict the metabolism of a xenobiotic by microorganisms in the GIT of an animal on the basis of information obtained with microorganisms from the GIT of another species. They observed large differences in the amount of azo-reductase, β-glucosidase, β-glucuronidase, nitratereductase, and nitroreductase activities in the microflora isolated from the hind gut of the rat, mouse, hamster, guinea pig and man. The correlation between the activities in the GIT of man and the GIT of all rodents was especially poor.

MULTIPLE LABELING AND ISOTOPE RATIOS

Xenobiotics isotopically labeled in two or more strategic locations can be used to determine some precursor-product relationships. For instance Mulford et al. (ref. 11, 12) compared the disposition of two different preparations of pentachlorothioanisole which were labeled as follows: (1) $^{14}CH_3$-^{32}S-$^{12}C_6Cl_5$; and; (2) $^{13}CH_3$-^{35}S-$^{14}C_6Cl_5$. When the first preparation was given to rats, a large percentage of the ^{14}C-activity given to the rats was converted to respiratory $^{14}CO_2$ and urinary $^{14}CH_3SO_2H$. When the second preparation was given to rats much of the ^{35}S-activity was excreted in the urine as $^{35}SO_4$ and $^{13}CH_3$-$^{35}SO_2H$. The ^{13}C in the ^{35}S-labeled methanesulfinic acid was not diluted with ^{12}C indicating that cleavage occurred at the bond between the sulfur and the aromatic carbon rather than the bond between the methyl group and the sulfur (probably after oxidation of CH_3S- to CH_3SO_2-).

Stillwell et al. (ref. 36) used a pool labeling technique to determine the source of the methyl group in certain xenobiotic metabolites which contained a methylthio group. Rats were given a methionine free diet for 9 days and then given a diet containing CD_3-methionine for 9 days prior to the administration of naphthalene. Several naphthalene metabolites containing a SCD_3 moiety were isolated from the urine and identified by GC-MS analysis providing evidence that the methyl in these methylthio containing metabolites was derived from methionine.

INDIRECT EVIDENCE FOR REACTIVE INTERMEDIATES

Some intermediates of xenobiotic metabolism are so reactive that their formation _in vivo_ cannot be verified by classical isolation and identification techniques. For instance the sulfate ester of 7,12-dihydroxymethylbenz[a]anthracene (DHBA-SO_4), was estimated to have a half-life of 8 minutes in water at pH 7.4 and 37°C (ref. 37). Nevertheless Watabe et al. (ref. 37) were able to obtain strong circumstantial evidence for the formation of this intermediate by comparative _in vitro_ studies. They demonstrated that DHBA was not mutagenic when incubated with a rat liver cytosol in the absence of 3′-phosphoadenosine-5′-phosphosulfate (PAPS) or when it was fortified with a NAPH-generating system. However DHBA was mutagenic when the _in vitro_ system was fortified with PAPS thereby providing evidence that DHBA was activated by sulfate conjugation. Because of this information, DHBA-SO_4 was synthesized and was shown to be strongly mutagenic. When the properties of the synthetic DHBA-SO_4 were known, these workers were able to devise special procedures to verify the formation of DHBA-SO_4 in a PAPS fortified tissue homogenate. Further studies demonstrated that DHBA-SO_4 was an obligatory intermediate in the formation of the glutathione conjugate of DHBA. The latter compound was non mutagenic. Therefore if only this relatively stable secondary metabolite would have been identified, the formation of the short lived, mutagenic intermediate would have been overlooked.

The mutagenicity of 5-hydroxymethylchrysene (5-HMC) was much higher when a PAPS generating system was included in the assay system (Ames test) as compared to the inclusion of a NADP-generating system (ref. 38) suggesting, that the sulfate ester of 5-HMC was the reactive metabolite. In subsequent studies the sulfate ester of 5-HMC was isolated from a hepatic cytosol fraction fortified with a PAPS generating system. The sulfate ester of 5-HMC had potent intrinsic mutagenic activity without activation and was found to bind to calf thymus DNA _in vitro_ (ref. 38). _In vitro_ studies have also been used to demonstrate that the sulfate esters of N-hydroxy-N-acetyl-2-aminofluorene (ref. 39-41) and 1′-hydroxysafrole (ref. 42) were reactive intermediates formed from 2-aminofluorene and safrole respectively.

Enzyme inhibitors such as pentachlorophenol, which inhibit hepatic sulfotransferase (ref. 43), have been used to demonstrate the formation of transient reactive intermediates _in vivo_. For instance, the induction of hepatic tumors in mice by both safrole and 1′-hydroxysafrole was inhibited by chronic administration of pentachlorophenol (ref. 44). Subsequent studies by Randerath et al. (ref. 45) demonstrated that the formation of DNA-safrole adducts in the livers of mice was greatly inhibited by administration of pentachlorophenol one hour before dosing with safrole. These results and related information indicated safrole was sequentially metabolized to 1′-hydroxysafrole (proximate carcinogen) and then to the sulfate ester of 1′-hydroxysafrole (ultimate carcinogen) and that the latter compound was the precursor to the observed DNA-safrole adducts.

Piperine reportedly depresses the formation of glucuronic acid conjugates (ref. 46) _in vitro_ and _in vivo_ and thereby alters the bioavailability of certain drugs (ref. 47). Piperine modifies the rate of glucuronic acid conjugation in two ways: (1) by decreasing the endogenous UDP-glucuronic acid concentration and; (2) by inhibiting UDP-glucuronyl transferase activity (ref. 48). Thus, piperine may have utility as a blocking agent in metabolism studies.

Comparative studies with normal animals and animals with unusually high or unusually low levels of a key xenobiotic metabolizing enzyme have been used to indicate the nature of important intermediates. For example, "rapid acetylators" converted isoniazid to acetylisoniazid which in turn was converted to acetylhydrazine (ref. 49). Acetylisoniazid and acetylhydrazine were more hepatotoxic than isoniazid and hydrazine; thus it was concluded that acetylation of isoniazid was the crucial step in the activation of this compound. Epidemiological studies indicate that "fast acetylators" are more susceptible than "slow acetylators" to poisoning by isoniazide (ref. 49). Boberg et al. (ref. 44) used brachymorphic mice, which are congenitally deficient in the hepatic synthesis of PAPS, to determine the involvement of sulfate ester conjugation in the activation of the carcinogen 1′-hydroxysafrole. When the brachymorphic mice were treated with 1′-hydroxysafrole the incidence of hepatic tumors and the formation of hepatic DNA-safrole adducts was much less than when 1′-hydroxysafrole was given to normal rats of the same age and sex. These data and other supporting information led to the conclusion that sulfate ester conjugation of 1′-hydroxysafrole yielded the tumor initiating metabolite.

Trapping agents can sometimes be used *in vivo* to provide evidence for the formation of a short lived, intermediate. For instance, the conversion of sulfamethazine to desaminosulfamethazine and insoluble residues in rats was greatly accelerated by high dietary nitrite. It appeared that the 4-amino moiety was diazotized under the acid conditions in the stomach and the reactive diazonium cation was the precursor to the observed end products (ref. 50, 51). A "trapping experiment" was conducted in which a rat was simultaneously orally dosed with ^{14}C-sulfamethazine, nitrite and dimethylaniline. One of the ^{14}C-labeled products formed in the GIT was isolated and identified as the diazo coupling product (4-dimethylaminophenyl){4-[4-6-dimethylpyrimidinyl)benzenesulphamido]} diazene, thereby demonstrating that the suspected precursor (sulfamethazinediazonium cation) was formed *in vivo* (ref. 51). Subsequently ^{14}C-labeled sulfamethazinediazonium tetrafluoroborate (stable when maintained as salt form) was synthesized and orally administered to rats. Radiolabeled products formed in these animals (primarily desaminosulfamethazine and insoluble residues) were the same as those formed in animals orally administered ^{14}C-sulfamethazine and high dietary nitrite (ref. 51).

CONCLUDING REMARKS

This brief discussion does not provide a complete review of all of the different experimental approaches which have been used to study intermediary xenobiotic metabolism in animals. However, the examples that were discussed illustrate that many techniques are available which can be used directly or with minor modification to study intermediary metabolism and to determine the organ or organ systems responsible for xenobiotic metabolism *in vivo*.

Knowledge of the intermediary xenobiotic metabolites and the sites of their formation is important as it relates to: (1) the efficacious use of drugs and other xenobiotics that animals are intentionally exposed to; (2) evaluation of results obtained from toxicological tests; (3) how factors such as the age, sex, nutritional status, health status and inherited metabolic deficiencies may affect the disposition and biological activity of a xenobiotic and; (4) how animal exposure to antibiotics, drugs or other xenobiotics may alter the disposition and biological activity of a second xenobiotic.

REFERENCES

1. J.W. Finley and D.E. Schwass, *Xenobiotic Metabolism: Nutritional Effects*, ACS Symposium Series 277, American Chemical Society, Washington, DC (1985).
2. J. Caldwell and G.D. Paulson, *Foreign Compound Metabolism*, Taylor and Francis, London (1984).
3. D.V. Parke, *Xenobiotica* *16*, 359-509 (1986).
4. R.T. Williams, *Detoxication Mechanisms*, 2nd ed., Chapman and Hall, London (1959).
5. D.H. Hutson and T.R. Roberts, *Progress in Pesticide Biochemistry*, Wiley, London (1982).
6. G.D. Paulson, D.S. Frear and E.P. Marks, *Xenobiotic Metabolism: In vitro Methods*, ACS Symposium Series 97, American Chemical Society, Washington, DC (1979).
7. P.G. Dayton, Z.H. Israili and J.D. Henderson, *Drug Metab. Reviews* *14*, 1193-1206 (1983).
8. K.L. Davison, L.D. Rowe and D.D. Witzel, *J. Animal Sci.* *62*, 173-178 (1986).
9. M. Nussbaum, E.S. Fiala, B. Kulkarni, K. El-Bayoumy and J.H. Weisburger, *Environ. Hlth. Persp.* *49*, 223-231 (1983).
10. J.E. Bakke, *Chemosphere* *12*, 793-798 (1983).
11. D.J. Mulford, J.E. Bakke, V.J. Feil and A. Bergman, Proceedings of the 191st ACS National Meeting NY, NY April 13-18, 1986 (Agrochemicals Division, paper 37).
12. D.J. Mulford "Metabolic Fate of methylsulfide-, methylsulfoxyl-, and methylsulfonyl-

containing xenobiotics in Mammals" PhD Thesis, Biochemistry Dept, North Dakota State University (1986).
13. J.L. Wood, Biochemistry of Mercapturic Acid Formation, In Metabolic Conjugation and Metabolic Hydrolysis, Vol. 2, W.H. Fishman, Ed., Academic Press, New York (1970).
14. J. Bakke, C. Struble, J. Gustafsson and B. Gustafsson, Proc. Natl. Acad. Sci. 82, 668-671 (1985).
15. K.R. Huckle, D.H. Chipman, D.H. Hutson, and P. Millburn, Drug Metab. Disp. 9, 360-367 (1981).
16. S. Brown and L.A. Griffiths, Experientia 39, 198-200 (1983).
17. C.B. Struble, G.L. Larsen, V.J. Feil and J.E. Bakke, Proc. 9th Intr. Symp. Polynuclear Aromatic Hydrocarbons, Battelle Press, 893-900 (1986).
18. E.H. Weyand and D.R. Bevan, Fed. Proc. 45, 1526 Abstract 262 (1986).
19. S.G. Bowes and A.G. Renwick, Xenobiotica 16, 531-523 (1986).
20. T.R.D. Elmhirst, J.K. Chipman, O. Ribeiro, P.C. Hirom and P. Millburn, Xenobiotica 15, 899-906 (1985).
21. J.K. Chipman, N.A. Bhave, P.C. Hirom and P. Millburn, Xenobiotica 12, 397-404 (1982).
22. J.K. Chipman, P. Millburn and T.M. Brooks, Toxicology Letters 17, 233-240 (1983).
23. N. Watari, M. Hanawa, M. Iwai and N. Kaneniwa, J. Pharm. Dyn. 7, 811-819 (1984).
24. H.W. Ruelius, E.M. Young, S.K. Kirkman, R.T. Schillings, S.F. Sisenwine and F.W. Janssen, Biochem. Pharmacol. 34, 451-452 (1985).
25. M. Inoue, M. Okazima and Y. Morino, Hepatology 2, 311-316 (1982).
26. K.L. Davison, J.D. Bakke and G.L. Larsen, Abstract 394, 78th Annual Meeting of American Society of Animal Science (1986) (In press).
27. R.F. Wideman and E.J. Braun, Laboratory Animal Sci. 298-301 (1982).
28. I. Sperber, Zool. Bidr. Uppsala 27, 429-448 (1948).
29. J.E. Bakke, P.W. Aschbacher and V.J. Feil, Xenobiotica 11, 173-178 (1981).
30. J.E. Bakke, J. Gustafsson and B.E. Gustafsson, Science 210, 433-435 (1980).
31. R. Gingell and J.W. Bridges, Xenobiotica 3, 599-604 (1973).
32. R.G. Dickinson, M.J. Eadie and W.D. Hooper, Biochem. Pharmacol. 34, 407-408 (1985).
33. G.L. Larsen, Xenobiotica 15, 199-209 (1985).
34. G.L. Larsen, J.D. Larsen and J-Å. Gustafsson, Xenobiotica 13, 689-700 (1983).
35. I.R. Rowland A.K. Mallett, C.A. Bearne and M.J.G. Farthing, Xenobiotica 16, 519-523 (1986).
36. W.G. Stillwell, O.J. Bouwsma and M.G. Horning, Res. Commun. Chem. Pathol. Pharmacol. 22, 329-343 (1978).
37. T. Watabe, A. Hiratsuka, K. Ogura and K. Endoh, Biochem. Biophys. Res. Comm. 131, 694-699 (1985).
38. H. Okuda, A. Hiratsuka, H. Nozima and T. Watabe, Biochem. Pharmacol. 35, 535-538 (1986).
39. J.R. DeBaun, J.Y. Rowley, E.C. Miller and J.A. Miller, Proc. Soc. Exptl. Biol. Med. 129, 268-273 (1967).
40. J.R. DeBaun, E.C. Miller and J.A. Miller, Cancer Res. 30, 577-595 (1970).
41. C.C. Irving, Xenobiotica 1, 387-398 (1971).
42. P.G. Wislocki, P. Borchert, J.A. Miller and E.C. Miller, Cancer Res. 36, 1686-1695 (1976).
43. G.J. Mulder and J.H.N. Meerman in "Conjugation Reactions in Drug Biotransformation" A. Aitio Editor, Elsevier/North-Holland Biomedical Press (1978).
44. W. Boberg, E.C. Miller, J.A. Miller, A. Poland and A. Liem, Cancer Res. 43, 5163-5173 (1983).
45. K. Randerath, R.E. Haglund, D.H. Philips and M.V. Reddy, Carcinogenis 12, 1613-1622 (1984).
46. C.K. Atal, R.K. Dubey and J. Singh, J. Pharmacol. Exp. Ther. 232, 258-262 (1985).
47. C.K. Atal, U. Zutshi and P.G. Rao, J. Ethanopharmacol. 4, 229-232 (1981).
48. J. Singh, R.K. Dubey and C.K. Atal, J. Pharmacol. Exp. Ther. 236, 488-493 (1986).
49. J.R. Mitchell, H.J. Zimmerman, K.G. Ishak, V.P. Thorgeirsson and J.A. Timbrell, Annals Intern. Med. 84, 181-192 (1976).
50. G.D. Paulson, Xenobiotica 16, 53-61 (1986).
51. G.D. Paulson, V.J. Feil and J.T. MacGregor, Xenobiotica (1986) (In Press).

Interpretation of animal metabolism data in terms of significance to man

D. H. Hutson

Shell Research Ltd., Sittingbourne Research Centre,
Sittingbourne, Kent, ME9 8AG, England.

Abstract - The extrapolation of the results of animal toxicity studies to man is often complicated by species differences in response. Metabolic and kinetic studies can aid the extrapolation by providing a different view of the interaction between subject and chemical. At best this view will include an understanding of the mechanism of toxicity, or the lack of toxic action. Some of the techniques of metabolism studies, particularly those *in vitro*, are readily applicable to man and can provide a useful link between the test species and man. Extrapolation cannot be a purely theoretical activity relying on background information from drug research. The direct approach - metabolic studies in man - has been used, but decreasingly over the last decade. Human studies clearly have their limitations and difficulties but they have provided some valuable information and should not be discouraged. As our understanding of the biochemistry of toxic action improves, however, it should be increasingly possible to design specific studies which avoid the classical human volunteer study.

INTRODUCTION

Metabolic studies are carried out on crop protection chemicals for two purposes. Studies on the fate in target organisms help to define the mode of action and, by feedback, aid in the design process. The purpose of studies in animals is to complement toxicological studies and, at their best, provide explanations in molecular terms for observed toxic effects. Species differences amongst animal models in the manifestation of toxic effects confound the accurate extrapolation of the results of animal studies to man. Many of these differences are due to species differences in the metabolism of the pesticide. Metabolism studies (studies of *fate* of the chemical) complement classical toxicological studies (studies of *effect* of the chemical) because they provide a qualitative and quantitative view of the compound - animal interaction in terms of the chemical as opposed to its observed action. The qualitative aspects are fundamentally important in that biotransformation of a chemical exposes the test species (or specific organs within the animal) to a small or large number of new chemicals (metabolites). Quantitative aspects are also important. Whilst the options open to a chemical for biotransformation are now largely predictable, the choice which even the well-studied rat will make between these options is not. Similarly it is not possible to predict with any certainty, from data in animals, the choice which man will make. Attempts at interspecies predictions are frequently invalidated by small variations in chemical structure.

Nevertheless, as animal metabolism data will always be our major source of information on the biological fate of a chemical, they must be interpreted in relation to man if animal studies have revealed significant toxic effects attributable to the chemical. The purpose of this paper, implicit in its title, is to describe exactly how this should be done or, failing that, to suggest in general terms how it might be done. The first question to pose is: can animal metabolism data be interpreted in isolation - without further experimentation involving some type of study of man? This may be possible, in which case we have two types of activity to consider: that without human experimentation (relying on existing knowledge of human xenobiochemistry and information derived from the study of therapeutic agents) and that involving some type of investigation in man.

INTERPRETATION WITHOUT FURTHER STUDY (IN MAN)

Design of metabolism studies

A variety of data emerges from a metabolism study in rodents, depending on the complexity of its design. The minimum requirement is for information on the persistence of the chemical. High persistence signals a problem that cannot be ignored. If a chemical is rapidly eliminated, its mechanism of elimination (e.g. metabolism, conjugation and renal clearance) must be defined. At this point analogies can be made with available information in man. For example, the wild oat herbicide flamprop methyl (Fig. 1) is rapidly hydrolysed and eliminated by the rat and is similarly disposed of in the mouse and rabbit (ref. 1). This herbicide is a methyl ester of a carboxylic acid. If we accept that (i) methyl (cf isopropyl) esters are particularly easily hydrolysed and (ii) man possesses an active carboxylesterase in the liver and other organs (ref. 2), it is inconceivable that man should not dispose of the herbicide in the same way. Given also that this compound did not elicit any significant adverse toxicity in the experimental animals studied (i.e. there is no target for biochemical toxicological studies), a reasonable judgement is that further study is unjustified. Where other biotransforming enzymes are crucial to rapid elimination, e.g. cytochrome P450 (etc.) - catalyzed hydroxylation, we should be more circumspect as it is known that there is considerable individual variation in the activity of this enzyme(s) in man. Furthermore, genetic polymorphism in man (ref. 3) complicates the issue. As an example of hydroxylation-initiated clearance, the organochlorine dieldrin is eliminated from the rat (ref. 4) relatively more rapidly than from man (ref. 5).

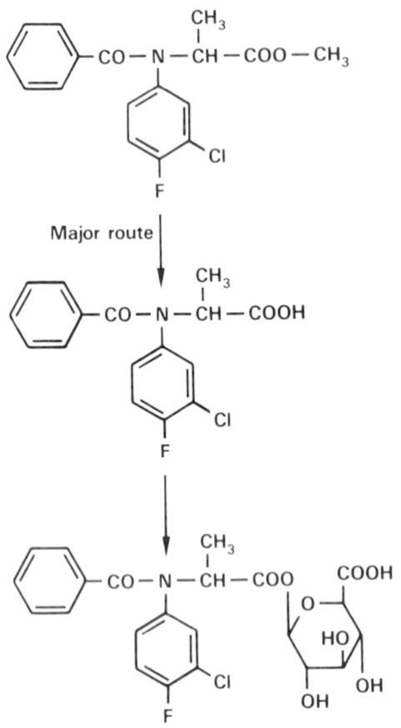

Fig. 1. The metabolism of flamprop-methyl in mammals

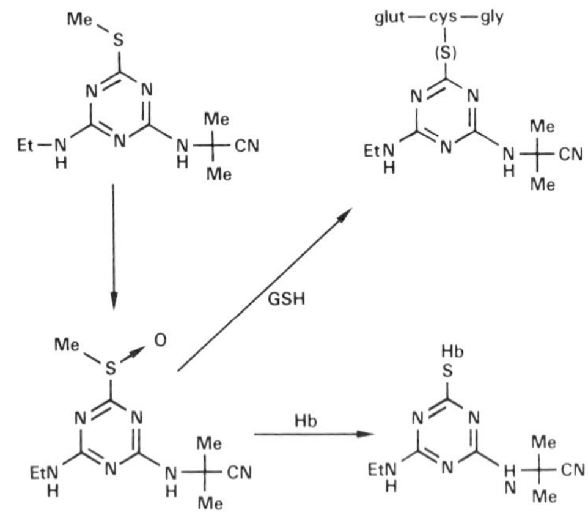

Fig. 2. The metabolism of cyanatryn in mammals

Another requirement from metabolism studies is the identification of all of the major (>5%?) excreted metabolites. The kidney and bladder as a minimum are exposed to these chemicals, but more important, this information allows the definition of most of the metabolic pathways for the chemical, including the putative intermediary metabolites which do not appear as terminal products. The existence of these can often be, and should be, confirmed by in vitro studies using rodent organs, cells or enzymes. Where these intermediates are obviously reactive they can be subjected directly to in vitro tests, e.g. for mutagenicity, to ameliorate (or otherwise) concern about their biological effects. One of the primary metabolites of the organophosphate insecticide chlorfenvinphos, 2,4-dichlorophenacyl chloride, is a case in point (ref. 6). The compound is a reactive chemical which combines spontaneously with glutathione (its major route of detoxification) and has the potential to react with other biological molecules. It appears, however, to have negligible reactivity towards DNA and is non-mutagenic when itself tested for mutagenicity in the Salmonalla/± microsome assay. Thus this intermediate, almost certainly formed in man, can be judged non-genotoxic. Another example is derived from the herbicidally active 2-thiomethyl-s-triazine cyanatryn (Fig. 2). This compound is oxygenated at sulphur in rodents to form a reactive intermediate which spontaneously triazinylates glutathione and, furthermore, reacts avidly with a thiol group in haemoglobin (ref. 7). This reactivity signalled caution which was allayed by (i) the failure of the metabolite (synthetic sample, radiolabelled) to react with DNA in vitro and (ii) its failure to afford a mutagenic response when tested for mutagenicity against bacteria at concentrations up to 5000 µg/ml. Whether the interaction of this intermediate with haemoglobin actually occurred in man is another question and one which cannot be answered without some sort of experimentation using a human system (see later).

The examples given above are few and selective but they serve to illustrate the sort of judgements which we have to make in relating animal metabolism results with human toxicology. These judgements have a sound basis and are valuable. It is obvious, however, that without recourse to some type of study of man one can neither confirm such judgements nor discover the unexpected or interactions unique to man. We hope that our animal data provides proof of rapid elimination via innocuous metabolites, support for toxicology in the form of explanations of observed events and - a more ambitious objective - data of value in predicting toxicity. If data of predictive value emerges from animal metabolism studies, the case for transposing the same experimental techniques (which may be very selective, and *in vitro* techniques as well as *in vivo* studies) to man must be a strong one.

METABOLIC STUDIES IN MAN - GENERAL COMMENTS

It is interesting, but on reflection not surprising, that when the need was perceived for studies on the metabolism and persistence of pesticides in man, the problem was addressed very directly with a full frontal attack. Only in the last ten years are more selective, less invasive techniques (based on biochemical mechanisms) being elaborated and applied. A relationship betweeen the fate of a chemical in the body and the intensity of its biological effect was assumed by pharmacologists decades ago and it is generally considered that metabolism studies in man should provide useful information enabling us:

(i) to validate (retrospectively) extensive studies in animals
(ii) to obtain a direct view of the man-chemical interaction
(iii) to choose the appropriate species for full toxicological studies.

Problems appear immediately. There is a dichotomy between (i) and (iii). The concept of using metabolism data to select the species to be used for full toxicological studies is unrealistic for at least two reasons. The enormous background knowledge derived from rats, mice and, to a lesser extent, primates ensures that these species will continue to be the main source of toxicological data. Furthermore, the full toxicological studies referred to in (iii) are required *before* it is ethically acceptable to study the fate of a non-therapeutic agent in man. Thus (iii) above would be better re-worded: (iii) to choose the appropriate species for *further selective* toxicological studies.

Should metabolism studies *invalidate* rodent toxicology then, of course, further toxicological studies become important.

The importance of (ii) above is that whilst it is ethically unacceptable to seek and study a toxic *effect* in man, it is acceptable, and possible, to derive much qualitative and quantitative information from man in terms of the *fate* of the chemical. This information is directly comparable with that derived from rodents.

There are several approaches to the effective study of the fate of pesticides in man:

(i) metabolism studies, including the acquisition of kinetic data, in man and in rodents under strictly comparable conditions
(ii) the follow-up of adventitious exposures
(iii) comparative metabolic studies using isolated tissues from man and experimental animals
(iv) specific measurements of a chemical lesion in target organs or target molecules. The problem of assessing this indirectly in man is being actively studied (see below).

Each of these approaches has its difficulties and limitations and it is probable that no single approach will suffice alone.

HUMAN METABOLISM STUDIES *IN VIVO*

The direct approach to the study of the fate of pesticides in man is the human volunteer study. Some examples of such studies, some from our own experience, are given below which serve to illustrate their usefulness and limitations. It is significant that most of these examples are now quite old.

Wayland Hayes, speaking at the last IUPAC Congress in Kyoto in 1982 (ref. 8), made the observations that, although about 100 pesticides had been studied in man, most of the results were published between 1960 and 1977, peaking at about 1970 and that the current rate of publication was very low. This is still the case in 1986.

Dieldrin

This organochlorine is one of the most thoroughly studied compounds in pesticide toxicology. As well as being well studied in mammals, dieldrin has been the subject of one of the more ambitious pesticide investigations in man. This research, a kinetic study, was initiated at a time when metabolism studies were non-routine. What little work had been done led researchers to the uneasy conclusion that dieldrin was not metabolized in mammals. The presence of this and of other organochlorines in man's diet stimulated concern that man's body burden of organochlorines would increase continually. The study (refs. 9 and 10) was designed with two objectives:

(i) to define the quantitative relationship between the oral intake of dieldrin and its concentration in blood and fat (thereby allowing meaningful monitoring of manufacturing, formulating and spraying operatives via blood sampling) and

(ii) to measure the kinetics of uptake and elimination in these tissues. The study also provided an opportunity for monitoring the health of human subjects under defined exposure conditions using full clinical and physiological examination. Metabolic studies could not be incorporated because neither the technology nor the knowledge of dieldrin metabolism was available at the time.

Twelve male volunteers, in groups of three, received 0, 10, 50, and 211 µg of dieldrin per day for up to 2 years. The study was continued for a further 8 months after cessation of treatment. Venous blood samples were taken monthly and samples of adipose tissue were taken by needle or open biopsy at 18 and 24 months and at 8 months post-exposure. No effects on the health of any of the volunteers were found; a further encouraging discovery was that, under continuous intake, the rate of increase of dieldrin concentration in the blood declined progressively and at about 15 months a plateau value was approached (0.014 µg/ml at an intake of 211 µg/day). The ratio of the concentration of dieldrin in adipose tissue to that in blood was 136. Both values were dependent on the amount of dieldrin ingested daily. When dieldrin administration was terminated, its concentration in blood fell exponentially with a half-life of 141-592 days. It was of interest that both the body burden of dieldrin (blood and fat values) and its rate of elimination (post-exposure blood values) were characteristic not only of the amount ingested, but also of the individual. This suggested that individual variations in the rate of metabolism of dieldrin had a profound effect on its kinetics.

Later metabolic studies in rodents and studies on human excretion products suggested that, whilst metabolism was slower in man, the major pathways of biotransformation were probably the same as those in rodents.

<u>Chlorfenvinphos</u>

This organophosphate insecticide has been extensively studied in a number of species in an attempt to relate acute toxicity and metabolic detoxification (ref. 10 and references therein). The biotransformation of ^{14}C-chlorfenvinphos was then studied in a human volunteer (ref. 11). An oral dose of 0.18 mg/kg was almost totally eliminated in the urine (72% in 4.5 hours and 95% in 26.5 hours). The yields of metabolites compared with those from rat, dog and rabbit are shown in Table 1.

These results are useful in that they demonstrate that the fate of chlorfenvinphos in man is similar to that in mammals. However, no judgement could be made on whether man was more like rat (sensitive) or dog (resistant) in his response to chlorfenvinphos. This aspect is discussed below in the section on <u>in vitro</u> studies. A further volunteer study was carried out (ref. 12) using non-radioactive chlorfenvinphos in order to relate the yield of a specific urinary metabolite (de-ethylchlorfenvinphos), to the amount of pesticide ingested. Fourteen male volunteers were each given 3 mg of chlorfenvinphos (ca 0.04 mg/kg) daily for 53 days. The yield of metabolite, however was only 5% (cf 22% in a single subject at the higher dose). This suggested a dose-dependency in the metabolism and highlighted the need to match rodent and human doses very closely.

TABLE 1. Comparative metabolism of ^{14}C-chlorfenvinphos

Metabolite	Per cent dose in following species: Rat (2.0)[a]	Dog (0.3)[a]	Rabbit (4.0)[a]	Man (0.18)[a]
Chlorfenvinphos	0	0	0	0
De-ethylchlorfenvinphos	27[c]	65[c]	47	22[c]
2,4-dichloromandelic acid	5	12	3	22[c]
2,4-Dichlorophenylethanol[b]	30	3	28	21
2,4-Dichlorophenylethandiol[b]	2	2	3	14
2,4-Dichlorobenzoylglycine	3	1	1	4

[a]Dose in mg/kg.
[b]As glucuronide conjugates.
[c]Isotope dilution analysis on whole urine.

Other organophosphates

The metabolism of ethyl parathion and methyl parathion has been studied in man (ref. 13) and the hydrolytic route known to occur in animals has been confirmed. Similarly, the fate of dichlorvos, thoroughly studied in rats, mice and hamsters, has been studied (using ^{14}C-label) in a human volunteer (ref. 14).

TABLE 2. Yields of radioactivity and metabolites 24 hours after the ingestion of ^{14}C-vinyl-dichlorvos

Metabolite	Elimination route	Per cent dose in following species: Rat (3.6)[a]	Mouse (7.3)[a]	Hamster (1.5-3.7)[a]	Man (0.07)[a]
^{14}C	Urine	10	27	15	8
^{14}C	Faeces	2	3	3	–[c]
$^{14}CO_2$	Respiration	29	23	34	27[b]
Hippuric acid	Urine	2	1	1	1
Demethyl dichlorvos	Urine	2 <1	19 <1	–[c]	<1
Urea	Urine	0.6	0.6	–[c]	<1

[a]Dose in mg/kg.
[b]Eight hours only.
[c]Not measured

The similarity between man and rat, in particular, was striking (see Table 2).

Bendiocarb

This carbamate insecticide, developed specifically for the control of insects in industry and public health has been studied relatively recently in man and rat. Rats were dosed at 0.125 and 2.5 mg/kg with ^{14}C-bendiocarb and a human volunteer was given the lower dose. Again, the rates of metabolism and the major pathways were common to both species, but the rat afforded minor amounts of ring-hydroxylated products not found in man (ref. 15).

Cypermethrin

The fate of cypermethrin in four male subjects given oral doses of 0.25-1.5 mg has been reported (ref. 16). The major objective of this study was to relate the amount ingested to the yield of a specific metabolite, to provide the basis of a biomedical monitoring procedure. This experiment succeeded and it is of interest that the yield of the metabolite was very similar to that found for rats and mice.

Conclusions

In the cases discussed above, the major pathways of biotransformation have been confirmed as being common to rat and man. Thus they provide evidence that the intermediary and terminal xenobiotic metabolites, as well as the parent molecule, have been accounted for in rodent toxicology. Rapid metabolism of ester pesticides has been demonstrated. Dieldrin is metabolized more slowly by man than by rat and the metabolism is subject to considerable inter-individual variation. The studies are also of value in confirming the basis of biomedical monitoring methods.

The weaknesses of many studies of pesticides in man are either inherent in the situation or due to poor design. Numbers of subjects are usually very low and clearly data from one individual are of limited value. Doses are low, correctly, for safety reasons and because low doses are closer to occupational exposures. However, the comparative dose in the rodent has often not been studied. Attention to the kinetic parameters of non-therapeutic agents has been scant. Clearance values and volumes of distribution of pesticides may be of limited value in isolation but they do offer widely accepted numerical values which should form the basis for comparison with other xenobiotics (e.g. drugs) in a variety of species. It is interesting to note, however, that drug research in this area also has its problems. Clark and Smith (ref. 17) have collected more than 900 published values for the clearance (CL) of various drugs administered at various doses via a variety of routes in the hope that some useful generalizations about species would emerge. They were disappointed by the variation in experimental techniques and some extreme differences of design between human and animal studies. Comparison of data for rat, dog, rhesus monkey, mouse and man revealed that the experimental species tended to afford higher clearance values and that values for the mouse were highly scattered. The conclusion was that prediction about an individual compound was impossible. These authors also highlight differences between man and experimental animals in urine flow and bile flow.

Perhaps the most serious indictment of most studies of biotransformation in man is that they confirm the known. Very rarely are novel reactions discovered. For this to be possible, reasonable amounts of radioactivity and weight of compound must be dosed. The use of double labelling with a second (stable) isotope should be encouraged. The most sensitive methodology available must be used. Immunochemical assays should be increasingly available for the analysis of parent pesticides and selected metabolites and these can replace the use of radiolabelling for certain purposes. Metabolism studies in man need the highest degree of co-operation between clinical research units and high-calibre analytical chemical facilities if the most is to be gained. It is a pity that, whilst analytical procedures have undergone spectacular improvements in 20 years. The ability of Societies to arrange controlled studies in man has declined.

HUMAN METABOLISM STUDIES *IN VITRO*

General studies

This term refers to studies designed to provide the maximum amount of information using an *in vitro* technique, i.e. to mimic *in vivo* metabolism using some form of tissue preparation. Thus Sullivan and co-workers (ref. 18) have used liver explants from a variety of species including man to study the metabolism of carbaryl. When the preparations were used under strictly controlled conditions, the variation between the *in vitro* results and *in vivo* results were much less than the variation between species. This technique could prove useful in the screening of species for the selection of one similar to man to be used in further toxicological studies.

Various levels of organisation may be used in these studies, e.g. isolated perfused organs, explants, slices, isolated cells, homogenates, subcellular fractions and individual enzymes. Generally speaking, the more disturbed or fragmented the tissue, the less realistic are the results. The higher the level of cellular organization used, however, usually the more difficult it is to obtain the human tissue preparations and this limits the applicability of the approach. However, even given ready access to tissue, the *in vivo*/*in vitro* correlation is difficult to maintain in practice for a generalized picture of metabolism. It is very important, therefore, that the *in vitro* preparation be confirmed as reflecting *in vivo* metabolism; i.e. the model must be validated in animals.

Specific studies

When a specific reaction has been identified in mammals as being important in metabolism and toxicity - possibly a critical determinant of toxicity - this reaction can be studied *in vitro* in human tissue and the result extrapolated to the human situation *in vivo*. The reaction may involve detoxification, bioactivation or reaction with a target molecule or receptor. Clearly the closer one can work to the target, the better. It is most important that the reaction to be studied *in vitro* be verified as important *in vivo* and this validation exercise must be done in an appropriate animal species (perhaps several) under strictly comparable conditions.

We have studied the detoxification of chlorfenvinphos in this way. Oxidative metabolism is critically important in limiting the acute toxicity in rats, mice, dogs and rabbits (ref. 10). The high rate of metabolism by human hepatic microsomal mono-oxygenase suggests that the acute toxicity of this insecticide in man will be much lower than it is in the sensitive rat (ref. 19).

Similarly, the rate of hydrolysis of the pyrethroid cypermethrin by hepatic microsomal carboxylesterase is important in limiting the action of the potentially toxic *cis*- and *trans*-isomers of this insecticide. The activity of the enzyme has been studied in rat, rabbit and man *in vitro* (after solubilization from the microsomal membrane) and the human enzyme has been found to be much more active than the rat enzyme towards the toxic *cis*-isomer (ref. 20). The limitations of such correlations must be stressed however, because the sensitivity of the target (the nervous system) has not been addressed. The triazine cyanazine was referred to above. This was a case of concern about the electrophilic reactivity of the *S*-oxide as evidenced by its facile reaction *in vivo* with glutathione and with haemoglobin. Studies with human liver (ref. 21) showed that the *S*-oxide was rapidly formed. However, studies *in vitro* of the reaction between the *S*-oxide and human and rat erythrocytes showed that rat haemoglobin was much more reactive (ref. 7) and gave a greatly exaggerated impression of the electrophilic reactivity of the intermediate.

If the target for a particular toxicity is known and accessible it can be studied directly. My colleagues are currently validating an *in vitro* model for skin carcinogenesis initiated by the action of benzo(a)pyrene. The profile of DNA adducts derived from radiolabelled benzo(a)-pyrene-treated mouse skin *in vivo* and in explants in *in vitro* culture have been shown to be quantitatively similar and qualitatively nearly identical (ref. 22). The *in vitro* technique has now been applied to human skin explants thus allowing a direct comparison between mouse and man. In principle, this approach may be applied to any human tissue. Furthermore, by studying the molecular interactions of reactive intermediates with the target molecule, it is possible to avoid the uncertainties of predicting or extrapolating differences in bioactivation and detoxification reactions.

TARGET DOSIMETRY

The example given above leads on naturally to a consideration of the target dose approach. Toxic effects are usually defined in relation to exposure dose, i.e. the dose of chemical absorbed by the test subject. Target dosimetry seeks to measure the amount of chemical (or active metabolite) that reaches the target and reacts with it. The two major problems with this approach are (i) defining the target and (ii) gaining access to it in man. Obviously the latter must be achieved by indirect methods.

Such an approach was widely attempted when the organophosphate insecticides were in their peak research and development phases. The inhibition of plasma and erythrocyte "cholinesterase" was (and still is) used as an indirect measure of the inhibition of the target cholinesterases. The approach has not been wholly successful for a variety of reasons, not the least of which is the uncertainty of the location of the target enzyme. This, with competing esterase inhibitions, and the method used (measurement of remaining active enzyme as opposed to inhibited enzyme) have confounded the proper validation of animal models.

A more encouraging prospect is the measurement of alkylating agents reaching their target. The target for the initiation of carcinogenesis is widely accepted to be DNA. There will be other targets for the promotional factors operating in carcinogenesis but, in assessing risk associated with low doses of alkylating agents, the promoting activity of the latter will be insignificant in comparison with environmental and genetic factors. An important aspect of dosimetry is its monitoring function - and when applied to man a non-invasive approach is needed. Ehrenberg and co-workers (ref. 23) chose erythrocyte haemoglobin (Hb) as an indirect target for dose monitoring for ethylene oxide (EO) exposure. The technique used was the isolation and assay of *N*-3-(2-hydroxyethyl)histidine by mass fragmentography. Circulating Hb in man has a biological life-time of approximately four months. If this time holds also for alkylated Hb it can be used to monitor chronic as well as acute exposures. It should be stressed that this approach must be validated in laboratory animals to ensure that there is a proportional relationship between interaction with target organ DNA and with Hb. EO proved a good initial choice because protein adduct measurements indicate that it is rapidly distributed throughout the body in several species and gives a similar tissue dose in all organs. More recently *N*-alkylated *N*-terminal valine residues of Hb have been used for monitoring. These adducts can be selectively and gently cleaved off and measured (ref. 24). The interpretation of the results of these measurements in terms of carcinogenic risk is difficult but, at least for EO, the application of the "rad-equivalence" concept by Ehrenberg and co-workers appears to be successful (ref. 24).

Hb will not always be a useful target-dose monitor. Some highly reactive alkylating agents generated *in vivo* by metabolism may never leave the cell in which they are formed. To cover such cases, Wright (ref. 25) has suggested that the use of specific secreted proteins of potential target organs may prove useful.

CONCLUSIONS

It seems impossible to recommend a simple, routine approach to the interpretation of animal metabolism data in relation to man. A number of problems are apparent: the elusive species which routinely metabolizes xenobiotics similarly to man has not been found; for non-therapeutic agents, extensive toxicological studies in laboratory animals are required before studies in man are sanctioned; primates, regarded as suitable on evolutionary grounds, are too scarce and expensive for use in large enough numbers for statistical analysis of results; the enormous background available for the laboratory rodent (effectively the rat and mouse) dictates its continued use.

The way forward would appear to be via the following sequence:

(i) the understanding of the mechanism of toxicity of the pesticide in the commonly used laboratory species

(ii) improved awareness of, and more information about, the differences between these species and man in their metabolism of xenobiotics generally

(iii) the comparison of the rate and routes of metabolism of the pesticide between laboratory species and man using *in vitro* models which have been carefully validated *in vivo* in animals

(iv) *in vivo* studies (including kinetics) in man at low doses which have also been studied (in addition to high doses) in laboratory animals

(v) specific and selective studies, e.g. target dosimetry, in man by carefully validated indirect methods to address specific types of toxicity identified in animals.

These various approaches, including detailed compilations of species differences in the metabolism of xenobiotics in experimental animals and man, and of the fate of pesticides in man, have been reviewed in a recent IUPAC Pesticide Chemistry Commission report entitled 'Recommended approaches to the validation of animal metabolism in relation to man' (ref. 26).

Clearly we cannot predict the unpredictable. Surprises will continue to present themselves but these are more likely to be derived from drug therapy where substantial numbers of people ingest relatively high doses of chemicals.

The understanding of particular forms of chemical toxicity at the molecular level is of paramount importance. Armed with knowledge, we can design targeted incursions into man rather than produce general data which provide little more than a comfort factor.

REFERENCES

1. J.V. Crayford and D.H. Hutson, Pestic. Sci., 11, 573-590 (1980).
2. W. Junge, E. Heymann, K. Krisch and H. Hollandt, Arch. Biochem. Biophys., 165, 749-763 (1974).
3. R.L. Smith, Xenobiotica: Special issue - human genetic variation in oxidative drug metabolism, 16, 361-365 (1986) and eleven following papers (Eds. R. A. Franklin and D. V. Parke).
4. J. Robinson, M. Roberts, M.K. Baldwin and A.I.T. Walker, Food Cosmet. Toxicol., 7, 317-332 (1969).
5. C.G. Hunter, J. Robinson and M. Roberts, Arch. Environ. Health, 18, 12-21 (1969).
6. D.H. Hutson, D.S. Holmes and M.J. Crawford, Chemosphere, 5, 79-84 (1976).
7. M.J. Crawford, D.H. Hutson and G. Stoydin, Xenobiotica, 10, 169-185 (1980).
8. W.J. Hayes, Jr., Pesticide Chemistry, Human Welfare and the Environment, Vol. 3 (Eds. J. Miyamoto and P.C. Kearney), Pergamon Press, Oxford, pp. 387-394 (1983).
9. C.G. Hunter and J. Robinson, Arch. Environ. Health, 15, 614-626 (1967).
10. D.H. Hutson and A.S. Wright, Chem.-Biol. Interact., 93-101 (1980).
11. D.H. Hutson, Progress in Pesticide Biochemistry, Vol. 1 (Eds. D.H. Hutson and T.R. Roberts), Wiley, Chichester, pp. 287-333 (1981).
12. C.G. Hunter, J. Robinson, C.T. Bedford and J.M. Lawson, J. Occup. Med., 14, 119-122 (1972).
13. D.P. Morgan, H.L. Hetzler, E.F. Slach and L.I. Lin, Arch. Environ. Contam. Toxicol., 6, 159-173 (1977).
14. D.H. Hutson and E.C. Hoadley, Arch. Toxicol., 30, 9-18 (1972).
15. I.R. Challis and J.W. Adcock, Pestic Sci., 12, 638-644 (1981).
16. C.V. Eadsforth and M.K. Baldwin, Xenobiotica, 13, 67-72 (1983).
17. B. Clark and D.A. Smith, CRC Critical Reviews in Toxicology, 12, 343-385 (1984).
18. L.J. Sullivan, B.H. Chin and C.P. Carpenter, Toxicol. Appld. Pharmacol., 22, 161-174 (1972).
19. D.H. Hutson and C.J. Logan, Xenobiotica, 16, 87-93 (1986).
20. A. Croucher, D.H. Hutson and C.J. Logan, Vth Int. Congr. Pesticide Chemistry, Kyoto, Abstr. Ve 4(1982).
21. M.J. Crawford and D.H. Hutson, Xenobiotica, 10, 187-192 (1980).
22. K.R. Huckle, R.J. Smith, W.P. Watson and A.S. Wright, Carcinogenesis, 7, 965-970 (1986).
23. C.J. Calleman, L. Ehrenberg, B. Jansson, S. Osterman-Golkar, D. Segerback, K. Svensson and C.A. Wachtmeister, J. Environ. Pathol. Toxicol., 2, 427-442 (1978).
24. M. Tornqvist, J. Mowrer, S. Jenson and L. Ehrenberg, Anal. Biochem., in press.
25. A.S. Wright, ACS Symposium Series, No. 160, 285-304 (1981).
26. J. Miyamoto, H. Kaneko, D. H. Hutson, H. O. Esser, S. Gorbach and E. Dorn, Pure. Appld. Chem., in press.

Interpretation of animal metabolism studies

Christian Schlatter

Institute of Toxicology of the Swiss Federal Institute of Technology and the University of Zurich, CH-8603 Schwerzenbach, Switzerland

Abstract - The investigation of animal metabolism plays an important role in the safety studies of a new pesticide. A key issue is the recognition of possible species differences. Also important is the comparison of plant and animal metabolites. Although the cell biochemistry shows many differences between plants and animals, surprisingly enough the spectrum of pesticide metabolites found in plants is usually quite similar to that observed in mammals. For this reason a toxicological study performed on the parent compound will also be applicable to a situation where only metabolites are present since the toxicity of both the parent compound as well as the metabolites could be expressed in the animal study. There are, however, situations where a compound becomes activated during metabolism. These intermediates are usually unstable, but due to their reactivity they are often responsible for toxic reactions including alkylation of DNA leading possibly to cell transformation. One of the best known examples is the epoxidation of double bonds. In cases where tumours are found in animal studies a careful search for such reactive metabolites becomes mandatory in order to obtain insights into the mechanism of tumour induction. Since carcinogenicity studies are performed in animals at rather high dose levels it will be important to determine whether the reactive intermediates are formed only at high doses and not at the trace levels to which the applicator and consumer are exposed. By the very fact of being metabolized the compound often induces an adaptive response in the mammalian organism: The responsible enzymes are produced in higher amounts. Such a longlasting change in the cellular biochemistry has been shown especially in rodents to be toxicologically important, since malignancies can develop in such induced organs. Such tumours do not, however, seem to be relevant for man. At low dose levels no enzyme induction exists and, in addition, it seems that tumour formation as a reaction to life-long enzyme induction is a peculiarity of the rodent liver. A similar situation exists in peroxisome proliferation which is induced by several chemicals known to be hepatocarcinogens in rats or mice. Again this phenomenon does not occur in man and so there is no tumour risk to man from these compounds.

INTRODUCTION

Pesticide-toxicology differs mainly in two aspects from other areas of toxicology such as drug-toxicology. Toxicity in its broadest sense, i.e. damage of biological materials, is the key property of a pesticide whereas in other chemicals toxicity only occurs occasionally as an unwanted side effect. Although the toxicity of pesticides is usually directed toward non-mammalian organisms such as plants, fungi or insects, one must also expect toxicity to mammals due to the close similarities of the fundamental features of all living cells. This situation is the reason for the widespread fear of pesticides. However, the danger of pesticides for mammals is usually rather overestimated since the second important aspect of pesticide toxicology has not been taken into account: the exposure of man to only low levels, which usually are biologically ineffective. This absence of mammalian effects at low dose levels is indeed one of the predominant requirements for a pesticide since any effects in non-target organisms would not be acceptable. Only in the relatively rare situations of acute accidental or suicidal intoxications with undiluted pesticide formulations overt toxicity symptoms may occur. Pesticide toxicology has to focus therefore mainly on questions of dose dependency of toxic symptoms and on effects possibly still occurring at low levels. The vast area of toxicology is thus restricted to some few key points. The most important effects possible at low-level exposure are carcinogenicity and mutagenicity or accumulation.

BASIC STUDIES

The basic studies of metabolic and kinetic behaviour of a pesticide are needed together with the basic toxicological data to investigate the possibility of low dose effects. The metabolic data are very often used to corroborate toxicological findings. The following examinations should be carried out in this respect:

Investigation of kinetics

Usually metabolic studies are performed with radiolabelled compounds. It is logical that before elucidation of the qualitative and quantitative metabolite pattern, the time course of the overall radioactivity in the body is followed. Important information regarding possible adverse effects is obtained already from these simple examinations: The determination of the absorption kinetics after one single administration by the oral, i.p. or dermal route shows if a compound principally may permeate through cell membranes and organ barriers and secondly how rapidly it is eliminated again.
From these first results important predictions regarding toxicity can be deduced or puzzling toxicity results may be explained:

A low absorption rate is suggested by the detection of most of the radioactivity in the feces; however, a predominant elimination of radioactivity via feces also occurs after absorption and subsequent biliary excretion. Such a situation can be detected either by - technically rather difficult - biliary canulation or by i.v. or i.p. injection: Elimination via feces indicates biliary secretion whereas appearance in urine will point to a lack of gastrointestinal absorption after oral administration. Compounds with a low absorption rate are not expected to be of much toxicological concern since the lack of penetration through cell walls will not be confined to cells of the gastrointestinal tract but to all other cells of the body. There are however exceptions possible. For instance, when a special transport system exists which is only available in certain cell types, these cells will take up the compound. This is the case for the specific uptake and accumulation into lung cells of paraquat (ref. 1), a polar herbicide with a very low absorption rate from the gastrointestinal tract.

In cases of rapid absorption and elimination the body burden and peak levels in the organs will strongly depend on the time schedule of dose administration. One has to expect a marked difference in toxicity if the compound is given as a single oral dose by gavage or when mixed into the diet. A typical example is metaldehyde, an effective snail and slug poison: The oral lethal dose for rats is approx. 150 mg/kg, however 5 % in the diet corresponding to a daily dose of 5000 mg/kg do not show any effects at all, even if given for a prolonged period of time. The primary distribution is mainly given by the lipophilicity of the compound and blood supply to the organs. Secondarily, the distribution might become rearranged by metabolic reactions rendering the compound more polar and excretion. The aim of the kinetic investigation is to examine, if indeed the theoretical predictions are fulfilled. If not then such a special situation must be carefully evaluated concerning possible adverse effects. A typical example is again the active uptake of paraquat into the lung cells due to the similarities between paraquat and some polyamines which are normal cell constituents (ref. 1). This accumulation finally leads to a high local paraquat concentration resulting in toxicity.

For the elimination of a compound the renal route plays the most important role. Only compounds reasonably water-soluble can be excreted renally. Often, however, such polar compounds or metabolites are bound to serum proteins. Such complexes cannot be excreted through the kidney. The degree of protein binding is markedly species dependent. Consequently, there will also be a marked species difference in the elimination half-life.

Studies in man

In the extremely rare cases where a considerable amount of a pesticide in the range of the ADI is likely to be taken in by man as a residue in foodstuff or where high exposure during application occurs it seems advisable to perform some kinetic studies in man. There are many examples showing important differences in kinetics between laboratory animals and man, so that it is not regarded as acceptable to rely only on results from animal studies in such exceptional critical cases. Not only renal excretion of a compound may vary according to the species but also differences in organ distribution and metabolism rate might lead to markedly different kinetic parameters:

We recently observed a striking prolongation of the half-life of the chlorinated dioxin 2,3,7,8-TCDD due to a relatively low deposition in the liver and a very slow metabolism rate in man. Whereas in many animal species tested the elimination half-life was in the order of 2 to 3 weeks, it was in man approximately 5 years (ref. 5). The toxicological consequence of this difference is important, since for reaching a given steady state in the body burden in man, a daily dose of only 1/80 of that in the animal is required. Another example refers to the biocide pentachlorophenol: This compound has an elimination half-life in rats of 30 hours whereas in man it is 17 days due to very slow renal elimination (ref.2).

In man PCP is excreted in the urine as 30 % unchanged and 70 % as glucuronide. Based on a protein binding of 96 % a renal clearance of 1.25 mL was expected; however, the measured renal clearance was only 0.07 ml/min. showing that in addition to protein binding in man a tubular reabsorption of 99 % occurred. There is widespread reluctance to perform investigations in man especially in non-drug-areas for ethical reasons. This reluctance however seems not to be justified under the condition that certain prerequisites are fulfilled: In a typical case of a pesticide the NOEL (for instance determined in a 90 day study in rats) is in the order of some mg/kg bw. Usually for a kinetic study a single dose of approximately 1 mg is needed. Provided that the compound in question is not mutagenic nor does show otherwise extraordinary toxic effects, such a study seems to be acceptable since there is a safety factor of at least 100 between the human dose and the animal NOEL. The radiation health hazard from the use of a labelled compound in a human study likewise is negligible since generally only 1 - 10 uCi of a ^{3}H or ^{14}C-labelled compound is administered. Assuming an even distribution in the body and an elimination half-life of about 10 days the total radiation burden is in the order of 5 mrem per study. The natural radiation burden/year ranges between 100 and 300 mrem or even more depending on geology and height above sea level of the location.

Metabolite pattern

During the elucidation of the metabolite pattern again one must ensure whether all goes according to expectation. Usually the metabolic pathways are rather uniform and can be predicted quite satisfactorily. Because of their higher polarity the metabolites are usually of lower toxicity than the parent compound. An important exception, however, is the formation of reactive intermediates. Therefore, one has to give special attention to the occurrence of unexpected types of metabolites which may have transient reactive metabolites as their precursors. In such a case further investigations on the possible implications have to follow. Reactive metabolites not only lead to cellular toxicity which is mostly reversible in nature but, of more importance, also by interaction with nuclear DNA, to DNA-damage, somatic mutation and eventually also to tumours.

The formation of unexpected metabolites is rare; its existence, however, can explain toxic effects of otherwise harmless chemicals as in the case of diallate. Schuphan et al. (ref. 3) were able to identify the in vitro formation of the strong bacterial mutagen of 2-chloroacrolein from the herbicide diallate after biological S-oxidation, a sigmatropic rearrangement and 1,2-elimination. This reaction is strongly inhibited by the important cellular detoxifying agent glutathione.

It remains to be shown if the same reactions also occur in the mammalian organism. Another example for an unexpected metabolic behaviour is the phase-II-reaction of phenoxyacids with alcohols (ref. 4) leading to some accumulation of these acids in glycerides, e.g. of cell membranes. Once such an unique reaction is discovered the possible biochemical and toxicological consequences must be assessed. In this case, no adverse effects at normal environmental levels have been found so far.

Another important aspect of metabolic studies are comparisons between plants and animals: Usually the same phase I metabolites (chemical changes of the parent compounds) are formed in plant and animals. In such cases toxicological study of the parent compound in animals can be used for the evaluation of residues occurring on plants either as parent compounds or as metabolites. Only in the very rare cases of plant metabolites which are not formed in mammals, a specific toxicological study of these metabolites might become necessary. Usually the differences in metabolism refer only to phase II reactions (conjugations). Since these conjugations represent bound or highly polar residues they usually do not show any essential biological activity at all nor an activity which differs from that of the mammalian conjugates. A special study of such metabolites, therefore, is not needed. In a recent example, it was shown that from several triazole-fungicides triazolylalanine is formed as a common metabolite. This conjugate which also occurs naturally indeed did not show any biological activity such as organ toxicity, mutagenicity or prenatal toxicity (ref. 6).

SPECIAL STUDIES RELATED TO CARCINOGENICITY

In addition to these metabolic studies which are done on a routine basis according to a more or less standardized scheme, there is a growing importance for studies of metabolism in cases where specific problems with a compound arose. Since trace level toxicology is the main feature of pesticide toxicology and since the main - and almost only - possible effect at trace levels is carcinogenicity, metabolism studies are required more and more in cases where a carcinogenicity issue is raised. Metabolism studies primarily are needed to clarify if a suspected compound exhibits some genotoxic activity. Typical chemical carcinogens are activated through metabolic reactions within the body to reactive electrophilic intermediates which can react with nucleophilic compounds such as water, thiols, nitrogen- and oxygen-atoms in proteins or DNA. Although alkylation of proteins can induce severe cytotoxicity or even cell death, such effects are usually reversible, whereas even an extremely low degree of

DNA-alkylation can lead to an irreversible change of the genetic material. Such mutations can manifest themselves among other effects finally as a malignant transformation of the cells.

A possible activation of a pesticide to an electrophilic compound can be studied in chemical or biochemical model systems where biological oxidation systems are simulated by mild chemical oxidants such as peroxyacids or by use of cell homogenates or enzyme mixtures. A possible activation can also be suspected simply by the way of discussing structure - activity relationships. It must be pointed out quite clearly that although such preliminary studies can give useful hints for possible biological activation of a compound, they never should be used for final decisions. There are too many intervening factors possible which can completely change the behaviour of the compound within the mammalian body, so that an activation in a model system always has to be shown also to occur in vivo. Since the reactive intermediates usually cannot be detected chemically in vivo due to their very short half-life and consequently their extremely low concentration, one has to examine the resulting reaction products. By use of radiolabelled compounds it is possible to determine the alkylation of DNA in different mammalian organs at very low levels, so that even very weak carcinogens can be detected (ref. 7). Again, however, it must be kept in mind that there is still a long way from a DNA alkylation to a visible tumour. A perfect correlation between DNA binding and carcinogenic potency is therefore not to be expected.

A further important aspect regarding formation of active intermediates is the dose dependence of that process (ref. 8). Since an extrapolation of high dose levels used in animal experiments to the very low exposure levels of man has to be made, any information on dose dependent phenomena is important and must be used in deriving quantitative extrapolations.

Tumours can also be induced by chemicals or their metabolites which do not directly interact with DNA, but which considerably disturb certain cell functions related to DNA synthesis, radical formation, cell growth and cell replication. This usually called "epigenetic" carcinogenicity is a typical high dose phenomenon, occurring only after a sustained period of application and being at least partially reversible. The investigation of biochemical changes induced by many so called "epigenetic" carcinogens such as induction of cytochrome P-450 containing enzymes or peroxisomes, will help to explain the mechanisms by which tumours were formed in experimental animals (9). Again the crucial step will be the extrapolation of the effects seen at high dose levels to the low level situation in man. Since it is unlikely that these epigenetic mechanisms of tumour induction are also effective at the very low dose levels of pesticides to which man is exposed , pesticides with some epigenetically induced tumorigenicity are not regarded to represent a significant health risk.

Metabolism studies can supply important information about the actions of pesticides at cellular and molecular levels and are therefore indispensable for a sound interpretation of toxic symptoms with regard to their relevance for man.

REFERENCES

1. L.L. Smith, Pesticide Chemistry, Human Welfare and the Environment edited by J. Miyamoto and P.C. Kearney, pp. 505-510, Pergamon Press (1983).
2. S. Uhl, P. Schmid and Ch. Schlatter, Arch. Toxicol. 58, 182-186 (1986).
3. J. Schuphan, J.D. Rosen and J.E. Casida, Science 205, 1013-1015 (1979).
4. J.V. Crayford and D.H. Hutson, Xenobiotica 10, 349-354 (1980).
5. H. Poiger and Ch. Schlatter, Chemosphere 15 (1986), in press.
6. Properties and Safety Assessment of "Triazolyl Alanine", submitted by Bayer, Ciba-Geigy, ICI and Rohm & Haas (1985).
7. W.K. Lutz, Mutat. Res. 65, 289-356 (1979).
8. M.E. Andersen, Critical Reviews in Toxicology 9, 105-150 (1981).
9. ECETOC: Hepatocarcinogenesis in Laboratory Rodents: Relevance for Man. Monograph No. 4, Brussels (1982).

Metabolism of pesticides—an industry view

Dexter B. Sharp

Monsanto Agricultural Company, 800 N. Lindbergh, St. Louis, MO, 63167

Abstract - For a pesticide manufacturer, metabolism data constitute a major part of the technical foundation that permits meaningful assessment of potential impacts of a pesticide upon humans as consumers of food from pesticide-treated crops and from animals eating such crops, and as applicators and workers handling pesticides. Such assessments are key features of proper product stewardship and successful registration. The pesticide industry applies very substantial resources to pesticide metabolism studies in a variety of plant, animal and environmental systems. Identification of major metabolites is possible now via state-of-the-art spectrometric instrumentation. Metabolism data are pre-requisites for development of adequate residue assay methodology. Studies of pesticide metabolism include crops, food-producing animals, and animal species used in toxicology tests. Critical planning and selection of type and sequence of studies is essential for optimal use of resources to obtain necessary data within the shortest possible time. The extent and detail of studies presently required exceed what is needed for sound assessment of risks. Examples include the elaborate extent of rotation crop studies, identification of metabolites below 0.05 ppm where only characterization is feasible, and suggested exaggerated dosing which can lead to metabolite patterns not relevant to expected exposure situations. Rules of reason should be applied to limit the extent and content of some studies required in ways which still provide necessary and sufficient answers for risk assessment. Tier and bench-mark approaches offer means to gather adequate data which provide realistic risk assessments for potential exposures to pesticides.

Industry is committed to the development of a necessary and sufficient technical information base which will permit meaningful assessments of the potential impacts of a pesticide upon humans and the environment. An important part of the technical data base is an understanding of the nature of pesticide degradation products. Pesticide metabolism refers to the chemical changes in the pesticide molecule caused by metabolic processes in plants, animals and micro-organisms. The transformed pesticides are new compounds called metabolites. The structures of major metabolites must be known in order to develop analytical methods which can quantify as much as possible of the total residue of the pesticide and its metabolites that remain in the treated crop. Ultimately, residue levels are compared to results of toxicology tests to estimate relative degrees of safety, or hazard, posed by the presence of and potential exposure to the pesticide residue. These safety/hazard evaluations are key features of proper product stewardship and successful registration. The pesticide industry applies very substantial resources to pesticide metabolism (Chart 1), although the metabolism area appears to be buried within the mass of other parts of total pesticide development from initial discovery to commercial sales.

This Chart 1 is a good news - bad news chart.

GOOD NEWS: "We have discovered a new compound that is phytotoxic and has market potential."

BAD NEWS: "We can sell our first kilogram of new herbicide in 1997."

It should be noted that the commercialization decision (line 7) is four years after the initial discovery of biological activity (line 3), and that first registration is seven years after the decision to commercialize (lines 14, 15). Metabolism and environmental chemistry (lines 8, 9) encompass the aforementioned area of metabolism. The results of metabolism studies in biological systems form the essential foundation for residue method development. For biodegradable pesticides, residue methodology can pose tough problems.

GOOD NEWS: "The new candidate pesticide degrades to 30 metabolites."

BAD NEWS: "Five different residue methods are required to measure total residues."

Chart 1

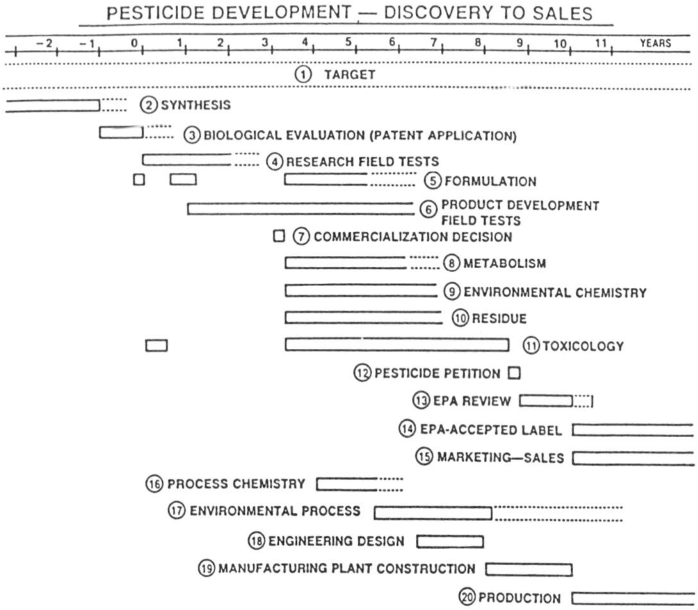

The Discovery to Sales costs for pesticide development range from 50 to 100 million U. S. dollars. An expanded scale for metabolism studies represented by lines 8 and 9 is shown in Chart 2 below. Residue studies are included to show the precursory position of metabolism, i. e., metabolite identification, to the initiation and progression of residue measurements. The residue and metabolism costs are 5 to 10 million dollars, roughly split equally between the two areas. In general, the cost of development of registration data is now in the 15 to 20 million dollar range, including Metabolism, Residue, Toxicology, Formulation, Product Chemistry, Process Chemistry (TOSCA) and Field Efficacy. To achieve manufacturing production status, another 25 to 75 million dollars investment is required.

GOOD NEWS: "The company has 100 million dollars to invest in the new herbicide."

BAD NEWS: "The first profit dollar will arrive in the year 2001."

Chart 2

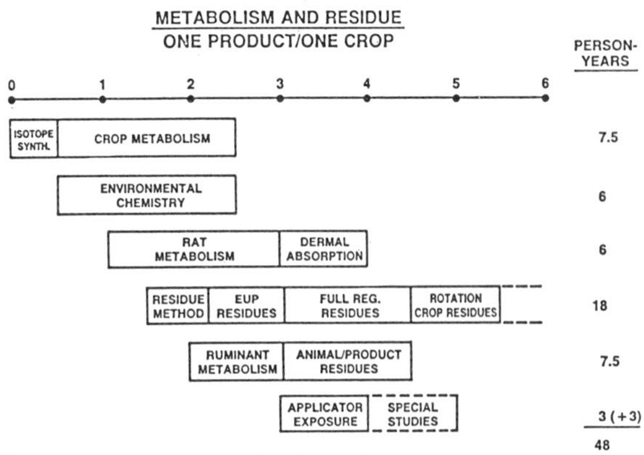

In Chart 2, the estimated person-years required to develope adequate information in each sub-area are shown in the column to the right. Crop metabolism and environmental chemistry are started together; residue method for parent pesticide can also begin. When major crop metabolites have been identified, residue method development keyed to measurement of these metabolites is started; normally, this is about one year after the start of crop metabolism programs. Rat metabolism follows later to take advantage of separations knowledge gained in the metabolism studies. The same rat species used in chronic-oncogenic toxicology studies is used for rat metabolism. Comparison of crop and rat metabolite patterns can be useful if the patterns are similar; also, a biochemical rationale for toxicological effects may be developed.

GOOD NEWS: "We have massive data on metabolism in the rat."

BAD NEWS: "Rat metabolism is different from that of any other animal species."

Crop metabolite identification is a precursor to ruminant and chicken metabolism studies. Ruminant/chicken metabolism with synthetic, isotopically labeled crop metabolites (and parent, if present) is a precursor to the large animal tissue/product residue programs, the so-called "meat, milk and eggs" studies. Residues of pesticide in food from crops or animals represent indirect exposure for humans. For the metabolism phases alone, a total of almost 25 person-years is required. In general, overall costs for metabolism are in the 2.5-5.0 million dollar (US) range. Further breakdown is illustrated in the following summary (Table 1), identifying studies with reference to U. S. EPA Registration Guideline numbers (ref. 1).

TABLE 1. Metabolism Studies - Estimated Resources

EPA RGL No.	Nature of residue, metabolite identification	Elapsed Time, Yrs.	Person No.	Person-Yrs.
171-4	Plants	2.5	3	7.5
162-1	Aerobic soil	0.75	2	1.5
162-2	Anaerobic soil	0.5	2	1.0
85-1	Rat	2.0	2.5	5.0
171-4	Goat and chicken	2.0	2.5	5.0
85-2	Dermal absorption	0.5	2	1.0
133-3	Applicator exposure	0.5	1	0.5
165-1	Rotation crops	2.0	2	4.0

Dermal absorption is a necessary precursor for studies of human or applicator direct exposure. These can be carried out <u>in vivo</u> with rats, rabbits or monkeys. The rhesus monkey, a primate species, is generally recognized as the more suitable surrogate for extrapolation to estimate human dermal absorption. Distribution patterns and excretion kinetics for parent and metabolites are important facets of such studies.

GOOD NEWS: "Our applicator exposure gauze patch test data are excellent."

BAD NEWS: "There is absolutely <u>NO</u> correlation between patch data and dermally absorbed dose."

Considering Table 1 and Chart 2, about 12 metabolism professionals are required in the years of maximum activity. For most efficient data generation in metabolism, one HPLC per professional and one GLC per two professionals are recommended. Computer data processing is mandatory. For the metabolism group, optimal instrumentation includes several radioactive scintillation counters, two GC-MS instruments and one 500 megaherz, or equivalent, NMR spectrometer.

GOOD NEWS: "Our scientist says that instruments and methods exist to identify most metabolites."

BAD NEWS: "Management says we cannot afford to buy the instruments - use TLC."

Major metabolite identifications are now possible with present-day technology. Enriched carbon-13 labeling of pesticide samples provides valuable information by ^{13}C-NMR. Wide chemical shifts accompany substituent changes at the labeled carbon position. Single position labeling with 50:50 ^{12}C/^{13}C gives in mass spectra ion fragment doublets of equal intensity separated by one atomic mass unit. These are valuable leads to structures of metabolites retaining the ^{12}C/^{13}C labeled position. Carbon-14 labeled sample added to stable isotope-labeled samples permits detection of metabolites in separation fractions, and provides the very important total radioactive carbon balance for the experiment.

These advances in technology have imposed a severe penalty upon biodegradable pesticides. The greater the extent of degradation, and the longer the time from application to harvest, the more difficult become the tasks of metabolite identification and total residue method development and quantitation. Rotation crop radiolabeled metabolism studies are very difficult because of the progressive dilution of multiple metabolites to very low levels of the individual metabolites, and in some cases because of the great increase in the number of metabolites with time. Rotation crop metabolism studies with radiolabeled pesticide could be dropped entirely from registration requirements. The primary crop residue method should be adequate to define rotation crop residue levels with significant savings in cost and resources. The four person-years in Table 1 could be applied to other programs.

GOOD NEWS: "The ^{14}C was taken up by the rotation crop, but no peak is greater than one percent of the total radioactivity."

BAD NEWS: "Carbon-14 is measurable in the crop and, therefore, residues are present. Increase application rate 10-50 fold and identify rotation crop metabolites; then do the field residue trials."

Residues in processed food components from pesticide-treated crops are receiving more attention by regulatory authorities. In cases where residues in the raw agricultural commodity are at undetectable sub-part per million levels, some regulatory authorities are suggesting repeat studies using exaggerated pesticide application rates. Such studies will be very costly in time, money and resources, will delay registration by two years, and the high rates could produce metabolite patterns which are not the same as patterns arising from use rates. In fact, exaggerated rates for herbicides are infeasible because of excessive phytotoxicity and crop death. Residues in processed foods should be required only when raw agricultural commodities have residues at detectable finite levels.

GOOD NEWS: "We can identify a metabolite at 0.025 parts per million."

BAD NEWS: "A known metabolite at 0.025 ppm is viewed with greater alarm than a bound residue at 25 ppm."

As technology continues to improve, some of these questions may be answerable. Meanwhile, efforts are continuing to establish tier or benchmark approaches which may provide acceptable answers at a lower cost in time, money and resources. An example is a general approach to pesticide metabolism recommended by the IUPAC Commission on Pesticide Chemistry (ref. 2). This approach uses two means to guide testing of metabolism of pesticides in soil, water, plants and animals: (1) a text outlining the rationale for selecting types and designs of studies; and (2), a flow-scheme supplemented with brief explanations, i. e., a "road map" with commentary. The flow-scheme for pesticide behavior in crop plants shown in Figure 1 is a representative example; a prototype version is available as a handout at the Congress presentation. Symbols used to present the information in easy-to-read form are as follows:

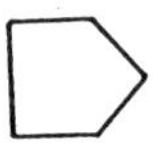	Input information prerequisite to design/performance of test/assessment		Description of test or calculation/assessment to be performed
	Description of test result, usually in two parts if one leads out of test scheme or both require further testing	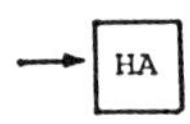	Output information for use as an element of hazard assessment directly or after verification in other tests. No further tests are needed.

This Figure and the preceding Charts are glaring examples of terrible slides for a presentation. In part, they have been used purposely to emphasize the extraordinarily complex problems intrinsic to development of adequate metabolism data for a pesticide. The industry, however, knows of no other way to assure proper product stewardship and to achieve registration. Some final GOOD NEWS - BAD NEWS:

BAD NEWS: "We have found a pesticide candidate which does not biodegrade, is persistent, and is very toxic by acute dosing."

GOOD NEWS: "The costs of registration data development will be reduced substantially, and with that high acute toxicity, a carcinogenic response is most unlikely. However, keep it out of the environment."

Or, GOOD NEWS: "We all know that 0.1 part per million is a trace residue."

BAD NEWS: "The media and the public believe that 100,000 parts per trillion is much larger than 0.1 part per million."

Finally, GOOD NEWS: "We can now analyze a residue at 1.0 part per trillion."

BAD NEWS: "One part per trillion equals one homogenized flea evenly distributed inside of 30 elephants."

REFERENCES

1. Pesticide Assessment Guidelines, (U.S.) Environmental Protection Agency, U. S. Dept. of Commerce, National Technical Information Service (NTIS), November, 1982.

2. H. O. Esser, R. J. Hemingway, W. Klein, D. B. Sharp and H. Vonk, "Recommended Approach to the Evaluation of the Environmental Behaviour of Pesticides", IUPAC Commission (VI.5) on Pesticide Chemistry. Manuscript in preparation.

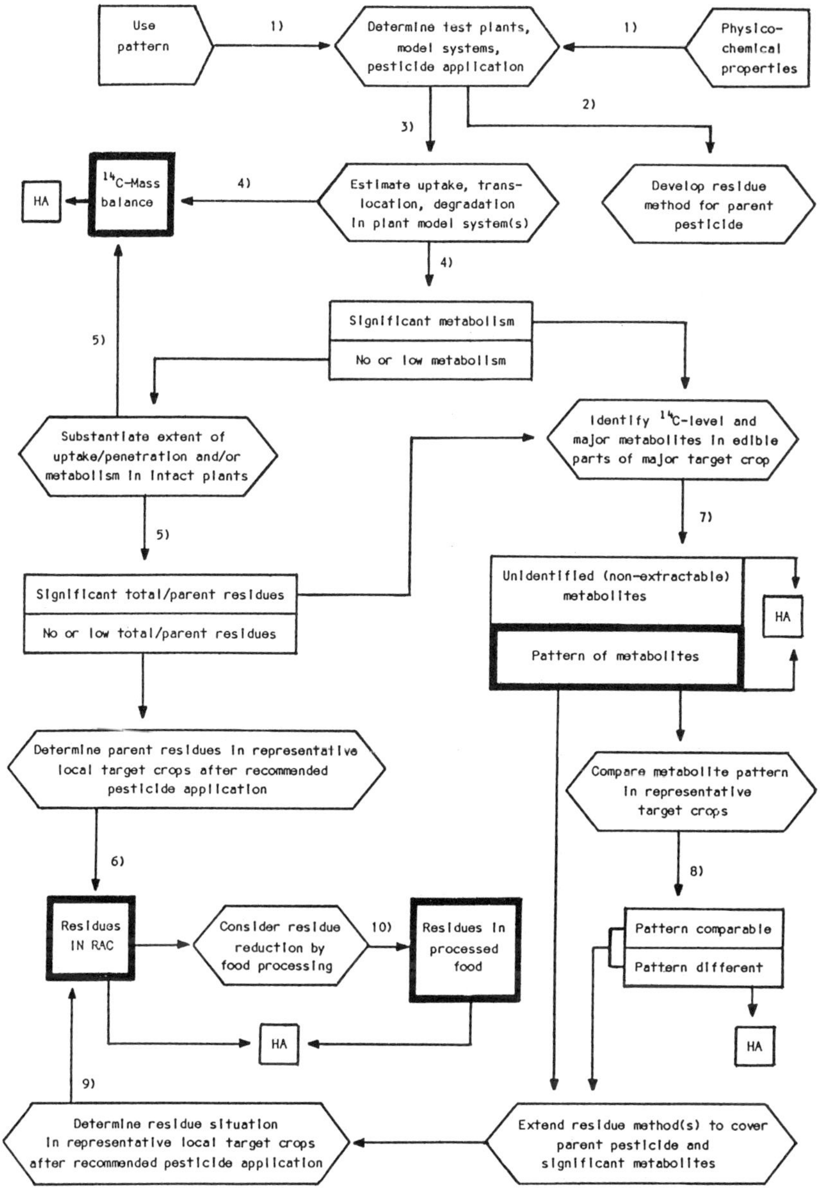

Fig. 1. Testing pesticide behaviour in crop plants

Commentary to scheme "Testing Pesticide Behaviour in Crop Plants"

1) Consider use pattern. Select one representative crop for each major crop group and the pertinent application procedure if pesticide finds broad use. Give attention to those physico-chemical properties which have impact on uptake and translocation such as Sw, Pow and pK.

2) Develop specific analytical method for parent pesticide; validate it with those crops for which tolerances are to be set.

3) Use small screening tests such as immature plants, stem injection, excised leaves and cell cultures of major target crop for this estimate. If relevant, consider influence of volatility and photo-degradation on parent dissipation.

4) If intact plants (in closed systems) are used, a ^{14}C-mass balance can be obtained which may allow the assessment of the extent of bio-availability and sources of loss of pesticide during and after its application. If no or low metabolism is observed in plant model systems, check in intact target plants for uptake/penetration and/or metabolism of parent. In the case of significant metabolism, identify major metabolites produced in the model system to define primary pathways and to facilitate identification of significant residues (e.g., >0.025 ppm extractable ^{14}C) in edible parts of the main target plant which had been treated with the pesticide as recommended (greenhouse or field).

5) If not obtained in the screening system and uptake/penetration behaviour indicates a need, a ^{14}C-mass balance can be established in this study. In case of significant parent (and few metabolite) residues follow as indicated in 4), Significant Metabolism. A low parent residue situation needs confirmation in target field crops by residue analysis.

6) Determine parent residue levels in the "raw agricultural commodity" (RAC) of individual target crops from a restricted number of representative locations (including re-entry situations, if appropriate) using residue method 2). Use data for the tolerance setting process in HA.

7) Evaluate toxicological relevance of metabolites in HA by comparing the metabolite pattern in edible plant parts with that found in the animal species used in the toxicity studies, preferably the rat. Major plant metabolites not found in this animal species may require metabolism testing in animals. Bioavailability of radiolabelled unidentified/non-extractable metabolites may be checked in "transfer studies" by direct feeding. Consider major metabolites of biological or toxicological significance in recommended residue method(s); if pattern is highly complex development of a "common moiety" approach is indicated.

8) If the pesticide is applied to several non-related target crops, comparison of the metabolite pattern from (maximum 3) species is indicated; if possible, use screening systems of 3). If patterns contain (structurally) different major metabolite(s) of biological or toxicological significance, adapt residue method(s) accordingly; follow tolerance setting considerations individually in these target crops.

9) Determine residue levels of parent pesticide plus relevant metabolites in the RAC of individual target crops from a representative number of typical locations (including re-entry situations, if appropriate) for use in the tolerance setting process in HA.

10) Check reduction of significant residues in RAC (e.g., >0.025 ppm ^{14}C in form of parent and/or metabolites) by food processing, if appropriate; use residue level in processed food as a factor in HA.

Pesticide metabolism studies—a regulatory viewpoint

J.A.R. Bates

Pesticides Registration & Surveillance Department, Ministry of Agriculture, Fisheries and Food, Hatching Green, Harpenden, Herts, UK

INTRODUCTION

Pesticide registration implies the acceptance, by a competent authority, of evidence submitted in support of all claims for safety and efficacy for the proposed product/use. A registered product may then be marketed with restrictions or control of packaging, labelling, supply and use recommended by the registration authority. The registration authority must be rational, fair and efficient in the interests of the public and the pesticide industry. All pest-control chemicals are not equally hazardous and registration authorities should be flexible and not apply the data requirements of a maximum-hazard testing approach to low-hazard chemicals. Scientific assessment is the core of the registration process in a generally administrative structure of control of pesticides and unreasonable demands for data when the product shows little hazard potential could damage the credibility of scientific evaluations. Flexibility is essential to a successful registration procedure and, in the general interest of reducing costs of development, the fullest use should be made of all data and a registration authority should ask for tests to be repeated within its territory only if there are valid scientific (not administrative) reasons for doing so. The registration process has become steadily more expensive to both industry and governments and several bodies have considered and investigated procedures whereby total costs could be reduced and unnecessary delays and disadvantages could be avoided. A number of countries already base their registration process, wholly or in part, on data and evaluations from international agencies or other countries. Futher discussion between interested countries with a view to expansion of this facility of data sharing would make a valuable contribution towards control of rising costs and the consequences of avoidable duplication.

However, the risk/benefit judgement involved in registering a pesticide often differs from country to country and it would be wrong to establish arbitrarily the same value judgement for the use of a pesticide for all countries. Countries should, if at all possible while developing their value judgements, bear in mind the consequences of such judgements beyond their national boundaries.

It is important to recognise that countries can only achieve a level of reassurance commensurate with their needs, technical resources and achievements and finance.

In order to assess the risks of using a pesticide, its chemical and toxicological properties must be investigated thoroughly and the overall design of the test programme depends to a large extent on the manner in which man is likely to be exposed. The rationale of testing and the techniques used are described in detail in publications by WHO, Council of Europe and in many national registration procedures and guidelines.

Although one tends to consider safety to man, domestic animals, wildlife and the environment under separate headings the problem essentially is how to minimise pesticide risks to organisms other than the pest. It is obviously impossible to test all the conditions under which the pesticide may be used in practice but experience has shown that, in most instances, a sound prediction can be made of the probable effects of the use of a compound from a consideration of certain basic studies. Data developed prior to registration are essentially predictive and test programmes have to be designed according to the characteristics and conditions of use of the product and the test species carefully selected in order to justify broad predictions on the basis of results from a feasible test programme.

In earlier years of pesticide registration, data requirements tended to concentrate on the measurable or extrapolated effects of the use of a pesticide with less emphasis on the mechanisms by which these effects came about. Uncertainty was reflected in requests by registration authorities for additional or repeated studies in toxicology or for the provision of more residues data. Gradually the real need to fully understand the mechanisms involved in both plant and animal metabolism has been emphasised although even now the full potential of skilful predictions from metabolism studies has not been realised.

A pesticide application results in deposits of the pesticide in desirable and undesirable places and it is the size, nature, characteristics and fate of these deposits which concern registration officials and studies on metabolism clearly have a very important role to play in evaluations carried out by registration authorities.

METABOLISM STUDIES IN PLANTS

The term metabolism in this context is often used to include purely chemical transformations such as oxidation and hydrolysis outside the plant as well as the effects of metabolic processes within the plant although 'fate' more properly describes the overall changes in the pesticide deposit on plants and in the environment. (Ref.1).

Adequate plant metabolism studies fulfil at least four purposes: (1) they provide an estimate of the quantity at the complexity of <u>total</u> residues in the treated crops; (2) they identify the major components of the terminal residue, thus indicating the components to be looked for in residue quantification studies; (3) they indicate the distribution of residues, eg whether the pesticide is absorbed through roots or foliage, whether translocation occurs, or whether the residues are entirely surface residues; and (4) they show the efficiency of analytical extraction procedures for various components of the residue.

The minimum application rate in a plant metabolism study should approximate the proposed label rates; however, exaggerated rate studies may be necessary in order to obtain sufficient activity for identification. The metabolism studies should be carried to cropping or plant maturity whenever possible, so that the composition of the residues found is indicative of that in the terminal residues at harvest.

The study should include characterisation of the residue in all plant parts used for food or feed and carried out for each type of plant for which use is proposed. For example, metabolism studies in wheat could be regarded as representative of all cereals but could not be extrapolated to root crops or fruit. Generally if three metabolism studies in unrelated crops indicate a similar metabolic route in the three crops, then further metabolism studies will not be required.

METABOLIC STUDIES TO PROVIDE DATA ON THE NATURE AND AMOUNT OF PESTICIDE RESIDUES IN FOOD OF ANIMAL ORIGIN

The purpose of these studies is to identify the nature and quantity of the residues in the edible tissue of livestock, milk and eggs. Animal metabolism studies are required whenever a pesticide is applied directly to livestock, animal premises are to be treated, or measurable residues occur in crops or crop parts used for feed. Data on the metabolism of a pesticide in laboratory animals cannot generally be extrapolated to predict residues in food of animal origin although in some cases laboratory animal metabolism data may be used to supplement livestock metabolism studies in which complete characterisation of the residue is not achieved. Initial rat studies using radiolabelled chemicals are an essential first step to identifying the metabolites that may be present in animal tissues but further radiolabelled metabolism studies on livestock animals are required to identify the nature of residues in animal products. A livestock metabolism study should primarily identify the compounds for which analytical methods and residue data must be generated. It should also indicate the distribution of residues in tissues, eggs and milk, and whether residues are stored and accumulated. A livestock metabolism study should also result in the elucidation of the efficiency of extraction of the various components of the residue so that extraction/residue release procedures can be developed as part of the analytical methods.

Usually the most important studies are those involving ruminants and poultry. Normally lactating cows are used but, for the ruminant radiolabelled study, a goat is an acceptable alternative for economic reasons. In the case of poultry, chickens are the animals of choice. If residues have been identified in animal products in the radiochemical studies, and, if the crop on which it is proposed to use the pesticide is an important part of an animal's diet, non-radioactive feeding studies should be carried out. Similarly, cold treatment studies will be necessary where a pesticide is applied to the skin of an animal. These studies will provide the quantitative data on which maximum residue limit (MRL) can be set for foods of animal origin.

Separate feeding studies are required for a ruminant (lactating cows) and poultry (laying chickens) whenever residues occur in items of feed for these animals.

Where feeding studies are required, they are often carried out using the parent pesticide which is normally the residue of greatest toxicological significance on the treated crop. In cases where this is not so, the animal transfer studies may be carried out using a metabolite or a realistic crop residue mixture.

The Codex Committee on Pesticide Residues has prepared guidelines (ref.2) on supervised studies to provide data on the nature and amount of pesticide residues in food of animal origin and these are appropriate to the needs of registration authorities.

CHARACTERIZATION OF THE TOTAL TERMINAL RESIDUE

Before complete residue detection methodology and residue quantification data can be developed it is necessary to know the composition of the terminal residues and the use of radiolabelled pesticide compounds is usually the only satisfactory way of providing such data. While *in vitro* data are useful to show if the pesticide is likely to undergo hydrolysis, oxidation or reduction, photolysis, or other changes, additional data must usually be submitted to show the fate in plants and animals. These metabolism studies are required whenever a pesticide is used on a food or food crop. Based on the results of the characterisation studies, the chemical definition of the total toxic residue should be proposed. The term "total toxic residue" is used to describe the sum of the parent pesticide and its degradation products, metabolites (free or bound) and impurities which are of toxicological concern.

On occasions residues in foodstuffs can consist of a number of metabolites as well as parent pesticide and this can present practical problems, both in the legal definition of the residue and in subsequent monitoring and enforcement of maximum residues limits. Some countries adopt the attitude that the total toxic residue must always be measured even during routine enforcement analyses. However practical difficulties can occur when an MRL is expressed in terms which include metabolites for which analytical standards are not readily available in food control laboratories. Where such difficulties arise it is sometimes permissible to base an estimate of total residues on the measurement of a specifically determined compound eg the parent, a metabolite or a derivative produced in an analytical procedure which must have been shown to bear a known relationship to the concentration of the total toxicologically significant residue in the foodstuff. Clearly the decision of regulatory authorities to include specific metabolites/ degradation products in the definition of the residue or the MRL must be based on adequate metabolism studies including both qualitative and quantitative aspects. Occasionally metabolites of toxicological concern are known to be present in significant amounts and may be determined and evaluated separately.

METABOLISM IN MAMMALS

Because man, his animals and wildlife may come into contact with pesticides directly or indirectly it is imperative to study the fate of pesticides in animals, the latter representing either models for man or direct target organisms.

Pesticide metabolism studies in animals should supply information about

- rates and routes of uptake, distribution and excretion
- biodegradation and metabolic pathways
- retention in specific organs and tissues
- variation of metabolic behaviour depending on the sex of the test animals, dose and (if necessary) route of pesticide administration (oral or dermal) and, in certain cases, with animal species.

This information is needed to interpret the nature and selectivity of observed toxic effects with regard to

- pharmacokinetic behaviour (origin, duration, possible accumulation of effects)
- molecular basis of the toxic effects observed (comparison with chemicals of similar structure/behaviour/effects)
- transferability of data between animal species and possibly from animals to man

to assess the toxicological potential of plant metabolites; and to provide, if needed, the basis for monitoring exposure to pesticides by establishing a relationship between the amount of specific metabolites excreted and the dose or exposure

Although some information should be available on the dependence of the pharmacokinetic and metabolic behaviour of a pesticide on the dose administered, studies in animals which are in a state of poisoning by massive overdosing of the pesticide should be avoided. Unlike in toxicology where biological effects on the animals are observed, pharmacokinetic studies focus upon the fate of the chemical itself. Statistical considerations in pharmacokinetic studies are irrelevant not only because a small number of animals are used but also because the excellent sensitivity levels inherent in radiolabelled studies provide adequate information needed with regard to accuracy of metabolic pathway definition and variations among individual animals. Pharmacokinetic investigations provide information on the absorption, distribution, excretion and retention of the pesticide in test animal species. It is important that they are conducted in such a way that they provide an accurate total balance for the radioactivity administered.

In a single dose mass-balance study the excretion of radioactivity after a single oral dose of labelled test pesticide to rats and the retention of radioactivity in their organs at the end of the excretion phase are determined. Other routes of application, eg dermal, and other test animal species, eg farm animals, may be required if the use pattern of the result of toxicity studies indicate a need. In studies with rats, generally male and female animals are exposed to two dose levels. One dose level should be in the order of potential residues supplied with a reasonable safety factor eg 0.5 mg/kg body weight corresponding to about 5 ppm in food the other dose level should be significantly higher (eg at 100 times the low dose but not exceeding $\frac{1}{4}$ of the LD 50) to recognise a potential dependency of the metabolism pathways on dose level.

When single-dose experiments show that significant retention of radioactivity in certain organs takes place, and possibly, when there are concerns about cumulative toxicological effects, information is needed on the concentration of radioactivity in organs and tissues when the pesticide is dosed, generally orally, over a certain period of time and on the depletion of radioactivity from the body after termination of dosing.

The results from such a multiple dose mass-balance study will provide information about the rate of buildup and the magnitude of "steady-state" (plateau) concentrations of total radioactivity and parent pesticide in:

- major organs and tissues

- milk and/or meat and/or eggs if farm animals are investigated

This type of study can also be used when the significance of generally unidentified radiolabelled residues in plants or relevance to human or animal nutrition is to be assessed. Where plant metabolites are identified and are different from those found in animals, synthetic samples of labelled metabolites can be used in "transfer studies".

Special kinetic studies may be needed to provide information on special pharmacokinetic aspects such as blood levels, biliary excretion, enterohepatic circulation if the results of the balance and/or toxicity studies require such additional background data. Pharmacokinetic studies may require elucidation of significant organ residues. The results will also show whether the metabolites found in plants and animal products are the same as or similar to those found in the animal species used in toxicological evaluation studies. Plant metabolites which occur in these animal species need no further toxicity testing.

In addition special metabolism studies may be required to explain

- the nature of non-extractable radioactivity

- any reaction of pesticide or its metabolites with cell constituents including biopolymers

- the molecular basis of a toxic effect of special interest

The species currently used in animal tests are generally satisfactory for predicting human metabolism, particularly where a foreign compound shows no species differences. However, their use raises problems when compounds showing remarkable species differences are encountered. A few pesticides are metabolised differently in different animal species with respect either to their excretion patterns or to the structures and relative quantities of the metabolites. Miyamoto et al.,(ref.3) concluded that man has virtually all of the capabilities that have been demonstrated in experimental animals for the metabolism of xenobiotics, ie man, in general, does not suffer from any particular defects in his ability to metabolise pesticides. However, whilst the metabolic options for a given compound are now largely predictable from chemical structures, the metabolic route which occurs even the rat is not always predictable. Similarly, it is not always possible to predict accurately the route in man from data in animals. Even within quite a narrow range of structural analogues, generalisations may break down when they are extrapolated from one species to another. Thus, foreign compound metabolism is not only compound-specific but may be species-specific in some cases.

PHARMOKINETIC STUDIES IN MAN

Immediate reasons for studying the fate of a pesticide in man are to ensure that it is removed from the human body by metabolic processes and that its fate is similar to that in the experimental animals used in toxicological studies. Metabolism and toxicity are usually inextricably linked and certain biotransformation reactions signal possible toxicity and it is necessary to know if these reactions occur and cause toxicity (in animals) and whether they occur in man. Methodology developed for the analysis of metabolites in test animals should be applied to samples (eg blood and urine) derived from (potentially) exposed personnel at the work-place. If any of the metabolites are found in these samples, exposure to the pesticide is demonstrated and further studies may be required. The main objective of such studies would be to relate the output of a particular metabolite or group of metabolites with a known exposure and then to develop a quantitative exposure monitoring method.

Any data which indicate that human metabolism is different from that in rats or mice indicate the need for a metabolism study in man. Volunteer studies with changes in clincial chemistry as end points to indicate toxic effects must be designed in the expectation of negative results. They are otherwise ethically unacceptable.
The pre-requisites and techniques for studies in man are discussed by Miyamoto et al. (ref.3).

SUMMARY

In conclusion it should be emphasised that:

1. Registration authorities are concerned with the fate of pesticide deposits on man, his food and his animals and their feed and the environment.

2. A full knowledge of the degradation/metabolism of a pesticide in plants and animals is essential to a complete understanding of the fate of a pesticide deposit.

3. The routes of degradation or metabolism in plants and animals should be well presented and mass balances provided to define quantitative aspects of metabolism.

4. The relevance, scope and quality of metabolism studies are of vital importance to their acceptability (GLP).

5. It is advisable to discuss the scope of metabolism studies with the registration authority.

6. The capability of analytical methods for residues should be clearly defined.

REFERENCES

1. H.O. Esser, R. Hemingway, W. Klein, D. Sharp and H. Vank. IUPAC Report on Pesticides No 23. J. Appld. Chem., 1986 (to be published).

2. Codex Committee on Pesticide Residues. Guidelines on Supervised studies to provide data on the Nature and Amount of Pesticides Residues in Food of Animal origin. Alinorm 87/24 Appendix IV, Annex I. (1986).

3. J. Miyamoto, H. Kaneko, D.H. Hutson, H.O. Esser, S. Gorbach and E. Dorn. IUPAC Report on Pesticides No 22. J. Appld. Chem., 1986 (to be published).

Metabolism of Phenmedipham in the sugar beet *(Beta vulgaris* L.*)*

J.-I. Celorio, G.-A. Hoyer, J. Iwan, L. Kölsch

Agrochemical Division, SCHERING AG, Federal Republic of Germany

ABSTRACT

Phenmedipham was metabolised by sugar beet plants to a range of metabolites. Of particular interest was the formation of sodium 1-(3-methoxycarbonylaminophenyl)-β-D-glucopyranos-2-yl sulphate. Reports on compounds isolated from plants and containing sugar sulphates have been reported but this is the first time that sulphate is associated with the metabolism of xenobiotics in plants.

Other conjugates included those with the unusual C-O-N linkage between exocon and endocon.

INTRODUCTION

Phenmedipham, 3-methoxycarbonylaminophenyl 3-methylphenylcarbamate, is the active ingredient of Betanal[R], a post-emergence herbicide for use in beet crops. It acts via the leaves and has no effect when applied to soil (ref. 1, 7). Phenmedipham is a strong inhibitor of the Hill reaction: 10^{-7} mol/l suffice to produce 50% inhibition in isolated chloroplasts (ref. 12). Although its metabolism in sugar beet has been studies before, the structure of the main metabolites could not be elucidated. Nonetheless, the reason for phenmedipham (PMP) selectivity toward sugar beet seems to depend on two factors: the speed with which the compound is absorbed by the plant, and the rate of metabolism. Resistant plants, such as sugar beet, inactivated the compound so fast that they recovered rapidly after an initial period in which the fixation of CO_2 was reduced (ref. 1, 3, 5, 6). Polar hydrophilic metabolites are produced which partially release unchanged parent compound following acid hydrolysis (ref. 5). Clearly little has been accomplished to establish the metabolism of PMP in sugar beet; it was the aim of this study to fill this gap (ref. 4).

TIME-COURSE STUDY

Beet grown in hydrophonic culture were treated on the leaves with radiolabelled phenmedipham (3-methoxycarbonylamino-[U-^{14}C]-phenyl 3-methylphenylcarbamate). The plants were harvested 12, 24, 48 hours, 4, 7, 15, 30, and 60 days after treatment. The plants were separated into leaves and roots. The leaves were washed and extracted following the Bligh/Dyer method (ref. 2, 10, 11), then subjected to Soxhlet extraction and/or combustion. The roots were freeze-dried and combusted.

The radiolabel in the washes declined from 72.5% (12 h) to 2.0% (60 days); phenmedipham equivalents in the Bligh/Dyer organic phase amounted to 25% after 12 h, rose to 49.6% (4 days), then declined to 0.3% (60 days); the radioactivity content in the aqueous phase rose from 6.2% (12 h) to 70.5% (60 days); finally, the bound residues in the leaves were in the range of 2.1% (12 h) to 28.1% (60 days). No radioactivity was detected in the roots.

Thin-layer chromatography of the apolar phase showed that the bulk of radiolabel was attributable to PMP and to its main hydrolysis product, methyl 3-hydroxy-phenylcarbamate (MHPC). In the aqueous phase up to 6 radioactive compounds were detected, plus PMP and MHPC. Also, two of the compounds migrated towards the anode when subjected to thin-layer electrophoresis at pH 2.

ISOLATION OF POLAR METABOLITES

For the isolation of metabolic products, over 2000 sugar beet plants were treated by injecting a solution containing ^{14}C-PMP into the hypocotyl. This technique produced the same metabolic pattern as the leaf application technique. After harvesting 8 and 10 weeks after treatment, the plants were separated into leaves and roots. Only the leaves were analysed and worked-up, since the amounts of polar conversion products in the roots were very low. The extracted radiolabel from the leaves was again separated by the Bligh/Dyer technique into apolar and polar materials. The apolar phase was shown to contain, as before, PMP and MHPC only, and was discarded.

The aqueous phase was worked-up in batches using the following techniques for the isolation of metabolites: partitioning into 1-butanol, column chromatography using DEAE-Sephadex A-25, thin-layer and reversed-phase high-performance liquid chromatography, and in special cases also thin-layer electrophoresis. Four polar metabolites (0.2 - 2 mg) were isolated and identified by chemical, enzymatic, and spectrometric techniques (IR, MS, NMR).

Metabolites 1A (cf. Figure 3)
Sodium 1-(3-methoxycarbonylaminophenyl)-β-D-glucopyranos-2-yl sulfate
$C_{14}H_{18}NO_{11}SNa$ Molecular weight: 431.4
Acid hydrolysis released MHPC. Enzymatic hydrolysis with β-glucosidase had no effect.

Metabolite 2 (cf. Figure 3)
Sodium 1-[N-(3-methoxycarbonylaminophenoxycarbonyl)-3-methylphenylaminooxy]-β-D-glucopyranos-2-yl sulfate
$C_{22}H_{25}N_2O_{13}SNa$ Molecular weight: 580.5
Acid hydrolysis released PMP. Enzymatic hydrolysis with β-glucosidase had no effect.

Figure 1 shows the FT-NMR spectrum of metabolite 2 in D_2O; the paramagnetic shift of the sugar's 2-H proton at δ4.03 ppm (usually located in the region of δ3.4 ppm) indicates the presence of a polar group (sulfate) at C-2; this effect was also observed in the NMR spectra of metabolite 1A. Figure 2 shows the FAB mass spectrum.

Metabolite 3 (cf. Figure 3)
Methyl 3-(β-D-glucopyranosyloxy)phenylcarbamate
$C_{14}H_{19}NO_8$ Molecular weight: 329.3
Acid hydrolysis and enzymatic hydrolysis with β-glucosidase released MHPC.

Metabolite 4 (cf. Figure 3)
3-Methoxycarbonylaminophenyl N-(β-D-glucopyranosyloxy)-3-methylphenylcarbamate
$C_{22}H_{26}N_2O_{10}$ Molecular weight: 478.5
Acid hydrolysis and enzymatic hydrolysis with β-glucosidase released PMP.

The position of the binding site in the exocon of metabolites 2 and 4 was established by comparing the chemical shift of the remaining free NH with those of phenmedipham's two mono-N-methyl derivatives.

A fifth metabolite (1B) containing phenmedipham was detected but could not be identified.

DISCUSSION

Although a C-O-N linkage between exocon and endocon is not very common in the metabolism of xenobiotics in plants, it is the presence of sulfate groups at the sugar's C-2 which make the compounds particularly unusual. Reports on compounds isolated from plants and containing sugar sulfates have appeared in the literature (ref. 8, 9); however, to our knowledge this is the first time that sulfate is associated with the metabolism of xenobiotics in plants.

Phenmedipham is thus deactivated in sugar beet by hydrolysis and/or conjugation of the parent compound (cf. Figure 3). The polar metabolites formed are stored in the leaves since no basipetal translocation was observed. Therefore no residues should be expected in the roots under normal conditions.

With regard to inhibition of the Hill reaction in isolated chloroplasts, the polar metabolites were 600 to 1700 times less active than PMP, confirming conjugation in this case as an effective detoxification mechanism.

ACKNOWLEDGMENTS

The authors wish to express their thanks to the following persons: Prof. Dr. W. Baltes, Dr. G. Hohne and Dr. J. Jakupovic, TU Berlin; and Dr. H.-R. Schulten, University of Bonn, FRG.

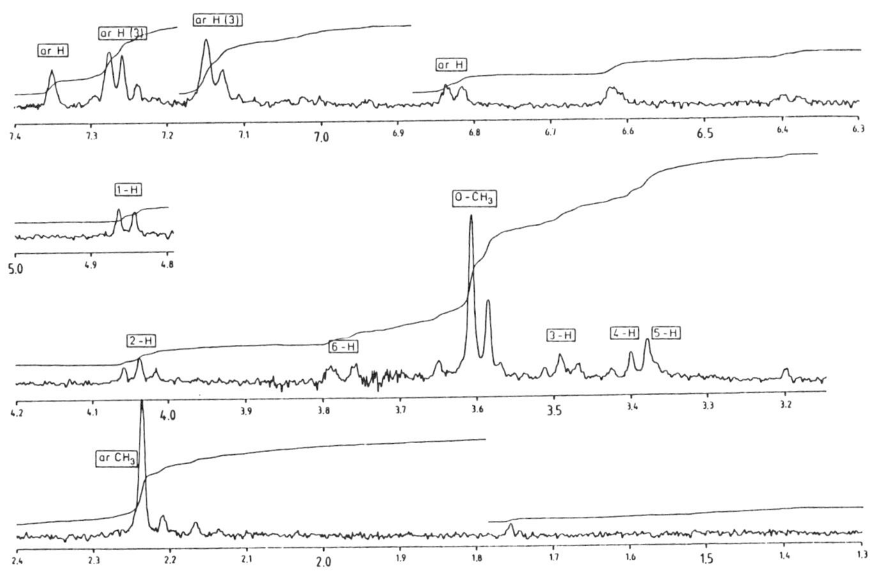

FIGURE 1: FT-NMR (400 MHz) spectrum in D_2O of metabolite 2

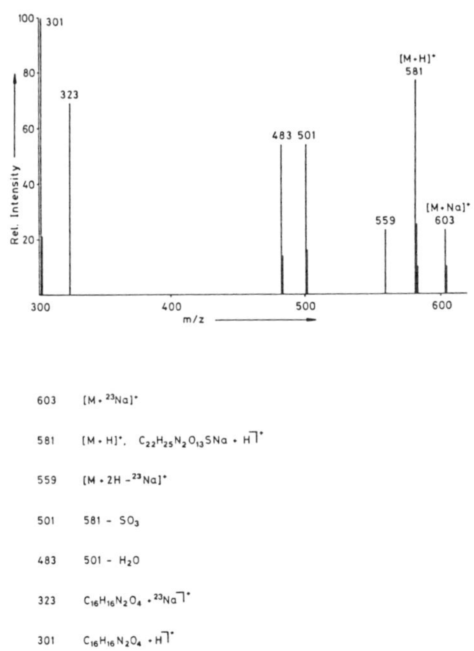

FIGURE 2: FAB mass spectrum of metabolite 2

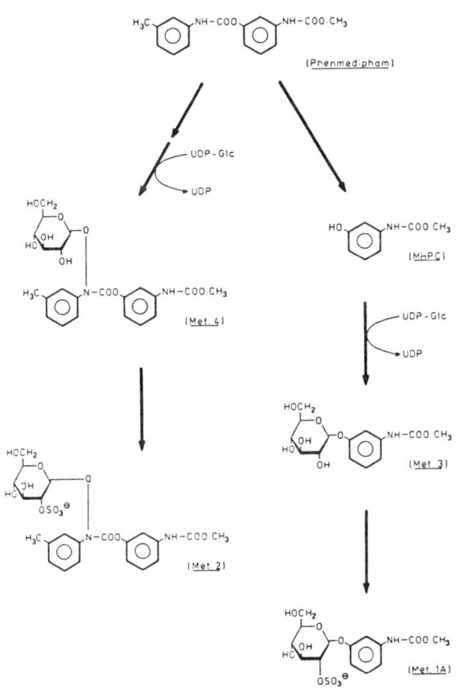

FIGURE 3: Proposed metabolic pathway of phenmedipham in sugar beets

REFERENCES

1. F. Arndt and C. Kötter. Weed Res., 8, 259-271 (1968).
2. E.G. Bligh and W.J. Dyer. Can. J. Biochem., 37, 911-917 (1959).
3. H. Börner. Z. Naturforsch., 34 c, 926-930 (1979).
4. J.-I. Celorio. Ph. D. Thesis, D 83/146, TU Berlin, FRG (1983).
5. H. Kassebeer. Z. PflKrankh. PflSchutz., 78, 158-174 (1971).
6. C. Kötter and F. Arndt. Z. PflKrankh. PflSchutz., Sonderheft V, 81-88 (1970).
7. Schering AG. Betanal[R], Berlin, FRG (1973).
8. H. Schildknecht. Endeavour, 8, 113-117 (1984).
9. H. Schildknecht and K. Schumacher. Pure and Appl. Chem., 54, 2501-2514 (1982).
10. G.G. Still and E.R. Mansager. Phytochemistry, 11, 515-520 (1972).
11. G.G. Still and E.R. Mansager. Pestic. Biochem. Physiol., 3, 87-95 (1973).
12. A. Trebst, E. Pistorius, G. Borochewski and H. Schulz. Z. Naturforsch., 23 b, 342-348.

1,8-naphthalic anhydride/auxin protection against chlorsulfuron inhibition of corn seedling growth

D. S. Frear*, H. R. Swanson and E. R. Mansager

U. S. Department of Agriculture, Agricultural Research Service
Metabolism and Radiation Research Laboratory, State University Station
Fargo, North Dakota 58105 U.S.A

Abstract - One week after corn seedlings (Northrup King PX 9144) were treated with chlorsulfuron (0.25-0.50 µg/seedling), shoot and root growth were inhibited 50-70%. Growth inhibition was long lasting and sometimes persisted to maturity. Chlorsulfuron, however, was metabolized rapidly. Within 24 hr, 80% of the applied [^{14}C]chlorsulfuron was metabolized to polar products. The major metabolite, the O-glucoside of 1-(5-hydroxy-2-chlorophenylsulfonyl)-3-(4-methoxy-6-methyl-1,3,5-triazin-2-yl)urea, was isolated from leaf tissues and identified by fast atom bombardment (FAB) and ^{252}Cf plasma desorption mass spectrometry. Pretreatment with 1,8-naphthalic anhydride (NAN), either as a seed dressing (0.33% w/w) or as a leaf application (10 µg/seedling), protected corn seedlings from subsequent chlorsulfuron inhibition of growth. Absorbed [^{14}C]NAN was degraded to 1,8-naphthalic acid (NA). Both NAN and NA prevented chlorsulfuron inhibition of seedling growth and also stimulated a positive gravitropic reaction in seedling roots. No significant differences in the rate of chlorsulfuron metabolism were observed between control and NAN-protected seedlings. Acetolactate synthase (ALS) preparations from control and NAN-protected seedling shoot tissues had similar specific activities, kinetic properties and Sephacryl$^{(R)}$ S-400 gel filtration patterns. In contrast, the specific activities of ALS preparations from chlorsulfuron-inhibited shoot tissues were reduced 30-60%. In vitro enzyme studies also showed that chlorsulfuron inhibition was reversible and that NAN and NA did not protect against chlorsulfuron inhibition. Substantial levels of branched-chain amino acids (valine/isoleucine) and ALS activity were present in apical meristems isolated from chlorsulfuron-treated seedlings. Several auxins; including, β-naphthaleneacetic acid, indolebutyric acid, phenylacetic acid and indoleacetic acid also protected seedlings from growth inhibition when applied to leaves 24 hr prior to chlorsulfuron treatment.

INTRODUCTION

Chlorsulfuron [1-(2-chlorophenylsulfonyl)-3-(4-methoxy-6-methyl-1,3,5-triazin-2-yl)urea] is a potent inhibitor of plant growth (ref. 1). As a member of a new class of highly active sulfonylurea herbicides, chlorsulfuron is registered for use as a selective preemergence and early postemergence herbicide in small grains. Selectivity has been attributed to differences in metabolism by sensitive vs. tolerant plant species (ref. 2 and 3). In sensitive dicot species (pea, tobacco and soybean), chlorsulfuron action has been correlated with the inhibition of acetolactate synthase (ALS) and the biosynthesis of branched-chain amino acids (ref. 4, 5 and 6). In corn and other sensitive monocot species, reports of chlorsulfuron metabolism and mechanism of action studies have been limited. A number of herbicide antidotes or safeners protect corn and other cereals from chlorsulfuron injury (ref. 7 and 8). As a seed dressing or a postemergence treatment, 1,8-naphthalic anhydride (NAN) is the most effective of these protectants, but its mechanism of action is unknown. In corn, NAN is degraded to naphthalic acid (NA), but subsequent metabolism does not appear to occur (ref. 9). The objectives of this study were to determine the metabolic fate and interactions of chlorsulfuron and NAN in corn seedlings, and the mechanism(s) of action of NAN as a protectant against chlorsulfuron inhibition of corn seedling growth.

RESULTS AND DISCUSSION

Chlorsulfuron inhibition of corn seedling growth

Chlorsulfuron is a potent inhibitor of corn seedling growth. Treatment with as little as 0.5 µg of chlorsulfuron/seedling typically results in a 50-60% inhibition of shoot growth 7 days after herbicide application (Table 1). Root growth is also inhibited. The inhibition of plant growth is long lasting and may persist to maturity. A characteristic response of corn seedlings to chlorsulfuron treatment is a pronounced inhibition of adventitious root development and an enlargement of the apical node with its associated shoot meristem.

TABLE 1. NAN Protection against chlorsulfuron inhibition of corn seeding growth.

Treatment[a]	Shoot height[b] (cm)	Shoot fr.wt.[b] (gm)	Adventitious root dev.[b]
Control	36.5 (100)	2.93 (100)	++++
NAN	32.2 (88)	2.28 (78)	++++
Chlorsulfuron(0.5 µg)	16.6 (45)	1.27 (43)	++
Chlorsulfuron(0.5 µg) + NAN	29.0 (79)	2.30 (78)	++++
Chlorsulfuron(1.0 µg)	12.6 (35)	0.87 (30)	+
Chlorsulfuron(1.0 µg) + NAN	28.2 (77)	2.28 (78)	+++

[a]Chlorsulfuron solution (10 µl) in 0.02 M TRIS buffer pH 7.5 with 0.2% Tween 80 applied to leaf whorl of 8-day-old seedling NAN applied as seed dressing (0.33% w/w).

[b]Shoot measurements and observations of adventitious root development made 7 days after chlorsulfuron treatment. Average values for 4 seedlings/treatment. Parentheses indicates % of control treatment.

Chlorsulfuron metabolism

Procedures for the isolation of chlorsulfuron metabolites are summarized in Fig. 1. Fifty seedlings (5-day-old) were each treated via the leaves with 50 µg of [phenyl-^{14}C]chlorsulfuron (1.23×10^3 dpm/µg). Three days after treatment, leaf tissues were harvested for metabolite isolation and identification.

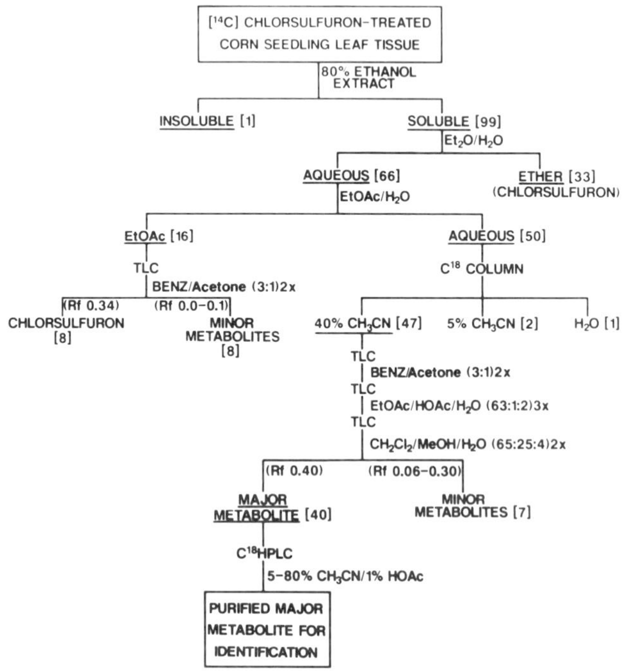

Fig. 1. Extraction and isolation of the major chlorsulfuron metabolite. Numbers in brackets indicate percent ^{14}C distribution.

Acid hydrolysis of the purified major metabolite resulted in the formation of 5-hydroxy-2-chlorobenzenesulfonamide. Hydrolysis with β-glucosidase yielded a product that cochromatographed (TLC and HPLC) with the A-1 aglycone reported in wheat (ref. 2). Direct mass spectral analysis confirmed that the isolated major metabolite was the O-glucoside of 1-(5-

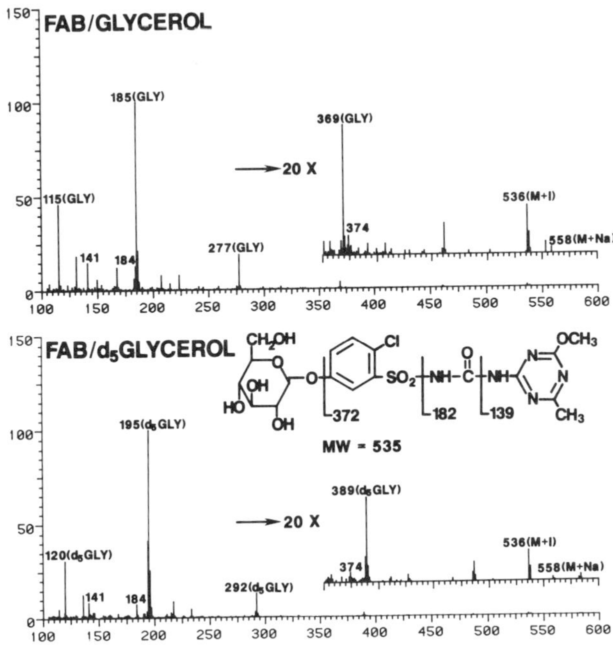

Fig. 2. Mass spectral identification of major chlorsulfuron metabolite.

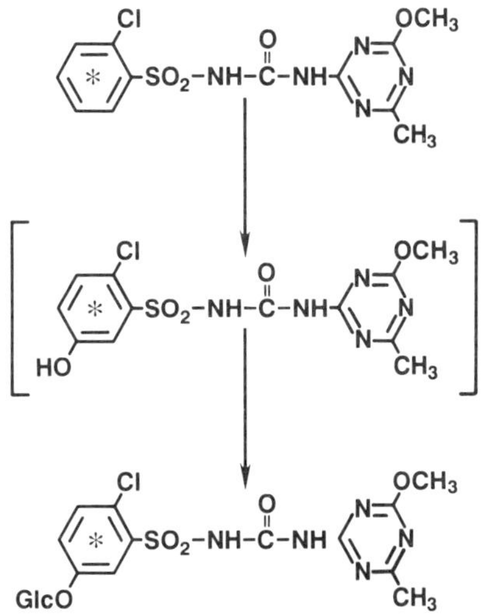

Fig. 3. Major pathway of chlorsulfuron metabolism in sensitive corn.

hydroxy-2-chlorophenylsulfonyl)-3-(4-methoxy-6-methyl-1,3,5-triazin-2-yl)urea (Fig. 2). Comparative positive ion FAB mass spectra in glycerol and glycerol-d_5 matrixes facilitated the identification of metabolite ion fragments. Weak molecular ions were observed at m/z 558 $(M+Na)^+$ and 536 $(M+1)^+$. Weak protonated aglycone, urea-triazine and amino-triazine ion fragments were also observed at m/z 374, 184, and 141, respectively. In addition, positive ion ^{252}Cf plasma desorption mass spectra of the tetraacetylated derivative showed weak molecular ions at m/z 750 $(M+2Na-H)^+$, 727 $(M+Na)^+$, and 704 $(M+H)^+$ together with a strong protonated amino-atrazine ion fragment at m/z 141 and weak glucose tetraacetate ion fragments at m/z 331, 271, 169 and 109.

Chlorsulfuron is oxidized first to a reactive phenolic intermediate and then rapidly conjugated as an O-glucoside (Fig. 3). The free phenol intermediate (brackets) was not detected as a metabolite. A similar metabolic pathway has been reported in tolerant wheat and other tolerant grasses (ref. 2). The phenolic intermediate (20 µg) was not phytotoxic when applied directly to 4-day-old seedling leaf tissues. In wheat and corn, the rapid in vivo conjugation and detoxication of the phenolic intermediate as the O-glucoside has been reported (ref. 10). Differential metabolism may not be a major factor responsible for the selective action of chlorsulfuron in tolerant wheat and sensitive corn.

Effects of NAN/NA on chlorsulfuron metabolism and action in corn

Chlorsulfuron was rapidly absorbed and metabolized (<<12 hr half-life) by both control and NAN-protected seedlings (Fig. 4). As seed dressings (Table 1) or foliar treatments (Fig. 5), NAN and NA provided similar protection against chlorsulfuron growth inhibition. NAN seed treatments increased 1-chloro-2,4-dinitrobenzene (CDNB):glutathione S-transferase (GST) activities 40-60%, but did not affect the glutathione (GSH) content of 4-day-old seedling shoot tissues. Similar results have been reported by others (ref. 11, 12 and 13). In contrast, chlorsulfuron treatments (0.5 µg/seedling) did not affect either GSH content or CDNB: GST activity in 3-day-old seedlings. There appears to be little direct evidence for the involvement of GSH or GST in chlorsulfuron metabolism or NAN protection in corn seedlings.

The effects of NAN and chlorsulfuron treatments on branched-chain amino acid levels in target meristem tissues are shown in Table 2. Amino acid levels varied substantially between experiments, but chlorsulfuron inhibition of seedling growth did not appear to reduce total valine/leucine-isoleucine concentrations in affected tissue sections. Generally, NAN-treated tissues contained higher levels of valine/leucine-isoleucine than control or chlorsulfuron-treated tissues.

Preparations of corn seedling ALS were unstable. Only limited purification by differential

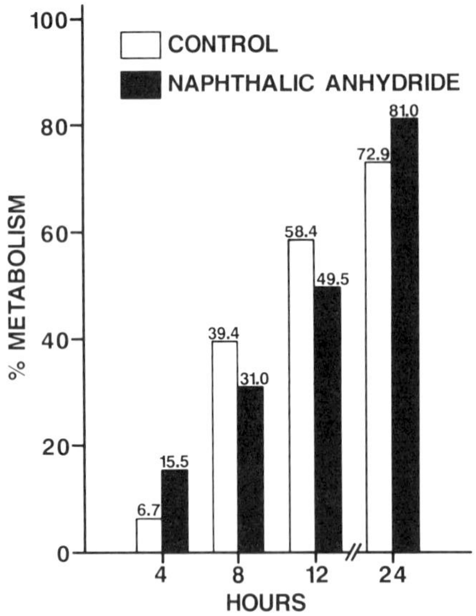

Fig. 4. Rate of [^{14}C]chlorsulfuron metabolism in control and NAN protected corn seedlings. Average of duplicate studies with foliar-applied [^{14}C]chlorsulfuron (2.3 µg or 4.4 µg/4-day-old seedling). Five seedlings harvested for each treatment and time period.

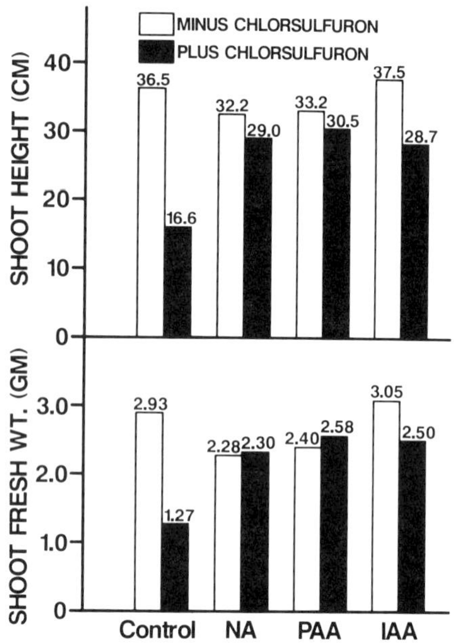

Fig. 5. Auxin protection against chlorsulfuron inhibition of corn seedling growth. Six-day-old seedling shoot tissues treated with 10 µg of NA, PAA or IAA 24 hr prior to treatment with 0.5 µg chlorsulfuron. Shoot height and fresh wt. measurements (aver. of 4 seedlings) taken 7 days after chlorsulfuron treatment.

TABLE 2. Amino acid concentrations in excised (apical) meristem tissue sections from NAN-protected and chlorsulfuron-inhibited corn seedlings

Treatments[a]	Valine/leucine-isoleucine conc. (nmoles/apical section)[b] Experiments[c] (1)	(2)	(3)
Control	12.2/ 5.7	42.8/17.5	16.4/ 8.6
NAN	14.9/ 9.1	49.0/22.8	28.3/13.7
Chlorsulfuron	18.4/ 6.5	45.6/22.5	35.3/13.0
NAN + Chlorsulfuron	20.6/12.3	60.5/28.3	36.2/13.0

[a]NAN applied as a seed dressing (1 gm/300 gm seed) and chlorsulfuron (0.5 µg/seedling) applied either as a foliar treatment (1) and (2) or by direct injection (3) into apical node (meristem).
[b]HPLC analysis of phenylisothiocyanate derivatives (ref. 14) with [^{14}C]valine as the internal standard. Leucine and isoleucine peaks not completely resolved.
[c]Excised apical (meristem) tissue sections (0.4 cm) from 6-day-old corn seedlings (9-15 sections/treatment). Chlorsulfuron treatments made 48 hr (1), 60 hr (2) and 24 hr (3) prior to excision and amino acid analysis.

centrifugation, $(NH_4)_2SO_4$ precipitation and gel filtration was possible. NAN seed treatments increased *in vitro* ALS I stability but did not appear to affect ALS I and II activities in seedling shoot tissues (Table 3). Chlorsulfuron treatment inhibited seedling growth and reduced the specific activity of ALS II by approximately 50%. Chlorsulfuron also appeared to stabilize ALS II, but did not seem to affect ALS I. No significant differences in the apparent Ki values for partially purified ALS preparations from control and NAN-treated seedlings were observed. Additional studies with corn seedling ALS preparations showed that chlorsulfuron inhibition was reversible and that NAN did not provide protection against chlorsufuron inhibition *in vitro*.

Chlorsulfuron/auxin interactions in corn

Protection against chlorsulfuron inhibition of seedling growth by phenylacetic acid (PAA) and indoleacetic acid (IAA) is shown in Fig. 5. Indolebutyric acid and β-naphthaleneacetic acid also provided protection. In other studies, NA and NAN seed treatment (0.3% w/w)

TABLE 3. Effect of NAN and chlorsulfuron on corn seedling ALS activity.

Treatment[a]	Enzyme Fraction[b]	Specific Activity (μmoles/mg protein/30 min)	Total Units[c]	Relative Stability[d]	Ki (nM)
Control	I	0.107	1.50	++	1.4
	II	0.170	5.35	+	3.1
NAN	I	0.105	1.47	++++	1.5
	II	0.193	5.98	+	3.0
Chlorsulfuron	I	0.121	1.57	++	---
	II	0.099	2.57	++++	---

[a]Seed treatments: NAN (0.33% w/w); chlorsulfuron (0.5 g/100 ml acetone/250 g seed for 2 min).
[b]4.5- to 6-day-old seedling shoots (12.0-13.0 g fresh wt). Modified (ref. 5) extraction and assay procedures. Partial ALS purification: 78,000 x g supernatant; 0-60% $(NH_4)_2SO_4$ ppt; and Sephacryl S-400 fraction I (7×10^5-2×10^6 kd) and fraction II (2×10^5-4×10^5 kd).
[c]Unit = μmole acetoin/30 min.
[d]When stored 20 hr at 4°C, losses in enzyme activity ranged from 0-80%.

stimulated a strong positive gravitropic reaction in dark germinated (72 hr) seedling roots. Auxin regulation of protein synthesis has been suggested as a requirement for the transduction of a gravitational stimulus into a graviresponse (ref. 15 and 16). In corn roots, gravistimulation induces a physiological asymmetry in auxin and calcium transport and/or gradients (ref. 17).

CONCLUSIONS

Chlorsulfuron is rapidly metabolized and detoxified *via* the same metabolic pathway in sensitive corn and tolerant wheat. Differential metabolism may not be a major factor responsible for chlorsulfuron selectivity in these crops. NAN and NA are effective protectants against chlorsulfuron inhibition of corn seedling growth, but do not appear to directly affect chlorsulfuron metabolism or ALS activity. NAN does, however, appear to elevate valine/leucine-isoleucine levels in target meristem tissues and may provide transient protection against chlorsulfuron inhibition of ALS while chlorsulfuron is rapidly detoxified as an O-glucoside. Evidence suggests that GSH and glutathione conjugation may not be important factors in chlorsulfuron metabolism or NAN protection against corn seedling growth inhibition. Substantial levels of ALS activity remained in shoot tissues from chlorsulfuron inhibited corn seedlings. Within 1-2 days after chlorsulfuron treatment, ALS activity in target apical meristem tissues was apparently sufficient to maintain total valine/isoleucine concentrations at control levels. The amelioration of chlorsulfuron effects by exogenous auxins and the enhanced graviresponsiveness of roots following NAN or NA seed-treatment suggest that NAN and exogenous auxins may be acting in a similar manner. Further studies on possible relationships between NAN and auxins as chlorsulfuron protectants are in progress.

REFERENCES

1. T.B. Ray, Pest. Biochem. Physiol. 17, 10-17 (1982).
2. P.B. Sweetster, G.S. Schow, and J.M. Hutchinson, Pest. Biochem. Physiol. 17, 18-23 (1982).
3. J.M. Hutchison, R. Shapiro, and P.B. Sweetser, Pest. Biochem. Physiol. 22, 243-247 (1984).
4. T.B. Ray, Plant Physiol. 75, 827-831 (1984).
5. R.S. Chaleff and C.J. Mauvais, Science 224, 1443-1445 (1984).
6. D. Scheel and J.E. Casida, Pest. Biochem. Physiol. 23, 398-412 (1985).
7. P.B. Sweetser, U.S. Pat. No. 4,343,649 (1980).
8. C. Parker, W.G. Richardson, and T.M. West, Proc. Brit. Crop Protection Conf. 15-22 (1980).
9. J.R. Riden and W.J. Asbell, Environ. Qual. Saf. Suppl. 3, 175-179 (1975).
10. D.L. Erbes, Pest. Biochem. Physiol. (In Press).
11. M.M. Lay and J.E. Casida, Pestic. Biochem. Physiol. 6, 442-446 (1976).
12. T.J. Mozer, D.C. Tiemeier, and E.G. Jaworski, Biochemistry 22, 106-1072 (1983).
13. T. Komives, M. Balays, V. Komives, and F. Dutka, Proc. 5th International Conf. on Cytochrome P-450, pp. 451-454, Budapest, Hungard (1985).
14. B.A. Bidlingmeyer, S.A. Cohen and T.L. Tarvin, J. Chromatogr. 336, 93-104 (1984).
15. L.J. Feldman, Physiol. Plant. 65, 341-344 (1985).
16. R. Moore, Amer. J. Bot. 75, 733-736 (1985).
17. J.S. Lee and M.L. Evans, Plant Physiol. 77, 824-827 (1985).

Metabolism of fenthion by aquatic microbial communities

*P.H. Pritchard[1], D.R. Cripe[1], L.H. Mueller[1], and E.J. O'Neill[2]

[1]U.S. Environmental Protection Agency and [2] Technology Applications Inc., Gulf Breeze, Florida 32561

Abstract - The microbial metabolism of the mosquito control agent, fenthion, has been studied in shake flask systems containing water and sediment from a salt marsh. The usefulness of this information in describing the fate of fenthion in microcosms and in a field dosing experiment was detemined. Our results show that microbial communities associated with the sediment, the presence of invertebrate animals in the sediment bed, and the anaerobic conditions of the sediment contribute significantly to the fate of fenthion under natural conditions.

INTRODUCTION

Recent developments in hazard assessment procedures have centered on improvements in exposure concentration estimations, particularly in those that extrapolate laboratory information to field situations. Metabolism by microorganisms is frequently one of the major processes that control the fate of a pesticide in the environment. Although a number of standardized tests for establishing susceptibility to microbial metabolism exist, few investigations (ref. 1 and 2) have attempted to relate the resulting information to the complex environmental conditions found in the field.

The objective of our study was to further clarify these relationships. Specifically, we proposed to characterize the environmental factors that control the microbial metabolism of the organophosphate pesticide, fenthion (Baytex), and then determine the relevance of this information in describing the fate of fenthion in microcosms (laboratory surrogates to the field) and field dosing experiments. Because fenthion is extensively applied to coastal areas to control mosquito populations, our studies dealt with the fate of this pesticide in an estuarine salt marsh.

FLASK STUDIES

The ability of microorganisms to metabolize fenthion was established in laboratory shake flask studies. Inocula for the flask studies consisted of water and detrital sediment collected from a local salt marsh. The detrital sediment was light, fluffy, high organic particulates composed largely of decaying plant material and heavily colonized with bacteria, algae, and protozoa; sand was excluded. Flasks containing water or water-sediment slurries (500 mg/L dry weight) were spiked with fenthion to a final concentration of approximately 200 µg/l, and incubated on a rotary shaker, in the dark, at 25° C. Changes in the concentration of fenthion (Fig. 1) were followed by periodically removing samples from the flasks, extracting the samples with hexane or acetonitrile, and analyzing the extracts by gas chromatography.

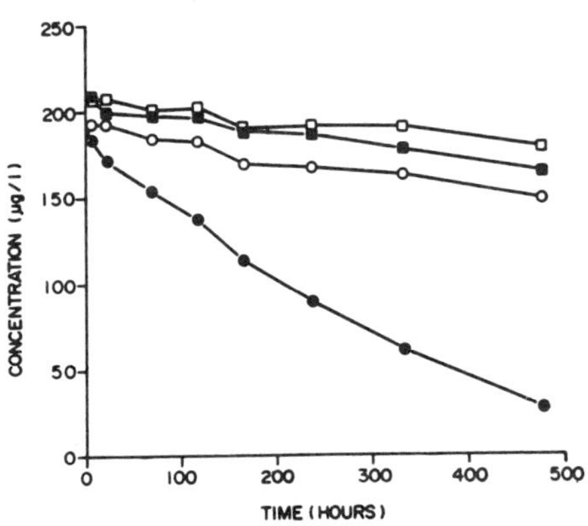

Figure 1. Disappearance rates of fenthion in formalin-sterile (open symbols) and nonsterile (closed symbols) flasks containing water (squares) and sediment (circles).

Fenthion disappeared slowly (significantly different [α = 0.05] from a slope of zero; half life = > 30 days) in sterile flasks, indicating little susceptibility to hydrolysis. The slow loss in nonsterile water (not significantly different [α = 0.05] from the sterile water; half life = > 30 days) suggested that microbial populations contained therein either were not able to metabolize fenthion or the biomass of degrading microorganisms was not sufficient to cause significant loss during the time course of the experiment. The rapid loss of fenthion in the presence of nonsterile sediment (half life = 7.4 days) indicates that the microbial populations associated with the sediment surface were capable of degrading the pesticide.

Additional flask experiments with radiolabeled fenthion showed the production of polar (relative to fenthion) products, the production of bound residues (radioactivity associated with sediment that could not be extracted with conventional solvents), and the production of very small amounts of radiolabeled carbon dioxide. These results are consistent with the idea that the disappearance of fenthion in the presence of sediments is due to the metabolic activity of bacteria associated with the sediment.

Several other environmental factors affected the fate of fenthion in saltmarsh environments, as shown by additional flask studies. Biodegradation rates were proportional to detrital sediment concentration and the microorganisms responsible for the degradation activity associated with the sediments could not be released into the water if vigorously shaken with the sediment. Degradation was much slower when oxygen was removed by sparging the sediment-water slurries in the flasks with nitrogen gas. The pesticide was also photolabile, having a half life in water of 6.2 days when exposed directly to sunlight.

MICROCOSM STUDIES

As a means of verifying the importance of sediment in the degradation of fenthion under more natural conditions, the fate of fenthion in microcosms was determined. Microcosms consisted of 3-L vessels (15 cm in diameter) containing an intact core of sediment (detrital sediment mixed with sand) from the salt marsh; 2 1/2 L of seawater from the marsh provided the water column. Complete mixing of the water column without resuspension of the sediment was accomplished with a bent glass rod attached to a constant rpm motor. The microcosms were exposed to a 12:12 hour light:dark cycle, using fluorescent lighting. Experiments commenced by spiking the water column with fenthion to a final concentration of 200 µg/l. Water samples were removed periodically and the concentration of the pesticide determined. After 170 hours of incubation, the water in the microcosms was siphoned off and the sediment bed was sliced into 0.5 cm layers (by frozen core fractionation), extracted, and the total concentration (dissolved and sorbed) of fenthion determined. Porosity and the concentration of detrital sediment were also determined for each layer. Microcosms treated with formalin were used to separate biotic form abiotic effects.

The changes in concentration of fenthion in the water column and the distribution of fenthion in the sediment bed after 170 hours of incubation are shown in figures 2a and b respectively.

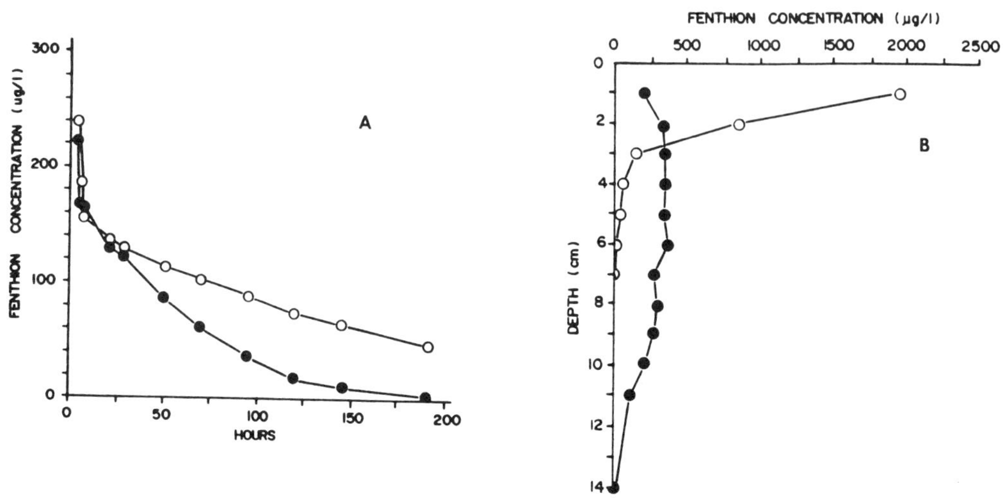

Figure 2. Concentration of fenthion in the water (A) and sediment at 170 hours (B) (open circles) and nonsterile (closed circles) saltmarsh microcosms.

Loss of the fenthion from the water column was faster in the nonsterile microcosm than in the sterile microcosm, indicating that a biotic component was controlling the fate of the pesticide. Substantial amounts of fenthion were found in the sediment bed of the nonsterile microcosms at greater depths than in the sterile microcosms. This distribution further suggests the involvement of biotic factors.

To more carefully define these biotic contributions and separate them from the other fate processes occurring simultaneously, a mathematical model, developed specifically for microcosm studies (ref. 3) was used to help interpret the data. Running the mathematical model with the appropriate kinetic parameters for diffusion (based on the movement of tritiated water), partitioning to the detrital sediment (from batch sorption tests), hydrolysis and photolysis (flask studies), showed that the fate of fenthion in the sterile microcosms could be adequately explained; i.e., good agreement between predicted and actual concentrations of the pesticide.

Similar use of the model, however, with kinetic parameters for the biodegradation of fenthion in water and sediment from the flask studies, failed to explain the fate of the pesticide in the active microcosms. This was due to a much slower rate of biodegradation and a much deeper penetration of the fenthion into the sediment bed than expected. The presence of numerous sediment-dwelling invertebrate animals in the microcosms, which are capable of extensive mixing of interstitial water and reworking of sediments (ref. 4 and 5), led to the hypothesis that such activities were responsible for the observed distribution of fenthion in the sediment bed. This bioturbation activity, coupled with the lack of anaerobic degradation of fenthion (anaerobic conditions in the microcosms occurred in the sediment at 2 cm depth and below) would explain the observed results. Rerunning the model with higher diffusion rates to simulate the bioturbation and biodegradation only in the top cm of the sediment bed projected a curve that reasonably fitted, based on chi squared analysis, the observed data. This fit is shown in figure 3 (curve D).

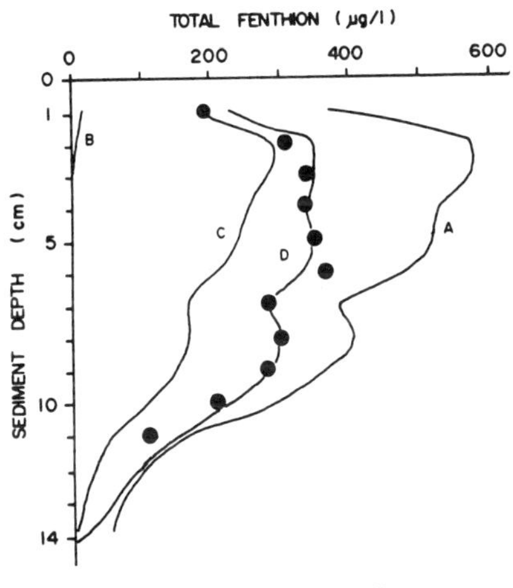

Figure 3. Concentration of fenthion with depth in the sediment of microcosms (data points) and predicted concentrations from a mathematical model (solid line).

Using biodegradation rates equal to zero (curve A), to the rate obtained from the flask studies (curve B), or to the flask reduced 40x and assumed to be equal at all depths in the sediment (curve C), gave relatively poor fits to the data. From these experiments, we have shown that a much different picture of the fate of fenthion was obtained by using experimental systems that simulate the complexity of natural systems; i.e., a variety of environmental factors can simultaneously affect the fate of a pesticide. Our experiments also suggest that biodegradation rate information from laboratory shake flask tests may not accurately reflect the rates that occur in nature.

FIELD STUDIES

To further test the importance of this information from the microcosm studies, the fate of fenthion in a field dosing experiment was considered. Fenthion, in acetone carrier, was sprayed on the water surface of the local salt marsh from which the water and sediment samples were taken for the above studies. Pesticide concentration was then monitored over time. No mechanical mixing of the marsh water was undertaken to avoid resuspending the sediments. A water-soluble dye, rhodamine WT, was also sprayed at the same time as the fenthion to measure losses due to dilution from tidal exchange. Dye concentrations were monitored by fluorescence spectrometry. Disappearance of fenthion from the marsh water was faster than the disappearance of the dye, indicating that one or more fate processes were operating at a rate which exceeded losses due to dilution. As a rough means of explaining the factors affecting the disappearance of fenthion in the field, the microcosm model was applied to the field data. This was accomplished by assuming that the microcosm although a scaled down version of the field, replicated the essential features of that environment in all important details (i.e., that the conditions of the sediment core in the microcosm were typical of all areas of the marsh) with a flow-through condition to the washout rate of the dye in the marsh. A predicted rate of disappearance of fenthion was thus estimated from the model and compared with the field data (Fig. 4).

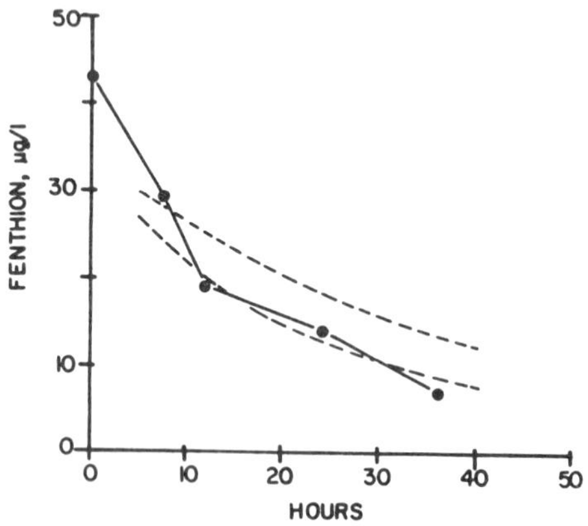

Figure 4. Actual (solid line) and predicted (dotted line) disappearance of fenthion following dosing of the water column in a salt marsh. Predicted values were based on a dilution rate calculated from dye disappearance and the kinetic parameters from the microcosm study.

As can be seen, agreement between the observed and computed disappearance curves was not close. However, in performing a similar modeling exercise using disappearance data from field conditions which more closely approximated the microcosm conditions, i.e., an open-ended plexiglass box placed in the field after the dosing to isolate a defined section of water and sediment, good agreement was obtained between observed and computed disappearance of fenthion. This result suggested that the microcosm conditions were reflective of conditions in the field: i.e., sorption, hydrolysis, photolysis and biodegradation were similar. The only major difference between the inside and the outside of the plexiglass box was the degree of water turbulence, the water mixing in the marsh being heavily influenced by wind and groundwater intrusion. Accordingly, the microcosm model was rerun with faster diffusion rates to simulate the additional water turbulence in the field. A good fit of the field data was obtained with diffusion rates that were 10x those obvserved in the microcosm results (figure 4). Changes in the other kinetic parameters of the model did not produce as good a fit. Thus, the scaling of diffusion and turbulent mixing processes between laboratory and field studies can be a crucial limitation in extrapolation efforts.

Although further field data are necessary, our results are consistent with the idea that microcosms are good surrogates for the field. We can conclude, therefore, that the fate of fenthion in a salt marsh is determined by complex factors: being affected significantly by sorption, biodegradation and bioturbation in the sediments and photolysis in the water column. Our experiments further emphasize the importance of mathematical models in studying environmental fate and the caution that should be exercised in extrapolating laboratory data to the field, particularly in the context of estimating exposure concentrations for hazard assessment procedures.

REFERENCES

1. L.M. Games, Modeling the Fate of Chemicals in the Aquatic Environment, Ann Arbor Science, Ann Arbor, Michigan, 325-346 (1982).
2. W.K. Marshall and J.R. Roberts, Scientific Criteria for Environment Quality, Nat. Res. Counc. Rev. Canada, 2, 252-278 (1971).
3. P.H. Pritchard, C.A. Monti, E.J. O'Neill, A.W. Bourquin and D.G. Ahearn, Environ. Toxicol. Chem. 5, 647-657 (1986).
4. D.C. Rhodes, J. Sediment. Petrol., 33, 723-727 (1983).
5. E.J. O'Neill, C.A. Monti, P.H. Pritchard, A.W. Bourquin and D.G. Ahearn, Environ. Toxicol. Chem., 4, 453-458 (1985).

The disposition of bound ^{14}C agrochemical residues in plants and animal tissues when consumed by animals

M. Humayoun Akhtar

Animal Research Centre, Research Branch, Agriculture Canada, Ottawa, Ontario, K1A 0C6

Abstract: Agrochemicals can be transported and deposited in various plant and animal tissues as the parent compound, metabolites, conjugated metabolites and bound residues. In plants insoluble pesticide residues are usually incorporated into lignin; while in animal tissues drugs and their metabolites are covalently bound to macromolecules. The significance of bound residues in plants and animal products is directly related to their bioavailability to human and animals when consumed. The bioavailability of plant bound residues to animals is dependent upon the type of material (starch, cellulose, lignin) to which the residues are bound. In most cases, plant bound residues are very poorly absorbed from the gastrointestinal tract of the rat and are excreted unchanged. Thus, most insoluble agrochemical related residues in plant tissues are of minimal toxicological concern.

INTRODUCTION

Agrochemicals such as pesticides and animal drugs, are routinely employed as part of the integrated farming practices to protect crops and animals from insects, weeds, diseases. Because of their potential toxicological properties, residues of these compounds in edible portion of crops and animal products are viewed with great concern by society. Xenobiotics may be deposited into edible portion as the parent compound, as breakdown products (chemical and enzymatic, including conjugates) or occasionally in a bound form. Experiments have shown that part of the ^{14}C-activity administered to plants and animals as ^{14}C-labeled pesticides and drugs was incorporated into non-endogenous products which could not be removed from the biological matrix by exhaustive extraction with commonly used solvents. This unextractable component was generally described as bound (unextractable) residues. Since the 1975 ACS Conference in Vail/Colorado considerable progress has been made towards isolating and determining the nature and quantity of bound residues (ref. 1-5).

The significance of bound residues in plant (crops) and animal tissues is a function of bioavailability to humans and animals when they consume a food containing bound residues. Recent studies have expanded our knowledge of the nature of bound pesticides residues in plants; but very little attention has been paid to the disposition of bound residues in plants when consumed by animals. Available data suggests that bioavailability of pesticide bound residues in plants to animals is highly dependent on the nature of binding sites.

DEFINITION

Prior to the Vail/Colorado conference, the radioactive residues which could not be extracted from the plant matrix were referred to by various scientific and quasi-scientific terms including 'residual', 'insoluble', 'non-extractable', 'un-extractable', and 'bound'. In recent years both 'bound' and 'unextractable' have been used interchangeably to identify the radioactivity which could not be isolated. The IUPAC commission on pesticides (ref. 6) proposed that these materials be defined as follows:

> 'non-extractable residues (sometimes referred to as 'bound' or non-extractable residues) in plants are defined as chemical species originating from pesticides used according to good agricultural practices, that are unextracted by methods which do not significantly change the chemical nature of these residues. These non-extractable residues are considered to exclude fragments recycled through pathways leading to natural products'.
>
> Chemical residues in this context refers either to the parent material or to derivatives or fragments of it. Methods in this context refer to any procedures, such as solvent extraction and distillation, used to

exhaustively remove chemical species from a plant matrix. In each reference to a non-extractable residue, the extraction procedure must be given.

Bound animal drug residues are defined as (ref. 7):

'those residues (parent compound or metabolites) which are covalently bound to biological macromolecules, are not extractable from the macromolecules by exhaustive extraction, denaturation or solubilization techniques, and are not the result of endogenous incorporation'.

BIOAVAILABILITY

Bound (non-extractable) residues of pesticides and drugs in edible crops and animal tissues are of toxicological significance only if they become systemically bioavailable to the species which are ingesting the bound residues. Bioavailable residues are defined as (ref. 7)

'compounds (parent or metabolites) which are absorbed from the gastro-intestinal tract and which may be found in the cells of gastrointestinal tract or which may be further absorbed and found in the tissues or fluids of the ingesting species as well as in the expired carbon dioxide'.

METHODOLOGY FOR THE DETERMINATION OF BIOAVAILABILITY

The use of ^{14}C-labeled xenobiotics has provided samples of ^{14}C-residues bound in plants and animal tissues; the bioavailability of some of these residues in animals - mainly in the rat has been studied. Plant portions and animal tissues containing ^{14}C residues are extracted with organic solvents and water in order to remove soluble (extractable) ^{14}C. The insoluble material is dried at 25-30°C for 24-48 h and the radioactivity in the residues is determined by combustion. The radiocarbon content in the insoluble material is generally referred to as bound ^{14}C-residues. In some cases the chemical identity of ^{14}C in the insoluble material has been determined by gas chromatography (GC) and gas chromatography-mass spectrometry (GC-MS) of thermally labile ^{14}C components when insoluble residues were subjected to high temperature distillation (HTD) (ref. 4) or treated with supercritical methanol (ref. 5). This insoluble material was then used as the bound ^{14}C-residues for bioavailability studies.

The bioavailability of bound residues has generally been determined by studying the intestinal absorption in the animal. The bound ^{14}C residues are mixed with regular diet and the contaminated feed is compressed into pellets of adequate size, where possible. Animals may be fed the fortified diet for 2 or 3 consecutive days in order to allow sufficient accumulation of radiocarbon in various tissues and organs, when absorbed. Feces, bile and urine are usually collected during the treatment period and 2 or 3 days after the treatment has stopped. Major organs such as kidney, liver, intestine are removed and analyzed for radioactivity. The presence of radioactivity in urine, bile, carcass, liver, kidney and intestinal wall tissues indicate absorption and bioavailability. Usually complete elimination of ingested radiocarbon in the feces provide evidence for no absorption and thus non-bioavailability. Presence of radioactivity in the feces may also be due to excretion of radiocarbon (i) in bile, pancreatic and gastric juice, (ii) by lungs and into saliva and swallowing and (iii) into intestinal lumen, in addition to incomplete absorption of ingested radiocarbon (ref. 8). The amount of radiocarbon eliminated in the urine, respired as CO_2 and retained in tissues are the guidelines for absorption. The toxicological significance of bound residues is directly proportional to the concentration of radioactivity in the liver, kidney, bile, intestinal wall and other tissues. Characterization of the ^{14}C-labeled compound(s) is important when the ^{14}C content in urine and tissues is high. The fate of bound residues of selected pesticides and drugs in animals are reviewed.

BIOAVAILABILITY OF ^{14}C-RESIDUES FROM PLANTS

One of the earliest experiments to determine the fate of ^{14}C-residues from plants in rats was reported by Paulson et al. (ref. 9). These workers observed that when the insoluble fraction of both alfalfa roots and shoots (containing ^{14}C-residues from ^{14}C-propham) was fed to rats, more than 86% of the ^{14}C was excreted in the feces, about 2% was found in the urine and less than 3% was retained in carcass. Thus most of the propham-related bound residues in alfalfa was shown to be non-bioavailable and of no toxicological concern.

When Sutherland (ref. 10) fed ^{14}C-propanil-related residues (containing 36% extractable ^{14}C residues) in 12 wk rice plant to albino rats, the amount of ^{14}C excreted in the urine was 11% and in the feces 89%. Radiocarbon in the urine may have been mainly due to absorption of the soluble ^{14}C. Similarly, dogs and mice failed to digest unextractable propanil rice plant residues (ref. 10). Thus, it can be concluded that bound propanil residues should be considered of little toxicological concern.

Stratton and Wheeler (ref. 11) reported that 20-30% of the total ^{14}C in early maturing radishes grown in ^{14}C-dieldrin-treated soil was bound; and a major portion of the bound residues was in the lignin fraction. The bioavailability of the bound dieldrin residues was tested by feeding male Sprague-Dawley rats a diet containing extracted radish pulp for 2 days. Approximately 16% of the ingested ^{14}C was eliminated in the urine; while 74% was voided in the feces over a 4-day period. On the other hand, rats excreted about 32% of the ingested radiocarbon in the urine, and 47% in the feces within 4 days after they were fed unextracted (containing free, conjugated and bound residues) radish pulp for two days (ref. 12). These data indicated that bound dieldrin residues in radishes were absorbed but at a considerably slower rate than free dieldrin and its conjugates.

It has been shown that the compound present in the form of bound ^{14}C residues in radishes was dieldrin (ref. 13). Dieldrin was the only compound identified in the urine and feces of the rats fed ^{14}C dieldrin residues in radishes (ref. 13). These observations were not unexpected because of the properties of this compound. Dieldrin is a relatively persistent chemical that does not undergo extensive metabolism in plants and animals (ref. 14).

Khan et al. (ref. 15) fed bound ^{14}C-atrazine-related residues in corn plants (mixed in with regular diet) to rats for 2 days and collected the feces and urine during the treatment and 2 days thereafter. Approximately 88% of the total ingested ^{14}C was eliminated in feces, and 10% in the urine after 4 days. Almost all of the ^{14}C in feces was bound. Only trace levels of deethylatrazine, its 2-OH analogue, and 2-OH ammeline were identified in the urine. When rats were fed extracted corn material fortified with ^{14}C-atrazine there were appreciable amounts of the ^{14}C in the liver (2.1%) and kidney (0.3%) 4 days after the dosing (ref. 15), whereas extremely small amounts (<0.1% of the consumed ^{14}C) of ^{14}C were found in the liver and kidney of animals treated with bound residues. The data show that bound atrazine residues in corn plants have very low degree of bioavailability and as such their toxicological concern should be minimal.

Rats, when fed bound ^{14}C-carbofuran related residues in radishes, excreted about 90% of the ^{14}C-activity in feces and less than 2% in the urine during a 4 day period. Only 2% of the ^{14}C in the feces from rats fed the bound ^{14}C carbofuran-related residues in radishes could be extracted (ref. 12). This indicated that almost all of the ^{14}C in the feces was still in the bound form. Carbofuran, 3-hydroxycarbofuran and 3-ketocarbofuran were detected in the urine and feces of the treated rats. The ^{14}C content in various tissues and organs of treated rats was very low (<0.1%) (ref. 12).

Mostafa et al. (ref. 16) observed that sweet potato tubers absorbed about 28% of the total ^{14}C when soaked in a 50 ppm solution of ^{14}C-carbofuran for one week. Approximately 23% of the ^{14}C in the tubers was in bound form. Rats when fed the bound ^{14}C-residues in tubers eliminated ^{14}C as follows: feces (66%), urine (9%) and carbon dioxide (11%). Small amounts of radioactivity were also detected in brain (0.03%), fat (0.02%), blood (0.09%), liver (0.23%) and kidney (0.08%). The chemical nature of the ^{14}C was not known in either bound residues or in the biological samples.

The above studies shed light on the factors that may affect the disposition of bound ^{14}C-residues (crops) when consumed by rats (monogastric). In radish most of the bound ^{14}C was in the lignin fraction; whereas in the tuber it was probably in the starch fraction (tubers contain starch or fructans as the main component). In animals the bioavailability of bound ^{14}C-residues can be expected to be greatly dependent on the digestibility of the food portion to which the residues are bound. The digestibility of a food is closely related to its chemical composition. The cell walls consist mainly of cellulose and hemicellulose and are poorly absorbed if highly lignified. Cellulose and hemicellulose are not attacked by digestive enzyme secreted by animals, but they are efficiently broken down by microorganisms in the rumen of these animals. The rumen micro-organisms degrade starch to glucose, and ultimately to CO_2, fatty acids and other low molecular weight products. However, lignin is not digested to any appreciable extent by rumen microorganism.

In view of the varying rates of degree of digestibility of lignin and starch, it was not surprising to observe a great difference in the rate of absorption between the bound ^{14}C residues in radishes and tubers in the rats. Although data are not available from studies with ruminants, a similar pattern would be expected. In human (monogastric) ^{14}C-residues bound to a lignin fraction would be expected to be less bioavailable than those bound to a starch fraction. Thus, bound ^{14}C-residues in radishes may be of less toxicological concern than those in the tubers.

Treatment of goats with a single or multiple oral dose of ^{14}C-dichloran produced protein bound ^{14}C-residues in liver (ref. 17). When the ^{14}C-bound residues in goat liver derived from ^{14}C-dichloran were incorporated in feed and fed to rats, about 8% of the ^{14}C-activity was excreted in the urine. The major portion of the ingested ^{14}C was excreted in the feces, of which 20% could be extracted with a methanol:water mixture. The data indicated low bioavailability of the ^{14}C-activity in the liver, only trace amounts were found in the liver and none in kidney, muscle and fat (ref. 17). It is thus apparent that although part of the bound residues of ^{14}C-residues were available, they were of minimal toxicological significance since they were neither retained nor reacted with cellular macromolecules.

BIOAVAILABILITY OF BOUND ^{14}C DRUG RESIDUES FROM ANIMAL TISSUES

As with pesticides, animal drugs are subjected to thorough investigation with respect to human safety. The edible tissues of the meat, milk and egg producing animals are examined to determine the nature and level of drug residues - parent, metabolites, conjugates and bound residues. The regulatory agencies are faced with a challenging task of establishing protocols for characterization and toxicological assessment of bound residues. Since 1975, a few studies on characterization and fate of the bound residues of some important animal drugs have been reported.

Oral administration of (phenyl-U-^{14}C)p-toluoyl chloride phenylhydrazone (TCPH) to sheep produced persistent residues in blood. A detailed analysis of the whole blood revealed that 86% of the radioactivity in the blood was associated with erythrocytes and was covalently bound to both globin (49%) and heme (20%). Rats fed the bound ^{14}C-residues in blood (added to diet for 7 days) eliminated 89% of the ^{14}C in the feces and only 8% in the urine. Small amounts of radiocarbon were present in various organs during the treatment period, but rapidly declined when the feeding of the ^{14}C-blood was discontinued (ref. 18). Feeding dried blood from TCPH treated sheep to rats at a level 2000 times the potential exposure, for a period of 90 days did not produce any observable toxic symptom (ref. 18). Thus the bound residues of TCPH in sheep tissues should be of little toxicological concern.

CONCLUSION

A critical evaluation of the data presented here indicated that, in most cases the bound agrochemical related residues that have been examined to date were not bioavailable and are therefore assumed to be non-toxic. Residues bound to lignin fraction were not bioavailable (e.g. dieldrin residues, in radish), while those bound to starch were more bioavailable. Further research on bioavailability of bound residues of other chemicals is needed to evaluate their possible effects on animals including man.

ACKNOWLEDGEMENT:

I thank Drs. Khan, and Mostafa and their co-workers for making their unpublished data available.

REFERENCES

1. S.U. Khan, Res. Rev. 84, 1-25 (1982)
2. R. Huber, S. Otto, 'Bound Pesticide Residues in Plants' In Pesticide Chemistry: Human Welfare and Environment (eds. J. Miyamoto and P.C. Kearney), Vol. 3. p 357-362. Pergamon Press, Oxford (1983).
3. J.B. Pillmoor, T.R. Roberts, 'Approaches to the study of non-extractable (bound) residues in plants' In Progress in Pesticide Biochemistry and Toxicology (eds. D.H. Hutson and T.R. Roberts) Vol. 4. p. 85-101, Wiley, Chichester, England (1985).
4. S.U. Khan, H.A. Hamilton, J. Agric. Food Chem. 28, 126-132 (1980).
5. P. Capriel, A. Haisch, S.U. Khan, J. Agric. Food Chem. 34, 70-73 (1986).
6. P.C. Kearney, J. Assoc. Off. Anal. Chem. 65, 1031-1032 (1982).
7. N.E. Weber, Toxicol. Europ. Res. 5, 271-275 (1982).
8. P.G. Dayton, Z.H. Israili, J.D. Henderson, Drug Metab. Rev. 14, 1193-1206 (1983).
9. G.D. Paulson, A.M. Jacobsen, G.G. Still, Pest. Biochem. Physiol. 5, 523-525 (1975).
10. M.L. Sutherland, 'Nature of propanil bound residues in rice plants as measured by plant fractionation and animal bioavailability experiments' In Bound and Conjugated Pesticide Residues (eds. D.D. Kaufman, G.G. Still, G.D. Paulson and S.K. Bandal) ACS Symposium Series No. 29, p 153-155, Washington, D.C. (1976).
11. G.D. Stratton, Jr., J.B. Wheeler, J. Agric. Food Chem. 31, 1076-1079. (1983).
12. S.U. Khan, S. Kacew, G.D. Stratton, Jr., W.B. Wheeler, Private communication.
13. S.U. Khan, G.D. Stratton, Jr., W.B. Wheeler, J. Agric. Food Chem. 32, 1189-1191 (1984).
14. F. Matsumaura, Toxicology of Insecticides, Plenum Press: New York, 1975, p. 332.
15. S.U. Khan, S. Kacew, S.J. Molnar, J. Agric. Food Chem. 33, 712-717 (1985).
16. I.Y. Mostafa, S.M.A.D. Zayed, M. Farghaly, Private communication.
17. P.S. Jaglan, T.S. Arnold, 'Abstracts of Papers'. 190th American Chemical Society Meeting, Chicago, Ill. Agrochem. paper No. 61 (1985).
18. P.S. Jaglan, M.W. Glenn, A.W. Neff, J. Toxicol. & Environ. Health 2, 815-826 (1977).

Contribution No: 1415

Intermediary metabolism of mercapturic acid pathway metabolites

*J.E. Bakke[1], D.J. Mulford[2], V.J. Feil[1] and G.L. Larsen[1]

[1]US Department of Agriculture, ARS, Metabolism and Radiation Research Laboratory, and [2]Department of Biochemistry, North Dakota State University, Fargo, ND 58105

A summary of the catabolic processes that have been shown to occur in vivo with mercapturic acid pathway (MAP) metabolites is given in Fig. 1 along with references to reviews wherein specific citations can be located. Here only unpublished and previously unreviewed information on the intermediary metabolism and interorgan translocations involved in MAP metabolites will be discussed. Some of these, such as the formation of an acetylated glutathione conjugate and a glutamylcysteine conjugate in rats dosed with 1,2,4-trichlorobenzene (unpublished) need only be mentioned because the biological significance of their formation in animal systems is not known. Glutamylcysteine conjugates are more usual metabolites in plants (ref. 5). The discovery of these new metabolites demonstrates the utility of fast atom bombardment mass spectrometry in the elucidation of metabolic pathways. Many of the arrows in Fig. 1 represent overall processes demonstrated to occur in vivo rather than defined step by step biochemical transformations. Factors which influence the diversion of MAP metabolites into these various pathways are, for the most part, unknown. It must also be remembered, as will be exemplified below, that interorgan translocation of metabolic intermediates has an important influence on the intermediary metabolism of MAP metabolites of xenobiotics.

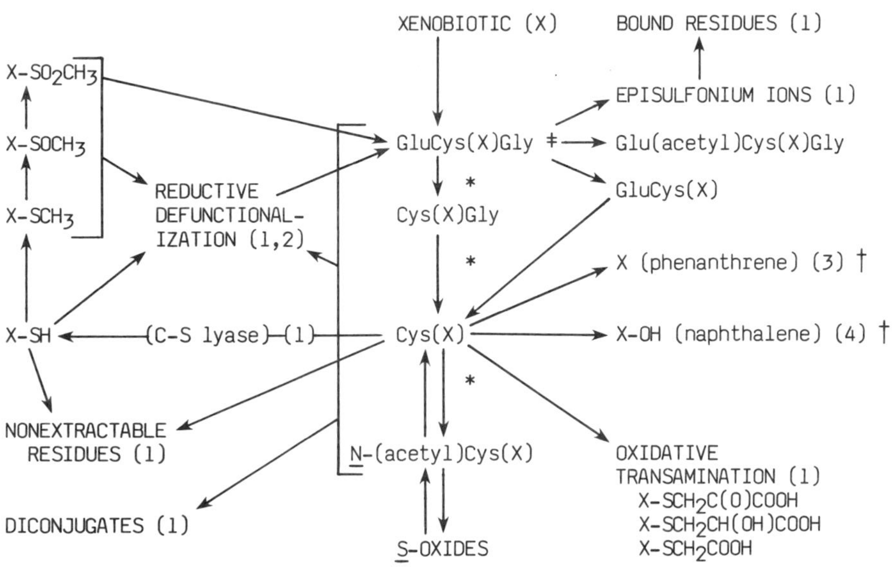

Fig. 1. Catabolic pathways known to occur in vivo for mercapturic acid pathway metabolites. The mercapturic acid pathway is denoted by asterisks. Numbers in parenthesis are references. † These biotransformations have been observed only with the compounds shown. ǂ GluCysGly = glutathione; GluCys(X)Gly = glutathione conjugates.

It is currently accepted that cells conjugate xenobiotics (or intermediates) with glutathione (GSH) and then excrete the GSH conjugate from the cell (ref. 6) for disposal by the organism in much the same fashion that GSH is synthesized in the cell and then translocated from the cell to supply GSH for other organs. Much of the complexity observed in the intermediary metabolism of MAP metabolites occurs because organisms attempt to excrete these GSH conjugates by routes (bile and kidney) that contain efficient systems for the conservation of the component amino acids of GSH; these systems also function to hydrolyze the peptide bonds in GSH conjugates. This conservation system results, with GSH conjugates, in the formation of S-cysteine conjugates. Cysteine conjugates, which appear to occupy a pivotal position in the intermediary metabolism of MAP metabolites, are rarely excreted without further biotransformation. These biotransformations may occur to negate the biological activity attributed to some cysteine conjugates (ref. 7), for example, pathways involving cysteine conjugate β-lyase (C-S lyase) (Fig. 1). Kidney damage can result from C-S lyase cleavage of cysteine conjugates when the appropriate GSH conjugates enter the renal GSH-conservation system (ref. 8). These observations would indicate that some of the pathways open to cysteine conjugate biotransformation, e.g. N-acetylation, have evolved to prevent them from entering toxic routes. The biliary/intestinal GSH-conservation system appears to add the most complexity to intermediary metabolism of xenobiotics through enterohepatic circulation of either intact MAP metabolites and/or microfloral metabolites. The microfloral metabolites can be thiols, phenols, parent aromatic hydrocarbons, or reductive defunctionalization products.

C-S lyase mediated methylthiolation is a well known metabolic route which essentially proceeds biochemically as outlined in Fig. 1 for the formation of X-SCH_3, however, biliary excretion of MAP metabolites complicates the pathway beyond that shown. Metabolic forces operating in the biliary tree and the intestinal lumen convert MAP metabolites and their sulfoxides into cysteine conjugates which then become substrates for microfloral C-S lyases. The resulting thiols are usually methylated after absorption into the tissues (S-methyltransferase activity has not been detected in the intestinal lumen). The S-methyl compounds, because of their low polarity, then become substrates for further detoxication processes which can include, among other phase I processes, displacement of the methylthio-group with glutathione; this results in methylthio-turnover which appears to be a futile process except that the displacement with GSH aids in mobilizing the xenobiotic for eventual excretion (see below).

The mechanism for reductive dechlorination of 2,4-dichlorophenacyl chloride (ref. 9) has been shown to be a glutathione dependent process as outlined here:

$$\text{R-Cl} \xrightarrow{\text{GSH}} \text{R-SG} \xrightarrow{\text{GSH}} \text{RH + GSSG}$$

No extension of this mechanism to include reductive replacement of other functional groups from either aliphatic or aromatic nucleii has been studied, however, reductive defunctionalizations have been observed in the metabolism of pentachloronitrobenzene (PCNB) and hexachlorobenzene (HCB) in rats (ref. 2) and in the metabolism of propachlor in chickens and soil (ref. 10,11). In these examples, all of the intermediates of the methylthio-turnover pathway have served as precursors for reductive defunctionalization in vivo. Whether all the sulfur-containing groups involved in the cycle can be displaced or if they must be first transformed to the GSH conjugate, as in the above example, needs to be determined.

Studies of the metabolism of the methylthio groups in pentachlorothioanisole (PCTA) and bis-methylthiotetrachlorobenzene (bis-MTTCB) exemplify complexities which result from methylthiolation, methylthio-turnover, reductive defunctionalizations and enterohepatic circulation. PCTA and bis-MTTCB along with reductive defunctionalization products (esp. tetrachlorothioanisole, TCTA) are rat metabolites from HCB and PCNB.

The intermediary metabolism shown in Fig. 2 for PCTA was deduced from published data (ref. 2,12,13) and recent metabolism studies with $CH_3S\text{-}^{14}C_6Cl_5$, $^{14}CH_3S\text{-}C_6Cl_5$, $(^{14}CH_3S)_2\text{-}C_6Cl_4$ and $^{13}CH_3{}^{35}S\text{-}^{14}C_6Cl_5$ (ref. 14) done in control, bile-duct-cannulated and germfree rats. These studies showed that both tissue and microfloral C-S lyases were involved in methylthiolation and methylthio-turnover and that some intermediates underwent at least two passes in enterohepatic circulation before excretion. Details of these studies will be reported elsewhere.

On each of the first two passes of PCTA and associated intermediates through the tissues, methylthio-displacement (mostly methylthio-turnover) occured which resulted in excretion of the elements of the methylthio-group as respiratory CO_2, urinary sulfate and urinary methane sulfinate. Studies with $^{13}CH_3{}^{35}S\text{-}^{14}C_6Cl_5$ (Table 1) indicated that the $^{13}CH_3{}^{35}S$-group was removed intact and not stepwise because the decrease in ^{35}S specific activity and the dilution of ^{13}C with ^{12}C in all the fecal metabolites were about equal. Had there been measurable S-demethylase activity, we would have expected methyl-turnover to have occurred and therefore a dilution of the ^{13}C that was greater than the decrease in ^{35}S-specific activity. This conclusion is presented with reservation because ^{13}C dilution could not be measured in the nonextractable residues, the source of which is unknown. First pass methylthio-displacement involved at least 15% of the dose; overall turnover or displacement in control rats involved 50% of the dose. As indicated in both Figs. 1 and 2, the oxidation state of the methylated sulfur is involved in the displacement process, but which oxidation states are actually displaced is not known. PCTA and the sulfoxide gave about equivalent amounts of CO_2 and methane sulfinate; the sulfone gave mostly methane sulfinate.

TABLE 1. Specific activities of ^{35}S and percent dilution of ^{13}C in metabolites and excreta from control and bile-duct cannulated rats dosed with $^{13}CH_3{}^{35}S$-$^{14}C_6Cl_5$

source	specific activity		dilution of ^{13}C
	dpm $^{35}S/\mu m$ C_6	(% decrease)	%
CH_3S-C_6Cl_5			
dose	9.8 X 10^5		-
feces	8.9 X 10^5	(9)	11
$(CH_3S)_2{=}C_6Cl_4$	5.3 X 10^5	(46)	46
CH_3S-C_6Cl_4H	5.1 X 10^5	(48)	49
nonextractable residues (feces)	5.9 X 10^5		-

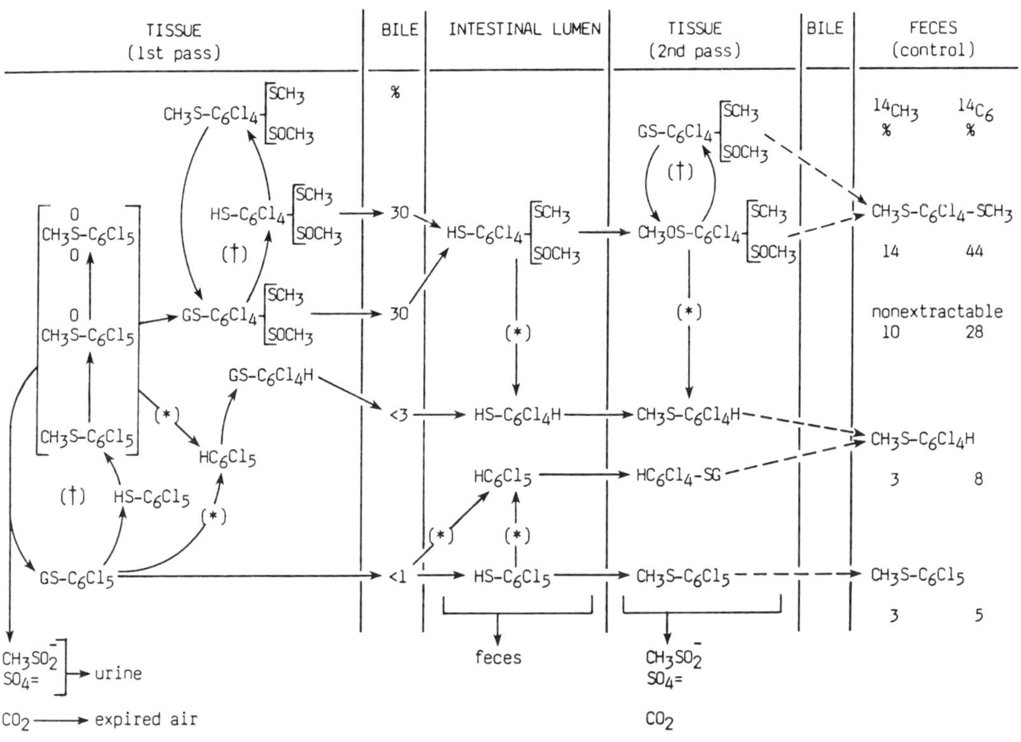

Fig. 2. Intermediary metabolism of pentachlorothioanisole in male rats. (*) Reductive defunctionalization (see text). (†) Methylthio-turnover as defined in Fig. 1. All % are % of dose.

Structures of the first pass biliary metabolites demonstrate the displacement of a chlorine in PCTA by GSH and the function of the other enzymes of the methylthiolation pathway in the tissue. Tissue reductive displacement of the methylthio group by animal tissues was indicated from the detection of small amounts of $^{14}C_6Cl_5H$-MAP metabolites in first-pass bile. In the control rats a portion of these biliary metabolites was apparently excreted with the feces in the forms listed in Fig. 2. The rest of the biliary and intestinal metabolites were reabsorbed from the intestinal lumen for subsequent passes in enterohepatic circulation with tissue and microfloral metabolism until they were excreted with the feces. It is of interest to note that nearly all the biliary metabolites contained an oxidized sulfur but none of the fecal metabolites of PCTA contained oxidized sulfur, therefore, all sulfoxides and sulfones were either reduced or displaced before excretion.

Some of the PCTA (3.6%) passes through the gastrointestinal tract and is excreted with the feces unmetabolized, but a small portion (0.4% of the dose) of the fecal PCTA from control rats has undergone methylthio-turnover. This was discovered when the fecal PCTA from rats dosed with $^{13}CH_3{}^{35}S\text{-}^{14}C_6Cl_5$ was found to be diluted with $CH_3S\text{-}^{14}C_6Cl_5$ (10% dilution) (Table 1). This dilution was not observed in the feces from bile-duct-cannulated rats dosed with triply-labeled PCTA, and therefore the turnover was probably effected by tissue systems. The corresponding C_6Cl_5-MAP metabolites were probably excreted with first-pass bile but were present in quantities too small to be detected.

Studies with triply-labeled PCTA also revealed information about the formation of bis-MTTCB, TCTA and the insoluble residues. The similarities in the specific activities of the ^{35}S in these fecal metabolites (Table 1) led to the conclusion that all three were formed by a common route. The slightly greater decrease in the specific activity of the ^{35}S in TCTA could be due to dilution of the TCTA formed in the route common with bis-MTTCB formation with that formed from GS-C_6Cl_4H and/or GS-C_6Cl_5; these latter metabolites would not contain any ^{35}S. All the reductive defuntionalization pathways given in Fig. 2 have been demonstrated in vivo or in vitro (ref. 2); the incubations with intestinal contents, however, need to be studied further because the incubations were carried out under aerobic conditions (ref. 2).

The reductive defunctionalizations leading to the tetrachlorobenzene formations observed by Renner (ref. 2) could readily fit into the reaction sequences of Fig. 2 by reductive defunctionalization of any of the R-C_6Cl_4H metabolites (R = Cl-, HS-, CH_3S-, GS-). We have not detected any of these in our studies.

In a previous study of the metabolism of ring-labeled PCTA in germfree rats, we concluded that the intestinal microflora must be present for the formation of bis-MTTCB because germfree rats excreted only water soluble fecal metabolites which were mainly the same MAP metabolites that were found in bile (ref. 13). Metabolism studies with methyl-labeled PCTA in germfree rats revealed extensive excretion of $^{14}CO_2$ (16%) and $^{14}CH_3SO_2$ (24%). This indicates turnover of at least 40% of the methylthio-group in the germfree rat and that the intestinal microflora mediate the form in which the biliary metabolites are eventually excreted as fecal metabolites.

The results of the studies outlined above demonstrate how hidden intermediate metabolites were detected using combinations of labeled compounds along with surgically modified and germfree animals. These intermediate metabolites would not have been undetected if only $CH_3\text{-}S\text{-}^{14}C_6Cl_5$ would have been used in these studies.

REFERENCES

1. J.E. Bakke, Xenobiotic Conjugation Chemistry, p. 301-321, American Chemical Society, Washington, DC (1986).
2. G. Renner and P.-T. Nguyen, Xenobiotica 14, 693-704 (1984).
3. C.B. Struble, G.L. Larsen, V.J. Feil and J.E. Bakke, Polynuclear Aromatic Hydrocarbons: Chemistry, Characterization and Carcinogenesis, Battelle Press, Columbus, OH (in press)
4. J.E. Bakke, C. Struble, J.-Å. Gustafsson and B.E. Gustafsson, Proc. Nat. Acad. Sci. U. S., 82, 668-671 (1985).
5. G.L. Lamoureux and D.G. Rusness, Sulfur in Pesticide Metabolism and Function, p. 133, American Chemical Society, Washington, D.C. (1981).
6. D.J. Reed and M.J. Meredith, Drugs and Nutrients, p. 179-224, Marcel Dekker, New York and Basel (1984).
7. R.H. Smith, Reports of the Rowett Inst. 30, 112-131 (1974).
8. J.A. Nash, L.J. King, E.A. Lock and T. Green, Toxicol. Appl. Pharmacol. 73, 124-137 (1984).
9. D.H. Hutson, D.S. Holmes and M.J. Crawford, Chemosphere 5, 79 (1976).
10. J.E. Bakke and G.L. Larsen, Chemosphere 14, 1749-1754 (1985).
11. J-K. Lee, R.D. Minard and J-M. Bollag, J. Korean Agr. Chem. Soc. 20, 44 (1982).
12. J.E. Bakke, Chemosphere 12, 793-798 (1983).
13. J.E. Bakke, P.W. Aschbacher, V.J. Feil, and B.E. Gustafsson, Xenobiotica 11, 173-178 (1981).
14. D.J. Mulford, J.E. Bakke, V.J. Feil and Bergman, Å., NY Abst No. 37 Agrochemicals, 191st ACS National Meeting, New York City, NY (1986).

In vitro systems in the study of mechanisms of pesticide metabolism in plants

Ralph O. Mumma

Pesticide Research Laboratory and Graduate Study Center, Department of Entomology, The Pennsylvania State University, University Park, PA 16802 USA

Abstract - In vitro plant systems have played a major role in furthering our knowledge of pesticide metabolism by plants and have contributed to our understanding of mode of action of agrochemicals. These plant systems are also playing an increasing role by industry in the search for, and the evaluation of, new synthetic chemicals with potential biological activity. In vitro plant systems include isolated plant parts (e.g., excised leaves, roots, and leaf disks), plant cells in culture (both suspension and callus), subcellular fractions and isolated enzymes. These model systems have been used to study the fate of agrochemicals (metabolism), their mode of action and biological activity.

By far the most used in vitro system is plant tissue culture. Most investigators have used suspension cultures, but callus tissues also have many desirable features. Metabolism studies with plant tissue cultures offer a number of advantages over those of intact plants. Metabolites are more easily isolated, purified and identified. The culture conditions can be controlled such that metabolism conditions mimic those of a whole plant or may be used to favor the mass production of certain metabolites. The expensive ^{14}C-labelled compounds can be efficiently used and the metabolism of specific metabolites or reactive intermediates can be studied. They can be used to determine the limit and scope of the active centers (structure activity relationships) of newly synthesized compounds without the complications of absorption and translocation processes. Plant tissue cultures are being used to compare interspecies metabolism with the objective of finding species specific metabolic reactions of target crops or weed species which can be used in the development of new agrochemicals.

Mode of action studies often require special conditions. Efflux and chase experiments, initial reactions, rates of uptake, pool sizes, cellular localization, and some enzymatic activity can more easily be studied with plant tissue cultures than with whole plants.

INTRODUCTION

How have in vitro systems contributed to our knowledge of metabolism and mechanisms of action of pesticides? What is currently being done and what are the future thrusts in this area? These are some of the questions I will try to answer from my own prejudiced perspective. I could summarize all the literature dealing with in vitro systems, but that would be a major effort requiring more space than permitted in this article. Also, recent reviews have adequately covered the subject (refs. 1-5). I will try to be a little more philosophical and critically evaluate accomplishments and predict the future. The agrochemical industries' concerns are different from those of academics and government, but industry is actively pursuing this area. Unfortunately, most of their contributions have not yet reached the public literature. Even so, I will still try to summarize the industrial efforts as I perceive them. I will concentrate on contributions made through the use of plant tissue culture to our knowledge of metabolism and mechanism of action since this is the area I know best. I will briefly summarize what has been done (prior to 1981), what is being done now (in the last five years), and what the future holds. I will point out some of the significant accomplishments with an example and cite some of the problems of this area.

PAST STUDIES

In vitro studies have either used differentiated plant tissue such as excised leaves, roots, and leaf disks (ref. 6), or undifferentiated plant tissue such as plant suspension cultures and callus cultures (refs. 2,5). The latter two techniques have been the most commonly used, but a hybrid of these two was used when treatments were conducted with callus pieces placed in liquid medium under shake conditions (ref. 7). Most studies dealt with identification of the metabolites and elucidating the pathways of degradation of pesticides. Rigorous identification of metabolites was not always achieved. Structures of sugar conjugates were usually not confirmed by spectroscopic techniques. In fact, usually the glycoside fraction was treated with β-emulsin and the fraction which hydrolyzed was considered to be glucosides. The fraction that did not hydrolyze was often not examined further. Other metabolites were often identified by co-chromatography and sometimes only proposed from the available data. The metabolism of only a small number of pesticides had been examined prior to 1981. These studies primarily focused on herbicides, e.g., amides, carbamates, ureas, phenoxy's and halogenated hydrocarbons (insecticides) (refs. 3, 8-14). Most laboratories utilized plant suspension cultures while we tended to use the callus tissue-liquid treatment technique for metabolism studies (refs. 3,7). The former is the method of choice of most laboratories because the growth of cells can be more easily defined (lag, log and stationary phases), and the cells are readily separated from the media and harvested. Callus tissue with localized regions of cell division may be more typical of whole plants where only the growing meristems are actively dividing.

Most early studies used only one or two species of plants in cultures for metabolism studies and they usually did not simultaneously compare the results with whole plant experiments. A great variety of plants were used including the agronomic plants such as corn, soybean, rice, potato, cotton, tobacco, and the not so common plants such as bedstraw, bindweed, carrot, lettuce, parsley, purple cockle, and others. Because of the lack of use of the same species, variations in growth conditions and nonrigid structural proof, it was difficult to make comparisons or draw firm conclusions from many of these early metabolism studies.

PRESENT STUDIES

Experimentation in the last five years has also centered on identification of structures and metabolic pathways. However, these studies have been increasingly concerned with the more important agronomic crops such as soybean, wheat, corn, oats, cotton, tobacco, peanut, and alfalfa. Most studies have involved the use of plant suspension cultures and structures have been determined more rigidly. These results more frequently have been compared with results of whole plants. Some investigators have compared the metabolic capabilties of different plant species and of resistant and susceptible plants (refs. 15-17). Also, there has been considerable interest in determining pool size of metabolites versus age of culture and cellular localization (refs. 18-19). Recent efflux studies, which can only be performed with plant suspension cultures, determined membrane permeability, both cytoplasmic and vacuolar, of amino acid and glycoside conjugates (ref. 18).

In vitro systems offer many advantages for mechanism of action studies. Reaction times of seconds or a few minutes can readily be achieved with plant suspension cultures. Gressel studied membrane permeability of amino acids relative to the influence of an agrochemical (ref. 20). Efflux and chase experiments, which contribute to our understanding of mode of action and metabolisms are more readily conducted with plant suspension cultures (refs. 18,21).

There is considerable interest in studying the enzymes involved in detoxification reactions (Phase I reaction) and conjugation reactions (Phase II reactions) (ref. 1). For example, glutathione *S*-transferase is an enzyme which is involved in detoxification mechanism of plants and animals. It catalyzes the reaction of glutathione with a nucleophilic sink, resulting in a polar metabolite. The enzyme has been chemically and physically characterized, but titers of this enzyme in major agronomic and weed plants remain to be determined. One reason for the current focus on this enzyme is that specific chemicals (antidotes, safeners) can affect the level of this enzyme in plants providing specific plant species resistance to certain herbicides.

In a major achievement, Schmitt et al. (ref. 22) have been able to isolate, from soybean tissue cultures, enzymes that catalyze the formation of the glycoside conjugates and the amino acid conjugates. They have been able to purify and chemically characterize these enzymes. Bound residues are important terminal deposits for many agrochemicals. Studies with plant tissue cultures have contributed much to our knowledge on how residues are copolymerized with lignin and polysaccharides (refs. 23-27). Also, the bioavailability of these bound residues has been studied and these results suggest that in some instances, following enzymatic degradation, the parent pesticide can be again liberated.

The agrochemical industry is actively using in vitro systems, particularly plant tissue cultures, for metabolism and mechanism studies. Although companies do not usually publish their accomplishments in a timely manner, if at all, a number of generalizations can be made. Agrochemical companies are routinely using plant tissue cultures as a complement to metabolism studies with whole plants. They are using tissue cultures to mass produce hard-to-obtain metabolites such as aglycones and polar conjugates. The companies are under increased pressure to identify minor metabolites and in vitro systems offer one method to mass produce these metabolites. They have established plant tissue cultures of the normal agronomic crops and of the major weed species, e.g., velvet leaf, bindweed, Johnson grass, etc. They are using these cultures for bioassays in the search for new bioactive molecules as well as for metabolism studies.

Another area of intense activity for the agrochemical industry is to develop biorational approaches to new biologically active compounds. Companies can no longer afford to synthesize 10,000-20,000 structures before they find a new useful compound. The biorational approach involves the derivation of structures based upon biochemical data and computer-assisted analysis, and plant tissue cultures are being used to complement this approach. Metabolism studies with tissue cultures usually do not depend upon cuticular penetration and translocation. Since many chemicals under development are sensitive in minute amounts and cause inhibition of growth, plant tissue cultures are being used to more closely define the molecular active centers devoid of the absorption and translocation phenomena. Once the active molecular structures have been defined, the molecules can be modified to add functional groups to assist in absorption and translocation and to take advantage of species specific metabolic detoxification reactions. In this manner a biologically active center can be modified to be useful on all major agronomic crops. This whole procedure depends upon knowing the biochemcial capabilities of the target and weed crops.

CITED EXAMPLE

We could select a number of pesticides to illustrate the advances in our knowledge concerning metabolism in plants by utilizing in vitro techniques, but for the sake of this article I will use 2,4-D as a model compound, since this is one area which we have contributed to this knowledge. In 1968 it was recognized from whole plant studies that 2,4-D was metabolized to several polar metabolites (refs. 28-30). Direct conjugation of the carboxyl group gave the glucoside ester and the aspartic acid conjugate. The aromatic ring of 2,4-D was hydroxylated by plants producing two major products, 4-hydroxy-2,3-dichloro- and 4-hydroxy-2,5-dichloro-phenoxyacetic acid (4-OH-2,3,D and 4-OH-2,5-D), which existed primarily as β-glucosides. There also was a report of possible chain elongation products (ref. 31). Ojima and Gamborg were the first to study 2,4-D metabolism with plant suspension cultures and they showed that polar metabolites were produced (ref. 32). Feung et al. isolated the glucoside conjugates of 2,4-D, 4-OH-2,3-D and 4-OH-2,5-D, and the glutamic acid conjugate of 2,4-D from soybean callus cultures (ref. 33). Subsequently, he isolated five additional amino acid conjugates of 2,4-D (alanine, valine, leucine, phenylalanine and tryptophan) from soybean tissue cultures and demonstrated that the concentration of metabolites varied with age of the culture (refs. 7,34). Subsequently, other investigators identified amino acid conjugates of 2,4-dichlorophenoxypropionic acid, 2,4,5-trichlorophenoxyacetic acid, and indoleacetic acid (refs. 35,38) from alfalfa, soybean, geranium, and crown gall callus.

In vitro techniques have made a significant contribution in understanding species variability of metabolism. Monocots seem to form mainly the β-glucoside of 2,4-D (carboxylic ester) or the β-glucoside of ring hydroxylated 2,4-D (refs. 39-42). Rice and Acer pseudoplantanus (a dicot exception) formed almost exclusively the β-glucoside ester of 2,4-D (refs. 40,43). On the other hand, corn metabolized 2,4-D to ring hydroxylated products in addition to the carboxylic glycoside (ref. 39). Dicots metabolized 2,4-D to amino acid conjugates, hydroxylated products, and to the carboxylic glycoside; the relative percentage of metabolites varied greatly with the plant species. Plant varieties also showed considerable metabolism differences (refs. 44,45) as well as resistant and susceptible plants (refs. 15-17). Even soybean tissue cultures derived from cotyledon and leaves exhibited metabolism differences (ref. 44).

In some interesting studies with soybean callus tissue, the age of the tissue and the concentration of 2,4-D influenced the formation of the amino acid conjugates of 2,4-D (ref. 46). Young callus tissue (<3 weeks old) formed only small amounts of amino acid conjugates, but older tissue (6 to 9 weeks old) converted most of the applied 2,4-D to amino acid conjugates and seemed to regulate the amount of free 2,4-D in the tissue at a constant level. Soybean roots in culture exhibited this same phenomenon of regulating the level of 2,4-D in the tissue by converting the excess 2,4-D into amino acid conjugates. Using efflux and chase experiments, Davidonis et al. (ref. 18) developed data suggesting that the glycosides and amino acid conjugates of 2,4-D were located in the plant vacuoles.

At about the same time, Schmitt and Sandermann (ref. 19) isolated plant vacuoles and showed that they contained glycoside conjugates, but not amino acid conjugates. Whether the latter conjugates leaked out of the vacuoles or were not there originally remains yet to be proven. These latter experiments would have been difficult to perform with whole plants and illustrate the advantage of working with in vitro systems.

The amino acid conjugates seem to be a reservoir of potential free 2,4-D. Davidonis et al. (ref. 21) has shown that amino acid conjugates can be enzymatically hydrolyzed to free 2,4-D. In this manner the pool of amino acid conjugates can be a reservoir and regulate the amount of free 2,4-D in the cells. Amino acid conjugates are not dead end products but are biologically active; they stimulate cell division and growth. With Acer pseudoplantanus cells in culture Leguay and Guern (ref. 43) clearly demonstrated that the carboxylic glycoside of 2,4-D is not available as a pool of potential free 2,4-D. However, studies with diclofop-methyl in oat suspension cultures indicated there was some inter-conversion of the carboxylic glycoside metabolite to free diclofop (ref. 17,47-49). These data suggest that in some plants the glycoside fraction may be a complete dead end product, not biologically active, while in other species it may be a potential source of the biologically active compound.

Recently Schmitt et al. (refs. 22,50) isolated the O-malonyl-β-D-glucosyl conjugates from soybean and wheat. The result is particularly interesting since early workers did not, in general, look for the malonyl derivatives of the glucoside metabolites of 2,4-D (unstable) and the published literature will have to be reconsidered with the possibility of the malonyl derivatives being present. Schmitt et al. (ref. 22) have also isolated, purified, and characterized the O-glucosyl- and O-malonyltransferases of soybean and wheat cells in culture. They also have partially purified an enzyme which catalyzes the formation of amino acid conjugates of 2,4-D from the coenzyme A thioester of 2,4-D. The isolation and characterization of these enzymes, obtained with in vitro techniques, will provide the basis for understanding the glycoside and amino acid conjugate metabolism pathways of plants.

We have come a long way in our understanding of the metabolism of 2,4-D in the last 18 years. Most of these data have been generated by in vitro techniques, primarily with plant tissue cultures, and these results will have wide meaning for all plant metabolism studies. Unfortunately, we still do not understand how a molecule, such as 2,4-D, actually exerts its biological activity. The chain elimination pathway is minor in the in vitro studies since good recovery of ^{14}C-label was usually obtained. Perhaps the chain elimination pathway is important in some species of plants not yet studied in tissue culture. Since plants can metabolize 2,4-D in four major directions: 1) to the carboxylic glycoside, 2) to the amino acid conjugates, 3) to the hydroxylated derivatives, and 4) chain elimination, there seems to be opportunities for species specific metabolic degradation of 2,4-D.

FUTURE THRUSTS

Future uses of in vitro techniques in metabolism and mechanism studies are only speculative and my prediction of what will happen is influenced by my prejudicial perspective. The academic and governmental laboratories will probably decrease their number of metabolism studies as industry increases its efforts to publish its metabolic fate studies. The public is becoming more concerned about the environmental fate of agrochemicals and some industries are responding by increasingly making their environmental fate data public. Still, metabolism data for many pesticides have not been examined in in vitro systems. These chemicals will be targets for future studies. More emphasis will be placed on mechanisms of detoxification, the enzymes involved, their species specific titer, titer changes with age of the plant, titer changes as influenced by other chemicals, the enzymes' subcellular localization, and their physical and chemical properties. The amino acid sequence of the enzymes will be determined and this will permit the localization of the genes responsible for these enzymes. Biotechnology techniques can then be utilized to enhance the expression of the gene or transfer the gene responsible for detoxification to important sensitive economic crops.

A greater emphasis will be placed on comparing the detoxification capabilities of the major agronomic crops and troublesome weed plants. Resistant and susceptible plant lines will be compared to further our understanding of the metabolic basis of resistance. These studies will be important for developing the background data for critically evaluating the risk assessment of biotechnology-derived new plant strains proposed for introduction as an agronomic crop.

Industry will continue to use in vitro systems as a complement to whole plant systems because it saves them time and ultimately money. The agrochemical industry operates on tight schedules, and if they can bring a product to the market faster, they will have financial benefits. With plant tissue cultures, metabolites often can be more easily isolated and identified. By choosing the proper conditions they can mimic conditions of the whole plant or they can choose conditions that favor the mass production of specific metabolites. With the demands for more thorough understanding of the fate of chemicals there will be greater demands for the identification of minor metabolites. Plant tissue cultures will play a major role in the production of these minor metabolites.

As mentioned before, the agrochemical industry is using *in vitro* techniques to support their biorational approach to find new biologically active chemicals. I envision this use of *in vitro* systems to increase dramatically in the future, but in general, to remain unknown to the public.

In vitro systems offer many advantages over whole plants. The metabolites can be more easily isolated and identified because of the lack of pigments, starches and other secondary substances. Only a small amount of radiolabeled compound needs to be used with *in vitro* systems. This is particularly important with academic research groups where cost is critical. Culture conditions can be chosen to mimic that of the whole plant or conditions can be selected which favor the production of certain desired metabolites. A series of plant species, such as weeds or resistant cultivars, can readily be screened for their metabolic capability. Reactive metabolite intermediates possessing stabilities of very short times (seconds or minutes) can be followed and isolated with plant tissue cultures. Reaction mechanisms can be easily studied with cell suspension cultures, especially short-term reactions. Chase and efflux experiments are readily performed with cells in culture where they would be difficult to conduct meaningfully with whole plants.

Although it may seem that *in vitro* systems are a panacea for all occasions, this is incorrect. There are still many problems, apparent failures, and unexplained results. Each research group seems to use its own system of culture and culture type. It is time we standardize methods and species types so that extrapolation of data would be more reliable. Since a high degree of variation of metabolism has been shown between plant species, we need to know more concerning the metabolism of non-targeted plants. At times there has not been good correlation of results with those from whole plants. Whether this difference is dependent upon culture conditions, my personal belief, remains yet to be proven. As new biotechnologically derived plant species are introduced, their metabolic capabilities need to be thoroughly characterized for risk-assessment purposes. Standardized procedures for *in vitro* systems should be established by our regulatory agencies so that metabolic capabilities of varieties, species and new biotypes can be properly compared. Several German research groups, in a collaborative effort, have proposed standard guidelines for test procedures to evaluate metabolism and degradation of chemicals by plant cell cultures (refs. 51-53).

ACKNOWLEDGMENTS

Supported in part by Northeastern Regional Research Project NE-115. Paper No. 7461 in the Journal Series of the Pennsylvania Agricultural Experiment Station.

REFERENCES

1. G. L. Lamoureux and D. G. Rusness, Xenobiotic Conjugation Chemistry, p. 62, Am. Chem. Soc. Sym. Series No. 299 (1986).
2. R. O. Mumma and G. H. Davidonis, Progress in Pesticide Biochemistry and Toxicology, p. 255, John Wiley and Sons Ltd. (1983).
3. R. O. Mumma and R. H. Hamilton, Xenobiotic Metabolism: In Vitro Methods, p. 35, Am. Chem. Soc. Sym. Series No. 97 (1979).
4. R. O. Mumma and R. H. Hamilton, J. Toxicol.-Clin. Toxicol., 19(6&7), 535-555 (1982-83).
5. H. Sandermann, H. Diesperger and D. Scheel, Plant Tissue Culture and Its Bio-technological Application, p. 178, Springer-Verlag, Berlin, Heidelberg, New York (1977).
6. R. H. Shimabukuro and W. C. Walsh, Xenobiotic Metabolism: In Vitro Methods, p. 1, Am. Chem. Soc. Sym. Ser. No. 97 (1979).
7. C. S. Feung, R. H. Hamilton, F. H. Whitham and R. O. Mumma, Plant Physiol. 50, 80-86 (1972).
8. D. G. Davis, R. H. Hodgson, K. E. Dusbabek and B. L. Hoffer, Physiol. Plant. 44, 87-91 (1978).
9. D. S. Frear and H. R. Swanson, Pest. Biochem. Physiol. 5, 73-80 (1975).
10. T. B. Ray and C. C. Still, Pest. Biochem. Physiol. 5, 171-177 (1975).
11. R. K. Locke, J. T. Chen, J. N. Damico, L. R. Dusold and J. A. Sphon, Arch. of Environ. Contam. and Toxicol. 4, 60-99 (1976).

12. J. Stockigt and B. Ries, Plant Tissue Culture and Its Biotechnological Applications, p. 204, Springer-Verlag, Berlin, Heidelberg, New York (1977).
13. K. R. Brain and D. S. Lines, Plant Tissue Culture and Its Biotechnological Applications, p. 197, Springer-Verlag, Berlin, Heidelberg, New York (1977).
14. D. Scheel and H. Sandermann, Planta 133, 315-320 (1977).
15. G. H. Davidonis, R. H. Hamilton and R. O. Mumma, Plant Physiol. 70, 104-107 (1982).
16. T. H. Oswald, A. E. Smith and D. V. Phillips, Pestic. Biochem. Physiol. 8, 73-78 (1978).
17. R. H. Shimabukuro, W. C. Walsh and R. A. Hoerauf, J. Agric. Food Chem. 27, 615-623 (1979).
18. G. H. Davidonis, R. H. Hamilton and R. O. Mumma, Plant Physiol. 70, 939-942 (1982).
19. R. Schmitt and H. Sandermann, Jr., Z. Naturforsch. 37, 772-778 (1982).
20. S. Zilkah and J. Gressel, Planta 145, 272-278 (1979).
21. G. H. Davidonis, R. H. Hamilton and R. O. Mumma, Plant Physiol. 66, 537-540 (1980).
22. R. Schmitt, R. Winkler, B. Santo, T. Bauknecht and H. Sandermann, Jr., Sixth International Congress of Pesticide Chemistry, poster.
23. D. Scheel and H. Sandermann, Jr., Planta 152, 253-258 (1981).
24. H. Sandermann, Jr., D. Scheel and Th. v.d. Trenck, J. Appl. Poly Sci. 37, 407-420 (1983).
25. C. Langebartels and H. Harms, Ecotox. Env. Saf. 10, 268-279 (1985).
26. D. Scheel, W. Schafer and H. Sandermann, Jr., J. Agric. Food Chem. 32, 1237-1241 (1984).
27. M. Arjmand and H. Sandermann, Jr., Verlag der Z. Naturforsch. 41c, 206-214 (1986).
28. E. W. Thomas, B. C. Loughman and R. G. Powell, Nature (London) 204, 286 (1964).
29. H. D. Klambt, Planta 57 , 339-353 (1961).
30. R. H. Hamilton, J. Hurter, J. K. Hall and C. D. Ercegovich, J. Agric. Food Chem. 19, 480-483 (1971).
31. D. L. Linscott, R. D. Hagin and J. E. Dawson, J. Agric. Food Chem. 16, 844-848 (1968).
32. K. Ojima and O. L. Gamborg, Proc. Sixth Int. Conf. Plant Growth Subst. P.-857 (1968)
33. C. S. Feung, R. H. Hamilton and F. H. Witham, J. Agric. Food Chem. 19, 475-479 (1971).
34. C. S. Feung, R. H. Hamilton and R. O. Mumma, J. Agric. Food Chem. 21, 637-640 (1973).
35. A. E. Smith, Weed Sci. 27, 392-396 (1979).
36. M. Arjmand, R. H. Hamilton and R. O. Mumma, J. Agric. Food Chem. 26, 1125-1128 (1978).
37. C. S. Feung, R. H. Hamilton and R. O. Mumma, Plant Physiol. 58, 666-669 (1976).
38. R. F. Troxler and R. H. Hamilton, Plant Physiol. 40, 400-440 (1965).
39. C. S. Feung, R. H. Hamilton and R. O. Mumma, J. Agric. Food Chem. 23, 373-376 (1975).
40. C. S. Feung, R. H. Hamilton and R. O. Mumma, J. Agric. Food Chem. 24, 1013-1017 (1976).
41. D. W. Bristol, A. M. Ghanuni and A. E. Oleson. J. Agric. Food Chem. 25, 1308-1314 (1977).
42. C. S. Feung, S. L. Loerch, R. H. Hamilton and R. O. Mumma, J. Agric. Food Chem. 26, 1064-1067 (1978).
43. J.-J. Lequay and J. Guern, Plant Physiol. 60, 265-270 (1977).
44. G. H. Davidonis, R. H. Hamilton and R. O. Mumma, Plant Physiol. 65, 94-97 (1980).
45. M. J. Montague, R. K. Enns, N. R. Siegel and E. G. Jaworski, Plant Physiol. 67, 603-607 (1981).
46. G. H. Davidonis, R. H. Hamilton and R. O. Mumma, Plant Physiol. 62, 80-83 (1978).
47. W. W. Donald and R. H. Shimabukuro, Physiol. Plant. 49, 459-464 (1980).
48. J. A. Dusky, D. G. Davis and R. H. Shimabukuro, Physiol. Plant. 49, 151-156 (1980).
49. J. A. Dusky, D. G. Davis and R. H. Shimabukuro, Physiol. Plant. 54, 490-494 (1982).
50. R. Schmitt, J. Kaul, Th. v.d. Trenck, E. Schaller and H. Sandermann, Jr., Pest. Biochem. Physiol. 24, 77-85 (1985).
51. I. Schuphan, A. Haque and W. Ebing, Chemosphere 13, 301-313 (1984).
52. A. Haque, W. Ebing and I. Schuphan, Chemosphere 13, 315-320 (1984)
53. W. Ebing, A. Haque, I Schuphan, H. Harms, C. Langebartels, D. Scheel, K. T. von der Trenck and H. Sandermann, Jr., Chemosphere 13, 947-957 (1984).

In vitro systems in the study of mechanisms of pesticide metabolism in animals

T. Nakatsugawa* and J.M. Becker

State University of New York
College of Environmental Science and Forestry, Syracuse, NY 13210, U.S.A.

Abstract - Although biotransformation is important as a determinant of toxicities of pesticides in live animals, many of its key products are unstable, transient and difficult to capture and range widely in toxicological properties. Thus, many metabolic steps were initially suspected in in vivo experiments, but their proof had to depend upon studies of each step in isolation. Others could only be shown in vitro. Characterization of enzymes also necessitated demonstration of cofactors, ionic requirements and other necessary conditions in vitro using cell-free systems. Finally the best-defined in vitro metabolism systems, i.e., purified enzymes have furnished detailed characterization of each enzyme and, in the case of multi-enzyme complexes, a reconstituted system has been a powerful tool to investigate molecular mechanisms involved as well as substrate specificity.

Recent interest in isolated organs and cells as model systems reflects the need to synthesize the qualitative characterization of enzymatic steps into quantitative interpretation of phenomena occurring in vivo. One of the advantages of in vitro systems, especially cell-free preparations, is that unnatural conditions can be used without the constraints of "live" systems. Precisely because of this, however, extrapolation of results obtained in model systems to metabolism in vivo requires clear understanding of the differences between the in vitro models and live animals. Mechanisms of penetration, distribution and uptake that influence biotransformation of both parent chemicals and metabolites are inseparable from the studies of metabolism in isolated organs, tissues and cells and in cell-free preparations, but are often ignored. Optimum conditions determined in vitro may not always hold in vivo. Availability of cofactors, oxygen or lack thereof, pH, binding to proteins or lipids, blood supply and the residence time of the substrate in the cell are among the factors that must be addressed to assess the relative contributions of various enzymatic metabolic mechanisms when they work together in a whole organism. Presence of isozymes and their heterogeneous distribution will further complicate the task of understanding the metabolic mechanisms in the integrated system. The value as well as shortcomings and difficulties of various in vitro systems are discussed using examples from past studies, and future research needs are considered.

AN OVERVIEW

The fact that only a very small fraction of a given dose of most toxicants reaches the target macromolecules reflects the importance of metabolic activities that modify toxicant molecules. Although the term detoxication reveals earlier emphasis of the overall effect of metabolic changes leading to elimination of toxicants, the more general term biotransformation better reflects the unpredictable nature of the toxicological consequences of metabolism. Aside from the classical "lethal synthesis" of fluoroacetate, earlier cases of bioactivation were almost entirely found among microsomal oxidation products (ref. 1). It has turned out, however, that the unpredictability is not the exclusive property of hydroxylation systems. Even the conjugation enzymes, the classical phase II enzymes to facilitate excretion by conferring the ultimate water-solubility, are now known to produce reactive toxicants, e.g., sulfate conjugation of the acetylaminofluorene metabolite (ref. 2) and glutathione conjugates of bromobenzene metabolites (ref. 3). While the tendency of such system to facilitate excretion is still true, intermediate metabolites are often reactive, more toxic than parent compounds, transient and difficult to capture. Thus, we owe much of our understanding of metabolic steps to the use of experimental model systems that permitted characterization of the biochemical machinery in isolation from rapid alteration of products that occurs in vivo.

Cell-free systems such as tissue homogenates or a subcellular fraction like microsomes and cytosol fraction are the only means by which one can define individual metabolic steps and define cofactor requirement of enzymes involved. For instance, mercapturic acids were recognized very early in the study of biotransformation (ref. 4), but the formation of initial products, i.e., glutathione conjugates were discovered and characterized only in in vitro studies (ref. 5). The recognition of the omnipotence of mixed-function oxidase system in the metabolism of the majority of insecticides, while benefited from the beacon offered by in vivo data (ref. 6), came after extensive characterization of metabolism in vitro (ref. 7). Bioactivation of phophorothioate insecticides to their phosphate analogs was originally suspected in a study in vivo (see ref. 8), but definitive characterization of the metabolites and the enzyme system involved required in vitro systems (see ref. 7). Generation of reactive sulfur species in the same reaction was also originally suspected and demonstrated in in vitro experiments (ref. 9).

While the history amply illustrates the utility of model systems in the study of pesticide metabolism in animals, it is important to bear in mind that these systems are means to understand metabolic events occurring in vivo, and not the end in themselves. There is definitely a need to synthesize information obtained through these efforts using in vitro system. To this goal, the use of perfused livers either in situ or in isolation has become popular. Other organs such as the lung and kidney have also been studied in perfusion systems or tissue fragments (ref. 10, 11) as a means of filling the gap between the cell-free system and the live animals. It is important, however, that even these "model systems" are double-edged swords and careful design and interpretation of results are essential.

ADVANTAGES AND DISADVANTAGES OF VARIOUS MODEL SYSTEMS

The perfused organ system preserves structural integrity and spacial relationship between various cell types. Xenobiotic uptake and exchange, supply of oxygen, nutrients and hormones as well as neural influence can be controlled in a mechanical setting similar to that in vivo. With each organ preparation, however, one or at most a few controlled experiments can be performed because the organ may not be "reusable" for a particular experiment. Viability of such preparations normally holds for a few hours at best and limits the utility of this model system. With large animals, the cost of animals as well as reagents may be prohibitive.

Isolated cell or tissue slice systems are therefore a model of choice where organ models cannot fulfill the goal. A variety of conditions can be examined in detail with an abundance of cellular material. Cellular uptake and excretion can be readily quantitized. Viability may sometimes be maintained over several days or more, allowing studies of long-term modulation of biotransformation enzymes, e.g. enzyme induction. Perhaps the most popular recent approach is the use of isolated hepatocytes which has become practical since the critical improvement of the collagenase perfusion procedure by Berry and Friend (ref. 12). Systematic studies and improvement of the procedures have been made by a number of investigators as reviewed by Seglen (ref. 13). Thus, isolated hepatocytes have become a useful tool in the study of biotransformation of xenobiotics with the hope of bridging the gap between the cell-free systems and in vivo conditions. Perifusion (ref. 14, 15), which allows continuous exchange and sampling of media for suspended cell systems may offer a useful modification. Fragility of isolated cells, however, require continual attention in maintaining their conditions as close to in vivo as possible.

Homogenates and subcellular fractions are often the first step towards enzyme characterization. They are the most convenient means of establishing enzyme cofactors, oxygen and cation requirements, subcellular localization, and enzyme-substrate interaction. In these systems, rather unnatural conditions can be employed to define enzyme parameters. For instance, experiments may be run under nitrogen atmosphere to determine oxygen requirement. Caution must be applied, however, in interpreting data from these systems. For instance, DDD has been found in the livers of rats fed DDT (ref. 16). Several investigators reported reductive dechlorination in vitro by a cytochrome P-450-linked system, by FAD-requiring enzyme or non-enzymatic systems. As discussed by Esaac and Matsumura (ref. 17), the in vitro systems require anaerobic conditions and therein lies the root of controversy. Reductive conversion of DDT to DDD (TDE) is the critical step for biodegradation of DDT (ref. 17). A crucial question pertains to the in vivo significance of these systems (ref. 18). Since the liver, a highly vascular aerobic organ, rapidly turns anaerobic upon ischemia, DDD found in the DDT-fed rat may represent post-mortem enzyme action. A flavoprotein-linked reductive system (ref. 17) may also have its role in vivo, but further investigation is needed to understand the significance of this and other reductive metabolism in animals.

Reconstituted purified enzyme systems are the ultimate tool in characterization of enzymes, their subgroups and individual enzymological parameters to allow detailed interpretation of modulation of biotransformation in vivo. One of the most studied enzyme complexes is the mixed-function oxidases involving multiple molecular species of cytochrome P-450. The elucidation and characterization of these enzymes well illustrate the utility of this model system. Guengerich (ref. 19) purified several cytochromes P-450 from rat and rabbit liver, reconstituted the enzyme system with highly purified preparations of NADPH-cytochrome P-450 reductase, synthetic phospholipid and deoxycholate and then compared turnover numbers for various metabolites of environmentally important toxicants including parathion. Varied ratios of parathion metabolites produced by different cytochromes P-450 partially answer the question of whether activation to paraoxon and oxidative hydrolysis to diethylphosphorothioic acid are catalyzed by the same or different enzyme species. In view of the purity of the cytochromes used, each molecular species seems to catalyze both pathways in different proportions. Such studies open further question as to how each cytochrome P-450 is regulated in vivo. For example, DDE is known to induce paraoxon deethylase of the rat over 15 fold while the total cytochrome P-450 level is increased only 2 to 3 fold (ref. 20). Even more dramatically, chlorfenvinphos deethylase is reported to be increased by a factor of 600 by dieldrin feeding in the rat (ref. 21). It should be kept in mind, however, the process of enzyme purification may not equally preserve all isozymes. As pointed out by Kurihara et al. (ref. 22), preparation of cell-free fractions may lead to differential loss of cytochrome P-450 species to give potentially artifactual results.

PROBLEMS AND FUTURE RESEARCH NEEDS

Often it is assumed that preservation of structural organization constitute a better model for in vivo conditions. This cannot, however, be taken as prima facie evidence. It is too easy to ignore or forget that each step of simplification from live animals transfers the burden of defining the missing conditions on the investigator. Perfusion systems certainly allow study of a particular organ in isolation. The use of artificial perfusates, however, involves a number of critical assumptions that are too often ignored. Isolated cell systems potentially suffer from similar problems. Cell to cell communication is also lost in suspension cultures and the relative location of cells within the organ no longer exists to impart a characteristic role of each cell. In broken cell preparations, while a number of "natural" components may be present, other conditions are far from those in vivo. Purified enzyme systems cannot by themselves furnish information regarding levels of the substrate, cofactors, cations, pH and oxygen that may prevail in vivo.

In the study of xenobiotic biotransformation, one of the most critical and least investigated problems that pervade modeling efforts in general is the composition of the media that surround enzymes in each system. Generally high lipophilicity of substrates is a unique feature of metabolic studies of pesticide and related toxicants of environmental concern. These chemicals cannot stay in any appreciable concentration in water or salt solutions and their fugacity drives them to bind to hydrophobic sites. Yet, nearly always homogenization and suspension media for cell-free systems are salt or sucrose solutions with little else in the media to facilitate dissolution of xenobiotic substrates. Because of this, many substrates in in vitro model systems are often introduced as emulsions or in organic solvents such as alcohol, acetone, dimethylsulfoxide or acetonitrile. While the small amount of these materials does not generally affect enzyme or cellular functions, the influence of solvent as well as protein components on xenobiotic solubilities needs to be carefully evaluated. Moss and Hathway pointed out long ago (ref. 23) that rabbit serum increases solubilities of dieldrin and Telodrin 4000 times compared with water solubility. Lewis et al (ref. 24) noted also that essentially all the aldrin substrate is bound to microsomes so that the epoxidation rate is a function of the ratio of the amount of substrate and microsomal protein and not of the overall substrate concentration. These early findings seem to have received much less attention than they deserve. If, for instance, the K_m or apparent K_m is to be determined in a purified enzyme system, subcellular fraction or an isolated cell system, the way the substrate is made available is a decisive factor. Importance of protein concentration in the partitioning of xenobiotics in suspended isolated hepatocytes has been noted (ref. 25).

If one is to simulate conditions in vivo, one needs to look carefully at the way xenobiotics are carried in perfusion systems as well. The choice of perfusion media is often dictated by the need to maintain the viability of the organ while the influence of media on xenobiotic partitioning is relegated to secondary considerations. Binding of xenobiotics in the blood, however, greatly influences xenobiotic disposition, and this issue has just begun to be seriously investigated. In rabbit lung perfusion, for instance, uptake of an anorexigenic drug chlorphentermine was greater from an artificial perfusate containing 4.5% bovine serum albumin than from whole autologous blood (ref. 26). As Wiersma et al. (ref. 27) stated, many xenobiotics are bound to albumins, lipoproteins and blood cells. They have demonstrated that clearance of benzo(a)pyrene by perfused rat liver is greatly influenced by the protein composition of perfusate.

Finally, partitioning characteristics of xenobiotic substrates may have serious consequence on their uptake in the cells of the perfused organ. In the liver, properties of xenobiotics such as the partitioning between blood and hepatocytes may influence hepatic uptake pattern within the lobule. Absorption may be relatively uniform throughout the lobule or may be centrilobularly or periportally skewed. In the latter case, the chemical may migrate in a chromatographic fashion repeating uptake and release along the sinusoid (ref. 28, 29). Such periportally absorbed xenobiotics may be totally or primarily metabolized by periportal hepatocytes so that metabolic activities expressed as mole/mg liver protein/min, for instance, may not compare fairly with data obtained with whole liver preparations such as usual isolated hepatocytes and cell-free systems. Significant influence of bovine serum albumin on the uptake and metabolism of parathion in the perfused mouse liver has been observed (ref. 30). Much remains to be learned in these areas of investigation.

REFERENCES

1. P. A. Dahm and T. Nakatsugawa, Enzymatic Oxidations of Toxicants, 89-112 (1968).
2. J. A. Miller and E. C. Miller, In Biological Reactive Intermediates, Formation, Toxicity, and Inactivation, D. J. Jollow, J. J. Kocsis, R. Snyder and H. Vainio, Eds., Plenum Press, New York, 6-24 (1977).
3. T. J. Monks, S. S. Lau, R. J. Highet and J. R. Gillette, Drug Metab. Dispos. 13, 553-559 (1985).
4. E. Boyland and L. F. Chasseaud, Adv. Enzymol. 32, 173-219 (1969).
5. H. G. Bray, T. J. Franklin and S. P. James, Biochem. J. 71, 690-696 (1959).
6. Y. P. Sun and E. R. Johnson, J. Agric. Food Chem. 8, 261-266 (1960).
7. T. Nakatsugawa and M. A. Morelli, In Insecticide Biochemistry and Physiology, C. F. Wilkinson, Ed., Plenum Press, New York, 61-114 (1976).
8. A. N. Davison, Biochem. J. 61, 203-209 (1955).
9. T. Nakatsugawa and P. A. Dahm, Biochem. Pharmacol. 16, 25-38 (1967).
10. H. M. Mehendale, L. S. Angevine and Y. Ohmiya, Toxicology 21, 1-36 (1981).
11. J. M. Tager, H. D. Söling and J. R. Williamson, Editors, Use of Isolated Liver Cells and Kidney Tubules in Metabolic Studies, North-Holland, Amsterdam, (1976).
12. M. N. Berry and D. S. Friend, J. Cell Biol. 43, 506-520 (1969).
13. P. O. Seglen, Methods Cell Biol. 13, 29-83 (1976).
14. R. van der Meer, E. J. de Haan and J. M. Tager, In Use of Isolated Liver Cells and Kidney Tubules in Metabolic Studies, J. M. Tager, H. D. Söling and J. R. Williamson, Eds., 159-164, North-Holland, Amsterdam, (1976).
15. P. Walter, F. Nyfeler, P. Fasel, G. von Glutz, U. K. Moser and M. R. de Sagarra, In Use of Isolated Liver Cells and Kidney Tubules in Metabolic Studies, J. M. Tager, H. D. Söling and J. R. Williamson, Eds., 167-174, North-Holland, Amsterdam, (1976).
16. A. K. Klein, E. P. Laug, P. R. Datta, J. O. Watts and J. T. Chen, J. Assoc. Off. Anal. Chem. 47, 1129-1145 (1964).
17. E. G. Esaac and F. Matsumura, Bull. Environ. Contam. Toxicol. 19, 15-22 (1978).
18. E. G. Esaac and F. Matsumura, Pestic. Biochem. Physiol. 13, 81-93 (1980).
19. F. P. Guengerich, J. Biol. Chem. 252, 3970-3979 (1977).
20. H. T. Appleton and T. Nakatsugawa, Pestic. Biochem. Physiol. 7, 451-465 (1977).
21. D. H. Hutson and A. S. Wright, Chem.-Biol. Interact. 31, 93-101 (1980).
22. N. Kurihara, N. Hori and R. Ichinose, Pestic. Biochem. Physiol. 21, 63-73 (1984).
23. J. A. Moss and D. E. Hathway, Biochem. J. 91, 384-393 (1964).
24. S. E. Lewis, C. F. Wilkinson and J. W. Ray, Biochem. Pharmacol. 16, 1195-1210 (1967).
25. T. Nakatsugawa and S. Tsuda, in IUPAC Pesticide Chemistry, Human Welfare and the Environment, J. Miyamoto, Ed. Vol. 3, 395-400, Pergamon Press, New York, (1983).
26. L. S. Angevine and H. M. Mehendale, Toxicol. Appl. Pharmacol. 52, 336-346 (1980).
27. D. A. Wiersma, P. M. Stemmer and R. A. Roth, Biochem. Pharmacol. 33, 3433-3438 (1984).
28. T. Nakatsugawa, W. L. Bradford and K. Usui, Pestic. Biochem. Physiol. 14, 13-25 (1980).
29. T. Nakatsugawa, S. Tsuda and W. K. Sherman, In IUPAC Pesticide Chemistry, Human Welfare and the Environment, J. Miyamoto, Ed. Vol. 3, 469-474, Pergamon Press, New York, (1983).
30. L. G. Sultatos and L. D. Minor, Drug Metab. Dispos. 14, 214-220 (1986).

Pesticide toxicology and human health

The complexities of interpreting carcinogenicity data

Robert A. Squire

Division of Comparative Medicine, Johns Hopkins School of Medicine

Baltimore, Maryland 21205

INTRODUCTION

Early this Summer, a symposium was held in Chapel Hill, N.C. entitled "Interdisciplinary Discussion Group on Carcinogenicity Studies". Most of the experts in the carcinogenesis testing field were there. If anyone had doubted that there are complexities and, in fact, uncertainties involved in the interpretation of animal cancer tests - they would have been amazed to have listened to the discourse and debate among the most knowledgeable biologists and statisticians in the field. In fact, a lay observer might have drawn the conclusion that animal tests are so complex and controversial, that they serve no useful purpose other than to provoke debate, and that would have been an erroneous conclusion.

It is true that animal studies are often misinterpreted, and there are many problems involved in their execution and evaluation. Nonetheless, they remain the most practical means generally available to identify potential human carcinogens. Human epidemiological studies are equally complex and controversial, and since levels of chemical exposures are often so difficult to define and latent periods so long - they are considered relatively insensitive. And earlier expectations that *in vitro* short-term tests would provide a rapid and less complex means of identifying carcinogens have not been fulfilled. In fact, the overall correlations between mutagens and carcinogens have not been much better than one could attribute to mere chance (ref. 1). It may be that further attention towards short-term *in vivo* tests will provide more accurate predictors of carcinogenicity in the future.

Unfortunately, since we do not yet know the exact mechanisms of cancer development, we cannot test for the mechanistic process itself. Current efforts directed towards the process of oncogene activation may soon change this. However, it is very possible that even when we do understand carcinogenic mechanisms, we still may not be able to simulate the process *in vitro* to the extent of quantifying risk. The present use of animal surrogates occurs amid debates over interspecies and other unproven assumptions necessary to extrapolate to humans. Imagine the potential uncertainties in quantifying human risk from a petri dish!

The evidence of carcinogenicity derived from animal tests has, in the past, yielded only 2 possible outcomes: negative or positive. More recently it has been properly recognized that many different levels of evidence can result from an animal experiment. Consequently, we see government agencies and other classifying compounds as showing "limited", "suggestive", "sufficient" or "clear" evidence of carcinogenicity. Such terms as *limited* or *suggestive* or even *equivocal* often cause frustration among regulators and other interested parties, because they may cast a shadow, but they do not provide a clear answer.

Ultimately, of course, of greater importance than whether a compound is an animal carcinogen is whether it is a potential human carcinogen. In other words, are the animal data in a given case relevant to humans? Again, government agencies and others have classified the evidence concerning potential human carcinogenicity into different categories such as "positive", "probable" or "possible". As presently applied, however, these are over-simplifications, mainly because the frequent use of the term "probable" does not consider differences in species susceptibility, or the conditions or levels of compound exposure. A chemical may be a probable human carcinogen if exposure is long-term and very high. But at actual exposure levels, there may be only a small possibility of cancer development. The fatalistic disregard among the lay community that can be created by routine overstatement of risk may be as detrimental to the public health as routinely understating risk. In fact, I believe a degree of such disregard has already developed.

There are several factors involved in, first, assessing the evidence for carcinogenicity in animals and, secondly, determining whether a positive (or negative) finding is relevant to humans. Today, I would like to discuss a few of the more important of these factors. They are:

1) Interpretation of pathological findings in animals.
2) Statistical analysis of animal test data.
3) Comparative metabolism and pharmacokinetics.
4) Relative tumor susceptibility in the test species.
5) Possible carcinogenic mechanisms.

I. INTERPRETATION OF PATHOLOGICAL FINDINGS IN ANIMALS

Cancer development in most animal and human tissues studied to date is a multi-stage process beginning with cell proliferation and progression, over time, through stages of hyperplasia, benign and then malignant tumors. However, all except the late stages of the process are reversible if the carcinogenic stimulus is removed. And there are no specific pathological markers indicating the point at which autonomous, irreversible growth is reached - i.e. the stage at which a lesion becomes "cancer". Pathologists primarily rely upon the collective, historical experience which demonstrates general correlations between morphological features of lesions and their biological or clinical behaviour and there is little disagreement over the late stages of cancer development, one the morphological features of severe cellular atypia or, especially, the biological properties of local invasion or distant metastases are present. However, there are often "gray" areas of uncertainty between the early stages of hyperplasia and benign tumors, or between benign and malignant tumors. And, of course, in animal testing, the somewhat arbitrary classification or terminology applied to such controversial stages may determine whether or not a chemical is considered a carcinogen or a non-carcinogen. The reported tumor incidences generated by an animal study are often interpreted as if they are hard data. In fact, they may be quite soft. There are several instances where only the early stages of tumor development are induced in animals - even following very high, toxic doses of a chemical. The "infamous" liver foci and nodules in mice and rats are examples of such early lesions. Is a chemical that induces only early changes a carcinogen? This is both a scientific and a legal question, and there may be little objective basis for answering it. Instead, the opinion and classification of one pathologist may determine the outcome. Of course this process is not unique to animal testing; reliance on medical judgement is the common practice in most medical disciplines and is the basis for second and third opinions in clinical areas. Similarly, in the case of controversial pathological and end-points, it is important that additional opinions be sought. This may require a panel of pathologists who examine the tissues "blind", i.e. without knowledge of the experimental group. Such a procedure is routinely used by the National Toxicology Program (NTP) in the U.S.

II. STATISTICAL ANALYSIS OF ANIMAL TEST DATA

Tumors occur naturally in control rodents just as they do in all animals. In fact, the incidences of some tumors are unusually high, as I will mention later. A carcinogenic effect of a chemical is usually expressed as an increase in a specific tumor type or types in treated animals over that present in appropriate controls. It is also expressed as increased tumor progression to malignancy, or as a decrease in time to tumor development, i.e. tumor latency. Determination that an increased incidence is compound-related rather than a result of normal variation, requires a full biological, pathological and statistical evaluation. This is primarily because certain tumors have not only a high incidence, but the incidence is highly variable among groups of control animals. For instance, the incidences of specific tumors among control animals in the NTP have varied from 0 to 65% among different groups, and the differences between contemporary control groups from the same laboratory may be statistically significant (ref. 2). It is, therefore, very important to recognize that statistics assist in evaluating the biological conclusion, but a biological conclusion is not determined by the statistical results alone.

III. COMPARATIVE METABOLISM AND PHARMACOKINETICS

Most chemical carcinogens are not active until they undergo metabolic activation. Consequently, the nature, extent and sites of biotransformation often determine the actual target dose of the ultimate carcinogen and the potential cancer risk, and it is well recognized that there may be major differences among species in the metabolic profiles of chemicals. These are most often quantitative differences. However, they may also include qualitative differences where specific metabolites are produced in some animal species, but not in others. Such differences may also occur within a species at different dose levels, particularly as a result of metabolic over-load and depletion of enzymatic detoxification pathways.

There are also differences in the rates of absorption and excretion or in tissue distribution among species, or following different routes of exposure. Actual target doses may, therefore, also vary for chemicals that do not require metabolic activation. It is obvious, then, that comparative metabolism may be a major factor in determining potential human risk, and attention has recently focused on this area as part of risk assessment. If it is demonstrated that a given test species metabolizes a chemical qualitatively different than humans, or, if metabolic products vary with dose levels or routes of exposure that do not simulate human exposure, the test findings may be irrelevant to human risk. Such data would certainly be inappropriate for quantitative risk assessment - unless metabolic differences were incorporated into the mathematical procedures.

The simplistic assumptions of the past that test animals are necessarily quantitative and qualitative human surrogates are no longer tenable. This is recognized by regulators and other scientists, and there are efforts to factor in metabolic and other differences. The necessary comparisons between animal and human data may require <u>in vitro</u> tests with human tissues using appropriate metabolic activation. In other instances, <u>in vivo</u> observations in humans may be possible. As a result of such efforts, on a case-by-case basis, risk assessment will certainly be a more valid and accurate process.

IV. RELATIVE TUMOR SUSCEPTIBILITY OF TEST SPECIES

Another factor that should be considered in the interpretation of animal carcinogenicity tests and risk assessment, but generally is not, is the genetically determined susceptibility to tumor formation. I have seen all kinds of interspecies comparisons of carcinogenic potency based upon empirical or theoretical data. However, they overlook one major factor. Laboratory rodents, because of their genetic composition, have very high spontaneous tumor rates in certain tissues - rates enormously higher than occur in humans. Consider, for example, liver and endocrine tumor or leukemia incidences of 20, 30, or 50% and even higher. In several other tissues, e.g. kidneys, mammary glands, lungs and even brains, tumor rates may exceed 5% - still a very high spontaneous incidence compared to human cancer. These rates vary among animal strains and may be influenced by breeding practices. In fact, many of the laboratory rodents used today were selectively bred decades ago for high tumor rates for use in carcinogenesis and cancer treatment research.

Irrespective of the basis for this susceptibility, whether it be a tendency towards oncogene expression, latent viruses or other genetic factors, the spontaneous background of tumors contributes to the susceptibility of a specific tissue to experimental cancer induction. Applying the operational terms presently in use, it may be observed that a tumor background indicates a population of naturally or spontaneously "initiated" cells that are more responsive to the effects of "promotion" or, perhaps, further "initiation".

Since many of the carcinogenic end-points in animals, as the result of chemical testing, consist of enhancement of tumors with very high backgrounds in control animals, one may question the applicability of these findings to quantitative risk assessment in humans. It is certainly true that any increase in tumors provides information that may be relevant to humans under certain conditions of human exposure. However, the use of such data to quantitate the levels of potential human risk on the assumption of comparable susceptibilities seems to be very unreasonable.

The counter argument to this view has stated that carcinogenic sites do not necessarily correlate between test animals and humans. For example, a mouse liver carcinogen may be a bladder carcinogen in humans. Well, this is true, but apparently to a very limited extent. Where comparable data on known human and animal carcinogens exist, the correlations are almost 90%. Of approximately 17 such chemicals, there are only 2 where human tumor sites were not also tumor sites in rats or mice (ref. 3).

Of course the reciprocal to over-sensitive test animals may be under-sensitive animals if one is considering a tissue where tumor rates are high in humans. Thus, a chemical which induces colon, prostate or breast cancer in rodents may pose an even greater risk for humans in the western hemisphere than one would predict from the animal data.

Such biological issues are generally not included in carcinogenesis risk assessment, though they may certainly affect potential human risk. Again, this points to the need for interpreting carcinogenicity data on a case-by-case basis.

V. POSSIBLE MECHANISMS OF CARCINOGENICITY

The final factor I would like to discuss may ultimately be the most important, i.e. the possible mechanisms involved in cancer development. If we were able to classify chemicals on the basis of mechanism of action, it would probably allow a more precise approach to risk estimation, including, perhaps, the identification of no-effects or threshold levels. However, it is generally acknowledged that the present state-of-the-science does not yet allow us to do this. It does, however, allow us to give full consideration to the circumstantial evidence relating to carcinogenic mechanisms and how such evidence impacts upon risk assessment.

It is obvious, for example, that all carcinogens do not operate in the same manner. Some induce rare forms of cancer in animals, even at relatively low level exposures. Many such chemicals, or their metabolites, also have the ability to directly alter DNA, i.e. they are genotoxic. At the other extreme are chemicals that only increase the incidence of tumors that are already common in control animals, and do so only at high, toxic exposure levels. Most of these chemicals are not genotoxic in most short-term tests. If they do damage DNA, it appears to be only indirectly - for example, only at doses that result in chronic cytotoxicity. There is apparently a prerequisite level of cell toxicity and resulting cell proliferation required before neoplastic transformation occurs following exposure to some chemicals. There are several theories to explain how such indirect DNA damage may occur as the result of tissue damage and regeneration. The answer probably lies in the process by which cellular oncogene function is deranged. The fact is we do not know how this happens, but if there is no direct genotoxic potential for a chemical or its metabolite, then the process must be an indirect one.

We do know that cell toxicity and proliferation are threshold phenomena. There are clearly no-effect levels in limited test systems. Although it is difficult, if not impossible, to establish a population threshold for any effect, neoplastic or non-neoplastic, such mechanistic observations do indicate that cancer risk at low-level exposure would not be directly proportional to that at high level exposure. In other words, dose-response relationships would not be linear, and risk estimates are invariably lower with non-linear extrapolations.

Current regulatory policies in the U.S. generally assume linear dose-response when assessing carcinogens. They utilize mathematical models which assume low-dose linearity, regardless of the empirical evidence. But, in light of the accumulating data on many of these so-called secondary carcinogens, I don't believe such conservative policies can be maintained across the board. Quantitative risk assessment procedures will have to be tailored to the scientific evidence on a case-by-case basis. There is already interest in developing separate risk assessment procedures for so-called "promoters", so I am optimistic that there will soon be substantial progress in this area.

In summary then, the interpretation of carcinogenicity data is indeed complex. The simplistic assumptions of the past that test animals are always valid human surrogates no longer holds. The factors I have discussed today, and others, must be considered on a case-by-case basis. Animals may be over-sensitive or under-sensitive indicator of human risk, or the animal data may be totally irrelevant. As stated by the Interdisciplinary Panel on Carcinogenicity in their 1984 report entitled "Criteria for Evidence of Chemical Carcinogenicity", "... the characterization of human risk always requires inter-disciplinary evaluation of the entire array of data on a case-by-case basis. Animal studies, whenever possible, should be augmented by studies of mechanisms, metabolism, and pharmacodynamics. Such studies may assist in assessing risk to man. Recognizing the utility of such data should point the way for better assessment in the future" (ref. 4).

REFERENCES

1. M.D. Shelby and S. Staslenvicz, Environ. Mutagenesis 6: 871-879 (1984).
2. J.M. Ward, Progress Exp. Tumor Res. 26: 241-258 (1983).
3. International Agency for Research in Cancer (IARC) Monographs Supplement 4. Chemicals, Industrial Processes and Industries Associated with Cancer in Humans. IARC, Lyon, October 1982.
4. Interdisciplinary Panel on Carcinogenicity. Science 225: 682-687 (1984).

Changing profiles in human health effects of pesticides

John E. Davies and James A. Lee

Department of Epidemiology and Public Health, University of Miami School of Medicine, 12500 S.W. 152nd Street, Miami, Florida 33177-1411

Abstract - Concern for the possible adverse human health effects of pesticides began in 1962. Subsequently President Kennedy's Science Advisory Committee on Pesticide translated this concern into action. Initial research started in the United States where the health effects of acute, chronic and incidental exposures were studied. Acute poisoning was the first essentially preventable disease, and was occurring as a result of inappropriate use of pesticide by an uninformed infrastructure. Though cases of poisoning are still not reliably quantified in the United States, training and technological advances have lessened this serious medical problem. This does not hold true for the developing countries because the transfer of safety has not paralleled the technological transfer of chemicals. Human poisonings are epidemic in certain countries and currently, newer chemicals, solvents and contaminants, present diagnostic challenges. Recently health effects from chronic exposure have also been recognized and compounds such as Kepone, leptophos, dibromochloropropane and inorganic arsenic have resulted in neurologic, behavioral, reproductive and carcinogenic effects, emphasizing the need for occupational surveillance. Incidental human exposure from widespread use of organochlorines is also of public health concern prompting regulatory action in the U.S. Significant declines in human body burdens have occurred during the last 15 years. In developing countries where human pesticide monitoring is in its infancy, remarkable geographical differences exist; recognition of these differences could have a significant impact on agricultural use of these chemicals. Currently, from a global perspective the major public health problems of pesticide use are poisoning, persistence and resistance, and disposal of chemicals and their containers. The relative public health significance of these problems varies greatly throughout the world; and could have a serious impact on international trade.

INTRODUCTION

In the United States the concerns for the possible adverse human health effects of pesticides really began in 1962 with the publication of Rachel Carson's book Silent Spring (ref. 1). Political action soon followed this publication and a plethora of epidemiologic and public health research programs were designed to answer three areas of major concern. These were 1) the human pesticide poisoning problem, 2) chronic long-term health effects of pesticide exposures, 3) the magnitude and significance of incidental pesticide exposures and human monitoring studies.

HUMAN PESTICIDE POISONINGS

In the U.S. pesticide poisonings were the earliest end-points to be studied and they were first evaluated through the organization of 19 community pesticide studies. Suspect cases of reported pesticide poisonings were subjected to field investigation and cholinesterase determinations, pesticide residue studies of gastric contents or urinary metabolites. It soon became apparent that human pesticide poisonings were of real public health concern and that the organophosphate and carbamate insecticides were the major group of chemicals involved. Poisonings were recognized in the occupationally exposed, particularly in those employed in mixing pesticides. Another group at special risk were children in the home where numerous reports emerged of serious poisoning and deaths resulting from the ready availability of highly toxic chemicals in appropriate containers in the home scene. Regulatory action involving the certification and training of applicators, improved packaging and labeling of pesticides, classification of pesticides into restricted and non-restricted categories, and general education and training contributed significantly to a lessening of this problem in the United States. Of major concern, however, was the absence of valid data on the magnitude of poisoning and even today, no accurate, reliable estimate of the number of pesticide poisonings is available.

Between 1966 and 1979 the Environmental Protection Agency (EPA) sought to tackle this problem through a Pesticide Incident Monitoring System (PIMS)(ref. 2), wherein it invited people to report on a voluntary basis, information on human and/or environmental poisonings. Between 1966 and 1979 PIMS had recorded 30,708 incidents, including poisoning by pesticides used in the home and garden. This, however, did not quantify all sources and mechanisms of poisoning. Mortality statistics was another approach and in 1974, Hayes and Vaughn (ref. 3), documented the occurrence of 52 accidental deaths. They concluded that the number of suicides from deliberate ingestion or exposure to pesticides was three times greater than the number of accidental deaths, so that the total number of deaths from pesticides was about 200 per year.

The most comprehensive data available that were found were in the State of California where the physician is reimbursed for his treatment of the poisoned victim only when the case has been reported to the Workmens Compensation Bureau (ref. 4). An average of 1,468 poisonings were thought to have occurred between 1971 and 1973 (ref. 5). These cases were not verified and were almost certainly mild since only a few were hospitalized. Pimental (ref. 6) has extrapolated from regional poisoning statistics from California (ref. 7) and from South Carolina (ref. 8) and it was estimated that 109,280 human pesticide poisonings occurred annually in the U.S., and in 1974, the U.S. EPA estimated 200 deaths from pesticides (ref. 9). These should be compared to the World Health Organization (WHO) which estimated that approximately 500,000 cases occurred annually with about a 1 percent fatality rate (ref. 10). It is apparent that in both the developed and developing world there are numerous data gaps on the actual number of poisonings and information is especially needed on such issues as the number of accidental poisonings versus suicidal poisoning, the sources of data, the degree of verification of cases by field studies and analytical methods, and some classification with respect to the severity of poisonings. For, it makes little sense to include minor skin lesions with fulminating system poisoning cases in the same statistic. In order to try to obtain more information on the number of poisonings in the developing countries, and especially the role of suicide, a pilot questionnaire was sent out to experts in the field of 67 countries. Replies were received from 27 countries and morbidity and mortality data from ten of those countries are shown in Table 1.

TABLE 1

Some Morbidity and Mortality Data of Pesticide Poisonings

Country	1980 Cases Recorded	1980 Deaths	1981 Cases Recorded	1981 Deaths	1982 Cases Recorded	1982 Deaths	1983 Cases Recorded	1983 Deaths
Barbados*(ref.11)	2	0	8	2	23	8	17	3
Canada**(ref.12)	-	4	-	3	-	4	3,404	2
Egypt#(ref.13)	129	-	352	-	70	-	36	-
Israel**(ref.14)	-	-	781	10	722	-	-	-
Kenya*(ref.15)	-	-	-	-	-	-	1,000	-
Philippines#(16)	650	93	576	68	-	-	-	-
Sri Lanka*(17)	11,372	1,045	11,811	1,112	-	-	-	-
Thailand**(18)	1,857	-	2,170	-	2,187	-	2,353	-
Trinidad/ Tobago*(ref.19)	-	-	-	-	-	-	232	94
U.K.(1)##(ref.20)	30	-	9	-	18	-	8	-
U.K.(2)##(ref.21)	-	-	112	-	167	-	58	-

* Hospital data; ** Poison Control data; # Survey data; ## Governmental Agency data.

Of necessity the WHO estimates have been based on models. Models are only as predictive as are the hypotheses which have been used to make them; the results of global pesticide poisoning estimates are obviously tenuous and the WHO has acknowledged this with regards to their estimates. The same holds true for poison control statistics for they tend to overestimate the magnitude of the problem. Hospitalization data are probably the best, for these sources of data would include only serious cases and would have the benefit of medical and laboratory verification. Such data, however, are usually not informative with regards to the number of suicides versus the number of accidents. The percentage of suicides reported in response to the questionnaire were 60, 73 and 90 percent in Philippines, Sri Lanka, and Trinidad-Tobago, respectively.

The causes of accidental pesticide poisonings might be rectified by involving local public health clinics through making a field visit to the scene of the accident. This has been the way in which we have verified alleged pesticide poisoning in south Florida. During the period April 1, 1976 through December 31, 1977, 121 suspected cases were investigated in south Florida. Of these suspect poisoning episodes, 51 were confirmed based on clinical and laboratory data and field visits.

The problem of pesticide poisoning verification has been recognized in other countries. In response to our questionnaire, one of the respondents from the United Kingdom described their poisoning verification process; he stated, "anyone who suspected that they had sustained a poisoning, reported this to the government Health and Safety Commission (HSC). This could be done either directly or through the victim's doctor. The incident would then be investigated by HSC inspector and the surrounding events reported to a central panel called APAP (Agricultural Poison Appeal Panel). This panel consists of representatives of the medical profession, the Health and Safety Commission, and representatives from a body representing labor. This panel decides whether a case is confirmed or unconfirmed. In 1981, 46.3% of incidents were confirmed, whereas, in 1983 only 20% of cases were confirmed" (ref. 21).

CHRONIC PESTICIDE EXPOSURE STUDIES--HEALTH CONCERNS OF LONG-TERM EXPOSURE

During the last 20 years or so, several epidemiologic studies have highlighted different diseases in occupationally exposed workers. These diseases, some of which may persist from a few months to years, have been recognized in a variety of different target organisms. Thus, in the reproductive area the association of occupational exposure to dibromochloropropane with impaired fertility was an important observation (ref. 22). Azoospermia and oligospermia with elevations of follicular-stimulating hormones and luteinizing hormones in 14 persons occupationally exposed workers.

Similar male reproductive disorders were noted in chlordecone workers exposed in a chlordecone manufacturing plant in Hopewell, Virginia. In addition, central nervous system (CNS) involvement manifested by the occurrence of tremors, stuttering speech, opsiclonus, were observed in highly exposed workers (ref. 23). Serious derangement of the central nervous system was also noted in workers manufacturing leptophos and a variety of different neurologic lesions were found similar to those in patients with multiple sclerosis (ref. 24) Leptophos will produce delayed neurotoxicity a CNS effect which was first observed in Egypt (ref. 25) in 1,300 water buffaloes that were noted to have hind leg paralysis from exposure to this insecticide.

In the cardiovascular area, hypertension and hyperlipidemia were observed more frequently in workers exposed to organochlorine pesticides (ref. 26). These observations have not been reconfirmed in other studies.

Insofar as the liver is concerned, as might be expected, a variety of different effects have been noted in the pesticide exposed worker. The clinical and biochemical effects have ranged from liver microsomal enzyme induction (DDT and dieldrin) to angiosarcoma of the liver in agricultural workers spraying the Moselle Vineyards with arsenic (ref. 27).

In the genitourinary system, hemorrhagic cystitis has been linked to occupational exposure to chlordimeform (ref. 28) and renal tubular abnormalities to organophosphate exposures (ref. 29).

Dermatologic lesions are the commonest of all pesticide related illnesses. Skin lesions with an allergic mechanism and others with a toxic mechanism, have followed occupational exposure to a variety of different pesticides and the lesions ranging in severity from toxic dermatitis, chloracne and necrotizing folicular dermatitis have been described (ref. 30).

No disease, however, is of more universal concern and interest than is cancer, yet on the whole, with the exception of inorganic arsenic and lung cancer the picture is somewhat reassuring. Several studies have focused on cancer risk factors in the occupationally exposed. The only positive study was by Mabuchi, et al. (ref. 31) who found an excess risk of lung cancer due to occupational exposure to inorganic arsenic. Their study identified 23 deaths from lung cancer in the studies' 1,050 participants. Rates increased with dosage and in the evaluation of arsenic and arsenic compounds including inorganic arsenic pesticides IARC (ref. 32) concluded that inorganic arsenic compounds were skin and lung carcinogens.

Retrospective mortality studies of pest control applicators (ref. 31) and of four organochlorine manufacturing plants (ref. 33) showed no overall excess rates of cancer, although small excesses of cancers of the lungs, skin, bladder, stomach, oesophagus and lymphatic system were noted in individual studies. In every study, insecticide exposure were multiple and no information was available on cigarette smoking.

Insofar as the herbicides are concerned studies of railroad workers, forestry and agricultural workers have been linked to an increased risk of soft tissue sarcoma following occupational exposure to phenoxyacetic acids and the chlorophenols. In a review of these studies by Coggon and Acheson (ref. 34) concluded that further research was urgently needed.

The findings have not been observed in other parts of the world, and the area is one which is currently under study in the Ranch Hand studies on morbidity and mortality of Agent Orange (refs. 35, 36).

HUMAN MONITORING STUDIES

The third area of public concern stemmed from the recognition of human body burdens of pesticides. These were acquired from trace amounts of organochlorine pesticides in the environment. These incidental exposures prompted annual blood and adipose tissue human monitoring studies (ref. 37, 38). In the twenty-five year period covered by this review the only significant documentation of temporal change has been noted in the United States where until recently there had been a regular, ongoing measurement of body burdens of human adipose tissue, blood and mother's milk (ref. 39). During the period 1970 to 1978 several organochlorine pesticides including DDT, chlordane, dieldrin and heptachlor have been cancelled. The impact of these cancellations on general population body burdens during this period are shown in Table 2. It will be noted that there has been a steady decline in the adipose pesticide levels in the U.S. general population following the cancellation orders in 1972. Levels of DDT in blacks were consistently higher than in whites a difference which was found to be only partially related to socioeconomic differences (ref. 40).

TABLE 2
Temporal Changes of Geometric Means of Residues in Adipose Tissue
Selected Organochlorine Pesticides in Black and White U.S. Populations
During the Period - 1970-1978. Geometric $\pm$ (ppm)

		U.S. White Population			U.S. Black Population		
Year	No.	Total DDT	Beta-BHC	Dieldrin	Total DDT	Beta-BHC	Dieldrin
1970	1,373	7.46	0.44	0.22	12.97	0.69	0.35
1971	1,545	7.04	0.37	0.24	16.21	0.65	0.35
1972	1,877	6.23	0.30	0.31	12.16	0.47	0.27
1973	1,085	5.70	0.32	0.21	9.66	0.37	0.27
1974	892	4.96	0.23	0.18	8.44	0.31	0.19
1975	771	4.67	0.24	0.16	7.04	0.27	0.16
1976	667	4.38	0.25	0.14	6.70	0.30	0.16
1977	775	3.35	0.20	0.12	4.47	0.19	0.11
1978	802	3.64	0.19	0.12	6.07	0.23	0.14

Our laboratory conducted several small serological surveys of organochlorine pesticides in different Caribbean populations during the period 1972 to 1978, some results are shown in Table 3.

TABLE 3

Comparison of Serum Organochlorine Levels (PPB) in Caribbean Countries and the United States

			Total DDT			Dieldrin			Lindane		
Year	Country & Pop.	No.	%Pos.	Geo.Mean	Range	%Pos.	Geo.Mean	Range	%Pos.	Geo.Mean	Range
1971	Bahamas (General Pop.)	98	100	58.5	33.6-381.1	11	---	1-3.2	-0-	-0-	-0-
1981-1982	Haiti (Migrants)	54	100	106.2	24.7-655.9	N.D.	N.D.	N.D.	-0-	-0-	-0-
1981	Caribbean Is. (Occup/Exposed)	108	100	38.2	16.0-72.0	83	3.5	1-44.4	-0-	-0-	-0-
1981-1982	Caribbean Is. (General Pop.)	51	82	23.4	5.4-236.1	82	2.0	1-12.4	31.0	<1.0	<1-16.2
1976-1980	United States (General Pop.)	3441	98.8	13.9	0 - 447	9.1	<1.0	0-16.1	0.2	<1.0	0-6.2

Striking differences were noted with regards to the frequency of pesticide identification. Thus, in the Bahamas survey which took place in 1972, this excessive contamination was attributed to household use of DDT and subsequent contamination of house dust (ref. 41).

DDT contamination observed in a small number of Haitian migrants reflected the use of DDT on that island for malaria control and residues were positively correlated with serum triglycerides (ref. 42).

In one Caribbean Island, two small serological surveys were conducted on public health workers and the general population. In both, dieldrin was identified in 82-83 percent of the samples studied. Dieldrin had not been used by the public health workers for malaria control for at least ten years, but the farmers used cyclodiene insecticide for the control of mole crickets, cutworms and ants (ref. 43). Local farmers were observed applying the cyclodiene insecticides by pouring the compound into the irrigation ditches between rows of vegetables, and then scooping buckets full of the water/insecticide mixture directly from the ditch onto the vegetables. Pesticide management changes on the island have taken place as a result of these findings. The findings are illustrative because they show how human monitoring studies can identify excessive and undesirable agricultural pest management practices.

Insofar as health effects of incidental pesticide exposures are concerned apart from liver microsomal induction (ref. 44) and hyperlipidemia with triglyceridemia (ref. 45) adverse health effects from incidental exposures of these orders of magnitude have not been demonstrated.

SUMMARY

Although pesticide management practices differ enormously in different parts of the world, surveys have shown that they are really all pesticide management problems which stem from current pesticide use patterns. These are 1) poisoning, 2) the problem of persistence, 3) the problem of resistance and the resurgence of vector borne disease, and 4) the problem of inappropriate disposal and the secondary effects of environmental degradation and surface and groundwater contamination.

While only poisonings represent a direct adverse human health, the other problem areas involving human and environmental contamination, will sooner or later impinge on human health.

The relative significance of these problems is different in the developed and developing world. In the former the disposal problem is currently dominant and has greatly contributed to the public's health concerns which stem from the recognition of hazardous waste sites. Local clusters of hot spots and exposed people are increasingly being recognized and cleanup is a major area of concern. In contrast in the developing countries, pesticide poisoning is probably the most important pesticide management problem, because examples of gross misuse of pesticide are only too apparent. As the amount of pesticide being used by the developing countries increases, so too will the poisoning problems grow unless there is a concerted attempt to transfer the technology of safety.

At the same time as the problem of poisoning is being recognized, the disposal problem has hardly been addressed, and it is probably one of the most pressing problems. In a recent pesticide management symposium in Nairobi, Kenya sponsored by U.S. Agency for International Development in March, 1985, one of the participants showed that there were sizable collections of unwanted deteriorating supplies of pesticides scattered throughout the countryside (ref. 46). Some of these reflect exports of donated material from donor countries to developing countries where not all of the materials are used up and now, as they begin to break down, pose a serious threat to rural communities. Pesticide disposal, therefore, is a demanding and vitally important problem which should be of concern for the developed and developing world.

REFERENCES

1. R. Carson, Silent Spring, Fawcett, New York, (1978).
2. National Study of Hospital Admitted Pesticide Poisonings. Epidemiologic Studies Program, Human Effects Monitoring Branch, Tech. Serv. Div., Off. Of Pest. Program. U.S.EPA, Washington, DC, (1976).
3. W.J. Hayes, Jr. and W.K. Vaughn, Toxi. and Appl. Pharmac. 42, 235-252 (1977).
4. K. Maddy, Unpublished data, Sacramento, CA, Department of Food and Agriculture (1978).
5. Anon., Occupational Disease in California, Attributed to Pesticides and Other Agricultural Chemicals, 1970-1973, Occcupational Health Section and Center for Health Statistics, State of California, Dept. of Health, Berkeley, CA (1974).
6. D. Pimentel, personal communication (1979).
7. E.P. Savage, T. Keefe and G. Johnson, Agrichem Age (1976).
8. N. Whitlock, J.E. Keil and S.H. Sandifer, J.S.C. Med. Assoc. 68, 109 (1972).
9. Strategy of the Environment Protection Agency for Controlling the Adverse Effects of

Pesticides, Off. of Pest. Programs, Off. of Water and Hazardous Materials., Washington, DC (1974).
10. J.F. Copplestone, Pesticide Management and Pesticide Resistance, pp. 147-153, Academic Press, New York (1977).
11. P.P. Santanam, Sr. Med. Off., Min. of Health, St. Michael, Barbados, World Resources Institute (WRI) Questionnaire (based on hospital data) (1984).
12. J. Graham, Poison Control, Health & Welfare, Bureau of Epidemiology, Ottawa, Ontario WRI Questionnaire (based on poison control data) (1984).
13. M.M. Amr, Prof., Indust. Med. & Occup. Dis., Cairo University, Egypt WRI Questionnaire (based on survey data) (1984).
14. Dr. Teitelman, Israel Poisoning Info. Ctr., Haifa, WRI Questionnaire (based on poison control data) (1984).
15. A.O.K. Obel, Sr. Head, Div. of Pharmac & Therapeutics, University of Nairobi, WRI Questionnaire (based on hospital data) (1984).
16. C.C. Zapatos, Agromedical Monitoring of Pesticides in Relations to Human Health in the Philippines (Thesis), University of Miami, FL (1982).
17. (a)J. Jeyaratnam, R.S. de Alwis Seneviratne, & J.F. Copplestone, Bull. of the WHO, 60 (4):615-619 (1982); (b) N. Senanayake, Head, Dept. Med., University of Peradeniya, Sri Lanka, WRI Questionnaire (based on hospital data) (1984).
18. K. Snidvongs, Sec. Gen., Off. of the Natl. Env. Bd., Bangkok, Thailand, WRI Questionnaire (based on poison control data) (1984).
19. R. Doon, Med. Off., Occup. Hlth. Unit, Min. of Hlth, Trinidad, WI, WRI Questionnaire (based on hospital records) (1984).
20. D.H. Wright, Hlth. & Safety Exec., H.M. Agri. Insp., Merseyside, UK, WRI Questionnaire (based on governmental agency records) (1984).
21. C. Clutterbuck, Union Dele., Hlth & Safety Comm., Lancashire, UK, WRI Questionnaire (based on governmental agency records) (1984).
22. D. Whorton, R.M. Krauss, S. Marshall and T.H. Milby, Lancet 2(8051), 2359 (1977).
23. W.J. Cohn, J.T. Boyland, R.V. Blanke, et al. N.E. J. of Med. 298, 243-248 (1978).
24. American Occupational Medicine Association Report, January (1978).
25. Near East News Roundup, FAO, RNEA., Cairo, Egypt (1971).
26. D.P. Morgan, Residue Reviews 75, 97-102 (1980).
27. Intl. Agcy. Res. Ca. (IARC) Monograph 2, 48-73 (1973).
28. D.S. Folland, R.D. Kimbrough, R.E. Cline, et al. J.A.M.A. 239(11), 105-1062 (1978).
29. J.B. Mann, J.E. Davies and R.W. Shane, Acute Glomerulonephritis, Little, Brown & Co. Boston, MA (1967).
30. J.E. Davies, V.H. Freed and F.W. Whittemore, An Agromedical Approach to Pesticide Management, UM Printing, Miami, FL (1982).
31. K. Mabuchi, A.M. Lillienfeld and L.M. Snell, Prev. Med. 9, 51-77 (1980).
32. Intl. Agcy. Res. Ca. (IARC) Monograph 30, 37-56 (1983).
33. D. Ditraglia, D.P. Brown, T. Namekata and N. Iverson, Scand. J. Work Envi. Hlth. 7(4), 140-146 (1981).
34. D. Coggon and E.D. Acheson, Lancet 8280(1), 1057-1059 (1982).
35. Project Ranch Hand II, "An Epidemiologic Investigation of Health Effects in A.F. Personnel Following Exposure to Herbicides," Baseline Mortality Study Results, June (1983).
36. Project Ranch Hand II, "An Epidemiologic Investigation of Health Effects in A.F. Personnel Following Exposure to Herbicides," Baseline Morbidity Study Results, February (1984).
37. F.W. Kutz, S.C. Strassman and J.D. Sperling, Fiscal Years 1970-1975, Ecological Monitoring Branch (TS-768), U.S. E.P.A., Washington, DC (1979).
38. F.W. Kutz, G.W. Sovocool, S.C. Strassman and R.G. Lewis, Bull. Environ. Contamin. & Toxicol, 16 (1), (1976).
39. S.C. Strassman and F.W. Kutz, Insecticide Residues in Human Milk from Arkansas and Mississippi, 1973-1974, 10 (4) U.S. E.P.A., Washington, DC (1977).
40. J.E. Davies, W.F. Edmundson, A. Raffonelli and C. Morgade, Amer. J. Epid. 96(5), 334 (1972).
41. J.E. Davies, W.F. Edmundson and A. Raffonelli, Amer. J. Pub. Hlth, 65(1), 53-57 (1975).
42. J.E. Davies, R.S. Levine and M. Fordyce, B.M.J. 282 2054 (1981).
43. R.M. Barrow, Proceedings of Seminar/Workshop, Country Papers, III, 178-192 (1980).
44. M.A. Campbell, J. Gyorkos, B. Leece, K. Homonko and S. Safe, Gen. Pharmac. 14(4) 445-454 (1983).
45. K. Kreiss, M.M. Zack, R.D. Kimbrough, L.L. Needham, A.L. Smrek and B.T. Jones, J.A.M.A. 245(19) 1926-1930 (1981).
46. H.G. Alford, Univ. of CA at Davis, "Management of Pesticides in Countries of East and South Africa," Presented at the 101st Natl. Mtg. Am. Chem. Soc., New York (1986).

Risk management and regulatory decisions on pesticides

John A. Moore

Assistant Administrator, Office of Pesticides and Toxic Substances
United States Environmental Protection Agency, Washington, D.C. 20460

I. INTRODUCTION

Over the past three years, the United States Environmental Protection Agency (USEPA), has framed many of its regulatory decisions on pesticides in the terms of "Risk Assessment/ Risk Management" considerations. This approach to decisionmaking traces its roots to the Agency's annoucement in 1976 of interim procedures for assessing the risks associated with exposure to chemicals which pose a carcinogenic hazard. Prior to 1976, there was little guidance to aid decisionmakers when faced with the prospect of such a hazard. Risk assessment was a crude tool which could be used in estimating the order of magnitude of the problem.

In 1983, the U.S. National Academy of Sciences issued an important report (Managing Risk in the Federal Government) which cogently presented the currently accepted conceptual framework for Risk Assessment/Risk Management decisionmaking. By providing a common lexicon of terms, this book has greatly facilitated communication both within and outside of the regulatory agencies.

In fact, the USEPA Administrator at the time, William Ruckelshaus, adopted Risk Assessment/Risk Management analysis as a hallmark of his tenure at the Agency, a practice which continues today. The paradigm has been adopted by several other organizations as well. For example, the Office of Science and Technology Policy in the White House utilized this approach in its discussion of "cancer principles". The Department of Health and Human Services has published an entire book on the use of Risk Assessment/Risk Management within its operations. And the OECD has discussed many of its concerns/actions in terms of assessing and managing risks associated with toxic chemical substances.

Today, I would like to briefly review the Risk Assessment/Risk Management concepts. I will then use the cases of ethylene dibromide, alachlor and pentachlorophenol to provide examples of how these concepts are applied in pesticide regulation.

II. CONCEPTUAL FRAMEWORK OF RISK ASSESSMENT/RISK MANAGEMENT DECISIONMAKING

Figure I illustrates the relationship between the elements of Risk Assessment (RA). First, Hazard Identification involves the determination of the qualitative, inherent toxic properties of the material in question. This anlaysis deals with information such as the results of animal, human, and short term studies. Second, in the Dose-Response Assessment segment one quantitatively estimates the degree of that toxicity, selecting the most appropriate material from the available scientific data. In the third step, Exposure Assessment, one estimates the degree of exposure to the substance likely to be encountered by people involved in the particular activities under discussion; for example, during pesticide application or consumption of pesticide treated food. Finally, in the Risk Characterization step, the results of the first three steps are combined, melding together the total weight of the evidence, in order to generate some estimate of the concern one has for the substance under specific use or exposure conditions.

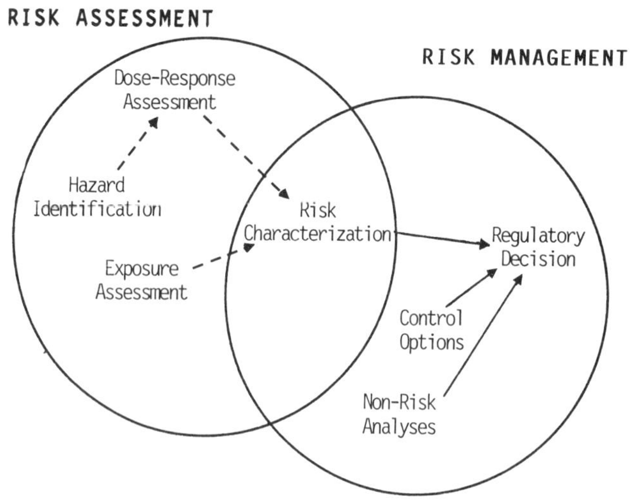

FIGURE I Risk Assessment and Risk Management

Figure I illustrates the relationship between the elements involved in Risk Management (RM); i.e., determining what to do about a specific risk once it has been identified and characterized. Note that the last step of Risk Assessment -- Risk Characterization -- is the first step of Risk Management. In addition, a number of "non risk" factors are brought into consideration, including:

1. The constraints of the statutory authorities under which the decision is being made (e.g., risk/benefit balancing vs. risk-driven mandates).

2. The range and effectiveness of control options which are possible (e.g., tolerances, use restrictions, and/or banning).

3. The availability of alternative substances and the degree to which the risks associated with those alternatives are known.

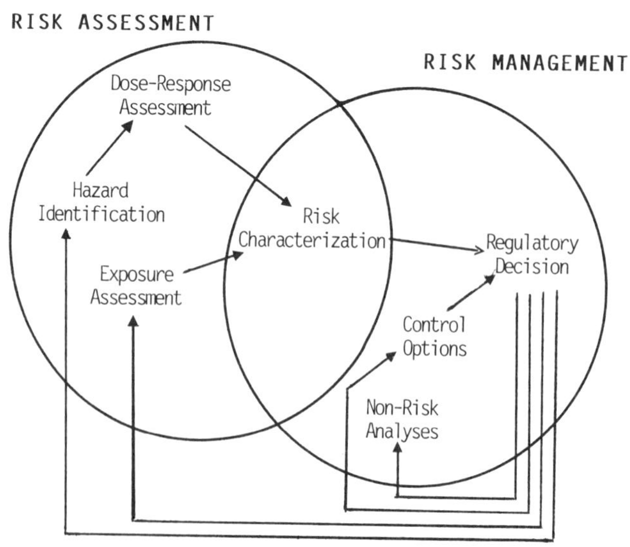

FIGURE II Intersection of Risk Assessment and Risk Management

Finally, Figure II illustrates how all of these parts fit together and interact with each other. The results of the Risk Assessment provide input to Risk Management analysis, which leads to specific regulatory decision. The impact of the proposed decision feeds back, in general, to reduce uses, exposure, and/or the inherent toxicity itself to such an extent that the risk and non-risk factors are placed into a more appropriate balance.

III. RISK ASSESSMENT/RISK MANAGEMENT APPLIED TO REGULATORY PESTICIDE DECISIONMAKING

In this Section, I will use specific cases to illustrate how the Agency has adopted the philosophical concepts discussed in Section II in "real world" situations involving pesticide regulation.

A. Ethylene Dibromide

Ethylene dibromide (EDB) burst upon the national scene in the United States in late 1983 with a flurry of news reports and telecasts covering the removal of "killer muffin mixes" from grocery store shelves in many States. This risk management action had been taken by certain States, based on relatively strong Hazard Identification information; i.e., cancer studies which demonstrated that administration of EDB to laboratory animals was associated with early onset of malignant, life shortening tumors, in multiple sites, in both sexes, of two species. Unfortunately, due to limitations in the data set, the same experiments did not lend themselves easily to a Dose-Response Assessment. However, the discovery of residues of EDB in grain products (that is, an Exposure Assessment), coupled with the potential cancer hazard, constituted a sufficient basis to prompt some State regulators to take action; hence the removal of certain grain products from grocery store shelves.

Sometimes lost in the re-telling of the EDB story is the fact that several months prior to the grain products saga, USEPA had taken steps to limit the use of EDB as a soil fumigant, based upon the discovery of EDB residues in ground water near treated fields. Therefore, two different exposure triggers -- food and water -- raised principal concerns about the risks posed to humans by EDB.

As noted above, however, the Dose-Response step of the Risk Assessment was made difficult by certain aspects of the available data. In fact, the technical limitations of the quantitative aspects of the animal studies favored making a more subjective judgment regarding the significance of the levels observed. Further, the Exposure Assessment data were limited in the case of the ground water residues and were somewhat problematic in the case of the food residues, since the latter data were being generated rather quickly by different scientists, in different labs, in different States, and using different methods of analysis.

At this point, the non-risk factors came into play in the Risk Management portion of the decisonmaking process. Specifically, attention turned to the possible alternatives to EDB for grain fumigation. These chemicals are not without their own potential risks, including cancer concerns and the possibility of explosions. Also, the fact that various States were taking action independently was leading to confusion for both consumers and grain product producers. As the story of EDB continued to find the spotlight in the news media, the need for definitive, action by the Agency became increasingly clear.

The Agency's annoucement of a decision to suspended some uses of EDB immediately and phrase out others over time did much to calm the public's concern. Within a few months, most indicators of worry had returned to near normal, including the sales of grain products. Therefore, while we would have preferred to have had better data to guide us in our decisionmaking, the very fact that a decision was made did much to alleviate the situation. To paraphrase the Administrator: We probably did the right thing, but we didn't necessarily do the thing right.

EDB is illustrative of a further point: the need for a broader range of residue data. As you know, pesticide residues are established for raw agricultural commodities. Analyzing grain as it was sold for milling, revealed moderately high EDB values -- sometimes a strikingly high value. Since our concern is with human health, a more relevant value for risk assessment purposes proved to be residues of EDB in commodity food products as consumed. We found that processing procedures commonly reduced residue levels by 1000-fold, which had a profound impact on quantitative estimates of risk. The need for correct exposure values is critical to the acceptance of a credible risk assessment.

It should be noted that the Agency has continued to pursue its original concern about EDB; i.e., ground water contamination. We are actively taking action on EDB and other pesticides which we have reason to believe pose similar threats to ground water. Further, we are preparing a major study of ground water contamination which will define more accurately the extent of the problem. Finally, we are devising an Agricultural Chemicals Strategy which will describe our approach to dealing with both surface and ground water contamination by both pesticides and fertilizers. Therefore, "final" decisions are seldom final. More often than not, they lead to consideration of separate, but related, matters.

B. Alachlor

In the case of alachlor, we again have Hazard Identification data on carcinogenicity resulting from long term studies in animals. In comparison to the case of EDB, the age at tumor onset, the degree of malignancy and the extent of the response is less. However, the data set for alachlor is more amenable to quantitative Dose-Response Assessment than in the case of EDB.

As expressed in the Agency's initial position document, the Exposure Assessment, which includes both field measurements and mathematical modeling, indicates that the greatest human exposure to alachlor is likely to be to applicators of the pesticides. Combining all of this information into a Risk Characterization, one estimates the upper limit of excess cancer risk to applicators to range from one-in-a-million to one-in-a-thousand. then, turns on the availability of substitutes and what, if any, hazards they may pose. Also, since a major concern is potential risk to applicators, one should consider those factors that may reduce the exposure to such applicators by providing specialized training to these individuals and altering the formulation and/or the application practices. In the case of alachlor some steps in this direction have been taken in the U.S. unilaterally by the registrant.

My Canadian colleagues have made their initial Risk Management decision regarding alachlor: they prohibited its use during the 1986 growing season. In the United States, we anticipate a decision well in advance of the 1987 growing season. The added risk, if any, during this additional year's use of alachlor is roughly 1-2% of that estimated for a lifetime's use of the substance. We believe that acceptance of this comparatively small incremental risk has allowed us to conduct a much improved Risk Assessment, based, for example, on more definitive information on human exposure.

Therefore, the description of the ideal Risk Assessment/Risk Management approach set forth in Section II does not reflect all of the vagaries associated with the "real world" situation of data base limitations. In reality, difficult judgments have to be made on when to make a Risk Management decision and when to await better data.

C. Pentachlorophenol (Penta)

EPA began work on its review of the three major wood preservative pesticides in the late-1970s. The strategy was to complete all of the Risk Assessment/ Risk Management reviews at the same time, in order to minimize any possible dislocation in the market as users shifted from one product to another and the Agency went through a protracted sequential consideration of the alternatives. In 1984, the Agency issued its position on inorganic arsenicals, creosote and pentachlorophenol wood preservatives. I would like to focus this discussion on penta, at this time, the most widely used wood preservative in the United States.

While there are a number of concerns within the area of Hazard Identification associated with penta, one of the major factors in our regulatory action was the presence of hexachlorodibenzo-p-dioxin (HxCDD) as a contaminant in the product. This material, at concentrations in the parts per million range in penta, has been shown to be carcinogenic in laboratory animals. The similarity of its chemical structure and biological activity to 2,3,7,8-tetrachlordibenzo-p-dioxin (2,3,7,8-TCDD or "dioxin") reinforces this qualitative concern about potential hazards.
An Exposure Assessment of the uses of penta, coupled with the Dose-Response Assessment for the substance, indicated cause for concern when examined in the Risk Characterization step.

As you might imagine, the Agency's Risk Assessment/ Risk Management examination of the wood preservatives has been quite involved, as reflected by the nearly 1000 pages required to present our rationale. To summarize in part, the Agency has proposed severe limitations on certain uses of penta (e.g., indoor uses), restrictions on who may purchase the substance (e.g., limited to persons willing to take special training in its use), and a mandatory reduction in the permissible levels of HxCDD by more than one order of magnitude.

The Federal Insecticide, Fungicide, and Rodenticide Act (FIFRA) under which pesticides are regulated in the United States, permits adversely affected parties, including registrants of existing pesticide products, to challenge an action the government might take to restrict the sale or use of those products. Therefore, shortly after the Agency issued its final decision, makers and users of penta asked for an administrative hearing to review the Agency's decision on the chemical. At the present time, the use restriction issues have been resolved by settlement; however, there is still a proceeding on the reduction of the HxCDD levels. This administrative hearing is now recessed pending the outcome of settlement discussions among the Agency and other parties.

As is always the case, however, as time passes, new data come to light which have a bearing on the Risk Assessment/Risk Management decisions which have been made. The case of penta and HxCDD is a good example. In recent years, the public has become increasingly aware of "dioxin" issues. This public concern has been heightened by a debate over the emission of chlorinated dibenzo-p-dioxins (CDDs) and the related chlorinated dibenzofurans (CDFs) from combustion sources (e.g., municipal waste combustors) and their relevance to human exposure and human health. In addition, an increasing number of reports are being published which suggest that a significant portion of the general population may be accumulating small quantities of the CDDs and CDFs, including HxCDD, in their tissues. The exact consequences for human health of these observations cannot be determined at the present time. The World Health Organization has acknowledged a contribution of combustors to the environmental loadings of CDDs and CDFs, but they have gone on to say that the combustors cannot account for the levels seen in humans. Therefore, other sources must be considered. Penta is a candidate source since it contains several of the more highly chlorinated CDDs and CDFs, which are often found in relatively high concentrations in human tissues.

This experience with pentachlorophenol is a clear example of a situation in which a Risk Assessment/Risk Management decision is not static, but evolutionary. In the light of new information, new assessments and new regulatory decisions may be called for.

IV. Conclusion

So how has the conceptual framwork of Risk Assessment/ Risk Management descisionmaking served us "in the real world"? In the cases that I have discussed and many others, I would say that Risk Assessment/Risk Management decisionmaking has proved to be an effective tool in getting us to gather the appropriate data, to look at it carefully, and to reach reasoned decisions.

In practice, however, there is a danger that the technique of Risk Assessment becomes so formalized that it takes on a totally unmerited patina of precision. Under such circumstances, an unwary, myopic decisionmaker could be led to inappropriate conclusions. The risk manager must remain conscious of the fact that Risk Assessment does not provde precise predictions of "cancer cases avoided" or "body counts".

Rather, in most instances, the technique supplies some rough estimate of the magnitude of the risks involved. Therefore, the risk manager must be sure that the other half of the decisionmaking process (Risk Management) is just as carefully constructed and considered in light of the uncertainties and limitations of risk assessment. While not generally the subject of as much attention and scrutiny as Risk Assessment, the Risk Management process should also be examined for possibilities of innovation and improvement.

Through a complete and balanced approach to assessment and management of risk, we are more likely to reach the types of decisions that address the most important environmental problems and efficiently reduce the most important risks to acceptable levels.

The percutaneous absorption of oxadiazon in the adult female rat

J. B. Knaak[1]*; J. P. Schreider[2]; O. G. Raabe[3], M.A. Al-Bayati[3]; and B. W. Wilson[4]

[1] California Department of Health Services, Sacramento, California 95814;
[2] California Department of Food and Agriculture, Sacramento, California 95814;
[3] Laboratory for Energy-related Health Effects; and
[4] Department of Avian Sciences, University of California-Davis, Davis, CA 95616

Abstract—The percutaneous and intravenous absorption of oxadiazon was studied in the adult female Sprague-Dawley rat. Parental dosages (118, 236, and 354 ug/kg) were rapidly eliminated in urine (50 to 75%) and feces (48.9 to 19%) over a period of 120 hours. The back skin of 36 rats was treated (51.7 ug cm^{-2}, 11.6 cm^2) and three were killed at 12 time intervals between 0.5 and 168 hours. The treated skin, plasma, heart, liver, kidneys, urine, feces, and the carcass from each rat were analyzed for ^{14}C. Recovery data indicated [^{14}C] oxadiazon penetrated skin, was absorbed by blood and other tissues, and eliminated in urine (26.8%) and feces (34.0%) over a period of 168 hours. The half-lives for retention on skin (penetrated and surface residues), absorption from skin and elimination from blood plasma were 40, 14, and 40 hours, respectively. Oxadiazon was absorbed through skin at a maximum rate of 0.5 ug hr^{-1} cm^{-2} (0.9% of the applied dose hr^{-1}).

INTRODUCTION

Agricultural chemical companies are presently required by state and federal regulatory agencies to determine the toxicological risk to workers applying pesticides. In order to perform a risk assessment, data on exposure, percutaneous absorption and the adverse health effect in animal models are needed.

Oxadiazon (2-tert-butyl-4-(2,4 dichloro-5-isopropoxyphenyl) 1,3,4-oxadiazolin-5-one) the active ingredient in Chipco Ronstar G Herbicide was recently shown to produce liver cancer in the rat and mouse. This adverse health effect necessitated the development of percutaneous data in an animal model and mixer/loader-applicator exposure data for quantitative risk assessment (ref. 1, 2). A mixer/loader-applicator study was conducted by Hamburger (ref. 3). The present percutaneous absorption study was conducted in the adult female Sprague-Dawley rat to determine the pharmacokinetics of absorption.

RESULTS

[^{14}C] Oxadiazon with a specific activity of 4.0 mCi/mmole was used (Fig. 1) in these studies. the radiopurity was 98% as determined by TLC on silica gel F-254 plates.

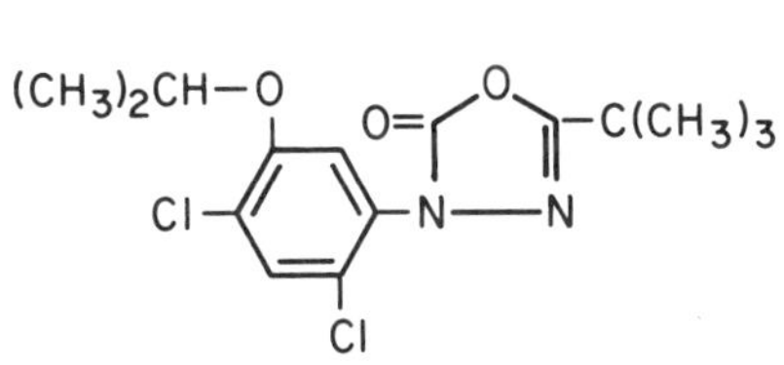

Fig. 1. Structure of oxadiazon. Pesticide was uniformly labeled with [^{14}C] in the phenyl ring.

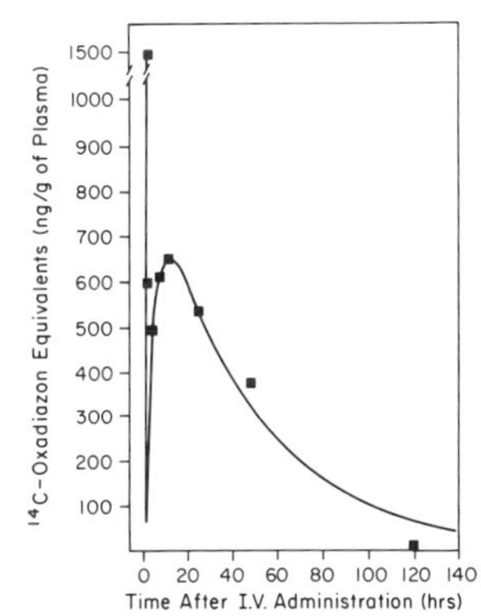

Fig. 2. Time-concentration curve for the elimination of [^{14}C] oxadiazon from plasma after i.v. administration.

Intravenous Studies

Intravenous dosages of 118, 236, and 354 ug/kg of [^{14}C] oxadiazon were individually administered in 25 ul of alcohol to three 280 g female rats. Blood samples (0.5 ml) were taken from each rat via a jugular vein cathether at 0.25, 0.5, 1, 2, 4, 8, 12, 24, 48, 72, and 120 hours post administration and analyzed for radioactivity. Oxadiazon was rapidly eliminated from plasma by distribution to tissues in the main compartment. The data points did not fit a single or biphasic declining exponential function, but fitted, in the case of 118 and 354 ug/kg dosages, a biphasic concentration-time function described by Nonlin (Metzler et al., ref. 4) for the simultaneous absorption and elimination of a drug. Figure 2 displays the time-concentration curve for the 354 ug/kg dose.

The absorption-elimination half-lives are given in Table 1. The administered radioactivity was completely recovered in urine and feces over a five day period. A threefold increase in dose resulted in an increase, from 50 to 75, in the percentage of [^{14}C] metabolities in urine and a decrease, from 48.5 to 19, in the percentage found in feces.

TABLE 1. Pharmacokinetic Constants[a] for the absorption-elimination of [^{14}C] oxadiazon equivalents by plasma after intravenous administration of [^{14}C] oxadiazon

Constants	Dosages ug/kg			Units
	118	236[b]	354	
Initial Conc.	389	–	848	ng ml^{-1}
Absorption, t½	3.8	–	3.1	hr
Elimination, t½	26.3	–	31.3	hr

a Nonlin function: $y(t) = yo\ k_1/k_1\text{-}k_2\ (e^{-k2t}\text{-}e^{-k1t})$

b. Values could not be fitted to the function

Percutaneous absorption studies

The percutaneous absorption of oxadiazon was studied in the adult 280 g female Sprague-Dawley rat using the methods and procedures of Knaak, et al. (ref. 5). [^{14}C] Oxadiazon was topically applied to the back skin (11.6 cm^2) at a concentration of 51.7 ug cm^{-2}. Thirty-six rats were treated and groups of three were killed at 0.5, 2, 4, 8, 12, 24, 48, 72, 96, 120, 144, and 168 hours.

Material balance. [^{14}C] Oxadiazon was recovered during the course of the study in feces, urine, carcasses, skin (penetrated and surface residues), cage washings and G.I. tracts, and contents. The recovery data in Fig. 3 is given in percentage of the dose except for urine, feces, and cage washings, which are expressed in cumulative percent of dose.

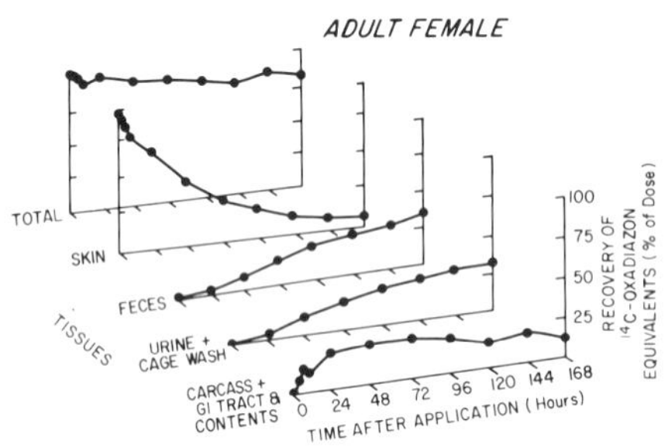

Fig. 3. Time-course recovery of topically applied [^{14}C] oxadiazon.

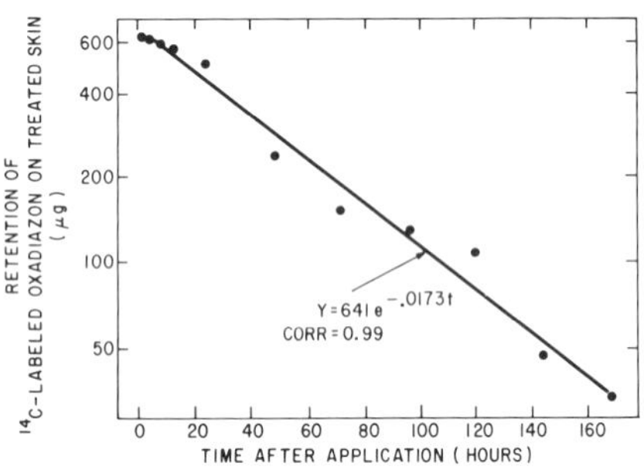

Fig. 4. Skin retention of surface and penetrated [^{14}C] oxadiazon equivalents

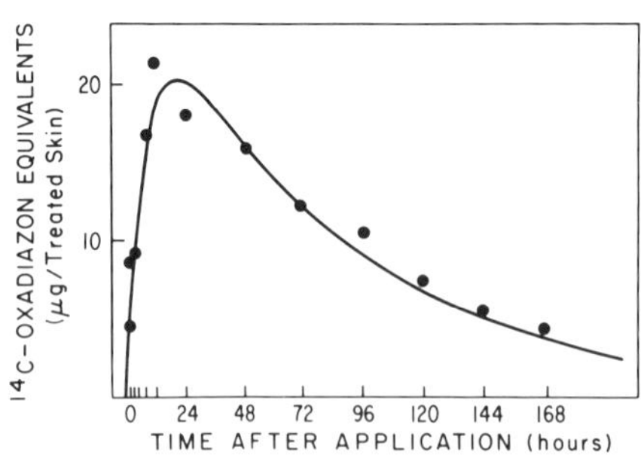

Fig. 5. Absorption-elimination profile of penetrated and bound [^{14}C] oxadiazon equivalents in skin

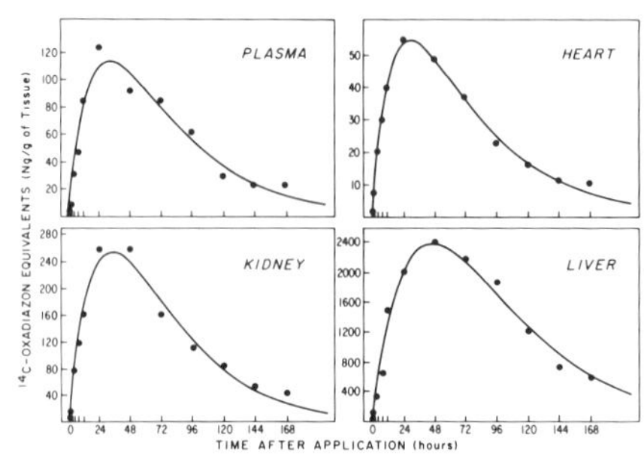

Fig. 6. Time-concentration curves for the simultaneous absorption-elimination of [^{14}C] oxadiazon equivalents in plasma, heart, liver, and kidney.

Recovery varied from 100 percent of the dose at the beginning of the study to 83.5 percent after 168 hours. Oxadiazon was absorbed and excreted in feces (34.0%) and urine (26.8%) as indicated in Fig. 3. A substantial quantity of the [^{14}C] oxadiazon equivalents were found in the carcass during the study (1.9 to 18.0% of dose). Approximately 80 percent of the applied dose was absorbed over the course of the study. Data on the disappearance of the topically applied dose correlated well with the recovered dose in urine, feces, and carcass.

Retention on skin, surface, penetrated and bound residues. The half-life for the retention of the dose (surface and penetrated residues) on skin was determined to be 40 hours in the adult female rat as shown in Fig. 4. [^{14}C] oxadiazon was largely found on the surface (Table 2) of the skin (98.6 to 86.9%) with smaller amounts being present within skin (penetrated residues) as soluble (0.7 to 8.9%) and bound (0.1 to 4.2%) residues. The time-course for absorption-elimination (Fig. 5) was prepared from the penetrated and bound skin residue data. The half-lives for absorption and elimination were 3.9 and 70 hours, respectively.

Simultaneous absorption-elimination. The penetrated dose was absorbed by blood and transferred to tissues in the main compartment. The time-concentration curves for blood, heart, kidney, and liver are shown in Fig. 6. The half-lives for absorption were 14.3, 12.0, 20.5, and 32.1 hours for plasma, heart, kidney, and liver while the elimination half-lives for these same tissues were 40. , 40.9, 28.1, and 32.5 hours, respectively. The rate constants and other additional pharmacokinetic data are listed in Table 3. The 1:1 ratios for the liver and kidney constants resulted in large standard deviations and poor estimates for these constants. The peak concentration in the liver and the area under the curve (AUC) were more than 20 times the corresponding plasma values.

Skin absorption rates. The rate of absorption through skin was determined to be 0.5 ug hr^{-1} cm^{-2} according to the procedure shown in Table 4. The skin permeability constant, permeability relative to triadimefon (ref. 5), and effective concentration on skin are also given in this Table.

TABLE 2. Time-Course Recovery of [^{14}C] Labeled Oxadiazon Equivalents From Skin of Adult Female Rats

	Percentage of residual [^{14}C] Oxadiazon Equivalents[a]		
Sacrifice Time (hr)	Washings from Intact Skin[b]	Washings from Ground Skin[c]	Combusted Skin[d]
0.5	97.2	0.7	0.1
2.0	98.6	1.2	0.3
4.0	98.4	1.7	0.4
8.0	97.0	2.5	0.5
12.0	96.1	3.2	0.7
24.0	96.3	3.2	0.5
48.0	93.2	5.4	1.4
72.0	92.2	5.1	2.7
96.0	91.5	6.2	2.3
120.0	93.1	4.9	2.0
144.0	87.9	7.6	4.5
168.0	86.9	8.9	4.2

a Total [^{14}C] oxadiazon equivalents at sacrifice.
b Surface residues removed by washing excised treated skin with acetone.
c Penetrated residues extracted using methanol followed by acetone.
d Bound residues.

TABLE 3. Pharmacokinetic Constants for the Simultaneous Absorption-Elimination of [^{14}C] Labeled Oxadiazon Equivalents in Plasma and Organs of Adult Female Rats

Constants	Plasma	Heart	Kidney	Liver	Units
Theoretical Initial Concentration	199.0[a]	90.4[d]	602[g]	6425[j]	ng ml^{-1}
Absorption Constant	0.048[b]	0.0577[e]	0.0338[h]	0.0216[k]	h^{-1}
Corresponding, t½	14.3	12.0	20.5	32.1	h
Elimination Constant	0.0172[c]	0.0169[f]	0.0147[i]	0.0213[l]	h^{-1}
Corresponding, t½	40.3	40.9	28.1	32.5	h
Position of Peak Time	34	24	36	48	h
Height	112.0	53.2	260	2374	ng ml^{-1}
R-Squared	0.94	0.99	0.97	0.99	
Absorption-Elimination Ratio k_{12}:k_{20}	2.8:1	3.4:1	1.3:1	.1:1	
Area Under Curve	10.6	4.9	23.2	263.2	mg h ml^{-1}

a ±44.8 d ±6.5 g ±337 j ±162,592
b ±0.015 e ±0.006 h ±0.0210 k ±0.547
c ±0.005 f ±0.0015 i ±0.0147 l ±0.542

Nonlin was used to determine the constants (ref. 4).

TABLE 4. Absorption of [^{14}C] Oxadiazon Through Adult Female Rat Skin in Vivo

Skin Absorption Rate[a] (ug hr^{-1} cm^{-2})	Skin Permeability Constant p[b] (cm hr^{-1})	Relative[c] Permeability	Concentration in skin C[d] (ug cm^{-3})
0.52	10 x 10^{-3}	0.9	41.4

a $$r = \frac{1/2 \text{ Absorbed Dose (ug)}}{t½ \text{ (hr)}} \times \frac{1}{\text{area treated in cm}^2}$$

1/2 absorbed dose, 240 ug; t½ for plasma elimination and retention on skin, 40 hr.

b p = r/applied dose in ug cm^{-3}

c Permeability relative to triadimefon in adult females, Knaak et al. (ref. 5).

d C = Applied dose in ug cm^{-3} x percentage absorbed in decimal form.

DISCUSSION

The i.v. studies were performed to determine the plasma elimination half-lives for increasing doages of [^{14}C] oxadiazon and to obtain information on the elimination of oxadiazon in urine and feces. The results obtained with plasma were not anticipated or easily explained until the results from the percutaneous absorption studies were available. The uptake and metabolism of the dose by liver appears to be largely responsible for the plasma elimination curve. Many investigators have used i.v. urinary elimination data to correct percutaneous absorption data for incomplete elimination (Wester and Maibach, ref. 6). If this procedure were carried out using the i.v. and percutaneous urinary elimination data, the calculated absorbed dose would be smaller (52 vs 80%) than the actual dose absorbed in the percutaneous studies.

A dose (51.7 ug cm^{-2}, 11.6 cm^2 treated) likely to be received by workers using oxadiazon was topically applied to the back skin of adult female rats. The skin retention half-life of this finite dose correlated well with its absorption from skin and elimination from plasma. The multi-step process was characterized in part by the simultaneous absorption elimination of [^{14}C] oxadiazon by skin, blood, and tissues. In this process, small but continuous quantities of oxadiazon were absorbed by skin and transferred to blood, distributed to tissues, metabolized, and eliminated in urine and feces. The half-life required to eliminate the absorbed dose from the main compartment (40 hours) was equivalent to the retention half-life of the applied dose on skin. We believe that the elimination half-life for plasma and the half-life for the retention of the dose on skin are good estimates of the time required to absorb the topically applied dose. A modification of the procedure of Knaak, et al., (ref. 5) was used to calculate the rate oxadiazon was absorbed through skin. The finding of a twentyfold increase in the concentration of [^{14}C] oxadiazon or equivalents in liver over plasma supports the results of the oncogenicity studies (ref. 1 and 2) conducted with oxadiazon. The percutaneous absorption data obtained in this study was used by Crump et al., (ref. 7) to quantitate the risk to workers in conjunction with animal tumor (ref. 1 and 2) and exposure data (ref. 3). The risk was less than one in a million of getting cancer from oxadiazon.

CONCLUSIONS

Oxadiazon (51.7 ug cm^{-2}, 11.6 cm^2) was absorbed through the back skin of adult female rats at the rate of 0.5 ug hr^{-1} cm^{-2}. This rate was equivalent to 0.9% of the applied dose per hour. The urinary values obtained from the intravenous study underestimated the amount absorbed via the percutaneous route.

The uptake of large quantities of [^{14}C] oxadiazon by the liver supported the results of cancer studies in which the liver was shown to be the target organ.

The results of this percutaneous absorption study, the exposure study of Hamburger (ref. 3), and a cancer study (ref. 1 and 2) were used by Crump et al., (ref. 7) to assess the risk to workers applying oxadiazon. The risk was less than one in a million of developing cancer.

Acknowledgement–The authors wish to thank Fiorella Gielow, Moire Robertson, and Gene Lee for their technical assistance.

REFERENCES

1. G. M. Wang, Regul. Toxicol. and Pharmacol. 4, 355–360 (1984).
2. G. M. Wang, Regul. Toxicol. and Pharmacol. 4, 361–371 (1984).
3. S. J. Hamburger, Internal Report, Rhone-Poulenc, Inc. (1984).
4. C. M. Metzler, D. L. Elfring, A. I. McEwen, Biometrics 30, 562–563 (1974).
5. J. B. Knaak, Karin Yee, C. R. Ackerman, G. Zweig, and B. W. Wilson, Toxicol. Appl. Pharmacol. 72, 406–416 (1984).
6. R. C. Wester and H. I. Maibach, Drug Metabolism Reviews 14(2), 169–205 (1983).
7. K. S. Crump, G. M. Wang, and K. T. Maddy, Report HS-1330, California Department of Food and Agriculture, Sacramento, CA (1985).

An investigation into the desirability of the inclusion of a second litter per generation in rat multigeneration studies

David J. Clegg* and Frank Wandelmaier

Health Protection Branch, Health and Welfare Canada, Ottawa, K1A 0L2

Abstract - Data on 159 studies submitted for regulatory purposes were screened to determine whether inclusion of a second (b) litter would improve the appraisal of rat multigeneration studies. A total of 41 studies showing some adverse effect remained after the initial screening. Examination of median values for measured parameters indicated 13 compounds which appeared to show an increased sensitivity of "b" litters, compared to "a" litters in the same generation. Full evaluation of all available data for the reproduction studies on these 13 compounds indicated evidence of increased "b" litter sensitivity in 8 of the studies although in only 2 cases were these data the limiting factor in determining the NOEL for the study.

INTRODUCTION

The scientific data base (on protocols generally utilized in the rat multigeneration studies required for regulatory purposes) has been examined with respect to the optimum number of generations required in such a study (ref. 1). However, there are disparities in the number of litters required in each generation (refs. 2 & 3). The present study was therefore undertaken to determine the necessity for inclusion of second litters in rat multigeneration studies.

METHOD

The material used in this survey comprised 159 valid rat multigeneration studies submitted to Food Directorate of the Canadian Health Protection Branch, to support of the registration of pesticides. To obtain studies that were uniformly designed and conducted, all studies were screened and those that failed to conform to the following criteria were rejected: 1. Dosing was continuous throughout the study. 2. Dosing was for at least 65 days prior to pairing, for the F_0 generation. 3. At least 2 litters/generation were produced. 4. There were no changes in dose levels during the study. 5. No litters were killed at birth, or utilized for teratology studies. 6. Data were reported separately for "a" and "b" litters. 7. Studies encompassed at least 2 generations. 8. The "rest" period for parent animals, between termination of lactation and pairing for the second litter was at least 7 days. 9. Parent animals subsequent to the F_0 generation were derived from "b" litters. 10. Both "a" and "b" litters were treated similarly. 11. Age of parents at second mating was less than 200 days. 12. Selection of animals at all stages was random. 13. Group size/ dose level for parent animals exceeded 10/group. 14. Parent animals were not replaced during the course of the study. Finally, the study design and results had to be adequate for regulatory purposes. As a result of this screening 106 studies were rejected, leaving 53 studies for further analysis.

A superficial evaluation of the remaining 53 studies (based on mean values for the parameters investigated) resulted in a further 12 being eliminated since these 12 studies did not show any evidence of toxicity to either maternal animals or offspring. (By present standards, these studies would probably be rejected as being inadequate for regulatory purposes but at the time of their original evaluation, were accepted for this purpose.).

RESULTS

A total of 41 studies remained for the comparative examination of the sensitivity of "b" (second) versus "a" (first) litters. This examination indicated 13 compounds (14 studies) with apparently increased sensitivity of "b" over "a" litters. All 14 studies were then evaluated in depth.

Compound 1: Increased "b" litter incidence (as compared to "a" litter incidence) of 5-21 day post-natal mortality was apparent in all generations at the 3 upper dose levels and in litters in the second and third generations at the fourth (low) dose level. The "b" litter effects at the 2 lower doses were the definitive effects precluding determination of a NOEL (no observable effect level).

Compound 2: Increased "b" litter incidence was apparent at the mid- and high-dose levels with respect to 1-5 day and 5-21 day post natal mortality in all generations. There was also an indication of increased "b" litter sensitivity with respect to decreased litter size, which occurred in 2/3 generations at the mid-dose, and 2/2 generations at the high dose level. However, the limiting parameter (i.e. the parameter which determined the NOEL) in the study was reduced fertility which was of comparable severity in matings for both "a" and "b" litters.

Compound 3: Significantly reduced pup weight at weaning was observed in F_{1b}, F_{2a}, F_{2b} and F_{3b} litters at the top dose level. In all generations, the degree of weanling weight reduction on "b" litters was greater than that of "a" litters. At the mid-dose level, pup weight gain (days 1-21) was also decreased to a greater degree in "b" litters but did not attain statistical significance (P 0.05). The limiting factor in this study was parental body weight at the mid-dose level. These data were only available on "b" litter derived parents, since "a" litters were sacrificed at weaning, thus no direct comparison was possible. However, since sub-acute (90 day) and chronic studies did not affect body weight of test animals, it is likely that the effects initiated on body weight during lactation were directly related to the reduced parental weight, in which case "b" litter data would be the limiting parameter.

Compound 4: Effects on post-natal mortality from day 5-21 post-partum showed an increased incidence in "b" litters. However, the converse was true in the 1-5 day period, "a" litters showing increased incidence with respect to mortality. Thus the diminished population during the period of "b" litter increased sensitivity renders definitive evaluation difficult. Further, the test material, being an organo-chlorine compound, which is not only cumulative, but also acts as a microsomal enzyme stimulator further complicates the evaluation of early versus late post-natal mortality. The NOEL in this study was based on post-natal mortality and, over the full pre-weaning period, the reduced survival is comparable in both "a" and "b" litters.

Supplementary data from a published 6 generation mouse study on this compound (aldrin)[4] does show increased "b" litter incidence with respect to 4-21 day post partum mortality, with significant increases in all 6 "b" litters, but only in 1 "a" litter. This study also shows increased mortality over the 1-4 day period is greater in "a" and "b" litters, with significant increases in 6 "a" litters, but only in 4 "b" litters.

Compound 5: Only 1 dose level was utilized in this study. In the F_1 and F_2 litters, the post-natal survival was reduced (days 1-21) in both "a" and "b" litters, the reduction being greater in "b" litters. In the F_3 generation, post-natal survival was not affected in either litter. The NOEL for this study was based on significantly increased liver weights of F_{3b} pups. Data on this parameter were not available for any other litters.

Compound 6: Increased post natal mortality occurred in all generations at the top dose level, with increased effects in F_1, F_2 and F_4 "b" litters. In the F_3 generation, the effect was greater in "a" litters. At low- and mid-dose levels, effects were sporadic, non-dose related, and random with respect to sensitivity of "a" and "b" litters.

Compound 7: Increased incidence in "b" litters was observed at the top dose level, with respect to 1-28 day mortality. At lower doses, mortality was generally comparable to controls. Data on the third generation are not comparable since F_{3b} pups were sacrificed after 21 days rather than at the termination of the 28 day weaning period. The NOEL was based on litter size at birth. With respect to this parameter, "a" litters were more sensitive than "b" litters.

Compound 8: Increased post natal mortality was noted at the top dose level, the effect being greater in "b" litters in the F_1 generation. This could not be confirmed since the top dose level was discontinued due to the lack of sufficient "b" litter survivors to provide a parental group. At lower doses, in the F_1 generation, post natal mortality in "b" litters is greater than in "a" litters at all dose levels. However, the converse is true in the F_2 generation.

Compound 9: There were no marked differences between "a" and "b" litters. At most, there is a slightly increased sensitivity of "b" litters with respect to post-natal mortality in the F_2 and F_3 generations.

Compound 10: Two studies, performed 2 years apart in the same laboratory, using the same protocol, were reviewed. The first study, following culling on day 5, shows an increased incidence in "b" litters, with respect to post-natal mortality in all 3 generations at the only dose level used. The second study did not show any effects on post-natal survival, and thus apparently failed to confirm the earlier results. However, information is available indicating a change in the manufacturing process during the 2 year period between studies. Unfortunately without specifications on the test material (which could not be ascertained) there is no certainty that the same test material was used in both studies.

Compounds 11 to 13: Problems with wide variations between generations in the control values, lack of consistency in results especially with respect to dose or markedly decreased survival in individual groups resulted in detailed evaluation of the studies on these compounds being incapable of determining the relative sensitivity of "b" litters with respect to "a" litters.

Discussion

The above examination, based on proprietary data submitted to the Health Protection Branch (thus precluding identification of the specific compounds) is, in reality, a preliminary study in that, because of the design of the investigation, not all studies received equal attention. The investigation design was intended to ensure a homogeneous sample, which resulted in the rejection of 106 studies which failed to meet one or more of the listed criteria. It should be emphasized that the applied criteria do not necessarily indicate that a study would be rejected for regulatory purposes. Further, it should also be remembered that the time frame in which these studies were performed covers at least two decades.

In selecting the screening criteria, the major consideration was that of obtaining a uniform data base. Some of these criteria may be obsolete, such as the minimum pre-treatment period of the F_0 parents, which, in terms of effects other than bio-accumulation may be strain dependent, rather than generalized. No data appear to be available on the effects of utilizing "a" litters rather than "b" litters for selection of parental animals. Neither could any data be obtained on the effect of the duration of the "rest" period between weaning of "a" litters and re-pairing for mating to obtain "b" litters. These criteria were included as a precaution, rather than because of known variations resulting from change. The possibility of such variations remains to be assessed.

The superficial review of the 53 studies was performed to determine where the possibility of increased "b" litter sensitivity might occur. Tabulated mean value data only were considered. If effects on "b" litters in the upper dose level exceeded the incidence of effects on "a" litters the studies were scheduled for full evaluation. No attempt was made to determine the incidence of increased sensitivity of "a" litters.

Because of the variation in the time of performance of the studies the quality and quantity of available data for each of the studies varied. In some cases, data were inadequate to permit statistical examination, and observations were therefore reliant upon the experience of the authors.

Of the 41 studies which formed the positive homogeneous sample utilized, 14 studies on 13 compounds were reviewed "in depth". Of these the NOEL for the reproduction study was based on the "b" litters in the case of 2 compounds (compounds 1 and 3). In a third case (compound 2) although clear indication of increased "b" litter sensitivity was observed the parameter affected was not that which determined the NOEL in the study.

Compound 4 through 8 appear to indicate an increased sensitivity in "b" litters, but in no case were the "b" litters the limiting consideration. Compound 10 showed equivocal results, based on the 2 studies performed. Although one study did show increased "b" litter sensitivity to post-natal mortality, confirmation of this effect in the second study was not obtained. However, it is impossible to ascertain purity or identity of contaminants for the test material in these studies. It should be noted that now 2 generation studies (as opposed to 3 or more) are the norm, the concern over "b" litter data would be increased with respect to compounds 6 and 7.

In the detailed evaluation the parameters in those studies where "b" litter incidence is increased are post-natal mortality and decreased body weight gain during lactation. It should be noted that all reported parameters were examined, but that "b" litter increased sensitivity was only noted in post-natal parameters.

CONCLUSION

The regulatory standpoint is normally to consider the "worst-case" situation. The present study indicates that there is some evidence that increased "b" litter sensitivity does occur and that it may affect the NOEL for multigeneration studies. Consequently, it was concluded that 1 litter/generation could not be universally accepted. However, it would be adequate if:

1. No adverse effects are induced on reproductive parameters, at dose levels up to either a dose level causing minimal toxic effects in short or long term feeding studies in the same species, or an adequate multiple of maximum anticipated exposure levels, whichever is higher.

2. Effects are restricted to clear effects on pre-natal data (such as fertility), where the test material is non-cumulative, or where blood and tissue levels have stabilized prior to mating, as demonstrated in adequate experimental studies.

3. Effects are limited to induction of non-lethal congenital malformations.

REFERENCES

1. D.J. Clegg, "Animal reproduction and carcinogenicity studies in relation to human safety evaluation." Toxicology and Occupational Medicine, p.45-59, Elsevar/North Holland, (1979)
2. Organization for Economic Co-operation and Development, OECD Guidelines for Testing of Chemicals. Section 4, Health Effects, Guideline 416 (1983).
3. Toxicological Principles for the Safety Assessment of Direct Food Additives and Color Additives used in Food, p. 80-107, Food and Drug Administration, USA (1982).
4. M. Keplinger, W.B. Duchman and F. Sala "Effects of combinations of Pesticides on Reproduction in Mice", Pesticide Symposia", p.125-138, Hales & Assoc.

Reactive nitroso, hydroxylamino, epoxide and ozonide intermediates in pesticide degradation

Luis O. Ruzo*, William M. Draper, Ella C. Kimmel and John E. Casida

Pesticide Chemistry and Toxicology Laboratory, Department of Entomological Sciences, University of California, Berkeley, California 94720, U.S.A.

Abstract - Peracid oxidation is used to generate reactive intermediates as follows: nitroso derivatives of anilines formed from formamidine insecticides and chloroacetanilide and nitrodiphenyl ether herbicides; hydroxylamino derivatives of these compounds and of glyphosate herbicide and the aniline derived from fluvalinate insecticide; epoxy derivatives of resmethrin insecticide and methfuroxam fungicide; ozonolysis products of pyrethroids.

INTRODUCTION

Most organic pesticides undergo activation or detoxification by oxidative metabolism and photodegradation. An understanding of these processes helps to define the mechanism of pesticidal action and to anticipate potential hazards. Biomimetic reactions using model oxidants can assist in the generation of intermediates of toxicological interest in relatively simple environments where the reactions can be monitored spectroscopically and additional reactants or derivatizing agents can be introduced as desired (ref. 1). This review considers our recent studies on the oxidation of pesticides yielding toxicologically-significant products identical to those obtained on metabolism or photodecomposition and of potential environmental relevance.

NITROSO AND HYDROXYLAMINO DERIVATIVES

The formamidine insecticides chlordimeform (R_1=Cl, R_2=CH_3) and amitraz (R_1=CH_3, R_2=bis-amidine) and the chloroacetanilide herbicides alachlor (R_1=R_2=Et, R_3=CH_2Cl, R_4=CH_2OCH_3), metolachlor (R_1=Me, R_2=Et, R_3=CH_2Cl, R_4=$CH(CH_3)CH_2OCH_3$) and butachlor (R_1=R_2=Et, R_3=CH_2Cl, R_4=$CH_2OC_4H_9$) revert in part to the corresponding disubstituted-anilines on mammalian metabolism and environmental degradation. Some of these pesticides and their anilines are promutagens in bacterial assays and carcinogens in rats or mice. m-Chloroperbenzoic acid (MCPBA) converts the anilines to the same hydroxylamino- and nitroso-derivatives observed (GC-MS) in the liver of rats treated with the pesticides or their aniline derivatives and in microsomal mixed-function oxidase reaction mixtures. These hydroxylamino and nitroso derivatives are mutagens in the Ames Salmonella typhimurium strain TA 100 assay (Ref. 2).

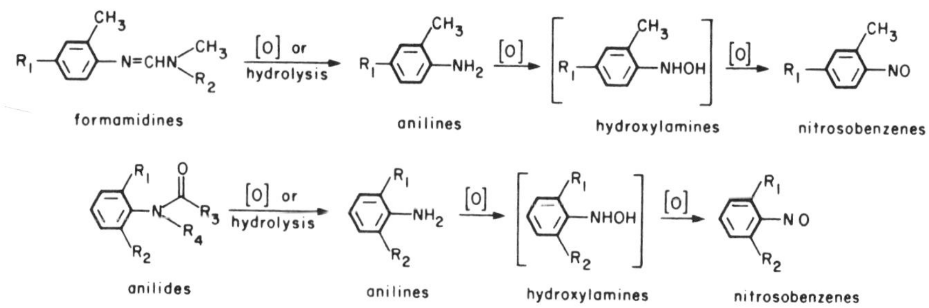

The nitrodiphenyl ether herbicides nitrofen, acifluorfen and oxyfluorfen undergo reduction in rats to the corresponding anilines which are excreted in the feces. The presumed intermediates are the nitroso and hydroxylamino derivatives (ref. 3). The overall reduction sequence and the reverse oxidation are also observed on photodegradation. MCPBA converts these aminodiphenyl ethers to nitroso and hydroxylamino derivatives also formed on metabolism. These materials are mutagens in bacterial assays. Nitroso-nitrofen adds to alkenes to yield N-hydroxy derivatives which are oxidized to relatively stable free radicals observable by ESR (ref. 4). The type of free radicals involved in the herbicidal action of nitrodiphenyl ethers is not conclusively defined. Nitrosoaryls are also reactive with thiols, including glutathione.

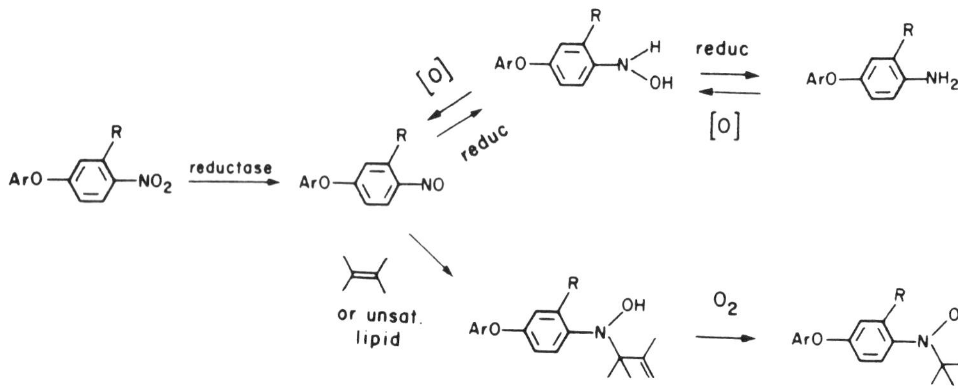

Glyphosate and the acid moiety of fluvalinate undergo carbon-nitrogen bond cleavage on reaction with MCPBA leading ultimately to many of the same products formed on metabolism and environmental degradation (ref. 5). N-Oxidation followed by dehydration of the hydroxylamines and hydrolysis of the resulting imines are plausible steps to explain the conversion of glyphosate to aminomethylphosphonic acid, glycine and formaldehyde and of the fluvalinate anilino acid to 2-chloro-4-trifluoromethylaniline. The anilino acid from fluvalinate is converted to the intermediate imine and the corresponding aniline on incubation with liver homogenate.

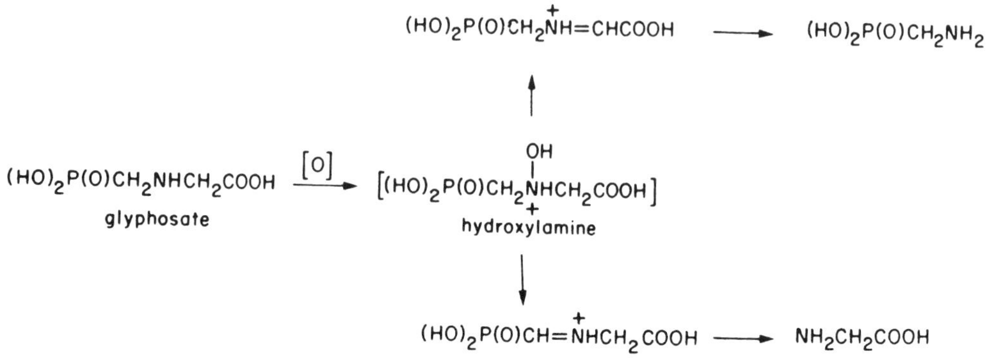

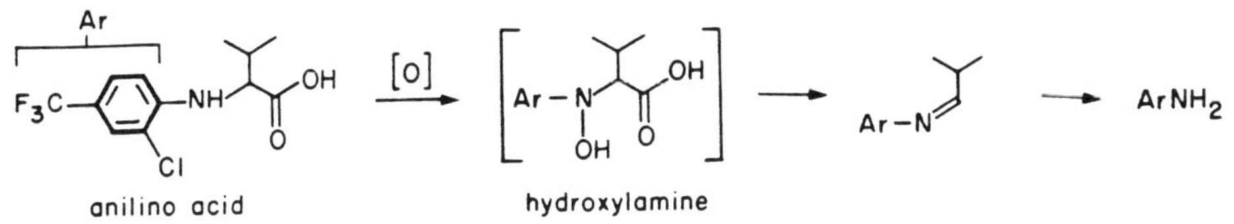

EPOXYFURAN INTERMEDIATES

The furan-containing alcohol moiety of the pyrethroid resmethrin binds to microsomal proteins on oxidative metabolism. Resmethrin and its analogs yield α,β-unsaturated ketoaldehydes on rat microsomal metabolism, identical to those obtained on treatment with MCPBA (ref. 6). The position for initial oxidation is established by studies with [^{18}O]-MCPBA. The ketoaldehyde reacts readily with bovine serum albumin and with thiols.

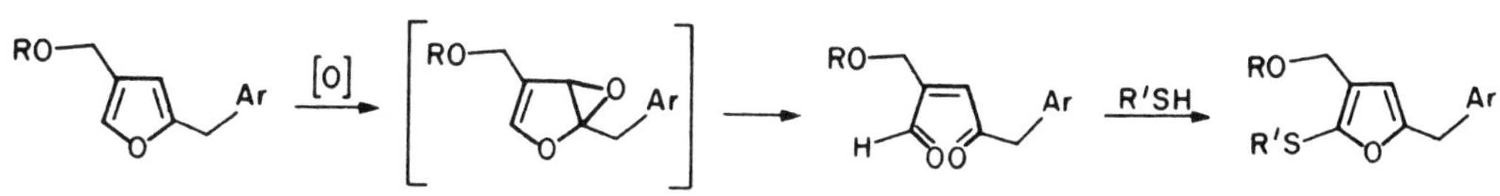

Peracid oxidation of the fungicide methfuroxam at low temperature allows direct ^1H and ^{13}C NMR observation of the epoxyfuran intermediate (Ref. 7). At room temperature the epoxyfuran decomposes to the unsaturated diketone. It remains to be determined if one of these products is responsible for the adverse toxicological properties of methfuroxam.

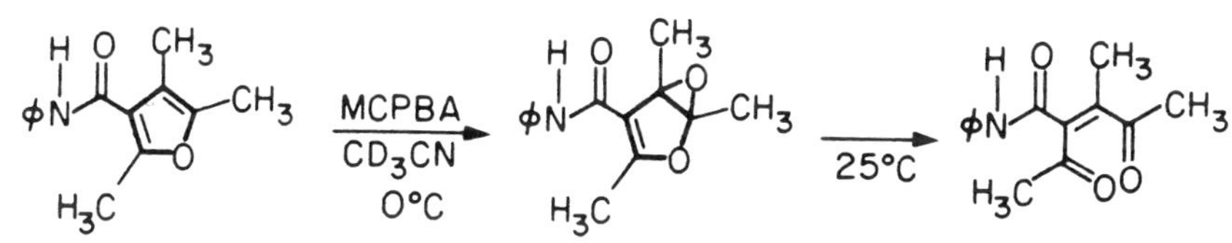

EPOXIDES AND OZONIDES FROM PYRETHROIDS

Ozonization of isobutenyl ($X_1=X_2=CH_3$) and chlorotrifluoromethylvinyl ($X_1=CF_3,X_2=Cl$)-substituted pyrethroids such as phenothrin and halothrin as solutions and as thin films on glass yields predominantly ozonides, while dihalovinyl-substituted compounds such as deltamethrin ($X_1=X_2=Br$) and permethrin ($X_1=X_2=Cl$) form dihaloepoxides as major products (ref. 8). The ozonization reaction mixtures contain unstable bacterial mutagens and degrade to a variety of products identified by a combination of CI-MS and ^{19}F-, ^{13}C- and ^1H-NMR.

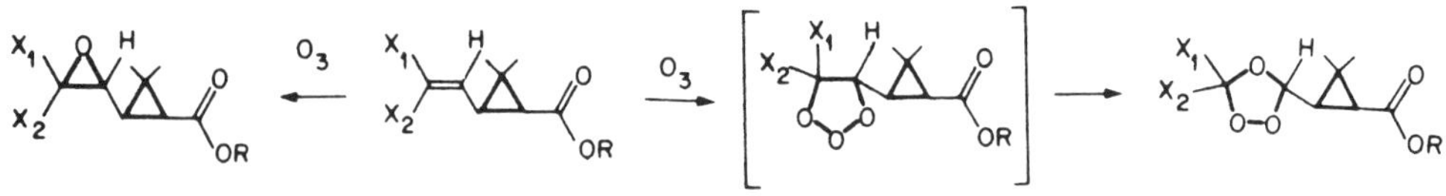

ACKNOWLEDGEMENT

This work was supported in part by National Institutes of Health Grant P01 ES00049.

REFERENCES

1. J.E. Casida and L.O. Ruzo, Xenobiotica, in press (1986).
2. E.C. Kimmel, J. E. Casida and L.O. Ruzo, J. Agric. Food Chem. 34, 157-162 (1986).
3. W. M. Draper and J. E. Casida, J. Agric. Food Chem. 31, 227-231 (1983).
4. W. M. Draper and J. E. Casida, J. Agric. Food Chem. 33, 103-108 (1985).
5. K. Gohre, J. E. Casida and L.O. Ruzo, J. Agric. Food Chem., submitted (1986).
6. L.O. Ruzo, J. E. Casida and I. Holden, J. Agric. Food Chem. 33, 622-625 (1985).
7. L.O. Ruzo, J. E. Casida and I. Holden, J. Chem. Soc. Chem. Comm., 1642-1643 (1985).
8. L.O. Ruzo, E. C. Kimmel and J.E. Casida, J. Agric. Food Chem., in press (1986).

Supernumerary ribs: a transient developmental defect in the rat

G.A. de S. Wickramaratne, M.E. Killick and H. Appleby

Imperial Chemical Industries PLC, Central Toxicology Laboratory, Alderley Park, Macclesfield, Cheshire, SK10 4TJ, U.K.

Abstract - Supernumerary ribs are a frequent occurrence in rat teratology studies and have been considered to indicate a teratogenic potential by some authors but not to be so indicative by others. In an attempt to determine whether the change was permanent and hence by definition teratogenic a study to follow the fate of supernumerary ribs in the post-natal development of the rat was undertaken. Aspirin, an established teratogen in the rat, was used to increase the frequency of supernumerary ribs. The results show that over the first 60 days *post-partum* the frequency of supernumerary ribs in both control and treated groups decline to essentially zero even though aspirin treatment of the dams doubled the initial frequency. This decline in frequency is accompanied by an increase in the proportion of foetuses with a fully developed transverse process on the first lumbar vertebra. It is suggested that supernumerary ribs in the rat are a reflection of developmental delays in a labile region of the axial skeleton and not a manifestation of a teratogenic event.

INTRODUCTION

Supernumerary ribs (extra-ribs, 14th ribs) are a frequent observation in rat teratology studies. Most strains of rat show a spontaneous frequency of occurrence and many compounds or treatments such as Aspirin (ref. 1) and stress (ref. 2) have been shown to increase their frequency. They have been considered to represent a teratogenic event by some authors (ref. 3) and not to be so by others (ref. 4,5). It is generally accepted that teratogenicity leads to a permanent morphological change in the offspring when the mother is exposed to extrinsic agents during pregnancy. Changes in the offspring such as retarded ossification are usually transient and not considered to be teratogenic.

In an attempt to provide evidence to support one or other view a study to follow the *post-natal* prevalance of supernumerary ribs after treatment with a chemical known to increase their frequency was designed. This paper reports the outcome of the study and discusses the implications on the interpretation of supernumerary ribs in teratogenicity studies.

MATERIALS AND METHODS

Primigravid rats of the Alpk:AP (Wistar derived) strain were housed in a barrier-maintained unit. This strain is known to have a spontaneous frequency of supernumerary ribs of the order of 20%, some of which are vestigial and others of normal length. It is also known from studies in this laboratory to be susceptible to an increased frequency of extra ribs when dams are treated with aspirin.

Time mated rats (day of sperm positive vaginal smear=day 1) were dosed by gavage on days 7-16 (inclusive) of gestation with 100 mg/kg/day of aspirin (Aspirin BP) as an aqueous suspension. Concurrent controls received an equivalent volume of 10 ml/kg/day of distilled water.

On day 19 of gestation the cages were provided with solid bottoms and bedding. The dams were allowed to litter and rear their offspring, the day of birth being day 1. Thirty control litters and 30 treated litters were available for the study. On *post-natal* days 12, 21, 40 and 60 two pups per litter were selected at random for evaluation of their skeletons. After euthanasia they were eviscerated and stained with alizarin (ref. 6) for examination. The number of ribs on each side of the thoracic cage was counted and recorded.

RESULTS

The frequencies of supernumerary ribs at each time point for control and treated groups is summarised in TABLE 1. The frequency of a fully ossified transverse process on the first lumbar vertebra (21st vertebra) is also shown in the table.

TABLE 1. Percentage occurrence of supernumerary ribs and transverse processes

Observation	Treatment	Age (days)			
		12	21	40	60
Supernumerary ribs					
Combined	Control	21	9	0	0
	Aspirin	47	38	0	2
Right	Control	5	3	0	0
	Aspirin	8	2	0	0
Left	Control	10	6	0	0
	Aspirin	8	15	0	0
Bilateral	Control	6	0	0	0
	Aspirin	32	21	0	2
Transverse processes					
Right	Control	0	0	8	11
	Aspirin	0	3	5	6
Left	Control	0	0	10	6
	Aspirin	0	0	11	14
Bilateral	Control	0	0	35	44
	Aspirin	0	6	39	55
No. of pups	Control	63	64	64	63
	Aspirin	66	66	65	66

The spontaneous frequency of supernumerary ribs shortly after birth is 20% and declined to zero by day 60 post-partum. In parallel with this decline the frequency of transverse processes on the first lumbar vertebra increased.

Treatment with aspirin at 100 mg/kg/day produced no adverse effects on the dams. There were no significant effects on the body weight of the dams nor were there any overt malformations in the surviving offspring. This is in keeping with previous experience in this laboratory where a dose of 250 mg/kg/day of aspirin is required to elicit dysmorphogenesis such as cranio-rachischisis. However at the dose used (100 mg/kg/day) the frequency of supernumerary ribs shortly after birth is approximately double that of the control group, again in line with previous experience with the strain. By day 60 post-partum the frequency of supernumerary ribs in the aspirin treated groups has declined essentially to zero. As with the controls this decline is accompanied by an increase in the prevalence of a fully ossified transverse process on the first lumbar vertebra. Some of the supernumerary ribs seen shortly after birth in the aspirin treated groups were of normal length others were of the vestigial type.

DISCUSSION

In an earlier presentation the pattern of transformation of spontaneously occurring supernumerary ribs was demonstrated (ref. 7). This study shows that even when their frequency of occurrence is increased by treatment of the dams with a chemical such as aspirin, the same pattern of transformation is maintained.

The thoracico-lumbar border is a particularly labile region during development (ref. 8) and readily affected by slight disturbances. As a result it is likely that low levels of stress would interfere with pattern formation under the direction of inductive forces during differentiation of this region. This could lead to the formation of supernumerary ribs by the segment that should form the 21st vertebra as a lumbar (the 33rd metameric segment) turning it into an apparent thoracic vertebra. Subsequently the normal structure of this vertebra is restored by remodelling of the bony elements involved via the normal processes of erosion and accretion that shape bone (ref. 9). Thus it is unlikely that changes in frequency of supernumerary ribs seen in rat teratology studies, unless forming an integral part of a syndrome of events, could be indicative of a teratogenic event. Even if it were to occur as a part of a collection of more serious malformations it is still likely that the supernumerary ribs reflect only stress related disturbances at the thoracico-lumbar border and are not _de facto_ a part of a teratogenic syndrome.

It is hence considered that the frequency changes in the incidence of supernumerary ribs observed in teratogenicity studies should not be considered as indicative of foetal dysmorphogenesis but as a manifestation of the non-specific stress the dam is placed under in such studies inducing transient changes at the unstable border between the thorax and lumbar region. An increase in frequency of supernumerary ribs could therefore be evidence that a minimal toxic dose had been achieved in the absence of more overt signs such as a depression in body weight gain.

REFERENCES

1. J. Wardany and E. Takacs. _Am. J. Pathol. 35,_ 315-331, (1959).
2. N. Chernoff, P. Parson and R.J. Kavlock. _The Toxicologist, 5,_ 185, (1985).
3. C.A. Kimmel and J.G. Wilson. _Teratology, 8_, 309-315, (1973).
4. A. Hudak and G. Ungvary. _Toxicology, 11,_ 55-63, (1978).
5. K.S. Khera. _Teratology, 31,_ 129-153, (1985).
6. R.E. Staples and V.L. Schnell. _Stain Technol. 39,_ 62-63, (1964).
7. G.A. de S. Wickramaratne, D. Richards, D. Kinsey and M. Kilmartin. _Teratology, 32,_ 36A, (1985).
8. E.L. Green. _Genetics, 47,_ 1085-1096, (1962).
9. N. Bateman. _J. Anat. 88,_ 212-261, (1954).

Postnatal alterations in development resulting from prenatal exposure to pesticides

Robert J. Kavlock*, John M. Rogers, L. Earl Gray, Jr., and Neil Chernoff

Perinatal and Reproductive Toxicology Branches, Developmental Biology Division, U.S. Environmental Protection Agency, Research Triangle Park, North Carolina, USA 27711

Abstract - Alterations in the developmental processes of embryos resulting from exposure to chemicals are not limited to morphological abnormalities that can be observed in the near-term fetus. In our research on the developmental toxicology of pesticides we have noted morphological defects (cataracts, ablation of Harderian gland development, malformation of the otoliths) in postnatal offspring exposed prenatally that could not be detected in fetal examinations because of the immature nature of these structures at that stage of developement. In addition, other effects on physiological functions have been detected (locomotor hyperactivity and decreased renal concentrating ability) that have no obvious morphological basis. Thus, a postnatal component of a standard teratology bioassay can yield critical new information in the overall assessment of developmental toxicity.

INTRODUCTION

In the assessment of developmental toxicity, four major endpoints of adverse effects are considered: 1) death of the conceptus, 2) structural abnormalities such as cleft palate, 3) altered growth such as fetal growth retardation, and 4) functional deficiencies of organ systems (ref. 1). Generally accepted guidelines for the assessment of teratogenic potential (e.g. the FIFRA Guidelines) basically address only the first three manifestations of developmental toxicity. The fourth currently receives little attention in such studies because their focus is on the morphology of the fetus rather than its physiology; in addition, many organ systems are not yet functional (e.g. the lungs and reproductive organs) or fully differentiated (e.g. the central nervous system, the reproductive tract, and the excretory system) in the fetal rodent. Yet it has become increasingly clear over the last decade that subtle alterations in the performance of a variety of key organ systems may result from exposure during critical prenatal periods. Much of the early work in this field occurred in the area of central nervous system dysfunction, and the study of 'behavioral teratology' has been widespread since the mid 1970's. That pesticides may induce such changes is exemplified by the use of methyl mercury as a prototype behavioral teratogen (ref. 2). Less work has been done on the effects of in utero exposure to chemicals on the function of other organ systems, but sufficient information exists to indicate that the cardiopulmonary, immune, endocrine, hepatic, renal, and reproductive systems are also subject to alterations in functional competence. Often times these functional changes are seen at dose levels levels below those inducing overt structural malformations. The following examples of such observations have been derived from work performed in our laboratory on approaches to evaluating such effects in the offspring of rodents exposed to xenobiotics in utero.

NITROFEN INDUCED DEVELOPMENTAL TOXICITY

Nitrofen (2,4-dichloro-p-nitrophenyl ether) is a herbicide with extremely low toxicity to adult mammals, yet has a remarkable ability to adversely affect the development of embryos. While extensive research has indicated that the most sensitive measure of nitrofen teratogenicity are the diaphragmatic hernias (which occur at doses of less than 1 mg/kg/d), a variety of effects have been observed at slightly higher dose levels. Some of the nitrofen-related effects occur in tissues and organ systems that are extremely difficult to evaluate in the rodent fetus. Even the magnitiude of neonatal mortality following in utero exposure to nitrofen cannot be predicted by routine methodologies, primarily due to its impact on cardiopulmonary development and function (refs. 3 and 4). Other effects (e.g. ablation of Harderian gland development) do not become obvious until later in development (refs. 5 and 6). The Harderian gland is an exocrine gland located in the orbit of reptiles, birds, and mammals that have a nictitating membrane. This teratogenic effect is not identifiable in the fetus because the glands are small and inactive until the animals are 10-12 days old. At that time the gland grows rapidly and begins its secretory process under the partial influence of the thyroid gland. Other nitrofen-induced effects that become evident in the postnatal period include altered renal function (refs. 7 and 8), hyperactivity (refs. 9 and 10), and altered organ growth (ref. 9). In renal function studies, pups exposed to 75 mg/kg on day 11 of gestation had a diminished capacity to concentrate their urine regardless of the presence or absence of hydronephrosis. Although only offspring with hydronephrosis had altered renal function after the neonatal period, the techniques applied to the neonates were predictive of long-term alterations in renal physiology. Delayed development of the renal papilla is a relatively common finding in teratology studies, and the procedures used in this study may provide an efficient means of determining whether the delay is a transient developmental phenomenon or whether progressive damage to renal parenchyma is likely to occur.

DINOCAP INDUCED TORTICOLLIS

Dinocap (a mixture of 2,4- and 2,6-dinitrooctylphenyl crotonates) is a fungicide used in the control of powdery mildew on a variety of crops. Using standard teratology bioassay techniques, we have demonstrated that exposure of mice to dinocap during organogenesis induces fetal growth retardation, cleft palate, and supernumerary ribs in the fetus at dosages not associated with maternal toxicity (ref. 11). While fetal growth was inhibited at the lowest dose tested (5 mg/kg/d), malformations were seen only at 20 mg/kg/d and higher. When other dams were allowed to litter and their offspring evaluated during postnatal maturation, a peculiar behavioral abnormality was observed in the home cage (ref. 12). After postnatal day 21 a number of mice exposed to 12 and 25 mg/kg/d dinocap displayed a twisting of the neck with resultant head tilt which we referred to as torticollis. These mice held their heads tilted constantly to one side. As these animals aged, the behavior of some became more aberrant; some circled repeatedly in one direction, others bobbed their heads and did backflips, while others would roll over constantly. At 30 days of age, the incidence of affected individuals was 6% at 12 mg/kg/d and 24% at 25 mg/kg/d. (Similar effects were not observed in either the rat or the hamster.) This particular behavior has been observed previously in a number of mutant strains of mice and in offspring of manganese-deficient mice. In those two examples, the behavioral deficit has been attributed to defects in the development of the otoliths of the inner ear. Recently (ref. 13) we have demonstrated a similar etiology for the dinocap-induced torticollis. The otoliths of the mammalian ear are gravity receptors located in the utricular and sacular regions and are responsible for sensing static equilibrium and linear acceleration. Each is composed of a sensory epithelium of hair cells over which lies the otoconial membrane. The membrane itself is composed of crystals of calcium carbonate (the otoconia) embedded in a mucopolysacchride matrix. The examination of these very small, highly specialized sense organs requires a particular dissection and clearing procedure which is not typical of a standard fetal skeletal examination. While malformations of the otoliths may be difficult to routinely detect in the fetus, observations of home cage behaviors can readily identify the lesion. Similar to the example with the herbicide nitrofen, the utility of a postnatal component of the standard teratology examination is clearly demonstrated in the effects of dinocap on the embryonic mouse. Unlike the example of nitrofen, however, the dinocap induced functional effects are detected at a dose level below that which causes defects that can be detected by standard procedures.

ENDRIN INDUCED HYPERACTIVITY

Endrin (1,2,3,4,10,10-hexachloro-6,7-epoxy-1,4,4a,5,6,7,8,8a,octahydro-1,4-endo-5,8-dimethanonapthalene) is one of the few cyclodiene insecticides still registered for use in the United States. In teratology studies (ref. 14), endrin induced fetal growth retardation in the mouse, but not the rat, at dose levels that were also toxic to the dams. In the hamster (ref. 15), exposure to endrin during the entire period of organogenesis resulted in increased prenatal mortality and fetal growth retardation at maternally toxic doses (1.5 mg/kg/d). However, when exposure was limited to a single dose on day 8 of gestation, fused ribs and meningoencephaloceles were observed at a nonmaternally toxic dose (5 mg/kg). In a subsequent study, we followed the behavioral development of rats and hamsters exposed perinatally to endrin (ref. 16). Open field behaviors and Figure 8 maze activities were recorded on multiple days between the onset of activity (postnatal day 15) and adulthood. The Figure 8 maze is an automated activity monitor that records the photocell interruptions which occur as an animal travels through the alleys. Because it is an automated device, longer-term observations of activity are possible using this apparatus. In the hamster, exposure to 1.5 mg/kg/d on days 5-14 of gestation produced a persistent elevation in postnatal locomotor activity when measured either in an open field or the Figure 8 maze. The increase in activity amounted to 75% on day 15, 45% on day 20, and 23% at 125 days of age. Non-locomotor behaviors such as rearing, running wheel activity, and sexual behaviors were unaffected by endrin treatment, indicating the specificity of the behavioral effect. Rats exposed to 0.15 or 0.30 mg/kg/d from time of implantation to beginning of solid food intake (day 16 of lactation) were 30% more active than controls prior to weaning, but not as adults. These doses did not result in maternal toxicity or affect postnatal growth or viability. The similarity of behavioral changes noted in the young of both species is suggestive of similar alterations of the central nervous system function even though endrin produces morphological defects only in the hamster.

MIREX INDUCED CATARACTS

Mirex (dodecachloro-octahydro-1,3,4-metheno-2H-cyclobuta(c,d)-pentalene) is a pesticide once used extensively for fire ant control in the southeastern regions of the U.S. That mirex could cause cataracts in developing rodents was first noted in pups nursing mothers exposed to the pesticide (refs. 17 and 18). Subsequently it has been demonstrated that cataracts also are present in fetuses exposed transplacentally (refs. 19-21). Similar to the case with nitrofen-induced ablation of Harderian gland development and dinocap-induced malformations of the otoliths, the cataracts in the fetus were difficult to detect. Not only are the routine fixation and examination procedures not well suited for detecting cataracts in the fetus, but the cataracts occcured in edematous fetuses, thus possibly obscuring the more subtle effect. A systemic hypoproteinemia was correlated with the formation of cataracts, once again pointing to alterations in perinatal physiology as an important toxicological endpoint. While relatively sophisticated techniques are necessary to pinpoint the etiology of this lesion, the adverse outcome of chemical exposure during the developmental period manifested itself in an obvious fashion when the postnatal growth and development of the offspring was monitored.

CONCLUSION

The examples presented above clearly demonstrate that teratogenic effects which may seriously influence the quality of life of the affected individuals can go undetected in standard teratology bioassays. Some of these effects might be expected to be detected in a well designed and executed multigeneration study, but are more likely to be detected in a teratology type bioassay due to the usually higher dosages, route of administration (oral gavage versus dietary), as well as shorter duration of treatment (less chance for induction of detoxification pathways). However, in order to take advantage of the factors of the teratology bioassay that favor the induction of adverse developmental outcomes, it is necessary to extend the period of observation into the postnatal period. Once the developmental effects are observed, further experiments can be designed to fully characterize the nature of the effect and their position on the overall dose-response curve. The types of observations to include in the initial investigation need not be highly sophisticated because many of the lesions that can be missed in the standard fetal visceral and skeletal examinations express themselves quite visibly when the appropriate phase of development is reached.

DISCLAIMER

The research described in this article has been reviewed by the Health Effects Research Laboratory, U.S. Environmental Protection Agency and approved for publication. Approval does not signify that the contents necessarily reflect the views and policies of the Agency nor does mention of trade names or commercial products constitute endorsement or recommendation for use.

REFERENCES

1. U.S.E.P.A., Federal Register, 49, 46324-46331 (1984).
2. L.W. Chang and Z. Annau, In: Neurobehavioral Teratology, J. Yanai, editor, pp. 405-432, Elsevier Science Publishers, Amsterdam (1984).
3. R.D. Kimbrough, T.B. Gaines, and R.E. Linder, Arch. Environ. Hlth., 28, 316-320 (1974).
4. C. Lau, A.M. Cameron, O. Irsula, and K.S. Robinson, Toxicol. Appl. Pharm., in press (1986).
5. L.E. Gray, Jr., R.J. Kavlock, N. Chernoff, J. Ferrell, J. McLamb, and J. Ostby, Science, 215, 293-295 (1982).
6. L.E.Gray, Jr., R.J. Kavlock, N. Chernoff, J. Ostby, and J. Ferrell, Toxicol. Appl. Pharm., 67, 1-14 (1983).
7. R.J. Kavlock and J.A. Gray, J. Toxicol. Environ. Hlth., 11, 113 (1983).
8. R.J. Kavlock and L.E. Gray, Jr., J. Toxicol. Environ. Hlth., 11, 679-690 (1983).
9. Ostby, J., L.E. Gray, R.J. Kavlock, and J.M. Ferrell, Toxicology, 34, 285-297 (1985).
10. L.E. Gray, Jr., R.J. Kavlock, J.Ostby, J. Ferrell, J. Rogers, and K. Gray, Neurotoxicology, 7, 447-460 (1986).
11. J.M. Rogers, B. Carver, L.E. Gray, Jr., J.A. Gray, and R.J. Kavlock, Terato. Carcino. Muta. 6, 375-381 (1986).
12. L.E. Gray, Jr., J.M. Rogers, R.J. Kavlock, J.S. Ostby, J.M. Ferrell and K. Gray, Terato. Carcino. Muta., 6, 33-44 (1986).
13. L.E. Gray, Jr., J.M. Rogers, J.S. Ostby, K. Gray, and J.M. Ferrell, Teratology, 33, 86C (1986).
14. R.J. Kavlock, N. Chernoff, R.C. Hanisch, J.A. Gray, E. Rogers, and L.E. Gray, Jr., Toxicology, 21, 141-150 (1981).
15. N. Chernoff, R.J. Kavlock, R.C. Hanisch, D.A. Whitehouse, J.A. Gray, L.E. Gray, Jr., and W. Sovocol, Toxicology, 13, 155-165 (1979).

16. L.E. Gray, Jr., R.J. Kavlock, N. Chernoff, J.A. Gray, and J. McLamb, Toxicology, 21, 187-202 (1981).
17. T.B. Gaines and R.D. Kimbrough, Arch. Environ. Hlth., 21, 7-14 (1970).
18. N. Chernoff, R.E. Linder, T.M. Scotti, E.H. Rogers, B.D. Carver and R.J. Kavlock, Environ. Res., 18, 257-269 (1979).
19. J.M. Rogers and C.T. Grabowski, Teratology, 27, 343-349 (1983).
20. J.M. Rogers, L. Morelli and C.T. Grabowski, Environ. Res., 155-161 (1984).
21. J.M. Rogers and C.T. Grabowski, Exp. Eye Res., 39, 563-573 (1984).

Pesticide exposure assessment

*Wm. Popendorf[1] and C.A. Franklin[2]

The University of Iowa,[1] Institute of Agricultural Medicine and Environmental Health, Iowa City, Iowa, U.S.A., 52242; and Department of National Health and Welfare[2] Ottawa, Ontario, CANADA K1A 0L2.

Abstract - Only recently has the importance of accurately estimating worker and user exposure to pesticides been fully recognized. The traditional method for estimating inhalation exposure has been through the use of personal monitoring pumps to estimate the concentration of pesticide in the ambient air in the breathing zone. Dermal contact exposure has been estimated through the use of patches placed over or under the clothing and more recently emphasis has been placed on estimating hand exposure through the use of cotton sampling gloves or by washing the hands in suitable solvents. These environmental monitoring techniques attempt to estimate the deposited dosage but give no indication of the actual absorbed dosage. Biological monitoring or, more precisely, measurement of biological parameters such as urinary metabolites, is being used more frequently as an alternative for estimating worker exposure. Techniques for biological monitoring are improving rapidly; however, more research is needed to correlate metabolite levels to toxic effects and to understand the relationship between measured metabolite values and worker exposure.

INTRODUCTION

Although the first worker exposure assessments were reported over 30 years ago, only recently has it been recognized that there were few, if any accurate estimates of the amount of pesticide a user was exposed to during application. As a result, there has been an rise in the last 10 to 15 years of the number of applicator exposure studies. One of the primary objectives of obtaining an accurate estimate of exposure is to enable assessment of the potential human health risks associated with use of the pesticide (ref. 1). Risk assessment entails identification of the toxicological hazard as determined through extensive animal testing (acute, short term, long term studies, including reproduction, teratogenicity, oncogenicity, mutagenicity and neurotoxicity tests) which is required by most countries prior to registration and estimation of the probability of that hazard occurring in humans. A comparison can be made of the dose at which no effects are observed in the test animals with the estimated exposure (dose) to humans. If the resultant margin is wide enough, the product may then be considered for registration.

There are many different reasons for conducting worker or applicator exposure studies:

(1) to quantify exposure under a given set of use conditions to enable risk assessment prior to registration of new chemicals or new uses of previously registered chemicals;
(2) to estimate exposure under special conditions, e.g. ultra-low volume application, application using newly designed machinery, and applications using personal protective equipment to reduce exposure;
(3) to estimate exposure that might have occurred in a cohort of people studied in short-term epidemiologic studies (hypothesis testing);
(4) to monitor exposure to people manufacturing, formulating or using pesticides (policy enforcement).

The majority of the exposure assessment studies are conducted by researchers who emphasize one of three disciplines:

(1) chemists emphasize analytical methodologies;
(2) agricultural engineers emphasize field practices; or
(3) industrial hygienists emphasize health risk management.

The best studies should involve all three disciplines: an interdisciplinary approach.

How exposure assessments are done should be determined by the why, rather than by whom. Two broad strategies of assessing exposure utilize environmental and biological monitoring. These two strategies have in the past been referred to by some as direct and indirect methods, respectively. Environmental monitoring is a broad term which includes the measurement of the concentration of the chemical in the ambient air as well as more precise measurement of the concentration of the chemical actually impinging on body surfaces. Patches have been used to estimate the amount of pesticide in the breathing zone and contacting the lungs and estimates have been made of oral exposure both through swallowing of large airborne droplets and as a result of poor hygiene. The term passive dosimetry has been used to describe these estimates of contact exposure. Biological monitoring estimates the absorbed dose through measurement of the concentration of the chemical or its metabolites in body fluids. In the past, it has also included measurements of parameters such as cholinesterase enzyme levels, but these are now more appropriately referred to as health effects monitoring.

Such strategies for assessing exposure evolve through characteristic stages by which one can judge their scientific growth and/or administrative maturity:
(1) Review: a consolidation of current knowledge;
(2) Laboratory study: basic science in controlled tests;
(3) Field study: applied science in actual (or uncontrolled) practice;
(4) Management: transferral of knowledge into policy.

By categorizing the admittedly selective nature of the papers presented at this session into the following matrix of environmental or biological assessment monitoring (or both) and the above stages of maturity, it seems that environmental assessments have reached a stage of greater administrative or management maturity, but scientific interest in biological assessments is more active (see Table 1).

TABLE 1. Numbers of papers presented

Stage	Biological	Both	Environmental
Review	1		1
Laboratory	4		2
Field	5	3	4
Management		1	3

ENVIRONMENTAL MONITORING

Environmental monitoring in the workplace initially focused on the measurement of levels of the chemical in ambient air and control of exposure was achieved though the setting of guidelines such as Threshold Limit Values (TLV's) for allowable concentrations of the chemical in air. This approach has a number of weaknesses even for the industrial workplace, since the airborne concentration of a chemical is not necessarily linearly correlated with the amount absorbed (ref. 2), and the dermal route of exposure is ignored.

It was because of these problems that techniques were developed to estimate the amount of pesticide that actually impinged on the workers' skin (dermal contact exposure) including the use of various dermal patch or wash techniques to either trap dermal deposits or extract unabsorbed deposits from exposed skin (ref. 3). As recently as 15 years ago, it was first recognized that both wet and dry pesticides could penetrate most clothing. The quantification of penetration data in the field involves measuring dermal exposures outside the clothing and dermal dose inside the clothing. Although a wash-extraction technique can be used to evaluate penetration through clothing, most such studies have used the inside and outside patch technique.

TABLE 2. Examples of the range of reported clothing field penetration data

30 - 40% for dry residues through light clothing
2 - 10% for wet sprays through cotton coveralls
less than 1% for wet sprays through Tyvek & Gortex (lab research)

Such inside-outside patches should not overlap, but if they are separated, one runs the risk that spatial variability of such deposits will affect the apparent penetration data. Spatial variability in exposure can affect individual data pairs, but the average of multiple samples will rapidly yield consistent and reliable results. The importance of clothing motion perpendicular to the skin as a mechanism inducing penetration has not yet been evaluated in regard to the applicability of laboratory methods to predict pesticide penetration in field usage.

The accurate assessment of hand exposure represents a critical and somewhat unique challenge: critical because it can represent the majority of the dermal contact deposit and unique because it can involve a liquid versus dry dust deposition mechanism. The use of gloves for hand assessments has been criticized because of reports that a glove or cloth pad dosimeter can retain 5 times more wet pesticide than can be washed from the skin following short exposures. However, after 1.5 hours this difference is usually less than 2 times and the importance of the rate of dermal penetration may not yet be appreciated in such comparisons. As a generalization, the precision or variability (geometric deviation) among environmental measures of exposure can be approximated as shown in Table 3. Typical ranges of dermal contact exposure clearly vary predominantly with task within the reported range shown in Table 4.

TABLE 3. Variability (geometric deviation) of environmental measures of exposure.

intralaboratory procedures	≤5% or ≤1.05X
intraperson, intratask	10-100% or 1.1 -2X
interperson, intratask	25-100% or 1.25-2X
interperson, intertask	700%[a]

[a] range of means, see Table 4.

TABLE 4. Ranges for dermal contact exposure.

mixer-loader	10-100 mg/hr
applicator	2-20 mg/hr
flagger	3-300 mg/hr
harvester	0.5-30 mg/hr

As shown, the use of patches and handwashes to estimate exposure have a number of limitations. Perhaps one of the more serious is the manner in which high values are treated. Since many exposure studies involve only a few workers, high values frequently are statistically treated as outliers. The development of the generic pesticide database (ref. 4) in which existing data on exposure will be included, may help in establishing the upper limits from numerous studies. Some of the inherent difficulties in placement of patches and the estimation of facial and hand exposure may be alleviated by new approaches for estimating dermal contact exposure, such as the use of fluorescent tracers and video imaging (ref. 5). This new technique may also clarify the issue with respect to extreme values since much less extrapolation of data will have to be done.

BIOLOGICAL MONITORING

In the industrial setting, the uncertainties inherent in correlating ambient concentration of the chemical with exposure led to the development of biological monitoring, where an integrated estimate of absorbed dose can be derived by measuring the concentration of the chemical or metabolite in biological media such as blood, urine, saliva, sweat, expired air, hair and adipose tissue. Routine monitoring of a biological response or of the chemical in question can determine whether the worker is receiving an unacceptably high exposure, usually based on experience gained in the workplace (ref. 6). Biological monitoring is actually one aspect of health surveillance monitoring and although depression of cholinesterase enzyme levels (ChE) in workers exposed to organophosphorus compounds (OP's) has been included by some authors in biological monitoring, it should more appropriately be considered as health effects monitoring. It has also been suggested that ChE monitoring might be useful in quantitatively estimating exposure to ChE inhibitors of known dermal toxicity (ref. 7, 8). It must be emphasized that although biological monitoring as currently practiced was never developed to provide an accurate quantitative estimate of actual exposure, it can provide useful information in situations where workers are receiving continuous exposure (manufacturing and re-entry) and there is sufficient information to enable correlation of levels measured and health outcome.

In situations where it is essential that an accurate estimate of exposure be obtained, e.g. for risk assessment for registration purposes, it is feasible that measurement of urinary metabolites in workers could be used to quantify actual exposure. It has been shown in controlled laboratory studies that there is a linear relationship over a wide dose range between the amount of azinphosmethyl applied dermally and the amount of the DMTP metabolite excreted in the urine of both rats (ref. 9) and humans (ref. 10). Knowledge of this relationship enables a standard curve approach to be used and if urinary DMTP levels were measured in workers, an estimate of dermal dose could be made using this standard curve.

Exposure studies conducted on orchardists in Ontario showed that they excreted an average of 688 ug of DMTP (ref. 10). In this same study, their dermal exposure, as estimated using patches and cotton sampling gloves, was 5200 ug. If the standard curve which was developed to correlate dermal dose with level of metabolite excreted is used, urinary output of 688 ug DMTP would be equivalent to 6880 ug of azinphosmethyl on the skin. These two estimates of dermal contact exposure (5200 ug and 6880 ug) are extremely close. In another study conducted in Nova Scotia, the standard curve approach provided a more reliable estimate of exposure than did the patch estimate (ref. 10).

Another completely different way in which urinary metabolites could be used to estimate risk would be to measure urinary metabolite levels in animals dosed with the chemical and then correlate their metabolite levels with the observed toxic effects. The NOEL derived from the study could be expressed in terms of dose or of metabolite level. If the chemical was metabolized and excreted in a similar manner in man, then the urinary metabolite levels measured in workers could be directly correlated with toxic effects. This approach would eliminate the need to estimate contact exposure as currently done through environmental monitoring. It is unlikely that this could be done for all products since some will have extremely complicated metabolism, but it could prove very useful for products with known metabolic pathways.

Comparable interperson variability in biological measures of exposure can be estimated from urinary excretion data from field exposure that varies ±45%; even excretion data from controlled laboratory exposure varies ±50%. The Unified Field Model enzyme response coefficient for organophosphorus compounds based on laboratory data is ±25% (ref. 7,8); and that from field studies varied ±45% (ref. 8). The statistical variability in extrapolating chronic effects from animal studies to humans is uncertain.

Thus overall, it appears that with more research, we may expect to reduce the variability in our field study data from circa ±100% for environmental assessments to ±50% for biological assessments; but the uncertainty in either strategy is minor in comparison to our ignorance of actual field practices. As a result we must make managerial decisions based on incomplete knowledge of such risk elements as the size of the exposed population, their exposure frequency and duration, and the use and integrity of available controls.

The selection of an assessment strategy depends not only on the goal of an assessment study, upon its desired accuracy and precision, but also upon less tangible factors. Examples of other advantages attributable to each include the following:

Environmental monitoring:
(1) can isolate exposure sources
(2) less invasive
(3) applicable to chemicals which lack measurable biological response

Biological monitoring:
(1) simultaneously assess all sources
(2) can assess cumulative responses
(3) may permit fewer samples

SUMMARY

Whereas scientific benefits can be gained by refining either biological or environmental assessment methods, our emphasis in the occupational (field use) setting should be to assess the factors underlying exposure variability (especially those associated with high exposure), to identify those factors most economically controlled, and to implement policies which encourage or enforce their use. Among the factors affecting variability are:
(1) pesticide formulation (e.g. liquid, paste, powder, granule)
(2) pesticide concentration (product and mixture)
(3) application process and equipment (enclosures and failures)
(4) clothing (including gloves)
(5) personal techniques (those amenable to education)
(6) uncontrolled conditions (weather, field foliage)

Exposure assessment is an approach to pesticide safety which is rapidly evolving, having developed a scientific basis and for the first time implementing policies to utilize that science. In our enthusiasm for scientific accuracy and precision, let us not lose sight of the question of why we are on the quest.

REFERENCES

1. C.A. Franklin, Dermal Exposure Related to Pesticide Use, p. 429-444, ACS Symposium Series, No. 273, American Chemical Society (1985).
2. E. King, A. Conchie, D. Hiett and B. Milligan, Ann. Occup. Hyg. 22, 213-239 (1979).
3. W.F. Durham and H.R. Wolfe, Bull. WHO 26, 75-92 (1962).
4. R.C. Honeycutt, Toxicology Letters, in press (1986).
5. R.A. Fenske, J.T. Leffingwell and R.C. Spear, Dermal Exposure Related to Pesticide Use, p. 377-393, ACS Symposium Series, No. 273, American Chemical Society (1985).
6. R.R. Lauwerys, Industrial Chemical Exposure: Guidelines for Biological Monitoring, Biomedical Publications, Davis, California, 150 pp. (1983).
7. J.B. Knaak, P. Schlocker, C.R. Ackerman and J.N. Seiber, Bull. Environ. Contam. Toxicol. 24, 796-804 (1980).
8. W. Popendorf and J.T. Leffingwell, Residue Rev. 82, 125-201 (1982).
9. C.A. Franklin, R. Greenhalgh and H.I. Maibach, IUPAC Pesticide Chemistry, Human Welfare and the Environment, p. 221-226, Pergamon Press (1983).
10. C.A. Franklin, N.I. Muir, and R.P. Moody, Toxicology Letters, in press (1986).

Application of pharmacokinetics in risk assessment for pesticides

Jim R. Withey* and Duncan Murdoch

Environmental Health Directorate, Tunney's Pasture, Ottawa, Ontario, Canada

Abstract - The safe use of pesticides in agriculture relies on an assessment of the potential hazard arising from their intrinsic toxicity and exposure. Pharmacokinetics can greatly assist in understanding the mechanisms involved in generating the toxic response and in setting acceptable use patterns. Toxicologists and regulators should be aware of the need to distinguish between the applied dose and the dose delivered to the target tissue.

INTRODUCTION

Pesticides have been regulated on the basis of their intrinsic toxicity and potential hazard. The acute and chronic human toxicity are usually determined by extrapolation from animal studies, in two or more species and by all potential exposure routes, i.e. dermal, inhalation and oral. The potential risk, posed during formulation or application depends not only on the intrinsic toxicity of the pesticide but also on the magnitude and duration of exposure as well as the route (ref. 1). Classifications of pesticide toxicity are available to accommodate the former; use patterns and field re-entry safety intervals (for a limited number of pesticides) have been established to accommodate the latter (ref. 2). For a given dose, the oral route is usually more responsive than the inhalation route which, in turn, is greater than the dermal route (ref. 3). In practice, applicators and formulators are most likely to be exposed via the dermal route (refs. 4, 5 and 6).

HAZARD ASSESSMENT

In assessing the potential hazard arising from continuous or repeated exposures, the amount absorbed may not adequately reflect the nature or magnitude of the toxic response. In Fig. 1, illustrating the temporal relationship of blood levels, three exposure scenarios are considered in which the absorbed dose is identical. Curve (A) represents the profile obtained during an 8 hr continuous dermal (or inhalation) exposure. In this case, the pharmacokinetics of the pesticide is assumed to follow a one compartment model with a zeroth-order rate of uptake of

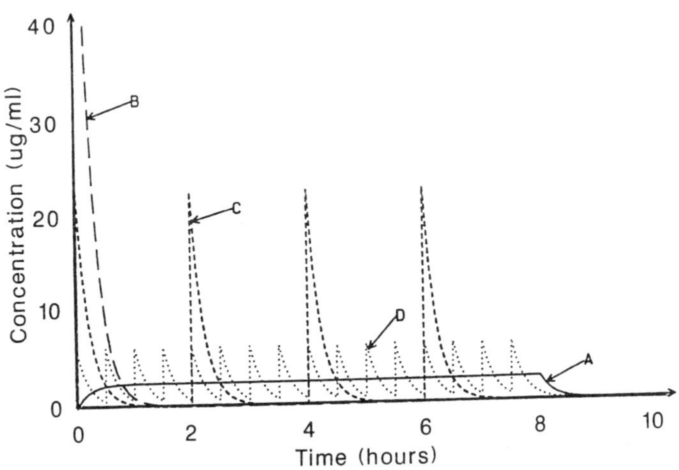

Fig. 1. Temporal blood concentration relationships for uptake by dermal and oral routes.

1 mg/min and a first-order elimination rate of 4.61 hr^{-1}. A plateau or steady-state level of 2.43 μg/ml is reached after 30 min of exposure and complete clearance from the blood occurs within 1.5 hr of termination. The area under this curve, sometimes termed the systemic availability (ref. 7) is 19.3 μghr/ml.

Curve (B) is derived by assuming that the same systemic availability as that obtained from the dermal exposure is achieved following the administration of a single oral dose of 480 mg. Absorption is assumed to be instantaneous and the first order elimination rate is 4.61 hr^{-1}. Curve (C) is the blood level profile after the administration of the same total dose administered as four equal doses of 120 mg spaced two hours apart. In this case it is worth noting that transient peak levels of 21.7 µg/ml are observed, which is nearly ten times the steady-state level reached during the dermal exposure. Curve (D) is derived by assuming that the same total dose is administered as sixteen equal doses of 30 mg ingested every half hour, and the same elimination rate of 4.61 hr^{-1}.

A number of important conclusions may be derived from this illustration. First, there is no potential for bioaccummulation if this scenario is repeated daily. Bioaccumulation occurs when the elimination rate is at least ten times slower than that assumed in this case, which it may well be for some of the chlorinated aromatic hydrocarbon pesticides. Secondly, although the systemic availability (the area under the blood level-time curve) is the same for all four exposure scenarios, it is not improbable that the acute effects arising from the levels depicted in curves (B) and (C) will be very much more severe than those from cases (A) and (D). Indeed, there are examples which reflect this conclusion. Rats tolerate more than one half of an LD50 daily dose of parathion for months when it is administered in their diet (ref. 4). In this case, the daily dosage is distributed somewhat throughout the day, similar to case (D) in Fig. 1. Guinea-pigs are known to survive twenty times the acute lethal dose of hydrogen sulfide when it is administered at a steady rate over a 67 min period (ref. 8).

The use pattern for pesticides varies considerably, from repetitive daily applications in greenhouses to seasonal application in fields and orchards. Additional factors, which have a high impact on the magnitude of exposure are the type of activity, the method and rate of application, wind velocity and direction, duration of exposure, route of exposure and attitude of workmen. The reports of Wolfe et al. (refs. 5 and 6) review the nature and magnitude of exposure for workers to 21 pesticides in 31 work activities. It is evident from these reports, and the principles illustrated in Fig. 1., that the effects of high acute dose received over a few days, could yield unanticipated effects although, when amortized over a season or converted to an annual time weighted average, would appear to be acceptable in terms of the apparent exposure.

PHARMACOKINETICS AND RATIONAL RISK ASSESSMENT

Recent applications involving the principles of physiologically-based pharmacokinetic modelling (ref. 9) show great promise as a rational method of extrapolating animal data to man, in calculating the toxicological effects when the dose is administered by other routes and distinguishing between the response at high acute exposures and those anticipated at low chronic exposures by calculating the "delivered" or surrogate dose of the active species. The physiological model, used in this role, groups individual organs and tissues having similar perfusion-volume ratios into physiological compartments arranged as a series of parallel shunts between the arterial and venous blood, as shown in Fig. 2. Thus, the well perfused organs like the adrenals, thyroid, brain and heart might be grouped into a compartment termed the vessel-rich group (VRG). The muscles and skin may be considered to comprise the muscle group (MG) while adipose tissue and bone marrow make up the vessel poor or fat group (VPG).

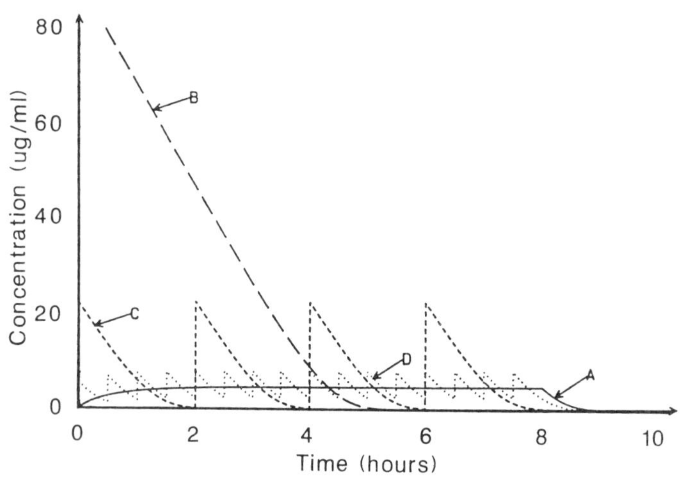

Fig. 2 Physiologically-based pharmacokinetic model.

Three other physiological compartments, especially concerned with metabolism, uptake or excretion are added to allow a description of these important processes. Where percutaneous absorption is considered, the lung compartment may be replaced by the skin.

Three types of data are required to implement a physiological analysis: partition coefficients, giving the relative solubilities of the pesticide in blood and various tissues; physiological constants, for blood flows and tissue volumes and biochemical constants, defining the rate, extent and mechanisms involved in the principal metabolic pathways. Partition coefficients for a specific compound are usually determined in the laboratory, physiological constants are generally available from the literature or calculated from allometric relationships and the biochemical rate parameters are determined from pharmacokinetic studies over appropriate dose ranges and routes.

The physiologically-based pharmacokinetic method allows a more rational determination of the dose of the toxicant actually delivered to the target organs. This calculated "internal" or "delivered" dose is then used as a surrogate for the exposure or administered dose in the dose- response relationship. In calculating the surrogate dose it is thus possible to incorporate metabolic thresholds, arising from saturable metabolic pathways, since fundamental metabolic parameters are used. The beauty of this approach is that it avoids the errors that are frequently involved by assuming that the "internal" dose is linearly related to the exposure. Thus in the case where high acute blood levels are induced as a consequence of an exposure scenario similar to cases (B) and (C) in Fig. 1, a saturable metabolic threshold may be exceeded and evoke an entirely different toxic end point to that resulting from an "equivalent" dose which does not saturate a metabolic pathway. A recent publication has illustrated how these principles can be applied in the risk assessment for methylene chloride (ref. 10).

Saturable excretion processes can also affect the systemic availability of "equivalent" doses delivered by different exposure scenarios. For example, if the linear elimination process in the kinetic model used in deriving Fig. 1 is replaced by a saturable process, with a maximum velocity of 23 μg/ml/hr and a Michaelis constant of 5 μg/ml, the elimination rate at low blood concentrations will approximate the elimination rate originally assumed of 4.61 hr^{-1}. As the blood levels increase, the concentration curves will assume the profiles, shown in Fig. 3 which vary with the exposure scenario. Thus, the single oral dose will result in a systemic availability that is 5 times greater than the scenario involving dermal exposure to the same dose. The calculated systemic availabilities are shown in Table 1.

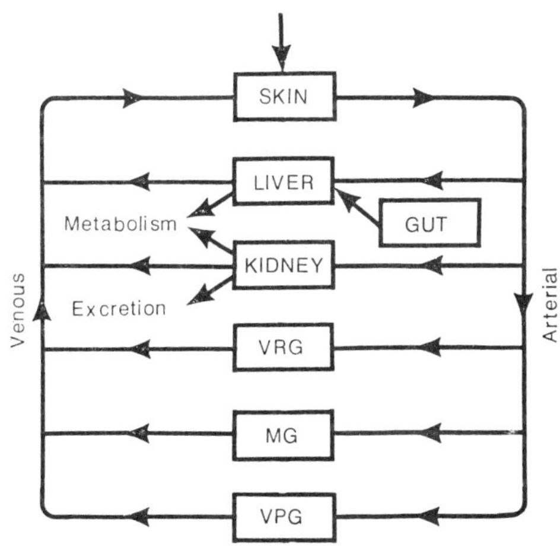

Fig. 3. Blood concentration over time (saturable kinetics).

Table 1. Systemic availability for the same dose administered by four different scenarios and involving saturable kinetics.

Exposure Scenario (Number of Doses)	Systemic Availability (μghr/ml)
Continuous	36.4
16	38.9
4	63.6
1	193.7

REFERENCES

1. G.H. Ware, Pesticides Theory and Application Chapter 21, W.H. Freeman and Co., San Francisco (1983)
2. Federal Register, Part II 40, 28279 (1976).
3. C.D. Klaassen and J. Doull, Casarett and Doull's Toxicology, 2nd Edit, p. 14, Macmillan Publishing Co., Inc., New York, (1980).
4. H.R. Wolfe, W.F. Durham and J.F. Armstrong, Arch. Environ. Health 6, 458-464 (1963).
5. H.R. Wolfe, W.F. Durham and J.F. Armstrong, Arch. Environ. Health 14, 672-683 (1967).
6. H.R. Wolfe, J.F. Armstrong, D.C. Staiff and S.W. Comer, Arch. Environ. Health 25, 29-31 (1972).
7. G. Levy and M. Gibaldi, Handbook of Experimental Pharmacology 28, Part 3, 1-34, Springer Verlag, New York (1975).
8. C.L. Evans, Quat. J. Exp. Physiol. 52, 231-248 (1967).
9. National Academy of Sciences, Drinking Water and Health 6, National Research Council, Washington, 183-233 (1986).
10. M.E. Andersen, H.J. Clewell III, M.L. Gargas, F.A. Smith, and R.H. Reitz, Toxicol. Appl. Pharmacol. (In Press).

Study of applicator exposure to alachlor by biological monitoring technology

John E. Cowell and Samuel Dubelman*

Monsanto Agricultural Company, Technology Division, Environmental Science Department, Life Sciences Research Center, 700 Chesterfield Village Parkway, Chesterfield, Missouri 63198

Abstract - Biological monitoring by chemical analysis of the operator's urine, in contrast to passive dosimetry on the operator or ambient monitoring of the environment, directly evaluates the amount of chemical that is absorbed by the body as an internal dose. When sufficient metabolic and pharmacokinetic data is known, biological monitoring provides a quantitative measure of the absorbed chemical resulting from exposure integrating all routes - pulmonary, cutaneous and oral.

For alachlor, mammalian metabolites belonging to two major classes were identified from oral, intravenous (IV) and dermal studies. Analytical methodology was developed which converts all metabolites to two chemophores for quantitation from a urine matrix. From an intravenous administration of ^{14}C-labeled alachlor to rhesus monkeys, it was determined that alachlor was rapidly and quantitatively cleared from the body with 96.4% of the systemic dose recovered in urine and feces within 10 days of administration. Additionally 80% was excreted in the urine in a 48 hour period, thus demonstrating urine as a very good monitoring matrix. A percutaneous absorption study indicated that 8.5% of a topically administered dose was systemically absorbed and again rapidly eliminated in the urine (88%) and feces. In applicator exposure studies, total body dose was measured by summing all the urine metabolite residues and by normalizing for the applicator's body weight and the quantity of the active ingredient applied. Applicator exposure studies showed body dose averages of 7.6×10^{-6} mg/kg bw/lb applied from open pour tank-fill and applications, and 1.6×10^{-7} mg/kg bw/lb applied from closed system tank-fill and applications.

INTRODUCTION

The objective of this presentation is to outline the prerequisite studies and the monitoring procedures which are utilized in conducting applicator exposure studies designed to measure actual body dose. Biological monitoring technology used for measurement of applicator exposure to alachlor from typical agricultural operations will be described as an example.

The optimal preparation for a biological monitoring study is a human pharmacokinetic study. Most of the time, this is not a viable option because of ethical considerations. In lieu of human data, three separate animal studies can be used to demonstrate the probable pharmacokinetic behavior of the chemical under study. First, an oral ^{14}C-labeled pesticide dosing study with rodents provides sufficient amounts of excretion products for metabolite identification purposes. A study of this type is typically required for pesticide registration. This data then serves as a basis to develop analytical methods to monitor the biological specimens for pesticide residues. The second study involves an IV injection of ^{14}C-labeled pesticide to monkeys, which have been shown to be a good model for man, to determine excretory recovery and distribution of the total amount of chemical between urine and feces. This data is then used to provide the necessary correction factors for interpreting the data developed in subsequent dermal monkey studies. The third study is a dermal application of ^{14}C-labeled pesticide to monkeys, which provides percutaneous absorption data regarding the percent of residue in the monitored matrix from a dermal dose. It also confirms the IV excretion profile if enough pesticide is absorbed through the skin. These three studies form the basis for an accurate biological monitoring approach to applicator exposure.

RESULTS

Mammalian metabolites of alachlor were identified from oral dosing of rodents, and IV and dermal dosing of monkeys with ^{14}C-labeled alachlor. No single metabolite was present in large amount (% of dose), but the metabolites could be classified into two distinct types - those which can be hydrolyzed to 2,6-diethylaniline, and those yielding hydroxyethylethylaniline upon hydrolysis. Analytical residue methodology was developed to hydrolyze the numerous metabolites to these two chemophores. This methodology employed base pressure hydrolysis of the urine residue followed by high pressure liquid chromatorgraphy (HPLC) separation and electrochemical detection. A GC/MS procedure was also used for confirmation or, in some cases, as the primary analytical technique. The GC/MS method was based on negative chemical ionization quantitation of two mass fragments which correspond to the derivatized anilines. Analysis of these two chemophores and subsequent conversion to alachlor equivalents produced the total alachlor residue picture. From the IV administration of alachlor to rhesus monkeys, it was determined that alachlor was rapidly and quantitatively cleared from the body with 96.4% of the systemic dose recovered in urine and feces within 10 days of administration. Additionally 80% was excreted in the urine in a 48 hour period and the total urine recovery was about 87%, thus demonstrating urine as a very good monitoring matrix. The percutaneous absorption study indicated that approximately 8.5% of the topical alachlor dose was systemically absorbed following an undiluted or diluted application to the abdomen of rhesus monkeys. Urinary excretion accounted for 88% of this absorbed dose and was 91% complete within 48 hours of topical application. Thus this study showed a profile and urinary recovery very similar to the IV study. With the completion of these studies and their interpretation, there was a firm foundation for a biological monitoring approach to an alachlor applicator exposure measurement. Results from the dermal monkey study were then used to correct measured applicator urine values for % dose excretion.

Planning, organization, site selection, requisition of equipment and materials, personnel recruitment and training on the proper collection of samples, were all important logistics that needed to be carried out well in advance of the study. Next analytical considerations needed to be addressed to ensure quality data. For biological monitoring, it was important that a liter of control urine from all applicators and mixer/loaders was collected as close to the beginning of the study as practical. This was very important as individual diets present different background levels sometimes prohibiting accurate low level quantitation of the intended analytes. Limit of Detection (LOD) and Limit of Quantitation (LOQ) were determined by evaluating this background data. Method validation was also performed on the actual study control samples. Control urine was fortified in the field with representative metabolites and the spiked urine shipped frozen to the laboratory. Analyses of these field fortified samples were performed at the end of the analytical work, and correction of all data for these recoveries compensated for field transport, storage stability, and laboratory analytical recovery. In general analytical recoveries were 94% for DEA and 67% for HEEA (N = 90).

Field information was gathered during the conduct of the study. Survey forms were a valuable aid in recording weather conditions, agricultural application details, and personnel clothing and hygiene information. Weather readings such as temperature, relative humidity, wind speed and wind direction were montiored during each operation of the study. The method of agricultural operation was described in detail to include such items as type and model of equipment, amount and rate of chemical diluted and applied, and time required for performance of the operation. The applicators wore typical pesticide application clothing. Protective equipment such as rubber gloves, rubber boots, and goggles was provided as specified on the label. The applicators and mixer/loaders were advised to avoid contact with the chemical and related chemicals during a period prior to the experiment. This was to ensure that the exposure measured after the experiment was directly attributable to the experiment. They were instructed to only perform the intended operation(s) with the pesticide in their usual manner. The operator's body weight and all details of the experiment were recorded. Photographs and videotapes of the operations proved very useful for presentation and documentation. After performance of the operations, the exposed personnel were required to collect all urine specimens in individual glass bottles. The bottles were labeled with name, date and time of specimen collection and immediately frozen. All samples were stored frozen until composited for analysis. The time of removal of clothing worn during experiment and showering were recorded in the 1985 studies to determine if any correlation of body dose with showering intervals existed.

Urine samples were composited beginning with the completion of the operation on a 12 hour basis. A total of 10 composites (12 hr each; 120 hr total collection) were then analyzed for each applicator. After analysis of the urine composites, calculation was accomplished by multiplying the amount (μg/mL) of alachlor equivalents found by the volume (mL) of urine excreted in the composite interval. Total excreted dose was calculated by summing all the composite amounts (mg). Total excreted dose was then normalized for body weight (kg) and weight (lb) of alachlor applied. This normalized excretion was corrected for the amount of dermal dose (88%) which was excreted in the urine from the monkey studies, giving the "Total Body Dose" (last column, Table 1).

CONCLUSIONS

Biological monitoring via urinary excretion of metabolites with adequate knowledge of pharmacokinetics is an effective technique for estimating actual body dose. The technique takes into account inhalation and oral exposure, clothing protection and percutaneous absorption. The Total Body Dose of alachlor during mixing/loading and application was found to be minimal (10^{-5} - 10^{-7} mg/kg bw/lb applied) as shown in Table 1. Elimination of alachlor dose is rapid (less than 3 days as shown in Fig. 1) and mostly via urinary excretion (88%). Correlation of metabolic patterns have indicated that the monkey is a good model for man and Fig. 2 demonstrates that both monkey and human applicators excrete the same metabolite class composition. Fig. 3 shows a comparison of the dose measured in the human applicators' urine to the doses given the monkeys in the pharmacokinetic studies. Finally, sensitive state-of-the-art analytical techniques were used to detect low ppb levels of alachlor metabolites in urine. The detection limit was 5.0 ppb (2.5 ppb for each metabolite class).

TABLE 1. BIOLOGICAL MONITORING RESULTS

Applicator Exposure Studies	N	Urinary Excretion (mg/kg/lb)	Total Body Dose (mg/kg/lb) +
1984 Ontario, Canada Open pour tank-fill Small containers (2.5-5.0 gal.) Small open cab tractor No soil incorporation MSL-4424	4	4.4×10^{-5}	5.0×10^{-5}
1984 Indiana, US Open pour tank-fill Small containers (2.5 gal.) Large closed cab tractor Soil incorporated application MSL-4207	8	6.9×10^{-6}	7.8×10^{-6}
1985 Missouri, US Open pour tank-fill Small containers (2.5 gal.) Large closed cab tractor Soil incorporated application Farmer education program MSL-5398	20	7.1×10^{-7}	8.1×10^{-7}
1985 Missouri, US Closed system tank-fill Mini-bulk containers (60-100 gal.) Large closed cab tractor Soil incorporated application MSL-5320	16	1.6×10^{-7}	1.8×10^{-7}

* Total Body Dose = Amount Excreted/Urinary Excretion Factor (monkey studies).

+ Experiments were performed in 20 acre (US) or 40 acre (Canada) plots, at maximum label rate of 4 lb/acre active ingredient. The "mg/kg/lb" unit allows extrapolation to any size plot. For example, the total dose that an applicator treating 20 acres would receive, based on the 1984 US Indiana study, is: 7.8×10^{-6} mg/kg/lb X 20 acres X 4 lb/acre = 6.2×10^{-4} mg/kg

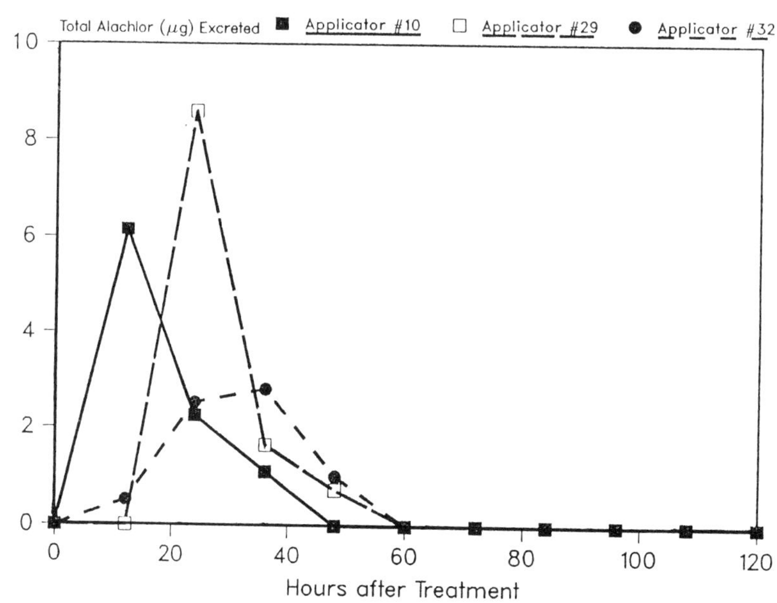

Fig. 1. Examples of applicator urinary excretion profiles.

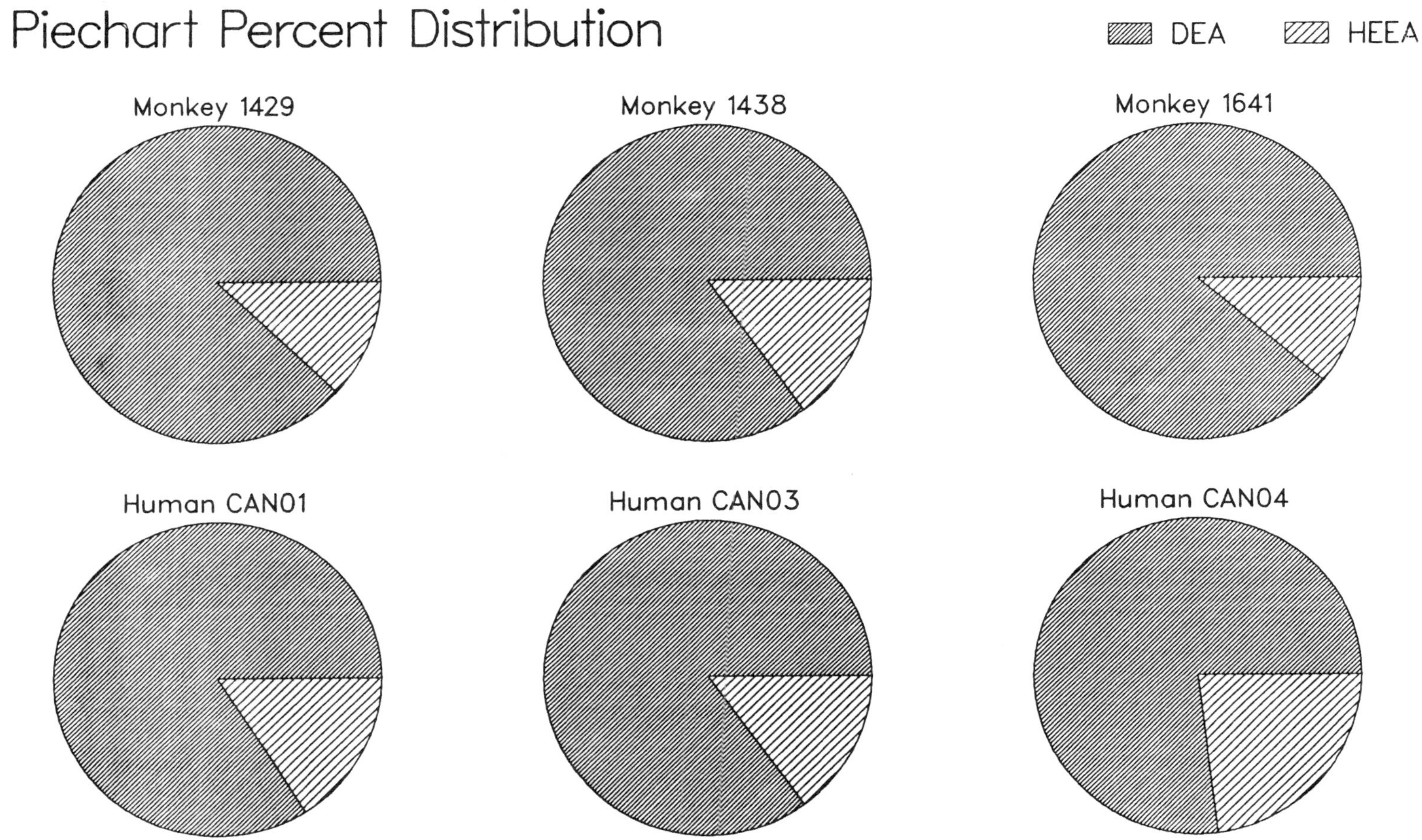

Fig. 2. Metabolite composition of urinary excretion of monkeys and humans

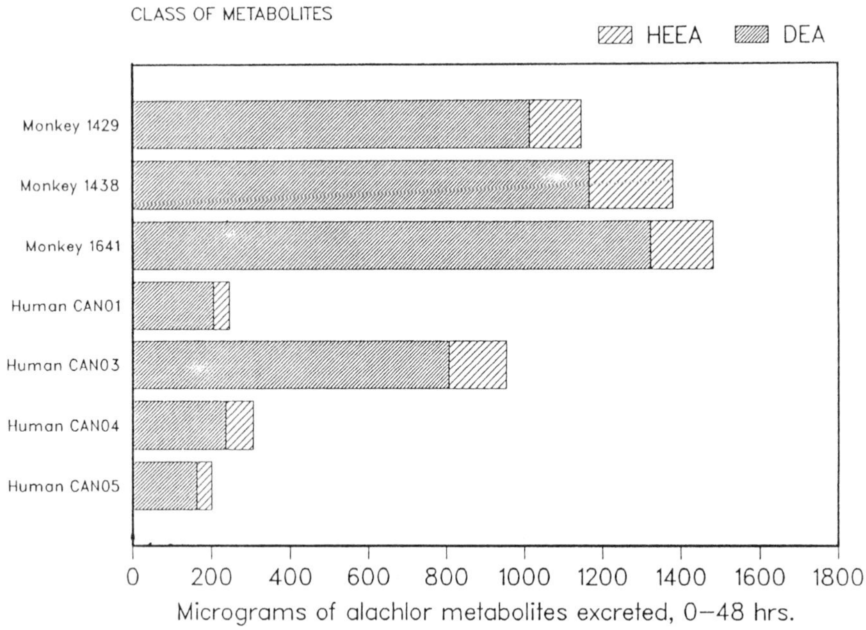

Fig. 3. Total quantity of metabolites excreted by monkeys and humans 48 hours after exposure.

Assessment of dermal exposure to pesticides: a comparison of the patch technique and the video imaging/fluorescent tracer technique

Richard A. Fenske

Department of Environmental Science, NJ Agric. Experiment Station, Rutgers University, New Brunswick, NJ, USA, 08903

Abstract - Two dermal exposure methodologies were compared in studies of orchard/airblast and greenhouse/backpack pesticide applications. Dermal deposition of a fluorescent tracer was quantitated by a television camera interfaced to a microcomputer with video digitizing capabilities. Extrapolation of pesticide levels recovered from patches to body surface areas provided a traditional estimate of exposure. Patch exposure levels exhibited systematic rather than random variability. The location of the patches greatly affects the exposure estimate derived from extrapolation. The fluorescent tracer - video imaging technique revealed nonuniform patterns of exposure, and exposures which escape detection by the patch technique. It also demonstrated that substantial exposure occurs beneath protective clothing. Patch technique estimates for the head were much higher than fluorescent tracer estimates. The actual exposure level is probably between these estimates.

INTRODUCTION

There is an increasing need for the accurate quantitation of dermal exposure among agricultural workers. Such measurements are an essential component of total exposure and risk assessment. Knowledge of the extent and magnitude of dermal exposure serves as the foundation for effective strategies to minimize exposure. In the case of agricultural workers it is well established that the dermal route of exposure is of primary importance during the mixing and application of pesticides. This paper summarizes data from two studies in which dermal exposure was evaluated by the traditional patch technique and by the fluorescent tracer technique. The first study involved airblast application of malathion to citrus in the Central Valley of California. The second study monitored backpack sprayer application of the fungicide Fosetyl-Al in greenhouses in southern New Jersey. The strengths and weaknesses of the two techniques are discussed in light of these data.

PATCH TECHNIQUE

The development of the patch technique over 30 years ago was prompted by the potential for acute intoxication among handlers of organophosphate compounds (ref 1). This technique has proven to be an extremely useful surveillance tool in characterizing the major variables in the exposure process, and in allowing comparisons among application procedures (ref 2). However, it appears to have inherent limitations as a quantitative method.

First, the method has never been standardized among investigators in regard to specific patch location and procedures for extrapolating regional exposure from the patch deposition values. Second, use of the patch technique has traditionally required two major assumptions: 1) exposure is uniform over each body region; 2) minimal exposure occurs beneath protective clothing. The second assumption has been questioned recently (ref. 3-4), and many investigators have modified the method to include patches beneath clothing. However, both the standard method and this modification still rely on the assumption of uniform exposure. If exposure is not uniform, two critical questions emerge: where should the patches be placed? Which patches should be used for extrapolation to regional exposure?

This paper will review the recommendations for estimating head exposure made in the proposed U.S. Environmental Protection Agency Pesticide Assessment Guidelines (ref. 5). These guidelines incorporate what have long been standard procedures for studies in the United States, and are very similar to the standard protocol recommended by the World Health Organization (ref. 6). The guidelines provide general rules for patch location, but specific placement is left to the investigator. The guidelines also recommend three different procedures for extrapolating patch values to estimate exposure to the head/face: 1) the mean value of all torso patches (two shoulder, one chest and one back patch); 2) if shoulder, chest and back patches are not used a head patch can be substituted; 3) the chest patch can be used alone as an estimate of face exposure.

FLUORESCENT TRACER - VIDEO IMAGING TECHNIQUE

The use of fluorescent tracers to visualize the deposition pattern of pesticides on the skin has a number of obvious strengths. As a qualitative tool this approach can provide immediate feedback concerning protective clothing performance, work practices, and worker hygiene. This method is now in use by investigators in Canada and the United Kingdom, as well as in the United States. The quantitative use of fluorescent tracers in conjunction with video imaging technology has been described in detail in recent publications (ref. 7-8). A fluorescent tracer is added to the spray mixture and workers are asked to conduct normal application activities. The workers are examined with the imaging system both prior to and immediately following exposure. This system consists of a television camera interfaced to a microcomputer with analog-to-digital capabilities. Subjects are illuminated by longwave ultraviolet in a darkened room, and a series of images of the skin surface are recorded onto disk. These images are later analyzed to determine the magnitude and extent of fluorescent tracer patterns.

The quantitation procedure involves several important steps. First, it must be established that the pesticide and tracer are deposited on the skin surface in a predictable ratio. Second, it must be demonstrated that pesticide and tracer penetration of protective clothing is similar and predictable. Finally, it must be assumed that the amount of fluorescence detected by the video imaging system is proportional to the amount of tracer deposited on the skin surface. Substantial progress has been made toward establishing the validity of the tracer technique in regard to aerosol deposition and protective clothing penetration. The most problematic issue in the method at this point is the assumption of a quantitative relationship between tracer deposition and fluorescence. Laboratory studies have shown this relationship to hold true over a wide range of tracer densities on the skin, but accurate quantitation cannot be achieved when very high depositions occur. Cases of high deposition can generally be avoided by adjustment of tracer concentration in the spray system. However, it cannot be known a priori if a very heavy exposure will take place through spills, wind effects, etc. Thus, there may be cases where exposure will be underestimated by this technique.

Preliminary results from the airblast study indicate a high correlation (r = .92) between 24 hour excretion of malathion metabolites in the urine and fluorescent tracer deposition. The correlation appears to be consistent for both mixer and applicator exposure. These results are an encouraging step in the validation of this method, and have led to further studies which combine surface deposition measurements with biological indicators of exposure. This paper will present evidence based on the fluorescent tracer - video imaging technique which challenge the basic assumptions of the patch technique.

VARIABILITY DUE TO PATCH LOCATION

In the study of air blast mixers and applicators two patches were attached to the baseball cap worn be each worker. One patch faced forward above the bill of the cap, while the other was attached to the hat above the right ear. The distribution of exposure on the head exhibited a systematic rather than random pattern, as evidenced by the magnitude of pesticide deposition on the two patches. Furthermore, the relative magnitude of front and side head patches was reversed depending on work activity. The front head patch value was 2 times higher than the side head patch value for applicators, but the side head patch value for mixers was 4.4 times higher than the front head patch value. Thus, the magnitude of exposure for the face or head calculated from a head patch is dependent on the location of the patch.

The relative magnitude of head, chest and torso patch values varied from study to study and between work activities. Estimates of exposure will vary substantially depending on the patch location chosen for extrapolation. Again, this variability is systematic rather than random. These patterns of exposure are illustrated by data from both the airblast study and the greenhouse study. For example, calculation of head/face exposure based on the head patch will be 4 times higher than the estimate based on the chest patch for airblast applicators, but will be 2 times lower for greenhouse applicators. Similarly, head/face exposure based on the head patch will be 4 times lower than the estimate based on the chest patch for airblast mixers, and will be 3 times lower for greenhouse mixers. Finally, use of average torso patch values will result in an exposure calculation which falls between those derived from head and chest patch values for both airblast mixers and applicators. However, the same calculation for greenhouse mixers and applicators will produce results which are 2-5 times higher than those based on either the head or chest patch values.

NONUNIFORMITY OF DERMAL EXPOSURE

The deposition of aerosols during mixing and application is not uniform over individual body regions. Fluorescent tracer evaluation demonstrates that substantial exposure may occur on one portion of the body part while other portions remain unexposed. Exposure to the head of greenhouse applicators illustrates the problem of patch location. Since application is downward onto a 75 cm bench, most exposure comes from below and is deposited under the chin and at the top of the neck. Little exposure occurs on the cheeks and almost none is found on the forehead. What patch location best represents this type of exposure? The tracer technique also reveals substantial splashes to mixers that go undetected by the patch technique. Similarly, tracer studies demonstrate that the pesticide aerosol can be drawn in at the neck and wrists and deposited under clothing.

EXPOSURE BENEATH PROTECTIVE CLOTHING

A substantial proportion of total dermal exposure occurred on regions protected by clothing. Table 1 presents the regional distribution of tracer exposure for airblast mixers and applicators. For mixers 42% of total exposure was to the hands, despite the fact that chemical-resistant gloves were worn during the mixing process. An additional 37% of total exposure was found on body regions protected by coveralls. The hands accounted for only 12% of total tracer exposure among applicators. For these workers exposure to regions protected by clothing was 64%, or nearly two-thirds the total dermal exposure. This value was higher (70%) for applicators wearing workshirts than for those wearing coveralls (59%).

COMPARISON OF HEAD EXPOSURE ESTIMATES

It is not possible to compare total exposure by these two methods, since the airblast study did not include a handwash procedure or patches beneath the clothing. Comparative analysis will focus instead on head exposure, employing the direct estimates from the fluorescent tracer technique and the extrapolation from the mean value of torso patches discussed earlier. Table 2 presents the mean estimates of head exposure for mixers and applicators. In each case the tracer estimate is lower than the patch estimate. For mixers the patch technique estimates a total of 2161 ug of malathion exposure to the head. The tracer technique measured 762 ug of malathion on the skin, or only about 35% of the patch estimate. If only the face (650 cm^2) is considered, the patch estimate is reduced to 1300 ug. The tracer estimate is then about 60% of the patch value. For applicators the difference between the two methods is much larger. The patch technique estimates a head exposure of 63,514 ug (63.6 mg), while the tracer technique estimates 1213 ug malathion reaching the skin surface. The tracer value is thus less than 2% of the patch value for the head, and less than 4% for the face only.

TABLE 1. Percent Tracer Exposure by Body Region

Worker Type	N	Hand %	Head %	Body[a] %
MIXERS	7	42	21	37
APPLICATORS	14	12	24	64
(Workshirt)	6	10	20	70
(Coveralls)	8	14	27	59

[a]Body = regions under coveralls

TABLE 2. Comparative Estimates of Head Exposure (micrograms)

Worker Type	N	Tracer Estimate	Patch Estimate[a]	%[b]
Mixer	7	762	2161	35.3
Applicator	14	1213	63514	1.9

[a]Mean of torso patches x 1080 cm^2
[b]Tracer estimate as % of patch estimate

DISCUSSION

A fundamental question is raised by these studies. Can a standardized patch location protocol be developed that is appropriate for all mixer/applicator exposure studies? In general, the evidence suggests that the exposure pattern is dependent on worker activity (i.e., mixer or applicator), and on the application process (e.g., airblast, backpack). The variability of deposition on head patches and the discrepancy between torso and head patches appear to be systematic rather than random. Is it the goal of patch studies to monitor the maximum exposure to a given region ("worst case" exposure)? If so, it may be necessary to carefully characterize exposure patterns by worker activity and application process to determine appropriate patch locations. However, if the goal of such studies is the accurate assessment of dermal exposure, there may be no "standard" patch method which can overcome the fact that exposure is not uniform. Fluorescent tracer analysis also demonstrates that substantial exposure occurs beneath protective clothing. Much of this exposure occurs at the openings of protective garments (i.e. sleeves and neck), rather than as actual penetration through the garments. Thus, the question of field performance of protective clothing is not simply a question of penetration resistance.

Several possible explanations for the discrepancy between the patch and tracer estimates for head exposure can be put forth. First, the surface area quantitated by the video imaging system is much smaller than the head surface area estimate of 1080 cm^2 employed with the patch technique. In these studies all workers wore hats, and much of the face was covered by a half-mask respirator and in some cases by glasses or goggles. Second, the tracer does not tend to fluoresce on the hair unless high levels of exposure occur. The method also does not assess exposure to the top of the head. It is not clear to what extent exposure to the hair results in percutaneous absorption. Third, exposure to the head was very high and in somes cases may have resulted in tracer deposition beyond the quantitative range of the technique. In conclusion, it is likely that an accurate measure of dermal exposure lies somewhere between these two estimates. Results from further studies comparing these techniques should provide greater insight into the sources of variability which currently limit our understanding of dermal exposure processes. [New Jersey Agricultural Experiment Station, Publication No. J-07905-1-86, supported by State Funds.]

REFERENCES

1. W.F. Durham and H.R. Wolfe, Bull World Health Organ. 26, 75-91 (1962).
2. J.E. Davis, Residue Review 75, 33-50 (1980).
3. C.A. Franklin, R.A. Fenske, R. Greenhalgh, L. Mathieu, H.V. Denley, J.T. Leffingwell and R.C. Spear, J. Toxicol. Environ. Health 7, 715-731 (1981).
4. R.A. Fenske, J.T. Leffingwell and R.C. Spear: Dermal Exposure Related to Pesticide Use, R.C. Honeycutt, G. Zweig, N.N. Ragsdale, eds., American Chemical Society Symposium Series 273, pp. 377-393 (1985).
5. U.S. Environmental Protection Agency, Pesticide Assessment Guidelines Subdivision U: Applicator Exposure Monitoring, Office of Pesticide Programs, Washington, D.C.
6. W.F. Tordoir and E.A.H. Van Heemstra-Lequin, eds., Field Worker Exposure During Pesticide Application, Studies in Environmental Science 7, (1980).
7. R.A. Fenske, J.T. Leffingwell and R.C. Spear, Am. Ind. Hygiene Assoc. J. 47, December (1986).
8. R.A. Fenske, S.M. Wong, J.T. Leffingwell and R.C. Spear, Am. Ind. Hyg. Assoc. J. 47, December (1986).

Biotechnology, pesticides and pest control

Impacts of biotechnology on agricultural pesticides

Ernest G. Jaworski

Biological Sciences Department, Monsanto Company,
700 Chesterfield Village Parkway, St. Louis, MO 63198, USA

Abstract - The applications of biological organisms, systems and processes to agricultural needs has created a new and exciting wave of biotechnology. These technologies applied to increasing agricultural productivity and to the development of new pest control measures should provide a complimentary synergistic addition to the more traditional chemical and genetic approaches. Successful transformation of plant cells has been obtained utilizing vectors and DNA delivery methods derived from the plant pathogen, *Agrobacterium tumefaciens*. This soil bacterium is capable of transferring a DNA segment (T-DNA), located between specific nucleotide border sequences from its large tumor inducing (Ti) plasmid into the nuclear DNA of infected plant cells. The exploitation of the *Agrobacterium*/Ti plasmid system for plant cell transformation has been facilitated by 1) the construction of modified *Agrobacterium* strains in which the genes responsible for pathogenecity have been deleted, 2) the design of intermediate vectors containing selectable drug markers for introducing foreign genes into the Ti plasmid and subsequently into plant cells, and 3) the development of efficient *in vitro* methods for transforming plant cells and tissues with engineered *Agrobacterium* strains. These modifications have led to the development of simple, efficient and reproducible transformation systems from which morphologically normal transformed plants can be readily regenerated. The foreign genes are stably maintained and expressed in the resulting plants and are inherited by progeny as typical Mendelian traits. The availability of plant transformation systems has now facilitated the study of gene expression and regulation in plants and will be extremely useful for the modification of various crop species in an agronomically significant manner. In addition, similar techniques utilized in the genetic modification of microbes will find significant utility for the development of microbial pesticides.

INTRODUCTION

Within the last fifteen years we have seen a dramatic advancement in biology. With reference to plant biology, enormous advances have been made in what was called the new biology which led through meristem culture to the clonal propagation and increase of materials of a particular genotype. The meristem culture procedures led to tissue culture techniques which were utilized to grow cells in suspension culture and in the form of callus. Such tissue culture, with its subsequent ability to regenerate whole plants, created a whole new era of plant biology investigations and found numerous scientific and practical applications. The cell culture procedures were subsequently utilized to produce protoplasts which could then be fused to produce somatic cell hybrids. This technology, unfortunately, has not been as successful in leading to regeneration of plant materials and a number of hurdles remain to be solved before we will see routine application of protoplast technology to agriculture.

The second major development during this period was the construction of biologically functional bacterial plasmids *in vitro*. This work demonstrated that one could clone pieces of functional DNA in bacterial systems by the use of restriction endonuclease generated fragments ligated into bacterial plasmids capable of autonomous replication within a bacterial species. This technology subsequently was utilized in the cloning of bacterial plasmids derived from the tumor inducing soil bacterium, *Agrobacterium tumefaciens*.

Agrobacterium is a naturally occurring microbe capable of inserting its DNA stably into the chromosome of plant cells resulting in the production of tumors or *Crown* gall. This system has now been effectively utilized in conjunction with plant tissue culture technology to successfully insert foreign genes into plant cells and regenerate new whole plant genotypes.

PLANT CELL TRANSFORMATION

In order to transform plants, variant strains of the *Agrobacterium tumefaciens* that contain the large Ti, or tumor inducing plasmid, have been utilized to infect plant tissue. The Ti plasmid contains a segment of DNA, termed the T-DNA, which is transferred and stably incorporated into the nuclear DNA of transformed cells. This T-DNA is transcribed within the cells and expresses a number of genes responsible for phytohormone synthesis and the synthesis of novel metabolites called opines. The transfer and insertion process for T-DNA transformation involves repeated nucleotide sequences ("T-DNA borders") which define the ends of the T-DNA segment as well as other genes of unknown function located in a specific region (VIR region) outside of the T-DNA.

These properties of *Agrobacterium* and its Ti plasmid system have been used to design a large number of new transforming vectors. Intermediate vectors containing selectable antibiotic resistant markers for introduction of foreign genes into the Ti plasmid have been constructed. The tumor forming properties of the *Agrobacterium* have been deleted by removing certain phytohormone coding genes, and a new efficient *in vitro* plant cell transformation technique has been perfected.

Chimeric genes were constructed that contained the neomycin phosphotransferase (NPT) coding sequences from bacterial transposon Tn-5 joined to the 5' and 3' regulatory sequences of the nopaline synthase gene derived from the Ti plasmid. This gene confers resistance to the amino glycoside antibiotic, kanamycin, to which plant cells are sensitive. The 5' and 3' regulatory regions of the nopaline synthase gene have been well characterized, and it was known that the gene was expressed constitutively in all plant tissues transformed with the Ti plasmid.

Since direct cloning approaches using the Ti plasmid were not practical, it was necessary to create shuttle and intermediate vectors. Vectors were created either to integrate with a resident Ti plasmid by recombination or to replicate independently of the Ti plasmid as trans vectors. The integrated vector was used for the transfer of a number of foreign genes into *Agrobacterium* cells. An essential characteristic of this plasmid include a segment of pBR322 DNA for replication in the *E. coli*, a portion of a Ti plasmid (pTiT37) containing the functional nopaline synthase gene for ease in scoring transformed plant cells, a *Streptomycin/spectinomycin* resistance determinant from Tn7 for selection in *Agrobacterium*, a portion of DNA from another Ti plasmid (pTiA6) to provide homology for recombination with a resident octopine type plasmid in *A. tumefaciens*, a synthetic multi-linker containing unique restriction sites for gene insertion and the chimeric kanamycin resistance gene (NOS/NPTII/NOS). This plasmid and its derivatives were introduced into *A. tumefaciens* by conjugation procedures and homologous recombination between the plasmid and the wild type octopine Ti plasmid to produce cointegrates. The system was then useful for the study of gene expression and inheritance of traits but was not highly efficient for routine production of transformed plants. A derivative of a virulent Ti plasmid was therefore constructed to form a cointegrate with the intermediate vector or derivative. By double cross-over events between the wide host-range plasmid, pRK290 derivative, and pTIB6S3, the phytohormone biosynthetic tumor genes were deleted. Subsequent cointegrate formations therefore resulted in the generation of an avirulent and selectable T-DNA system.

An *in vitro* transformation procedure was developed using plant protoplasts incubated directly with an *Agrobacterium tumefaciens* bacterial cell suspension. Protoplasts were prepared from plant leaf tissue by conventional enzyme digestion. The bacteria attached to the protoplast during cell wall regeneration and by an unknown mechanism, transferred the T-DNA into the plant cells. Such cells were easily identified within three weeks by selection for kanamycin resistance.

An improved alternative procedure has been developed to obviate some of the problems of protoplast isolation and regeneration. In this procedure whole leaf punches are infected with *Agrobacterium*, and the subsequent callus formation around the circumference of the disk results in regenerated plantlets under appropriate conditions. This simple reproducible system produces transformed plantlets within 3-4 weeks.

Using either the protoplast cocultivation system or the leaf disk method, it was possible to demonstrate the stable maintenance and expression of foreign genes (kanamycin resistance). In addition, progeny seed derived from selfed plants were shown to inherit the kanamycin resistance in a simple Mendelian pattern.

GENE EXPRESSION

With the development of a transformation system which allowed for the stable inheritable introduction and expression of a foreign gene, a tool was available for the analysis of gene expression in general. One of the first studies involved a tissue-specific light regulated gene coding for the small subunit of ribulose-1,5-bisphosphate-carboxylase (RuBP-ss) from peas. The genomic clone encoding the pea RuBP-ss gene was inserted into an appropriate vector and after transfer into *Agrobacterium tumefaciens* was introduced into petunia cells by co-cultivation (collaborative studies with Dr. Nam-Hai Chua and Dr. Richard Broglie). Northern and S1 analyses of transformed petunia cells revealed that the pea RuBP-ss gene was expressed in petunia under the control of its own promoter. Furthermore, the gene expression was regulated by light in a manner identical to that seen for the endogenous RuBP-ss gene in pea. The pea RuBP-ss retained its tissue specific expression pattern in leaves derived from regenerated transformed petunia plants. In subsequent studies using *in vivo* radio labelling followed by immunoprecipitation of ribulose-1,5-bisphosphate-carboxylase, it was demonstrated on 2-D gels that the heterologeous RuBP-ss protein of pea could be separated from the endogenous petunia RuBP-ss. This indicated that the pea RuBP-ss protein was correctly processed *in vivo* by petunia chloroplast. *In vivo* labelled pea RuBP-ss was also recovered from whole enzymes that were immuno-selected with the petunia anti-large subunit antibody indicating that the pea small subunit could form a hybrid holo-enzyme assembly with petunia large subunits.

In collaborative studies with Dr. Roger Beachy, Washington University, the transfer of legume storage protein genes into solanaceous plants was shown to result in tissue specific accumulation of storage protein in the seeds of transformed plants. The regulated expression of soybean α'-conglycinin protein in petunia seeds, was demonstrated. This model system for storage protein expression will permit the determination of the structural sequences required for protein translocation, glycosylation and assembly as well as the regulatory sequences essential for seed specific expression. Such systems should prove useful in analyzing the processing, assembly, stability and functioning of storage proteins which have been modified through mutagenesis to alter their amino acid composition and nutritional value.

The expression of two mammalian genes in plant cells has also been demonstrated using a cDNA clone encoding α-human choriono-gonadotropin (α-hCG) under the control of the *Cauliflower mosaic* virus 35s promoter and the expression of a mouse cDNA clone encoding a methotrexate insensitive dihydrofolic reductase (DHFR) gene also under the control of the CaMV 35s promoter. These results with α-hCG and mouse DHFR suggest the broad utility of the *A. tumefaciens* transformation system for the study of gene regulation and expression in plants.

APPLICATIONS

Plants

Within the past three years the availability of gene transfer systems has led to several important new insights into the regulation of gene expression and protein transport in plants. In addition, it has now been demonstrated using single gene traits that herbicide, insect and viral disease tolerance can be produced within plants.

Glyphosate, N-phosphonomethylglycine, is the active herbicidal ingredient in ROUNDUP® and inhibits the aromatic biosynthetic pathway; specifically, enolpyruvyl-shikimate-3-phosphate (EPSP) synthase. This enzyme in the aromatic pathway is involved in the synthesis of phenylalanine, tyrosine and tryptophane. It has been demonstrated that glyphosate specifically inhibits this enzyme and that overproduction of this enzyme or of mutants of this enzyme in transformed tissue result in herbicide tolerance. Utilizing a chimeric gene construct consisting of a petunia EPSP synthase cDNA clone flanked by the CaMV 35s 5' promoter and the NOS 3' regulatory regions, transformed petunia cells have been demonstrated to overproduce EPSP synthase by 30-60 fold. Plants regenerated from these cell lines were tolerant when sprayed with 0.8 lbs/acre of glyphosate in contrast to control plants which were killed when sprayed with 0.2 lbs/acre of the herbicide. Similar strategies are now being utilized to create cell lines tolerant to many other herbicides. Since crop plants such as rape, tomato, potato, tobacco and lettuce are currently amenable to transformation using the *Agrobacterium* system, we can expect to see considerable practical demonstrations in the near future. It should not be forgotten, however, that this technology will also be extremely powerful for studies in gene regulation and the developmental biology of higher eukaryotics due to the relative ease by which the technology can now be applied to model systems such as tobacco and petunia.

Collaborative studies with R. Beachy, Washington University, have also demonstrated the ability to utilize the transformation system for creating viral resistant plants. The *tobacco mosaic* virus (TMV) coat protein gene was engineered into a plasmid similar to the one used for the glyphosate tolerance studies. Tobacco and tomato plants transformed with the coat protein gene have been demonstrated to either be resistant to TMV infection or to demonstrate significantly (weeks) delay in symptomology following infection. Comparable control plants have always demonstrated severe symptoms within days following inoculation with the TMV virus. The mechanism for this cross protection of plants from viral infection is currently under study.

Most recently it has been demonstrated by a Belgium firm that the *Bacillus thuringiensis* (B.t.) toxin gene can be inserted into the Ti plasmid and be expressed in tobacco to protect regenerated tobacco plants from attack by lepidopterous larvae such as the tobacco hornworm. This demonstration clearly illustrates the ability to protect a plant from insect damage utilizing the transformation technologies now available.

Microbes

It may be worth touching on one final example which involves the transformation of microbes rather than plants but in a fashion which is useful in enhancing the productivity of plants. This example relates to the development of transformed *Pseudomonas fluorescens* microbes that are natural root colonizers on major crops such as corn and soybeans. These *Pseudomonads* have now been engineered to carry and express the *Bacillus thuringienses* toxin gene or the *E. coli* lacZY genes which code for β-galactosidase and lactose permease. In both instances, these genes and their products are well characterized and are not toxic to man, animals or plants. Also, they have been permanently inserted into the chromosome using a disarmed Tn7 system. This system, as well as the disarmed Tn5 procedure used for *B.t.* toxin genes, insures an extremely low frequency of gene loss or movement from the chromosome. While these systems have not yet been tested in the field, laboratory and greenhouse studies have clearly indicated that these genes are effectively expressed without loss of root colonizing ability and competitiveness with other soil microbes. Much further work is needed to develop performance characteristics of these microbes under normal field environments and soil types, particularly to generate more concrete data on the ecology of these microbes. The lacZY marked *Pseudomonas fluorescens* provides a particularly useful organism for this purpose since it contains four characteristics that make it possible to isolate this organism from soil samples and detect one bacterium per gram of soil. The diagnostic characteristics of the engineered *Pseudomonas fluorescens* include its fluorescent properties, natural rifampicin resistance, ability to grow on a simple lactose media and detection by the X-gal chromogenic dye. The organism provides us with a sensitive, selectable tracking system which would be extremely useful for microbial ecology studies under natural environmental conditions.

CONCLUSIONS AND FUTURE NEEDS

The next decade promises to bring with it significant advances in the genetic modification of plants including many with direct value to plant breeders. Most practical benefits will initially focus on such single gene traits as herbicide, insect and viral disease tolerance.

The use of modern molecular biology to insert new genetic traits into plants will be especially interesting in developing herbicide and pesticide tolerant plants. Examples now include the development of oilseed rape resistant to the atrazine herbicide, development of imidazolinone-resistant maize, studies on the development of sulfonylurea resistant crop plants, on the development of phosphonotrycine resistance and the developments of glyphosate tolerant plants. Over 180 herbicides are currently being used worldwide, and it is not difficult to see that it will become more difficult and costly to develop new broad-spectrum products. Herbicide resistant crops could therefore be an effective way to combine useful weed control practices with existing products in a productive and novel way. Advantages should acrue for those products which are broad spectrum, non-selective and which have been demonstrated to be environmentally and toxicologically safe. The basic unknown question is whether the introduction of genes will in some say reduce the productivity of the crop and such information can only be collected under wide-scale field trials, that is to say, under natural stress conditions in different geographic locations.

One area of concern which will require considerable attention, particularly with regard to public perception is the concern over developing weed species which are resistant to the herbicide. Attention to the possible cross breeding of herbicide tolerant weed species will be required along with the evaluation of options, should such an unlikely event actually occur. It would appear that there are adequate number of herbicides in our current arsenal of compounds to obviate such a problem even if it could occur.

Additional advances will be made in the combination of traits mentioned above as well as in the broadening to other characteristics as our knowledge base of plant biochemistry improves. Protein engineering may well see its first application in simple amino acid codon shifts in seed storage proteins to enhance their content of lysine (corn) or methionine (soybeans). Gene amplification techniques may also be useful in improving the yield of protein levels in forage crops as well as in the development of systems for the production of valuable rare proteins utilizing this technology.

Other methods for the tranfer of DNA into plant cells are currently in their infancy but display potential. Such procedures as electroporation, microinjection, laser treatment and viral vectors are receiving extensive study and while still relatively unexplored, will certainly impact the entire field.

Some of the technical hurdles that remain will include cell culture procedures to broaden the base of crop species that can be regenerated with particular emphasis on the development of regeneration procedures from protoplast culture. New transformation systems, particularly those related to monocot transformation, are sorely needed and while receiving attention, still present a challenging problem for scientists. Gene structure and function will continue to receive attention with particular regard to the regulatory sequences which affect the expression of the gene in a tissue or developmentally specific manner. And finally, the identification of agronomically important genes lies at the heart of true exploitation of this technology in order to maximize its impact on the breeding of improved crops.

What is clear at the present time is the lack of basic biochemical knowledge in plant system. This knowledge is vital for our understanding of the genetic components involved in achieving such desirable traits as frost tolerance, heat tolerance, drought tolerance, heavy metal tolerance, disease resistance, insect tolerance, etc. While new initiatives are being made to improve this situation, the social and economic impacts of modern molecular biology will only be filled when more resources are directed toward improvements in plant tissue culture and regeneration, novel transformation systems (transposable elements, gemini virus vectors, direct DNA insertion, etc.), understanding of gene structure, organization and function, the selection, isolation and characterization of agronomically important genes and the development of unique plant breeding techniques.

REFERENCES

1. Recombinant DNA, a Short Course, J. D. Watson, J. Tooze, D. T. Kirks, Scientific American Books, W. H. Freeman and Co., New York. (1983).

2. Genetic Transformation in Higher Plants, R. T. Fraley, S. G. Rogers, R. B. Horsch. CRC Critical Reviews in Plant Sciences, Vol.4, Issue 1 (1986).

ACKNOWLEDGEMENTS

I would like to express my appreciation to the numerous colleagues who have contributed to this exciting technology. These include Monsanto associates Robb Fraley, Rob Horsch, Steve Rogers, David Drahos, Harry Klee, Dilip Shah, Ganesh Kishore, Steve Padgette, Gary Bild, Gerard Barry, Guy della-Cioppa, Mark Obukowicz, and to our colleagues and external collaborators including Roger Beachy (Washington University), Jeff Schell (Max-Planck Institute, Köln, Germany), Marc Van Montagu (University of Gent), Nam-Hai Chua (The Rockefeller University), Richard Broglie (The Rockefeller University), Ferenc Nagy (The Rockefeller University) and Mary-Dell Chilton (Ciba-Geigy).

Engineering soil microorganisms for pesticide degradation

P. C. Kearney, J. S. Karns, and W. W. Mulbry

Pesticide Degradation Laboratory, U.S. Department of Agriculture
Agricultural Research Service, Beltsville, Maryland 20705, U.S.A.

Abstract - The use of native and genetically engineered microorganisms holds considerable promise for biological degradation of pesticide waste waters. At this time only a limited number of useful genes encoding degradative enzyme synthesis have been isolated and characterized. Among these are the genes which encode parathion hydrolase, two haloacid dehalogenases and the enzymes involved in 2,4-D degradation. The properties of these genes and their related enzymes are reviewed from the perspective of their utility in pesticide degradation systems.

INTRODUCTION

The advent of modern molecular biology offers the pesticide biochemist some very promising opportunities to increase our understanding of the genetic processes that lead to degradative enzyme synthesis. It also provides us with an opportunity to manipulate genes encoding useful degradative enzymes in more useful host organisms to solve a number of important agricultural problems. One current problem of considerable magnitude is the safe and economical disposal of pesticide wastes to avoid groundwater pollution. It is estimated that during 1981 approximately 25,835 metric tons of 51 listed pesticide wastes were generated in the United States (Ref. 1). According to this U.S. Environmental Protection Agency survey, 38% were pesticide manufacturing residuals and the rest discarded commercial products. At the user level, one of the most difficult problems has been to safely dispose of fairly dilute volumes of wastewater generated from rinsing containers and application equipment. Wittaker (Ref. 2) estimates that the typical agricultural applicator produces between 100 to 400 L of pesticide-contaminated wash water each time a piece of equipment is cleaned. For the aerial applicator, the pesticide concentration in rinsewater is estimated to be about 500 ppm. Thus, the potential annual volume may run as high as 226 million L containing 120,000 kg of pesticide from the approximate 6000 commercial spray planes (Ref. 3). Some of these residues have been discarded in unlined soil evaporation pits which may have contributed to groundwater contamination. Clearly a need exists for a simple, economical, on-site method for degrading waste pesticide chemicals. Native and engineered soil microorganisms offer considerable promise for solving this specific agricultural problem.

PESTICIDE DEGRADATION ENZYMES AND GENES

There is a wealth of literature concerning microbial transformations of pesticides. In many instances, the responsible bacterial enzymes have been identified and at least partially characterized. In a relatively few cases, the genes encoding these degradative enzymes have been described. A comprehensive review of the biochemical and genetic biodegradation literature is outside of the scope of this paper (for an excellent review see Ref. 4). Instead, we have attempted to outline the most recent research in these areas of pesticide biodegradation which have pertinence to waste disposal applications.

Although the first report of bacterial genes specifying pesticide degrading enzymes/pathways appeared more than ten years ago, there are still relatively few degradative genes which are well characterized. Still fewer are the known degradation genes which seem amenable to being incorporated into pesticide waste disposal schemes.

In many cases, xenobiotic degradation genes, including those specifying pesticide degradation pathways, have been localized to bacterial plasmids. However, there are several instances where such pathways are partially encoded by chromosomal genes. In some cases, degradation genes may be on transposons, discrete pieces of DNA that are mobile and can move between bacterial plasmids and chromosomes. Although the original location of these genes may not be of consequence in future genetic engineering of these genes in novel bacterial hosts, the localization of these genes has been important in attempting to unravel the origins of these genes. Comparisons between independently isolated degradation genes which encode functionally similar enzymes have been pursued to trace the dissemination of these genes among the microbial community.

As yet, only natural isolates of bacteria have been employed to degrade pesticide wastes. For the immediate future, such strains will be used rather than recombinant/engineered strains for a variety of reasons. First, as noted above, there are at present many more strains known at the biochemical level that possess useful activities toward pesticides than have been characterized genetically. Hence, exploitation of these activities will necessarily be constrained to the use of native strains until the pertinent genes can be isolated. Second, where degradative genes have been cloned into novel bacterial hosts, expression of these genes is frequently much lower in novel hosts than it is in the native hosts. Thus, to employ these genes in recombinant organisms will also require that synthetic regulatory elements be inserted to improve expression. And finally, use of native organisms in disposal schemes obviates the potential problems of containment that would be of great concern in working with recombinant organisms.

0,0 Dialkylthiophosphates- Over the last five years, the enzyme parathion hydrolase E. 3.2.3) and the gene which encodes it (termed opd for organophophosphate degradation) have been studied extensively in several laboratories. The characteristics of parathion hydrolase make it an attractive prospect for organophosphate waste disposal applications. Parathion hydrolase requires no cofactors for activity and is quite stable in crude extracts. Moreover, the enzyme has a broad substrate range within the dialkylthiophosphates, shows broad pH and temperature ranges, and is not inhibited by some organic solvents or reaction end products. Important insecticides hydrolyzed by this enzyme include methyl parathion, diazinon, fenitrothion, cyanophos, chlorpyrifos, and coumaphos. The opd gene has been clone from the 70 kilobase (kb) *Pseudomonas diminuta* plasmid (pCMS1) and expressed in other bacterial strains (Ref. 5). More recently, an opd gene from a 39 kb C *Flavobacterium* sp. (ATC 27551) plasmid (pPDL2) has been cloned (Ref. 6). Hybridization experiments using a restriction fragment containing the *Pseudomonas* opd gene against Southern blots of restriction enzyme digested *Flavobacterium* plasmid DNA revealed that the opd genes from thes two sources shared extensive homology (Ref. 6). Comparison of the restriction maps of pPDL2 and pCMS1 suggests that the opd genes in these two plasmids are both contained within a highly conserved region of approximately 2.3 kb. Within this region the restriction sites for six endonucleases were found to be identical, while outside of this region no comparable sites were evident (Fig. 1).

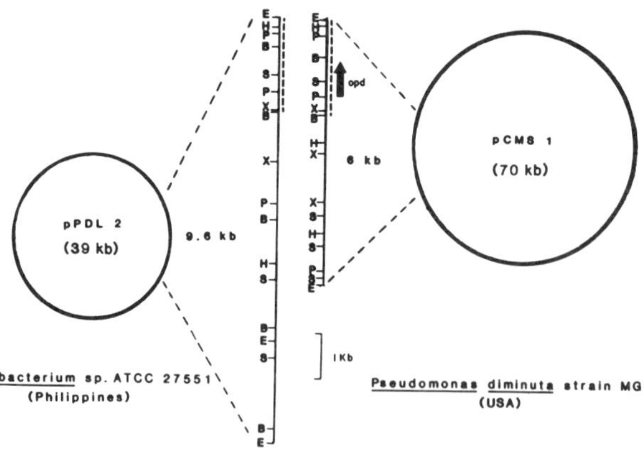

Fig. 1. Comparative restriction enzyme maps of the opd regions of plasmids from *Flavobacterium* sp. ATCC 27551 (pPDL2) and from *Pseudomonas diminuta* (pCMS1). The linear maps represent a short region of each plasmid (9.6 kb from pPDL2 and 6 kb from pCMS1). The lettered lines are sites where the DNA was cleaved by specific restriction endonucleases. Restriction endonucleases used were: E-EcoRI; B-BamHl; H-HindIII; P-PstI; S-SalI; X-XhoI.

In their native hosts, both opd genes are constitutively expressed under varying growth conditions. However, in foreign bacterial hosts, expression of these genes is completely dependent on promoter sequences located on the cloning vector. For example, Fig. 2 shows that in *E. coli*, the opd gene from *Flavobacterium* is not expressed when it is located on the vector pBR325.

However, when a restriction fragment containing the gene is subcloned into the expression vector pUC19, opd is expressed, but only under conditions where the lac promoter is induced. Increases in parathion hydrolase production in novel bacterial hosts should be easily attainable both by bringing the opd coding sequences closer to strong synthetic promoters and by increasing the number of opd genes contained within each host cell.

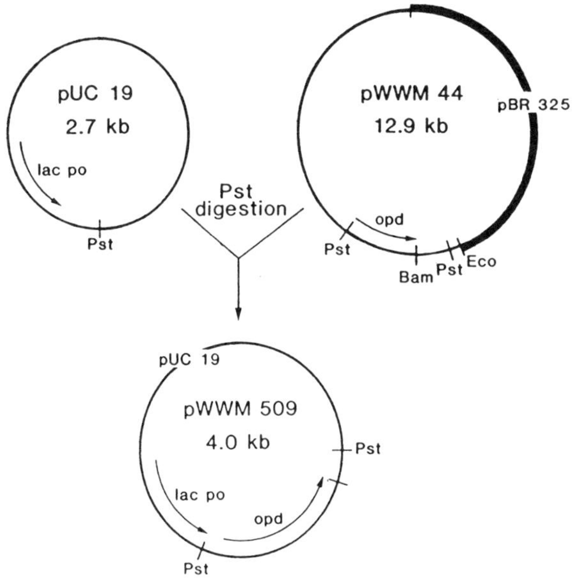

Host/plasmid	Hydrolase Activity (umoles/min-mg protein)	Relative Activity
Flavobacterium sp. 27551	4.3810	1.0
HB 101/pWWM 44	< 0.0001	-
JM 105/pWWM 509	0.0008	0.0002
JM 105/pWWM 509 + IPTG	0.0868	0.0198

Fig. 2. Expression of opd in E. coli. A 7.3 kb Eco RI fragment containing the opd gene from Flavobacterium sp. ATCC 27551 was cloned into the vector plasmid pBR325 to yield plasmid pWWM44. There was no measurable expression of parathion hydrolase when this plasmid was placed into E. coli strain HB101. When a 1.3 kb PstI fragment containing opd was removed from pWWM44 and inserted into the plasmid pUC19 [which contains the regulatory sequences of the lactose operon (lac po)], expression of parathion hydrolase was obtained in E. coli strain JM105 only when isopropylthio-B-D-galactoside (IPTG, a non-metabolizable inducer of expression of the lactose operon) was present. Such a result demonstrates that the regulatory sequences of the opd gene cannot be recognized by E. coli.

Haloacids - Genes associated with enzymatic dehalogenation of the short chain chlorinated or fluorinated aliphatic acids may play a role in waste disposal, although admittedly these compounds constitute a small number of biocides in the family of compounds that comprise pesticides. The important pesticides in this class include dalapon (2,2-dichloropropionate), TCA (trichloroacetic acid), and 1080 (fluoroacetate). Two different dehalogenases have been studied in some detail, haloacetate dehalogenase (EC 3.8.1.3) and 2-haloacetate dehalogenase (EC 3.8.1.2). The latter enzyme class is further subdivided into two types, those that act on fluoroacetate and those that do not. Most dehalogenases exhibit a high degree of stereospecificity and replace halogens with a hydroxyl group only on L-2-haloacids. An enzyme has been isolated from a Pseudomonas sp., however, that catalyzed a non-stereospecific dehalogenation of both optical isomers of 2-chloropropionate through an SN_2 type reaction (Ref. 7). L- and D-lactates were formed from D- and L-chloropropionates, respectively. This enzyme differs, therefore, from all previously reported dehalogenases that react on the chiral carbon of exclusive enantiomers.

A 73 kb plasmid (pUO1) has been isolated from a Moraxella strain B that possesses genes-encoding two haloacetate dehalogenases (EC 3.8.1.3) (Ref. 8). This organism can dehalogenate both fluoroacetate and chloroacetate. A mutant possessing a 67 kb plasmid (pUO11) was isolated that lacked activity on fluoroacetate but retained chloroacetate activity. Analysis of restriction endonuclease digests suggested that pUO11 was a deletion mutant derived from pUO1. The authors concluded that the 6 kb of DNA that is missing in pUO1, must harbor the gene encoding the fluoroacetate dehalogenase.

Research from the Warwick group (Ref. 9) has shown that Pseudomonas putida PP3 contained two independent pairs of chromosomal dehalogenase and permease genes (designated dehalogenase I, permease I, dehalogenase II and permease II) that permit growth on several haloacids. Recently it was observed that when strain PP3 was grown under conditions of carbon limitation, one or more of the gene functions were lost at extremely high rates (Ref. 10). At least one pair of genes (dehalogenase I and permease I) were located on a transposable element. Spontaneous mutations which yielded mutants unable to degrade halocids involved excision of this element. The observation that the frequency of this excision was strongly influenced by environmental stress has important implications for use of highly mobile genes in any disposal operation.

Microbial Degradation of Phenoxyalkanoic Acid Herbicides - Bacterial metabolism of the herbicides 2,4-dichlorophenoxyacetic acid (2,4-D), 2,4,5-trichlorophenoxy acetic acid (2,4,5-T), 4-chloro-2-methylphenoxyacetic acid (MCPA), and 2,4,5- trichlorophenoxypropionic acid (silvex) have been studied extensively (Refs. 11-15). Metabolism of these compounds begins with an oxidative attack on the ether bond that links the aliphatic acid portion of the molecules to chlorinated aromatic moieties (Refs. 13, 16). The resulting short chain organic acids and substituted phenol metabolites are further degraded and used by the organism as carbon and energy sources (Refs. 16, 17).

The genetic basis of 2,4-D degradation has been determined in Alcaligenes eutrophus. The genes encoding 2,4-D degradation in this bacterium were completely encoded on a single 93 kb conjugal plasmid (Ref. 18).

The biophysical and genetic properties of six independently isolated plasmids encoding the degradation of 2,4-D and MCPA from A. paradoxus and A. eutrophus have been described (Ref. 19). Four of the conjugative plasmids pJP3, pJP4, pJP5 and pJP7 had molecular masses of 82 kb. Transposon mutagenesis of pJP4 by Tn5 and Tn1771 and DNA cloning techniques were employed to establish the plasmid locations of genes encoding five catabolic enzymes involved in the degradation of 2,4-D (Ref.20).

While still not completely understood, the complete degradation of 2,4,5-T by Pseudomonas cepacia AC1100 involves both chromosomal and plasmid encoded genes (Ref. 21). DNA/DNA hybridization has demonstrated that plasmid DNA from strain AC1100 shares homology with plasmid DNA sequences from the 2,4-D degrading Alcaligenes (Ref. 21).

Microbial Degradation of Carbamate Pesticides - The carbamates are a broad class of pesticides which include N-phenyl-, thio-, dithio-, and N-methylcarbamate herbicides as well as N-methylcarbamate insecticides. All carbamate pesticides seem to be susceptible to some form of microbial degradation, with hydrolysis being the major route of inactivation. Three important bacterial hydrolases have been reported with activities toward the different classes of carbamates. Kearney (Ref. 22) reported the isolation and purification of an enzyme that could hydrolyze a number of N-phenylcarbamates from a chloropropham degrading Pseudomonas striata. Derbyshire et al. (Ref. 23) have reported the isolation of a single enzyme from Achromobacter strain WM111 (Ref. 24) that catalyzes the hydrolysis of several N-methylcarbamate insecticides. Little is known of the genetics of carbamate hydrolysis. Tam et al. report in this volume on the isolation of a bacterial plasmid that encodes the degradation of the thiocarbamate herbicide EPTC (Ref. 25).

USE OF MICROORGANISMS IN THE DISPOSAL OF PESTICIDE WASTES

Pesticide degrading microorganisms offer the potential for the development of low-cost methods for the disposal of aqueous pesticide wastes, or for the cleanup of heavily contaminated soils. Kilbane et al. (Ref. 26) used a 2,4,5-T degrading Pseudomonas cepacia to clean up soil containing 20,000 ppm 2,4,5-T. The compound was removed to the point that lettuce plants could actually grow in the treated soils. We have developed a system for the elimination of the organophosphate insecticide coumaphos from waste cattle dips. In the laboratory, resting cells of the parathion hydrolase producing Flavobacterium sp. ATCC 27551 were used to rapidly hydrolyze the coumaphos to less toxic products (Ref. 27). UV-ozonation was then used to further oxidize the hydrolysis products. These oxidized hydrolysis products were rapidly degraded when they were added to ordinary soils (Ref. 27). In a field trial on 2285 L of material at a dipping station in Texas, we were able to successfully degrade the waste through microbial hydrolysis and ozonation. On such a large volume, resting cell preparations would require such large amounts of bacterial culture that this method was impractical. To overcome this problem, we inoculated the vat material with 20 L of active culture and supplemented it with carbon and nitrogen sources so that the organisms would grow in the waste dip. Under these conditions, the organism required only 48 h to completely hydrolyze the coumaphos contained in the waste (Fig. 3).

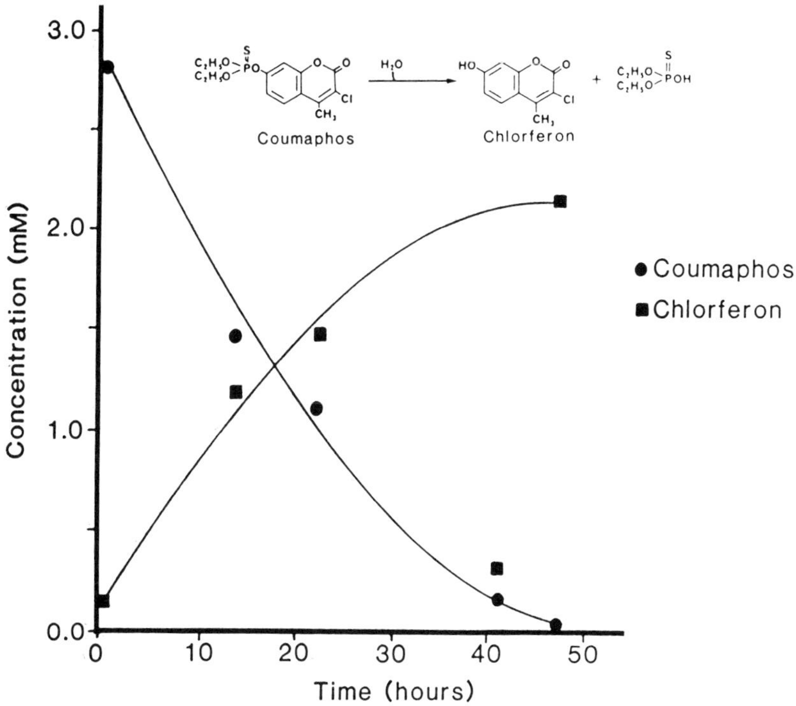

Fig. 3 Hydrolysis of coumaphos in 2285 L of waste cattle dip by growing cells of the parathion hydrolase producing Flavobacterium sp. ATCC 27551. The vat material was supplemented with 10 kg of xylose and 5 kg of ammonium sulfate fertilizer as well as 1.5 kg of monobasic- and 2 kg of dibasic potassium phosphate. The material was then inoculated with 20 L of growing Flavobacterium cells and was vigorously aerated throughout the course of the experiments. The reaction catalyzed by parathion hydrolase with coumaphos as substrate is shown in the inset.

The enzyme parathion hydrolase has also been used to degrade diazinon in modeled spill situations. When added to soil, the enzyme degraded 1000 ppm diazinon by 97% in 24 h (Ref. 28). In an expanded study, parathion hydrolase was added to soils containing 500, 1000, 2000 and 5000 ppm diazinon. The half-lives of diazinon were 1, 1.2, 5.6 and 128 h respectively (Ref. 29). At 500 ppm, the half-life for diazinon was 9.6 days without enzyme. The authors emphasize that much research remains to be done on the soil parameters that influence the stability of the enzyme. Munnecke (Ref. 30) demonstrated that immobilized preparations of parathion hydrolase could be used to reduce parathion concentrations in industrial effluents to below 500 ppb.

Biotechnology can play an important role in two areas of pesticide waste disposal. First, should the use of enzymes like parathion hydrolase prove to be a feasible method of decontamination at spill sites, then large quantities of the biocatalyst would be required for commercial purposes. Increased enzyme production through gene manipulation would be an obvious route to achieve this goal. Second, the direct addition of microorganisms into aqueous wastes has considerable merit from the standpoint of economy, safety, and efficiency. Insertion of a degradative gene into a host microorganism indigenous to the waste environment has the advantage of providing the engineered organism with resistance to the solvent, thermal, saline, and other stresses present in the waste. Hybrid systems using physical methods such as UV and/or ozonation combined with native or engineered organisms are being actively pursued in our laboratory.

REFERENCES

1. D. Friedman, Treatment and Disposal of Pesticide Wastes p. 18, Amer. Chem. Soc. Symp. Series 259, Washington (1984).

2. K. F. Whittaker, J. C. Nye, K. F. Wukasch, R. G. Squires, A. C. York and h. Kazimier, PB 82-255 365, Oil and Hazardous Materials Spill Branch, MERL-Cincinnati, USEPA, Edison, N.Y. (1982).

3. J. N. Seiber, Treatment and Disposal of Pesticide Wastes p XI, Amer. Chem. Soc. Symp. Series 259, Washington (1984).

4. A. M. Chakrabarty, Biodegradation and Detoxification of Environmental Pollutants, p 1-147 CRC Press Inc., Boca Raton, Fla. (1982).

5. C. M. Serdar, D. T. Gibson, D. M. Munnecke and J. H. Lanchaster, Appl. Environ. Microbiol. 44, 246-249 (1982).

6. W. M. Mulbry, J. S. Karns, P. C. Kearney, J. O. Nelson, S. C. McDaniel and J. R. Wild, Appl. Environ. Microbiol. 51, 926-930 (1986).

7. K. Motosugi, N. Eaki and K. Soda, J. Bacteriol. 150, 522-527 (1982).

8. H. Kawaski,. H. Yohara, K. Tonomura, Agric. Biol. Chem. 45, 1477-1481 (1981).

9. A. J. Weightman, H. L. Weightman and J. H. Slater, J. Gen. Microbiol. 128, 1755-1762 (1982).

10. J. H. Slater, A. J. Weightman and B. G. Hall, Mol. Biol. Evol. 2, 557-567, (1985).

11. W. C. Evans, B. S. W. Smith, P. Moss and H. N. Fernley. Biochem.J. 122, 509-517 (1971).

12. G. R. Bell, Can. J. Microbiol. 66, 325-337 (1960).

13. A. Rosenberg and M. Alexander, J. Agric. Food Chem. 28, 297-302 (1980).

14. L. T. Ou and H. C. Sikka, J. Agric. Food Chem. 25, 1336-1339 (1977).

15. J. J. Kilbane, D. K. Chatterjee, J. S. Karns, S. T. Kellogg and A. M. Chakrabarty, Appl. Environ. Microbiol. 44, 72-78 (1982).

16. J. S. Karns, J. J. Kilbane, S. Du Hag Upta and A. M. Chakrabarty, Appl. Environ. Microbiol. 45, 1697-1700 (1983).

17. J. N. Tiedje, J. M. Duxbury, M. Alexander and J. E. Dawson, J. Agric. Food Chem. 17, 1021-1026 (1969).

18. P. R. Fisher, J. Appleton and J. M. Pemberton, J. Bacteriol. 135, 798-804 (1978).

19. R. H. Don and J. M. Pemberton, J. Bacteriol. 145, 681-685 (1981).

20. R. H. Don, A. J. Weightman, H. J. Knockmuss, K. N. Timmis, J. Bacteriol. 161, 85-90 (1986).

21. P. H. Tomasek, B. Frantz, D. K. Chatterjee and A. M. Chakrabarty, Beltsville Symposia in Agricultural Research 10, 355-368, Martinus Nijhoff Publishers, Dordrecht (1985).

22. P. C. Kearney, J. Agric. Food. Chem. 13, 561-564 (1965).

23. M. K. Derbeyshire, J. S. Karns, J. O. Nelson and P. C. Kearney, (Abstracts) Ann. Mtg. Amer. Soc. Microbiol. pp. 206 (1986).

24. J. S. Karns, W. W. Mulbry, J. O. Nelson and P. C. Kearney, Pesticide Biochem. and Physiol. 25, 211-217 (1986). 25. A. C. Tam, R. M. Behki and S. U. Khan, (Abstracts) Sixth International Congress of Pesticide Chemistry, 66-68 (1986).

25. A. C. Tam, R. M. Behki and S. U. Khan, (Abstracts) Sixth International Congress of Pesticide Chemistry, 66-68 (1986).

26. J. J. Kilbane, D. K. Chatterjee and A. M. Chakrabarty, Appl. Environ. Microbiol. 45, 1697-1700 (1983).

27. P. C. Kearney, J. S. Karns, M. T. Muldoon and J. M. Ruth, J. Agric. Food Chem. 34, 702-706 (1986).

28. D. M. Munnecke, J. Appl. Environ. Microbiol. 32, 7-13 (1976).

29. R. Honeycutt, L. Ballantine, H. LeBaron, D. Paulson, V. Seim, C. Ganz, G. Milad, Treatment and Disposal of Pesticide Wastes p. 343, Amer. Chem. Soc. Symp. Series 259, Washington (1984).

30. D. M. Munnecke, Biotech. Bioengineer. 21, 2247-2261 (1979).

Monoclonal antibodies for the detection of trace chemicals

M. Vanderlaan*, J. Van Emon, B. Watkins, and L. Stanker

Biomedical Sciences Division, Lawrence Livermore National Laboratory, Livermore, CA, USA 94550

Abstract - Problems in analytical chemistry may limit monitoring for trace organic residues by traditional chromatographic methods. For example, the cost and analysis time per sample may preclude adequate sampling, making the development of alternative technologies desirable. Immunoassays are one such alternative, with the potential for cost reduction by automation and parallel sample processing. A particularly significant advance in the past decade has been the development of monoclonal antibodies, which offer greater selectivity and reproducibility than conventional antisera. Immunoassays can be developed that use simple, field-portable instrumentation, give rapid results, and have detection limits of less than a part-per-billion. This paper reviews the general technology for developing monoclonal antibodies to small organic molecules using the immunoassay of 2,3,7,8-tetrachlorodibenzo-dioxin (2,3,7,8-TCDD) as an example.

INTRODUCTION

An increasing number of immunoassays for pesticides have been reported (ref. 1-3). These studies demonstrate that antibodies can be made that selectively bind to a wide range of small organic molecules (haptens). Immunoassays incorporating these antibodies offer the potential for high sensitivity, low cost, and rapid detection of contaminants. Immunoassays are particularly appropriate to quantify residues when chromatographic methods are difficult (e.g. paraquat and cypermethrin), when the compounds are complex biologically-derived products (e.g. Aflatoxins, Ivermectin, or Bacillus thuringiensis israelensis toxin), or when only specific isomers are of interest (e.g. dioxin and S-bioallethrin) (ref.4-9). This paper reviews the significant characteristics of monoclonal antibodies (Mabs) for their application in residue analysis, using our anti-dioxin antibodies as an example (ref.9). General immunoassay methodology and examples of specific applications are beyond the scope of this paper, and readers are referred elsewhere (ref.10,11).

The last ten years have seen an explosive growth in the generation and use of Mabs. Mabs and the technology for their development have many distinctions from conventional serology which are significant in the development of immunoassays for environmental monitoring. Those not familiar with the field are often struck with the high initial time and labor associated with Mab production, and question whether monoclonal antibodies are really different from polyclonal antisera, and worthy of the additional efforts and resources. Our view is that Mabs are not only worth the cost, but are essential for standardized, hapten-specific immunoassays that can be used to analyze large numbers of samples in many different laboratories.

Antibody assays exploit the natural capacity of the immune system to respond to an apparently limitless number of different antigens. Figure 1 illustrates several features of the immune response. Small molecules, such as dioxins, do not elicit an antibody response if injected alone. Rather they must be conjugated to a carrier molecule, usually a protein. The choice of linkage site and chemistry is significant, since it will influence the spectrum of antibodies produced. In the case illustrated, a reactive group was introduced into the dioxin molecule by synthesizing 1-amino-3,7,8-tri-chloro-dibenzo-dioxin (A-triCDD) and this chemical was then conjugated to the protein bovine serum albumin (BSA) via a linker arm by published methods (ref. 5).

CLONAL NATURE OF THE IMMUNE RESPONSE

A central feature of the antibody immune response is its clonal basis (ref.12). All vertebrates have B-lymphocytes which possess immunoglobulin receptors on their external membranes and each B-cell possesses a single receptor whose combining site is characteristic of that cell. Following immunization, those B-cells able to bind immunogen, proliferate and secrete antibodies with the same combining site as their cell membrane receptor. Thus the multiplicity of antibodies in serum reflects the corresponding multiplicity of cells, each secreting one and only one, antibody.

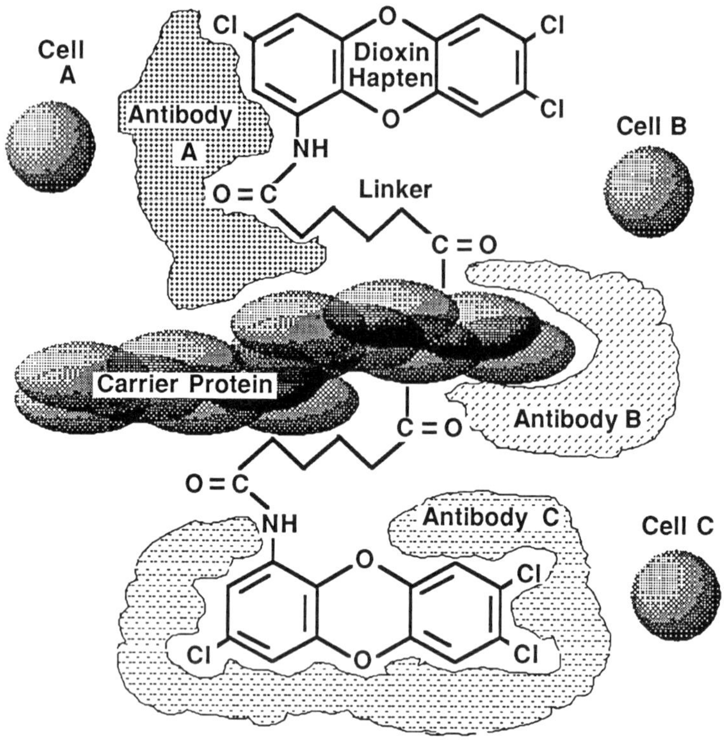

Fig. 1. Illustration of the immune response to dioxin haptens linked to a carrier protein. Three different antibodies produced by three B-cells are shown. Antibodies A and B are unsuitable for immunoassays since they react with the carrier protein and/or the linker arm. Antibody C is suitable since it reacts primarily with the dioxin hapten. For an immunoassay based on polyclonal antisera, antibody C must be purified from other serum proteins; For one based on monoclonal antibodies, cell C must be cloned and cultured to produce antibody C.

Figure 1 illustrates three B-cells, and their corresponding antibodies secreted in response to the hapten-protein immunogen. Clones of all of these cells will be present in an immunized animal. The serum will contain all of the antibodies, and hence is termed "polyclonal antiserum." Some antibodies (e.g. C) react with hapten alone, others (e.g. B) react with the carrier protein, and still others (e.g. A) react with some combination of hapten, linker, and/or protein. There will be numerous other B-cells secreting antibodies to such things as contaminants in the immunogen. Only antibodies such as C are useful in immunoassays for the hapten and separating the desired antibodies from the undesirable ones is often difficult. Separation methods usually rely on affinity chromatography which may damage or change the specificity of the desired antibodies. Antibodies such as A, having complex combining sites, far outnumber hapten-specific antibodies, when the hapten has a molecular weight less than 500 (ref.12). Failure to purify the specific anti-hapten antibodies may compromise the immunoassay.

HYBRIDOMAS AND MONOCLONAL ANTIBODIES

In contrast to polyclonal antiserum, the preparation of monoclonal antibodies does not involve the isolation of hapten-specific antibodies from other proteins in the serum, but rather the isolation of the corresponding cells that secrete the hapten-specific antibodies. Following immunization, the B-cells are removed and induced to grow in culture by fusing them with tumor cells which already grow in vitro. These fused cells are called "hybridomas" and can grow in culture and secrete antibodies. The details of culturing the cells have been reviewed elsewhere (ref.14,15). What is important from the stand point of environmental monitoring is that each colony is descended from a single cell, and produces its own unique antibody reagent with invariant affinity for the specific hapten and invariant selectivity among related haptens. The hybridoma cells can be grown indefinitely in culture and the antibody produced is a uniform, standardized reagent which can be distributed to many laboratories.

Following a fusion, literally thousands of hybridomas are produced which need to be evaluated to identify the small number of clones suitable for future development. The importance of this screening step cannot be over emphasized. Screening must be rapid, allowing all the clones to be evaluated in a single day, and it must also be accurate, since errors in screening result in wasted time on clones that ultimately prove unsatisfactory. Therefore, an early screening assay should closely mimic the eventual application of the antibody. For example, immunoassays for small organics almost always require that the antibody recognize the free hapten in solution. Thus, an early screen testing the Mab's ability to recognize the free hapten in the absence of carrier protein is essential in identifying the best clones. Subtle changes in the assay conditions may alter the test by which the most desirable antibody is selected. An appropriately designed screening assay will identify the clones of cells secreting antibodies such as C in Fig. 1. Cultures of these clones may then be expanded and the characteristics of the antibodies determined.

We use an enzyme-linked immunosorbent assay (ELISA) to screen and characterize the binding specificity of our Mabs, and for the eventual immunoassay. In a competition ELISA the antibody partitions between free compound in the sample and a hapten-protein conjugate adsorbed to a plastic microtiter plate. The amount of antibody bound to the immobilized hapten is then determined by detecting enzymatic activity that is linked to the antibody. The competition ELISA format is both simple and rapid. It is a quantitative and sensitive procedure that can be used for evaluating an antibody's binding specificity and in the eventual application of the antibody to unknown samples.

Once a clone has been isolated, the selectivity of its antibody must be characterized in detail. Antigen-antibody binding is based primarily on steric factors influencing the fit of the antigen into the antibody receptor site. One of the advantages of Mabs, is that antibodies with differing selectivities can be chosen for specific applications. To illustrate, some anti-dioxin antibodies may distinguish dioxin from 2,4,5-trichlorophenoxy) acetic acid, but may cross react with polyclorinated biphenyls (PCBs). These antibodies would be suitable for immunoassays to detect dioxins in 2,4,5-T treated areas. Antibodies with the reverse specificity would be suitable for detecting dioxins in PCB-contaminated sites. The characterization of a Mab's selectivity defines the nature of its combining site, while the specificity of polyclonal antisera reflects the aggregate sum of combining sites in all the antibodies produced by the set of clones in the animal. This set of clones will differ from animal to animal, and change with time in any given animal.

MONOCLONAL ANTIBODIES TO DIOXIN

Spleen cells from seven dioxin-BSA immunized mice were fused and the resulting hybriodmas screened. The first tier in our screening process selected antibodies distinguishing both A-triCDD-BSA and A-triCDD-rabbit serum albumin (A-triCDD-RSA) from BSA and RSA. This identified clones whose antibodies reacted with the dioxin independently of protein, and did not react with the protein. This test limited us to the 25 clones shown in table 1. The next screening assay required that the antibodies reacted with the free analogs: linker-A-triCDD, A-triCDD, and 2,3,7,8-tetrachloro-dibenzodioxin (TCDD). These compounds range from the hapten used in the immunogen, (linker-A-triCDD), to the dioxin congener of greatest toxicological concern, (TCDD). The values listed in Table 1 are the 50% inhibition points for the competition ELISA-the lower the $I50$, the greater the affinity for that compound. All ranges of selectivity were encountered. Four clones (11-1a-71,74,75, and 76) did not compete with any dioxin tested. Three clones from mouse 3-7 required the presence of the linker arm. The majority of clones (e.g. clone 11-1b-47) recognized dioxins having either the linker arm or an amino group in the #1 position, but did not react with TCDD. Five clones (DD-1,3,4,5, and 6) were identified that do react with TCDD.

Several features of the clonal nature of the immune response are apparent from the results in Table 1. For example, the spectrum of clones present in each animal differed. This is a common finding and indicates how varied the immune response may be, even between individuals of the same inbred strain. One strategy to take advantage of these differences is to screen the serum of many mice for rare individuals producing the potentially suitable antibodies. The antibody-producing cells in these mice can then be "immortalized" as hybridomas. Such a screening of serum would have rejected mouse 3-7 from being fused. Relative to mice, it is harder to screen many rabbits, goats, or other large animals for the individuals that are generating the best set of polyclonal antibodies.

Consider also the likely properties of polyclonal serum present in these mice. Mouse 11-1a, for example, produced many antibodies with differing specificities, but only one hybridoma in nine recognized TCDD. Assuming that there is the same relative concentration of antibodies in serum as hybridomas produced, dioxin would only inhibit the

TABLE 1. Characteristics of monoclonal antibodies produced by 28 hybridoma clones. The concentrations required for 50% inhibition (I_{50}) of the ELISA are listed for 2,3,7,8-tetrachloro-dibenzodioxin (TCDD), 1-amino-3,7,8-trichloro-dibenzodioxin (A-triCDD), and the hapten used for immunization (linker-DD).

Mouse Number	Clone Number	I_{50} (ng) Linker-DD	A-TriCDD	TCDD
11-1a	61	5	10	>10
	71	>10	>10	>10
	74	>10	>10	>10
	75	>10	>10	>10
	96	>10	>10	>10
	102	0.3	0.3	>10
	108	5	5	>10
	110	1	0.6	>10
	DD-1	0.15	<.02	1
11-1b	47	0.6	0.03	>10
	DD-3	0.15	<.02	7
11-8	75	0.3	.07	>10
	DD-5	0.15	<.02	2.5
12-6a	42	7	2.5	>10
12-6b	52	0.6	0.3	>10
	DD-4	0.3	<.02	1
2-4	9	1	1	>10
	19	0.6	<.02	>10
	22	1	.3	>10
	23	0.6	0.6	>10
	24	0.6	<.02	>10
	26	0.6	<.02	>10
	32	1	0.7	>10
	DD-6	0.5	<.02	1
3-7	12	0.6	>20	>20
	14	>20	>20	>20
	24	1	>20	>20
	25	1	>20	>20

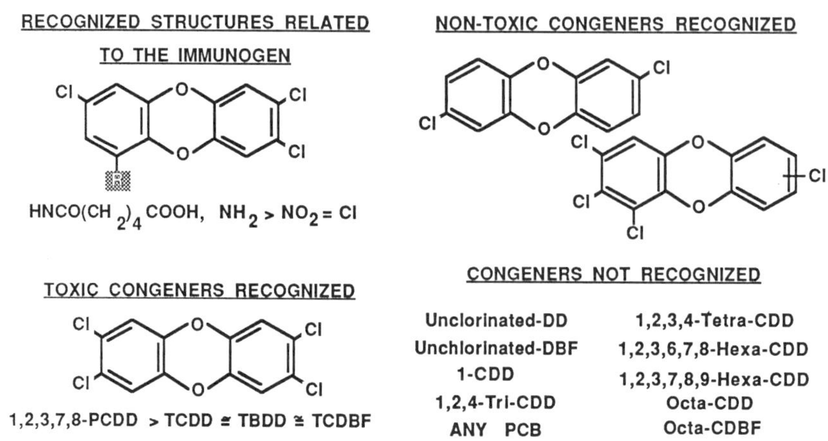

Fig. 2. Binding specificity for Mab DD-3, which is representative of the group of anti-dioxin antibodies. The antibody binds dioxins related to the immunizing hapten, with a preference for either the linker or an amino group in the #1 position. Toxic dioxins containing substituents in the 2,3,7, and 8 positions are recognized by the Mab. Some less toxic congeners are also recognized, providing there is at least one chlorine on each ring, and no more than five chlorines total.

binding of 1/9th of the antibodies in 11-1a antiserum in a competition ELISA. By selecting the clone, we are able to obtain large quantities of a pure antibody which was present as a very small minority of the total antibodies present in the population of mice. Hybridomas allow the immortalization of rare antibody-producing cells that might have valuable selectivity properties, but are very small components in polyclonal serum.

The five anti-TCDD antibodies have been tested for their reactivity with many compounds and results are summarized in Fig. 2 for Mab DD-3, which is representative of the set of antibodies. The Mabs react with only dioxin congeners having intermediate levels of chlorination, including the highly toxic congeners 2,3,7,8-TCDD, 1,2,3,7,8-penta-CDD, 2,3,7,8-tetrachlorodibenzofuran (TCDBF), and 2,3,7,8-tetrabromo-dibenzodioxin (TBDD). They do not react with less toxic congeners, such as the octa-CDD or unchlorinated dioxin, nor do they react with a large number of PCBs, chlorinated herbicides or chlorinated hydrocarbons. These antibodies therefore are ideal candidates for testing for the presence of TCDD in a variety of contexts.

PROSPECTIVES ON IMMUNOASSAYS FOR ENVIRONMENTAL STUDIES

Environmental and human monitoring for potentially hazardous compounds by immunoassay is a challenging new field with broad potential for aiding regulatory agencies and basic research in toxicology. How should Mab-based immunoassays complement more established analytical methods, such as GC/MS or HPLC? The high specificity of a given immunoassay contrasts with the general applicability of chromatography. If the contaminant is unknown, or many contaminants must be determined in a few samples, then GC/MS or HPLC are the appropriate methods. Conversely, if there are many samples that must be analyzed for a limited number of compounds then immunoassays are appropriate. For example, one might screen many soil samples for TCDD contamination and then do detailed GC/MS analysis on the positive samples to determine the nature of the contamination. This two tiered approach allows the immunoassay to guide the field collection of samples, and direct the GC/MS to the most interesting samples. One should also consider hybrid systems, where antibodies concentrate a sample before chromatography, or where the immunoassay is performed on fractions collected from a chromatographic separation.

For regulatory and health agencies, the employment of immunoassays offers a cost effective means of screening food and environmental sites for residues, and can be used to quantify worker exposure. For example, an ELISA for paraquat was applied to air filters, cloth exposure patches, and biological samples to determine worker exposure during an aerial spraying of cotton (ref.11). Since assays can be portable and require little in the way of sophisticated equipment, they can be done on site, providing timely indications of contamination. Automation of immunoassays permits parallel sample processing, greatly increasing the number of samples that can be analyzed. Monoclonal antibodies contribute to these advantages by being invariant reagents, facilitating quality controls for the assay. Multiple laboratories can perform assays using the same standardized reagents to make valid comparisons.

For the research community immunoassays permit experiments that are not now possible. By reducing the task of chemical analysis, more extensive sampling can be done to address questions related to uptake, metabolism, and the environmental fate of pesticides. For example, the specificity of Mabs allows the selection of antibodies that either cross react with several metabolites, or discriminate between them. One of the best examples of the potential for Mab-based immunoassays in research is the detection of carcinogen-DNA adducts. Studies with ethylating agents allow the quantitation of dose at the single-cell level using histochemical methods. Differences in the initial distribution and repair of DNA adducts can be seen between organs in the whole animal, and between cell types in the same organ (ref.16). Studies using antibodies to benzo-a-phrene-DNA and methyldeoxyguanosine adducts in humans provide a chemical basis for estimating dose in epidemiological studies (ref.17,18). In these studies Mabs provide the critically needed selectivity and sensitivity. Similar studies on the uptake, distribution, and retention of pesticides in insects, plants, and mammals including humans should contribute significantly to basic research in toxicology.

ACKNOWLEDGMENTS

This project was funded by the USEPA under IAG #DW89931433-01-0 and performed under the auspices of the US Dept. of Energy by Lawrence Livermore National Laboratory under contract #W-7405-ENG-48.

REFERENCES

1. B.D. Hammock and R.O. Mumma, In: Recent Advances in Pesticide Analytical Methodology p. 321, J. Harvey Jr. and G. Zweig, Eds., American Chemical Society Symposium Series, Washington, D.C. (1980).
2. W.H. Newsome and J. B. Shields, J. Agric. Food Chem. 29, 220-222 (1981).

3. D.F. Rinder and J.R. Fleeker, Bull. Environm. Contam. Toxicol. 26, 375-380 (1981).

4. K.D. Wing and B.D. Hammock, Experientia 35, 1619-1620 (1979).

5. P.W. Albro, M.I. Luster, K. Chae, G. Clark and J.D. Mckinney, Toxicol Applied Pharmacol 50, 137-146 (1979).

6. J.D. Groopman, L.J. Trudel, P.R. Donahue, A. Marshak-Rothstein, and G.N. Wogan, Proc Natl Acad Sci, USA 81, 7728-7731 (1984).

7. S.I. Wie, R.E. Andrews, B.D. Hammock, R.M. Faust and L.A., Bull. Appl. Environ. Microbiol. 43, 891-894 (1982).

8. J.M. Van Emon, B. Hammock , and J.N. Seiber, Analytical Chemistry 58, 1866-1873, (1986).

9. L.Stanker, B. Watkins, N. Rogers, and M. Vanderlaan, Chemosphere (submitted).

10. W.R. Butt, Practical Immunoassay: The State of the Art., Marcel Dekker, New York,(1984).

11. J.M. Van Emon, J.N. Seiber, and B.D. Hammock, p 307, In: Bioregulators for Pest Control, American Chemical Society Symposium Series #276, P.A. Hedin, Ed., Washington, D.C. (1985).

12. M.Z. Atassi, C.J. Van Oss, and D.R. Absolom, Molecular Immunology: a Textbook, Marcel Dekker, New York (1984).

13. M. Delaage, B. de Veyrac, A. Morel, H. Cailla, and J.L. Drocourt, In: Monoclonal Antibodies and New Trends in Immunoassays, p 53-58, C.A. Bizollon, Ed., Elsevier, New York (1984).

14. J.W. Goding, Monoclonal Antibodies: Principles and Practice, Academic Press, New York (1983).

15. V.T. Oi and L.A. Herzenberg, In: Selected Methods in Cellular Immunology, p 351-372, B.B. Mishell and S.M. Shiigi, Eds., Freeman, San Francisco (1980).

16. G.J. Menkveld, C.J. Van Der Laken, T. Hermsen, E. Kriek, E. Scherer, and L. Den Engelse, Carcinogenesis 6, 263-270 (1985).

17. F.P. Perera, R.M. Santella, and M.C. Poirier, In: Banbury Report 19: Risk Quantitation and Regulatory Policy, p 211-229, D.G. Hoel, R.A. Merrill and F.P. Perera, Eds., Cold Spring Harbor Press, New York (1985).

18. C.P. Wild, D. Umbenhauer, B. Chapot, and R. Montesano, J. Cellular Biochem. 30, 171-179 (1986).

Author Index